Polypharmacology

Zhiguo Wang • Baofeng Yang

Polypharmacology

Principles and Methodologies

Springer

Zhiguo Wang
Department of Pharmacology
Harbin Medical University
Harbin, Heilongjiang, China

Baofeng Yang
Department of Pharmacology
Harbin Medical University
Harbin, Heilongjiang, China

ISBN 978-3-031-04997-2 ISBN 978-3-031-04998-9 (eBook)
https://doi.org/10.1007/978-3-031-04998-9

This Springer imprint is published by the registered company Springer Nature Switzerland AG
The registered company address is: Gewerbestrasse 11, 6330 Cham, Switzerland

Preface

Polypharmacology is frequently referred to as the design or use of pharmaceutical agents that act on multiple molecular targets or multiple signaling pathways for multi-factorial human disease. Yet, such a definition is at its best incomplete and inaccurate. This is because polypharmacology on its birth possesses its own unique theoretical basis and principles of modern pharmacology and practical approaches and methodologies for drug discovery and pharmacological treatment of human disease, based on its own unique philosophy—multitarget drug therapy and multitarget drug development. Since its advent, polypharmacology has grown rapidly and become a new subdiscipline of general pharmacology and a new research field as well. Nowadays, polypharmacology is gaining its predominance in the therapy of multifactorial complex human disease, and polypharmacology-based multitarget drugs are considered the new paradigm for drug discovery and development. The rapid advance and evolution of polypharmacology has revolutionized pharmacology and the pharmaceutical industry with new thoughts and innovative technologies, which inspired and prompted us to write this book summarizing the most up-to-date concepts, research progress, translational medicine, and clinical applications of polypharmacology.

This book aims to target a broad range of readership, which includes university graduate students, postdoctoral fellows, teachers/professors, and scientists and researchers in academic and research institutions involved in pharmacological studies of any fields, as well as investigators and entrepreneurs involved in the drug research and development (R&D) in pharmaceutical industries. The contents of the book are also suitable for medical practitioners from residents to professors in various medical divisions, who are interested in utilizing, testing, and developing new medications for the treatment of human disease. The book can be used as a teaching material and reference book as it provides novel concepts and ideas, proof of concepts, state-of-the-art approaches, cutting-edge methodologies, and practical protocols on polypharmacology.

The contents in *Polypharmacology: Principles and Methodologies* is roughly divided into five parts, including (a) "The Basics of Polypharmacology"; (b) "Polypharmacology in Clinical Applications"; (c) "Polypharmacology in Drug Discovery and Development"; (d) "Polypharmacokinetics Studies"; and (e) "Process Rational Design and Discovery of Multitarget Drugs." Part I begins with Chap. 1 on the concept and definition of polypharmacology, followed by historical account of pharmacology on how polypharmacology evolved into a new sub-discipline of general pharmacology, and ends with Chap. 2 on detailed comparisons among the three major forms of polypharmacology, combination drug therapy (CDT), fixed-dose single pill combination therapy (FDC), and multitarget drug (MTD), in terms of their applicability, strengths and weaknesses, and types. The second part, "Polypharmacology in Clinical Applications," includes nine chapters discussing the three forms of polypharmacology for the treatment of cancer, cardiovascular disease, metabolic disorders, neurological disorders, respiratory disease, gastrointestinal disease, renal disease, infectious disease, and inflammatory disease, respectively. This part also tentatively categorizes polypharmacology into anti-cancer polypharmacology, cardiovascular polypharmacology, and Neurological polypharmacology. The third part has four chapters covering several aspects of the applications of polypharmacology to drug design and discovery. The major topics include the basis for rational design of multitarget drugs, drug repurposing for old drug rediscovery, polypharmacology-based prediction of drug toxicity, and natural products in polypharmacology-based drug discovery. The fourth part focuses on polypharmacokinetics by discussing the differences between regular pharmacokinetics and polypharmacokinetics and describing the major methods for polypharmacokinetics studies, as well as the influence of drug-drug interactions on polypharmacokinetics. The final part of this book aims to provide a general guideline for polypharmacology-based rational design and discovery of multitarget drugs. This part begins with an introduction to the general process of drug discovery and development, followed by formulating a dozen of most frequently referred strategies for rational design and discovery of multitarget drugs. Next, the Internet resources for rational design and discovery of multitarget drugs are presented, followed by descriptions of several selected methods for rational design and discovery of multitarget drugs. Finally, a few exemplary protocols of rational design of multitarget drugs are provided.

To our knowledge, this book is the first of its kind. The major distinction to published books on topics related polypharmacology is that we do not limit polypharmacology as a new strategy for drug development; instead, we take it as a new sub-discipline of general pharmacology for it possesses its own unique theoretical system and technological system as well. By presenting this book to the audience, we hope it be a source of education on this new sub-discipline of pharmacology and a preliminary guide for rational design and discovery of multitarget drugs, so as to promote polypharmacology. Yet, we are fully aware that due to our limitations, the book may not cover all aspects of polypharmacology, and some of the contents may not be exactly precise because of the rapid development of this new sub-discipline. Therefore, feedbacks and criticisms are more than welcome and highly encouraged.

Harbin, Heilongjiang, China

Zhiguo Wang
Baofeng Yang

Acknowledgement

Writing this book was a "mission impossible" for us who specialize in cardiovascular pharmacology to cover all subjects involving polypharmacology, just like a single conifer trying to cover the whole vast field to create a panoramic view. But we finally had it done, solely because we were provided with a wealth of information from researchers and clinicians all over the world.

We entirely give credit to the actual contributors and inventors of the techniques and methodologies described in this book; this book is full of citations and quotes from them. Since the beginning of the writing process, we have been immediately inspired by the actual contributors with their ingenious, creative thoughts, strongly amazed by their serious efforts of searing scientific doctrine, and deeply touched by their non-conserved sharing-out of every detail of their experimental protocols. We were not only gathering the information from their studies for our book writing but are also learning a great deal from these investigators, for "great works are performed not by strength but by perseverance (Samuel Johnson)" and "the talent of success is nothing more than doing well whatever you do without a thought of fame (Longfellow)." We wish to give our most sincere and respectful thanks to all these people, whether we knew them in person or not; we know their names from their papers and their works by heart. Without their sweat and intellect, completion of this book would have been truly impossible.

We wish to convey our hearty gratitude to our wives for standing behind us and selflessly supporting us, while sharing with us the harsh times and the joyful times as well. We also want to thank our kids for understanding our long unavailability and our lack of care of them as a result of the writing task.

Contents

About the Authors

Zhiguo Wang PhD, received his PhD in cardiovascular pharmacology from the Faculty of Medicine and Health Sciences, McGill University, Canada. He is presently a professor in the Department of Pharmacology, Harbin Medical University, Harbin, China. He is a ChangJiang Scholar Endowed Professor (China) and Longjiang Scholar Endowed Professor in Heilongjiang (China). He is also Fellow of the American Heart Association (FAHA) and Fellow of the European Society of Cardiology (FESC). His current research interests include cardiovascular pharmacology and epigenetic polypharmacology related to cardiovascular disease.

Baofeng Yang MD, PhD, received his PhD in pharmacology from Tongji Medical University, China. He is presently Professor of Pharmacology and the president of Harbin Medical University, and an academician of the Chinese Academy of Engineering. He is a chief scientist for the National Program on Key Basic Research Project of China; the vice president of the Chinese Pharmacologic Society; a visiting professor at West Virginia University, the University of Missouri-Kansas City, and Nippon Medical University; and an honorable professor at Perm Pharmaceutical Academy and Shiga-University of Medical Science. His current research interests include cardiovascular and molecular pharmacology involving non-coding RNAs and epigenetics.

1 Basics of Polypharmacology

Abstract

Polypharmacology is frequently referred to the design or use of pharmaceutical agents that act on multiple molecular targets or multiple signaling pathways for multifactorial human disease. Yet, such a definition is at its best incomplete and inaccurate. This is simply because polypharmacology has evolved and emerged as a new subdiscipline of Pharmacology and a unique research field, which in addition to retaining all areas of general Pharmacology, possesses its own theoretical basis and principles of modern Pharmacology and practical approaches and methodologies for drug discovery and pharmaceutical treatment of human disease, based on its own philosophy—multitarget drug therapy and drug development. Polypharmacology revolutionizes traditional and general Pharmacology. This chapter aims to answer the following questions: (1) What is Polypharmacology; (2) What relationship of Polypharmacology to Pharmacology is; (3) Why we need Polypharmacology; (4) What are the mission and tasks of Polypharmacology; (5) What opportunities and challenges Polypharmacology can offer. To this end, the historical account of Polypharmacology is given, followed by the outlines of principles and other aspects of Polypharmacology. At the end, network pharmacology and epigenetic polypharmacology as two of the branches of Polypharmacology are also briefly discussed. Some new concepts of Polypharmacology we proposed are presented in this chapter.

Keywords

Pharmacology · Polypharmacology · Combination drug therapy (CDT) · Fixed-dose single pill combination (FDC) · Multitarget drug (MTD)

1.1 The Concept and Definition of Polypharmacology

Before discussing Polypharmacology, we need to know the basics of Pharmacology. Pharmacology is a branch of medical and pharmaceutical sciences with a broad scope involving drugs or medications and their biological actions and clinical applications [1–5]. A drug is defined as any artificial/synthetic, natural, or endogenous (from within the body) molecule (chemical substance, compound, protein, antibody, RNA, etc.) which exerts biochemical or physiological effects on the cell, tissue, organ, or organism. Sometimes, the term "Pharmacon" is used to encompass the endogenous and exogenous bioactive species. The Pharmacology discipline encompasses drug composition and properties, synthesis and drug design, molecular and cellular mechanisms, organ/systems mechanisms, signal transduction/cellular communication, molecular diagnostics,

Z. Wang, B. Yang, *Polypharmacology*, https://doi.org/10.1007/978-3-031-04998-9_1

interactions, chemical biology, therapy, and medical applications and antipathogenic capabilities.

There are many subspecialties within the general discipline of pharmacology, such as basic pharmacology, clinical pharmacology, receptor pharmacology, cellular pharmacology, molecular pharmacology, genetic pharmacology, epigenetic pharmacology, antiaging pharmacology, and polypharmacology [6–15]. Each of these subspecialties involves two main areas—pharmacodynamics and pharmacokinetics, which together investigate the interactions that occur between a living organism and chemicals affecting normal or abnormal biochemical function [4, 5]. Specifically, pharmacodynamics focuses on the effects of drugs on biological systems and specifically addresses the chemical properties and physiological and behavioral effects of drugs arising from their interaction with molecular targets such as receptor proteins or enzyme systems and therapeutic effectiveness/clinical applications. On the other hand, pharmacokinetics deals with what biological systems do to the drug and encompasses elucidations of drug liberation, absorption, distribution, metabolism/biotransformation, and excretion (L-ADME) in the body, essential information for the design of drug treatment schedules in different patient populations and experimental animals, and for the prediction of drug-drug interactions that may enhance or compromise the effectiveness and safety of therapeutic agents.

At a deeper level, pharmacology is also the study of mechanisms of drug actions at various levels including molecular, subcellular, cellular, organ, and whole-body levels [1–4]. If substances have medicinal properties, they are considered pharmaceuticals.

In a broader scope, pharmacology also involves drug discovery with identification and validation of new targets for drug action and the design, synthesis, composition, and property analyses, and development of new drugs to prevent, treat, and cure disease [16].

In the long run, pharmacology research is also a critical component in the development of modern "personalized medicine" or "precision medicine" [17–19].

Pharmacology is not synonymous with pharmacy and the two terms are frequently confused. Pharmacology is the science-oriented research field, whereas pharmacy is the direct-patient care practice [20]. Pharmacology deals with the research, discovery, and characterization of chemicals which show biological effects and the elucidation of cellular and organismal function in relation to these chemicals. In contrast, pharmacy, a health services profession, is concerned with the application of the principles learned from pharmacology in its clinical settings.

Polypharmacology is frequently referred to the design or use of pharmaceutical agents that act on multiple molecular targets or multiple signaling pathways for multifactorial human disease. Yet, such a definition is at its best incomplete and inaccurate. This is simply because polypharmacology has evolved and emerged as a new subdiscipline of pharmacology and a unique research field, which in addition to retaining all areas of general pharmacology, possesses its own theoretical basis and principles of modern pharmacology and practical approaches and methodologies for drug discovery and pharmaceutical treatment of human disease, based on its own philosophy—multitarget drug therapy and drug development [21–23]. This multitarget approach can come in different forms, combination of drugs, integration of drugs or a single drug with multiple ligands, all intended for multiple targets.

The term polypharmacology was initially coined to describe the phenomenon of unintended off-target effects of therapeutic drugs, which confers drugs with adverse actions and toxicities. It has now evolved to describe the ability of a drug to act on multiple molecular targets and thereby affect multiple biological processes pertaining to a single disease [24–27]. Even though polypharmacology is still used to refer to the fact that most drugs unintendedly bind to multiple proteins, its meaning has shifted recently to reflect exploitation of this characteristic for the design or use of pharmaceutical agents to simultaneously interact with multiple functionally related targets that act together to enhance the drug effectiveness in disease therapy and reduce the side effects. Polypharmacology is also used to identify the drug effects that differ functionally from its primary target to produce additional relevant effects—thus repurposing the drug for new effects [22, 28–31]. Polypharmacology is different from "drug promiscuity" which is typically characterized by a broad spectrum of biological activities directed against targets often not related to the specific disorder of interest, thus leading to a plethora of adverse reactions [26, 27]. Drug promiscuity is often responsible for the occurrence of drug toxicity at the preclinical stage.

Polypharmacology is a conceptual and technological innovation and advancement of Pharmacology, representing both opportunities and challenges in drug discovery and research. It is a more holistic discipline that looks at the use and development of therapeutic compounds from a systems perspective. Polypharmacology opens novel avenues to rationally design the next generation of more effective, but less toxic, therapeutic agents. Polypharmacology suggests that more effective drugs can be developed by specifically modulating multiple targets [21–23].

Polypharmacology has also revolutionized our comprehension and appreciation of drug targeting and efficacy and safety of therapeutic drugs from at least three layers. First, many of the highly effective therapeutic agents in the market are multitarget compounds although they were initially developed as single-target drugs. Second, the concept of multitarget drugs is not limited to a single disease but may be expanded to one drug acting on more than one disease entity or simultaneously acting on a disease and its comorbidity. Third, a multitarget drug does not mean nonselective actions; instead, it refers to "selectively nonselective" property, or it targets a right group of molecules for a specific disease entity or related disease entities. These novel conceptual advances point to the future direction of pharmacology as the guidance of new drug discovery and development.

1.2 The Historical Account of Polypharmacology

The approaches and methodologies for drug discovery and therapy have never been static, but continually more dynamic in response to accumulating knowledge about the pathological mechanisms and therapeutic targets, as well as increasing global morbidity and mortality rates caused by the so-called incurable diseases.

1.2.1 Traditional Pharmacology: The "Natural Multicomponents Medicine"

Traditional medicine (TM) has a long history and is often regarded as ancient medicine indigenous to different cultures, though it has nowadays been considered part of complementary and alternative medicine (CAM) [32–35]. In general, TM refers to health practices used in the maintenance of health as well as in the prevention, diagnosis, improvement, or treatment of physical and mental illness, which is based on experiences, skills, approaches, knowledge, theories, and beliefs, whether explicable or not. It addresses holistic and systems approaches and measures. Traditional drug therapy (TDT), as the main component of TM, incorporates plant, animal, and mineral-based medicines with spiritual therapies, manual techniques and exercises, applied singularly or in combination to treat, diagnose, and prevent illnesses or maintain well-being. In ancient times, traditional drug discovery was primarily and essentially derived from firsthand individual experience in reactions after intake of certain natural materials (e.g., herbs) selected based on field study and subsequent modifications (varying combinations of different natural products, different forms of preparations of natural products, etc.) and word of mouth or written records, in conjunction with certain degrees of personal intuition and hunch [36–39]. Li Shizhen, regarded as the "Saint of Medicine" and one of the greatest physicians and pharmacologists in Chinese history, is the best-exemplified figure of such a practice. Li risked his own life by testing the effect of hundreds of herbs on himself (in many cases by tasting poisonous herbs). His monumental medical work, "Bencao Gangmu" or "Compendium of Materia Medica," [39] is the most comprehensive medical book ever written about traditional Chinese medicine (TCM).

A single herbal formula or even a single herb contains multiple ingredients with innumerous chemical molecules and is used directly or in various forms or preparations [36–39]. The multiple components contained in an herbal formula or a single herb act on multiple targets through multiple mechanisms, exerting additive or synergistic and holistic or systematic advantages of therapeutic effects. While properly prepared and quality-controlled natural products may not be as effective in the treatment of human disease as the target-specific modern medicine, they generally have better safety profiles with less side effects and toxicities [40–46]. For example, Daming capsule (DMC), a combination of six herbs, is a multifunctional Chinese herbal medicine used as a lipid-lowering medicine to treat patients with hyperlipidemia in China, as well as other diseases including diabetes, cardiac and cerebral ischemia, arrhythmia, and atherosclerosis [46, 47]. A multicenter randomized clinical trial reported that DMC significantly decreases the serum levels of triglyceride and low-density lipoprotein cholesterol in patients with hypercholesterolemia, exhibiting a clinical efficacy comparable to pravastatin, good tolerance, and safety profile with minimal side-effects [48]. Analyses revealed that 53 compounds are absorbed into the circulation after oral administration of DMC, of which 6 anthraquinones were identified as the main bioactive components of DMC, namely emodin, emodin-*O*-glucoside, chrysophanol, rhein, aurantio-obtusin, and aloe-emodin, which largely account for the therapeutic efficacies of DMC [46].

It is generally believed that modern pharmaceuticals may not be within the reach of many people and "health for all" may only be realized with the help of TDT using natural products. Mankind has been using herbal medicine for healing and well-being from the beginning of human civilization. With the advent of "modern medicine," TDT has been looked down upon, especially by western societies because of the serious concerns regarding their safety, claimed efficacy, and quality [49, 50]. First, the mechanisms of action of TDT are largely unclear, despite that there has been much empirical evidence for their efficacy in various disorders. The lack of mechanistic understanding of mode of action of TDT creates a doubt on claimed efficacy and a constrain for its clinic application [51]. Second, the potential herb-drug interactions of multiple compositions are uncertain making the outcomes of herbal medicine less predictable as to whether it produces synergistic actions or amplified adverse side effects [51]? Third, although herbal medicine is generally considered safe due to their age-old usage, significant side effects have been reported for many herbal products, including herbal medicine, presumably due to the lack of standardized guidelines and measures for controlling and inspecting the quality of medicinal preparations [51]. Finally, accidental contamination and intentional adulteration are considered another issue associated with the side effects of herbal medicine [52].

Yet, since the past two decades, there has been a clear upward trend of popularity of TM as alternative medicine—natural products have revived interest in traditional remedies and use of herbal medicine for healthcare has gained its momentum of steady increase all over the world [49, 50]. This could be partially ascribed to the rapid development of alternative more acceptable forms of herbal medicine preparation by extracting, isolating, or even synthesizing the herbal medicinal products or other natural bioactive substances or nutraceuticals, instead of using the raw materials and crude extractions. The leaf extract of Ginkgo biloba for the treatment of cognitive impairment and dementia [53–55] and artemisinin, an endoperoxide sesquiterpene lactone isolated from *Artemisia annua L.*, as an antimalarial drug with anticancer property [56–58] are such two successful examples. The emerging of polypharmacology as a novel drug discovery strategy contributed and will continue to contribute to the boom of TDT as polypharmacology and TDT share the same philosophy—multitargets drug therapy.

However, for bringing more objectivity and to confirm traditional claims of the therapeutic effectiveness and safety profile, more systematic, well-designed animal and randomized clinical studies with sufficient sample sizes are necessary

[59–62]. Multidisciplinary basic research and preclinical studies are required to exploit the vast potential of herbs and other resources of TDT and to delineate their active compositions and mechanisms of action, as well as their potential adverse side effects.

1.2.2 Modern Pharmacology—Single-Target Drugs (STD): The "Single Agent on Single Target for Single Disease" Strategy

The advent of modern medicine (or western medicine) brought in research-based approaches to drug therapy and discovery with laboratory research and clinical trial replacing experience-based ones in traditional medicine. This was made possible owing to the ever-broadening and deepening of our understanding of pathology for the causes and mechanisms of human disease, physiology for the function and characteristics of therapeutic targets, and pharmacology for drug actions and metabolism. Before the 1980s, drug discovery based on modern pharmacology began by using animal models to test compounds isolated or synthesized by medicinal chemists. Drugs were deemed successful by virtue of their effects rather than the number of molecular targets to which they bound or acted. For every safe and effective promiscuous drug, there were good treatments that caused major side effects, and plenty of other drugs that were just plain unsafe. At the time, trying to predict side effects and understand them was almost impossible as in most cases no one knew exactly how the drugs worked.

The selective approach for drug discovery was made possible once biochemical and genetic studies began to reveal the molecular mechanisms that underlie common illnesses such as cancer and cardiovascular disease. Companies were able to pick a protein that they thought would make a good target, design compounds that interact with this protein, and test these compounds to find potential drugs. This revolution gave birth to the "single-target drug (STD)" or "one drug–one target–one disease" strategy. STDs were quickly accepted by researchers and pharmaceuticals as the most rational approach for drug discovery and drug therapy and became the common practice in modern medicine [63–65]. More specifically, STD is a concept of drug action and therapeutic application and an approach for drug discovery and development based on the design of highly selective chemical entities to target a single biological entity, considered as a dominant player in a certain pathology. The aims and advantages of STDs are primarily to simplify compound screening, reduce unwanted side effects, and simplify registration [63–65]. Using STD strategy, researchers have expected to avoid any undesired side effects and guarantee a more drug-like property of drug candidates.

Why and how did modern medicinal chemists come up with the concept of STD? The answer is the "Lock and Key" theory. This theory has been used as a doctrine for figuratively explaining and accurately modeling protein-ligand binding (e.g., receptor-agonist binding) mechanism. It is a useful starting point for finding new lead compounds or drug candidates and an important step in structure-based drug design. STD resorts to this traditional lock and key model to describe the interaction between a ligand and its molecular target (or an updated version of this paradigm that contemplates the ligand and target flexibility, such as the hand-in-glove analogy). The general idea is that the ligand (an agonist or the key) and the target (e.g., receptor or the lock) should have complementary features to efficiently interact and trigger some biological response: the agonist activates the receptor (the key turns the lock open) [66–70].

It is undeniable that the single-target drug strategy has had, over a long time, countless successes for the Pharma industry and the treatment of diseases, paving the current stage of evolution in therapeutics [71–76]. Since the isolation of quinine from Cinchona bark tree, in 1820, the first chemically purified effective treatment for malaria, several other natural and synthetic compounds have been developed. For example, drugs such as artemisinin and chloroquine have been traditionally used for the treatment of malaria. On the other hand, while insulin continues to be the most utilized form of treatment for type-1 diabetes, several oral hypoglycemic drugs are utilized in the management of type-2 diabetes. Carboplatin, adriamycin, and fluorouracil are among the most popular drugs for cancer treatment. Probably the best example of accurate, causal therapy is molecular-targeted therapy with the disease-causing molecular defect as a target of a drug [77–80]. A therapeutic drug of molecular-targeted therapy is often a small molecule drug that targets markers inside the cell or an antibody that attaches to specific targets on the outer surface of cells. Most molecularly targeted therapies are used in the treatment of cancer and this STD strategy is one of the cornerstones of personalized medicine [77–80].

Despite a massive progress in many scientific and technological developments in pharmaceutical research and considerable progress in genome- and proteome-based high-throughput screening methods and rational drug design over the past two decades, the number of new single-target drugs approved or successfully translated into clinical use has significantly declined or at least failed to increase appreciably. Moreover, despite massive investment on drug research and development (R&D) by the global biotechnology and pharmaceutical industries, STDs are frequently withdrawn from markets. The decline in pharmaceutical efficiency can be attributed to many factors, including lack of efficacy due to drug resistance and individual variation in treatment responses, as well as clinical safety or toxicology observed for the candidate drug compounds in preclinical or clinical studies in vivo, and the increase in withdrawal is primarily due to their side effects or toxicities. Additionally, highly selective or specific therapeutic STD developed based on the STD concept to direct to single molecular targets have proven to lack in efficacy, especially for complex human diseases [63–65]. Not surprisingly, this issue has rendered drug discovery increasingly inefficient [81]. Furthermore, STD has also been associated with the increasing incidences of drug resistance. In addition, STD might be of benefit or they might in contrast raise severe safety concerns due to their side effects or toxicities. For example, even the most specific agents for molecularly targeted therapy can result in various toxicities in patients [82–86].

Then, what is wrong with STD?

In theory, drugs are designed based on the STD concept with the aim of targeting a single biological entity, usually a protein (the so-called "on-target"), with high selectivity to avoid any unwanted effects arising from mis-targeting other biological targets ("off-targets"). In reality, however, drug molecules often unintendedly interact with multiple targets or the binding of a ligand to multiple protein targets (a promiscuous ligand), causing side effects or toxic actions. This is because the STD approach has some inherent limitations that could hardly be avoided. The possible explanations for this point are multiple. First, STDs are mostly nonselective or nonspecific agents due to drug promiscuity, because truly specific drugs are hard, if not impossible to produce. The promiscuity could elicit some unwanted effects or side effects or adverse effects. For example, a single agonist or antagonist binds to a specific membrane receptor; a single activator or inhibitor acts on a specific enzyme; a single blocker or opener on a specific transmembrane ion channel. On this basis, the concept of drugs interacting with multiple targets has long been flagged as undesirable, as it was inherently associated with adverse side effects. However, the complexity of the current incurable pathologies has clearly demonstrated that such single-target drugs are inadequate to achieve a therapeutic effect.

Second, even a nonpromiscuous ligand developed based on the STD strategy may activate different targets due to the following possibilities. (i) It is frequent for a given STD to act on several isoforms of the same protein. For instance, xilocaine (lidocaine) can produce anesthetic, antiarrhythmic, and anticonvulsant effects by blocking the peripheral nervous system, heart, and central nervous system sodium channels [87]. (ii) Alternatively, different members of a given biochemical pathway targeted by a given STD might share, to some extent, ligand specificity due to co-evolution. (iii) Finally, a given STD might display affinity to two or more unrelated targets by combining different pharmacophores in the same molecule leading to molecules that are either enthalpically or entropically unfavorable, which conspires against the specificity of action [88].

Third, except for monogenic diseases caused by mutation or elimination of specific genes, many other human diseases including cancer, metabolic disorders, cardiovascular disease, and CNS diseases are mostly multietiologic multifactorial physiological disorders with complex subnetworks within the interactome, referred to polypathology, and the STD approach oversimplifies disease mechanisms. Moreover, disease definitions are mostly symptom—rather than mechanism-based, and hence the therapeutics are likewise.

In contrast, network models suggest that partial inhibition of a surprisingly small number of targets can be more efficient than the complete inhibition of a single target, suggesting that systematic drug-design strategies should be directed against multiple targets. In this context, despite that chemical biology continues to be a reductionist discipline regarding chemical probes as highly selective small molecules enabling specific targeting, medicinal chemistry has searched for new approaches and tools or other alternatives capable of reaching more agility, security, and a more efficient direction in the design and prospection of drug candidates.

It should be noted that an STD is originally designed for a single specific target, but it often can act on multiple unwanted targets and elicit multiple undesirable effects or off-target effects or toxicities. Targeting multiple unwanted molecules is the major distinction from multitarget drugs to be introduced in a following subsection.

1.2.3 Modern Polypharmacology v1.0—Combination Drug Therapy (CDT): The "Multiple Agents on Multiple Targets for Single Disease" Strategy

"Drug Combination Therapy (CDT)" refers to the strategy of "Multiple Agents on Multiple Targets for Single Disease" and the combination of two or more drugs for co-administration or sequential administration (within a timeframe so that the combining drugs could exert their respective effects in an overlapping manner). The major distinction of CDT from other formats of polypharmacology is that it is a joining or merging of different parts or qualities in which the individual component elements are separate and distinct. With CDT, the doses of the combining drugs can be "freely" determined based on the specific applications; thus, CDT is also called free-dose combinations.

The single target-based STD, as a reductionistic paradigm, begins with the premise that one drug should be developed for one molecular target for one disease. Yet, 20 years down the road of drug discovery and development, this premise has been challenged by the proponents of network pharmacology, who argue that networks of signaling pathways are more relevant drug targets than single molecules. It turns out that the single target-based STD does not always guarantee success. Some of these selective drugs work in only a select population of patients. AstraZeneca's Iressa (gefitinib), for example, is designed to treat lung cancer by targeting the epidermal growth factor receptor. The drug does give an incredibly potent response, but only in about one-tenth of the patients who receive it [89]. On the other hand, Roth and his team found that the drugs that bound to the most receptors were the most successful in the clinic in their efforts to understand how antipsychotic drugs work by investigating into the nerve-cell receptors that are being targeted by a range of antipsychotic drugs [90–92]. The idea that promiscuous drugs might be more effective than targeted ones has also been emerging from that time being. The schizophrenia drug Clozaril (clozapine), for example, works well in patients because it targets a large number of receptor proteins [93]. Variations designed to bind to fewer targets and reduce the adverse side effects of Clozaril did not lead to any improvement. These and other similar observations are now not difficult to understand considering that human diseases are mostly, if not all, multifactorial and multimechanism entities and targeting one of the factors or mechanisms may yield only minimum effectiveness because cells have the mechanisms to compensate for a molecule whose gene expression and functional activity are affected by a drug, a phenomenon known as redundancy. One example of this concept is synthetic lethality, i.e., the fact that two individual gene deletions may not produce a phenotype, but their simultaneous deletion is lethal, suggesting that many signaling networks have redundancies that can compensate for loss of a single component, but not a loss of two components. This principle may be applied to drug discovery, where two drugs that are individually ineffective (because of redundancies in signaling networks) could be effective in combination.

Given the fact that STD drugs are less effective for multifactorial human diseases but more toxic than expected, alternative strategies of drug therapy were in urgent demand. CDT refers to a combination of multiple theoretically single-targeted drugs, which is also coined combinational drug therapy, multidrug combination therapy or multicompound medication (the association of multiple active principles in the same pharmaceutical formulation), and quickly emerged and came into being as a new approach alternative to STD. The concept of CDT implies the concurrent use of two or more therapeutic agents or CDT in the form of free-dose combination to treat a particular disease entity, and it has become a universally accepted fact in medical practice. Pharmacology has from this point ushered in a new era, the polypharmacology v1.0 era. It is interesting to note that traditional drug therapy (TDT) or natural multicomponents medicine is an analog approach to CDT; or in other words, CDT is the return of TDT. While CDT belongs to modern polypharmacology, TDT can be viewed as a traditional polypharmacology. CDT initiates the era of polypharmacological therapeutic strategies and modalities.

CDT is best exemplified by the HIV/AIDS treatment with drug cocktail, some types of cancer, tuberculosis, and hypertension [94–111]. The benefits of combinational antiretroviral therapy for HIV replication are obvious; Combinational antiretroviral therapy affords durable (likely lifelong) suppression of viral replication and effective control of viral transmission, optimized tolerability, and long-term safety. The success and benefits of CDT for the HIV/AIDS treatment have led to universal recommendation. CDT for the treatment of cancer is also becoming more popular because it generates synergistic anticancer effects, reduces individual drug-related toxicity, and suppresses multidrug resistance through different mechanisms of action. In recent years, nanotechnology-based combination drug delivery to tumor tissues has emerged as an effective strategy by overcoming many biological, biophysical, and biomedical barriers that the body stages against successful delivery of anticancer drugs. The sustained, controlled, and targeted delivery of CDT chemotherapeutic drugs enhances therapeutic anticancer effects with reduced drug-associated side effects. Combination treatment of hypertension has been introduced almost 50 years ago, and accumulating evidence supports CDT for hypertension with respect to efficacy, safety, and adherence to treatment. Clinical data indicate that three-drug combinations can control hypertension in about 90% of patients.

The advantages of CDT over STD may be conferred by a number of elements. First, combining drugs with complementary mechanisms of action can target multiple factors and mechanisms of a disease, which is expected to yield more complete and radical therapeutic outcomes. Second, the CDT approach allows for reduction of doses of each of the multiple agents so as to minimize adverse events and improve tolerability relative to the use of high doses of monotherapy. Third, CDT might generate synergistic or additive actions against a disease, amplifying the therapeutic effectiveness [112]. Fourth, CDT can mitigate drug resistance because the development of resistance to multiple drugs is much less likely. In fact, CDT is used as a tool to prevent emergence of bacterial resistance [113]. In addition, CDT also allows for personalized therapy with varying combinations of different agents, which can be predetermined and readjusted according to patients' conditions and dynamic disease progression.

According to the reasonings described above, CDT should be superior to STD monotherapy, providing better therapeutic effects without additional toxicity. Nevertheless, clinical outcome of CDT is not always as good as anticipated; rather is often associated with higher toxicities, although nonoverlapping toxicity is a major consideration for selecting drug combination candidates [114–118]. In addition to the unwanted escalation of toxicity, the success of CDT is hampered by inability of component drugs to achieve desired spatiotemporal distribution when they are delivered in free molecule form. The inefficiency and deleterious effects of CDT could be ascribed to undesirable drug–drug interactions, distinct pharmacokinetics of different agents in combination with differing bioavailability, differing times to therapeutic or peak effect, and differing metabolism routes and rates, different physiochemical characteristics such as solubility. A key challenge in the design of CDT is attaining a balanced activity of multiple agents at each target of interest while simultaneously achieving a wider selectivity (multiple targets) and a suitable pharmacokinetic profile.

The drawbacks of CDT have become a driving force for further exploitation of newer approaches for more effective and safer drug therapy of human disease, which let us enter a new era of Polypharmacology v2.0.

1.2.4 Modern Polypharmacology v2.0—Fixed-Dose Combination (FDC): The "Single Pill with Multiple Agents on Multiple Targets for Single Disease" Strategy

"Fixed-dose combination" follows the strategy of "single pill with multiple agents on multiple targets for a single disease." In this format, multiple drugs are mixed or blended together to form a single pill that can act on multiple therapeutic molecular targets pertinent to a specific disease entity. In other words, FDC refers to a single pill/tablet made by mixing or blending two or more preselected drugs with each at a predetermined dose, within which the individual drugs are no longer separable and distinguishable.

One of the problems associated with the free-dose combination or the DCS strategy is higher patient pill burden that can negatively affect adherence and persistence, which in turn can lead to poorer efficacy outcomes (most patients with cardiovascular disease do not have the adherence and persistence to take recommended medications long-term). One way to bypass this obstacle is to encapsulate multiple active pharmaceutical ingredients into a single pharmaceutical formulation to produce a final medicinal product, usually in pill form (i.e., tablet or capsule). Such a co-formulated drug product was initially coined "polypill" nearly two decades ago [119–121] and is nowadays also referred to single-pill combination (SPC) therapy [122–134]. Or occasionally, a synonym "combopill" is used. In this book, FDC will be used throughout as it is the most frequently used expression in the literature.

FDC drugs are commonly manufactured as mass-produced fixed-dose combination drug product for the treatment or prevention of chronic disease. They may be aimed to be consumed by healthy people as a means of preventive medicine, and/or treating actual pathophysiological condition(s), the former typically involving lower dosages than the latter. One of the advantages of the FDC strategy is that it can reduce the number of tablets or capsules that need to be taken, which in turn may facilitate handling and administration of pharmaceuticals as well as alleviate patient pill-burden. Often, the multidrugs in a given polypill might all be aimed at a single underlying condition (or, group of related conditions). In addition to the noted fixed-dose types of polypills, polypills can also be custom-made for specific patients through a process called pharmacy compounding.

FDC therapy could be traced back to hundreds of years ago, but the concept received a setback in the 1950s due to the "medical accidence" caused by the fixed-dose combination of potassium chloride (KCl) and diuretic for the treatment of hypertension and other illnesses as well. The idea behind the combination is simple and reasonable: the use of diuretics is to remove excessive sodium from the body, but these agents also tend to remove potassium simultaneously. One way to counter lost potassium is to be supplement potassium via KCl tablets. Unfortunately, the KCl in the diuretic/KCl combination is lying in the gut at one spot punctured the stomach lining, requiring surgery in many instances [133]. Subsequently, a review of the claims put forward for the use of penicillin-sulfonamides, penicillin-streptomycin, and certain other fixed combinations of antimicrobial agents has convinced five panels organized under the auspices of the National Academy of Sciences—National Research Council that such combinations are "ineffective as fixed-dose combinations" [134]. The authorities at the time believed that although the individual active ingredients may be useful in specific entities, no greater effectiveness can be expected for the combination than for any one ingredient, and the use of a proper dose of one ingredient may require excessive or inadequate doses of the other. The panels reached the judgment "the use of these fixed combinations should be discontinued and that the physician should use the individual components according to his best clinical judgment" [134]. Since then, "Fixed-dose Combination" has gained its bad reputation, and medical students were taught that prescribing fixed-dose combination drugs is a serious error!

Such a misconception was hold for some time until "Polypill" (an alternative term for fixed-dose drug combination) was first conceptualized by Wald and Law in the year 2003 [119]. They proposed that the polypill comprising of six drugs (aspirin, statin, beta-blocker, angiotensin-converting enzyme (ACE) inhibitor, diuretic, and folic acid) could reduce the burden of cardiovascular disease (CVD) by 80% when taken by all above the age of 55 years and those with CVD. The resurgence of FDC is also ascribed to its adventitious success for the treatment of AIDS/HIV and for oral contraceptives. Moreover, the era of evidence-based medicine has aided the stature of fixed-dose combination products. Since it is known up to 50% of hypertensive patients cannot be controlled by monotherapy [135], the concept of simultaneous multitherapy has gained momentum with the obvious advantages of combining ingredients into a single dosage form. For instance, angiotensin-converting enzyme inhibitor/calcium channel blocker combination has a superior therapeutic ratio, as a class, than either of their components used as monotherapies [136]. Indeed, the "Polypill" concept was based on research accumulated over years on the benefit of blood pressure lowering, cholesterol lowering, and antiplatelet drugs in reducing CVD morbidity and mortality [119, 133]. Today, multiple fixed-dose combination polypills are available with differing active components, applications (primary vs. secondary prevention), and doses.

One of the first recommended applications of FDC is to patients with hypertension [122–131], heart disease [132, 133, 137], ischemic stroke [138], type 2 diabetes mellitus [139–141], and HIV [142, 143]. The polypill makes prescribing easier by reducing the need for complex treatment algorithms and dose titration. It also appears to be cost-effective. The challenge today is to produce polypills containing drugs with established efficacy and complementary actions. However, there are several issues that need to be addressed before the polypill can be used routinely. A single polypill formulation may not be suitable for all patients. It may be necessary to develop several types of polypill to meet the needs of different patient groups. With respect to clinical outcomes, Machnicki et al. showed a significant advantage of FDC treatment regarding hospitalizations, emergency room visits, and hypertension-related emergency room visits with the positive effects on adherence and persistence being translated into clinical outcome and effectiveness parameters as well [128]. These findings are supported by a recently published study by Verma et al. [129].

An FDC containing a fixed dose of aspirin, a statin, and one or two antihypertensives has enormous potential in treating and preventing cardiovascular disease based on the results of relevant clinical trials [144–146]. Yet, though holding a great promise, no drugs based on the fixed-dose FDC principle has been widely marketed anywhere in the world to date, and the full benefits of such a pill remain unclear until it can be put to the test in more large-scale clinical trials.

1.2.5 Modern Polypharmacology v3.0—Multitarget Drugs (MTD): The "Single Agent on Multiple Targets for Single Disease" Strategy

MTD refers to a single agent that is naturally formed or rationally designed to contain multiple (more than one) binding sites toward, or multiple pharmacophores for, multiple disease-related therapeutic targets. It follows the "Single Agent on Multiple Targets for Single Disease" strategy. MTD is relative to STD introduced in a previous subsection. The major distinction of MTD from other formats of polypharmacology is that it integrates two or more target-binding motifs or sites or pharmacophores (a pharmacophore refers to a part of a molecular structure that is responsible for a particular biological or pharmacological interaction) into one by fusion or hybridization.

Despite that the employment of FDC/Polypill can help relieve the issue of pill-taking burden associated with CDT and offers improved adherence and persistence thereby the therapeutic outcomes in patients, it retains the inherent drawbacks of the latter. That is, both CDT and FDC/Polypill are associated with undesirable drug–drug interactions and different bioavailability, which often render compromised therapeutic efficacy and unexpected (even unpredictable) adverse effects, due to distinct physiochemical characteristics and pharmacokinetics of multiple agents in combination. Apparently, an upgraded strategy is in urgent need for better drug therapy. Polypharmacology v3.0—the single-agent multitargets drugs (MTD) approach—comes into play and quickly becomes a new generation of drug therapy and discovery strategy and an overwhelmingly hot research field.

MTD-SD is a "single-agent—multiple targets—one disease" paradigm for pharmacological therapy and drug development, which is the cornerstone of the modern polypharmacology [21, 24, 25, 88, 147–152]. It aims to afford manifold effects with multitarget treatments of multifactorial diseases. The key to successful MTD therapy is the designed or targeted multitarget single agent that "nonselectively" acts on selected multiple key and sensitive nodes (molecules) belonging to a network of interacting targets pertinent to a specific pathological entity. The philosophy behind such a selectively nonselective MTD strategy is that complex diseases such as cancer, cardiovascular disease, and central nervous system diseases may require complex therapeutic interventions [24, 25, 88]. In this respect, a drug that "hits" multiple related targets offers the potential for higher efficacy and may limit drawbacks generally arising from the use of STD, CDT, or FDC. The MTD strategy advocates the combination, in a single scaffold, of different structural subunits that allow the molecular recognition by more than one bioreceptor or bioeffector, acting in a simultaneous mode in multiple targets interconnected to biochemical networks responsible for the pathophysiology of multifactorial diseases. Most likely, a multitarget drug will be one that produces moderate influences on several targets, either positively or negatively, at the appropriate concentrations. In other words, a balanced modulation of several related targets in a disease network can provide a superior therapeutic effect to side effect profile compared to the use of a selective ligand.

In this sense, drug "promiscuity" can be a virtue. In theory, the MTD strategy is superior to any other approaches with greater therapeutic efficacy and fewer/lower adverse or off-target side effects. According to the "Lock and Key" concept, a multitarget drug/ligand might be conceived as a skeleton or master key capable of unlocking several locks (target receptors or effector molecules). Additional advantages of MTD multitarget drugs vs CDT/FDC drug combinations are the more predictable and simpler pharmacokinetic and pharmacodynamic relationship of a single medicine should improve the bioavailability to yield the greater efficacy against advanced-stage diseases with the lower toxicities and better patient compliance [24, 88, 153]. To benefit from those effects, it is required that the multitarget drug exhibits balanced in vitro and in vivo activities to match potency for the corresponding targets, as well as optimized pharmacokinetic and safety profiles. A combination of drugs (CDT or FDC) faces the problem of the different solubilities that may modify the bloodstream uptake, which requires fine-tuning the formulation in order to ensure the required blood level of each drug. MTD is therefore the ultimate solution for these issues. Moreover, MTD is associated with better adherence and persistence of medication taking by patients than CDT, which can be translated into better clinical outcomes and effectiveness parameters. Furthermore, the regulatory requirements are more complex when the agents are used in combination, since the safety profile of each drug needs to be demonstrated before clinical trials, and this can be further delayed due to regulatory and intellectual property issues, in particular if the two drugs are being developed by different companies. The MTD strategy is a solution to such a problem.

The first virtual MTD strategy can be dated back to 1998 when a few pioneering groups reported the possibility of efficiently exploiting the multitarget profiles of small organic molecules [154]. These researchers demonstrated that purposely designed compounds could be endowed with a multitarget mechanism of action. Nevertheless, the primordial idea of Polypharmacology v3.0 was first proposed in 2004 by Roth et al. [24]. The authors reasoned that most common central nervous system disorders are polygenic in origin and attempts to develop more effective treatments for diseases such as schizophrenia and depression by discovering drugs selective for single molecular targets ("magic bullets") have been largely unsuccessful. They proposed a proof of concept that designing selectively nonselective drugs (that is, "magic shotguns") that interact with several molecular targets will lead to new and more effective medications for a variety of central nervous system disorders. A similar concept was independently proposed in the year of 2006 by Wang's research team [25] who used the term "one drug, multiple targets" or "single agent–multiple targets" (MTD) to describe the same principle as "magic shotguns," and his research provided the very first experimental evidence for the feasibility, effectiveness, and advantages of epigenetic polypharmacology [147], specifically the "complex decoy oligodeoxynucleotides technology cdODN" attacking multiple target transcription factors, in the treatment of xenograft breast cancer in mice. Subsequently, Wang's team extended the MTD to designing single agent that can act on multiple miRNAs targeting cancer cells and cardiac pacemaker channel genes and calcium channel genes as a new therapeutic approach [147–151]. In 2008, Shokat and his colleagues described a single compound that blocks the proliferation of tumor cells by direct inhibition of oncogenic tyrosine kinases and phosphatidylinositol-3-OH kinases and termed it "multitargeted drug" along with the concept of "Targeted Polypharmacology" [152]. Since then, the MTD drug discovery and e-related research fields have been growing at a remarkable pace, offering new paradigms with the potential to overcome some of the major limitations of classic STD strategy and CDT and FDC approaches as well, and polypharmacology has become a new branch of pharmacology discipline and research field as well as one of the new directions and strategies for drug development [21]. MTD, or the "magic shotguns" concept, has triggered a recent shift in efforts to create drugs that hit more than one target simultaneously by purposely engineering promiscuous drugs for the treatment of polygenic or polypathologic diseases with complex pathomechanisms [153].

1.2.6 Modern Polypharmacology v3.1—Multitarget Drugs (MTD): The "Single Agent on Multiple Targets for Multiple Diseases" Strategy (MTD-MD)

Drug repurposing, also known as drug repositioning, drug reprofiling, drug redirecting, retasking, drug rediscovering, or rescue of drugs, is defined as developing new uses for an existing drug (including approved or investigational drugs) beyond its original use or initial approved indication [155–160]. In other words, drug repurposing aims to uncover the potential multitargeting for multiple diseases or to identify the good side of a-coin-of-two-sides drug promiscuity [161, 162]. Thus, the MTD-MD approach for drug repurposing to expand the therapeutic spectrum of an existing drug and extract the therapeutic element from adverse effects of a promiscuous drug holds the potential to complement traditional drug discovery and offers various advantages over developing an entirely new drug for a given indication.

First, the whole process of the traditional or current model of new drug discovery and development is in general substantially monetary costly, research- and labor demanding, time-consuming with slow pace, and highly risky with high attrition rates or failure rate of ~45% associated with safety or toxicity issues [157]. These facts imply not only a high investment by the pharmaceutical industry, but also ultimately high drug prices to patient customers. The rediscovery of a generic drug is also a challenging pursuit because there is no formal regulatory approach and a lack of economic interest by pharmaceutical companies. The MTD-MD approach of drug repositioning confers an attractive prospect for drug developers and patients alike because it was developed to bypass much of the costs and timelines involved with bringing a new drug to market saving of up to 5–7 years in average drug development time and mitigate the safety risk by using de-risked compounds, particularly in those cases where preclinical safety studies have already been completed [155–160]. Thus, MTD-MD drug repositioning has attracted increasing attention in recent years as drug companies seek potentially inexpensive alternatives to compensate for the high costs and disappointing success rate associated with the drug discovery pipeline.

Second, drug repositioning helps find "old" drugs to treat rare or orphan diseases in addition to common ailments [163–168]. While nearly 8000 rare diseases have been identified in humans to date, only 5% have licensed treatments. As most of these rare diseases are life-threatening, it underscores the urgent need for new drugs. The MTD-MD approach for drug repurposing consists in identifying new uses for approved or investigational drugs that are outside the scope of the original medical indication. It represents an opportunity for rare diseases and patients with unmet needs. It is an alternative option in drug development and is often presented as being a viable, risk-managed strategy for pharmaceutical companies developing orphan drugs [163–168].

Third, since the beginning of the novel coronavirus (SARS-CoV-2) disease outbreak, there has been a rapidly increasing interest in finding a potential therapeutic agent for the disease. However, outbreaks of novel emerging infections such as coronavirus disease 2019 (COVID-19) have unique challenges in front of the health professionals to select appropriate therapeutics/pharmacological treatments in the clinical setup with very little time available for the new drug discovery, and there are currently no treatment options available for the deadly contagious disease, COVID-19. In response to such an emergent crisis, the MTD-MD approach for drug repositioning has quickly and timely become an overwhelming strategy for discovering or rediscovering drugs for fighting COVID-19 [169–173]. Using the basic knowledge of viral pathogenesis and pharmacodynamics of drugs as well as using computational tools, many drugs are currently in pipeline to be repurposed. In the current scenario, repositioning of the drugs could be considered the new avenue for the treatment of COVID-19. Number of drugs such as remdesivir, favipiravir, ribavirin, lopinavir, ritonavir, darunavir, arbidol, chloroquine, hydroxychloroquine, tocilizumab, and interferons have shown inhibitory effects against the SARS-CoV2 in-vitro as well as in clinical conditions [171]. These drugs either act through virus-related targets such as RNA genome, polypeptide packing, and uptake pathways or target host-related pathways involving angiotensin-converting enzyme-2 (ACE2) receptors and inflammatory pathways.

Paradigm shifts in science are often coupled to technological advances. Three specific approaches have been widely used in drug repositioning: computational approaches, biological experimental approaches, and mixed approaches [168, 173–176]. Data such as gene expression, drug-target interactions, protein networks, electronic health records, clinical trial reports, and drug adverse event reports have become accessible in standardized forms. The repository of knowledge and omics data available in pharmaceutical research leads to the rise of some computational methods which are novel and exciting in the field of drug repositioning. These computational methods are capable to make a high-level integration of all the knowledge and data and help in understanding the new signaling pathways and generate novel insights into drug mechanisms, side effects, and interactions which further speed up drug discovery. A recent study presented integrative network-based systems pharmacology, methodology which quantifies the interplay between the coronavirus–human cell interactome and drug targets in the human protein–protein interaction network which help in rapid identification of repurposable drugs against SARS-CoV2. A study using this kind of approach was able to identify 30 potential repurposable drugs against COVID-19 [174, 175].

In a bibliometric review of drug repurposing by scanning >25 million papers in PubMed and using text-mining methods to gather, count, and analyze chemical-disease therapeutic relationships [177], the authors found that >60% of the ~35,000 drugs or drug candidates identified in their study have been tried in more than one disease, including 189 drugs that have been tried in >300 diseases each. Whereas in the majority of cases, these drugs were applied in therapeutic areas close to their original use, there have been striking, and perhaps instructive, successful attempts of drug repurposing for unexpected, novel therapeutic areas. There have been notable successes for drug repurposing, though repurposing does not always succeed. The MTD-MD drug repositioning approach is considered as a very effective strategy for drug discovery as it involves less time and cost to find a therapeutic agent in comparison to the de novo drug discovery process. Especially, given the estimation that 75% of known drugs could be repositioned for various diseases [178], the MTD-MD drug repositioning approach is of great promising and attractive. While some pharmaceutical companies now consider exploring repositioning alternatives for drugs in the pipeline, the MTD-MD approach could be taken much further, by designing multipurpose drugs to treat different conditions; prominently, frequently comorbid disorders (e.g., diabetes and cardiac disease; anxiety and peptic ulcer disease, epilepsy, and depression) or, alternatively, underlying pathologies plus disease symptoms.

Collectively, while both STD and MTD belong to the single-drug strategy, the former presumably acts on a single therapeutic target whereas the latter binds to multiple molecular targets related to a specific disease entity thereby being an integration of multiple pharmacophores compounded into a single molecule. Both CDT and FDC contain two or more different therapeutics, but CDT is a combination or co-administration of multiple separate drugs to acquire multitargeting capability, whereas FDC is a single pill or tablet with mixture of two or more drug components that are no longer separable. By comparison, traditional medicine (TM) could be a mix of CDT, FDC, and MTD; for example, herbal medicine contains many phytochemicals/natural products (like CDT), and some of these phytochemicals might be naturally formed STDs or MTDs and some might be the mix of multiple difficult-to-separate bioactive components.

Intriguingly, the human medicine began with traditional pharmacology based on natural multicomponent multitarget medication, then evolved into sophisticated modern pharmacology advocating single-target drug therapy, and ends up with polypharmacology back to multitarget concept but rationally designed. Figure 1.1 summarizes the evolution of pharmacology in regard to the targeting concepts.

1.3 The Principles of Polypharmacology

The concept of polypharmacology involves the interaction of drug molecules with multiple targets, which may interfere with a single or multiple disease-causing signaling pathways.

Polypharmacology refers to the ability of drugs to interact simultaneously and specifically with multiple targets (multitarget drugs). Although polypharmacology might be associated with compound promiscuity, it should be considered a different category since the so-called promiscuous drugs are instead a class of compound that show a wide spectrum of biological activities and adverse reactions [15].

The principles of polypharmacology predict that phenotypic screening should have certain advantages over target-based screening. First, the best phenotypic screens have the potential to elucidate the broad spectrum of biochemical interactions and biological effects elicited by a compound. Second, network pharmacology reintroduces an old, but intuitively obvious concept: that the broad biological profile of a drug is more important than any particular molecular interaction.

1.3.1 Designed Multitargeticity Is the Core of Polypharmacology

Multitargeticity, referring to the capability of a given drug to exhibit activity with respect to multiple molecular (binding site) targets as already described in previous sections [15, 22, 26–30], is the most fundamental element and core connotation of polypharmacology [26, 27, 179, 180]. Or in other words, by connotation the term "polypharmacology" was originally derived from drug "multitargeticity."

The "multitargeticity" includes two different categories: designed multitargeticity and unintended multitargeticity, both of which are within the scope of polypharmacology for drug research

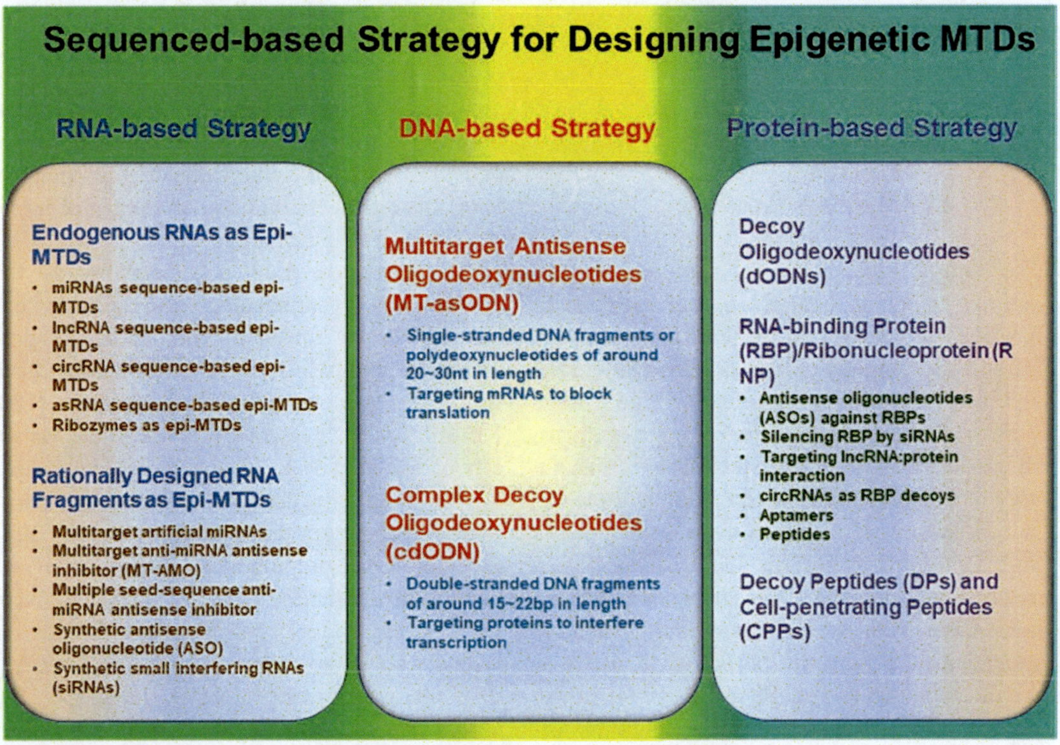

Fig. 1.1 Diagram showing the evolution of pharmacology from traditional pharmacology to modern pharmacology of various stages in regard to therapeutic targeting. Arrows indicate the direction of development and lines represent inclusion

and development. The former is about the discovery and development of drugs as more effective and safer therapeutics, whereas the latter aims to identify and characterize drug promiscuity that essentially forms the source of off-target side effects or beneficial actions (such as the naturally occurring products and the synthetic compounds that happen to give rise to net beneficial effects). According to these definitions, designed multitargeticity is a highly desirable property of therapeutics that polypharmacology aims to achieve and unintended multitargeticity or nondesigned promiscuity is an unwanted attribute for drug treatment of disease, which polypharmacology is applied to avoid and eliminate. Strictly speaking, central to polypharmacology is designed multitargeticity. The designed multitargets are not necessarily in the same functional class or the same signaling pathway but are normally for a specific disease or pathologic entity. In fact, many of the highly effective therapeutic agents in the market are multitarget compounds although they were initially developed as single-target drugs.

1.3.2 Designed Polyspecificity Is the Essence of Polypharmacology

The term polyspecificity was primarily used to describe antibody recognition, and the concept has been around for more than three decades [181, 182]. Antibody polyspecificity refers to the ability of an antibody molecule to bind a large variety of diverse epitopes in different antigens, which is analogous to drug promiscuity, a single drug with the capability of acting on multiple targets. Likewise, if an antibody is a therapeutic drug, then antibody polyspecificity can be viewed as drug polyspecificity. On the other hand, the connotation of the term polyspecificity was later expanded to explain the phenomenon of target promiscuity; that is, a single molecular target can bind or interact with multiple structurally dissimilar drugs (or ligands, or compounds). Thus, polyspecificity can either refers to drug polyspecificity or target polyspecificity, depending on specific context.

"Drug polyspecificity" implies the designed "selectively nonselective" property or combinatorial selectivity, of drugs with a single agent being able to act on multiple targets related to a disease entity, whereas "target polyspecificity" refers to a single molecular target that can be modulated by multiple unrelated drugs. The concept of "selective nonselectivity" for multiple targets or "drug polyspecificity" with combinatorial selectivity was initially proposed by Roth et al. [24] and vividly described as "magic shotguns" to describe the ability of a drug to engage in specific interactions with multiple targets from the polypharmacology perspective. This concept is also expressed as "target set-based promiscuity," "meaningful promiscuity," "a specific multitarget profile," or "pseudo-specificity" [162, 180, 183–188]. Kinase inhibitors may be the best analog to antibody in terms of their drug polyspecificity rationalized as a form of "selectively nonselective" interactions or kinases to antigens for target polyspecificity [183]. Thus, the key point that distinguishes drug polyspecificity from drug multitargeticity is that drug polyspecificity emphasizes the specificity of interactions between a given drug and its multiple targets related to a disease entity, while drug multitargeticity stresses the multiplicity or number of target molecules for a given drug.

The designed drug polyspecificity is the essence of polypharmacology because it is essential for drugs to elicit polypharmacological effects in vivo, which are of critical importance for the treatment of multifactorial diseases [180, 183–187]. The "selectively nonselective" property of designed drug polyspecificity implies that a given multitarget drug is not one that nonselectively acts on extensive targets; instead, it is expected to act only on a set of selected targets that are not limited to closely related molecules from the same family but extends to distantly or virtually unrelated molecules that are all relevant to a disease of concern [180, 184–186]. The design of multitarget drugs with predefined activities has become a topical issue in pharmaceutical research. Insights into the molecular origins of drug promiscuity would aid in the design of small molecules with predefined multitarget activities for polypharmacology [162, 188–196]. Thus far, however, such a design is mainly driven by combining pharmacophore elements of known single-target compounds, also taking target (binding site) knowledge into account, and to a lesser extent by machine learning [196].

In a word, while drug multitargeticity is a prerequisite of polypharmacology, drug polyspecificity is a warranty for the efficacy and safety of multitarget drugs, because multitargeticity in polypharmacology is based upon the ability of drugs to form "specifically nonspecific" interactions with different targets.

Therefore, polypharmacology can be redefined as the ability of drugs to interact simultaneously (multitargeticity) and specifically (polyspecificity) with multiple targets (multitarget drugs).

1.3.3 Polypharmacology Is Network-Based Pharmacology

A system is defined as an entity which maintains its existence through the interactions between its parts. In systems biology, a system is commonly described as a network of nodes representing functional elements or biological factors (such as biomolecules, diseases, phenotypes, etc.) connected by "edges" describing their relationships or functional interactions (physical interactions, shared metabolic pathway, shared gene, shared trait, etc.) [197]. Systems or network biology, a recent trend in bioscience research which focuses on the complex interactions in biological systems from a holistic perspective, rather than altering the single molecular component [198, 199], has proven useful for deciphering fundamental research questions, such as how perturbations in the cellular networks lead to certain phenotypes, including human diseases [200–205]. Biological networks, such as protein-protein interactions and metabolic pathways, is built upon the recent development of network theory [206], the organizing principles of which can be comprehensively analyzed by representing systems as complex networks, which are collections of nodes linked together by a particular relationship.

More recent field of network medicine aims at applying network science toward identifying, preventing, and treating diseases, and is a medically focused area of systems biology. The term "network medicine" was coined and popularized in a scientific article by Albert-László Barabási entitled "Network medicine – From obesity to the "diseasome" [207]. This field focuses on using network topology and network dynamics toward identifying diseases and developing medical drugs. Thus, the network medicine is based on the idea that understanding complexity of gene regulation, metabolic reactions, and protein-protein interactions and that representing these as complex networks will shed light on the causes and mechanisms of diseases. Four key networks for understanding human disease include the molecular network, metabolic network, the disease network, and the social network. (a) The molecular network refers to the whole set of molecular interactions in the human cell, also known as the interactome, and is being used for disease identification and prevention [208]. These networks have been technically classified as scale-free, disassortative, small-world networks, having a high betweenness centrality [209]. Using interactome networks, one can discover and classify diseases, as well as develop treatments through knowledge of its associations and their role in the networks. (b) The metabolic network is the complete set of metabolic and physical processes and pathways or biochemical reactions that determine the physiological and biochemical properties of a cell [210]. As such, these networks comprise the chemical reactions of metabolism, the metabolic pathways, as well as the regulatory interactions that guide these reactions. With the sequencing of complete genomes, it is now possible to reconstruct the network of biochemical reactions in many organisms, from bacteria to human. Several of these networks are available online: Kyoto Encyclopedia of Genes and Genomes (KEGG), EcoCyc, BioCyc and metaTIGER. Metabolic networks are powerful tools for studying and modeling metabolism. Metabolic networks can be used to detect comorbidity patterns in diseased patients and to determine if two disorders are connected due to their correlated reactions. Researchers have used databases such as KEGG to map these networks. Other networks include cell signaling networks, gene regulatory networks, and RNA networks. (c) The human disease network, also called the diseasome, is a network in which the nodes are diseases and the links, the strength of correlation between them and are used to map relationships between diseases and biological factors, also play an important role in the field. This correlation is commonly quantified based on associated cellular components that two diseases share. The human disease network is also a network of human disorders and diseases with reference to their genetic origins or other features or the map of human disease associations referring mostly to disease genes [211]. The first-published human disease network looked at genes, finding that many of the disease-associated genes are nonessential genes, but are the genes that do not completely disrupt the network and are able to be passed down generations [212]. Metabolic disease networks, in which two diseases are connected by a shared metabolite or metabolic pathway, have also been extensively studied and are especially relevant in the case of metabolic disorders [213]. (d) The social networks and transportation networks can be used to study epidemiology by modeling the spreading of disease across, e.g., the role of social ties in the spread of obesity in populations [214].

Network pharmacology is a developing field based in systems pharmacology that looks at the effect of drugs on biological networks, especially the interactome and the diseasome [14, 215–229]. The drug-target network can play an important role in understanding the mechanisms of action of approved and experimental drugs [217]. Network-based studies are becoming an increasingly important tool in understanding the relationships between drug action and disease susceptibility genes. Systems pharmacology is an emerging area of pharmacology which utilizes network analysis of drug action as one of its approaches. Systems pharmacology can provide new approaches for drug discovery for complex diseases. The integrated approach used in systems pharmacology can allow for drug action to be considered in the context of the whole genome. The network theory view of pharmaceuticals is based on the effect of the drug in the interactome, especially the region that the drug target occupies. The concept of disease modules can be used to aid in drug discovery, drug design, and the development of biomarkers for disease detection [206]. There are a variety of ways for drug discovery with network pharmacology; a simple example of this is the "guilt by association" method, which implies if two diseases are treated by the same drug, a drug that treats one disease may treat the other [217]. Drug repurposing, drug-drug interactions, and drug side-effects have also been studied in this field. In such network context, the efficacy of multitarget therapy can be understood from a robustness of disease networks to deal with single node perturbations, due to inherent diversity and redundancy of compensatory signaling pathways that result in highly resilient network architecture with modular and interconnected topology [212]. Therefore, network-based drug discovery seeks for drug target combinations to perturb a specific subset of nodes in the disease-associated networks to inhibit the bypass mechanisms at systems level [218]. These recent developments have resulted in moving away from the traditional "one target" strategy toward subnetwork targets and systems or network pharmacology, a novel paradigm with potential to provide more global understanding of the mechanism behind drug action, resistance, and side effects by considering drugs and their targets in the context of biological networks and interconnected pathways [206, 215–229].

Classical pharmacology considers a single transduction pathway, connecting the processes on the causal path between drug administration and response, as the basis of drug action. Recently systems pharmacology has been introduced as a novel approach to predict in vivo drug effects, in which biological networks rather than single transduction pathways are considered as the basis of drug action and disease progression. This is because the majority of diseases are associated with alterations in multiple molecular pathways and complex interactions at the cellular and organ levels, and single-target monotherapies therefore have intrinsic limitations with respect to their maximum therapeutic benefits.

Network biology has revealed that the deletion of individual nodes has little effect on the disease networks [228, 230–239]. The majority of diseases are associated with alterations in multiple molecular pathways and complex interactions at the cellular and organ levels. Single-target monotherapies therefore have intrinsic limitations with respect to their maximum therapeutic benefits. In drug discovery, network-based approaches are expected to spotlight our understanding of drug action across multiple layers of information. On one hand, network pharmacology considers the drug response in the context of a cellular or phenotypic network. On the other hand, a chemical-based network is a promising alternative for characterizing the chemical space. Both can provide complementary support for the development of rational drug design and better knowledge of the mechanisms underlying the multiple actions of drugs.

Systems pharmacology is an emerging field that uses both experiments and computation to develop an understanding of drug action across multiple scales of complexity ranging from molecular and cellular levels to tissue and organism levels. Systems pharmacology is the application of systems biology principles to the field of pharmacology. It seeks to understand how drugs affect the human body as a single complex biological system. By integrating multifaceted approaches, systems pharmacology can provide mechanistic understanding of both the therapeutic and adverse effects of drugs. This includes understanding of how drugs act in different tissues and cell types, as well as the issues of multiple actions within a single cell type due to the presence of several interacting pathways. Such studies are important from a translational perspective because they help identify new drug targets, predict adverse events, and can improve the safety and efficacy of existing drugs. The application of computational and experimental systems biology approaches to pharmacology allows us to expand the definition of systems pharmacology to include network analyses at multiple scales of biological organization and to explain both therapeutic and adverse effects of drugs. Systems pharmacology analyses rely on experimental "omics" technologies that are capable of measuring changes in large numbers of variables, often at a genome-wide level, to build networks for analyzing drug action. A major use of omics technologies is to relate the genomic status of an individual to the therapeutic efficacy of a drug of interest. Combining pathway and network analyses, pharmacokinetic and pharmacodynamic models, and a knowledge of polymorphisms in the genome will enable the development of predictive models of therapeutic efficacy.

1.3.4 Polypharmacology Is Constellation-Directed Pharmacology

The central concept of Constellation Pharmacology is to differentiate between cell types by their cell-specific constellations of key signaling proteins, using a set of pharmacological agents, including both agonists and antagonists. Constellation pharmacology is a novel cell-based high-content phenotypic screening platform that not only interrogates numerous molecular mechanisms simultaneously, but also ties phenotypes elicited by pharmacological compounds to molecular targets and mechanisms in native cell types.

As defined by Teichert et al. [240–242], a constellation refers to a group of functionally linked ion channels, receptors, or other signaling proteins that coordinate to elicit a physiological function. For example, for skeletal muscle contraction, a neuromuscular signaling must be completed by a constellation of ion channels and receptors including voltage-gated calcium channels at the presynaptic terminus, the postsynaptic nicotinic acetylcholine receptor (nAChR), and voltage-gated sodium channels in muscles. Specifically, voltage-gated calcium channels at the presynaptic terminus must open before neurotransmitter is released, which then elicits opening of the postsynaptic nAChR. This in turn causes depolarization at the postsynaptic terminus, which triggers activation of voltage-gated sodium channels resulting in the muscle action potential that ultimately triggers muscle contraction. Inhibiting any one of these sequentially acting signaling components can cause an inhibition of muscle contraction. Most importantly, these receptors and ion channels are functionally linked to each other and comprise the target "constellation" of the motor cabal.

While the network biology/medicine/pharmacology primarily weights on intracellular signaling pathways or networks, constellation pharmacology pays a greater attention to intercellular signaling pathways or networks. With skeletal muscle contraction as an example again, the whole signaling pathway involves one ion channel protein and one receptor protein expressed in two neuronal subclasses and another ion channel protein in one muscle fiber.

In order to extend the concepts of cabals and constellations to drug discovery, the first step is to define cell-specific constellations. The cell, as the fundamental unit of life, is also the first level of significant integration of molecular components in biological systems. A specific cell type, with a specific physiological function, is therefore a potential entity for drug targeting, and therefore also a meaningful unit for phenotypic screening. Differences in cell-specific gene expression create distinct cell types with divergent physiological properties, functions, and roles. In neurons, the ion channels and receptors expressed within a particular neuronal cell type are functionally linked to each other in order to integrate signaling inputs that result in a physiologically relevant output, such as action-potential firing and neurotransmitter release. The function of each cell-specific constellation is to produce the integrated physiological signaling of that neuronal cell type. Therefore, constellation pharmacology is a term used to describe the approach of uncovering constellations (combinations) of ion channels and receptors found in different cell types that are revealed by the application of target selective pharmacology. This approach is based on the principle that different cell types uniquely express cell-specific constellations of ion channels and receptors that define their functional phenotypes. Constellation pharmacology aims at uncovering the combinations of ion channels and receptors that are unique to each cell type. In addition, it has the potential to comprehensively elucidate the cell-specific constellations of every cell type in a heterogeneous cell population and to evaluate changes that occur in cell-specific constellations as a function of disease progression. As a screening platform, constellation pharmacology can be used to screen and deconvolute complex biological mixtures (e.g., venoms or bacterial extracts) or synthetic small molecules in order to identify compounds that exhibit unique molecular and cellular targeting-selectivity profiles [240–247].

However, it should be noted that the concept of constellation pharmacology has not been commonly accepted [248]. Though by the multitargeticity and polyspecificity principles, constellation pharmacology can be considered organic part of polypharmacology, its potential application to drug discovery awaits more rigorous investigation and validation.

1.3.5 Holistic and Integrative Approach Is the Future of Polypharmacology

As an ancient theme concept, "holism" has resurfaced in recent literature and become the new catchwords both in academic circles and in the general public. Holism is the belief that various systems (e.g., physical, biological, cellular) should be viewed as wholes, not merely as a collection of parts because parts of a whole are in intimate interconnection, such that they cannot exist independently of the whole, or cannot be understood without reference to the whole [249, 250]. The exact meaning of "holism" depends on context. However, as a common usage, "holism" usually refers to the idea that a whole is greater than the sum of its parts [251]. In this sense, "holism" may also be spelled "wholism" (although the two are not etymologically related), and it may be contrasted with reductionism.

The rise of molecular biology and genetics in the late twentieth century shifted the focus from the anatomical to the molecular, nudging the underlying philosophy from holism to reductionism. Developmental biology is currently experiencing yet another transformation back to a holistic approach, which was triggered by "-omics" technology and propelled forward by CRISPR genome editing. Emergence of the epigenetic landscape as an "antireductionist systemic view of the organism" emphasizes the complex interrelatedness. Reductionist and holistic science are not mutually exclusive; rather, the findings derived from each methodology are complementary [252–254]. The most complete understanding of a system as complex as the development of multicellular organisms will best be achieved by merging the two philosophies.

While the holistic view of polypharmacology might be rooted in ancient philosophy, it is based on holistic medicine, a whole-body approach to healthcare aiming to improve health and wellness through the body, mind, and soul [255–258]. Holistic medicine combines conventional treatments guided by modern medicine with traditional medicine and complementary and alternative medicine, and so will do polypharmacology.

Holistic polypharmacology resembles systems pharmacology in that it aims to understand drug actions and pathophysiological mechanisms at all levels (gene, molecular, subcellular, cellular, tissue, organ, and organismal levels) of an organism. On the other hand, polypharmacology is also an integrative medicine-based approach integrating multiple disciplines and transdisciplinary methodologies. Specifically, polypharmacology integrates the transdisciplinary knowledge, different pharmacokinetic and biological processes as well as multiple computational and experimental techniques [259–265]. In addition, there exists various approaches to uncover polypharmacological relationships: (a) structure-based; (b) similarity-based; (c) function-based; (d) expression-based; (e) sequence-based; (f) network-based, etc. Holistic and integrative polypharmacology embraces and integrate all these different approaches for drug discovery and development.

Therefore, the holistic philosophy and view and integrative transdisciplinary knowledge and methodology together form the conceptual strategy and practice approach of polypharmacology.

1.4 The Mission and Task of Polypharmacology

Polypharmacology is the design or use of pharmaceutical agents that act on multiple targets in a disease pathway or in multiple disease pathways. The mission of polypharmacology is to employ the most innovative concepts, the most updated knowledge and information, the most upgraded methods, the most advanced state-of-the-art technologies to design and produce therapeutics that are most effective and least toxicant in the treatment of human disease and to restore and maintain the highest standard of health status of global populations.

The self-imposed mission of polypharmacology determines it to carry on multiple tasks that are accompanied by strong conviction, commitment, and dedication. The first of its tasks is multitarget-drug discovery,

1.4.1 Multitarget Drug Discovery—Rational Design of New Drugs

The complexity of diseases has led to recent interest in polypharmacology, which suggests that many effective drugs specially modulate multiple targets and shift the new drug discovery from "single drug on single target (STD)" to "single drug on multiple targets (MTD)" paradigm. Designed multitargeticity and polyspecificity are the core and essence of polypharmacology, and the MTD approach is therefore to discover and develop drugs with multitargeticity and polyspecificity through rational design and rigorous validation [24–30, 162, 179–196]. In addition to MTD, multiple agents on multiple targets (CDT) and single pill with multiple agents on multiple targets (FDC) are alternative approaches of multitarget-drug discovery.

Drugs with multiple targets can provide a superior therapeutic effect and decrease in side effect profile compared to ligands with single target, especially in the treatment of complex diseases, such as tumors, nervous system diseases, and inflammatory diseases. The network-based holistic and integrative polypharmacology holds the promise of expanding the opportunity for novel targets and drug identification. In this regard, computational methods to predict the potential novel multitarget drugs with lower toxicity and higher efficacy are a powerful approach for multitarget-drug discovery and development [167, 169, 188, 223]. High-level data curation/integration and methodology development from various drug discovery disciplines would be needed for accurate prediction of polypharmacology prediction and rational design of multitargeting agents. Computational approaches for polypharmacology modeling will witness rapid growth and wide application in drug discovery.

The procedures for the task involve at least four major steps: (a) Target discovery and validation and lead identification and optimization; (b) Preclinical research on polypharmacodynamics and polypharmacokinetics and drug delivery methods; (c) Clinical development; (d) FDA (or other equivalent organizations) review; (e) post-market monitoring.

1.4.2 Drug Repurposing—Rediscovery of Old Drugs for New Uses

Although not designed on purpose, numerous drugs are known for their multitargeting activities. The polypharmacology approaches aim to discover the unknown off-targets for the existing drugs [157–178]. Off-target effects of drugs could either be beneficial or harmful, a face-changing game. To the face of angel, off-target effects of existing drugs could be utilized for drug repurposing or repositioning for their possible new therapeutic uses (i.e., finding a second or further medical use of an existing drug in the market beyond its original or primary therapeutic application, including approved, discontinued, shelved, and experimental drugs). To the face of evil, however, off-target effects could impose adverse side-effects or toxic actions that need to be avoided and minimized. The great advantages of re-using drugs with safe pharmacological profiles with favorable pharmacokinetic properties highlight drug repositioning as a valid alternative to rational drug design, massive drug development efforts, and high-throughput screening, especially when supported by in silico techniques. Moreover, existing multitarget agents are a pool of natural candidates for more innovative, off-target drug repositioning.

Drug repositioning is therefore the second task of polypharmacology. Yet, in the past, most of the successful drug repositioning cases have been found by serendipity or through exploitation of the original action mechanism of a drug for new indications. The advent of polypharmacology has made drug repositioning an intended and rational strategy for the redevelopment of old drugs for their new uses/indications. The approach needs the systematic integration of the data derived from different disciplines including computational modeling, synthetic chemistry, in vitro/in vivo pharmacological testing, and clinical studies. Computational strategies have demonstrated great potential in predicting polypharmacology and in facilitating drug repurposing [159, 174–176, 266, 267]. Various computational approaches enable comprehensive prediction and analysis of in vitro and in vivo drug-response phenotypes and outline their potential for drug discovery. Thus far, computational approaches to drug repositioning have mainly focused on retrospective drug repositioning: screening known drugs collections/libraries to find novel indications for already known therapeutic agents. In the future, however, prospective drug repositioning in the context of designed polypharmacology will explore drug repositioning possibilities much earlier in the drug discovery process [268].

1.4.3 Identifying Drug Promiscuity—Predicting Off-Target Effects

It is known that therapeutic drugs generally act with multiple molecular targets and exhibit similar or distinct pharmacological effects, which is mostly unintended or nondesigned. Such a property is designated as "drug promiscuity." If a drug interacts with unwanted or undesirable targets, the "drug promiscuity" is considered bad because it is the cause for off-target effects or adverse side-effects or toxic effects. In case if the unintended multiple-molecule targeting generates beneficial effects either by enhancing the potency and efficacy or by acting on other targets for new indication, the "drug promiscuity" is regarded as good because it forms the basis for drug repurposing as a task of polypharmacology. In this sense, "drug promiscuity" can be considered a double-edged sword and cause clinical problems when it is not fully understood [161, 162, 248, 269]. Clearly, "drug promiscuity" is an inaccurate, and could be misleading, term to give exact and precise definition or description of the nonspecificity of therapeutic drugs. To avoid such an embarrassment and confusion, we define "drug promiscuity" for its bad side in this book.

One of the primary tasks of polypharmacology is to uncover "drug promiscuity" or new off-targets for the existing drugs, thereby predicting the drug side-effects and disastrous toxicities. Such an effort should greatly help implement measures to avoid or minimize adverse effects on patients receiving the therapeutics for disease treatment.

1.4.4 Characterizing Drug Interactions

Drug interactions can be categorized into three groups: interactions of drug with disease condition (drug-disease interaction), drugs with other drugs (drug-drug interaction), and drugs with food (drug-food interaction). Most of the important drug interactions result from a change in the absorption, metabolism, or elimination of a drug.

Drug-disease interaction is an event in which a drug that is intended for therapeutic use to treat a disease causes worsening of another disease in a patient because of a disease or condition that the patient has [270–272]. Drug-disease interactions are relatively uncommon with the exception of interactions when a patient also has chronic kidney disease, cardiovascular disease, digestive disease, or other diseases that drastically alter the body's ability to metabolize a drug.

Drug-drug interaction (DDI) occurs when two or more drugs react with each other and may cause unexpected side effects, particularly when two drugs that have similar (additive) effects or opposite (canceling) effects on the body are administered together [273–275]. Such situations are seen with two major therapeutic approaches in clinics. One is associated with polypharmacology regimens for a disease with CDT (or combinational or drug cocktail therapy)-based therapy or FDC-based therapy. The other is associated with polypharmacy with the use of five or more medications daily by an individual for the treatment of two or more conditions [276]. For example, mixing a sleep-aid drug (a sedative) and an antiallergic drug (an antihistamine) can slow patient's reactions and make driving a car or operating machinery dangerous. Another source of DDI is when one drug alters the concentration of an endogenous substance in the body, which in turn influences (reduces or enhances) the effect of another drug that is being taken. The DDI between warfarin (Coumadin) and vitamin K-containing products is a good example of this type of interaction. Warfarin acts by reducing the concentration of the active form of vitamin K in the body. Therefore, when vitamin K is taken, it reduces the effect of warfarin.

Drug-food/beverage interactions result from drugs reacting with certain substances in foods or beverages one takes [277]. Interactions between drugs and foods may inadvertently reduce or increase the drug effect. Drug-food interactions can be clinically relevant, especially when they must be avoided to prevent undesirable effects or exploited to optimize medication therapy. The majority of clinically relevant drug-food interactions are caused by food-induced changes in the bioavailability of the drug.

Drug interactions, especially DDIs pertaining to polypharmacology, could be beneficial (enhancing the effectiveness and safety of drug treatment due to additive or synergetic actions among multiple drugs) or adverse (compromising the effectiveness and safety of therapeutic agents by mitigating or canceling targeting actions, or even producing toxic effects), depending on specific situations of a patient [278–281]. An important task of polypharmacology is to prospectively predict and retrospectively clarify the possible drug interactions of various types (particularly drug-drug interactions) so that clinicians could optimize the drug treatment based on polypharmacology with CDT or FDC strategy and/or polypharmacy.

1.4.5 Elucidating Polypharmacodynamics and Polypharmacokinetics

Like general pharmacology, polypharmacology involves two main areas: polypharmacodynamics and polypharmacokinetics [282–286]. Polypharmacodynamics studies the effects (efficacy and potency) and therapeutic effectiveness/clinical applications of drugs on biological systems. This is particularly important for CDT- or FDC-based drugs. Polypharmacodynamics addresses the chemical properties and physiological and behavioral effects of drugs arising from their interaction with molecular targets (drug-target interactions).

Polypharmacokinetics, on the other hand, focuses on what and how biological systems react to drug treatment. It elucidates the liberation, absorption, distribution, metabolism, and excretion (L-ADME) of multiple drugs and drug-interactions in the body to provide clinicians with the essential information for the design of drug treatment schedules in different patient populations.

To maximize the therapeutic effectiveness (efficacy and potency) and meanwhile minimize the potential toxic effects, it is essential for polypharmacological sciences to elucidate as thoroughly as possible the polypharmacodynamical and polypharmacokinetic properties of each component of polypharmacology-based drugs (mainly CDT- or FDC-based drugs) and the combined drugs in a whole as well [CDT- or FDC- or MTD-based drugs].

1.4.6 Delineating Mechanisms of Action of Polypharmacological Therapeutics

Polypharmacology-based therapeutics in the form of CDT, FDC, or MTD may involve complex multiple mechanisms at the molecular and cellular levels. This is particularly true with drug combination (CDT and FDC) as each individual drug could have its unique mechanism of action, as well as with MTD drugs designed to target multiple disease pathways with multiple mechanisms [287]. In theory, the mechanisms of action of MTD drugs designed to target a single disease pathway should be known because they are supposed to be predetermined at design. Nonetheless, in reality such designed MTD drugs can still elicit certain unintended off-target effects through unknown mechanisms due to unavoidable drug promiscuity. Therefore, delineating the mechanisms of action of polypharmacological therapeutics is unshirkable responsibility and duty of polypharmacology. Moreover, the mechanistic link between drug-target, drug-drug, and drug-disease interactions will in no doubt provide invaluable information for deepening our understanding of target reaction to drugs and improving the rational drug design and discovery.

1.5 Opportunities and Challenges of Polypharmacology

Polypharmacology has emerged as an essential paradigm for modern drug discovery and development [15, 30]. It can be employed for characterizing drug promiscuity to identify and predict the off-targets and the associated adverse side effects of existing drugs in clinical use and new drugs in development, respectively [288, 289]. Moreover, polypharmacology may also help identify new indications and novel applications of existing drugs through drug repositioning as an alternative strategy for drug discovery [20, 290–293].

1.5.1 Improved Drug Efficacy

Numerous lines of evidence suggest that agents capable of modulating multiple disease-relevant targets in a selective manner may offer improved therapeutic efficacy and improved balance between therapeutic efficacy and safety compared to single-targeted agents [15, 30, 294, 295]. This is because human diseases are mostly complex multifactorial entities, and polypharmacological treatment of human diseases with CDT-, FDC-, or MTD-based drugs acts through simultaneous blockade of multiple targets or pathways relevant to a disease and therefore is deemed to confer increased effectiveness. For example, the polypharmacological treatment with the current antineoplastic agents suggests that drug clinical efficacy in oncology can be achieved only via modulation of multiple cellular mechanisms [295]. Moreover, polypharmacological treatment of human disease also benefits from additive or synergistic actions of CDT/FDC or multitarget drugs. As a matter of fact, polypharmacology is often a key contributor to the efficacy of a drug [296].

1.5.2 Enhanced Drug Potency

It is essential for a successful drug to possess two basic characteristics: satisfactory pharmacological action with sufficient potency and efficacy. It is generally believed that the potency of a given drug can be increased through the engagement of multiple targets involved in a certain disease [31, 297–299]. Many newly identified compounds have low potency; this limits their immediate clinical applications because the known, tolerated plasma drug concentrations are lower than the required therapeutic drug concentrations. Drug combinations of two or more compounds with different mechanisms of action are an alternative approach to increase the potency of drug actions.

1.5.3 Increased Drug Safety

The relationship between polypharmacology and drug safety could be understood from the following arguments. First, multiple lines of evidence suggest that agents capable of modulating multiple targets in a selective manner may by nature afford improved balance between therapeutic efficacy and safety compared to single-targeted agents [30, 195, 294, 297]. Additionally, combinational drug therapy based on either CDT or FDC strategy offers an opportunity to cut the dose of each drug of the combination so as to mitigate the overall toxic effects. Even for MTD-based drugs, the dose for disease treatment may be reduced to the range where the maximum efficacy-to-safety ratio can be achieved. Moreover, polypharmacology approach can be applied to predicting drug promiscuity allowing for tailored dosages and regimens to avoid or minimize the potential safety risks of drugs. Furthermore, drug repurposing based on polypharmacological approaches can be channeled to the collection of approved drugs that have been demonstrated to be innocuous or less toxic or have excellent safety profiles following long-term clinical applications. Finally, it has also been proposed that one way to reduce the risk of idiopathic toxicity with kinase inhibitors or PARP-1 inhibitors as promising anticancer agents is to design compounds that target different protein families based on so-called "interfamily polypharmacology," as opposed to the "intrafamily polypharmacology" strategy with compounds acting on multiple target proteins belonging to a same family [300, 301].

1.5.4 Targeting Complex Disorders

Human diseases are mostly complex disorders that are multifactorial health conditions triggered by a number of intrinsic and/or environmental factors acting together on an organism [299–302]. This is particularly true with literally all chronic diseases including cancer, neurodegenerative diseases, cardiovascular diseases, and chronic inflammation. Complex diseases are due to deviations of fundamental physiological systems, with different pathologies being characterized by similar malfunctioning biological networks. The ensuing compensatory mechanisms may weaken the body's dynamic ability to respond to further insults and reduce the efficacy of conventional single target treatments. These are the major reason why despite the advances on the comprehension of the biological basis of these conditions and the huge investments made by the pharmaceutical sector, pharmaceutical solutions remain elusive. The continued drug discovery failures in complex diseases have raised questions about the classical paradigm "Single Agent on Single Target for Single Disease" (STD) strategy. For example, chronic pain with complex etiologies is unlikely to respond to single, target-specific therapeutics but rather require intervention at multiple points within a perturbed disease system. Polypharmacology, which broadly embodies the use of pharmaceutical agents acting on multiple targets, seems to be the best way to restore the complex diseased network and to provide disease-modifying effects. Although in some cases complex human disorders can be or are approached through combined therapies with CDT or FDC, multitarget ligands based on MTD strategy present clear advantages, among them more predictive pharmacokinetics, better patient compliance, and reduced risk of drug interactions. It is generally thought that complex diseases may require complex therapeutic approaches. In this respect, a drug that "hits" multiple sensitive nodes belonging to a network of interacting targets offers the potential for higher efficacy and may limit drawbacks generally arising from the use of a single-target drug or a combination of multiple drugs. Polypharmacology offers a model for the way drug discovery must evolve to develop therapies most suited to treating currently incurable diseases with safer and more effective profiles and affordable medicines against the most complex diseases [26].

1.5.5 Mitigating Drug Resistance

Drug resistance of either inherent or adaptive nature, the reduction in effectiveness of a medication in treating a disease or condition, is a common problem in pharmaceutical therapy of various human diseases, though it is more frequently manifested by antimicrobial, anticancer, or antineoplastic drugs [280, 285, 303–306]. It is widely accepted that drug discovery often requires a systems-level polypharmacology approach to tackle problems such as lack of efficacy and emerging resistance of single-targeted compounds. Simultaneously impacting different targets could be advantageous to approach individuals expressing intrinsic or induced variability in drug response due to modifications in key disease-relevant biological pathways and activation of compensatory mechanisms [279]. Apart from the obvious applications in the field of antimicrobial chemotherapy, the multitargeting strategy could also be pertinent to treat noninfectious conditions characterized by high incidence of the drug resistance phenomena, e.g., epilepsy [280]. One-third of the epileptic patients suffer from refractory epilepsy. One of the prevalent hypotheses to explain refractory epilepsy cases proposes that at least part of the nonresponsive patients might express variations in molecular targets of antiepileptic drugs [282]. The use of multitarget drugs is promising as it lowers the possibility of the disease to evolve into a drug-resistant phenotype [307].

An analysis revealed that over the past two decades, 52% (14 out of 27) of the FDA-approved antibiotics have demonstrated synergistic, multitarget mechanisms of action, of which are three second-generation lipoglycopeptides, five new-generation quinolones, and six modernized β-lactams. The polypharmacological drug action of these antibiotics demonstrates how these multitarget antibiotics successfully achieve the dual objectives of maximizing bactericidal or bacteriostatic efficacy and minimizing antibiotic resistance [308].

The polypharmacology approach was born with the aim to overcome the limitations of the single-targeted therapy, as manifested by drug resistance caused by redundancy and robustness of biological pathways [286]. Polypharmacology not only provides great opportunities for drug repurposing to exploit off-target effects in a new single-target indication but through simultaneous blockade of multiple targets or pathways offers exciting opportunities to slow, overcome or even prevent inherent or adaptive drug resistance [296].

However, polypharmacology is and will be in many years to come facing many challenges. First, the design of small molecules that have the property to modulate at the same time several molecular targets with contrasting or totally unrelated mechanisms of action is a difficult task though not a mission impossible. The identification of the most appropriate targets and of the relative hit compound(s), as well as the optimization of its (their) structure activity relationship represent further difficulties. Moreover, the development of multitarget drugs from leads is more complex than that of single drugs. Likewise, it is not easy to optimize drug-like molecular properties for multiple pharmacological activities. Another challenge is to avoid or minimize unintended interactions with additional unwanted targets. Finally, balancing the pharmacological activities is another complication.

Table 1.1 provides a succinct summary of the definition, principles, mission, advantages and weaknesses, and opportunities and challenges of polypharmacology for better understanding of the subject.

1.6 Polypharmacology and Network Pharmacology

Three years after Roth et al. first proposed the concept of multitarget-drug polypharmacology in 2004 [24, 195], Hopkins formulated another related concept in 2007—network pharmacology [14, 215]. Network pharmacology and polypharmacology share the same is essentially the same core connotation—innovative discovery and development of multitarget drugs aiming to maximize drug efficacy and potency and minimize off-target side effects and increase the success rate of clinical trials and decrease the costs of new drug development. Both network pharmacology and polypharmacology are based in systems pharmacology that looks at the effect of drugs on biological networks, especially the interactome and the diseasome [14, 215–229]. However, the two are not exactly the same; literally speaking, network pharmacology is part of polypharmacology or a branch of polypharmacology subdiscipline. More specifically, network pharmacology is one of the strategies of polypharmacology for drug discovery and the key technical approach for efficient identification and rational design of disease network-related multitarget drugs.

An approach to drug design that encompasses systems biology, network analysis, connectivity, redundancy, and pleiotropy, which is capable of describing complex interactions among biological systems, drugs, and diseases from network and holistic perspectives. Specifically, network pharmacology considers the drug response in the context of a cellular or phenotypic network. Often, network pharmacology is taken as a synonym of systems pharmacology [29], which is

Table 1.1 Polypharmacology: definition, principle, and mission

Definition	A new subdiscipline of general Pharmacology and a new branch of medicine, biology and pharmaceutical sciences concerned with the therapeutic effects and molecular mechanisms and discovery and development of new or reused drugs that act on multiple molecular targets and thereby affect multiple biological processes pertaining to a single disease.
Principles	Rationally designed multitargeticity.
	Rationally designed polyspecificity.
	Network-based systems pharmacology focusing on intracellular signaling pathways or networks.
	Constellation-directed pharmacology focusing to intercellular signaling pathways or networks.
	Holistic and integrative pharmacology.
Missions/ tasks	Multitarget drug discovery—Rational design of new drugs.
	Drug repurposing—Rediscovery of old drugs for new uses.
	Identifying drug promiscuity—Predicting off-target effects.
	Characterizing drug interactions.
	Elucidating polypharmacodynamics and polypharmacokinetics.
	Delineating mechanisms of action of polypharmacological therapeutics.
Strengths	Improved drug efficacy by minimizing physiological redundancy.
	Enhanced drug potency with synergetic effects.
	Increased drug safety profile.
	Targeting complex disorders via multitargeticity.
	Mitigating drug resistance by minimizing compensatory mechanisms.
Limitations	The effectiveness may not be as expected due to the inappropriate selection of molecular targets with present incomplete understanding of the disease pathways and mechanisms.
	We do not comprehend the full polypharmacological effect on the body at a molecular level.
	The technology and research currently available are often not sufficient to guarantee safety to patients receiving treatments with high confidence.
	The costs to conduct necessary studies can be high.

the application of systems biology principles to the field of pharmacology. It seeks to understand how drugs affect the human body as a single complex biological system. Instead of considering the effect of a drug to be the result of one specific drug-protein interaction, systems pharmacology considers the effect of a drug to be the outcome of the network of interactions a drug may have. Networks of interaction may include chemical-protein, protein–protein, genetic, signaling, and physiological (at cellular, tissue, organ, and whole body levels). Systems pharmacology uses bioinformatics and statistics techniques to integrate and interpret these networks.

The biological networks that network pharmacology is based upon include three major areas, as already described in the previous section entitled "The Principles of Polypharmacology": the molecular network covering the whole set of molecular interactions in the human cell including cell signaling networks, gene regulatory networks, and RNA networks [208]; the metabolic network encompassing the biochemical reactions in metabolic pathways [210]; the human disease networks (diseasome) that maps relationships between diseases and biological factors [212]. Network pharmacology can be reconstructed with these biological networks that integrate multidisciplinary concepts including biochemical, bioinformatics, and systems biology [14, 215]. Many complex diseases involve the interactions of multiple genes and functional proteins [309]. Network pharmacology models aim at addressing questions such as how and where in the disease network one target inhibits or activates the disease phenotypes. This ideally leads to therapies that are less vulnerable to drug resistance and lesser side effects by means of attacking the disease network at the systems level through synergistic and lethal interactions. Network pharmacology focuses on systematic characterization of drug targets and explores the interaction between multitarget drugs and disease-related networks [310–313].

The drug development based on the network pharmacology approaches involves three major consecutive steps: network construction, network analysis, and network verification. The aim of constructing a network is to achieve the interaction between the bioactive compounds and targets and the interaction between various targets, and then find out and validate the key nodes via network analysis and network verification.

1.6.1 Network Construction and Pragmatic Network Model

The first step is to establish a pragmatic network model for predicting the drug targets [205, 207, 209, 314, 315]. The networks could be based on public databases or available data of earlier researches or constructed anew according to the need. Alternatively, a "drug-target-disease" network prediction model could be reconstructed using the high-throughput screen technology and bioinformatics methods.

1.6.2 Network Analysis and Visualization

The goal of network analysis and visualization is to predict the mechanism of functional drug by the network equilibrium principle based on the established network model.

(a) For network analysis, three types of network analysis can be used. The first one involves the elucidation of the optimal topological structure and statistical properties of network following extraction of specific network data [316, 317]. The second method of network analysis is the generation and comparison of random networks for ensuring the reliability of the existing network by inducing acceptable modulation [318]. Another method is comparison and evaluation of the hierarchical network clustering algorithms to predigest the complicated genetic interaction network [319, 320].

(b) Network visualization aims to extract the interaction information from inter-associated data and subsequently transform into a visual network using visualization tools [321]. This process can be done by enriching network attributes, adding network nodes, and increasing connection power of network and describing the network and the architectural feature that clearly and intuitively represents the network.

1.6.3 Data Collection and Network Validation

The third step for the drug development based on the network pharmacology approaches is data collection and validation. Network pharmacology takes into account the aforementioned principles to optimize the efficacy and safety of a candidate drug and their potent combinations. To this end, original experimental data need to be collected and selected to build a biological network, followed by the experimental validation for the predicted network model. The validated data can be quantified using integrated methods including genomics, proteomics, metabolomics, and high-throughput screen/high-content screening (HCS) technologies [320]. These technologies allow for rapid detection of millions of data sets, identification of substances, and modulatory mechanisms of a particular molecular pathway or alter the phenotype of a cell.

Integrating network biology and polypharmacology holds the promise of expanding the current opportunity space for druggable targets. However, the rational design of polypharmacology faces considerable challenges in the need for new methods to validate target combinations and optimize multiple structure-activity relationships while maintaining drug-like properties. Advances in these areas are creating the foundation of the next paradigm in drug discovery: network pharmacology.

The rationale for network pharmacology as a major component of drug discovery and development in polypharmacology is that a disease can be caused by perturbation of the disease-causing network and a drug may be designed to interact with multiple targets for modulation of such a network from the disease status toward normal status. Therefore, network pharmacology has been applied to guide and assist in drug repositioning [322]. Drugs exerting their therapeutic effects may directly target disease-associated proteins, but they may also modulate the pathways involved in the pathological process.

1.7 Epigenetic Polypharmacology

The concept of "epigenetics" is derived from "genetics" that is broadly defined as the study of genes and their heredity. At the molecular level, genetics involves the organism's DNA sequence or genome and how it influences the biological phenotype. Genetics has a profound influence on medicinal chemistry as mutations in single or multiple genes are directly linked to or affect the susceptibility to thousands of human diseases. The newer science of epigenetics is concerned not with the actual genes but how their expression is regulated, and it studies the heritable and reversible changes in gene expression without changes in the DNA sequence of bases [323–326]. Thus, epigenetics can be defined as covalent modifications of DNA and core histones that regulate gene activity without altering DNA sequence.

In eukaryotic organisms, epigenetics involves an elaborate pattern of structural modification of the DNA and histone proteins that constitute the nucleosome. These alterations are introduced by a set of enzymes (called "writers") and then serve as recognition elements for specific protein-protein and protein-nucleic acid interaction domains (the "readers"). Meanwhile, a separate set of enzymes (the "erasers") reverses the action of the writers. Together, the three families of epigenetic players serve to coordinate the complex transcription machinery in eukaryotes and determine if a specific gene is expressed or silenced. Since dysfunctional gene regulation lies at the heart of most human ailments, the modulation of epigenetic processes holds much promise as a new strategy for drug discovery. Over the last two decades, small molecules with high affinity and selectivity for epigenetic writers, readers, and erasers have been identified. These efforts have culminated in numerous clinical candidates and seven approved anticancer agents.

Since dysfunctional gene regulation is responsible for many human diseases, the modulation of epigenetic processes is presently considered an innovative and very interesting therapeutic strategy [327]. There are at least six major mechanisms for epigenetics as target for drug discovery and development:

(a) Targeting the methylation of cytosine residues within CpG dinucleotides in DNA via covalent modifications [327–330].
(b) Targeting chromatin remodeling through histone covalent chemical modifications such as acetylation, methylation, phosphorylation, etc. [327, 331–333].
(c) Targeting RNA methylation [334–336].
(d) Targeting mRNAs using siRNA [337–339], antisense oligodeoxynucleotides (ASO) [340–344], antisense RNAs (asRNAs), etc. [345–350].
(e) Targeting noncoding RNAs, such as microRNAs (miRNAs) [148, 351–354], long noncoding RNAs (lncRNAs) [355, 356], circular RNAs (circRNAs), etc. [357–360].
(f) Targeting transcription factors using decoy oligodeoxynucleotides (dODN) [25, 361–364].

The convergence of epigenetics and polypharmacology forms another branch of polypharmacology: Epigenetic Polypharmacology [147, 148, 286, 326–372]. Epigenetics has the potential to be a key element in a paradigm change of our understanding of aging, development, cancer, heart disease, psychological disorders, and other diseases. Indeed, search for small molecule epigenetic modulators or sequence-based molecules has become an increasingly attractive strategy for discovering and developing epigenetic drugs (epi-drugs) for disease therapy, especially multi-target epi-drugs or MTD epi-drugs. Epigenetic polypharmacology can be considered another branch of the polypharmacology discipline.

MTD epi-drugs include two major classes: (a) Function-based MTD epi-drugs mainly include "small-molecule drugs" and "macromolecular drugs" and (b) Sequence-based MTD epi-drugs including DNA/RNA sequence-based and amino acid sequence-based MTD drugs.

The sequence-based MTD approach offers remarkable opportunities for the easy and precise design of MTD drugs able to act on multiple targets (multitargeticity) with preselected specificities (polyspecificity), representing the core and essence of polypharmacology. The main challenge to this approach is efficient delivery of MTD epi-drugs into the target site.

For function-based MTD epi-drugs, the Epigenetics Compound Library by TargetMol, which contains 932 compounds related to epigenetic regulation, could be an abundant resource for epi-drug discovery—high throughput screening and high content screening for new drugs in epigenetic modification.

Detailed descriptions of epigenetic polypharmacology will be given in Chap. 13.

References

1. Vallance P, Smart TG. The future of pharmacology. Br J Pharmacol. 2006;147(S1):S304–7.
2. Flower RJ. Pharmacology 2.0. Br J Clin Pharmacol. 2013;76(5):625–9.
3. Barrett JE, Page C, Michel MC. Perspectives of pharmacology over the past 100 years. Handb Exp Pharmacol. 2019;260:3–16.
4. Currie G. Pharmacology, Part 1: Introduction to pharmacology and pharmacodynamics. J Nucl Med Technol. 2018;46:81–6.
5. Currie GM. Pharmacology, Part 2: Introduction to pharmacokinetics. J Nucl Med Technol. 2018;46(3):221–30.
6. Reidenberg MM. Clinical pharmacology. JAMA. 1991;265(23):3107–9.
7. Kenakin T. Emergent concepts of receptor pharmacology. Handb Exp Pharmacol. 2019;260:17–41.
8. Bittar E. Molecular and cellular pharmacology. Elsevier Science; 1997.
9. Dickenson J, Freeman F, Mills CL, et al. Molecular pharmacology: from DNA to drug discovery. Wiley; 2013.
10. Ylä-Herttuala S. The pharmacology of gene therapy. Mol Ther. 2017;25(8):1731–2.
11. Stefanska B, MacEwan DJ. Epigenetics and pharmacology. Br J Pharmacol. 2015;172(11):2701–4.
12. Vaiserman AM, Lushchak OV, Koliada AK. Anti-aging pharmacology: promises and pitfalls. Ageing Res Rev. 2016;31:9–35.
13. Wang Z, Yang B. Anti-aging pharmacology. Harbin: Scientific Press of Heilongjian; 2002.
14. Hopkins AL. Network pharmacology: the next paradigm in drug discovery. Nat Chem Biol. 2008;4(11):682–90.
15. Anighoro A, Bajorath J, Rastelli G. Polypharmacology: challenges and opportunities in drug discovery. J Med Chem. 2014;57(19):7874–87.
16. Fredholm BB, Fleming WW, Vanhoutte PM, Godfraind T. The role of pharmacology in drug discovery. Nat Rev Drug Discov. 2002;1(3):237–8.
17. Whirl-Carrillo M, McDonagh EM, Hebert JM, et al. Pharmacogenomics knowledge for personalized medicine. Clin Pharmacol Ther. 2012;92(4):414–7.
18. Baye J, Limberg C, Wall T. Pharmacology focus: pharmacogenomics: gene-directed prescribing in the era of personalized medicine. S D Med. 2019;72(6):280–1.
19. Cecchin E, Stocco G. Pharmacogenomics and personalized medicine. Genes (Basel). 2020;11(6):679.
20. Abood RR. Pharmacy practice and the law. 6th ed. Sudbury: Jones & Bartlett Learning; 2011.
21. Reddy AS, Zhang S. Polypharmacology: drug discovery for the future. Expert Rev Clin Pharmacol. 2013;6(1):41–7.
22. Bolognesi ML. Harnessing polypharmacology with medicinal chemistry. ACS Med Chem Lett. 2019;10(3):273–5.
23. Knight ZA, Lin H, Shokat KM. Targeting the cancer kinome through polypharmacology. Nat Rev Can. 2010;10:130–7.
24. Roth BL, Sheffler DJ, Kroeze WK. Magic shotguns versus magic bullets: selectively non-selective drugs for mood disorders and Schizophrenia. Nat Rev Drug Discov. 2004;3:353–9.
25. Gao H, Xiao J, Yang B, et al. A single decoy oligodeoxynucleotides targeting multiple oncoproteins produces strong anti-cancer effects. Mol Pharmacol. 2006;70:1621–9.
26. Bolognesi ML. Polypharmacology in a single drug: multitarget drugs. Curr Med Chem. 2013;20(13):1639–45.
27. Bolognesi ML, Cavalli A. Multitarget drug discovery and polypharmacology. ChemMedChem. 2016;11(12):1190–2.
28. Santos R, Ursu O, Gaulton A, et al. A comprehensive map of molecular drug targets. Nat Rev Drug Discov. 2017;16(1):19–34.
29. Zhang W, Bai Y, Wang Y, et al. Polypharmacology in drug discovery: a review from systems pharmacology perspective. Curr Pharm Des. 2016;22(21):3171–81.
30. Peters JU. Polypharmacology – foe or friend? J Med Chem. 2013;56(22):8955–71.
31. Sun W, Sanderson PE, Zheng W. Drug combination therapy increases successful drug repositioning. Drug Discov Today. 2016;21(7):1189–95.
32. Redvers N, Blondin B. Traditional Indigenous medicine in North America: a scoping review. PLoS One. 2020;15(8):e0237531.

33. Raja Ikram RR, Abd Ghani MK, Abdullah N. An analysis of application of health informatics in traditional medicine: a review of four traditional medicine systems. Int J Med Inform. 2015;84(11):988–96.
34. Herman PM, Craig BM, Caspi O. Is complementary and alternative medicine (CAM) cost-effective? A systematic review. BMC Complement Altern Med. 2005;5:11.
35. Arji G, Safdari R, Rezaeizadeh H, et al. A systematic literature review and classification of knowledge discovery in traditional medicine. Comput Methods Prog Biomed. 2019;168:39–57.
36. Tang JL, Liu BY, Ma KW. Traditional Chinese medicine. Lancet. 2008;372(9654):1938–40.
37. Lock O, Perez E, Villar M, et al. Bioactive compounds from plants used in Peruvian traditional medicine. Nat Prod Commun. 2016;11(3):315–37.
38. Nugraha RV, Ridwansyah H, Ghozali M, et al. Traditional herbal medicine candidates as complementary treatments for COVID-19: a review of their mechanisms, pros and cons. Evid Based Complement Alternat Med. 2020;2020:2560645.
39. Li Shizhen. The Compendium of Materia Medica (Bencao Gangmu). 1578 (first draft), 1596 (print) in Nanjing.
40. Wang J, Wong YK, Liao F. What has traditional Chinese medicine delivered for modern medicine? Expert Rev Mol Med. 2018;20:e4.
41. Liu SH, Chuang WC, Lam W, et al. Safety surveillance of traditional Chinese medicine: current and future. Drug Saf. 2015;38(2):117–28.
42. Hao P, Jiang F, Cheng J, et al. Traditional Chinese medicine for cardiovascular disease: evidence and potential mechanisms. J Am Coll Cardiol. 2017;69(24):2952–66.
43. Hu Q, Calduch RM. On traditional Chinese medicine regulation in China: how quality and safety of use are insured. Pharmacol Res. 2017;119:371–2.
44. Pferschy-Wenzig EM, Bauer R. The relevance of pharmacognosy in pharmacological research on herbal medicinal products. Epilepsy Behav. 2015;52(Pt B):344–62.
45. Hasani-Ranjbar S, Nayebi N, Larijani B, et al. A systematic review of the efficacy and safety of herbal medicines used in the treatment of obesity. World J Gastroenterol. 2009;15(25):3073–85.
46. Ju JM, Liu YY, Du WJ. et al. Therapeutic values of Daming capsule for cardiovascular diseases.
47. Hasani-Ranjbar S, Nayebi N, Moradi L, et al. The efficacy and safety of herbal medicines used in the treatment of hyperlipidemia; a systematic review. Curr Pharm Des. 2010;16(26):2935–47.
48. Ai J, Zhao LM, Lu YJ, et al. A randomized, multicentre, open-label, parallel-group trial to compare the efficacy and safety profile of Daming capsule in patients with hypercholesterolemia. Phytother Res. 2009;23(7):1039–42.
49. Mosihuzzaman M. Herbal medicine in healthcare – an overview. Nat Prod Commun. 2012;7(6):807–12.
50. Ernst E. The efficacy of herbal medicine – an overview. Fundam Clin Pharmacol. 2005;19(4):405–9.
51. Steinhoff B. Review: quality of herbal medicinal products: state of the art of purity assessment. Phytomedicine. 2019;60:153003.
52. Calahan J, Howard D, Almalki AJ, et al. Chemical adulterants in herbal medicinal products: a review. Planta Med. 2016;82(6):505–15.
53. Tan MS, Yu JT, Tan CC, et al. Efficacy and adverse effects of ginkgo biloba for cognitive impairment and dementia: a systematic review and meta-analysis. J Alzheimers Dis. 2015;43(2):589–603.
54. Ude C, Schubert-Zsilavecz M, Wurglics M. Ginkgo biloba extracts: a review of the pharmacokinetics of the active ingredients. Clin Pharmacokinet. 2013;52(9):727–49.
55. Sierpina VS, Wollschlaeger B, Blumenthal M. Ginkgo biloba. Am Fam Physician. 2003;68(5):923–6.
56. Tu Y. Artemisinin – A gift from traditional Chinese medicine to the world (Nobel lecture). Angew Chem Int Ed Engl. 2016;55(35):10210–26.
57. Miller LH, Su X. Artemisinin: discovery from the Chinese herbal garden. Cell. 2011;146(6):855–8.
58. Slezakova S, Ruda-Kucerova J. Anticancer activity of artemisinin and its derivatives. Anticancer Res. 2017;37(11):5995–26003.
59. Rastogi S, Pandey MM, Rawat AK. Traditional herbs: a remedy for cardiovascular disorders. Phytomedicine. 2016;23(11):1082–9.
60. Carresi C, Scicchitano M, Scarano F, et al. The potential properties of natural compounds in cardiac Stem cell activation: their role in myocardial regeneration. Nutrients. 2021;13(1):275.
61. Santana FP, Pinheiro NM, Mernak MI, et al. Evidences of herbal medicine-derived natural products effects in inflammatory lung diseases. Mediat Inflamm. 2016;2016:2348968.
62. Gurib-Fakim A. Medicinal plants: traditions of yesterday and drugs of tomorrow. Mol Asp Med. 2006;27(1):1–93.
63. Medina-Franco JL, Giulianotti MA, Welmaker GS, et al. Shifting from the single to the multitarget paradigm in drug discovery. Drug Discov Today. 2013;18(9–10):495–501.
64. Makhoba XH, Viegas C Jr, Mosa RA, et al. Potential impact of the multi-target drug approach in the treatment of some complex diseases. Drug Des Devel Ther. 2020;14:3235–49.
65. Ramsay RR, Popovic-Nikolic MR, Nikolic K, et al. A perspective on multi-target drug discovery and design for complex diseases. Clin Transl Med. 2018;7(1):3.
66. Talevi A. Multi-target pharmacology: possibilities and limitations of the "skeleton key approach" from a medicinal chemist perspective. Front Pharmacol. 2015;6:205.
67. Gouin SG. Multivalent inhibitors for carbohydrate-processing enzymes: beyond the "lock-and-key" concept. Chemistry. 2014;20(37):11616–28.

68. Tripathi A, Bankaitis VA. Molecular docking: from lock and key to combination lock. J Mol Med Clin Appl. 2017;2(1) https://doi.org/10.16966/2575-0305.106.
69. Schwyzer R. 100 years lock-and-key concept: are peptide keys shaped and guided to their receptors by the target cell membrane? Biopolymers. 1995;37(1):5–16.
70. Schneider HJ. Limitations and extensions of the lock-and-key principle: differences between gas state, solution and solid state structures. Int J Mol Sci. 2015;16(4):6694–717.
71. Colín-Lozano B, Estrada-Soto S, Chávez-Silva F, et al. Design, synthesis and in combo antidiabetic bioevaluation of multitarget phenylpropanoic acids. Molecules. 2018;23(2):1–16.
72. Qaseem A, Barry MJ, Humphrey LL, et al. Oral pharmacologic treatment of type 2 diabetes mellitus: a clinical practice guideline update from the American College of Physicians. Ann Intern Med. 2017;166(4):279–90.
73. Hussein Z, Wentworth JM, Nankervis AJ, et al. Effectiveness and side effects of thiazolidinediones for type 2 diabetes: real-life experience from a tertiary hospital. Med J Aust. 2004;181(10):536–9.
74. Vuylsteke V, Chastain LM, Maggu GA, et al. Imeglimin: a potential new multi-target drug for type 2 diabetes. Drugs R D. 2015;15(3):227–32.
75. Detaille D, Vial G, Borel A-L, et al. Imeglimin prevents human endothelial cell death by inhibiting mitochondrial permeability transition without inhibiting mitochondrial respiration. Cell Death Discov. 2016;2(1):15072.
76. Munhoz ACM, Frode TS. Isolated compounds from natural products with potential antidiabetic activity – a systematic review. Curr Diabetes Rev. 2018;14(1):36–106.
77. Zagouri F, Sergentanis TN, Chrysikos D, et al. Molecularly targeted therapies in metastatic pancreatic cancer: a systematic review. Pancreas. 2013;42(5):760–73.
78. Kayser S, Levis MJ. Advances in targeted therapy for acute myeloid leukaemia. Br J Haematol. 2018;180(4):484–500.
79. Zagouri F, Papadimitriou CA, Dimopoulos MA, et al. Molecularly targeted therapies in unresectable-metastatic gastric cancer: a systematic review. Cancer Treat Rev. 2011;37(8):599–610.
80. Huang M, Shen A, Ding J, et al. Molecularly targeted cancer therapy: some lessons from the past decade. Trends Pharmacol Sci. 2014;35(1):41–50.
81. Scannell JW, Blanckley A, Boldon H, et al. Diagnosing the decline in pharmaceutical R&D efficiency. Nat Rev Drug Discov. 2012;11:191–200.
82. Hedhli N, Russell KS. Cardiotoxicity of molecularly targeted agents. Curr Cardiol Rev. 2011;7(4):221–33.
83. Altzerinakou MA, Collette L, Paoletti X. Cumulative toxicity in targeted therapies: what to expect at the recommended phase ii dose. J Natl Cancer Inst. 2019;111(11):1179–85.
84. Renouf DJ, Velazquez-Martin JP, Simpson R, et al. Ocular toxicity of targeted therapies. J Clin Oncol. 2012;30(26):3277–86.
85. Ribecco AS, Pino MS, Cipriani G, et al. Molecularly targeted therapy: toxicity and quality of life considerations in advanced colorectal cancer. Expert Rev Anticancer Ther. 2013;13(10):1181–91.
86. Basu B, Vitfell-Pedersen J, Moreno Garcia V, et al. Creatinine clearance is associated with toxicity from molecularly targeted agents in phase I trials. Oncology. 2012;83(4):177–82.
87. Catterall WA. From ionic currents to molecular mechanisms: the structure and function of voltage-gated sodium channels. Neuron. 2000;26:13–25.
88. Morphy R, Kay C, Rankovic Z. From magic bullets to designed multiple ligands. Drug Discov Today. 2004;9:641–51.
89. Cohen MH, Williams GA, Sridhara R, et al. United States Food and Drug Administration Drug Approval summary: Gefitinib (ZD1839; Iressa) tablets. Clin Cancer Res. 2004;10(4):1212–8.
90. Roth BL, Ciaranello RD, Meltzer HY. Binding of typical and atypical antipsychotic agents to transiently expressed 5-HT1C receptors. J Pharmacol Exp Ther. 1992;260(3):1361–5.
91. Roth BL, Meltzer HY, Khan N. Binding of typical and atypical antipsychotic drugs to multiple neurotransmitter receptors. Adv Pharmacol. 1998;42:482–5.
92. Roth BL, Willins DL, Kristiansen K, et al. 5-Hydroxytryptamine2-family receptors (5-hydroxytryptamine2A, 5-hydroxytryptamine2B, 5-hydroxytryptamine2C): where structure meets function. Pharmacol Ther. 1998;79(3):231–57.
93. Saenz-Méndez P, Eriksson LA. Exploring polypharmacology in drug design. Methods Mol Biol. 2018;1824:229–43.
94. Ascierto PA, Marincola FM. Combination therapy: the next opportunity and challenge of medicine. J Transl Med. 2011;9:115.
95. Phougat N, Khatri S, Singh A, et al. Combination therapy: the propitious rationale for drug development. Comb Chem High Throughput Screen. 2014;17(1):53–67.
96. Ghosn J, Taiwo B, Seedat S, et al. HIV. Lancet. 2018;392(10148):685–97.
97. Deeks SG, Overbaugh J, Phillips A, Buchbinder S. HIV infection. Nat Rev Dis Primers. 2015;1:15035.
98. Atta MG, De Seigneux S, Lucas GM. Clinical pharmacology in HIV therapy. Clin J Am Soc Nephrol. 2019;14(3):435–44.
99. Lu DY, Lu TR, Yarla NS, et al. Drug combination in clinical cancer treatments. Rev Recent Clin Trials. 2017;12(3):202–11.
100. Correia AS, Gärtner F, Vale N. Drug combination and repurposing for cancer therapy: the example of breast cancer. Heliyon. 2021;7(1):e05948.
101. Parhi P, Mohanty C, Sahoo SK. Nanotechnology-based combinational drug delivery: an emerging approach for cancer therapy. Drug Discov Today. 2012;17(17–18):1044–52.

102. Düsing R, Waeber B, Destro M, et al. Triple-combination therapy in the treatment of hypertension: a review of the evidence. J Hum Hypertens. 2017;31(8):501–10.
103. Poulter NR, Prabhakaran D, Caulfield M. Hypertension. Lancet. 2015;386(9995):801–12.
104. Tsioufis C, Thomopoulos C. Combination drug treatment in hypertension. Pharmacol Res. 2017;125(Pt B):266–71.
105. Lee JH, Nan A. Combination drug delivery approaches in metastatic breast cancer. J Drug Deliv. 2012;2012:915375.
106. Waterhouse DN, Gelmon KA, Klasa R, et al. Development and assessment of conventional and targeted drug combinations for use in the treatment of aggressive breast cancers. Curr Cancer Drug Targets. 2006;6:455–89.
107. Yhee JY, Son S, Lee H, et al. Nanoparticle-based combination therapy for cancer treatment. Curr Pharm Des. 2015;21:3158–66.
108. Kerantzas CA, Jacobs WR Jr. Origins of combination therapy for tuberculosis: lessons for future antimicrobial development and application. MBio. 2017;8(2):e01586–16.
109. Patel L, Grossberg GT. Combination therapy for Alzheimer's disease. Drugs Aging. 2011;28(7):539–46.
110. Prabhakar PK, Kumar A, Doble M. Combination therapy: a new strategy to manage diabetes and its complications. Phytomedicine. 2014;21(2):123–30.
111. Van Gaal LF, De Leeuw IH. Rationale and options for combination therapy in the treatment of Type 2 diabetes. Diabetologia. 2003;46(Suppl 1):M44–50.
112. Li X, Xu Y, Cui H, et al. Prediction of synergistic anti-cancer drug combinations based on drug target network and drug induced gene expression profiles. Artif Intell Med. 2017;83:35–43.
113. Mouton JW. Combination therapy as a tool to prevent emergence of bacterial resistance. Infection. 1999;27(Suppl 2):S24–8.
114. Scully M, Hassoun A. Increasing evidence of potential toxicity of a common antibiotic combination. J Infect Public Health. 2018;11(4):594–5.
115. Zerkak D, Dougados M. Benefit/risk of combination therapies. Clin Exp Rheumatol. 2004;22(5 Suppl 35):S71–6.
116. Welz T, Wyen C, Hensel M. Drug interactions in the treatment of malignancy in HIV-infected patients. Oncol Res Treat. 2017;40(3):120–7.
117. Aronson J. Serious drug interactions. Practitioner. 1993;237(1531):789–91.
118. Sharma M, Vadhariya A, Chikermane S, et al. Clinical outcomes associated with drug-drug interactions of oral chemotherapeutic agents: a comprehensive evidence-based literature review. Drugs Aging. 2019;36(4):341–54.
119. Wald NJ, Law MR. A strategy to reduce cardiovascular disease by more than 80%. BMJ. 2003;326(7404):1419.
120. Webster R, Castellano JM, Onuma OK. Putting polypills into practice: challenges and lessons learned. Lancet. 2017;389(10073):1066–74.
121. Basson M. Prevention with a polypill. Nat Med. 2019;25(11):1647.
122. Chrysant SG. Single-pill triple-combination therapy: an alternative to multiple-drug treatment of hypertension. Postgrad Med. 2011;123(6):21–31.
123. Huan Y, Townsend R. The single pill triple combination of aliskiren, amlodipine, and hydrochlorothiazide in the treatment of hypertension. Expert Opin Pharmacother. 2012;13(16):2409–15.
124. Tsioufis K, Kreutz R, Sykara G, et al. Impact of single-pill combination therapy on adherence, blood pressure control, and clinical outcomes: a rapid evidence assessment of recent literature. J Hypertens. 2020;38(6):1016–28.
125. Weisser B, Predel HG, Gillessen A, et al. Single pill regimen leads to better adherence and clinical outcome in daily practice in patients suffering from hypertension and/or dyslipidemia: results of a meta-analysis. High Blood Press Cardiovasc Prev. 2020;27(2):157–64.
126. Angeli F, Reboldi G, Mazzotta G, et al. Fixed-dose combination therapy in hypertension: cons. High Blood Press Cardiovasc Prev. 2012;19:51–4.
127. Moriarty F, Bennett K, Fahey T. Fixed-dose combination antihypertensives and risk of medication errors. Heart. 2019;105:204–9.
128. Machnicki G, Ong SH, Chen W, et al. Comparison of amlodipine/valsartan/hydrochlorothiazide single pill combination and free combination: adherence, persistence, healthcare utilization and costs. Curr Med Res Opin. 2015;31(12):2287–96.
129. Verma AA, Khuu W, Tadrous M, et al. Fixed-dose combination antihypertensive medications, adherence, and clinical outcomes: a population-based retrospective cohort study. PLoS Med. 2018;15(6):e1002584.
130. Gupta AK, Arshad S, Poulter NR. Compliance, safety, and effectiveness of fixed-dose combinations of antihypertensive agents: a meta-analysis. Hypertension. 2010;55:399–407.
131. Sica DA. Rationale for fixed-dose combinations in the treatment of hypertension: the cycle repeats. Drugs. 2002;62:443–62.
132. Muñoz D, Uzoije P, Reynolds C, et al. Polypill for cardiovascular disease prevention in an underserved population. N Engl J Med. 2019;381(12):1114–23.
133. Roy A, Naik N, Srinath RK. Strengths and limitations of using the polypill in cardiovascular prevention. Curr Cardiol Rep. 2017;19(5):45.
134. National Academy of Sciences – National Research Council Division of Medical Sciences Drug Efficacy Study. Fixed combinations of antimicrobial agents. N Engl J Med. 1969;280(21):1149–54.
135. Koshy S, Balaris GL. Therapeutic approaches to achieve desired blood pressure goals: focus on calcium channel blockers. Cardiovasc Drugs Ther. 2000;14(3):295–301.

136. Kuschnir E, Acuña E, Sevilla D, et al. Treatment of patients with essential hypertension: amlodipine 5 mg/benazepril 20 mg, compared with amlodipine 5 mg, benazepril 20 mg and placebo. Clin Ther. 1996;18(6):1213–24.
137. Franczyk B, Gluba-Brzózka A, Jurkiewicz L, et al. Embracing the polypill as a cardiovascular therapeutic: is this the best strategy? Expert Opin Pharmacother. 2018;19(17):1857–65.
138. Ho WY, Abdul-Rahim AH, Dawson J, et al. The role of single pill combination therapy in the prevention of ischaemic stroke. Scott Med J. 2019;64(4):126–32.
139. Aronson R. Single-pill combination therapy for type 2 diabetes mellitus: linagliptin plus empagliflozin. Curr Med Res Opin. 2015;31(5):901–11.
140. Janssen VE, Visseren FL, de Boer A, et al. SMART Study Group. Combined use of polypill components in patients with type 2 diabetes mellitus. Eur J Prev Cardiol. 2018;25(14):1523–31.
141. Park JH, Lee YH, Ko SK, et al. Cost-effectiveness analysis of low density lipoprotein cholesterol-lowering therapy in hypertensive patients with type 2 diabetes in Korea: single-pill regimen (amlodipine/atorvastatin) versus double-pill regimen (amlodipine+atorvastatin). Epidemiol Health. 2015;37:e2015010.
142. Gandhi M, Gandhi RT. Single-pill combination regimens for treatment of HIV-1 infection. N Engl J Med. 2014;371(3):248–59.
143. Johnson LB, Saravolatz LD. The quad pill, a once-daily combination therapy for HIV infection. Clin Infect Dis. 2014;58(1):93–8.
144. Thom S, Poulter N, Field J, UMPIRE Collaborative Group, et al. Effects of a fixed-dose combination strategy on adherence and risk factors in patients with or at high risk of CVD: the UMPIRE randomized clinical trial. JAMA. 2013;310(9):918–29.
145. Lafeber M, Spiering W, van der Graaf Y, et al. The combined use of aspirin, a statin, and blood pressure-lowering agents (polypill components) and the risk of vascular morbidity and mortality in patients with coronary artery disease. Am Heart J. 2013;166(2):282–9.
146. Lenz TL, Hilleman DE. Aggrenox: a fixed-dose combination of aspirin and dipyridamole. Ann Pharmacother. 2000;34(11):1283–90.
147. Tomaselli D, Lucidi A, Rotili D, et al. Epigenetic polypharmacology: a new frontier for epi-drug discovery. Med Res Rev. 2020;40(1):190–244.
148. Lu Y, Xiao J, Lin H, et al. A single anti-microRNA antisense oligodeoxyribonucleotide (AMO) targeting multiple microRNAs offers an improved approach for microRNA interference. Nucleic Acids Res. 2009;37:e24–33.
149. Wang Z. New concepts of miRNAi technologies. In: Wang Z, editor. MicroRNA interference technologies. Heidelberg: Springer; 2009.
150. Wang Z. The concept of multiple-target anti-miRNA antisense oligonucleotides technology. Methods Mol Biol. 2011;676:51–7.
151. Wang Z. The concept of multiple-target anti-miRNA antisense oligonucleotides technology. In: Wu W, editor. MicroRNA and cancer. Totowa: Humana Press; 2011.
152. Apsel B, Blair JA, Gonzalez B, et al. Targeted polypharmacology: discovery of dual inhibitors of tyrosine and phosphoinositide kinases. Nat Chem Biol. 2008;4(11):691–9.
153. Morphy R, Rankovic Z. Designed multiple ligands. An emerging drug discovery paradigm. J Med Chem. 2005;48(21):6523–43.
154. Melchiorre C, Andrisano V, Bolognesi ML, et al. Acetylcholinesterase noncovalent inhibitors based on a polyamine backbone for potential use against Alzheimer's disease. J Med Chem. 1998;41(22):4186–9.
155. Pushpakom S, Iorio F, Eyers PA, et al. Drug repurposing: progress, challenges and recommendations. Nat Rev Drug Discov. 2019;18(1):41–58.
156. Simsek M, et al. Finding hidden treasures in old drugs: the challenges and importance of licensing generics. Drug Discov Today. 2018;23(1):17–21.
157. Ashburn TT, Thor KB. Drug repositioning: identifying and developing new uses for existing drugs. Nat Rev Drug Discov. 2004;3:673–83.
158. Schcolnik-Cabrera A, Juárez-López D, Duenas-Gonzalez A. Perspectives on drug repurposing. Curr Med Chem. 2021;28(11):2085–99.
159. Cha Y, Erez T, Reynolds IJ, et al. Drug repurposing from the perspective of pharmaceutical companies. Br J Pharmacol. 2018;175(2):168–80.
160. Pantziarka P, Pirmohamed M, Mirza N. New uses for old drugs. BMJ. 2018;361:k2701.
161. Mencher SK, Wang LG. Promiscuous drugs compared to selective drugs (promiscuity can be a virtue). BMC Clin Pharmacol. 2005;5:3.
162. Mei Y, Yang B. Rational application of drug promiscuity in medicinal chemistry. Future Med Chem. 2018;10(15):1835–51.
163. Fetro C, Scherman D. Drug repurposing in rare diseases: myths and reality. Therapie. 2020;75(2):157–60.
164. Sardana D, Zhu C, Zhang M, et al. Drug repositioning for orphan diseases. Brief Bioinform. 2011;12:346–56.
165. Scherman D, Fetro C. Drug repositioning for rare diseases: knowledge-based success stories. Therapie. 2020;75(2):161–7.
166. Sun W, Zheng W, Simeonov A. Drug discovery and development for rare genetic disorders. Am J Med Genet A. 2017;173(9):2307–22.
167. Delavan B, Roberts R, Huang R, et al. Computational drug repositioning for rare diseases in the era of precision medicine. Drug Discov Today. 2018;23(2):382–94.
168. Ekins S, Williams AJ, Krasowski MD, et al. *In silico* repositioning of approved drugs for rare and neglected diseases. Drug Discov Today. 2011;16:298–310.

169. Low ZY, Farouk IA, Lal SK. Drug repositioning: new approaches and future prospects for life-debilitating diseases and the COVID-19 pandemic outbreak. Viruses. 2020;12(9):1058.
170. Yousefi H, Mashouri L, Okpechi SC, et al. Repurposing existing drugs for the treatment of COVID-19/SARS-CoV-2 infection: a review describing drug mechanisms of action. Biochem Pharmacol. 2021;183:114296.
171. Singh TU, Parida S, Lingaraju MC, et al. Drug repurposing approach to fight COVID-19. Pharmacol Rep. 2020;72(6):1479–508.
172. Zhou Y, Hou Y, Shen J, et al. Network-based drug repurposing for novel coronavirus 2019-nCoV/SARS-CoV-2. Cell Discov. 2020;6:14.
173. Li X, Yu J, Zhang Z, et al. Network bioinformatics analysis provides insight into drug repurposing for COVID-2019. Med Drug Discov. 2021;10:100090.
174. Dudley JT, Deshpande T, Butte AJ. Exploiting drug–disease relationships for computational drug repositioning. Brief Bioinform. 2011;12:303–11.
175. Masoudi-Sobhanzadeh Y, Omidi Y, Amanlou M, et al. Drug databases and their contributions to drug repurposing. Genomics. 2020;112(2):1087–95.
176. Brown AS, Patel CJ. A standard database for drug repositioning. Sci Data. 2017;4:170029.
177. Baker NC, Ekins S, Williams AJ, et al. A bibliometric review of drug repurposing. Drug Discov Today. 2018;23(3):661–72.
178. Huang F, Zhang C, Liu Q, et al. Identification of amitriptyline HCl, flavin adenine dinucleotide, azacitidine and calcitriol as repurposing drugs for influenza A H5N1 virus-induced lung injury. PLoS Pathog. 2020;16(3):e1008341.
179. Sánchez-Tejeda JF, Sánchez-Ruiz JF, Salazar JR, et al. A definition of "multitargeticity": identifying potential multitarget and selective ligands through a vector analysis. Front Chem. 2020;8:176.
180. Overington JP, Al-Lazikani B, Hopkins AL. How many drug targets are there? Nat Rev Drug Discov. 2006;5(12):993–6.
181. Sawant MS, Streu CN, Wu L, et al. Toward drug-like multispecific antibodies by design. Int J Mol Sci. 2020;21(20):7496.
182. Dimitrov JD, Pashov AD, Vassilev TL. Antibody polyspecificity: what does it matter? Adv Exp Med Biol. 2012;750:213–26.
183. Morphy R. Selectively nonselective kinase inhibition: striking the right balance. J Med Chem. 2010;53(4):1413–37.
184. Feldmann C, Miljković F, Yonchev D, et al. Identifying promiscuous compounds with activity against different target classes. Molecules. 2019;24:e4185.
185. Maggiora G, Gokhale V. A simple mathematical approach to the analysis of polypharmacology and polyspecificity data. F1000Res. 2017;6:788.
186. Van Regenmortel MH. Specificity, polyspecificity, and heterospecificity of antibody-antigen recognition. J Mol Recognit. 2014;27(11):627–39.
187. Bansal Y, Silakari O. Multifunctional compounds: smart molecules for multifactorial diseases. Eur J Med Chem. 2014;76:31–42.
188. Zhang W, Pei J, Lai L. Computational multitarget drug design. J Chem Inf Model. 2017;57:403–12.
189. Proschak E, Stark H, Merk D. Polypharmacology by design: a medicinal chemist's perspective on multitargeting compounds. J Med Chem. 2019;62:420–44.
190. Zhou J, Jiang X, He S, et al. Rational design of multitarget-directed ligands: strategies and emerging paradigms. J Med Chem. 2019;62:8881–914.
191. Maruca A, Moraca F, Rocca R, et al. Chemoinformatic database building and in silico hit-identification of potential multi-targeting bioactive compounds extracted from mushroom species. Molecules. 2017;22:1571.
192. Maruca A, Lanzillotta D, Rocca R, et al. Multi-targeting bioactive compounds extracted from essential oils as kinase inhibitors. Molecules. 2020;25:2174.
193. Morales JF, Alberca LN, Chuguransky S, et al. Molecular topology and other promiscuity determinants as predictors of therapeutic class – a theoretical framework to guide drug repositioning? Curr Top Med Chem. 2018;18(13):1110–22.
194. Feldmann C, Yonchev D, Bajorath J. Structured data sets of compounds with multi-target and corresponding single-target activity from biological assays. Future Sci OA. 2021;7(5):FSO685.
195. Besnard J, Ruda GF, Setola V, et al. Automated design of ligands to polypharmacological profiles. Nature. 2012;492:215–20.
196. Feldmann C, Yonchev D, Bajorath J. Analysis of biological screening compounds with single- or multi-target activity via diagnostic machine learning. Biomol Ther. 2020;10(12):1605.
197. Chan SY, Loscalzo J. The emerging paradigm of network medicine in the study of human disease. Circ Res. 2012;111(3):359–74.
198. Kitano H. Systems biology: a brief overview. Science. 2002;295(5560):1662–4.
199. Sauer U, Heinemann M, Zamboni N. Getting closer to the whole picture. Science. 2007;316(5824):550–1.
200. Vidal M, Cusick ME, Barabási A-L. Interactome networks and human disease. Cell. 2011;144:986–98.
201. Schadt EE. Molecular networks as sensors and drivers of common human diseases. Nature. 2009;461:218–23.
202. Del Sol A, Balling R, Hood L, et al. Diseases as network perturbations. Curr Opin Biotechnol. 2010;21:566–71.
203. Chuang H-Y, Hofree M, Ideker T. A decade of systems biology. Annu Rev Cell Dev Biol. 2010;26:721–44.
204. Liu Z-P, Wang Y, Zhang X-S, et al. Network-based analysis of complex diseases. IET Syst Biol. 2012;6:22–33.
205. Barabási A-L, Oltvai ZN. Network biology: understanding the cell's functional organization. Nat Rev Genet. 2004;5:101–13.

206. Caldarelli G. Scale-free networks. Oxford University Press; 2007.
207. Barabási AL, Gulbahce N, Loscalzo J. Network medicine: a network-based approach to human disease. Nat Rev Genet. 2011;12(1):56–68.
208. Loscalzo J, Barabasi AL. Systems biology and the future of medicine. Wiley Interdiscip Rev Syst Biol Med. 2011;3(6):619–27.
209. Barabási AL. Network medicine – from obesity to the "diseasome". N Engl J Med. 2007;357(4):404–7.
210. Ravasz E, Somera AL, Mongru DA, et al. Hierarchical organization of modularity in metabolic networks. Science. 2002;297(5586):1551–5.
211. Boezio B, Audouze K, Ducrot P, et al. Network-based approaches in pharmacology. Mol Inform. 2017;36:10.
212. Goh KI, Cusick ME, Valle D, et al. The human disease network. Proc Natl Acad Sci. 2007;104(21):8685–90.
213. Braun P, Rietman E, Vidal M. Networking metabolites and diseases. Proc Natl Acad Sci. 2008;105(29):9849–50.
214. Christakis NA, Fowler JH. The spread of obesity in a large social network over 32 years. N Engl J Med. 2007;357(4):370–9.
215. Hopkins AL. Network pharmacology. Nat Biotechnol. 2007;25(10):1110–1.
216. Yıldırım MA, Goh KI, Cusick ME, et al. Drug – target network. Nat Biotech. 2007;25(10):1119–6.
217. Chiang AP, Butte AJ. Systematic evaluation of drug–disease relationships to identify leads for novel drug uses. Clin Pharmacol Therapeutics. 2009;86(5):507–10.
218. Sonawane AR, Weiss ST, Glass K, et al. Network medicine in the age of biomedical big data. Front Genet. 2019;10:294.
219. Schadt EE, Friend SH, Shaywitz DA. A network view of disease and compound screening. Nat Rev Drug Discov. 2009;8:286–95.
220. Zanzoni A, Soler-López M, Aloy P. A network medicine approach to human disease. FEBS Lett. 2009;583:1759–65.
221. Pawson T, Linding R. Network medicine. FEBS Lett. 2008;582:1266–70.
222. Creixell P, Schoof EM, Erler JT, et al. Navigating cancer network attractors for tumor-specific therapy. Nat Biotechnol. 2012;30:842–8.
223. Leung EL, Cao Z-W, Jiang Z-H, et al. Network-based drug discovery by integrating systems biology and computational technologies. Brief Bioinform. 2013;14(4):491–505.
224. Pujol A, Mosca R, Farrés J, et al. Unveiling the role of network and systems biology in drug discovery. Trends Pharmacol Sci. 2010;31:115–23.
225. Wist AD, Berger SI, Iyengar R. Systems pharmacology and genome medicine: a future perspective. Genome Med. 2009;1:11.
226. Arrell DK, Terzic A. Network systems biology for drug discovery. Clin Pharmacol Ther. 2010;88:120–5.
227. Boran ADW, Iyengar R. Systems approaches to polypharmacology and drug discovery. Curr Opin Drug Discov Devel. 2010;13:297–309.
228. Berger SI, Iyengar R. Network analyses in systems pharmacology. Bioinformatics. 2009;25:2466–72.
229. Zhao S, Iyengar R. Systems pharmacology: network analysis to identify multiscale mechanisms of drug action. Annu Rev Pharmacol Toxicol. 2012;52:505–21.
230. Csermely P, Ágoston V, Pongor S. The efficiency of multi-target drugs: the network approach might help drug design. Trends Pharmacol Sci. 2005;26(4):178–82.
231. Wermuth CG. Multitargeted drugs: the end of the "one-target-one-disease" philosophy? Drug Discov Today. 2004;9(19):826–7.
232. Keith CT, Borisy AA, Stockwell BR. Multicomponent therapeutics for networked systems. Nat Rev Drug Discov. 2005;4(1):71–8.
233. Sidders B, Karlsson A, Kitching L, et al. Network-based drug discovery: coupling network pharmacology with phenotypic screening for neuronal excitability. J Mol Biol. 2018;430(18 Pt A):3005–15.
234. Isgut M, Rao M, Yang C, et al. Application of combination high-throughput phenotypic screening and target identification methods for the discovery of natural product-based combination drugs. Med Res Rev. 2018;38(2):504–24.
235. Berger SI, Iyengar R. Role of systems pharmacology in understanding drug adverse events. Syst Biol Med. 2011;3:129–35.
236. Zhou W, Wang Y, Lu A, Zhang G. Systems pharmacology in small molecular drug discovery. Int J Mol Sci. 2016;17(2):246.
237. Stéphanou A, Fanchon E, Innominato PF, et al. Systems biology, systems medicine, systems pharmacology: the what and the why. Acta Biotheor. 2018;66(4):345–65.
238. van Hasselt JGC, Iyengar R. Systems pharmacology: defining the interactions of drug combinations. Annu Rev Pharmacol Toxicol. 2019;59:21–40.
239. Fliri AF, Loging WT, Volkmann RA. Drug effects viewed from a signal transduction network perspective. J Med Chem. 2009;52(24):8038–46.
240. Teichert RW, Raghuraman S, Memon T, et al. Characterization of two neuronal subclasses through constellation pharmacology. Proc Natl Acad Sci U S A. 2012;2012(109):12758–63.
241. Teichert RW, Memon T, Aman JW, et al. Using constellation pharmacology to define comprehensively a somatosensory neuronal subclass. Proc Natl Acad Sci U S A. 2014;111:2319–24.
242. Teichert RW, Schmidt EW, Olivera BM. Constellation pharmacology: a new paradigm for drug discovery. Annu Rev Pharmacol Toxicol. 2015;55:573–89.
243. Inagaki RT, Raghuraman S, Chase K, et al. Molecular characterization of frog vocal neurons using constellation pharmacology. J Neurophysiol. 2020;123(6):2297–310.

244. Curtice KJ, Leavitt LS, Chase K, et al. Classifying neuronal subclasses of the cerebellum through constellation pharmacology. J Neurophysiol. 2016;115(2):1031–42.
245. Raghuraman S, Xie JY, Giacobassi MJ, et al. Chronicling changes in the somatosensory neurons after peripheral nerve injury. Proc Natl Acad Sci U S A. 2020;117(42):26414–21.
246. Moutal A, Chew LA, Yang X, et al. (S)-lacosamide inhibition of CRMP2 phosphorylation reduces postoperative and neuropathic pain behaviors through distinct classes of sensory neurons identified by constellation pharmacology. Pain. 2016;157(7):1448–63.
247. Raghuraman S, Garcia AJ, Anderson TM, et al. Defining modulatory inputs into CNS neuronal subclasses by functional pharmacological profiling. Proc Natl Acad Sci U S A. 2014;111(17):6449–54.
248. Voets T, Talavera K, Nilius B. Transient receptor potential channel promiscuity frustrates constellation pharmacology. Proc Natl Acad Sci U S A. 2012;109(49):E3338.
249. Oshry B. Seeing systems: unlocking the mysteries of organizational life. Berrett-Koehler; 2008.
250. Auyang SY. Foundations of complex-system theories: in economics, evolutionary biology, and statistical physics. Cambridge University Press; 1999.
251. Kellenberger E. The evolution of molecular biology. EMBO Rep. 2004;5(6):546–9.
252. Delker RK, Mann RS. From reductionism to holism: toward a more complete view of development through genome engineering. Adv Exp Med Biol. 2017;1016:45–74.
253. Fang FC, Casadevall A. Reductionistic and holistic science. Infect Immun. 2011;79(4):1401–4.
254. Segall M. HLA and genetics of IDDM. Holism vs. reductionism? Diabetes. 1988;37(8):1005–8.
255. Torday JS, Miller WB Jr. The resolution of ambiguity as the basis for life: a cellular bridge between Western reductionism and Eastern holism. Prog Biophys Mol Biol. 2017;131:288–97.
256. Michaelson V, Pickett W, Davison C. The history and promise of holism in health promotion. Health Promot Int. 2019;34(4):824–32.
257. Povlsen L, Borup IK. Holism in nursing and health promotion: distinct or related perspectives? – A literature review. Scand J Caring Sci. 2011;25(4):798–805.
258. Frisch NC, Rabinowitsch D. What's in a definition? Holistic nursing, integrative health care, and integrative nursing: aeport of an integrated literature review. J Holist Nurs. 2019;37(3):260–72.
259. Fan D. Holistic integrative medicine: toward a new era of medical advancement. Front Med. 2017;11(1):152–9.
260. Madabushi R, Wang Y, Zineh I. A holistic and integrative approach for advancing model-informed drug development. CPT Pharmacometrics Syst Pharmacol. 2019;8(1):9–11.
261. Collis MC. Integrative pharmacology and drug discovery – is the tide finally turning? Nat Rev Drug Discov. 2006;5(5):377–9.
262. Sperber G, Flaws B. Integrative pharmacology (2nd edition of integrated pharmacology): combining modern pharmacology with integrative medicine. Blue Poppy Press; 2016.
263. Ma C, Wang X, Xu T, et al. An integrative pharmacology-based analysis of refined qingkailing injection against cerebral ischemic stroke: a novel combination of baicalin, geniposide, cholic acid, and hyodeoxycholic acid. Front Pharmacol. 2020;11:519.
264. Mehta D. Integrative medicine and cardiovascular disorders. Prim Care. 2017;44(2):351–67.
265. Tabrizi M, Zhang D, Ganti V, et al. Integrative pharmacology: advancing development of effective immunotherapies. AAPS J. 2018;20(4):66.
266. Park K. A review of computational drug repurposing. Transl Clin Pharmacol. 2019;27(2):59–63.
267. Karaman B, Sippl W. Computational drug repurposing: current trends. Curr Med Chem. 2019;26(28):5389–409.
268. Carrieri A, Pérez-Nueno VI, Lentini G, et al. Recent trends and future prospects in computational GPCR drug discovery: from virtual screening to polypharmacology. Curr Top Med Chem. 2013;13(9):1069–97.
269. Hopkins AL. Drug discovery: predicting promiscuity. Nature. 2009;462(7270):167–8.
270. van Tongeren JMZ, Harkes-Idzinga SF, van der Sijs H, et al. The development of practice recommendations for drug-disease interactions by literature review and expert opinion. Front Pharmacol. 2020;11:707.
271. Hanlon JT, Perera S, Newman AB, et al. Health ABC Study. Potential drug-drug and drug-disease interactions in well-functioning community-dwelling older adults. J Clin Pharm Ther. 2017;42(2):228–33.
272. Dumbreck S, Flynn A, Nairn M, et al. Drug-disease and drug-drug interactions: systematic examination of recommendations in 12 UK national clinical guidelines. BMJ. 2015;350:h949.
273. Niu J, Straubinger RM, Mager DE. Pharmacodynamic drug-drug interactions. Clin Pharmacol Ther. 2019;105(6):1395–406.
274. Carpenter M, Berry H, Pelletier AL. Clinically relevant drug-drug interactions in primary care. Am Fam Physician. 2019;99(9):558–64.
275. Roblek T, Vaupotic T, Mrhar A, et al. Drug-drug interaction software in clinical practice: a systematic review. Eur J Clin Pharmacol. 2015;71(2):131–42.
276. Masnoon N, Shakib S, Kalisch-Ellett L, et al. What is polypharmacy? A systematic review of definitions. BMC Geriatr. 2017;17(1):230.
277. Ased S, Wells J, Morrow LE, Malesker MA. Clinically significant food-drug interactions. Consult Pharm. 2018;33(11):649–57.

278. Lehár J, Krueger AS, Avery W, et al. Synergistic drug combinations tend to improve therapeutically relevant selectivity. Nat Biotechnol. 2009;27(7):659–66.
279. Zimmermann GR, Lehár J, Keith CT. Multi-target therapeutics: when the whole is greater than the sum of the parts. Drug Discov Today. 2007;12(1–2):34–42.
280. Bianchi MT, Pathmanathan J, Cash SS. From ion channels to complex networks: magic bullet versus magic shotgun approaches to anticonvulsant pharmacotherapy. Med Hypotheses. 2009;72(3):297–305.
281. Ding P, Yin R, Luo J, et al. Ensemble prediction of synergistic drug combinations incorporating biological, chemical, pharmacological, and network knowledge. IEEE J Biomed Health Inform. 2019;23(3):1336–45.
282. Loisios-Konstantinidis I, Paraiso RLM, Fotaki N, et al. Application of the relationship between pharmacokinetics and pharmacodynamics in drug development and therapeutic equivalence: a PEARRL review. J Pharm Pharmacol. 2019;71(4):699–723.
283. Chen B, Dong JQ, Pan WJ, et al. Pharmacokinetics/pharmacodynamics model-supported early drug development. Curr Pharm Biotechnol. 2012;13(7):1360–75.
284. Chaikin P, Rhodes GR, Bruno R, et al. Pharmacokinetics/pharmacodynamics in drug development: an industrial perspective. J Clin Pharmacol. 2000;40(12 Pt 2):1428–38.
285. Talevi A, Bruno-Blanch LE. On the development of new antiepileptic drugs for the treatment of pharmacoresistant epilepsy: different approaches to different hypothesis. In: Rocha L, Cavalheiro EA, editors. Pharmacoresistance in epilepsy: from genes and molecules to promising therapies. New York: Springer; 2013. p. 207–24.
286. Benedetti R, Conte M, Iside C, et al. Epigenetic-based therapy: from single- to multi-target approaches. Int J Biochem Cell Biol. 2015;69:121–31.
287. Zhao S, Iyengar R. Systems pharmacology: network analysis to identify multiscale mechanisms of drug action. Ann Rev Pharmacol Toxicol. 2012;52(52):505–21.
288. Lounkine E, Keiser MJ, Whitebread S, et al. Large-scale prediction and testing of drug activity on side-effect targets. Nature. 2012;486(7403):361–7.
289. Campillos M, Kuhn M, Gavin AC, et al. Drug target identification using side-effect similarity. Science. 2008;321(5886):263–6.
290. Oprea TI, Nielsen SK, Ursu O, et al. Associating drugs, targets and clinical outcomes into an integrated network affords a new platform for computer-aided drug repurposing. Mol Inform. 2011;30(2–3):100–11.
291. Ravot E, Lisziewicz J, Lori F. New uses for old drugs in HIV infection: the role of hydroxyurea, cyclosporin and thalidomide. Drugs. 1999;58(6):953–63.
292. Chong CR, Sullivan DJ Jr. New uses for old drugs. Nature. 2007;448(7154):645–6.
293. Vazquez-Martin A, López-Bonetc E, Cufí S, et al. Repositioning chloroquine and metformin to eliminate cancer stem cell traits in pre-malignant lesions. Drug Resist Updat. 2010;14(4–5):212–23.
294. Ravikumar B, Aittokallio T. Improving the efficacy-safety balance of polypharmacology in multi-target drug discovery. Expert Opin Drug Discov. 2018;13(2):179–92.
295. Amelio I, Lisitsa A, Knight RA, et al. Polypharmacology of approved anticancer drugs. Curr Drug Targets. 2017;18(5):534–43.
296. Antolin AA, Workman P, Mestres J, et al. Polypharmacology in precision oncology: current applications and future prospects. Curr Pharm Des. 2016;22(46):6935–45.
297. Schierle S, Flauaus C, Heitel P, et al. Boosting anti-inflammatory potency of zafirlukast by designed polypharmacology. J Med Chem. 2018;61(13):5758–64.
298. Tan Z, Chaudhai R, Zhang S. Polypharmacology in drug development: a minireview of current technologies. ChemMedChem. 2016;11(12):1211–8.
299. Senn N, Ott M, Lanz J, Riedl R. Targeted polypharmacology: discovery of a highly potent non-hydroxamate dual matrix metalloproteinase (MMP)-10/-13 inhibitor. J Med Chem. 2017;60(23):9585–98.
300. Meyers J, Chessum NEA, Ali S, et al. Privileged structures and polypharmacology within and between protein families. ACS Med Chem Lett. 2018;9(12):1199–204.
301. Passeri D, Camaioni E, Liscio P, et al. Concepts and molecular aspects in the polypharmacology of PARP-1 inhibitors. ChemMedChem. 2016;11(12):1219–26.
302. Brodie JS, Di Marzo V, Guy GW. Polypharmacology shakes hands with complex aetiopathology. Trends Pharmacol Sci. 2015;36(12):802–21.
303. Hayes JD, Wolf CR. Molecular mechanisms of drug resistance. Biochem J. 1990;272(2):281–95.
304. Levy SB, Marshall B. Antibacterial resistance worldwide: causes, challenges and responses. Nat Med. 2004;10(12 Suppl):S122–9.
305. Szakács G, Paterson JK, Ludwig JA, et al. Targeting multidrug resistance in cancer. Nat Rev Drug Discov. 2006;5(3):219–34.
306. He J, McLaughlin RP, van der Noord V, et al. Multi-targeted kinase inhibition alleviates mTOR inhibitor resistance in triple-negative breast cancer. Breast Cancer Res Treat. 2019;178(2):263–74.
307. Xie L, Bourne PE. Developing multi-target therapeutics to fine-tune the evolutionary dynamics of the cancer ecosystem. Front Pharmacol. 2015;6:209.
308. Wetzel C, Lonneman M, Wu C. Polypharmacological drug actions of recently FDA approved antibiotics. Eur J Med Chem. 2021;209:112931.
309. Wu X, Jiang R, Zhang MQ, et al. Network-based global inference of human disease genes. Mol Syst Biol. 2008;4:189.

310. Masoudi-Nejad A, Mousavian Z, Bozorgmehr JH. Drug-target and disease networks: polypharmacology in the post-genomic era. In Silico Pharmacol. 2013;1:17.
311. Benek O, Korabecny J, Soukup O. A perspective on multi-target drugs for Alzheimer's disease. Trends Pharmacol Sci. 2020;41(7):434–45.
312. Muhammad J, Khan A, Ali A, et al. Network pharmacology: exploring the resources and methodologies. Curr Top Med Chem. 2018;18(12):949–64.
313. Chen Y, Zhu J, Lum PY, et al. Variations in DNA elucidate molecular networks that cause disease. Nature. 2008;452(7186):429–35.
314. Boran ADW, Iyengar R. Systems approaches to polypharmacology and drug discovery. Curr Opin Drug Disc Dev. 2010;13(3):297–309.
315. Kibble M, Saarinen N, Tang J, Wennerberg K, Mäkelä S, Aittokallio T. Network pharmacology applications to map the unexplored target space and therapeutic potential of natural products. Nat Prod Rep. 2015;32(8):1249–66.
316. Zhang GB, Li QY, Chen QL, et al. Network pharmacology: a new approach for Chinese herbal medicine research. Evid Based Complement Alternat Med. 2013;2013:621423.
317. Gu J, Zhang H, Chen L, et al. Drug-target network and polypharmacology studies of a Traditional Chinese Medicine for type II diabetes mellitus. Comput Biol Chem. 2011;35(5):293–7.
318. Brinda KV, Vishveshwara S, Vishveshwara S. Random network behaviour of protein structures. Mol BioSyst. 2010;6(2):391–8.
319. Hou L, Wang L, Berg A, et al. Comparison and evaluation of network clustering algorithms applied to genetic interaction networks. Front Biosci. 2012;4:2150–61.
320. Cheng XR, Zhou WX, Zhang YX. Experimental techniques in network pharmacology. Chin J Pharmacol Toxicol. 2012;26(2):131–7.
321. Wu LH, Wang Y, Fan X. Tools for network pharmacology study: network visualization and network analysis. China J Chin Mater Medica. 2011;36(21):2923–5.
322. Ye H, Wei J, Tang K, et al. Drug repositioning through network pharmacology. Curr Top Med Chem. 2016;16(30):3646–56.
323. Pachaiyappan B, Woster PM. Design of small molecule epigenetic modulators. Bioorg Med Chem Lett. 2014;24(1):21–32.
324. Lu Q, Quinn AM, Patel MP, et al. Perspectives on the discovery of small-molecule modulators for epigenetic processes. J Biomol Screen. 2012;17(5):555–71.
325. Mohammad HP, Barbash O, Creasy CL. Targeting epigenetic modifications in cancer therapy: erasing the roadmap to cancer. Nat Med. 2019;25(3):403–18.
326. Janssens Y, Wynendaele E, Vanden Berghe W, et al. Peptides as epigenetic modulators: therapeutic implications. Clin Epigenet. 2019;11(1):101.
327. de Lera AR, Ganesan A. Epigenetic polypharmacology: from combination therapy to multitarget drugs. Clin Epigenetics. 2016;8:105.
328. Levine AJ. The p53 protein plays a central role in the mechanism of action of epigenetic drugs that alter the methylation of cytosine residues in DNA. Oncotarget. 2017;8(5):7228–30.
329. Ferguson LR, Tatham AL, Lin Z, et al. Epigenetic regulation of gene expression as an anticancer drug target. Curr Cancer Drug Targets. 2011;11(2):199–212.
330. Evison BJ, Bilardi RA, Chiu FC, et al. CpG methylation potentiates pixantrone and doxorubicin-induced DNA damage and is a marker of drug sensitivity. Nucleic Acids Res. 2009;37(19):6355–70.
331. Dawson MA, Kouzarides T. Cancer epigenetics: from mechanism to therapy. Cell. 2012;150(1):12–27.
332. Atlante S, Mongelli A, Barbi V, et al. The epigenetic implication in coronavirus infection and therapy. Clin Epigenetics. 2020;12(1):156.
333. Costantino S, Mohammed SA, Ambrosini S, et al. Epigenetic processing in cardiometabolic disease. Atherosclerosis. 2019;281:150–8.
334. Fustin JM, Doi M, Yamaguchi Y, Hida H, Nishimura S, Yoshida M, Isagawa T, Morioka MS, Kakeya H, Manabe I, Okamura H. RNA-methylation-dependent RNA processing controls the speed of the circadian clock. Cell. 2013;155(4):793–806.
335. Boriack-Sjodin PA, Ribich S, Copeland RA. RNA-modifying proteins as anticancer drug targets. Nat Rev Drug Discov. 2018;17(6):435–53.
336. Selberg S, Blokhina D, Aatonen M, et al. Discovery of small molecules that activate RNA methylation through cooperative binding to the METTL3-14-WTAP complex active site. Cell Rep. 2019;26(13):3762–71.e5.
337. Kanasty R, Dorkin JR, Vegas A, Anderson D. Delivery materials for siRNA therapeutics. Nat Mater. 2013;12(11):967–77.
338. Setten RL, Rossi JJ, Han SP. The current state and future directions of RNAi-based therapeutics. Nat Rev Drug Discov. 2019;18(6):421–46.
339. Gavrilov K, Saltzman WM. Therapeutic siRNA: principles, challenges, and strategies. Yale J Biol Med. 2012;85(2):187–200.
340. Stein CA, Narayanan R. Antisense oligodeoxynucleotides. Curr Opin Oncol. 1994;6(6):587–94.
341. Chan JH, Lim S, Wong WS. Antisense oligonucleotides: from design to therapeutic application. Clin Exp Pharmacol Physiol. 2006;33(5–6):533–40.
342. Chirila TV, Rakoczy PE, Garrett KL, et al. The use of synthetic polymers for delivery of therapeutic antisense oligodeoxynucleotides. Biomaterials. 2002;23(2):321–42.
343. Mahato RI, Cheng K, Guntaka RV. Modulation of gene expression by antisense and antigene oligodeoxynucleotides and small interfering RNA. Expert Opin Drug Deliv. 2005;2(1):3–28.
344. Holmlund JT. Applying antisense technology: Affinitak and other antisense oligonucleotides

in clinical development. Ann N Y Acad Sci. 2003;1002:244–51.
345. Dolnick BJ. Naturally occurring antisense RNA. Pharmacol Ther. 1997;75(3):179–84.
346. Faghihi MA, Wahlestedt C. Regulatory roles of natural antisense transcripts. Nat Rev Mol Cell Biol. 2009;10(9):637–43.
347. Wang Z. Antisense RNA and cancer. In: Chakrabarti J, Mitra S, editors. Cancer and noncoding RNAs, Translational epigenetics series. London: Elsevier AP Academic Press; 2017.
348. Yu AM, Choi YH, Tu MJ. RNA drugs and RNA targets for small molecules: principles, progress, and challenges. Pharmacol Rev. 2020;72(4):862–98.
349. Khorkova O, Myers AJ, Hsiao J, Wahlestedt C. Natural antisense transcripts. Hum Mol Genet. 2014;23(R1):R54–63.
350. Wahlestedt C. Natural antisense and noncoding RNA transcripts as potential drug targets. Drug Discov Today. 2006;11(11–12):503–8.
351. Wang Z. MicroRNA-interference technologies. Heidelberg/New York: Springer; 2009.
352. Wang Z. MicroRNA mimetics: design, validation and applications. New York: Springer/Humana Press; 2012.
353. Wang Z. The principle of miRNA-masking antisense oligonucleotides technology. In: Wu W, editor. MicroRNA and cancer: methods and protocols. New York: Springer/Humana Press; 2011. p. 43–50.
354. Wang Z. The concept of multiple-target anti-miRNA antisense oligonucleotides technology. In: Wu W, editor. MicroRNA and cancer: methods and protocols. New York: Springer/Humana Press; 2011. p. 51–8.
355. Zhang Y, Jiao L, Sun L, et al. LncRNA ZFAS1 as a SERCA2a inhibitor to cause intracellular Ca^{2+} overload and contractile dysfunction in a mouse model of myocardial infarction. Circ Res. 2018;122(10):1354–68.
356. Matsui M, Corey DR. Non-coding RNAs as drug targets. Nat Rev Drug Discov. 2017;16(3):167–79.
357. Mishra S, Verma SS, Rai V, et al. Long non-coding RNAs are emerging targets of phytochemicals for cancer and other chronic diseases. Cell Mol Life Sci. 2019;76(10):1947–66.
358. Wang Y, Chen S, Chen L, et al. Associating lncRNAs with small molecules via bilevel optimization reveals cancer-related lncRNAs. PLoS Comput Biol. 2019;15(12):e1007540.
359. Mäkinen P, Ruotsalainen AK, Ylä-Herttuala S. Nucleic acid-based therapies for atherosclerosis. Curr Atheroscler Rep. 2020;22(2):10.
360. Hansen TB, Wiklund ED, Bramsen JB, et al. miRNA-dependent gene silencing involving Ago2-mediated cleavage of a circular antisense RNA. EMBO J. 2011;30(21):4414–22.
361. Xu T, Wang M, Jiang L, et al. CircRNAs in anticancer drug resistance: recent advances and future potential. Mol Cancer. 2020;19(1):127.
362. Huang CK, Kafert-Kasting S, Thum T. Preclinical and clinical development of noncoding RNA therapeutics for cardiovascular disease. Circ Res. 2020;126(5):663–78.
363. Ruan H, Xiang Y, Ko J, et al. Comprehensive characterization of circular RNAs in ~ 1000 human cancer cell lines. Genome Med. 2019;11(1):55.
364. Hecker M, Wagner AH. Transcription factor decoy technology: a therapeutic update. Biochem Pharmacol. 2017;144:29–34.
365. Osako MK, Nakagami H, Morishita R. Modification of decoy oligodeoxynucleotides to achieve the stability and therapeutic efficacy. Curr Top Med Chem. 2012;12(15):1603–7.
366. Morishita R, Aoki M, Kaneda Y. Decoy oligodeoxynucleotides as novel cardiovascular drugs for cardiovascular disease. Ann N Y Acad Sci. 2001;947:294–301.
367. Nakagami H, Tomita N, Kaneda Y, et al. Anti-oxidant gene therapy by NF kappa B decoy oligodeoxynucleotide. Curr Pharm Biotechnol. 2006;7(2):95–100.
368. Ganesan A. Multitarget drugs: an epigenetic epiphany. ChemMedChem. 2016;11(12):1227–41.
369. Franci G, Miceli M, Altucci L. Targeting epigenetic networks with polypharmacology: a new avenue to tackle cancer. Epigenomics. 2010;2(6):731–42.
370. Singh AN, Sharma N. Epigenetic modulators as potential multi-targeted drugs against hedgehog pathway for treatment of cancer. Protein J. 2019;38(5):537–50.
371. Ranganna K, Selvam C, Shivachar A, et al. Histone deacetylase Inhibitors as multitarget-directed epi-drugs in blocking PI3K oncogenic signaling: a polypharmacology approach. Int J Mol Sci. 2020;21(21):8198.
372. Álvarez R, Altucci L, Gronemeyer H, et al. Epigenetic multiple modulators. Curr Top Med Chem. 2011;11(22):2749–87.

2 Strategies of Polypharmacology

Abstract

In Chap. 1, we have learned that the modern pharmacological therapy of human disease and new drug discovery are based primarily on four major strategies/modality: the STD strategy, the CDT strategy, the FDC strategy, and the MTD strategy. In addition, the "Natural Multicomponent Medicine" is in its widespread use either as a CAM (complementary and alternative medicine) or a mainstay medicine in some countries and regions. Among these pharmacological strategies for disease therapy and drug development, STD belongs to the Basic and Clinical Pharmacology, and CDT, FDC, and MTD fall into more advanced pharmacology—Polypharmacology and are essentially the applications of polypharmacology to fundamental research and clinical practice. Polypharmacology-based drugs aim to achieve a synergistic effect or target multiple disease-related signaling pathways aiming to ensure more complete and radical therapeutic effects. While the CDT and the FDC strategies are based on the combination of two or more active principles, each with its own specific target, the MTD strategy is focused on the development of a single molecule able to simultaneously interact with multiple different targets. To help better acquire fundamental knowledge and understand clinical applications of Polypharmacology, this chapter provides in-depth interpretations on the three major polypharmacology strategies. Specifically, comparisons of concepts, characteristics, strengths, and weaknesses, and types among CDT, FDC, and MTD are summarized. We proposed some new concepts related to each of the three Polypharmacological approaches and present them in this chapter.

Keywords

Polypharmacology-based drugs · Combination drug therapy (CDT) · Fixed-dose single pill combination (FDC) · Multitarget drug (MTD) · Comparison

2.1 CDT Strategy—The "Combination" Strategy

2.1.1 Concept of CDT

CDT, also termed "drug combination therapy/combination drug therapy (CDT)" [1–3], "drug cocktail" [4–6], "multiple-medication therapy (MMT)" [7–9], "multicomponent therapeutics" [10, 11], or "free-equivalent combination (FEC)" [12, 13], refers to Combination drug therapy (CDT) which is a "Combination" strategy with the concurrent or sequential administration of two or more separate therapeutic agents in the form of free-dose combination (FEC) for the treatment of human disease. It is the methodical

Z. Wang, B. Yang, *Polypharmacology*, https://doi.org/10.1007/978-3-031-04998-9_2

combination of pharmacophores from selective, single-target agents. The first clinical success of the CDT approach dates back to 1980 and consists in the treatment of childhood acute lymphoblastic leukemia (ALL), based on the co-administration of methotrexate, vincristine, prednisone, and 6-mercaptopurine [14]. Ever since then, CDT, the use of cocktails or multi-component drugs to address several targets simultaneously, has become the most widely employed polypharmacology-based therapeutic approach to treat multifactorial diseases such as cancer, cardiovascular diseases, and neurological disorders. We have gained tremendous experience and knowledge about the CDT strategy from intensive fundamental research and clinical practice.

According to Mihich and Grindey [15], in the CDT strategy, the type of drug interactions can be divided into three broad categories:

(a) Combinations based on cooperative effects of active drugs.
(b) Combinations in which the effectiveness of an active drug is increased by the concurrent administration of an inactive agent.
(c) Combination of an active drug with an agent capable of selectively reversing the toxicity of the first drug.

2.1.2 Basic Requirement for CDT

The CDT strategy is not a simple or an arbitrary co-administration of different drugs; instead, it follows some specific rules in order to warrant better therapeutic efficacy and safety profile. The following principles are commonly applied to the CDT strategy.

2.1.2.1 Multitargeting with Complementary Mechanisms

Combining drugs should target one or more disease-related pathways with complementary mechanisms of action to yield more complete and radical therapeutic outcomes or synergetic effects and optimize the therapeutic efficacy [3, 16].

2.1.2.2 Nonoverlapping Toxicity

Nonoverlapping toxicity should be a major consideration for selecting drug combination candidates [17, 18]. One of the goals of combination drug therapy is to minimize the toxicity occurring with conventional monotherapy, and any combinations that could potentially enhance existing or create new adverse side effects must be avoided.

2.1.2.3 Pharmacokinetics Comparability or Physicochemical Compatibility

Combining drugs should have comparable pharmacokinetics for synchronizing the effects of all active components to maximize the synergetic action and for rationally formulating therapeutic regimens [19].

2.1.2.4 Minimizing the Number of Combining Drugs

Unless with proven benefits (optimized efficacy and minimized toxicity) from higher-order of drug combination, the number of drugs in combination should always be kept minimal to reduce patients' economic burden (the cost of paying medicine bills) and pill burden (the difficulty of keeping medication compliance).

2.1.3 Strengths of CDT

Drugs have both therapeutic and adverse effects. A general goal in polypharmacology is to optimize the therapeutic efficacy while reducing the adverse event risks. Traditionally, this is done through medicinal chemistry by altering drug structure or by tailoring the choice of drug or dose to an individual patient's specific health conditions or pharmacogenomic status [20]. The CDT strategy has become a universally accepted pharmacological therapy in clinical practice owing to its following advantages [2, 3, 21].

2.1.3.1 Minimizing Compensatory Effect Due to Redundancies in Signaling Networks

Cellular function in living cells is delicately regulated by a multitude of signaling pathways that form complex signaling networks. Within the networks, different signaling pathways are in constant compensatory interactions conferring functional redundancies. The ensuing compensatory mechanisms due to robustness and redundancy of biological pathways may weaken the body's dynamic ability to respond to further insults and reduce the efficacy of conventional single target treatments. Many disease-related signaling networks have redundancies that can compensate for loss of a single component produced by a single drug, but not for loss of two or more components produced by combining drugs. That is, two or more drugs that are individually ineffective because of redundancies in signaling networks could be effective in combination that prevents the compensatory mechanism.

2.1.3.2 Creating Synergistic Actions

Combining drugs with complementary mechanisms of action can target multiple factors and/or multiple mechanisms of a disease, which might generate synergistic or additive actions against a disease, amplifying the therapeutic effectiveness and yielding more complete and radical therapeutic outcomes. Many concepts have been proposed to explain the synergistic interaction between two active drugs at the level of the target cell. These include multiple inhibition of a single enzyme, enhanced activation, decreased inactivation, increased drug uptake, sequential blockade, concurrent inhibition, complimentary inhibition, and concerted inhibition [15].

2.1.3.3 Combination of Infinite Number of Therapeutical Agents

CDT allows for combinations of theoretically and virtually infinite/unlimited number of different therapeutical agents (chemical substances, small-molecule compounds, peptides, proteins, antibodies, etc.) as needed. Combination of two or three drugs is now a common strategy for the treatment of a variety of human diseases; however, for effective therapy of cancer, higher-order drug combinations are sometimes required [22]. One such example is using a cocktail of drugs including a protease inhibitor, a nucleoside-based reverse transcriptase inhibitor, a nonnucleoside reverse transcriptase inhibitor, and/or an integrase inhibitor to treat patients infected with HIV [23]. Similar to these approaches, mixtures of rMAbs are now being developed as human therapeutics [24]. Nonetheless, unnecessary addition of more than the needed number of drugs should be avoided.

2.1.3.4 Multiple Options of Combination

CDT allows for combinations of two or more completely different forms of chemical/compound, for example, the small molecule/antibody combinations, and the traditional chemotherapeutics/epigenetic drugs (epi-drugs) combinations [25]. For several human therapies, therapeutic recombinant monoclonal antibodies may be used either alone or in combination with small molecule-based drugs to treat patients. One classic example of this paradigm is the treatment of rheumatoid arthritis patients with the small molecule, methotrexate, and an anti-TNF-α antibody such as infliximab, or adalimumab, or golimumab [26].

2.1.3.5 Offering the Possibility of Personalized Treatment

One of the central aspects of the CDT approach consists in the possibility to choose different dosages, establishing a prolonged administration of drugs and increasing the number and type of potential therapeutic schedules, paving the way to a personalized treatment. The CDT approach as a free-dose combination allows to modulate the dose of each single drug within the cocktail, according to the patient's needs [27]. CDT can also allow for personalized therapy with varying combinations of different agents, which can be predetermined and readjusted according to patients' conditions and dynamic disease progression.

2.1.3.6 Minimizing Drug Toxicity or Side Effects

The CDT approach can minimize adverse events due to drug toxicity or side effects through at least two mechanisms.

First, CDT allows for reduction of doses of the components to minimize the possible adverse events and intolerability that higher therapeutic doses of monotherapy could cause.

Second, addition of another drug to conventional monotherapy or existing combination therapy offers an opportunity to mitigate serious adverse events produced by the first drug or drug combination, for example, the use of atropinics to minimize the muscarinic adverse effects of cholinesterase inhibitors that are used for expedited recovery from nondepolarizing neuromuscular blockers [28]. Another example was presented by Zhao et al. [29] who tried to test the CDT approach for reducing the risk of myocardial infarction induced by rosiglitazone, an efficacious antidiabetic drug. Their analyses indicate that addition of exenatide, another antidiabetic agent of a different class, significantly reduces rosiglitazone-associated MI.

2.1.3.7 Mitigating Drug Resistance

CDT can mitigate drug resistance because the development of resistance to multiple drugs is much less likely. In fact, CDT has been used as a tool to prevent emergence of bacterial resistance [30].

2.1.4 Limitations of CDT

Despite the benefits, the CDT approach also has its disadvantages. In fact, clinical outcome of CDT is not always as good as anticipated; rather is often associated with unexpected higher toxicities.

2.1.4.1 Poor Patient Compliance

One of the main disadvantages is higher patient pill burden that can negatively impact adherence and persistence or patient compliance (the extent to which the patient follows medical instructions), which in turn can lead to poorer therapeutic outcomes [31]. Worldwide, an estimated 50% of CVD patients employing therapeutic agents as a primary prevention measure and 43% as a secondary prevention measure fail to adhere to treatment programs in the first 2 years [32]. Furthermore, when stratified by drug class, a recent meta-analysis of 34 studies showed that 46% of patients on statins, 41% on antihypertensive agents and 30% on aspirin fail to adhere to their medication [33].

2.1.4.2 Possible Different Bioavailability

Another potential limitation of CDT is the possible different bioavailability among combing agents due to their distinct and unpredictable pharmacokinetics. Such a distinction could well result in different times to therapeutic or peak effect, differing metabolism routs and rates, and different physiochemical characteristics such as solubility of the active substances that may modify the bloodstream uptake, not guaranteeing the simultaneous presence of the required drugs at the desired concentrations in the target tissue or desired spatiotemporal distribution. These differences could render appropriate dosages and regimens difficult.

2.1.4.3 Disturbing Drug Interactions

Undesirable drug-drug interactions (DDIs) might happen between or among the combining drugs, which could cause unpredictable adverse events. In addition, CDT may create more chances for off-target DDIs, which could also increase the probability for adverse events to occur.

2.1.4.4 Complicating Drug Development Process

For drug development, when the agents are used in combination or association, regulatory agencies generally require the demonstration of safety of each individual drug in addition to the combination before clinical trials. This could tremendously complicate and lengthen the development process. If a combination product comprises drugs that are already available, evidence of its likely efficacy and safety can largely be obtained from existing large-scale clinical trials and meta-analyses. Although the regulatory authorities will require, at the very least, a demonstration of pharmacodynamic/pharmacokinetic efficacy using surrogate markers, they will hopefully not require hard endpoint data. Large-scale morbidity and mortality trials would be both prohibitively expensive and ethically challenging.

2.1.4.5 Lengthening Drug Approval Procedure

If the drugs of interest for combination are owned by different pharmaceutical companies, the procedure for development can require an extended time.

2.1.5 Types of CDT

In clinical applications, CDT may be applied at different forms according to a patient's specific conditions (age, drug tolerance, overall health status, etc.), disease entities, and pathological stages. Different forms of CDT offer an array of options to meet different needs of different patients with different diseases. The decision on a specific form of CDT is made to improve the therapeutic outcomes with maximized efficacy and minimized toxicity. The choice of a specific form of CDT may also aim to enhance a patient's adherence to medication.

2.1.5.1 Classification Based on Chronological Order of Drug Administration

Concurrent-Administration Combination

Concurrent-administration combination or co-administration combination is the situation that two or more drugs are administered together and simultaneously [34–36]. This approach is appropriate for the combining drugs that have comparable pharmacokinetics. It can help patients keep good compliance.

Sequential-Administration Combination

Sequential-administration combination refers to the regimen with the second medication being administered with a certain time lag following administration of the first drug or drug combination [37–40]. This method can be used for combining drugs with distinct pharmacokinetics, and it can also help minimize overlapping toxicity that the combining agents could cause.

2.1.5.2 Classification Based on the Design of Drug Combination

Designed CDT

Designed CDT drugs are the drug combinations composed based on the knowledge about the chemical structure, physiochemical characteristics, and pharmacological properties of the combining compounds and the relevant basic research data and clinical trial outcomes. Hence, it is a rational approach of combination drug therapy.

Nondesigned CDT or Unintended CDT

Unintended CDT drugs refer to the nonartificially designed or naturally occurring drug mix prepared based on empirical information without the knowledge of chemical ingredients. Herbal medicine is an example of unintended CDT, in which while the mixture of multiple herbs is rationally chosen based on certain knowledge acquired from observation or experience, the exact bioactive substances or therapeutic components contained in the preparation are not known.

Partially Designed CDT

Partially designed CDT drugs are those that are designed or prepared based on known or at least partially informed facts related to their application as medicine or complementary medicine commonly used in the treatment of early-stage Alzheimer's disease, vascular dementia, peripheral claudication, and tinnitus of vascular origin [41, 42]. For example, Ginkgo biloba extract, an herbal extract derived from the leaves of the Maidenhair Tree, contains the following chemical compositions: Ginkgo biloba leaf extract contains flavonol and flavone glycosides, lactone derivatives (ginkgolides), bilobalide, ascorbic acid, catechin, iron-based superoxide, 6-hydroxykinuretic acid, protocatechuic acid, shikimic acid, sterols and allium acid [41, 42]. The major classes of active ingredients are the ginkgolides and bilobalides (also known as terpenes) and the flavonoids. Ginkgo biloba extract is usually standardized for 24% ginkgo flavone glycosides, 6% total terpene lactones, and more.

Recombinant Antibody Mixture (RAM)

Antibody mixtures are considered the next generation of antibody therapeutics [26, 43]. An antibody mixture usually contains two or more antibodies directed against either a single antigen or different antigens [26, 43]. Antibody mixtures against a single antigen may recognize different binding epitopes and may exhibit synergies in neutralizing and/or cell killing, whereas antibody mixtures against different antigens contain well-defined and well-characterized antibodies that target well-defined antigens. There are at least eight different examples of antibody mixtures that are currently in clinical development [26]. These antibody mixtures may provide advantages over antibody monotherapy because the individual antibody components contained in the mixture may complement each other to target the diseases [44]. Recombinant antibody mixtures offering targeting of more than one antigen is one of the new promising antibody technologies resulting in higher therapeutic effectiveness and/or broader reactivity.

Several strategies have been developed to achieve multiple targeting, including the administration of two or more mAbs to the patient, bispecific antibodies, and antibody mixtures [45, 46]. Recombinant antibody mixtures (RAM) can in principle be manufactured by different approaches but two main strategies are often applied, either individual manufacturing of the constituent antibodies or single batch manufacturing of the recombinant antibody mixture [45]. Symphogen has developed an expression platform, Sympress™, allowing single batch manufacturing of recombinant antibody mixtures, while other companies are currently using a manufacturing strategy based on production of the individual constituent monoclonal antibodies [46]. It is believed that the single batch manufacturing approach expressing a mixture of full-length IgG provides a robust and reproducible platform that can be used for cost-effective manufacturing of recombinant antibody mixtures [45]. However, RAM increases the complexity of development and manufacture. To address this issue, a method to efficiently express multiple monoclonal antibodies from a single cell has been developed which might be applicable to industrial-scale production of RAM [47].

2.1.5.3 Classification Based on the Number of Combining Drugs

Dual Combination

Two-drugs combination is the most commonly used CDT strategy. For example, dual antiplatelet therapy is commonly used for secondary stroke prevention [48].

Triple Combination

Three-drugs combination is also a frequently used therapeutic strategy, especially for the treatment of hypertension and cancer. It is estimated that triple-combination therapies are required to achieve blood pressure control in approximately one-fourth to one-third of patients [49]. Both the 5-fluorouracil/oxaliplati/leucovorin triple combination and the 5-fluorouracil/irinotecan/leucovorin triple combination are recommended as the preferred cytotoxic treatment options for first-line treatment of metastatic colorectal cancer [22].

Higher-Order Combinations

Combinations of more than 3 drugs for intensive therapy of certain severe or critical conditions. For example, for more advanced metastatic colorectal cancer, the quadruple combination with 5-FU + oxaliplatin + irinotecan + leucovorin is often employed [22].

In addition, there is another combinational approach with one of the agents at a fixed dose and the other one (or more) agent(s) at a series of escalating doses: "half fixed dose and half free dose" combination. Such a method is frequently used in testing anticancer drugs for optimal dosages that produced optimal efficacy with minimal toxicity.

2.2 FDC Strategy—The "Admixture" Strategy

2.2.1 Concept of FDC

The FDC strategy, "fixed-dose combination (FDC)" [53–55], or "multicompound medication (MCM)" [7], or "Polypill" [50–52], is a "Mix" strategy with multiple active agents (two or more) mixed or blended inseparable within the same pharmaceutical formulation or in a single pill (tablet or capsule) for disease treatment and/or risk factor control. The concept of FDC was initially proposed by Wald and Law to reduce cardiovascular disease (CVD) [50] and was later rapidly applied to a wide variety of other diseases [56–62]. The World Health Organization (WHO) reported that FDCs are becoming increasingly important from a public health perspective and are particularly useful in the treatment of many chronic diseases such as CVDs, diabetes, and obesity, as well as in the management of human immunodeficiency virus conditions such as HIV and AIDs, malaria and tuberculosis, which are considered to be the foremost infectious disease threats in the world today [56–64]. The development of fixed-dose combinations is becoming increasingly important from a public health perspective. Probably, while the most compelling application of FDC medications is in the secondary prevention of CVD [65], the most commonly used FDC is the multivitamin or multivitamin plus mineral products seen in any drug stores in the world. Often these tablets or capsules contain up to 20 different active ingredients though this is an extreme example and unusual.

The main differences between CDT and FDC are: (a) CDT involves multiple uses of a single agent, whereas FDC involves a single use of multiple agents simultaneously; (b) a CDT medication is a free-dose combination of drugs in which the dose of each active agent can be varied according to the need, while an FDC medication is a fixed-dose combination of drugs in which the doses of all active agents are fixed in a product though different products with different doses; (c) CDT drugs can be administered together concurrently or separately within a planned time interval.

2.2.2 Basic Requirement for FDC

The main requirements for the design and application of CDT drugs also apply to the FDC strategy (Table 2.1): multitargeting with complementary mechanisms; nonoverlapping toxicity; pharmacokinetics comparability or physicochemical compatibility; and minimizing the number of combining drugs. Justification for combining various active components in a single pill must be based on valid therapeutic principles, which take into account the potential advantages and disadvantages. International regulatory agencies and most major drug regulatory agencies have established guidelines along with regulations concerning preparation, labeling, and marketing for FDC products. Thus, FDC has to comply with additional requirements by the guidelines set by the FDA [66] and CPMP [67].

Three types of product labeling are available according to whether an FDC product is indicated in first-line therapy, second-line therapy, or as a convenience for patients who require simplified therapy.

(a) *First-line therapy*: For an FDC product to be indicated in "first-line therapy," a low-dose combination of Drugs A plus B must be more effective than the respective high-dose monotherapies and have a similar side-effect profile or be similarly effective but have fewer side effects.
(b) *Second-line therapy*: FDC drugs can be indicated in "second-line therapy" in patients who fail to achieve a satisfactory benefit/risk ratio with monotherapy and require additional agents or dose titration [68]. In such cases, FDC products are recommended if one or more of the components diminishes the dose-related side effects of the other(s).
(c) *Convenience therapy*: An FDC product can be indicated for "convenience therapy" if it has the potential to simplify therapy by improving the dosing strategy or pill burden.

Table 2.1 Comprehensive Comparisons among CDT, FDC, and MTD

	CDT	FDC	MTD
Concept	Combination drug therapy (CDT) is a "combination" strategy with concurrent or sequential administration of two or more separate therapeutic agents in the form of free-dose combination	Single-pill fixed-dose combination (FDC) or Polypill is a "mix" strategy with multiple active agents (two or more) mixed inseparable within the same pharmaceutical formulation or in a single pill (tablet or capsule)	The MTD strategy is an integrative strategy with the combination, in a single scaffold, of different structural subunits that allow the molecular recognition by more than one bioreceptor or bio-effector, acting in a concordant mode in multiple therapeutic targets responsible for multifactorial diseases.
Basic requirement for combination	Multitargeting with complementary mechanisms.	Apply the basic requirement for CDT.	Choosing the right combination of targets.
	Nonoverlapping toxicity.	Additionally, comply with additional requirements by the guidelines and regulations concerning preparation, labeling, and marketing for FDC products set by the FDA and CPMP.	Balancing activities toward targets.
	Pharmacokinetics comparability or physicochemical compatibility.		Excluding undesired targets.
	Minimizing the number of combining drugs.		
Strengths	Minimizing compensatory effect due to redundancies in signaling networks.	Improved medication compliance.	Optimized therapeutic efficacy and improved safety profile.
	Creating synergistic actions.	Superior disease control.	Lack of drug interactions.
	Combination of infinite number of therapeutical agents.	Better tolerability and safety.	Predictable pharmacokinetics.
	Multiple options of combination.	Increased cost-effectiveness.	Less prone to drug resistance.
	Offering the possibility of personalized treatment.	Simplified drug regimes.	Simplified therapeutic regimens.
	Minimizing drug toxicity or side effects.		Simplified regulatory requirements.
	Mitigating drug resistance.		Providing unprecedented opportunities for drug discovery.
Limitations	Poor patient compliance.	Difficulty in titration of doses.	Challenges of validating target combinations.
	Possible different bioavailability.	Impact of drug intolerance and side effects.	Difficulties in optimizing multiple structure-activity relationships.
	Disturbing drug interactions.	Undesirable drug interactions.	Requiring more advanced technologies.
	Complicating drug development process.	Incompatible pharmacokinetics.	
	Lengthening drug approval procedure.	Physician acceptability.	
		Pharmaceutical/formulation issues.	

(continued)

Table 2.1 (continued)

	CDT	FDC	MTD
Types	Classification based on chronological order of drug administration.	Rationally designed FDC.	Classification based on the design of drugs.
	1. Concurrent-administration combination.	1. Conventional form of Polypills.	1. Unintended MTD or nondesigned MTD.
	2. Sequential-administration combination.	2. Nanoparticles-bearing Polypills.	2. Rationally designed MTD or targeted MTD.
	Classification based on the design of drug combination.	Partly designed FDC.	Classification based on the disease pathway.
	1. Designed CDT.		1. Disease pathway MTD.
	2. Nondesigned or unintended CDT.		2. Disease network MTD.
	3. Partially designed CDT.		Classification based on the origins of drugs.
	4. Recombinant antibody mixture (RAM).		1. Naturally occurring MTD.
	Classification based on the number of combining drugs.		2. Artificially designed MTD.
	1. Dual combination.		Classification based on the mechanisms of action.
	2. Triple combination.		1. Function-based MTD.
	3. Higher-order combinations.		2. Sequence-based MTD.
			(a) DNA/RNA sequence-based MTD.
			(b) Protein sequence-based MTD.
			Classification based on approaches of drug discovery.
			1. New molecular entity MTD.
			2. Privileged scaffold-based MTD.
			3. Hybrid molecule MTD.
			(a) Similar pharmacophores-fused MTD.
			(b) Distinct pharmacophores-conjugated MTD.
			4. Repurposed MTD.

If a product achieves this without providing additional health benefits, the convenience claim should be restricted to nonprescription (over-the-counter) products, which are fraught with problems. Furthermore, as reducing the pill burden has the potential to improve compliance to therapy, it may be necessary for regulatory authorities to relax their former guidelines and rethink their current restrictions on product labeling.

The burden of proving the safety and efficacy of a combination product lies with the applicants rather than with the regulatory authorities. Applicants must provide proof that each product contains the appropriate drug ratios and minimal doses, that the product is safe and effective in a significant proportion of the target population, that the benefit/risk ratio is at least as good as that of the individual agents, and pharmacodynamic and pharmacokinetic data to assess drug–drug interactions. The applicant is also required to assess any clinically beneficial pharmacokinetic interactions in healthy subjects, patients, and high-risk subgroups (such as the elderly and those with diabetes and/or renal disease). Thus, although drug–drug interactions are a major concern for the development of combination products, thorough assessments have the potential to improve the safety of these medications when compared with more traditional single therapies, which are often used in combinations that have not been formally tested.

For products that combine established drugs, often used concomitantly (such as antihypertensive agents and statins), well-founded bibliographic data can be used to provide a rationale for the proposed doses, thereby minimizing the number of new studies that must be carried out.

2.2.3 Strengths of FDC

FDC drugs containing two or more agents with complementary mechanisms of action represent an example of incremental innovation that extends the range of therapeutic options in the treatment of numerous diseases. Several short and intermediate to long-term studies with different brands of polypills have analyzed the impact of polypill in phase II and III trials. Overall, the FDC has been demonstrated to have several advantages over the separate use of several medications (CDT) or monotherapy, including enhanced long-term patients' acceptability and adherence, better risk factor control and quality of life, lower cost, easier accessibility, etc. The specific strengths of FDC are discussed below.

2.2.3.1 Improved Medication Compliance

It has recently been suggested that by improving adherence by 10%, the incidence of cardiovascular events (both fatal and nonfatal) can be reduced by 6.7% [69]. Indeed, expert panels such as the WHO advocate addressing the factors contributing to nonadherence as a fundamental priority over the development of specific evidence-based drugs for improving cardiovascular health outcomes [70]. The combination of individual agents into one "polypill" has been shown to reduce pill burden [71–73], and therefore offers a potential method for improving adherence with high acceptability to patients [74]. Studies have reported a range of values for good adherence (generally defined as taking practitioner-prescribed medication >80% of the time) to indicate CVD medication in polypill form, measured using various methodologies over differing durations. Overall, the polypill has been conclusively shown to improve patient adherence to CVD preventative medication [75]. In addition, a systematic review on 67 original peer-reviewed articles documented that adherence to polypill regimens was significantly higher in 56 articles (84%) with reduction in the frequency of administration compared to multiple pill regimens, which was also supported by the results of 13 out of 17 selected previously published systematic reviews and meta-analyses dealing with this topic [73]. Clearly, adherence can be improved through the formulation of polypills, which is presumably why the interest in researching the FDC strategy and products has been kept growing rapidly and steadily.

2.2.3.2 Greater Disease Control

In general, FDC could yield better risk factor control and greater therapeutic efficacy in the treatment of human disease as well as increased quality of life following the FDC treatment [50, 53, 76–78]. For CVD, each polypill is generally composed of a minimum of a statin, an antiplatelet drug (commonly aspirin), and an antihypertensive agent (ACEI/ARB/thiazide/β-blocker/calcium-channel blocker), each selected based on clinical trial efficacy and pharmacological interaction data [52, 76]. An open prospective cohort study assessing the relative reductions in cardiovascular risk for a range of medicinal permutations found that a combination of statins, aspirin, and β-blockers was the most effective (reductions of 83%), followed by a combination of statins, aspirin, and ACEIs (reductions of 71%) [76]. In terms of quality of life (QoL), the UMPIRE study reported significantly higher scores on a visual analog scale for the polypill group compared to the standard treatment group, indicating greater levels of patient-perceived health [53].

In addition to reducing the pill burden, combination products have the potential to improve compliance by allowing some patients to achieve their therapeutic goals more quickly than they might with monotherapy, which is translated into improved therapeutic efficacy. This is especially important in patients with asymptomatic conditions, such as hypertension and dyslipidemia, who may be tempted to discontinue their medication unless a measurable benefit is achieved within the first few months of treatment.

Additional strengths may include more predictable and simpler pharmacokinetic and pharmacodynamic relationship of the components in a single medicine, the greater efficacy against advanced-stage diseases, the lower toxicities, and the simultaneous presence of the chemical entities in multiple tissues. Moreover, FDC may also include the counteracting by one agent of an adverse reaction produced by another one [78, 79].

2.2.3.3 Better Tolerability and Safety

A meta-analysis of 44 selected randomized controlled trials (RCTs) concludes that the tolerability was lower among those on "polypills" than those on placebo or one component, but differences were moderate [80]. This conclusion is supported by another analysis of reported safety data showing no safety concerns of FDC [81].

2.2.3.4 Increased Cost-Effectiveness

On the financial side, the usual advantages of FDC products are to reduce the number of individual medications that patients must remember to take, which often reduces costs to the patient and insurer due to lower costs of manufacturing compared to the costs of producing separate products administered concurrently, simpler logistics of distribution, a decreased number of dispensing fees, the use of fewer bottles, labels, etc., and from the possible situation where the price of the FDC is less than the medication price of the two separate ingredients dispensed individually [82]. Therefore, FDC is generally considered to yield better cost-effectiveness than monotherapy and CDT strategy.

2.2.3.5 Simplifying Drug Regimes

Another advantage of FDC drugs is said to be in the clinical realm where simplified therapy regimens enhance patient's medication taking compliance [78]. This is quite straightforward to understand: many patients, especially those with a high pill burden and multiple chronic illnesses, are less adherent to medication, and in medication treatments utilizing polypills, this problem is diminished since multiple drugs are fused into one formulation and, therefore, the therapy regimen is simplified.

2.2.4 Limitations of FDC

2.2.4.1 Difficulty in Titration of Doses

The lurking limitations include difficulty with dose adjustment to targets or inflexible fixed dose ratio, fear of mass medicalization, and low acceptability among physicians, of which the inconvenience of dose adjustments with a polypill is the key worry for most practitioners [64, 72, 78, 83]. The lack of ability to fine tune one component of the polypill if the therapeutic goals are not reached is a constant concern.

2.2.4.2 Impact of Drug Intolerance and Side Effects

The other concern with polypill has been that a side-effect of any one of the components can lead to discontinuation of all drugs. A Cochrane review revealed that discontinuation rates were 26% higher in participants who were randomized to polypill [54] than in control subjects receiving usual care and placebo. However, the long-term studies showing better risk factor control suggest that this increased discontinuation is offset by other advantages of polypill including better adherence. Overall, the common side effects associated with the polypill are the same as those associated with the individual component parts. Studies suggest that the polypill neither reduces nor enhances the prevalence of serious adverse events associated with usual CVD treatment [83].

2.2.4.3 Undesirable Drug Interactions

The probability of undesirable drug-drug interactions and off-target drug-target interactions is often increased by the number of components included in the polypill, further limiting feasibility due to possible cross-toxic effects [78, 83]. Both of the CDT and FDC approaches show potential pharmacodynamic and pharmacokinetic limitations, such as the possible different solubility of the active substances that may modify the bloodstream uptake, not guaranteeing the simultaneous presence of the required drugs at the desired concentrations in the target tissue. If an adverse drug reaction occurs from using an FDC drug, it may be difficult to identify the active ingredient responsible for causing the reaction [84].

2.2.4.4 Incompatible Pharmacokinetics

The compounds mixed in an FDC pill may well have incompatible pharmacokinetics and different bioactivities, which can well confront and/or weaken the efficacies of therapy.

2.2.4.5 Physician Acceptability

Physicians like to have the flexibility to titrate individual agents and, as a result, are often hostile to the concept of polypills [78, 83]. In practice, however, physicians often manage unresponsive patients by prescribing an alternative agent rather than titrating the dose. Considerable re-education of both primary care and specialist physicians will, therefore, be required to improve physician acceptance of the polypill.

On the other hand, physician and pharmacist who accept the application of FDC may unintendedly ignore the content of the pill and the associated multi-side effects.

2.2.4.6 Pharmaceutical/Formulation Issues

As pointed out by Sleight et al. [78], not all drugs are suitable for use in combination products. Candidate agents must be proven safe, well tolerated, effective in submaximal doses, and physicochemically compatible with the other components of the pill. The potential for drug-drug interactions that affect the bioavailability and/or efficacy of the individual agents must also be considered.

2.2.5 Types of FDC

FDC can be classified into two general groups: rationally designed FDC and partly designed FDC.

2.2.5.1 Rationally Designed FDC

Rationally designed FDC drugs refer to those developed entirely based upon proof-of-concept and complete understanding of the main active substances to be contained in the materials and their pharmacological properties. They can be further divided into two subcategories.

Conventional Form of Polypills

Regular form of polypills refers to the pharmaceutical drug products manufactured with a specific mixture of multiple active ingredients and inactive components (excipients) in a particular configuration (such as a capsule shell or tablet), which are apportioned into a particular dose. This formulation is currently the most common dosage form of FDC drugs.

Nanoparticles-Bearing Polypills

Nanoparticles offer several advantages in drug delivery for combination therapy due to their unique characteristics. Some of these advantages include: potential for functionalization for enhanced drug-carrying capacity [85], tissue or organ-specific transport and delivery [85, 86], reduction in administered dose and toxicity [87], the ability to carry and deliver multiple classes of diagnostics and therapeutic agents loaded within a nanoparticle, which would then exert their various effects in a controlled manner [85, 88] and reduction in the frequency of administration [89].

FDC using nanoparticle formulations provides certain advantages over regularly formulated tablets or capsules. The controlled release feature offered by nanoparticle systems can normalize the pharmacokinetics, biodistribution, and stability of drugs that possess very different chemical properties that would independently have produced contrasting pharmacological behaviors [89, 90]. These long-circulating formulations are capable of continuous release of drugs at controlled ratios or permit independent modification of release rates of each drug in ways that would not be achievable with conventional formulations of FDC which are rapidly cleared from the system.

Multidrug-loaded nanoparticle polypills comprising different classes of therapeutic agents have been developed, such as polymer-lipid hybrid nanoparticles [89, 90], multidrug-loaded nanoparticle micellar formulation [91]. Combination therapy with a co-delivered chemotherapeutic agent and nucleic acid using a nanoparticle system is also a promising strategy for the effective treatment of disease.

Varying biodistribution/pharmacokinetics of combination drugs through cocktail administration has been attributed to their ineffectiveness in the clinic. The problem is being solved by a nanotechnology platform for drug delivery. The unique ability of multifunctional therapeutic nanoparticles to provide site-specific lesion targeting (e.g., tumor), improve the solubility of combining drugs, synchronize the disposition (pharmacokinetics) of encapsulated drugs (drug combination), overcome drug resistance, and enhance the effectiveness of therapeutic drugs represents an important innovation in drug delivery [90].

2.2.5.2 Partly Designed FDC

Partly designed FDC drugs refer to those that are prepared by combining two or more materials based on partial understanding of the main active substances contained in the materials and their pharmacological properties. For example, flavonoids-enriched natural products are often used to produce FDC drugs as flavonoids are powerful antioxidants with anti-inflammatory and immune system benefits. Polyphenols have historically been used in Chinese and Ayurvedic medicine, according to the Global Healing Center, and they are associated with skin protection, brain function, blood sugar, and blood pressure regulation, in addition to antioxidant and anti-inflammatory activity. Chinese patent drugs, herbal medicines generally consisting of extracted condensed pills (ready-to-use forms of formulation such as tablets, oral solutions, or dry suspensions), as opposed to crude herbal preparations that require cooking (hot water extraction), all belong to the partly designed FDC category as they are formulated based on the existing knowledge about the main bioactive ingredients contained in several herbs or other natural origins without knowing the nature of other substances contained in the preparations. Danshen dripping pill is a typical example of a partly designed FDC compound that is widely used in the clinical treatment of cardiovascular and cerebrovascular diseases [92–94]. It is prepared from the root of *Salvia miltiorrhiza* containing two major groups of bioactive substances, i.e., liposoluble tanshinones and water-soluble phenolics (mainly salvianolate) [92, 93]. To date, there have been more than 230 randomized controlled trials (RCTs) providing evidence indicating the superior efficacy of the Danshen dripping pill in treating angina pectoris [94].

2.3 MTD Strategy—The "Integration" Strategy

2.3.1 Concept of MTD

MTD is defined as the "single agent on multiple targets for single disease" strategy for disease therapy and drug discovery. It is frequently termed "multitarget drugs (MTDs)" [95–98], "multitarget-directed ligands (MTDLs)" [99–102] or "one drug, multiple targets" [103–105].

While both CDT and FDC are combinations or mix of multiple separate pharmacotherapeutics or nutrients (complementary medicine), MTD is a proper combination of multiple targeting activities (Table 2.1). The MTD strategy advocates the combination, in a single scaffold, of different structural subunits that allow the molecular recognition by more than one bioreceptor or bioeffector, acting in a simultaneous mode in multiple targets interconnected to biochemical networks responsible for the pathophysiology of multifactorial diseases.

Traditionally drugs have been designed with the aim of targeting a single biological entity, usually a protein (the so-called "on-target"), with high selectivity to avoid any unwanted effects arising from mis-targeting other biological targets ("off-targets"). On this basis, the concept of drugs interacting with multiple targets has long been flagged as undesirable, as it was inherently associated with adverse side effects. However, the complexity of the current incurable pathologies has clearly demonstrated that such single-target drugs are inadequate to achieve a therapeutic effect [106, 107]. In parallel, we have learned that molecules hitting more than one target may possess in principle a safer profile compared to single-targeted ones. Building on such accumulating evidence, the concept of multitarget drugs has made rapid and spectacular progress from being an emerging paradigm when first enunciated at the beginning of 2000 [108–110], to one of the hottest topics in drug discovery today.

In the last two decades, the number of MTDs on the market has been rising. From 2015 to 2017, 21% of the drugs approved by the Food and Drug Administration (FDA) belong to the MTD category, primarily antineoplastic agents [91]. This trend may indicate that the number of multi-target drugs will continue to rise since they present advantages over single-target drugs (or STD), CDT, and FDC [111]. Although the number of single-target small molecules (34%) is still greater than that of multitarget drugs (21%), the latter keeps increasing compared to previous years (16%) [112, 113]. Together with the newly approved therapeutic combinations (10%), polypharmacological drugs add up to a total of 31%.

The MTD strategy is the core of polypharmacology, and rational drug design is the core issue of the MTD strategy. Thus, multitarget polypharmacology is also called "Rational Polypharmacology" [114–118]. MTDs developed based on "Rational Polypharmacology" possess several advantages over CDT and FDC and is the cornerstone of the modern polypharmacology.

2.3.2 Basic Requirement for MTD

2.3.2.1 Choosing the Right Combination of Targets

Choosing the right combination of multiple targets for the disease of interest is a prerequisite for rational design of MTDs and an ultimate of polypharmacology. It requires good understanding of target-disease associations, disease-related pathways and networks, target-drug-disease-pathway/network relationships, and adverse events profiling [119]. Furthermore, the selection should be based on whether or not modulating the selected targets could lead to additive or synergistic effects [120]. Only with a proper selection of appropriate molecular targets can an MTD acquire optimized therapeutic efficacy and minimized toxicity. It is important to note that additive effects can be obtained when the targets belong to the same pathway, whereas synergism can be achieved only if the selected targets are located on functionally complementary pathways. With either additive or synergetic action, lower doses, thereby lower toxicities, of MTDs can be attained with respect to STDs [120].

2.3.2.2 Balancing Activities Toward Targets

Establishing the same strength range of modulation for each of the selected multiple targets or achieving balanced affinities and activities toward the selected multiple targets is a critical factor to be addressed for rational design of an MTD. Additionally, achieving balanced in vitro and in vivo activities is also crucial for matching potency for the corresponding targets. These are another requisite for an MTD to acquire predictable synergism of pharmacological actions [121]. Probably a more favorable MTD will be one that produces moderate influences on several targets, either positively or negatively, at lower concentrations.

2.3.2.3 Excluding Undesired Targets

Designing out unwanted drug-target interactions to exclude undesired target thereby off-target effects is still another critical factor to be addressed for rational design of MTDs. This is of particular importance when designing MTDs directed to the molecular targets belonging to the same family (e.g., multikinases inhibitors), as well as in the case of shared functional domains and/or binding sites across target families [121]. MTD should be the optimal combination and integration of multitargeticity and polyspecificity (the "selectively nonselective" property) (For more details, see Chap. 1, Sect. 1.3 "The Principles of Polypharmacology").

2.3.3 Strengths of MTD

MTD possesses all the advantages of STD, CDT, and FDC drugs and beyond. It solves the concerns about complex therapeutic regimens, difficulty in including numerous drugs in a single formulation and drug interactions at the different pharmacokinetics levels: absorption, distribution, metabolism, and elimination [122]. On the other hand, MTD is devoid of the limitations of CDT and FDC drugs in terms of unpredictable and unfavorable pharmacokinetics and drug interactions. MTD is therefore the ultimate solution of the issues associated with single target drugs and drug combinations.

2.3.3.1 Optimized Therapeutic Efficacy and Improved Safety Profile

It is well recognized that with appropriate combination and integration of multitargeticity and polyspecificity, MTDs in theory can afford optimized therapeutic efficacy for complex diseases and minimized possibility of toxic effects. Despite that in reality there have not been any ideal MTDs in the market, this approach has resulted in a number of multitarget drugs superior to STD, CDT, and FDC drugs [91, 111–113]. The strength of MTDs is conferred in part by the following advantages.

2.3.3.2 Lack of Drug Interactions

As a single chemical entity, there is no concern about adverse drug interactions in the use of MTDs, thereby minimal adverse events as a result of drug interactions. This property should manifest the effectiveness of MTDs in disease therapy.

2.3.3.3 Predictable Pharmacokinetics

A combination of drugs (CDT or FDC) faces the problem of the different solubilities that may modify the bloodstream uptake, which requires fine-tuning the formulation in order to ensure the required blood level of each drug. By comparison, MTD as a single medicine has more predictable and uniform pharmacokinetic and pharmacodynamic relationship, which should yield greater bioavailability thereby greater efficacy. Additionally, an MTD is devoid of drug-drug interactions and the associated cross-toxic effects [27, 109].

2.3.3.4 Less Prone to Drug Resistance

Drug resistance is the reduction in the effectiveness of a medication to cure a disease or condition. Though drug resistance more frequently refers to the situations of antimicrobial or antineoplastic therapy, it could also be used, in a broader sense, to describe the decreased reaction of targeted molecules to drug actions due to desensitization. In any case, by acting on multiple molecular targets, an MTD less likely encounters drug resistance as the odds for all targeted molecules to simultaneously develop resistance to this MTD is small.

2.3.3.5 Simplified Therapeutic Regimens

MTD is associated with better compliance/adherence and persistence of medication taking by patients due to a less complicated treatment regimen in comparison to CDT, which can be translated into better clinical outcomes and effectiveness parameters owing to synergy and a diminished number of adverse reactions and toxicity.

2.3.3.6 Simplified Regulatory Requirements

The regulatory requirements are more complex when the agents are used in combination (as for CDT and FDC drugs), since the safety profile of each drug needs to be demonstrated before clinical trials, and this can be further delayed due to regulatory and intellectual property issues, especially if the two drugs are being developed by different companies. The MTD strategy is a solution to such a problem.

2.3.3.7 Providing Unprecedented Opportunities for Drug Discovery

The strengths of the MTD strategy explain why it is rapidly emerging as the main paradigm of drug discovery and therapy. Of note, the MTD strategy opens up the opportunities for the discovery and development of new drugs with optimized therapeutic efficacy and minimized off-target toxicity, as well as for re-discovery of new uses of old drugs by their property of polyspecificity or nonselective selectivity identified through drug repurposing. Additionally, the MTD strategy also enables the prediction and clarification of potential and known off-target side effects of drugs due to drug promiscuity thereby helping improve the safety profiles by laying the groundwork for modifying chemical structure and physicochemical properties and/or optimizing therapeutic regimens.

2.3.4 Limitations of MTD

2.3.4.1 Challenges of Validating Target Combinations

The challenge in MTD discovery is the design of small molecules that have the property to modulate at the same time several of the molecular targets with contrasting or totally unrelated mechanisms of action [123]. Despite that the available technologies are able to allow the design of clinically useful multitarget agents [119, 120, 124–126], but these procedures have to be integrated with phenotype screening [127] and a proper selection of targets relevant for a given pathology [119, 120, 124–126].

2.3.4.2 Difficulties in Optimizing Multiple Structure-Activity Relationships

While balancing the activities toward multiple targets in a quantitative manner, particularly those associated with different disease-related pathways or mechanisms, is one of the principles of the MTD strategy for drug discovery and therapy, it is not easy to achieve such a task [119]. On the contrary, to optimize multiple structure-activity relationships we first need to have adequate understanding of disease-related networks, key target molecules, drug-target interactions, cooperativities of different pharmacophores in an MTD, and phenotypic alterations upon interfering the disease-related networks by targeting the key molecules in the networks.

2.3.4.3 Requiring More Advanced Technologies

It is technically more demanding to produce new MTDs regardless of whether they are derived from brand new structure, privileged scaffold, molecular hybrid, antibody conjugate, or even drug repositioning. Combined computational approaches with target-based screening, cell-based screening, phenotypic screening (combining the advantages of animal models and cell cultures to allow drug developers to identify problems with drug candidate leads at early stages), and/or combined targeted phenotypic screening approaches are required [128–131]. Functional cell-based assays can enable high-throughput compound screening (HTS) to advance drug discovery success, in which they measure the functional behavior of the cell of interest in response to the candidate compound. Cell-based assays, because of their peculiar advantages of predictability, possibility of automation, multiplexing, and miniaturization, seem the most appealing tool for the high demands of the early stages of the drug-discovery process. Nevertheless, cellular screening, relying on different strategies ranging from reporter gene technology to protein fragment complementation assays, still presents a variety of challenges. The relatively recent introduction of target-based approaches is increasingly accepted as an equally important technology in the context of the long-time frames of drug discovery and development [132, 133]. While phenotypic screening is considered an empirical approach, target-based screening is believed to be a molecular approach which is predominantly a hypothesis-driven drug discovery process. In general, phenotypic screening has a slight advantage when it comes to identifying first-in-class drugs, whereas target-based screening has yielded more best-in-class drugs [133].

2.3.5 Types of MTD

2.3.5.1 Classification Based on the Design of Drugs

MTDs can be categorized into two major classes based on whether they are rationally designed multitargeting agents or are single-target agents with off-target actions.

Unintended MTD or Nondesigned MTD

Unintended MTD refers to a given drug which is initially designed based on "single-agent, single-target" (STD) principle, but may elicit unintended off-target actions which could be beneficial as a multitarget drug or detrimental due to drug promiscuity. Aspirin is one example of a beneficial unintended MTD. It was originally designed as an analgesic to relieve minor pains or as an antipyretic to reduce fever [134]. However, it also acts as an anti-inflammatory medication to treat rheumatoid arthritis [135], pericarditis [136], and Kawasaki diseases [137]. Additionally, it has been used in the prevention of transient ischemic attacks [138], strokes, heart attacks [139], pregnancy loss [140], and even cancer [141]. In fact, there have not been any true, pure STDs in the market except for molecularly targeted monoclonal antibodies; all therapeutics in clinical use have certain degrees of multitarget effects.

A screening study of the 3665 FDA-approved and investigational drugs was conducted using databases containing the chemical structures of hundreds of thousands of biologically active compounds for which the binding characteristics to a panel of 1400 target proteins were known [142]. A massive network of interactions (nearly 7000 of them with high probability) for the studied compounds with off-targets were predicted, which indicates that polypharmacology is, perhaps unintentionally, a feature intrinsic to the therapeutic efficacies of drugs. It is becoming clear that many drugs may derive their therapeutic action (and/or their off-target activities) from the interactions with multiple proteins; for non-cancerous diseases, the average number of interaction partners for protein targets of marketed drugs is 4.24, and for cancer targets it is 7.82 [125, 126].

Rationally Designed MTD or Targeted MTD

Designed MTDs are those generated on purpose based on the MTD strategy, which ideally only act on the selected targets. Targeted MTD is the central concept of polypharmacology and the principal strategy for drug therapy and discovery. Designed MTDs could be new molecular entities, structurally modified compounds based on privileged scaffolds, or molecular hybrids derived from fusion or conjugation of multiple pharmacophores.

Imeglimin, a drug currently in the development as the first in a new class of oral agents for the treatment of type 2 diabetes mellitus (T2DM), is an excellent example of rationally designed MTD [143]. Although the precise mechanism of action has not been fully understood, it appears, based on pre-clinical results, to target several components of the pathology of T2DM. Studies indicate that imeglimin is associated with stimulation of skeletal muscle glucose uptake, protection against β-cell apoptosis, and reduction of liver steatosis [144–150]. Furthermore, imeglimin has been shown to enhance glucose-dependent insulin secretion [143].

2.3.5.2 Classification Based on the Disease Pathway

According to the disease pathway, the targeted MTDs can be divided into two groups: single disease pathway (disease pathway MTD) and multiple disease pathways (or disease network MTD).

Disease Pathway MTD

"Disease pathway MTDs" are primarily designed to target multiple signaling molecules along a selected signaling pathway or a unique disease pathway which plays the key role in the development of a disease under concern.

Disease Network MTD

"Disease network MTDs" are designed to target multiple signaling molecules belonging to several different but related signaling pathways that interplay or even are interconnected to form a disease-related signaling network or multiple disease pathways that control a specific pathological process. In addition, polypharmacology for complex diseases is likely to employ multiple drugs acting on distinct functionally related targets, that are part of networks regulating various physiological responses [119].

2.3.5.3 Classification Based on the Origins of Drugs

Naturally Existing MTD

Naturally existing MTD agents can be roughly divided into two subclasses: natural products and endogenous molecules.

Natural products used as traditional medicines or supplements can be classified into the category of unintended MTDs [151]. Numerous examples of naturally occurring MTD exist within nature, such as phytochemicals. A representative agent of such naturally occurring MTD is Berberine, a natural pentacyclic isoquinoline alkaloid as the principal component of many popular medicinal plants such as the genus Berberis, Coptis, and Hydrastis. Numerous basic studies and clinical trials have documented the multifunctional nature of berberine as a therapeutic agent as an attribute of its diverse effects on enzymes, receptors and cell signaling pathways and the efficacies against cardiovascular disease, metabolic diseases, gastrointestinal disease, neurological disease, cancer, etc. [152–154]. Resveratrol, a natural polyphenolic stilbene isolated from various plants, foods, and beverages with a broad spectrum of biological and pharmacological properties through modulating diverse targets and signaling pathways is another imminent example of naturally occurring MTD [155–157].

Endogenous molecules existing in human cells that act like "MTDs" include noncoding RNAs such as microRNAs (miRNAs), long noncoding RNAs (lncRNAs), circular RNAs (circRNAs), etc. An important characteristic and mode of action of these regulatory RNAs is that each molecule can act simultaneously on multiple signaling pathways to elicit multiple cellular functions through targeting multiple genes by multiple mechanisms [158–161].

Artificially Designed MTD

Artificially designed MTDs are rationally designed MTDs based on the principles of polypharmacology, as already described above.

2.3.5.4 Classification Based on the Mechanisms of Action

MTDs can also be divided into two major classes according to their mechanisms of actions: function-based MTD and sequence-based MTD.

Function-Based MTD

Function-based MTDs mainly include "small-molecule drugs" and "macromolecular drugs."

The term "small-molecule drugs" is relative to macromolecular drugs. Compared to biologics, small molecule drugs are relatively simple chemical compounds and can be manufactured by chemical synthesis [162–165]. These compounds are typically comprised of 20 to 100 atoms and have a molecular mass of less than 1000 g/mol or 1 kilodalton [kDa]. The pharmacological activity, stability, and permeability of small molecules largely depend on chemistry as opposed to structural interactions. Because of their properties, small-molecule drugs can be administered by a variety of routes, including orally. Their oral bioavailability is indeed the most obvious advantage over biologics in drug development.

In contrast, macromolecular drugs (also called bioactive molecules) including protein, DNA, and RNA offer a newer class of drugs that are in general large molecules with complex structure and function [165–167]. For example, therapeutic proteins or amino acid-based molecules such as peptides, proteins, and antibodies range in size from smaller peptides (1 to <10 kDa) to much larger proteins (>10 kDa) like monoclonal antibodies. The examples of therapeutic proteins include antibody-based drugs, Fc fusion proteins, anticoagulants, blood factors, bone morphogenetic proteins, engineered protein scaffolds, enzymes, growth factors, hormones, interferons, interleukins, and thrombolytics. Even the smallest of these molecules are larger than most small molecule drugs, and they typically contain from 5000 to 50,000 atoms per molecule.

Traditionally, small-molecule drugs are STD-based compounds designed to target functional proteins (e.g., ligands, agonists, or antagonists, for receptor proteins) based on the pathophysiological functions of the proteins [161–164]. Their target spectrum has now been expanded to regulatory molecules and protein-protein interaction other than specific functional proteins, such as RNA including mRNA and noncoding RNA (e.g., miRNA) and epigenetic modulators that play key roles in regulating the expression and function of proteins [157–161, 168, 169], modifying chromatin remodeling to alter the expression and function of protein-coding genes and their protein products [170, 171], and other epigenetic processes [25, 123, 172–177]. Yet, no matter what targets are, small-molecule drug discovery is still a function-based approach. In this sense, the "small-molecule MTDs" refer to small-molecule drugs that are designed to possess multitargeticity and polyspecificity properties and that target functional proteins or epigenetic regulators or both. In other words, small-molecule MTDs are single multitargeting compounds [2, 25, 109, 159–183].

Sequence-Based MTD

Sequence-based drug molecules are designed to base-pair the sequence domain of target molecules (DNA or RNA) or to bind proteins such as transcriptional factors or functional proteins [184–202]. When sequence-based drugs are designed to possess multitargeticity and polyspecificity, they become sequence-based MTDs. This class of drugs essentially belongs to epigenetics-based therapeutics. Based on their binding targets, sequence-based drugs can be categorized into two subclasses: DNA/RNA sequence-based drugs and protein sequence-based drugs. (For more details, see Chap. 13 in Section I)

(a) *DNA/RNA Sequence-based MTD*: DNA/RNA sequence-based drugs are considered the most strictly target-specific and hold the potential to cause a major revolution in the pharmaceutical industry because they can be used for precise and individualized therapy. This subclass of sequence-based drugs can be further divided into several groups: antisense oligonucleotides (ASO) that are sequence commentary to DNA or mRNA [187–191], mRNA-based therapeutics [192, 193], small interference RNA (siRNA) directed to silence mRNAs [194–196], anti-miRNA antisense inhibitors (AMO) [197, 198, 203–206], etc. DNA sequence-based drugs include decoy oligodeoxynucleotides (dODN) that can be recognized and bond by target transcription factors [197–201] and complex dODN (cdODN) that can bind and target multiple transcription factors [207–213].

(b) *Protein sequence-based MTD*: Protein sequence-based drugs include drugs targeting RNA-protein interaction, which are directed to interrupt RNA-protein binding based on either the sequence of protein-binding domain in RNA or that of RNA binding region (amino acid sequence) in protein [202, 214–218].

The first experimental validation of sequence-based MTD approach was conducted by our research team [202]. In a study, we developed an innovative dODN technology that we term complex decoy oligodeoxynucleotide (cdODN) in which multiple cis-elements are engineered into single dODNs attacking multiple target transcription factors. Specifically, we designed dODNs targeting NF-κB, E2F, and Stat3 separately and a cdODN targeting NF-κB, E2F, and Stat3 concomitantly. We evaluated the effects of the cdODN on the expression of cancer-related genes, viability of human cancer cell lines, and in vivo tumor growth in nude mice. The cdODN targeting all NF-κB, E2F, and Stat3 together demonstrated markedly enhanced efficacy and increased potency by approximately two orders of magnitude compared with each of the dODNs or the combination of all three dODNs. The cdODN also showed earlier onset and longer-lasting action. Most strikingly, the cdODN has the ability to attack multiple molecules critical to cancer progression via multiple mechanisms, leading to the elimination of regression. The findings suggest that the cdODN knocked down the expression of the genes regulated by the target transcription factors. The cdODN strategy offers resourceful combinations of varying cis-elements for concomitantly targeting multiple molecules in cancer biological processes and opens the door to "one-drug, multiple-target" therapy for a broad range of human cancers. In this sense, a cdODN is an MTD agent. The cdODN techniques have been employed by other researchers in subsequent studies [208–213].

Ebert et al. [198] invented a miRNA inhibitor AMO containing multiple, tandem binding sites to a specific miRNA of interest, which they termed "microRNA sponges." When vectors encoding this sponge are transiently transfected into cultured cells, sponges derepress the targeted miRNA strongly. Notably, the sponge specifically inhibits miRNAs with the same seed sequence (5′ end 2–8 nucleotides), such that a single sponge can be used to block an entire miRNA seed family. Such a mode of action can be viewed as a special example of MTD as one AMO can target multiple miRNAs of an identical seed site.

We innovated the above approach by engineering multiple AMO units into a single sequence that are designed to simultaneously knockdown multiple miRNAs of different seed sites, which we named "the multiple-target AMO or MT-AMO" [203]. We validated the technique with two separate MT-AMOs: anti-miR-21/anti-miR-155/anti-miR-17-5p and anti-miR-1/anti-miR-133. We first verified the ability of the MT-AMOs to antagonize the repressive actions of their target miRNAs using luciferase reporter activity assays and to specifically knock down the levels of their target miRNAs using real-time RT-PCR methods. We then used the MT-AMO approach to identify several tumor suppressors-TGFBI, APC, and BCL2L11 as the target genes for oncogenic miR-21, miR-155, and miR-17-5p, respectively, and two cardiac ion channel genes HCN2 (encoding a subunit of cardiac pacemaker channel) and CACNA1C (encoding the alpha-subunit of cardiac L-type Ca^{2+} channel) for the muscle-specific miR-1 and miR-133. We further demonstrated that the MT-AMO targeting miR-21, miR-155, and miR-17-5p produced a greater inhibitory effect on cancer cell growth, compared with the regular single-target AMOs. Similarly, the MT-AMO for miR-1 and miR-133 knocks down both miR-1 and miR-133. Therefore, MT-AMO as another example of sequence-based MTDs may be used for its broad applications to diseases involving multiple miRNAs and multiple genes.

2.3.5.5 Classification Based on Approaches of Drug Discovery

MTDs can also be divided into the following four classes according to the approaches and methods by which they are discovered and developed.

New Molecular Entity (NME) MTD

NMEs represent novel chemical structures or biologics (proteins, peptides, antibodies) [91]. These structures or molecules consist of unique backbones and chemical groups and even conformation that confer their ability to interact specifically with the selected molecular targets so as to elicit desired polypharmacological effects.

Privileged Scaffold-Based MTD

The term "privileged scaffold" was coined by Evans in 1988 [219] as a single molecule that can be adapted to interact with diverse targets by functionalization, early examples of which were benzodiazepines, indoles, and 1,4-dihydropyridines [220]. This concept has been explored widely for chemically diverse scaffolds as a strategy for constructing high-affinity ligands from core structures that can bind more than one receptor [221, 222]. Since then, the privileged scaffold-based design has evolved from a stand-alone technology to an integral component of various lead generation platforms, particularly polypharmacology. For example, imeglimin is synthesized from metformin as a precursor or privileged scaffold via a single-step chemical reaction [223]; Dronedarone is a benzofuran derivative developed with amiodarone as a privileged scaffold, and is a "multichannel blocker" as a new antiarrhythmic [224].

Hybrid Molecule MTD

Hybrid molecules-based drug design is the combination of two or more structural domains or bioactive molecules having different biological functions and multiple activities into a single chemical entity or an MTD, indicating that a hybrid molecule acts as multiple distinct pharmacophores [225, 226]. Hybrid molecules, also named "chimeric molecules" [227], are therefore the engineered constructs as new molecules with multitargeting capabilities. Hybrid molecules [228], in contrast to pro-drugs, contain two or more domains with different biological functions and dual/multiple activities that ideally act as distinct pharmacophores, although not necessarily on the same biological target. Multiple ligands usually consist of the combination of pharmacophores of selective ligands (either existing drugs or drug candidates).

The hybrid molecules strategy has been used to confer the multitargeticity and polyspecificity properties on a single chemical entity, aiming to achieve better affinity and efficacy or improved properties compared with the parent molecules, to interact with two or multiple targets, to reduce undesirable side effects, to decrease drug-drug interactions, or to reduce the emergence of drug resistance [225, 226]. From the last 1–2 decades, the synthesis of hybrid molecules by the combination of different biologically relevant moieties has been under constant escalation along with their evaluation as diverse range of pharmacological agents and as potent drugs.

(a) *Similar pharmacophores-fused MTD*: Pharmacophores that are similar and share common substructures, usually hydrophobic or basic ring systems, can be synthetically fused or merged.
(b) *Distinct pharmacophores-conjugated MTD*: Pharmacophores that are dissimilar can be joined as conjugates with cleavable or non-cleavable linkers, although this strategy often leads to structures of high molecular weight and lipophilicity [228].

Alternatively, when two or more pharmacophores are connected by a linker that is labile or can be easily cleaved in vivo, they are called pro-drugs since each of them uses another pharmacologically active compound instead of some inert molecule as carrier. Being released simultaneously inside the cells, they might act synergistically and affect distinctive cellular targets. This latter strategy can be viewed as a hybrid MTD-CDT approach: the conjugate molecule is absorbed into the cells in the form of MTD, but it acts on the target in the form of CDT.

New Use of Old Drugs—Repurposed MTD

New drug discovery and development are a process of slow pace, high costs, high attrition rates, high failure risks, and high levels of intensive research and labor [229, 230]. According to the Eastern Research Group [231], it usually takes 10–15 years to develop a new drug clinical use.

Moreover, the success rate of developing a new molecular entity is only 2.01% on average [232]. The number of drugs approved by the FDA keeps declining since 1995 [233]. Furthermore, investment in drug research and development (R&D) has been rapidly climbing up year-by-year [234]. On the other hand, repurposing "old" drugs for their new uses in indications (for the treatment of both common and rare diseases) other than the ones for which they were initially marketed is increasingly becoming an attractive proposition. This is because drug repositioning involves the use of de-risked compounds, with potentially lower overall development costs and shorter development timelines. Traditional drug development procedures usually take five obligatory steps: discovery and preclinical, safety review, clinical research, FDA review, and FDA post-market safety monitoring [234, 235]. For drug repositioning, however, the procedures are simplified to only four steps: compound identification, compound acquisition, development, and FDA post-market safety monitoring. Due to the rapid growth of bioinformatics knowledge and biology big data, various data-driven and experimental approaches have been established for the identification of repurposable drug candidates. The R&D investment required for drug repositioning is tremendously reduced compared to traditional approaches, and the timeline is also shortened to a total of 9–10 years [219, 232–235]. The basis for drug repurposing is the potential multitargeticity and polyspecificity of existing "single-target" drugs, the properties defined by the principles of polypharmacology.

Building on the accumulating evidence, the concept of multitarget drugs has made rapid and spectacular progress from being an emerging paradigm when first enunciated at the beginning of 2000 [91, 108–110], to one of the hottest topics in drug discovery nowadays. The concepts related to polypharmacology have triggered the interest of the drug discovery community both in academia and pharmaceutical companies to such a point that a plethora of multitarget drugs is already available on the market. The documentation by Ramsay and colleagues [91] provided an overview on the FDA-approved new molecular entities (NMEs) from 2015 to 2017 (status September 2017) as a clear proof of such translational success of multitarget drugs into the clinic. According to their report [91], the 101 new NMEs on FDA.gov approved over this triennium were classified into NME categories (small molecules, biologics, therapeutic combinations, and diagnostics). By using the DrugBank database, which compiles information on approved drugs, together with their target(s) and mechanism(s) of action (MoA), the small molecules have been further analyzed and subdivided into single-target and multitarget drugs. Biotech drugs (proteins, peptides, and monoclonal antibodies) represent 31% of the novel NMEs, nearly approaching the number of single-target drugs (34%). This clearly reflects the recent increased pharma interest in the discovery of biologics, which build on the premise of a personalized treatment.

Collectively, as already described in Chap. 1, the three forms of polypharmacological medication differ in many perspectives. Both CDT and FDC regimens contain two or more different therapeutics; however, while CDT is a combination or co-administration of multiple separable drugs, FDC as a single pill or tablet is a mix of two or more drug components that are unable to be separated. By comparison, MTD is an integration (hybrid or fusion) of two or more pharmacophores in the form of a single compound (Fig. 2.1 and Table 2.1). In respect to STD, though MTD also belongs to the single-drug strategy, it is comprised of multiple molecular targets related to a specific disease entity as an integration of multiple drugs, whereas STD presumably acts on a single therapeutic target.

It should be noted that CDT could be the combination of an STD and an FDC or an STD and an MTD (Fig. 2.1), in addition to STD + STD combination as decided in Sect. 2.1.5. STD + FDC and STD + MTD combinations are frequently used in cancer therapy (to be discussed in Chap. 3) and in the treatment of neurological disorders (to be discussed in Chap. 6).

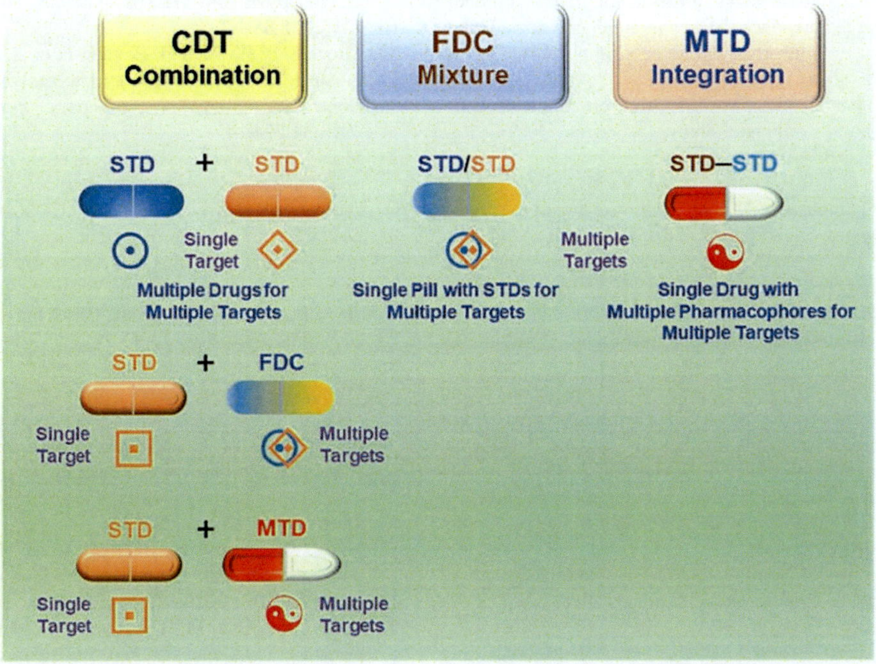

Fig. 2.1 Schematic illustration of three major strategies of Polypharmacology. *CDT* combination drug therapy or multiple drugs for multiple therapeutic targets, *FDC* fixed-dose combination or a single pill mixed with multiple drugs for multiple targets, *MTD* multitarget drug or a single drug constructed with multiple pharmacophores for multiple targets, *STD* single-target drug or a single drug for a single target

References

1. Sun W, Sanderson PE, Zheng W. Drug combination therapy increases successful drug repositioning. Drug Discov Today. 2016;21(7):1189–95.
2. Ascierto PA, Marincola FM. Combination therapy: the next opportunity and challenge of medicine. J Transl Med. 2011;9:115.
3. Phougat N, Khatri S, Singh A, et al. Combination therapy: the propitious rationale for drug development. Comb Chem High Throughput Screen. 2014;17(1):53–67.
4. Xu KJ, Song J, Zhao XM. The drug cocktail network. BMC Syst Biol. 2012;6(Suppl 1):S5.
5. Zhou H, Tong Z, McLeod JF. "Cocktail" approaches and strategies in drug development: valuable tool or flawed science? J Clin Pharmacol. 2004;44(2):120–34.
6. Preissner S, Dunkel M, Hoffmann MF, et al. Drug cocktail optimization in chemotherapy of cancer. PLoS One. 2012;7(12):e51020.
7. Tomaselli D, Lucidi A, Rotili D, et al. Epigenetic polypharmacology: a new frontier for epi-drug discovery. Med Res Rev. 2020;40(1):190–244.
8. Steinman MA, Hanlon JT. Managing medications in clinically complex elders: "There's got to be a happy medium". JAMA. 2010;304(14):1592–601.
9. Sutaria A, Liu L, Ahmed Z. Multiple medication (polypharmacy) and chronic kidney disease in patients aged 60 and older: a pharmacoepidemiologic perspective. Ther Adv Cardiovasc Dis. 2016;10(4):242–50.
10. Keith CT, Borisy AA, Stockwell BR. Multicomponent therapeutics for networked systems. Nat Rev Drug Discov. 2005;4(1):71–8.
11. McNeil SE. Nanoparticle therapeutics: a personal perspective. Wiley Interdiscip Rev Nanomed Nanobiotechnol. 2009;1(3):264–71.
12. Gupta AK, Arshad S, Poulter NR. Compliance, safety, and effectiveness of fixed-dose combinations of antihypertensive agents: a meta-analysis. Hypertension. 2010;55(2):399–407.
13. Sherrill B, Halpern M, Khan S, et al. Single-pill *vs* free-equivalent combination therapies for hypertension: a meta-analysis of health care costs and adherence. J Clin Hypertens (Greenwich). 2011;13(12):898–909.
14. Berry DH, Fernbach DJ, Herson J, et al. Comparison of prednisolone, vincristine, methotrexate and

6-mercaptopurine vs. 6-mercaptopurine and prednisone maintenance therapy in childhood acute leukemia: A Southwest Oncology Group Study. Cancer. 1980;46(5):1098–103.

15. Mihich E, Grindey GB. Multiple basis of combination chemotherapy. Cancer. 1977;40(1 Suppl):534–43.
16. Van Gaal LF, De Leeuw IH. Rationale and options for combination therapy in the treatment of Type 2 diabetes. Diabetologia. 2003;46(Suppl 1):M44–50.
17. Welz T, Wyen C, Hensel M. Drug interactions in the treatment of malignancy in HIV-infected patients. Oncol Res Treat. 2017;40(3):120–7.
18. Aronson J. Serious drug interactions. Practitioner. 1993;237(1531):789–91.
19. Scheen AJ. Pharmacokinetic characteristics and clinical efficacy of an SGLT2 inhibitor plus DPP-4 Inhibitor combination therapy in type 2 diabetes. Clin Pharmacokinet. 2017 Jul;56(7):703–18.
20. Evans WE, Relling MV. Moving towards individualized medicine with pharmacogenomics. Nature. 2004;429:464–8.
21. Tsioufis C, Thomopoulos C. Combination drug treatment in hypertension. Pharmacol Res. 2017;125(Pt B):266–71.
22. Horn T, Ferretti S, Ebel N, et al. High-order drug combinations are required to effectively kill colorectal cancer cells. Cancer Res. 2016;76(23):6950–63.
23. Olender S, Wilkin TJ, Taylor BS, Hammer SM. Advances in antiretroviral therapy. Top Antivir Med. 2012;20:61–86.
24. Meng Q, Garcia-Rodriguez C, Manzanarez G, et al. Engineered domain-based assays to identify individual antibodies in oligoclonal combinations targeting the same protein. Anal Biochem. 2012;430:141–50.
25. Benedetti R, Conte M, Iside C, et al. Epigenetic-based therapy: from single- to multi-target approaches. Int J Biochem Cell Biol. 2015;69:121–31.
26. Raju TS, Strohl WR. Potential therapeutic roles for antibody mixtures. Expert Opin Biol Ther. 2013;13(10):1347–52.
27. Morphy R, Rankovic Z. Designed multiple ligands. An emerging drug discovery paradigm. J Med Chem. 2005;48(21):6523–43.
28. Ostheimer GW. A comparison of glycopyrrolate and atropine during reversal of nondepolarizing neuromuscular block with neostigmine. Anesth Analg. 1977;56:182–6.
29. Zhao S, Nishimura T, Chen Y, et al. Systems pharmacology of adverse event mitigation by drug combinations. Sci Transl Med. 2013;5(206):206ra140.
30. Mouton JW. Combination therapy as a tool to prevent emergence of bacterial resistance. Infection. 1999;27(Suppl 2):S24–8.
31. Ho PM, Bryson CL, Rumsfeld JS. Medication adherence: its importance in cardiovascular outcomes. Circulation. 2009;119:3028–35.
32. Naderi SH, Bestwick JP, Wald DS. Adherence to drugs that prevent cardiovascular disease: meta-analysis on 376,162 patients. Am J Med. 2012;125:882–7.
33. Chowdhury R, Khan H, Heydon E, et al. Adherence to cardiovascular therapy: a meta-analysis of prevalence and clinical consequences. Eur Heart J. 2013;34:2940–8.
34. Moja L, Tagliabue L, Balduzzi S, et al. Trastuzumab containing regimens for early breast cancer. Cochrane Database Syst Rev. 2012;2012(4):CD006243.
35. Bellos I, Karageorgiou V, Pergialiotis V, et al. Acute kidney injury following the concurrent administration of antipseudomonal beta-lactams and vancomycin: a network meta-analysis. Clin Microbiol Infect. 2020;26(6):696–705.
36. Rosenstock J, Handelsman Y, Vidal J, et al. Propensity-score-matched comparative analyses of simultaneously administered fixed-ratio insulin glargine 100 U and lixisenatide (iGlarLixi) vs sequential administration of insulin glargine and lixisenatide in uncontrolled type 2 diabetes. Diabetes Obes Metab. 2018;20(12):2821–9.
37. Niquet J, Lumley L, Baldwin R, et al. Rational polytherapy in the treatment of cholinergic seizures. Neurobiol Dis. 2020;133:104537.
38. Piñero F, Silva M, Iavarone M. Sequencing of systemic treatment for hepatocellular carcinoma: second line competitors. World J Gastroenterol. 2020;26(16):1888–900.
39. Soni KS, Thomas D, Caffrey T, et al. A polymeric nanogel-based treatment regimen for enhanced efficacy and sequential administration of synergistic drug combination in pancreatic cancer. J Pharmacol Exp Ther. 2019;370(3):894–901.
40. Chou TC. Theoretical basis, experimental design, and computerized simulation of synergism and antagonism in drug combination studies. Pharmacol Rev. 2006;58(3):621–81.
41. Liu H, Ye M, Guo H. An updated review of randomized clinical trials testing the improvement of cognitive function of Ginkgo biloba extract in healthy people and Alzheimer's patients. Front Pharmacol. 2020;10:1688.
42. Sierpina VS, Wollschlaeger B, Blumenthal M. Ginkgo biloba. Am Fam Physician. 2003;68(5):923–6.
43. Haurum JS. Recombinant polyclonal antibodies: the next generation of antibody therapeutics? Drug Discov Today. 2006;11:655–60.
44. Skartved NJ, Jacobsen HJ, Pedersen MW, et al. Preclinical pharmacokinetics and safety of Sym004: a synergistic antibody mixture directed against epidermal growth factor receptor. Clin Cancer Res. 2011;17:5962–72.
45. Rasmussen SK, Næsted H, Müller C, et al. Recombinant antibody mixtures: production strategies and cost considerations. Arch Biochem Biophys. 2012;526(2):139–45.
46. Robak T. The emerging therapeutic role of antibody mixtures. Expert Opin Biol Ther. 2013;13(7):953–8.

47. de Kruif J, Kramer A, Nijhuis R, et al. Generation of stable cell clones expressing mixtures of human antibodies. Biotechnol Bioeng. 2010;106(5):741–50.
48. Stringberg A, Camden R, Qualls K, et al. Update on dual antiplatelet therapy for secondary stroke prevention. Mo Med. 2019;116(4):303–7.
49. Van Cutsem E, Cervantes A, Nordlinger B, et al. ESMO Guidelines Working Group. Metastatic colorectal cancer: ESMO Clinical Practice Guidelines for diagnosis, treatment and follow-up. Ann Oncol. 2014;25(Suppl 3):iii1–9.
50. Wald NJ, Law MR. A strategy to reduce cardiovascular disease by more than 80%. BMJ. 2003;326(7404):1419.
51. Wiley B, Fuster V. The concept of the polypill in the prevention of cardiovascular disease. Ann Glob Health. 2014;80:24–34.
52. Wald DS, Morris JK, Wald NJ. Randomized polypill crossover trial in people aged 50 and over. PLoS One. 2012;7:e41297.
53. Thom S, Poulter N, Field J, et al. Effects of a fixed-dose combination strategy on adherence and risk factors in patients with or at high risk of CVD: the UMPIRE randomized clinical trial. JAMA. 2013;310:918–29.
54. de Cates AN, Farr MRB, Wright N, et al. Fixed-dose combination therapy for the prevention of cardiovascular disease. Cochrane Database Syst Rev. 2014;4(4):CD009868.
55. Sanz G, Fuster V. Fixed-dose combination therapy and secondary cardiovascular prevention: rationale, selection of drugs and target population. Nat Clin Pract Cardiovasc Med. 2009;6:101–10.
56. Dhillon S. Decitabine/cedazuridine: first approval. Drugs. 2020;80(13):1373–8.
57. Kang C, Dhillon S, Deeks ED. Trifluridine/tipiracil: a review in metastatic gastric cancer. Drugs. 2019;79(14):1583–90.
58. Albanna AS, Smith BM, Cowan D, et al. Fixed-dose combination antituberculosis therapy: a systematic review and meta-analysis. Eur Respir J. 2013;42(3):721–32.
59. Zappulo E, Scotto R, Buonomo AR, et al. Efficacy and safety of a fixed dose combination tablet of asunaprevir + beclabuvir + daclatasvir for the treatment of Hepatitis C. Expert Opin Pharmacother. 2020;21(3):261–73.
60. Hu J, Zou P, Zhang S, et al. Empagliflozin/metformin fixed-dose combination: a review in patients with type 2 diabetes. Expert Opin Pharmacother. 2016;17(18):2471–7.
61. Naltrexone/bupropion for obesity. Drug Ther Bull. 2017;55(11):126–9.
62. Farooq S, Singh SP. Fixed dose-combination products in psychiatry: systematic review and meta-analysis. J Psychopharmacol. 2015;29(5):556–64.
63. WHO. WHO Expert Committee on specifications for pharmaceutical preparations. WHO technical report series 929. Geneva: 39th Report; 2005. p. 95.
64. Webster R, Castellano JM, Onuma OK. Putting polypills into practice: challenges and lessons learned. Lancet. 2017;389(10073):1066–74.
65. World Health Organization. Secondary prevention of non-communicable disease in low and middle income countries through community-based and health service interventions. Wellcome Trust meeting report, 1–3 August 2001. Geneva: World Health Organization. p. 2002.
66. Food and Drug Administration (FDA). Guidance for industry and FDA current good manufacturing practice for combination products. http://www.fda.gov/cder/guidance/OCLove1dft.htm#_ftn2. 20 March 2006.
67. Committee for Proprietary Medicinal Products (CPMP). Note for guidance on fixed-combination medicinal products. http://www.emea.eu.int/pdfs/human/ewp/024095en.pdf#search='note%20for%20guidance%20on%20fixed%20combination%20medicinal%20products'. 20 March 2006.
68. Giles TD. Rationale for combination therapy as initial treatment for hypertension. J Clin Hypertens. 2003;5:4–11.
69. Becerra V, Gracia A, Desai K, et al. Cost-effectiveness and public health benefit of secondary cardiovascular disease prevention from improved adherence using a polypill in the UK. BMJ Open. 2015;5:e007111.
70. WHO. Adherence to long-term therapies: evidence for action. 2003.
71. Truelove M, Patel A, Bompoint S, et al. The effect of a cardiovascular polypill strategy on pill burden. Cardiovasc Ther. 2015;33:347–52.
72. Roy A, Naik N, Srinath RK. Strengths and limitations of using the polypill in cardiovascular prevention. Curr Cardiol Rep. 2017;19(5):45.
73. Baumgartner A, Drame K, Geutjens S, et al. Does the polypill improve patient adherence compared to its individual formulations? A systematic review. Pharmaceutics. 2020;12(2):190.
74. Soliman EZ, Mendis S, Dissanayake WP, et al. A polypill for primary prevention of cardiovascular disease: a feasibility study of the World Health Organization. Trials. 2011;12:3.
75. Bramlage P, Sims H, Minguet J, et al. The polypill: an effective approach to increasing adherence and reducing cardiovascular event risk. Eur J Prev Cardiol. 2017;24(3):297–310.
76. Hippisley-Cox J, Coupland C. Effect of combinations of drugs on all cause mortality in patients with ischaemic heart disease: nested case–control analysis. BMJ. 2005;330:1059–63.
77. Brainin M, Feigin V, Martins S, et al. Cut stroke in half: polypill for primary prevention in stroke. Int J Stroke. 2018;13(6):633–47.
78. Sleight P, Pouleur H, Zannad F. Benefits, challenges, and registerability of the polypill. Eur Heart J. 2006;27(14):1651–6.

79. Rosenthal T. Can a polypill one single tablet combat different cardiovascular risk factors? J Am Soc Hypertens. 2018;12(5):335–9.
80. Elley CR, Gupta AK, Webster R, et al. The efficacy and tolerability of 'polypills': meta-analysis of randomized controlled trials. PLoS One. 2012;7(12):e52145.
81. Selak V, Webster R. Polypills for the secondary prevention of cardiovascular disease: effective in improving adherence but are they safe? Ther Adv Drug Saf. 2018;9(2):157–62.
82. Wertheimer AI. The economics of polypharmacology: fixed dose combinations and drug cocktails. Curr Med Chem. 2013;20(13):1635–8.
83. Teo KK, Liang Y. Polypill: lights and shadows. Curr Hypertens Rep. 2010;12(4):276–81.
84. Jayasheel BG. Perspect. Clin Res. 2010;1(4):120–3.
85. Bhaskar S, Tian F, Stoeger T, et al. Multifunctional nanocarriers for diagnostics, drug delivery and targeted treatment across blood-brain barrier: perspectives on tracking and neuroimaging. Part Fibre Toxicol. 2010;7(1):3.
86. Drbohlavova J, Chomoucka J, Adam V, et al. Nanocarriers for anticancer drugs-new trends in nanomedicine. Curr Drug Metab. 2013;14(5):547–64.
87. Su H, Wang Y, Gu Y, et al. Potential applications and human biosafety of nanomaterials used in nanomedicine. J Appl Toxicol. 2018;38(1):3–24.
88. Davis ME, Chen Z, Shin DM. Nanoparticle therapeutics: an emerging treatment modality for cancer. Nat Rev Drug Discov. 2008;7(9):771–82.
89. Ma L, Kohli M, Smith A. Nanoparticles for combination drug therapy. ACS Nano. 2013;7(11):9518–25.
90. Fisusia FA, Akalaa EO. Drug combinations in breast cancer therapy. Pharm Nanotechnol. 2019;7(3):3–23.
91. Ramsay RR, Popovic-Nikolic MR, Nikolic K, et al. A perspective on multi-target drug discovery and design for complex diseases. Clin Transl Med. 2018;7:3.
92. Ren J, Fu L, Nile SH, et al. Salvia miltiorrhiza in treating cardiovascular diseases: a review on its pharmacological and clinical applications. Front Pharmacol. 2019;10:753.
93. Li ZM, Xu SW, Liu PQ. Salvia miltiorrhiza Burge (Danshen): a golden herbal medicine in cardiovascular therapeutics. Acta Pharmacol Sin. 2018;39(5):802–24.
94. Jia Y, Leung SW. How efficacious is Danshen (Salvia miltiorrhiza) dripping pill in treating angina pectoris? Evidence assessment for meta-analysis of randomized controlled trials. J Altern Complement Med. 2017;23(9):676–84.
95. Jalili-Baleh L, Babaei E, Abdpour S, et al. A review on flavonoid-based scaffolds as multi-target-directed ligands (MTDLs) for Alzheimer's disease. Eur J Med Chem. 2018;152:570–89.
96. Sameem B, Saeedi M, Mahdavi M, et al. A review on tacrine-based scaffolds as multi-target drugs (MTDLs) for Alzheimer's disease. Eur J Med Chem. 2017;128:332–45.
97. Scotti L, Monteiro AFM, de Oliveira VJ, et al. Multi-target drugs against metabolic disorders. Endocr Metab Immune Disord Drug Targets. 2019;19(4):402–18.
98. Talevi A. Tailored multi-target agents. Applications and design considerations. Curr Pharm Des. 2016;22(21):3164–70.
99. Makhoba XH, Viegas C Jr, Mosa RA, et al. Potential impact of the multi-target drug approach in the treatment of some complex diseases. Drug Des Devel Ther. 2020;14:3235–49.
100. Rodríguez-Soacha DA, Scheiner M, Decker M. Multi-target-directed-ligands acting as enzyme inhibitors and receptor ligands. Eur J Med Chem. 2019;180:690–706.
101. Zhang X, He X, Chen Q, et al. A review on the hybrids of hydroxycinnamic acid as multi-target-directed ligands against Alzheimer's disease. Bioorg Med Chem. 2018;26(3):543–50.
102. Gontijo VS, Viegas FPD, Ortiz CJC, et al. Molecular hybridization as a tool in the design of multi-target directed drug candidates for neurodegenerative diseases. Curr Neuropharmacol. 2020;18(5):348–407.
103. Zhang HY. One-compound-multiple-targets strategy to combat Alzheimer's disease. FEBS Lett. 2005;579(24):5260–4.
104. Antoszczak M, Huczyński A. Salinomycin and its derivatives – a new class of multiple-targeted "magic bullets". Eur J Med Chem. 2019;176:208–27.
105. Pereira DM, Rodrigues PM, Borralho PM, et al. Delivering the promise of miRNA cancer therapeutics. Drug Discov Today. 2013;18(5–6):282–9.
106. Bolognesi ML. Polypharmacology in a single drug: multitarget drugs. Curr Med Chem. 2013;20(13):1639–45.
107. Bolognesi ML, Cavalli A. Multitarget drug discovery and polypharmacology. ChemMedChem. 2016;11(12):1190–2.
108. Roth BL, Sheffler DJ, Kroeze WK. Magic shotguns versus magic bullets: selectively non-selective drugs for mood disorders and schizophrenia. Nat Rev Drug Discov. 2004;3(4):353–9.
109. Morphy R, Kay C, Rankovic Z. From magic bullets to designed multiple ligands. Drug Discov Today. 2004;9(15):641–51.
110. Youdim MBH, Buccafusco JJ. Multi-functional drugs for various CNS targets in the treatment of neurodegenerative disorders. Trends Pharm Sci. 2005;26(1):27–35.
111. Talevi A. Multi-target pharmacology: possibilities and limitations of the "skeleton key approach" from a medicinal chemist perspective. Front Pharmacol. 2015;6:205.
112. Lin HH, Zhang LL, Yan R, et al. Network analysis of drug-target interactions: a study on FDA-approved new molecular entities between 2000 to 2015. Sci Rep. 2017;7:12230.
113. Lu JJ, Pan W, Hu YJ, et al. Multi-target drugs: the trend of drug research and development. PLoS One. 2012;7(6):e40262.

114. Zhou J, Li Q, Wu M, et al. Progress in the rational design for polypharmacology drug. Curr Pharm Des. 2016;22(21):3182–9.
115. McKie SA. Polypharmacology: in silico methods of ligand design and development. Future Med Chem. 2016;8(5):579–602.
116. Reddy AS, Zhang S. Polypharmacology: drug discovery for the future. Expert Rev Clin Pharmacol. 2013;6(1):41–7.
117. Al-Ali H, Lee DH, Danzi MC, et al. Rational polypharmacology: systematically identifying and engaging multiple drug targets to promote axon growth. ACS Chem Biol. 2015;10(8):1939–51.
118. Hopkins AL, Mason JS, Overington JP. Can we rationally design promiscuous drugs? Curr Opin Struct Biol. 2006;16(1):127–36.
119. Hopkins AL. Network pharmacology: the next paradigm in drug discovery. Nat Chem Biol. 2008;4(11):682–90.
120. Bottegoni G, Favia AD, Recanatini M, et al. The role of fragment-based and computational methods in polypharmacology. Drug Discov Today. 2012;17(1–2):23–34.
121. Decker M. Design of hybrid molecules for drug development. Amsterdam: Elsevier; 2017.
122. Rosini M. Polypharmacology: the rise of multitarget drugs over combination therapies. Future Med Chem. 2014;6:485–7.
123. de Lera AR, Ganesan A. Epigenetic polypharmacology: from combination therapy to multitarget drugs. Clin Epigenetics. 2016;8:105.
124. Costantino L, Barlocco D. Challenges in the design of multitarget drugs against multifactorial pathologies: a new life for medicinal chemistry? Future Med Chem. 2013;5(1):5–7.
125. Costantino L, Barlocco D. Designed multiple ligands: basic research vs clinical outcomes. Curr Med Chem. 2012;19(20):3353–87.
126. Hase T, Tanaka H, Suzuki Y, et al. Structure of protein interaction networks and their implications on drug design. PLoS Comput Biol. 2009;5:e1000550.
127. Lee JA, Uhlik MT, Moxham CM, et al. Modern phenotypic drug discovery is a viable, neoclassic pharma strategy. J Med Chem. 2012;55(10):4527–38.
128. Michelini E, Cevenini L, Mezzanotte L, et al. Cell-based assays: fueling drug discovery. Anal Bioanal Chem. 2010;398(1):227–38.
129. Moffat JG, Rudolph J, Bailey D. Phenotypic screening in cancer drug discovery – past, present and future. Nat Rev Drug Discov. 2014;13(8):588–602.
130. Hurko O. Target-based drug discovery, genetic diseases, and biologics. Neurochem Int. 2012;61(6):892–8.
131. Nishio K, Arao T, Shimoyama T, et al. Translational studies for target-based drugs. Cancer Chemother Pharmacol. 2005;56(Suppl 1):90–3.
132. Eder J, Sedrani R, Wiesmann C. The discovery of first-in-class drugs: origins and evolution. Nat Rev Drug Discov. 2014;13(8):577–87.
133. Swinney DC. Phenotypic vs. target-based drug discovery for first-in-class medicines. Clin Pharmacol Ther. 2013;93(4):299–301.
134. Vane JR, Botting RM. The mechanism of action of aspirin. Thromb Res. 2003;110(5–6):255–8.
135. Simpson MR, Simpson NR, Masheter HC. New drugs. 8. Flufenamic acid in rheumatoid arthritis. Comparison with aspirin and the results of extended treatment. Ann Phys Med. 1966;8(6):208–13.
136. Berman J, Haffajee CI, Alpert JS. Therapy of symptomatic pericarditis after myocardial infarction: retrospective and prospective studies of aspirin, indomethacin, prednisone, and spontaneous resolution. Am Heart J. 1981;101(6):750–3.
137. Durongpisitkul K, Gururaj VJ, Park JM, et al. The prevention of coronary artery aneurysm in Kawasaki disease: a meta-analysis on the efficacy of aspirin and immunoglobulin treatment. Pediatrics. 1995;96(6):1057–61.
138. Grundmann K, Jaschonek K, Kleine B, et al. Aspirin non-responder status in patients with recurrent cerebral ischemic attacks. J Neurol. 2003;250(1):63–6.
139. Amory JK, Amory DW. Dosing frequency of aspirin and prevention of heart attacks and strokes. Am J Med. 2007;120(4):e5.
140. Daya S. Recurrent spontaneous early pregnancy loss and low dose aspirin. Minerva Ginecol. 2003;55(5):441–9.
141. Baron JA. Aspirin and cancer: trials and observational studies. J Natl Cancer Inst. 2012;104(16):1199–200.
142. Keiser MJ, Setola V, Irwin JJ, et al. Predicting new molecular targets for known drugs. Nature. 2009;462(7270):175–81.
143. Fouqueray P, Leverve X, Fontaine E, et al. Imeglimin – a new oral anti-diabetic that targets the three key defects of type 2 diabetes. J Diabetes Metab. 2011;2:126.
144. Vuylsteke V, Chastain LM, Maggu GA, et al. Imeglimin: a potential new multi-target drug for type 2 diabetes. Drugs R&D. 2015;15(3):227–32.
145. Pacini G, Mari A, Fouqueray P, et al. Imeglimin increases glucose-dependent insulin secretion and improves β-cell function in patients with type 2 diabetes. Diabetes Obes Metab. 2015;17(6):541–5.
146. Pirags V, Lebovitz H, Fouqueray P. Imeglimin, a novel oral antidiabetic, exhibits a good efficacy and safety profile in type 2 diabetic patients. Diabetes Obes Metab. 2012;14:852–8.
147. Vial G, Chauvin M-A, Bendridi N, et al. Imeglimin normalizes glucose tolerance and insulin sensitivity and improves mitochondrial function in liver of a high-fat, high-sucrose diet mice model. Diabetes. 2015;64(6):2254–64.
148. Fouqueray P, Leverve X, Fontaine E, et al. Imeglimin – a new oral anti-diabetic that targets the three key defects of type 2 diabetes. J Diabet Metab. 2011;2(4):126.
149. Lablanche S, Tubbs E, Cottet-Rousselle C, et al. Imeglimin protects INS-1 cells and human islets against high glucose–and high fructose–induced cell

death by inhibiting the mitochondrial PTP opening. Diabetes. 2018;67(Suppl1):81.
150. Bolze S, Fouqueray P, Hallakou-Bozec S. Imeglimin, a new mitochondria-targeted agent for type 2 diabetes treatment. J World Mitochondria Soc. 2015;1(2)
151. Li J, Yu H, Wang S, et al. Natural products, an important resource for discovery of multitarget drugs and functional food for regulation of hepatic glucose metabolism. Drug Des Devel Ther. 2018;12:121–35.
152. Habtemariam S. Berberine pharmacology and the gut microbiota: a hidden therapeutic link. Pharmacol Res. 2020;155:104722.
153. Feng X, Sureda A, Jafari S, et al. Berberine in cardiovascular and metabolic diseases: from mechanisms to therapeutics. Theranostics. 2019;9(7):1923–51.
154. Imenshahidi M, Hosseinzadeh H. Berberine and barberry (Berberis vulgaris): a clinical review. Phytother Res. 2019;33(3):504–23.
155. Dyck GJB, Raj P, Zieroth S, et al. The effects of resveratrol in patients with cardiovascular disease and heart failure: a narrative review. Int J Mol Sci. 2019;20(4):904.
156. Ko JH, Sethi G, Um JY, et al. The role of resveratrol in cancer therapy. Int J Mol Sci. 2017;18(12):2589.
157. Ahmadi R, Ebrahimzadeh MA. Resveratrol – a comprehensive review of recent advances in anticancer drug design and development. Eur J Med Chem. 2020;200:112356.
158. Wang Z. MicroRNAs and cardiovascular disease. Potomac: Bentham Science Publishers; 2010. eISBN:978-1-60805-184-7
159. Wang Z. MicroRNA-interference technologies. Heidelberg/New York: Springer; 2009. ISBN-13:978-3-642-00488-9
160. Zhang Y, Du W, Yang B. Long non-coding RNAs as new regulators of cardiac electrophysiology and arrhythmias: molecular mechanisms, therapeutic implications and challenges. Pharmacol Ther. 2019;203:107389.
161. Wang Y, Liu B. Circular RNA in diseased heart. Cells. 2020;9(5):1240.
162. Arkin MR, Wells JA. Small-molecule inhibitors of protein-protein interactions: progressing towards the dream. Nat Rev Drug Discov. 2004;3(4):301–17.
163. Arkin MR, Tang Y, Wells JA. Small-molecule inhibitors of protein-protein interactions: progressing toward the reality. Chem Biol. 2014;21(9):1102–14.
164. Raina K, Crews CM. Targeted protein knockdown using small molecule degraders. Curr Opin Chem Biol. 2017;39:46–53.
165. Su XD, Shuai Y. Macromolecules and antibody-based drugs. Adv Exp Med Biol. 2020;1248:485–530.
166. Tyagi P, Santos JL. Macromolecule nanotherapeutics: approaches and challenges. Drug Discov Today. 2018;23(5):1053–61.
167. Rizvi NF, Smith GF. RNA as a small molecule druggable target. Bioorg Med Chem Lett. 2017;27(23):5083–8.
168. Xia T, Li J, Cheng H, et al. Small-molecule regulators of microRNAs in biomedicine. Drug Dev Res. 2015;76(7):375–81.
169. Van Meter EN, Onyango JA, Teske KA. A review of currently identified small molecule modulators of microRNA function. Eur J Med Chem. 2020;188:112008.
170. Hauser AT, Robaa D, Jung M. Epigenetic small molecule modulators of histone and DNA methylation. Curr Opin Chem Biol. 2018;45:73–85.
171. Finley A, Copeland RA. Small molecule control of chromatin remodeling. Chem Biol. 2014;21(9):1196–210.
172. Ganesan A. Multitarget drugs: an epigenetic epiphany. ChemMedChem. 2016;11(12):1227–41.
173. Franci G, Miceli M, Altucci L. Targeting epigenetic networks with polypharmacology: a new avenue to tackle cancer. Epigenomics. 2010;2(6):731–42.
174. Mohammad HP, Barbash O, Creasy CL. Targeting epigenetic modifications in cancer therapy: erasing the roadmap to cancer. Nat Med. 2019;25(3):403–18.
175. Janssens Y, Wynendaele E, Vanden Berghe W, et al. Peptides as epigenetic modulators: therapeutic implications. Clin Epigenetics. 2019;11(1):101.
176. Lu Q, Quinn AM, Patel MP, et al. Perspectives on the discovery of small-molecule modulators for epigenetic processes. J Biomol Screen. 2012;17(5):555–71.
177. Pachaiyappan B, Woster PM. Design of small molecule epigenetic modulators. Bioorg Med Chem Lett. 2014;24(1):21–32.
178. Catterall WA. From ionic currents to molecular mechanisms: the structure and function of voltage-gated sodium channels. Neuron. 2000;26:13–25.
179. Cohen MH, Williams GA, Sridhara R, et al. United States Food and Drug Administration drug approval summary: Gefitinib (ZD1839; Iressa) tablets. Clin Cancer Res. 2004;10(4):1212–8.
180. Roth BL, Ciaranello RD, Meltzer HY. Binding of typical and atypical antipsychotic agents to transiently expressed 5-HT1C receptors. J Pharmacol Exp Ther. 1992;260(3):1361–5.
181. Roth BL, Meltzer HY, Khan N. Binding of typical and atypical antipsychotic drugs to multiple neurotransmitter receptors. Adv Pharmacol. 1998;42:482–5.
182. Roth BL, Willins DL, Kristiansen K, et al. 5-Hydroxytryptamine2-family receptors (5-hydroxytryptamine2A, 5-hydroxytryptamine2B, 5-hydroxytryptamine2C): where structure meets function. Pharmacol Ther. 1998;79(3):231–57.
183. Saenz-Méndez P, Eriksson LA. Exploring polypharmacology in drug design. Methods Mol Biol. 2018;1824:229–43.
184. Niv MY, Rubin H, Cohen J, et al. Sequence-based design of kinase inhibitors applicable for therapeutics and target identification. J Biol Chem. 2004;279(2):1242–55.
185. Mitra A, Sarkar N. Sequence and structure-based peptides as potent amyloid inhibitors: a review. Arch Biochem Biophys. 2020;695:108614.

186. Li Q, Lai L. Prediction of potential drug targets based on simple sequence properties. BMC Bioinformatics. 2007;8:353.
187. Bennett CF, Baker BF, Pham N, et al. Pharmacology of antisense drugs. Annu Rev Pharmacol Toxicol. 2017;57:81–105.
188. Bennett CF, Krainer AR, Cleveland DW. Antisense oligonucleotide therapies for neurodegenerative diseases. Annu Rev Neurosci. 2019;42:385–406.
189. Bennett CF. Therapeutic antisense oligonucleotides are coming of age. Annu Rev Med. 2019;70:307–21.
190. Dowdy SF. Overcoming cellular barriers for RNA therapeutics. Nat Biotechnol. 2017;35(3):222–9.
191. Zhou LY, Qin Z, Zhu YH, et al. Current RNA-based therapeutics in clinical trials. Curr Gene Ther. 2019;19(3):172–496.
192. Sergeeva OV, Koteliansky VE, Zatsepin TS. mRNA-based therapeutics – advances and perspectives. Biochemistry (Mosc). 2016;81(7):709–22.
193. Weng Y, Li C, Yang T, et al. The challenge and prospect of mRNA therapeutics landscape. Biotechnol Adv. 2020;40:107534.
194. Kanasty R, Dorkin JR, Vegas A, et al. Delivery materials for siRNA therapeutics. Nat Mater. 2013;12(11):967–77.
195. Setten RL, Rossi JJ, Han SP. The current state and future directions of RNAi-based therapeutics. Nat Rev Drug Discov. 2019;18(6):421–46.
196. Gavrilov K, Saltzman WM. Therapeutic siRNA: principles, challenges, and strategies. Yale J Biol Med. 2012;85(2):187–200.
197. Velagapudi SP, Gallo SM, Disney MD. Sequence-based design of bioactive small molecules that target precursor microRNAs. Nat Chem Biol. 2014;10(4):291–7.
198. Ebert MS, Neilson JR, Sharp PA. MicroRNA sponges: competitive inhibitors of small RNAs in mammalian cells. Nat Methods. 2007;4(9):721–6.
199. Smith CIE, Zain R. Therapeutic oligonucleotides: state of the art. Annu Rev Pharmacol Toxicol. 2019;59:605–30.
200. Goodchild J. Therapeutic oligonucleotides. Methods Mol Biol. 2011;764:1–15.
201. Hecker M, Wagner AH. Transcription factor decoy technology: a therapeutic update. Biochem Pharmacol. 2017;144:29–34.
202. Mohibi S, Chen X, Zhang J. Cancer the 'RBP' eutics-RNA-binding proteins as therapeutic targets for cancer. Pharmacol Ther. 2019;203:107390.
203. Lu Y, Xiao J, Lin H, et al. A single anti-microRNA antisense oligodeoxyribonucleotide (AMO) targeting multiple microRNAs offers an improved approach for microRNA interference. Nucleic Acids Res. 2009;37:e24–33.
204. Wang Z. New concepts of miRNAi technologies. In: Wang Z, editor. MicroRNA interference technologies. Heidelberg: Springer; 2009.
205. Wang Z. The concept of multiple-target anti-miRNA antisense oligonucleotides technology. Methods Mol Biol. 2011;676:51–7.
206. Wang Z. The concept of multiple-target anti-miRNA antisense oligonucleotides technology. In: Wu W, editor. MicroRNA and cancer. Totowa: Humana Press; 2011.
207. Gao H, Xiao J, Yang B, et al. A single decoy oligodeoxynucleotides targeting multiple oncoproteins produces strong anti-cancer effects. Mol Pharmacol. 2006;70:1621–9.
208. Wang X, Liu Q, Hou B, et al. Concomitant targeting of multiple key transcription factors effectively disrupts cancer stem cells enriched in side population of human pancreatic cancer cells. PLoS One. 2013;8(9):e73942.
209. Rad SM, Bamdad T, Sadeghizadeh M, et al. Transcription factor decoy against stem cells master regulators, Nanog and Oct-4: a possible approach for differentiation therapy. Tumour Biol. 2015;36(4):2621–9.
210. Swindell WR, Sarkar MK, Stuart PE, et al. Psoriasis drug development and GWAS interpretation through in silico analysis of transcription factor binding sites. Clin Transl Med. 2015;4:13.
211. Johari B, Zargan J. Simultaneous targeted inhibition of Sox2-Oct4 transcription factors using decoy oligodeoxynucleotides to repress stemness properties in mouse embryonic stem cells. Cell Biol Int. 2017;41(12):1335–44.
212. Bigdelou Z, Mortazavi Y, Saltanatpour Z, et al. Role of Oct4-Sox2 complex decoy oligodeoxynucleotides strategy on reverse epithelial to mesenchymal transition (EMT) induction in HT29-ShE encompassing enriched cancer stem-like cells. Mol Biol Rep. 2020;47(3):1859–69.
213. Johari B, Rezaeejam H, Moradi M, et al. Increasing the colon cancer cells sensitivity toward radiation therapy via application of Oct4-Sox2 complex decoy oligodeoxynucleotides. Mol Biol Rep. 2020;47(9):6793–805.
214. Cheng MS, Su MX, Wang MX, et al. Probes and drugs that interfere with protein translation via targeting to the RNAs or RNA-protein interactions. Methods. 2019;167:124–33.
215. Baker JD, Uhrich RL, Strovas TJ, et al. Targeting pathological tau by small molecule inhibition of the poly(A): MSUT2 RNA-protein interaction. ACS Chem Neurosci. 2020;11(15):2277–85.
216. DeJong ES, Luy B, Marino JP. RNA and RNA-protein complexes as targets for therapeutic intervention. Curr Top Med Chem. 2002;2(3):289–302.
217. Hermann T. Strategies for the design of drugs targeting RNA and RNA-protein complexes. Angew Chem Int Ed Engl. 2000;39(11):1890–904.
218. Zhang Y, Jiao L, Sun L, et al. LncRNA ZFAS1 as a SERCA2a inhibitor to cause intracellular Ca^{2+} overload and contractile dysfunction in a mouse model of myocardial infarction. Circ Res. 2018;122(10):1354–68.
219. Evans BE, Rittle KE, Bock MG, et al. Methods for drug discovery: development of potent, selective,

orally effective cholecystokinin antagonists. J Med Chem. 1988;31(12):2235–46.
220. Horton DA, Bourne GT, Smythe ML. The combinatorial synthesis of bicyclic privileged structures or privileged substructures. Chem Rev. 2003;103(3):893–930.
221. Zhao H, Dietrich J. Privileged scaffolds in lead generation. Expert Opin Drug Discov. 2015;10(7):781–90.
222. Welsch ME, Snyder SA, Stockwell BR. Privileged scaffolds for library design and drug discovery. Curr Opin Chem Biol. 2010;14(3):347–61.
223. Yendapally R, Sikazwe D, Kim SS, et al. A review of phenformin, metformin, and imeglimin. Drug Dev Res. 2020;81(4):390–401.
224. Vamos M, Hohnloser SH. Amiodarone and dronedarone: an update. Trends Cardiovasc Med. 2016;26(7):597–602.
225. Shaveta S, Mishra S, Singh P. Hybrid molecules: the privileged scaffolds for various pharmaceuticals. Eur J Med Chem. 2016;124:500–36.
226. Li Y, Zhao C, Zhang J, et al. HybridMolDB: a manually curated database dedicated to hybrid molecules for chemical biology and drug discovery. J Chem Inf Model. 2019;59(10):4063–9.
227. Borsari C, Trader DJ, Tait A, et al. Designing chimeric molecules for drug discovery by leveraging chemical biology. J Med Chem. 2020;63(5):1908–28.
228. Meunier B. Hybrid molecules with a dual mode of action: dream or reality? Acc Chem Res. 2008;41(1):69–77.
229. Xue H, Li J, Xie H, et al. Review of drug repositioning approaches and resources. Int J Biol Sci. 2018;14(10):1232–44.
230. Pushpakom S, Iorio F, Eyers PA, et al. Drug repurposing: progress, challenges and recommendations. Nat Rev Drug Discov. 2019;18(1):41–58.
231. Sertkaya A, Birkenbach A, Berlind A, et al. Examination of clinical trial costs and barriers for drug development, vol. 1. US Department of Health and Human Services, Office of the Assistant Secretary for Planning and Evaluation Report; 2014. p. 1–92.
232. Yeu Y, Yoon Y, Park S. Protein localization vector propagation: a method for improving the accuracy of drug repositioning. Mol BioSyst. 2015;11(7):2096–102.
233. Pharmaceutical Research and Manufacturers of America. http://www.phrma.org/sites/default/files/pdf/2015_phrma_profile.pdf.2015
234. Drug Development Process. https://www.fda.gov/Drugs/default.htm
235. Drug Approval Process. https://www.fda.gov/Drugs/DevelopmentApprovalProcess/default.htm

3 Polypharmacology in Clinical Applications—Anticancer Polypharmacology

Abstract

Polypharmacology is not only a theoretical discipline but also an experimental and application discipline. The ultimate goal of Polypharmacology is to translate the new concept and fundamental knowledge into drug discovery and therapeutics development for clinical uses. The clinical applications of Polypharmacology have been realized in many disease entities though it is still in its early stage of development. From this chapter to Chap. 11, the status and advancement of clinical applications of Polypharmacology in disease therapy will be discussed. The present chapter focuses on the application of polypharmacology to cancer therapy. In this respect, an introduction to the conventional pharmacology therapy of cancer is given as preknowledge for better understanding of the contents in the subsequent sections. Then, the therapeutic applications of Polypharmacology in the forms of combination drug therapy (CDT), fixed-dose single-pill combination (FDC), and multitarget drug (MTD) will be detailed with their uses in the five major types of cancer, including lung cancer, colorectal cancer, hepatocellular carcinoma, stomach cancer, and female breast cancer. The efficacy and safety issues of Polypharmacological treatment will be described. Balanced discussion on both advantages and limitations of Polypharmacological regimens will be provided.

Keywords

Cancer · Anticancer polypharmacology · Multitargeted tyrosine kinase inhibitors · Multitargeting HDAC inhibitors · Multitarget therapy

3.1 Pharmacological Therapy of Cancer

Cancer is a group of diseases characterized by the uncontrolled growth and spread of abnormal cells, which affects 1 in 3 people in the United States alone. If the spread is not controlled, it can result in death. Cancer can develop anywhere in the body and is named for the part of the body where it started. There are two main categories of cancer: (1) hematologic (blood) cancers are cancers of the blood cells, including leukemia, lymphoma, and multiple myeloma and (2) solid tumor cancers are cancers of any of the other body organs or tissues. According to the Global Cancer Statistics 2020 by the International Agency for Research on Cancer [1], an estimated 19.3 million new cancer cases (18.1 million excluding nonmelanoma skin cancer) and almost 10.0 million cancer deaths (9.9 million excluding nonmelanoma skin cancer) occurred in 2020 worldwide. Female breast cancer has surpassed lung cancer as the most commonly diagnosed cancer, with an estimated 2.3 million new cases (11.7%), followed by lung (11.4%), colorectal

Z. Wang, B. Yang, *Polypharmacology*, https://doi.org/10.1007/978-3-031-04998-9_3

(10.0%), prostate (7.3%), and stomach (5.6%) cancers. Lung cancer remained the leading cause of cancer death, with an estimated 1.8 million deaths (18%), followed by colorectal (9.4%), liver (8.3%), stomach (7.7%), and female breast (6.9%) cancers. Overall incidence was from 2-fold to 3-fold higher in transitioned versus transitioning countries for both sexes. The global cancer burden is expected to be 28.4 million cases in 2040, a 47% rise from 2020 [1].

In the past decades, significant advances in cancer biology have led to the identification of key factors contributing to tumorigenesis and elucidation of a variety of molecular pathways underlying cell growth, the cell cycle, apoptosis, angiogenesis, and invasion providing molecular targets for cancer treatment. We now know that new vessel formation (angiogenesis) is an essential physiological process for embryologic development, normal growth, and tissue repair and it is tightly regulated at the molecular level; however, this process is dysregulated in several pathological conditions such as cancer. Accordingly, many molecular-targeted agents have been developed and exhibited significant antitumor activity in a variety of tumor types such as hematologic malignancies, colorectal cancer, breast cancer, renal cell carcinoma, and gastrointestinal stromal tumors.

The vascular endothelial growth factor (VEGF) and its tyrosine kinase receptors (VEGFR-1 and VEGFR-2) are important signaling proteins involved in both vasculogenesis (the formation of the circulatory system) and angiogenesis (the growth of blood vessels from preexisting vasculature) [2, 3] and play major roles not only in physiological but also in most pathological angiogenesis, such as cancer [4, 5]. The discovery of VEGF/VEGFR as major drivers of tumor angiogenesis and growth suggests VEGF/VEGFR2 pathway as the central therapeutic target in anti-angiogenic treatment in multiple cancers and has led to efforts to develop novel therapeutics aimed at inhibiting its activity. Anti-VEGF therapy (anti-VEGF monoclonal antibody and VEGFR tyrosine kinase inhibitors (VEGFR-TKI)) has become an important option for the management of several human malignancies; however, a significant number of patients do not respond to anti-VEGF therapy when used either as single agent or in combination with chemotherapy. In addition, the benefit of antiangiogenic therapy is relatively short lived, and the majority of patients relapse and progress.

In addition, the epidermal growth factor (EGF) is a protein that stimulates cell growth and differentiation by binding to its receptor, EGFR. EGFR is a multifunctional receptor transmembrane glycoprotein and a member of the tyrosine kinase family of growth factor receptors. Epidermal growth factor (EGF) is the specific ligand of the EGFR, and it activates the receptor by binding and phosphorylating the tyrosine kinase receptor [6, 7]. Receptor activation stimulates a number of intracellular signal transduction pathways and thereby promotes tumor cell division, migration, and angiogenesis. Therefore, EGFR signal transduction can be targeted and blocked to inhibit tumor proliferation, invasion, and distant metastasis during the molecular-targeted treatment of gastric cancer [8, 9]. The main anti-EGFR therapeutic agents include anti-EGFR monoclonal antibodies and EGFR tyrosine kinase inhibitors (EGFR-TKI). These therapeutic strategies include epidermal growth factor receptor (EGFR) inhibitors, angiogenesis inhibitors, cell-cycle inhibitors, and matrix metalloproteinase inhibitors.

According to their mechanisms of action, currently available anticancer drugs can be classified into chemotherapeutic agents, hormone therapy/endocrine drugs, and immunological agents.

3.1.1 Chemotherapeutic Agents

Chemotherapeutic agents, also referred to as antineoplastic agents, are used to directly or indirectly inhibit the uncontrolled growth and proliferation of cancer cells. They can be further classified into several subgroups including DNA alkylating agents, antimetabolites, topoisomerase inhibitors, antibiotics, mitotic inhibitors, and protein kinase inhibitors. Chemotherapy is associated with a range of adverse effects (e.g., nausea, vomiting, immunosuppression, and impaired growth of healthy cells), and some agents increase the risk of secondary neoplasm development [10].

3.1.1.1 DNA Alkylating Agents

The DNA alkylating agents used to treat breast cancer include cyclophosphamide and doxorubicin.

Cyclophosphamide is an oxazaphosphorine, interferes with the duplication of DNA and RNA transcription, and its most common undesired effects are myelosuppression, hepatotoxicity, pulmonary fibrosis, nephrotoxicity, mucosis, megaloblastic anemia, birth defects, and neurotoxicity [11, 12].

Doxorubicin, which is an anthracycline, is a topoisomerase II inhibitor that increases DNA degradation, preventing cell replication. Additionally, it can intercalate into DNA and promote the formation of free radicals, which in turn causes the fragmentation of DNA strands. Unfortunately, this agent can also cause cardiotoxicity, myelosuppression, alopecia, extravasation, infertility, and urine discoloration [11, 12].

3.1.1.2 Antimetabolites

Antimetabolites used in breast cancer can be divided into antifolates (such as methotrexate) and pyrimidine antagonists (such as 5-FU, capecitabine, or gemcitabine). Methotrexate is a competitive inhibitor of the dihydrofolate reductase which causes a decrease in pyrimidine and DNA synthesis. On the other hand, 5-FU or capecitabine can form a complex with the thymidylate synthase-folic acid, inhibiting its activity and reducing thymidine and DNA synthesis, whereas gemcitabine can incorporate a pyrimidine analog into DNA, decreasing its synthesis. The most common collateral effects of these drugs are myelosuppression, cardiotoxicity, hepatotoxicity, pulmonary toxicity, hemolytic uremic syndrome, and hyperammonemic encephalopathy [11, 12].

3.1.1.3 Antimitotic Drugs

The antimitotic drugs used in breast cancer treatment can be divided into three types: vinca alkaloids (such as vinorelbine), taxanes (such as paclitaxel and docetaxel), and non-taxane microtubule inhibitors (such as eribulin, ixabepilone and epothilone). Vinorelbine binds to B tubulin, inhibiting its polymerization into microtubules; it also prevents the formation of the mitotic spindle and arrests cells in the M phase. Taxanes produce hyper-stabilization in polymerized microtubules, thus inhibiting the degradation of the mitotic spindle, and arrest cells in the M phase. Non-taxane microtubule inhibitors can either prevent the formation of the mitotic spindle (eribulin) or bind to B tubulin, hyper-stabilizing microtubules and producing cell arrest. Undesirable actions of these drugs include myelosuppression, peripheral neuropathy, QT prolongation, and hypersensitivity [11, 13].

3.1.1.4 Topoisomerase Inhibitors

Topoisomerase inhibitors are chemical compounds that block the action of topoisomerases, which are broken into two broad subtypes: type I topoisomerases (TopI) and type II topoisomerases (TopII) [14–16]. Topoisomerase plays important roles in cellular reproduction and DNA organization, as they mediate the cleavage of single- and double-stranded DNA to relax supercoils, untangle catenanes, and condense chromosomes in eukaryotic cells. Topoisomerase inhibitors influence these essential cellular processes. Some topoisomerase inhibitors prevent topoisomerases from performing DNA strand breaks while others, deemed topoisomerase poisons, associate with topoisomerase-DNA complexes, and prevent the re-ligation step of the topoisomerase mechanism [16]. These topoisomerase-DNA-inhibitor complexes are cytotoxic agents, as the un-repaired single- and double-stranded DNA breaks that they cause can lead to apoptosis and cell death [15, 16]. Because of this ability to induce apoptosis, topoisomerase inhibitors have gained interest as therapeutics against infectious and cancerous cells.

3.1.1.5 Monoclonal Antibody Drugs—Targeted Cancer Therapy

Monoclonal antibody therapy is a form of immunotherapy that uses monoclonal antibodies (mAbs) to bind monospecifically to certain cells or proteins. The objective is that this treatment will stimulate the patient's immune system to attack those cells. Alternatively, in radioimmunotherapy a radioactive dose localizes a target cell line, delivering lethal chemical doses [17]. More recently antibodies have been used to bind to molecules involved in T-cell regulation to remove inhibitory pathways that block T-cell responses. This is known as immune checkpoint therapy [18]. It is possible to create a mAb that is specific to almost any extracellular/cell surface target. R&D is underway to create antibodies for diseases. Many monoclonal antibodies have been approved for treating many different types of cancer. Monoclonal antibodies are administered through a vein (intravenously). Monoclonal antibodies are designed to function in different ways. A particular drug may actually function by more than one means. Examples include:

Flagging cancer cells: Some immune system cells depend on antibodies to locate the target of an attack. Cancer cells that are coated in monoclonal antibodies may be more easily detected and targeted for destruction.

Triggering cell-membrane destruction: Some monoclonal antibodies can trigger an immune system response that can destroy the outer wall (membrane) of a cancer cell.

Blocking cell growth: Some monoclonal antibodies block the connection between a cancer cell and proteins that promote cell growth—an activity that is necessary for cancer growth and survival.

Preventing blood vessel growth: In order for a cancerous tumor to grow and survive, it needs a blood supply. Some monoclonal antibody drugs block protein-cell interactions necessary for the development of new blood vessels.

Blocking immune system inhibitors: Human body keeps the immune system from being overactive by making proteins that control the activity of the immune system cells. Monoclonal antibodies can interfere with that process so that your immune system cells are allowed to work without controls against cancer cells.

Directly attacking cancer cells: Certain monoclonal antibodies may attack the cell more directly. When some of these antibodies attach to a cell, a series of events inside the cell may cause it to self-destruct.

Delivering radiation treatment: Because of a monoclonal antibody's ability to connect with a cancer cell, the antibody can be engineered as a delivery vehicle for other treatments. When a monoclonal antibody is combined with a small radioactive particle, it transports the radiation treatment directly to cancer cells and may minimize the effect of radiation on healthy cells.

Delivering chemotherapy: Similarly, some monoclonal antibodies are combined with a chemotherapy drug in order to deliver the treatment directly to the cancer cells while avoiding healthy cells.

Binding cancer and immune cells: Some drugs combine two monoclonal antibodies, one that attaches to a cancer cell and one that attaches to a specific immune system cell. This connection may promote immune system attacks on the cancer cells.

3.1.1.6 Mitotic Inhibitors

Mitotic inhibitors are drugs derived from natural plant alkaloids. They inhibit cell division or mitosis, where a single cell divides into two genetically identical daughter cells. Mitotic inhibitors bind to tubulin and inhibit its polymerization into microtubules. Microtubules are structures responsible for pulling the cell apart when it divides. Mitotic inhibitors affect cancer cells more than normal cells because cancer cells divide (mitotic cell division) more rapidly therefore are more susceptible to mitotic inhibition. Different mitotic inhibitors are used to treat particular types of cancers, such as leukemia, lymphoma, breast cancer, lung cancer, and other types of cancers. Examples of mitotic inhibitors frequently used in the treatment of cancer include paclitaxel, docetaxel, vinblastine, vincristine, and vinorelbine [19–21]. Colchicine and griseofulvin are mitotic inhibitors used in the treatment of gout and toenail fungus, respectively.

3.1.1.7 Protein Kinase Inhibitors

Protein kinases are ubiquitous intracellular and cell surface proteins that play critical roles in cell signaling pathways involved in metabolism, injury responses, adaption, growth, and differentiation [22]. They act by adding a phosphate group to a protein (phosphorylation), usually on a specific amino acid which often makes the protein or enzyme "active" [22]. The human genome has more than 500 protein kinases and they can be classified as (1) tyrosine, (2) serine-theonine, or (3) nonspecific (both), based upon their amino acid specificity [22]. Many protein kinases are cell surface receptors and act to initiate an intracellular pathway of activation, after the receptor is engaged by its ligand, typically a cytokine or growth factor. Inhibitors of these kinases are called protein kinase receptor inhibitors. Other kinases are intracellular and take part in cell signaling [22]. Protein kinases can be specifically involved in cell growth, proliferation, and differentiation, and mutations may lead to unregulated growth and proliferation that is typical of cancerous cells [22]. These mutated protein kinases represent an attractive target for anticancer agents.

The kinase inhibitors are a large group of unique and potent antineoplastic agents which specifically target protein kinases that are altered in cancer cells and that account for some of their abnormal growth [22]. These kinases can be targeted by "non-receptor" protein kinases. Finally, some kinase inhibitors have specificity for multiple kinases and are called "multikinase inhibitors" [22]. The potent activity and lack of generalized toxicity of the kinase inhibitors relate to the specificity of antagonist for the mutated protein. In like manner, their toxicity often relates to off-target activity, either to the unmutated kinase or to closely related, normal kinases [22–24].

The protein kinases can be categorized based upon the amino acid that they phosphorylate: either serine, threonine, or tyrosine. The tyrosine kinase receptor inhibitors were the initial and are perhaps the best-characterized kinase inhibitors. The protein kinase inhibitors are relatively recently developed agents, all having been introduced since 2001. They are unique and represent a major advance in cancer chemotherapy, away from broadly cytotoxic agents and toward drugs that specifically target the molecular abnormalities of cancer cells [22–24]. The first tyrosine kinase inhibitor approved for use in the United States was imatinib (Gleevec: 2001) which is used to treat Philadelphia chromosome-positive chronic lymphocytic leukemia, which has a mutated kinase receptor (BCR-ABL) that is created by the specific translocation that creates the Philadelphia chromosome. Imatinib is a specific inhibitor of the BCR-ABL kinase. The introduction of this first protein kinase inhibitor was followed by more than a dozen others within the next 10 years.

3.1.2 Hormone Therapy/Endocrine Therapy (ET)

Hormone therapy/endocrine therapy (ET) is a cancer treatment that slows or stops the growth of cancer that uses hormones to grow and is considered a systemic treatment because the hormones they target circulate in the body. Different types of ET work in different ways. They can stop the body from making the hormone, block the hormone from attaching to cancer cells, and/or antagonize the hormone from binding to their respective receptors. They can be used to treat a certain kind of cancer by stopping or slowing its growth and/or to lessen symptoms related to a certain type of cancer. Hormone therapeutics are generally classified according to their applications to specific endocrine organs (glands). For example, endocrine drugs for BC are used to treat estrogen receptor (ER)- or progesterone receptor (PR)-positive breast tumors. These drugs are synthetic analogs of the anti-gonadotropin-releasing hormone, anti-progestins, or anti-estrogens. The latter can be divided into aromatase inhibitors and ER antagonists. The most common side effects of these drugs are flushing sensation, nightly sweats, vaginal dryness, high blood clot risk, apoplexy, cataracts, endometrial cancer, uterus cancer, bone-mass decrease, and gastrointestinal symptoms [11, 25, 26].

3.1.2.1 Breast Cancer

Aromatase inhibitors (AIs): anastrozole, exemestane, and letrozole.

Selective estrogen receptor modulators (SERMs): tamoxifen and raloxifene.

Estrogen receptor antagonists: fulvestrant and toremifene.

Luteinizing hormone-releasing hormone (LHRH) agonists: goserelin, leuprolide, and triptorelin.

3.1.2.2 Prostate Cancer

Anti-androgens: apalutamide, enzalutamide, darolutamide, bicalutamide, flutamide, and nilutamide (also called androgen deprivation therapy or ADT).

CYP17 inhibitors: abiraterone and ketoconazole.

Luteinizing hormone-releasing hormone (LHRH) agonists and antagonists: goserelin, leuprolide, triptorelin, and degarelix.

3.1.2.3 Endometrial Cancer

Progestins: medroxyprogesterone acetate or megestrol acetate.

Selective estrogen receptor modulators (SERMs): tamoxifen and raloxifene.

Luteinizing hormone-releasing hormone (LHRH) agonists: goserelin and leuprolide.

Aromatase inhibitors (AIs): letrozole, anastrozole, and exemestane.

3.1.2.4 Adrenal Cancer

Adrenolytics: mitotane.

Estrogen receptor antagonists: fulvestrant and toremifene.

Selective estrogen receptor modulators (SERMs): tamoxifen and raloxifene.

3.1.3 Immunological Agents

Immunological agents are drugs that can stimulate the immune system to detect and eliminate breast cancer cells. They also include monoclonal antibodies against specific tumor cell proteins, and cancer treatment vaccines. Different types of immunological agents can be administered to inhibit check-points, block the suppression of the immune response, and/or over-activate T cells from patients. The most common side effects of these drugs are fatigue, fever, shivers, weakness, nausea, vomits, dizziness, body aches, and high or low blood pressure [11, 27–29].

Immune checkpoint inhibitors (ICIs): monoclonal antibodies targeting cytotoxic T lymphocyte antigen-4 (CTLA-4), programmed cell death protein-1 (PD-1), or PD-1 ligand (PD-L1).

3.1.4 Targeted Cancer Therapies

In addition to the therapeutics introduced above, targeted therapies are evolving as the focus of much anticancer drug development. They are a cornerstone of precision medicine, a form of medicine that uses information about a person's genes and proteins to prevent, diagnose, and treat cancer. Targeted cancer therapies are drugs or other substances that block the growth and spread of cancer by interfering with specific molecules ("molecular targets") that are involved in the growth, progression, and spread of cancer [30–40]. Targeted cancer therapies are sometimes called "molecularly targeted drugs," "molecularly targeted therapies," "precision medicines," or similar names.

Targeted therapies differ from standard chemotherapy in several ways:

(a) Targeted therapies act on specific molecular targets that are associated with cancer, whereas most standard chemotherapies act on all rapidly dividing normal and cancerous cells.
(b) Targeted therapies are deliberately chosen or designed to interact with their target, whereas many standard chemotherapies were identified because they kill cells.
(c) Targeted therapies are often cytostatic (that is, they block tumor cell proliferation), whereas standard chemotherapy agents are cytotoxic (that is, they kill tumor cells).

Many targeted cancer therapies have been approved by the FDA to treat specific types of cancer. Others are being studied in clinical trials (research studies with people), and many more are in preclinical testing (research studies with animals).

One approach to identify potential targets is to compare the amounts of individual proteins in cancer cells with those in normal cells. Proteins that are present in cancer cells but not normal cells or that are more abundant in cancer cells would be potential targets, especially if they are known to be involved in cell growth or survival. An example of such a differentially expressed target is the human epidermal growth factor receptor 2 protein (HER-2). HER-2 is expressed at high levels on the surface of some cancer cells. Several targeted therapies are directed against HER-2, including trastuzumab (Herceptin), which is approved to treat certain breast and stomach cancers that overexpress HER-2.

Another approach to identify potential targets is to determine whether cancer cells produce mutant (altered) proteins that drive cancer progression. For example, the cell growth signaling protein BRAF is present in an altered form (known as BRAF V600E) in many melanomas. Vemurafenib (Zelboraf) targets this mutant form of the BRAF protein and is approved to treat patients with inoperable or metastatic melanoma that contains this altered BRAF protein.

Researchers also look for abnormalities in chromosomes that are present in cancer cells but not in normal cells. Sometimes these chromosome abnormalities result in the creation of a fusion gene (a gene that incorporates parts of two different genes) whose product, called a fusion protein, may drive cancer development. Such fusion proteins are potential targets for targeted cancer therapies. For example, imatinib mesylate (Gleevec) targets the BCR-ABL fusion protein, which is made from pieces of two genes that get joined together in some leukemia cells and promotes the growth of leukemic cells.

Once a candidate target has been identified, the next step is to develop a therapy that affects the target in a way that interferes with its ability to promote cancer cell growth or survival. For example, a targeted therapy could reduce the activity of the target or prevent it from binding to a receptor that it normally activates, among other possible mechanisms.

Most targeted therapies are either small molecules or monoclonal antibodies. Small-molecule compounds are typically developed for targets that are located inside the cell because such agents are able to enter cells relatively easily. Monoclonal antibodies are relatively large and generally cannot enter cells, so they are used only for targets that are outside cells or on the cell surface.

Candidate small molecules are usually identified in what are known as "high-throughput screens," in which the effects of thousands of test compounds on a specific target protein are examined. Compounds that affect the target (sometimes called "lead compounds") are then chemically modified to produce numerous closely related versions of the lead compound. These related compounds are then tested to determine which are most effective and have the fewest effects on nontarget molecules.

Monoclonal antibodies are developed by injecting animals (usually mice) with purified target proteins, causing the animals to make many different types of antibodies against the target. These antibodies are then tested to find the ones that bind best to the target without binding to nontarget proteins.

Before monoclonal antibodies are used in humans, they are "humanized" by replacing as much of the mouse antibody molecule as possible with corresponding portions of human antibodies. Humanizing is necessary to prevent the human immune system from recognizing the monoclonal antibody as "foreign" and destroying it before it has a chance to bind to its target protein. Humanization is not an issue for small-molecule compounds because they are not typically recognized by the body as foreign.

Many different targeted therapies have been approved for use in cancer treatment. These therapies include hormone therapies, signal transduction inhibitors, gene expression modulators, apoptosis inducers, angiogenesis inhibitors, immunotherapies, and toxin delivery molecules.

Like classical chemotherapy regimens used to treat cancer, targeted therapies also rely upon polypharmacology, but it remains elusive as to what combinations of molecularly targeted drugs may be most efficacious. A study used image-based proliferation and apoptosis assays in colorectal cancer cell lines to systematically investigate the efficacy of combinations of two to six drugs that target critical oncogenic pathways [40]. Drug pairs targeting key signaling pathways resulted in synergies across a broad spectrum of genetic backgrounds but often yielded only cytostatic responses. Enhanced cytotoxicity was observed when additional processes including apoptosis and cell cycle were targeted as part of the combination. In some cases, where cell lines were resistant to paired and tripled drugs, increased expression of antiapoptotic proteins was observed, requiring a fourth-order combination to induce cytotoxicity [40].

3.1.4.1 Hormone Therapies

Hormone therapies slow or stop the growth of hormone-sensitive tumors, which require certain hormones to grow. Hormone therapies act by preventing the body from producing the hormones or by interfering with the action of the hormones. Hormone therapies have been approved for both breast cancer and prostate cancer.

3.1.4.2 Signal Transduction Inhibitors

Signal transduction inhibitors block the activities of molecules that participate in signal transduction, the process by which a cell responds to signals from its environment. During this process, once a cell has received a specific signal, the signal is relayed within the cell through a series of biochemical reactions that ultimately produce the appropriate response(s). In some cancers, the malignant cells are stimulated to divide continuously without being prompted to do so by external growth factors. Signal transduction inhibitors interfere with this inappropriate signaling.

3.1.4.3 Gene Expression Modulators

Gene expression modulators modify the function of proteins that play a role in controlling gene expression.

3.1.4.4 Apoptosis Inducers

Apoptosis inducers cause cancer cells to undergo a process of controlled cell death called apoptosis. Apoptosis is one method the body uses to get rid of unneeded or abnormal cells, but cancer cells have strategies to avoid apoptosis. Apoptosis inducers can get around these strategies to cause the death of cancer cells.

3.1.4.5 Angiogenesis Inhibitors

Angiogenesis inhibitors block the growth of new blood vessels to tumors (a process called tumor angiogenesis). A blood supply is necessary for tumors to grow beyond a certain size because blood provides the oxygen and nutrients that tumors need for continued growth. Treatments that interfere with angiogenesis may block tumor growth. Some targeted therapies that inhibit angiogenesis interfere with the action of vascular endothelial growth factor (VEGF), a substance that stimulates new blood vessel formation. Other angiogenesis inhibitors target other molecules that stimulate new blood vessel growth.

3.1.4.6 Immunotherapies

Immunotherapies trigger the immune system to destroy cancer cells. Some immunotherapies are monoclonal antibodies that recognize specific molecules on the surface of cancer cells. Binding of the monoclonal antibody to the target molecule results in the immune destruction of cells that express that target molecule. Other monoclonal antibodies bind to certain immune cells to help these cells better kill cancer cells.

3.1.4.7 Monoclonal Antibodies

Monoclonal antibodies that deliver toxic molecules can cause the death of cancer cells specifically. Once the antibody has bound to its target cell, the toxic molecule that is linked to the antibody—such as a radioactive substance or a poisonous chemical—is taken up by the cell, ultimately killing that cell. The toxin will not affect cells that lack the target for the antibody—i.e., the vast majority of cells in the body.

The FDA has approved targeted therapies for the treatment of some patients with the following types of cancer (some targeted therapies have been approved to treat more than one type of cancer):

Bladder Cancer

Atezolizumab (Tecentriq), nivolumab (Opdivo), avelumab (Bavencio), pembrolizumab

(Keytruda), erdafitinib (Balversa), enfortumab vedotin-ejfv (Padcev), sacituzumab govitecan-hziy (Trodelvy).

Brain Cancer
Bevacizumab (Avastin), everolimus (Afinitor).

Breast Cancer
Everolimus (Afinitor), tamoxifen (Nolvadex), toremifene (Fareston), trastuzumab (Herceptin), fulvestrant (Faslodex), anastrozole (Arimidex), exemestane (Aromasin), lapatinib (Tykerb), letrozole (Femara), pertuzumab (Perjeta), ado-trastuzumab emtansine (Kadcyla), palbociclib (Ibrance), ribociclib (Kisqali), neratinib maleate (Nerlynx), abemaciclib (Verzenio), olaparib (Lynparza), talazoparib tosylate (Talzenna), atezolizumab (Tecentriq), alpelisib (Piqray), fam-trastuzumab deruxtecan-nxki (Enhertu), tucatinib (Tukysa), sacituzumab govitecan-hziy (Trodelvy), pertuzumab, trastuzumab, and hyaluronidase-zzxf (Phesgo), pembrolizumab (Keytruda), margetuximab-cmkb (Margenza).

Cervical Cancer
Bevacizumab (Avastin), pembrolizumab (Keytruda).

Colorectal Cancer
Cetuximab (Erbitux), panitumumab (Vectibix), bevacizumab (Avastin), ziv-aflibercept (Zaltrap), regorafenib (Stivarga), ramucirumab (Cyramza), nivolumab (Opdivo), ipilimumab (Yervoy), encorafenib (Braftovi), pembrolizumab (Keytruda).

Dermatofibrosarcoma Protuberans
Imatinib mesylate (Gleevec).

Endocrine/Neuroendocrine Tumors
Lanreotide acetate (Somatuline Depot), avelumab (Bavencio), lutetium Lu 177-dotatate (Lutathera), iobenguane I 131 (Azedra).

Endometrial Cancer
Pembrolizumab (Keytruda), lenvatinib mesylate (Lenvima), dostarlimab-gxly (Jemperli).

Esophageal Cancer
Trastuzumab (Herceptin), ramucirumab (Cyramza), pembrolizumab (Keytruda), nivolumab (Opdivo), fam-trastuzumab deruxtecan-nxki (Enhertu).

Head and Neck Cancer
Cetuximab (Erbitux), pembrolizumab (Keytruda), nivolumab (Opdivo).

Gastrointestinal Stromal Tumor
Imatinib mesylate (Gleevec), sunitinib (Sutent), regorafenib (Stivarga), avapritinib (Ayvakit), ripretinib (Qinlock).

Giant Cell Tumor
Denosumab (Xgeva), pexidartinib hydrochloride (Turalio).

Kidney Cancer
Bevacizumab (Avastin), sorafenib (Nexavar), sunitinib (Sutent), pazopanib (Votrient), temsirolimus (Torisel), everolimus (Afinitor), axitinib (Inlyta), nivolumab (Opdivo), cabozantinib (Cabometyx), lenvatinib mesylate (Lenvima), ipilimumab (Yervoy), pembrolizumab (Keytruda), avelumab (Bavencio), tivozanib hydrochloride (Fotivda).

Leukemia
Tretinoin (Vesanoid), imatinib mesylate (Gleevec), dasatinib (Sprycel), nilotinib (Tasigna), bosutinib (Bosulif), rituximab (Rituxan), alemtuzumab (Campath), ofatumumab (Arzerra), obinutuzumab (Gazyva), ibrutinib (Imbruvica), idelalisib (Zydelig), blinatumomab (Blincyto), venetoclax (Venclexta), ponatinib hydrochloride (Iclusig), midostaurin (Rydapt), enasidenib mesylate (Idhifa), inotuzumab ozogamicin (Besponsa), tisagenlecleucel (Kymriah), gemtuzumab ozogamicin (Mylotarg), rituximab and hyaluronidase human (Rituxan Hycela), ivosidenib (Tibsovo), duvelisib (Copiktra), moxetumomab pasudotox-tdfk (Lumoxiti), glasdegib maleate (Daurismo), gilteritinib (Xospata), tagraxofusp-erzs (Elzonris), acalabrutinib (Calquence).

Liver and Bile Duct Cancer
Sorafenib (Nexavar), regorafenib (Stivarga), nivolumab (Opdivo), lenvatinib mesylate (Lenvima), pembrolizumab (Keytruda), cabozantinib (Cabometyx), ramucirumab (Cyramza), ipilimumab (Yervoy), pemigatinib (Pemazyre), atezolizumab (Tecentriq), bevacizumab (Avastin).

Lung Cancer
Bevacizumab (Avastin), crizotinib (Xalkori), erlotinib (Tarceva), gefitinib (Iressa), afatinib dimaleate (Gilotrif), ceritinib (LDK378/Zykadia), ramucirumab (Cyramza), nivolumab (Opdivo), pembrolizumab (Keytruda), osimertinib (Tagrisso), necitumumab (Portrazza), alectinib (Alecensa), atezolizumab (Tecentriq), brigatinib (Alunbrig), trametinib (Mekinist), dabrafenib (Tafinlar), durvalumab (Imfinzi), dacomitinib (Vizimpro), lorlatinib (Lorbrena), entrectinib (Rozlytrek), capmatinib hydrochloride (Tabrecta), ipilimumab (Yervoy), selpercatinib (Retevmo), pralsetinib (Gavreto), cemiplimab-rwlc (Libtayo), tepotinib hydrochloride (Tepmetko).

Lymphoma
Ibritumomab tiuxetan (Zevalin), denileukin diftitox (Ontak), brentuximab vedotin (Adcetris), rituximab (Rituxan), vorinostat (Zolinza), romidepsin (Istodax), bexarotene (Targretin), bortezomib (Velcade), pralatrexate (Folotyn), ibrutinib (Imbruvica), siltuximab (Sylvant), idelalisib (Zydelig), belinostat (Beleodaq), obinutuzumab (Gazyva), nivolumab (Opdivo), pembrolizumab (Keytruda), rituximab and hyaluronidase human (Rituxan Hycela), copanlisib hydrochloride (Aliqopa), axicabtagene ciloleucel (Yescarta), acalabrutinib (Calquence), tisagenlecleucel (Kymriah), venetoclax (Venclexta), mogamulizumab-kpkc (Poteligeo), duvelisib (Copiktra), polatuzumab vedotin-piiq (Polivy), zanubrutinib (Brukinsa), tazemetostat hydrobromide (Tazverik), selinexor (Xpovio), tafasitamab-cxix (Monjuvi), brexucabtagene autoleucel (Tecartus), crizotinib (Xalkori), umbralisib tosylate (Ukoniq), lisocabtagene maraleucel (Breyanzi), loncastuximab tesirine-lpyl (Zynlonta).

Malignant Mesothelioma
Ipilimumab (Yervoy), nivolumab (Opdivo).

Microsatellite Instability-High or Mismatch Repair-Deficient Solid Tumors
Pembrolizumab (Keytruda).

Multiple Myeloma
Bortezomib (Velcade), carfilzomib (Kyprolis), panobinostat (Farydak), daratumumab (Darzalex), ixazomib citrate (Ninlaro), elotuzumab (Empliciti), selinexor (Xpovio), isatuximab-irfc (Sarclisa), daratumumab and hyaluronidase-fihj (Darzalex Faspro), belantamab mafodotin-blmf (Blenrep), melphalan flufenamide hydrochloride (Pepaxto), idecabtagene vicleucel (Abecma).

Myelodysplastic/Myeloproliferative Disorders
Imatinib mesylate (Gleevec), ruxolitinib phosphate (Jakafi), fedratinib hydrochloride (Inrebic).

Neuroblastoma
Dinutuximab (Unituxin), naxitamab-gqgk (Danyelza).

Ovarian Epithelial/Fallopian Tube/Primary Peritoneal Cancers
Bevacizumab (Avastin), olaparib (Lynparza), rucaparib camsylate (Rubraca), niraparib tosylate monohydrate (Zejula).

Pancreatic Cancer
Erlotinib (Tarceva), everolimus (Afinitor), sunitinib (Sutent), olaparib (Lynparza).

Plexiform Neurofibroma
Selumetinib sulfate (Koselugo).

Prostate Cancer
Cabazitaxel (Jevtana), enzalutamide (Xtandi), abiraterone acetate (Zytiga), radium 223 dichloride (Xofigo), apalutamide (Erleada), darolu-

tamide (Nubeqa), rucaparib camsylate (Rubraca), olaparib (Lynparza).

Skin Cancer
Vismodegib (Erivedge), sonidegib (Odomzo), ipilimumab (Yervoy), vemurafenib (Zelboraf), trametinib (Mekinist), dabrafenib (Tafinlar), pembrolizumab (Keytruda), nivolumab (Opdivo), cobimetinib (Cotellic), alitretinoin (Panretin), avelumab (Bavencio), encorafenib (Braftovi), binimetinib (Mektovi), cemiplimab-rwlc (Libtayo), atezolizumab (Tecentriq).

Soft Tissue Sarcoma
Pazopanib (Votrient), alitretinoin (Panretin), tazemetostat hydrobromide (Tazverik).

Solid Tumors That Are Tumor Mutational Burden-High (TMB-H)
Pembrolizumab (Keytruda).

Solid Tumors with an NTRK Gene Fusion
Larotrectinib sulfate (Vitrakvi), entrectinib (Rozlytrek).

Stomach (Gastric) Cancer
Pembrolizumab (Keytruda), trastuzumab (Herceptin), ramucirumab (Cyramza), fam-trastuzumab deruxtecan-nxki (Enhertu), nivolumab (Opdivo).

Systemic Mastocytosis
Imatinib mesylate (Gleevec), midostaurin (Rydapt).

Thyroid Cancer
Cabozantinib (Cometriq), vandetanib (Caprelsa), sorafenib (Nexavar), lenvatinib mesylate (Lenvima), trametinib (Mekinist), dabrafenib (Tafinlar), selpercatinib (Retevmo), pralsetinib (Gavreto).

Often, the combinations of molecular-targeted agents and cytotoxic agents (chemotherapeutics) or TKIs, combined immunotherapy and vascular-targeted regimen or varying combinations among chemotherapeutics or as CDT approaches yield superior effects in anticancer therapy. There is no doubt that the CDT strategy for the treatment of cancer has gained popularity worldwide, and for certain cancers, CDT has become the first-line treatment. Meanwhile, there is a need for a comprehensive anticancer strategy that simultaneously targets abnormal proliferation, angiogenesis rates, and development of chemotherapy resistance. There have been tremendous efforts from both fundamentalists and clinicians in establishing the FDC- and MTD-based approaches for the discovery or rational design and development of multitarget compounds as anticancer agents. To date, however, while the CDT strategy has been intensively and extensively tested and used as anticancer therapeutics, the application of FDC is still relatively scanty and rational MTD is largely in preclinical development or in laboratory investigation except for multitarget TKIs and some natural products or natural herbs and plant formulations laden with a variety of phytochemicals.

3.2 CDT in Anticancer Therapy

3.2.1 CDT in Lung Cancer

The impact of lung cancer is enormous worldwide making it the leading cause of cancer-related mortality with an estimated 2.2 million new cases (11.4) and 1.8 million deaths (18%) in 2020 [1]. Non-small cell lung cancer (NSCLC) accounts for approximately 85% of all cases of lung cancer [1, 40], with the rest being small cell lung cancer (SCLC). Lung cancer is a complex disease caused by a multitude of genetic and environmental factors. The progression of lung cancer involves dynamic changes in the genome and a complex network of interactions between cancer cells with multiple, distinct cell types that form tumors [41].

Chemotherapy continues to play an essential role in the treatment of most stages of NSCLC. Standard first-line treatment options for NSCLC depend on disease and patient characteristics and may include surgery, platinum-based doublet chemotherapy, and targeted therapies [42]. Although surgical resection is curative if diagnosis occurs at early stage I or stage II disease, almost half of all newly diagnosed patients are with advanced-stage disease and candidates for palliative systemic therapies. To date, less than 25% of patients affected by NSCLC are optimal candidates to radical surgery and unfortunately most of them subsequently relapse. Mono-chemotherapy for NSCLC has shown modest improvements in patients with stages II and IIIA NSCLC with an improvement in both overall survival (OS) and quality of life [43, 44].

On the other hand, the polychemotherapy or combination chemotherapy approach has been proven highly effective due to the ability to affect multiple cellular pathways involved in the disease progression and established as the cornerstone of systemic treatment for advanced lung cancer in the last ten to fifteen years. However, improvements with new drug combinations in recent years have been rather small and a general outcome plateau has been reached with one-year survival rates of about 40% and two-year survival rates of less than 15%. Survival over three to four years is still a rare event in this disease, and more and more efforts are being made to develop innovative systemic treatment strategies with mechanisms of action different from conventional cytotoxic drugs. These molecularly targeted agents have made a strong move forward in the management of this disease since Gefitinib, a small molecule epidermal growth factor receptor (EGFR) tyrosine kinase inhibitor, was registered in 2003 by the US Food and Drug Administration (FDA) and a number of further countries for the third-line treatment of NSCLC. Since then, every month findings have been reported about new cellular targets on lung-cancer cells and, consequently, new agents aiming at these molecular targets are being developed, preclinically. Some of these agents have already been tested in the clinics within phase I, phase II, and some even within randomized phase III trials. The most promising strategies have been aiming at the EGFR family, serum vascular endothelial growth factor (VEGF), and the VEGF receptor (VEGFR) family (VEGFR-1 and -2, respectively). Results from pivotal registration trials are expected within the next one or two years for a number of these new drugs, and the standards of care for advanced lung cancer may change dramatically, comparable to what we have seen in other solid tumors such as metastasized breast and colon cancer [45].

Treatment of patients with advanced NSCLC remains a vexing problem and long-term survival beyond 5 years is extremely rare. Five new agents, paclitaxel, docetaxel, vinorelbine, gemcitabine, and irinotecan, have been introduced for the treatment of NSCLC and investigated extensively both preclinically and clinically. Monotherapy with one of these agents has produced survival benefits over the best supportive care in phase III studies. Combination chemotherapy with a new agent and platinum produced a higher response rate and superior quality of life and/or survival compared with the single agents or conventional cisplatin-based chemotherapy in some randomized trials [46, 47]. There was little difference in efficacy and toxicity between the chemotherapeutic regimens with a new agent and a platinum in phase III trials, suggesting the clinical utility of these regimens is similar. Many trials have focused on regimens containing two new agents, with or without platinum. Preliminary results of phase III trials of three drug combinations versus two drug combinations suggested the former to be more promising, in terms of response rates and survival. Doublet chemotherapy consisting of a new agent combined with platinum, cisplatin by preference where tolerated, has become the standard of care for advanced disease.

In a large, multinational, randomized phase III trial, 1,220 patients with advanced and/or metastatic non-small cell lung cancer were treated with docetaxel/cisplatin, docetaxel/carboplatin, or vinorelbine/cisplatin combination [48]. Preliminary results of the phase III trial showed that OS among patients randomized to receive docetaxel/cisplatin was significantly better than that those treated with vinorelbine/cisplatin and similar to docetaxel/versus control arm. The three arms were similar for toxicity data, except the use of vinorelbine/cisplatin was associated with a significantly greater incidence of grade (3/4) anemia and nausea and vomiting than the use of docetaxel plus either cisplatin or carboplatin.

A literature-based meta-analysis of 12 randomized studies provides evidence supporting the use of combination chemotherapy in patients with NSCLC [49]. The use of combination chemotherapy resulted in a statistically significant better overall survival compared to single-agent chemotherapy (11 trials, 1114 patients). The survival benefit was pronounced when platinum-based combination was used, but no survival benefit was observed in non-platinum-based combinations. Grade 3/4 anemia, thrombocytopenia, and neutropenia, but not febrile neutropenia, were significantly more frequent with combination chemotherapy.

3.2.1.1 Taxane-Platinum Combinations

Platinum-based chemotherapy is the treatment of choice for patients with NSCLC, and taxane-platinum combinations are often used as first-line therapy for this disease, owing to their single-agent activities and synergistic effects. Four large, multicenter, randomized phase III clinical trials (the TAX 326 trial, the Southwest Oncology Group 9509 trial, the Italian Lung Cancer Project, and the Eastern Cooperative Oncology Group 1594 trial) have compared taxane-platinum combinations (docetaxel and paclitaxel) with other platinum combinations (vinorelbine and gemcitabine) in chemotherapy-naïve patients with good performance status scores and advanced disease [26]. The end points for these large randomized clinical trials were survival, response rate, adverse events, and quality of life.

Docetaxel/Cisplatin Combination

Of the taxane-platinum combinations tested, docetaxel-cisplatin was the only platinum combination to yield survival and response rates superior to another platinum combination. In terms of adverse events, the taxane-platinum combination of paclitaxel-carboplatin demonstrated less grade 3 or 4 neutropenia and lower rates of febrile neutropenia than other taxane-platinum combinations but higher rates of irreversible grade 3 or 4 peripheral neuropathy than any of the other taxane-platinum combinations. The docetaxel-platinum combination demonstrated broad benefits to quality of life for patients receiving this combination, and this benefit was not observed with the other platinum or taxane-platinum combinations [50].

Irinotecan/Cisplatin Combination

The DNA topoisomerase inhibitor irinotecan (CPT-11, Camptosar) is being evaluated as a novel chemotherapeutic agent for small-cell lung cancer that may complement other agents and treatment modalities. In a recent Japanese phase III study, irinotecan in combination with cisplatin significantly improved survival of previously untreated patients with extensive SCLC compared with standard etoposide/cisplatin therapy. The results demonstrate better outcomes with irinotecan/cisplatin combination than with etoposide/cisplatin combination [51].

Carboplatin/Gemcitabine Combination

The treatment of advanced NSCLC has evolved rapidly over the past few years. Systemic chemotherapy is associated with both quality of life and modest survival benefit for patients with advanced NSCLC. Platinum-based doublet combinations are the "standard of care" [52]. The FDA has approved gemcitabine (Gemzar), a pyrimidine analog, to be used in combination with cisplatin for the treatment of advanced NSCLC in the first-line setting. Randomized clinical trials have established comparable efficacy with improved therapeutic index for the carboplatin/gemcitabine regimen when compared with cisplatin/gemcitabine and other platinum doublets. Nonhematologic toxicities occur at a lower frequency with carboplatin/gemcitabine combinations compared with other "standard" platinum-based doublets, whereas dose-limiting thrombocytopenia, the most common toxicity, rarely requires therapeutic intervention. Both the 3- and 4-week schedules of carboplatin/gemcitabine result in similar efficacy and toxicity profiles, but the 3-week regimen is preferred. The combination of carboplatin and gemcitabine is an effective regimen with an acceptable toxicity profile for the treatment of advanced NSCLC. This regimen can also be used as a foundation for the development of innovative combinations with molecularly targeted agents.

Gemcitabine/Nedaplatin Combination

Four clinical studies on the efficacy and safety of gemcitabine/nedaplatin combination-based regimens including 112 patients with NSCLC were evaluated [53]. The analysis revealed that the pooled response rate was 40.2% (45/112). Main side effects included grade 3-4 neutropenia, thrombocytopenia, and anemia. Grade 3-4 nonhematological toxicity included nausea and vomiting, diarrhea, and hepatic dysfunction. There were no treatment-related deaths. This led to a conclusion that the combination of gemcitabine and nedaplatin is associated with good response rate and accepted toxicity for treating patients with non-small cell lung cancer.

Carboplatin/Etoposide Combination

There is a global trend of an increasing incidence of elderly population, and this demographic shift is a main reason for an increase of cancer incidence because age is the greatest risk factor for cancer. Lung cancer is a typical disease of the elderly patients. The number of randomized phase III trials concerning the treatment of SCLC in elderly patients is very limited. Current treatment standards for limited disease (LD)-SCLC (polychemotherapy plus local thoracic irradiation and additional prophylactic cranial irradiation in case of complete remission) seem to be feasible for elderly (>70 years) LD-SCLC patients with a good performance and full functional capacities. When standard treatment is not feasible due to co-morbidity or loss of functional capacity, several alternative combination regimens are available, which appear to be slightly superior to single-agent treatment, although randomized data for elderly on that issue are sparse. Carboplatin and etoposide may be the most appropriate two-drug combination in elderly patients at the time being, but there are a variety of active and low toxic third-generation agents like taxanes, gemcitabine, and vinorelbine which are active in both NSCLC and SCLC [54].

S-1: Fluoropyrimidine/5-Fluorouracil/Gimeracil/Oteracil Combination

S-1 is an oral fluoropyrimidine agent in a fixed formulation of a prodrug of the thymidylate synthase inhibitor 5-fluorouracil (5-FU), and two biochemical and pharmacological modulators of 5-FU to improve its tumor-selective cytotoxicity. They comprise gimeracil, an inhibitor of dihydropyrimidine dehydrogenase (DPD), the rate-limiting enzyme in 5-FU catabolism, and oteracil, an inhibitor of orotate phosphoribosyl transferase, the enzyme responsible for the conversion of 5-FU to the ribonucleotide fluorouridylate monophosphate [55–57]. S-1 has been approved for patients with gastric cancer in 7 Asian countries and 15 European countries. In Japan, S-1 has also been approved for eight types of cancers, including NSCLC. A study evaluated the efficacy and toxicity of S-1 and paclitaxel (PTX) combination therapy in 40 patients with previously treated NSCLC [39]. Their data indicate that S-1/PTX combination elicits satisfactory efficacy without severe toxicities and treatment-related deaths in patients with previously treated advanced NSCLC. On the other hand, PTX-based combination chemotherapy, particularly with carboplatin, has become a very popular combination in the United States for patients with advanced NSCLC [58].

Inoue et al investigated the feasibility of post-operative adjuvant chemotherapy using S-1/carboplatin combination in patients with completely NSCLC in a prospective, phase II, multicenter trial conducted in five Japanese institutes [56, 57]. The primary endpoint of the study was the completion rate of four courses of adjuvant chemotherapy. The secondary endpoints included safety and 2-year OS and disease-free rates (DFS). A total of 35 patients were enrolled and 34 were eligible for the analysis. The authors reported a completion of protocol adjuvant chemotherapy in 29/34 (85.3%) patients and the 50% of patients completed 4 courses without dose reduction. Among secondary end points, there were no treatment-related deaths and two-year OS and DFS were 96.3% and 53.3%, respectively. The authors concluded that adjuvant chemotherapy with carboplatin plus S-1 is safe and feasible and the combination of carboplatin and S-1 appears to be a valid treatment option in patients with advanced NSCLC. It is also well known that in the adjuvant setting of NSCLCs, the best results have been observed with the combination of cisplatin and vinorelbine [59].

It should be noted that the currently used drug combination designs are primarily based on empirical clinical studies, and little attention has been given to dosage regimens, i.e., how administration routes, onsets, and durations of the combinations influence the therapeutic outcome [40]. Combination chemotherapy is recognized as the most effective means of improving survival in patients with extensive-stage small-cell lung cancer, but until recently, no one combination had emerged as superior. This first-line, platinum-based doublet chemotherapy regimen is associated with modest clinical benefits, but it has significant toxicities. Furthermore, chemotherapy combinations for more advanced disease have shown to convey no benefit on overall survival or quality of life beyond 4–6 cycles [60, 61]. Agents targeting the epidermal growth factor receptor (EGFR), particularly the tyrosine kinase inhibitors gefitinib and erlotinib, have proven to be clinically active in patients with advanced-stage NSCLC. Because of these findings, it is logical to expect that combinations of these 2 classes of antineoplastic agents would prove more efficacious than either one alone. Yet 4 large randomized phase III trials of chemotherapy with or without an EGFR tyrosine kinase inhibitor in unselected patients with advanced-stage NSCLC, altogether totaling > 4000 patients, did not demonstrate improvement in clinical outcomes with the combination. Whether these negative results will be reproduced in ongoing combination studies of chemotherapy plus monoclonal antibodies directed against EGFR remain to be determined [62]. This is partly because combination therapy is challenged by distinct physicochemical properties and in vivo pharmacokinetics/pharmacodynamics of the individual pharmaceuticals, including small molecule drugs and biopharmaceuticals, which make the optimization of dosing and administration schedule challenging.

Although treatments, which target individual pathway, have shown clinical successes, the ability of tumor cell to develop resistance to circumvent inhibition of a single signaling pathway drives the urge to target multiple signaling pathways. Therapeutic approaches to inhibit multiple pathways using multiple single-targeted agents may help to maximize the suppression of oncogenic processes involved in disease progression.

Diabetes mellitus (DM), one of the most common life-threatening illnesses worldwide, is a group of metabolic diseases, characterized by sustained hyperglycemia and associated with dysfunction of different organs such as heart, eyes, blood vessels, nerves, and kidneys. Notably, there is also a strong association between DM (especially type 2 DM, T2DM) and carcinogenesis [63–65]. Metformin is one of the most commonly prescribed oral antidiabetic medicines and it is suggested as the first-line therapy due to its comparatively safe, inexpensive, effective, and well-tolerated. Some of the in vitro and in vivo investigations proved that metformin may have a direct anticancer action by preventing the proliferation of malignant cells and formations of the colony, inducing arrest of cell cycle and apoptosis and suppressing tumor growth. Observational data suggest that the use of the standard antidiabetic agent metformin decreases lung cancer incidence and mortality. Synergistic actions of metformin with antitumoral agents in preclinical NSCLC models have then been highlighted. Recent retrospective studies advocated improved outcomes in NSCLC diabetic patients receiving metformin in combination with chemotherapeutic compounds (including conventional platinum-based chemotherapy and EGFR tyrosine kinase inhibitors) [66–69]. The advantages of combination therapy with metformin and antitumor agents have also been reported in non-diabetic advanced NSCLC patients [70]. The antiproliferative mechanism of metformin alone or in combination with various chemotherapeutic agents is complex and involves several beneficial roles in clinical trials investigating the effects of metformin in lung cancer. Several prospective randomized trials are therefore currently assessing the addition of metformin to standard therapy in non-diabetic lung cancer patients [66].

3.2.1.2 VEGFR/PDGFR Dual Inhibitor Combinations

Sorafenib/Erlotinib Combination

Sorafenib (Bay 43-9006, Nexavar, Bayer; Leverkusen, Germany) acts by blocking the phosphorylation of VEGFR, PDGFR, Raf, and stem cell factor receptor (KIT) [71]. In phase II studies, single-agent sorafenib prolonged progression-free survival (PFS) [71, 72]. Several studies about the combination of sorafenib with chemotherapeutic agents reported no beneficial effects. Placebo-controlled phase III trials of carboplatin/paclitaxel [73] and gemcitabine/cisplatin [74, 75] alone or with sorafenib have demonstrated no additional benefit of response rate, PFS, or OS. Patients with squamous histology were found to be more vulnerable in mortality index. But the combination of sorafenib with another EGFR tyrosine kinase inhibitor erlotinib proved promising clinical activity in patients with advanced NSCLC with an acceptable safety profile. A multicenter phase II study of erlotinib and sorafenib in patients with adenocarcinoma (72%), squamous (10%), large cell (12%), and NSCLC NOS (6%) histology reported an OS of 10.9 months with its adverse events which included fatigue (16%), hand-foot skin reaction (16%), rash (16%), diarrhea (14%), and hypophosphatemia (42%) [76]. Another phase II study (patients with squamous histology were excluded) demonstrated the higher activity of sorafenib/erlotinib combination than sorafenib/gemcitabine combination with a higher 1-year survival rate for first group patients [77]. Benefit for the combination of erlotinib/sorafenib over single-agent erlotinib with respect to PFS (3.38 months for sorafenib/erlotinib versus 1.94 months for placebo/erlotinib) and OS (8 months for sorafenib/erlotinib versus 4.5 months for placebo/erlotinib) has also been reported [78]. According to Adjei et al [79], in patients with advanced NSCLC who achieve a prolonged response or stable disease with sorafenib given as a single agent or as part of a combination regimen, sorafenib treatment could be continued until disease progression without major long-term safety or tolerability problems. These studies indicate that the erlotinib/sorafenib combination can take NSCLC therapy into a plateau [71].

Sunitinib/Erlotinib Combination

Sunitinib (SU11248, Sutent, Pfizer; New London, CT, USA) is an oral, small-molecule, and multi-targeted receptor tyrosine kinase (RTK) inhibitor of VEGFR-2 (Flk-1), PDGFR-β, and KIT, rearranged during transfection (RET), colony-stimulating factor 1 receptor (CSF-1R), and FMS-like tyrosine kinase-3 receptor (FLT3) [71, 80]. Phase II studies of sunitinib on a continuous daily dosing (CDD) schedule suggested sunitinib as a safe and potential therapeutic agent for patients with previously treated advanced NSCLC of all histologies, though single-agent sunitinib was associated with a few manageable hematologic and neurologic abnormalities [81, 82]. Combinations of sunitinib with other agents were studied in order to gain high efficacy. But the results were not so satisfactory because of toxicities. A phase III trial revealed sunitinib plus erlotinib did not improve OS compared with erlotinib alone (9.0 months versus 8.5 months) in patients with NSCLC of nonsquamous (about 82% adenocarcinoma) histology, but the combination was associated with a statistically significant longer PFS (3.6 months versus 2.0 months) and greater overall response rate (ORR) (10.6% versus 6.9%) [83]. That trial also revealed a greater grade 3 or higher toxicities with combined therapy. Earlier a study about sunitinib with bevacizumab plus paclitaxel/carboplatin was discontinued because of toxicities including neutropenia, thrombocytopenia, leukopenia, and fatigue on its first phase [84]. However, Blumenschein Jr et al [85] suggested on their randomized, double-blind, and multicenter study that a dosage of sunitinib concurrently with erlotinib was tolerable in subjects with platinum-refractory NSCLC of adenocarcinoma (70.0%), squamous cell (26.7%), and large cell (3.3%) histology, but the adverse effects were not totally removed with diarrhea (76.9%), fatigue (61.5%), and decreased appetite (53.8%) being the most frequent adverse events.

Cediranib/Gemcitabine/Carboplatin Combination

Cediranib (AZD2171, Recentin, AstraZeneca; Wilmington, DE, USA) is an oral TKI that targets VEGFR-1, VEGFR-2, VEGFR-3, PDGFRs, and c-kit [71, 86]. Cediranib as a monotherapy was generally well tolerated and showed encouraging antitumor activity in a phase I, dose-escalation, and pharmacokinetic study in patients with advanced solid tumor [87]. Phase I data suggested that cediranib could be combined with standard doses of cisplatin and gemcitabine in patients with advanced NSCLC with promising activity but predictable and manageable toxicity [88]. Based on these data, a randomized phase II/III trial of carboplatin/paclitaxel plus cediranib or placebo as first-line therapy for advanced NSCLC was initiated [89]. Although PFS favored the combination of chemotherapy plus cediranib regardless of patient's histology, the study encountered excessive toxicities including hypertension, hypothyroidism, hand-foot syndrome, hypoalbuminemia, and an imbalance in treatment-related deaths (10 and 2 for chemotherapy plus cediranib and chemotherapy plus placebo, respectively). A randomized phase II study of gemcitabine and carboplatin with or without cediranib as first-line therapy in advanced NSCLC of squamous (15.5% versus 27.6%), adenocarcinoma (37.9% versus 55.2%), and other (46.6% versus 17.2%) histologies did not meet its primary endpoint of overall response rate (ORR) (19% versus 20%) but met its secondary endpoint of 6-month progression-free survival (48% versus protocol-specified threshold of at least 40%) with more grade 3 plus nonhematologic adverse events experienced by cediranib group [90].

Axitinib/Paclitaxel/Carboplatin or Axitinib/Gemcitabine/Cisplatin

Axitinib (AG-013736, Pfizer; New London, CT, USA) is an oral TKI with activity against VEGFR, PDGFR, and c-kit [71, 91]. In vitro, axitinib enhances the efficacy of chemotherapeutic drugs (topotecan and mitoxantrone) via inhibiting the drug transport function of adenosine triphosphate- (ATP-) binding cassette subfamily G member 2 (ABCG2) and reversed ABCG2-mediated multidrug resistance (MDR) [92]. In a phase I trial study, Kozloff et al [93] demonstrated that administration of axitinib with paclitaxel/carboplatin or gemcitabine/cisplatin regimens was well tolerated with improved clinical efficacy. They have also reported no grade ≥3 hemoptysis among 12 patients with squamous cell NSCLC out of total 49 patients suggesting clinically working further with this drug.

Motesanib/Carboplatin/Paclitaxel Combination

Motesanib (AMG 706; Amgen, Thousand Oaks, CA, USA) is an orally administered potent small molecule, which targets VEGFR-1, VEGFR-2, VEGFR-3, PDGFR, and KIT [71, 94]. In human NSCLC xenograft models, motesanib has been proved as a potent antitumor agent and that it had enhanced activity when combined with cisplatin or docetaxel [95]. But clinical trials of combination have different results. Several phase III trials have failed to achieve their primary endpoint of OS. A randomized, double-blind phase III study of motesanib plus carboplatin/paclitaxel in 1090 patients (of those, 890 had adenocarcinoma) with advanced NSCLC demonstrated the combination did not significantly improve OS over carboplatin/paclitaxel alone, with a median OS of 13 months and 11 months and median PFS of 5.6 months and 5.4 months for carboplatin/paclitaxel plus motesanib and carboplatin/paclitaxel alone, respectively [83]. The grade ≥3 and grade 5 adverse effects were higher with motesanib treatment. Similarly, another MONET1 (Motesanib NSCLC Efficacy and Tolerability) study reported improvement on PFS but no OS benefit with increased toxicity [71, 96].

3.2.1.3 VEGFR/EGFR Dual Inhibitor Combinations

Vandetanib/Chemotherapeutic Agent Combinations

Vandetanib (AZD6474, Zactima, AstraZeneca; Wilmington, DE, USA) is a multitargeted tyrosine kinase inhibitor (TKI) that inhibits VEGFR, EGFR, and RET [71, 97]. Single-agent vandetanib has antitumor activity with an acceptable safety and tolerability profile [98]. Combinations of vandetanib with chemotherapeutic agents are being evaluated for better efficacy as an expected combination of vandetanib with paclitaxel and/or carboplatin has increased PFS [99]. But the adverse effects of combining therapy are still in concern. Three phase III studies also provide contradictory concepts. The ZEAL trial (N = 534), which compared vandetanib plus pemetrexed versus placebo plus pemetrexed, failed to meet its primary endpoint of prolonged PFS [100]. Whereas the ZODIAC trial, which compared vandetanib plus docetaxel (N = 694; of those 412 had adenocarcinoma, 184 had squamous, and 98 had other histology) versus placebo plus docetaxel (697 patients; of those 417 had adenocarcinoma, 160 had squamous, and 120 had other histologies) [101], the ZEPHYR trial (924 patients; of those 726 had adenocarcinoma, 102 had squamous, and 96 had other histologies) evaluated vandetanib in patients with advanced NSCLC, and prior EGFR TKI therapy demonstrated significantly improved PFS for the combination but did not significantly improve OS [102]. In addition, both studies showed higher incidence of adverse effects for the vandetanib group [71].

3.2.1.4 VEGFR/FGFR Dual Inhibition

Brivanib/Ixabepilone Combination

Brivanib (BMS-540215, Bristol-Myers Squibb; New York, NY, USA) is a small-molecule TKI, formulated as an orally administered l-alanine ester prodrug, brivanib alaninate [71, 103]. It is the first-selective dual inhibitor of FGF and VEGF signaling [104], and preclinical studies have demonstrated that dual inhibition of FGF and VEGF signaling by brivanib has strong anti-angiogenic effects [105]. In phase I study in patients with advanced or metastatic solid tumors, brivanib demonstrated promising antiangiogenic and antitumor activity and manageable toxicity with most frequent serious toxic effects recorded being nausea, pyrexia, aspartate aminotransferase (AST) or alanine aminotransferase (ALT) elevations, and thrombocytopenia [106, 107]. Synergistic activity of ixabepilone plus brivanib has suggested a promising anticancer activity both in vitro and in vivo [108].

3.2.1.5 VEGFR/PDGFR/FGFR Triple-Target Inhibition

Nintedanib/Pemetrexed Combination

Nintedanib or BIBF 1120 (Boehringer Ingelheim; Ingelheim, Germany) is an oral small-molecule triple kinase inhibitor which inhibits VEGFR, PDGFR, and FGFRs and members of the Src family and flt-3 [71, 109]. Single-agent treatment of nintedanib (BIBF 1120) in advanced NCSLC was well tolerated, with median PFS of 6.9 weeks and median overall survival (OS) of 21.9 weeks [110]. This drug also has proved its safety and well tolerability on its phase I trials in combination with chemotherapy. Phase I trial of 26 patients (7 with adenocarcinoma, 3 with squamous, 1 with large cell carcinoma, 1 with adenosquamous, and 14 with unspecified histology) with recurrent advanced-stage NSCLC receiving BIBF 1120 in combination with standard-dose pemetrexed was tolerable, with promising signs of efficacy [111]. For combination with paclitaxel and carboplatin, nintedanib demonstrated an acceptable safety profile in a phase I trial [112]. These findings indicate nintedanib as a potential triple kinase inhibitor and as a candidate for further study [71].

Pazopanib/Gemcitabine Combination

Pazopanib (GW786034, GlaxoSmithKline; London, UK) is another triple-targeted TKI targeting VEGF, PDGFR, and FGFR [71, 113]. Phase II study of pazopanib monotherapy in 35 patients (23 with adenocarcinoma, 4 with squamous, and 8 with other histology) with stage I/II NSCLC demonstrated well tolerability with 86% of patients who achieved tumor volume reduction after pazopanib treatment [114]. Clinical data with pazopanib activity in advanced disease stage and in combination with other drugs are strikingly limited. A phase I study of pazopanib plus gemcitabine combination therapy in patients with advanced solid tumors explored that the combination was tolerable, and there was no apparent pharmacokinetic interaction [115].

3.2.2 CDT in Colorectal Cancer

Colorectal cancer (CRC) is ranked as the third most commonly diagnosed cancer with 1.9 million new cases (10.0%) and the second most deadly cancer with 0.94 million deaths (9.4%) in the year of 2020 [1]. Localized colorectal cancer (LCRC) is often successfully treated with curative surgery followed by chemotherapy in patients at high risk for recurrence [116]. Overall prognosis for patients with LCRC is favorable, with a 5-year survival rate of up to 90% [117]. Because early-stage CRC can be asymptomatic, only a minority (~40%) of CRC cases are diagnosed early, leaving a large number of patients initially diagnosed with metastatic CRC (mCRC), for whom the 5-year survival rate is poor (~13%). Approximately 20% of patients with CRC will have metastatic disease at the time of initial diagnosis, and approximately 30–50% of patients with primary colon cancer will relapse and die of metastatic cancer.

mCRC remains incurable in most cases, but survival has improved with advances in cytotoxic chemotherapy and targeted agents. Currently recommended treatment approaches are primarily the combination of different cytotoxic compounds or combinations of targeted agents and cytotoxic therapeutics [118]. The feasibility of combining antiangiogenic agents, such as bevacizumab, with established cytotoxic chemotherapy regimens in mCRC was initially demonstrated in 2004 with the pivotal randomized phase III study investigating the combination of bevacizumab and IFL (folinic acid, 5-FU, and irinotecan) chemotherapy for the first-line treatment of mCRC [119]. This was the first randomized phase III study to document the clinical benefit of bevacizumab with respect to significantly improving progression-free survival (PFS), response rate (RR), and OS associated with IFL chemotherapy. The NO16966 randomized phase III study confirmed the clinical benefits of combining bevacizumab with oxaliplatin-based chemotherapy (XELOX or FOLFOX) in the first-line setting, with a significant improvement in PFS [120]. The recommended combinations among the backbone chemotherapies (cytotoxic agents) and combinations between targeted agents and cytotoxic therapeutics are provided below.

An experimental study using image-based proliferation and apoptosis assays in CRC cell lines to systematically investigate the efficacy of combinations of two to six drugs that target critical oncogenic pathways revealed that high-order drug combinations are needed to kill drug-resistant cancer cells [39]. Horn et al [121] believed that to achieve optimal efficacy, it is critical to target essential growth mechanisms, like RAS/MAPK signaling. They showed that pairwise combinations of drugs targeting the RAS/MAPK and PI3K/AKT pathways resulted in synergies broadly across cancer cell lines of a broad spectrum of genetic backgrounds but often yielded only cytostatic responses. The resulting maximum effects were largely independent of the targeted node in either of the pathways, which could be of interest in clinical settings where targeting alternative nodes could provide better tolerability. Importantly, most combinations targeting growth signaling could not adequately kill cells, and enhanced cytotoxicity was observed when additional processes including apoptosis and cell cycle were targeted as part of the combination. This suggests that other cellular processes need to be targeted in combination and that dual inhibition therapy, even when targeting the tumor's major genetic alterations, might not be sufficient for adequate killing. Triple combinations could enhance both growth inhibition and apoptosis when compared with their underlying drug pairs and in some cell lines were required to tip the balance from cytostatic to cytotoxic effects. Dual growth pathway inhibition provided a good "base" for most of the third agents tested, including CDK4/6 and MDM2 inhibitors. In some cases, where cell lines were resistant to paired and tripled drugs, increased expression of antiapoptotic proteins was observed, requiring a fourth-order combination to induce cytotoxicity. However, this study did not address the potential synergistic toxicity that high-order drug combinations could cause [39].

3.2.2.1 CDT Cytotoxic Therapy (Chemotherapy)

The National Comprehensive Cancer Network (NCCN) guidelines recommend several CDT regimens as cytotoxic therapy of mCRC [30]. The cytotoxic agents include 5-fluorouracil (5-FU), irinotecan, a topoisomerase I inhibitor, oxaliplatin, a third-generation platinum analog, and capecitabine, an oral fluoropyrimidine.

(a) FOLFOX: 5-FU+oxaliplatin+leucovorin.
(b) FOLFIRI: 5-FU+irinotecan+ leucovorin.
(c) XELOX: capecitabine+oxaliplatin.
(d) FOLFOXIRI: 5-FU+oxaliplatin+irinotecan+leucovorin for patients with mCRC who are appropriate for intensive therapy [30].

The European Society for Medical Oncology guidelines similarly recommend a first-line backbone chemotherapy of a fluoropyrimidine in various schedules and combinations [122], and FOLFOX and FOLFIRI are currently considered the preferred cytotoxic treatment options for first-line treatment of mCRC [122, 123].

3.2.2.2 CDT Targeted Therapy

Angiogenesis plays a significant role in tumor growth and metastasis in colorectal carcinoma [124, 125]. Since the mid-1990s, six targeted antiangiogenic agents have been approved by the FDA for the treatment of mCRC, including bevacizumab (a monoclonal antibody mAb to VEGF) [126], cetuximab (EGFR mAb) [127], panitumumab (EGFR mAb) [128], ziv-aflibercept (a recombinant fusion protein that binds diffusible VEGF) [129], regorafenib (a multikinase inhibitor that inhibits EGFR and VEGF receptor) [130], and ramucirumab (VEGFR-2 mAb) [131]. Current treatment approaches and optimal treatment sequencing across the first-, second-, and third-line settings in mCRC [116, 123]. Results from several clinical trials proved that targeting tumor-mediated angiogenesis improves overall survival for patients with mCRC when used in combination with first-line and second-line chemotherapy regimens [132].

First-Line Treatment of mCRC

Effective first-line therapy is a key determinant of treatment outcomes and should be selected after considering both clinical factors and biological markers. Bevacizumab, cetuximab, and panitumumab are the only targeted therapies currently indicated as first-line treatment of mCRC in combination with backbone chemotherapies, with approvals based on findings from their respective pivotal phase III trials [119, 133–135]. For example, the phase III Cancer and Leukemia Group B (CALGB)/Southwest Oncology Group (SWOG) 80405 [136] and FIRE-3 [137] studies compared bevacizumab with cetuximab to identify the optimal first-line targeted agent for use in combination with FOLFIRI (examined CALGB/SWOG 80405 and FIRE-3) or modified FOLFOX6 (mFOLFOX6; examined in CALGB/SWOG 80405 only) in KRAS (KRAS proto-oncogene, GTPase, a guanosine-nucleotide-binding protein involved in cell growth, differentiation, and survival, thereby carcinogenesis)–wild-type mCRC. Based on the collective findings from these studies, the NCCN panel currently recommends cetuximab and panitumumab for the first-line treatment of left-sided tumors only, in combination with cytotoxic agents in KRAS–wild-type mCRC [134, 138–140]. Bevacizumab may be preferred for right-sided tumors in this setting because these tumors are unlikely to respond to anti-EGFR antibodies [116, 123]. A randomized phase III study (PRIME) of panitumumab plus FOLFOX4 versus FOLFOX4 alone in first-line treatment of KRAS–wild-type mCRC showed a significant median PFS benefit and a trend toward median OS benefit [134, 138]. The addition of bevacizumab to combination chemotherapy in the first- and second-line treatment of mCRC has resulted in meaningful improvement in overall and progression-free survival.

Second-Line Treatment of mCRC

The second-line regimen choice depends on the systemic therapies given in first-line. Both cetuximab and panitumumab are appropriate to use in the second-line setting in patients with KRAS–wild-type mCRC following progression on chemotherapy [125, 126], but are not recommended for patients who previously received an EGFR inhibitor [39]. The phase III 181 study demonstrated a significant improvement in PFS with panitumumab plus FOLFIRI versus FOLFIRI alone in patients with KRAS–wild-type mCRC following progression on chemotherapy [139]. Objective response rate was 35% versus 10% for the combination versus FOLFIRI alone, and there was a trend toward improved OS with the combination in patients with wild-type KRAS. Efficacy was not improved with panitumumab in patients with mutant KRAS [139]. The pivotal phase III trial (VELOUR) of ziv-aflibercept plus FOLFIRI versus placebo plus FOLFIRI in the second-line setting showed OS and PFS benefits with ziv-aflibercept [140]. Another phase III trial (RAISE) evaluated the role of ramucirumab plus FOLFIRI versus placebo plus FOLFIRI and likewise showed OS and PFS benefits with the addition of ramucirumab to FOLFIRI in patients who progressed on first-line therapy with bevacizumab, fluoropyrimidine, and oxaliplatin [141]. In a phase II randomized trial (BOND) assessing the role of cetuximab plus irinotecan versus cetuximab monotherapy in mCRC patients refractory to irinotecan, higher responses were observed in the combination treatment arm, with significantly longer PFS and a trend toward longer OS [142].

Based on the clinical trials, Ciombor and Bekaii-Saab [116] proposed the following recommendations for second-line therapy of mCRC:

(a) If a patient was previously treated with FOLFOX- or XELOX-based regimen with bevacizumab, then cetuximab/panitumumab (RAS–wild-type only), bevacizumab, ramucirumab, or ziv-aflibercept in combination with FOLFIRI is recommended [123, 143].
(b) If the patient was previously treated with FOLFIRI in combination with bevacizumab, then FOLFOX or XELOX with bevacizumab or with cetuximab/panitumumab (KRAS–wild-type only) is recommended; alternatively, cetuximab/panitumumab can be combined with FOLFIRI or single-agent irinotecan [144–146].
(c) If the patient was previously treated with FOLFOXIRI, then cetuximab/panitumumab with or without irinotecan are recommended for patients with KRAS–wild-type mCRC.

Third-Line Treatment of mCRC

Molecular profiling is important in third-line treatment, with options in KRAS wild-type patients including EGFR inhibitors (cetuximab or panitumumab), regorafenib and trifluridine/tipiracil [116, 118, 147]. Immunotherapy with pembrolizumab or nivolumab in combination with ipilimumab may be considered for patients with high microsatellite instability disease.

Collectively, combination regimens including a fluoropyrimidine (5-FU or capecitabine) with either oxaliplatin (FOLFOX or XELOX) or irinotecan (FOLFIRI) have been widely accepted as standard first-line treatment for patients with mCRC [133–138]. The addition of bevacizumab to any of these combination regimens is reasonable for patients who are appropriate candidates for aggressive therapy. The addition of an anti-EGFR antibody (e.g., cetuximab or panitumumab) might also be indicated in selected patients with wild-type KRAS status.

3.2.3 CDT in Hepatocellular Carcinoma

Liver cancer is the sixth most common cancer globally, accounting for 4.7% of all new cancer cases in 2020 and represents the third most common cause of cancer-related death (0.83 million death in 2020, 8.3%) worldwide [1, 116, 148]. Of the primary liver cancers, hepatocellular carcinoma (HCC) is the most prevalent histological subtype and accounts for >80% of primary liver cancers [117, 148, 149]. Surgical resection and liver transplant are associated with the best survival outcomes for patients with HCC and are potentially curative treatments [150]. Locoregional therapies including arterially directed therapies, ablation, and radiotherapy are also associated with good survival outcomes in patients with unresectable disease confined to the liver. However, over 50% of patients with HCC are diagnosed at an advanced stage or with other characteristics that preclude surgical or locoregional treatment [151]. For these patients, systemic therapy is usually the recommended treatment option [152]. Treatment options for patients with very early, early, or intermediate stage (0-B) Barcelona Clinic Liver Cancer (BCLC) stage HCC include curative resection, liver transplantation, radiofrequency ablation, and chemoembolization. Pharmacological treatment of human disease is a mostly systemic treatment that is any medication that travels through the body in the bloodstream to find, damage, or destroy cancer cells. It includes chemotherapy, immunotherapy, hormone therapy, or targeted therapy. Although early-stage HCC may be curable by resection, liver transplantation, or ablation, most patients present with unresectable disease and have a poor prognosis.

3.2.3.1 Anti-PD-L1/Anti-VEGF Combination: Atezolizumab/Bevacizumab Combination

Over the past 3 years, the number of approved first-line systemic therapies for patients with HCC has expanded greatly, and numerous drugs and drug combinations have been evaluated in this setting [153]. Between 2007 and 2018, sorafenib was the only approved systemic treatment for HCC based on the results of the phase III SHARP trial, which showed a survival benefit for sorafenib *vs* placebo [154]. In the decade following the approval of sorafenib, numerous unsuccessful trials of systemic therapies in advanced HCC were conducted until the approval of lenvatinib in 2018 [155]. Lenvatinib was approved for first-line use in advanced HCC following the successful outcome of the phase III REFLECT trial. In this trial, lenvatinib showed a non-inferior overall survival (OS) *vs* sorafenib for the treatment of advanced HCC [156]. Since the approval of lenvatinib, the immunotherapy drugs nivolumab and pembrolizumab, as well as other tyrosine kinase inhibitors (TKI), have been approved for the second-line treatment of HCC. Most recently, combination therapy with the anti-PD-L1 agent atezolizumab (a monoclonal antibody against the protein programmed cell death-ligand 1) plus bevacizumab (a monoclonal antibody against vascular endothelial growth factor A, VEGF-A) demonstrated better OS and progression-free survival (PFS) than sorafenib in the phase III IMbrave 150 trial [157]. Atezolizumab/bevacizumab combination, as an immune-checkpoint inhibitor (ICI) + humanized anti-VEGF monoclonal antibody pair, is now considered the standard of care in the first-line setting in patients with advanced HCC and is also the first combined immunotherapy and vascular-targeted regimen to be recommended as a first-line treatment option in the National Comprehensive Cancer Network HCC guidelines [158–162].

Systemic therapeutic agents currently approved for use or are currently being tested in advanced HCC (aHCC) can be divided into three classes [163]:

(i) Tyrosine kinase inhibitors (TKI) with broad affinity to various receptor tyrosine kinases including VEGFR. The agents in this multi-target TKI class of drugs include sorafenib, lenvatinib, cabozantinib, and regorafenib.
(ii) Targeted agents that inhibit the VEGF/VEGFR pathway, either TKIs targeting the kinase domain of VEGFR or monoclonal antibodies against VEGF or VEGFR, including bevacizumab (anti-VEGF mAb), apatinib (VEGFR2 TKI), and ramucirumab (anti-VEGFR2 mAb).
(iii) Immune checkpoint inhibitors against either the B7/CTLA-4 or PD-1/L1 axis to reinvigorate antitumor immune responses. These include the anti-PD-1 antibodies, nivolumab, pembrolizumab, and camrelizumab, the anti-PD-L1 antibodies, durvalumab, atezolizumab, and avelumab, and the anti-CTLA-4 antibodies, ipilimumab and tremelimumab.

A network meta-analysis evaluated OS and PFS [164]. Fourteen trials with 8 in the first-line setting and 6 in the second-line setting were included for the analysis. The 8 trials in the first-line setting encompassed a total of 6290 patients, with an age range of 18–89 years. The 5 trials included in the second-line analysis encompassed a total of 2653 patients, with an age range of 18 to 91 years. The analysis revealed the combination of atezolizumab and bevacizumab yielded superior outcomes in patients with HCC treated in the first-line setting compared with lenvatinib, sorafenib, or nivolumab alone. In the refractory setting, NMA showed that all studied drugs had PFS benefit compared with placebo.

In another network meta-analysis of randomized controlled trials including 27 qualified trials [165], survival rate, response rate, and safety profile were evaluated with comparisons among atezolizumab plus bevacizumab, brivanib, donafenib, dovitinib, FOLFOX4, lenvatinib, linifanib, nintedanib, nivolumab, sorafenib, sunitinib, vandetanib, 11 sorafenib combination therapies, and three other combination therapies. It was found that lenvatinib ranked 1/19 in terms of overall response rate, followed by atezolizumab/bevacizumab combination and nivolumab. For progression-free survival (PFS), atezolizumab/bevacizumab combination was ranked 1/15, followed by lenvatinib. For overall survival, atezolizumab/bevacizumab combination was ranked 1/25, followed by vandetanib or donafinib, with lenvatinib ranked 6/25. Atezolizumab/bevacizumab combination was associated with a lower risk of death *vs* lenvatinib. However, atezolizumab/bevacizumab combination was ranked 1/13 for the incidence of treatment terminations due to adverse events. These results are supported by findings from a recent meta-analysis showing that atezolizumab/bevacizumab combination was associated with the best OS and PFS outcomes of all therapies (including lenvatinib, nivolumab, and sorafenib) included in the analysis [166]. Although bevacizumab monotherapy failed phase II trials in unresectable HCC, in combination with atezolizumab it led to superior efficacy compared with bevacizumab monotherapy. However, consideration of treatment safety and tolerability is also an important factor in clinical decision-making. Their analysis revealed that atezolizumab/bevacizumab combination was associated with the highest incidence of discontinuation due to AEs. It appears therefore there is no single first-line treatment for advanced HCC associated with superior outcomes across all outcome measurements and first-line systemic treatment should be selected based on individualized treatment goals.

Similar conclusions were reached by a meta-analysis with first-line studies comprised of 8335 patients (13 studies) and second-line studies comprised of 4612 patients (11 studies) [167]. In the first-line setting, atezolizumab/bevacizumab combination was associated with the highest OS benefit over sorafenib. Additionally, lenvatinib was associated with the greatest objective response rate benefit in the first-line setting. In the second-line setting, cabozantinib was associated with the highest PFS benefit over placebo as well as the highest ORR benefit. Thus, atezolizumab plus bevacizumab appears to have superior efficacy among first-line agents whereas cabozantinib appears to be superior in the second-line setting.

3.2.3.2 Anti-PD-L1/Anti-VEGFR2 Combinations

Atezolizumab/Cabozantinib Combination

The success of this anti-VEGF and ICI combination is in keeping with the move toward combination therapy evaluations in clinical trials where this specific combination is currently the most frequently evaluated regimen [168]. For instance, the COSMIC-021 trials have demonstrated promising efficacy data for atezolizumab and cabozantinib (a small molecule inhibitor of the tyrosine kinases c-Met, VEGFR2, AXL, and RET) in prostate cancer, non-small cell lung cancer, renal cell carcinoma, and urothelial carcinoma [168–171] and the CAMILLA study has demonstrated the potential efficacy of durvalumab with cabozantinib in advanced GI cancers [172].

Camrelizumab/Apatinib Combination

Furthermore, a recent, multicenter, phase II trial tested the safety and efficacy of a camrelizumab (anti-PD-1 antibody) plus apatinib (anti-VEGFR-2) in aHCC patients who were either treatment-naïve or refractory to first-line therapy [173]. The data showed promising efficacy in the second-line settings. The toxicity was high as expected with an ICI plus anti-VEGF combination, with grade 3 or higher adverse events occurring in 77.4% of patients [165]. Thus, targeted agents against the VEGF/VEGFR pathway are powerful partners to ICIs in combination regimens and more such combination therapies will likely demonstrate efficacy in aHCC.

Nivolumab/Ipilimumab Combination
In the CheckMate-040 trial, which evaluated nivolumab with or without ipilimumab, nivolumab monotherapy demonstrated an ORR of 14% and a median OS of 15.1 months in aHCC patients who progressed or were intolerant of sorafenib [174]. Also, nivolumab plus ipilimumab demonstrated an ORR of 31–32%, a median duration of response (mDOR) of 16.6–22.2 months, and an mOS of 12.5–22.8 months [175].

Tremelimumab/Durvalumab Combination
In a recent trial presented at GI-ASCO 2020 (NCT02519348), tremelimumab plus durvalumab demonstrated an ORR of 9.5–22.7%, mDOR of 13.2 months to not reached, and mOS of 11.3–18.7 months after 11.7 months of median follow up [176]. In comparison, nivolumab monotherapy appears comparable while nivolumab with ipilimumab and tremelimumab with durvalumab appear superior to pembrolizumab monotherapy, although the indirect comparison was not conducted.

These findings suggest that atezolizumab/bevacizumab combination should be the initial choice for first-line therapy in aHCC patients who are candidates for receiving immune checkpoint inhibitor therapy [116]. On the other hand, patients who are ineligible for immune checkpoint inhibitor therapy due to underlying conditions, such as active autoimmune disorders, should be considered for either sorafenib or lenvatinib [116].

3.2.3.3 Anti-PD-L1/Tyrosine Kinase Inhibitor Combination: Pembrolizumab/Lenvatinib Combination

Lenvatinib combined with the anti-PD-1 antibody pembrolizumab showed promising antitumor activity with a tolerable safety profile in patients with aHCC [13]. In the open-label phase IIb study (NCT03006926), 104 patients with BCLC stage B or C, CPA and ECOG performance status ≤1 received lenvatinib (body weight ≥ 60 kg: 12 mg/day; < 60 kg: 8 mg/day) and pembrolizumab (200 mg IV Q3W). Updated results of expansion part indicate that the ORR was as high as 46% in 100 patients that could be evaluated, with tumors completely disappearing in 11% of them and achieving a complete response (CR) [13]. Compared with lenvatinib (ORR 24.1%) in the REFLECT trial [8] or pembrolizumab (ORR 17%) in the KEYNOTE-224 trial [12], the combination has an outstanding advantage over single-drug treatment. Furthermore, mDOR was 8.6 months, median OS was 22.0 months, and median PFS was 9.3 months [13]. Investigators performed tumor assessments using mRECIST for HCC. The most common grade ≥ 3 treatment-emergent adverse events (TEAEs) was hypertension; no new safety signals emerged beyond the established safety profile for each agent [13].

The impressive result has pushed this therapeutic combination to the forefront. Last year, the FDA granted the BTD for lenvatinib plus pembrolizumab, which is used for the first-line treatment of patients with aHCC. The phase III multicenter, randomized, double-blinded, active-controlled, LEAP-002 trial (NCT03713593) is being evaluated, mainly with regard to assessment of safety and effectiveness as first-line therapy for aHCC patients. To be inspired, the combination of lenvatinib plus other ICIs, which are effective for HCC, are being evaluated or designed, such as nivolumab, sintilimab, and toripalima.

3.2.4 CDT in Stomach Cancer

Gastric cancer (GC) is a disease in which malignant (cancer) cells form in the lining of the stomach, and most cases of stomach cancers are gastric carcinomas. It is the fifth most common cancer (1.1 million new cases in 2020, 5.6% of total) and the fourth most common cause (0.77

million death in 2020, 7.7% of total) of cancer death globally [1]. GC is a multifactorial disease, where both environmental and genetic factors can have an impact on its occurrence and development. Of all people with gastric cancer, in countries where screening is not routinely performed, 80% to 90% are either diagnosed at an advanced stage when the tumor is inoperable or develop a recurrence within five years after surgery. The incidence rate of GC rises progressively with age; the median age at diagnosis is 70 years [177, 178]. However, approximately 10% of gastric carcinomas are detected at the age of 45 or younger. GC is usually diagnosed histologically after endoscopic biopsy and staged using CT, endoscopic ultrasound, PET, and laparoscopy. It is a molecularly and phenotypically highly heterogeneous and pathologically highly aggressive malignance. The gastric carcinomas affect mostly men, who are two times more likely to develop them than women [179, 180]. Patients with gastric cancer exhibit "three high and three low" characteristics, whereby the incidence, metastasis rate, and mortality rate are high, and the early diagnosis rate, radical resection rate, and 5-year survival rate are low [181, 182].

Early-stage GC (eGC) is limited primarily to the mucosa or submucosa, regardless of the size of the lesion and the presence of lymph node metastasis. GC that extends beyond the submucosa to invade the gastric muscular layer is diagnosed as middle GC (mGC), whereas tumors that infiltrate into or beyond the subserosa to nearby organs or metastasize belong to advanced GC (aGC). Advanced GC includes intermediate and advanced tumors. The stage of the tumor determines the treatment effectiveness thereby treatment strategy. The main treatment for eGC patients is radical endoscopic resection followed by chemotherapy with postoperative 5-year survival rate of 90%. However, the detection rate of eGC is low because of the lack of specific signs of early gastric cancer, and therefore, most patients (>70%) develop into aGC with high metastatic potential and overall poor prognosis. Non-early operable gastric cancer is treated with surgery, which includes D2 lymphadenectomy (including lymph node stations in the perigastric mesentery and along the celiac arterial branches). Perioperative (neoadjuvant) or adjuvant chemotherapy improves survival in patients with stage 1B or higher cancers. More than 50 years since its development, infusional 5-FU remains the backbone of most combination chemotherapy regimens in aGC [177].

3.2.4.1 Platinum/Fluoropyrimidine Combinations

5-Fluorouracil (5-FU)/Epirubicin/Mitomycin Combination

Advanced GC is treated with sequential lines of systemic chemotherapy that remains the mainstay of treatment for the human epidermal growth factor receptor 2 (HER2)-negative patients, starting with a combined platinum and fluoropyrimidine doublet in the first line [177]. Neoadjuvant chemotherapy is a new strategy for the treatment of aGC, and there have been no standard guidelines on the selection of agents; instead, the decision is based mainly on the results of computed tomography, barium meal, the accurate application of a gastroscope, and laparoscopic staging. GC is relatively sensitive to chemotherapy, and the focal application of neoadjuvant chemotherapy can significantly reduce tumor staging, increase the surgical success rate, and prolong patient survival time [182]. Wilke et al [183] first reported the application of neoadjuvant chemotherapy for the treatment of GC. They reported the laparoscopic exploration of 34 cases of unresectable aGC, given etoposide, adriamycin, and cisplatin chemotherapy. Of these, 33 cases required a reoperation followed by two cycles of postoperative chemotherapy; the remission rate was 70%. In 1994, Mai et al [184] reported 24 cases of aGC patients who were given a combinational chemotherapy regimen with 5-fluorouracil (5-FU) + epirubicin + mitomycin or mtx/5-FU. Malignant ascites disappeared in 82% of the patients, 68% of the patients underwent radical resection, and the postoperative median survival time was 14 months.

5-FU/Calcium Folinate/Cisplatin (FLP) Combination

In 1997, Crookes et al [185] reported 56 cases of aGC who received preoperative chemotherapy with 5-FU + calcium folinate + cisplatin, of whom 40 patients received radical resection. There were 5 cases of complete remission, 12 cases were reduced to stage I, and 13 cases dropped to stage II. However, some other reports have suggested that neoadjuvant chemotherapy can improve the R0 resection rate and reduce tumor staging, but it had no obvious advantage on the long-term survival rate [177, 186–188].

5-FU/Cetuximab/Irinotecan/Leucovorin Combination

Another study reported that combination treatment with cetuximab and irinotecan, leucovorin, and 5-FU in aGC patients achieved an objective remission rate of 44% and a median time to disease progression (TTP) of 8 months [189]. Wagner et al [190] reported a meta-analysis of overall survival with 60 RCTs (11,698 participants). The results led the authors to conclude that chemotherapy improves survival (by an additional 6.7 months) in comparison to best supportive care (BSC), and combination chemotherapy improves survival (by an additional month) compared to single-agent 5-FU. For HER-2 negative patients, all different two- and three-drug combinations including irinotecan, docetaxel, oxaliplatin, or oral 5-FU prodrugs are valid treatment options for aGC, but consideration of the side effects of each regimen is essential in the treatment decision. Irinotecan-containing combinations and docetaxel-containing combinations (docetaxel + single-agent or two-drug platinum/5-FU combination) show significant survival benefits [177]. Furthermore, docetaxel-containing three-drug regimens increase response rates, but the advantages are counterbalanced by increased toxicity [177].

Cisplatin/S-1 Combination

In recent years, two oral fluoropyrimidines: capecitabine and S-1, were shown to be at least equal in efficacy to 5-FU [177]. S-1 itself is a combination of tegafur with two enzyme inhibitors: 5-chloro-2,4-dihydroxypyridine (CDHP), a reversible inhibitor of dihydropyrimidine dehydrogenase, and potassium oxonate (Oxo). CDHP enhances the anticancer activity of tegafur by increasing its half-life, and Oxo reduces the gastrointestinal toxicity of tegafur. The pivotal trial, which evaluated S-1 in a Western population, is the randomized phase III "FLAGS" trial. This trial compared a regimen of cisplatin + S-1 with cisplatin + 5-FU over a 28-d cycle. It was observed that the cisplatin/S-1 combination was as effective as the cisplatin/5-FU combination, and patients receiving 5-FU experienced significantly more side effects than did patients treated with S-1 [191, 192].

Targeted therapies licensed to treat gastric cancer include trastuzumab (HER2-positive patients first line), ramucirumab (anti-angiogenic second line), and nivolumab or pembrolizumab (anti-PD-1 third line). For HER-2-positive aGC, significant benefits from targeted therapies are seen with trastuzumab in combination with capecitabine or other chemotherapeutics as first-line treatment. In second line, ramucirumab, alone or in combination with paclitaxel, demonstrated significant survival benefits [190].

Collectively, while meaningful advances have been made in the treatment of other solid tumors in the past years, numerous phase III studies on GC therapy have had negative outcomes [177]. Successes of targeted therapies so far include the introduction of trastuzumab in the first-line treatment of HER2-positive GC, and second-line antiangiogenic treatment with the anti-VEGF-2 receptor antibody ramucirumab [177]. Taxanes have become established in the perioperative setting and in second-line treatment, but taxanes in combination with a platinum- and 5-fluorouracil-based regimen are generally not recommended because they lack a survival benefit and confer high toxicity [193].

3.2.5 CDT in Female Breast Cancer

Breast cancer (BC) is the most frequently diagnosed cancer in women and ranks second among causes for cancer-related death in women. In 2020, female BC has surpassed lung cancer as the most commonly diagnosed cancer with an estimated 2.3 million new cases (11.7%) globally and 0.69 million death (6.9%) ranked number five in overall cancer death [1]. Like most cancers, BC is a heterogeneous disease with different molecular subtypes.

BC therapy involves multidisciplinary multimodal approaches with local therapies (including surgery and radiotherapy) and systemic therapies (including chemotherapy, hormonal/endocrine therapy, and immunotherapy/targeted therapy, as well as mitomycin therapy) [194–200]. The choice of systemic therapies is made depending upon the recurrence risk established through the prognosis score for each patient. DNA studies led to the identification of an "intrinsic list" of over 500 genes which define 5 biological subtypes of BC: **(a)** luminal A; **(b)** luminal B Her2^{-}; **(c)** luminal B Her2^{+}; **(d)** Non luminal Her2^{+}; and **(e)** basal-like.

Applications of systemic therapies can also be viewed as neoadjuvant chemotherapy and adjuvant chemotherapy, hormonal therapy, and targeted therapy [194–200]. Currently, neoadjuvant chemotherapy is extensively used in early-stage BC and locally advanced BC (aBC) as it helps to provide greater chances for breast-conserving surgery by downstaging the disease and to determine tumor response to therapy [201, 202]. Adjuvant therapy for BC may involve local irradiation, systemic therapy with cytotoxic agents/chemotherapeutics, molecular targeted agents or any combination of these [195, 196, 203, 204]. Unlike early-stage BC, metastatic BC (mBC) is considered largely incurable with currently available therapies [205, 206]. In mBC, the role of surgery for removal of primary tumor is traditionally seen as that of palliation rather than for survival benefits [207]. Systemic chemotherapy with cytotoxic agents or endocrine therapy has been the core treatment strategy for mBC for several decades [207] and it remains a crucial component of treatment regimens [208]. Chemotherapeutic agents are appropriate for most patients with mBC including those with hormone receptor-positive (HR^{+}) disease with extensive visceral involvement, HR^{+} disease after failure of hormone-directed therapy, HER2^{+} disease (systemic chemotherapeutic agents used in combination with HER2-directed therapy with trastuzumab and/or pertuzumab), and HR^{-}/HER2^{-} disease (triple-negative disease) [208]. Nevertheless, there is no generally accepted first-line chemotherapy protocol or approach for the treatment of mBC [209].

Unfortunately, systemic therapies have demonstrated limited efficacy in aBC. With aggressive systemic therapies, patients often experience significant toxicity, while still only achieving a 50% or lower response rate [210]. These toxicities can persist as long-term ailments, affecting the cardiac and neurological systems and overall quality of life, as well as leading to the development of new primary cancers [211]. Combination therapies or the CDT approach have therefore been utilized to increase treatment efficiency, and are in extensive use today.

3.2.5.1 Chemotherapeutics Combinations

Chemotherapy of BC generally involves the use of initial neoadjuvant chemotherapy [212–214] before the initiation of adjuvant chemotherapy subsequent to appropriate surgical therapy and/or radiotherapy [215]. Neoadjuvant chemotherapy frequently uses a drug combination modality [212–219], so does adjuvant therapy [220–226]. It is generally accepted that combination chemotherapy affords several advantages such as higher response rates than single-agent chemotherapy, chances for better efficacy and dose reduction while increasing or maintaining efficacy, decreased toxicity, and reduced or delayed development of drug resistance. Due to these advantages, combination chemotherapy has now become the conventionally applied strategy in clinical practice. Advances in the areas of isolation technology and chemical synthetic capability, as well as omics and cell biology, have also played an important role in increasing the application of drug combinations in modern medical practice.

Docetaxel/Epirubicin Combination

In a phase II neoadjuvant trial, docetaxel/epirubicin combination was evaluated for activity and toxicity in women with large, operable, or locally advanced (Stage III) breast carcinoma, as well as patients with inflammatory breast carcinoma [216]. The results of the trial showed an observed response rate (ORR) of 76.7%. More than 25% of the patients experienced clinically significant diarrhea and 80% experienced grade 4 neutropenia with one-third experiencing febrile neutropenia. The subsequent use of prophylactic hemopoietic growth factor support allowed patients to complete the planned treatment. The investigators concluded that neoadjuvant epirubicin plus docetaxel was active and feasible for patients with breast carcinoma including patients with unfavorable disease presentations such as locally advanced breast carcinoma and inflammatory breast carcinoma [216].

Docetaxel/Doxorubicin/Cyclophosphamide Combination

The effects of adding sequential preoperative docetaxel to preoperative doxorubicin and cyclophosphamide were reported in the National Surgical Adjuvant Breast and Bowel Project Protocol B-27 [217]. The study reported that the addition of four cycles of preoperative docetaxel after four cycles of preoperative doxorubicin + cyclophosphamide resulted in a significant increase in clinical and pathologic response rates for operable breast cancer.

Docetaxel/Doxorubicin/Cyclophosphamide Combination

Earl and co-investigators in the Neo-tAnGo (phase III neoadjuvant trial) study assessed the advantage of adding gemcitabine to accelerated paclitaxel with epirubicin and cyclophosphamide [218]. They also evaluated the effect of sequencing the blocks of the epirubicin/cyclophosphamide/paclitaxel (with or without gemcitabine) combination. Their findings showed that the addition of gemcitabine to the combination chemotherapy does not improve pathological complete response (pCR). However, sequencing chemotherapy such that taxanes are received prior to anthracyclines could improve pCR in standard BC neoadjuvant chemotherapy [218].

Carboplatin/Docetaxel/Pegfilgrastim Combination

Another phase II neoadjuvant trial assessed the efficacy and toxicity of the docetaxel/carboplatin combination as neoadjuvant therapy for stage II or III BC in 57 patients (9 with triple-negative BC, TNBC) [219]. The results showed that 9 of the 57 patients (16%) had pCR, and 4 of the 9 (44%) patients with TNBC achieved pCR. The investigators concluded that 4 cycles of 2-weekly neoadjuvant carboplatin/docetaxel combination followed by pegfilgrastim is an active regimen for BC resulting in pCR rates similar to other anthracycline-containing regimens, with an acceptable toxicity profile [219].

On the other hand, therapeutic combinations in adjuvant therapy of BC may take the form of chemotherapeutic combinations (with at least two chemotherapeutic agents) or of chemotherapeutic agents/hormonal therapy or chemotherapeutic agents/immunotherapy combinations. The chemotherapeutic combination regimen for adjuvant chemotherapy was first tested in 1976 [220] in a prospective clinical trial comprised an alkylating agent (cyclophosphamide) and two antimetabolites (methotrexate and 5-fluorouracil) (CMF regimen). There were clear improvement of disease-free survival (DFS) and OS and significant reduction in risk of recurrence for the triple combination compared with the control (no additional treatment) population. Based on the accumulating evidence in favor of combination chemotherapy, the National Institute of Health consensus panel in the US in 2001 recommended the use of multiple agents rather than single agents in most women with local BC, irrespective of lymph node, menopausal, or hormone receptor status [221].

Tamoxifen/5-FU/Cyclophosphamide/ Methotrexate Combination

The National Surgical Adjuvant Breast and Bowel Project (NSABP) B-20 trial found that the addition of the CMF combination to tamoxifen improved 5-year DFS and OS for patients with ER-positive, lymph node-negative disease at lower risk for recurrence [222]. The NSABP B-11 trial evaluated melphalan and 5-fluorouracil with or without doxorubicin in 697 (non-tamoxifen responsive) patients and found an improved 5-year DFS as well as a trend for improved OS for patients that received doxorubicin [223].

Gemcitabine/Capecitabine/Vinorelbine Combination

The CDT comprising gemcitabine, capecitabine, and vinorelbine has also been used in the therapy of BC [193]. A phase II randomized trial evaluated lapatinib combination with capecitabine, vinorelbine, or gemcitabine in HER2-positive MBC patients with progression after treatment with a taxane (Latin American Cooperative Oncology Group (LACOG) 0801 Study). The trial enrolled a total of 142 women 18 years and older, with histologically confirmed locally advanced or metastatic invasive HER2-positive adenocarcinoma of the breast. The primary endpoint of the trial was an overall response (percentage of patients experiencing a confirmed complete response or partial response) and secondary endpoints included progression-free survival, overall survival, and duration of response. The results suggest that these combinations represent potential alternatives to be further explored in the sequence of regimens for patients with HER2-positive BC [193].

Other Drug Combinations

Other drug combinations that were employed in the earlier chemotherapy combinations include: 5-FU/adriamycin (doxorubicin)/cyclophosphamide (FAC) [221]; 5-FU/epirubicin/cyclophosphamide (FEC) [222, 224]. Combination regimens involving the use of platinum-based drugs such as cisplatin have been evaluated in combination with adriamycin (doxorubicin) and cyclophosphamide [225], 5-FU and etoposide [226], and other agents such as vinblastine [227], epirubicin [228], methotrexate, cyclophosphamide, mitomycin C, and vincristine [229].

The taxanes form an important class of cytotoxic agents commonly used in chemotherapy combinations for the treatment of BC. A phase II multicenter randomized study involving 142 patients with at least one measurable lesion compared the efficacy and safety of docetaxel/epirubicin with the 5-FU/epirubicin/cyclophosphamide CDT as first-line chemotherapy for mBC [230]. The results showed an overall response rate of 59% for docetaxel/epirubicin and 32% for 5-FU/epirubicin/cyclophosphamide after a median seven and six cycles, respectively. Combinations of paclitaxel with cisplatin, an agent with less myelosuppressive effects than many other cytotoxic agents used for the treatment of mBC, had been employed in clinical practice. A preliminary study showed an overall response rate of 85% with an 11% complete response rate and there was an acceptable tolerance of the therapy [230]. The paclitaxel/cisplatin combination has also been assessed by other investigators at different dose combinations [231, 232]. Combinations of paclitaxel with other agents such as cyclophosphamide, 5-FU, and mitoxantrone have also been used in the treatment of mBC [233]. Docetaxel has been used in combination with agents like doxorubicin and this combination is considered to be very effective in the treatment of locally advanced and mBC [234]. In phase III multicenter study, the docetaxel/doxorubicin combination was compared with the doxorubicin/cyclophosphamide CDT as first-line chemotherapy for mBC. The results showed that although there was no difference in overall survival, the overall response rate in patients that received docetaxel/doxorubicin was significantly greater than for those that received doxorubicin/cyclophosphamide. In another clinical trial [235], the epirubicin/docetaxel combination chemotherapy can be administered in locally aBC patients with acceptable toxicity. Furthermore, the combination of docetaxel with cisplatin was reported to be effective in patients with anthracycline-resistant mBC [236].

Mitomycins as a class of chemotherapeutics have been in clinical use against a broad range of cancers since the 1960s [237], and mitomycin combinations are another strategy used in the treatment of BC. A study reported the efficacy of low-dose mitomycin and low-dose weekly doxorubicin in mBC patients previously treated with the CMF-CDT (cyclophosphamide/methotrexate/5-fluorouracil combination) [238]. Forty-six biopsy-proven BC patients previously treated with the CMF-CDT were entered in the study. Patients received a chemotherapy regimen consisting of doxorubicin and mitomycin. The results revealed an overall response rate of 43% (11% complete remission and 32% partial response), 11.5 months median duration of survival, and median duration of response of 8 months for responders (complete response and partial response). The toxicity of the regimen was moderate, comprising mainly neutropenia, thrombocytopenia, pneumonitis, and cardiomyopathy. Thus, the low-dose mitomycin and low-dose weekly doxorubicin combination chemotherapy were considered effective for treating patients previously treated with CMF [238]. Another study reported that the mitomycin C/vinblastine combination is active and well tolerated in heavily pre-treated BC patients [239]. This combination chemotherapy regimen resulted in 35% partial response (14 patients) with additional 10 patients (25%) having stable disease lasting for 4 months or longer, giving a total clinical benefit of 60%. The median TTP and OS durations were 4 months and 11 months, respectively, for a median follow-up period of 11 months. Also, 1- and 2-year overall survival rates were 39.4% and 15.7%, respectively.

In addition, the therapeutic benefits of the mitomycin C/methotrexate, mitomycin C/mitoxantrone, and mitomycin C/capecitabine combinations have also been documented based on the results from clinical studies [240, 241]. And the retrospective reviews also favor the use of the mitomycin C/methotrexate combination for BC therapy [209, 240–242].

Heat shock protein 90 (Hsp90) inhibitors have also been used in combination therapy with other anticancer drugs such as taxanes, cisplatin, etoposide, and trastuzumab [243–245]. For example, a combination of the Hsp90 inhibitor, 17-allylamino, 17-demethoxygel-danamycin (17-AAG), a geldanamycin analog with trastuzumab produced encouraging results in the treatment of HER2-positive mBC progressing on trastuzumab [243].

(***Docetaxel:*** a taxoid antineoplastic chemotherapeutic agent used in the treatment of various cancers, such as locally advanced or metastatic breast cancer, metastatic prostate; **Mitomycin C:** an alkylating, antitumor antibiotic used as a chemotherapeutic agent; **Doxorubicin (adriamycin):** a DNA alkylating agent and an anthracycline, is a topoisomerase II inhibitor that increases DNA degradation, preventing cell replication; **Cyclophosphamide**: a DNA alkylating agent and an oxazaphosphorine, interferes with the duplication of DNA and RNA transcription; **Methotrexate**: a competitive inhibitor of the dihydrofolate reductase which causes a decrease in pyrimidine and DNA synthesis; **Taxanes:** a class of mitotic inhibitors that block cell growth; **Cisplatin:** a chemotherapy drug that contains platinum; **Carboplatin:** an analog of cisplatin; **Etoposide**: an antineoplastic agent and an epipodophyllotoxin (a semisynthetic derivative of the podophyllotoxins) that works by slowing the growth of cancer cells; **17-Allylamino:** a Hsp90 inhibitor; **17-Demethoxygel-danamycin:** a Hsp90 inhibitor; **Geldanamycin:** a natural existing HSP90 inhibitor and a benzoquinone antineoplastic antibiotic; **Paclitaxel:** a tetracyclic diterpenoid (taxane) mitotic inhibitor isolated originally from the bark of the Pacific yew tree, *Taxus brevifolia*; **Epirubicin:** a 4′-epi-isomer of the anthracycline antineoplastic antibiotic doxorubicin; **Vinblastine:** a vinca alkaloid antineoplastic agent; **Fluorouracil (5-FU):** a pyrimidine analogue used as an antineoplastic agent to treat multiple solid tumors including colon, rectal, breast, gastric, pancreatic, ovarian, bladder, and liver cancer; **Gemcitabine:** a broad-spectrum antimetabolite and deoxycytidine analogue with antineoplastic activity; **Capecitabine:** an antimetabolite chemotherapeutic agent; **Vinorelbine (Navelbine):** a semisynthetic vinca alkaloid derived from vinblastine and an antimitotic chemotherapy drug; **Filgrastim:** a human granulocyte colony-stimulating factor (G-CSF) produced by recombinant DNA technology; **Pegfilgrastim:** a PEGylated form of the recombinant human granulocyte colony-stimulating factor (GCSF) analog filgrastim; **Mitoxantrone:** a type II topoisomerase inhibitor that disrupts DNA synthesis and DNA repair in both healthy cells and cancer cells).

3.2.5.2 Hormonal/Endocrine Therapeutics Combinations

Approximately 60% of BC among premenopausal women express the nuclear estrogen receptor (ER^+ breast cancer). Adjuvant hormonal therapy or endocrine therapy (ET) is an integral component of care for ER^+ breast cancer, exerting its effect by reducing the availability of estrogen to micrometastatic tumor cells. Endocrine strategies in premenopausal women include estrogen receptor blockade with tamoxifen, temporary suppression of ovarian estrogen synthesis by luteinizing hormone-releasing hormone (LHRH) agonists, or permanent interruption of ovarian estrogen synthesis with oophorectomy or radiotherapy [246]. ET is mandatory for all patients with hormone receptor-positive breast neoplasms. It is active both in adjuvant and metastatic disease. The only active adjuvant hormonal therapy in pre- and post-menopause is tamoxifen. The purpose of hormonal therapy is to prevent the interaction between estrogens and estrogen-dependent pathways of stimulating neoplastic cells [247].

Anastrazole/Fulvestrant Combination

New ET strategies include combining ET agents as well as adding inhibitors targeting growth factors, angiogenesis, the mechanistic target of rapamycin, phosphoinositide 3-kinase (PI3K), or cyclin-dependent kinase 4/6 (CDK4/6) to ET [248]. Emerging data show that ET combinations can improve progression-free and overall survival outcomes in patients with hormone receptor-positive, HER2-negative (HR^+/HER^-) aBC. Level 1 evidence supports consideration of dual ET regimens (e.g., fulvestrant/anastrozole combination or palbociclib/letrozole combination), particularly in ET-naïve patients, or palbociclib/letrozole combination as the first-line therapy, and the addition of mTOR (the everolimus/exemestane combination) or CDK4/6 inhibitors (the palbociclib/fulvestrant combination) to established ET in the second-line setting and in select first-line patients [248].

Some combinations are associated with increased risk of class-specific toxicities that will require individualized risk stratification, earlier and more rigorous agent-specific monitoring, and patient education [248]. Yet, although the safety profiles of these combinations are generally predictable and manageable, both everolimus and palbociclib are associated with an increased risk of potentially serious or early-onset toxicities requiring individualized a priori adverse event risk stratification, earlier and more rigorous agent-specific monitoring, and patient education [247, 248]. Moreover, although each of these combinations improves progression-free survival, none with the exception of the anastrazole/fulvestrant combination have demonstrated improved overall survival [247, 248]. Therapeutic ratios of select ET combinations support their use in first- and second-line settings, but optimal sequencing has yet to be determined. PI3K catalytic-α mutations assessed from circulating tumor DNA represent the first potentially viable serum biomarker for the selection of ET combinations, and new data demonstrate the feasibility of this minimally invasive technique as an alternative to traditional tissue analysis [247, 248].

LHRH Agonists/Tamoxifen or LHRH Agonists/Fulvestrant Combination

The LHRH agonists/tamoxifen combination seems effective and safe and is considered as being superior to either monotherapy [249]. Due to their mechanism of action, aromatase inhibitors alone are not suitable for the treatment of premenopausal patients, but the combination with LHRH agonists may result in excellent disease control. Fulvestrant, in conjunction with LHRH agonists, also yields interesting results regarding clinical benefit rate and time to progression [249]. Currently, other orally available selective estrogen receptor downregulators are under clinical evaluation. Recently, targeted drugs have been added to ET in order to reverse endocrine resistance, but only limited information regarding their activity in premenopausal patients is available. The CDK4/6 inhibitor palbociclib when combined with fulvestrant and LHRH agonists was shown to prolong progression-free survival over endocrine therapy alone in pretreated patients [249]; similar results were obtained with the addition of abemacicilib or ribociclib to endocrine therapy. Currently, the activity of the mammalian target of rapamycin inhibitor everolimus in combination with letrozole and goserelin is under assessment in premenopausal patients after progression on tamoxifen (MIRACLE trial).

Palbociclib/Letrozole or Palbociclib/ Fulvestrant Combination

A systematic review and network meta-analysis with a systematic literature search on PubMed, Embase, Cochrane Central Register of Clinical Trials, Web of Science, and online archives of the most relevant international oncology conferences [250]. The study included all phase II and III randomized controlled trials investigating chemotherapy with or without targeted therapies and hormone therapies with or without targeted therapies as first-line or second-line treatments, or both, in postmenopausal women with hormone-receptor-positive (HR^+), HER2-negative metastatic breast cancer. The investigators identified 2689 published results and 140 studies (comprising 50,029 patients) for inclusion in their analysis. They found that the combination pairs of palbociclib/letrozole, ribociclib/letrozole, abemaciclib/anastrozole or letrozole, palbociclib/ fulvestrant, ribociclib/fulvestrant, abemaciclib/ fulvestrant, everolimus/exemestane, and in patients with a PIK3CA mutation, alpelisib/fulvestrant, and several chemotherapy-based regimens, including anthracycline and taxane-containing regimens, are associated with better progression-free survival than was anastrozole alone [250]. No chemotherapy or hormone therapy regimen was significantly better than the palbociclib/letrozole combination for progression-free survival. Paclitaxel/bevacizumab was the only clinically relevant regimen that was significantly better than palbociclib/ letrozole in terms of the proportion of patients achieving an overall response. The investigators therefore believe that in the first-line or second-line setting, the combinations of CDK4/6 inhibitors and hormone therapies are better than standard hormone therapies in terms of progression-free survival [250]. Moreover, no chemotherapy regimen with or without targeted therapy is significantly better than CDK4/6 inhibitors plus hormone therapies in terms of progression-free survival. Their findings support treatment guideline recommendations involving the new combinations of hormone therapies plus targeted therapies as first-line or second-line treatments, or in both settings, in women with hormone-receptor-positive, $HER2^+$ metastatic breast cancer [250]. Another meta-analysis verified that the palbociclib/letrozole and palbociclib/fulvestrant combinations yielded significantly greater PFS than endocrine therapy in treatment-naïve and previously treated patients with advanced/metastatic BC. And palbociclib/ fulvestrant was associated with significantly less toxicity than the everolimus/exemestane combination [196]. Still another network meta-analysis indicates that combinations of CDK4/6 inhibitors and aromatase inhibitors, including palbociclib/ letrozole, ribociclib/letrozole, and abemaciclib/ letrozole or anastrozole combination pairs, are all superior to fulvestrant alone as first-line treatment of HR^+ aBC.

(**Palbociclib** is the first cyclin-dependent kinase 4/6 (CDK4/6) inhibitor approved in the United States for $HR^+/HER2^-$ advanced/metastatic BC; **Abemaciclib** is another CDK inhibitor selective for CDK4 and CDK6; **Ribociclib:** a small-molecule cyclin-dependent kinase (CDK) inhibitor; **Letrozole (Femara)**, **anastrozole** and **exemestane** are aromatase (estrogen synthase) inhibitor used in the treatment of hormonally responsive breast cancer after surgery; **Fulvestrant** and **tamoxifen** are estrogen receptor antagonists; **Everolimus** is an inhibitor of mTOR which blocks cancer cell proliferation and an immunosuppressive agent; **Alpelisib:** a targeted therapy medicine called a PI3K inhibitor that blocks the PI3K pathway; **Goserelin:** is a luteinizing hormone-releasing hormone (LHRH) agonist).

3.2.5.3 Immunotherapeutics Combinations

Immunotherapy, revolutionizing the management of multiple solid tumors, is emerging as a new treatment modality in BC [27, 251–254]. After long-standing use of endocrine therapy and targeted biological therapy, improved understanding of immune evasion by cancer cells and the discovery of selective immune checkpoint inhibitors have created novel opportunities for treatment. Single-drug therapies with monoclonal antibodies against programmed death-1 (PD-1) and programmed death ligand-1 (PD-L1) have shown little efficacy in patients with mBC. There is growing interest in the development of combinations of immunotherapy and molecularly targeted therapies for mBC. Current clinical efforts are focused on developing immunotherapy combinations that convert nonresponders to responders, deepen those responses that do occur, and surmount acquired resistance to immunotherapy. Identifying biomarkers that can predict the potential for response to single-agent immunotherapy, identify the best immunotherapy combinations for a particular patient, and guide salvage immunotherapy in patients with progressive disease are high priorities for clinical development [27]. It is anticipated that combination therapy strategies will be the way forward for immunotherapy in breast cancer, with an improved understanding of tumor, microenvironment, and host factors informing treatment combination decisions [254].

Pertuzumab/Trastuzumab/Docetaxel Combination

Combination therapy with monoclonal antibodies such as trastuzumab and pertuzumab has been reported [255–259]. A phase III randomized, double-blind, placebo-controlled trial demonstrated the efficacy of first-line therapy with pertuzumab, trastuzumab, and docetaxel combination therapy in patients with $HER2^+$ mBC (The Clinical Evaluation of Pertuzumab and Trastuzumab CLEOPATRA trial). Adjuvant or neoadjuvant chemotherapy with or without trastuzumab was permitted. Primary end point analysis revealed that patients in the pertuzumab/trastuzumab/docetaxel combination treatment had a significantly longer median PFS, compared to those in the placebo/trastuzumab/docetaxel combination group. Furthermore, the second interim analysis of OS confirmed the significantly longer survival with the pertuzumab/trastuzumab/docetaxel combination but similar safety profiles between the two groups. The final prespecified OS results with a median follow up of 50 months showed a median OS of 56.5 months in the pertuzumab/trastuzumab/docetaxel combination, compared to 40.8 months in the placebo/trastuzumab/docetaxel combination, revealing a difference of 15.7 months in median survival between the two groups. Thus, it was concluded that the addition of pertuzumab to trastuzumab and docetaxel significantly improved the median overall survival in patients with HER2-positive mBC [255–259].

(**Pertuzumab:** a monoclonal antibody which targets the surface of the cells human epidermal growth factor receptor 2 protein (HER2) on the cancer cell, interfering with HER2 causing cancer cell death; **Trastuzumab:** a monoclonal anti-HER2 antibody used to treat $HER2^+$ breast, gastroesophageal, and gastric cancers.)

Taken together, combination therapy offers the potential of improving therapeutic efficacy and efficiency in BC treatment. Various classes of agents have been used in combination chemotherapy regimens over the years. Maximum benefit from therapeutic combinations would emanate from proper identification of molecular subclass of disease to permit the use of agents that would provide the maximum therapeutic advantage. In other words, a rational combination of therapeutic agents based on disease profile would provide the greatest benefit to BC patients. Thus, therapeutics to be employed in combination therapy need to be carefully selected to ensure that patients derive maximum benefit from the therapy. The number of therapeutic agents must be kept to a minimum to avoid unnecessarily increasing the agents to be administered unless there is evidence of additional benefits to be derived from the inclusion of more agents in the treatment regimen. Moreover, the addition of some agents to existing regimens may result in increased toxicity without significant improvement in therapeutic efficacy. The application of nanoparticles in combination treatment may further extend the benefits derivable from such a treatment strategy based on the unique advantages presented by the nanoparticle platform.

3.2.5.4 Combinations of Muscarinic Acetylcholine Receptor Agonists with Other Classes of Anti-BC Drugs

Most recently, muscarinic acetylcholine receptors (mAChRs) are emerging as new targets for tumor therapy including BC [260]. mAChRs are G protein-coupled receptors (GPCRs), which belong to the superfamily of seven-transmembrane domain receptors and can modulate many functions in normal and tumor cell biology [260–262]. Sales' research team described for the first time the over-expression of mAChRs of the GPCR family in breast tumor tissues and in cancer cell lines from murine and human origin [263, 264]. The expression of mAChRs in malignant tissues is different from that in normal tissues. These differences are manifested by an increase in the expression of mAChRs and/or a modulation in the expression pattern of mAChR subtypes. In human colon cancer, for example, the expression of the M_3 receptor subtype is increased more than 100-fold in respect to normal tissue and its activation modulates cell proliferation, progression, and invasion of this neoplasia [265]. In small cell lung cancer, the expression of M_3 is upregulated, promoting cell migration and invasion [266]. In human bladder, while normal tissue expresses M_1, M_2, and M_3 subtypes, tumor tissue expresses only the M_2 receptor, and its activation induces a decrease in cell proliferation and migration [260, 267]. Regarding BC, Sales' group demonstrated the presence of different mAChR subtypes in BC cells of murine and human origins and the absence of these receptors in normal mammary cells [268–271]. Cell proliferation has been found to be induced by the M_5 receptor subtype in melanoma cells [272], by the M_1 and M_4 receptor subtypes in glioblastoma cells [273], by the M_3 receptor subtype in stomach and colon cancer [274, 275], by the M_2 receptor subtype in non-small cell lung cancer [276], and by all receptor subtypes in the esophagus [277]. The increases of M_1 and M_3 in hepatocellular carcinoma enhance cell proliferation, cell invasiveness, and migration ability [266, 278]. Activation of mAChRs induces proliferation and angiogenesis in MCF-7 cells [279].

Collectively, Español et al [260] proposed that mAChRs should be considered as new targets in BC therapy since they are expressed in breast tumor tissue but not in normal tissue. The conventional treatment for BC involves surgery, radiotherapy, and/or chemotherapy. The latter presents disadvantages such as limited specificity, the appearance of resistance to treatment, and other side effects. To prevent these side effects, several schedules of drug administration, like metronomic therapy, have been developed. Metronomic therapy is a type of chemotherapy in which one or more drugs are administered at low concentrations repetitively. Recently, two chemotherapeutic agents usually used to treat BC have been shown to activate mAChRs [260]. The combination of low concentrations of these chemotherapeutic agents with mAChRs agonists could be a useful option to be applied in BC treatment, since this combination not only reduces tumor cell survival without affecting normal cells, but also decreases pathological neoangiogenesis, the expression of drug extrusion proteins and the cancer stem cell fraction. The addition of mAChRs agonists at low concentrations combined with traditional chemotherapeutic drugs presents promising results. The administration of these combinations in a metronomic schedule is effective to kill tumor cells, reduce tumor invasion and neoangiogenesis. Promising results have also been reported regarding the expression of drug extrusion pumps and the decrease of the cancer stem cell fraction, which would be useful to reduce resistance to traditional chemotherapy and the relapse of breast cancer [260].

3.2.6 Potential Combinations of Oncolytic Virotherapy and Chemotherapy for Cancer Treatment in the Future

Oncolytic viruses (OV) are uniquely capable of targeting cancer cells through intrinsic or engineered means [280–282]. They are derived or engineered from naturally occurring viruses to target and specifically kill cancer cells, and they come in many forms, mainly from four major virus groups which are classified by their genome structure and modes of replication and transcription as defined by the Baltimore classification system. These viruses are subsequently engineered to utilize transcriptional and transductional targeting strategies that restrict replication of oncolytic vector constructs to cancer cells, thus sparing normal cells. In this way, oncolytic viruses target and kill cancer cells, and even stimulate immunotherapeutic effects in patients.

Research in oncolytic virotherapy has been ongoing for decades but only recently has the approach advanced to investigations at the clinical level. In recent years, oncolytic viruses have been moving toward clinical application at an accelerated pace. One example is T-VEC, an oncolytic herpes simplex virus (oHSV), which has been approved by the FDA for clinical use [283]. Oncolytic viruses are particularly attractive due to the myriad of targeting strategies they can utilize, thus extending them into the burgeoning field of personalized medicine. In BC research, various viruses have already been extensively tested preclinically to assess their oncolytic efficacy. Of the seven groups in the Baltimore classification system, viruses from group I (double-stranded DNA viruses), group III (double-stranded RNA viruses), group IV (single-stranded RNA viruses—positive-sense), and group V (single-stranded RNA viruses—negative-sense), have been extensively investigated as candidates for BC therapies, based on their previous use as vaccines or ease of handling and genetic manipulation. As with current treatments in the clinic, oncolytic viruses were initially being explored as single-agents and later in combination with existing therapies.

According to Martin and Bell [281], the combination of OVs with approved chemotherapeutics holds several benefits, foremost the known safety profile and indications of an approved drug. Several preclinical and clinical studies have been conducted investigating the combination of OVs with approved chemotherapeutics [284–288]. For example, Binz et al [285] demonstrated the successful combination of the two chemotherapeutics nab-paclitaxel and gemcitabine with oncolytic VACV GLV-1h68 for the treatment of pancreatic cancer *in vitro*, Tanaka et al [286] investigated a combination therapy with oncolytic HSV and dacarbazine in a mouse melanoma model, and Bourgeois-Daigneault et al [287] were able to show synergistic cytopathic activity, and therefore reduced tumor growth and prolonged survival, in a murine breast cancer model treated with a combination of Maraba MG1 and paclitaxel.

The trans combination of OV therapy with organic or inorganic molecules (so-called viral sensitizers) to potentiate virus infection and spreading within the tumor as well as antitumor immune response has been successfully demonstrated in several syngeneic and xenograft tumor models [289–291]. For example, Selman et al [292] demonstrated that dimethyl fumarate inhibits type I interferon (IFN) production and response by blocking nuclear translocation of NF-κB, thus potentiating oncolytic virotherapy. Similarly, Arulanandam et al [293] showed an improved bystander killing effect caused by reduction of IFN expression and secretion when combining VSV with microtubule-destabilizing agents. VSV as well as VACV have also been synergistically combined with histone deacetylase inhibitors [294, 295] as well as second mitochondrial activator of caspase (Smac)-mimetics [296]. The chemotherapeutic cyclophosphamide has been shown to enhance oncolytic HSV efficacy by inhibiting the innate immune response when combined in *trans* [297]; therefore, Currier et al [298] have successfully constructed an oncolytic HSV encoding a prodrug converting enzyme for cyclophosphamide and demonstrated the safety and efficacy of this *cis* combination in xenograft models. Another important aspect of the innate immunity is the activation of macrophages and microglia, causing the expression and secretion of tumor necrosis factor alpha (TNF-α), which in turn induces both intrinsic and extrinsic apoptosis and necroptosis as demonstrated for Newcastle disease virus (NDV)-infected (tumor) cells [299]. The effect of TNF-α on OV therapy is being investigated because although it enhances the bystander effect, especially when encoded under a late promoter in HSV [300], it can also inhibit virus replication by inducing apoptosis of infected cells. This effect, however, can be counteracted by transient TNF-α blockade [301]. The innate antiviral defense is not the only factor diminishing OV efficacy: as already discussed, a tumor is usually highly heterogenic, comprising fibrotic barriers, extracellular matrix, and other physical barriers inhibiting OV spread and infection of tumor cells [302]. Several groups have pursued a variety of *cis* combinations to enhance virus replication, e.g., encoding fusogenic glycoproteins to enhance virus spread via syncytia formation as demonstrated for HSV [303, 304], as well as VSV and measles virus [305], or by encoding enzymes actively remodeling the extracellular matrix as demonstrated by AdV encoding relaxin [306] or hyaluronidase [307], the latter demonstrating improved efficacy over the parental virus in orthotopic murine models for osteosarcoma [308], as well as glioma [309]. Several groups have also successfully combined antiangiogenic therapies with OVs, both in *trans* as well as in *cis*. Tan et al [310], for example, showed synergistic effects, i.e., improved viral distribution throughout the tumor, as well as enhanced tumor hypoxia of HSV *trans*-combined with the anti-vascular endothelial growth factor (VEGF) antibody bevacizumab in a human breast cancer xenograft mouse model. Similarly, Buckel et al [311] successfully *cis*-combined VACV with a single-chain variant of bevacizumab together with radiotherapy for the treatment of a subcutaneous patient-derived glioma xenograft model. Using a comparable model of glioblastoma, Zhang et al [312] demonstrated that treatment with bevacizumab is augmented by *trans* combination with an HSV armed with angiostatin, an antiangiogenic polypeptide. The observed synergistic effects of this combination are only partly caused by the antiangiogenic effect, but also by modulation of the innate immune response as demonstrated for an HSV encoding different anti-VEGF antibodies [313]. Not pursuing an antibody-based approach, Bolyard et al [314] showed significant antitumor efficacy of a *cis*-combination of HSV and vasculostatin, a secreted angiogenesis inhibitor, in an ovarian cancer xenograft mouse model, additionally combined in *trans* with the approved chemotherapeutic doxorubicin.

As pointed out by Martin and Bell [281], OV therapy, and even more so the combination of OVs and immunotherapy, is a rapidly expanding field. There is considerable reason for optimism because combinations of OVs with innovative immune modulators like immune checkpoint inhibitors (ICIs), Chimeric antigen receptor T-cell (CAR-T cells), and Bi-specific T-cell engagers (BiTEs) are accelerating the field toward our long-term goal of achieving lasting cures for cancer patients. At the molecular level, unraveling the details of the interactions between OVs, the TME, and the immune system is crucial to the optimization of combination strategies. Coupling this with a thorough analysis of the many ways cancer cells evade immune surveillance will help in the design of multiplexed immunotherapeutic approaches to target the heterogeneity of cancer. Yet translating this knowledge into the clinic remains a challenge. Preclinical animal models are of limited value when using therapeutics that have species-specific interactions. The field needs more rationally designed, high-content clinical trials exploring thoughtful therapeutic combinations. As multiple studies demonstrate, for the majority of patients with systemic disease, the only way to fight cancer is through combinations with multiple therapeutics.

The multikinase inhibitors sorafenib and lenvatinib are the approved first-line systemic treatments for unresectable hepatocellular carcinoma on the basis of studies showing modestly longer survival with sorafenib than with placebo and noninferiority of lenvatinib to sorafenib. Both are associated with considerable side effects that impair quality of life.

3.3 FDC in Anticancer Therapy

Application of the FDC strategy to anticancer therapy is relatively rare compared to the CDT approach or free combinations. Below are examples of the FDC approach in anticancer therapy.

3.3.1 FDC in Lung Cancer

3.3.1.1 Pioglitazone/Metformin FDC

Combination treatment with pioglitazone and metformin is utilized clinically in the treatment of type II diabetes. Treatment with this drug combination reduced the development of aerodigestive cancers in this patient population. This anticancer property is ascribed to the off-target effect [315–325]. Pioglitazone is a thiazolidinedione (TZD) peroxisome proliferator-activated receptor gamma (PPARγ) activator with chemoprevention capacity preclinically and clinically in both head and neck and lung carcinoma [317, 323–330]. This drug class has been demonstrated to have anti-inflammatory effects targeting NF-κB in both diabetes and cancer. Observational studies derived from the Veterans Administration Veterans Integrated Service Network (VA VISN) database have shown greater than 35% reductions in both head and neck and lung cancer incidence in diabetics treated with TZD drugs [331, 332].

Seabloom et al [315] expanded this treatment into clinical lung cancer chemoprevention. They hypothesized that dietary delivery of metformin/pioglitazone would prevent lung adenoma formation in A/J mice in a benzo[a]pyrene (B[a]P)-induced carcinogenesis model while modulating chemoprevention and anti-inflammatory biomarkers in residual adenomas [315]. They found that metformin (500 and 850 mg/kg/d) and pioglitazone (15 mg/kg/d) produced statistically significant decreases in lung adenoma formation both as single-agent treatments and in combination, compared with untreated controls, after 15 weeks [315]. Treatment with metformin alone and in combination with pioglitazone resulted in statistically significant decreases in lung adenoma formation at both early- and late-stage interventions. Pioglitazone alone resulted in significant decreases in adenoma formation only at early treatment intervention. We conclude that oral metformin is a viable chemopreventive treatment at doses ranging from 500 to 1,000 mg/kg/d. Pioglitazone at 15 mg/kg/d is a viable chemopreventive agent at early-stage interventions. Combination metformin and pioglitazone performed equal to metformin alone and better than pioglitazone at 15 mg/kg/d. The investigators of this study believe that because the drugs are already FDA-approved, rapid movement to human clinical studies is possible [315].

Initial interest in metformin as an anticancer agent has come from clinical and epidemiologic research which has shown, for type II diabetes patients prescribed metformin, reduced cancer incidence and/or mortality [333–342]. This has provided an empiric basis for its evaluation in the clinical setting. One of the effects of metformin is reducing cell growth and proliferation via attenuation of the insulin/IGF-1R pathway, which inhibits PI3K/Akt/mTOR signaling [315]. In one recent study by Dennis et al., dietary metformin reduced lung cancer burden in the NNK model of mouse carcinogenesis by over 70%, with a potential mechanism being downregulation of IGF-1 receptor phosphorylation [316]. Additionally, several recent studies using metformin point to NFκB attenuation by this agent with respect to both angiogenic and matrix metalloproteinase activity in both cancer and atherogenesis [329, 343–346]. To summarize, both pioglitazone and metformin have been demonstrated to prevent tobacco smoke-induced lung tumor development in preclinical models; however, the pharmacological effectiveness of each agent individually in preventing lung tumor development has been deemed modest. Therefore, it is imperative to elucidate potential additive effects on lung cancer of these agents in combination in a prevention setting [347, 348].

3.3.1.2 Erlotinib/Pertuzumab FDC

Pertuzumab, a dimerization inhibitor of human epidermal growth factor receptor 2 (HER2), has demonstrated pharmacodynamic activity, with stable disease in NSCLC. Combining erlotinib and pertuzumab may enhance antitumor activity. The study conducted by Felip et al [349] aimed to establish the recommended dosing of the erlotinib and pertuzumab combination; assess safety, preliminary efficacy, and pharmacokinetics; and analyze biomarkers. Fifteen patients with stage IIIb/IV NSCLC who failed chemotherapy were recruited. The patients received erlotinib, then combination therapy. Pertuzumab was given intravenously at 840 mg, then 420 mg once every three weeks, with erlotinib given daily (100 or 150 mg). In this study, no dose-limiting toxicities were observed. Adverse events were generally grade 1/2 and manageable. The objective response rate was 20%, median overall progression-free survival was 9.3 weeks. High HER1, HER2, and HER3 messenger RNA expression correlated with increased progression-free survival. Combination therapy did not affect erlotinib's pharmacokinetics [349]. The authors concluded that erlotinib/pertuzumab FDC was well tolerated in patients with good performance status, with encouraging efficacy. A loading dose of pertuzumab 840 mg followed by 420 mg once every three weeks plus daily erlotinib 150 mg appears to be the most appropriate regimen for this FDC.

3.3.2 FDC in Stomach Cancer

3.3.2.1 Dual Combinations of S-1 FDC with Other Agents

As already introduced in an earlier section of this chapter, S-1 is a novel oral fluorouracil antitumor drug that contains a combination of 3 pharmacological agents: tegafur (FT), a 5-fluorouracil (5-FU) prodrug, 5-chloro-2,4-dihydroxypyridine (CDHP), which inhibits the activity of dihydropyrimidine dehydrogenase (DPD), and potassium oxonate (Oxo) to reduce the gastrointestinal toxicity of 5-FU [350]. S-1 and docetaxel (a chemotherapy medication) have both been identified as effective agents for the treatment of gastric cancer [350]. However, little is known about the effects and/or adverse effects of a combination of these drugs in the treatment of gastric cancer. A phase I study determined the maximum-tolerated dose and the recommended dose of docetaxel with a fixed dose of S-1 in patients with advanced or recurrent gastric cancer [350]. The results led to a conclusion that the S-1/docetaxel drug combination showed a good safety profile, with diarrhea and general fatigue being common, but manageable, adverse reactions. Moreover, the responses observed in this study suggest that the drug combination shows a high degree of efficacy in patients with advanced and/or recurrent gastric cancer.

3.3.2.2 Triple Combination of S-1 FDC with Docetaxel and Cisplatin

A dose-escalation study of a triple combination of docetaxel, cisplatin, and S-1 investigated the dose-limiting toxicity, maximum-tolerated dose, recommended dose, and antitumor activity in patients with advanced gastric cancer [351]. The results showed that this regimen yields a high rate of tumor response and can be administered safely [351].

3.3.3 FDC in Female Breast Cancer

In general, combining chemotherapy agents for treating women with metastatic breast cancer results in regimens that are more active, offer superior tumor response rates with more time before progression and improve overall survival [352]. To compare the use of single chemotherapy agents with regimens containing a combination of agents for the treatment of metastatic breast cancer (mBC), Carrick et al [352] performed a meta-analysis of 37 eligible RCTs, based on an estimated 4220 deaths in 5707 women. Their data indicate a modest advantage for combination chemotherapy regimens compared with single agents with a hazard ratio (HR) for overall survival of 0.88 (95% CI=0.83-0.94, P<0.0001) and no evident heterogeneity. Results are similar if the analysis is limited to trials in women receiving first-line chemotherapy. Combination regimens are favorably associated with time to progression and tumor response rates although significant heterogeneity was observed. Meanwhile, women receiving combination regimens experienced a significantly higher toxicity level for leukopenia, hair loss and nausea and vomiting compared with those receiving a single agent. Thus, compared with single-chemotherapy agents, combination regimens show a statistically significant advantage for tumor response and time to progression in women with mBC, a modest improvement in overall survival and significantly worse toxicities [352].

3.3.3.1 Pertuzumab/Trastuzumab FDC

A subcutaneous formulation of pertuzumab and trastuzumab with recombinant human hyaluronidase in one ready-to-use, fixed-dose combination vial (pertuzumab, trastuzumab, and hyaluronidase-zzxf) was approved by the US Food and Drug Administration (FDA) on June 29, 2020 [353]. A primary analysis of the FeDeriCa study (a randomized, open-label, international, multicenter, non-inferiority, phase III study) was designed to assess the pharmacokinetics, efficacy, and safety of the fixed-dose subcutaneous formulation compared to intravenous pertuzumab plus trastuzumab in patients with HER2-positive early breast cancer in the neoadjuvant-adjuvant setting [353]. The study met its primary endpoint: the pertuzumab/trastuzumab FDC for subcutaneous injection provides non-inferior cycle 7 pertuzumab serum Ctrough concentrations to intravenous pertuzumab plus trastuzumab in the neoadjuvant setting with comparable total pathological complete response rates, supporting the FDA approval. Safety was found similar between treatment groups, and in line with other pertuzumab, trastuzumab, and chemotherapy trials [353].

Pertuzumab plus trastuzumab, administered intravenously (iv) with chemotherapy, is the standard treatment for HER2-positive metastatic or high-risk early breast cancer. Pertuzumab and trastuzumab are administered over 1-2.5 h traditionally; however, the need for iv infusions places a strain on medical centers with respect to scheduling, preparation, and administration [354]. A novel fixed-dose combination of pertuzumab and trastuzumab for subcutaneous injection (PHESGO®, PH FDC SC) can be administered in approximately 5-8 min. PH FDC SC was non-inferior to IV pertuzumab plus trastuzumab in terms of pertuzumab and trastuzumab serum levels in the phase III FeDeriCa study, which enrolled 500 patients with HER2-positive early breast cancer. Total pathologic complete response rates were comparable after PH FDC SC (59.7%) or IV pertuzumab plus trastuzumab (59.5%), as was the incidence of grade ≥3 and serious adverse events. The results of a phase II clinical trial (PHranceSCa) showed that the majority of patients (85%) preferred PH FDC SC treatment over IV pertuzumab plus trastuzumab. This new product takes much less time to administer than IV pertuzumab-trastuzumab and has the potential to alleviate time constraints for patients and busy clinics [354].

3.4 MTD in Anticancer Therapy

3.4.1 Multitargeted Tyrosine Kinase Inhibitors (MT-TKIs) as Anticancer MTDs

Targeted therapies by means of compounds that inhibit a specific target molecule represent a new perspective in the treatment of cancer. In contrast to conventional chemotherapy which acts on all dividing cells generating toxic effects and damage of normal tissues, targeted drugs allow to hit, in a more specific manner, subpopulations of cells directly involved in tumor progression [355]. Molecules controlling cell proliferation and death, such as tyrosine kinase receptors (RTKs) for growth factors, are among the best targets for this type of therapeutic approach [355]. Two classes of compounds targeting RTKs are currently used in clinical practice: monoclonal antibodies and tyrosine kinase inhibitors (TKIs) [355]. The era of targeted therapy began with the approval of trastuzumab, a monoclonal antibody against HER2, for the treatment of metastatic breast cancer, and imatinib, a small TKI targeting BCR-Abl, in chronic myeloid leukemia [355]. While monoclonal antibodies can be considered a "single agent for single target" (STD) strategy, TKIs might be multitarget drugs. At present, there are only few rationally designed multitarget drugs as anticancer agent in clinical use except for multitargeted tyrosine kinase inhibitors (MT-TKIs). Tyrosine kinases (TKs) are enzymes responsible for the activation of many proteins by signal transduction cascades [356]. The proteins are activated by adding a phosphate group to the protein (phosphorylation), a step that TKIs inhibit. TKs can be categorized into receptor tyrosine kinases (RTKs), non-receptor tyrosine kinases (NRTKs), and a small group of dual-specificity kinases (DSK) which can phosphorylate serine, threonine, and tyrosine residues [356]. RTKs are transmembrane receptor that includes VEGFR, platelet-derived growth factor receptors (PDGFR), insulin receptor (InsR) family, and the ErbB receptor family which includes EGFR and HER2 [356]. NRTKs are cytoplasmic proteins that consist of nine families, including Abl, Ack, Csk, Fak, Fes/Fer, Jak, Src, Syk/Zap70, and Tec with the addition of Brl/Sik, Rak/Frk, Rlk/Txk, and Srm, which fall outside the nine defined families [356]. The most notable example of DSKs is the mitogen-activated protein kinase kinases (MEKs), which are principally involved in the MAP pathways [23, 357, 358]. As of now, there are over 50 FDA-approved TKIs [356]. As a whole, TKs phosphorylate specific amino acids on substrate enzymes, which subsequently alter signal transduction leading to downstream changes in cellular biology. The downstream signal transduction set off by TKs can modify cell growth, migration, differentiation, apoptosis, and death. Constitutive activation or inhibition, either by mutations or other means, can lead to dysregulated signal cascades, potentially resulting in malignancy and other pathologies [357, 359]. Therefore, blocking these initial signals via TKIs can prevent the aberrant action of the mutated or dysfunctional TKs.

Despite the diverse primary amino acid sequences, human kinases share similar 3D structures, particularly when it comes to the ATP-binding pocket located in the catalytically active region. The starting amino acid sequence (ASP-Phe-Gly or DFG) of the flexible activation loop that controls access to the activation site is also typically conserved [360]. TKIs are either irreversible or reversible. The irreversible kinase inhibitors tend to covalently bind and block the ATP site resulting in irreversible inhibition. The reversible kinase inhibitors can further subdivide into four major subtypes based on the confirmation of the binding pocket as well as the DFG motif [361, 362]. These structural features form the basis for the multitargeticity of MT-TKIs.

Despite the initial enthusiasm for the efficacy of monoclonal antibodies and TKIs, clinicians had to face soon the problem of relapse, as almost invariably cancer patients developed drug resistance, often due to the activation of alternative RTKs pathways [355]. In this view, the rationale at the basis of targeting drugs is radically shifting. In the past, the main effort was aimed at developing highly specific inhibitors acting on single RTKs. Now, there is a general agreement that molecules interfering simultaneously with multiple RTKs might be more effective than single-target agents [355]. With the recent approval by FDA of sorafenib and sunitinib—targeting VEGFR, PDGFR, FLT-3, and c-Kit—a different scenario has been emerging, where a new generation of anticancer drugs, able to inhibit more than one pathway, would probably play a major role [355].

Efforts to identify drugs that inhibit key pathways involved in the pathogenesis of cancer have led to the development of multitargeted agents. Small-molecule TKIs that inhibit receptors such as VEGFR, PDGFR, Raf, and KIT simultaneously may offer advantages over agents with single targets, and therefore a higher likelihood of single-agent activity [363]. In addition, because multitargeted TKIs are often available orally, they may be more convenient for patients [363]. Conversely, a potential disadvantage is the potential for toxicity resulting from off-target kinase inhibition, and the additive toxicity that may be particularly relevant when the agents are combined with chemotherapy.

TKIs hold promising efficacy in anticancer therapy. However, their multitargeting property is primarily limited to the inhibition of angiogenesis that is merely one driving force for the development and progression of cancers. Such a limitation often makes these agents less effective in cancer therapy when used as a stand-alone regimen; moreover, the lack of target specificity of TKIs often leads to unexpected toxicity (hypertension, hemorrhage, and cavitation are common toxicities amongst this class of agents). These issues stimulated major effort on combination with other categories of anticancer drugs in numerous clinical trials. The outcomes of these studies in general support that TKIs can be synergistic or additive with traditional cytotoxic chemotherapy drugs and other novel compounds, even with reduced dosages: enhanced efficacy and mitigated toxicity.

Several multitargeted, antiangiogenic agents have been approved for clinical use or being studied in clinical trials. Most of these agents inhibit VEGFR, and some also inhibit PDGFR, Raf, and EGFR [363, 364] (see Table 3.1 for summary of approved and experimental anticancer MTDs).

3.4.1.1 Vandetanib

Vandetanib (CAPRELSA® or AZD6474, Zactima, AstraZeneca; Wilmington, DE, USA) is an MT-TKI that inhibits VEGFR, EGFR, and rearranged during transfection (RET) [71, 364, 365]. Single-agent vandetanib has antitumor activity with an acceptable safety and tolerability profile [98]. It is an approved orphan drug vandetanib for the treatment of medullary thyroid cancer that cannot be removed by surgery or that has spread to other parts of the body. Combinations of vandetanib with chemotherapeutic agents are being evaluated for better efficacy as expected combination of vandetanib with paclitaxel and/or carboplatin has increased PFS [99]. But the adverse effects of combining therapy are still in concern.

3.4.1.2 Sunitinib

Sunitinib (SU11248, Sutent, Pfizer; New London, CT, USA) is an oral, small-molecule, and multitargeted receptor tyrosine kinase (RTK) inhibitor of VEGFR-2 (Flk-1), PDGFR-β, stem cell factor receptor (KIT), RET, colony-stimulating factor 1 receptor (CSF-1R), and FMS-like tyrosine kinase-3 receptor (FLT3) [71, 364, 366]. It is approved multinationally for the treatment of advanced renal cell carcinoma (aRCC) and imatinib-resistant or imatinib-intolerant gastrointestinal stromal tumor [364]. Phase II studies of sunitinib on a continuous daily dosing (CDD) schedule suggested sunitinib as a safe and potential therapeutic agent for patients with previously treated advanced NSCLC of all histologies, though single-agent sunitinib was associated with a few manageable hematologic and neurologic abnormalities [82, 367]. Combinations of sunitinib with other agents were studied in order to gain high efficacy. But the results were not so satisfactory because of toxicities.

3.4.1.3 Axitinib

Axitinib (AG-013736; Pfizer Inc., New York) is a TKI with activity against all three VEGFRs (VEGFR-1, VEGFR-2, and VEGFR-3) at clinical doses and is currently being studied in multiple solid tumor settings [71, 364, 368]. FDA approves pembrolizumab plus axitinib for advanced renal cell carcinoma (aRCC). In a phase II study of patients with advanced NSCLC treated with single-agent axitinib (in the first-line, second-line, or third-line setting), the median PFS interval was 4.9 months overall (95% CI, 3.6–7.0 months), and the median OS time of 14.6 months was greater than expected in such a mixed patient population [364, 369]. Treatment was also well tolerated.

3.4.1.4 Sorafenib

Sorafenib (Bay 43-9006, Nexavar, Bayer; Leverkusen, Germany) can block the phosphorylation of VEGFR, PDGFR, Raf, and KIT [71, 364]. It is licensed in many countries for the treatment of aRCC and HCC [370]. According to Adjei et al [79], in patients with advanced NSCLC who achieve a prolonged response or stable disease with sorafenib given as a single agent or as part of a combination regimen, sorafenib treatment could be continued until disease progression without major long-term safety or tolerability problems. These studies can conclude that combination of erlotinib and sorafenib would take NSCLC therapy into a plateau.

Table 3.1 Anticancer MTDs

Target category	Anticancer MTD
Multitargeted Tyrosine Kinase Inhibitors	*Vandetanib*: inhibiting VEGFR, EGFR, and RET; FDA approved for the treatment of medullary thyroid cancer; maybe used in combination with chemotherapeutic agents. *Sunitinib*: inhibiting VEGFR-2 (Flk-1), PDGFR-β, KIT, RET, CSF-1R, and FLT3; FDA approved for the treatment of aRCC and imatinib-resistant or imatinib-intolerant GIST; maybe used for aNSCLC. *Axitinib*: inhibiting VEGFR-1, VEGFR-2, and VEGFR-3; FDA approved pembrolizumab plus axitinib for the first-line treatment of patients with aRCC; maybe used for aNSCLC. *Sorafenib*: inhibiting the phosphorylation of VEGFR, PDGFR, Raf, and KIT; approved for the treatment of aRCC and HCC/aNSCLC; given as a single agent or as part of a combination regimen. *Pazopanib*: inhibiting VEGFR, PDGFR, and FGFR; FDA approved for the treatment of advanced soft tissue sarcoma such as aNSCLC; may be used as monotherapy or combination with gemcitabine. *Lenvatinib*: inhibiting VEGFR1, VEGFR2, VEGFR3, FGFR, PDGFRα, RET, and KIT; FDA approved as the combination with pembrolizumab for first-line treatment of adult patients with advanced aRCC. *Vatalanib*: inhibiting VEGFR, PDGFR, and KIT; an investigational drug not approved yet; being studied as a possible treatment for several types of cancer, particularly cancer that is at an advanced stage. *Cediranib*: inhibiting VEGFR-1, VEGFR-2, VEGFR-3, PDGFRs, and c-kit; an investigational drug not yet approved; maybe used as a monotherapy for advanced solid tumors such as aNSCLC. *Motesanib*: inhibiting VEGFR-1, VEGFR-2, VEGFR-3, PDGFR, and KIT; an investigational drug not yet approved; maybe used as a monotherapy for NSCLC. *Linifanib*: inhibiting VEGFR-1, VEGFR-2, VEGFR-3, and PDGFR; an investigational drug not yet approved; maybe used for advanced NSCLC. *Tesevatinib*: inhibiting EGFR, HER2, VEGFR, and EphB4; an investigational drug not yet approved; optimized as an inhibitor of a spectrum of growth-promoting and angiogenic receptor tyrosine kinases (RTKs) to simultaneously block tumor growth and vascularization; maybe used for MSCLC. *Brivanib*: inhibiting FGF and VEGF; an investigational drug not yet approved. *Nintedanib*: inhibiting VEGFR, PDGFR, and FGFRs and members of the Src family and FLT3; FDA approved for idiopathic pulmonary fibrosis; maybe used for advanced-stage NSCLC.
Multitargeting HDAC Inhibitors	HDACs are epigenetic enzymes that can synergistically improve the efficacy of other anticancer drugs; Four multi-isoform HDAC inhibitors Vorinosat (SAHA), Romidepsin (FK-228), Belinostat (PXD101), and Panobinostat (LBH589) have been approved by the FDA for the treatment of specific lymphoma and myeloma diseases; Four HDAC inhibitor hybrids, CUDC-101, CUDC-907, Tinostamustine, and Domatinostat, have entered different phases of clinical trials for the treatment of different types of blood and solid tumors; Two HDAC inhibitor–kinase inhibitor hybrid compounds: (a) Erlotinib-based hybrid with activity in HDACs, EGFR, and HER2 (b) Fimepinostat, an orally active pan PI3K/HDAC inhibitor.

aNSCLC advanced non-small cell lung cancer, *aRCC* advanced renal cell carcinoma, *CSF-1R* colony-stimulating factor 1 receptor, *EGFR* epidermal growth factor receptor, *EphB4* Ephrin type-B receptor 4, *FGFR* fibroblast growth factor receptor, *FLT3* FMS-like tyrosine kinase-3 receptor, *GIST* gastrointestinal stromal tumor, *HCC* hepatocellular carcinoma, *HDAC* histone deacetylase, *HER2* human epidermal growth factor receptor 2, *KIT* stem cell factor receptor, *RET* rearranged during transfection, *PDGFR* platelet-derived growth factor receptor, *VEGFR* vascular endothelial growth factor receptor

3.4.1.5 Vatalanib

Vatalanib (PTK787; Novartis/Schering AG, Berlin, Germany) is a VEGFR, PDGFR, and KIT inhibitor that is currently being studied in phase II/III trials. Data from a phase II trial examining vatalanib monotherapy administered once or twice daily in previously treated patients with NSCLC have been reported [364, 371]. Single-agent treatment appeared active, with a trend toward greater efficacy with twice-daily treatment (11% of evaluable patients had a partial response in this cohort). Additionally, treatment was well tolerated, with no apparent differences between once- and twice-daily dosing [364].

3.4.1.6 Cediranib

Cediranib (AZD2171, Recentin®, AstraZeneca; Wilmington, DE, USA) is an oral TKI that targets VEGFR-1, VEGFR-2, VEGFR-3, PDGFRs, and c-kit [71, 86, 364]. Cediranib as a monotherapy was generally well tolerated and showed encouraging antitumor activity in a phase I, dose escalation, and pharmacokinetic study in patients with advanced solid tumor [87]. Phase I data suggested that cediranib could be combined with standard doses of cisplatin and gemcitabine in patients with advanced NSCLC with promising activity but predictable and manageable toxicity [88, 372]. Based on these data, a randomized phase II/III trial of carboplatin/paclitaxel plus cediranib or placebo as first-line therapy for advanced NSCLC was initiated [89].

3.4.1.7 Motesanib

Motesanib (AMG 706; Amgen, Thousand Oaks, CA, USA) is an orally administered potent small molecule, which targets VEGFR-1, VEGFR-2, VEGFR-3, PDGFR, and KIT [71, 364, 373]. In human NSCLC xenograft models, motesanib has been proved as a potent antitumor agent and that it had enhanced activity when combined with cisplatin or docetaxel [374]. But clinical trials of combination have different results. Several phase III trials have failed to achieve their primary endpoint of OS. A randomized, double-blind phase III study of motesanib plus carboplatin/paclitaxel in 1090 patients (of those, 890 had adenocarcinoma) with advanced NSCLC demonstrated the combination did not significantly improve OS over carboplatin/paclitaxel alone, with a median OS of 13 months and 11 months and median PFS of 5.6 months and 5.4 months for carboplatin/paclitaxel plus motesanib and carboplatin/paclitaxel alone, respectively [83]. The grade ≥3 and grade 5 adverse effects were higher with motesanib treatment. Similarly, another MONET1 (Motesanib NSCLC Efficacy and Tolerability) study reported improvement on PFS but no OS benefit with increased toxicity [96].

3.4.1.8 Pazopanib

Pazopanib (GW786034; GlaxoSmithKline, Philadelphia) is a VEGFR, PDGFR, and FGFR inhibitor approved by FDA for the treatment of advanced soft tissue sarcoma and also currently in phase III development for advanced RCC and in phase II development in advanced NSCLC [71, 364, 375]. Initial data from a phase II trial in early (stage I/II) NSCLC have been presented; 87% of patients had a reduction in tumor volume, and the tolerability profile in this setting was favorable [376]. A phase II study of pazopanib monotherapy in 35 patients (23 with adenocarcinoma, 4 with squamous, and 8 with other histology) with stage I/II NSCLC demonstrated well tolerability with 86% patients who achieved tumor volume reduction after pazopanib treatment [114]. Clinical data with pazopanib activity in advanced disease stage and in combination with other drugs are strikingly limited. A phase I study of pazopanib plus gemcitabine combination therapy in patients with advanced solid tumors explored that the combination was tolerable, and there was no apparent pharmacokinetic interaction [377]. Numerous phase II studies of pazopanib in patients with advanced NSCLC have either been completed or are ongoing, and data are keenly awaited. These include pazopanib as monotherapy, as well as combination studies with paclitaxel and pemetrexed.

3.4.1.9 Linifanib

Linifanib (ABT-869, Abbott; Abbott Park, IL, USA) targets VEGFR-1, VEGFR-2, VEGFR-3, and PDGFR [71, 378]. Phase I study showed that linifanib had a favorable safety profile with minimal dose-limiting toxicities (DLTs) and median PFS of 5.7 months [379]. Another phase II trial of linifanib in 139 (122 nonsquamous and 17 squamous histology) patients with advanced NSCLC showed that it was active in advanced NSCLC as second- or third-line therapy. Comparatively, low dose linifanib can be a potent small-molecule inhibitor against NSCLC though study about its combination with other targeting agents is limited [71].

3.4.1.10 Tesevatinib

Tesevatinib (XL647 or EXEL 7647) is an oral TKI with activity against EGFR, HER2, VEGFR, and EphB4 [71, 380]. XL647 inhibits EGFR phosphorylation as well as phosphorylation of downstream signals (AKT, ERK). XL647 demonstrated antitumor activity in patients with EGFR-activating mutations and adenocarcinoma histology (41 patients) in a phase II study with most common adverse effects being diarrhea, nausea, and fatigue [381]. XL647 has also shown to be active against cells containing the T790M mutation, which have been associated with emerging resistance to first-generation EGFR TKIs such as gefitinib or erlotinib [382]. Further studies are required to develop this agent as a potential second-generation TKI to delay or overcome acquired resistance to first-generation EGFR TKIs [71].

3.4.1.11 Brivanib

Brivanib (BMS-540215, Bristol-Myers Squibb; New York, NY, USA) is a small-molecule TKI, formulated as an orally administered l-alanine ester prodrug, brivanib alaninate [71, 103]. It is the first-selective dual inhibitor of fibroblast growth factor (FGF) and VEGF signaling [104], and preclinical studies have demonstrated that dual inhibition of FGF and VEGF signaling by brivanib has strong antiangiogenic effects [105]. In phase I study in patients with advanced or metastatic solid tumors, brivanib demonstrated promising antiangiogenic and antitumor activity and manageable toxicity with most frequent serious toxic effects recorded being nausea, pyrexia, aspartate aminotransferase (AST) or alanine aminotransferase (ALT) elevations, and thrombocytopenia [106, 107]. Synergistic activity of ixabepilone plus brivanib has suggested a promising anticancer activity both in vitro and in vivo [108].

3.4.1.12 Nintedanib

Nintedanib or BIBF 1120 (Boehringer Ingelheim; Ingelheim, Germany) is an oral small-molecule triple kinase inhibitor which inhibits VEGFR, PDGFR, and fibroblast growth factor receptors (FGFRs) and members of the Src family and flt-3 [71, 109, 364]. Single-agent treatment of nintedanib (BIBF 1120) in advanced NCSLC was well tolerated, with median PFS of 6.9 weeks and median OS of 21.9 weeks [110]. This drug also has proved its safety and well tolerability on its phase I trials in combination with chemotherapy. Phase I trial of 26 patients (7 with adenocarcinoma, 3 with squamous, 1 with large cell carcinoma, 1 with adenosquamous, and 14 with unspecified histology) with recurrent advanced-stage NSCLC receiving BIBF 1120 in combination with standard-dose pemetrexed was tolerable, with promising signs of efficacy [111]. For combination with paclitaxel and carboplatin, nintedanib demonstrated an acceptable safety profile in a phase I trial [112]. These findings indicate nintedanib as a potential triple kinase inhibitor and as a candidate for further study [71].

3.4.1.13 Lenvatinib

Lenvatinib is a multitargeted TKI against VEGFR1, VEGFR2 and VEGFR3, FGFR, PDGFRα, as well as proto-oncogenes RET and KIT [383–385]. The combination of lenvatinib (Lenvima, Eisai) plus pembrolizumab (Keytruda, Merck) was approved by FDA for first-line treatment of adult patients with advanced renal cell carcinoma (RCC). Preclinical studies have shown that lenvatinib has potent antiangiogenic activity, mainly through the inhibition of VEGF and FGF signaling pathways [386, 387]. The results of the REFLECT trial for advanced hepatocellular carcinoma (aHCC) patients reported by Masatoshi Kudo et al. showed lenvatinib to be non-inferior to sorafenib with regard to OS [388], and all secondary efficacy endpoints in the lenvatinib group indicated a statistically significant improvement compared with sorafenib. Although all patients experienced at least one adverse event (AE) in both phase II and REFLECT trial of lenvatinib, most of the AEs were class effects associated with TKIs, which can be manageable with appropriate and prompt treatment [388, 389]. Based on the REFLECT trial, lenvatinib has been approved in the USA, EU, China, Japan, and other countries as an alternative first-line treatment option to sorafenib for aHCC.

The application of lenvatinib has changed the status of sorafenib as the only first-line TKI treatment for HCC. Evidence has shown that lenvatinib possesses antitumor proliferation and immunomodulatory activity in preclinical studies. In comparison, lenvatinib was non-inferior to sorafenib in OS, and even shows superiority with regard to all the secondary efficacy endpoints. ICIs are now being incorporated into HCC treatment. Positive outcomes have been achieved in the combination of lenvatinib plus ICIs, bringing broader prospects for HCC.

3.4.2 Multitargeting HDAC Inhibitors as Anticancer MTDs

Despite the encouraging clinical progress of chemotherapeutic agents in cancer treatment, innovation and development of new effective anticancer candidates still represents a challenging endeavor [390]. With 15 million death every year in 2030 according to the estimates, cancer has increased rising of an alarm as a real crisis for public health and health systems worldwide. Therefore, scientist began to introduce innovative solutions to control the cancer global health problem [390]. One of the promising strategies in this issue is the multitarget or smart hybrids having two or more pharmacophores targeting cancer [390]. These rationalized hybrid molecules have gained great interests in cancer treatment as they are capable to simultaneously inhibit more than one cancer pathway or target without drug-drug interactions and with less side effects. A prime important example of these hybrids, the histone deacetylase (HDAC) hybrid inhibitors or referred as multitargeting HDAC inhibitors by Bass et al [390] or MT-HDACIs.

3.4.2.1 HDAC Inhibitor Hybrids

HDACs are a type of epigenetic enzymes that in recent years have aroused great interest in oncology [391]. Its aberrant activity has been related to cell proliferation in a variety of human pathologies, especially solid tumors and myeloid neoplasia [392–398]. In humans, there are 18 HDAC isoforms that are divided into four different classes (I–IV). Class I includes HDAC1, 2, 3, and 8; class II includes 4, 5, 6, 7, 9, and 10; class III is made up of members of the sirtuin family and class IV is represented by HDAC11 [391]. Except class III, they are Zn-dependent enzymes. Class I HDACs are ubiquitously expressed and play a critical role in proliferation, while classes II and IV have tissue-specific roles [399]. These enzymes catalyze the deacetylation of the amino group of lysine residues in the N-terminal tails of histone and non-histone proteins and are involved in cell cycle progression, transcriptional regulation, and apoptosis [400, 401]. For a normal cell growth, a balance between acetylation by histone acetyl transferases (HATs) and deacetylation by HDACs is essential. HDAC inhibitors (HDACIs), by inducing hyperacetylation of histones, alter gene transcription and exert antitumor effects through differentiation, growth arrest, or/and apoptosis [402, 403]. HDAC inhibition has become an important strategy in epigenetic drug discovery, and many HDAC inhibitors are currently under clinical investigation for cancer therapy [404, 405]. Four multi-isoform HDAC inhibitors Vorinosat (SAHA), Romidepsin (FK-228), Belinostat (PXD101), and Panobinostat (LBH589) have been approved by the FDA for the treatment of specific lymphoma and myeloma diseases [406].

The ability of HDAC inhibitors to synergistically improve the efficacy of other anticancer drugs and moreover, the ease of HDAC inhibitors cap group modification prompt many medicinal chemists to innovate and develop new generation of HDAC hybrid inhibitors. Notably, and during this short period, there are four HDAC inhibitor hybrids have entered different phases of clinical trials for the treatment of different types of blood and solid tumors, namely, CUDC-101, CUDC-907, Tinostamustine, and Domatinostat. The review article by Bass et al [390] shed light on the hybrids of HDACIs with one or more other cancer target pharmacophore including topoisomerase inhibitors, kinase inhibitors, nitric oxide releasers, antiandrogens, FLT3 and JAC-2 inhibitors, PDE5 inhibitors, NAMPT inhibitors, Protease inhibitors, BRD4 inhibitors, and other targets.

3.4.2.2 HDAC Inhibitor–Kinase Inhibitor Hybrids

Some examples of multitarget inhibitors that combine structural features of HDACI and other anticancer agents have been published in the recent years [390, 407]. A combination of kinase inhibitors with the HDAC pharmacophore has been recently reported [408, 409]. Two hybrid compounds are currently in clinical trials: an Erlotinib-based hybrid with activity in HDACs, EGFR, and HER2 (IC_{50} = 4.4, 2.4, and 15.7 nM, respectively), and Fimepinostat (CUDC-907), which is an orally active pan PI3K/HDAC inhibitor [410].

As described by Martínez et al [391], protein kinase CK2 constitutes an attractive option for a secondary target, owing not only to its role in the regulation of HDACs by post-translational modification [410, 411], but also to its capacity to activate HDAC1 in tumors associated with hypoxia [412, 413]. By using tetrabromobenzotriazole (TBB) and 2-dimethylamino-4,5,6,7-tetrabromo-benzimidazole (DMAT) as scaffolds for CK2 inhibition, and a hydroxamate to coordinate the zinc atom present in the active site of HDAC (zinc binding group, ZBG), Martínez and colleagues designed and synthesized new multitarget inhibitors [391]. Their *in vitro* assays demonstrated that *N*-Hydroxy-6-(4,5,6,7-tetrabromo-2-(dimethylamino)-1H-benzo[d]imidazol-1-yl)hexanamide is the most interesting compound, with IC_{50} values of 0.66; 1.46 and 3.67 μM. for HDAC6; HDAC1 and CK2; respectively. Cellular assays on different cancer cell lines rendered promising results for *N*-Hydroxy-8-(4,5,6,7-tetrabromo-2-(dimethylamino)-1H-benzo[d]imidazol-1-yl)octanamide [391]. This inhibitor presented the highest cytotoxic activity, proapoptotic capability, and the best mitochondria-targeting and multidrug-circumventing properties, thus being the most promising drug candidate for further in vivo studies [391].

References

1. Sung H, Ferlay J, Siegel RL, et al. Global cancer statistics 2020: GLOBOCAN estimates of incidence and mortality worldwide for 36 cancers in 185 countries. CA Cancer J Clin. 2021;71(3):209–49.
2. Holmes K, Roberts OL, Thomas AM, et al. Vascular endothelial growth factor receptor-2: structure, function, intracellular signaling and therapeutic inhibition. Cell Signal. 2007;19(10):2003–12.
3. Shibuya M. Vascular endothelial growth factor (VEGF) and its receptor (VEGFR) signaling in angiogenesis: A crucial target for anti- and pro-angiogenic therapies. Genes Cancer. 2011;2(12):1097–105.
4. Rapisarda A, Melillo G. Role of the VEGF/VEGFR axis in cancer biology and therapy. Adv Cancer Res. 2012;114:237–67.
5. Carmeliet P. VEGF as a key mediator of angiogenesis in cancer. Oncology. 2005;69(Suppl 3):4–10.
6. Becker JC, Muller-Tidow C, Serve H. Role of receptor tyrosine kinases in gastric cancer: new targets for a selective therapy. World J Gastroenterol. 2006;12:3297–305.
7. Johnston JB, Navaratnam S, Pitz MW. Targeting the EGFR pathway for cancer therapy. Curr Med Chem. 2006;13:3483–92.
8. Arteaga CL. Overview of epidermal growth factor receptor biology and its role as a therapeutic target in human neoplasia. Semin Oncol. 2002;29:3–9.
9. Krozely P. Epidermal growth factor receptor tyrosine kinase inhibitors: evolving role in the treatment of solid tumors. Clin J Oncol Nurs. 2004;8:163–8.
10. Peart O. Breast intervention and breast cancer treatment options. Radiol Technol. 2015;86:535M–58M. quiz 559-62
11. International Agency for Research on Cancer. Data visualization tools for exploring the global cancer burden in 2020. In: International Agency for Research on Cancer. Available from: https://gco.iarc.fr/today/home.
12. Falzone L, Salomone S, Libra M. Evolution of cancer pharmacological treatments at the turn of the third millennium. Front Pharmacol. 2018;9:1300.
13. Shetty N, Gupta S. Eribulin drug review. South Asian J Cancer. 2014;3:57–9.
14. Cooper, Geoffrey M. The Cell: A Molecular Approach Eighth Edition. Oxford University Press. 2019; p222. ISBN 9781605357072.
15. Nelson DL. Cox MM. Lehninger Principles of Biochemistry Seventh Edition. W. H. Freeman and Company. 2017; pp963–71. ISBN 9781464126116.
16. Delgado JL, Hsieh CM, Chan NL, Hiasa H. Topoisomerases as anticancer targets. Biochem J. 2018;475(2):373–98.
17. Waldmann TA. Immunotherapy: past, present and future. Nat Med. 2003;9(3):269–77.
18. Sharma P, Allison JP. The future of immune checkpoint therapy. Science. 2015;348(6230):56–61.

19. What Are the Different Types of Chemotherapy Drugs? American Cancer Society. Archived from the original on 2007-07-17. Retrieved 2007-08-05.
20. Definition of mitotic inhibitor. National Cancer Institute. Retrieved 2007-08-05.
21. Treatment Options: Mitotic Inhibitors. Drug Digest. Archived from the original on 2007-02-16. Retrieved 2007-08-05.
22. Protein Kinase Inhibitors. In: LiverTox: Clinical and Research Information on Drug-Induced Liver Injury [Internet]. Bethesda: National Institute of Diabetes and Digestive and Kidney Diseases; 2012–. 2021 Nov 30.
23. Roskoski R Jr. Properties of FDA-approved small molecule protein kinase inhibitors: A 2020 update. Pharmacol Res. 2020 Feb;152:104609.
24. Gross S, Rahal R, Stransky N, et al. Targeting cancer with kinase inhibitors. J Clin Invest. 2015;125(5):1780–9.
25. Abotaleb M, Kubatka P, Caprnda M, et al. Chemotherapeutic agents for the treatment of metastatic breast cancer: An update. Biomed Pharmacother. 2018;101:458–77.
26. Harbeck N, Penault-Llorca F, Cortes J, et al. Breast cancer. Nat Rev Dis Primers. 2019;5:66.
27. Esteva FJ, Hubbard-Lucey VM, Tang J, et al. Immunotherapy and targeted therapy combinations in metastatic breast cancer. Lancet Oncol. 2019;20:e175–86.
28. Guo ZS, Lu B, Guo Z, et al. Vaccinia virus-mediated cancer immunotherapy: cancer vaccines and oncolytics. J Immunother Cancer. 2019;7:6.
29. Greten TF, Lai CW, Li G, et al. Targeted and immune-based therapies for hepatocellular carcinoma. Gastroenterology. 2019;156(2):510–24.
30. Santos R, Ursu O, Gaulton A, et al. A comprehensive map of molecular drug targets. Nat Rev Drug Discov. 2017;16(1):19–34.
31. Hasan M, Alam S, Poddar SK. Antibody-drug conjugates: A review on the Epitome of Targeted anti- cancer therapy. Curr Clin Pharmacol. 2018;13(4):236–51.
32. Gasser M, Waaga-Gasser AM. Therapeutic antibodies in cancer therapy. Adv Exp Med Biol. 2016;917:95–120.
33. Lee YT, Tan YJ, Oon CE. Molecular targeted therapy: Treating cancer with specificity. Eur J Pharmacol. 2018;834:188–96.
34. Gotwals P, Cameron S, Cipolletta D, et al. Prospects for combining targeted and conventional cancer therapy with immunotherapy. Nat Rev Cancer. 2017;17(5):286–301.
35. Mayekar MK, Bivona TG. Current landscape of targeted therapy in lung cancer. Clin Pharmacol Ther. 2017;102(5):757–64.
36. Bashraheel SS, Domling A, Goda SK. Update on targeted cancer therapies, single or in combination, and their fine tuning for precision medicine. Biomed Pharmacother. 2020;125:110009.
37. Parameswaran S, Kundapur D, Vizeacoumar FS, et al. A road map to personalizing targeted cancer therapies using synthetic lethality. Trends Cancer. 2019;5(1):11–29.
38. Meric-Bernstam F, Johnson AM, Dumbrava EEI, et al. Advances in HER2-targeted therapy: Novel agents and opportunities beyond breast and gastric cancer. Clin Cancer Res. 2019;25(7):2033–41.
39. Horn T, Ferretti S, Ebel N, et al. High-order drug combinations are required to effectively kill colorectal cancer cells. Cancer Res. 2016;76(23):6950–63.
40. Wu L, Leng D, Cun D, et al. Advances in combination therapy of lung cancer: Rationales, delivery technologies and dosage regimens. J Control Release. 2017;260:78–91.
41. Sause WT. Combination chemotherapy and radiation therapy in lung cancer. Semin Oncol. 1994;21(3 Suppl 6):72–8.
42. Ettinger DS, Akerley W, Borghaei H, et al. Non-small cell lung cancer. J Nat Comp Cancer Netw. 2012;10(10):1236–71.
43. Fossella F, Pereira JR, von Pawel J, et al. Randomized, multinational, phase III study of docetaxel plus platinum combinations versus vinorelbine plus cisplatin for advanced non-small-cell lung cancer: the TAX 326 Study Group. J Clin Oncol. 2003;21(16):3016–24.
44. Scagliotti GV, Parikh P, von Pawel J, et al. Phase III study comparing cisplatin plus gemcitabine with cisplatin plus pemetrexed in chemotherapy-naive patients with advanced-stage non-small-cell lung cancer. J Clin Oncol. 2008;26(21):3543–51.
45. Korfee S, Gauler T, Hepp R, et al. New targeted treatments in lung cancer—overview of clinical trials. Lung Cancer. 2004;45(Suppl 2):S199–208.
46. Cheong K, Spicer J, Chowdhury S, et al. Combination therapy versus single agent chemotherapy in non-small cell lung cancer. Expert Opin Pharmacother. 2005;6(10):1693–700.
47. Sekine I, Saijo N. Novel combination chemotherapy in the treatment of non-small cell lung cancer. Expert Opin Pharmacother. 2000;1(6):1131–61.
48. Belani CP. TAX 326 Study Group. Docetaxel in combination with platinums (cisplatin or carboplatin) in advanced and metastatic non-small cell lung cancer. Semin Oncol. 2002;29(3 Suppl 12):4–9.
49. Mörth C, Valachis A. Single-agent versus combination chemotherapy as first-line treatment for patients with advanced non-small cell lung cancer and performance status 2: a literature-based meta-analysis of randomized studies. Lung Cancer. 2014;84(3):209–14.
50. Rigas JR. Taxane-platinum combinations in advanced non-small cell lung cancer: a review. Oncologist. 2004;9(Suppl 2):16–23.
51. Sandler A. Irinotecan plus cisplatin in small-cell lung cancer. Oncology (Williston Park). 2002;16(9 Suppl 9):39–43.

52. Ramalingam S, Belani CP. Carboplatin/gemcitabine combination in advanced NSCLC. Oncology (Williston Park). 2004;18(8 Suppl 5):21–6.
53. Yang S. A pooled study on combination of gemcitabine and nedaplatin for treating patients with non-small cell lung cancer. Asian Pac J Cancer Prev. 2015;16(14):5963–6.
54. Weinmann M, Jeremic B, Bamberg M, et al. Treatment of lung cancer in elderly part II: small cell lung cancer. Lung Cancer. 2003;40(1):1–16.
55. Chihara Y, Yoshimura A, Date K, et al. Phase II study of S-1 and paclitaxel combination therapy in patients with previously treated non-small cell lung cancer. Oncologist. 2019;24(8):1033–e617.
56. Inoue M, Kusumoto H, Shiono H, et al. Feasibility of postoperative adjuvant chemotherapy using carboplatin plus S-1 in completely resected non-small cell lung cancer patients. J Chemother. 2019;31(6):343–8.
57. Roviello G, Nobili S, Mini E. A potential novel drug combination for adjuvant therapy in lung cancer: reflections on an early phase of clinical development. J Chemother. 2019;31(6):295–6.
58. Greco FA. Paclitaxel-based combination chemotherapy in advanced non-small cell lung cancer. Lung Cancer. 2001;34(Suppl 4):S53–6.
59. Douillard JY, Rosell R, De Lena M, et al. Adjuvant vinorelbine plus cisplatin versus observation in patients with completely resected stage IB-IIIA non-small-cell lung cancer (Adjuvant Navelbine International Trialist Association [ANITA]): a randomized controlled trial. Lancet Oncol. 2006;7:719–27.
60. von Plessen C, Bergman B, Andresen O, et al. Palliative chemotherapy beyond three courses conveys no survival or consistent quality-of-life benefits in advanced non-small-cell lung cancer. Br J Cancer. 2006;95(8):966–73.
61. Joon OP, Kim SW, Jin SA, et al. Phase III trial of two versus four additional cycles in patients who are nonprogressive after two cycles of platinum-based chemotherapy in non-small-cell lung cancer. J Clinical Oncol. 2007;25(33):5233–9.
62. Gandara DR, Davies AM, Gautschi O, et al. Epidermal growth factor receptor inhibitors plus chemotherapy in non-small-cell lung cancer: biologic rationale for combination strategies. Clin Lung Cancer. 2007;8(Suppl2):S61–7.
63. Wojciechowska J, Krajewski W, Bolanowski M, et al. Diabetes and cancer: A review of current knowledge. Exp Clin Endocrinol Diabetes. 2016;124(5):263–75.
64. Shlomai G, Neel B, LeRoith D, et al. Type 2 diabetes mellitus and cancer: The role of pharmacotherapy. J Clin Oncol. 2016;34(35):4261–9.
65. Gallagher EJ, LeRoith D. Obesity and diabetes: The increased risk of cancer and cancer-related mortality. Physiol Rev. 2015;95(3):727–48.
66. Gupta G, de Jesus Andreoli Pinto T, Chellappan DK, et al. A clinical update on metformin and lung cancer in diabetic patients Panminerva Med. 2018; 60(2): 70–75.
67. Levy A, Doyen J. Metformin for non-small cell lung cancer patients: Opportunities and pitfalls. Crit Rev Oncol Hematol. 2018;125:41–7.
68. Parikh AB, Marrone KA, Becker DJ, et al. A pooled analysis of two phase II trials evaluating metformin plus platinum-based chemotherapy in advanced non-small cell lung cancer. Cancer Treat Res Commun. 2019;20:100150.
69. Zhang HH, Guo XL. Combinational strategies of metformin and chemotherapy in cancers. Cancer Chemother Pharmacol. 2016;78(1):13–26.
70. Li L, Jiang L, Wang Y, et al. Combination of metformin and gefitinib as first-line therapy for nondiabetic advanced NSCLC patients with EGFR mutations: A randomized, double-blind phase II trial. Clin Cancer Res. 2019;25(23):6967–75.
71. Roy M, Luo YH, Ye M, et al. Nonsmall cell lung cancer therapy: insight into multitargeted small-molecule growth factor receptor inhibitors. Biomed Res Int. 2013;2013:964743.
72. Schiller JH, Lee J, Hanna N, et al. A randomized discontinuation phase II study of sorafenib versus placebo in patients with non-small cell lung cancer who have failed at least two prior chemotherapy regimens. J Clin Oncol. 2008;26(suppl5):S427.
73. Scagliotti G, Novello S, von Pawel J, et al. Phase III study of carboplatin and paclitaxel alone or with sorafenib in advanced non-small-cell lung cancer. J Clin Oncol. 2010;28(11):1835–42.
74. Wang Y, Wang L, Liu Y, et al. Randomize trial of cisplatin plus gemcitabine with either sorafenib or placebo as first-line therapy for non-small cell lung cancer. Chin J Lung Can. 2011;14(3):239–44.
75. Paz-Ares LG, Biesma B, Heigener D, et al. Phase III, randomized, double-blind, placebo-controlled trial of gemcitabine/cisplatin alone or with sorafenib for the first-line treatment of advanced, nonsquamous non-small-cell lung cancer. J Clin Oncol. 2012;30(25):3084–92.
76. Lind JSW, Dingemans AMC, Groen HJM, et al. A multicenter phase II study of erlotinib and sorafenib in chemotherapy-naïve patients with advanced non-small cell lung cancer. Clinical Cancer Research. 2010;16(11):3078–87.
77. Gridelli C, Morgillo F, Favaretto A, et al. Sorafenib in combination with erlotinib or with gemcitabine in elderly patients with advanced non-small-cell lung cancer: a randomized phase II study. Ann Oncol. 2011;22(7):1528–34.
78. Spigel DR, Burris HA III, Greco FA, et al. Randomized, double-blind, placebo-controlled, phase II trial of sorafenib and erlotinib or erlotinib alone in previously treated advanced non-small-cell lung cancer. J Clin Oncol. 2011;29(18):2582–9.
79. Adjei A, Blumenschein G Jr, Mandrekar S, et al. Long-term safety and tolerability of sorafenib in patients with advanced non-small-cell lung cancer: a case-based review. Clin Lung Cancer. 2011;12(4):212–7.
80. Mendel DB, Douglas Laird A, et al. In vivo antitumor activity of SU11248, a novel tyrosine kinase

inhibitor targeting vascular endothelial growth factor and platelet-derived growth factor receptors: determination of a pharmacokinetic/pharmacodynamic relationship. Clin Can Res. 2003;9(1):327–37.
81. Novello S, Scagliotti GV, Rosell R, et al. Phase II study of continuous daily sunitinib dosing in patients with previously treated advanced non-small cell lung cancer. Br J Can. 2009;101(9):1543–8.
82. Novello S, Camps C, Grossi F, et al. Phase II study of sunitinib in patients with non-small cell lung cancer and irradiated brain metastases. J Thoracic Oncol. 2011;6(7):1260–6.
83. Scagliotti GV, Vynnychenko I, Park K, et al. International, randomized, placebo-controlled, double-blind phase III study of motesanib plus carboplatin/paclitaxel in patients with advanced nonsquamous non-small-cell lung cancer: MONET1. J Clin Oncol. 2012;30(23):2829–36.
84. Socinski MA, Scappaticci FA, Samant M, et al. Safety and efficacy of combining sunitinib with bevacizumab + paclitaxel/carboplatin in non-small cell lung cancer. J Thoracic Oncol. 2010;5(3):354–60.
85. Blumenschein GR Jr, Ciuleanu T, Robert F, et al. Sunitinib plus erlotinib for the treatment of advanced/metastatic non-small-cell lung cancer: a lead-in study. J Thoracic Oncol. 2012;7(9):1406–16.
86. Nikolinakos P, Heymach JV. The tyrosine kinase inhibitor cediranib for non-small cell lung cancer and other thoracic malignancies. J Thoracic Oncol. 2008;3(6):S131–4.
87. Yamamoto N, Tamura T, Yamamoto N, et al. Phase I, dose escalation and pharmacokinetic study of cediranib (RECENTIN™), a highly potent and selective VEGFR signaling inhibitor, in Japanese patients with advanced solid tumors. Cancer Chemother Pharmacol. 2009;64(6):1165–72.
88. Goss G, Shepherd FA, Laurie S, et al. A phase I and pharmacokinetic study of daily oral cediranib, an inhibitor of vascular endothelial growth factor tyrosine kinases, in combination with cisplatin and gemcitabine in patients with advanced non-small cell lung cancer: a study of the National Cancer Institute of Canada Clinical Trials Group. Eur J Cancer. 2009;45(5):782–8.
89. Goss GD, Arnold A, Shepherd FA, et al. Randomized, double-blind trial of carboplatin and paclitaxel with either daily oral cediranib or placebo in advanced non-small-cell lung cancer: NCIC clinical trials group BR24 study. J Clin Oncol. 2010;28(1):49–55.
90. Dy GK, Mandrekar SJ, Nelson GD, et al. A randomized phase II study of gemcitabine and carboplatin with or without cediranib as first-line therapy in advanced non-small-cell lung cancer: North Central Cancer Treatment Group Study N0528. J Thoracic Oncol. 2013;8(1):79–88.
91. Choueiri TK. Axitinib, a novel anti-angiogenic drug with promising activity in various solid tumors. Curr Opin Investig Drugs. 2008;9(6):658–71.
92. Wang F, Mi YJ, Chen SG, et al. Axitinib targeted cancer stemlike cells to enhance efficacy of chemotherapeutic drugs via inhibiting the drug transport function of ABCG2. Mol Medicine. 2012;18:887–98.
93. Kozloff MF, Martin LP, Krzakowski M, et al. Phase I trial of axitinib combined with platinum doublets in patients with advanced non-small cell lung cancer and other solid tumours. Br J Can. 2012;107(8):1277–85.
94. Raghav KPS, Blumenschein GR. Motesanib and advanced NSCLC: experiences and expectations. Expert Opinion Investig Drugs. 2011;20(6):859–69.
95. Coxon A, Ziegler B, Kaufman S, et al. Antitumor activity of motesanib alone and in combination with cisplatin or docetaxel in multiple human non-small-cell lung cancer xenograft models. Mol Can. 2012;11:70.
96. Das M, Wakelee H. Anti-angiogenic agents in Non-Small-Cell Lung Cancer (NSCLC): a perspective on the MONET1 (Motesanib NSCLC Efficacy and Tolerability) study. J Thoracic Dis. 2012;4(6):558–61.
97. Wedge SR, Ogilvie DJ, Dukes M, et al. ZD6474 inhibits vascular endothelial growth factor signaling, angiogenesis, and tumor growth following oral administration. Cancer Research. 2002;62(16):4645–55.
98. Zhang L, Li S, Zhang Y, et al. Pharmacokinetics and tolerability of vandetanib in Chinese patients with solid, malignant tumors: an open-label, phase I, rising multiple-dose study. Clin Therapeutics. 2011;33(3):315–27.
99. Heymach JV, Paz-Ares L, De Braud F, et al. Randomized phase II study of vandetanib alone or with paclitaxel and carboplatin as first-line treatment for advanced non-small-cell lung cancer. J Clin Oncol. 2008;26(33):5407–15.
100. De Boer RH, Arrieta Ó, Yang CH, et al. Vandetanib plus pemetrexed for the second-line treatment of advanced non-small-cell lung cancer: a randomized, double-blind phase III trial. J Clin Oncol. 2011;29(8):1067–74.
101. Herbst RS, Sun Y, Eberhardt WEE, et al. Vandetanib plus docetaxel versus docetaxel as second-line treatment for patients with advanced non-small-cell lung cancer (ZODIAC): a double-blind, randomised, phase 3 trial. Lancet Oncol. 2010;11(7):619–26.
102. Lee JS, Hirsh V, Park K, et al. Vandetanib Versus placebo in patients with advanced non-small-cell lung cancer after prior therapy with an epidermal growth factor receptor tyrosine kinase inhibitor: a randomized, double-blind phase III trial (ZEPHYR). J Clin Oncol. 2012;30(10):1114–21.
103. Marathe PH, Kamath AV, Zhang Y, et al. Preclinical pharmacokinetics and in vitro metabolism of brivanib (BMS-540215), a potent VEGFR2 inhibitor and its alanine ester prodrug brivanib alaninate. Can Chemother Pharmacol. 2009;65(1):55–66.
104. Bhide RS, Cai ZW, Zhang YZ, et al. Discovery and preclinical studies of (R)-1-(4-(4-fluoro-2-methyl-1H-indol-5-yloxy)-5-methylpyrrolo[2,1-f][1,2,4]triazin-6-yloxy)propan-2-ol (BMS-540215), an

in vivo active potent VEGFR-2 inhibitor. J Medicinal Chem. 2006;49(7):2143–6.
105. Bhide RS, Lombardo LJ, Hunt JT, et al. The antiangiogenic activity in xenograft models of brivanib, a dual inhibitor of vascular endothelial growth factor receptor-2 and fibroblast growth factor receptor-1 kinases. Mol Can Therapeutics. 2010;9(2):369–78.
106. Mekhail T, Masson E, Fischer BS, et al. Metabolism, excretion, and pharmacokinetics of oral brivanib in patients with advanced or metastatic solid tumors. Drug Metab Disposition. 2010;38(11):1962–6.
107. Jonker DJ, Rosen LS, Sawyer MB, et al. A phase I study to determine the safety, pharmacokinetics and pharmacodynamics of a dual VEGFR and FGFR inhibitor, brivanib, in patients with advanced or metastatic solid tumors. Ann Oncol. 2011;22(6):1413–9.
108. Lee F, Jure-Kunkel MN, Salvati ME. Synergistic activity of ixabepilone plus other anticancer agents: preclinical and clinical evidence. Therapeutic Adv Med Oncol. 2011;3(1):11–25.
109. Hilberg F, Roth GJ, Krssak M, et al. BIBF 1120: triple angiokinase inhibitor with sustained receptor blockade and good antitumor efficacy. Can Res. 2008;68(12):4774–82.
110. Reck M, Kaiser R, Eschbach C, et al. A phase II double-blind study to investigate efficacy and safety of two doses of the triple angiokinase inhibitor BIBF 1120 in patients with relapsed advanced non-small-cell lung cancer. Ann Oncol. 2011;22(6):1374–81.
111. Ellis PM, Kaiser R, Zhao Y, et al. Phase I open-label study of continuous treatment with BIBF 1120, a triple angiokinase inhibitor, and pemetrexed in pretreated non-small cell lung cancer patients. Clin CanRes. 2010;16(10):2881–9.
112. Doebele RC, Conkling P, Traynor AM, et al. A phase I, open-label dose-escalation study of continuous treatment with BIBF, 1120 in combination with paclitaxel and carboplatin as first-line treatment in patients with advanced non-small-cell lung cancer. Ann Oncol. 2012;23(8):2094–102.
113. Kumar R, Knick VB, Rudolph SK, et al. Pharmacokinetic-pharmacodynamic correlation from mouse to human with pazopanib, a multikinase angiogenesis inhibitor with potent antitumor and antiangiogenic activity. Mol Can Therapeutics. 2007;6(7):2012–21.
114. Altorki N, Lane ME, Bauer T, et al. Phase II proof-of-concept study of pazopanib monotherapy in treatment-naive patients with stage I/II resectable non-small-cell lung cancer. J Clin Oncol. 2010;28(19):3131–7.
115. Plummer R, Madi A, Jeffels M, et al. A Phase I study of pazopanib in combination with gemcitabine in patients with advanced solid tumors. Can Chemother Pharmacol. 2013;71(1):93–101.
116. Ciombor KK, Bekaii-Saab T. A comprehensive review of sequencing and combination strategies of targeted agents in metastatic colorectal cancer. Oncologist. 2018;23(1):25–34.
117. American Cancer Society. Cancer Facts & Figures. 2021. Accessed June 28, 2021. https://www.cancer.org/research/cancer-facts-statistics/all--cancer-facts-figures/cancer-facts-figures-2021.html
118. Modest DP, Pant S, Sartore-Bianchi A. Treatment sequencing in metastatic colorectal cancer. Eur J Cancer. 2019;109:70–83.
119. Hurwitz H, Fehrenbacher L, Novotny W, et al. Bevacizumab plus irinotecan, fluorouracil, and leucovorin for metastatic colorectal cancer. N Engl J Med. 2004;350:2335–42.
120. Saltz LB, Clarke S, Diaz-Rubio E, et al. Bevacizumab in combination with oxaliplatin-based chemotherapy as first-line therapy in metastatic colorectal cancer: a randomized phase III study. J Clin Oncol. 2008;26:2013–9.
121. National Comprehensive Cancer Network. NCCN Clinical Practice Guidelines in Oncology (NCCN Guidelines). Colon Cancer. Version 2.2021. Accessed June 28, 2021. https://www.nccn.org/guidelines/guidelines-detail?category=1&id=1428
122. Van Cutsem E, Cervantes A, Nordlinger B, et al. ESMO Guidelines Working Group. Metastatic colorectal cancer: ESMO Clinical Practice Guidelines for diagnosis, treatment and follow-up. Ann Oncol. 2014;25(suppl 3):iii1–9.
123. Tournigand C, Andre T, Achille E, et al. FOLFIRI followed by FOLFOX6 or the reverse sequence in advanced colorectal cancer: A randomized GERCOR study. J Clin Oncol. 2004;22:229–37.
124. Mody K, Baldeo C, Bekaii-Saab T. Antiangiogenic therapy in colorectal cancer. Cancer J. 2018;24(4):165–70.
125. Lee JJ, Chu E. Sequencing of antiangiogenic agents in the treatment of metastatic colorectal cancer. Clin Colorectal Cancer. 2014;13(3):135–44.
126. AVASTIN. (bevacizumab) solution for intravenous infusion. South San Francisco: Genentech, Inc; 2015.
127. ERBITUX. (cetuximab) injection, for intravenous infusion. Branchburg: ImClone LLC; 2016.
128. VECTIBIX. (panitumumab) injection for intravenous infusion. Thousand Oaks: Amgen, Inc.; 2015.
129. ZALTRAP. (ziv-aflibercept) injection for intravenous infusion. Bridgewater: Sanofi-Aventis U.S. LLC; 2016.
130. Wilhelm SM, Dumas J, Adnane L, et al. Regorafenib (BAY 73-4506): A new oral multikinase inhibitor of angiogenic, stromal and oncogenic receptor tyrosine kinases with potent preclinical antitumor activity. Int J Cancer. 2011;129:245–55.
131. CYRAMZA (ramucirumab) injection, for intravenous use [package insert]. Indianapolis: Eli Lilly and Company; 2015.
132. Jenab-Wolcott J, Giantonio BJ. Antiangiogenic therapy in colorectal cancer: where are we 5 years later? Clin Colorectal Cancer. 2010;9(Suppl 1):S7–15.
133. Van Cutsem E, Kohne CH, Hitre E, et al. Cetuximab and chemotherapy as initial treatment for metastatic colorectal cancer. N Engl J Med. 2009;360:1408–17.

134. Van Cutsem E, Kohne CH, Lang I, et al. Cetuximab plus irinotecan, fluorouracil, and leucovorin as first-line treatment for metastatic colorectal cancer: Updated analysis of overall survival according to tumor KRAS and BRAF mutation status. J Clin Oncol. 2011;29:2011–9.
135. Douillard JY, Siena S, Cassidy J, et al. Randomized, phase III trial of panitumumab with infusional fluorouracil, leucovorin, and oxaliplatin (FOLFOX4) versus FOLFOX4 alone as first-line treatment in patients with previously untreated metastatic colorectal cancer: The PRIME study. J Clin Oncol. 2010;28:4697–705.
136. Venook AP, Niedzwiecki D, Lenz HJ, et al. CALGB/SWOG 80405: Phase III trial of irinotecan/5-FU/leucovorin (FOLFIRI) or oxaliplatin/5-FU/leucovorin (mFOLFOX6) with bevacizumab (BV) or cetuximab (CET) for patients (pts) with KRAS wild-type (wt) untreated metastatic adenocarcinoma of the colon or rectum (MCRC). J Clin Oncol. 2014;32(suppl:5s): LBA3a.
137. Heinemann V, von Weikersthal LF, Decker T, et al. FOLFIRI plus cetuximab versus FOLFIRI plus bevacizumab as first-line treatment for patients with metastatic colorectal cancer (FIRE-3): A randomised, open-label, phase 3 trial. Lancet Oncol. 2014;15:1065–75.
138. Douillard JY, Siena S, Cassidy J, et al. Final results from PRIME: Randomized phase III study of panitumumab with FOLFOX4 for first-line treatment of metastatic colorectal cancer. Ann Oncol. 2014;25:1346–55.
139. Peeters M, Price TJ, Cervantes A, et al. Randomized phase III study of panitumumab with fluorouracil, leucovorin, and irinotecan (FOLFIRI) compared with FOLFIRI alone as second-line treatment in patients with metastatic colorectal cancer. J Clin Oncol. 2010;28:4706–13.
140. Van Cutsem E, Tabernero J, Lakomy R, et al. Addition of aflibercept to fluorouracil, leucovorin, and irinotecan improves survival in a phase III randomized trial in patients with metastatic colorectal cancer previously treated with an oxaliplatin-based regimen. J Clin Oncol. 2012;30:3499–506.
141. Tabernero J, Yoshino T, Cohn AL, et al. Ramucirumab versus placebo in combination with second-line FOLFIRI in patients with metastatic colorectal carcinoma that progressed during or after first-line therapy with bevacizumab, oxaliplatin, and a fluoropyrimidine (RAISE): A randomised, double-blind, multicentre, phase 3 study. Lancet Oncol. 2015;16:499–508.
142. Cunningham D, Humblet Y, Siena S, et al. Cetuximab monotherapy and cetuximab plus irinotecan in irinotecan-refractory metastatic colorectal cancer. N Engl J Med. 2004;351:337–45.
143. Colucci G, Gebbia V, Paoletti G, et al. Phase III randomized trial of FOLFIRI versus FOLFOX4 in the treatment of advanced colorectal cancer: a multicenter study of the Gruppo Oncologico Dell'Italia Meridionale. J Clin Oncol. 2005;23:4866–75.
144. Goldberg RM, Sargent DJ, Morton RF, et al. A randomized controlled trial of fluorouracil plus leucovorin, irinotecan, and oxaliplatin combinations in patients with previously untreated metastatic colorectal cancer. J Clin Oncol. 2004;22:23–30.
145. Saltz LB, Cox JV, Blanke C, et al. Irinotecan plus fluorouracil and leucovorin for metastatic colorectal cancer. Irinotecan Study Group. N Engl J Med. 2000;343:905–14.
146. Douillard JY, Cunningham D, Roth AD, et al. Irinotecan combined with fluorouracil compared with fluorouracil alone as first-line treatment for metastatic colorectal cancer: a multicenter randomized trial. Lancet. 2000;355:1041–7.
147. Foubert F, Matysiak-Budnik T, Touchefeu Y. Options for metastatic colorectal cancer beyond the second line of treatment. Dig Liver Dis. 2014;46:105–12.
148. El-Serag HB, Rudolph KL. Hepatocellular carcinoma: epidemiology and molecular carcinogenesis. Gastroenterology. 2007;132(7):2557–76.
149. Zhou M, Wang H, Zhu J, et al. Cause-specific mortality for 240 causes in China during 1990-2013: a systematic subnational analysis for the Global Burden of Disease Study 2013. Lancet. 2016;387:251–72.
150. Allemann P, Demartines N, Bouzourene H, et al. Long-term outcome after liver resection for hepatocellular carcinoma larger than 10 cm. World J Surg. 2013;7:452–8.
151. Park JW, Chen M, Colombo M, et al. Global patterns of hepatocellular carcinoma management from diagnosis to death: the BRIDGE Study. Liver Int. 2015;35:2155–66.
152. Forner A, Reig M, Bruix J. Hepatocellular carcinoma. Lancet. 2018;391:1301–14.
153. Li D, Sedano S, Allen R, et al. Current Treatment Landscape for Advanced Hepatocellular Carcinoma: Patient Outcomes and the Impact on Quality of Life. Cancers (Basel). 2019;11(6):841.
154. Bruix J, Raoul JL, Sherman M, et al. Efficacy and safety of sorafenib in patients with advanced hepatocellular carcinoma: subanalyses of a phase III trial. J Hepatol. 2012;57:821–9.
155. Bruix J, da Fonseca LG, Reig M. Insights into the success and failure of systemic therapy for hepatocellular carcinoma. Nat Rev Gastroenterol Hepatol. 2019;16:617–30.
156. Kudo M, Finn RS, Qin S, et al. Lenvatinib *vs* sorafenib in first-line treatment of patients with unresectable hepatocellular carcinoma: a randomised phase 3 non-inferiority trial. Lancet. 2018;391:1163–73.
157. Finn RS, Qin S, Ikeda M, et al. IMbrave150 Investigators. Atezolizumab plus Bevacizumab in Unresectable Hepatocellular Carcinoma. N Engl J Med. 2020;382:1894–905.
158. Upadhaya S, Neftelino ST, Hodge JP, et al. Combinations take centre stage in PD1/PDL1 inhibitor clinical trials. Nat Rev Drug Discov. 2020;11:020–00204.

159. National Comprehensive Cancer Network. NCCN Clinical Practice Guidelines in Oncology (NCCN Guidelines). Hepatobiliary Cancers. Version 3.2021. Accessed June 30, 2021. https://www.nccn.org/guidelines/guidelines-detail?category=1&id=1438
160. Heimbach JK. Overview of the Updated AASLD Guidelines for the Management of HCC. Gastroenterol Hepatol (N Y). 2017;13:751–3.
161. Vogel A, Cervantes A, Chau I, et al. ESMO Guidelines Committee. Hepatocellular carcinoma: ESMO Clinical Practice Guidelines for diagnosis, treatment and follow-up. Ann Oncol. 2018;29:iv238–55.
162. Kudo M. Systemic therapy for hepatocellular carcinoma: Latest advances. Cancers (Basel). 2018;10
163. Pinter M, Peck-Radosavljevic M. Review article: systemic treatment of hepatocellular carcinoma. Aliment Pharmacol Ther. 2018;48(6):598–609.
164. Sonbol MB, Riaz IB, Naqvi SAA, et al. Systemic therapy and sequencing options in advanced hepatocellular carcinoma: A systematic review and network meta-analysis. JAMA Oncol. 2020;6(12):e204930.
165. Han Y, Zhi WH, Xu F, et al. Selection of first-line systemic therapies for advanced hepatocellular carcinoma: A network meta-analysis of randomized controlled trials. World J Gastroenterol. 2021;27(19):2415–33.
166. Lee MS, Ryoo BY, Hsu CH, et al. GO30140 investigators. Atezolizumab with or without bevacizumab in unresectable hepatocellular carcinoma (GO30140): an open-label, multicentre, phase 1b study. Lancet Oncol. 2020;21:808–20.
167. Park R, Lopes da Silva L, Nissaisorakarn V, et al. Comparison of efficacy of systemic therapies in advanced hepatocellular carcinoma: updated systematic review and frequentist network meta-analysis of randomized controlled trials. J Hepatocell Carcinoma. 2021;8:145–54.
168. Agarwal N, Loriot Y, McGregor BA, et al. Cabozantinib in combination with atezolizumab in patients with metastatic castration-resistant prostate cancer: results of cohort 6 of the COSMIC-021 study. J Clin Oncol. 2020;38:5564.
169. Choueiri TK, Albiges L, Powles T, et al. A phase III study (COSMIC-313) of cabozantinib (C) in combination with nivolumab (N) and ipilimumab (I) in patients (pts) with previously untreated advanced renal cell carcinoma (aRCC) of intermediate or poor risk. J Clin Oncol. 2020;38:TPS767–TPS767.
170. Neal JW, Lim FL, Felip E, et al. Cabozantinib in combination with atezolizumab in non-small cell lung cancer (NSCLC) patients previously treated with an immune checkpoint inhibitor: results from cohort 7 of the COSMIC-021 study. J Clin Oncol. 2020;38:9610.
171. Pal SK, Agarwal N, Loriot Y, et al. Cabozantinib in combination with atezolizumab in urothelial carcinoma previously treated with platinum-containing chemotherapy: results from cohort 2 of the COSMIC-021 study. J Clin Oncol. 2020;38:5013.
172. Saeed A, Phadnis M, Park R, et al. Cabozantinib (cabo) combined with durvalumab (durva) in gastroesophageal (GE) cancer and other gastrointestinal (GI) malignancies: preliminary phase Ib CAMILLA study results. J Clin Oncol. 2020;38:4563.
173. El-Khoueiry AB, Melero I, Yau TC, et al. Impact of antitumor activity on survival outcomes, and nonconventional benefit, with nivolumab (NIVO) in patients with advanced hepatocellular carcinoma (aHCC): subanalyses of CheckMate-040. J Clin Oncol. 2018;36:475.
174. Yau T, Kang YK, Kim TY, et al. Efficacy and safety of nivolumab plus ipilimumab in patients with advanced hepatocellular carcinoma previously treated with sorafenib: the check mate 040 randomized clinical trial. JAMA Oncol. 2020;6(11):e204564.
175. Kelley RK, Sangro B, Harris WP, et al. Efficacy, tolerability, and biologic activity of a novel regimen of tremelimumab (T) in combination with durvalumab (D) for patients (pts) with advanced hepatocellular carcinoma (aHCC). J Clin Oncol. 2020;38:4508.
176. Yau T, Park JW, Finn RS, et al. CheckMate 459: a randomized, multi-center phase III study of nivolumab (NIVO) vs sorafenib (SOR) as first-line (1L) treatment in patients (pts) with advanced hepatocellular carcinoma (aHCC). Ann Oncol. 2019;30:v874–5.
177. Smyth EC, Nilsson M, Grabsch HI, et al. Gastric Cancer. Lancet. 2020;396(10251):635–48.
178. Forman D, Burley V. Gastric cancer: Global pattern of the disease and an overview of environmental risk factors. Best Pract Res Clin Gastroenterol. 2006;20:633–49.
179. Correa P. Human gastric carcinogenesis: A multistep and multifactorial process—First American Cancer Society Award Lecture on Cancer Epidemiology and Prevention. Cancer Res. 1992;52:6735–40.
180. Reddavid R, Sofia S, Chiaro P, et al. Neoadjuvant chemotherapy for gastric cancer. Is it a must or a fake? World J Gastroenterol. 2018;24(2):274–89.
181. Song Z, Wu Y, Yang J, et al. Progress in the treatment of advanced gastric cancer. Tumour Biol. 2017;39(7):10.
182. Wu H, Wang W, Tong S. Nucleostemin regulates proliferation and migration of gastric cancer and correlates with its malignancy. Int J Clin Exp Med. 2015;8:17634–43.
183. Wilke H, Preusser P, Fink U. Preoperative chemotherapy in locally advanced and nonresectable gastric cancer: a phase II study with etoposide, doxorubicin, and cisplatin. J Clin Oncol. 1989;7:1318–26.
184. Mai M, Takahashi Y, Fujimoto T. Neoadjuvant chemotherapy for far-advanced gastric carcinoma. Gan To Kagaku Ryoho. 1994;21:431–9.
185. Crookes P, Leichman CG, Leichman L. Systemic chemotherapy for gastric carcinoma followed by postoperative intraperitoneal therapy: a final report. Cancer. 1997;79:1767–75.
186. Schuhmacher CP, Fink U, Becker K. Neoadjuvant therapy for patients with locally advanced gastric carcinoma with etoposide, doxorubicin, and cis-

platinum. Closing results after 5 years of follow-up. Cancer. 2001;91:918–27.
187. Gianni L, Panzini I, Tassinari D. Meta-analyses of randomized trials of adjuvant chemotherapy in gastric cancer. Ann Oncol. 2001;12:1178–80.
188. Mirza A, Pritchard S, Welch I. The postoperative component of MAGIC chemotherapy is associated with improved prognosis following surgical resection in gastric and gastrooesophageal junction adenocarcinomas. Int J Surg Oncol. 2013;2013:781742.
189. Lordick F, Luber B, Lorenzen S. Cetuximab plus oxaliplatin/leucovorin/5-fluorouracil in first-line metastatic gastric cancer: a phase II study of the Arbeitsgemeinschaft Internistische Onkologie (AIO). Br J Cancer. 2010;102:500–5.
190. Wagner AD, Syn NL, Moehler M, et al. Chemotherapy for advanced gastric cancer. Cochrane Database Syst Rev. 2017;8(8):CD004064.
191. Ajani JA, Rodriguez W, Bodoky G, et al. Multicenter phase III comparison of cisplatin/S-1 with cisplatin/infusional fluorouracil in advanced gastric or gastroesophageal adenocarcinoma study: the FLAGS trial. J Clin Oncol. 2010;28:1547–53.
192. Digklia A, Wagner AD. Advanced gastric cancer: Current treatment landscape and future perspectives. World J Gastroenterol. 2016;22(8):2403–14.
193. Lorenzen S, Stahl M, Hofheinz RD, et al. Influence of taxanes on treatment sequence in gastric cancer. Oncol Res Treat. 2020;43(1-2):42–7.
194. Fisusi FA, Akala EO. Drug combinations in breast cancer therapy. Pharm Nanotechnol. 2019;7(1):3–23.
195. Chew HK. Adjuvant therapy for breast cancer: who should get what? West J Med. 2001;174(4):284–7.
196. de Matteis A, Nuzzo F, D'Aiuto G, et al. Docetaxel plus epidoxorubicin as neoadjuvant treatment in patients with large operable or locally advanced carcinoma of the breast: a single-center, phase II study. Cancer. 2002;94(4):895–901.
197. Drăgănescu M, Carmocan C. Hormone therapy in breast cancer. Chirurgia (Bucur). 2017;112(4):413–7.
198. Diaby V, Tawk R, Sanogo V, et al. A review of systematic reviews of the cost-effectiveness of hormone therapy, chemotherapy, and targeted therapy for breast cancer. Breast Cancer Res Treat. 2015;151(1):27–40.
199. Williams AD, Payne KK, Posey AD Jr, et al. Immunotherapy for breast cancer: Current and future strategies. Curr Surg Rep. 2017;5:31.
200. Lenaz L. Mitomycin C in advanced breast cancer. Cancer Treat Rev. 1985;12(4):235–49.
201. Tanaka S, Iwamoto M, Kimura K, et al. Phase ii study of neoadjuvant anthracycline-based regimens combined with nanoparticle albumin-bound paclitaxel and trastuzumab for human epidermal growth factor receptor 2-positive operable breast cancer. Clin. Breast Cancer. 2015;15(3):191–6.
202. Kummel S, Holtschmidt J, Loibl S. Surgical treatment of primary breast cancer in the neoadjuvant setting. Br J Surg. 2014;101(8):912–24.
203. Neuman HB, Morrogh M, Gonen M, et al. Stage IV breast cancer in the era of targeted therapy: does surgery of the primary tumor matter? Cancer. 2010;116(5):1226–33.
204. Hennigs A, Riedel F, Marmé F, et al. Changes in chemotherapy usage and outcome of early breast cancer patients in the last decade. Breast Cancer Res. Treat. 2016;160(3):491–9.
205. Harbeck N, Gnant M. Breast cancer. Lancet. 2017;389(10074):1134–50.
206. Rashid OM, Takabe K. Does removal of the primary tumor in metastatic breast cancer improve survival? J Womens Health Wellness. 2014;23(2):184–8.
207. Hortobagyi GN. Treatment of breast cancer. N. Engl. J. Med. 1998;339(14):974–84.
208. Sachdev JC, Jahanzeb M. Use of cytotoxic chemotherapy in metastatic breast cancer: putting taxanes in perspective. Clin Breast Cancer. 2016;16(2):73–81.
209. Vrdoljak E, Boban M, Omrcen T, et al. Combination of capecitabine and mitomycin C as first-line treatment in patients with metastatic breast cancer. Neoplasma. 2011;58(2):172–8.
210. Gonzalez-Angulo AM, Morales-Vasquez F, Hortobagyi GN. Overview of resistance to systemic therapy in patients with breast cancer. Adv Exp Med Biol. 2007;608:1–22.
211. Azim HA Jr, de Azambuja E, Colozza M, et al. Long-term toxic effects of adjuvant chemotherapy in breast cancer. Ann Oncol. 2011;22(9):1939–47.
212. Heys SD, Hutcheon AW, Sarkar TK, et al. Neoadjuvant docetaxel in breast cancer: 3-Year Survival Results from the Aberdeen trial. Clin Breast Cancer. 2002;3:S69–74.
213. Malhotra V, Dorr VJ, Lyss AP, et al. Neoadjuvant and adjuvant chemotherapy with doxorubicin and docetaxel in locally advanced breast cancer. Clin Breast Cancer. 2004;5(5):377–84.
214. Colleoni M, Goldhirsch A. Neoadjuvant chemotherapy for breast cancer: any progress? Lancet Oncol. 2014;15(2):131–2.
215. Kaufmann M, von Minckwitz G, Mamounas EP, et al. Recommendations from an international consensus conference on the current status and future of neoadjuvant systemic therapy in primary breast cancer. Ann Surg Oncol. 2012;19(5):1508–16.
216. De Matteis A, Nuzzo F, D'Aiuto G, et al. Docetaxel plus epidoxorubicin as neoadjuvant treatment in patients with large operable or locally advanced carcinoma of the breast. Cancer. 2002;94(4):895–901.
217. Bear HD, Anderson S, Brown A, et al. The effect on tumor response of adding sequential preoperative docetaxel to preoperative doxorubicin and cyclophosphamide: preliminary results from national surgical adjuvant breast and bowel project protocol B-27. J Clin Oncol. 2003;21(22):4165–74.
218. Earl HM, Vallier AL, Hiller L, et al. Effects of the addition of gemcitabine, and paclitaxel-first sequencing, in neoadjuvant sequential epirubicin, cyclophosphamide, and paclitaxel for women with high-risk early breast cancer (neo-tango): an open-

label, 2x2 factorial randomised phase 3 trial. Lancet Oncol. 2014;15(2):201–12.
219. Roy V, Pockaj BA, Allred JB, et al. A Phase II trial of docetaxel and carboplatin administered every 2 weeks as preoperative therapy for stage II or III breast cancer: NCCTG study N0338. J Clin Oncol. 2013;36(6):540–4.
220. Bonadonna G, Brusamolino E, Valagussa P, et al. Combination chemotherapy as an adjuvant treatment in operable breast cancer. N Engl J Med. 1976;294(8):405–10.
221. Abrams JS. Adjuvant therapy for breast cancer--results from the USA consensus conference. Breast Cancer. 2001; 8(4): 298–304.
222. Fisher B, Dignam J, Wolmark N, et al. Tamoxifen and chemotherapy for lymph node-negative, estrogen receptor-positive breast cancer. J Natl Cancer Inst. 1997;89(22):1673–82.
223. Fisher B, Redmond C, Wickerham DL, et al. Doxorubicin-containing regimens for the treatment of stage ii breast cancer: the national surgical adjuvant breast and bowel project experience. J Clin Oncol. 1989;7(5):572–82.
224. Hortobagyi GN, Blumenschein GR, Spanos W, et al. Multimodal treatment of locoregionally advanced breast cancer. Cancer. 1983;51(5):763–8.
225. French Epirubicin Study Group A prospective randomized phase Iii trial comparing combination chemotherapy with cyclophosphamide, fluorouracil, and either doxorubicin or epirubicin. French epirubicin study group. J Clin Oncol. 1988;6(4):679–88.
226. French Adjuvant Study Group. Benefit of a high-dose epirubicin regimen in adjuvant chemotherapy for node-positive breast cancer patients with poor prognostic factors: 5-year follow-up results of French adjuvant study group 05 randomized trial. J Clin Oncol. 2001;19(3):602–11.
227. Creagan ET, Green SJ, Ahmann DL, et al. A phase III clinical trial comparing the combination cyclophosphamide, adriamycin, cisplatin with cyclophosphamide, 5-fluorouracil, prednisone in patients with advanced breast cancer. J Clin Oncol. 1984;2(11):1260–5.
228. Saphner T, Tormey DC, Albertini M. Continuous infusion 5-fluorouracil with escalating doses of intermittent cisplatin and etoposide. A phase I study. Cancer. 1991;68(11):2359–62.
229. Trump DL, Ettinger DS, Abeloff MD. Doxorubicin, vincristine, and cis-diamminedichloroplatinum (II) therapy in patients with advanced breast cancer. Med Pediatr Oncol. 1981;9:1–3.
230. Bonneterre J, Dieras V, Tubiana-Hulin M, et al. Phase II multicentre randomised study of docetaxel plus epirubicin *vs.* 5-fluorouracil plus epirubicin and cyclophosphamide in metastatic breast cancer. Br J Cancer. 2004;91(8):1466–71.
231. Wasserheit C, Frazein A, Oratz R, et al. Phase II trial of paclitaxel and cisplatin in women with advanced breast cancer: an active regimen with limiting neurotoxicity. J Clin Oncol. 1996;14(7):1993–9.
232. Gelmon KA, O'Reilly SE, Tolcher AW, et al. Phase I/II trial of biweekly paclitaxel and cisplatin in the treatment of metastatic breast cancer. J Clin Oncol. 1996;14(4):1185–91.
233. Hortobagyi GN. Paclitaxel-based combination chemotherapy for breast cancer. Oncology. 1997;11:29–37.
234. Esteva FJ, Valero V, Pusztai L, et al. Chemotherapy of metastatic breast cancer: what to expect in 2001 and beyond. Oncologist. 2001;6(2):133–46.
235. Nabholtz J-M, Falkson C, Campos D, et al. Docetaxel and doxorubicin compared with doxorubicin and cyclophosphamide as first-line chemotherapy for metastatic breast cancer: results of a randomized, multicenter, phase III Trial. J Clin Oncol. 2003;21(6):968–75.
236. Smith RE, Anderson SJ, Lembersky BC, et al. Phase ii trial of a doxorubicin/docetaxel doublet for locally advanced and metastatic breast cancer: results from national surgical adjuvant breast and bowel project trial Bp-57. Clin Breast Cancer. 2004;5(3):208–15.
237. Andrez J-C. Mitomycins syntheses: a recent update. Beilstein J Org Chem. 2009;5:33.
238. Colozza M, Tonato M, Grignani F, et al. Low-dose mitomycin and weekly low-dose doxorubicin combination chemotherapy for patients with metastatic breast carcinoma previously treated with cyclophosphamide, methotrexate, and 5-fluorouracil. Cancer. 1988;62(2):262–5.
239. Ospovat I, Siegelmann-Danieli N, et al. Mitomycin C and vinblastine: an active regimen in previously treated breast cancer patients. Tumori. 2009;95(6):683–6.
240. Panasci L, Shenouda G, Begin L, et al. Mitomycin C and mitoxantrone chemotherapy for advanced breast cancer: efficacy with minimal gastrointestinal toxicity and alopecia. Cancer Chemother. Pharmacol. 1990;26(6):457–60.
241. Tanabe M, Ito Y, Tokudome N, et al. Possible use of combination chemotherapy with mitomycin C and methotrexate for metastatic breast cancer pretreated with anthracycline and taxanes. Breast Cancer. 2009;16(4):301.
242. Fukuda T, Tanabe M, Kobayashi K, et al. Combination chemotherapy with mitomycin C and methotrexate is active against metastatic HER2-negative breast cancer even after treatment with anthracycline, taxane, capecitabine, and vinorelbine. Springerplus. 2015;4(1):376.
243. Lu X, Xiao L, Wang L, Ruden DM. Hsp90 inhibitors and drug resistance in cancer: the potential benefits of combination therapies of Hsp90 inhibitors and other anti-cancer drugs. Biochem. Pharmacol. 2012;83(8):995–1004.
244. Modi S, Stopeck AT, Gordon MS, et al. Combination of trastuzumab and tanespimycin (17-aag, kos-953) is safe and active in trastuzumab-refractory her-2-overexpressing breast cancer: a phase I dose-escalation study. J Clin Oncol. 2007;25(34):5410–7.

245. Modi S, Stopeck A, Linden H, et al. Hsp90 inhibition is effective in breast cancer: a phase ii trial of tanespimycin (17-aag) plus trastuzumab in patients with HER2-positive metastatic breast cancer progressing on trastuzumab. Clin Cancer Res. 2011;17(15):5132–9.
246. Goel S, Sharma R, Hamilton A, et al. LHRH agonists for adjuvant therapy of early breast cancer in premenopausal women. Cochrane Database Syst Rev. 2009;2009(4):CD004562.
247. Tancredi R, Furlanetto J, Loibl S. Endocrine therapy in premenopausal hormone receptor positive/human epidermal growth receptor 2 negative metastatic breast cancer: between guidelines and literature. Oncologist. 2018;23(8):974–81.
248. Pritchard KI, Chia SK, Simmons C, et al. Enhancing endocrine therapy combination strategies for the treatment of postmenopausal $HR^+/HER2^-$ advanced breast cancer. Oncologist. 2017;22(1):12–24.
249. Freedman RA, Tolaney SM. Efficacy and safety in older patient subsets in studies of endocrine monotherapy versus combination therapy in patients with $HR^+/HER2^-$ advanced breast cancer: a review. Breast Cancer Res Treat. 2018;167(3):607–14.
250. Giuliano M, Schettini F, Rognoni C, et al. Endocrine treatment versus chemotherapy in postmenopausal women with hormone receptor-positive, HER2-negative, metastatic breast cancer: a systematic review and network meta-analysis. Lancet Oncol. 2019;20(10):1360–9.
251. Chirila C, Mitra D, Colosia A, et al. Comparison of palbociclib in combination with letrozole or fulvestrant with endocrine therapies for advanced/metastatic breast cancer: network meta-analysis. Curr Med Res Opin. 2017;33(8):1457–66.
252. Emens LA. Breast cancer immunotherapy: Facts and hopes. Clin Cancer Res. 2018;24(3):511–20.
253. Marra A, Viale G, Curigliano G. Recent advances in triple negative breast cancer: the immunotherapy era. BMC Med. 2019;17(1):90.
254. Adams S, Gatti-Mays ME, Kalinsky K, et al. Current landscape of immunotherapy in breast cancer: A review. JAMA Oncol. 2019;5(8):1205–14.
255. Baselga J, Cortés J, Kim S-B, et al. Pertuzumab plus trastuzumab plus docetaxel for metastatic breast cancer. N Engl J Med. 2012;366(2):109–19.
256. Swain SM, Kim S-B, Cortés J, et al. Pertuzumab, trastuzumab, and docetaxel for HER2-positive metastatic breast cancer (Cleopatra study): overall survival results from a randomised, double-blind, placebo-controlled, phase 3 study. Lancet Oncol. 2013;14(6):461–71.
257. Swain SM, Ewer MS, Cortés J, et al. Cardiac tolerability of pertuzumab plus trastuzumab plus docetaxel in patients with HER2-positive metastatic breast cancer in Cleopatra: a randomized, double-blind, placebo-controlled phase III Study. Oncologist. 2013;18(3):257–64.
258. Swain SM, Baselga J, Kim SB. Pertuzumab, trastuzumab, and docetaxel in HER2-positive metastatic breast cancer. N Engl J Med. 2015;372(8):724–34.
259. Baselga J, Swain SM. CLEOPATRA: a phase III evaluation of pertuzumab and trastuzumab for HER2-positive metastatic breast cancer. Clin Breast Cancer. 2010;10(6):489–91.
260. Español A, Salem A, Sanchez Y, et al. Breast cancer: Muscarinic receptors as new targets for tumor therapy. World J Clin Oncol. 2021;12(6):404–28.
261. Zhang N, Sun P, Xu Y, et al. The GPER1/SPOP axis mediates ubiquitination-dependent degradation of ERα to inhibit the growth of breast cancer induced by oestrogen. Cancer Lett. 2021;498:54–69.
262. Hsu LH, Chu NM, Lin YF, et al. G-Protein Coupled Estrogen Receptor in Breast Cancer. Int J Mol Sci. 2019;20
263. Español A, Eiján AM, Mazzoni E, et al. Nitric oxide synthase, arginase and cyclooxygenase are involved in muscarinic receptor activation in different murine mammary adenocarcinoma cell lines. Int J Mol Med. 2002;9:651–7.
264. Rimmaudo LE, de la Torre E, Sacerdote de Lustig E, et al. Muscarinic receptors are involved in LMM3 tumor cells proliferation and angiogenesis. Biochem Biophys Res Commun. 2005;334:1359–64.
265. Cheng K, Shang AC, Drachenberg CB, et al. Differential expression of M3 muscarinic receptors in progressive colon neoplasia and metastasis. Oncotarget. 2017;8:21106–14.
266. Lin G, Sun L, Wang R, et al. Overexpression of muscarinic receptor 3 promotes metastasis and predicts poor prognosis in non-small-cell lung cancer. J Thorac Oncol. 2014;9:170–8.
267. Pacini L, De Falco E, Di Bari M, et al. M2muscarinic receptors inhibit cell proliferation and migration in urothelial bladder cancer cells. Cancer Biol Ther. 2014;15:1489–98.
268. Español AJ, Jacob G, Dmytrenko G, et al. Muscarinic activation enhances the anti-proliferative effect of paclitaxel in murine breast tumor cells. Anticancer Agents Med Chem. 2013;13:1273–9.
269. Español AJ, Salem A, Rojo D, et al. Participation of non-neuronal muscarinic receptors in the effect of carbachol with paclitaxel on human breast adenocarcinoma cells. Roles of nitric oxide synthase and arginase. Int Immunopharmacol. 2015;29:87–92.
270. Español AJ, Salem A, Di Bari M, et al. The metronomic combination of paclitaxel with cholinergic agonists inhibits triple negative breast tumor progression. Participation of M_2 receptor subtype. PLoS One. 2020;15:e0226450.
271. Salem AR, Martínez Pulido P, Sanchez F, et al. Effect of low dose metronomic therapy on MCF-7 tumor cells growth and angiogenesis. Role of muscarinic acetylcholine receptors. Int Immunopharmacol. 2020;84:106514.
272. Kohn EC, Alessandro R, Probst J, et al. Identification and molecular characterization of a m5 muscarinic receptor in A2058 human melanoma cells.

Coupling to inhibition of adenylyl cyclase and stimulation of phospholipase A2. J Biol Chem. 1996;271:17476–84.
273. Liu H, Xia J, Wang T, et al. Differentiation of human glioblastoma U87 cells into cholinergic neuron. Neurosci Lett. 2019;704:1–7.
274. Yu H, Xia H, Tang Q, et al. Acetylcholine acts through M_3 muscarinic receptor to activate the EGFR signaling and promotes gastric cancer cell proliferation. Sci Rep. 2017;7:40802.
275. Tolaymat M, Larabee SM, Hu S, et al. The role of M_3 muscarinic receptor ligand-induced kinase signaling in colon cancer progression. Cancers (Basel). 2019;11(3):308.
276. Zhao Q, Yue J, Zhang C, et al. Inactivation of M_2 AChR/NF-κB signaling axis reverses epithelial-mesenchymal transition (EMT) and suppresses migration and invasion in non-small cell lung cancer (NSCLC). Oncotarget. 2015;6:29335–46.
277. Wang J, Krysiak PS, Laurier LG, et al. Human esophageal smooth muscle cells express muscarinic receptor subtypes M_1 through M_5. Am J Physiol Gastrointest Liver Physiol. 2000;279:G1059–69.
278. Zhang L, Wu LL, Huan HB, et al. Sympathetic and parasympathetic innervation in hepatocellular carcinoma. Neoplasma. 2017;64:840–6.
279. Fiszman GL, Middonno MC, de la Torre E, et al. Activation of muscarinic cholinergic receptors induces MCF-7 cells proliferation and angiogenesis by stimulating nitric oxide synthase activity. Cancer Biol Ther. 2007;6:1106–13.
280. Russell SJ, Peng K-W, Bell JC. Oncolytic virotherapy. Nat Biotechnol. 2012;30:658–70.
281. Martin NT, Bell JC. Oncolytic virus combination therapy: Killing one bird with two stones. Mol Ther. 2018;26(6):1414–22.
282. Bommareddy PK, Shettigar M, Kaufman HL. Integrating oncolytic viruses in combination cancer immunotherapy. Nat Rev Immunol. 2018;18(8):498–513.
283. Bommareddy PK, Patel A, Hossain S, et al. Talimogene laherparepvec (T-VEC) and other oncolytic viruses for the treatment of melanoma. Am J Clin Dermatol. 2017;18(1):1–15.
284. Simpson GR, Relph K, Harrington K, et al. Cancer immunotherapy via combining oncolytic virotherapy with chemotherapy: recent advances. Oncolytic Virother. 2016;5:1–13.
285. Binz E, Berchtold S, Beil J, et al. Chemovirotherapy of pancreatic adenocarcinoma by combining oncolytic vaccinia virus GLV-1h68 with *nab*-paclitaxel plus gemcitabine. Mol Ther Oncolytics. 2017;6:10–21.
286. Tanaka R, Goshima F, Esaki S, et al. The efficacy of combination therapy with oncolytic herpes simplex virus HF10 and dacarbazine in a mouse melanoma model. Am. J. Cancer Res. 2017;7:1693–703.
287. Bourgeois-Daigneault M-C, St-Germain LE, Roy DG, et al. Combination of Paclitaxel and MG1 oncolytic virus as a successful strategy for breast cancer treatment. Breast Cancer Res. 2016;18:83.
288. Bourgeois-Daigneault MC, St-Germain LE, et al. Combination of Paclitaxel and MG1 oncolytic virus as a successful strategy for breast cancer treatment. Breast Cancer Res. 2016;18(1):83.
289. Dornan MH, Krishnan R, Macklin AM, et al. First-in-class small molecule potentiators of cancer virotherapy. Sci. Rep. 2016;6:26786.
290. Selman M, Rousso C, Bergeron A, et al. Multimodal potentiation of oncolytic virotherapy by vanadium compounds. Mol Ther. 2018;26:56–69.
291. Diallo JS, Le Boeuf F, Lai F, et al. A high-throughput pharmacoviral approach identifies novel oncolytic virus sensitizers. Mol Ther. 2010;18:1123–9.
292. Selman M, Ou P, Rousso C, et al. Dimethyl fumarate potentiates oncolytic virotherapy through NF-κB inhibition. Sci Transl Med. 2018;10 eaao1613
293. Arulanandam R, Batenchuk C, Varette O, et al. Microtubule disruption synergizes with oncolytic virotherapy by inhibiting interferon translation and potentiating bystander killing. Nat Commun. 2015;6:6410.
294. MacTavish H, Diallo J-S, Huang B, et al. Enhancement of vaccinia virus based oncolysis with histone deacetylase inhibitors. PLoS ONE. 2010;5:e14462.
295. Nguyên TL-A, Abdelbary H, Arguello M, et al. Chemical targeting of the innate antiviral response by histone deacetylase inhibitors renders refractory cancers sensitive to viral oncolysis. Proc Natl Acad Sci USA. 2008;105:14981–6.
296. Kim DS, Dastidar H, Zhang C, et al. Smac mimetics and oncolytic viruses synergize in driving anticancer T-cell responses through complementary mechanisms. Nat Commun. 2017;8:344.
297. Fulci G, Breymann L, Gianni D, et al. Cyclophosphamide enhances glioma virotherapy by inhibiting innate immune responses. Proc Natl Acad Sci USA. 2006;103:12873–8.
298. Currier MA, Gillespie RA, Sawtell NM, et al. Efficacy and safety of the oncolytic herpes simplex virus rRp450 alone and combined with cyclophosphamide. Mol Ther. 2008;16:879–85.
299. Liao Y, Wang H-X, Mao X, et al. RIP1 is a central signaling protein in regulation of TNF-α/TRAIL mediated apoptosis and necroptosis during Newcastle disease virus infection. Oncotarget. 2017;8:43201–17.
300. Han ZQ, Assenberg M, Liu BL, et al. Development of a second-generation oncolytic Herpes simplex virus expressing TNFalpha for cancer therapy. J Gene Med. 2007;9:99–106.
301. Meisen WH, Wohleb ES, Jaime-Ramirez AC, et al. The impact of macrophage- and microglia-secreted TNFα on oncolytic HSV-1 therapy in the glioblastoma tumor microenvironment. Clin Cancer Res. 2015;21:3274–85.
302. Vähä-Koskela M, Hinkkanen A. Tumor restrictions to oncolytic virus. Biomedicines. 2014;2:163–94.

303. Simpson GR, Han Z, Liu B, et al. Combination of a fusogenic glycoprotein, prodrug activation, and oncolytic herpes simplex virus for enhanced local tumor control. Cancer Res. 2006;66:4835–42.
304. Fu X, Tao L, Jin A, et al. Expression of a fusogenic membrane glycoprotein by an oncolytic herpes simplex virus potentiates the viral antitumor effect. Mol Ther. 2003;7:748–54.
305. Ayala-Breton C, Russell LOJ, Russell SJ, et al. Faster replication and higher expression levels of viral glycoproteins give the vesicular stomatitis virus/measles virus hybrid VSV-FH a growth advantage over measles virus. J Virol. 2014;88:8332–9.
306. Kim JH, Lee YS, Kim H, et al. Relaxin expression from tumor-targeting adenoviruses and its intratumoral spread, apoptosis induction, and efficacy. J Natl Cancer Inst. 2006;98:1482–93.
307. Rodríguez-García A, Giménez-Alejandre M, Rojas JJ, et al. Safety and efficacy of VCN-01, an oncolytic adenovirus combining fiber HSG-binding domain replacement with RGD and hyaluronidase expression. Clin Cancer Res. 2015;21:1406–18.
308. Martínez-Vélez N, Xipell E, Vera B, et al. The oncolytic adenovirus VCN-01 as therapeutic approach against pediatric osteosarcoma. Clin Cancer Res. 2016;22:2217–25.
309. Vera B, Martínez-Vélez N, Xipell E, et al. Characterization of the antiglioma effect of the oncolytic adenovirus VCN-01. PLoS ONE. 2016;11:e0147211.
310. Tan G, Kasuya H, Sahin TT, et al. Combination therapy of oncolytic herpes simplex virus HF10 and bevacizumab against experimental model of human breast carcinoma xenograft. Int J Cancer. 2015;136:1718–30.
311. Buckel L, Advani SJ, Frentzen A, et al. Combination of fractionated irradiation with anti-VEGF expressing vaccinia virus therapy enhances tumor control by simultaneous radiosensitization of tumor associated endothelium. Int J Cancer. 2013;133:2989–99.
312. Zhang W, Fulci G, Buhrman JS, et al. Bevacizumab with angiostatin-armed oHSV increases antiangiogenesis and decreases bevacizumab-induced invasion in U87 glioma. Mol Ther. 2012;20:37–45.
313. Currier MA, Eshun FK, Sholl A, et al. VEGF blockade enables oncolytic cancer virotherapy in part by modulating intratumoral myeloid cells. Mol Ther. 2013;21:1014–23.
314. Bolyard C, Yoo JY, Wang P-Y, et al. Doxorubicin synergizes with 34.5ENVE to enhance antitumor efficacy against metastatic ovarian cancer. Clin Cancer Res. 2014;20:6479–94.
315. Seabloom DE, Galbraith AR, Haynes AM, et al. Fixed-dose combinations of pioglitazone and metformin for lung cancer prevention. Cancer Prev Res (Phila). 2017;10(2):116–23.
316. Memmott RM, Mercado JR, Maier CR, et al. Metformin prevents tobacco carcinogen--induced lung tumorigenesis. Cancer Prev Res (Phila) 2010; 3(9): 1066–1076.
317. Fu H, Zhang J, Pan J, et al. Chemoprevention of lung carcinogenesis by the combination of aerosolized budesonide and oral pioglitazone in A/J mice. Mol Carcinog. 2011;50(12):913–21.
318. Ihle NT, Lemos R, Schwartz D, et al. Peroxisome proliferator-activated receptor gamma agonist pioglitazone prevents the hyperglycemia caused by phosphatidylinositol 3-kinase pathway inhibition by PX-866 without affecting antitumor activity. Mol Cancer Ther. 2009;8(1):94–100.
319. Nemenoff RA, Weiser-Evans M, Winn RA. Activation and Molecular Targets of Peroxisome Proliferator-Activated Receptor-gamma Ligands in Lung Cancer. PPAR Res. 2008;2008:156875.
320. Wang Y, James M, Wen W, et al. Chemopreventive effects of pioglitazone on chemically induced lung carcinogenesis in mice. Mol Cancer Ther. 2010;9(11):3074–82.
321. Ondrey F. Peroxisome proliferator-activated receptor gamma pathway targeting in carcinogenesis: implications for chemoprevention. Clin Cancer Res. 2009;15(1):2–8.
322. Harris G, Ghazallah RA, Nascene D, et al. PPAR activation and decreased proliferation in oral carcinoma cells with 4-HPR. Otolaryngol Head Neck Surg. 2005;133(5):695–701.
323. Chang TH, Szabo E. Induction of differentiation and apoptosis by ligands of peroxisome proliferator-activated receptor gamma in non-small cell lung cancer. Cancer Res. 2000;60(4):1129–38.
324. Chang TH, Szabo E. Enhanced growth inhibition by combination differentiation therapy with ligands of peroxisome proliferator-activated receptor-gamma and inhibitors of histone deacetylase in adenocarcinoma of the lung. Clin Cancer Res. 2002;8(4):1206–12.
325. Avis I, Martínez A, Tauler J, et al. Inhibitors of the arachidonic acid pathway and peroxisome proliferator-activated receptor ligands have super-additive effects on lung cancer growth inhibition. Cancer Res. 2005;65(10):4181–90.
326. Wright SK, Wuertz BR, Harris G, et al. Functional activation of PPARγ in human upper aerodigestive cancer cell lines. Mol Carcinog. 2017;56(1):149–62.
327. Han EJ, Im CN, Park SH, et al. Combined treatment with peroxisome proliferator-activated receptor (PPAR) gamma ligands and gamma radiation induces apoptosis by PPARγ-independent up-regulation of reactive oxygen species-induced deoxyribonucleic acid damage signals in non-small cell lung cancer cells. Int J Radiat Oncol Biol Phys. 2013;85(5):e239–48.
328. Bren-Mattison Y, Meyer AM, Van Putten V, et al. Antitumorigenic effects of peroxisome proliferator-activated receptor-gamma in non-small-cell lung cancer cells are mediated by suppression of cyclooxygenase-2 via inhibition of nuclear factor-kappaB. Mol Pharmacol. 2008;73(3):709–17.
329. Hazra S, Batra RK, Tai HH, et al. Pioglitazone and rosiglitazone decrease prostaglandin E2 in

non-small-cell lung cancer cells by up-regulating 15-hydroxyprostaglandin dehydrogenase. Mol Pharmacol. 2007;71(6):1715–20.
330. Han S, Sidell N, Fisher PB, et al. Up-regulation of p21 gene expression by peroxisome proliferator-activated receptor gamma in human lung carcinoma cells. Clin Cancer Res. 2004;10(6):1911–9.
331. Govindarajan R, Ratnasinghe L, Simmons DL, et al. Thiazolidinediones and the risk of lung, prostate, and colon cancer in patients with diabetes. J Clin Oncol. 2007;25(12):1476–81.
332. Govindarajan R, Siegel ER, Simmons DL, et al. Thiazolidinedione (TZD) exposure and risk of squamous cell carcinoma of head and neck (SCCHN). J Clin Oncol 2007 ASCO Ann Meet eedings Part I. 2007; 25(Suppl18): 1511–2007.
333. Decensi A, Puntoni M, Goodwin P, et al. Metformin and cancer risk in diabetic patients: a systematic review and meta-analysis. Cancer Prev Res. 2010;3(11):1451–61.
334. Franciosi M, Lucisano G, Lapice E, et al. Metformin therapy and risk of cancer in patients with type 2 diabetes: systematic review. PLoS One. 2013;8(8):e71583.
335. Goodwin PJ, Stambolic V. Obesity and insulin resistance in breast cancer--chemoprevention strategies with a focus on metformin. Breast. 2011; 20(Suppl 3): S31–S35.
336. He X, Esteva FJ, Ensor J, et al. Metformin and thiazolidinediones are associated with improved breast cancer-specific survival of diabetic women with HER2+ breast cancer. Ann Oncol. 2012;23(7):1771–80.
337. Lai SW, Liao KF, Chen PC, et al. Antidiabetes drugs correlate with decreased risk of lung cancer: a population-based observation in Taiwan. Clin Lung Cancer. 2012;13(2):143–8.
338. Rozengurt E, Sinnett-Smith J, Kisfalvi K. Crosstalk between insulin/insulin-like growth factor-1 receptors and G protein-coupled receptor signaling systems: a novel target for the antidiabetic drug metformin in pancreatic cancer. Clin Cancer Res. 2010;16(9):2505–11.
339. Tan BK, Adya R, Chen J, et al. Metformin treatment exerts antiinvasive and antimetastatic effects in human endometrial carcinoma cells. J Clin Endocrinol Metab. 2011;96(3):808–16.
340. Vaccaro O, Masulli M, Bonora E, et al. Addition of either pioglitazone or a sulfonylurea in type 2 diabetic patients inadequately controlled with metformin alone: impact on cardiovascular events. A randomized controlled trial. Nutr Metab Cardiovasc Dis. 2012;22(11):997–1006.
341. Xie Y, Wang YL, Yu L, et al. Metformin promotes progesterone receptor expression via inhibition of mammalian target of rapamycin (mTOR) in endometrial cancer cells. J Steroid Biochem Mol Biol. 2011;126(3–5):113–20.
342. Zhang ZJ, Bi Y, Li S, et al. Reduced risk of lung cancer with metformin therapy in diabetic patients: a systematic review and meta-analysis. Am J Epidemiol. 2014;180(1):11–4.
343. Cacicedo JM, Yagihashi N, Keaney JF Jr, et al. AMPK inhibits fatty acid-induced increases in NF-kappaB transactivation in cultured human umbilical vein endothelial cells. Biochem Biophys Res Commun. 2004;324(4):1204–9.
344. Hattori Y, Suzuki K, Hattori S, et al. Metformin inhibits cytokine-induced nuclear factor kappaB activation via AMP-activated protein kinase activation in vascular endothelial cells. Hypertension. 2006;47(6):1183–8.
345. Huang NL, Chiang SH, Hsueh CH, et al. Metformin inhibits TNF-alpha-induced IkappaB kinase phosphorylation, IkappaB-alpha degradation and IL-6 production in endothelial cells through PI3K-dependent AMPK phosphorylation. Int J Cardiol. 2009;134(2):169–75.
346. Isoda K, Young JL, Zirlik A, et al. Metformin inhibits proinflammatory responses and nuclear factor-kappaB in human vascular wall cells. Arterioscler Thromb Vasc Biol. 2006;26(3):611–7.
347. Li SN, Wang X, Zeng QT, et al. Metformin inhibits nuclear factor kappaB activation and decreases serum high-sensitivity C-reactive protein level in experimental atherogenesis of rabbits. Heart Vessels. 2009;24(6):446–53.
348. Tan BK, Adya R, Chen J, et al. Metformin decreases angiogenesis via NF-kappaB and Erk1/2/Erk5 pathways by increasing the antiangiogenic thrombospondin-1. Cardiovasc Res. 2009;83(3):566–74.
349. Felip E, Ranson M, Cedrés S, et al. A phase Ib, dose-finding study of erlotinib in combination with a fixed dose of pertuzumab in patients with advanced non-small-cell lung cancer. Clin Lung Cancer. 2012;13(6):432–41.
350. Rino Y, Takanashi Y, Yukawa N, et al. A phase I study of bi-weekly combination therapy with S-1 and docetaxel for advanced or recurrent gastric cancer. Anticancer Res. 2006;26(2B):1455–62.
351. Nakayama N, Koizumi W, Sasaki T, et al. A multicenter, phase I dose-escalating study of docetaxel, cisplatin and S-1 for advanced gastric cancer (KDOG0601). Oncology. 2008;75(1-2):1–7.
352. Carrick S, Parker S, Wilcken N, et al. Single agent versus combination chemotherapy for metastatic breast cancer. Cochrane Database Syst Rev. 2005;(2):CD003372.
353. Tan AR, Im SA, Mattar A, et al. FeDeriCa study group. Fixed-dose combination of pertuzumab and trastuzumab for subcutaneous injection plus chemotherapy in HER2-positive early breast cancer (FeDeriCa): a randomised, open-label, multicentre, non-inferiority, phase 3 study. Lancet Oncol. 2021;22(1):85–97.
354. DuMond B, Patel V, Gross A, et al. Fixed-dose combination of pertuzumab and trastuzumab for subcutaneous injection in patients with HER2-positive breast cancer: A multidisciplinary approach. J Oncol Pharm Pract. 2021;27(5):1214–21.

355. Petrelli A, Giordano S. From single- to multi-target drugs in cancer therapy: when aspecificity becomes an advantage. Curr Med Chem. 2008;15(5):422–32.
356. Thomson RJ, Moshirfar M, Ronquillo Y. Tyrosine Kinase Inhibitors. In: StatPearls [Internet]. Treasure Island (FL): StatPearls Publishing; 2021.
357. Jiao Q, Bi L, Ren Y, et al. Advances in studies of tyrosine kinase inhibitors and their acquired resistance. Mol Cancer. 2018;17(1):36.
358. Patterson KI, Brummer T, O'Brien PM, et al. Dual-specificity phosphatases: critical regulators with diverse cellular targets. Biochem J. 2009;418(3):475–89.
359. Kolibaba KS, Druker BJ. Protein tyrosine kinases and cancer. Biochim Biophys Acta. 1997;1333(3):F217–48.
360. Tong M, Seeliger MA. Targeting conformational plasticity of protein kinases. ACS Chem Biol. 2015;10(1):190–200.
361. Norman RA, Toader D, Ferguson AD. Structural approaches to obtain kinase selectivity. Trends Pharmacol Sci. 2012;33(5):273–8.
362. Wu P, Nielsen TE, Clausen MH. FDA-approved small-molecule kinase inhibitors. Trends Pharmacol Sci. 2015;36(7):422–39.
363. Gentile C, Martorana A, Lauria A, et al. Kinase inhibitors in multitargeted cancer therapy. Curr Med Chem. 2017;24(16):1671–86.
364. Scagliotti G, Govindan R. Targeting angiogenesis with multitargeted tyrosine kinase inhibitors in the treatment of non-small cell lung cancer. Oncologist. 2010;15(5):436–46.
365. Wedge SR, Ogilvie DJ, Dukes M, et al. ZD6474 inhibits vascular endothelial growth factor signaling, angiogenesis, and tumor growth following oral administration. Cancer Res. 2002;62(16):4645–55.
366. Mendel DB, Douglas Laird A, Xin X, et al. In vivo antitumor activity of SU11248, a novel tyrosine kinase inhibitor targeting vascular endothelial growth factor and platelet-derived growth factor receptors: determination of a pharmacokinetic/pharmacodynamic relationship. Clin Cancer Res. 2003;9(1):327–37.
367. Novello S, Scagliotti GV, Rosell R, et al. Phase II study of continuous daily sunitinib dosing in patients with previously treated advanced non-small cell lung cancer. Br J Cancer. 2009;101(9):1543–8.
368. Hu-Lowe DD, Zou HY, Grazzini ML, et al. Nonclinical antiangiogenesis and antitumor activities of axitinib (AG-013736), an oral, potent, and selective inhibitor of vascular endothelial growth factor receptor tyrosine kinases 1, 2, 3. Clin Cancer Res. 2008;14:7272–83.
369. Schiller JH, Larson T, Ou SH, et al. Efficacy and safety of axitinib in patients with advanced non-small-cell lung cancer: Results from a phase II study. J Clin Oncol. 2009;27:3836–41.
370. Escudier B, Eisen T, Stadler WM, et al. Sorafenib in advanced clear-cell renal-cell carcinoma. N Engl J Med. 2007;356:125–34.
371. Gauler TC, Besse B, Gounant V, et al. Phase II open-label study to investigate efficacy and safety of PTK787/ZK 222584 (PTK/ZK) orally administered once daily or twice daily at 1,250 mg as second-line monotherapy in patients (pts) with stage IIIB/IV non-small cell lung cancer (NSCLC). J Clin Oncol. 2007; 25(18 suppl): Abstract 7541
372. Wedge SR, Kendrew J, Hennequin LF, et al. AZD2171: A highly potent, orally bioavailable, vascular endothelial growth factor receptor-2 tyrosine kinase inhibitor for the treatment of cancer. Cancer Res. 2005;65:4389–400.
373. Raghav KPS, Blumenschein GR. Motesanib and advanced NSCLC: experiences and expectations. Expert Opin Investig Drugs. 2011;20(6):859–69.
374. Coxon A, Ziegler B, Kaufman S, et al. Antitumor activity of motesanib alone and in combination with cisplatin or docetaxel in multiple human non-small-cell lung cancer xenograft models. Mol Cancer. 2012;11:p70.
375. Kumar R, Knick VB, Rudolph SK, et al. Pharmacokinetic-pharmacodynamic correlation from mouse to human with pazopanib, a multikinase angiogenesis inhibitor with potent antitumor and antiangiogenic activity. Mol Cancer Therapeut. 2007;6(7):2012–21.
376. Altorki N, Guarino M, Lee P, et al. Preoperative treatment with pazopanib (GW786034), a multikinase angiogenesis inhibitor in early-stage non-small cell lung cancer (NSCLC): A proof-of-concept phase II study. J Clin Oncol. 2008;26(15 suppl) Abstract 7557
377. Plummer R, Madi A, Jeffels M, et al. A Phase I study of pazopanib in combination with gemcitabine in patients with advanced solid tumors. Cancer Chemother Pharmacol. 2013;71(1):93–101.
378. Albert DH, Tapang P, Magoc TJ, et al. Preclinical activity of ABT-869, a multitargeted receptor tyrosine kinase inhibitor. Mol Cancer Therapeutics. 2006;5(4):995–1006.
379. Asahina H, Tamura Y, Nokihara H, et al. An open-label, phase 1 study evaluating safety, tolerability, and pharmacokinetics of linifanib (ABT-869) in Japanese patients with solid tumors. Cancer Chemother Pharmacol. 2012;69(6):1477–86.
380. Yu HA, Riely GJ. Second-generation epidermal growth factor receptor tyrosine kinase inhibitors in lung cancers. J Nat Compreh Cancer Network. 2013;11(2):161–9.
381. Pietanza MC, Gadgeel SM, Dowlati A, et al. Phase II study of the multitargeted tyrosine kinase inhibitor XL647 in patients with non-small-cell lung cancer. J Thoracic Oncol. 2012;7(5):856–65.
382. Gendreau SB, Ventura R, Keast P, et al. Inhibition of the T790M gatekeeper mutant of the epidermal growth factor receptor by EXEL-7647. Clin Cancer Res. 2007;13(12):3713–23.
383. Zhao Y, Zhang YN, Wang KT, et al. Lenvatinib for hepatocellular carcinoma: From preclinical mecha-

nisms to anti-cancer therapy. Biochim Biophys Acta Rev Cancer. 2020;1874(1):188391.
384. Matsui J, Funahashi Y, Uenaka T, et al. Multi-kinase inhibitor E7080 suppresses lymph node and lung metastases of human mammary breast tumor MDA-MB-231 via inhibition of vascular endothelial growth factor-receptor (VEGF-R) 2 and VEGF-R3 kinase. Clin Cancer Res. 2008;14(17):5459–65.
385. Tohyama O, Matsui J, Kodama K, et al. Antitumor activity of lenvatinib (e7080): an angiogenesis inhibitor that targets multiple receptor tyrosine kinases in preclinical human thyroid cancer models. J Thyroid Res. 2014;2014:638747.
386. Ichikawa K, Miyano SW, Adachi Y, et al. Abstract 1374: Lenvatinib, tri-specific targeted therapy to VEGFR/FGFR/RET, suppresses angiogenesis through the inhibition of both VEGFR and FGFR signaling pathways. Cancer Res. 2015;75:1374.
387. Matsuki M, Hoshi T, Yamamoto Y, et al. Lenvatinib inhibits angiogenesis and tumor fibroblast growth factor signaling pathways in human hepatocellular carcinoma models. Cancer Med. 2018;7:2641–53.
388. Kudo M, Finn RS, Qin S, et al. Lenvatinib versus sorafenib in first-line treatment of patients with unresectable hepatocellular carcinoma: a randomised phase 3 non-inferiority trial. Lancet. 2018;391:1163–73.
389. Ikeda K, Kudo K, Kawazoe S, et al. Phase 2 study of lenvatinib in patients with advanced hepatocellular carcinoma. J Gastroenterol. 2017;52:512–9.
390. Bass AKA, El-Zoghbi MS, Nageeb EM, et al. Comprehensive review for anticancer hybridized multitargeting HDAC inhibitors. Eur J Med Chem. 2021;209:112904.
391. Martínez R, Geronimo BD, Pastor M, et al. Multitarget anticancer agents based on histone deacetylase and protein kinase CK2 inhibitors. Molecules. 2020;25(7):1497.
392. Witt O, Deubzer HE, Milde T, et al. HDAC family: What are the cancer relevant targets? Cancer Lett. 2009;277:8–21.
393. Marks PA, Rifkind RA, Richon VM, et al. Histone deacetylases and cancer: Causes and therapies. Nat Rev Cancer. 2001;1:194–202.
394. Gräff J, Tsai L-H. The potential of HDAC inhibitors as cognitive enhancers. Annu Rev Pharmacol. 2013;53:311–30.
395. Andrews KT, Haque A, Jones MK. HDAC inhibitors in parasitic diseases. Immunol Cell Biol. 2012;90:66–77.
396. Rotili D, Simonetti G, Savarino A, et al. Non-Cancer Uses of Histone Deacetylase Inhibitors: Effects on Infectious Diseases and β-Hemoglobinopathies+. Curr Top Med Chem. 2009;9:272–91.
397. Gluckman PD, Hanson MA, Buklijas T, et al. Epigenetic mechanisms that underpin metabolic and cardiovascular diseases. Nat Rev Endocrinol. 2009;5:401–8.
398. Gregoretti I, Lee Y-M, Goodson HV. Molecular evolution of the histone deacetylase family: functional implications of phylogenetic analysis. J Mol Biol. 2004;338:17–31.
399. Haberland M, Montgomery RL, Olson EN. The many roles of histone deacetylases in development and physiology: Implications for disease and therapy. Nat Rev Genet. 2009;10:32–42.
400. Lehrmann H, Pritchard LL, Harel-Bellan A. Histone acetyltransferases and deacetylases in the control of cell proliferation and differentiation. Adv Cancer Res. 2002;86:41–65.
401. Lane AA, Chabner BA. Histone Deacetylase Inhibitors in Cancer Therapy. J Clin Oncol. 2009;27:5459–68.
402. Minucci S, Pelicci PG. Histone deacetylase inhibitors and the promise of epigenetic (and more) treatments for cancer. Nat. Rev. Cancer. 2006;6:38–51.
403. Pontiki E, Hadjipavlou-Litina D. Histone deacetylase inhibitors (HDACIs). Structure—Activity relationships: History and new QSAR perspectives. Med Res Rev. 2012;32:1–165.
404. Paris M, Porcelloni M, Binaschi M, et al. Histone deacetylase inhibitors: From bench to clinic. J. Med. Chem. 2008;51:1505–29.
405. Schobert R, Biersack B. Multimodal HDAC inhibitors with improved anticancer activity. Curr Cancer Drug Targets. 2018;18:39–56.
406. Seo SY. Multi-targeted hybrids based on HDAC inhibitors for anti-cancer drug discovery. Arch Pharm Res. 2012;35(2):197–200.
407. Ling Y, Liu J, Qian J, et al. Recent advances in multi-target drugs targeting protein kinases and histone deacetylases in cancer therapy. Curr Med Chem. 2020;27:1–24.
408. Chu-Farseeva YY, Mustafa N, Poulsen A, et al. Design and synthesis of potent dual inhibitors of JAK2 and HDAC based on fusing the pharmacophores of XL019 and vorinostat. Eur J Med Chem. 2018;158:593–619.
409. Luan Y, Li J, Bernatchez JA, et al. Kinase and histone deacetylase hybrid inhibitors for cancer therapy. J Med Chem. 2018;62:3171–83.
410. Cai X, Zhai H-X, Wang J, et al. Discovery of 7-(4-(3-Ethynylphenylamino)-7-methoxyquinazolin-6-yloxy)-N-hydroxyheptanamide (CUDC-101) as a potent multi-acting HDAC, EGFR, and HER2 inhibitor for the treatment of cancer. J Med Chem. 2010;53:2000–9.
411. Mahboobi S, Dove S, Sellmer A, et al. Design of chimeric histone deacetylase- and tyrosine kinase-inhibitors: A series of imatinib hybrides as potent inhibitors of wild-type and mutant BCR-ABL, PDGF-Rβ, and histone deacetylases. J Med Chem. 2009;52:2265–79.
412. Khan DH, He S, Yu J, et al. Protein kinase CK2 regulates the dimerization of histone deacetylase 1 (HDAC1) and HDAC2 during mitosis. J Biol Chem. 2013;288:16518–28.
413. Pluemsampant S, Safronova OS, Nakahama K-I, et al. Protein kinase CK2 is a key activator of histone deacetylase in hypoxia-associated tumors. Int J Cancer. 2008;122:333–41.

4 Polypharmacology in Clinical Applications: Cardiovascular Polypharmacology

Abstract

Cardiovascular disease (CVD) is a class of disorders of the heart and blood vessels and the leading cause of death worldwide. Yet, treatment of CVDs is the major challenge in medicine because they are mostly multifactorial and multi-mechanism conditions that require simultaneous targeting on multitargets to produce optimal results. Even monogenic CVDs are influenced by multiple cellular regulators and environmental factors. Because of these traits of CVDs, it has recently come to a common consensus that combination therapies overall produce superior outcomes for many types of CVDs compared to monotherapy. In this chapter, we will introduce the polypharmacology strategy, particularly free-dose drug combination (CDT), fixed-dose single-pill drug combination (FDC), and multitarget drug, for the treatment of some most common CVDs including hypertension, hyperlipidemia, coronary artery disease (CAD), heart failure (HF), cardiac arrhythmias, and stroke. At present, while CDT and FDC have been utilized in nearly all major type of CVDs, the application of MTD is primarily restricted to hypertension and arrhythmias.

Keywords

Cardiovascular polypharmacology · Hypertension · Coronary artery disease (CAD) · Heart failure (HF) · Cardiac arrhythmias

4.1 CDT in the Treatment of CVD

Cardiovascular disease (CVD) is a class of disorders of the heart and blood vessels and the leading cause of death worldwide except Africa where the contribution of noncommunicable diseases led by CVD has significantly increased in recent years [1–5]. In 2019, approximately 18.6 million deaths were attributed to CVD globally, and the highest mortality rates attributable to CVD in 2019 were in Eastern Europe and Central Asia. CVD includes coronary artery diseases (CAD) such as angina and myocardial infarction (commonly known as a heart attack), heart failure (HF), hypertensive heart disease, rheumatic heart disease, cardiomyopathy, abnormal heart rhythms (arrhythmias), congenital heart disease, valvular heart disease, carditis, aortic aneurysms, cerebrovascular disease (e.g., stroke), peripheral artery disease, thromboembolic disease, and venous thrombosis. Globally, it was estimated that in 2019, 197.2 million people were living with CAD, and the mortality rates of CAD were 118.0 per 100,000. Global prevalence of stroke in 2019 was 101.5 million people, whereas that of ischemic stroke was 77.2 million, that of intracerebral hemorrhage was 20.7 million, and that of subarachnoid hemorrhage was 8.4 million, and a total of 3.3 million individuals died of ischemic

Z. Wang, B. Yang, *Polypharmacology*, https://doi.org/10.1007/978-3-031-04998-9_4

stroke. Between 1990 and 2015, the number of deaths related to systolic blood pressure (SBP) ≥140 mm Hg was around two million in high-income countries, high-middle-income countries, and middle-income countries. Based on 2015 data, there were 3.47 billion adults worldwide with SBP of 110–115 mm Hg or higher, of whom 874 million had SBP ≥140 mm Hg. Hypertension is one of the most important risk factors for CAD, stroke, other CVDs, chronic kidney disease, and dementia. Elevated BP is a leading preventable cause of CVD mortality and disease burden, globally and in most regions of the world [5, 6]. The underlying mechanisms for CVD vary depending on the disease. CAD, stroke, and peripheral artery disease involve atherosclerosis [4]. This may be caused by hypertension, smoking, diabetes mellitus, lack of exercise, obesity, high blood cholesterol, poor diet, excessive alcohol consumption, and poor sleep, among others. Hypercholesterolemia and hypertension constitute major physiological risk factors of CAD. Tremendous advances have been made in the use of drugs to treat many heart conditions. These drugs can help prevent certain heart conditions from getting worse, prolong life, and reduce the effect of symptoms on the ability to perform daily activities. Drug therapy can lower BP, lower cholesterol levels, prevent the blood from creating life-threatening clots, eliminate chest pain, regulate the heartbeat, and suppress arrhythmias.

Recently, it has come to a common consensus that combination therapies of CVDs, either in the form of fixed-dose combinations (FDCs) [7, 8] or free-dose combinations (CDTs) [9, 10], produce superior outcomes for many types of CVDs compared to monotherapy. This is because CVDs are multifactorial and multi-mechanism conditions that require simultaneous targeting on multitargets to produce optimal results.

4.1.1 CDT in Hypertension

Hypertension or high blood pressure (BP) is recognized as a major risk factor for cardiovascular and renal diseases and represents the leading cause of mortality worldwide [11–16]. Among the risk factors for CVD, hypertension is associated with the strongest evidence for causation [5]. Large cohort studies have demonstrated that high BP is an important risk factor for heart failure, atrial fibrillation, chronic kidney disease, heart valve diseases, aortic syndromes, and dementia, in addition to coronary heart disease and stroke. In multivariate modeling, the presumed attributable risk of high BP for stroke and coronary heart disease has increased steadily with progressive use of lower values for normal BP. Meta-analysis of BP-lowering randomized controlled trials has demonstrated a benefit which is almost identical to that predicted from BP risk relationships in cohort studies. Prevention of age-related increases in BP would, in large part, reduce the vascular consequences usually attributed to aging and, together with intensive treatment of established hypertension, would eliminate a large proportion of the population burden of BP-related CVD. However, in spite of proven benefits of hypertension treatment, BP control (BPC) rates are poor, even in economically well-developed countries with virtually full access to therapies. Nearly 75% of hypertensive patients do not achieve adequate control with monotherapy [6]. More than 70% of adults treated for primary hypertension will eventually require at least two antihypertensive agents, either initially as combination therapy or as add-on therapy if monotherapy and lifestyle modifications do not achieve adequate blood pressure control.

4.1.1.1 Antihypertensives

Antihypertensives are a class of drugs that are used to treat hypertension (high blood pressure), and they comprise several classes of compound with the therapeutic intention of preventing, controlling, or treating hypertension and of preventing the complications of high blood pressure, such as stroke and myocardial infarction [6]. Evidence suggests that reduction of the blood pressure by 5 mmHg can decrease the risk of stroke by 34% and of ischemic heart disease by 21% and reduce the likelihood of dementia, heart failure, and mortality from cardiovascular disease [5, 6]. The classes of antihypertensive drugs can be categorized into several classes by mechanism or site of action, which differ both structurally and functionally and act by different mechanisms. Among the most important and most widely used medications are thiazide diuretics, calcium channel blockers (CCBs), angiotensin-converting enzyme inhibitors (ACEIs), angiotensin II receptor blockers (ARBs), and adrenergic receptor antagonists (ARAs).

The fundamental goal of treatment should be the prevention of the important endpoints of hypertension, such as heart attack, stroke, and heart failure. Patient age, associated clinical conditions, and end-organ damage also play a part in determining dosage and type of medication administered [13–16]. Antihypertensive drugs are also used frequently in other unrelated conditions, for example, beta-blockers (β-blockers) in thyrotoxicosis and anxiety or ACEIs in HF.

Diuretics

Diuretics help the kidneys eliminate excess salt and water from the body's tissues and blood to achieve BPC. They are often used in combination with additional prescription therapies. This class of antihypertensives includes the following subclasses and compounds.

Loop diuretics: bumetanide, ethacrynic acid, furosemide, torsemide

Thiazide diuretics: epitizide, hydrochlorothiazide and chlorothiazide, bendroflumethiazide, methyclothiazide, polythiazide

Thiazide-like diuretics: indapamide, chlorthalidone, metolazone, xipamide, clopamide

Potassium-sparing diuretics: amiloride, triamterene, spironolactone, eplerenone

Combination diuretics: amiloride hydrochloride/hydrochlorothiazide, spironolactone/hydrochlorothiazide, triamterene/hydrochlorothiazide.

In the United States, the JNC-8 (Eighth Joint National Committee on the Prevention, Detection, Evaluation, and Treatment of High Blood Pressure) recommends thiazide-type diuretics to be one of the first-line drug treatments for hypertension, either as monotherapy or in combination with calcium channel blockers, ACE inhibitors, or angiotensin II receptor antagonists [16]. There are fixed-dose combination drugs, such as ACE inhibitor and thiazide combinations. Despite thiazides being cheap and effective, they are not prescribed as often as some newer drugs. This is because they have been associated with increased risk of new-onset diabetes and as such are recommended for use in patients over 65 where the risk of new-onset diabetes is outweighed by the benefits of controlling systolic blood pressure [17]. Another theory is that they are off-patent and thus rarely promoted by the drug industry [18].

Calcium Channel Blockers (CCBs)

CCBs prevent calcium from entering the smooth muscle cells of the heart and arteries. When calcium enters these cells, it causes a stronger and harder contraction, so by decreasing the calcium, the hearts' contraction is not as forceful. CCBs relax and open up narrowed blood vessels, reduce heart rate, and lower blood pressure. They include the following agents.

Dihydropyridines: amlodipine, barnidipine, bepridil, cilnidipine, clevidipine, felodipine, isradipine, lercanidipine, levamlodipine, nicardipine, nifedipine, nimodipine, nisoldipine, nitrendipine

Non-dihydropyridines: diltiazem, verapamil

The eighth Joint National Committee (JNC-8) recommends CCBs to be a first-line treatment either as monotherapy or in combination with thiazide-type diuretics, ACEIs, or ARBs for all patients regardless of age or race [19].

Angiotensin-Converting Enzyme Inhibitors (ACEIs)

ACEIs inhibit the activity of angiotensin-converting enzyme, an enzyme responsible for the conversion of angiotensin I into angiotensin II, a potent vasoconstrictor. Angiotensin is a chemical that causes the arteries to become narrow, especially in the kidneys but also throughout the body. ACEIs help the body produce less angiotensin, which helps the blood vessels relax and open up, which, in turn, lowers BP.

ACEIs include benazepril, captopril, enalapril, fosinopril, lisinopril, moexipril, perindopril, quinapril, ramipril, and trandolapril.

A systematic review of 63 trials with over 35,000 participants indicated ACEIs significantly reduced doubling of serum creatinine levels compared to other drugs (ARBs, α-blockers, β-blockers, etc.), and the authors suggested this as a first line of defense [20]. The AASK trial showed that ACEIs are more effective at slowing down the decline of kidney function compared to CCBs and β-blockers [21]. As such, ACEIs should be the drug treatment of choice for patients with chronic kidney disease regardless of race or diabetic status [16]. However, ACEIs (and ARBs) should not be a first-line treatment for black hypertensives without chronic kidney disease [16]. Results from the Antihypertensive and Lipid-Lowering Treatment to Prevent Heart Attack Trial (ALLHAT) showed that thiazide-type diuretics and CCBs are both more effective as monotherapy in improving cardiovascular outcomes compared to ACEIs for this subgroup [22]. Furthermore, ACEIs are less effective in reducing blood pressure and had a 51% higher risk of stroke in black hypertensives when used as initial therapy compared to a CCB [23].

Angiotensin II Receptor Blockers (ARBs)

ARBs work by antagonizing the activation of angiotensin receptors. These drugs block the effects of angiotensin, a chemical that causes the arteries to become narrow. Angiotensin needs a receptor—like a chemical "slot" to fit into or bind with—in order to constrict the blood vessel. ARBs block the receptors to render a failure of angiotensin to constrict the blood vessel and so as to lower BP.

ARBs include azilsartan, candesartan, eprosartan, fimasartan, irbesartan, losartan, olmesartan, telmisartan, and valsartan.

ARBs happen to be the favorable alternative to ACEIs if the hypertensive patients with the heart failure type of reduced ejection fraction treated with ACEIs was intolerant of cough, angioedema other than hyperkalemia, or chronic kidney disease [24, 25].

Adrenergic Receptor Antagonists (ARAs)

ARAs include β-blockers and α-blockers. β-Blockers reduce the heart rate, the heart's workload, and the heart's output of blood to lower BP, and α-blockers reduce the arterial resistance, relaxing the muscle tone of the vascular walls to produce antihypertensive effect.

β-Blockers: acebutolol, atenolol, bisoprolol, betaxolol, carteolol, carvedilol, labetalol, metoprolol, nadolol, nebivolol, oxprenolol, penbutolol, pindolol, propranolol, sotalol, timolol

α-Blockers: doxazosin, phentolamine, indoramin, phenoxybenzamine, prazosin, terazosin, tolazoline, urapidil

Mixed α-blockers + β-blockers: bucindolol, carvedilol, labetalol

α_2-Receptor Agonists

These drugs reduce blood pressure by decreasing the activity of the sympathetic (adrenaline-producing) portion of the involuntary nervous system. Methyldopa is considered a first-line antihypertensive during pregnancy because adverse effects are infrequent for the pregnant woman or the developing fetus.

Central Agonists

Central agonists help decrease the blood vessels' ability to tense up or contract. The central agonists follow a different nerve pathway than the α- and β-blockers but accomplish the same goal of BP reduction. This subclass includes alpha methyldopa, clonidine, guanabenz, and guanfacine.

Peripheral Adrenergic Inhibitors

These medications reduce BP by blocking neurotransmitters in the brain. This blocks the smooth muscles from getting the "message" to constrict. These drugs include guanadrel, guanethidine, and reserpine, but they are rarely used unless other medications do not help.

Blood Vessel Dilators (Vasodilators)

Vasodilators include hydralazine, minoxidil, and nitroprusside. These drugs act directly on the smooth muscle of arteries to relax their walls so blood can move more easily through them. They are only used in hypertensive emergencies or when other antihypertensive drugs have failed and even so are rarely given alone. Sodium nitroprusside, a very potent, short-acting vasodilator, is most commonly used for the quick, temporary reduction of blood pressure in emergencies (such as malignant hypertension or aortic dissection) [26, 27]. Hydralazine and its derivatives are also used in the treatment of severe hypertension, although they should be avoided in emergencies [27]. These drugs are no longer indicated as first-line therapy for high blood pressure due to side effects and safety concerns, but hydralazine remains a drug of choice in gestational hypertension [26].

Endothelin Receptor Antagonists (ERAs)

ERA is another class of vasodilators for targeted therapy of pulmonary artery hypertension (PAH). Endothelin is made in the layer of cells that line the heart and blood vessels. It causes the blood vessels to constrict. In people with PAH, the body produces too much endothelin. This causes the blood vessels in the lungs to become narrow, increasing the blood pressure in the pulmonary arteries. Targeted therapies with ERA slow the progression of PAH and may even reverse some of the damage to the heart and lungs [28, 29]. Currently, three ERAs are available for clinical use: ambrisentan, bosentan, and macitentan.

4.1.1.2 Combination Therapy of Hypertension

Combination therapy of hypertension has been introduced more than 50 years ago, because of the marked BP elevation of recruited patients in the early randomized controlled trials of BP lowering. However, in all subsequent trials, combination treatment was per protocol anticipated irrespectively of the initial randomized treatment to ensure either a desirable BP lowering or a comparable level of BP reduction among arms. Beyond clinical trials, combination treatment is mainly used in the clinical practice to reinforce ongoing single-agent treatment to achieve hypertension control. Renin-angiotensin system inhibiting drugs are the cornerstone of combination treatment of hypertension because they have been repeatedly tested in clinical trials in combination with other agents either from the beginning or during the follow-up. Effective BP lowering following combination treatment depends on the activation of complementary pathophysiological pathways, but different agents can stimulate a common mode of action more effectively. The rate of adverse events following combination treatment may be reduced because effects of each agent are reciprocally counterbalanced.

Although β-blockers lower blood pressure, they do not have a positive benefit on endpoints as some other antihypertensives [30]. In particular, β-blockers are no longer recommended as first-line treatment due to relative adverse risk of stroke and new-onset of type 2 diabetes mellitus (T2DM) when compared to other medications [15], while certain specific β-blockers such as atenolol appear to be less useful in overall treatment of hypertension than several other agents [31]. A systematic review of 63 trials with over 35,000 participants indicated β-blockers increased the risk of mortality, compared to other antihypertensive therapies [20]. They do, however, have an important role in the prevention of heart attacks in people who have already had a heart attack [32].

Despite lowering BP, α-blockers have significantly poorer endpoint outcomes than other antihypertensives and are no longer recommended as a first-line choice in the treatment of hypertension [33]. However, they may be useful for some men with symptoms of prostate disease.

Currently appreciated strategies to improve BP control include the prompt shift from monotherapy to combination therapy [34, 35], the initial treatment with a two-drug combination, and the use of fixed-dose combinations in a single pill. Preferred combinations include a renin-angiotensin blocker, either an ACEI or an ARB plus a CCB or a diuretic. Some patients will also require a triple combination to achieve blood pressure control. Hypertension is a major, if not the most important, contributor to the disease burden and premature death globally which is largely related to cardiovascular disease. In both the primary and the secondary preventions of CVD, BP targets are often not achieved which is similar to achievement of cholesterol goals.

Combination pharmacotherapy with two or more drugs is required in order to reach the currently recommended blood pressure goals in the majority of hypertensive patients, particularly those with a goal of <130/80 mm Hg [36–41]. Further to the potentiation of the antihypertensive effects, benefits of combination therapy include the potential of fewer adverse effects and of improvement of patients' compliance. Current guidelines recommend that combination pharmacotherapy might also be considered as initial treatment in patients with significant elevation of blood pressure and evidence of complications. Several effective and well-tolerated antihypertensive drug classes available today offer multiple options for combination therapy. The choice of antihypertensive agents should be made on the basis of current recommendations regarding first-line drugs and compelling indications. Specific drug combinations might have additional beneficial or detrimental long-term metabolic effects, beyond their effects on blood pressure [42].

An ATOM systematic review and meta-analysis of randomized clinical trials according to PRISMA statement evaluated 208 trials including 94,305 patients [43]. The results demonstrated that in monotherapy, the drugs (olmesartan, amlodipine, olmesartan, hydrochlorothiazide, felodipine, metoprolol, valsartan, and hydrochlorothiazide) result in the decrease of SBP by 10–15 mmHg and DBP by 8–10 mmHg. By comparison, the olmesartan/amlodipine, olmesartan/hydrochlorothiazide, felodipine/metoprolol, and valsartan/hydrochlorothiazide combinations achieved >20 mmHg reductions in mean SBP.

The aim of combination therapy should always be both to improve BP control and to reduce cardiovascular events. Antihypertensive drugs can be effectively combined if they have different and complementary mechanisms of action [44]. This is crucial to obtain additive BP-lowering effects without impacting on tolerability. One typical combination is the association of drugs blocking and stimulating the renin-angiotensin system (RAS) (ACEI or ARB and CCB or diuretic, respectively). In contrast, some combinations (e.g., calcium antagonists plus diuretics or beta-blockers plus RAS blockers) have no additive BP-lowering effects, while other combinations (e.g., clonidine plus alpha-1 receptor blockers) can have a negative interaction. Regardless, BP reduction is not the only mechanism that reduces cardiovascular risk. Scientific evidence indicates that some drug classes are better than others in this respect, and therefore some drug combinations are also better than others. The results of the ASCOT-BPLA and ACCOMPLISH trials suggested that an ACE inhibitor/calcium antagonist combination had better cardioprotective effects than β-blocker/diuretic or ACE inhibitor/diuretic combinations. It is worth noting that no controlled clinical trials have used hard endpoints when investigating the effects of an angiotensin receptor blocker/calcium antagonist combination. In conclusion, combination therapy is needed for optimal antihypertensive management, with the first choice being an ACE inhibitor plus a calcium antagonist. This approach should improve BP control and provide better cardiovascular protection.

Hypertension is a serious public health concern with inadequate control of BP worldwide [40]. Contributing factors include low efficacy of drugs, underuse of combination therapies, irrational combinations, physicians' therapeutic inertia, and poor adherence to treatment. Current guidelines recommend the use of initial (dual) combination therapy in high-risk patients for immediate BP response, better short- and long-term BP control, and continued/improved patient adherence. This article aims to review the existing evidence of triple-combination therapies with respect to efficacy, safety, and adherence to treatment. It is estimated that three drugs are required to achieve BP control in approximately one-fourth to one-third of patients. RCTs have shown that triple combinations of amlodipine/valsartan/hydrochlorothiazide, amlodipine/olmesartan/hydrochlorothiazide, and amlodipine/telmisartan/hydrochlorothiazide produce greater BP reductions, with greater proportions of patients achieving BP control compared with dual therapies. Further evidence also demonstrates that triple-combination therapy is efficacious for moderate to severe hypertension, with substantial additional BP reduction over dual regimens. Both RCTs and post-marketing observational studies have shown consistent and comparable efficacy in both the general population and high-risk hypertensive subgroups. Triple therapies are generally well tolerated with adverse event profiles similar to dual regimens. In addition, fixed-dose combinations used as single pill improve patient adherence leading to better long-term BP control. Depending on regional circumstances, they may also be cost-effective. Thus, single-pill triple combinations of different classes of drugs with complementary mechanisms of action help to treat patients to goal with improved efficacy and better adherence to treatment.

4.1.1.3 Principles of Antihypertensive Combination

Four main classes of medications are used in combination therapy for the treatment of hypertension: thiazide diuretics, CCBs, ACEIs, and ARBs [41].

1. ACEIs and ARBs should not be used simultaneously.
2. In black patients, at least one agent should be a thiazide diuretic or a CCB.
3. Patients with HF with reduced ejection fraction should be treated initially with a beta-blocker and an ACEI or ARB, followed by add-on therapy with a mineralocorticoid receptor antagonist and a diuretic based on volume status.
4. Treatment for patients with CKD and proteinuria should include an ACEI or ARB plus a thiazide diuretic or a calcium channel blocker.
5. Patients with diabetes mellitus should be treated similarly to those without diabetes unless proteinuria is present, in which case combination therapy should include an ACEI or ARB.
6. Inadequate control with monotherapy is the clearest indication for adding another medication, which can be done before or after titrating the first agent to the maximal dosage. If a patient does not achieve adequate control with a lower initial dose of a single agent, it is reasonable to titrate that medication or to add a second agent [16, 41]. Initiating a second agent before titration of the first may result in a larger reduction in BP compared with increasing the dose of the first agent [45]. Response to initial monotherapy varies significantly with individual plasma renin levels, so a second mechanism of action may more appropriately address the patient's individual physiology rather than increasing the dosage of a relatively ineffective first agent [35].

Approximately 45% of patients with hypertension and 84% of those with uncontrolled hypertension are not adherent to their antihypertensive regimen [39, 44]. Multiple studies report that combination pills containing two or three agents increase medication adherence and lower overall costs [47–49]. Although improved adherence is expected to decrease morbidity and mortality, a large systematic review found that the effects of fixed-dose combination therapy on all-cause mortality or atherosclerotic cardiovascular disease events are unclear [50].

Available data demonstrate that the ACEI/diuretic or CCB combinations produce synergistic effects on BP, and the β-blocker/diuretic or CCB combinations produce additive effects, as does the rarer combination of diltiazem plus a dihydropyridine CCB [51]. Ineffective combinations include the β-blocker/ACEI combinations and the dihydropyridine CCB/diuretic combinations, although there are specific clinical circumstances where these combinations may be used [51].

4.1.1.4 Initial Combination Therapy (ICT)

Initial combination therapy (ICT) of BP is being increasingly investigated and recommended. Low-dose combinations of BP lowering provide more effective BP lowering, with fewer adverse events [52]. Recent advances include listing of four dual combinations on the WHO Essential Medicines List, completion of a triple half-dose combination trial, and a pilot of quadruple quarter-dose combination, and recent cardiovascular polypill trials have included two BP-lowering medicines at low dose. These trials all demonstrated improvements in achieving BP targets with low-dose combination therapy [52].

Low-dose combination therapy is a promising option for initial treatment of hypertension that appears to be safe and effective. Larger trials of triple and quadruple low-dose combination therapy in multiple locations are underway and should provide stronger evidence of efficacy as well as information on the side effect profile [52].

Compared with monotherapy, ICT improves the average decrease in BP and achieves BPC faster, with similar tolerability [35, 46, 52–58]. There is general agreement that ICT is safe and more effective than monotherapy in patients with SBP >160 mmHg or > 20 mmHg above goal or with DBP >100 mmHg or > 10 mmHg above goal [16, 59–61]. Whichever treatment strategy is chosen, escalation of therapy should occur within 1 month, if needed, to achieve target BP. However, no randomized controlled trials have shown decreased cardiac risk with ICT, although some observational data have shown decreased risk [34, 43]. This may be because a significant number of patients are not escalated to combination therapy after titration of monotherapy as directed by guidelines, a phenomenon termed therapeutic inertia [54]. Furthermore, delays of even a few months in therapy escalation are associated with an increased risk of cardiac events or death [62]. In addition, in a randomized controlled trial, patients who started on monotherapy eventually achieved BPC similar to that of patients who started on combination therapy [35].

Guidelines recommend the addition of a third agent for patients whose BP is not controlled with dual therapy [16]. Randomized controlled trials have shown significantly higher rates of BPC in patients using a combination of an ARB, CCB, and thiazide diuretic compared with those on a dual regimen of an ARB and CCB [63, 64]. A recent meta-analysis suggests that adding a third agent to dual therapy is more effective for lowering blood pressure than increasing dosages of dual therapy and is equally safe [65]. Because of the lack of mortality data from randomized controlled trials, it is reasonable to either titrate individual agents to maximal dosages before adding an additional agent or to add an additional agent before reaching the maximal dosage of the current agent.

Recommended CDTs exhibit complimentary pharmacology producing an additional BP-lowering effect over monotherapy and additive BP reduction, are well tolerated, and include components with demonstrated endpoint reduction in long-term clinical trials. Thoughtful use of the CDT strategy is critical for achieving therapeutic objectives in hypertensive individuals and populations—more rapid BPC and more effective endpoint reduction. Practical strategies for the optimal use of CDT continue to evolve from the older stepped-care approach to more recent recommendations favoring the use of low-dose combinations, and ICT, particularly in patients with stage 2 hypertension [39].

CDT should be based on the understanding of the mechanisms of action of the different classes of drugs used and of their interaction. Rational combinations include an ACEI or a β-blocker together with moderate salt restriction or a thiazide diuretic or a CCB. Less rational CDTs include the ACEI/β-blocker combination and CCB/thiazide diuretic combination [38].

4.1.1.5 Combination Therapy of Pulmonary Arterial Hypertension (PAH)

Pulmonary arterial hypertension (PAH) is a rare, progressive disorder typified by occlusion of the pulmonary arterioles owing to endothelial dysfunction and uncontrolled proliferation of pulmonary artery smooth muscle cells and fibroblasts. PAH carries a poor prognosis if not promptly diagnosed and appropriately treated associated with substantial morbidity and mortality and can lead to increases in pulmonary vascular resistance and thereby increased pressure in the pulmonary arteries, often resulting in right ventricular failure with shortness of breath and syncope and death if left untreated [66–68]. The development and approval of 14 medications over the last several decades have led to a rapidly evolving approach to therapy and have necessitated periodic updating of evidence-based treatment guidelines [69, 70]. Therapeutic options for the patient with PAH continue to expand through basic discovery, translational science, and clinical trials. Over the last 25 years, there has been a significant evolution in the therapies to treat PAH. An important change of PAH therapy is the focus on initial upfront combination therapy for most patients [69, 70].

Combination therapy is now regarded as the standard of care in PAH and is becoming widely used in clinical practice [71–75]. Despite their differing modes of action, all of the currently available classes of PAH therapy induce systemic vasodilation. In ICT regimens in particular, this may lead to additive side effects and reduced tolerability compared with monotherapy. However, approaches such as staggered treatment initiation and careful up-titration may reduce the risk of additive side effects and have been used successfully in clinical practice, as well as in clinical trials and registry studies. At present, compelling evidence has been obtained that combined therapy of PAH with specific drugs can significantly slow the progression of PAH. Combination therapy, administered as either a sequential or an initial regimen, can improve long-term outcomes in PAH [73]. The dual combination—phosphodiesterase-5 inhibitor (PDE-5i)/ERA—is the most widely utilized CDT regimen. However, some patients fail to achieve their treatment goals on dual combination and may benefit from the addition of a third drug. The use of triple combination in clinical practice was previously reserved for patients with severe disease due to the need for parenteral administration of prostanoids. Although triple therapy with parenteral prostanoids plays a key role in the management of severe PAH, the approval of oral therapies that target the prostacyclin pathway means that all three pathways can now be targeted with oral drugs at an earlier disease stage. Furthermore, there is evidence demonstrating that this approach can delay disease progression. Based on the evidence available, it is becoming increasingly clear that all PAH patients should be offered the benefits of combination therapy [73].

Although dual-combination therapy targeting two pathogenic pathways at once, either with initial or sequential therapy, is now widespread in clinical practice [75], the results from early short-term trials were inconsistent [73]. Three of these studies demonstrated that this treatment strategy delayed disease progression compared with monotherapy, regardless of whether an initial or sequential strategy was employed [76–78]. For example, the AMBITION study has provided data supporting the use of ICT in PAH [76]. AMBITION was a double-blind RCT which was designed to evaluate the treatment strategy of ICT with the ERA ambrisentan and PDE-5i tadalafil vs monotherapy with either agent in 500 newly diagnosed treatment-naïve PAH patients. The two drugs were initiated at half of their maximum approved doses and then up-titrated to their maximum approved doses (tadalafil 40 mg and ambrisentan 10 mg) over a period of 8 weeks. There was a 50% risk reduction in clinical failure events (composite primary endpoint) for patients on combination therapy compared with the pooled monotherapy group. At week 24, significant differences in favor of combination therapy were also observed in the percentage of patients with a satisfactory clinical response and in the change from baseline in 6MWD (+49 m in the combination therapy arm vs + 24 m in the pooled monotherapy arm) and N-terminal pro-brain natriuretic peptide (NT-proBNP, a biomarker for HF) (−67.2% in the combination therapy arm vs −50.4% in the pooled monotherapy arm) [77].

For example, the long-term effects of sequential dual-combination therapy with PDE-5i as a background PAH-specific therapy and subsequent macitentan (10 mg; an antagonist/blocker of endothelin receptors on blood vessels and smooth muscle, and it blocks the stimulation of vasculature hypertrophy, inflammation, fibrosis, proliferation, and vasoconstriction) were examined in the SERAPHIN RCT, a double-blind, placebo-controlled, phase 3 study in 742 PAH patients [78]. In the overall study population, macitentan significantly reduced the risk of a morbidity/mortality event (composite primary endpoint) by 45% vs placebo. Furthermore, macitentan significantly improved 6-min walking distance (6MWD) by 25.9 m from baseline to month 6 vs placebo in patients on PDE-5i background therapy [74, 78]. A hemodynamic substudy enrolling 187 patients from 38 centers indicated that macitentan also significantly improved pulmonary vascular resistance (PVR) treatment effect and cardiac index from baseline to month 6 in patients on PDE-5i background therapy [77]. The benefits of combining an ERA and a PDE-5i sequentially were also indicated in COMPASS-2, a double-blind, placebo-controlled, phase 4 RCT in 334 PAH patients [78, 79].

4.1.2 CDT in Hyperlipidemia

Hyperlipidemia is a chronic condition with abnormally elevated levels of any or all lipids (fats, cholesterol, or triglycerides) or lipoproteins in the blood [80–83]. It is used as an umbrella term covering any of various acquired or genetic disorders that result in that finding. Hyperlipidemia represents a subset of dyslipidemia and a superset of hypercholesterolemia, and the two major types of lipids found in the blood are triglycerides and cholesterol. Hyperlipidemias are divided into primary and secondary subtypes. Primary hyperlipidemia is usually due to genetic causes, while secondary hyperlipidemia arises due to other underlying causes such as diabetes. Lipid and lipoprotein abnormalities are common in the general population and are regarded as modifiable risk factors for cardiovascular disease due to their influence on atherosclerosis. Triglycerides are made when the body stores the extra calories and come directly from diet in foods such as red meat and whole-fat dairy. Cholesterols are produced naturally in the liver and also come in fatty foods like eggs, red meat, and cheese.

Hyperlipidemia predisposes a person to atherosclerosis. Atherosclerosis is the accumulation of lipids, cholesterol, calcium, and fibrous plaques within the walls of arteries [80–83]. This accumulation narrows the blood vessel and reduces blood flow and oxygen to muscles of the heart. Over time fatty deposits can build up, hardening and narrowing the arteries until organs and tissues do not receive enough blood to properly function. If arteries that supply your heart with blood are affected, a person might have angina (chest pain). Complete blockage of the artery causes infarction of the myocardial cells, also known as heart attack. Fatty buildup in the arteries can also lead to stroke, if a blood clot blocks blood flow to the brain. In the United States, more than 100 million, or roughly 53% of adults, have elevated low-density lipoprotein cholesterol (LDL-C) levels [84]. Yet, fewer than 50% of patients with high LDL-C receive treatment to reduce their levels, and among those receiving treatment, fewer than 35% achieve adequate control.

People with hyperlipidemia are at roughly twice the risk of developing CVD as compared to those with normal total cholesterol levels. Patients with familial hypercholesterolemia (FH) have an even greater risk of developing CVD at an earlier age; therefore, early detection and treatment are imperative to reduce cardiovascular events and premature death. The approach to determining the appropriate treatment options has also undergone important changes. Guidelines for the management of patients with hyperlipidemia vary in their recommendations, with the American College of Cardiology/American Heart Association recommending that treatment decisions be based on the intensity of response associated with various statins, while multiple other guidelines (e.g., National Lipid Association (NLA) and the American Association of Clinical Endocrinologists and American College of Endocrinology) still support attaining prespecified lipid values to reduce cardiovascular risk [85–88]. Some evidence supports adding ezetimibe to statin therapy in patients with acute coronary syndrome or chronic kidney disease.

4.1.2.1 Hypolipidemic Drugs

There are several classes of drugs used to decrease cholesterol. These include:

Statins

Statins are HMG-CoA reductase inhibitors and the mainstay treatment for hyperlipidemia. Seven statins are currently approved by the FDA for the treatment of hyperlipidemia in combination with a heart-healthy diet [88, 89], including atorvastatin (Lipitor), fluvastatin (Lescol XL), lovastatin (Altoprev), pitavastatin (Livalo), pravastatin

(Pravachol), rosuvastatin (Crestor), and simvastatin (Zocor). These agents decrease LDL-C and triglycerides and slightly increase high-density lipoprotein cholesterol (HDL-C). However, the limitations of statins include treatment resistance, intolerance due to adverse events, and a lack of adherence which contribute to poor outcomes. As such, many patients require adjunct therapies to properly control hyperlipidemia including niacin, bile acid sequestrants, fibric acids, and ezetimibe.

Proprotein Convertase Subtilisin/Kexin Type 9 (PCSK9) Inhibitors
These drugs include alirocumab (Praluent) and evolocumab (Repatha), which decrease LDL-C. They are usually reserved for people who have a genetic condition that causes very high LD-CL levels or people with heart disease who cannot tolerate statins or other cholesterol-lowering drugs.

Cholesterol Absorption Inhibitor
Ezetimibe (Zetia) decreases LDL-C, slightly decreases triglycerides, and slightly increases HDL-C.

Adenosine Triphosphate-Citrate Lyase (ACL) Inhibitors
Bempedoic acid (Nexletol) and bempedoic acid-ezetimibe (Nexlizet) decrease LDL-C.

Bile Acid Sequestrants
Cholestyramine (Prevalite), colesevelam (Welchol), and colestipol (Colestid) decrease LDL-C and may slightly increase HDL-C.

Combination Cholesterol Absorption Inhibitor and Statin
Ezetimibe-simvastatin (Vytorin) decreases LDL-C and triglycerides and increases HDL-C.

Combination Calcium Channel Blocker and Statin
Amlodipine-atorvastatin (Caduet). The statin component decreases LDL-C and triglycerides; the CCB lowers BP.

Fibrates (Fibric Acid Derivatives)
Fenofibrate (Antara, Lipofen, others) and gemfibrozil (Lopid) decrease triglycerides, modestly decrease LDL-C, and increase HDL-C.

Nicotinic Acid
Niacin (Niacor, Niaspan) decreases LDL-C and triglycerides and increases HDL-C.

Long-Chain Omega-3 Fatty Acids
Prescription versions include Lovaza, Omacor, and Vascepa, which decrease triglycerides and may increase HDL-C.

4.1.2.2 Combination Therapy of Hyperlipidemia

Statins are considered first-line pharmacological therapy for lowering LDL-C levels [90, 91], although bile acid sequestrants were the first class of drugs to show a mortality benefit related to LDL-C lowering. The West of Scotland Coronary Prevention Study demonstrated that the reduction in LDL-C with pravastatin (mean reduction 26% vs no change with placebo) was accompanied by a significant reduction in cardiovascular risk with 31% reduction in CAD death or nonfatal myocardial infarction (MI) over an average follow-up period of 4.9 years among men with hypercholesterolemia and no history of MI [90, 92]. A similar reduction in cardiovascular risk in the secondary prevention setting was demonstrated by the Scandinavian Simvastatin Survival Study; a reduction in LDL-C of 35% (vs + 1% with placebo) was accompanied by a 30% reduction in total mortality and a 34% reduction in risk of major coronary events over a median of 5.4 years of follow-up [90, 93]. Statins also reduce triglyceride levels (7–30%) and increase HDL-C levels (5–15%) [90, 91]. However, a substantial proportion of patients (up to 50%) does not achieve LDL-C goals with statin monotherapy, especially in high-risk patients, and in some patient subgroups with atherogenic dyslipidemia [90, 94]. Another consideration is that some patients may be unwilling to use or unable to tolerate statin therapy because of adverse effects. The most well-known type of adverse effect associated with statin therapy is myopathy, the effects of which can range in severity from asymptomatic increases in creatine kinase, through myalgias and muscle weakness, up to the rare but potentially fatal rhabdomyolysis [90, 95]. Also, statins may increase risk of new-onset diabetes in patients at high risk for diabetes [90]. Thus, there remains a need for other lipid-lowering drugs to be used in combination with or in place of statins. Indeed, clinical trials have shown that these hypolipidemic drug classes can be safely combined, either in the form of fix-dose combination or free-dose combination, in order to augment the lipid-lowering efficacy [90, 96–98]. Indeed, a combination of a statin with other hypolipidemic drugs has been reported to afford additional benefits. In this regard, Jones and Nwose [90] presented an excellent review on the available data from the literature on the use of colesevelam, a second-generation bile acid sequestrant, in combination with other lipid-lowering agents.

Statin/Colesevelam Combination

The combinations of varying statins including atorvastatin, lovastatin, rosuvastatin, or simvastatin and colesevelam (a bile acid sequestrant) have been investigated in patients with hypercholesterolemia [90]. The statin/colesevelam combination was more effective in lowering LDL-C than the corresponding monotherapies [99–101]. Notably, in a randomized, double-blind, placebo-controlled, parallel-design, multicenter, 8-week study evaluating the efficacy of colesevelam 3.75 g/day when coadministered with atorvastatin in 94 patients with moderate hypercholesterolemia, the LDL-C reduction achieved (48%) was not significantly different from that observed with atorvastatin (53%) [100]. A prospective, randomized, open-label, blinded endpoint study that assessed changes in lipid profile as a secondary endpoint in 40 patients assigned to the rosuvastatin/colesevelam combination therapy or rosuvastatin monotherapy for 3 months showed that similar reductions in LDL-C were observed [102]. It should be noted that the small sample size and open-label design were listed as study limitations and could have potentially introduced bias into the study's findings.

On the other hand, the addition of colesevelam to existing statin therapy has been shown to have an incremental benefit on LDL-C levels. A pooled analysis of three randomized, double-blind, placebo-controlled, parallel, 6-week clinical trials evaluated the efficacy of adding colesevelam or placebo to stable existing statin therapy (atorvastatin, pravastatin, or simvastatin) in 204 patients with primary hypercholesterolemia. Study results showed a significant incremental reduction in LDL-C with the addition of colesevelam to statins [103]. Similarly, in a post hoc analysis of pooled data from three double-blind, placebo-controlled studies in which colesevelam or placebo was added to existing metformin-, sulfonylurea-, or insulin-based therapy in patients with T2DM, among patients receiving statins at baseline, the addition of colesevelam led to a significant reduction in LDL-C [368]. Levels of HDL-C significantly increased with both colesevelam plus atorvastatin 10 mg/day and with atorvastatin 80 mg/day alone, but the increase was numerically greater for the combination (11 vs 5%) [100]. Significant reductions from baseline in apolipoprotein B (ApoB, a predictor of heart disease risk) were seen with colesevelam or atorvastatin monotherapy, and greater reduction was seen with the atorvastatin/colesevelam combination. Similarly, in a multicenter, randomized, double-blind, placebo-controlled trial, the lipid-lowering efficacy of the simvastatin/colesevelam combination in 251 patients with primary hypercholesterolemia was significantly greater than either agent as monotherapy [101]. In addition, the lovastatin/colesevelam achieved reductions in ApoB that exceeded those of the individual components [99]. On the other hand, the atorvastatin/colesevelam combination produced greater increases in apolipoprotein A-I (ApoA-I the major protein component of HDL) than colesevelam monotherapy [100, 101]. Triglyceride levels generally decreased from baseline with statin monotherapies, increased with colesevelam monotherapy, and remained relatively unchanged with the combination of colesevelam and a statin

[99–101]. Triglyceride levels remained relatively unchanged when colesevelam was combined with statins, fibrates, ezetimibe, or statin plus ezetimibe and decreased with the triple combination of colesevelam, statin, and niacin. Colesevelam offset the negative glycemic effects of statins and niacin in subjects with insulin resistance or impaired glucose tolerance [90].

Fibrates/Bile Acid Sequestrant Combination
Recent trials have investigated the usefulness of fenofibrate, alone and in combination with other lipid-lowering therapies, in the treatment of hyperlipidemia. Studies of the fenofibrate/bile acid sequestrants demonstrate that these two therapies may have an additive effect in reducing total cholesterol (TC), LDL-C, very-low-density lipoprotein cholesterol (VLDL-C), and triglyceride levels in patients with hyperlipoproteinemia or familial hypercholesterolemia. These lipoprotein changes have been associated with a regression of tendon xanthoma. Pharmacokinetic studies have shown that bile acid sequestrants do not alter the absorption or the plasma levels of fenofibrate. The combined use of fenofibrate with bile acid sequestrants has been found to be comparably effective with the new 3-hydroxy-3-methylglutaryl coenzyme A reductase inhibitor, synvinolin, with respect to the reduction of TC and LDL-C. Although synvinolin was more effective in lowering LDL-C, VLDL-C and triglycerides were reduced to a greater extent with fenofibrate. Another notable difference was that the fenofibrate/bile acid sequestrant combinations more markedly increased HDL-C levels. The combination fenofibrate/nicotinic acid combination also appears to have a beneficial effect on lipoproteins [98].

The primary effects of fibrates include reduction of triglycerides (20–50%) and increases in HDL-C levels (10–20%); moderate increases in LDL-C (5–20%) may also be observed, but the effects vary depending on the patient's lipid profile [91, 104, 105]. One study has evaluated the dual combination of colesevelam and a fibrate [106]. This randomized, double-blind, placebo-controlled, parallel-group study investigates the efficacy of the fenofibrate/colesevelam combination vs fenofibrate monotherapy in 129 patients with mixed hyperlipidemia. Patients were randomized to receive fenofibrate in combination with colesevelam 3.75 g/day or placebo for 6 weeks. The combination therapy produced significantly greater reductions in LDL-C levels than fenofibrate monotherapy, and effects on ApoB were similar between combination therapy and monotherapy. No significant differences were observed in HDL-C and triglycerides levels between the combination therapy and the fenofibrate monotherapy. Colesevelam and fenofibrate had additive effects on LDL-C levels, regardless of whether patients' triglyceride levels were > 200 mg/dL [106].

Cholesterol Absorption Inhibitor/Bile Acid Sequestrant Combination
Ezetimibe (a cholesterol absorption inhibitor) primarily reduces LDL-C levels, with only modest reductions in triglycerides and increases in HDL-C in patients with primary hypercholesterolemia, as reported by Pandor et al. [107] in a systematic review and meta-analysis of randomized controlled trials comprising 2722 patients administered with ezetimibe monotherapy for a minimum treatment period of 12 weeks. In a multicenter, randomized, double-blind, placebo-controlled, parallel-group study in 86 adults with primary hypercholesterolemia that was designed to determine the superiority of colesevelam plus ezetimibe vs ezetimibe alone, a significantly greater reduction in LDL-C was seen with the combination therapy compared with ezetimibe monotherapy [108]. In addition, subjects receiving colesevelam plus ezetimibe, compared with ezetimibe monotherapy, had a significantly greater reduction in ApoB and increase in ApoA-I. Thus, it is possible to achieve LDL-C reductions with this combination that are similar to those seen with low-dose statin monotherapy.

Statin/Bile Acid Sequestrant/Nicotinic Acid Triple Combination
Two studies have evaluated the efficacy of colesevelam-based, triple-drug combination lipid-lowering regimens. A randomized, double-blind, partially placebo-controlled, 1-year study

in adults with atherosclerotic disease and dyslipidemia evaluated the incremental effects of adding extended-release niacin and colesevelam to atorvastatin [109]. Compared with atorvastatin monotherapy, there was no further reduction in LDL-C with atorvastatin plus niacin (−47% for both), but a significantly greater reduction was achieved with atorvastatin plus niacin plus colesevelam (−57%); a lower average atorvastatin dose was required to meet LDL-C targets with dual and triple therapy. In addition, the triple-combination group, compared with the atorvastatin monotherapy group, had significantly greater reductions in ApoB (−51% vs −40%) and triglycerides (−42% vs −25%) and a significantly greater increase in ApoA-I (+15% vs + 7%; $p = 0.01$). Changes in ApoB (−45%), triglycerides (−33%), and ApoA-I (+11%) in the dual-combination group were not significantly different from those in the atorvastatin monotherapy group [109].

The addition of colesevelam to stable, maximally tolerated therapy with a statin plus ezetimibe for 12 weeks in adults with familial hypercholesterolemia resulted in significantly greater reductions in LDL-C (−11.4% vs + 0.6%) and apo B (−6.0 vs + 4.6%) compared with placebo [110]. Triglyceride levels increased in both groups, resulting in no significant between-group difference. An unexpected finding was that hemoglobin A1c was significantly reduced in patients receiving colesevelam as compared with placebo (least-squares mean treatment difference −0.12%). The patients' mean baseline hemoglobin A1c of 5.7 ~ 5.8% suggests that many were at high risk for development of new-onset T2DM [110].

4.1.2.3 Limitations of Combination Therapy

A meta-analysis indicated that the combination of statins with fibrates or ezetimibe (a cholesterol absorption inhibitor) was not associated with better clinical outcomes than high-dose statin monotherapy; limited outcomes data were available for combinations of statins with other lipid-lowering agents [106, 111]. Although colesevelam/niacin combination therapy demonstrated lipid-lowering and metabolic benefits vs the placebo/niacin combination, results of the AIM-HIGH and HPS2-THRIVE studies suggest that the use of niacin in lipid-lowering combination therapy may be limited [107]. These studies demonstrated that the addition of niacin to a statin had no incremental benefit in reducing cardiovascular events (AIM-HIGH) or the risk of heart attacks, strokes, or operations to open blocked arteries (HPS2-THRIVE) in patients with well-controlled LDL-C levels. Furthermore, the unwanted side effects associated with the addition of niacin appeared to outweigh the benefits of its use [112].

Recent American Diabetes Association guidelines indicate that combination lipid-lowering therapy is not broadly recommended because of the lack of demonstrated incremental cardiovascular benefit over that achieved with statins [113]. Similarly, the current European Society of Cardiology/European Atherosclerosis Society guidelines for the management of dyslipidemia recommend considering combination therapy in high-risk individuals, those with very high LDL-C levels, or in patients who are either statin-intolerant or unable to tolerate higher statin doses [114]. However, a systematic review of agents used to treat dyslipidemias in clinical practice concluded that complementary effects of agents other than statins might substantially provide clinical benefits in not only individuals with severe dyslipidemias but in other patients as well. It is recommended that multiple drug combinations should be more widely used to potentially get more people to lipoprotein goals, even in light of the fact that cardiovascular outcomes data are lacking for such combination therapies [115].

4.1.3 CDT in Coronary Artery Disease (CAD)

Cardiovascular diseases (CVD) account for 30% of all deaths worldwide with coronary artery disease (CAD) accounting for almost half of all cardiovascular deaths [116–118]. CAD, also called coronary heart disease (CHD) or ischemic heart disease (IHD) or atherosclerotic heart disease (AHD), is a type of heart disease that develops when the arteries of the heart cannot deliver enough oxygen-rich blood to the heart and a leading cause of death worldwide [116–118]. It is the leading cause of death in the United States and many other countries as well. It is a condition of recurring chest pain or discomfort that occurs when a part of the heart does not receive enough blood. This condition occurs most often during exertion or excitement when the heart requires greater blood flow. Ischemic heart disease, also called coronary heart disease, is common in the United States and is a leading cause of death worldwide.

Atherosclerosis is the common pathophysiological process underlying both CAD and peripheral artery disease (PAD). According to the "injury-repair" hypothesis initially proposed by Ross [119], platelet adhesion, activation, and aggregation (both platelet-platelet and platelet-monocyte) play pivotal roles in the initiation and progression of atherosclerotic lesions and the final manifestation of thrombotic complications (atherothrombosis) following atherosclerotic plaque erosion or rupture [119, 120]. CAD develops when cholesterol particles in the blood begin to accumulate on the walls of the arteries that supply blood to the heart. Eventually, deposits called plaques may form. These deposits narrow the arteries and eventually block the flow of blood. This decrease in blood flow reduces the amount of oxygen supplied to the heart muscle. CAD progression from a stable to an unstable state has been characterized as a pathophysiological condition associated with heightened platelet activation, hypercoagulability, and inflammation, a thromboinflammatory syndrome [121, 122].

4.1.3.1 Drug Treatment of CAD

In acute coronary syndrome (ACS), sudden blockage in a coronary artery greatly reduces or cuts off the blood supply to an area of the heart muscle (myocardium). The lack of blood supply to any tissue is termed ischemia. If the supply is greatly reduced or cut off for more than a few minutes, heart tissue dies. A heart attack, also termed myocardial infarction (MI), is death of heart tissue due to ischemia.

The overall goals of drug treatment of CAD are to control symptoms and to stop or slow the progression of the disease [123]. The specific aims are **(a)** to relieve chest pain by reducing the heart's workload and widening arteries (nitrates); **(b)** to prevent angina and acute coronary symptoms from occurring (beta-blockers, calcium channel blockers, and sometimes ranolazine); **(c)** to prevent and reverse coronary artery narrowing from atherosclerosis; and **(d)** to open a blocked artery.

Nitrates

Most people are given nitroglycerin, which relieves pain by lowering BP to reduce the workload of the heart and possibly by dilating arteries.

Morphine

Morphine has a calming effect and reduces the workload of the heart. It is given when nitroglycerin cannot be used or is not effective; however, recent data suggest it may interact with antiplatelet drugs and reduce their effectiveness and may increase the risk of death.

β-Blockers

Because decreasing the heart's workload also helps limit tissue damage, a β-blocker is usually given to slow the heart rate (HR). Slowing HR enables the heart to work less hard and reduces the area of damaged tissue and the symptoms of angina by improving microcirculation and coronary flow and is a recognized strategy to prevent ischemic episodes [117]. Elevated HR is an established risk factor for cardiovascular events in patients with CAD and in those with chronic HF. Data from SIGNIFY indicate that HR is not a modifiable risk factor in patients with CAD who do not also have HF [117]. HR is also an important determinant of cardiac arrhythmias; a low HR can be associated with atrial fibrillation, and a high heart rate after exercise can be associated with sudden cardiac death (SCD).

Calcium Channel Blockers

Calcium channel blockers prevent blood vessels from narrowing (constricting) and can counter coronary artery spasm. All calcium channel blockers reduce blood pressure. Some of these drugs, such as verapamil and diltiazem, may also reduce HR. This effect can be useful to many people, especially those who cannot take beta-blockers or who do not get enough relief from nitrates.

Heart Rate (HR)-Lowering Drug Ivabradine
Ivabradine is a unique agent that is distinct from beta-blockers and calcium channel blockers as it reduces heart rate without affecting myocardial contractility or vascular tone. Ivabradine is a use-dependent inhibitor targeting the sinoatrial node [124]. Ivabradine is recommended in the European Society of Cardiology guidelines on acute and chronic heart failure for stable, symptomatic patients with left ventricular ejection fraction below or equal to 35% in sinus rhythm with a resting heart rate above or equal to 70 beats per minute despite therapy with beta-blocker, angiotensin-converting enzyme inhibitor (or angiotensin receptor blocker), or a mineralocorticoid receptor antagonist or for people in whom beta-blocker therapy is contraindicated [125]. Ivabradine is also recommended as the second-line therapy for angina relief in the European Society of Cardiology guidelines on stable coronary artery disease [126]. It is approved for use in the United States as an adjunct therapy for heart rate reduction in patients with heart failure with reduced ejection fraction [124].

Late Sodium Channel Blocker Ranolazine
Ranolazine, a piperazine derivative sold under the trade name Ranexa, is a well-tolerated medication that selectively inhibits the late sodium current. It has beneficial metabolic properties and does not affect HR or BP. Ranolazine is currently approved in the United States and Europe as a second-line agent in the management of chronic stable angina pectoris in people who continue to have symptoms despite taking all other antianginal therapy [127].

Angiotensin-Converting Enzyme Inhibitors and Angiotensin II Receptor Blockers
Angiotensin-converting enzyme (ACE) inhibitors and angiotensin II receptor blockers (ARBs) can reduce heart enlargement and increase the chance of survival for many people. Therefore, these drugs are usually given in the first few days after a heart attack and prescribed indefinitely.

Statins
Statins have long been used to help prevent coronary artery disease, but doctors have recently found that they also have short-term benefit for people with acute coronary syndrome (for more detail, see "**(2) Hyperlipidemia**").

Antiplatelet Drugs
In patients with CAD, there is a central role played by platelet inhibition in both treatment and prevention of short- as well as long-term atherothrombotic events. Oral antiplatelet agents recommended for secondary prevention include the cyclooxygenase-1 inhibitor aspirin (also named ASA for acetylsalicylic acid) and the platelet adenosine diphosphate P2Y12 receptor inhibitors clopidogrel, prasugrel, and ticagrelor [128]. Daily aspirin or other blood thinner reduces the tendency of blood to clot, which may help prevent obstruction of coronary arteries. In case of a heart attack, aspirin can help prevent future attacks. But aspirin can be dangerous to people who have a bleeding disorder or who are already taking another blood thinner. People who think they may be having a heart attack should chew an aspirin tablet immediately after calling an ambulance. If aspirin is not taken at home or given by emergency personnel, it is immediately given at the hospital. This therapy improves the chances of survival by reducing the size of the clot (if present) in the coronary artery. Other types of antiplatelet drugs are the platelet P2Y12 receptor antagonists including clopidogrel, ticlopidine, or ticagrelor taken by mouth or glycoprotein IIb/IIIa inhibitors given by vein (intravenously) [129].

Antithrombotic Drugs (or Thrombolytic Drugs or Clot-Dissolving Drugs)
Clot-dissolving drugs are given intravenously to open the arteries if percutaneous coronary interventions cannot be done within 90 min after the person arrives at the hospital.

Anticoagulants

Anticoagulant drugs, such as heparin and warfarin (a vitamin K antagonist), help prevent the formation of additional blood clots. Recently, a group of drugs collectively termed as novel oral anticoagulants were discovered, which consists of the drugs ximelagatran, darexaban, dabigatran, rivaroxaban, and apixaban, of which apixaban (Eliquis), dabigatran (Pradaxa), edoxaban (Savaysa), and rivaroxaban (Xarelto) are approved for clinical use [130].

Oxygen

Often, oxygen is given through nasal prongs or a face mask. Providing more oxygen to the heart helps keep heart tissue damage to a minimum.

4.1.3.2 Combination Therapy

Dual Antiplatelet Therapy

Among several potential combinations of antiplatelet agents or dual antiplatelet therapy (DAPT), the aspirin/P2Y12 receptor inhibitor combination has been shown to reduce recurrent major ischemic events in patients with ACS [128]. Numerous studies have evaluated the safety and efficacy of DAPT with clopidogrel, prasugrel, or ticagrelor in combination with aspirin (ASA) to prevent secondary ischemic events including stroke and TIA. The 2018 Guidelines for Management of Acute Ischemic Stroke state that clopidogrel and aspirin could be beneficial for secondary stroke prevention if initiated within 24 h of the stroke and then continued for 21–90 days [131]. DAPT with aspirin and clopidogrel was associated with lower thrombotic event rates than aspirin monotherapy in patients with acute coronary syndromes, and those undergoing percutaneous coronary intervention changed the treatment paradigm. Moreover, the demonstration that more pharmacodynamically potent P2Y12 inhibitors than clopidogrel were associated with fewer thrombotic event occurrences further solidified the dual antiplatelet therapy approach. According to Rollini et al. [129], dual antiplatelet therapy with the aspirin/P2Y12-receptor inhibitor combination is the cornerstone of treatment of patients with ACS and of those undergoing percutaneous coronary intervention. Prasugrel and ticagrelor have more prompt, potent, and predictable antiplatelet effects than those of clopidogrel and result in reduced ischemic outcomes in patients with ACS, albeit at the expense of an increased risk of bleeding.

The CURE study demonstrated a significant reduction of the primary composite endpoint (cardiovascular death, nonfatal MI, or stroke) in the clopidogrel/aspirin combination arm compared with aspirin alone at 12 months, with an increased risk of major bleeding in 12,262 patients with non-ST elevation (NSTE)-ACS. However, there was no significant difference in the mortality between treatment groups [131, 132].

CHARISMA was one of the first trials that evaluated the long-term use of clopidogrel plus aspirin as compared with aspirin alone in 15,630 patients at high risk for a cardiovascular event. The trial failed to demonstrate any significant benefits of prolonged DAPT in reducing the rate of MI, stroke, or death from cardiovascular causes at a median duration of 28 months and suggested harm in terms of bleeding in patients with multiple risk factors [133].

The PEGASUS-TIMI 54 was a randomized trial that evaluated ticagrelor plus aspirin to aspirin alone in 21,162 stable ACS patients at high risk and followed up for a median of 33 months. The results demonstrated that prolonged DAPT with ticagrelor on top of aspirin reduced ischemic events compared to aspirin alone. However, major bleeding increased, though there was no significant difference between fatal bleeding and nonfatal intracranial hemorrhage in either ticagrelor or placebo [134].

In the DAPT study, extension of DAPT using thienopyridines (either clopidogrel or prasugrel) with aspirin was evaluated up to 30 months after percutaneous coronary intervention (PCI) in 9961 patients who received a drug-eluting stent (DES). It was the first study that showed a significant reduction in MI from both target and nontarget lesions, suggesting a secondary preventive effect of prolonging DAPT. It also showed a significant reduction in stent thrombosis, MI, and major adverse cardiovascular events (MACE including cardiac death, nonfatal myocardial infarction, and recurrent ischemia requiring hospitalization) but no significant benefit in reduction of all-cause mortality and stroke. Prolonging DAPT was also associated with an increase in moderate-to-severe bleeding [135].

In the THEMIS trial, 19,220 patients with T2DM and stable CAD and no history of previous MI or stroke were randomly assigned to receive either ticagrelor plus aspirin or placebo plus aspirin, with a median follow-up of 39.9 months. The primary endpoint, a composite of CV death MI or stroke, was lower in the ticagrelor group than in the placebo group. The incidence of primary safety endpoint thrombolysis in myocardial infarction (TIMI) major bleeding was higher in the ticagrelor group than in placebo group. Incidence of ICH was higher in the ticagrelor group, but there was no significant difference in incidence of fatal bleeding. The study concluded that in patients with stable CAD and diabetes without a history of MI or stroke, using ticagrelor with aspirin lowered the incidence of ischemic cardiovascular events but had a higher incidence of major bleeding than those who received placebo plus aspirin [136].

In the THEMIS PCI trial, a prespecified subgroup of 11,154 patients having a history of previous PCI was analyzed. The primary endpoint, composite of CV death, MI, or stroke was 7.3% in the ticagrelor group compared to 8.6% in the placebo group. TIMI major bleed occurred in 2% of patients in the ticagrelor arm vs 1.1% in the placebo arm. Fatal bleed and ICH occurrence were similar in both groups. Patients with PCI had a better net clinical benefit compared to those without PCI, and benefit was present irrespective of time from most recent PCI [137].

Antiplatelet/Anticoagulant Combination

Antiplatelet therapy is the mainstay for the treatment of acute and chronic arterial disease involving the coronary and peripheral beds. However, recurrent thrombotic events occur in nearly one out of ten patients in the first year following an acute coronary syndrome (ACS) event. Moreover, despite evidence of clinical event reduction associated with potent P2Y12 inhibitors plus acetylsalicylic acid (ASA) in large-scale trials, residual ischemic risk persists along with risks for bleeding that are proportional to the intensity of the antiplatelet regimen [134, 138–140]. Such a limitation stimulated the interest in exploring the efficacy and safety of the antiplatelet/anticoagulant combination therapy with non-vitamin K antagonist direct oral anticoagulants [141]. These investigations have included treatment with very-low-dose oral anticoagulation, and even its replacement of acetylsalicylic acid in the presence of a P2Y12 inhibitor, in patients stabilized after an ACS event. Recent basic and translational studies have suggested noncanonical effects of coagulation factor inhibition that may further modulate clinical benefits. This in-depth review will discuss developments in our understanding of the roles that platelets and coagulation factors play in atherothrombosis and review the rationale and clinical evidence for combining antiplatelet and oral anticoagulant therapy in patients with coronary and peripheral artery disease. In patients with prior MI, the ticagrelor/aspirin combination modestly reduced MACE but increased bleeding in comparison with aspirin monotherapy, without an effect on cardiovascular or all-cause mortality [142]. Blockade of PAR-1 with vorapaxar in patients with stable cardiovascular disease on top of single or dual antiplatelet therapy reduced MACE, but increased bleeding and did not reduce mortality [143]. However, a benefit of vorapaxar on reducing the rates of hospitalization for acute limb ischemia and peripheral artery revascularization in patients with stable cardiovascular disease has been reported [144].

Earlier studies used vitamin K antagonists as an anticoagulant in an attempt to improve clinical outcomes in patients with stable CAD and PAD. When administered in doses sufficient to produce a therapeutic international normalized ratio, vitamin K antagonists reduced cardiovascular events in patients with stable CAD but increased bleeding [145]. In contrast, no benefit was observed for low-dose vitamin K antagonists that did not prolong the international normalized ratio to therapeutic levels, but increased bleeding in comparison with ASA alone [145]. There is no documented evidence of the long-term benefit of therapeutic levels of vitamin K antagonist therapy in patients with PAD [146].

Smaller studies have demonstrated the potential benefits of warfarin plus aspirin vs aspirin alone in patients with ACS [144, 147]. A meta-analysis including 25,307 patients with ACS reported that combination therapy of ASA plus warfarin therapy (irrespective of international normalized ratio) versus ASA therapy alone was not associated with less MACE (all-cause death, nonfatal MI, and nonfatal thromboembolic stroke), but was associated with more major bleeding [148]. In an analysis of studies with international normalized ratio values between 2 and 3, long-term warfarin plus ASA therapy was associated with fewer MACE and an increased risk of major bleeding in comparison with ASA alone [148].

As pointed out by Gurbel et al. [141], the success of combined antiplatelet and anticoagulant therapy is critically dependent on the safety profile of the anticoagulant. For example, therapeutic doses of warfarin reduced recurrent ischemic events after MI, but this effect was offset by an increase in serious bleeding. In the chronic disease phase, a combination of very-low-dose rivaroxaban plus ASA to elicit dual pathway inhibition reduces MACE in patients with CAD. Both DAPT and antiplatelet/anticoagulant combinations reduce the risk of adverse cardiovascular events in patients with CAD [149, 150]. However, the addition of P2Y12 inhibitor to aspirin is associated with increased risk of significant bleeding [151, 152]. The European Society of Cardiology (ESC) guidelines endorse prescription of proton pump inhibitors (PPIs reduce the production of acid by the stomach by irreversibly blocking an enzyme called H^+/K^+ ATPase which controls acid production) with DAPT for all CAD patients to reduce gastrointestinal (GI) bleeding [153], whereas the 2016 American College of Cardiology/American Heart Association focused update recommends the concomitant use of PPI with DAPT in the following patients: (a) prior history of GI bleeding and (b) higher risk of GI bleeding (i.e., advanced age, concomitant use of warfarin, steroids, or nonsteroidal inflammatory drugs. The routine use of PPIs is not recommended for patients at low risk of GI bleeding [154]. A meta-analysis of 39 studies [4 RCTs and 35 observational studies) with 221,204 patients supports and endorses the use of PPI with DAPT for reducing GI bleeding without worsening cardiovascular outcomes [155]. At present, however, it is recommended that combination antiplatelet therapy should be reserved only for selected cases and following thoughtful consideration of the associated risk/benefit ratio [156].

Triple-Combination Therapy

Current guidelines recommend a global approach to risk management, using agents that address as many risk factors as possible. Statins, ACEIs, and CCBs have markedly changed the clinical progression of patients with CAD. A review article provides the rationale and evidence for combining these three drug classes in hypertensive patients with hypercholesterolemia or CAD. Dual-combination therapies are an important component of guideline-recommended therapy in hypertension. Studies indicate that triple therapy with a statin, ACEI, and CCB is associated with a significant reduction in MACE [157]. The benefits of the combination of an ACEI, CCB, and statin, as well as the synergism among these active substances, were confirmed by the post hoc analyses of the Anglo-Scandinavian Cardiac Outcomes Trial-Lipid Lowering Arm (ASCOT-LLA) study [158]. The beneficial impact of these agents on clinical events may be attributed largely to the synergistic, vascular protective, and metabolic effects common to perindopril, amlodipine, and atorvastatin [159, 160]. A post hoc analysis in 1056 patients with stable CAD participating in the EUROPA trial indicated that the addition of perindopril to a CCB and a lipid-lowering agent was associated with a 46% reduction in the composite of cardiovascular death, MI, and resuscitated cardiac arrest [159]. In addition, single-pill formulations are known to result in better adherence to the treatment. Single-pill formulations that combine a statin, an ACE inhibitor, and a CCB appear to offer an effective approach to the management of global cardiovascular risk, which will be introduced in the next chapter.

4.1.4 CDT in Heart Failure (HF)

Heart failure (HF) is considered an epidemic and a deadly disease in the modern world affecting approximately 1–2% of adult population and is more common in the elderly, with around 6–10% of patients over 65 suffering from the condition [161]. Chronic congestive heart failure is a common condition that, if untreated, markedly impairs the quality of life and is associated with a high risk of recurrent hospitalization and death. HF presents a multifactorial, systemic disease with a constellation of symptoms and signs that are commonly attributed to the inability of the heart to produce a cardiac output that meets the demands of the body. The coordinated, complex processes in HF lead to excessive volume overload, increased sympathetic activity, and circulation redistribution and result in different, parallel developing clinical signs and symptoms. Hence, HF is also considered a heterogeneous clinical syndrome stemming from cardiac overload and injury that leads to considerable morbidity and mortality [162]. Yet, at present, evidence-based treatment options are available only for congestive heart failure with a low ejection fraction.

Modern implantable defibrillators and special pacemakers for cardiac resynchronization are well-established device treatment of HF, and alternative devices (baroreflex modulation or cardiac contractility modulation) are under investigation for their clinical applications. It was recently shown that the catheter-based treatment of secondary mitral regurgitation with a MitraClip improves the outcome of selected patients [163].

Pharmacotherapy is mainly based on neurohumoral inhibition of the renin-angiotensin-aldosterone (RAS) system and the adrenergic system. Recent introduction of combined angiotensin receptor antagonists and neprilysin inhibitors has greatly improved the prognosis of HF patients [163]. Yet, despite recent therapeutic gains, HF remains a major cause of mortality and hospitalizations worldwide.

4.1.4.1 Drugs for HF Treatment

HF patients may need multiple medications. Each one treats a different symptom or contributing factor. Some medicines for HF relieve cardiac workload by lowering HR and BP, removing excess fluid from the body, and dilating blood vessels so blood can flow more easily to the heart and improve how well your heart pumps. A combination of medicines is often used to manage your condition and help you feel better.

β-Blockers (Carvedilol, Metoprolol, and Bisoprolol)

These medicines lower HR and BP. β-Blockers are often started at a very low dose and then are increased until a dose is reached that works best (for more detail, see "**(1) Hypertension**").

ACEIs (Lisinopril and Captopril)

ACEIs reduce the amount of heart-damaging hormones the body produces and dilate blood vessels and lower BP to lessen the workload of the heart (for more detail, see "**(1) Hypertension**").

ARBs (Losartan)

ARBs are similar to ACEIs in their cardiovascular effects.

Angiotensin-Receptor Neprilysin Inhibitors (ARNIs) or Combination Therapy

ARNIs are a new drug combination of a neprilysin inhibitor and an ARB, and the sacubitril/valsartan combination is the most frequently used ARNI (neprilysin is an enzyme that degrades natriuretic and other vasoactive peptides) [164].

Mineralocorticoid Receptor Antagonists (MRAs) or Aldosterone Antagonists

Commonly prescribed MRAs include spironolactone (Aldactone) and eplerenone (Inspra). These medicines prevent the body from producing too much of the hormone aldosterone that can damage the heart [165]. MRAs are recommended as standard of care in management of HF patients.

I_f Channel Blocker

This drug class reduces HR to minimize cardiac workload, similar to **another** class of drugs called β-blockers. At present, ivabradine (Corlanor) is the only clinically available I_f

blocker (for more detail, see "**(3) Coronary artery disease (CAD)**").

Digoxin (Lanoxin)

This medicine slows HR and might improve the pumping ability of your heart.

Hydralazine and Nitrates (Apresoline, Nitro-Bid, Imdur, and Isordil)

Hydralazine and nitrates (or isosorbide dinitrate) are often used together to treat heart failure. They dilate blood vessels so it is easier for your heart to receive and pump blood.

Diuretics [Furosemide (Lasix), Bumetanide (Bumex), Torsemide (Demadex), and Metolazone (Zaroxolyn)]

These medicines are called water pills because they remove excess fluid from the body to help you breathe easier and feel better (for more detail, see "**(1) Hypertension**").

Other Medications

Other less commonly used drugs may also be prescribed depending on the additional health problems in addition to HF. These drugs include:

(a) Anticoagulants in HF patients with atrial fibrillation (AF).
(b) Cholesterol-lowering drugs (statins) in HF patients who also have high cholesterol or have had a heart attack.

4.1.4.2 Combination Therapy of HF

In July 2015, the US FDA approved the first of a new class of drugs for the treatment of HF: the valsartan/sacubitril (formerly known as LCZ696 and currently marketed by Novartis as Entresto) combination with ARB valsartan and the neprilysin inhibitor prodrug sacubitril in a 1:1 ratio in a sodium supramolecular complex [166]. Sacubitril is converted by esterases to LBQ657, which inhibits neprilysin, the enzyme responsible for the degradation of the natriuretic peptides and many other vasoactive peptides. Thus, this combination addresses two of the pathophysiological mechanisms of HF: activation of the RAAS system and decreased sensitivity to natriuretic peptides [166]. In the Prospective Comparison of ARNI With ACEI to Determine Impact on Global Mortality and Morbidity in Heart Failure (PARADIGM-HF) trial, valsartan/sacubitril significantly reduced mortality and hospitalization for HF, as well as blood pressure, compared with enalapril in patients with heart failure, reduced ejection fraction, and an elevated circulating level of brain natriuretic peptide or N-terminal pro-brain natriuretic peptide [166–169].

Solomon et al. [170] studied the effects of valsartan/sacubitril in patients with heart failure and preserved ejection fraction (HFpEF) in the Prospective comparison of ARni with ARB on Management Of heart failUre with preserved ejectioN fracTion (PARAMOUNT) trial. NT-proBNP was significantly reduced after 12 weeks in the valsartan/sacubitril combination compared to the valsartan alone. The difference in NT-proBNP was not sustained at 36 weeks. Left atrial size at 36 weeks was significantly decreased by valsartan/sacubitril, which was postulated to reflect improved left ventricular filling pressures in patients receiving valsartan/sacubitril. Notably, BP was reduced to a significantly greater degree in the valsartan/sacubitril combination vs in the valsartan group. There were no significant differences between treatment groups in other echocardiographic endpoints, in a clinical composite assessment, or in quality of life. The incidence of adverse events was similar in the two groups [170].

On the other hand, the therapeutic management of chronic HF, particularly for patients with reduced ejection fraction (EF), is built mainly around drug combinations including angiotensin-converting enzyme inhibitors (ACEIs), angiotensin receptor blockers (ARBs), beta-blockers (BBs), mineralocorticoid receptor antagonists (MRAs), and diuretics. Recently, the international guidelines in Europe [125] and in the United States [171] for the treatment of HF with reduced EF (HFrEF) have been updated to include the use of sacubitril/valsartan (the ARNI/ARB combination) and ivabradine (a sinus node I_f channel inhibitor). According to Komajda et al. [172] and Laribi et al. [173], over the last three decades, the implementation of guideline-recommended drug treatments for HFrEF has resulted in a decline of HF mortality.

In a large, phase III, randomized, controlled clinical trial (PARADIGM-HF), HF patients with reduced EF (HFrEF) who were treated with sacubitril/valsartan had 20% less risk for cardiovascular death or hospitalization for heart failure (the primary endpoint), 20% less risk for cardiovascular death, 21% less risk for first hospitalization for heart failure, and 16% less risk for death from any cause, compared with enalapril [174]. Concerning tolerability, the sacubitril/valsartan group had higher proportions of patients with hypotension and nonserious angioedema but lower proportions with renal impairment, hyperkalemia, and cough, compared with the enalapril group.

A recently published network meta-analysis of drug treatment for HFrEF has emphasized the benefits of certain drug combinations over the last 30 years [165, 175]. A more comprehensive network meta-analysis of 58 relevant randomized controlled trials (RCTs) on all recommended drug groups for the treatment of HF with reduced ejection fraction was performed to assess the relative efficacy and incremental benefit [172]. The relative efficacy of each treatment group (or combination therapy) in terms of all-cause mortality, cardiovascular mortality, all-cause hospitalizations, and hospitalizations for HF, per patient-year of follow-up, was included in their analysis. The pairwise comparison between each regimen and for each outcome was estimated. Combinations of drug groups showed incremental benefits on outcomes over single groups. Among the different possible combinations, those including the most recently developed drugs, IVA and sacubitril/valsartan are among the most efficacious: ARNI/β-block/MRA and ACEI/β-blocker/MRA/ivabradine tended to be the combinations associated with the greatest reductions of the mortality endpoints and for hospitalizations. There was a relative risk reduction of 62% and 59%, respectively, on all-cause death, 64% and 59% on cardiovascular deaths, 42% and 42% on all-cause hospitalizations, and 73% and 75% on hospitalizations for worsening HF. However, the investigators emphasized that none of these estimates could be taken as strong evidence of superiority over other combinations; instead, options need to be carefully weighed given the characteristics of the individual patient. Numerous factors including rhythm (sinus or AF), HR at baseline, a low systolic blood pressure (SBP), and severe kidney disease can influence the choice of ARNI or ivabradine as third-line therapy in HFrEF based on the respective indication and contraindication of each drug. However, this combination of the two drugs is recommended by the European Society of Cardiology (ESC) guidelines if the patient's profile fits. Their results therefore firmly established the role of combination therapy in HFrEF and reaffirms the validity of international guidelines.

4.1.5 CDT in Cardiac Arrhythmia

A significant fraction of the cardiovascular disease-related deaths is a direct or indirect consequence of cardiac arrhythmias [176, 177]. For example, approximately 20% of all deaths occur suddenly, often due to ventricular tachycardias (VT) degenerating into ventricular fibrillation (VF) and sudden cardiac arrest (SCA), leading to sudden cardiac death (SCD) [178]. Similarly, approximately half of the patients with chronic heart failure (HF) die from ventricular arrhythmias [179]. In addition, atrial fibrillation (AF) is the most common clinically relevant cardiac arrhythmia, affecting 30 million people worldwide. AF is also strongly associated with increased morbidity and mortality and is designated the underlying cause of death in more than 15% of deaths [180]. Despite significant technological advances and applications, including implantable cardioverter defibrillator (ICD), cardiac resynchronization therapy (CRT), pacemaker, and cardiac ablation, antiarrhythmic drugs (AADs) remain the most commonly used treatment for patients with cardiac arrhythmias, which help reduce the arrhythmia burden in a wide range of patients [181–183]. However, currently available AADs have suboptimal efficacy and a relatively high risk of adverse effects, notably drug-induced proarrhythmia or organ toxicity

[184, 185]. Thus, there is a clear unmet clinical need for safer and more effective AADs.

Cardiac electrical activity can be reflected in a whole heart level by electrocardiogram (ECG) recording from the surface of human body, or by action potential (AP) recording in a multicell myocardial preparation and in a single cardiac cell, or even by ion current recording from a patch of cell membrane. The key to understanding cardiac electrophysiology is to understand AP, a cyclic change in voltage (membrane potential) across the cell membrane of heart cells. This is caused by the movement of ions with charges between the inside and outside of the cell, through proteins called ion channels.

A cardiac AP in human and many other species is characterized by a relatively long plateau phase after the initial rapid membrane depolarization, which distinguishes itself from non-cardiac APs, such as nerves [186–190]. A cardiac AP typically contains 5 phases, with phase 0 representing the rapid upstroke evoked by the initial depolarization, then phase 1 displaying a brief and rapid repolarization, followed by characteristic phase 2 manifesting a long isopotential plateau due to the secondary membrane depolarization which is terminated by phase 3 turning into an "all-or-none" repolarization, leading to quick returning back to the phase 4 resting state or phase 4 spontaneous depolarization for the next excitation.

The cardiac AP represents the rapid flux of ions across the cell membrane of excitable tissue. The orderly changes of transmembrane potential during an AP are underlain by finely orchestrated sequential activation and inactivation of ion channels and transporters (the transmembrane proteins embedded across the cytoplasmic membrane of cardiomyocytes) as well as intracellular Ca^{2+} handling proteins [191]. In cardiomyocytes, there are four major categories of ion channels with each of which being responsible for letting one of the four major electrolytes or ions including three cations (Na^+, Ca^{2+}, and K^+) and one anion (Cl^-) to flow across the cell membrane [188, 189]. Inward cation currents (Na^+ and Ca^{2+} influx) or inward anion current (Cl^- outflux) tends to depolarize membrane—conversely, outward cation currents (K^+ influx) or outward anion current (Cl^- influx). In resting state, cardiac cells are normally sitting at a negative transmembrane potential (intracellular ~ − 80 mV). Upon activation, cells are depolarized by rapid entry of Na^+ through Na^+ channels to generate a large fast inward-flowing (depolarizing) Na^+ current (I_{Na}). The maximum rate of voltage upstroke during phase 0 is mainly determined by the size of I_{Na}. Ca^{2+} entry into the cell through the L-type Ca^{2+} current (I_{CaL}) also contributes to phase 0 depolarization. The resultant depolarization travels across the heart and produces a wavefront of excitation. This initial membrane depolarization is ended by Na^+ channel inactivation together with subsequent rapid repolarization due to K^+ efflux carried by a rapidly activating and inactivating transient outward K^+ current (I_{to}). In human atrial myocytes, another K^+ current named ultrarapid delayed rectifier (I_{Kur}) is activated in accompany with I_{to}. Then cardiac cells enter the characteristic plateau phase (phase 2) during which there is a delicate balance between inward currents I_{CaL} together with late Na^+ current ($I_{Na,L}$) and rapid delayed rectifier K^+ current (I_{Kr}) as well as I_{Kur} in atrial myocytes. During this phase, entry of Ca^{2+} through I_{CaL} triggers Ca^{2+}-induced Ca^{2+} release from sarcoplasmic reticulum (SR) Ca^{2+} store through ryanodine receptor 2 (Ryr2), a SR Ca^{2+} channel primarily found in cardiac cells, which facilitate the electro-mechanical coupling for muscle contraction. The plateau phase is then terminated by the time-dependent sequential activation of several outward K^+ currents with initial I_{Kr}, subsequent slow delayed rectifier K^+ current (I_{Ks}), and finally inward rectifier K^+ current (I_{K1}) to bring the membrane to a complete repolarization [190, 191]. During this electrically resting state and mechanically diastolic period, cytoplasmic Ca^{2+} is up-taken back into SR mainly by SERCA2a, a SR Ca^{2+}-ATPase (Ca^{2+} pump), as well as through Ca^{2+} extrusion to extracellular space via the Na^+/Ca^{2+}-exchanger type-1 (NCX1), which stops contraction rendering myocardial relaxation and preventing intracellular Ca^{2+} overload to maintain intracellular Ca^{2+} homeostasis. Ca^{2+} homeostasis in cardiac cells is critically determined by the balances between intracellular and extracellular concentrations and cytoplasmic and SR distributions of Ca^{2+}. In addition, Na^+ and K^+ homeostasis or intracellular/extracellular balance is recovered and maintained primarily through the Na^+/K^+-ATPase (Na^+ pump). Clearly, the morphology and properties of a cardiac AP are determined by the orderly activation and inactivation/deactivation of multiple ion currents and the dynamic balance between inward and outward currents.

4.1.5.1 Antiarrhythmic Drug (AAD) Treatment of Cardiac Arrhythmia

According to the classical Vaughan Williams classification system [192–194], AADs include the following four classes:

Class I AADs

Na^+ channel blockers (NCBs) slowing cardiac conduction velocity (CV) and mainly increasing effective refractory period (ERP) in fast-channel tissues (working atrial and ventricular myocytes, His-Purkinje system). They are further subclassified into drugs with intermediate (Ia), fast (Ib), and slow (Ic) offset kinetics of the Na^+ channel blockade. The primary indications are supraventricular tachycardia (SVT) for Class Ia and Ic drugs and VT for all Class I drugs. In the electrocardiogram (ECG), the effects of Class I AADs may or may not be reflected, dependent on HR, as widening of the P wave and QRS complex, prolongation of the PR interval, or a combination. Class I drugs mainly include:

(a) *Class Ia*: procainamide, disopyramide (Rythmodan), quinidine.
(b) *Class Ib*: lidocaine, tocainide, mexiletine (Teva-Mexiletine).
(c) *Class 1c*: flecainide (Apo-Flecainide, Mar-Flecainide), propafenone (Rythmol), moricizine.

Class II AADs

β-Adrenoceptor (β-AR) antagonists, which affect predominantly slow-channel tissues, mainly including sinoatrial node (SAN) and atrioventricular node (AVN), where they decrease automaticity and heart rate, slow CV, and prolong ERP. These effects are manifested as the PR interval lengthening in ECG recordings. Class II drugs are used primarily to treat SVT, including sinus tachycardia (ST), AVN reentry, AF, and atrial flutter, as well as VT by raising the threshold for VF and reducing the ventricular proarrhythmic effects of β-AR stimulation. Examples of beta-blockers include:

(a) *β_1-Selective*: acebutolol (Sectral), atenolol (Tenormin), betaxolol, bisoprolol (Zebeta), esmolol (Brevibloc), metoprolol (Lopressor, Toprol-XL), nebivolol,
(b) *Non-selective β_1/β_2*: carteolol, carvedilol, labetalol, nadolol (Corgard), penbutolol, pindolol, propranolol (Inderal LA, InnoPran XL), sotalol, timolol.

Class III AADs

Drugs that predominantly block K^+ channels (KCBs) to prolong action potential duration (APD) and ERP and reduce rate of automaticity without affecting CV in both slow- and fast-channel tissues. The predominant effect on the ECG is QT interval prolongation. These AADs are used to treat SVT and VT. Drugs in this class include amiodarone (the most commonly used drug in this class), dronedarone (Multaq), bretylium, dofetilide, ibutilide, and sotalol (Pro-Sotalol, PMS-Sotalol).

Class IV AADs

Non-dihydropyridine L-type Ca^{2+} channel blockers (CCBs) depressing HR, automaticity, and CV and lengthening refractoriness, particularly in slow-response tissues SAN and AVN. These are reflected by lengthening of the PR interval in ECG. They are used primarily to treat SVT and may also be used to slow rapid AF or atrial flutter. Examples of common calcium channel blockers include amlodipine (Norvasc), diltiazem (Cardizem, Tiazac), felodipine, isradipine, nicardipine (Cardene SR), nifedipine (Procardia), nimodipine, nisoldipine (Sular), nitrendipine, and verapamil (Calan, Verelan, Covera-HS).

In addition, some new drugs that have antiarrhythmic efficacy have been recently added to the ADDs. These include the following:

New Class 0 AADs Acting on Sinoatrial Automaticity [195]
Ivabradine is the only one compound assigned to this new class based on its pharmacological efficacy in suppressing the pacemaker current I_f in SAN cells [196] and clinical application to reducing HF in situations of inappropriate ST [197, 198] or when ST accompanies HF [199], by blocking hyperpolarization-activated cyclic nucleotide-gated channel (HCN) [200].

Class Id AADs (Extension of Vaughan Williams Classification)
The Modernized classification [195] adds a new subclass to the original Vaughan Williams Class I, Class Id, by introducing new compounds ranolazine, GS-458967, and F15845 with their actions distinct from those belonging to Class Ia through 1c in that Class Id agents inhibit the relatively small but persistent late Na^+ current ($I_{Na,L}$) that follows the principal rapidly inactivating I_{Na} decay and influences APD and plateau level and secondary membrane depolarization, thereby affecting the generation of early depolarization (EAD). Both clinical and experimental evidence has proven that they have potential antiarrhythmic effects in $I_{Na,L}$-related arrhythmia [201, 202].

Class IIc AADs: Blockers of Receptors Other Than Adrenergic Receptors (Extension of Vaughan Williams Class II)
In addition to sympathetic innervation and adrenergic receptors in the heart, G_i proteins mediate parasympathetic cholinergic muscarinic (M_2) [203] or adenosine (A_1) receptor [204] activation which is also important in determining the vulnerability and propensity of cardiac arrhythmias by affecting membrane excitation and chronotropic and conduction function. Therefore, subclasses are introduced into the original Vaughan Williams Class II. Specifically:

(a) *New Class IIc*: M_2 inhibitors, exemplified by atropine, indicated for relieving sinus bradycardia and supra-His, atrioventricular block at or below the His bundle level.
(b) *New Class IId*: M_2 activator carbachol.
(c) *New Class IIe*: adenosine (A_1) receptor activator adenosine [204]. The major therapeutic use of adenosine is as an antiarrhythmic drug for the rapid treatment of supraventricular tachycardias. Its suppression of atrioventricular conduction makes it very useful in treating paroxysmal supraventricular tachycardia in which the AV node is part of the reentry pathway (as in Wolff-Parkinson-White syndrome). For these indications, adenosine is administered either as bolus intravenous injection or as an intravenous infusion.

Cardiac Glycosides
Cardiac glycosides represent a family of compounds that are derived from the foxglove plant (*Digitalis purpurea*). They act as inhibitors of cellular Na^+/K^+-ATPase.

Electrolyte Supplements (Magnesium and Potassium)
Magnesium is an important ion in many enzymatic reactions, including cardiac Na^+-K^+-ATPase. Hypomagnesemia can inhibit this vital ion transport system and lead to cellular depolarization and associated arrhythmia. Potassium ion plays an important role in membrane potentials, particularly in the resting membrane potential. It is also very important in the repolarization phase 3 of cardiac pacemaker and non-pacemaker APs to participate arrhythmias which include ventricular tachycardia and fibrillation, premature ventricular complexes, supraventricular tachycardias (e.g., Wolff-Parkinson-White syndrome), atrial tachycardias including flutter and fibrillation, and arrhythmias associated with digitalis toxicity.

New Class VII AADs Acting on Upstream Modulatory Targets
The introduction of Class VII aims to encompass cardiac remodeling processes and their consequently longer-term changes that are missing in the original Vaughan Williams classification. Cardiac remodeling involves at least two mechanisms. One is the molecular remodeling involving gene expression deregulation and gene mutation (channelopathies) that influences longer-term changes upstream of the electrophysiological processes, and another is myocardial structural anomaly leading to chronic

scar-related arrhythmogenesis. Many non-AADs including renin-angiotensin-aldosterone inhibitors, omega-3 fatty acids, and statins have demonstrated their antiarrhythmic efficacies via their "upstream therapy" to prevent cardiac electrical and/or structural remodeling [205–207]. The Modernized classification system includes these non-AADs as new AADs in the new Class VII.

Anticoagulants

In addition, for some people, an abnormal heart rhythm changes how the blood flows through their system. For instance, AF may cause blood to pool in the heart, which may result in blood clots leading to stroke. In this case, anticoagulants, though do not fix arrhythmias, help reduce the risk of blood clots. Warfarin (Coumadin) is one of the most common anticoagulants. However, non-vitamin K oral anticoagulants (NOACs) are now recommended over warfarin. NOACs include dabigatran (Pradaxa), rivaroxaban (Xarelto), apixaban (Eliquis), and edoxaban (Savaysa) (for more detail, see "**(3) Coronary artery disease (CAD)**").

4.1.5.2 Combination AAD Therapy

AAD therapy was used to be the mainstay of arrhythmia treatment. Most AADs directly act at the level of the individual ion channel, altering total membrane current and/or intracellular ion fluxes, thereby affecting cardiac electrophysiological properties that modulate the risk of ectopic activity and reentry, as well as other types of rhythm disorders. In addition, many AADs indirectly regulate cardiac electrophysiology by blocking the regulation of the heart by the autonomic nervous system (ANS) through inhibition of various G-protein-coupled receptor signaling pathways. While ion channel blockers primarily produce short-term inhibitory effects on the excitability and electrical activity of cardiomyocytes, ANS antagonists might impose long-term regulatory effects on adverse cardiac electrical and structural remodeling processes. The remodeling processes involve deregulated expression of genes and production of proteins and adverse alterations of the size and configuration of the sarcomeres present in the myocytes, primarily impeding impulse propagation and hence decreases CV. Remodeling is the root cause that precipitates sustain atrial and ventricular arrhythmias, the latter of which often leads to sudden cardiac death (SCD). Short-term acting ion channel blockers, while helping relieve the arrhythmias of concern, often elicit severe to lethal cardiac and extracardiac adverse effects, as revealed by the Cardiac Arrhythmia Suppression Trial (CAST) [208, 209], and because of this, many of them have been withdrawn from clinical use.

Antiarrhythmic treatment with single agents is often ineffective and can be limited by dose-dependent side effects. Therefore, combinations of antiarrhythmic drugs in smaller and well-tolerated doses are advocated in cases refractory to single antiarrhythmic drugs. The combination of two antiarrhythmic agents with complimentary electrophysiological activities and differing toxicities might offer significant advantages in overcoming the drawbacks of standard antiarrhythmic drug therapy [210]. A number of combination therapies have been tested for suppression of complex ventricular ectopy, nonsustained ventricular tachycardia (nsVT) and sustained ventricular tachycardia (sVT), and ventricular fibrillation (VF). Moreover, combinations of antiarrhythmic drugs likely have synergistic antiarrhythmic effects and may also allow the use of lower dosages with reduced side effects and better balance the inward and the outward currents in normalizing AP [211, 212]. The most successful combinations have been the Class Ia/Ib combinations and the Class Ia/Class II combinations. In general, these combinations tend to show higher efficacy rates in suppressing all forms of ventricular ectopy and ventricular tachyarrhythmias and usually have a lower incidence of toxic side effects compared with individual agents alone.

β-Blocker/Class III KCB Combination [213]

Class III AADs (e.g., amiodarone, dronedarone, ibutilide) are frequently used for conversion of AF and atrial flutter (AFL). Postoperative atrial fibrillation (POAF) is associated with a higher rate of postoperative complications and mortality, as well as with longer hospitalization and increased treatment costs. A randomized trial of pharmacological prophylaxis on POAF was conducted to evaluate whether combination therapy is more useful than monotherapy. The study included 240 stable patients with CAD, who were referred to elective surgical revascularization of the myocardium. The patients were assigned into 3 groups of 80 patients each: β-blocker (comparator), β-blocker/amiodarone (a K^+ channel blocker KCB) combination, and β-blocker/rosuvastatin (a statin) combination. Their results showed that the number of new POAF cases is the lowest in β-blocker/amiodarone combination (14; 17.5%) compared to β-blocker alone (25; 31.25%) and β-blocker/rosuvastatin (27; 33.75%). Absolute risk reduction was ≈14% less POAF in β-blocker/amiodarone compared to β-blocker alone, and relative risk reduction was 56%. Hence, prophylaxis of POAF with the β-blocker/amiodarone combination therapy was the most efficient one relative to other therapies. Therefore, amiodarone combined with beta-blockers is often used clinically to maintain sinus rhythm of AF patients [214].

β-Blocker/Class Ic NCB Combination [215]

Recurrence of catecholaminergic polymorphic ventricular tachycardia (CPVT) in patients following β-blocker treatment is a frequent event, and flecainide (a Class Ic Na^+ channel blocker NCB) is an alternative adjunctive antiarrhythmic agent known to provide incomplete protection to CPVT patients. A meta-analysis investigated the efficacy and tolerability of flecainide for CPVT with 7 retrospective cohort studies (RCs in 333 CPVT patients) and 1 RCT involving 333 CPVT patients with 152 treated with flecainide. Their results revealed that flecainide monotherapy was superior to standard therapy in alleviating the risk of arrhythmic events and exercise-induced arrhythmia scores. The β-blocker/flecainide combination was superior to β-blocker monotherapy in reducing the risk of arrhythmic and symptomatic events, peak heart rate, and exercise-induced arrhythmia scores. Flecainide did not increase the risk of all side effects compared to that with β-blockers alone. Yet, due to the limited number of RCTs and patients, the conclusion needs further investigation.

The efficacy of a different form of β-blocker/Class Ic NCB combination with bisoprolol plus propafenone in the treatment of recent-onset AF was evaluated in 164 patients [216]. The results showed that in the initial stages of regularization, the combined therapy restored the sinus rhythm in a greater number of patients in comparison with the monotherapy.

β-Blocker/Class IV CCB Combination [217]

A study involving 144 patients with persistent AF investigated the efficacy of carvedilol (β-blocker) for preventing the recurrence of AF and reducing QT prolongation induced by bepridil (a calcium channel blocker CCB). The pharmacological conversion rate by the carvedilol/bepridil combination was 77%, the candesartan (an antihypertensive)/bepridil combination was 63%, and bepridil alone was 57% (significantly smaller than the carvedilol/bepridil combination). The maintenance of sinus rhythm (SR) at 3 years was 60% in carvedilol plus bepridil, 59% in candesartan plus bepridil, and 40% in bepridil alone (significantly smaller than the carvedilol/bepridil combination). QTc and QT interval were significantly prolonged in candesartan plus bepridil and bepridil alone but not in carvedilol plus bepridil. Hence, the combination therapy with carvedilol plus bepridil is more effective for maintaining SR than bepridil alone therapy.

β-Blocker/Digoxin Combination [218]

A RCT involving 47 patients with persistent AF and HFrEF examined the relative effects of digoxin (Na^+ pump inhibitor), carvedilol (β-blocker), and their combination as these ADDs are known to reduce the ventricular rate and improve symptoms. In the initial phase, digoxin was compared with the combination of digoxin and carvedilol (4 months); in the second phase, digoxin was withdrawn in the carvedilol-treated arm, thus allowing a comparison between digoxin and carvedilol (6 months). Compared with digoxin alone, the carvedilol/digoxin combination significantly lowered the ventricular rate and during submaximal exercise and improved left ventricular ejection fraction (LVEF) and symptom score. There was no significant difference between digoxin and carvedilol monotherapies. The mean ventricular rate increased, and LVEF decreased following a switch from combination therapy to carvedilol monotherapy. Hence, the β-blocker/digoxin combination appears generally superior to either carvedilol or digoxin alone in the management of AF in patients with HF.

A small-scale prospective randomized crossover study in 35 patients with chronic AF including 15 patients with left ventricular dysfunction was conducted with comparison between the two combinations: low-dose betaxolol (β-blocker)/digoxin and low-dose diltiazem (a Ca^{2+} channel blocker)/digoxin [219]. The data suggest that the low-dose betaxolol/digoxin combination was superior to the low-dose diltiazem/digoxin combination in improving ventricular rate, reducing rate-pressure products and exercise capacity, and the beneficial effects were maintained for 7 months.

Atrioventricular (AV) tachycardia includes both AV nodal reentrant tachycardia (AVNRT) and AV reentrant tachycardia (AVRT) using an accessory pathway. The treatment of the acute attack is different from the long-term treatment of both AVNRT and AVRT. Verapamil and adenosine, by prolonging the refractory period of the AV node, are highly effective in terminating acute attacks of AVNRT and orthodromic AVRT. Conversion to sinus rhythm is achieved in approximately 90% of the episodes of tachycardias with both agents given intravenously. Other agents producing lengthening of the refractory period of the accessory pathway in AVRT or of the fast pathway in AVNRT often terminate reentry tachycardia. Such agents are Class IC antiarrhythmic drugs such as flecainide or propafenone and Class IA drugs such as procainamide. In patients with accessory pathways and antidromic tachycardia or AF conducting via an accessory pathway, treatment with verapamil or digoxin should be avoided because these agents may enhance the conduction properties of the accessory pathway, thereby leading to an increase of the ventricular rate or even to ventricular fibrillation. Prevention of AVNRT episodes can be obtained with various antiarrhythmic drugs. Digoxin alone or in combination with beta-blockers is effective in approximately 50% of the cases [220].

Nonetheless, the lack of advantage or even the appearance of intolerability with the β-block/digoxin combination has also been reported [221–223]. The discrepancies are attributable to different treatment regimens (including doses, drug delivery routes, and treatment courses) in different studies.

Class Ia/Class Ib Combination [224]

Programmed ventricular stimulation was used to assess the effect of oral disopyramide, mexiletine, and a combination of both drugs in a randomized crossover study on 12 patients with documented ventricular tachycardia (VT) and/or VF. In two cases, all three types of treatment were ineffective; in four cases, disopyramide and mexiletine monotherapies were ineffective. However, when both drugs were administered in combination, VT was more difficult to induce or was no longer inducible. In two other cases, disopyramide was the more effective drug. In the remaining four patients, all three types of treatment were similarly effective. Seven patients received both disopyramide and mexiletine and tolerated this therapy well during a follow-up of 42 weeks. The conclusion was that the disopyramide/mexiletine combination (the Class Ia/Class Ib combination) is effective and safe in patients with VT, and it is tolerated without major side effects even during long-term therapy.

This conclusion was supported by another study evaluating the efficacies and side effects of disopyramide and mexiletine used alone and in combination in 29 patients with chronic ventricular arrhythmias [225]. A combination of disopyramide and mexiletine in smaller doses (one-half or two-thirds of the conventional doses of each drug) may provide almost the same or enhanced antiarrhythmic effects, no aggravation of ECG parameters, and less incidence of side effects when compared to the conventional dose of each drug alone. The same author also performed a ventricular arrhythmia suppression trials to compare the efficacies and side effects of disopyramide and mexiletine used alone and in combination and of mexiletine and propranolol used alone and in combination in patients with chronic ventricular premature contractions (VPCs, greater than or equal to 3000 beats/day) [226]. Unexpectedly, this study demonstrated that the number of patients with significant reduction in VPCs (≥75%) and the elimination of VT did not differ significantly among the combination and monotherapies. However, while the prematurity index (PI), vulnerability index (VI), and QTc tended to be aggravated by disopyramide monotherapy, these detrimental changes were corrected by combination therapy. No patients withdrew from the study due to side effects during combination therapy, although three patients withdrew from the study due to severe side effects during single-drug therapy.

Another group assessed the efficacy and tolerance of disopyramide and mexiletine used alone and in combination in 21 patients with frequent (≥30/h) PVCs [227]. Additive antiarrhythmic effects of the disopyramide/mexiletine combination were observed with an effective regimen (>83% reduction in VPCs and abolition of nsVT) in 5 (24%) of 21 patients during disopyramide monotherapy, in 3 (14%) receiving mexiletine alone, and in 13 (62%) receiving combination therapy. Similar results were obtained by another research team [228].

A dose escalation crossover study on an alternative Class Ia/Class Ib combination was documented [229]. Specifically, the low-dose quinidine/mexiletine combination therapy was compared with quinidine monotherapy in 15 patients with frequent ventricular premature complexes (VPCs) and nonsustained VT. The investigators observed that the quinidine/mexiletine combination suppressed 80% of VPCs in 13 of 14 patients and suppressed 100% of VT episodes in 6 of 8 patients. By comparison, monotherapy was less effective with 80% suppression of VPCs in 5 of 15 patients and 100% suppression of VT in only 2 of 9 patients. Adverse systemic effects occurred in 3 patients on the combination therapy and in 11 on quinidine monotherapy.

Class Ic/Class III Combination [230]

The safety and efficacy of ibutilide (Class III KCB) when added to propafenone (Class Ic NCB) in treating both paroxysmal and chronic AF and atrial flutter (AFL) were assessed in 202 patients with AF/AFL without left ventricular dysfunction. The sequential administration of ibutilide following initial propafenone offered an overall conversion efficacy of 66.3% (69 of 104 patients), 70.8% for patients with paroxysmal AF/AFL, and 62.5% for patients with chronic AF/AFL. The investigators believed that the propafenone/ibutilide combination is a relatively safe and effective alternative for the treatment of paroxysmal and chronic AF/AFL to both rapidly restore SR in nonresponders to propafenone monotherapy and prevent immediate recurrences of the arrhythmia.

Intriguingly, a similar sequential combination of propafenone and ibutilide in a reverse sequence (propafenone added to ibutilide) was assessed for its therapeutic value in persistent AF [231]. This study unraveled that success rates for cardioversion of AF to SR were 41.1% (21 of 51 patients) for ibutilide alone and 71.4% (35 of 49 patients) for propafenone plus ibutilide. A comparable increase in the QTc interval was observed in both groups, but one case of sustained torsade de pointes, requiring electrical cardioversion, was observed in the propafenone plus ibutilide group. No other complications were noted during the hospitalization period. The results therefore advocate that the Class III/Class Ic combination for pharmacological cardioversion of persistent AF is safe and more effective than ibutilide alone.

Class Id/Class III Combination [232]

Amiodarone (AMIO) is for many years effectively used to control ventricular rate during AF and to convert it into sinus rhythm. However, due to its delayed onset of action, ranolazine (a late Na^+ channel blocker), a new Class Id AAD and antianginal agent with atrial-selective electrophysiological properties, has recently been attempted as add-on therapy with AMIO to facilitate AF conversion. A study in 173 consecutive patients with recent-onset (<48-h duration) AF reported faster sinus rhythm restoration and enhanced conversion rate of AF after the amiodarone/ranolazine combination in patients with preserved EF and left atrial size without serious clinically evident adverse effects.

Another study examined the same combination therapy (the amiodarone/ranolazine combination) and amiodarone monotherapy on conversion rate of postoperative atrial fibrillation (POAF). The results showed that either of the combination or monotherapy converted POAF faster in patients with reduced EF than in those with preserved EF and suggested the amiodarone/ranolazine combination as a valuable alternative treatment for terminating POAF in HFrEF patients [233].

In fact, multiple earlier studies have demonstrated the accelerated termination of AF by the amiodarone/ranolazine combination therapy and addition of ranolazine to standard amiodarone therapy is equally safe and appears to be more effective compared to amiodarone alone for conversion of recent-onset AF [232, 234–236]. While the evidence for the combination is based upon the results from relatively small RCTs, a recent meta-analysis involving ten studies (eight RCTs and two nonrandomized observational studies) came up with the conclusion that ranolazine given orally at appropriate doses showed the property to significantly quicken the conversion of AF to sinus rhythm when combined with amiodarone, compared to amiodarone alone [237].

In a manner similar to in the human studies with the ranolazine/amiodarone combination, ranolazine/dronedarone (an amiodarone derivative) has also been studied, where the combination has again been shown to have enhanced efficacy in terminating AF [212, 238]. The small-scale HARMONY trial reported that a combination of ranolazine with low-dose dronedarone has more pronounced antiarrhythmic effects than either ranolazine monotherapy or dronedarone monotherapy [212]. Thus, targeting a combination of ion channels likely has synergistic antiarrhythmic effects, and it may be desirable to achieve multichannel block through a specific combination of selective compounds, whereby the exact antiarrhythmic mechanism remains unknown. Specifically, midrange ranolazine (750 mg twice daily), midrange dronedarone (225 mg twice daily), ranolazine (750 mg twice daily) combined with low-dose dronedarone (150 mg twice daily), and ranolazine 750 mg twice daily combined with dronedarone 225 mg twice daily vs placebo in the suppression of paroxysmal AF were examined. Study participants were 134 patients with paroxysmal AF demonstrated prior to the initiation of drug administration and with previously implanted permanent pacemakers capable of being interrogated to demonstrate AF burden. The study revealed no efficacy of 225 mg of dronedarone twice daily vs placebo in reducing the AF burden but did report a progressive, absolute decrease in AF burden with 750 mg of ranolazine twice daily, 750 mg of ranolazine twice daily plus 150 mg of dronedarone twice daily, and 750 mg of ranolazine twice daily plus 225 mg of dronedarone twice daily. The tolerance and safety of the combination were excellent. Several experimental studies using in vivo animal models and in vitro cellular models have been performed to confirm and help explain the results of the HARMONY trial [239–243]. The results indicate that targeting a combination of ion channels likely has synergistic antiarrhythmic effects, and it may be desirable to achieve multichannel block through a specific combination of selective compounds, whereby the exact antiarrhythmic mechanism remains unknown. For example, recent work has shown that additional inhibition of specific K^+ currents (I_{SK}, I_{Kr}, or I_{Kur}) also potentiated the AF-selective antiarrhythmic effects of Na^+ channel inhibition with specific state-dependent blockers [241–243], providing a proof of concept for mechanism-based AAD combinations. Most notably, like amiodarone and dronedarone, ranolazine also inhibits multiple cardiac ion channels that carry $I_{Na.L}$ along with the I_{Kr}, I_{Ks}, and I_{CaL} currents, all belonging to MTD.

Class III/Electrolyte Combination

Intravenous magnesium as an adjunctive therapy has been shown to reduce toxicity and enhance efficacy of Class III AADs [244, 245]. Specifically, adjunctive use of intravenous magnesium offers the following advantages:

(a) Magnesium attenuates the corrected QT interval (QTc) prolongation associated with ibutilide.
(b) Concurrent administration of magnesium with ibutilide was associated with a 78% increased odds of successful chemical conversion of AF and AFL to SR compared to ibutilide monotherapy, and 4 g magnesium appeared to provide the greatest benefit [246].
(c) Pre-injection of magnesium significantly enhances the efficacy of ibutilide for the conversion of typical AFL but not of atypical AF [247].
(d) Combined prophylactic therapy with amiodarone and magnesium at the early postoperative period without a maintenance phase is an effective, simple, well-tolerated, and possibly cost-effective regimen to prevent POAF in normomagnesemic, high-risk patients [248].
(e) Sotalol or magnesium monotherapy significantly reduces the incidence of POAF, and this effect is magnified by combining sotalol or magnesium [249].

Digoxin/Electrolyte Combination

A meta-analysis including 10 RCTs containing a total of 515 patients with acute-onset AF (AOAF) indicates that intravenous magnesium was not effective in converting AOAF to SR when compared to placebo or an alternative AAD [250]. When compared to placebo, adding intravenous magnesium to digoxin decreases the proportion of patients with fast ventricular response in AOAF. The use of intravenous magnesium was associated with transient minor symptoms of flushing, tingling, and dizziness in about 17% of the patients.

AAD/Non-AAD Combination

It has been mentioned above that many non-AADs including renin-angiotensin-aldosterone inhibitors, omega-3 fatty acids, and statins have demonstrated their antiarrhythmic efficacies via their "upstream therapeutic effects" to prevent cardiac electrical and/or structural remodeling thereby arrhythmias [205–207]. Some clinical studies provide evidence in support of this idea. **(a)** In patients with paroxysmal AF (PAF), ACEI enhances the efficacy of amiodarone therapy in maintaining sinus rhythm and preventing the development of structural remodeling in atria [251]; **(b)** similarly, the addition of both pravastatin (a lipid-lowering statin) and enalapril (an antihypertensive ACEI) to AAD therapy is more effective for maintaining SR in patients with PAF [252]; **(c)** in patients with post-rheumatic valve replacement, the combination of amiodarone and irbesartan (an antihypertensive ARB) demonstrated a lower rate of AF recurrence after cardioversion than amiodarone alone [253]; **(d)** addition of ascorbic acid to β-blocker for the prophylaxis of POAF is effective, in addition to being well tolerated and relatively safe [254].

Ni et al. [255] presented a novel systems pharmacology framework to quantitatively evaluate synergistic anti-AF effects of combined block of multiple atrial-predominant K^+ currents (ultrarapid delayed rectifier K^+ current, I_{Kur}, small conductance Ca^{2+}-activated K^+ current, I_{KCa}, K2P3.1 2-pore-domain K^+ current, I_{K2P}) in AF. The rationale behind the study is that AF is characterized by rapid and irregular atrial activations, and thus compounds displaying potent antiarrhythmic effects at fast and minimal effects at slow rates are highly desirable. Their results reveal that in both virtual atrial myocytes and tissues, multiple atrial-predominant K^+-current block promotes favorable rate-dependent APD prolongation with increasing synergistic antiarrhythmic effects at fast pacing vs slow rates. The finding suggests that simultaneous block of multiple atrial-predominant K^+ currents may be a valuable antiarrhythmic pharmacotherapeutic strategy for AF.

4.1.5.3 Principles of Combination AAD Therapy

Both experimental and clinical studies have generated evidence for the favorable efficacy and toxicity profiles of AAD/AAD combinations and AAD/non-AAD combinations for preventing and terminating cardiac arrhythmias [256]. Nonetheless, there have not been any standardized protocols or guidelines for the combination therapy. In this regard, Lüderitz et al. [257] provided a useful summary on the principles of combination AAD therapy of cardiac arrhythmias.

In general, agents belonging to the same AAD class should not be combined. However, drugs of different subsets of Class I may be combined.

Agents that have pharmacokinetic interactions, such as quinidine and amiodarone, should not be given together because this combination may be associated with a considerable proarrhythmic effect.

A combination of β-blockers with Class I AADs may be effective, mainly in cases in which the arrhythmia is dependent on adrenergic stimulation.

The combination of Class Ib NCBs and III KCBs can be useful in some cases, from the electrophysiological and clinical point of view. Among the successful combinations of this type are the mexiletine/amiodarone combination, mexiletine/sotalol combination, and tocainide/sotalol combination. For example, in 34 patients, the reduction of ventricular premature beats by sotalol alone was 28% and by sotalol plus mexiletine or tocainide was 79%. Complex ventricular arrhythmias were suppressed by sotalol alone by <40% and by sotalol plus mexiletine or tocainide by >80%. There was no difference in the effectiveness of mexiletine and tocainide (both of them being Class Ib drugs) in this combination.

(v) In patients refractory to amiodarone alone or to a combination with mexiletine, the Class Ic (flecainide or encainide)/amiodarone prolongs the cycle length of VT, but does not suppress induction of VT during programmed stimulation. Combination therapy with amiodarone and encainide is associated with a remarkable incidence of proarrhythmic effects. Nevertheless, a combination of antiarrhythmic drugs improves efficacy of therapy in selected patients.

Collectively, as stated by Greenspan et al. [258], the combination of two antiarrhythmic agents with complimentary electrophysiological activities and differing toxicities might offer significant advantages in overcoming the drawbacks of standard antiarrhythmic drug therapy. A knowledge of the pharmacodynamics of the major classes of antiarrhythmic agents will allow informed choices of drugs for use in combination therapy. Judging antiarrhythmic drug efficacy is a complex problem, requiring an understanding of the influence of the arrhythmia monitoring technique, arrhythmia morphology, and the response to previous drugs on drug efficacy rates, so that accurate comparisons of drug effectiveness can be made among different agents or combinations [258].

4.1.6 CDT in Stroke

A stroke is a medical emergency and requires prompt treatment. With more than 795,000 cases occurring every year, stroke has become a major problem in the United States across all demographics [259]. Stroke is the leading cause of long-term disability and is the fifth leading cause of death in the United States. Ischemic stroke represents 87% of total strokes in the United States and is currently the main focus of stroke research. There are two main causes of stroke: a blocked artery (ischemic stroke) or leaking or bursting of a blood vessel (hemorrhagic stroke).

Ischemic stroke (IS) is the most common type of stroke. It happens when the brain's blood vessels become narrowed or blocked, causing severely reduced blood flow (ischemia). Blocked or narrowed blood vessels are caused by fatty deposits that build up in blood vessels or by blood clots or other debris that travel through your bloodstream and lodge in the blood vessels in the brain. Some people may have only a temporary disruption of blood flow to the brain, known as a transient ischemic attack (TIA) or minor stroke (MS), that does not cause lasting symptoms.

Hemorrhagic stroke (HS) occurs when a blood vessel in your brain leaks or ruptures. Brain hemorrhages can result from many conditions that affect your blood vessels. Factors related to hemorrhagic stroke include uncontrolled high blood pressure, overtreatment with blood thinners (anticoagulants), bulges at weak spots in your blood vessel walls (aneurysms), trauma (such as a car accident), protein deposits in blood vessel walls that lead to weakness in the vessel wall (cerebral amyloid angiopathy), and ischemic stroke leading to hemorrhage.

Stroke is a highly complex process which involves cerebrovasculature and parenchymal tissues through the interaction of multiple mechanisms [260]. Given the complexity of the cellular-biological mechanisms in the pathogenesis of neurovascular dysfunction following stroke, adjuvant pharmacological agents with multi-complementary mechanisms on neurovascular protection may enhance the efficacy of thrombolytic therapy without adversely compromising the integrity of the BBB [261].

4.1.6.1 Prevention and Treatment of Acute Stroke

Overall, the available treatments for ischemic stroke are extremely limited. Some stroke drugs actually break up existing blood clots [259]. Others help prevent blood clots from forming in your blood vessels. Some work to adjust high blood pressure and cholesterol levels to help prevent blood flow blockages. Preventive medications for stroke are mainly to help reduce the risk of having another stroke [262–265]. The common agents used for stroke include:

Antiplatelet Drugs

Antiplatelet drugs make platelets less sticky and less likely to clot [264]. The most commonly used antiplatelet medication is aspirin. Another frequently prescribed drug is Aggrenox, a combination of low-dose aspirin and the antiplatelet drug dipyridamole. For TIA, the aspirin/clopidogrel (Plavix) combination is used for a period of time to reduce the risk of secondary stroke.

Anticoagulants

These drugs reduce blood clotting. Heparin is fast acting and may be used short term in the hospital [263]. Slower-acting warfarin (Coumadin, Jantoven) may be used over a longer term. Several newer blood-thinning medications (anticoagulants) are available for preventing strokes in people who have a high risk. These medications include dabigatran (Pradaxa), rivaroxaban (Xarelto), apixaban (Eliquis), and edoxaban (Savaysa). They are shorter acting than warfarin and usually do not require regular blood tests or monitoring during the medication. These drugs are also associated with a lower risk of bleeding complications (for more detail, see "**(3) Coronary artery disease (CAD)**").

Tissue Plasminogen Activator (tPA)

tPA is the only stroke drug that actually breaks up a blood clot. It is used as a common emergency treatment during a stroke, for which tPA is injected into a vein so it can get to the blood clot quickly. Yet, thrombolysis with tPA carries a higher risk of symptomatic hemorrhagic complications, which hampers the clinical use of tPA; hence, tPA is not used for people at high risk of bleeding into the brain.

Statins

Statins help lower high cholesterol levels. When cholesterol levels are too high, cholesterol can start to build up along the walls of the arteries to build up plaque. Statins block HMG-CoA reductase, an enzyme that is needed to make cholesterol. This helps reduce the risk of plaque and prevent ministrokes and heart attacks caused by clogged arteries (for more detail, see "**(2) Hyperlipidemia**").

Antihypertensives

High blood pressure can play a major role in stroke. It can contribute to chunks of plaque breaking off, which can lead to the formation of a blood clot (for more detail, see "**(1) Hypertension**").

4.1.6.2 Recovery Medications After Stroke

In contrast to acute stroke treatments such as tPA and thrombectomy (surgical removal of a thrombus or blood clot under image guidance), for which the target is a clot and the goal is to salvage threatened brain tissue to limit injury, recovery treatments target surviving neural tissue with the goal of promoting neural repair. Motor deficits are common after stroke (82% of patients) and are linked with reduced quality of life [266, 267].

Drugs that enhance recovery are separate from those that promote neuroprotection or reperfusion in patients with stroke. Recovery-based drugs have distinct therapeutic targets that are related to plasticity and growth following stroke, and in general, improvements in behavioral outcome are not accompanied by a reduction in infarct volume. Apparently, stroke recovery trials target brain tissue to promote neural repair in remaining tissue not vasculature to rescue dying brain tissue as in acute stroke treatments. Conventional therapies for patients recovering from stroke include physical, occupational, and speech therapy, which are delivered across numerous care settings. The two classes of drugs that have been most investigated as stroke recovery treatments to date are serotonergic and dopaminergic.

Serotonergic Drugs

Serotonin normally plays in a role in modulating multiple cognitive functions, particularly response inhibition and memory consolidation, and modulates the impact of punishment-related signals on learning and emotion [268]. Recent reports suggest potential clinical utility of selective serotonin reuptake inhibitor (SSRI) drugs for promoting improved motor outcome after stroke [269, 270]. SSRIs include antidepressants such as citalopram (Celexa), fluoxetine (Prozac, Sarafem), fluvoxamine, paroxetine (Paxil, Pexeva, Brisdelle), and sertraline (Zoloft).

Dopaminergic Drugs

Dopamine regulates many aspects of neural functioning, including excitability, synaptic transmission, plasticity, protein trafficking, and gene transcription and thus has a key role in wide-ranging brain processes such as movement, reward, learning, and plasticity [266, 271]. Dopaminergic agents include dopamine receptor agonists such as apomorphine, bromocriptine, cabergoline, dihydrexidine (LS-186,899), dopamine, fenoldopam, piribedil, lisuride, pergolide, pramipexole, ropinirole, and rotigotine. These drugs are used in the treatment of Parkinson's disease and to treat depression and anxiety.

Noradrenergic Drugs

Noradrenergic neurotransmission broadly amplifies neuronal activity, increases the general level of excitability, and selectively amplifies activities evoked by unexpected inputs. This effect of norepinephrine on regulating overall arousal levels has a modulatory effect on executive function [272]. To date there has been only a handful of studies of noradrenergic drugs to promote stroke recovery.

Cholinergic Drugs

In the cortex, acetylcholine inputs positively enable plasticity by selectively amplifying only anticipated and selectively weakening non-anticipated inputs. Modulation of nicotinic cholinergic neurotransmission alters attention, while muscarinic receptors play a greater role in cognitive flexibility. It has been advocated for cholinergic therapies to enhance recovery, yet very few controlled studies in humans with stroke have been published to date, and limited data in non-motor aspects of stroke recovery are promising [273].

Amphetamine

Amphetamines increase neurotransmission in several monoamine systems. Initial studies of amphetamine to enhance post-stroke motor or language recovery were small but promising [274, 275].

4.1.6.3 Combination Treatment of Stroke

Dual Antiplatelet Therapy (DAPT): Clopidogrel/Aspirin Combination

Numerous studies have evaluated the safety and efficacy of dual antiplatelet therapy (DAPT) with clopidogrel, prasugrel, or ticagrelor in combination with aspirin (ASA) to prevent secondary ischemic events including stroke and TIA. The 2018 Guidelines for Management of Acute Ischemic Stroke state that the clopidogrel/aspirin combination could be beneficial for secondary stroke prevention if initiated within 24 h of the stroke and then continued for 21–90 days [131, 276]. However, several clinical trials have generated unfavorable results on DAPT for prevention of secondary stroke and TIA [277–279].

The Clopidogrel with Aspirin in Acute Minor Stroke or Transient Ischemic Attack (CHANCE) trial was a randomized, double-blind, placebo-controlled trial in China that tested the hypothesis that 3 weeks of dual antiplatelet therapy with clopidogrel and ASA would decrease the risk of a recurrent stroke in comparison to ASA alone in patients with acute high-risk TIA or minor ischemic stroke [280]. This is the first demonstration that administration of aspirin and clopidogrel within 24 h of symptom onset of TIA reduced 90-day stroke recurrence without an increased rate of bleeding. Subsequently, in a one-year outcome follow-up study, the benefit of treatment with clopidogrel and aspirin for 21 days followed by clopidogrel alone for 90 days reduced the risk of subsequent stroke for up to 1 year [281].

The North Americans the Clopidogrel and Aspirin in Acute Ischemic Stroke and High-Risk TIA (POINT) trial was a randomized, double-blind, placebo-controlled trial at 269 sites in 10 countries [282]. Results of this study demonstrated dual treatment patients experiencing fewer events compared to aspirin monotherapy. And statistically significant difference was also seen in the individual secondary efficacy outcomes of IS and HS. Although more efficacious, the clopidogrel and aspirin group also had a statistically significant increase in major hemorrhages and minor hemorrhages compared to aspirin [282].

The Trial to Assess Improvement in Therapeutic Outcomes by Optimizing Platelet Inhibition with Prasugrel-Thrombolysis in Myocardial Infarction 38 (TRITON-TIMI 38) was a phase 3, randomized clinical trial that evaluated clopidogrel versus prasugrel in patients with an acute coronary syndrome awaiting PCI. This study found a benefit of prasugrel over clopidogrel for the composite primary endpoint of MI, stroke, or death from cardiovascular causes (9.9% vs 12.1%). Although prasugrel had fewer cardiovascular complications, the rate of non-CABG TIMI major bleeding was higher in the prasugrel vs clopidogrel group (2.4% vs 1.8%). Importantly, post hoc analysis also revealed a lack of benefit and harm to patients in the prasugrel group that had experienced a previous ischemic stroke or TIA, weighed <60 kilograms, or were > 75 years old. Based on this data, prasugrel is contraindicated in patients with a history of stroke or TIA, and therapy should be discontinued if a patient currently receiving prasugrel experiences one of these events.

Dual Antiplatelet Therapy (DAPT): Ticagrelor/Aspirin Combination

Clinical studies suggest that ticagrelor/aspirin is superior to clopidogrel/aspirin in inhibiting platelet reactivity measured by the PL platelet function analyzer among patients with TIA [283, 284]. This is attributed to that ticagrelor/aspirin therapy possessed greater platelet inhibition and more rapid onset in platelet inhibition compared with clopidogrel/aspirin therapy both in carriers and non-carriers of CYP2C19 lose-of-function alleles with TIA (the polymorphisms of the CYP2C19 gene have been identified to play a significant role in the metabolism of clopidogrel) [285, 286].

Another RCT involving a total of 11,016 patients (5523 in the ticagrelor/aspirin group and 5493 in the aspirin monotherapy) showed that among patients with a mild-to-moderate acute noncardioembolic IS or TIA who were not undergoing intravenous or endovascular thrombolysis, the risk of the composite of stroke or death within 30 days was lower with ticagrelor/aspirin than with aspirin alone, but the incidence of disability did not differ significantly between the two groups [287].

Collectively, dual antiplatelet therapy post-acute minor stroke or high-risk TIA has proved to be efficacious in preventing a future ischemic event, but also puts patients at a higher risk for bleeding complications. More potent P2Y12 inhibitors including prasugrel and ticagrelor as well as intensive therapy with three antiplatelet agents may exacerbate bleeding risk and should be avoided until more data become available. Aspirin is recommended as initial treatment to prevent recurrent ischemic stroke. Clopidogrel is recommended as an alternative monotherapy and in patients allergic to aspirin. The combination of clopidogrel and aspirin is not recommended for long-term use (more than two to three years) because of increased bleeding risk. Aspirin/dipyridamole is at least as effective as aspirin alone, but it is not as well tolerated. Warfarin should not be used for prevention of recurrent ischemic stroke [288]. Further studies are needed to determine the optimal duration for DAPT as benefit appears to be most pronounced in the short-term period post-stroke or TIA (within 7 days), while bleeding risk remains elevated throughout the extended duration of therapy. Lastly, additional clinical trials are needed to provide evidence for treatment preferences in patients with more severe ischemic strokes who are also not candidates for thrombolysis.

Dual Antiplatelet Therapy (DAPT): Cilostazol/Aspirin Combination

A systematic review and meta-analysis to explore the efficacy and safety of cilostazol (an antiplatelet agent) as a mono or combination (plus aspirin or clopidogrel) treatments compared to conventional single antiplatelet therapy (aspirin) for secondary stroke prevention was reported [289]. Based on their analysis, the authors concluded that cilostazol monotherapy is a more effective and safer treatment option than aspirin for IS and HS and the cilostazol/aspirin combination is more effective in reducing recurrence.

Preclinical studies have indicated that combination therapy with normobaric oxygenation (NBO) and ethanol enhances the neuroprotective effect produced by each therapy alone. The mechanism behind this synergistic action is related to changes in cellular metabolism after ischemia reperfusion [290, 291].

4.2 FDC in the Treatment of CVD

Cardiovascular disease (CVD) is the major cause of death in developed countries and is rapidly becoming the major cause of death in the developing world. The increasing rates of obesity and type 2 diabetes, however, may slow the current favorable trends for deaths attributable to CVD in many developed countries. To improve control of risk factors for CVD, Wald and Law proposed the concept of "polypill" in which five commonly used preventative medications were combined into a single pill [292], a typical example of fixed-dose single-pill combination (FDC) or "multiple agents for multiple targets" FDC strategy, for complementary care in both the primary and secondary preventions of cardiovascular disease. Specifically, in their controversial article published in 2003, Wald and Law advocated that a single daily pill combining half-doses (to minimize toxicity) of a β-blocker, thiazide diuretic, and an ACE inhibitor, together with a statin, folic acid, and aspirin, and taken by everyone aged >55, could reduce the incidence of CVD by over 80%. Furthermore, the authors claimed that approximately one in three people would gain an average of 11–12 years of heart attack- and/or stroke-free life. The objective of the polypill was to improve four key CV risk factors simultaneously: LDL-C, BP, serum homocysteine levels, and platelet function. The proposed benefits are based on a series of meta-analyses, which suggest that statin lowers mean LDL-C levels by 1.8 mM (69.7 mg/dL), thereby decreasing the risk of IHD and stroke by 60 and 17%, respectively. Furthermore, half-doses of multiple antihypertensive drugs can significantly lower BP with fewer side effects than the full dose and can reduce the risk of IHD by 46% and stroke by 63%. Importantly, Wald and Law estimated that adverse events would only warrant discontinuation of the pill in 1–2% of patients and that fatal side effects would occur in less than 1 in 10,000 users [292–294]. The beauty of such combination pharmacotherapy is it could be made widely available without treating specific risk factors or individuals [294, 295]. Today, multiple fixed-dose combination polypills are available with differing active components, applications (primary vs secondary prevention), and doses.

In general, each cardiovascular polypill (CV-Polypill) is composed of a minimum of a hypolipidemic agent (e.g., statin), an antiplatelet drug (commonly aspirin), and an antihypertensive agent (ACEI, ARB, thiazide, β-blocker, or calcium channel blocker), each selected based on clinical trial efficacy and pharmacological interaction data [294–296]. A UK, open prospective cohort study assessing the relative reductions in cardiovascular risk for a range of medicinal permutations found that a combination of statins, aspirin, and β-blockers was the most effective (reductions of 83%), followed by a combination of statins, aspirin, and ACEIs (reductions of 71%) [294]. It therefore follows that these are the most common combinations for polypills, although aspirin is often omitted for primary prevention due to unjustifiably increased bleeding risk. In terms of quality of life (QoL), the UMPIRE study reported significantly higher scores on a visual analog scale for the polypill group compared to the standard treatment group, indicating greater levels of patient-perceived health [297, 298]. Estimates focusing mostly on primary prevention show that using such a combination drug, a reduction in the 5-year stroke incidence by 50% can be achieved—especially in low- and middle-income countries with a high prevalence of risk factors even among people at young ages [299]. The benefit from addition of aspirin to CV polypill, that is, a lower incidence of cardiovascular events, has been confirmed by Yusuf et al. [296].

4.2.1 FDC in Hypertension

Data from the National Health and Nutritional Examination Survey and from the World Health Organization have clearly demonstrated that, worldwide, less than 30% of hypertensive patients are adequately controlled by our currently accepted blood pressure goals. Although monotherapy is often unable to achieve blood pressure goals, the use of fixed low-dose combination drugs as alternative treatment seems to be related to a better antihypertensive efficacy and higher response rates in the low range of doses as the result of complementary mechanisms of antihypertensive effects [300]. Indeed, clinical trials have shown that initial low-dose combination therapy is superior as compared with treatment by the stepped-care and the sequential monotherapy approach, while recently, low-dose combination therapy for initial antihypertensive therapy instead of the stepped-care approach or of sequential monotherapy has been recommended [300].

Indeed, the evidence for benefit in polypill/FDCs/FDC for the treatment of hypertension has been well established, showing clear improvements in efficacy and adherence [301] as well as the reduction in cardiovascular events for patients using combination therapy vs individual tablets [302]. Probably the most popular application of FDC to medications of human disease is antihypertensive FDC. A number of international guidelines are recommending early combination therapy, such as CCB-RAS blocker combinations in the majority of patients based on the available evidence, with such combinations showing benefits in terms of compliance, BP lowering and control, and safety [303]. Guidelines such as the 2017 American Heart Association Hypertension guidelines [304] already endorse the use of FDCs in hypertension to improve adherence, with the initiation of treatment with multiple drugs recommended in patients with blood pressure > 20/10 mmHg above their target. The 2018 European Society of Hypertension guidelines have also moved to broaden their recommendations to acknowledge that combination therapy is now recognized as the most effective initial treatment strategy in most patients, with no specific threshold for initiation of combination therapy [305].

Additionally, recent studies have indicated that improved BP control may be achieved with simplified guidelines and the use of FDC, as earlier use of antihypertensive FDC may help to shrink the current gap between antihypertensive use and BP target control [303]. Most guidelines acknowledge that combination therapy is required in the majority of patients, and FDC is regarded as a suitable strategy, having demonstrated better compliance compared with free-dose combinations. Overall, these findings indicate that FDC could be used as first-line treatment [300, 303].

A non-systematic search of PubMed (from 2007 to 2012) performed for randomized, controlled trials in order to capture the evidence on FDC versus monotherapy and free-dose combinations as first-line therapy identified two emergent issues [47]. Firstly, there is a discord between antihypertensive use and actual blood pressure (BP) control achieved—despite an increase in the use of antihypertensives over the last 10 years, BP control rates remain low. Secondly, a greater association between BP and cardiovascular risk in Asians may magnify this discrepancy. These issues could be effectively resolved by using the FDC strategy [47].

Nine trials included in the systematic review by Gupta et al. [301] provided information on BP-lowering efficacy, and additionally, three of those trials reported the influence on both systolic and diastolic blood pressures. Apart from the findings reported by Gupta et al. [301] and described in his systematic review, Kawalec et al. [47] identified three additional clinical trials [306–308] referring to BP-lowering efficacy that were included in the narrative review.

In a cohort study performed by Bronsert et al. [306], patients receiving FDC treatment in comparison with those receiving free-dose combination (FEC) or CDT treatment showed a significantly greater reduction in SBP and DBP. The analysis was adjusted for the difference in confounder variables between the compared groups. A significantly higher rate of BP normalization on FDC treatment in comparison with CDT treatment was also found [306]. Kumagai et al. [307] found that home BP obtained by treatment was significantly lower in patients receiving FDC than those receiving CDT, and this was accompanied by increasing adherence to medication. On the other hand, Ulusoy et al. [308] demonstrated a reduction in all-day, daytime, night-time, and early morning SBP and DBP (BP variability) in the FEC group as compared with the FDC group. However, the conclusions should probably be attributed to the difference in the time of administration of FEC compounds in the control group (valsartan in the morning and amlodipine in the afternoon in the control group; FDC in the morning in the study group) [308].

In addition, based on the synthesis of direct-comparison studies, it was found that FDC treatment is associated with a significant improvement in adherence and persistence in comparison with CDT treatment [47]. Consistently, Sherrill et al. [309] also demonstrated a significantly higher adherence to treatment in the FDC group compared with the CDT group, based on a meta-analysis of studies reporting average MPR in the groups. Likewise, Gupta et al. [301] revealed that the adherence ratio in the FDC group was 1.2 times higher than that in the FEC group. Similarly, Bangalore et al. [310] found that FDC treatment decreased the risk of medication noncompliance in patients with hypertension, tuberculosis, diabetes, and human immunodeficiency virus.

Additionally, the recent TRIUMPH study [311], a randomized controlled trial of low-dose triple-combination therapy vs usual care, demonstrated a significant improvement in the proportion of patients reaching their BP targets within 6 weeks when a low-dose triple-combination pill was prescribed for patients either not on therapy or uncontrolled on monotherapy, thus supporting the early use of low-dose combination therapy within a simplified treatment regimen to improve early blood pressure control.

4.2.2 FDC in Hyperlipidemia

Cardiovascular disease is a major cause of morbidity and mortality throughout the world, and hypercholesterolemia, particularly elevated low-density lipoprotein cholesterol (LDL-C) and reduced high-density lipoprotein cholesterol (HDL-C) levels, is one of the key risk factors. Statins are the first-line treatment to reduce atherogenic lipids, and there is substantial and robust evidence with atorvastatin for reduction of cardiovascular events and mortality. LDL-C targets may not be achieved by statin monotherapy, especially in high-risk patients. Furthermore, in some patient subgroups, atherogenic dyslipidemia is observed. Despite guideline recommendations for intensive therapy among patients at high risk for CHD to lower LDL-C, such lowering has failed to prevent approximately two-thirds of cardiovascular events. As a result of new data, guidelines have begun to focus on non-HDL-C, HDL-C, and triglycerides as treatment targets, with the end result being a recommendation for combination therapy, such as niacin plus statin for the treatment of dyslipidemia. As a result, a combination of a statin with other hypolipidemic drugs may have additional benefits, especially in the form of a single tablet.

Statins, ezetimibe, and fibrates have established their efficacy and safety in the treatment of hypercholesterolemia and hypertriglyceridemia. Clinical trials have shown that these hypolipidemic drug classes can be safely combined in order to augment the lipid-lowering efficacy. Furthermore, novel hypolipidemic drugs such as bempedoic acid and berberine have shown some promising initial results. The combination of different hypolipidemic regimens in a fixed-dose formulation can enhance the adherence to hypolipidemic treatment leading to improved outcomes. Moreover, complementary mechanisms of action of the combined hypolipidemic drugs may also provide additional benefits such as improvement in carbohydrate metabolism. As a result, fixed-dose combinations of hypolipidemic agents may provide an attractive option for the effective and safe management of hypercholesterolemia.

4.2.2.1 Rosuvastatin/Ezetimibe FDC

In this combination, rosuvastatin is an HMG-CoA reductase inhibitor (i.e., statin) with particularly strong inhibitory effects on hepatic cholesterol synthesis, and ezetimibe inhibits the intestinal absorption of cholesterol. Rosuvastatin/ezetimibe FDC is indicated as an adjunctive therapy to diet for the management of primary hypercholesterolemia in adults in numerous countries worldwide [312]. In well-designed clinical trials evaluating the therapeutic efficacy of rosuvastatin/ezetimibe administered as either separate agents or as a FDC, this FDC was significantly more effective than rosuvastatin monotherapy (including at double the dose of rosuvastatin) or simvastatin/ezetimibe in reducing LDL-C and total cholesterol in adults with hypercholesterolemia [313]. Furthermore, rosuvastatin/ezetimibe enabled significantly higher proportions of patients to achieve recommended LDL-C levels than rosuvastatin monotherapy or simvastatin/ezetimibe. The rosuvastatin/ezetimibe FDC did not significantly differ from rosuvastatin monotherapy with respect to incidences of treatment-related or serious adverse events in these short-term trials and displayed a similar safety profile to simvastatin/ezetimibe [313].

4.2.2.2 Simvastatin/Ezetimibe FDC

A double-blind, randomized, 6-week study [314] assessed the lipid-lowering efficacy of simvastatin/ezetimibe 20/10 mg versus atorvastatin 10 or 20 mg and simvastatin/ezetimibe 40/10 mg versus atorvastatin 40 mg in 1128 patients with hypercholesterolemia and metabolic syndrome. The primary endpoint was the percentage of change from baseline in LDL-C. Additional endpoints included changes in other lipids, lipoprotein ratios, high-sensitivity C-reactive protein, and attainment of prespecified lipid levels. Significantly greater improvements in the levels of LDL-C, non-high-density lipoprotein cholesterol (non-HDL-C), apolipoprotein B, and lipid/lipoprotein ratios resulted with simvastatin/ezetimibe compared with atorvastatin at all specified dose comparisons. The attainment of prespecified LDL-C and non-HDL-C levels was also significantly greater with simvastatin/ezetimibe FDC than with atorvastatin at all dose comparisons. HDL-C increases were significantly greater with simvastatin/ezetimibe than with atorvastatin 10 mg and simvastatin/ezetimibe 40/10 mg than with atorvastatin 40 mg. The changes in triglycerides, very-low-density lipoprotein cholesterol (vLDL-C), and high-sensitivity C-reactive protein were similar for both treatments. The incidence of liver, muscle, and gastrointestinal-, hepatitis-, and allergic reaction/rash-related adverse events were low and generally similar to those in previous studies of simvastatin/ezetimibe and/or atorvastatin. The researchers concluded that simvastatin FDC was more likely to result in lipid treatment endpoints than atorvastatin and was generally well tolerated at the doses compared in our patients [314].

4.2.2.3 Simvastatin/Niacin FDC

Compared with statin monotherapy, a combination of niacin and statin therapy provides beneficial effects on a broad range of lipid particles, and some evidence suggests a further reduction in CHD risk. Recent studies have shown that the combination of a fixed dose of extended-release niacin plus simvastatin reduces non-HDL-C, LDL-C, triglycerides, and total cholesterol/HDL-C ratio by approximately 50% while increasing HDL-Cl by 25% [315]. The safety of this combination is consistent with the safety profiles of each individual component and is well tolerated [315].

4.2.2.4 Lovastatin/Niacin FDC

A controlled multicentric trial in India to evaluate the efficacy and safety of a fixed-dose combination of lovastatin and extended-release niacin (niacin-ER) formulation in patients with moderate-to-severe dyslipidemia was reported [316]. The lovastatin/niacin FDC was found to significantly improve cholesterol lipoprotein lipids as well as lipoprotein A and apoA1/apoB levels in Asian Indian dyslipidemic patients. Satisfactory safety and tolerability profile in this population was also demonstrated.

4.2.2.5 Ezetimibe/Simvastatin FDC Plus Extended-Release Niacin

Currently, niacin is the most potent lipid-lowering agent for raising HDL-C levels together with lowering of triglyceride and LDL-C [317–319]. Although the use of extended-release (ER) formulation can reduce flushing, the tolerability and compliance of niacin remain suboptimal. A 24-week multicenter, randomized, double-blind study including 1220 type IIa or IIb hyperlipidemic patients was reported, which involved treatment with ezetimibe/simvastatin FDC (10/20 mg/day) plus extended-release niacin (titrated to 2 g/day) or the FDC or niacin alone [317]. Changes from baseline in LDL-C (primary) and other secondary variables were assessed in the completers and modified intent-to-treat populations. The results demonstrated that the FDC/niacin combination resulted in significantly greater reductions in LDL-C, non-NDL-C, triglycerides, apolipoprotein B, and lipid/lipoprotein ratios, compared with either agent alone. The combination increased levels of apolipoprotein A-I and HDL-C significantly more than the FDC alone and reduced high-sensitivity C-reactive protein levels significantly more than niacin alone [317]. A significantly greater percentage of patients discontinued the study in the niacin alone (25.0%) and the FDC/niacin combination (23.3%) groups, compared with the FDC alone (9.6%, $p < 0.001$) because of clinical adverse experiences (primarily flushing). Incidences of other clinical and laboratory adverse experiences (liver-, muscle-, and gastrointestinal-related) were similar for all groups. The investigators concluded that the FDC/niacin combination showed superior lipid-altering efficacy compared with the FDC or niacin alone in type IIa or IIb hyperlipidemia patients and was generally well tolerated aside from N-associated flushing. This combination offers an effective, broad, lipid-altering therapy with improvements in lipid effects beyond LDL-C in these patients. The same results and conclusions were documented by another group in their study with 942 patients with type IIa/IIb hyperlipidemia for 64 weeks in a randomized, double-blind study [318].

Several randomized, placebo-controlled trials have now confirmed that FDCs/FDC improve blood pressure and cholesterol compared to usual care [320] and to the same degree and with similar rates of adverse events compared to individual component drugs [295, 321–323]. Most recently, the first FDC trial to look at CVD outcomes (PolyIran) found a reduction in major CVD events among the polypill group compared to minimal care, and this reduction was more pronounced when analyzing those with the highest adherence alone [324].

Four trials (UMPIRE, IMPACT, Kanyini-GAP, and FOCUS) have assessed the effects of FDCs in patients either with an established disease or at high calculated risk (≥15% over five years) on adherence, LDL-C, and SBP compared to either usual care or separate drug components [297, 325–327]. All four trials have shown improvements in adherence to CVD preventive medication with the use of a polypill with absolute increases ranging from 25% to 40% at 12 months. A meta-analysis of the Single Pill to Avert Cardiovascular Event (SPACE) Collaboration group of three trials (UMPIRE, IMPACT, and Kanyini-GAP) also showed an improvement in SBP and LDL-C [328].

4.2.3 FDC in Coronary Artery Disease (CAD)

Ischemic heart disease remains the leading cause of health loss in both sexes [329]. Angina pectoris (AP) due to CAD affects approximately 112 million people in the world [330]. Patients with established CAD are likely to receive a combination of aspirin, a statin, and BP-lowering agents. Combining these pharmacologic agents into a single pill (termed polypill or FDC) is recommended by guidelines for reducing prescription gaps and nonadherence in high-risk patients (Table 4.1).

4.2.3.1 Pacemaker Channel Blocker (PCB)/β-Blocker FDC

Heart rate (HR) is one of the main determinants of myocardial oxygen consumption. Elevated HR shortens the length of each cardiac cycle and reduces diastolic perfusion time, decreasing oxygen supply [331]. This imbalance leads to myocardial ischemia, which can trigger angina symptoms [332]. In patients with stable angina, most episodes of exercise-induced myocardial ischemia are preceded by an increase in HR [331, 333]. Development of ischemia is related to baseline HR as well as to the extent and duration of HR increase [334]. The frequency of ambulatory ischemic episodes in treated CAD patients is related to their mean HR, with patients with a HR of 80 beats per minute (bpm) experiencing ischemia almost twice as often as those with a HR of 70 bpm [335]. There is also strong evidence supporting the association between HR and mortality in patients with stable angina [336, 337].

Reducing resting HR is very important when treating patients with stable angina and CAD, and the latest guidelines for the treatment of stable angina recommend reaching target levels of ≤ 60 bpm [331, 338]. β-Blockers are currently recommended as the first-line antianginal treatment for reducing HR and angina symptoms [338]. In the CLARIFY registry, the majority of patients (79%) with angina symptoms were treated with a β-blocker; however, their average HR was above the target levels. Moreover, among the 7301 patients with angina symptoms, only 25% (n = 1818) had their HR controlled (≤60 bpm) [339].

Combining a β-blocker with ivabradine, a pacemaker channel blocker (PCB) has been shown to further reduce the HR of patients with stable angina, along with myocardial ischemia and symptoms of angina [331, 340, 341]. Ivabradine selectively inhibits the pacemaker current, which reduces HR, leading to a decrease in myocardial oxygen demand while increasing both coronary filling time and blood flow, thus protecting against ischemia and angina symptoms [342, 343].

The beneficial effects of combining ivabradine with a β-blocker have been shown in several previously published studies with a similar design [331]. The ASSOCIATE study [340] showed that combining ivabradine with atenolol 50 mg led to an improvement in exercise tolerance test parameters. In the ADDITIONS study, in which more than 2000 patients were prospectively followed for 4 months, combining ivabradine with a β-blocker in stable angina patients led to an additional reduction in HR, weekly angina attacks, and nitrate consumption [341]. An observational, prospective, open-label study in 2403 Greek patients with chronic stable angina receiving ivabradine in combination with β-blockers showed antianginal effectiveness and improvement in quality of life (QoL) [344].

The benefits of the ivabradine and β-blocker combination have also been demonstrated in specific patient subpopulations, including the elderly, post-PCI patients, and patients with comorbidities. In elderly patients (≥ 80 years old) with stable angina, the 4-month, open-label, observational REDUCTION study [345] reported a significant improvement in HR, angina attack frequency, and short-acting nitrate (SAN) consumption with ivabradine twice daily combined with a β-blocker. A subgroup analysis of the ADDITIONS study in elderly patients (≥75 years old) also found that the addition of ivabradine to a β-blocker effectively reduced HR, angina attacks, and nitrate consumption and improved Canadian Cardiovascular Society (CCS) symptom scores and QoL [346]. The antianginal effectiveness of ivabradine administration in addition to a β-blocker has also been shown in post hoc analyses of two observational studies in stable angina patients post-revascularization [331, 347, 348]. A pooled analysis of eight multicenter, randomized, double-blind studies in patients with stable angina found similar reductions in HR and angina attack frequency both in patients with and without diabetes mellitus [349]. A recent analysis of pooled data from 3 observational studies in 8555 stable angina patients receiving ivabradine 2.5 mg, 5 mg, or 7.5 mg twice daily for 4 months found a significant reduction in angina attack frequency and SAN consumption of 87% [350]. Ivabradine was effective and safe in all subpopulations of angina patients seen in clinical practice, independent of age, comorbidities, and use of β-blockers [331, 351].

4.2.3.2 ACEI/CCB FDC

A multicenter, prospective, observational, non-interventional, open-label, 6-month clinical study [352] investigated the fixed-dose combination of perindopril (an ACEI) and amlodipine (a CCB, Ca^{2+} channel blocker) which improved the exercise capacity in patients with stable coronary artery disease (SCAD) after 6 months of treatment. The perindopril/amlodipine FDC produced favorable effects on the cardiovascular system, not only by its BP-lowering effect and its effect on vascular resistance but also through its direct cardiovascular protective effects [352].

4.2.3.3 β-Blocker/ARB/Diuretic FDC

In a phase III, open-label, multicenter study, 254 adults with SCAD having uncontrolled hypertension despite being treated with FDC of metoprolol (25/50 mg) and telmisartan (40 mg) were included [353]. Triple FDC of metoprolol (a β-blocker), telmisartan (an ARB), and chlorthalidone (a diuretic) was considered effective and well tolerated in patients with SCAD and hypertension who respond inadequately to dual therapy [353].

4.2.3.4 Dual Antiplatelet Agent FDC

A FDC of aspirin and clopidogrel bisulfate may improve medication adherence. However, the absence of data on the relative antiplatelet efficacy of FDC and separate dual pills of aspirin and clopidogrel in real-world patients with SCAD is a major factor retarding clinical introduction of such a FDC [354]. A single-center, randomized, open-label, parallel-group, non-inferiority trial studied the outcome in patients who maintained a regimen of separate aspirin and clopidogrel pills for at least 1 year after drug-eluting stent implantation without adverse events were enrolled. Patients were randomly assigned to either the dual antiplatelet aspirin/clopidogrel FDC group or the aspirin/clopidogrel free-dose combination group. Antiplatelet efficacy and tolerability were assessed at baseline and at 4 weeks. The results indicated that in patients with prior percutaneous coronary intervention, the antiplatelet efficacy of the aspirin/clopidogrel FDC was non-inferior to that of free-dose combination, and the tolerability of the two regimens was similar after 4 weeks of treatment [354]. Qualitatively the same results were obtained, and the same conclusions were reached by another study documented by Oh et al. [355].

4.2.3.5 Statin/Antihypertensive FDC Plus Aspirin Combination

Lafeber et al. [356] evaluated the effect of the concomitant use of aspirin, a statin, and BP-lowering agent(s) in patients with CAD on vascular morbidity and mortality in an observational study to provide insights in the combination pill concept related to feasibility and applicability. In total, 2706 patients with CAD enrolled in the Second Manifestations of ARTerial disease study were followed for the occurrence of a subsequent vascular event (i.e., myocardial infarction, ischemic cerebrovascular accident, vascular death) and all-cause mortality. The relationship between combination therapy and cardiovascular events and all-cause mortality was assessed using Cox proportional hazards regression models to calculate hazards ratios with a 95% CI. Both covariate and propensity score adjusting methods were used to reduce confounding by indication [356]. Their results showed statin/antihypertensive FDC plus aspirin combination therapy was associated with a lower risk of myocardial infarction, ischemic cerebrovascular

Table 4.1 FDCs in the treatment of coronary artery disease and stroke

Pathology	FDCs in use or under development
Coronary artery disease	**PCB/β-blocker**: Combining ivabradine with a β-blocker for stable angina patients leading to an additional reduction in HR [345–348] **ACEI/CCB**: The perindopril/amlodipine FDC produced favorable effects on the cardiovascular system, not only by its BP-lowering effect and its effect on vascular resistance but also through its direct cardiovascular protective effects [352] **β-Blocker/ARB/diuretic**: Triple FDC metoprolol/telmisartan/chlorthalidone is effective and well tolerated in patients with stable CAD and hypertension who respond inadequately to dual therapy [353] **Dual antiplatelet agents**: The antiplatelet efficacy of the aspirin/clopidogrel FDC was non-inferior to that of free-dose combination, and the tolerability of the two regimens was similar [354, 355] **Statin/antihypertensive + aspirin**: Statin/antihypertensive FDC + aspirin combination therapy was associated with a lower risk of myocardial infarction, ischemic cerebrovascular accident, composite vascular endpoint, vascular mortality, and all-cause mortality compared with the absence of combination therapy [356] **Statin/β-blocker/diuretic/ACEI + aspirin**: Statin/β-blocker/diuretic/ACEI polypill + aspirin led to a lower incidence of cardiovascular events than did placebo among participants without cardiovascular disease [296]
Stroke	**Dual antiplatelet drug FDC (aspirin/dipyridamole)**: This FDC was found more favorable for the prevention of atherothrombotic cerebral ischemic events than aspirin monotherapy [369] **Anticoagulant/antiplatelet drug FDC (warfarin/aspirin)**: The concept that reducing the dose of warfarin to reduce the associated risk of bleeding in conjunction with use of low-dose aspirin to simultaneously reduce the risk of thromboembolism is a safer alternative needs more evidence **ARB/diuretic FDC (losartan/hydrochlorothiazide)**: Losartan was frequently administered with hydrochlorothiazide in the losartan intervention for endpoint reduction in hypertension (LIFE) study, in which there was a 25% risk reduction for stroke in the losartan-based compared with the atenolol-based treatment group [370]

ACEI angiotensin-converting enzyme inhibitor, *ARB* angiotensin II receptor blocker, *CAD* coronary artery disease, *CCB* calcium channel blocker, *HR* heart rate, *PCB* pacemaker channel blocker

accident, composite vascular endpoint, vascular mortality, and all-cause mortality compared with the absence of combination therapy, after adjusting for confounding covariates in a propensity score. The use of one or only two components of combination therapy was associated with a higher risk for cardiovascular events compared with the FDC combination therapy [356]. Although treatment effect in observational studies should be interpreted with caution, the results of this study support supposed benefits from combination therapy. However, the effect of fixed-dose combination pill on clinical outcome needs to be demonstrated in randomized clinical trials.

4.2.3.6 Statin/β-Blocker/Diuretic/ACEI FDC plus Aspirin Combination

A clinical study [296] randomly treated 5713 participants without cardiovascular disease who had an elevated INTERHEART Risk Score to receive a polypill (containing 40 mg of simvastatin, 100 mg of atenolol, 25 mg of hydrochlorothiazide, and 10 mg of ramipril) or placebo daily, aspirin (75 mg) or placebo daily, and vitamin D or placebo monthly, and the mean follow-up was 4.6 years. For the polypill-alone and polypill-plus-aspirin comparisons, the primary outcome was death from cardiovascular causes, myocardial infarction, stroke, resuscitated cardiac arrest, heart failure, or revascularization. For the aspirin comparison, the primary outcome was death from cardiovascular causes, myocardial infarction, or stroke. Safety was also assessed. The LDL-C level was lower by approximately 19 mg per deciliter, and SBP was lower by approximately 5.8 mm Hg with the polypill and with polypill/aspirin combination therapy than with placebo. Combined treatment with a polypill plus aspirin led to a lower incidence of cardiovascular events than did placebo among participants without cardiovascular disease who were at intermediate cardiovascular risk [296].

4.2.4 FDC in Heart Failure (HF)

Heart rate (HR) control is an effective approach for the management of patients with CAD and HF and their manifestations. Although in epidemiological and, in less degree, in clinical studies derived indications of a possible pathogenetic role of HR in major cardiac diseases, clinical trials did not provide any strong evidence. β-Blockers are the drugs most frequently used for HR control in patients with cardiac disease. However, recent studies constantly find insufficient effectiveness of β-blockers in HR control and the need for an additional therapy. Ivabradine, a pure HR inhibitor, added to classic β-blocker treatment represents the new therapeutic option in HF [357].

Indeed, with its pure negative chronotropic action, it is recommended by European Society of Cardiology and American College of Cardiology/American Heart Association guidelines in symptomatic heart failure patients (NYHA ≥2) with ejection fraction 35% or less, sinus rhythm and HR at least 70 bpm, despite maximally titrated β-blocker therapy [358]. Data supporting this indication mainly derive from the SHIFT study [359, 360], in which ivabradine reduced the combined endpoint of mortality and hospitalization, despite the fact that only 26% of patients enrolled were on optimal β-blocker doses.

Other than ivabradine/β-blocker FDC, other fixed-dose combinations have also been proposed [361, 362]. For example, polypills for heart failure with reduced ejection fraction (HFrEF) have been proposed as an implementation strategy to simplify the current treatment paradigm for undertreated patients with HFrEF [361, 363]. Nonetheless, reports on clinical evaluations on polypills for HF have been overall scanty, if any. In a secondary analysis of a prospective, interrupted times series study, the authors explored the eligibility for a potential HFrEF polypill in 937 participants with HFrEF (n = 494 control period, n = 443 intervention period), more than 80% of patients were eligible for a HFrEF polypill at discharge [363]. Polypill increased rates of hospital discharge from 28.8% in the control period to 42.3% in the intervention period, and routine administration of a HFrEF polypill could increase this rate and improve adherence by 44% [50, 364].

4.2.5 FDC in Stroke

A systematic review was performed for RCTs testing the effect of a fixed-dose combination compared to usual care, placebo, or an active drug comparator for primary or secondary stroke prevention [365]. The search was performed for RCTs published in the databases PubMed and CENTRAL (Cochrane library) before 1 May 2017. The search terms used were "stroke," "cardiovascular," "fixed-dose combination," and "polypill." The authors identified 170 RCTs and clinical trials in PubMed and 251 clinical trials in CENTRAL. None of these trials used stroke as primary outcome [365]. However, in four RCTs stroke was included in the outcome "major cardiovascular event" [50, 323, 327, 365–368]. The follow-up period of these four trials varied between 12 and 23 months with changes in LDL-C, BP levels, and adherence as primary outcomes. Overall, these 4 RCTs included 3615 individuals and reported 28 stroke events, 17 (0.9%) in the polypill group and 11 (0.6%) in the usual care group. Such a low number of events does not allow for a conclusion on the effect of a cardiovascular polypill for stroke prevention. Moreover, to date, no published or ongoing primary or secondary prevention trial has included stroke as a primary outcome. Thus, the benefits of a polypill for primary stroke prevention can only be projected from the reduction in risk factors found in these trials.

4.2.5.1 Dual Antiplatelet Drug FDC (Aspirin/Dipyridamole) for Stroke

The American College of Chest Physicians (ACCP) recommends aspirin 50–325 mg/d to be the initial antiplatelet of choice for the prevention of atherothrombotic cerebral ischemic events [369]. However, with the favorable results of the ESPS-2, it may be appropriate to substitute aspirin/extended-release (ER) dipyridamole for aspirin alone as the drug of choice. Dipyridamole is a nucleoside transport inhibitor and a PDE3 inhibitor that inhibits blood clot formation when given chronically and causes blood vessel dilation when given at high doses over a short time. It is used to dilate blood vessels in people with peripheral arterial disease and CAD. The aspirin/dipyridamole FDC appears to have a favorable adverse effect profile. The relative effectiveness of aspirin/ER dipyridamole compared with clopidogrel and ticlopidine has yet to be determined. If alternative antiplatelet therapy is needed, the ACCP recommends clopidogrel rather than ticlopidine because of its lower incidence of adverse effects. The ACCP further states that the aspirin/dipyridamole FDC may be more effective than clopidogrel; these agents have a similarly favorable adverse effect profile [369].

4.2.5.2 ARB/Diuretic FDC (Losartan/Hydrochlorothiazide) for Stroke

A fixed-dose combination of losartan/hydrochlorothiazide (HCTZ) therapy may be a logical choice for antihypertensive treatment, including for initial therapy in patients with blood pressure elevation >20/10 mmHg above treatment target. The renin-angiotensin-aldosterone-system-activating effect of hydrochlorothiazide augments the efficacy of blocking the angiotensin II type 1 (AT1) receptor with losartan. Some adverse effects associated with hydrochlorothiazide, including increased risk for new-onset diabetes mellitus, may be offset by losartan. Losartan was frequently administered with hydrochlorothiazide in the Losartan Intervention for Endpoint reduction in hypertension (LIFE) study, in which there was a 25% risk reduction for stroke in the losartan-based compared with the atenolol-based treatment group. The efficacy, tolerability, and convenience of losartan/HCTZ combination therapy may increase patient compliance and lower risk for stroke [370], a devastating outcome in patients with hypertension.

4.2.5.3 Anticoagulant/Antiplatelet FDC (Warfarin/Aspirin) for Stroke

Adjusted-dose warfarin is highly efficacious for prevention of ischemic stroke in patients with AF. However, this treatment carries a risk of bleeding and the need for frequent medical monitoring. A safer alternative would be to reduce the dose of warfarin to reduce the associated risk of bleeding in conjunction with use of low-dose aspirin to simultaneously reduce the risk of thromboembolism. To this end, fixed-dose combination with warfarin and aspirin was proposed [371]. However, the efficacy of such a regimen has been in debate. One study involving [372] 1044 patients with AF and with at least one thromboembolic risk factor (congestive heart failure or left ventricular fractional shortening < or = 25%, previous thromboembolism, systolic blood pressure of more than 160 mm Hg at study enrollment, or being a woman aged over 75 years) found that low-intensity, fixed-dose warfarin plus aspirin is insufficient for stroke prevention in patients with non-valvular AF at high risk for thromboembolism; by comparison, adjusted-dose warfarin importantly reduces stroke for high-risk patients. Similarly, another study concluded that adjusted-dose warfarin seemed superior to minidose warfarin and to warfarin plus aspirin after 1 year of treatment for prevention of stroke in AF patients [373].

The concept of a polypill for the prevention of CVD was already proposed for primary prevention as a population treatment strategy by Wald and Law in 2003 [298]. Since then a number of polypills have been developed for the prevention of CVD; however, their use is not widespread [365]. The major barriers are a low level of pharmaceutical investment, little patient advocacy, skepticism of clinicians, and hemorrhagic side effects from aspirin [365, 374]. The current WHO Model List of Essential Medicines 2017 recognizes for cardiovascular medicines "the potential value of fixed-dose combinations of currently listed essential medicines, with regulatory approval and demonstrated bioavailability for the management of chronic NCDs" [375].

Estimates of the initial model of Wald and Law suggest that a polypill strategy could reduce incidence of heart attack by 88% and stroke by 80% if used by everyone aged 55 and older. A more recent model using US national data suggests that with optimal prevention (intake of aspirin for high-risk individuals, control of prediabetes, weight reduction in obese individuals, BP reduction, reduction of LDL-C in patients with coronary artery disease), two out of three myocardial infarctions and one out of three strokes could be prevented [376].

4.3 MTD in the Treatment of CVD

4.3.1 MTD in Hypertension

Hypertension is recognized as one of the leading risk factors for human morbidity and mortality. It is a major risk factor for myocardial infarction, congestive heart failure, stroke, and end-stage renal disease [377]. The difficulty in controlling hypertension is related, at least in part, to the complex pathogenesis of hypertension. In spite of the availability of variety of antihypertensive agents, the control of blood pressure in the general population is at best inadequate. Despite the intensive and extensive studies and applications of the CDT and FDC in an attempt to overcome some of the problems associated with conventional single agent for single-target therapy and enhance the effectiveness in achieving BP control goals, the possibility of certain unpredictable side effects or toxicities associated with these combinational regimens remains because of the potential drug-drug interactions. The ultimate solution to these issues would naturally be multitarget antihypertensive agents or the MTD strategy. Nevertheless, at present, drugs by such a principle are still under development.

For example, MT-1207 is a novel multitarget blocker of α1 receptor, 5-HT2A receptor, and calcium channel that is currently under development for the treatment of hypertension [378] (Table 4.2). In the study documented by Zhang et al. [378], the tolerability, pharmacokinetics (PK), and pharmacodynamics (PD) of MT-1207 in healthy Chinese subjects were evaluated, specifically on the effects of a single-ascending dose (SAD) of MT-1207 (5–40 mg) and MT-1207 (40 mg) administered in combination with food in 56 healthy subjects. No serious adverse events or discontinuations due to adverse events (related to MT-1207) occurred in either study. MT-1207 was rapidly absorbed (median T_{max}: 0.5–1.25 h). The mean t1/2 of MT-1207 was approximately 4–7 h. Systemic exposure (C_{max} and AUC) to MT-1207 increased in proportion to dose. For 4–24 h after administration, the BP drop was greater in the MT-1207 group than in the placebo group. Unfortunately, no comparison with known antihypertensives was attempted in this study, making it impossible to judge whether this multitarget compound yields better efficacy than regular single-target drugs.

In another study, tylerdipine hydrochloride (KBP-5660), a novel L- and T-type dual calcium channel blocker developed for the treatment of hypertension, was examined for its pharmacokinetics, safety, and tolerability in healthy Chinese subjects [379]. However, no pharmacodynamic data were presented in this report [379].

4.3.2 MTD in Cardiac Arrhythmia

Cardiac arrhythmias are major causes of morbidity and mortality [380, 381], with atrial fibrillation, the most prevalent cardiac arrhythmia, being a major risk factor for stroke [382]. Antiarrhythmic drugs are literally the most problematic class of therapeutics, not only because of the uncertainty of antiarrhythmic efficacy but also due to their high profile of proarrhythmic actions [383]. Though the main mechanisms of cardiac arrhythmias are no more than disorders of impulse generation, conduction, and repolarization as described above, the actual mechanism of an arrhythmia in the clinical settings is not clear-cut but difficult, if not impossible, to be determined [186, 188]. Arrhythmias of apparently the same form may well have distinct mechanisms due to alterations of different ion channels. For instance, APD/QT prolongation could be a consequence of impairment of any one or combination of various outward K^+ currents or of enhancement of inward currents I_{CaL} and/or $I_{Na,L}$. Mapping techniques help identify and define the mechanisms of an ongoing arrhythmia, but it can be done only in the surgery room, and it does not lead to understanding of ion channel targets.

Moreover, arrhythmias are dynamic and can transform from one type to another or to others during evolving of the underlying pathological disease or remodeling process. Thus, antiarrhythmic targets are moving targets with which ion channels/transporters/receptors involved in the development of arrhythmias may be constantly changing. For example, ectopic beats or triggered activities can predispose into a sustained reentrant arrhythmia.

Furthermore, antiarrhythmic targets may be complex "moving targets" with overlap or coexistence of multiple mechanisms involving multiple ion channels/transporters/receptors for an arrhythmia in a patient under circumstances.

The currently available AADs were, except for β-blockers, originally developed as specific ion channel blockers or openers. Unfortunately, blocking a specific ion channel can exacerbate the existing arrhythmia and/or induce a new arrhythmia by creating or worsening the imbalance of electrolytes and ion currents and disturbance of electrophysiological homeostasis, called proarrhythmias. Clearly, the concept of specific ion channel blocking is an oversimplified strategy for a complex problem and addresses an incomplete link between the mechanism of AAD action and the clinical efficacy of AADs and the mechanisms that generate/maintain the arrhythmia. However, there are exceptions; Class III agents amiodarone and its derivative dronedarone have much fewer adverse effects without confirmed increase in mortality and less proarrhythmias but with overall superior antiarrhythmic efficacy over other AADs [384, 385]. These compounds are known to act on multiple ion channel targets plus adrenergic antagonizing effect. The possible explanation for this case may be that unlike other AADs, these two agents possess appropriate proportional effects on multiple ion channels and adrenergic receptors.

4.3.2.1 Amiodarone as a MTD Antiarrhythmic

Amiodarone is the most commonly prescribed antiarrhythmic drug owing to its efficacy and minimal proarrhythmic side effects [380, 386].

Table 4.2 MTDs in the treatment of hypertension and arrhythmias

Pathology	MTDs in use or under development
Hypertension	**MT-1207**: a novel multitarget blocker of α1 receptor, 5-HT2A receptor, and calcium channel that is currently under development for the treatment of hypertension [378] **Tylerdipine hydrochloride (KBP-5660)**: a novel L-/T-type dual calcium channel blocker is under studying [379]
Cardiac arrhythmias	**Amiodarone**: Blocking multiple cardiac K^+ currents, I_{Na} and I_{CaL}, as well as α- and β-adrenergic receptors being effective and relatively safe without proarrhythmic effects; an anti-AF agent [387–390] **Dronedarone**: an amiodarone-like ADD designed based on amiodarone as a scaffold; used for AF [385] **Vernakalant**: Blocking fast sodium current and late sodium current, it also inhibits the early activating K^+ channels (I_{Kur}; I_{to}) and I_{KACh}; its unique ion channel-blocking profile makes it an anti-AF ADD [388–390] **Ranolazine**: Blocking $I_{Na,L}$ and peak I_{Na} and I_{Kr} in the atrium and improving intracellular calcium handling; used both alone or in combination with other drugs such as amiodarone and dronedarone **NIP-142**: a novel benzopyrane derivative that blocks potassium (I_{Kur}, I_{KACh}, I_{to}, I_{Kr}, and I_{Ks}) [415–418], calcium, and sodium channels and shows atrial-specific APD prolongation [415, 416]; may be used for AF

AAD antiarrhythmic drug, *AF* atrial fibrillation, *APD* action potential duration, *5-HT2A* serotonin 5-HT2A receptors, I_{CaL} L-type Ca^{2+} current, I_{Kur} ultrarapid delayed rectifier K^+ current, I_{KACh} acetylcholine-sensitive K^+ current, I_{Kr} rapid delayed rectifier K^+ current, I_{Ks} slow delayed rectifier K^+ current, $I_{Na,L}$ late Na^+ current, I_{Na} peak Na^+ current, I_{to} transient outward K^+ current

Amiodarone acts through multiple mechanisms—prolongation of repolarization, reduction of excitability, and slowing of conduction—exerting these effects by altering the function of diverse membrane proteins: ion channels, ion exchangers, and adrenergic receptors [184]. Amiodarone has been shown to block several cardiac K^+ currents, as well as I_{Na} and $I_{Ca,L}$. It is also a noncompetitive antagonist of α- and β-adrenergic receptors. Amiodarone has been demonstrated to be effective and relatively safe without inducing TdP polymorphic ventricular tachycardia that is seen as a result of QT prolongation [384, 387].

Importantly, this multitarget therapeutic mechanism is a feature shared by other antiarrhythmics in current use, including dronedarone, vernakalant, and ranolazine, which have been shown to be effective clinically for AF with low incidence of proarrhythmias [388–390]. These facts suggest that a multipronged mechanism of action may be a desired feature of antiarrhythmic drugs, although the mechanism for such target promiscuity is unclear. In the case of membrane proteins, the concurrent regulation of many different proteins could be caused by a common mechanism arising from drug-dependent changes in lipid bilayer properties that alter the energetic coupling between membrane proteins and their host bilayer [391]. This bilayer-mediated mechanism results from hydrophobic coupling between membrane proteins and the surrounding lipid bilayer [392].

A hallmark of many current and in-development antiarrhythmics is their effects on numerous, diverse membrane proteins [380, 386, 389, 393], as is the case for amiodarone and dronedarone [386, 394, 395]. Although well established, the mechanism(s) underlying these multitarget effects remains unresolved. The prevailing paradigm attributes drug regulation of protein function to direct interactions (binding) of a drug to its target, but this paradigm does not readily explain the multitargeting behavior over a narrow effective concentration range. So, alternatively, amiodarone and dronedarone could act through a common, more general mechanism such as drug-induced changes in the energetic coupling between the bilayer-embedded proteins and their host bilayer [380, 396–399]. Amiodarone and dronedarone alter lipid bilayer properties at their clinical, pharmacologically relevant [184, 386, 394] concentrations. In contrast, propranolol and pindolol block β-adrenergic receptors [386] and have their clinical effects in the micromolar concentration range [400, 401], which does not overlap with the concentrations at which they alter lipid bilayer properties [380].

The concepts of drug promiscuity [402, 403] and polypharmacology [404, 405] denote the effects of a drug on multiple targets, usually with an assumption of direct binding to the target proteins. The study by Rusinova et al. [380], together with those in earlier studies [391, 404–407], suggests an alternative mechanism for drug promiscuity/polypharmacology for membrane proteins, namely, that the drug in question alters the physical properties of the common feature among all membrane proteins: their host lipid bilayer. Amiodarone and dronedarone alter bilayer properties at clinically relevant concentrations and alter the function of multiple membrane proteins [380], suggesting that their promiscuity, and maybe even their clinical effects, may be due, at least in part, to their effects on the lipid bilayer, thus representing a novel form of polypharmacology.

Many other current drugs similarly exhibit multichannel effects [380, 395], and some drug targets, such as Nav1.5, are mechanosensitive [408, 409]. For example, ranolazine, a novel multichannel antiarrhythmic [395], reduces the mechanosensitivity of Nav1.5 channels in mouse cardiomyocytes, which may be caused by changes in lipid bilayer properties [410], as well as the shear-stretch sensitivity of endogenous voltage-dependent currents in cell lines derived from human atrial myocytes [411]. Rusinova et al. [391] have previously found a correlation between drug off-target effects and bilayer-modifying potency in a family of insulin-sensitizing drugs—the thiazolidinediones—where they altered bilayer properties at the same concentration as their effects on the Nav. Moreover, the rank order of the bilayer-modifying potency coincided with severity of their side effects, with troglitazone being the most effective but also the most toxic and the most bilayer modifying [380].

Introduced as an antianginal compound in 1962, amiodarone emerged in the 1970s as a uniquely effective antiarrhythmic and antifibrillatory drug with a multiplicity of properties, the most striking being the lengthening of the repolarization in the atria and ventricles associated with bradycardia but without a significant propensity for inducing torsade de pointes. It was approved by the US FDA in 1985. Although amiodarone is considered a Class III antiarrhythmic agent, it also has Class I, II, and IV actions, making it a unique and effective antiarrhythmic agent. Because of its minimal negative inotropic activity and very low rate of proarrhythmia, it is considered safe in treating arrhythmias in patients with coronary artery disease and left ventricular systolic dysfunction [384, 385, 412–414]. Amiodarone may have clinical value in patients with left ventricular dysfunction and heart failure as first-line treatment for atrial fibrillation, though other agents are available. Amiodarone is useful in acute management of sustained ventricular tachyarrhythmias, regardless of hemodynamic stability [384, 385, 412–414]. Although amiodarone appears to be more effective than other current antiarrhythmics for a rhythm control approach in AF patients, many side effects limit its long-term use. Based on both clinical evidence and experimental data, the idea has arisen that amiodarone might serve as a model for the synthesis and development of newer and novel antiarrhythmic agents devoid of organ toxicity [384]. Amiodarone is an iodinated compound, and the possibility that its molecule may be modified with or without elimination of the iodine is being tested. The deletion of the iodine radical, substituting ethyl with butyl in the side chain of the aromatic ring, and adding a methylsulfonyl radical to the benzofuran component of the molecule lead to the loss of thyroid and pulmonary effects. The resulting compound, dronedarone, is in advanced clinical development [384]. It is likely the first amiodarone congener that will be used in the control of AF [384]. Others are in the earlier phases of development.

4.3.2.2 Dronedarone as an MTD Antiarrhythmic

Dronedarone, a benzofuran amiodarone derivative, is structurally modified to reduce toxicities often associated with chronic amiodarone therapy. With the addition of a methylsulfonyl group, dronedarone is less lipophilic and has a lower tissue accumulation and a much shorter serum half-life of 24 h compared with amiodarone [414]. Dronedarone is also designed without the iodine moieties that are responsible for thyroid dysfunctions associated with amiodarone. Similar to amiodarone, dronedarone exhibits electrophysiological characteristics of all four Vaughan Williams classifications [414]. Phase III clinical trials have shown dronedarone to be effective at reducing ventricular rate, reducing recurrence of AF, and reducing cardiovascular morbidity and mortality in patients with AF or atrial flutter (AFL) [414]. However, dronedarone was associated with increased mortality in one study that included patients with severe heart failure (HL) and left ventricular dysfunction. Overall, dronedarone appears to be well tolerated. The most common side effects are gastrointestinal in nature and include nausea, vomiting, and diarrhea. Because of its more favorable adverse effect profile, dronedarone is likely to be a useful addition to the therapeutic management of AF.

Dronedarone has amiodarone-like antiarrhythmic and electrophysiological properties, despite it having a modified structure and lacking an iodine moiety. Thus, dronedarone lacks amiodarone's organ toxicity (including adverse thyroid and pulmonary effects). The efficacy of dronedarone has been investigated in several clinical trials, proving its effect in the prevention of AF recurrence, rate control in paroxysmal/persistent and permanent AF, reduction of cardiovascular hospitalization or death from any cause, and others. Indirect comparisons with amiodarone, as well as one head-to-head study of the two drugs, indicate that the relative safety of dronedarone may be at a cost of its lower antiarrhythmic efficacy compared with amiodarone. Dronedarone has fewer adverse effects and is better tolerated than amiodarone. Nonetheless, in one head-to-head comparison of dronedarone and amiodarone, the latter drug was superior to dronedarone for maintenance of sinus rhythm post-cardioversion, but dronedarone was safer and better tolerated, with useful benefit to decrease hospitalizations and thus healthcare costs. However, dronedarone was less efficient than amiodarone in maintaining normal sinus rhythm (DIONYSOS trial) [413] and is contraindicated in severe or deteriorating heart failure (ANDROMEDA trial).

Electrophysiological studies show that dronedarone shares amiodarone's multichannel blocking effects, inhibiting transmembrane Na^+, K^+, Ca^{2+}, and slow L-type calcium channels, as well as its antiadrenergic effects. Unlike amiodarone, it has a little effect at thyroid receptors [385]. Possessing both rate- and rhythm-control properties, dronedarone has proved safe and effective in preventing recurrence of AF in patients with persistent AF in the Dronedarone Atrial Fibrillation Study After Electrical Cardioversion (DAFNE) trial, the first prospective randomized trial to evaluate its efficacy and safety. Dronedarone has since undergone further extensive evaluation in three pivotal phase III trials. In two sister studies, the European Trial in Atrial Fibrillation or Flutter Patients Receiving Dronedarone for the Maintenance of Sinus Rhythm (EURIDIS) and American-Australian-African Trial with Dronedarone in Atrial Fibrillation/Flutter Patients for the Maintenance of Sinus Rhythm (ADONIS), dronedarone 400 mg b.i.d. showed significant efficacy against placebo in prevention of AF recurrence [412]. Additionally, in patients with permanent AF, dronedarone was highly effective at controlling ventricular rate on top of standard rate-controlling therapies in the Efficacy and Safety of Dronedarone for the Control of Ventricular Rate during Atrial Fibrillation (ERATO) study [385].

4.3.2.3 Vernakalant as a MTD Antiarrhythmic

Vernakalant [388–390] is a novel aminocyclohexyl ether drug, which has a unique ion channel-blocking profile among drugs used clinically for cardioversion of AF. It exerts a frequency- and voltage-dependent I_{Na} block, including inhibition of $I_{Na,L}$, which is probably the most important of its electrophysiological effects with regard to termination of AF. The powerful effect on intra-atrial conduction at fast rates is due to blockade of this current. However, it also inhibits the early activating K^+ channels (I_{Kur}; I_{to}) and I_{KACh}. I_{Kur} and I_{KACh} are currents which are specific to the atrium and cause prolongation of atrial rather than ventricular refractoriness and may contribute to the efficacy of the drug. The effect of vernakalant on these currents was responsible for its original classification as an ARDA (atrial repolarization-delaying agent). However, vernakalant also blocks the rapidly activating potassium current I_{Kr} which accounts for mild QT prolongation. QRS widening due to I_{Na} blockade also contributes to QT prolongation.

4.3.2.4 Ranolazine as a MTD Antiarrhythmic

Ranolazine was developed as an antianginal agent and was approved by the Food and Drug Administration in 2006 for use in chronic stable angina pectoris. Experimental and clinical studies have shown that it also has antiarrhythmic properties based on the frequency-dependent blockade of peak sodium channel current (peak I_{Na}) and rapidly activating delayed rectifier potassium current (I_{Kr}) in the atria and blockade of late phase of the inward sodium current (late $I_{Na,L}$) and I_{CaL} in the ventricles. It suppresses the early and delayed after depolarizations and is effective in the suppression of atrial and ventricular arrhythmias (off-label use) without significant proarrhythmic effect. Recent clinical studies have revealed the efficacy of ranolazine in prevention of atrial fibrillation in patients with acute coronary syndromes, prevention as well as conversion of postoperative AF after cardiac surgery, conversion of recent-onset AF, and maintenance of sinus rhythm in recurrent AF. Ranolazine has also been shown to reduce ventricular tachycardia and drug-refractory implantable cardioverter defibrillator shocks. The antiarrhythmic effect of ranolazine is preserved in the setting of chronic heart failure, and clinical studies have demonstrated its safety in patients with heart failure.

4.3.2.5 NIP-14 as a MTD Antiarrhythmic

NIP-142 is a novel benzopyrane derivative, which blocks potassium, calcium, and sodium channels and shows atrial-specific action potential duration prolongation [415, 416]. NIP-142 preferentially blocks the ultrarapid delayed rectifier potassium current (I_{Kur}) and the acetylcholine-activated potassium current (I_{KACh}) [435, 436,]. In addition, it also blocks I_{to}, I_{Kr}, and I_{Ks} currents [417, 418] (Table 4.2). Since I_{Kur} and I_{KACh} have been shown to be expressed more abundantly in the atrial than in the ventricular myocardium, the atrial-specific repolarization prolonging effect of NIP-142 is considered to be due to the blockade of these potassium currents. In canine models, NIP-142 was shown to terminate the microreentry-type AF induced by vagal nerve stimulation and the macroreentry-type atrial flutter induced by an intercaval crush. These effects of NIP-142 have been attributed to the prolongation of atrial effective refractory period (ERP), because this compound prolonged atrial ERP without affecting intraatrial and interatrial conduction times in these models. The ERP prolongation by NIP-142 was greater in the atrium than in the ventricle. NIP-142 also terminated the focal activity-type AF induced by aconitine. In addition, NIP-142 reversed the atrial ERP shortening and the loss of rate adaptation induced by short-term rapid atrial pacing in anesthetized dogs. Thus, although clinical trials are required to provide evidence for its efficacy and safety, the novel multiple ion channel blocker, NIP-142, appears to be a useful agent for the treatment of several types of AF with a low risk of proarrhythmic activity. However, whether NIP-142 can ultimately get into market awaits more clinical studies.

4.3.2.6 Other MTD Antiarrhythmics

In a whole, novel antiarrhythmic agents that have been developed are mainly for the management of atrial fibrillation, targeting atrial-specific ion channels [419]. An example is vernakalant, a multichannel blocker of the ultrarapid (I_{Kur}) and acetylcholine-activated (I_{KACh}) potassium currents. This has minimal effects on the ventricles [420] and is thus less likely to induce malignant ventricular arrhythmias. However, there has been no recently licensed drug for the treatment of ventricular arrhythmias, although ranolazine, a late I_{Na} inhibitor licensed for angina, has demonstrated efficacy in experimental studies [421]. There is a need to develop new drugs for several reasons. First, implantable cardioverter defibrillators (ICDs) are not able to prevent the occurrence of ventricular arrhythmias, but can only terminate them, and are not without significant morbidity. Yet, they are the only form of treatment shown to increase life expectancy. Secondly, current pharmacological agents possess significant cardiac and extracardiac side effects, the most concerning of which are their proarrhythmic properties [422].

Recently, experimental data in human cardiomyocytes and the small-scale HARMONY trial have shown that a combination of ranolazine with low-dose dronedarone has more pronounced antiarrhythmic effects than each AAD alone [423, 424]. Thus, targeting a combination of ion channels likely has synergistic antiarrhythmic effects. However, it may be desirable to achieve multichannel block through a specific combination of selective compounds instead of the "shotgun approach" offered by amiodarone or dronedarone with or without additional AADs, whereby the exact antiarrhythmic mechanism remains unknown. For example, recent work has shown that additional inhibition of specific K^+-currents (I_{SK}, I_{Kr} or I_{Kur}) also potentiated the AF-selective antiarrhythmic effects of Na^+-channel inhibition with specific state-dependent blockers [211, 242, 243], providing a proof of concept for mechanism-based AAD combinations. These drug combinations may include compounds directed at the targets discussed in this review.

References

1. Joseph P, Leong D, McKee M, et al. Reducing the global burden of cardiovascular disease, part 1: the epidemiology and risk factors. Circ Res. 2017;121(6):677–94.
2. Leong DP, Joseph PG, McKee M, et al. Reducing the global burden of cardiovascular disease, part 2: prevention and treatment of cardiovascular disease. Circ Res. 2017;121(6):695–710.
3. Virani SS, Alonso A, Aparicio HJ, et al. American Heart Association Council on epidemiology and prevention statistics committee and stroke statistics subcommittee. Heart disease and stroke statistics-2021 update: a report from the American Heart Association. Circulation. 2021;143(8):e254–743.
4. Mendis S, Puska P, Norrving B. Global atlas on cardiovascular disease prevention and control (PDF). World Health Organization in collaboration with the World Heart Federation and the World Stroke Organization; 2011. p. 3–18.
5. Fuchs FD, Whelton PK. High blood pressure and cardiovascular disease. Hypertension. 2020;75(2):285–92.
6. Ettehad D, Emdin CA, Kiran A, et al. Blood pressure lowering for prevention of cardiovascular disease and death: a systematic review and meta-analysis. Lancet. 2016;387(10022):957–67.
7. Working Group on the Summit on Combination Therapy for CVD, Yusuf S, Attaran A, Bosch J, et al. Combination pharmacotherapy to prevent cardiovascular disease: present status and challenges. Eur Heart J. 2014;35(6):353–64.
8. Smith R, McCready T, Yusuf S. Combination therapy to prevent cardiovascular disease: slow progress. JAMA. 2013;309(15):1595–6.
9. Toth PP, Farnier M, Tomassini JE, et al. Statin combination therapy and cardiovascular risk reduction. Futur Cardiol. 2016;12(3):289–315.
10. Combination Pharmacotherapy and Public Health Research Working Group. Combination pharmacotherapy for cardiovascular disease. Ann Intern Med. 2005;143(8):593–9.
11. Elliott WJ. Systemic hypertension. Curr Probl Cardiol. 2007;32(4):201–59.
12. Doroszko A, Janus A, Szahidewicz-Krupska E, et al. Resistant hypertension. Adv Clin Exp Med. 2016;25(1):173–83.
13. Hui R. Hypertension drug therapy. Adv Exp Med Biol. 2020;1177:149–268.
14. Wright JM, Musini VM, Gill R, Wright JM, editors. First-line drugs for hypertension. Cochrane Database Syst Rev. 2018;4:CD001841.
15. Rahmawati R, Bajorek BV. Self-medication among people living with hypertension: a review. Fam Pract. 2017;34(2):147–53.
16. James PA, Oparil S, Carter BL, et al. 2014 evidence-based guideline for the management of high blood pressure in adults: report from the panel members appointed to the Eighth Joint National Committee (JNC 8). JAMA. 2014;311(5):507–20.
17. Zillich AJ, Garg J, Basu S, et al. Thiazide diuretics, potassium, and the development of diabetes: a quantitative review. Hypertension. 2006;48(2):219–24.
18. Wang TJ, Ausiello JC, Stafford RS. Trends in antihypertensive drug advertising, 1985–1996. Circulation. 1999;99(15):2055–7.
19. Handler J, Lackland DT. Translation of hypertension treatment guidelines into practice: a review of implementation. J Am Soc Hypertens. 2011;5(4):197–207.
20. Wu HY, Huang JW, Lin HJ, et al. Comparative effectiveness of renin-angiotensin system blockers and other antihypertensive drugs in patients with diabetes: systematic review and Bayesian network meta-analysis. BMJ. 2013;347:f6008.
21. Wright JT, Bakris G, Greene T, et al. Effect of blood pressure lowering and antihypertensive drug class on progression of hypertensive kidney disease: results from the AASK trial. JAMA. 2002;288(19):2421–31.
22. The Allhat Officers and Coordinators for The Allhat Collaborative Research Group. Major outcomes in high-risk hypertensive patients randomized to angiotensin-converting enzyme inhibitor or calcium channel blocker vs diuretic: The Antihypertensive and Lipid-Lowering Treatment to Prevent Heart Attack Trial (ALLHAT). JAMA. 2002;288(23):2981–97.
23. Leenen FH, Nwachuku CE, Black HR, et al. Clinical events in high-risk hypertensive patients randomly assigned to calcium channel blocker versus angiotensin-converting enzyme inhibitor in the antihypertensive and lipid-lowering treatment to prevent heart attack trial. Hypertension. 2006;48(3):374–84.
24. Yusuf S, Pitt B, Davis CE, et al. Effect of enalapril on survival in patients with reduced left ventricular ejection fractions and congestive heart failure. N Engl J Med. 1991;325(5):293–302.
25. Li ECK, Heran BS, et al. Angiotensin converting enzyme (ACE) inhibitors versus angiotensin receptor blockers for primary hypertension. Cochrane Database Syst Rev. 2014;2014(8):CD009096.
26. Brunton L, Parker K, Blumenthal D, et al. Therapy of hypertension. In: Goodman & Gilman's manual of pharmacology and therapeutics. New York: McGraw-Hill; 2007. p. 544–60.
27. Varon J, Marik PE. The diagnosis and management of hypertensive crises. Chest. 2000;118(1):214–27.
28. Correale M, Ferraretti A, Monaco I, et al. Endothelin-receptor antagonists in the management of pulmonary arterial hypertension: where do we stand? Vasc Health Risk Manag. 2018;14:253–64.
29. Pulido T, Adzerikho I, Channick RN, et al. SERAPHIN Investigators. Macitentan and morbidity and mortality in pulmonary arterial hypertension. N Engl J Med. 2013;369(9):809–18.
30. Lindholm LH, Carlberg B, Samuelsson O. Should beta blockers remain first choice in the treatment of primary hypertension? A meta-analysis. Lancet. 2005;366(9496):1545–53.

31. Carlberg B, Samuelsson O, Lindholm LH. Atenolol in hypertension: is it a wise choice? Lancet. 2004;364(9446):1684–9.
32. Freemantle N, Cleland J, Young P, et al. Beta Blockade after myocardial infarction: systematic review and meta regression analysis. BMJ. 1999;318(7200):1730–7.
33. LLHAT Officers and Coordinators for the ALLHAT Collaborative Research Group. Diuretic versus alpha-blocker as first-step antihypertensive therapy: final results from the Antihypertensive and Lipid-Lowering Treatment to Prevent Heart Attack Trial (ALLHAT). Hypertension. 2003;42(3):239–46.
34. Garjón J, Saiz LC, Azparren A, et al. First-line combination therapy versus first-line monotherapy for primary hypertension. Cochrane Database Syst Rev. 2020;2(2):CD010316.
35. MacDonald TM, Williams B, Webb DJ, et al. British Hypertension Society Programme of Prevention and Treatment of Hypertension with Algorithm-based Therapy (PATHWAY): combination therapy is superior to sequential monotherapy for the initial treatment of hypertension: a double-blind randomized controlled trial. J Am Heart Assoc. 2017;6(11):e006986.
36. Tsioufis C, Thomopoulos C. Combination drug treatment in hypertension. Pharmacol Res. 2017;125(Pt B):266–71.
37. Gorostidi M, de la Sierra A. Combination therapy in hypertension. Adv Ther. 2013;30(4):320–36.
38. Cappuccio FP, MacGregor GA. Combination therapy in hypertension. J Hum Hypertens. 1991;5(Suppl 2):9–14.
39. Gradman AH. Strategies for combination therapy in hypertension. Curr Opin Nephrol Hypertens. 2012;21(5):486–91.
40. Düsing R, Waeber B, Destro M, et al. Triple-combination therapy in the treatment of hypertension: a review of the evidence. J Hum Hypertens. 2017;31(8):501–10.
41. Smith DK, Lennon RP, Carlsgaard PB. Managing hypertension using combination therapy. Am Fam Physician. 2020;101(6):341–9.
42. Stergiou GS. Combination pharmacotherapy in hypertension. Int Urol Nephrol. 2006;38(3–4): 673–82.
43. Paz MA, de-La- Sierra A, Sáez M, et al. Treatment efficacy of anti-hypertensive drugs in monotherapy or combination: ATOM systematic review and meta-analysis of randomized clinical trials according to PRISMA statement. Medicine (Baltimore). 2016;95(30):e4071.
44. Taddei S. Combination therapy in hypertension: what are the best options according to clinical pharmacology principles and controlled clinical trial evidence? Am J Cardiovasc Drugs. 2015;15(3):185–94.
45. Wald DS, Law M, Morris JK, et al. Combination therapy versus monotherapy in reducing blood pressure: meta-analysis on 11,000 participants from 42 trials. Am J Med. 2009;122(3):290–300.
46. Abegaz TM, Shehab A, Gebreyohannes EA, et al. Nonadherence to anti-hypertensive drugs: a systematic review and meta-analysis. Medicine (Baltimore). 2017;96(4):e5641.
47. Kawalec P, Holko P, Gawin M, et al. Effectiveness of fixed-dose combination therapy in hypertension: systematic review and meta-analysis. Arch Med Sci. 2018;14(5):1125–36.
48. Ferrario CM, Panjabi S, Buzinec P, et al. Clinical and economic outcomes associated with amlodipine/renin-angiotensin system blocker combinations. Ther Adv Cardiovasc Dis. 2013;7(1):27–39.
49. Baser O, Andrews LM, Wang L, et al. Comparison of real-world adherence, healthcare resource utilization and costs for newly initiated valsartan/amlodipine single-pill combination versus angiotensin receptor blocker/calcium channel blocker free-combination therapy. J Med Econ. 2011;14(5):576–83.
50. Bahiru E, de Cates AN, Farr MR, et al. Fixed-dose combination therapy for the prevention of atherosclerotic cardiovascular diseases. Cochrane Database Syst Rev. 2017;3:CD009868.
51. Rodgers PT. Combination drug therapy in hypertension: a rational approach for the pharmacist. J Am Pharm Assoc (Wash). 1998;38(4):469–79.
52. Atkins ER, Chow CK. Low-dose combination therapy for initial treatment of hypertension. Curr Hypertens Rep. 2020;22(9):65.
53. Gradman AH, Parisé H, Lefebvre P, et al. Initial combination therapy reduces the risk of cardiovascular events in hypertensive patients: a matched cohort study. Hypertension. 2013;61(2):309–18.
54. Rea F, Corrao G, Merlino L, et al. Initial antihypertensive treatment strategies and therapeutic inertia. Hypertension. 2018;72(4):846–53.
55. Byrd JB, Zeng C, Tavel HM, et al. Combination therapy as initial treatment for newly diagnosed hypertension. Am Heart J. 2011;162(2):340–6.
56. Tedesco MA, Natale F, Calabrò R. Effects of monotherapy and combination therapy on blood pressure control and target organ damage: a randomized prospective intervention study in a large population of hypertensive patients. J Clin Hypertens (Greenwich). 2006;8(9):634–41.
57. Egan BM, Bandyopadhyay D, Shaftman SR, et al. Initial monotherapy and combination therapy and hypertension control the first year. Hypertension. 2012;59(6):1124–31.
58. Salam A, Kanukula R, Atkins E, et al. Efficacy and safety of dual combination therapy of blood pressure-lowering drugs as initial treatment for hypertension: a systematic review and meta-analysis of randomized controlled trials. J Hypertens. 2019;37(9):1768–74.
59. Weber MA, Schiffrin EL, White WB, et al. Clinical practice guidelines for the management of hypertension in the community: a statement by the American Society of Hypertension and the International Society of Hypertension. J Hypertens. 2014;32(1):3–15.

60. Cifu AS, Davis AM. Prevention, detection, evaluation, and management of high blood pressure in adults. JAMA. 2017;318(21):2132–4.
61. Williams B, Mancia G, Spiering W, ESC Scientific Document Group, et al. 2018 ESC/ESH guidelines for the management of arterial hypertension. Eur Heart J. 2018;39(33):3021–104.
62. Garjón J, Saiz LC, Azparren A, et al. First-line combination therapy versus first-line monotherapy for primary hypertension. Cochrane Database Syst Rev. 2017;1:CD010316.
63. Xu W, Goldberg SI, Shubina M, et al. Optimal systolic blood pressure target, time to intensification, and time to follow-up in treatment of hypertension: population based retrospective cohort study. BMJ. 2015;350:h158.
64. Weir MR, Hsueh WA, Nesbitt SD, et al. A titrate-to-goal study of switching patients uncontrolled on antihypertensive monotherapy to fixed-dose combinations of amlodipine and olmesartan medoxomil ± hydrochlorothiazide. J Clin Hypertens (Greenwich). 2011;13(6):404–12.
65. Volpe M, Christian Rump L, Ammentorp B, et al. Efficacy and safety of triple antihypertensive therapy with the olmesartan/amlodipine/hydrochlorothiazide combination. Clin Drug Investig. 2012;32(10):649–64.
66. Salam A, Atkins ER, Hsu B, et al. Efficacy and safety of triple versus dual combination blood pressure-lowering drug therapy: a systematic review and meta-analysis of randomized controlled trials. J Hypertens. 2019;37(8):1567–73.
67. Southgate L, Machado RD, Gräf S, et al. Molecular genetic framework underlying pulmonary arterial hypertension. Nat Rev Cardiol. 2020;17(2):85–95.
68. Dodson MW, Brown LM, Elliott CG. Pulmonary arterial hypertension. Heart Fail Clin. 2018;14(3):255–69.
69. Coons JC, Pogue K, Kolodziej AR, et al. Pulmonary arterial hypertension: a pharmacotherapeutic update. Curr Cardiol Rep. 2019;21(11):141.
70. Klinger JR, Elliott CG, Levine DJ, et al. Therapy for pulmonary arterial hypertension in adults: update of the CHEST guideline and expert panel report. Chest. 2019;155(3):565–86.
71. Parikh V, Bhardwaj A, Nair A. Pharmacotherapy for pulmonary arterial hypertension. J Thorac Dis. 2019;11(Suppl 14):S1767–81.
72. Burks M, Stickel S, Galiè N. Pulmonary arterial hypertension: combination therapy in practice. Am J Cardiovasc Drugs. 2018;18(4):249–57.
73. Avdeev SN, TsarEva NA, Gaisin IR. Combination therapy is a new standard for treatment of pulmonary arterial hypertension. Ter Arkh. 2018;90(3):72–80.
74. Sitbon O, Gaine S. Beyond a single pathway: combination therapy in pulmonary arterial hypertension. Eur Respir Rev. 2016;25(142):408–17.
75. Levinson AT, Klinger JR. Combination therapy for the treatment of pulmonary arterial hypertension. Ther Adv Respir Dis. 2011;5(6):419–30.
76. Johnson SR, Brode SK, Mielniczuk LM, et al. Dual therapy in IPAH and SSc-PAH. A qualitative systematic review. Respir Med. 2012;106:730–9.
77. Galiè N, Barberà JA, Frost AE, et al. Initial use of ambrisentan plus tadalafil in pulmonary arterial hypertension. N Engl J Med. 2015;373:834–44.
78. Sitbon O, Channick R, Delcroix M, et al. Effect of macitentan on haemodynamics in SERAPHIN, a randomised controlled trial in pulmonary arterial hypertension. Eur Respir J. 2013;42(Suppl 57):P4060.
79. Pulido T, Adzerikho I, Channick R, et al. Macitentan and morbidity and mortality in pulmonary arterial hypertension. N Engl J Med. 2013;369:809–18.
80. Karr S. Epidemiology and management of hyperlipidemia. Am J Manag Care. 2017;23(9 Suppl):S139–48.
81. Wierzbicki AS, Mikhailidis DP, Wray R, et al. Statin-fibrate combination: therapy for hyperlipidemia: a review. Curr Med Res Opin. 2003;19(3):155–68.
82. Bergheanu SC, Bodde MC, Jukema JW. Pathophysiology and treatment of atherosclerosis: current view and future perspective on lipoprotein modification treatment. Neth Hear J. 2017;25(4):231–42.
83. Last AR, Ference JD, Menzel ER. Hyperlipidemia: drugs for cardiovascular risk reduction in adults. Am Fam Physician. 2017;95(2):78–87.
84. Centers for Disease Control and Prevention (CDC). Vital signs: prevalence, treatment, and control of high levels of low-density lipoprotein cholesterol—United States, 1999–2002 and 2005–2008. MMWR Morb Mortal Wkly Rep. 2011;60(4):109–14.
85. Jacobson TA, Ito MK, Maki KC, et al. National Lipid Association recommendations for patient-centered management of dyslipidemia: part 1—full report. J Clin Lipidol. 2015;9(2):129–69.
86. Jellinger PS, Handelsman Y, Rosenblit PD, et al. American Association of Clinical Endocrinologists and American College of Endocrinology guidelines for management of dyslipidemia and prevention of cardiovascular disease. Endocr Pract. 2017;23(suppl 2):1–87.
87. Stone NJ, Robinson JG, Lichtenstein AH, et al. 2013 ACC/AHA guideline on the treatment of blood cholesterol to reduce atherosclerotic cardiovascular risk in adults. J Am Coll Cardiol. 2014;63(25):2889–934.
88. Ai C, Zhang S, He Q, Shi J. Comparing the combination therapy of ezetimibe and atorvastatin with atorvastatin monotherapy for regulating blood lipids: a systematic review and meta-analyse. Lipids Health Dis. 2018;17(1):239.
89. Davidson M, McKenney J, Stein E, et al. Comparison of one-year efficacy and safety of atorvastatin versus lovastatin in primary hypercholesterolemia. Atorvastatin Study Group I. Am J Cardiol. 1997;79(11):1475–81.
90. Jones MR, Nwose OM. Role of colesevelam in combination lipid-lowering therapy. Am J Cardiovasc Drugs. 2013;13(5):315–23.

91. National Cholesterol Education Program Expert Panel on Detection, Evaluation and Treatment of High Blood Cholesterol in Adults (Adult Treatment Panel III). Third Report of the National Cholesterol Education Program (NCEP) Expert Panel on Detection, Evaluation, and Treatment of High Blood Cholesterol in Adults (Adult Treatment Panel III) final report. Circulation. 2002;106(25):3143–421.
92. Shepherd J, Cobbe SM, Ford I, The West of Scotland Coronary Prevention Study Group, et al. Prevention of coronary heart disease with pravastatin in men with hypercholesterolemia. N Engl J Med. 1995;333(20):1301–7.
93. Scandinavian Simvastatin Survival Study Group. Randomised trial of cholesterol lowering in 4444 patients with coronary heart disease: the Scandinavian Simvastatin Survival Study (4S). Lancet. 1994;344(8934):1383–9.
94. Kotseva K, Wood D, De Backer G, The EUROASPIRE Study Group, et al. Cardiovascular prevention guidelines in daily practice: a comparison of EUROASPIRE I, II, and III surveys in eight European countries. Lancet. 2009;373(9667):929–40.
95. Egan A, Colman E. Weighing the benefits of high-dose simvastatin against the risk of myopathy. N Engl J Med. 2011;365(4):285–7.
96. Pappa E, Rizos CV, Filippatos TD, Elisaf MS. Emerging fixed-dose combination treatments for hyperlipidemia. J Cardiovasc Pharmacol Ther. 2019;24(4):315–22.
97. Ferri N, Corsini A. Efficacy and safety of bempedoic acid added to ezetimibe in statin-intolerant patients with hypercholesterolemia: a randomized, placebo-controlled study. Clinical pharmacology of statins: an update. Curr Atheroscler Rep. 2020;22(7):26.
98. Crismaru I, Pantea Stoian A, Bratu OG, et al. Low-density lipoprotein cholesterol lowering treatment: the current approach. Lipids Health Dis. 2020;19(1):85.
99. Davidson MH, Toth P, Weiss S, et al. Low-dose combination therapy with colesevelam hydrochloride and lovastatin effectively decreases low-density lipoprotein cholesterol in patients with primary hypercholesterolemia. Clin Cardiol. 2001;24(6):467–74.
100. Hunninghake D, Insull W Jr, Toth P, et al. Coadministration of colesevelam hydrochloride with atorvastatin lowers LDL cholesterol additively. Atherosclerosis. 2001;158(2):407–16.
101. Knapp HH, Schrott H, Ma P, et al. Efficacy and safety of combination simvastatin and colesevelam in patients with primary hypercholesterolemia. Am J Med. 2001;110(5):352–60.
102. Florentin M, Liberopoulos EN, Rizos CV, et al. Colesevelam plus rosuvastatin 5 mg/day versus rosuvastatin 10 mg/day alone on markers of insulin resistance in patients with hypercholesterolemia and impaired fasting glucose. Metab Syndr Relat Disord. 2013;11(3):152–6.
103. Bays HE, Davidson M, Jones MR, et al. Effects of colesevelam hydrochloride on low-density lipoprotein cholesterol and high-sensitivity C-reactive protein when added to statins in patients with hypercholesterolemia. Am J Cardiol. 2006;97(8):1198–205.
104. Jialal I, Abby SL, Misir S, et al. Concomitant reduction in low-density lipoprotein cholesterol and glycated hemoglobin with colesevelam hydrochloride in patients with type 2 diabetes: a pooled analysis. Metab Syndr Relat Disord. 2009;7(3):255–8.
105. Chapman MJ. Fibrates: therapeutic review. Br J Diab Vasc Dis. 2006;6(1):11–20.
106. McKenney J, Jones M, Abby S. Safety and efficacy of colesevelam hydrochloride in combination with fenofibrate for the treatment of mixed hyperlipidemia. Curr Med Res Opin. 2005;21(9):1403–12.
107. Pandor A, Ara RM, Tumur I, et al. Ezetimibe monotherapy for cholesterol lowering in 2,722 people: systematic review and meta-analysis of randomized controlled trials. J Intern Med. 2009;265(5):568–80.
108. Bays H, Rhyne J, Abby S, et al. Lipid-lowering effects of colesevelam HCl in combination with ezetimibe. Curr Med Res Opin. 2006;22(11):2191–200.
109. Moore A, Phan BAP, Challender C, et al. Effects of adding extended-release niacin and colesevelam to statin therapy on lipid levels in subjects with atherosclerotic disease. J Clin Lipidol. 2007;1(6):620–5.
110. Huijgen R, Abbink EJ, Bruckert E, The Triple Study Group, et al. Colesevelam added to combination therapy with a statin and ezetimibe in patients with familial hypercholesterolemia: a 12-week, multicenter, randomized, double-blind, controlled trial. Clin Ther. 2010;32(4):615–25.
111. Sharma M, Ansari MT, Abou-Setta AM, et al. Systematic review: comparative effectiveness and harms of combination therapy and monotherapy for dyslipidemia. Ann Intern Med. 2009;151(9):622–30.
112. The AIM-HIGH Investigators. Niacin in patients with low HDL cholesterol levels receiving intensive statin therapy. N Engl J Med. 2011;365(24):2255–67.
113. American Diabetes Association. Standards of medical care in diabetes—2013. Diabetes Care. 2013;36(Suppl 1):S11–66.
114. Reiner Ž, Catapano AL, De Backer G, et al. ESC/EAS guidelines for the management of dyslipidaemias. The Task Force for the management of dyslipidaemias of the European Society of Cardiology (ESC) and the European Atherosclerosis Society (EAS). Eur Heart J. 2011;32:1769–818.
115. Reiner Ž. Combined therapy in the treatment of dyslipidemia. Fundam Clin Pharmacol. 2010;24(1):19–28.
116. Libby P, Theroux P. Pathophysiology of coronary artery disease. Circulation. 2005;111(25):3481–8.
117. Ferrari R, Fox K. Heart rate reduction in coronary artery disease and heart failure. Nat Rev Cardiol. 2016;13(8):493–501.

118. Fox KAA, Metra M, Morais J, et al. The myth of 'stable' coronary artery disease. Nat Rev Cardiol. 2020;17(1):9–21.
119. Ross R, Glomset J, Harker L. Response to injury and atherogenesis. Am J Pathol. 1977;86:675–84.
120. Borissoff JI, Spronk HM, ten Cate H. The hemostatic system as a modulator of atherosclerosis. N Engl J Med. 2011;364:1746–60.
121. Tantry US, Bliden KP, Suarez TA, et al. Hypercoagulability, platelet function, inflammation and coronary artery disease acuity: Results of the Thrombotic Risk Progression (TRIP) study. Platelets. 2010;21:360–7.
122. Weitz JI. Insights into the role of thrombin in the pathogenesis of recurrent ischaemia after acute coronary syndrome. Thromb Haemost. 2014;112:924–31.
123. Kandaswamy E, Zuo L. Recent advances in treatment of coronary artery disease: role of science and technology. Int J Mol Sci. 2018;19(2):424.
124. Koruth JS, Lala A, Pinney S, et al. The clinical use of ivabradine. J Am Coll Cardiol. 2017;70(14):1777–84.
125. Ponikowski P, Voors AA, Anker SD, et al. 2016 ESC guidelines for the diagnosis and treatment of acute and chronic heart failure: the Task Force for the diagnosis and treatment of acute and chronic heart failure of the European Society of Cardiology (ESC). Developed with the special contribution of the Heart Failure Association (HFA) of the ESC. Eur J Heart Fail. 2016;18(8):891–975.
126. Task Force M, Montalescot G, Sechtem U, et al. 2013 ESC guidelines on the management of stable coronary artery disease: the Task Force on the management of stable coronary artery disease of the European Society of Cardiology. Eur Heart J. 2013;34(38):2949–3003.
127. Rayner-Hartley E, Sedlak T. Ranolazine: A contemporary review. J Am Heart Assoc. 2016;5(3):e003196.
128. Degrauwe S, Pilgrim T, Aminian A, et al. Dual antiplatelet therapy for secondary prevention of coronary artery disease. Open Heart. 2017;4(2):e000651.
129. Rollini F, Franchi F, Angiolillo DJ. Switching P2Y12-receptor inhibitors in patients with coronary artery disease. Nat Rev Cardiol. 2016;13(1):11–27.
130. McMahon SR, Brummel-Ziedins K, Schneider DJ. Novel oral anticoagulants in the management of coronary artery disease. Coron Artery Dis. 2016;27(5):412–9.
131. Powers WJ, Rabinstein AA, Ackerson T, et al. Guidelines for the early management of patients with acute ischemic stroke: a guideline for healthcare professionals from the American Heart Association/American Stroke Association. Stroke. 2018;49:e46–99.
132. Effient® (prasugrel) tablets. Indianapolis (IN): Eli Lilly. 2009. https://pi.lilly.com/us/effient.pdf. Accessed 8 July 2021.
133. Bhatt D, Fox K, Hacke W, et al. Clopidogrel and aspirin versus aspirin alone for the prevention of atherothrombotic events. Am J Ophthalmol. 2006;142(2):366.
134. Bonaca MP, Bhatt DL, Cohen M, PEGASUS-TIMI 54 Steering Committee and Investigators, et al. Long-term use of ticagrelor in patients with prior myocardial infarction. N Engl J Med. 2015;372(19):1791–800.
135. Mauri L, Kereiakes DJ, Yeh RW, DAPT Study Investigators, et al. Twelve or 30 months of dual antiplatelet therapy after drug-eluting stents. N Engl J Med. 2014;371(23):2155–66.
136. Steg P, Bhatt D, Simon T, et al. Ticagrelor in patients with stable coronary disease and diabetes. N Engl J Med. 2019;381(14):1309–20.
137. Bhatt D, Steg P, Mehta S, et al. Ticagrelor in patients with diabetes and stable coronary artery disease with a history of previous percutaneous coronary intervention (THEMIS-PCI): a phase 3, placebo-controlled, randomized trial. Lancet. 2019;394(10204):1169–80.
138. Gurbel PA, Tantry US. Combination antithrombotic therapies. Circulation. 2010;121:569–83.
139. Wiviott SD, Braunwald E, McCabe CH, et al. TRITON-TIMI 38 Investigators. Prasugrel versus clopidogrel in patients with acute coronary syndromes. N Engl J Med. 2007;357:2001–15.
140. Wallentin L, Becker RC, Budaj A, PLATO Investigators, et al. Ticagrelor versus clopidogrel in patients with acute coronary syndromes. N Engl J Med. 2009;361:1045–57.
141. Gurbel PA, Fox KAA, Tantry US, et al. Combination antiplatelet and oral anticoagulant therapy in patients with coronary and peripheral artery disease. Circulation. 2019;139(18):2170–85.
142. Morrow DA, Braunwald E, Bonaca MP, TRA 2P–TIMI 50 Steering Committee and Investigators, et al. Vorapaxar in the secondary prevention of atherothrombotic events. N Engl J Med. 2012;366:1404–13.
143. Bonaca MP, Scirica BM, Creager MA, et al. Vorapaxar in patients with peripheral artery disease: results from TRA2P-TIMI 50. Circulation. 2013;127:1522–1529, 1529e1–6.
144. Cohen M, Adams PC, Parry G, et al. Combination antithrombotic therapy in unstable rest angina and non-Q-wave infarction in nonprior aspirin users. Primary end points analysis from the ATACS trial. Antithrombotic Therapy in Acute Coronary Syndromes Research Group. Circulation. 1994;89:81–8.
145. Anand SS, Yusuf S. Oral anticoagulants in patients with coronary artery disease. J Am Coll Cardiol. 2003;41(4 suppl S):S62–9.
146. The Warfarin Antiplatelet Vascular Evaluation Trial Investigators. Oral anticoagulant and antiplatelet therapy and peripheral arterial disease. N Engl J Med. 2007;357:217–27.
147. Anand SS, Yusuf S, Pogue J, et al. Long-term oral anticoagulant therapy in patients with unstable angina or suspected non-Q-wave myocardial infarction: organization to assess strategies for ischemic syndromes (OASIS) pilot study results. Circulation. 1998;98:1064–70.

148. Andreotti F, Testa L, Biondi-Zoccai GG, et al. Aspirin plus warfarin compared to aspirin alone after acute coronary syndromes: an updated and comprehensive meta-analysis of 25,307 patients. Eur Heart J. 2006;27:519–26.
149. American College of Emergency P, Society for Cardiovascular A, Interventions, O'Gara PT, Kushner FG, Ascheim DD, et al. 2013 ACCF/AHA guideline for the management of ST-elevation myocardial infarction: a report of the American College of Cardiology Foundation/American Heart Association Task Force on Practice Guidelines. J Am Coll Cardiol. 2013;61(4):e78–140.
150. Amsterdam EA, Wenger NK, Brindis RG, et al. 2014 AHA/ACC Guideline for the Management of Patients with Non-ST-Elevation Acute Coronary Syndromes: a report of the American College of Cardiology/American Heart Association Task Force on Practice Guidelines. J Am Coll Cardiol. 2014;64(24):e139–228.
151. Yusuf S, Zhao F, Mehta SR, et al. Effects of clopidogrel in addition to aspirin in patients with acute coronary syndromes without ST-segment elevation. N Engl J Med. 2001;345(7):494–502.
152. Sabatine MS, Cannon CP, Gibson CM, et al. Addition of clopidogrel to aspirin and fibrinolytic therapy for myocardial infarction with ST-segment elevation. N Engl J Med. 2005;352(12):1179–89.
153. Valgimigli M, Bueno H, Byrne RA, et al. 2017 ESC focused update on dual antiplatelet therapy in coronary artery disease developed in collaboration with EACTS: the Task Force for dual antiplatelet therapy in coronary artery disease of the European Society of Cardiology (ESC) and of the European Association for Cardio-Thoracic Surgery (EACTS). Eur Heart J. 2018;39(3):213–60.
154. Levine GN, Bates ER, Bittl JA, et al. 2016 ACC/AHA Guideline Focused Update on Duration of Dual Antiplatelet Therapy in Patients With Coronary Artery Disease: A Report of the American College of Cardiology/American Heart Association Task Force on Clinical Practice Guidelines: An Update of the 2011 ACCF/AHA/SCAI Guideline for Percutaneous Coronary Intervention, 2011 ACCF/AHA Guideline for Coronary Artery Bypass Graft Surgery, 2012 ACC/AHA/ACP/AATS/PCNA/SCAI/STS Guideline for the Diagnosis and Management of Patients With Stable Ischemic Heart Disease, 2013 ACCF/AHA Guideline for the Management of ST-Elevation Myocardial Infarction, 2014 AHA/ACC Guideline for the Management of Patients With Non-ST-Elevation Acute Coronary Syndromes, and 2014 ACC/AHA Guideline on Perioperative Cardiovascular Evaluation and Management of Patients Undergoing Noncardiac Surgery. Circulation. 2016;134(10):e123–55.
155. Khan SU, Lone AN, Asad ZUA, et al. Meta-analysis of efficacy and safety of proton pump inhibitors with dual antiplatelet therapy for coronary artery disease. Cardiovasc Revasc Med. 2019;20(12):1125–33.
156. Alexopoulos D, Katogiannis K, Sfantou D, et al. Combination antiplatelet treatment in coronary artery disease patients: a necessary evil or an overzealous practice? Platelets. 2018;29(3):228–37.
157. Dézsi CA. Treatment with triple combination of atorvastatin, perindopril, and amlodipine in patients with stable coronary artery disease: a subgroup analysis from the PAPA-CAD study. J Int Med Res. 2018;46(5):1902–9.
158. Sever P, Dahlöf B, Poulter N, et al. Potential synergy between lipid-lowering and blood-pressure-lowering in the Anglo-Scandinavian Cardiac Outcomes Trial. Eur Heart J. 2006;27:2982–8.
159. Bertrand ME, Vlachopoulos C, Mourad JJ. Triple combination therapy for global cardiovascular risk: atorvastatin, perindopril, and amlodipine. Am J Cardiovasc Drugs. 2016;16(4):241–53.
160. Nádházi Z, Dézsi CA. The results of ACES (Antihypertensive Combinations' Long Term Efficacy Comparing Study): analysis of metabolic effects of antihypertensive combination therapies. Clin Drug Investig. 2016;36:819–27.
161. Tanai E, Frantz S. Pathophysiology of heart failure. Compr Physiol. 2015;6(1):187–214.
162. Snipelisky D, Chaudhry SP, Stewart GC. The many faces of heart failure. Card Electrophysiol Clin. 2019;11(1):11–20.
163. Berliner D, Hänselmann A, Bauersachs J. The treatment of heart failure with reduced ejection fraction. Dtsch Arztebl Int. 2020;117(21):376–86.
164. Jhund PS, McMurray JJV. The neprilysin pathway in heart failure: a review and guide on the use of sacubitril/valsartan. Heart. 2016;102(17):1342–7.
165. Burnett H, Earley A, Voors AA, et al. Thirty years of evidence on the efficacy of drug treatments for chronic heart failure with reduced ejection fraction: a network meta-analysis. Circ Heart Fail. 2017;10(1):e003529.
166. Hubers SA, Brown NJ. Combined angiotensin receptor antagonism and neprilysin inhibition. Circulation. 2016;133(11):1115–24.
167. McMurray JJ, Packer M, Desai AS, et al. Angiotensin-neprilysin inhibition versus enalapril in heart failure. N Engl J Med. 2014;371:993–1004.
168. Solomon SD, McMurray JJV, Anand IS, PARAGON-HF Investigators and Committees, et al. Angiotensin-neprilysin inhibition in heart failure with preserved ejection fraction. N Engl J Med. 2019;381(17):1609–20.
169. McMurray JJ, Packer M, Desai AS, et al. Dual angiotensin receptor and neprilysin inhibition as an alternative to angiotensin-converting enzyme inhibition in patients with chronic systolic heart failure: rationale for and design of the Prospective comparison of ARNI with ACEI to Determine Impact on Global Mortality and morbidity in Heart Failure trial (PARADIGM-HF). Eur J Heart Fail. 2013;15:1062–73.
170. Solomon SD, Zile M, Pieske B, et al. The angiotensin receptor neprilysin inhibitor LCZ696 in heart

failure with preserved ejection fraction: a phase 2 double-blind randomised controlled trial. Lancet. 2012;380:1387–95.

171. Yancy CW, Jessup M, Bozkurt B, et al. 2016 ACC/AHA/HFSA focused update on new pharmacological therapy for heart failure: an update of the 2013 ACCF/AHA Guideline for the Management of Heart Failure: a report of the American College of Cardiology/American Heart Association Task Force on Clinical Practice Guidelines and the Heart Failure Society of America. J Am Coll Cardiol. 2016;68:1476–88.
172. Komajda M, Böhm M, Borer JS, et al. Incremental benefit of drug therapies for chronic heart failure with reduced ejection fraction: a network meta-analysis. Eur J Heart Fail. 2018;20(9):1315–22.
173. Laribi S, Aouba A, Nikolaou M, GREAT Network, et al. Trends in death attributed to heart failure over the past two decades in Europe. Eur J Heart Fail. 2012;14:234–9.
174. Smith KR, Hsu CC, Berei TJ, et al. PARADIGM-HF trial: secondary analyses address unanswered questions. Pharmacotherapy. 2018;38(2):284–98.
175. Khder Y, Shi V, McMurray JJV, et al. Sacubitril/valsartan (LCZ696) in heart failure. Handb Exp Pharmacol. 2017;243:133–65.
176. Priori SG, Blomstrom-Lundqvist C, Mazzanti A, et al. 2015 ESC Guidelines for the management of patients with ventricular arrhythmias and the prevention of sudden cardiac death: The Task Force for the Management of Patients with Ventricular Arrhythmias and the Prevention of Sudden Cardiac Death of the European Society of Cardiology (ESC). Endorsed by: Association for European Paediatric and Congenital Cardiology (AEPC). Eur Heart J. 2015;36:2793–867.
177. Kirchhof P, Benussi S, Kotecha D, et al. 2016 ESC guidelines for the management of atrial fibrillation developed in collaboration with EACTS. Eur Heart J. 2016;37:2893–962.
178. Hayashi M, Shimizu W, Albert CM. The spectrum of epidemiology underlying sudden cardiac death. Circ Res. 2015;116:1887–906.
179. Kjekshus J. Arrhythmias and mortality in congestive heart failure. Am J Cardiol. 1990;65:42I–8I.
180. Fedeli U, Ferroni E, Pengo V. Mortality associated to atrial fibrillation still on the rise: United States, 1999 to 2014. Int J Cardiol. 2016;222:788–9.
181. Schleifer JW, Sorajja D, Shen WK. Advances in the pharmacologic treatment of ventricular arrhythmias. Expert Opin Pharmacother. 2015;16:2637–51.
182. Qin D, Leef G, Alam MB, et al. Comparative effectiveness of antiarrhythmic drugs for rhythm control of atrial fibrillation. J Cardiol. 2016;67:471–6.
183. Santangeli P, Muser D, Maeda S, et al. Comparative effectiveness of antiarrhythmic drugs and catheter ablation for the prevention of recurrent ventricular tachycardia in patients with implantable cardioverter-defibrillators: a systematic review and meta-analysis of randomized controlled trials. Heart Rhythm. 2016;13:1552–9.
184. Heijman J, Voigt N, Dobrev D. New directions in antiarrhythmic drug therapy for atrial fibrillation. Futur Cardiol. 2013;9:71–88.
185. Frommeyer G, Eckardt L. Drug-induced proarrhythmia: risk factors and electrophysiological mechanisms. Nat Rev Cardiol. 2016;13:36–47.
186. Wang Z. The role of microRNA in cardiac excitability. J Cardiovasc Pharmacol. 2010;56:460–70.
187. Klabunde RE. Cardiac electrophysiology: normal and ischemic ionic currents and the ECG. Adv Physiol Educ. 2017;41(1):29–37.
188. Yang B, Lu Y, Wang Z. Control of cardiac excitability by microRNAs. Cardiovasc Res. 2008;79:571–80.
189. Black N, D'Souza A, Wang Y, et al. Circadian rhythm of cardiac electrophysiology, arrhythmogenesis, and the underlying mechanisms. Heart Rhythm. 2019;16(2):298–307.
190. Grandi E, Sanguinetti MC, Bartos DC, et al. Potassium channels in the heart: structure, function and regulation. J Physiol. 2017;595(7):2209–28.
191. Bartos DC, Grandi E, Ripplinger CM. Ion channels in the heart. Compr Physiol. 2015;5:1423–64.
192. Vaughan WE. Classification of antiarrhythmic drugs. In: Sandoe E, Flensted-Jensen E, Olsen K, editors. Symposium on cardiac arrhythmias. Elsinore: Astra; 1970. 826pp.
193. Vaughan Williams EM. Classification of antidysrhythmic drugs. Pharmacol Ther B. 1975;1:115–38.
194. Vaughan Williams EM. A classification of antiarrhythmic actions reassessed after a decade of new drugs. J Clin Pharmacol. 1984;24:129–47.
195. Lei M, Wu L, Terrar DA, et al. Modernized classification of cardiac antiarrhythmic drugs. Circulation. 2018;138(17):1879–96.
196. DiFrancesco D. The role of the funny current in pacemaker activity. Circ Res. 2010;106:434–46.
197. Mathew ST, Po SS, Thadani U. Inappropriate sinus tachycardia-symptom and heart rate reduction with ivabradine: a pooled analysis of prospective studies. Heart Rhythm. 2018;15:240–7.
198. Zellerhoff S, Hinterseer M, Felix Krull B, et al. Ivabradine in patients with inappropriate sinus tachycardia. Naunyn Schmiedeberg's Arch Pharmacol. 2010;382:483–6.
199. Mulder P, Barbier S, Chagraoui A, et al. Long-term heart rate reduction induced by the selective I_f current inhibitor ivabradine improves left ventricular function and intrinsic myocardial structure in congestive heart failure. Circulation. 2004;109:1674–9.
200. Sartiani L, Mannaioni G, Masi A, et al. The hyperpolarization-activated cyclic nucleotide-gated channels: from biophysics to pharmacology of a unique family of ion channels. Pharmacol Rev. 2017;69(4):354–95.
201. Saad M, Mahmoud A, Elgendy IY, et al. Ranolazine in cardiac arrhythmia. Clin Cardiol. 2016;39(3):170–8.

202. Pulford BR, Kluger J. Ranolazine therapy in cardiac arrhythmias. Pacing Clin Electrophysiol. 2016;39(9):1006–15.
203. Shen MJ, Zipes DP. Role of the autonomic nervous system in modulating cardiac arrhythmias. Circ Res. 2014;114(6):1004–21.
204. Li N, Csepe TA, Hansen BJ, et al. Adenosine-induced atrial fibrillation. Circulation. 2016;134:486–98.
205. Jia Q, Han W, Shi S, et al. The effects of ACEI/ARB, aldosterone receptor antagonists and statins on preventing recurrence of atrial fibrillation: a protocol for systematic review and network meta-analysis. Medicine (Baltimore). 2021;100(1):e24280.
206. Williams EA, Russo V, Ceraso S, et al. Anti-arrhythmic properties of non-antiarrhythmic medications. Pharmacol Res. 2020;156:104762.
207. Das MK, Zipes DP. Antiarrhythmic and nonantiarrhythmic drugs for sudden cardiac death prevention. J Cardiovasc Pharmacol. 2010;55(5):438–49.
208. Cardiac Arrhythmia Suppression Trial (CAST) Investigators. Preliminary report: effect of encainide and flecainide on mortality in a randomized trial of arrhythmia suppression after myocardial infarction. N Engl J Med. 1989;321(6):406–12.
209. The Cardiac Arrhythmia Suppression Trial II (CAST) Investigators. Effect of the antiarrhythmic agent moricizine on survival after myocardial infarction. N Engl J Med. 1992;327:227–33.
210. Bigger JT Jr, Giardina EG. Rational use of anti-arrhythmic drugs alone and in combination. Cardiovasc Clin. 1974;6(2):103–17.
211. Aguilar M, Xiong F, Qi XY, et al. Potassium channel blockade enhances atrial fibrillation-selective antiarrhythmic effects of optimized state-dependent sodium channel blockade. Circulation. 2015;132:2203–11.
212. Reiffel JA, Camm AJ, Belardinelli L, HARMONY Investigators, et al. The HARMONY trial: combined ranolazine and dronedarone in the management of paroxysmal atrial fibrillation: mechanistic and therapeutic synergism. Circ Arrhythm Electrophysiol. 2015;8:1048–56.
213. Osmanovic E, Ostojic M, Avdic S, et al. Pharmacological prophylaxis of atrial fibrillation after surgical myocardial revascularization. Med Arch. 2019;73(1):19–22.
214. Schupp T, Behnes M, Reiser L, et al. Prognostic impact of beta-blocker compared to combined amiodarone therapy secondary to ventricular tachyarrhythmias. Int J Cardiol. 2019;277:118–24.
215. Wang G, Zhao N, Zhong S, et al. Safety and efficacy of flecainide for patients with catecholaminergic polymorphic ventricular tachycardia: a systematic review and meta-analysis. Medicine (Baltimore). 2019;98(34):e16961.
216. Negreva MN, Penev AP. Effect of selective beta-blockade with bisoprolol in the treatment of recent-onset atrial fibrillation. Folia Med (Plovdiv). 2012;54(2):27–31.
217. Kawamura M, Munetsugu Y, Tanno K, et al. Carvedilol is effective and safe in combination with Bepridil for persistent atrial fibrillation and decreases the QT prolongation induced by Bepridil therapy. J Cardiovasc Pharmacol. 2013;61(1):77–82.
218. Khand AU, Rankin AC, Martin W, et al. Carvedilol alone or in combination with digoxin for the management of atrial fibrillation in patients with heart failure? J Am Coll Cardiol. 2003;42(11):1944–51.
219. Koh KK, Song JH, Kwon KS, et al. Comparative study of efficacy and safety of low-dose diltiazem or betaxolol in combination with digoxin to control ventricular rate in chronic atrial fibrillation: randomized crossover study. Int J Cardiol. 1995;52(2):167–74.
220. Goy JJ, Fromer M. Antiarrhythmic treatment of atrioventricular tachycardias. J Cardiovasc Pharmacol. 1991;17(Suppl 6):S36–40.
221. Fauchier L, Grimard C, Pierre B, et al. Comparison of beta blocker and digoxin alone and in combination for management of patients with atrial fibrillation and heart failure. Am J Cardiol. 2009;103(2):248–54.
222. Wu X, Wang C, Zhu J, et al. Meta-analysis of randomized controlled trials on magnesium in addition to beta-blocker for prevention of postoperative atrial arrhythmias after coronary artery bypass grafting. BMC Cardiovasc Disord. 2013;13:5.
223. Munsif AN, Saksena S. Efficacy of nadolol alone or in combination with a type IA antiarrhythmic drug in sustained ventricular tachycardia: a prospective study. Pacing Clin Electrophysiol. 1989;12(11):1816–26.
224. Breithardt G, Seipel L, Abendroth RR. Comparison of the antiarrhythmic efficacy of disopyramide and mexiletine against stimulus-induced ventricular tachycardia. J Cardiovasc Pharmacol. 1981;3(5):1026–37.
225. Tanabe T, Takahashi K, Yoshioka K, et al. Evaluation of disopyramide and mexiletine used alone and in combination for ventricular arrhythmias in patients with and without overt heart disease. Int J Cardiol. 1991;32(3):303–12.
226. Tanabe T. Combination antiarrhythmic treatment among class Ia, Ib, and II agents for ventricular arrhythmias. Cardiovasc Drugs Ther. 1991;5(Suppl 4):827–34.
227. Kim SG, Mercando AD, Tam S, et al. Combination of disopyramide and mexiletine for better tolerance and additive effects for treatment of ventricular arrhythmias. J Am Coll Cardiol. 1989;13(3):659–64.
228. Sakurada H, Motomiya T, Hiraoka M. Efficacy of disopyramide and mexiletine used alone or in combination in the treatment of ventricular premature beats. Cardiovasc Drugs Ther. 1991;5(Suppl 4):835–41.
229. Giardina EG, Wechsler ME. Low dose quinidine-mexiletine combination therapy versus quinidine monotherapy for treatment of ventricular arrhythmias. J Am Coll Cardiol. 1990;15(5):1138–45.
230. Chiladakis JA, Kalogeropoulos A, Patsouras N, et al. Ibutilide added to propafenone for the conversion of

atrial fibrillation and atrial flutter. J Am Coll Cardiol. 2004;44(4):859–63.
231. Korantzopoulos P, Kolettis TM, Papathanasiou A, et al. Propafenone added to ibutilide increases conversion rates of persistent atrial fibrillation. Heart. 2006;92(5):631–4.
232. Tsanaxidis N, Aidonidis I, Hatziefthimiou A, et al. Ranolazine added to amiodarone facilitates earlier conversion of atrial fibrillation compared to amiodarone-only therapy. Pacing Clin Electrophysiol. 2017;40(4):372–8.
233. Simopoulos V, Hevas A, Hatziefthimiou A, et al. Amiodarone plus ranolazine for conversion of post-cardiac surgery atrial fibrillation: enhanced effectiveness in reduced versus preserved ejection fraction patients. Cardiovasc Drugs Ther. 2018;32(6):559–65.
234. Fragakis N, Koskinas KC, Katritsis DG, et al. Comparison of effectiveness of ranolazine plus amiodarone versus amiodarone alone for conversion of recent-onset atrial fibrillation. Am J Cardiol. 2012;110(5):673–7.
235. Koskinas KC, Fragakis N, Katritsis D, et al. Ranolazine enhances the efficacy of amiodarone for conversion of recent-onset atrial fibrillation. Europace. 2014;16(7):973–9.
236. Simopoulos V, Tagarakis GI, Daskalopoulou SS, et al. Ranolazine enhances the antiarrhythmic activity of amiodarone by accelerating conversion of new-onset atrial fibrillation after cardiac surgery. Angiology. 2014;65(4):294–7.
237. De Vecchis R, Ariano C, Giasi A, et al. Antiarrhythmic effects of ranolazine used both alone for prevention of atrial fibrillation and as an add-on to intravenous amiodarone for its pharmacological cardioversion: a meta-analysis. Minerva Cardioangiol. 2018;66(3):349–59.
238. Reiffel JA. Ranolaziodarone-A synergism you should not miss. J Innov Card Rhythm Manag. 2021;12(3):4429–31.
239. Verrier RL, Pagotto VPF, Kanas AF, et al. Low doses of ranolazine and dronedarone in combination exert potent protection against atrial fibrillation and vulnerability to ventricular arrhythmias during acute myocardial ischemia. Heart Rhythm. 2013;10(1):121–7.
240. Burashnikov A, Sicouri S, Di Diego JM, et al. Synergistic effect of the combination of ranolazine and dronedarone to suppress atrial fibrillation. J Am Coll Cardiol. 2010;56(15):1216–24.
241. Harrmann N, Mason FE, Braun I, et al. The combined effects of ranolazine and dronedarone on human atrial and ventricular electrophysiology. J Mol Cell Cardiol. 2016;94:95–106.
242. Burashnikov A, Belardinelli L, Antzelevitch C. Inhibition of I_{Kr} potentiates development of atrial-selective I_{Na} block leading to effective suppression of atrial fibrillation. Heart Rhythm. 2015;12:836–44.
243. Kirchhoff JE, Diness JG, Sheykhzade M, et al. Synergistic antiarrhythmic effect of combining inhibition of Ca^{2+}-activated K^+ (SK) channels and voltage-gated Na^+ channels in an isolated heart model of atrial fibrillation. Heart Rhythm. 2015;12:409–18.
244. Henyan N, White CM. Adjunctive intravenous magnesium to reduce toxicity and enhance efficacy of class III antiarrhythmic agents. Conn Med. 2004;68(10):627–9.
245. Wang A. Efficacy of class III antiarrhythmics and magnesium combination therapy for atrial fibrillation. Pharm Pract (Granada). 2012;10(2):65–71.
246. Tercius AJ, Kluger J, Coleman CI, et al. Intravenous magnesium sulfate enhances the ability of intravenous ibutilide to successfully convert atrial fibrillation or flutter. Pacing Clin Electrophysiol. 2007;30(11):1331–5.
247. Steinwender C, Hönig S, Kypta A, et al. Pre-injection of magnesium sulfate enhances the efficacy of ibutilide for the conversion of typical but not of atypical persistent atrial flutter. Int J Cardiol. 2010;141(3):260–5.
248. Cagli K, Ozeke O, Ergun K, et al. Effect of low-dose amiodarone and magnesium combination on atrial fibrillation after coronary artery surgery. J Card Surg. 2006;21(5):458–64.
249. Forlani S, Moscarelli M, Scafuri A, et al. Combination therapy for prevention of atrial fibrillation after coronary artery bypass surgery: a randomized trial of sotalol and magnesium. Card Electrophysiol Rev. 2003;7(2):168–71.
250. Ho KM, Sheridan DJ, Paterson T. Use of intravenous magnesium to treat acute onset atrial fibrillation: a meta-analysis. Heart. 2007;93(11):1433–40.
251. Komatsu T, Ozawa M, Tachibana H, et al. Combination therapy with amiodarone and enalapril in patients with paroxysmal atrial fibrillation prevents the development of structural atrial remodeling. Int Heart J. 2008;49(4):435–47.
252. Komatsu T, Tachibana H, Sato Y, et al. Long-term efficacy of upstream therapy using angiotensin-converting enzyme inhibitors and statins in combination with antiarrhythmic agents for the treatment of paroxysmal atrial fibrillation. Int Heart J. 2009;50(4):465–76.
253. Ji Q, Mei Y, Wang X, et al. Combination of irbesartan and amiodarone to maintain sinus rhythm in patients with persistent atrial fibrillation after rheumatic valve replacement. Circ J. 2010;74(9):1873–9.
254. Eslami M, Badkoubeh RS, Mousavi M, et al. Oral ascorbic acid in combination with beta-blockers is more effective than beta-blockers alone in the prevention of atrial fibrillation after coronary artery bypass grafting. Tex Heart Inst J. 2007;34(3):268–74.
255. Ni H, Fogli Iseppe A, Giles WR, et al. Populations of in silico myocytes and tissues reveal synergy of multiatrial-predominant K^+-current block in atrial fibrillation. Br J Pharmacol. 2020;177(19):4497–515.
256. Lüderitz B, Mletzko R, Jung W, et al. Combination of antiarrhythmic drugs. J Cardiovasc Pharmacol. 1991;17(Suppl 6):S48–52.

257. Lüderitz B, Jung W, Manz M. Combination antiarrhythmic drug therapy. Z Kardiol. 1992;81(Suppl 4):157–61.
258. Greenspan AM, Spielman SR, Horowitz LN. Combination antiarrhythmic drug therapy for ventricular tachyarrhythmias. Pacing Clin Electrophysiol. 1986;9(4):565–76.
259. Barthels D, Das H. Current advances in ischemic stroke research and therapies. Biochim Biophys Acta Mol basis Dis. 2020;1866(4):165260.
260. Lo EH, Dalkara T, Moskowitz MA. Mechanisms, challenges and opportunities in stroke. Nat Rev Neurosci. 2003;4(5):399–415.
261. Zhang L, Zhang ZG, Chopp M. The neurovascular unit and combination treatment strategies for stroke. Trends Pharmacol Sci. 2012;33(8):415–22.
262. Wardlaw JM, Murray V, Berge E, et al. Thrombolysis for acute ischaemic stroke. Cochrane Database Syst Rev. 2014;2014(7):CD000213.
263. Sandercock PA, Counsell C, Kane EJ. Anticoagulants for acute ischaemic stroke. Cochrane Database Syst Rev. 2015;2015(3):CD000024.
264. Milionis H, Liontos A, Vemmos K, et al. Antiplatelet treatment in stroke: new insights. Curr Pharm Des. 2016;22(29):4617–26.
265. Rabinstein AA. Treatment of acute ischemic stroke. Continuum (Minneap Minn). 2017;23(1, Cerebrovascular Disease):62–81.
266. Cramer SC. Recovery after stroke. Continuum (Minneap Minn). 2020;26(2):415–34.
267. Cramer SC. Drugs to enhance motor recovery after stroke. Stroke. 2015;46(10):2998–3005.
268. Cools R, Roberts AC, Robbins TW. Serotoninergic regulation of emotional and behavioural control processes. Trends Cogn Sci. 2008;12:31–40.
269. Dam M, Tonin P, De Boni A, et al. Effects of fluoxetine and maprotiline on functional recovery in poststroke hemiplegic patients undergoing rehabilitation therapy. Stroke. 1996;27:1211–4.
270. Perrier JF, Cotel F. Serotonergic modulation of spinal motor control. Curr Opin Neurobiol. 2014;33C:1–7.
271. Molina-Luna K, Pekanovic A, Rohrich S, et al. Dopamine in motor cortex is necessary for skill learning and synaptic plasticity. PLoS One. 2009;4:e7082.
272. Logue SF, Gould TJ. The neural and genetic basis of executive function: attention, cognitive flexibility, and response inhibition. Pharmacol Biochem Behav. 2014;123:45–54.
273. Whyte EM, Lenze EJ, Butters M, et al. An open-label pilot study of acetylcholinesterase inhibitors to promote functional recovery in elderly cognitively impaired stroke patients. Cerebrovasc Dis. 2008;26:317–21.
274. Crisostomo E, Duncan P, Propst M, et al. Evidence that amphetamine with physical therapy promotes recovery of motor function in stroke patients. Ann Neurol. 1988;23:94–7.
275. Walker-Batson D, Smith P, Curtis S, et al. Amphetamine paired with physical therapy accelerates motor recovery after stroke further evidence. Stroke. 1995;26:2254–9.
276. Stringberg A, Camden R, Qualls K, et al. Update on dual antiplatelet therapy for secondary stroke prevention. Mo Med. 2019;116(4):303–7.
277. Hankey GJ, Johnston SC, Easton JD, et al. Effect of clopidogrel plus ASA vs. ASA early after TIA and ischaemic stroke: a substudy of the CHARISMA trial. Int J Stroke. 2011;6(1):3–9.
278. Diener HC, Bogousslavsky J, Brass LM, et al. Aspirin and clopidogrel compared with clopidogrel alone after recent ischaemic stroke or transient ischaemic attack in high-risk patients (MATCH): randomized, double-blind, placebo-controlled trial. Lancet. 2004;364:331–7.
279. The SPS3 Investigators. Effects of clopidogrel added to aspirin in patients with recent lacunar stroke (SPS3 substudy). N Engl J Med. 2012;367(9):817–25.
280. Wang Y, Wang Y, Zhao X, et al. Clopidogrel with aspirin in acute minor stroke or transient ischemic attack. N Engl J Med. 2013;369(1):11–9.
281. Wang Y, Pan Y, Zhao X, et al. Clopidogrel with Aspirin in Acute Minor Stroke or Transient Ischemic Attack (CHANCE) trial one-year outcomes. Circulation. 2015;132:40–6.
282. Johnston SC, Easton JD, Farrant M, et al. Clopidogrel and aspirin in acute ischemic stroke and high-risk TIA (POINT) trial. N Engl J Med. 2018;379(3):215–25.
283. Wang Y, Chen W, Lin Y, PRINCE Protocol Steering Group, et al. Ticagrelor plus aspirin versus clopidogrel plus aspirin for platelet reactivity in patients with minor stroke or transient ischaemic attack: open label, blinded endpoint, randomised controlled phase II trial. BMJ. 2019;365:l2211.
284. Yang Y, Chen W, Pan Y, et al. Ticagrelor is superior to Clopidogrel in inhibiting platelet reactivity in patients with minor stroke or TIA. Front Neurol. 2020;11:534.
285. Zhou M, Chen W, Pan Y, et al. Antiplatelet effect of ticagrelor with aspirin in acute minor stroke and transient ischemic attack stratified by CYP2C19 metabolizer status: subgroup analysis of the PRINCE trial. Aging (Albany NY). 2020;13(3):3994–4006.
286. Mega JL, Close SL, Wiviott SD, et al. Cytochrome p-450 polymorphisms and response to clopidogrel. N Engl J Med. 2009;360(4):354–62.
287. Johnston SC, Amarenco P, Denison H, THALES Investigators, et al. Ticagrelor and aspirin or aspirin alone in acute ischemic stroke or TIA. N Engl J Med. 2020;383(3):207–17.
288. Oza R, Rundell K, Garcellano M. Recurrent ischemic stroke: strategies for prevention. Am Fam Physician. 2017;96(7):436–40.
289. Kim SM, Jung JM, Kim BJ, et al. Cilostazol mono and combination treatments in ischemic stroke: an updated systematic review and meta-analysis. Stroke. 2019;50(12):3503–11.
290. Geng X, Fu P, Ji X, et al. Synergetic neuroprotection of normobaric oxygenation and ethanol in isch-

emic stroke through improved oxidative mechanism. Stroke. 2013;44(5):1418–25.
291. Geng X, Elmadhoun O, Peng C, et al. Ethanol and normobaric oxygen: novel approach in modulating pyruvate dehydrogenase complex after severe transient and permanent ischemic stroke. Stroke. 2015;46(2):492–9.
292. Wald NJ, Law MR. A strategy to reduce cardiovascular disease by more than 80%. BMJ. 2003;326(7404):1419.
293. Nguyen C, Cheng-Lai A. The polypill: a potential global solution to cardiovascular disease. Cardiol Rev. 2013;21(1):49–54.
294. Wiley B, Fuster V. The concept of the polypill in the prevention of cardiovascular disease. Ann Glob Health. 2014;80:24–34.
295. Wald DS, Morris JK, Wald NJ. Randomized polypill crossover trial in people aged 50 and over. PLoS One. 2012;7:e41297.
296. Yusuf S, Joseph P, Dans A, International Polycap Study 3 Investigators, et al. Polypill with or without aspirin in persons without cardiovascular disease. N Engl J Med. 2021;384(3):216–28.
297. Thom S, Poulter N, Field J, et al. Effects of a fixed-dose combination strategy on adherence and risk factors in patients with or at high risk of CVD: the UMPIRE randomized clinical trial. JAMA. 2013;310:918–29.
298. Thom S, Field J, Poulter N, et al. Use of a Multidrug Pill In Reducing cardiovascular Events (UMPIRE): rationale and design of a randomised controlled trial of a cardiovascular preventive polypill-based strategy in India and Europe. Eur J Prev Cardiol. 2014;21(2):252–61.
299. Hippisley-Cox J, Coupland C. Effect of combinations of drugs on all cause mortality in patients with ischaemic heart disease: nested case–control analysis. BMJ. 2005;330:1059–63.
300. Papadopoulos DP, Papademetriou V. Low-dose fixed combination of bisoprolol/hydrochlorothiazide as first line for hypertension: a review of the rationale and clinical evidence. Angiology. 2009;60(5):601–7.
301. Gupta AK, Arshad S, Poulter NR. Compliance, safety, and effectiveness of fixed-dose combinations of antihypertensive agents: a meta-analysis. Hypertension. 2010;55(2):399–407.
302. Verma AA, Khuu W, Tadrous M, Gomes T, Mamdani MM. Fixed-dose combination antihypertensive medications, adherence, and clinical outcomes: a population-based retrospective cohort study. PLoS Med. 2018;15(6):e1002584.
303. Kjeldsen SE, Messerli FH, Chiang CE, et al. Are fixed-dose combination antihypertensives suitable as first-line therapy? Curr Med Res Opin. 2012;28(10):1685–97.
304. Whelton PK, Carey RM, Aronow WS, et al. ACC/AHA/AAPA/ABC/ACPM/AGS/APhA/ASH/ASPC/NMA/PCNA Guideline for the Prevention, Detection, Evaluation, and Management of High Blood Pressure in Adults: A Report of the American College of Cardiology/American Heart Association Task Force on Clinical Practice Guidelines. J Am Coll Cardiol. 2017;
305. Williams B, Mancia G, Spiering W, et al. 2018. ESC/ESH guidelines for the management of arterial hypertension. Eur Heart J. 2018;39(33):3021–104.
306. Bronsert MR, Henderson WG, Valuck R, et al. Comparative effectiveness of antihyperthensive therapeutic classes and treatment strategies in the initiation of therapy in primary care patients: a Distributed Ambulatory Research in Therapeutics Network (DARTNet) study. J Am Board Fam Med. 2013;26:529–38.
307. Kumagai N, Onishi K, Hoshino K, et al. Improving drug adherence using fixed combinations caused beneficial treatment outcomes and decreased health-care costs in patients with hypertension. Clin Exp Hypertens. 2013;35:355–60.
308. Ulusoy S, Ozkan G, Konca C, et al. A comparison of the effects of fixed dose vs. single-agent combinations on 24-h blood pressure variability. Hypertens Res. 2012;35:1111–7.
309. Sherrill B, Halpern M, Khan S, et al. Single-pill vs. free-equivalent combination therapies for hypertension: a meta-analysis of health care costs and adherence. J Clin Hypertens. 2011;13:898–909.
310. Bangalore S, Kamalakkannan G, Parkar S, et al. Fixed-dose combinations improve medication compliance: a meta-analysis. Am J Med. 2007;120(8):713–9.
311. Webster R, Salam A, de Silva HA, et al. Fixed low-dose triple combination antihypertensive medication vs. usual care for blood pressure control in patients with mild to moderate hypertension in Sri Lanka: a randomized clinical trial. JAMA. 2018;320(6):566–79.
312. Pappa E, Rizos CV, Filippatos TD, et al. Emerging fixed-dose combination treatments for hyperlipidemia. J Cardiovasc Pharmacol Ther. 2019;24(4):315–22.
313. Lamb YN. Rosuvastatin/ezetimibe: a review in hypercholesterolemia. Am J Cardiovasc Drugs. 2020;20(4):381–92.
314. Robinson JG, Ballantyne CM, Grundy SM, et al. Lipid-altering efficacy and safety of ezetimibe/simvastatin versus atorvastatin in patients with hypercholesterolemia and the metabolic syndrome (from the VYMET study). Am J Cardiol. 2009;103(12):1694–702.
315. Vo AN, Kashyap ML. Fixed-dose combination of extended-release niacin plus simvastatin for lipid disorders. Expert Rev Cardiovasc Ther. 2008;6(10):1303–10.
316. Sharma M, Sharma DR, Singh V, et al. Evaluation of efficacy and safety of fixed dose lovastatin and niacin (ER) combination in Asian Indian dyslipidemic patients: a multicentric study. Vasc Health Risk Manag. 2006;2(1):87–93.
317. Guyton JR, Brown BG, Fazio S, et al. Lipid-altering efficacy and safety of ezetimibe/simvastatin coad-

ministered with extended-release niacin in patients with type IIa or type IIb hyperlipidemia. J Am Coll Cardiol. 2008;51(16):1564–72.

318. Fazio S, Guyton JR, Polis AB, et al. Long-term safety and efficacy of triple combination ezetimibe/simvastatin plus extended-release niacin in patients with hyperlipidemia. Am J Cardiol. 2010;105(4):487–94.
319. Yiu KH, Cheung BM, Tse HF. A new paradigm for managing dyslipidemia with combination therapy: laropiprant + niacin + simvastatin. Expert Opin Investig Drugs. 2010;19(3):37–49.
320. Munoz D, Uzoije P, Reynolds C, et al. Polypill for cardiovascular disease prevention in an underserved population. N Engl J Med. 2019;381(12):1114–23.
321. Indian Polycap S, Yusuf S, Pais P, et al. Effects of a polypill (Polycap) on risk factors in middle-aged individuals without cardiovascular disease (TIPS): a phase II, double-blind, randomised trial. Lancet. 2009;373(9672):1341–51.
322. Pill Collaborative Group, Rodgers A, Patel A, et al. An international randomised placebo-controlled trial of a four-component combination pill ("polypill") in people with raised cardiovascular risk. PLoS One. 2011;6(5):e19857.
323. Malekzadeh F, Marshall T, Pourshams A, et al. A pilot double-blind randomised placebo-controlled trial of the effects of fixed-dose combination therapy ('polypill') on cardiovascular risk factors. Int J Clin Pract. 2010;64(9):1220–7.
324. Roshandel G, Khoshnia M, Poustchi H, et al. Effectiveness of polypill for primary and secondary prevention of cardiovascular diseases (PolyIran): a pragmatic, cluster-randomised trial. Lancet. 2019;394(10199):672–83.
325. Castellano JM, Sanz G, Penalvo JL, et al. A polypill strategy to improve adherence: results from FOCUS (Fixed-dose Combination Drug for Secondary Cardiovascular Prevention) Project. J Am Coll Cardiol. 2014;64:2071–82.
326. Patel A, Cass A, Peiris D, et al. A pragmatic randomized trial of a polypill-based strategy to improve use of indicated preventive treatments in people at high cardiovascular disease risk. Eur J Prev Cardiol. 2014;22:920–30.
327. Selak V, Elley CR, Bullen C, et al. Effect of fixed dose combination treatment on adherence and risk factor control among patients at high risk of cardiovascular disease: randomised controlled trial in primary care. BMJ. 2014;348:g3318.
328. Webster R, Patel A, Selak V, et al. Effectiveness of fixed dose combination medication ('polypills') compared with usual care in patients with cardiovascular disease or at high risk: a prospective, individual patient data meta-analysis of 3140 patients in six countries. Int J Cardiol. 2016;205:147–56.
329. Roth GA, Johnson C, Abajobir A, et al. Global, regional, and national burden of cardiovascular diseases for 10 causes, 1990 to 2015. J Am Coll Cardiol. 2017;70:1–25.
330. Vos T, Flaxman AD, Naghavi M, et al. Years lived with disability (YLDs) for 1160 sequelae of 289 diseases and injuries 1990–2010: a systematic analysis for the Global Burden of Disease Study 2010. Lancet. 2012;380:2163–96.
331. Divchev D, Stöckl G, study investigators. Effectiveness and impact on adherence of a new fixed-dose combination of ivabradine and metoprolol in a wide range of stable angina patients in real-life practice. Cardiol Ther. 2019;8(2):317–28.
332. Heusch G. Pleiotropic action(s) of the bradycardic agent ivabradine:cardiovascular protection beyond heart rate reduction. Br J Pharmacol. 2008;155:970–1.
333. Panza JA, Diodati JG, Callahan TS, et al. Role of increases in heart rate in determining the occurrence and frequency of myocardial ischemia during daily life in patients with coronary artery disease. J Am Coll Cardiol. 1992;20:1092–8.
334. Andrews TC, Fenton T, Toyosaki N, et al. Subsets of ambulatory myocardial ischemia based on heart rate activity. Circadian distribution and response to anti-ischemic medication. The Angina and Silent Ischemia Study Group (ASIS). Circulation. 1993;88(1):92–100.
335. Pratt CM, McMahon RP, Goldstein S, et al. Comparison of subgroups assigned to medical regimens used to suppress cardiac ischemia (the Asymptomatic Cardiac Ischemia Pilot [ACIP] study). Am J Cardiol. 1996;77:1302–9.
336. Palatini P. Heart rate: a strong predictor of mortality in subjects with coronary artery disease. Eur Heart J. 2005;26:943–5.
337. Diaz A, Bourassa MG, Guertin MC, Tardif JC. Long-term prognostic value of resting heart rate in patients with suspected or proven coronary artery disease. Eur Heart J. 2005;26:967–74.
338. Members Task Force, Montalescot G, Sechtem U, Achenbach S, et al. ESC guidelines on the management of stable coronary artery disease: the Task Force on the management of stable coronary artery disease of the European Society of Cardiology. Eur Heart J. 2013;2013(34):2949–3003.
339. Tendera M, Fox K, Ferrari R, CLARIFY Registry Investigators, et al. Inadequate heart rate control despite widespread use of beta-blockers in outpatients with stable CAD: findings from the international prospective CLARIFY registry. Int J Cardiol. 2014;176:119–24.
340. Tardif JC, Ponikowski P, Kahan T, ASSOCIATE Study Investigators. Efficacy of the I(f) current inhibitor ivabradine in patients with chronic stable angina receiving beta blocker therapy: a 4-month, randomized, placebo-controlled trial. Eur Heart J. 2009;30:540–8.
341. Werdan K, Ebelt H, Nuding S, et al. Ivabradine in combination with beta-blocker improves symptoms and quality of life in patients with stable angina pectoris: results from the ADDITIONS study. Clin Res Cardiol. 2012;101:365–73.

342. DiFrancesco D. Cardiac pacemaker I_f current and its inhibition by heart rate-reducing agents. Curr Med Res Opin. 2005;21:1115–22.
343. DiFrancesco D, Camm John A. Heart rate lowering by specific and selective if current inhibition with Ivabradine. Drugs. 2004;64(16):1757–65.
344. John Z, Violetta G, Manolis K, et al. Treatment of stable angina pectoris with ivabradine in everyday practice: a Pan-Hellenic, prospective. Noninterventional Study Clin Cardiol. 2015;38(12):725–32.
345. Koester R, Kaehler J, Meinertz T. Ivabradine for the treatment of stable angina pectoris in octogenarians. Clin Res Cardiol. 2011;100(2):121–8.
346. Ursula M-W, Georg S, Henning E, et al. Ivabradine in combination with beta-blocker reduces symptoms and improves quality of life in elderly patients with stable angina pectoris: age-related results from the ADDITIONS study. Exp Gerontol. 2014;59:34–41.
347. Werdan K, Ebelt H, Nuding S, et al. Ivabradine in combination with beta-blockers in patients with chronic stable angina after percutaneous coronary intervention. Adv Ther. 2015;32:120–37.
348. Zarifis J, Grammatikou V, Kallistratos M, et al. Antianginal efficacy of ivabradine in patients with history of coronary revascularization. Angiology. 2017;68:10–8.
349. Borer JS, Tardif JC. Efficacy of ivabradine, a selective I(f) inhibitor, in patients with chronic stable angina pectoris and diabetes mellitus. Am J Cardiol. 2010;105:29–35.
350. Werdan K, Perings S, Köster R, et al. Effectiveness of ivabradine treatment in different subpopulations with stable angina in clinical practice: a pooled analysis of observational studies. Cardiology. 2016;135:141–50.
351. Divchev D, Stöckl G, study investigators. Treatment of stable angina with a new fixed-dose combination of ivabradine and metoprolol: effectiveness and tolerability in routine clinical practice. Cardiol Ther. 2017;6(2):239–49.
352. Forster T, Dézsi CA. Short-term cardioprotective effects of the original perindopril/amlodipine fixed-dose combination in patients with stable coronary artery disease: results of the PAPA-CAD study. Adv Ther. 2016;33(10):1771–81.
353. Sarkar G, Gaikwad VB, Sharma A, et al. Fixed-dose combination of metoprolol, telmisartan, and chlorthalidone for essential hypertension in adults with stable coronary artery disease: phase III study. Adv Ther. 2021; https://doi.org/10.1007/s12325--021-01971-9. Online ahead of print
354. Lim S, Kim PJ, Baek C, et al. Antiplatelet efficacy of fixed-dose aspirin-clopidogrel combination in patients with stable coronary artery disease treated with drug-eluting stent implantation. Clin Drug Investig. 2015;35(12):833–42.
355. Oh PC, Ahn T, Kim DW, et al. Comparative effect on platelet function of a fixed-dose aspirin and clopidogrel combination versus separate formulations in patients with coronary artery disease: a phase IV, multicenter, prospective, 4-week non-inferiority trial. Int J Cardiol. 2016;202:331–5.
356. Lafeber M, Spiering W, van der Graaf Y, et al. The combined use of aspirin, a statin, and blood pressure-lowering agents (polypill components) and the risk of vascular morbidity and mortality in patients with coronary artery disease. Am Heart J. 2013;166(2):282–9.e1.
357. Katsi V, Skalis G, Kallistratos MS, et al. Ivabradine and metoprolol in fixed dose combination: when, why and how to use it. Pharmacol Res. 2019;146:104279.
358. Tondi L, Fragasso G, Spoladore R, et al. Real-life indications to ivabradine treatment for heart rate optimization in patients with chronic systolic heart failure. J Cardiovasc Med (Hagerstown). 2018;19(7):351–6.
359. Das D, Savarese G, Dahlström U, et al. Ivabradine in heart failure: the representativeness of SHIFT (Systolic Heart Failure Treatment with the I_F Inhibitor Ivabradine Trial) in a broad population of patients with chronic heart failure. Circ Heart Fail. 2017;10(9):e004112.
360. Swedberg K, Komajda M, Böhm M, SHIFT Investigators, et al. Effects on outcomes of heart rate reduction by ivabradine in patients with congestive heart failure: is there an influence of beta-blocker dose: findings from the SHIFT (Systolic Heart failure treatment with the I_f inhibitor ivabradine Trial) study. J Am Coll Cardiol. 2012;59(22):1938–45.
361. Agarwal A, Yancy CW, Huffman MD. Improving care for heart failure with reduced ejection fraction-a potential polypill-based strategy. JAMA. 2020;324(22):2259–60.
362. Khan SS, Huffman MD, Shah SJ. Could a low-dose diuretic polypill improve outcomes in heart failure with preserved ejection fraction? Circ Heart Fail. 2021;14(3):e008090.
363. Vijay A, Mohanan PP, Kondal D, et al. Polypill eligibility for patients with heart failure with reduced ejection fraction in South India: a secondary analysis of a prospective, interrupted time series study. J Am Heart Assoc. 2021;10(20):e021676.
364. Agarwal A, Mohanan PP, Kondal D, et al. Effect of a quality improvement intervention for acute heart failure in South India: an interrupted time series study. Int J Cardiol. 2021;329:123–9.
365. Brainin M, Feigin V, Martins S, et al. Cut stroke in half: Polypill for primary prevention in stroke. Int J Stroke. 2018;13(6):633–47.
366. Thom S, Poulte N, Field J. Effects of a fixed-dose combination strategy on adherence and risk factors in patients with or at high risk of CVD: the UMPIRE randomized clinical trial. JAMA. 2013;310:918–29.
367. Patel A, Cass A, Peiris D. A pragmatic randomized trial of a polypill-based strategy to improve use of indicated preventive treatments in people at high cardiovascular disease risk. Eur J Prev Cardiol. 2015;22:920–30.

368. Huffman MD, Xavier D, Perel P. Uses of polypills for cardiovascular disease and evidence to date. Lancet. 2017;389:1055–65.
369. Lenz TL, Hilleman DE. Aggrenox: a fixed-dose combination of aspirin and dipyridamole. Ann Pharmacother. 2000;34(11):1283–90.
370. Kjeldsen SE, Lyle PA, Kizer JR, et al. Fixed combination of losartan and hydrochlorothiazide and reduction of risk of stroke. Vasc Health Risk Manag. 2007;3(3):299–305.
371. Albers GW. Choice of antithrombotic therapy for stroke prevention in atrial fibrillation: warfarin, aspirin, or both? Arch Intern Med. 1998;158(14):1487–91.
372. Blackshear JL, et al. Adjusted-dose warfarin versus low-intensity, fixed-dose warfarin plus aspirin for high-risk patients with atrial fibrillation: Stroke Prevention in Atrial Fibrillation III randomised clinical trial. Lancet. 1996;348(9028):633–8.
373. Gulløv AL, Koefoed BG, Petersen P, et al. Fixed minidose warfarin and aspirin alone and in combination vs adjusted-dose warfarin for stroke prevention in atrial fibrillation: second Copenhagen Atrial Fibrillation, Aspirin, and Anticoagulation Study. Arch Intern Med. 1998;158(14):1513–21.
374. Webster R, Murphy A, Bygrave H, et al. Implementing fixed dose combination medications for the prevention and control of cardiovascular diseases. Glob Heart. 2020;15(1):57.
375. World Health Organization. WHO model list of essential medicines, 20th list. 2017. www.who.int/medicines/publications/essentialmedicines/20th_EML2017.pdf?ua=1. Accessed 1 Sept 2017.
376. Kahn R, Robertson RM, Smith R, et al. The impact of prevention on reducing the burden of cardiovascular disease. Circulation. 2008;118:576–85.
377. Naik P, Gandhi H, Pawar V, et al. Management of hypertension-journey from single drug therapy to multitargeted ligand therapy: a clinical overview. Curr Clin Pharmacol. 2015;10(4):321–46.
378. Zhang H, Wu M, Hu Y, et al. Pharmacokinetics and pharmacodynamics of MT-1207, a novel multitarget blocker of alpha1 receptor, 5-HT2A receptor, and calcium channel, in healthy subjects. Expert Opin Investig Drugs. 2021;30(3):271–8.
379. Zhou S, Wang Y, Wang L, et al. Pharmacokinetics, Safety and Tolerability of Tylerdipine Hydrochloride, a Novel Dihydropyridine Dual L/T-type Calcium Channel Blocker, after Single and Multiple Oral Doses in Healthy Chinese Subjects. Clin Drug Investig. 2019;39(1):85–96.
380. Rusinova R, Koeppe RE 2nd, Andersen OS. A general mechanism for drug promiscuity: studies with amiodarone and other antiarrhythmics. J Gen Physiol. 2015;146(6):463–75.
381. Benjamin EJ, Wolf PA, D'Agostino RB, et al. Impact of atrial fibrillation on the risk of death: The Framingham Heart Study. Circulation. 1998;98:946–52.
382. Mozaffarian D, Benjamin EJ, Go AS, et al. Heart disease and stroke statistics—2015 update: a report from the American Heart Association. Circulation. 2015;131:e29–322.
383. Nattel S, Maguy A, Le Bouter S, et al. Arrhythmogenic ion-channel remodeling in the heart: heart failure, myocardial infarction, and atrial fibrillation. Physiol Rev. 2007;87:425–56.
384. Singh BN. Amiodarone as paradigm for developing new drugs for atrial fibrillation. J Cardiovasc Pharmacol. 2008;52(4):300–5.
385. Kozlowski D, Budrejko S, Lip GY, et al. Dronedarone: an overview. Ann Med. 2012;44(1):60–72.
386. Zimetbaum P. Antiarrhythmic drug therapy for atrial fibrillation. Circulation. 2012;125:381–9.
387. Zimetbaum P. Amiodarone for atrial fibrillation. N Engl J Med. 2007;356:935–41.
388. Burashnikov A, Antzelevitch C. New developments in atrial antiarrhythmic drug therapy. Nat Rev Cardiol. 2010;7:139–48.
389. Dobrev D, Carlsson L, Nattel S. Novel molecular targets for atrial fibrillation therapy. Nat Rev Drug Discov. 2012;11:275–91.
390. Grunnet M, Bentzen BH, Sørensen US, et al. Cardiac ion channels and mechanisms for protection against atrial fibrillation. Rev Physiol Biochem Pharmacol. 2012;162:1–58.
391. Rusinova R, Herold KF, Sanford RL, et al. Thiazolidinedione insulin sensitizers alter lipid bilayer properties and voltage-dependent sodium channel function: implications for drug discovery. J Gen Physiol. 2011;138:249–70.
392. Lundbæk JA, Koeppe RE II, Andersen OS. Amphiphile regulation of ion channel function by changes in the bilayer spring constant. Proc Natl Acad Sci U S A. 2010;107:15427–30.
393. Singh BN. Amiodarone: historical development and pharmacologic profile. Am Heart J. 1983;106:788–97.
394. Kodama I, Kamiya K, Toyama J. Cellular electropharmacology of amiodarone. Cardiovasc Res. 1997;35:13–29.
395. Heijman J, Heusch G, Dobrev D. Pleiotropic effects of antiarrhythmic agents: dronedarone in the treatment of atrial fibrillation. Clin Med Insights Cardiol. 2013;7:127–40.
396. Sackmann E, Kotulla R, Heiszler F-J, On the role of lipid-bilayer elasticity for the lipid-protein interaction and the indirect protein-protein coupling. Can J Biochem Cell Biol. 1984;62:778–88.
397. Andersen OS, Sawyer DB, Koeppe RE II. Modulation of channel function by the host bilayer. In: Easwaran KRK, Gaber B, editors. Biomembrane structure and function. Schenectady: Adenine Press; 1992. p. 227–44.
398. Keller SL, Bezrukov SM, Gruner SM, et al. Probability of alamethicin conductance states varies with nonlamellar tendency of bilayer phospholipids. Biophys J. 1993;65:23–7.

399. Lundbæk JA, Andersen OS. Lysophospholipids modulate channel function by altering the mechanical properties of lipid bilayers. J Gen Physiol. 1994;104:645–73.
400. Anavekar S-N, Louis WJ, Morgan TO, et al. The relationship of plasma levels of pindolol in hypertensive patients to effects on blood pressure, plasma renin and plasma noradrenaline levels. Clin Exp Pharmacol Physiol. 1975;2:203–12.
401. Woosley RL, Kornhauser D, Smith R, et al. Suppression of chronic ventricular arrhythmias with propranolol. Circulation. 1979;60:819–27.
402. Insel PA. Phenylalkylamines are promiscuous receptor blockers. Trends Pharmacol Sci. 1988;9:10.
403. Mencher SK, Wang LG. Promiscuous drugs compared to selective drugs (promiscuity can be a virtue). BMC Clin Pharmacol. 2005;5:3.
404. Hwang TC, Koeppe REI II, Andersen OS. Genistein can modulate channel function by a phosphorylation-independent mechanism: importance of hydrophobic mismatch and bilayer mechanics. Biochemistry. 2003;42:13646–58.
405. Lundbæk JA, Birn P, Tape SE, et al. Capsaicin regulates voltage-dependent sodium channels by altering lipid bilayer elasticity. Mol Pharmacol. 2005;68:680–9.
406. Bruno MJ, Koeppe REI II, Andersen OS. Docosahexaenoic acid alters bilayer elastic properties. Proc Natl Acad Sci U S A. 2007;104:9638–43.
407. Ingólfsson HI, Koeppe REI II, Andersen OS. Curcumin is a modulator of bilayer material properties. Biochemistry. 2007;46:10384–91.
408. Morris CE, Juranka PF. Nav channel mechanosensitivity: activation and inactivation accelerate reversibly with stretch. Biophys J. 2007;93:822–33.
409. Beyder A, Rae JL, Bernard C, et al. Mechanosensitivity of Nav1.5, a voltage-sensitive sodium channel. J Physiol. 2010;588:4969–85.
410. Beyder A, Strege PR, Reyes S, et al. Ranolazine decreases mechanosensitivity of the voltage-gated sodium ion channel Nav1.5: a novel mechanism of drug action. Circulation. 2012;125:2698–706.
411. Strege P, Beyder A, Bernard C. Ranolazine inhibits shear sensitivity of endogenous Na+ current and spontaneous action potentials in HL-1 cells. Channels (Austin). 2012;6(6):457–62.
412. Wegener FT, Ehrlich JR, Hohnloser SH. Dronedarone: an emerging agent with rhythm- and rate-controlling effects. J Cardiovasc Electrophysiol. 2006;17(Suppl 2):S17–20.
413. Pamukcu B, Lip GY. Dronedarone or amiodarone for rhythm control for atrial fibrillation: implications from the DIONYSOS study. Expert Opin Pharmacother. 2010;11(17):2775–8.
414. Garcia D, Cheng-Lai A. Dronedarone: a new antiarrhythmic agent for the treatment of atrial fibrillation. Cardiol Rev. 2009;17(5):230–4.
415. Tanaka H, Hashimoto N. A multiple ion channel blocker, NIP-142, for the treatment of atrial fibrillation. Cardiovasc Drug Rev. 2007;25(4):342–56.
416. Matsuda T, Takeda K, Ito M, et al. Atria selective prolongation by NIP-142, an antiarrhythmic agent, of refractory period and action potential duration in guinea pig myocardium. J Pharmacol Sci. 2005;98(1):33–40.
417. Tanaka H, Namekata I, Hamaguchi S, et al. Effect of NIP-142 on potassium channel alpha-subunits Kv1.5, Kv4.2 and Kv4.3, and mouse atrial repolarization. Biol Pharm Bull. 2010;33(1):138–41.
418. Namekata I, Tsuruoka N, Tsuneoka Y, et al. Blocking effect of NIP-142 on the KCNQ1/KCNE1 channel current expressed in HEK293 cells. Biol Pharm Bull. 2011;34(1):153–5.
419. Mason PK, DiMarco JP. New pharmacological agents for arrhythmias. Circ Arrhythm Electrophysiol. 2009;2:588–97.
420. Dorian P, Pinter A, Mangat I. The effect of vernakalant (RSD1235), an investigational antiarrhythmic agent, on atrial electrophysiology in humans. J Cardiovasc Pharmacol. 2007;50:35–40.
421. Antzelevitch C, Belardinelli L, Zygmunt AC. Electrophysiological effects of ranolazine, a novel antianginal agent with antiarrhythmic properties. Circulation. 2004;110:904–10.
422. Williams ES, Viswanathan MN. Current and emerging antiarrhythmic drug therapy for ventricular tachycardia. Cardiol Ther. 2013;2:27–46.
423. Reiffel JA, Camm AJ, Belardinelli L, et al. The HARMONY Trial: Combined ranolazine and dronedarone in the management of paroxysmal atrial fibrillation: mechanistic and therapeutic synergism. Circ Arrhythm Electrophysiol. 2015;8:1048–56.
424. Hartmann N, Mason FE, Braun I, et al. The combined effects of ranolazine and dronedarone on human atrial and ventricular electrophysiology. J Mol Cell Cardiol. 2016;94:95–106.

5 Polypharmacology in Clinical Applications: Metabolic Disease Polypharmacology

Abstract

Metabolic diseases consisting of different groups of disorders are becoming increasingly prevalent and have a major impact on public health worldwide. The major metabolic diseases include metabolic syndrome (mainly characterized by insulin resistance), diabetes (particularly type 2 diabetes), overweight and obesity, and nonalcoholic fatty liver disease. Despite current therapies that offer some protection through tight serum glucose control, ultimately such treatments cannot block the progression of disability and death realized with metabolic disorders. Oppositely, current treatment options have proven insufficient to cope with obesity and diabetes. One of the critical issues is that metabolic diseases are in general multietiologic, multifaceted, and multifactorial complex disorders that may not respond well to single-target drug (STD) therapy. It has now been commonly accepted that polypharmacology will offer better effectiveness and safety profiles for the treatment of metabolic diseases. This chapter will focus on the applications of polypharmacology, at three different forms (CDT, FDC, and MTD), to the management of diabetes and obesity. For each form of polypharmacological approaches, an introduction to the basics of diabetes and obesity and the conventional approaches of drug treatment of these two disorders will precede for better understanding of the main theme.

Keywords

Metabolic diseases · Diabetes · Obesity · Polypharmacology · Multitarget therapy

5.1 CDT for Metabolic Disease

Metabolism is the means by which the body derives energy and synthesizes the other molecules it needs from the fats, carbohydrates, and proteins obtained from food, by enzymatic reactions helped by minerals and vitamins [1]. Disease related to metabolism will occur if abnormal chemical reactions in the body disrupt metabolism due to lack of or excess production of a metabolism-related enzyme or hormone. On the other hand, metabolic diseases or disorders, characterized by the inability to properly use and/or store energy, negatively impact all systems of the body by interrupting the normal processes of metabolism within the body, the many biochemical reactions responsible for breaking down food and either using, storing, or disposing of the energy it contains. There are different groups of disorders: some affect the breakdown of amino acids, carbohydrates, or lipids, and others like mitochondrial diseases affect the parts of the cells that produce the energy. The burdens of metabolic disease, such as obesity or diabetes mellitus (MD), are believed to arise through a complex interplay between genetics and epigenetics predisposition, environment, and nutrition.

Z. Wang, B. Yang, *Polypharmacology*, https://doi.org/10.1007/978-3-031-04998-9_5

Metabolic diseases are becoming increasingly prevalent and have a major impact on public health worldwide. Major metabolic diseases include metabolic syndrome (mainly characterized by insulin resistance), type 2 diabetes, obesity, and nonalcoholic fatty liver disease (NAFLD). The prevalence of metabolic diseases is taking on epidemic proportions and poses a serious threat to human health. Despite current therapies that offer some protection through tight serum glucose control, ultimately such treatments cannot block the progression of disability and death realized with metabolic disorders. Oppositely, current treatment options have proven insufficient to cope with obesity and DM because they rarely restore normal metabolism and thus leave patients exposed to life-threatening complications. Successful management of these diseases depends on novel, improved therapeutic strategies targeting early intervention in disease progression. Discovery of novel metabolic disease targets has been hampered by the complexity of contributing environmental and genetic factors, as well as the need for potent but safe treatments suitable for chronic diseases [2]. As a result, novel therapeutic avenues are critical for further development to address these concerns. Replacement therapies have revolutionized treatment paradigms in metabolic diseases by restoring defective enzymes and supplementing missing downstream metabolites [3]. Through most of the twentieth century, no targeted therapies existed for these conditions; the only treatment options available focused on symptoms rather than the underlying disorders. Improved understanding of the molecular pathways underlying metabolic disease has allowed not only supplementation of missing metabolites and reduction of upstream substrates but also replacement of defective or missing enzymes.

5.1.1 CDT in Diabetes Mellitus (DM)

5.1.1.1 Basics of Diabetic Mellitus (DM)

Diabetic mellitus (DM) is a common degenerative chronic metabolic disease characterized by high blood glucose (hyperglycemia) over a prolonged period of time, which causes considerable negative impacts on health and lifespan with diabetic cardiovascular complications (DCCs) as major causes of death in diabetic population [4]. If left untreated, diabetes can cause many health complications. Lipoproteins, blood pressure, weight control, mental health, and lifestyle are important factors that contribute to the frequency of diabetes-associated complications. Acute complications can include diabetic ketoacidosis, hyperosmolar hyperglycemic state, or death. Serious long-term complications include cardiovascular disease, stroke, chronic kidney disease, foot ulcers, damage to the nerves, damage to the eyes, and cognitive impairment. There are three main types of DM:

Type 1 Diabetes Mellitus (T1DM) results from failure of the pancreas to produce enough insulin due to loss of β-cells [4]. This form was previously referred to as "insulin-dependent diabetes mellitus" (IDDM) or "juvenile diabetes." The loss of β-cells is caused by an autoimmune response.

Type 2 Diabetes Mellitus (T2DM) begins with insulin resistance, a condition in which cells fail to respond to insulin properly [4]. As the disease progresses, a lack of insulin may also develop. This form was previously referred to as "non-insulin-dependent diabetes mellitus" (NIDDM) or "adult-onset diabetes." The most common cause is a combination of excessive body weight and insufficient exercise.

Gestational Diabetes is the third main form and occurs when pregnant women without a previous history of diabetes develop high blood sugar levels [4].

Impaired glucose tolerance (IGT) and impaired fasting glycemia (IFG) are intermediate conditions in the transition between normality and diabetes. People with IGT or IFG are at a high risk of progressing to T2DM, although this is not inevitable [4].

DM is associated with several major complications including **(a)** hypoglycemia that can be severe and life-threatening and **(b)** hyperglycemia that is a main factor for the development of cardiovascular disease, and macrovascular and microvascular disease, both of which are hallmark features of TM-associated constraints. Therefore, the treatment goals in DM patients include both glycemic control and prevention of complications to maintain and improve quality of life in DM patients [4–10].

There is no known preventive measure for T1DM [4]. T2DM, which accounts for 85–90% of all cases worldwide, can often be prevented or delayed by maintaining a normal body weight, engaging in physical activity, and eating a healthy diet [5]. There are a number of different classes of antidiabetic medications. T1DM requires treatment with insulin, ideally using a "basal bolus" regimen that most closely matches normal insulin release: long-acting insulin for the basal rate and short-acting insulin with meals [4]. T2DM is generally taken with oral medications (e.g., metformin) although some eventually require injectable treatment with insulin or GLP-1 agonists [5].

5.1.1.2 Medications for T1DM

Insulin

Replacement therapy with insulin is the most common type of medication used in T1DM treatment [4]. It is also used in T2DM treatment. There are several types of insulin medications for use to meet different needs depending on the severity of insulin depletion of patients.

(a) Short-acting insulin—regular insulin (Humulin and Novolin)
(b) Rapid-acting insulins—insulin aspart (NovoLog, FlexPen, Fiasp), insulin glulisine (Apidra), insulin lispro (Humalog)
(c) Intermediate-acting insulin—insulin isophane (Humulin N, Novolin N)
(d) Long-acting insulins—insulin degludec (Tresiba); insulin detemir (Levemir); insulin glargine (Lantus); insulin glargine (Toujeo)
(e) Combination insulins—NovoLog Mix 70/30 (insulin aspart protamine-insulin aspart), Humalog Mix 75/25 (insulin lispro protamine-insulin lispro), Humalog Mix 50/50 (insulin lispro protamine-insulin lispro), Humulin 70/30 (human insulin NPH-human insulin regular), Novolin 70/30 (human insulin NPH-human insulin regular), Ryzodeg (insulin degludec-insulin aspart)

Amylinomimetic Drug

Pramlintide (Symlin®, SymlinPen®) is the only amylinomimetic in the medication class and is FDA-approved as adjunct to treatment with insulin in patients with T1DM and T2DM who have failed to achieve desired glucose control despite optimal insulin therapy [4]. Pramlintide is an injectable drug used before meals and works by delaying the time the stomach takes to empty itself. It reduces glucagon secretion after meals.

5.1.1.3 Medications for T2DM: Hypoglycemic Agents

The primary physiologic defects of T2DM include excessive hepatic glucose production, diminished peripheral glucose uptake by insulin-sensitive tissues, and inadequate insulin secretion [11]. In general, pharmacologic therapy includes not only initial antihyperglycemic agents but also more intensive approaches to glycemic control over time, often requiring multiple medications with synergistic mechanisms of action [12]. At first glance it seems that there are sufficient numbers of therapeutic classes of agents to treat T2DM, with each class varying in mechanism, safety, efficacy, and route of administration; however, few drugs target all three of the main defects associated with T2DM, which is why complex medication regimens are often required to achieve glycemic control. Metformin is widely recognized as first-line therapy in multiple treatment algorithms and is known to work primarily by limiting hepatic glucose production, one of the key defects of T2DM [11–13]. Metformin has a long track record of efficacy, minimal weight gain, and low risk of hypoglycemia. Despite the potential benefits of metformin, some patients cannot tolerate the gastrointestinal (GI) side effects associated with this drug [11]. Insulin secretagogues, such as sulfonylureas and meglitinides, are also frequently used early in the treatment of a patient with diabetes as they generally elicit clinical efficacy in the short term; however, they have been noted to fail to maintain efficacy over time. While the secretagogues are targeting one of the component defects of T2DM by stimulating endogenous insulin secretion, they are often associated with hypoglycemia and weight gain, thus limiting their use. Thiazolidinediones target a third aspect of the physiologic defects of the diabetic patient by increasing insulin sensitivity in the periphery and have shown sustained durability of effect and infrequent incidence of hypoglycemia; however, they, too, have side effects that possibly limit their use, such as edema and weight gain [14].

Newer classes of antidiabetic agents that target the incretin hormones, such as glucagon-like peptide-1 (GLP-1) receptor agonists and dipeptidyl peptidase-4 (DPP-4) inhibitors, have recently achieved a presence in the landscape of antihyperglycemic drugs. GLP-1 receptor agonists, delivered via subcutaneous injection, promote glucose-dependent insulin secretion, inhibition of inappropriate postprandial glucagon secretion, and slowed gastric emptying to achieve glycemic control. While they are not often associated with hypoglycemia and generally induce some weight loss, the GLP-1 receptor agonists are often poorly tolerated because of nausea and increased risk of pancreatitis. Additionally, the FDA has issued a boxed warning identifying the risk of medullary thyroid cancer with the GLP-1 receptor agonists [15]. DPP-4 inhibitors also work on incretin hormones by blocking the enzyme that is responsible for degrading endogenous GLP-1, allowing for higher concentrations of GLP-1 to exert its action; however, it should be noted that glycemic effects are generally less significant with the DPP-4 inhibitors compared with the exogenous GLP-1 receptor agonists. DPP-4 inhibitors, unlike GLP-1 receptor agonists, are administered orally and generally have good patient tolerability and a low incidence of hypoglycemia; however, the overall efficacy in glucose reduction is minimal compared with other classes [16]. Selective sodium glucose cotransporter 2 (SGLT2) inhibitors are the newest agents to gain approval for the treatment of T2DM. This class has a unique mechanism of action that does not target the pathology of diabetes but works by increasing glucose excretion in the urine. These agents are not associated with weight gain or hypoglycemia but are known to cause an increase in genital and urinary tract infections for both men and women [17]. Insulin therapies are by far the most efficacious agents for glycemic control and directly target insulin deficiency/insufficiency. Insulin does not have dose limitations or other side effects seen with non-insulin therapies but does have a significant risk of hypoglycemia and weight gain, and patients are generally averse to starting insulin therapy early on in the course of their disease [11]. After reviewing these agents used for the treatment of T2DM, it is evident that no single agent targets the three primary defects of this widespread disease.

Most medications used to treat diabetes act by lowering blood sugar levels through different mechanisms. There is broad consensus that when people with DM maintain tight glucose control—keeping the glucose levels in their blood within normal ranges—they experience fewer complications [5–9]. T2DM is a syndrome characterized by insulin deficiency, insulin resistance, and increased hepatic glucose output [5]. Current therapeutic strategies used to treat patients with this disease are aimed at correcting one or more of these physiologic abnormalities using five distinct classes of hypoglycemic agents available, each class displaying unique pharmacologic properties. In patients for whom diet and exercise do not provide adequate glycemic control, therapy with a single oral agent can be tried. If adequate blood glucose control is not attained using a single oral agent, a combination of agents with different mechanisms of action may have additive therapeutic effects and result in better glycemic control.

Insulin Resistance Inhibitors

(a) *Biguanides* (metformin)—increase hepatic adenosine monophosphate-activated protein kinase activity, thus reducing hepatic gluconeogenesis and release, suppressing lipogenesis, and enhancing insulin-mediated uptake of glucose in muscles to lower hemoglobin A1c (HbA1c) by 1.5%. Metformin is generally recommended as a first-line treatment for T2DM or as the initial oral hypoglycemic agent unless contraindicated.

(b) *Thiazolidinediones* (TZD; rosiglitazone, pioglitazone)—activate peroxisome proliferator-activated receptor gamma (PPARγ), a nuclear receptor, which increases insulin sensitivity in muscle and fat, peripheral uptake of glucose, and the level of adiponectin, a fat tissue-secreted cytokine, that not only increases the number of insulin-sensitive adipocytes but also stimulates fatty acid oxidation. These agents can lower HbA1c by 1.0–1.5%.

Insulin Secretagogues

(a) *Sulfonylureas* (acetohexamide, chlorpropamide, tolazamide, tolbutamide; glipizide, glyburide, gliclazide, glimepiride)—bind to adenosine triphosphate-sensitive potassium channels (K_{ATP} channels) in the β-cells of the pancreas; this leads to the inhibition of those channels and alters the resting membrane potential of the cell, causing an influx of calcium and the stimulation of insulin secretion. These agents can lower HbA1c by 1.5%.

(b) *Meglitinides* (repaglinide, nateglinide)—exert their effects via different pancreatic β-cell receptors, but they act similarly to sulfonylureas by regulating adenosine triphosphate-sensitive potassium channels in pancreatic beta cells, thereby causing an increase in insulin secretion. These agents can lower HbA1c by 0.5–1.0%.

Glucose Absorption Inhibitors

(a) *Alpha-glucosidase inhibitors* (acarbose, miglitol, voglibose)—competitively inhibit α-glucosidase enzymes in the intestinal brush border cells that digest the dietary starch, thus inhibiting the polysaccharide reabsorption and the metabolism of sucrose to glucose and fructose. These agents can lower HbA1c by 0.5–1.0%.

(b) *Sodium glucose cotransporter 2 (SGLT2) inhibitors* (dapagliflozin, canagliflozin, empagliflozin)—inhibit sodium-glucose cotransporter 2 (SGLT2) in proximal tubules of renal glomeruli, causing inhibition of 90% glucose reabsorption and resulting in glycosuria in people with diabetes which in turn lowers the plasma glucose levels. These agents can lower HbA1c by 1.0%.

Incretin Agents

(a) *Dipeptidyl peptidase-4 (DPP-4) inhibitors* (sitagliptin, saxagliptin, vildagliptin, linagliptin, alogliptin)—inhibit the enzyme dipeptidyl peptidase-4 (DPP-4) to inhibit incretin degradation resulting in insulin secretion to lower HbA1c by 0.5–1.0%.

(b) *Glucagon-like peptide-1 (GLP-1) agonists* (liraglutide, exenatide, pramlintide)—incretin mimetics that bind GLP-1 receptors decreasing glucagon release and increasing glucose-dependent insulin release to lower HbA1c by 1.0–1.5% and facilitate weight loss.

Dopamine D2 Receptor Agonist (Cycloset or Bromocriptine)

A sympatholytic dopamine D2 receptor agonist resets the hypothalamic circadian rhythm, which might have been altered by obesity. This action results in the reversal of insulin resistance and a decrease in glucose production. It can be used in combination with metformin (as part of dual or triple therapy) and may lower HbA1C by 0.7–1.4%.

Insulin

When insulin is used in T2DM, a long-acting formulation is usually added initially while continuing oral medications [10]. Doses of insulin are then increased until glucose targets are reached.

Antihypertensive Drugs
Cardiovascular disease is a serious complication associated with diabetes, and many international guidelines recommend blood pressure treatment targets that are lower than 140/90 mmHg for people with diabetes [18] (see "4.1.2 CDT for Cardiovascular Disease (CAD)—(1) Hypertension" for more detail).

Aspirin
The use of aspirin to prevent cardiovascular disease in diabetes is controversial [19]. Aspirin is recommended by some in people at high risk of cardiovascular disease; however, routine use of aspirin has not been found to improve outcomes in uncomplicated diabetes. The 2015 American Diabetes Association recommendations for aspirin use (based on expert consensus or clinical experience) are that low-dose aspirin use is reasonable in adults with diabetes who are at intermediate risk of cardiovascular disease (10-year cardiovascular disease risk, 5–10%) [19]. National guidelines for England and Wales by the National Institute for Health and Care Excellence (NICE) recommend against the use of aspirin in people with T1DM or T2DM who do not have confirmed cardiovascular disease [4, 5].

Imeglimin
Imeglimin is a drug currently in development as the first in a new class of oral antidiabetic agents acting on both insulin-resistant organs and pancreatic beta cells. Although the precise mechanism of action has not been fully understood, it appears, based on preclinical results, to target several components of the pathology of T2DM. In rodent studies, it has been shown to reduce fasting plasma glucose (FPG), reduce hemoglobin A1c (A1c), and inhibit hepatic glucose production similar to metformin. Imeglimin has also been associated with stimulation of skeletal muscle glucose uptake, protection against beta-cell apoptosis, and reduction of liver steatosis. Lastly, imeglimin has been shown to perform similarly to DPP-4 inhibitors with respect to glucose-dependent insulin secretion [20]. Researchers hope this drug will add to the arsenal of existing antidiabetic agents as either an alternative or complementary option in any stage of the disease.

In summary, imeglimin has the following properties:

(a) Imeglimin can induce insulin sensitivity through several molecular pathways [21].
(b) Imeglimin has been used to reduce hepatic gluconeogenesis [20, 22].
(c) Evidence suggests that imeglimin preserves beta-cell function and increases glucose-induced insulin secretion [23]. Pacini and coworkers in 2015 demonstrated that imeglimin protected beta cells and promoted islet function by improving glucose homeostasis in the diabetic milieu [24]. Lablanche and colleagues in 2018 provided further data indicating imeglimin attenuated beta-cell apoptosis by lowering the glucotoxicity by a mitochondrial-dependent mechanism [25].
(d) Mitochondrial function: some evidence suggests that imeglimin can improve mitochondrial function in patients with diabetes [21, 26].
(e) Antioxidative effect: recent evidence indicates that imeglimin has antioxidative potentials which enable it to ameliorate free radical generation and readjust the redox state [21].

In a whole, imeglimin, a novel oral antidiabetic, exhibits a good efficacy and safety profile in T2DM patients [21, 27–29].

5.1.1.4 Combination Therapy of T2DM
If adequate control is not obtained with the use of a single agent, combination therapy is an option. Several of the available oral agents have been studied in combination and have been shown to further improve glycemic control when compared to monotherapy [30]. As with monotherapy, the choice of a second agent should be based on individual characteristics. Reasonable combinations of agents include a sulfonylurea plus metformin, a sulfonylurea plus an alpha-glucosidase inhibitor, a sulfonylurea plus a thiazolidinedione, metformin plus repaglinide, biguanide plus alpha-glucosidase inhibitor, and metformin plus a thiazolidinedione. A newer formulation

(Glucovance) combines glyburide and metformin in one tablet, allowing a more convenient dosing schedule.

Some physicians advocate therapy combining three oral agents (sulfonylurea, metformin, thiazolidinedione or sulfonylurea, metformin, alpha-glucosidase inhibitor), although this approach has not been extensively studied [31]. Significant data support the combination of bedtime insulin with daytime sulfonylurea therapy [32–34]. The combination can be quite effective in reducing fasting plasma glucose (FPG). There are in general two forms of combination: **(a)** stepwise administration of multiple agents by sequential additions and **(b)** coadministration of multiple agents by simultaneous application. Stepwise therapy may result in good patient adherence and allow greater individualization of therapy, and minimization of side effects and cost, and so may be appropriate for patients who are closer to goal; however, it may also lead to frequent delays in achieving glycemic goals and longer exposure to hyperglycemia.

Combination therapy, which is now emerging as an important therapy option, has a number of potential advantages over stepwise therapy, including reduction in clinical inertia and earlier and more frequent achievement of glycated hemoglobin goals by targeting multiple pathogenic mechanisms simultaneously, which may more effectively delay disease progression. Compared with stepwise therapy, the disadvantages of combination therapy include reduced patient adherence resulting from complex, multidrug regimens, difficulty determining the cause of poor efficacy and/or side effects, patient refusal to accept disease, and higher cost [35].

The Metformin-Based Combinations [19]

A meta-analysis of 15 randomized controlled trials involving 6693 T2DM patients assessed whether initial combination therapy with metformin plus another agent (TZD, sulfonylureas/glinides, DPP-4 inhibitors, or SGLT2 inhibitors) improved glycemic control in treatment-naïve patients with T2DM. When compared with metformin monotherapy, combination treatment reduced HbA1c but increased the risk of hypoglycemia [36].

(a) *The TZD/metformin combination—rosiglitaz one/metformin* [20]: recent clinical trial results have demonstrated that, in patients with T2DM, second-line treatment of rosiglitazone (a TZD) in combination with metformin can lead to significant improvements in the control of FPG/glycosylated HbA1c after the failure of metformin monotherapy [37]. Treatment with rosiglitazone in combination with metformin provides better glycemic control over the remaining lifetime of patients than metformin and sulfonylurea combination therapy. Patients treated with rosiglitazone combination therapy were predicted to have a longer life expectancy, gaining 123 and 140 additional life years per 1000 patients in the obese and overweight cohorts, respectively. Improvements in morbidity and a delay in the start of insulin therapy resulted in a projected improvement in quality of life. These effects combine with projected improved survival to yield 131 and 209 additional quality-adjusted life years (QALYs) per 1000 patients in the obese and overweight cohorts, respectively. Discounted incremental cost-effectiveness ratios were estimated at pounds 16,700 per QALY gained for the obese cohort and pounds 11,600 per QALY gained for the overweight cohort.

(b) *The SGLT2 inhibitor/metformin combination—canagliflozin/metformin*: a study has demonstrated the efficacy of canagliflozin (a SGLT2 inhibitor) in combination with metformin in significantly reducing HbA1c *vs* metformin monotherapy in treatment-naïve patients with T2DM [38].

(c) *The DPP-4 inhibitor/metformin combination—linagliptin/metformin*: the efficacy of initial combination of linagliptin (a DPP-4 inhibitor) and metformin in T2DM patients was assessed in a 24-week trial [39], followed by a 1-year extension study [40]. Results demonstrated lower mean standard deviation (SD) reductions in HbA1c levels in patients who continued to receive linagliptin plus metformin than with metformin alone, with a low risk of hypoglycemia. In treatment-naïve T2DM patients with baseline HbA1c ≥ 9.5%, initial therapy with linagliptin plus metformin showed marked improvements in adjusted mean standard error (SE) changes from baseline in HbA1c after 24 weeks *vs* linagliptin monotherapy [41].

(d) **The DPP-4 inhibitor/metformin combination—alogliptin/metformin**: the initial combination of alogliptin (a DPP-4 inhibitor) and metformin provided significantly greater reductions in HbA1c levels from baseline to week 26 versus metformin [42].

The DPP-4 Inhibitor/TZD Combination: Alogliptin/Pioglitazone

The efficacy of the initial combination of alogliptin and pioglitazone (a TZD) on β-cell function and glycemic control was evaluated in patients with T2DM [43]. After 16 weeks, mean (SD) improvements in HbA1c were significantly greater with combination therapy than with alogliptin monotherapy or placebo.

The DPP-4 Inhibitor/TZD Combination

DPP-4 inhibitor and TZD also improved β-cell glucose sensitivity versus placebo, supporting the concept of combination therapy to address multiple pathophysiological defects.

The SGLT2 Inhibitor/DPP-4 Inhibitor Combination: Empagliflozin/Linagliptin

Treatment-naïve patients were randomized to dual-inhibitor combination therapy (SGLT2/DPP-4) of empagliflozin/linagliptin or monotherapy with empagliflozin or linagliptin [44]. After 24 weeks, reductions in HbA1c were significantly greater for empagliflozin/linagliptin versus monotherapies. Efficacy was maintained over 52 weeks. Modest reductions in SBP from baseline were observed with empagliflozin/linagliptin combination therapies and empagliflozin alone at week 52, but these changes were not significantly different between the combination therapies and their individual components. The investigators concluded that initial combination therapy studies provide evidence of improved glycemic control but lack evidence regarding long-term macrovascular and microvascular outcomes [44].

Other combinations have also been tested; these combinations include the pioglitazone/metformin combination (Actoplus Met), the pioglitazone/glimepiride combination (Duetact), the sitagliptin/metformin combination (Janumet), and the glipizide/metformin combination (Metaglip).

In addition to the combination therapies for glycemic control, broader combinations have also been evaluated for their efficacy in both glycemic control and complication prevention. For example, the Steno-2 study evaluated multifactorial therapy with intensive antihyperglycemic therapy, antihypertensives, aspirin, and lipid-lowering agents in patients with T2DM and microalbuminuria [45–47]. Patients had a mean HbA1c of 8.4–8.8% and median disease duration of 5.5–6.0 years [46]. During the first 3.8 years of follow-up, intensive (combination therapy) versus conventional (monotherapy) therapy reduced the risk of nephropathy, retinopathy, and autonomic neuropathy [46]. After 7.8 years, intensive therapy decreased the risk of the primary composite macrovascular endpoint [29]. After a mean follow-up of 17 years, the original multifactorial combination therapy group had an absolute risk reduction of 2.8% for composite CVD endpoints [47]. The authors concluded that the use of statins and antihypertensive drugs has the largest effect on long-term CVD risk.

The Anglo-Danish-Dutch Study of Intensive Treatment in People with Screen Detected Diabetes in Primary Care (ADDITION) trials were conducted to further verify the multifactorial combination therapy involving patients with T2DM from three different countries [48, 49]. The ADDITION-Europe study randomized patients with T2DM to intensive multifactorial therapy (combination therapy) or conventional therapy (monotherapy) and assessed primary cardiovascular endpoints and non-traumatic amputation for a mean of 5.3 years. Patients had a mean HbA1c of 7.0%, and >90% had a history of cardiovascular events [49]. Intensive therapy over a 5-year period was associated with a modest improvement in HbA1c, cholesterol, and blood pressure and a non-significant 17% relative reduction in the incidence of composite cardiovascular endpoints versus conventional therapy. The ADDITION-Denmark study screened 1533 patients for T2DM and randomized them to intensive or conventional therapy, followed by 6 years of observation for diabetic peripheral neuropathy (DPN) or peripheral artery disease (PAD) [48]. Baseline HbA1c was 6.4%. Intensive treatment did not change the incidence of DPN or PAD. The authors postulated that an extended period of intensive therapy was required to observe a clinically significant effect on DPN or PAD even with early treatment initiation.

As stated by Milligan [50], based on our current knowledge, metformin-based combination therapy as soon as possible after diagnosis results in improved glycemic control [36]. Early-onset pathophysiological damage induced by hyperglycemia and worsening metabolic control, a consequence of this progressive disease, substantiates the use of early combination therapy. The current clinical trial evidence includes a number of combination therapies in patients with long-standing diabetes in terms of glycemic control and microvascular and macrovascular outcomes. Evidence from ADVANCE [51–53] and VADT [54, 55] indicate long-term (~10–15 years) benefits of intensive glycemic control on microvascular outcomes, similar to the findings of the UKPDS [56]. Collectively, the rationale for early combination treatment in adults with T2DM is straightforward, and studies clearly support a glycemic benefit. However, evidence of long-term improvements in macrovascular outcomes with early combination therapy is limited. Early intensive combination therapy may be a way to offset early pathophysiological damage resulting from hyperglycemia, therefore reducing the period in which patients are exposed to hyperglycemia. (For more details about combination therapies of DM, please read the excellent review article by Milligan [50].)

5.1.2 CDT in Obesity

5.1.2.1 Basics of Obesity

Obesity is generally defined by an excessive accumulation of adipose tissue or body fat prolonged in time causing prejudice to health, which can no longer be evaluated solely by the body mass index (expressed in kg/m^2) because it represents a heterogeneous entity [57]. It is a condition of rising prevalence worldwide, with important socioeconomic implications, being considered as a growing public health concern. According to the last updated records from the World Health Organization (WHO), approximately 40% of people around the world are overweight, whereas 13% are obese (>600 million individuals). Moreover, in recent decades, obesity-associated metabolic diseases have been become a global epidemic. Thus, obesity represents one of the greatest public health concerns worldwide, being considered as an important risk factor for the development of chronic or noncommunicable diseases (NCD).

Contrary to just being a medical condition or risk factor for other diseases, obesity is a complex disease of multifaceted etiology, with its own disabling capacities, pathophysiologies, and comorbidities [quote from Conway and Rene] [58]. It meets the medical definition of disease in that it is a physiological dysfunction of the human organism with environmental, genetic, and endocrinological etiologies. As a set of disabling and pathophysiological problems with inherent comorbidities, a result of the interaction of multiple obesogenic factors, obesity is a response to environmental stimuli, genetic predisposition, and abnormalities and has a characteristic set of signs and symptoms with consistent anatomical alterations. Excess adipose tissue increases the work of the heart and leads to anatomical changes in this organ. It alters pulmonary, endocrine, and immunological functions, all with adverse effects on health. Some of the complications of obesity include cardiovascular disease, T2DM, obstructive pulmonary disease, arthritis, and cancer [59–61]. Given the excess mortality, substantial morbidity, and the economic toll of obesity, this is a disease that warrants serious attention by the medical community [62]. Obesity's status and acceptance as a disease are pivotal in determining its treatment, reimbursement for treatment, and the development of widespread interventions [quote from Conway and Rene] [58].

Paradoxically, some patients with a higher BMI were related to a minor risk of mortality, a fact known as the obesity paradox [63, 64]. Hence, it is important to consider other factors such as fat redistribution, metabolic status, cardiorespiratory fitness, age, or ethnic factors for a more accurate stratification to gain further insights into this condition and adopt/develop better approaches for the treatment of obesity.

5.1.2.2 Pharmacological Treatment of Obesity

As a chronic disease, obesity could cause long-standing serious consequences. Although lifestyle modification is considered first-line treatment, it is often ineffective, especially in the long term. Relatively few people with obesity will undergo the most effective currently available treatment of bariatric surgery [58]. The primary role of anti-obesity medications or weight loss medications is to bridge the gap between lifestyle modification and surgery. Anti-obesity medications are pharmacological agents that reduce or control weight by altering either appetite or absorption of calories.

Pancreatic Lipase Inhibitors (Orlistat or Xenical, Cetilistat)

Pancreatic lipase is an enzyme that breaks down triglycerides in the intestine, and inhibition of this enzyme prevents triglycerides from the diet from being hydrolyzed into absorbable free fatty acids and forces them to be excreted undigested [65, 66]. These agents were originally available only by prescription, but they have now been approved by the FDA for over-the-counter sale. Xenical might cause severe liver injury that has been reported rarely with the use of this medication [66]. A published phase II trial found that cetilistat significantly reduced weight and was better tolerated than orlistat.

GLP-1 Activators

GLP-1 is a hormone that is found naturally in the body that has effects on appetite and food intake. Saxenda® (liraglutide, an antidiabetic agent) is a GLP-1 therapy that is similar to the natural GLP-1 found in human body and is thought to help patients lose weight by decreasing appetite and the amount you eat. It has been documented by studies that liraglutide can also improve other health factors beyond weight, including waist circumference, blood sugar, blood pressure, and cholesterol levels.

Metformin

In people with T2DM, metformin (Glucophage, an antidiabetic agent) can reduce weight. Metformin limits the amount of glucose that is produced by the liver as well as increases muscle consumption of glucose. It also helps in increasing the body's response to insulin.

5.1.2.3 Combination Therapies for Obesity

However, many monotherapies have only modest efficacy or require high doses with unacceptable side effects. As with many other areas of medicine, combination therapy is now becoming accepted as a way of optimizing efficacy for weight management while minimizing adverse effects. Combination therapies use different medications with complementary modes of action. The ultimate aim of pharmacotherapies is to match the efficacy of bariatric surgery with a combination of medicines [67].

Bupropion/Naltrexone Combination

Naltrexone is used to treat alcohol and opioid dependence, and bupropion is an antidepressant and quit-smoking aid. The naltrexone/bupropion combination is used for weight loss in those that are either obese or overweight with some weight-related illnesses. It combines low doses of bupropion and naltrexone. While both medications have individually shown some evidence of effectiveness in weight loss, the combination has been shown to have some synergistic effects on weight. A sustained release formulation of the medication was approved for marketing in the United

Sates under the brand name Contrave. The combination was also approved for use in the European Union under the brand name Mysimba. Like all antidepressants, bupropion carries a warning about suicide risk. Bupropion/naltrexone can raise blood pressure, and monitoring is necessary at the start of treatment.

Phentermine/Topiramate Combination (Qsymia)

The phentermine/topiramate is a combination of a weight loss drug (phentermine) and an anticonvulsant (topiramate) or a combination of low-dose phentermine and sustained release topiramate. Phentermine has the potential to be abused because it acts like an amphetamine. Phentermine (Adipex-P, Lomaira) is also used for monotherapy of obesity and overweight and is one of the four similar weight loss drugs approved for short-term use. The combination with the brand name Qsymia (formerly Qnexa) was approved by the FDA as an obesity treatment complementary to a diet and exercise regimen. The European Medicines Agency, by contrast, rejected the combination as a treatment for obesity, citing concerns about long-term effects on the heart and blood vessels (increases in HR and BP), mental health, and cognitive side effects. Topiramate could also increase the risk of birth defects.

Pramlintide/Phentermine Combination

The combined pharmacotherapy of pramlintide and phentermine was superior to pramlintide alone in humans [68, 69], a fourfold more effective for reduction in body weight with the combination than with pramlintide alone.

Exenatide/Dapagliflozin Combination

The effects on weight loss in patients with T2DM of exenatide once weekly plus daily dapagliflozin (a SGLT2 inhibitor in renal tubular epithelium) and exenatide once weekly or dapagliflozin daily added to metformin monotherapy were investigated [68]. Among the overall population, weight loss was significantly greater after 28 weeks for patients receiving exenatide once weekly plus dapagliflozin (average weight loss 3.55 kg) compared to the same treatments individually (average weight loss of 2.0 kg with daily dapagliflozin and 1.5 kg with exenatide once weekly) [70]. There were greater reductions in body mass index (BMI) observed in all BMI groups, including normal weight and overweight patients, with combination treatment compared with either treatment alone [71]. Exenatide once weekly plus dapagliflozin was superior to placebo in a separate 24-week trial [72], and results were sustained over 1 year of treatment [73].

Exenatide/Liraglutide/Semaglutide Combination

This is a triple GLP-1 agonist combination. Exenatide (Byetta) is a long-acting analog of the hormone GLP-1, which the intestines secrete in response to the presence of food. Among other effects, GLP-1 delays stomach emptying and promotes a feeling of fullness after eating. Some obese people are deficient in GLP-1, and dieting reduces GLP-1 further [74]. Liraglutide (Saxenda) and semaglutide (Ozempic) are GLP-1 analogs for daily and weekly administration, respectively.

Amylin/Pramlintide Combination

An analog of amylin (secreted by the beta cells of the pancreas in a fixed ratio when insulin is released and activated), pramlintide, originally developed by Amylin Pharmaceuticals, now owned by AstraZeneca Pharmaceuticals, is currently available for treating diabetes and is in testing for treating obesity in non-diabetics.

Incretin/Pancreatic Hormone Combination

The incretins are hormones that are released after eating and augment the secretion of insulin from pancreatic beta cells. Incretins and pancreatic hormones generally inhibit the upper gastrointestinal motor function or secretion. In addition, many of these hormones also exert central effects that reduce appetite, and some (*e.g.*, GLP-1 analogs or GLP-1 receptor agonists) are efficacious in the treatment of obesity [69]. Several combinations of incretins with pancreatic hormones have been tested.

Coadministration of GLP-1 agonist with glucagon hormone in humans resulted in increased energy expenditure [74] and a reduction in food intake by an average of 13% [75], whereas glucagon or GLP-1, given individually at subanorectic doses, did not significantly reduce food intake [76]. The effect of the combination on resting energy expenditure is attributable to the effects of glucagon; there was no significant change in resting energy expenditure during GLP-1 infusion [68].

The combination of GLP-1 and PYY_{3-36} elicits a synergistic effect in two human studies. In a double-blind, four-arm, crossover, randomized study, 25 male participants received placebo, GLP-1, peptide tyrosine tyrosine3-36 (PYY_{3-36}), or GLP-1 plus PYY_{3-36}. The coinfusion of GLP-1 and PYY_{3-36} produced a reduction of 30.4% of food intake compared to placebo and more than the sum of each independently, suggesting a synergistic effect [77]. A subsequent experimental study tested acute continuous subcutaneous infusion for 10.5 hours/day of GLP-1, peptide YY, and oxyntomodulin [summarized as GOP], at doses that replicated the postprandial concentrations observed after Roux-en-Y gastric bypass (RYGB). Oxyntomodulin analogs have significant potential due to the inherent affinity of oxyntomodulin to both glucagon and GLP-1 receptors [68].

5.2 FDC for the Treatment of Metabolic Disease

Polypharmacy for metabolic diseases includes lipid-lowering agents, antihypertensive agents, antidiabetic agents, heart failure drugs, and anti-obesity therapies. Reduction of low-density lipoprotein (LDL) cholesterol by statins is a common treatment for patients with metabolic diseases. Statins are 3-hydroxy-3-methyl-glutaryl--coenzyme A (HMG-CoA) reductase inhibitors that affect the rate-limiting step in cholesterol biosynthesis [78–80]. Another lipid-lowering approach in metabolic diseases is to inhibit intestinal cholesterol absorption with ezetimibe [78–80]. Angiotensin system inhibitors are commonly used to combat hypertension associated with metabolic diseases [80, 81]. Angiotensin-converting enzyme inhibitors or angiotensin receptor blockers not only lower blood pressure but can decrease heart and kidney disease progression in diabetes [80]. Heart failure in metabolic disease is treated with diuretics, beta-adrenergic blockers, or calcium channel blockers [80, 82]. Antidiabetic agents include peroxisome proliferator-activated receptor-γ (PPARγ) agonist insulin-sensitizing agents, glucagon-like peptide-1 (GLP-1) receptor agonists, dipeptidyl peptidase-4 (DPP-4) inhibitors, and sodium-glucose cotransporter 2 (SGLT2) inhibitors [80]. These antidiabetic agents are effective in lowering blood glucose and hemoglobin A1c (HbA1c) levels [78–80]. Obesity can be treated with gastrointestinal lipase inhibitors, central nervous system acting drugs like serotonin 2C receptor agonists, or by bariatric surgery [78, 80]. A primary therapeutic approach for metabolic diseases has been to identify and treat each symptom or condition separately. This approach leads to several drugs being prescribed to patients with metabolic diseases.

Patients with metabolic diseases have a high rate of polypharmacy, and this is a growing concern in the elderly population [78, 83–85]. For example, when polypharmacy is defined as at least five drugs being prescribed, the rate of polypharmacy reaches levels as high as 60% in patients over 65 years old with non-insulin-dependent type 2 diabetes [84, 85]. The majority of these diabetics are on at least one antidiabetic drug, not less than one antihypertensive drug, and a lipid-lowering statin drug [78, 79, 83–85]. A potential solution to polypharmacy in metabolic diseases is to combine multiple drugs or invent drugs with multiple actions to combat metabolic diseases. Consequently, there is growing interest in developing novel therapeutic strategies that could target multiple metabolic disease components to provide a comprehensive treatment approach [86, 87]. Developing a drug that lowers plasma lipids, decreases inflammation, decreases blood pressure, and lowers blood glucose levels would be ideal to treat metabolic diseases. Currently, there are no approved drugs that can reliably reduce multiple conditions associated with metabolic diseases over the long term [79, 81, 88].

5.2.1 FDC in Diabetes Mellitus (DM)

Achieving a glycated hemoglobin (HbA1c) level of <7.0% is the primary endpoint of diabetic care, with evidence supporting this target for the reduction of microvascular complications and providing potential benefit to the macrovascular complications of diabetes [89, 90]. The complex pathophysiology of diabetes, with a combination of peripheral insulin resistance, reduced insulin production secondary to beta-cell dysfunction, and increased hepatic gluconeogenesis, provides a number of potential treatment targets. The British National Institute for Health and Care Excellence (NICE), the American Diabetes Association (ADA), and the European Association for the Study of Diabetes (EASD) recommend early intervention with oral antidiabetic agents to obtain the target HbA1c [89–91]. Type 2 diabetes mellitus is a chronic and progressive disease, and although monotherapy may be adequate initially, >50% of patients do not achieve adequate glycemic control.3 Dose adjustment and progression to multiple diabetes therapies are often required to maintain target HbA1c levels. The complex pathophysiology of diabetes, with a combination of peripheral insulin resistance, reduced insulin production secondary to beta-cell dysfunction, and increased hepatic gluconeogenesis, provides a number of potential treatment targets. The British National Institute for Health and Care Excellence (NICE), the American Diabetes Association (ADA), and the European Association for the Study of Diabetes (EASD) recommend early intervention with oral antidiabetic agents to obtain the target HbA1c [89–91]. Type 2 diabetes mellitus is a chronic and progressive disease, and although monotherapy may be adequate initially, >50% of patients do not achieve adequate glycemic control [92]. Dose adjustment and progression to multiple diabetes therapies are often required to maintain target HbA1c levels.

5.2.1.1 Sitagliptin/Metformin FDC for T2DM

Metformin hydrochloride (*N,N*-dimethylimidodicarbinimidic diamide hydrochloride) is a biguanide antihyperglycemic, with a molecular formula of $C_4H_{11}N_5HCl$ [93]. According to Hayes et al. [89], metformin has saturable and incomplete oral bioavailability of 40%–60%. The absorption is reduced and delayed in the presence of food [94]. Metformin is distributed without binding to plasma proteins and primarily undergoes renal excretion unchanged [93] with the remaining 20%–30% recovered in the feces. A significant increase in plasma half-life is seen in patients with renal impairment, and it is not indicated for use in patients with an estimated glomerular filtration rate <30 mL/min/1.73 m [91, 94]. The mean plasma half-life is estimated to be 4–8.7 hours.

Sitagliptin, MK-0431 [95], is an orally active beta-amino acid-derived DPP-4 inhibitor, with molecular formula $C_{16}H_{20}F_6N_5O_6P$ and a molecular weight of 523.3 g/mol [96]. It is a potent inhibitor of DPP-4 (inhibition mean of 91.3% after 12 hours [92]), with a half-life of ~8–14 hours and an oral bioavailability of 87% with negligible effect on the time to reach maximum concentration (t_{max}) when administered with food. The mean plasma area under the curve of sitagliptin is 8.52 μM/h with a maximum concentration of 950 nM with ~38% plasma protein bound.7 Sitagliptin is primarily eliminated renally via active tubular secretion, 79% being excreted unchanged. Near maximal DPP-4 inhibition is seen at the recommended sitagliptin dose of 100 mg/d [97].

Several studies including RCTs and smaller and single-blind studies have examined the efficacy, safety, and tolerability of sitagliptin and metformin as either a FDC or a dual therapy against placebo, monotherapy, as add-on, or other active combinations. Smaller and single-blind studies have supported the findings from RCT [89, 98–100]. A number of studies have examined the effect of the combined use of metformin and sitagliptin in relation to beta-cell function, which show a greater positive effect on beta-cell function compared to respective monotherapy [89, 101, 102].

Reasner et al. [103] assessed the efficacy and safety of sitagliptin/metformin FDC therapy with metformin alone for >18 weeks and 26 weeks in treatment-naïve participants with type 2 diabetes who had an HbA1c ≥7.5% on a diet and exercise therapy. A total of 1450 patients with a mean HbA1c of 9.9% were randomized to sitagliptin/metformin 50/500 mg or metformin 500 mg monotherapy. A higher proportion of participants receiving sitagliptin/metformin FDC achieved an HbA1c <7.0% at 18 weeks (49.2% vs 34.2%, respectively, $P<0.001$). Similarly, a higher proportion of participants taking sitagliptin/metformin FDC achieved an HbA1c <6.5% at 18 weeks compared to metformin alone. For fasting plasma glucose (FPG) at week 18, sitagliptin/metformin FDC had a larger decrease from baseline compared to the monotherapy group, resulting in a significant difference between the groups of −0.9 mM. Significantly greater improvements in measures of β-cell function were observed in the sitagliptin/metformin group at 18 weeks, and reduction in body weight from baseline was comparable in both groups. The safety profile was similar in both groups with no meaningful differences observed in drug-related adverse experiences or discontinuation of treatment. The incidence of reported hypoglycemia was low and similar in the two groups (2.1% sitagliptin/metformin group and 1.8% metformin monotherapy) [103]. Similar results were observed by another study [97].

5.2.1.2 Empagliflozin/Metformin FDC for T2DM

Sodium glucose cotransporter 2 (SGLT2) inhibitors improve glycemic control through increasing urinary glucose excretion, which is independent of β-cell function. In addition, they are generally well tolerated and associated with a low risk of hypoglycemia. Available literature supports that empagliflozin/metformin combination therapy could significantly improve glycemic control, body weight, and blood pressure with a low risk of hypoglycemia [104, 105]. In addition, the empagliflozin/metformin fixed-dose tablets have been demonstrated to reduce pill burden to further achieve the improved patients' adherence, better glycemic control, and optimized cost-effectiveness [104, 105].

5.2.1.3 Empagliflozin/Linagliptin FDC for T2DM

Although various single-pill combinations of oral glucose-lowering agents are available, empagliflozin/linagliptin was the first approved combination of a SGLT2 inhibitor with a DPP-4 inhibitor in the United States. Two publications of the clinical trial investigating the efficacy and safety of single-pill combinations of empagliflozin/linagliptin FDC in treatment-naïve or metformin-treated patients with T2DM (NCT01422876) provided evidence for the efficacy and safety of empagliflozin/linagliptin single pills [106–108]. The addition of an empagliflozin/linagliptin single pill may be considered in patients with inadequate glycemic control on metformin or as an alternative to first-line treatment with empagliflozin or linagliptin when metformin is not suitable, particularly in patients with very poor glycemic control or those who need to achieve target more quickly [106, 107].

A meta-analysis evaluated the efficacy and safety of empagliflozin/linagliptin FDC compared with either monotherapy [n = 6 randomized controlled trials; 2857 adults with T2DM on diet + exercise ± metformin] [109]. The empagliflozin (10 mg)/linagliptin (5 mg) FDC led to significantly greater reductions in glycated hemoglobin (HbA1c) and in fasting plasma glucose (−1.60 mM) compared with either drug alone over 24 weeks. Patients with T2DM treated with the FDC had more than three times higher likelihood of achieving HbA1c <7% than those on either monotherapy. Weight reduction was significantly greater in the combination group only when compared with linagliptin monotherapy. Safety profile was similar between combination treatment and monotherapies. Overall, the empagliflozin/linagliptin FDC had superior efficacy and similar safety in achieving euglycemia compared with either monotherapy [109].

5.2.1.4 Canagliflozin/Metformin FDC for T2DM

The American Association of Clinical Endocrinologists algorithm for the treatment of T2DM recommends metformin plus a sodium glucose cotransporter 2 (SGLT2) inhibitor as the first oral combination in patients who present with HbA1c ≥7.5%. Canagliflozin, an SGLT2 inhibitor, lowers the renal threshold for glucose and increases urinary glucose excretion leading to a mild osmotic diuresis and a net caloric loss. The effect of canagliflozin is insulin-independent and complementary to other antihyperglycemic agents, including metformin. The canagliflozin/metformin FDC is also available with variable dosing, which may be attractive to some patients owing to the potential for reduced pill burden and costs [110]. As initial combination therapy in drug-naïve patients or as dual therapy with metformin or triple therapy in combination with metformin and other antihyperglycemic agents, canagliflozin 100 and 300 mg improved glycemic control and provided reductions in body weight and SBP that were sustained for up to 104 weeks. Canagliflozin was generally well tolerated across studies in combination with metformin. An increased incidence of adverse events (AEs) related to the mechanism of SGLT2 inhibition (i.e., genital mycotic infections, urinary tract infections, osmotic diuresis-related AEs) was observed with canagliflozin. Canagliflozin was associated with a low incidence of hypoglycemia when not used in conjunction with antihyperglycemic agents associated with hypoglycemia (i.e., insulin or sulfonylurea). Together, these results support the use of a canagliflozin/metformin FDC as a treatment approach for a broad range of patients with T2DM [110].

The canagliflozin and metformin IR FDC is indicated as an adjunct to diet and exercise to improve glycemic control in adults with T2DM. The canagliflozin and metformin IR FDC formulation is available in four dose strengths (50/500 mg, 50/1000 mg, 150/500 mg, 150/1000 mg) and should be taken twice daily with food [110, 111]. An oral FDC containing immediate-release canagliflozin and metformin XR in a bilayer tablet formulation was recently approved in the United States [110–112]; the XR FDC formulation was under development. The efficacy and safety of the canagliflozin/metformin FDC are supported by data from six phase III studies or substudies of once-daily canagliflozin 100 or 300 mg in patients with T2DM as add-on to metformin with or without other AHAs and a phase III initial combination therapy study of canagliflozin plus metformin XR, which also evaluated canagliflozin versus metformin monotherapy [110].

5.2.1.5 Linagliptin/Metformin FDC for T2DM

Recently, a fixed-dose, single-tablet, combined formulation of linagliptin (a dipeptidyl peptidase-4 inhibitor) and metformin has been approved for use in T2DM patients and is indicated as an adjunct to diet and exercise for those patients who remain inadequately controlled despite maximal tolerated doses of metformin, metformin and sulfonylurea, or linagliptin and metformin monotherapies [113]. The combination tablet is administered twice daily and can be used either alone or combined with sulfonylureas. Clinical trials suggest that this FDC provides significantly superior glycemic control compared to linagliptin and metformin monotherapy, in terms of improving key parameters of glucose homeostasis such as glycosylated hemoglobin, fasting, and postprandial glucose levels [113]. It also exhibits an excellent tolerability profile, without promoting weight gain and hypoglycemic episodes. The compounds of this formulation do not display clinically relevant pharmacokinetic interactions with each other and exert synergistic (complementary) pharmacodynamic effects, including an enhanced incretin effect, suppressed hepatic glucose production, and improved peripheral insulin sensitivity [113]. As a result, a linagliptin/metformin FDC offers the potential to address multiple defects of T2DM pathophysiology (pancreatic islet dysfunction, insulin resistance, increased hepatic glucose output) and, most importantly, in the context of a safe, efficacious, flexible, and convenient therapeutic regimen [113].

5.2.1.6 Pioglitazone/Metformin FDC for T2DM

Metformin and pioglitazone are well-known insulin-sensitizing agents that act through different mechanisms. The mechanism of action of both metformin and pioglitazone is still not completely understood. However, the large spectrum of effects of pioglitazone makes the combination of metformin and pioglitazone a promising treatment option not only for optimizing management of glycemic control but also for prevention of the cardiovascular complications of diabetes [114]. Bioequivalence between the fixed-dose tablets and pioglitazone and metformin given separately has been established in terms of absorption and bioavailability parameters in two studies in which healthy volunteers were treated with a fixed-dose combination of pioglitazone and metformin at two dose strengths, 15 mg/500 mg and 15 mg/850 mg, or with the coadministration of each drug separately [115]. In a 16-week trial, combination therapy with pioglitazone and metformin significantly improved HbA_{1c} and fasting plasma glucose in diabetic patients inadequately controlled with metformin [114, 116].

5.2.1.7 Dapagliflozin/Saxagliptin FDC for T2DM

A review article on the pharmacology, pharmacokinetics, efficacy, safety, and place in therapy of the FDC product, QTERN, the dapagliflozin (a SGLT2 inhibitor)/saxagliptin (a DPP-4 inhibitor) FDC, suggests it as a reasonable option for patients with T2DM not controlled on metformin [117].

5.2.1.8 Pioglitazone/Glimepiride FDC for T2DM

While glimepiride stimulates β-cell secretion and leads to a reduction of blood glucose levels, pioglitazone activates peroxisome proliferator-activated receptor-γ and improves insulin resistance. Combining these two modes of action has been shown to improve glucose and lipid metabolism and to improve the overall cardiovascular risk in patients with T2DM [118]. The combination of glimepiride and pioglitazone is generally well tolerated, and a fixed combination may lead to an improved compliance in patients [118].

5.2.1.9 Ertugliflozin/Metformin FDC for T2DM

Ertugliflozin and metformin hydrochloride (ertugliflozin/metformin, SEGLUROMET) is a recently approved fixed-dose combination tablet containing the SGLT2 inhibitor ertugliflozin and metformin for the treatment of T2DM [119]. This information comes from the ertugliflozin/metformin prescribing information as well as published clinical trials obtained through a PubMed search. Expert commentary: SGLT2 inhibitors are an important class of antihyperglycemic agents that are efficacious as monotherapy and in combination with other antihyperglycemic agents. Given their favorable effects on glycemia control as well as "extra-glycemic" parameters such as body weight and blood pressure, they are ideal agents for appropriate patients with type 2 diabetes. The fixed-dose combination of ertugliflozin with metformin is an effective combination that is conveniently administered and may improve medication adherence and persistence.

5.2.1.10 Acarbose/Metformin FDC for T2DM

Compared with acarbose monotherapy, acarbose/metformin FDC has superior antihyperglycemic efficacy, brings proportionally more T2D patients to HbA1c goal, and further reduces body weight [120, 121]. Acarbose/metformin FDC is well tolerated without a significant risk of hypoglycemia and is a potentially advantageous therapy for T2D [120, 121].

5.2.1.11 Insulin Degludec/Liraglutide FDC for T2DM

Available data in the literature on the pharmacology, pharmacokinetics, efficacy, and safety of the fixed-dose combination of insulin degludec and the glucagon-like peptide-1 receptor agonist (GLP-1 RA), liraglutide (IDegLira), in the treatment of T2DM support the FDC as an additional glycemic control with fewer AEs for patients uncontrolled on a GLP-RA or basal insulin alone [122]. The FDC displayed pharmacokinetic and pharmacodynamic properties similar to that of the individual components. It has shown significant hemoglobin A1C (A1C) reductions of 1.3–1.9% and fasting plasma glucose reductions of 45–65 mg/dL when used in patients with T2DM previously receiving oral antihyperglycemic agents (AHAs), GLP-1 RAs, or basal insulin. Weight loss also occurred when the insulin degludec/liraglutide FDC was started in patients previously receiving oral AHAs or basal insulin. Adverse effects (AEs) tended to be mild and transient. The most common AEs were headache, nasopharyngitis, upper-respiratory infections, and gastrointestinal disorders. Hypoglycemia risk was lower with the FDC than basal insulin alone but higher than liraglutide alone.

Moreover, the recent approval of the insulin degludec/liraglutide FDC is unique not simply for the additive benefits of the two agents, but because it now permits adjustable dosing of liraglutide together with insulin, providing better glucose control than with either agent alone at lower dose levels [123]. A lower dosage of insulin degludec reduces the risk of hypoglycemia. Liraglutide combats the weight gain that accompanies the introduction of insulin therapy, and a reduced dose of liraglutide induces less GI intolerance [123].

5.2.1.12 Saxagliptin/Metformin FDC for T2DM

Saxagliptin/metformin FDC tablets can be used either alone or in combination with glyburide, thiazolidinediones, or insulin. It has been proven that the saxagliptin/metformin FDC leads to a significant improvement in glycemic control compared to placebo in patients with type 2 diabetes that is inadequately controlled with MET alone [124]. In addition, this FDC has been proven to be safe for people with diabetes mellitus and established cardiovascular disease, elderly patients, and patients with impaired renal function (>30 mL/minute), with dosage modification. Patient compliance, adherence, and persistence to the therapeutic regimen have been shown to be very good, while the titration of each compound according to the patient's profile is easy, given the availability of different formulations [124]. The saxagliptin/metformin FDC is a patient-friendly, dosage-flexible, and hypoglycemia-safe regimen with very few adverse events and a neutral or even favorable effect on body weight. It achieves significant glycosylated hemoglobin A1c reduction helping the patient to achieve his/her individual glycemic goals [124].

5.2.1.13 Dapagliflozin/Metformin FDC for T2DM

A review article on 14 studies reached a conclusion that the unique combined mechanism of action and favorable efficacy and safety profile of the dapagliflozin/metformin with extended-release support consideration of this fixed-dose combination as a treatment option for patients with T2DM [125] because the two compounds exert synergistic effects with better glycemic control and weight loss and reduced incidence of hypoglycemia, based on their complementary mechanisms of action [126].

Overall, it was found that metformin was the most widely used component of FDCs with other antihyperglycemic agents [127]. An increasing number of randomized trials on the safety and efficacy of newly emerging FDCs suggests that they would be better treatment options for T2DM patients [127].

5.2.2 FDC in Obesity

Obesity is a known health risk for the development of several preventable diseases. Obesity-related metabolic alterations negatively impact different physiological mechanisms, which supports the rationale for the use of combined drug therapy. Despite increasing public health concerns regarding obesity, few safe and effective drug treatments are available. Few studies on combination therapies for the treatment of obesity had been conducted until recently, when two fixed-dose combinations, bupropion-naltrexone ER fixed-dose combination and phentermine-topiramate ER titrated-dose combinations, were evaluated in clinical studies that ultimately led to FDA approval. In practice, many of the FDC drugs for T2DM could be applied to the polypharmacological treatment of obesity.

5.2.2.1 Naltrexone/Bupropion FDC for Obesity

Naltrexone/bupropion (Mysimba, Orexigen Therapeutics Ireland Limited) is a fixed-dose combination product for the treatment of adults who are obese or overweight with at least one weight-related comorbidity, as an adjunct to diet and lifestyle modifications [128–130]. The drug has been tested in four randomized, double-blind, placebo-controlled, phase III trials, and the co-primary endpoints were met in each case. Originally licensed by the European Medicines Agency (EMA) in 2015, it has recently been launched in the United Kingdom.

Naltrexone is an opioid antagonist for the treatment of opioid and alcohol dependency, whereas bupropion is a norepinephrine/dopamine reuptake inhibitor used to treat depression and smoking cessation. Although not effective as individual monotherapies for obesity, naltrexone and bupropion in combination produce weight loss and a metabolic profile beneficial for the potential treatment of obesity. The naltrexone/bupropion FDC has a greater weight loss efficacy than two FDA-approved medications, orlistat and lorcaserin [131]. Although the weight loss produced by phentermine/topiramate is superior to naltrexone/bupropion, the safety profile of naltrexone/bupropion has less severe adverse effects [131]. In addition, naltrexone/bupropion is well tolerated, with nausea being the most reported adverse event. Unlike other centrally acting medications, lorcaserin and phentermine/topiramate, naltrexone/bupropion has no abuse potential [131].

Combination treatment with sustained-release naltrexone and bupropion was developed to produce complementary actions in CNS pathways regulating body weight. The Contrave Obesity Research I (COR-I) study assessed the effect of such treatment on body weight in overweight and obese participants [132]. Men and women aged 18–65 years who had a body mass index (BMI) of 30–45 kg/m^2 and uncomplicated obesity or BMI 27–45 kg/m^2 with dyslipidemia or hypertension were eligible for enrolment in this randomized, double-blind, placebo-controlled, phase III trial undertaken at 34 sites in the United States [132]. Participants were prescribed mild hypocaloric diet and exercise and were randomly assigned in a 1:1:1 ratio to receive sustained-release naltrexone 32 mg per day plus sustained-release bupropion 360 mg per day combined in fixed-dose tablets (also known as NB32), sustained-release naltrexone 16 mg per day plus sustained-release bupropion 360 mg per day combined in fixed-dose tablets (also known as NB16), or matching placebo twice a day, given orally for 56 weeks [132]. The trial included a 3-week dose escalation. Randomization was done by use of a centralized, computer-generated, web-based system and was stratified by study center. Co-primary efficacy endpoints at 56 weeks were percentage change in body weight and proportion of participants who achieved a decrease in body weight of 5% or more [132]. The primary analysis included all randomized participants with a baseline weight measurement and a post-baseline weight measurement while on study drug (last observation carried forward). This study is registered with ClinicalTrials.gov, number NCT00532779. A total of 1742 participants were enrolled and randomized to double-blind treatment (naltrexone 32 mg plus bupropion, n = 583; naltrexone 16 mg plus bupropion, n = 578; placebo, n = 581). 870 (50%) participants completed 56 weeks of treatment, and 1453 were included in the primary analysis. Mean change in body weight was −1.3% (SE 0.3) in the placebo group, −6.1% in the naltrexone 32 mg plus bupropion group ($p < 0.0001$ *vs* placebo), and −5.0% in the naltrexone 16 mg plus bupropion group ($p < 0.0001$ *vs* placebo). 84 (16%) participants assigned to placebo had a decrease in body weight of 5% or more compared with 226 (48%) assigned to the naltrexone (32 mg)/bupropion group ($p < 0.0001$ *vs* placebo) and 186 (39%) assigned to the naltrexone (16 mg)/bupropion group ($p < 0.0001$ *vs* placebo) [132]. The most frequent adverse event in participants assigned to FDC treatment was nausea. Headache, constipation, dizziness, vomiting, and dry mouth were also more frequent in the naltrexone plus bupropion groups than in the placebo group

[132]. A transient increase of around 1.5 mm Hg in mean systolic and diastolic blood pressure was followed by a reduction of around 1 mm Hg below baseline in the naltrexone/bupropion FDC groups. Combination treatment was not associated with increased depression or suicidality events compared with placebo. The results indicate that a sustained-release combination of naltrexone plus bupropion could be a useful therapeutic option for treatment of obesity [132].

Binge-eating disorder (BED), the most prevalent eating disorder, is associated strongly with obesity and functional impairments [133]. Few evidence-based treatments for BED exist; a pharmacotherapy effective in reducing both binge eating and body weight is therefore urgently needed. A placebo-controlled double-blind pilot RCT was conducted to evaluate the acute effects of the naltrexone/bupropion FDC on BED with obesity and examined the longer-term effects through 6-month follow-up after the discontinuation of medication [133]. In this small-scale study, 22 adult patients with BED were randomized to receive 12 weeks of double-blind treatment with fixed-dose naltrexone (50 mg)/ bupropion ER (300 mg) or placebo. Independent (blinded) researcher-clinicians evaluated patients at major outcome time points (baseline, posttreatment, and 6-month follow-up after the treatment period); patients were also evaluated for the tracking of course/tolerability throughout treatments and at 3-month follow-up. Primary outcomes were changes from baseline in binge-eating frequency and percent body weight. Secondary outcomes were changes in eating-disorder psychopathology and depression. Their analyses revealed significant reductions from baseline in binge-eating, eating-disorder psychopathology, depression, and body weight during treatment, but these changes with the FDC did not differ significantly from those with placebo. The percentage of patients who attained 3% weight loss was significantly greater with the FDC than with placebo (45.5% vs 0%); weight loss and binge-eating reductions were significantly correlated in the group that received naltrexone/bupropion ER. At 6-month follow-up, outcomes remained improved relative to baseline, with no significant differences between the FDC and placebo [133]. The findings from this pilot RCT suggest that the FDC was well tolerated in these patients with BED and comorbid obesity. Most outcomes were not statistically different between the FDC and placebo [133]. A larger-scale, adequately powered RCT is needed for determining the efficacy of NB in the treatment of BED [133].

The naltrexone/bupropion FDC is one of the most promising new possibilities for the treatment of obesity in an era of increasing prevalence of this disease and decreasing options for its pharmacological management [134]. Although approved by FDA panel members, it was temporally rejected by the FDA afterward, who demanded more cardiovascular safety data for its commercialization.

5.2.2.2 Phentermine IR/Topiramate ER FDC for Obesity

The US Food and Drug Administration granted approval to a fixed-dose combination of phentermine immediate release and topiramate extended release in mid-2012 for treatment of obese patients or overweight patients with comorbid conditions [135]. Phentermine is a noradrenergic sympathomimetic amine approved for short-term treatment of obesity. Topiramate is a sulfamate-substituted monosaccharide derivative of the naturally occurring sugar monosaccharide D-fructose approved as a treatment for migraine headaches and seizure disorders. The new drug has shown significant weight loss compared with placebo for a period up to 2 years [135]. Obesity is spreading globally at an alarming speed. The management of obesity is multifaceted and includes lifestyle modifications as the cornerstone. The era of nonsurgical weight loss therapies or drug treatment of obesity was not begun until orlistat had been approved for long-term use in obesity [135, 136].

A multicenter, randomized, double-blind, parallel-design, placebo-controlled study assessed the pharmacokinetic (PK) and pharmacodynamic (PD) characteristics of VI-0521, a fixed-dose combination of immediate-release phentermine (phentermine IR) and extended-release topiramate (topiramate ER) (phentermine

IR/topiramate ER FDC) in adolescents aged 12–17 years with obesity [137]. A total of 42 adolescents were randomly assigned in a 1:1:1 ratio to placebo, or to a mid-dose (phentermine IR/topiramate ER 7.5 mg/46 mg), or to a top-dose (phentermine IR/topiramate ER 15 mg/92 mg) of VI-0521. A total of 26 adolescents were included in the PK analysis (14 from the mid-dose group and 12 from the top-dose group). A higher proportion of adolescents in both the mid- and top-dose groups (13.3% and 50.0%, respectively) compared with placebo (0.0%) reached ≥5% weight loss at day 56. Both phentermine IR/topiramate ER dose combinations were safe and well tolerated [137]. Thus, treatment of adolescents with obesity using a fixed-dose combination of phentermine IR/topiramate ER for 8 weeks resulted in statistically significant weight loss and a tolerable safety profile [137]. These data indicate that both mid- and top-dose levels are appropriate for longer-term safety and efficacy studies in adolescents [137].

Another study evaluated the long-term efficacy and safety of controlled release phentermine IR/topiramate in overweight and obese subjects with cardiometabolic disease [138]. This was a placebo-controlled, double-blind, 52-week extension study; volunteers at selected sites continued with original randomly assigned treatment with placebo, 7.5 mg phentermine/46 mg controlled-release topiramate, or 15 mg phentermine/92 mg controlled-release topiramate to complete a total of 108 weeks. Of 866 eligible subjects, 84.0% of subjects completed the study. At week 108, the phentermine/topiramate FDC was associated with significant, sustained weight loss [138]. Significantly more phentermine/topiramate FDC-treated subjects at each dose achieved ≥5%, ≥10%, ≥15%, and ≥20% weight loss compared with placebo. The phentermine/topiramate FDC improved cardiovascular and metabolic variables and decreased the rates of incident diabetes in comparison with placebo [138]. The phentermine/topiramate FDC was well tolerated over 108 weeks, with reduced rates of adverse events occurring between weeks 56 and 108 compared with rates between weeks 0 and 56 [138].

Positive caloric balance often causes pathologic adipocyte and adipose tissue anatomical and functional changes (termed adiposopathy or "sick fat"), which may lead to pathogenic adipocyte and adipose tissue responses and metabolic disease [139]. Fat weight loss may improve adiposopathy and thus improve metabolic disease in overweight patients. Unfortunately, the efficacy of nonsurgical weight loss therapies is often limited due to redundant physiological systems that help "protect" against starvation and/or negative caloric balance. One strategy to overcome these limitations is to combine weight loss drug therapies having complementary mechanisms of action, thereby affecting more than one physiologic process influencing body fat accumulation [139]. Although known to facilitate weight loss since its approval, topiramate monotherapy does not have a regulatory indication as an anti-obesity agent. Fixed-dose phentermine/topiramate controlled-release (phentermine/topiramate CR) is a combination agent containing immediate-release phentermine and controlled-release topiramate. Clinical trials involving thousands of patients demonstrate phentermine/topiramate CR to be effective in improving the weight of patients and also effective in improving adiposopathy-associated metabolic diseases [139].

5.2.2.3 Orlistat/Acarbose FDC for Obesity

Orlistat is an active ingredient in Alli, which promotes weight loss by decreasing the amount of dietary fat absorbed in the intestines. Orlistat (prescription and nonprescription) is used with an individualized low-calorie, low-fat diet and exercise program to help people lose weight. Prescription orlistat is used in overweight people who may also have high blood pressure, diabetes, high cholesterol, or heart disease. Orlistat is also used after weight loss to help people keep from gaining back that weight. Orlistat is in a class of medications called lipase inhibitors. It works by preventing some of the fat in foods eaten from being absorbed in the intestines. This unabsorbed fat is then removed from the body in the stool.

Acarbose is FDA approved for the treatment of adults with T2DM as an adjunct to diet only or diet and exercise, depending on the patient's health status. It is a complex oligosaccharide that acts as a competitive, reversible inhibitor of pancreatic alpha-amylase and membrane-bound intestinal α-glucoside hydrolase. Pancreatic α-amylase hydrolyzes complex carbohydrates to oligosaccharides in the small intestine. Intestinal α-glucosidase hydrolase breaks down oligosaccharides, trisaccharides, and disaccharides (sucrose, maltose) to monosaccharides (glucose, fructose) in the brush border of the small intestine. By delaying the digestion of carbohydrates, acarbose slows glucose absorption, resulting in a reduction of postprandial glucose blood concentrations.

A study compared the effect of a novel modified-release fixed-dose combination of orlistat and acarbose (the orlistat/acarbose FDC) with that of conventional orlistat regarding tolerability, appetite, and glucose metabolism [140]. Sixty-seven men with obesity, aged 24–60 years with body mass indexes (BMIs) 33–40 kg/m^2 or 30–32 kg/m^2, and waist circumference above 102 cm were included. They were randomized to either three different doses of the test formulation orlistat/acarbose (60 mg/20 mg, 90/30, and 120/40) or orlistat alone (120 mg) for 2 weeks. In the orlistat/acarbose FDC groups, reports of liquid and oily stools as well as fecal incontinence were fewer, whereas reports of gastric distension and flatulence were higher, compared with orlistat monotherapy [140]. More participants reported decreased hunger in the 90/30 and 120/40 orlistat, and postprandial plasma glucose concentration was reduced in all orlistat/acarbose FDC groups compared with orlistat. This study shows that by using a modified-release dosage form, orlistat and acarbose can be combined without compromising tolerability. Furthermore, the orlistat/acarbose FDC shows promising effects regarding reduction of appetite and reduces postprandial glucose [140]. Tolerability is coupled to compliance and thereby efficacy of a treatment; therefore, this novel orlistat/acarbose FDC could be an effective approach for weight loss treatment [140].

5.2.2.4 D-Norpseudoephedrine/Tri-iodothyronine/Atropine/Aloin/Diazepam FDC for Obesity

A FDC of d-norpseudoephedrine, tri-iodothyronine, atropine, aloin, and diazepam is used in Mexico for the short-term treatment of obesity; however, its efficacy and safety have been scarcely studied. A study was designed to analyze the efficacy and safety of this FDC in Mexican adult overweight and obese patients by a prospective, uncontrolled, multicenter, phase IV open-label study [141]. 3290 patients with a body mass index (BMI) ˃27 kg/m^2 were included in the current study. The primary outcome was the absolute body weight loss, while the secondary outcomes were the improvement of anthropometric and cardiometabolic parameters as well as the description of adverse events. The FDC decreased the body weight and BMI by −9.0 ± 5.6 kg and −3.4 ± 2.2 kg/m^2, respectively, at 6 months. In addition, 43.3% and 14.3% of subjects achieved at least 5% or 10% weight loss at 6 months, respectively [141]. The FDC also significantly improved waist circumference, hip circumference, body fat, visceral fat, SBP, DBP, diabetes risk, and mortality risk, at 3 and 6 months [141]. Moreover, the FDC seems to have better results in the following order: obese grade 3 ≈ obese grade 2 ˃ obese grade 1 ˃ overweight patients [141]. Mild mouth dryness, anxiety, and headache were the main reported adverse events. These data suggest that the d-norpseudoephedrine/tri-iodothyronine/atropine/aloin/diazepam FDC is effective and well tolerated for the short-term therapy of overweight and obesity in Mexican patients.

5.3 MTD for the Treatment of Metabolic Disease

There are several metabolic disease drugs that have actions at multiple targets or have multiple therapeutic actions that could serve as a starting point for developing novel bifunctional molecules to treat metabolic diseases. These drugs include statins, metformin, peroxisome proliferator-activated receptor (PPAR, a group of nuclear receptor proteins that function as transcription factors) agonists, and farnesoid X receptor (FXR, also known as bile acid receptor, a nuclear receptor) agonists.

Statins have a primary action to lower LDL-C levels in metabolic diseases [78, 81, 82]. These lipid-lowering actions are due to statins' ability to inhibit 3-hydroxy-3-methylglutaryl coenzyme A (HMG-CoA, an intermediate in the mevalonate and ketogenesis pathways) reductase [81, 82]. Atorvastatin in particular has also been demonstrated to enhance hepatic insulin sensitivity and signaling through undetermined mechanisms [78, 81, 82]. There is also evidence that statins have a direct antiinflammatory action since statins lower C-reactive protein levels [78, 82].

Another potential starting point is the biguanide metformin that is widely prescribed for T2DM and metabolic syndrome [78, 80, 82]. Metformin lowers blood glucose and HbA1c levels and prevents weight gain [80]. The primary action for metformin is to lower hepatic glucose output through adenosine monophosphate (AMP) kinase activation [78, 80]. Another positive aspect is that metformin decreases circulating plasminogen activator inhibitor-1 levels which could combat cardiovascular diseases associated with T2DM and metabolic syndrome [80, 82].

PPARγ agonists such as rosiglitazone are used as insulin-sensitizing agents to treat type 2 diabetes [78, 80]. Because PPARγ agonists act as a ligand-regulated transcription factor, these drugs regulate the expression of many genes that have a myriad of actions. One of the actions that these thiazolidinedione PPARγ agonists have is to decrease inflammatory genes to prevent atherosclerotic complications associated with T2DM [78, 80]. There have been attempts made to develop drugs that act on multiple PPARs including PPARα, PPARγ, and PPARδ for treating metabolic diseases [78, 80, 82, 142]. Regrettably, these dual-acting PPARs have failed to become approved for treating metabolic diseases. Drugs simultaneously targeting PPAR and other validated metabolic targets represent a promising multitarget approach to combine antihyperglycemic, antihyperlipidemic, and antihypertensive effects [143]. Nevertheless, the strategies to introduce PPARγ activity have been excessively used in various approaches to multitarget drugs for metabolic syndrome and were excellently reviewed by Ammazzalorso et al. [143].

Although not currently approved for humans, FXR agonists are being developed to treat metabolic diseases like nonalcoholic fatty liver disease (NAFLD) and fibrotic diseases [144–146]. FXR activation reduces hepatic fat accumulation and has been demonstrated to have anti-fibrotic and antiinflammatory actions [145, 146]. Obeticholic acid is an FXR agonist that has entered clinical trials for nonalcoholic steatohepatitis (NASH) and NAFLD; however, this FXR agonist has the unwanted effect of increasing total cholesterol levels and increasing the ratio of HDL-C to non-HDL-C [147, 148].

Nevertheless, a common denominator for this group of metabolic disease drugs that could serve as a starting point to develop multitarget drugs is their ability to combat inflammation [78]. Novel bifunctional molecules build on this antiinflammatory action and will have the capacity to impact multiple factors including BP, lipid and triglyceride levels, and insulin signaling in metabolic diseases like metabolic syndrome, T2DM, and NAFLD [78]. A multitarget pharmacologic approach could be advantageous for the therapy of metabolic multiple diseases, such as the metabolic syndrome, which is characterized by metabolic abnormalities associated with diabetes, obesity, hypertension, and increased cardiovascular risk [143].

5.3.1 MTD in Diabetes Mellitus (DM)

The treatment of patients with T2DM is challenging, since there is no cure available to date, but glycemia can be controlled by pharmacological therapy. Furthermore, a reliable glycemic control is not the only goal of a proper diabetic management, and other important objectives to be pursued should include reducing the body weight as well as alleviating the symptoms and preventing micro- and macrovascular damage.

5.3.1.1 PPARα/γ Dual Agonists as MTD Drugs for T2DM

As stated by Artasensi et al. in their excellent review article [149], the most promising efforts were made in the development of PPARα/γ dual agonist which revealed potent therapeutic effects for T2DM, cardioprotective effects, and dyslipidemia. Some recently investigated PPARα/γ dual agonists include naveglitazar, netoglitazone, muraglitazar, ragaglitazar, tesaglitazar, imiglitazar, MK-767, and LY-929 [149].

Clinical trials with dual PPARα/γ agonists demonstrate reduction in triglyceride concentration, an increase in cardioprotective HDL-C level with consequently improvements in insulin sensitivity [150].

Evidence reported in literature demonstrated the positive effects of the dual agonist ragaglitazar in decreasing cardiovascular risk factors by inducing an efficient hypotensive effect in spontaneously hypertensive rats and in the improvement of the endothelial function in Zucker diabetic fatty rats [151].

Ragaglitazar was also reported to have some carcinogenic effects in rodent, but there are not enough available data that support this, as well as the effects have been never observed in human. MK-767 has slightly less activity on PPARγ when compared with pioglitazone. MK-767 effectively normalizes hyperglycemia and hyperinsulinemia in the diabetic ab/ab mouse model. In healthy human subjects, it still reduces triglyceride, FFA, LDL, VLDL, and fasting plasma glucose level. Furthermore, it increases adiponectin levels in healthy subjects [152].

The clinical program of MK-767 was developed by Merck in collaboration with Kyorin Pharmaceutical Co., but it was halted at phase III, in a long-term safety assessment program, due to the identification of a rare form of malignant tumor in mice.

Unexpectedly, the developments of other dual PPARs agonist such as tesaglitazar have been withdrawn in phase III clinical trials because it may cause an increase in fibrosarcoma formation.

Muraglitazar has been discontinued as well because, although it is responsible for an increase in HDL-C levels and a decrease in total cholesterol, apolipoprotein B, triglycerides, and HbA_{1c}, this agent is associated with an increased risk of adverse cardiovascular events and heart failure.

These unsatisfactory results led to the investigation of the agonist ratio in order to avoid the exacerbation of adverse effects, and the gained lines of evidence show that supra-activation of PPARα or PPARγ may be associated with diabetes comorbidities such as renal dysfunction, fluid retention, heart failure, and carcinogenesis.

Tesaglitazar and muraglitazar also lack balance in binding affinity: the former has a high affinity toward PPARγ, while the latter is more active on PPARα.

Hence, one of the challenges is the design of a dual agonist with balanced PPARα/γ activity and good safety profile.

Further studies have been looking for combining the potential effects of PPARα/δ and PPARγ/δ, which may show interesting proprieties like the dual activation of PPARα/γ. For example, (R)-3-{2-ethyl-4-[3-(4-ethyl-2-pyridin-2-yl-phenoxy)-butoxy]-phenyl}propionic acid, a PPARγ/δ agonist, has been shown to lower the glucose level inducing less weight gain than rosiglitazone [153].

5.3.1.2 PPAR Pan-agonists as MTD Drugs for T2DM

The development of pan-agonists, which can act on more than two targets, is currently under investigation [149]. Pharmacodynamic tests on bezafibrate demonstrate that this lipid-lowering drug acts on all the three PPAR subtypes and induces lower LDL cholesterol and triglyceride levels while increasing HDL. Additional studies show that in patients with impaired glucose balance, bezafibrate may delay progress of diabetes by improving insulin sensitivity, inhibiting atherosclerosis, and preventing ischemic heart injury [154]. Furthermore, these agents are also expected to avoid weight gain. Bezafibrate analogs such as LY 465608 and BPR1H036 are under investigation as novel PPAR pan-agonist [155]. However, in vitro and in vivo studies have to be performed to collect more clinical evidence of their efficacy.

5.3.1.3 GLP-1/Glucagon Dual Agonists as MTD Drugs for T2DM

Glucagon is a hormone, which participates with insulin in glycemic homeostasis, and, while remaining unclear whether low glucose levels directly stimulate glucagon release, it is a fact that T2DM patients suffer from hyperglucagonemia [156], an excess of glucagon secretion, possibly as a result of alpha cell insulin resistance [149].

Glucagon can be utilized therapeutically as a satiety factor, which also increases energy expenditure and weight loss [157]. Since the incretin hormones and glucagon have some overlapping functions, their combined use could lead to synergistic effects on diabetes and related metabolic diseases [158].

Investigation of compounds simultaneously targeting the GLP-1R and glucagon receptor (GcgR) has been carried out. The research groups of Day et al. [159] and Pocai et al. [160] have first reported peptides that act as such dual drugs. In preclinical studies, they showed that the treatment with these co-agonists induces superior weight loss and lipid lowering, without causing hyperglycemia or any other adverse effects in diet-induced obese (DIO) mice. The relevance of the results reached with this approach stimulated big pharma companies such as Sanofi-Aventis and Eli Lilly to invest further in the field [149].

Oxyntomodulin (OXM) is a gut hormone that, similarly to GLP-1, is secreted in response to feeding. Its biological effects are attributed to a dual activation of GLP-1R and GcgR [149]. However, OXM is rapidly inactivated (around 12 minutes) in plasma by DPP-4 which cleavages the first two N-terminal amino acids of the peptide. Therefore, OXM mimetics should inevitably involve a modified N-terminus in order to protect the compound against DPP-4 inactivation and possibly to avoid decreasing of the efficacy. On these bases, investigation of OXM-analogs has been performed [161] to improve half-life. For example, the strategy proposed by Pocai [160] was to introduce a D-Ser^2 (S) substitution, while an enhanced metabolic stability was achieved through a cholesterol moiety (chol) at the C-terminus of these peptides. Furthermore, the polyethylene glycol spacer minimizes the loss in agonist potency because of plasma protein/lipid binding. Finally, the treatment with these compounds induced similar improvements in glycemic control compared to selective GLP-1R agonists, and additionally it has been shown to decrease body weight and food intake [149].

5.3.1.4 GLP-1/GIP Dual Agonists as MTD Drugs for T2DM

Incretin-based therapies exploit the actions of the glucose-dependent insulinotropic polypeptide hormone (GIP) and the glucagon-like peptide-1 (GLP-1) [162]. These enteroendocrine incretin hormones are released from the gut in response to intraluminal carbohydrates [163], and they act as important regulators of postprandial glycemic control. More specifically, they are involved in several beneficial pancreatic effects, including stimulation of insulin secretion and insulin gene expression, promoting β-cell survival, improving β-cell glucose sensitivity, and decreasing glucagon secretion. In addition, incretin hormones not only target pancreatic islet cells, but they also possess numerous extra-pancreatic actions that impart positive effects in terms of slowing the gastric emptying and reduction of food intake and weight loss [164]. In the metabolically disturbed state, T2DM subjects show a lacking glucose-lowering response mediated by incretin hormone [165]. This is due to reduced activity or desensitization of the GIP-receptors and reduced postprandial circulating level of GLP-1 [166, 167]. However, restoring GIP efficacy in T2DM patients is possible: preclinical and clinical studies show that this can be achieved with the enhancement of hyperglycemia due to drugs [168, 169] or weight loss [170]. Regarding the glucose-dependent insulinotropic hormone, patients treated with GLP-1 for six weeks exhibit an improved insulin sensitivity and reduced levels of glycated hemoglobin HbA1c [171].

Enhancing the incretin effects is a prominent approach to successfully treat diabetes and to control obesity.

The design process starts from the observation that GLP-1 and GIP share a high degree of sequence similarity, which can be modified in order to obtain potent and balanced co-agonism. More specifically this was achieved with the substitution of residues Glu^3 and Lys^{16} of GIP sequence with the obtained peptides having almost no in vitro glucagon activity. Furthermore, the modifications could include substitution with Lys^{40} that allows site-specific lipidation or PEGylation of the peptide in order to avoid DPP-4 inactivation and consequently to permit less frequent administrations [172–174].

Several in vivo studies show promising results in terms of decreasing glycated hemoglobin and weight loss in both healthy subjects and patients with T2DM [175].

5.3.2 MTD in Obesity

There have not been any published reports on multitarget agents for the management of obesity, though some ongoing efforts devote to identifying phytochemicals from natural products for helping obese or overweight population reversing to normal healthy conditions. In addition, the antidiabetic multitarget drugs, either in the market or under development, might also be applied to anti-obesity therapy as many of them have been shown to effectively reduce body weight. Furthermore, the drugs used for lowering blood lipid are also the choice for obesity depending on their specifications.

References

1. Henao-Mejia J, Elinav E, Thaiss CA, Flavell RA. Inflammasomes and metabolic disease. Annu Rev Physiol. 2014;76:57–78.
2. Keller AS, Keller Iv TCS, et al. Replacement therapies in metabolic disease. Curr Pharm Biotechnol. 2018;19(5):382–99.
3. Dohrmann CE. Target discovery in metabolic disease. Drug Discov Today. 2004;9(18):785–94.
4. National Institute for Health and Care Excellence. Type 1 diabetes in adults: diagnosis and management. www.nice.org.uk. 26 August 2015. Retrieved 07 July 2021.
5. WHO. Diabetes. 13 April 2021. Archived from the original on 14 July 2021. https://www.who.int/news-room/fact-sheets/detail/diabetes
6. Melmer A, Laimer M. Treatment goals in diabetes. Endocr Dev. 2016;31:1–27.
7. DeFronzo RA. Pharmacologic therapy for type 2 diabetes mellitus. Ann Intern Med. 1999;131:281–303.
8. Feinglos MN, Bethel MA. Treatment of type 2 diabetes mellitus. Med Clin North Am. 1998;82:757–90.
9. Luna B, Feinglos MN. Oral agents in the management of type 2 diabetes mellitus. Am Fam Physician. 2001;63(9):1747–56.
10. Ripsin CM, Kang H, Urban RJ. Management of blood glucose in type 2 diabetes mellitus. Am Family Physician. 2009;79(1):29–36.
11. Inzucchi SE, Bergenstal RM, Buse JB, et al. Management of hyperglycemia in type 2 diabetes: a patient-centered approach: position statement of the American Diabetes Association (ADA) and the European Association for the Study of Diabetes (EASD). Diabetes Care. 2012;35(6):1334–79.
12. Garber AJ, Abrahamson MJ, Barzilay JI, et al. AACE comprehensive diabetes management algorithm 2013 consensus statement. Endocr Pract. 2013;19(Suppl 2):1–48.
13. Qaseem A, Humphrey L, Sweet D, et al. Oral pharmacologic treatment of type 2 diabetes mellitus: a clinical practice guideline from the American College of Physicians. Ann Intern Med. 2012;156:218–31.
14. Actos. Deerfield: Takeda Pharmaceutical; 2013.
15. Byetta. San Diego: Amylin Pharmaceutical; 2011.
16. Januvia. Whitehouse Station: Merck & Co; 2014.
17. Farxiga. Princeton: Bristol-Myers Squibb; 2014.
18. Mitchell S, Malanda B, Damasceno A, et al. A roadmap on the prevention of cardiovascular disease among people living with diabetes. Global Heart. 2019;14(3):215–40.
19. Fox CS, Golden SH, Anderson C, et al. American Heart Association Diabetes Committee of the Council on Lifestyle and Cardiometabolic Health; Council on Clinical Cardiology, Council on Cardiovascular and Stroke Nursing, Council on Cardiovascular Surgery and Anesthesia, Council on Quality of Care and Outcomes Research; American Diabetes Association. Update on Prevention of cardiovascular disease in adults with type 2 diabetes mellitus in light of recent evidence: A scientific statement from the American Heart Association and the American Diabetes Association. Diabetes Care. 2015;38(9):1777–803.
20. Fouqueray P, Leverve X, Fontaine E, et al. Imeglimin—a new oral anti-diabetic that targets the three key defects of type 2 diabetes. J Diabetes Metab. 2011;2:126.

21. Vial G, Chauvin M-A, Bendridi N, et al. Imeglimin normalizes glucose tolerance and insulin sensitivity and improves mitochondrial function in liver of a high-fat, high-sucrose diet mice model. Diabetes. 2015;64(6):2254–64.
22. Vuylsteke V, Chastain LM, Maggu GA, et al. Imeglimin: a potential new multi-target drug for type 2 diabetes. Drugs R&D. 2015;15(3):227–32.
23. Perry RJ, Cardone RL, Petersen MC, et al. Imeglimin lowers glucose primarily by amplifying glucose-stimulated insulin secretion in high-fat-fed rodents. Am J Physiol-Endocrinol Metab. 2016;311(2):E461–70.
24. Pacini G, Mari A, Fouqueray P, et al. Imeglimin increases glucose-dependent insulin secretion and improves β-cell function in patients with type 2 diabetes. Diabetes Obes Metab. 2015;17(6):541–5.
25. Lablanche S, Tubbs E, Cottet-Rousselle C, et al. Imeglimin protects INS-1 cells and human islets against high glucose–and high fructose–induced cell death by inhibiting the mitochondrial PTP opening. Diabetes. 2018;67(Suppl1):81.
26. Hallakou-Bozec S, Vial G, Kergoat M, et al. Mechanism of action of Imeglimin: A novel therapeutic agent for type 2 diabetes. Diabetes Obes Metab. 2021;23(3):664-73
27. Pirags V, Lebovitz H, Fouqueray P. Imeglimin, a novel oral antidiabetic, exhibits a good efficacy and safety profile in type 2 diabetic patients. Diabetes Obes Metab. 2012;14:852–8.
28. Pacini G, Mari A, Fouqueray P, et al. Imeglimin increases glucose-dependent insulin secretion and improves beta-cell function in patients with type 2 diabetes. Diabetes Obes Metab. 2015;17:541–5.
29. Fouqueray P, Pirags V, Inzucchi SE, et al. The efficacy and safety of imeglimin as add-on therapy in patients with type 2 diabetes inadequately controlled with metformin monotherapy. Diabetes Care. 2013;36(3):565–8.
30. Riddle M. Combining sulfonylureas and other oral agents. Am J Med. 2000;108(suppl 6a):15S–22S.
31. Ovalle F, Bell DSH. Triple oral antidiabetic therapy in type 2 diabetes mellitus. Endocr Pract. 1998;4:146–7.
32. Riddle MC, Schneider J. Beginning insulin treatment of obese patients with evening 70/30 insulin plus glimepiride versus insulin alone. Glimepiride Combination Group. Diabetes Care. 1998;21:1052–7.
33. Johnson JL, Wolf SL, Kabadi UM. Efficacy of insulin and sulfonylurea combination therapy in type II diabetes. A meta-analysis of the randomized placebo-controlled trials. Arch Intern Med. 1996;156:259–64.
34. Buse J. Combining insulin and oral agents. Am J Med. 2000;108(suppl 6a):23S–32S.
35. Cersosimo E, Johnson EL, Chovanes C, et al. Initiating therapy in patients newly diagnosed with type 2 diabetes: combination therapy vs a stepwise approach. Diabetes Obes Metab. 2018;20(3):497–507.
36. Phung OJ, Sobieraj DM, Engel SS, et al. Early combination therapy for the treatment of type 2 diabetes mellitus: systematic review and meta-analysis. Diabetes Obes Metab. 2014;16(5):410–7.
37. Beale S, Bagust A, Shearer AT, et al. Cost-effectiveness of rosiglitazone combination therapy for the treatment of type 2 diabetes mellitus in the UK. Pharmacoeconomics. 2006;24(Suppl 1):21–34.
38. Rosenstock J, Chuck L, González-Ortiz M, et al. Initial combination therapy with canagliflozin plus metformin versus each component as monotherapy for drug-naive type 2 diabetes. Diabetes Care. 2016;39(3):353–62.
39. Haak T, Meinicke T, Jones R, et al. Initial combination of linagliptin and metformin improves glycaemic control in type 2 diabetes: a randomized, double-blind, placebo-controlled study. Diabetes Obes Metab. 2012;14(6):565–74.
40. Haak T, Meinicke T, Jones R, et al. Initial combination of linagliptin and metformin in patients with type 2 diabetes: efficacy and safety in a randomized, double-blind 1-year extension study. Intern J Clin Practice. 2013;67(12):1283–93.
41. Ross SA, Caballero AE, Del Prato S, et al. Initial combination of linagliptin and metformin compared with linagliptin monotherapy in patients with newly diagnosed type 2 diabetes and marked hyperglycemia: a randomized, double-blind, active-controlled, parallel group, multinational clinical trial. Diabetes Obes Metab. 2015;17(2):136–44.
42. Pratley RE, Fleck P, Wilson C. Efficacy and safety of initial combination therapy with alogliptin plus metformin versus either as monotherapy in drug-naive patients with type 2 diabetes: a randomized, double-blind, 6-month study. Diabetes Obes Metab. 2014;16(7):613–21.
43. Van Raalte DH, van Genugten RE, Eliasson B, et al. The effect of alogliptin and pioglitazone combination therapy on various aspects of beta-cell function in patients with recent-onset type 2 diabetes. Eur J Endocrinol. 2014;170(4):565–74.
44. Lewin A, DeFronzo RA, Patel S, et al. Initial combination of empagliflozin and linagliptin in subjects with type 2 diabetes. Diabetes Care. 2015;38(3):394–402.
45. Gaede P, Lund-Andersen H, Parving HH, et al. Effect of a multifactorial intervention on mortality in type 2 diabetes. N Engl J Med. 2008;358(6):580–91.
46. Gaede P, Vedel P, Parving HH, et al. Intensified multifactorial intervention in patients with type 2 diabetes mellitus and microalbuminuria: the Steno type 2 randomized study. Lancet. 1999;353(9153):617–22.
47. Gaede P, Vedel P, Larsen N, et al. Multifactorial intervention and cardiovascular disease in patients with type 2 diabetes. N Engl J Med. 2003;348(5):383–93.
48. Charles M, Ejskjaer N, Witte DR, et al. Prevalence of neuropathy and peripheral arterial disease and the impact of treatment in people with screen-detected type 2 diabetes: the ADDITION-Denmark study. Diabetes Care. 2011;34(10):2244–9.

49. Griffin SJ, Borch-Johnsen K, Davies MJ, et al. Effect of early intensive multifactorial therapy on 5-year cardiovascular outcomes in individuals with type 2 diabetes detected by screening (ADDITION-Europe): a cluster-randomized trial. Lancet. 2011;378(9786):156–67.
50. Milligan S. Combination therapy for the improvement of long-term macrovascular and microvascular outcomes in type 2 diabetes: rationale and evidence for early initiation. J Diabetes Complicat. 2016;30(6):1177–85.
51. ADVANCE Collaborative Group, Patel A, MacMahon S, Chalmers J, et al. Intensive blood glucose control and vascular outcomes in patients with type 2 diabetes. N Engl J Med. 2008;358(24):2560–72.
52. Zoungas S, Chalmers J, Neal B, et al. ADVANCE-ON Collaborative Group. Follow-up of blood-pressure lowering and glucose control in type 2 diabetes. N Engl J Med. 2014;371(15):1392–406.
53. Action to Control Cardiovascular Risk in Diabetes Study Group, Gerstein HC, Miller ME, Byington RP, et al. Effects of intensive glucose lowering in type 2 diabetes. N Engl J Med. 2008;358(24):2545–59.
54. Duckworth W, Abraira C, Moritz T, et al. VADT Investigators. Glucose control and vascular complications in veterans with type 2 diabetes. N Engl J Med. 2009;360(2):129–39.
55. Hayward RA, Reaven PD, Wiitala WL, et al. VADT Investigators. Follow-up of glycemic control and cardiovascular outcomes in type 2 diabetes. N Engl J Med. 2015;372(23):2197–206.
56. Holman RR, Paul SK, Bethel MA, et al. 10-year follow-up of intensive glucose control in type 2 diabetes. N Engl J Med. 2008;359(15):1577–89.
57. Moreno M. Definition and classification of obesity. Rev Méd Clín Condes. 2012;23:124–8.
58. Conway B, Rene A. Obesity as a disease: no lightweight matter. Obes Rev. 2004;5(3):145–51.
59. Fitzgerald FT. The problem of obesity. Annu Rev Med. 1981;32:221–31.
60. Piché ME, Tchernof A, Després JP. Obesity phenotypes, diabetes, and cardiovascular diseases. Circ Res. 2020;126(11):1477–500.
61. Neeland IJ, Poirier P, Després JP. Cardiovascular and metabolic heterogeneity of obesity: clinical challenges and implications for management. Circulation. 2018;137(13):1391–406.
62. Ortega MA, Fraile-Martínez O, Naya I, et al. Type 2 diabetes mellitus associated with obesity (diabesity). The central role of gut microbiota and its translational applications. Nutrients. 2020;12(9):2749.
63. Bosello O, Donataccio MP, Cuzzolaro M. Obesity or obesities? Controversies on the association between body mass index and premature mortality. Eat Weight Disord Stud Anorex Bulim Obes. 2016;21:165–74.
64. De Lorenzo A, Soldati L, Sarlo F, et al. New obesity classification criteria as a tool for bariatric surgery indication. World J Gastroenterol. 2016;22:681–703.
65. Yamada Y, Kato T, Ogino H, et al. Cetilistat (ATL-962), a novel pancreatic lipase inhibitor, ameliorates body weight gain and improves lipid profiles in rats. Hormone Metab Res. 2008;40(8):539–43.
66. Aronne LJ, Powell AG, Apovian CM. Emerging pharmacotherapy for obesity. Expert Opin Emerg Drugs. 2011;16(3):587–96.
67. Khera R, Murad MH, Chandar AK, et al. Association of pharmacological treatments for obesity with weight loss and adverse events: a systematic review and meta-analysis. JAMA. 2016;315(22):2424–34.
68. Camilleri M, Acosta A. Combination therapies for obesity. Metab Syndr Relat Disord. 2018;16(8):390–4.
69. Aronne LJ, Halseth AE, Burns CM, et al. Enhanced weight loss following coadministration of pramlintide with sibutramine or phentermine in a multicenter trial. Obesity (Silver Spring). 2010;18(9):1739–46.
70. Frías JP, Guja C, Hardy E, et al. Exenatide once weekly plus dapagliflozin once daily versus exenatide or dapagliflozin alone in patients with type 2 diabetes inadequately controlled with metformin monotherapy (DURATION-8): a 28 week, multicentre, double-blind, phase 3, randomised controlled trial. Lancet Diabetes Endocrinol. 2016;4:1004–16.
71. Jabbour SA, Frías JP, Guja C, et al. Effects of exenatide once weekly plus dapagliflozin, exenatide once weekly, or dapagliflozin, added to metformin monotherapy, on body weight, systolic blood pressure, and triglycerides in patients with type 2 diabetes in the DURATION-8 study. Diabetes Obes Metab. 2018;20:1515–9.
72. Lundkvist P, Sjöström CD, Amini S, et al. Dapagliflozin once-daily and exenatide once-weekly dual therapy: a 24-week randomized, placebo-controlled, phase II study examining effects on body weight and prediabetes in obese adults without diabetes. Diab Obes Metab. 2017;19:49–60.
73. Lundkvist P, Pereira MJ, Katsogiannos P, et al. Dapagliflozin once daily plus exenatide once weekly in obese adults without diabetes: sustained reductions in body weight, glycaemia and blood pressure over 1 year. Diabetes Obes Metab. 2017;19:1276–88.
74. de Luis DA, Gonzalez Sagrado M, Conde R, et al. Decreased basal levels of glucagon-like peptide-1 after weight loss in obese subjects. Ann Nutr Metab. 2007;51(2):134–8.
75. Tan TM, Field BC, McCullough KA, et al. Coadministration of glucagon-like peptide-1 during glucagon infusion in humans results in increased energy expenditure and amelioration of hyperglycemia. Diabetes. 2013;62:1131–8.
76. Cegla J, Troke RC, Jones B, et al. Coinfusion of low-dose GLP-1 and glucagon in man results in a reduction in food intake. Diabetes. 2014;63:3711–20.
77. Schmidt JB, Gregersen NT, Pedersen SD, et al. Effects of PYY3-36 and GLP-1 on energy intake, energy expenditure, and appetite in overweight men. Am J Physiol Endocrinol Metab. 2014;306:E1248–56.

78. Lillich FF, Imig JD, Proschak E. Multi-target approaches in metabolic syndrome. Front Pharmacol. 2021;11:554961.
79. Grundy SM. Drug therapy of the metabolic syndrome: minimizing the emerging crisis in polypharmacy. Nat Rev Drug Discov. 2006;5:295–309.
80. Bianchi C, Penno G, Romero F, et al. Treating the metabolic syndrome. Expet Rev Cardiovasc Ther. 2007;5:491–506.
81. Grundy SM, Cleeman JI, Daniels SR, et al. Diagnosis and management of the metabolic syndrome: an American heart association/national heart, lung, and blood Institute scientific statement. Circulation. 2005;112:2735–52.
82. Dandona P, Aljada A, Chaudhuri A, et al. Metabolic syndrome: a comprehensive perspective based on interactions between obesity, diabetes, and inflammation. Circulation. 2005;111:1448–54.
83. Grant RW, Devita NG, Singer DE, et al. Polypharmacy and medication adherence in patients with type 2 diabetes. Diabetes Care. 2003;26:1408–12.
84. Noale M, Veronese N, Perin PC, et al. Reply to letter to the editor Polypharmacy in elderly people with diabetes admitted to hospital. Acta Diabetol. 2016;53:859–60.
85. Alwhaibi M, Balkhi B, Alhawassi TM, et al. Polypharmacy among patients with diabetes: a cross-sectional retrospective study in a tertiary hospital in Saudi Arabia. BMJ Open. 2018;8:e020852.
86. Rollason V, Vogt N. Reduction of polypharmacy in the elderly: a systematic review of the role of the pharmacist. Drugs Aging. 2003;20:817–32.
87. Morphy R, Rankovic Z. Designed multiple ligands. An emerging drug discovery paradigm. J Med Chem. 2005;48:6523–43.
88. Handelsman Y, Jellinger PS. Overcoming obstacles in risk factor management in type 2 diabetes mellitus. J Clin Hypertens. 2011;13:613–20.
89. Hayes J, Anderson R, Stephens JW. Sitagliptin/metformin fixed-dose combination in type 2 diabetes mellitus: an evidence-based review of its place in therapy. Drug Des Devel Ther. 2016;10:2263–70.
90. National Institute for Health and Care Excellence. Type 2 Diabetes in Adults: Management. NICE guideline (NG28); 2015. [Accessed July 1, 2016]. Available from: http://www.nice.org.uk/guidance/ng28?unlid=9428993482016126103832
91. American Diabetes Association Standards of medical care in diabetes – 2016. Diabetes Care. 2016; 39(Suppl 1): S1–106.
92. Herman GA, Stevens C, Van Dyck K, et al. Pharmacokinetics and pharmacodynamics of sitagliptin, an inhibitor of dipeptidyl peptidase IV, in healthy subjects: results from two randomized, double-blind, placebo-controlled studies with single oral doses. Clin Pharmacol Ther. 2005;78(6):675–88.
93. National Center for Biotechnology Information PubChem Compound Database; CID=4091. [Accessed January 2016]. Available from: http://pubchem.ncbi.nlm.nih.gov/compound/4091.
94. Papanas N, Maltezis E. Metformin: a review of its use in the treatment of type 2 diabetes. Clin Med Ther. 2009;1:1367–81.
95. Perfetti R, Hui H. The role of GLP-1 in the life and death of pancreatic beta cells. Horm Metab Res. 2004;36(11–12):804–10.
96. National Center for Biotechnology Information PubChem Compound Database; CID=4369359. [Accessed January 2016]. Available from: http://pubchem.ncbi.nlm.nih.gov/compound/4369359
97. Chwieduk CM. Sitagliptin/metformin fixed-dose combination: in patients with type 2 diabetes mellitus. Drugs. 2011;71(3):349–61.
98. Raz I, Chen Y, Wu M, et al. Efficacy and safety of sitagliptin added to ongoing metformin in therapy in patients with type 2 diabetes. Curr Med Res Opin. 2008;24(2):537–50.
99. Brazg R, Xu L, Dalla MC, et al. Effect of adding sitagliptin, a dipeptidyl peptidase-4 inhibitor, to metformin on 24-hour glycemic control and B-cell function in patients with type 2 diabetes. Diabetes Obes Metab. 2007;9(2):186–93.
100. Aschner P, Chan J, Owens D, et al. EASIE investigators Insulin glargine versus sitagliptin in insulin-naive patients with type 2 diabetes mellitus uncontrolled on metformin (EASIE): a multicentre, randomised open-label trial. Lancet. 2012;379(9833):2262–9.
101. Campbell RK. Clarifying the role of incretin-based therapies in the treatment of type 2 diabetes mellitus. Clin Ther. 2011;33(5):511–27.
102. Williams-Herman D, Xu L, Teng R, et al. Effect of initial combination therapy with sitagliptin and metformin on B-cell function in patients with type 2 diabetes. Diabetes Obes Metab. 2012;14(1):67–76.
103. Reasner C, Olansky L, Seck T, et al. The effect of initial therapy with the fixed–dose combination of sitagliptin and metformin compared with metformin monotherapy in patients with type 2 diabetes mellitus. Diabetes Obes Metab. 2011;13(7):644–52.
104. Hu J, Zou P, Zhang S, et al. Empagliflozin/metformin fixed-dose combination: a review in patients with type 2 diabetes. Expert Opin Pharmacother. 2016;17(18):2471–7.
105. Empagliflozin/metformin (Synjardy) for type 2 diabetes. Med Lett Drugs Ther. 2015;57(1484):172–4.
106. Kedia R, Kulkarni S, Ross M, et al. Spotlight on empagliflozin/metformin fixed-dose combination for the treatment of type 2 diabetes: a systematic review. Patient Prefer Adher. 2016;10:1999–2006.
107. Woo V. Empagliflozin/linagliptin single-tablet combination: first-in-class treatment option. Int J Clin Pract. 2015;69(12):1427–37.
108. Jain RK. Empagliflozin/linagliptin single-pill combination therapy for patients with type 2 diabetes mellitus. Expert Opin Pharmacother. 2017;18(6):545–9.
109. Katsiki N, Ofori-Asenso R, Ferrannini E, et al. Fixed-dose combination of empagliflozin and linagliptin for the treatment of patients with type 2 diabetes mellitus: a systematic review and meta-analysis. Diabetes Obes Metab. 2020;22(6):1001–5.

110. Davidson JA, Sloan L. Fixed-dose combination of canagliflozin and metformin for the treatment of type 2 diabetes: an Overview. Adv Ther. 2017;34(1):41–59.
111. INVOKAMET® (canagliflozin and metformin hydrochloride) tablets for oral use. Titusville: Janssen Pharmaceuticals, Inc; 2016.
112. Rosenstock J, Chuck L, González-Ortiz M, et al. Initial combination therapy with canagliflozin plus metformin versus each component as monotherapy in drug-naïve type 2 diabetes. Diabetes Care. 2016;39:353–62.
113. Koliaki C, Doupis J. Linagliptin/Metformin fixed-dose combination treatment: a dual attack to type 2 diabetes pathophysiology. Adv Ther. 2012;29(12):993–1004.
114. Derosa G, Salvadeo SA. Pioglitazone and metformin fixed-dose combination in type 2 diabetes mellitus: an evidence-based review of its place in therapy. Core Evid. 2008;2(3):189–98.
115. Karim A, Slater M, Bradford D, et al. Oral antidiabetic drugs: effect of food on absorption of pioglitazone and metformin from a fixed-dose combination tablet. J Clin Pharmacol. 2007;47:48–55.
116. Einhorn D, Rendell M, Rosenzweig J, et al. Pioglitazone hydrochloride in combination with metformin in the treatment of type 2 diabetes mellitus: a randomized, placebo-controlled study. The Pioglitazone 027 Study Group. Clin Ther. 2000;22:1395–409.
117. Coppenrath VA, Hydery T. Dapagliflozin/Saxagliptin fixed-dose tablets: a new sodium-glucose cotransporter 2 and dipeptidyl peptidase 4 combination for the treatment of type 2 diabetes. Ann Pharmacother. 2018;52(1):78–85.
118. Forst T, Pfützner A. Fixed-dose combination of pioglitazone and glimepiride in the treatment of Type 2 diabetes mellitus. Expert Rev Endocrinol Metab. 2007;2(3):303–12.
119. Frias JP. Fixed-dose combination of ertugliflozin and metformin hydrochloride for the treatment of type 2 diabetes. Expert Rev Endocrinol Metab. 2019;14(2):75–83.
120. Wang JS, Huang CN, Hung YJ, et al. Acarbose/metformin fixed-dose combination study investigators. Acarbose plus metformin fixed-dose combination outperforms acarbose monotherapy for type 2 diabetes. Diabetes Res Clin Pract. 2013;102(1):16–24.
121. Joshi SR, Ramachandran A, Chadha M, et al. Acarbose plus metformin fixed-dose combination in the management of type 2 diabetes. Expert Opin Pharmacother. 2014;15(11):1611–20.
122. Harris K, Nealy KL. The clinical use of a fixed-dose combination of insulin degludec and liraglutide (Xultophy 100/3.6) for the treatment of type 2 diabetes. Ann Pharmacother. 2018;52(1):69–77.
123. Rendell M. First fixed-ratio combination of insulin degludec and liraglutide for the treatment of type 2 diabetes. Drugs Today (Barc). 2015;51(3):185–96.
124. Panagoulias GS, Doupis J. Clinical utility in the treatment of type 2 diabetes with the saxagliptin/metformin fixed combination. Patient Prefer Adherence. 2014;8:227–36.
125. Kuecker CM, Vivian EM. Patient considerations in type 2 diabetes – role of combination dapagliflozin-metformin XR. Diabetes Metab Syndr Obes. 2016;9:25–35.
126. Tan X, Hu J. Combination therapy for type 2 diabetes: dapagliflozin plus metformin. Expert Opin Pharmacother. 2016;17(1):117–26.
127. Vijayakumar TM, Jayram J, Meghana Cheekireddy V, et al. Safety, efficacy, and bioavailability of fixed-dose combinations in type 2 diabetes mellitus: a systematic updated review. Curr Ther Res Clin Exp. 2017;84:4–9.
128. Naltrexone/bupropion for obesity. Drug Ther Bull. 2017;55(11):126–129.
129. Ali KF, Shukla AP, Aronne LJ. Bupropion-SR plus naltrexone-SR for the treatment of mild-to-moderate obesity. Expert Rev Clin Pharmacol. 2016;9(1):27–34.
130. Halpern B, Mancini MC. Safety assessment of combination therapies in the treatment of obesity: focus on naltrexone/bupropion extended release and phentermine-topiramate extended release. Expert Opin Drug Saf. 2017;16(1):27–39.
131. Verpeut JL, Bello NT. Drug safety evaluation of naltrexone/bupropion for the treatment of obesity. Expert Opin Drug Saf. 2014;13(6):831–41.
132. Greenway FL, Fujioka K, Plodkowski RA, et al. COR-I Study Group. Effect of naltrexone plus bupropion on weight loss in overweight and obese adults (COR-I): a multicentre, randomised, double-blind, placebo-controlled, phase 3 trial. Lancet. 2010;376(9741):595–605.
133. Grilo CM, Lydecker JA, Morgan PT, et al. Naltrexone + bupropion combination for the treatment of binge-eating disorder with obesity: a randomized, controlled pilot study. Clin Ther. 2021;43(1):112–22.e1
134. Halpern B, Faria AM, Halpern A. Bupropion/naltrexone fixed-dose combination for the treatment of obesity. Drugs Today (Barc). 2011;47(8):575–81.
135. Singh J, Kumar R. Phentermine-topiramate: first combination drug for obesity. Int J Appl Basic Med Res. 2015;5(2):157–8.
136. Halpern B, Faria AM, Halpern A. Fixed-dose combination of phentermine-topiramate for the treatment of obesity. Expert Rev Clin Pharmacol. 2013;6(3):235–41.
137. Hsia DS, Gosselin NH, Williams J, et al. A randomized, double-blind, placebo-controlled, pharmacokinetic and pharmacodynamic study of a fixed-dose combination of phentermine/topiramate in adolescents with obesity. Diabetes Obes Metab. 2020;22(4):480–91.
138. Garvey WT, Ryan DH, Look M, et al. Two-year sustained weight loss and metabolic benefits with controlled-release phentermine/topiramate in obese and overweight adults (SEQUEL): a randomized,

placebo-controlled, phase 3 extension study. Am J Clin Nutr. 2012;95(2):297–308.
139. Bays H. Phentermine, topiramate and their combination for the treatment of adiposopathy ('sick fat') and metabolic disease. Expert Rev Cardiovasc Ther. 2010;8(12):1777–801.
140. Holmbäck U, Forslund A, Grudén S, et al. Effects of a novel combination of orlistat and acarbose on tolerability, appetite, and glucose metabolism in persons with obesity. Obes Sci Pract. 2020;6(3):313–23.
141. Del Valle-Laisequilla CF, Trejo-Jasso C, Huerta-Cruz JC, et al. Efficacy and safety of a fixed-dose combination of D-norpseudoephedrine, triiodothyronine, atropine, aloin, and diazepam in obese patients. Int J Clin Pharmacol Ther. 2018;56(11):531–8.
142. Jain MR, Giri SR, Bhoi B, et al. Dual PPARα/γ agonist saroglitazar improves liver histopathology and biochemistry in experimental NASH models. Liver Int. 2018;38:1084–94.
143. Ammazzalorso A, Maccallini C, Amoia P, et al. Multitarget PPARγ agonists as innovative modulators of the metabolic syndrome. Eur J Med Chem. 2019;173:261–73.
144. Ali AH, Carey EJ, Lindor KD. Recent advances in the development of farnesoid X receptor agonists. Ann Transl Med. 2015;3:5.
145. Oseini AM, Sanyal AJ. Therapies in non-alcoholic steatohepatitis (NASH). Liver Int. 2017;37:97–103.
146. Sumida Y, Yoneda M. Current and future pharmacological therapies for NAFLD/NASH. J Gastroenterol. 2018;53:362–76.
147. Mudaliar S, Henry RR, Sanyal AJ, et al. Efficacy and safety of the farnesoid X receptor agonist obeticholic acid in patients with type 2 diabetes and nonalcoholic fatty liver disease. Gastroenterology. 2013;145:574–82.
148. Han CY. Update on FXR biology: promising therapeutic target? Int J Mol Sci. 2018;19:2069.
149. Artasensi A, Pedretti A, Vistoli G, et al. Type 2 diabetes mellitus: a review of multi-target drugs. Molecules. 2020;25(8):1987.
150. Balakumar P, Rose M, Ganti SS, et al. PPAR dual agonists: are they opening Pandora's Box? Pharmacol Res. 2007;56:91–8.
151. Mamnoor PK, Hegde P, Datla SR, et al. Antihypertensive effect of ragaglitazar: a novel PPARα and γ dual activator. Pharmacol Res. 2006;54:129–35.
152. Pourcet B, Fruchart JC, Staels B, et al. Selective PPAR modulators, dual and pan PPAR agonists: multimodal drugs for the treatment of Type 2 diabetes and atherosclerosis. Expert Opin Emerg Drugs. 2006;11:379–401.
153. Gonzalez IC, Lamar J, Iradier F, et al. Design and synthesis of a novel class of dual PPARγ/δ agonists. Bioorganic Med Chem Lett. 2007;17:1052–5.
154. Tenenbaum A, Motro M, Fisman EZ. Dual and pan-peroxisome proliferator-activated receptors (PPAR) co-agonism: the bezafibrate lessons. Cardiovasc Diabetol. 2005;4:14.
155. Stefanski A, Majkowska L. Existing and potential therapeutic approaches targeting peroxisome proliferator-activated receptors in the management of Type 2 diabetes. Expert Opin Ther Pat. 2006;16:1713–33.
156. Lund A, Bagger JI, Christensen M, et al. Glucagon and type 2 diabetes: the return of the alpha cell. Curr Diabetes Rep. 2014;14:1–7.
157. Parker JA, McCullough KA, Field BCT, et al. Glucagon and GLP-1 inhibit food intake and increase c-fos expression in similar appetite regulating centres in the brainstem and amygdala. Int J Obes. 2013;37:1391–8.
158. Sadry SA, Drucker DJ. Emerging combinatorial hormone therapies for the treatment of obesity and T2DM. Nat Rev Endocrinol. 2013;9:425–33.
159. Day JW, Ottaway N, Patterson J, et al. A new glucagon and GLP-1 co-agonist eliminates obesity in rodents. Nat Chem Biol. 2009;5:749–57.
160. Pocai A, Carrington PE, Adams JR, et al. Glucagon-like peptide 1/glucagon receptor dual agonism reverses obesity in mice. Diabetes. 2009;58:2258–66.
161. Lynch AM, Pathak N, Flatt YE, et al. Comparison of stability, cellular, glucose-lowering and appetite suppressing effects of oxyntomodulin analogues modified at the N-terminus. Eur J Pharmacol. 2014;743:69–78.
162. Irwin N. New perspectives on exploitation of incretin peptides for the treatment of diabetes and related disorders. World J Diabetes. 2015;6:1285–95.
163. Drucker DJ. The biology of incretin hormones. Cell Metab. 2006;3:153–65.
164. Drucker DJ. Deciphering metabolic messages from the gut drives therapeutic innovation: the 2014 banting lecture. Diabetes. 2015;64:317–26.
165. Toft-Nielsen MB, Damholt MB, Madsbad S, et al. Determinants of the impaired secretion of glucagon-like peptide-1 in type 2 diabetic patients. J Clin Endocrinol Metab. 2001;86:3717–23.
166. Xu G, Kaneto H, Laybutt DR, et al. Downregulation of GLP-1 and GIP receptor expression by hyperglycemia: possible contribution to impaired incretin effects in diabetes. Diabetes. 2007;56:1551–8.
167. Pathak V, Vasu S, Flatt PR, et al. Effects of chronic exposure of clonal β-cells to elevated glucose and free fatty acids on incretin receptor gene expression and secretory responses to GIP and GLP-1. Diabetes Obes Metab. 2014;16:357–65.
168. Højberg PV, Vilsbøll T, Rabøl R, et al. Four weeks of near-normalisation of blood glucose improves the insulin response to glucagon-like peptide-1 and glucose-dependent insulinotropic polypeptide in patients with type 2 diabetes. Diabetologia. 2009;52:199–207.
169. Piteau S, Olver A, Kim SJ, et al. Reversal of islet GIP receptor down-regulation and resistance to GIP by reducing hyperglycemia in the Zucker rat. Biochem Biophys Res Commun. 2007;362:1007–12.

170. Asmar M, Tangaa W, Madsbad S, et al. On the role of glucose-dependent insulintropic polypeptide in postprandial metabolism in humans. Am J Physiol Endocrinol Metab. 2010;298:614–21.
171. Esposito K, Mosca C, Brancario C, Chiodini P, Ceriello A, Giugliano D. GLP-1 receptor agonists and HBA1c target of of >7% in type 2 diabetes: meta-analysis of randomized controlled trials. Curr Med Res Opin. 2011;27:1519–28.
172. Pocai A. Unraveling oxyntomodulin, GLP1's enigmatic brother. J Endocrinol. 2012;215:335–46.
173. Amori RE, Lau J, Pittas AG. Efficacy and safety of incretin therapy in type 2 diabetes: systematic review and meta-analysis. J Am Med Assoc. 2007;298:194–206.
174. Finan B, Tao M, Ottaway N, et al. Unimolecular dual incretins maximize metabolic benefits in rodents, monkeys, and humans. Sci Transl Med. 2013;5(209):209ra151.
175. Coskun T, Sloop KW, Loghin C, et al. LY3298176, a novel dual GIP and GLP-1 receptor agonist for the treatment of type 2 diabetes mellitus: from discovery to clinical proof of concept. Mol Metab. 2018;18:3–14.

6 Polypharmacology in Clinical Applications: Neurological Polypharmacology

Abstract

Neurological disorders include diverse subgroups, including neuromuscular diseases, genetic and metabolic disorders, developmental delay, traumatic brain disorders and injuries, and degenerative diseases. Polypharmacological treatment of neurological disorders is getting popular and, on some occasions, has become the mainstay therapeutic strategy. This chapter introduces the basic knowledge of polypharmacology-based therapeutic approaches for neurological disorders, as well as new multitarget drugs under development for the treatment of neurological disorders. Focus is given to CDT, FDC, and MTD approaches for the treatment of Alzheimer's disease, Parkinson's disease, epilepsy, multiple sclerosis, and schizophrenia, which represent some of the most difficult and challenging neurological disorders.

Keywords

Neurological polypharmacology · Alzheimer's disease · Parkinson's disease · Schizophrenia · Multitarget therapy

6.1 Basics of Neurological Disorders

The nervous system is a highly complex part of multicellular animals, which controls, regulates, communicates, and coordinates its actions and sensory information by transmitting signals to and from different parts of the body. It is the center of all mental activity including thought, learning, and memory. Together with the endocrine system, the nervous system is responsible for regulating and maintaining homeostasis. Through its receptors, the nervous system keeps us in touch with our environment, both external and internal. Like other systems in the body, the nervous system is composed of organs, principally the brain, spinal cord, nerves, and ganglia. These, in turn, consist of various tissues, including nerve, blood, and connective tissue. The nervous system is made up of the central nervous system (brain and spinal cord) and the peripheral nervous system (nerves) that connect the CNS to every other part of the body. The peripheral nervous system is divided into three separate subsystems, the somatic, autonomic, and enteric nervous systems. Any functional and/or physical damages

Z. Wang, B. Yang, *Polypharmacology*, https://doi.org/10.1007/978-3-031-04998-9_6

or degenerations can cause neurological disorders that are highly diverse, complex, and multifaceted. Neurological disorders include diverse subgroups, including neuromuscular diseases, genetic and metabolic disorders, developmental delay, traumatic brain disorders and injuries, and degenerative diseases.

The brain, spinal cord, and nerves make up the nervous system that is a complex, highly specialized network. The nervous system organizes, explains, and directs interactions between an individual and the surrounding environment. The nervous system controls voluntary and involuntary movements:

- Sight, hearing, taste, smell, and feeling (sensation).
- Voluntary and involuntary functions, such as movement, balance, and coordination. The nervous system also regulates the actions of most other body systems, such as blood flow and blood pressure.
- The ability to think and reason. The nervous system allows you to be conscious and have thoughts, memories, and language.

Neurological disorders or nervous system diseases are medically defined as disorders that affect the brain as well as the nerves found throughout the human body and the spinal cord [1, 2]. Neurological disorders can occur in and affect the central nervous system or the peripheral nervous systems and can impair the brain, spinal cord, peripheral nerve, or neuromuscular function.

The most common neurological disorders, including genetic defects (such as Huntington's disease and muscular dystrophy), neurodegenerative diseases (such as Parkinson's disease and Alzheimer's disease), psychiatric disorders, etc., have received recent attention with regard to pharmacogenomics and personalized medicine. Structural, biochemical, or electrical abnormalities in the brain, spinal cord, or other nerves can result in a range of symptoms. Examples of symptoms include paralysis, muscle weakness, poor coordination, loss of sensation, seizures, confusion, pain, and altered levels of consciousness.

Neurological disorders are the leading cause of disability and the second leading cause of death worldwide. In the past 30 years, the absolute numbers of deaths and people with disabilities owing to neurological diseases have risen substantially, particularly in low-income and middle-income countries, and further increases are expected globally as a result of population growth and aging. According to the US National Library of Medicine, there are more than 600 neurologic diseases. This rise in absolute numbers of people affected suggests that advances in prevention and management of major neurological disorders are not sufficiently effective to counter global demographic changes.

Neurological disabilities include a wide range of disorders, such as epilepsy, learning disabilities, neuromuscular disorders, autism, ADD, brain tumors, and cerebral palsy, just to name a few. Some neurological conditions are congenital, emerging before birth. Other conditions may be caused by tumors, degeneration, trauma, infections, or structural defects. Regardless of the cause, all neurological disabilities result from damage to the nervous system. Depending on where the damage takes place determines to what extent communication, vision, hearing, movement, and cognition are impacted.

Disorders of the nervous system may involve the following:

Vascular Disorders such as stroke, transient ischemic attack (TIA), subarachnoid hemorrhage, subdural hemorrhage and hematoma, and extradural hemorrhage

Infections such as meningitis, encephalitis, polio, and epidural abscess

Structural Disorders such as brain or spinal cord injury, Bell's palsy, cervical spondylosis, carpal tunnel syndrome, brain or spinal cord tumors, peripheral neuropathy, and Guillain-Barré syndrome

Functional Disorders such as headache, epilepsy, dizziness, and neuralgia

Degeneration such as Parkinson's disease, multiple sclerosis, amyotrophic lateral sclerosis (ALS), Huntington's chorea, and Alzheimer's disease.

The specific causes of neurological problems vary but can include genetic disorders, congenital abnormalities or disorders, infections, lifestyle or environmental health problems including malnutrition, and brain injury, spinal cord injury, or nerve injury. Thus, treatment of neurological disorders represents a tremendous challenge [3, 4]. Mental disorders, on the other hand, are "psychiatric illnesses" or diseases which appear primarily as abnormalities of thought, feeling, or behavior, producing either distress or impairment of function. Clinical studies support a bidirectional link between depression and neurological diseases [5–7]. "Psychiatric" and "neurologic" depression seem to share common abnormalities in specific brain areas, but sound brain imaging studies of the neural correlates of depression in neurological disorders are still lacking. Available treatments are efficacious, but no clear-cut guidelines about the best drugs and dosages can be defined because double-blind placebo-controlled studies are still scarce. Psychiatric disorders are common in many neurological disorders, including epilepsy, migraine, Alzheimer's disease, Parkinson's disease, essential tremor, and stroke. These comorbidities increase disease burden and may complicate the treatment of the combined disorders.

In general, the nervous system is vulnerable to various disorders and can be damaged by the following:

Trauma
Infections
Degeneration
Structural defects
Tumors
Blood flow disruption
Autoimmune disorders

6.2 CDT in Neurological Disease

6.2.1 CDT in Alzheimer's Disease (AD)

6.2.1.1 Basics of Alzheimer's Disease (AD)

Alzheimer's disease (AD) is a particular onset and course of cognitive and functional decline associated with age together with a particular neuropathology with an irreversible, progressive brain disorder that slowly destroys memory and thinking skills and, eventually, the ability to carry out the simplest tasks by causing the brain to shrink (atrophy) and brain cells to die. AD is the worldwide leading cause of dementia—the loss of cognitive functioning: thinking, remembering, and reasoning, and behavioral abilities to such an extent that it interferes with a person's daily life and activities. Approximately 5.8 million people in the United States age 65 and older live with AD. Out of the approximately 50 million people worldwide with dementia, between 60% and 70% are estimated to have AD. AD is currently ranked as the sixth leading cause of death in the United States, but recent estimates indicate the disorder may rank third, just behind heart disease and cancer, as a cause of death for older people [8, 9].

Changes in amyloid precursor protein (APP) cleavage and production of the APP fragment beta-amyloid (Aβ) along with hyperphosphorylated tau protein aggregation coalesce to cause reduction in synaptic strength, synaptic loss, and neurodegeneration. Metabolic, vascular, and inflammatory changes as well as comorbid pathologies are key components of the disease process. Symptomatic treatment offers a modest, clinically measurable effect in cognition, but disease-modifying therapies are desperately needed [9].

6.2.1.2 Drug Treatments in Alzheimer's Disease

Treating dementia has become a major challenge in clinical practice. Presently, acetylcholinesterase inhibitors are the first-line drugs in the treatment of AD. These options are now complemented by memantine, which is approved for the treatment of moderate to severe AD. Altogether, a minimum of six agent classes already exist, all of which are approved for clinical use and are either already being tested or ready for phase III clinical trials for the treatment of AD. These include cholinesterase inhibitors, blockers of the NMDA receptor, antioxidants or blockers of oxidative deamination (including *Gingko biloba*), anti-inflammatory agents, neurotrophic factors (including hormone replacement therapy and drugs acting on insulin signal transduction), and anti-amyloid agents (including cholesterol-lowering therapy) [10–12].

Cholinesterase (ChE) or Acetylcholinesterase (AChE) Inhibitors (ChEI or AChIs)

Reduction in the activity of the cholinergic neurons with declined levels of acetylcholine (ACh) that is important for alertness, memory, thought, and judgment is a well-known feature of AD [11]. Cholinesterase inhibitors can reduce the rate at which ACh is broken down, thereby boosting the concentration of ACh in the brain and combating the loss of ACh caused by the death of cholinergic neurons [11]. The following cholinesterase inhibitors are commonly prescribed for AD therapy:

(a) Tacrine was the first-generation ChEI but was limited by hepatotoxic side effects [12].
(b) Donepezil (Aricept) is approved to treat all stages of the disease [13, 14].
(c) Galantamine (Razadyne) is approved to treat mild to moderate Alzheimer's.
(d) Rivastigmine (Exelon) is approved for mild to moderate Alzheimer's disease.

Efficacy appears similar between these different agents so choice should be based on cost, individual patient tolerance, and physician experience.

NMDA Receptor Antagonist

The N-methyl-D-aspartate receptor (NMDAR) is a glutamate and ion channel protein receptor that is activated when glycine and glutamate bind to it. Glutamate is an excitatory neurotransmitter of the nervous system, although excessive amounts in the brain can lead to cell death [15]. Memantine (Namenda), an NMDAR antagonist/inhibitor (NMDARI), is approved by the FDA for treatment of moderate to severe Alzheimer's disease. It works by regulating the activity of glutamate, a messenger chemical widely involved in brain functions—including learning and memory. Memantine uncompetitively blocks the NMDAR [16, 17] and, thus, may be neuroprotective by preventing neuron loss, as well as improving symptoms by helping to restore function of damaged neurons. Memantine has been shown to have modest benefits in moderate to severe AD, with little evidence supporting its use in milder AD [17]. Additionally, the addition of memantine to donepezil monotherapy may be beneficial in those with mid-stage AD or who are deteriorating cognitively [18]. While these medications represent the best currently available pharmacological treatments in AD, they have relatively small average overall effect and do not alter the course of the underlying neurodegenerative process.

Aduhelm (Aducanumab)

Most recently, the FDA approved Aduhelm (aducanumab) to treat patients with AD using the accelerated approval pathway [19], under which the FDA approves a drug for a serious or life-threatening illness that may provide meaningful therapeutic benefit over existing treatments when the drug is shown to have an effect on a surrogate endpoint that is reasonably likely to predict a clinical benefit to patients and there remains some uncertainty about the drug's clinical benefit. Aducanumab, a human monoclonal antibody that selectively targets aggregated amyloid-β (Aβ), is the first treatment directed at the underlying pathophysiology of AD, the presence of Aβ plaques in the brain. AD is characterized by deposition of Aβ plaques and neurofibrillary tangles in the brain, accompanied by synaptic dysfunction and neurodegeneration [20]. Therapies targeting Aβ have been the focus for almost 30 years. The clinical trials for aducanumab were the first to show that a reduction in these plaques—a hallmark finding in the brain of patients with AD—is expected to lead to a reduction in the clinical decline of this devastating form of dementia [21, 22].

Ginkgo biloba Extract

An extract of *Ginkgo biloba* known as EGb761 has been widely used for treating AD and other neuropsychiatric disorders [23]. Its use is approved throughout Europe [24]. The World Federation of Societies of Biological Psychiatry guidelines list EGb761 with the same weight of evidence given to acetylcholinesterase inhibitors and memantine. EGb761 is the only one that showed improvement of symptoms in both AD and vascular dementia. EGb761 acts as a free radical scavenger to improve mitochondrial function and modulate serotonin and dopamine levels. Many studies of its use in mild to moderate dementia have shown it to significantly improve cognitive function, activities of daily living, and neuropsychiatric symptoms. However, its use has not been shown to prevent the progression to dementia [23].

Currently, researchers are also making their efforts to tau protein-targeting therapies, since tau protein appears to be better correlated with the severity of cognitive decline than Aβ. Currently, most anti-tau agents in clinical trials are immunotherapies and they are in the early stages of clinical research. Four monoclonal anti-tau antibodies (gosuranemab, tilavonemab, semorinemab, and zagotenemab) and one anti-tau vaccine (AADvac1) have reached phase II [25].

6.2.1.3 Combination Drug Therapy of Alzheimer's Disease

Even though the number of AD patients is rapidly growing, there is no effective treatment for this neurodegenerative disorder [26, 27]. Therefore, implementation of effective treatment approaches for AD is vital to meet clinical needs. In AD research, priorities concern the development of disease-modifying therapeutic agents to be used in the early phases of AD and the optimization of the symptomatic treatments predominantly dedicated to the more advanced AD stages. Until now, available therapeutic agents for AD treatment only provide symptomatic treatment. Since AD pathogenesis is multifactorial, use of a multimodal therapeutic intervention addressing several molecular targets of AD-related pathological processes seems to be the most practical approach to modify the course of AD progression. It has been demonstrated through numerous studies that the clinical efficacy of combination therapy is higher than that of monotherapy, the evidence for synergistic combination effects of symptomatic therapy, including delay or prevention of disease progression in AD patients. In case of AD, combination therapy is more effective, mostly when started early, at slowing the rate of cognitive impairment [27]. The efficacy, safety, and tolerability of combination therapy based on the NMDA receptor antagonist memantine with any of the ChEIs (donepezil, galantamine, or rivastigmine) for the treatment of AD have been evaluated [27].

ChEI/NMDARI (Galantamine/Memantine) CDT for AD [27]

The most widely studied combination drug therapy for AD treatment is the concomitant use of memantine and cholinesterase inhibitors. Furthermore, this treatment has proven clinical efficacy in AD treatment [26–28]. The effects of this combination therapy in AD have also been assessed in long-term observational studies, open-label trials, and randomized controlled trials (RCTs). In AD, RCTs primarily evaluate drug efficacy, and these trials involve the determination of four main criteria including neuropsychiatric symptoms, functioning in activities of daily living, cognition, and global clinical outcomes. The findings of these studies denote that combination drug therapies using memantine and cholinesterase inhibitors decrease the rate of functional and cognitive decline. Furthermore, as compared to no treatment or monotherapy with cholinesterase inhibitors, these combination therapies can reduce the emergence and the severity of neurobehavioral symptoms [26, 28–40]. It has also been demonstrated that combined therapies are more effective when started early [41].

Dysfunction in the activity of glutamatergic neurons along with the cholinergic system is vital in AD pathology [42]. Therefore, combination therapies with the ChEI/NMDARI combination or the galantamine/memantine combination that target glutamatergic and cholinergic systems concurrently are considered as the present standard of care for individuals with AD [28]. It is known that NMDAR is accountable for learning and memory. Furthermore, glutamate and glycine are the agonists that need to simultaneously bind with the NMDAR for activating this receptor. It has been shown clinically that NMDAR overactivation (i.e., through an excess of glutamate) can cause an uncontrolled influx of Ca^{2+} into the neuron, which can eventually lead to excitotoxicity. On the other hand, along with its inhibitory effects on AChE, galantamine exhibits neuroprotective activities against glutamate toxicity through the activation of nicotinic AChR [43]. Therefore, memantine and galantamine can act together in the same excitotoxic cascade to deliver a synergistic neuroprotective action [44]. A new series of hybrid compounds were designed by following this strategy that combined the pharmacological effects of memantine and galantamine [45]. On the other hand, following a dual-binding strategy, new hybrid compounds were designed by linking two drug molecules via a variable-length polymethylene linker. In this series, compounds that carry a hexamethylene spacer are considered to display the most auspicious properties. A 6-methylene spacer offered the optimal distance to allow a concurrent contact with both the catalytic active site and the peripheral anionic binding site on AChE [46–48].

ChEI/NMDARI (Nitroglycerin/Memantine) CDT for AD [27]

A second generation of memantine analogs was developed to improve the efficacy of memantine and to provide it with a substantial disease-modifying effect [15, 27]. An important pharmacophore (nitrooxy moiety, –ONO2) from nitroglycerin was linked with memantine to design these memantine analogs [49]. In a triple transgenic AD mouse model, the action of nitromemantine and memantine on extrasynaptic NMDARs was evaluated in vivo [50]. The triple transgenic AD mice showed impairments in cognition and synaptic function, which took place before extracellular Aβ deposition and tangle formation but which were linked with intracellular Aβ immunoreactivity [50]. Nitromemantine exhibited greater effects, as compared to memantine, on elevating synaptic and dendritic density in 9-month-old triple transgenic AD mice. Furthermore, only the nitromemantine-treated group exhibited considerably better functions on the location-novelty recognition test. In addition to this, these studies revealed that nitromemantine was able to reverse the loss of brain connections and bring the number of synapses all the way back to normal within a few months of treatment in mouse AD models [51, 52]. These aforesaid findings suggest that nitromemantine might have disease-modifying properties. Furthermore, nitromemantine did not show any negative effect on blood pressure [15, 53]. Collectively, these promising results in animal models have prompted the entry of nitromemantine in clinical trials. These initiatives are indeed raising hope in the treatment of early and later stages of AD.

ChEI/NMDARI (Donepezil/Clioquinol) CDT for AD [27]

Generally, the linking approach produces molecules with poor pharmacokinetic properties and high molecular weight. Therefore, smaller hybrid molecules might prove beneficial as they possess better pharmacokinetic properties. Thus, a linking strategy was utilized to design a new series of hybrid molecules that combine the pharmacophore of donepezil with the metal chelator of clioquinol [27, 54]. Amyloid plaques possess elevated concentrations of zinc and copper, which is also a common characteristic of inflammation [55]. In addition to this, amyloid toxicity can be mediated by the binding sites for both of these metal ions on Aβ oligomers [56]. Therefore, metal-chelating therapy may be a good approach to offset AD progression. On the other hand, combined antioxidant, Cu(II) complexing, and cholinergic properties have been revealed in a series of tacrine-8-hydroxyquinoline hybrids [57]. New series of hybrid compounds were thus rationally designed based on these aforesaid observations through fusing the 8-hydroxyquinoline part with benzylpiperidines (i.e., as a donepezil pharmacophore) to attain inhibition of Aβ aggregation, free radical scavenging property, chelation of Zn(II) and Cu(II), and inhibition of ChE [58]. Substitution of the donepezil's indanone core with the aromatic and planar 8-hydroxyquinoline preserves its affinity for the peripheral site of AChE without a rise in molecular weight. It was observed that most of the new hybrids obtained effectively suppressed the self-aggregation of Aβ and selectively targeted human butyrylcholinesterase (BuChE). Instead, the 7-((4-(2-methoxybenzyl)piperazin-1-yl)methyl)-8-hydroxyquinoline exhibited good in vitro antioxidant properties, Zn(II)- and Cu(II)-chelating properties, acceptable toxicity, well-balanced anti-aggregating, and AChE properties as well as displaying positive permeation of blood-brain barrier dysfunction in parallel artificial membrane permeability assays.

ChEI/NMDARI (Rivastigmine/Rasagiline) CDT for AD [27]

An example of successful molecular hybridization based on a drug combination strategy is the case of ladostigil, a neuroprotective agent currently undergoing clinical trials that contains the characteristics of old drugs such as rasagiline (i.e., anti-monoamine oxidase (MAO)-B) and rivastigmine (i.e., anti-AChE). Ladostigil was studied in the treatment of AD, Lewy body dementia, and PD [27, 59]. Moreover, this hybrid compound showed good reversible BuChE, AChE, and irreversible MAO-B inhibition properties and protected impairments of motor function and memory in mice. Ladostigil (1 μM) also exhibited antioxidant properties via its direct scavenging activity on free radicals and also indirectly through the induction of the expression of cellular antioxidant enzymes [60]. It has been shown upon in vivo studies that ladostigil decreased the concentration of the holo-amyloid precursor protein (APP). In addition to this, ladostigil also increased the concentrations of protein kinase C (PKC)a, PKCe, and p-PKC in the mice hippocampus [61].

ChEI/NMDARI (VK-28/Propargylamine) CDT for AD [27]

M30 was developed by combining the propargyl moiety of the anti-PD drug rasagiline with the skeleton of the neuroprotective iron chelator, VK-28. M30 possesses the vital pharmacophores (i.e., propargylamine moiety) from the Food and Drug Administration (FDA)-approved anti-PD rasagiline. The propargylamine moiety in the rasagiline's structure plays a crucial role in the disease-modifying potential of rasagiline since it can mediate the interaction of this compound with various neuroprotective/neurorescue pathways [27, 62]. PBT2 is a metal-protein-attenuating compound (MPAC) of second generation, which is now under phase II clinical trials for the treatment of AD. M30 possesses a metal-protein-attenuating/ionophore moiety similar to that of PBT2 [63]. MPACs typically show intermediate reversible affinity toward metals and have the capacity to reach certain intracellular compartments to target the harmful upstream metal-protein reactions [64]. M30 displays a selective inhibition for MAO-A and MAO-B action in the brain, with a comparatively poor inhibition of these enzymes in the small intestine and liver in vivo [65].

ChEI/PBT2 CDT for AD [27]

Chronic exposure to various metals is associated with AD [27, 66, 67]. PBT2 (i.e., a copper/zinc ionophore) is a metal-chelating agent that is used to decrease the toxicity of protein-metal complexes like those produced by Aβ oligomers with zinc and copper. Early phase II clinical trials have already been completed with PBT2. Interestingly, in individuals with early AD, the combination of PBT2 and ChEIs (i.e., galantamine, donepezil, or rivastigmine) enhanced executive functions in a dose-dependent manner and significantly decreased the levels of cerebrospinal fluid Aβ42 as compared to ChEIs alone [63, 68]. Although these results are encouraging, further studies are required in this domain.

ChEIs/NMDARI/Noradrenaline Reuptake Inhibitors CDT for AD [27]

Several studies have reported an intense noradrenergic dysfunction in AD [69, 70]. Atomoxetine is an inhibitor of noradrenaline reuptake. In mild to moderately severe AD, it has been observed that the combination of atomoxetine and ChEIs (i.e., galantamine, 16–24 mg per day; donepezil, 5–10 mg per day; or rivastigmine, 6–12 mg per day) or an NMDAR antagonist (i.e., memantine 5–20 mg per day) was normally well tolerated by patients but did not improve the monotherapy's clinical efficacy in mild to moderately severe AD [71].

6.2.2 CDT in Parkinson's Disease (PD)

6.2.2.1 Basics of Parkinson's Disease (PD)

Parkinson's disease (PD) is a progressive neurodegenerative disease characterized by tremor and bradykinesia and is a common neurologic ailment [72]. There are a range of associated non-motor symptoms, including cognitive impairment, depression, and sleep disturbance. PD affects 1–2 per 1000 of the population at any time with male sex and advancing age as independent risk factors. PD prevalence is increasing with age, and PD affects 1% of the population above 60 years, being the second most common neurodegenerative disease after Alzheimer's disease [73]. The main neuropathological finding is α-synuclein-containing Lewy bodies and loss of dopaminergic neurons in the substantia nigra, manifesting as reduced facilitation of voluntary movements. With progression of PD, Lewy body pathology spreads to neocortical and cortical regions. PD is regarded as a movement disorder with three cardinal signs: tremor, rigidity, and bradykinesia. A recent revision of the diagnostic criteria excludes postural instability as a fourth hallmark and defines supportive criteria, absolute exclusion criteria, and red flags. Non-motor symptoms in PD have gained increasing attention, and both motor and non-motor signs are now included among the supportive criteria. The cause of PD is unknown in most cases. Genetic risk factors have been identified, including monogenetic causes that are rare in unselected populations [74, 75]. Some genetic factor can be identified in 5–10% of the patients. Several environmental factors are associated with increased risk of PD. Autopsy studies show that the clinical diagnosis of PD is not confirmed at autopsy in a significant proportion of patients.

6.2.2.2 Drug Treatments in PD

PD seems to result from a complicated interplay of genetic and environmental factors affecting numerous fundamental cellular processes. The complexity of PD is accompanied by clinical challenges, including an inability to make a definitive diagnosis at the earliest stages of the disease and difficulties in the management of symptoms at later stages. Unfortunately, there are currently no treatments that slow the neurodegenerative process but ameliorate the symptoms [76–78].

Levodopa

The motor symptoms of PD are the result of reduced production of dopamine (a chemical the brain uses to send signals to the body) in the brain's basal ganglia. Dopamine does not cross the blood-brain barrier, so it cannot be taken as a medicine to boost the brain's depleted levels of dopamine. However, as a precursor of dopamine, levodopa (also called L-dopa) can pass through to the brain, where it is readily converted to dopamine. Administration of levodopa temporarily diminishes the motor symptoms of PD. Levodopa has been the most widely used PD treatment for over 40 years [79]. It is also the best at controlling the symptoms of the condition, particularly slow movements and stiff, rigid body parts. Yet only 5–10% of levodopa crosses the blood-brain barrier. Much of the remainder is metabolized to dopamine elsewhere in the body, causing a variety of side effects, including nausea, vomiting, and orthostatic hypotension.

Catechol-O-methyltransferase (COMT) Inhibitors

COMT inhibitors mainly include entacapone (Comtan), opicapone (Ongentys), and tolcapone (Tasmar). These medications mildly prolong the effect of levodopa therapy by blocking an enzyme that breaks down dopamine. During the course of PD, affected people can experience what is known as a "wearing off phenomenon," where they have a recurrence of symptoms after a dose of levodopa, but right before their next dose [80]. Catechol-O-methyltransferase (COMT) is a protein that degrades levodopa before it can cross the blood-brain barrier, and these inhibitors allow for more levodopa to cross [81]. They are normally not used in the management of early symptoms but can be used in conjunction with levodopa/carbidopa when a person is experiencing the "wearing off phenomenon" with their motor symptoms. Tolcapone has been available for several years, but its usefulness is limited by possible liver damage complications so requires liver-function monitoring. Entacapone and opicapone have not been shown to cause significant alterations to liver function.

Dopamine Agonists

Dopamine agonists include bromocriptine, pergolide, pramipexole, ropinirole, piribedil, cabergoline, apomorphine, and lisuride. These agents bind to dopamine receptors in the brain and have similar effects to levodopa, and unlike levodopa, dopamine agonists do not change into dopamine; instead, they mimic dopamine effects in your brain [79]. These were initially used as a complementary therapy to levodopa for individuals experiencing levodopa complications (on-off fluctuations and dyskinesias); they are now mainly used on their own as first therapy for the motor symptoms of PD with the aim of delaying the initiation of levodopa therapy, thus delaying the onset of levodopa's complications [79, 82]. Dopamine agonists are the preferred initial treatment for younger-onset PD, and levodopa is preferred for older-onset PD [76–78].

Monoamine Oxidase B (MAO-B) Inhibitors

MAO-B inhibitors (safinamide, selegiline, and rasagiline) increase the amount of dopamine in the basal ganglia by inhibiting the activity of monoamine oxidase B, an enzyme that breaks down dopamine [79]. They help alleviate motor symptoms when used as monotherapy (on their own); when used in conjunction with levodopa, they reduce the time spent in the "off" phase.

Amantadine

Amantadine works by raising the amount of dopamine in brain cells, which helps ameliorate PD symptoms [79, 83]. Recent studies have found that Symmetrel may help ease the involuntary movements that can happen with levodopa therapy. But it may cause side effects, such as confusion and memory problems.

Anticholinergics

Anticholinergics mainly include benztropine (Cogentin) and trihexyphenidyl (Artane). These drugs restore the balance between two brain chemicals, dopamine and acetylcholine, to ease tremors and muscle stiffness in people with PD [79, 83]. And they have been used for many years to help control the tremor associated with PD. But these medications can harm memory and thinking, especially in older people. Because of that, doctors rarely prescribe them today.

Other Drugs

In addition to motor symptoms, PD is accompanied by a diverse range of symptoms. Several drugs have been used to treat some of these problems [84]. Examples are the use of quetiapine for psychosis, cholinesterase inhibitors for dementia, and modafinil for excessive daytime sleepiness [84]. In 2016, pimavanserin was approved for the management of PD psychosis. Doxepin and rasagiline may reduce physical fatigue in PD [85].

6.2.2.3 Combination Drug Therapy in Parkinson's Disease

Dopa Decarboxylase Inhibitor (DDCI)/ Levodopa CDT for PD

L-3,4-dihydroxyphenylalanine (L-dopa) or levodopa is a dopamine precursor and is an effective and well-tolerated dopamine replacement agent as the gold standard for symptomatic treatment of PD, but long-term therapy is associated with the emergence of L-dopa-induced dyskinesia (LID). Oral levodopa has been widely used for over 40 years, often in combination with a dopa decarboxylase inhibitor (DDCI), which reduces many treatment complications, extending its half-life and increasing levodopa availability to the brain. Carbidopa and benserazide are dopa decarboxylase inhibitors that do not cross the blood-brain barrier and inhibit the conversion of levodopa to dopamine outside the brain, reducing side effects and improving the availability of levodopa for passage into the brain. One of these drugs is usually taken along with levodopa, often combined with levodopa in the same pill [86–88]. It is believed that levodopa in combination with benserazide or carbidopa constitutes the best available therapy for PD [89–91].

(a) *The Carbidopa/Levodopa Combination*

Sinemet is a mix of levodopa and carbidopa that makes the levodopa work better. This combination prevents many common side effects associated with levodopa monotherapy, such as nausea, vomiting, and irregular heart rhythms; Sinemet has the fewest short-term side effects, compared with other PD medications. But it does raise the odds for some long-term problems, such as involuntary movements [89–91].

(b) *The Benserazide/Levodopa Combination*

Levodopa and benserazide were loaded by poly (lactic-co-glycolic acid) microspheres (LBM), which can release levodopa and benserazide in a sustained manner in order to continuously stimulate dopaminergic receptors [89–91].

MAO-B Inhibitor/Levodopa CDT for PD

A systematic literature search and study including 27 publications eligible for network analyses compared the relative effects of three MAO-B inhibitors rasagiline, safinamide, and selegiline treatment given alone or in combination with levodopa [92]. Their results indicate that all three MAO-B inhibitors are efficient when given together with levodopa. Selegiline/levodopa was the most effective, and rasagiline/levodopa was the second best. All three MAO-B inhibitors were effective compared to placebo when given as monotherapy.

(a) *The Selegiline/Levodopa Combination*

A meta-analysis on 14 RCTs with 2008 participants showed that the selegiline/levodopa combination therapy is superior to levodopa monotherapy for the improvement of clinical symptoms in PD patients. Moreover, the safety profile of the pair is comparable with that of levodopa alone [93].

(b) *The Rasagiline/Levodopa Combination*

Another meta-analysis involving 14 RCTs with 2531 participants revealed that the rasagiline/levodopa combination therapy was superior to levodopa monotherapy for improvement of Unified Parkinson's Disease Rating Scale (UPDRS) scores and off-time in PD patients, and had similar safety and tolerability profiles [94, 95].

(c) *The Safinamide/Levodopa Combination*

Safinamide (Xadago) is an orally active, selective, reversible monoamine oxidase-B inhibitor with both dopaminergic and non-dopaminergic (glutamatergic) properties and an add-on medicine that may be prescribed when individuals taking levodopa and carbidopa have a breakthrough of PD symptoms that were previously under control. Studies show that adding this drug helps individuals experience longer times with reduced or no symptoms. The most common side effects are trouble falling or staying asleep, nausea, falls, and uncontrolled, involuntary movements.

Safinamide significantly increased mean on time (with no or non-troublesome dyskinesia) and reduced mean off time when used as first adjunct therapy in levodopa-treated patients and patients with mild motor fluctuations. Changes from baseline in mean daily off time, UPDRS Part II total score (off phase), UPDRS Part III total score (on phase), and UPDRS Part I also showed significant improvements, compared with placebo and levodopa monotherapy.

Additionally, safinamide improved bradykinesia, rigidity, tremor, and gait. Safinamide was a safe and effective first adjunct therapy in levodopa-treated patients and improved 4/5 cardinal symptoms of PD while providing benefits to mild and non-mild fluctuators and patients receiving other concomitant dopaminergic therapies [96, 97].

Dopamine Agonist (DA)/Levodopa CDT for PD

A meta-analysis of randomized placebo-controlled trials evaluated the use of dopamine agonist (DA) or placebo to preexisting levodopa therapy for the treatment of advanced PD. A total of 15 trials including 4380 subjects were included in the meta-analysis. Adjunctive DA use resulted in greater improvement as measured by the UPDRS ADL and motor score reduction as well as reduction in "off" time measured in hours/day and reduction in levodopa dose *vs* placebo. However, the incidence of dyskinesia and hallucinations was higher with DAs. The data indicate that adjunctive DA use to levodopa is superior to levodopa alone in reducing PD symptoms in patients not controlled with monotherapy. DAs seem especially useful among PD patients with wearing off phenomenon from levodopa therapy but can cause some adverse events [98].

DA/DDCI/Levodopa (Pramipexole/Benserazide/Levodopa) CDT for PD

Pramipexole is a DA that works on the nervous system to help treat the symptoms of PD. A meta-analysis of 29 RCTs with 3017 participants found that clinical efficacy and safety of pramipexole plus levodopa/benserazide combination therapy in the treatment of PD are significantly better than those of levodopa/benserazide dual combination, including better motor UPDRS score, activities of daily living UPDRS score, mental UPDRS score, and lower HAMD score (the Hamilton Depression Rating Scale score). In contrast, the triple combination therapy decreased the number of adverse events in PD patients. Thus, the pramipexole/levodopa/benserazide triple combination is superior to the levodopa/benserazide dual combination for improvement of clinical symptoms in PD patients. Moreover, the safety profile of pramipexole/levodopa/benserazide is also better than that of levodopa/benserazide [99].

COMTI/DDCI/Levodopa (Entacapone/Carbidopa/Levodopa) CDT for PD

Entacapone, a COMT inhibitor, is used to improve the bioavailability of levodopa, especially when used in conjunction with a DDCI. During early-stage PD, treatment will depend on the severity of symptoms; if greater symptomatic effect is required, then levodopa or dopamine agonists are usually the drugs of choice. The ability to remain employable or physically active is an important goal in younger patients; therefore, in some instances, levodopa initiation should be considered early on, either as a monotherapy or in combination with other drugs. The clinical use of levodopa may eventually be limited by the development of various treatment-related complications, including response fluctuations, dyskinesia, and psychiatric problems. Motor complications are related to the intermittent delivery of dopamine-replacing drugs to the brain. Triple combination of entacapone/carbidopa/levodopa available in a single tablet in multiple levodopa dose strengths offers flexibility and helps control response fluctuations [100, 101].

Mounting evidence indicates the short half-life of levodopa, and resultant pulsatile stimulation of striatal dopamine receptors leads to wearing off, motor fluctuations, and dyskinesias. Longer-acting dopaminergic agents, such as dopamine agonists, are less likely to cause motor fluctuations and dyskinesias but are not as efficacious for control of motor symptoms. Therefore, there is interest in exploring ways to deliver levodopa in a more continuous fashion, in an effort to maintain benefit through the day and reduce the development of motor fluctuations and dyskinesias. A dopa decarboxylase inhibitor (DDCI), such as carbidopa or benserazide, is administered with levodopa to attenuate its peripheral conversion to dopamine, reduce nausea, and increase central bioavailability. When levodopa is administered with a DDCI, its main route of peripheral metabolism is via catechol-O-methyl transferase (COMT). A COMT inhibitor can be added to the combination of levodopa and a DDCI to further extend the levodopa peripheral half-life and increase central bioavailability. Stalevo is a combination tablet comprised of levodopa, carbidopa, and the COMT inhibitor entacapone and is currently approved for use in Parkinson's disease patients with end-of-dose wearing off [101].

6.2.3 CDT in Epilepsy

6.2.3.1 Basics of Epilepsy

Epilepsy is a central nervous system (neurological) disorder in which brain activity becomes abnormal, causing seizures or periods of unusual behavior, sensations and sometimes loss of awareness [102–107]. It is primarily characterized by recurrent epileptic seizures.

The underlying mechanism of epileptic seizures is excessive and abnormal neuronal activity in the cortex of the brain [103] which can be observed in the electroencephalogram (EEG) of an individual. The reason this occurs in most cases of epilepsy is unknown (idiopathic) [104]; some cases occur as the result of brain injury, stroke, brain tumors, infections of the brain, or birth defects through a process known as epileptogenesis [104, 105]. Known genetic mutations are directly linked to a small proportion of cases [106, 107]. A better understanding of the pathogenesis of epilepsy unmasks various molecular targets for the treatment of epilepsy.

As of 2020, about 50 million people suffer from epilepsy [108]. Nearly 80% of cases occur in the developing world [104]. In 2015, it resulted in 125,000 deaths, an increase from 112,000 in 1990 [109]. Epilepsy is more common in older people [110, 111]. In the developed world, onset of new cases occurs most frequently in babies and the elderly [112]. In the developing world, onset is more common in older children and young adults due to differences in the frequency of the underlying causes [113]. About 5–10% of people will have an unprovoked seizure by the age of 80 [114], with the chance of experiencing a second seizure is between 40% and 50% [115]. In many areas of the world, those with epilepsy either have restrictions placed on their ability to drive or are not permitted to drive until they are free of seizures for a specific length of time [116].

6.2.3.2 Drug Treatments in Epilepsy

Treatment with medications or sometimes surgery can control seizures for the majority of people with epilepsy. Some people require lifelong treatment to control seizures, but for others, the seizures eventually go away. Some children with epilepsy may outgrow the condition with age.

The mainstay treatment of epilepsy is anticonvulsant medications, possibly for the person's entire life [108, 117]. The choice of anticonvulsant is based on seizure type, epilepsy syndrome, other medications used, other health problems, and the person's age and lifestyle [118]. A single medication is recommended initially; if this is not effective, switching to a single other medication is recommended [119]. Two medications at once is recommended only if a single medication does not work [119].

There are a number of medications available including phenytoin, carbamazepine [120], and valproate. Evidence suggests that phenytoin, carbamazepine, and valproate may be equally effective in both focal and generalized seizures [121, 122]. Controlled release carbamazepine appears to work as well as immediate release carbamazepine and may have fewer side effects [123]. In the United Kingdom, carbamazepine or lamotrigine are recommended as first-line treatment for focal seizures, with levetiracetam and valproate as second-line due to issues of cost and side effects [119]. Valproate is recommended first-line for generalized seizures with lamotrigine being second-line [119]. In those with absence seizures, ethosuximide or valproate is recommended; valproate is particularly effective in myoclonic seizures and tonic or atonic seizures [119]. If seizures are well-controlled on a particular treatment, it is not usually necessary to routinely check the medication levels in the blood [119].

The currently used antiepileptic drugs (AEDs) predominately target voltage-gated ion channels (Na^+, Ca^{2+}, and K^+), GABAA receptor, glutamate receptor, synaptic vesicle 2A (SV2A) protein, and carbonic anhydrase (CA) [102]. AEDs include carbamazepine, clobazam, lamotrigine, levetiracetam, oxcarbazepine, phenytoin, sodium valproate, topiramate, lacosamide, zonisamide, nitrazepam, ethosuximide, felbamate, and phenobarbitone.

6.2.3.3 Combination Drug Therapy in Epilepsy

Monotherapy remains the standard initial therapy of new-onset epilepsy, but when the first AED fails and control is not achieved, combination therapy may be considered. The choice of combination therapy should take into consideration pharmacokinetic interactions, as well as pharmacodynamic interactions related to mechanism of action. There is evidence that an AED combination with different mechanisms of action is more likely to be successful than a combination with the same mechanisms [124, 125]. Moreover, the introduction of over 14 new antiepileptic drugs (AEDs) for the adjunctive treatment in refractory epilepsy in the past twenty years has triggered a renewed interest in combination therapy. Nevertheless, the experimental and clinical evidence in support of "rational polytherapy" is sparse, with only the combination of sodium valproate with lamotrigine demonstrating synergism.

Lamotrigine/Valproate CDT for Epilepsy

The combination of lamotrigine (an antiepileptic drug belonging in the phenyltriazine class) and valproate (a commonly used antiepileptic drug for the treatment of generalized and partial seizures through affecting GABA levels, blocking voltage-gated sodium channels, and inhibiting histone deacetylases) has been demonstrated to be synergistic in its efficacy [126]. Early case series reported dramatic reductions in seizure frequency when lamotrigine was added to valproate in patients with absence seizures [127, 128], complex partial seizures [129], and myoclonic epilepsy [130]. In some of these cases, the patients had not responded to monotherapy with either drug.

In the study reported by Moeller and colleagues [126], almost two-thirds of patients with medically refractory epilepsy achieved a sustained improvement in seizure frequency with the combination of lamotrigine and valproate, and more than one-half of patients were completely seizure-free in long-term follow-up. This improvement occurred despite the fact that these patients had severe, medically refractory epilepsy (50% with at least one seizure weekly; median five previous AED failures).

As stared by these authors [126], there has been a debate whether combining AEDs with different mechanisms of action may be more beneficial than targeting the same cellular mechanisms by two different drugs [131, 132]. Although one can find examples supporting both scenarios, it appears that, overall, combining two drugs with similar mechanisms of action can provide greater therapeutic benefit than that of two drugs acting on different mechanisms [133–136]. Interestingly, combinations of drugs acting on the GABAergic system appear to be particularly beneficial in this regard [126].

On the basis of the numerous experimental findings, Moeller et al. [126] concluded that combinations of levetiracetam with other AEDs, particularly those enhancing GABAergic inhibition, lead to additive/synergistic effects on seizure protection, which are not associated with more pronounced side effects and pharmacokinetic interactions.

Levetiracetam/AEDs CDT for Epilepsy

Levetiracetam, the generic name (non-brand name) for the drug called Keppra (KEP-ruh), is an antiepileptic drug (AED) widely used for seizure medication and characterized by a novel mechanism of action, unique profile of activity in seizure models, and broad-spectrum clinical efficacy [137]. It has recently been approved in the European Union (EU) for monotherapy of partial onset seizures, with or without secondary generalization, in patients from 16 years or older with newly diagnosed epilepsy; however, it is frequently used in combination with other AEDs, which is often necessary in drug-resistant patients [138]. Several preclinical studies focused on combination therapy with levetiracetam, and other anticonvulsants in various seizure and epilepsy models have been reported [137]. Administration of levetiracetam together with many different clinically used AEDs or other anticonvulsants generally enhances their protective activity, and, among existing AEDs, this was particularly prevalent with valproate [137]. The protective activity of other AEDs was also enhanced by levetiracetam, which seems to be a universal finding that is independent of seizure model or drug combination studied [137]. However, particularly strong enhancement was observed when levetiracetam was combined with agents either enhancing GABAergic or reducing glutamatergic neurotransmission [137]. Importantly, these combinations were not associated with exacerbation of side effects or pharmacokinetic interactions [137]. Based on the available preclinical data, it appears that combination treatment with levetiracetam and other anticonvulsants provides additional therapeutic benefit that may be attributed to its novel and distinct mechanism of action [137]. Moreover, combinations of levetiracetam with clinically used AEDs that enhance GABAergic inhibition may be considered for rational polypharmacological therapy, which is often necessary in drug-resistant patients [137].

6.2.4 CDT in Multiple Sclerosis (MS)

6.2.4.1 Basics of Multiple Sclerosis (MS)

Multiple sclerosis (MS) is a chronic demyelinating disease of the central nervous system, driven by two main pathogenetic mechanisms: inflammation and degeneration [139–141]. The former is principally related to the occurrence of acute episodes of neurologic dysfunction (relapses) and the formation of focal demyelinating lesions in the brain and the spine. The latter is mainly responsible for progression of disability. The disease occurs in different clinical forms, possibly related to the varying interplay of these pathogenetic mechanisms. MS is the most common non-traumatic cause of neurological disability among young adults in Western Europe and North America and a potentially disabling disease of the brain and spinal cord (central nervous system). In MS, the immune system attacks the protective sheath (myelin) that covers nerve fibers and causes communication problems between your brain and the rest of your body. Eventually, the disease can cause permanent damage or deterioration of the nerves. This damage disrupts the ability of parts of the nervous system to transmit signals, resulting in a range of signs and symptoms, including physical, mental, and sometimes psychiatric problems [139–141].

While the cause is unclear, the underlying mechanism is thought to be either destruction by the immune system or failure of the myelin-producing cells [5]. Proposed causes for this include genetics and environmental factors, such as viral infections [140, 142]. MS is usually diagnosed based on the presenting signs and symptoms and the results of supporting medical tests [143].

Multiple sclerosis is the most common immune-mediated disorder affecting the central nervous system [144]. In 2015, about 2.3 million people were affected globally, with rates varying widely in different regions and among different populations [145, 146]. In that year, about 18,900 people died from MS, up from 12,000 in 1990 [109, 147]. The disease usually begins between the ages of 20 and 50 and is twice as common in women as in men [148]. MS was first described in 1868 by French neurologist Jean-Martin Charcot [149]. The name multiple sclerosis refers to the numerous glial scars (or sclerae—essentially plaques or lesions) that develop on the white matter of the brain and spinal cord [149]. As of 2009 a number of new treatments and diagnostic methods are under development [150].

6.2.4.2 Drug Treatments in Multiple Sclerosis (MS)

There is no known cure for multiple sclerosis [151]. Treatments attempt to improve function after an attack and prevent new attacks [140]. Medications used to treat MS, while modestly effective, can have side effects and be poorly tolerated [151]. Physical therapy and occupational therapy can help with people's ability to function [151]. Many people pursue alternative treatments, despite a lack of evidence of benefit [152].

The long-term outcome is difficult to predict; good outcomes are more often seen in women, those who develop the disease early in life, those with a relapsing course, and those who initially experienced few attacks [153]. Life expectancy is 5 to 10 years lower than that of the unaffected population [139].

Several classes of agents are being used for MS therapy [151]: **(a)** steroids inhibit lymphocyte activation and reduce production of proinflammatory cytokines and migration of immune cells into the central nervous system; **(b)** glatiramer acetate is a synthetic amino acid polymer resembling myelin basic protein. It acts by promoting a shift from a proinflammatory to an anti-inflammatory immune system setting, thereby reducing the number of active MRI lesions and the relapse rate; **(c)** interferon-β promotes the production of anti-inflammatory cytokines and decreases the production of proinflammatory ones by reducing the trafficking of inflammatory cells within the central nervous system; **(d)** monoclonal antibodies, natalizumab, the first monoclonal antibody approved for the treatment of MS, interacts with α4 integrin receptors on the surface of lymphocytes, preventing these cells from adhering to the blood-brain barrier and entering the central nervous system; **(e)** fingolimod is the first oral treatment available for MS through preventing T-cell migration from secondary lymphoid organs to circulating blood; and **(f)** traditional immunosuppressants.

During the 2000s and 2010s, there has been approval of several oral drugs that are expected to gain in popularity and frequency of use [150, 151, 154]. Several more oral drugs are under investigation, including ozanimod, laquinimod, and estriol. Laquinimod was announced in August 2012 and is in a third phase III trial after mixed results in the previous ones [155, 156]. Similarly, studies aimed to improve the efficacy and ease of use of already existing therapies are occurring. This includes the use of new preparations such as the PEGylated version of interferon-β-1a, which it is hoped may be given at less frequent doses with similar effects [157, 158]. Estriol, a female sex hormone found at high concentrations during late pregnancy, has been identified as a candidate therapy for women with relapsing-remitting MS and has progressed through phase II trials [159, 160]. Request for approval of peginterferon beta-1a is expected during 2013 [158].

Monoclonal antibodies have also raised high levels of interest. As of 2012 alemtuzumab, daclizumab, and CD20 monoclonal antibodies such as rituximab [161], ocrelizumab, and ofatumumab had all shown some benefit and were under study as potential treatments [162], and the FDA approved ocrelizumab for relapsing and primary MS in March 2017 [163]. Their use has also been accompanied by the appearance of potentially dangerous adverse effects, the most important of which being opportunistic infections [154]. Related to these investigations is the development of a test for JC virus antibodies, which might help to determine who is at greater risk of developing progressive multifocal leukoencephalopathy when taking natalizumab [154]. While monoclonal antibodies will probably have some role in the treatment of the disease in the future, it is believed that it will be small due to the risks associated with them [154].

Another research strategy is to evaluate the combined effectiveness of two or more drugs [164]. The main rationale for using a number of medications in MS is that the involved treatments target different mechanisms and, therefore, their use is not necessarily exclusive [164]. Synergies, in which one drug improves the effect of another, are also possible, but there can also be drawbacks such as the blocking of the action of the other or worsened side effects [164]. There have been several trials of combined therapy, yet none have shown positive enough results to be considered as a useful treatment for MS [164].

Research on neuroprotection and regenerative treatments, such as stem cell therapy, while of high importance, are in the early stages [165]. Likewise, there are not any effective treatments for the progressive variants of the disease. Many of the newest drugs as well as those under development are probably going to be evaluated as therapies for PPMS or SPMS [154].

Medications that influence voltage-gated sodium ion channels are under investigation as a potential neuroprotective strategy because of hypothesized role of sodium in the pathological process leading to axonal injury and accumulating disability. Currently, there is insufficient evidence of an effect of sodium channel blockers for people with MS [166].

6.2.4.3 Combination Drug Therapy in Multiple Sclerosis (MS)

MS is an inflammatory/demyelinating and neurodegenerative disease. Treatment of MS is currently based on various different therapeutic algorithms of a sequential or escalating therapy with immunomodulators or immunosuppressants, generated partly from evidence-based medicine and partly from expert's consensus [167]. However, these therapies are not always effective as monotherapies. An alternative would be the combination of agents which already have some proven efficacy in MS therapy, are directed against different mechanisms of the pathogenic chain, and ideally result in synergic effects and a profile of reduced toxicity [167]. Combination therapy in multiple sclerosis can be combination of two or more anti-inflammatory agents or combination of anti-inflammatory agents plus neuroprotective agents [167]. Many combinations of drugs have been or are being tested in MS. Clinical trials have included a low number of patients for short periods of time [167].

Moreover, MS pathogenesis is complex and probably heterogeneous among patients, suggesting that combination therapy strategies that target a range of disease mechanisms might be more effective than medications used as monotherapy [168]. Potential combinational regimens include interferons and glatiramer acetate with each other or with approved second-line drugs such as natalizumab and mitoxantrone. Disease-modifying therapies have also been used in combination with drugs approved for other indications, such as corticosteroids, methotrexate, azathioprine, and cyclophosphamide. Many preliminary studies have provided favorable results for various combination regimens [168]. However, several subsequent large, randomized, controlled trials have had negative or conflicting results. Therefore, the usefulness of combination therapy in MS remains uncertain [168].

Furthermore, combination of standard therapies with immunosuppressive agents or with new therapies may provide synergistic effects that will likely benefit patients with MS [164, 169].

Immunomodulators/Immunosuppressants CDT for MS

Immunomodulating agents have beneficial effects in the treatment of MS, decreasing the frequency of relapses, the progression of disability, and MRI measures of disease activity. Despite the efficacy of these agents, many patients continue to show progression of disability, breakthrough relapses, and active disease on MRI. Therefore, clinicians have employed a variety of combinations of agents in an attempt to decrease disease activity in those with active disease despite standard immunomodulatory therapy [170]. Although a variety of combination therapies have been used in clinical practice, there is a paucity of data available to guide clinical decision-making. A major pitfall in using combination therapy in the absence of data demonstrating safety is the possibility that the agent added to the primary therapy may have no effect or, worse, may antagonize the effect of the primary agent. The combination of mitoxantrone and interferon beta (IFNbeta) appears safe in short-term studies from a toxicity standpoint and is associated with a reduction in relapse rates, a decrease in the frequency of enhancing lesions, and a decrease in T2 lesion burden [170]. Other combinations that appear safe in preliminary studies include IFNbeta-1a and methotrexate, IFNbeta-1a and azathioprine, and mitoxantrone plus methylprednisolone [170].

Mycophenolate Mofetil (MMF)-Interferon β-1a (IFNB-1a) CDT for MS

Several studies have evaluated the efficacy and toxicity of the mycophenolate mofetil (MMF)/interferon β-1a (IFNB-1a) combination for MS [171–173]. The common findings are that this regimen is well tolerated and, despite the small sample size, therapeutic trends were observed in favor of the combination therapy. However, an adequately powered controlled trial is required to clarify the efficacy of this combination.

6.2.5 CDT in Schizophrenia

6.2.5.1 Basics of Schizophrenia

Schizophrenia is a chronic, severe, and debilitating mental (brain and behavior) disorder of a type involving a breakdown in the relation between thought, emotion, and acts, leading to faulty perception, inappropriate actions and feelings, withdrawal from reality and personal relationships into fantasy and delusion, and a sense of mental fragmentation [174–177]. People with schizophrenia can have trouble distinguishing reality from fantasy, expressing and managing normal emotions, and making decisions. Thought processes may also be disorganized, and the motivation to engage in life's activities may be blunted. Those with the condition may hear imaginary voices and believe others are reading their minds, controlling their thoughts, or plotting to harm them.

Most people with schizophrenia suffer from symptoms either continuously or intermittently throughout life and are often severely stigmatized by people who do not understand the disease. Contrary to popular perception, people with schizophrenia do not have "split" or multiple personalities, and most pose no danger to others. However, the symptoms are terrifying to those afflicted and can make them unresponsive, agitated, or withdrawn [174–177]. Although schizophrenia is not as common as other mental disorders, the symptoms can be very disabling [174–177]. People with schizophrenia attempt suicide more often than people in the general population, and estimates are that up to 10% of people with schizophrenia will complete a suicide in the first 10 years of the illness—particularly young men with schizophrenia, leading to an average decrease in life expectancy by 20 years [178]. In 2015, an estimated 17,000 deaths were linked to schizophrenia [147].

About 0.3–0.7% of people are diagnosed with schizophrenia during their lifetime [179]. In 2017, there were an estimated 1.1 million new cases and in 2019 a total of 20 million cases globally [180, 181]. Males are more often affected and on average have an earlier onset [180], although some large reviews have not found gender differences in the prevalence of the disorder [182]. The likely causes of schizophrenia include genetic and environmental factors [174–177]. Genetic factors include a variety of common and rare genetic variants [183]. Possible environmental factors include being raised in a city, cannabis use during adolescence, infections, the ages of a person's mother or father, and poor nutrition during pregnancy [177].

The mechanisms that underlie the development of schizophrenia are complex and not clearly understood. A number of hypotheses including the dopamine hypothesis and the glutamate hypothesis have been put forward in an attempt to explain the link between altered brain function and the symptoms and development of schizophrenia [174–177]. Proposed models are separate from the proposed causes, which deal with the risk factors that may lead to schizophrenia.

6.2.5.2 Drug Treatment of Schizophrenia

The mainstay of treatment is antipsychotic medication, along with counseling, job training, and social rehabilitation [174–177]. Up to a third of people do not respond to initial antipsychotics, in which case the atypical antipsychotic clozapine may be used [184]. In a network comparative meta-analysis of 15 antipsychotic drugs, clozapine was significantly more effective than all other drugs, although clozapine's heavily multimodal action may cause more side effects [185]. In situations where doctors judge that there is a risk of harm to self or others, they may impose short involuntary hospitalization [186]. Long-term hospitalization is used on a small number of people with severe schizophrenia. In some countries where supportive services are limited or unavailable, long-term hospital stays are more common [187].

The first-line treatment for schizophrenia is an antipsychotic. The first-generation antipsychotics, now called typical antipsychotics, are dopamine antagonists that block D2 receptors and affect the neurotransmission of dopamine. Those brought out later, the second-generation antipsychotics known as atypical antipsychotics, can also have effect on another neurotransmitter, serotonin. Antipsychotics can reduce the symptoms of anxiety within hours of their use, but for other symptoms, they may take several days or weeks to reach their full effect [188]. They have little effect on negative and cognitive symptoms, which may be helped by additional psychotherapies and medications [189].

There is no single antipsychotic suitable for first-line treatment for everyone, as responses and tolerances vary between people [190]. Stopping medication may be considered after a single psychotic episode where there has been a full recovery with no symptoms for 12 months. Repeated relapses worsen the long-term outlook, and the risk of relapse following a second episode is high, and long-term treatment is usually recommended [191].

Tobacco smoking increases the metabolism of some antipsychotics, by strongly activating CYP1A2, the enzyme that breaks them down, and a significant difference is found in these levels between smokers and nonsmokers [192]. It is recommended that the dosage for those smokers on clozapine be increased by 50% and for those on olanzapine by 30%. The result of stopping smoking can lead to an increased concentration of the antipsychotic that may result in toxicity, so that monitoring of effects would need to take place with a view to decreasing the dosage; many symptoms may be noticeably worsened, and extreme fatigue and seizures are also possible with a risk of relapse. Likewise, those who resume smoking may need their dosages adjusted accordingly [193]. The altering effects are due to compounds in tobacco smoke and not to nicotine; the use of nicotine replacement therapy therefore has the equivalent effect of stopping smoking, and monitoring would still be needed.

6.2.5.3 Combination Drug Therapy of Schizophrenia

Although the international guidelines for the treatment of schizophrenia recommend antipsychotic monotherapy as the treatment of choice, many schizophrenia patients receive two or more antipsychotics in clinical practice, while co-treatment with antidepressants, mood stabilizers, and benzodiazepines is also common [194]. Indeed, schizophrenia is a severe mental disorder characterized by a heterogeneous symptom profile which comprises a clinical platform for widespread use of polypharmacological therapy (or polypharmacy) even though antipsychotic monotherapy is the recommended treatment regimen [195].

Despite treatment with antipsychotic medication, approximately one-third of individuals with schizophrenia will fail to have an adequate response. To treat these patients, a commonly utilized approach is antipsychotic combination therapy. Antipsychotic polypharmacy is frequently prescribed instead of following international consensus of clozapine monotherapy in treatment-resistant patients [194–196]. Yet, antipsychotic combination therapy is controversial with mixed efficacy and tolerability results.

Antipsychotic/Antipsychotic Combination for Schizophrenia

Antipsychotic/antipsychotic combination is a CDT with concomitant administration of two or more different antipsychotic compounds. The frequency of antipsychotic polypharmacy prescribing varies geographically with higher incidences in Asia and Europe than in North America [195, 198]. Changes in prescribing practice also vary geographically with the prevalence of antipsychotic polypharmacy increasing insignificantly in North America from the 1980s (12.7%) to the 2000s (17.0%), decreasing significantly from 1980 (55.5%) to 2000 (19.2%) in Asia, and varying nonsignificantly in Europe.

According to Baandrup [195], clinical indications for antipsychotic polypharmacy typically include insufficient treatment response to monotherapy, intolerable side effects, and patients that are trapped on polypharmacy during a process of cross-titration from one antipsychotic to another. Theoretically, it is difficult to find a rationale for combining antipsychotics because of their similar mechanism of action, that is, via dopamine D2 receptor antagonism [195]. Thus, there is no reason to expect an increased antipsychotic efficacy when combining two antipsychotics above what is obtained when increasing the dosage of one single antipsychotic. Several reviews of clinical trials investigating different combinations of antipsychotic compounds were published from 2007 to 2009 all convincingly concluding the same: antipsychotic polypharmacy is not superior to monotherapy except in the group of partially clozapine-responsive patients where an additional antipsychotic effect may be obtained by adding another antipsychotic [199]. On the other hand, antipsychotic combination has not been found to be associated with increased risk of death when compared with antipsychotic monotherapy in patients with schizophrenia either [200]. The evidence comparing the impact of antipsychotic monotherapy vs combination therapy on rehospitalization is mixed [195]. Evidence is primarily retrospective in nature, and there is high heterogeneity between studies, which could partially explain the mixed results [195]. There is likely a subset of patients for whom antipsychotic combination therapy reduces the risk of hospitalization greater than antipsychotic monotherapy [195].

However, a study demonstrated that antipsychotic combination of low-dose antipsychotics (400 mg/day amisulpride plus 800 mg/day sulpiride, N=46) is as efficacious and safe as, but cheaper than, optimal-dose monotherapy (800 mg/day amisulpride, N=46) in the treatment of schizophrenia [201]. The same group conducted another study on antipsychotic combination with a pair of different agents and reached the same conclusion [202]. In this 6-week, double-blind, fixed-dose study, patients were randomized to receive 5 mg/day of olanzapine plus 5 mg/day of trifluoperazine or 10 mg/day of olanzapine for 6weeks. The results uncovered that the CDT is as efficacious and safe as, but cheaper than, monotherapy in the acute treatment of schizophrenia [202].

Antipsychotic/Benzodiazepine Combination for Schizophrenia

Antipsychotic drugs are class of psychotropic medication primarily used to manage psychosis, including atypical or second-generation antipsychotics (risperidone, quetiapine, olanzapine, ziprasidone, paliperidone, aripiprazole, and clozapine) and typical or first-generation antipsychotics (chlorpromazine, flupenthixol, fluphenazine, haloperidol, loxapine, perphenazine, pimozide, trifluoperazine, thiothixene, and zuclopenthixol) [190]. The main difference between the two types of antipsychotics is that the first-generation drugs block dopamine and the second-generation drugs block dopamine and also affect serotonin levels. Evidence suggests that some of the second-generation drugs have milder movement-related side effects than the first-generation drugs. The second-generation antipsychotics are usually the first choice for the treatment of schizophrenia.

Benzodiazepines are a class of psychoactive drugs or depressants used to treat conditions such as anxiety, insomnia, and seizures. Benzodiazepines enhance the effect of the neurotransmitter gamma-aminobutyric acid (GABA) at the GABAA receptor, resulting in sedative, hypnotic (sleep-inducing), anxiolytic (antianxiety), anticonvulsant, and muscle-relaxant properties. High doses of many shorter-acting benzodiazepines may also cause anterograde amnesia and dissociation [197]. These properties make benzodiazepines useful in treating anxiety, panic disorder, insomnia, agitation, seizures, muscle spasms, alcohol withdrawal, and as a premedication for medical or dental procedures. Benzodiazepines are categorized as short, intermediate, or long acting. Short- and intermediate-acting benzodiazepines are preferred for the treatment of insomnia; longer-acting benzodiazepines are recommended for the treatment of anxiety [197]. Approved benzodiazepines in the United States include alprazolam, chlordiazepoxide, clonazepam, clorazepate, diazepam, estazolam, flurazepam, lorazepam, midazolam, oxazepam, temazepam, triazolam, and quazepam. Although most benzodiazepines are used interchangeably, some are most commonly used for certain conditions.

Antipsychotic/benzodiazepine combination clearly has a role in the treatment of acute agitation, whereas there is no evidence to support an effect on core schizophrenia symptoms when chronically prescribed [195]. Oppositely, several larger observational studies have documented an increased risk of death with antipsychotic/benzodiazepine combination treatment in patients with schizophrenia [200, 203, 204]. This includes both increased mortality from suicide, which is most probably a result of confounding by indication, but it also includes increased risk of death from natural causes which is not readily explainable as confounding by indication [195].

Antipsychotic/Antidepressant Combination for Schizophrenia

Antidepressants are medications used to treat major depressive disorder, some anxiety disorders, and some chronic pain conditions and to help manage some addictions [205]. They include serotonin and noradrenaline reuptake inhibitors (SNRIs, such as duloxetine, venlafaxine, and desvenlafaxine), selective serotonin reuptake inhibitors (SSRIs; e.g., citalopram, escitalopram, fluoxetine, fluvoxamine, paroxetine, and sertraline), tricyclic antidepressants (TCAs; e.g., amitriptyline, amoxapine-clomipramine, desipramine, doxepin, imipramine, nortriptyline, protriptyline, and trimipramine), monoamine oxidase inhibitors (MAOIs; e.g., phenelzine, tranylcypromine, isocarboxazid, and selegiline), and noradrenaline and specific serotoninergic antidepressants (NASSAs; mianserin and mirtazapine). Clearly, antidepressants are a diverse group of agents, but a common trait is inhibition of neurotransmitter reuptake, specifically acting on the serotonergic and the noradrenergic systems.

Antidepressants are typically added to antipsychotic treatment in case of persistent negative symptoms. Available evidence suggests that antidepressants may improve negative symptom control in schizophrenia [195]. Combining an antipsychotic with an antiepileptic is not supported by any firm evidence, but individual mood stabilizers have come out positively in single trials. Generally, the evidence for such combination in schizophrenia maintenance treatment is sparse but may be warranted in certain clinical situations [195]. Recent meta-analyses indicate that there is some evidence of a small to moderate effect of combining an antidepressant with antipsychotic treatment for total symptom score improvement [206]. Selective serotonin reuptake inhibitors (SSRIs) and mirtazapine (a noradrenergic, specific serotonergic agent) are the compounds that have been most often investigated. As regards effect specifically on negative symptoms, the latest meta-regression found a small therapeutic effect of SSRIs and a larger effect of serotonergic and noradrenergic reuptake inhibitors (SNRIs) but no effect of mirtazapine [206]. Likewise, all-cause mortality has been found to be decreased in patients with schizophrenia treated with antidepressants compared with no exposure [204]. These results add to the overall view of a potentially beneficial effect of combining an antipsychotic with an antidepressant in patients with schizophrenia and depressive/negative symptoms [195].

Antipsychotic/Antiepileptic Combination for Schizophrenia

Augmenting antipsychotic treatment with an antiepileptic (or mood stabilizer) stems from clinical practice where this has been considered an option in a subgroup of patients with the aim of reducing aggression or stabilizing mood outside the licensed indication of bipolar disorder [195]. Theoretically, antiepileptic drugs may reduce aggression by reducing neuronal hyperexcitability associated with aggression, but the current evidence is not convincing [207]. The clinical practice of combining an antipsychotic with an antiepileptic in patients with schizophrenia has been the subject of several smaller clinical trials with the latest meta-regression results [206] indicating that lamotrigine might be associated with a therapeutic effect regarding total psychopathology score as well as psychotic and negative symptom score, whereas carbamazepine had no effect [195]. However, an effectiveness study based on US national Medicaid data [208] found that initiating use of a mood stabilizer in patients with schizophrenia on stable treatment with a single antipsychotic was associated with an increased risk of all-cause mortality. Overall, the results are mixed, and increased risk of death with use of antiepileptics in patients with schizophrenia represents a concern that cannot be completely ruled out based on the current data [195].

6.3 FDC in Neurological Disease

6.3.1 FDC in Alzheimer's Disease (AD)

Alzheimer's disease (AD) is the most common neurodegenerative disease worldwide and carries an immense societal burden. Unfortunately, no curative or disease-modifying treatment has yet been discovered. The currently approved medications are symptomatic. They include two classes: the cholinesterase inhibitors (ChEIs), such as donepezil, and the NMDA receptor antagonists (NMDARIs), e.g., memantine. Most evidence has shown that combining both classes is superior to monotherapy but may complicate the treatment regimen for patients and families [209, 210].

6.3.1.1 Memantine ER/Donepezil FDC for AD

In December 2014, the US Food and Drug Administration approved Namzaric™, a once-daily, fixed-dose combination (FDC) of memantine extended release (ER) and donepezil for patients with moderate to severe AD. The FDC capsule is bioequivalent to the coadministered individual drugs, and its bioavailability is similar when taken fasting, with food, or sprinkled onto applesauce. The combination of memantine and ChEIs in moderate to severe AD provides additional benefits to ChEI monotherapy across multiple domains and may delay the time to nursing home admission. A dedicated study of memantine ER compared to placebo in patients on a stable dose of a ChEI found statistically significant benefits on cognition and global status but not functioning [210]. Treatment with memantine ER and donepezil is generally well tolerated, although higher doses of ChEIs are associated with more serious adverse events such as vomiting, syncope, and weight loss [211, 212]. Potential advantages of the FDC include a simpler treatment regimen, reduction in pill burden, and the ability to sprinkle the capsule onto soft foods. Patients who may benefit from the FDC include those with significant dysphagia, a history of poor compliance, or limited caregiver interaction. Aside from its cost, FDC memantine ER/donepezil improves adherence to medication and reduces caregiver burden [209]. It allows patients to benefit from combination therapy as the disease progresses, especially in those with dysphagia, poor adherence, and limited caregiver support.

In a 24-week, phase III trial in patients with moderate to severe AD, addition of memantine ER 28 mg once daily to stable ChEI therapy was more effective than add-on placebo on measures of cognition, global clinical status, dementia behavior, and semantic processing ability, although between-group differences on a measure of daily function did not significantly differ [213, 214]. In subgroup analyses in donepezil-treated patients, add-on memantine ER was more effective than add-on placebo on measures of cognition, dementia behavior, and semantic processing, although there were no significant between-group differences on measures of global clinical status and daily function. Memantine ER plus ChEI combination therapy was generally well tolerated in the phase III trial [213], with diarrhea, dizziness, and influenza occurring at least twice as often with add-on memantine ER as add-on placebo in donepezil-treated patients. Thus, memantine ER plus donepezil combination therapy is an effective and well-tolerated treatment option for patients with moderate to severe AD. The fixed-dose combination is potentially more convenient than coadministration of the individual agents [213].

Meta-analyses also confirmed that the donepezil/memantine FDC produces greater improvement in cognitive functions, behavioral and psychological symptoms in dementia, and global functions than did donepezil alone in patients with moderate to severe AD.

6.3.1.2 Memantine/Rivastigmine FDC for AD

A study was conducted to determine if memantine has any effects on the steady-state pharmacokinetics of rivastigmine in patients with mild to moderate Alzheimer's disease [215]. Sixteen patients (seven males and nine females, age range 64–88 years, weight range 51.8–104 kg) were enrolled in this open-label, crossover study, which consisted of a 28-day screening period, a 36-hour baseline period, and a 35-day combination treatment phase. The combination of memantine (10 mg twice daily) with rivastigmine (1.5–6 mg twice daily) was found safe and well tolerated [215]. At each dose level of rivastigmine, the area under the concentration-time curve (AUC) values of rivastigmine and its metabolite as well as the metabolite to parent AUC ratios were unaffected by coadministration of memantine, confirming the absence of a meaningful pharmacokinetic drug-drug interaction.

In a multicenter, randomized, open-label study, patients entered an 8-week run-in period (a 5 cm^2 rivastigmine patch for 4 weeks, then a 10 cm^2 patch for 4 weeks) followed by 16 weeks of memantine plus rivastigmine patch or rivastigmine patch monotherapy [30]. The primary outcome measure was the retention rate at the end of the trial. Overall, 88 and 84 patients received rivastigmine patch with and without memantine, respectively, and of these, 77 (87.5%) and 70 (83.3%) patients completed the study. The difference in retention rate was not significant. The incidence of adverse events (AEs) (53.4 vs 50.6%) and discontinuation due to AEs (6.8 vs 4.8%) were not different between patients with and without memantine. The most frequent AEs were skin irritation in patients with and without memantine, but discontinuation due to skin irritation was rare. The incidence of gastrointestinal AEs was very low in patients with and without memantine. The Korean version of the Cohen-Mansfield Agitation Inventory scores favored rivastigmine patch monotherapy at the end of treatment. Changes in other efficacy measures were similar between the groups. Collectively, there were no significant differences in tolerability and safety between the treatment groups. The combination therapy of memantine plus rivastigmine patch did not show an advantage over rivastigmine patch monotherapy on efficacy analyses [30].

6.3.2 FDC in Parkinson's Disease (PD)

6.3.2.1 Melevodopa/Carbidopa FDC for PD

Stocchi et al. [216] performed a systematic literature review to evaluate pharmacokinetic (PK), efficacy, and safety profile of oral melevodopa (a dopaminergic agent)/carbidopa (a dopa decarboxylase inhibitor) fixed combination in the management of motor fluctuations in patients with PD. The results suggest that melevodopa/carbidopa FDC in a highly soluble formulation as effervescent oral tablets is effective in improving control of motor complications in PD patients compared to conventional levodopa [216].

6.3.2.2 Levodopa/Carbidopa/Entacapone FDC for PD

Levodopa is the most efficacious agent for the treatment of motor features of PD, but its chronic use is associated with the development of motor complications. Mounting evidence indicates the short half-life of levodopa, and resultant pulsatile stimulation of striatal dopamine receptors leads to wearing off, motor fluctuations, and dyskinesias. Longer-acting dopaminergic agents, such as dopamine agonists, are less likely to cause motor fluctuations and dyskinesias but are not as efficacious for control of motor symptoms. Therefore, there is interest in exploring ways to deliver levodopa in a more continuous fashion, in an effort to maintain benefit through the day and reduce the development of motor fluctuations and dyskinesias. A dopa decarboxylase inhibitor (DDCI), such as carbidopa or benserazide, is administered with levodopa to attenuate its peripheral conversion to dopamine, reduce nausea, and increase central bioavailability. When levodopa is administered with a DDCI, its main route of peripheral metabolism is via catechol-O-methyl transferase (COMT). A COMT inhibitor can be added to the combination of levodopa and a DDCI to further extend the levodopa peripheral half-life and increase central bioavailability. Stalevo is a combination tablet comprised of levodopa, carbidopa, and the COMT inhibitor entacapone. It is available in fixed-dose combinations of levodopa/carbidopa/entacapone, 50/12.5/200, 75/18.75/200, 100/25/200, 125/31.25/200, 150/37.5/200, and 200/50/200 mg [101]. Stalevo is currently approved for use in Parkinson's disease patients with end-of-dose wearing off [101].

6.3.3 FDC in Schizophrenia

ALKS 3831 is composed of a flexible dose of olanzapine (a second-generation antipsychotic) and a fixed dose of 10 mg samidorphan (a novel opioid system modulator), designed to provide the established antipsychotic efficacy of olanzapine and to mitigate olanzapine-induced weight gain. To support clinical development of ALKS 3831, Sun et al. [217] conducted a multicenter, randomized, open-label, phase I study to obtain steady-state exposure for olanzapine and samidorphan and short-term safety at the intended therapeutic dose range of ALKS 3831 10/10 (10 mg olanzapine/10 mg samidorphan) to ALKS 3831 20/10 (20 mg olanzapine/10 mg samidorphan) in subjects with schizophrenia. They found that samidorphan exposure was not affected by different dose levels of olanzapine in ALKS 3831. In addition, olanzapine exposure as a component of ALKS 3831 was comparable with previously published data for olanzapine monotherapy. These data indicate that combining olanzapine with samidorphan does not affect the pharmacokinetic profile of either drug and support continued clinical evaluation of ALKS 3831 [217].

6.4 MTD in Neurological Disease

Neurological disorders are in general multifactorial pathologies related to neuronal impairment and functional loss from different regions of the nervous system. In the last two decades, a new approach for the rational design of new drug candidates, the MTD strategy, has emerged and has been used in the design and for the development of a variety of hybrid compounds capable to act simultaneously in diverse biological targets. Nervous system application of polypharmacology has led to approval of four multitarget drugs [218]. Recently, an increased number of multitarget drugs have been developed to treat schizophrenia and major depressive disorders [219–221]. Concerning neurodegenerative disorders, the polypharmacological approach looks particularly promising, given the complexity and multifactorial nature of such diseases and their unknown etiopathogenesis [218]. For example, rasagiline, approved in 2006 for neurodegenerative diseases, displayed a multitarget profile. The first new chemical entity approved for neurodegeneration in over a decade is a multitarget drug called safinamide.

6.4.1 MTD in Alzheimer's Disease (AD)

Alzheimer's disease (AD) is a multifactorial neurodegenerative disorder. Several hallmarks such as β-amyloid (Aβ) aggregation underlying amyloid plaque formation, τ-hyperphosphorylation leading to production of neurofibrillary tangles, and decline in the number of cholinergic neurons appear to be fundamental in the pathophysiology of the disease. Other evidence points also to the involvement of oxidative stress, biometal dyshomeostasis, inflammation, and cell cycle regulatory failure. Taking into account such premises, many attractive targets for the development of anti-AD drugs have emerged. Specifically, the multifactorial nature of AD calls for multitarget-directed ligands (MTDLs) [222–224] or MTDs which can be beneficial by providing interactions with multiple targets. Consensus has recently been reached that MTDs are the ultimate solution for neurodegenerative diseases such as AD [222–227]. However, there has not been any rationally designed MTDs in the market, but many ongoing research and development projects are actively engaged in the discovery of AD MTDs. As presented by Marotta et al. [222] and by Gontijo et al. [223] in their excellent review articles, a number of approaches have been being extensively and intensively implemented in the development of MTDs for AD. Molecular hybridization is one of these approaches, which has demonstrated its particular value in this regard.

Memantine is an uncompetitive NMDARI approved to treat moderate to severe AD patients and is viewed as a privileged NMDAR-directed structure, thereby representing the driving motif in the design of a variety of multifunctional compounds. In the search for a disease-modifying drug for AD, the MDT approach has been applied by different research groups to obtain hybrid compounds able to hit simultaneously different pathways implicated in AD [222, 228]. In their review, Marotta et al. [222] present selected examples of small molecules designed by merging in a single entity the amantadine core of memantine with the chemical features of known neuroprotectants performing through different mechanism of action, such as antioxidant and anti-aggregating activity or AChE and monoaminoxidase inhibition.

Following the polypharmacological approach and the clinical relevance of both memantine and AChEIs, several MTDs have been developed by joining memantine to pharmacophores, inhibiting ChE with the aim to synergistically and more effectively tackle the AD pathological cascade.

6.4.1.1 Tacrine-Adamantanes Hybrids

Tacrine (THA) is a noncompetitive reversible AChEI active in the central nervous system (CNS) and slightly selective for BChE. It was the first AChEI drug approved by FDA to combat AD, but it was withdrawn from the market due to its hepatoxicity due to the frequent adverse effects including peripheral cholinergic effects and hepatotoxicity limiting its therapeutic potential [224, 229]. Its derivatives 7-methoxytacrine (7-MEOTA) and 6-chlorotacrine (6-Cl-THA) are potent anti-ChE compounds with lower side effects. An interesting strategy to obtain a multimodal profile toward ChE and NMDAR is to combine tacrine scaffold with the adamantane core through tethers of different length [222]. In this view, hybrids of 7-MEOTA and amantadine, linked by methylene-thiourea or methylene-urea tethers, were designed and synthesized [222]. The new compounds showed micromolar to submicromolar activities toward *h*AChE and human BChE (*h*BChE), NMDA receptor, and $A\beta_{1\text{-}40}$ aggregation [230, 231].

Huprine X, a reversible AChE inhibitor, designed by molecular hybridization of tacrine and huperzine A, has been shown to affect the amyloidogenic process in vitro and the AD-related neuropathology in vivo in mice models of the disease [232]. More recently, Giménez-Llort et al. [232] have shown that a group of donepezil-huprine heterodimers exerts a highly potent and selective inhibitory action on AChE both in vitro and ex vivo, simultaneously interacting with both peripheral and catalytic binding sites and inhibiting the β-amyloid aggregation, whereas some levetiracetam-huprine hybrids have been shown to reduce epileptiform activity, neuroinflammation, and amyloid burden in an animal model of AD.

6.4.1.2 Galantamine-Memantine Hybrids

Galantamine is an alkaloid-derived AChEI approved since 2001 for the treatment of mild to moderate forms of AD [222]. Furthermore, galantamine associates AChEI efficacy to potentiation of NMDAR transmission mediated by protein kinase C (PKC) and to the ability of allosterically modulated α-subunit of nicotinic receptors. Particularly, through the modulation of α7-nAChR on presynaptic glutamatergic neurons, it can enhance the beneficial effect of NMDAR's synaptic activation, potentiating at the same time memantine's neuroprotective effect [233].

6.4.1.3 Aminoadamantane-Carbazole/Tetrahydrocarbazole Hybrids

It has been reported that some carbazole derivatives are potential AD-modifying compounds, i.e., *N*-alkyl carbazoles are able to favor the formation of soluble form of Aβ, through different mechanisms and are also able to block apoptosis, thus exerting a neuroprotective effect [222, 234]. Moreover, recent studies suggested that some carbazole-based compounds are valuable anti-AD drug candidates, since they are able to inhibit both AChE and BChE activity as well as AChE- and self-induced Aβ aggregation. These findings support the rationale to develop novel hybrids with a multimodal profile by adding to biological properties of the carbazole core the neuroprotection derived by the memantine nucleus. Hybrids of aminoadamantanes with carbazole and tetrahydrocarbazole were synthesized and tested as inhibitors of *h*AChE and equine serum BChE, as antagonists of NMDARs and as stimulators of microtubule assembly [235]. In AD, microtubule structure disturbance and tubulin polymerization suppression are signs of the incoming neurodegeneration that lead to failures in the axonal transport; therefore, it is necessary to develop compounds that are able to stabilize microtubule structure and to enhance tubulin assembly [235, 236].

6.4.1.4 Memantine-Ferulic Acid Hybrids

Recently, Rosini and coworkers developed a series of hybrids between memantine and ferulic acid, envisioned by the concurrent ability to reduce Aβ-induced neurotoxicity and oxidative stress of this latter [237, 238]. The best derivatives of the series are those directly linked at the carboxylic function of ferulic acid and nitrogen atom of memantine through alkyl chain of different lengths, which were initially screened for their NMDAR antagonism efficacy.

6.4.1.5 Memantine-Glutathione/Lipoic Acid Hybrids

The possibility of connecting antioxidant agents with the memantine core has also been explored [238, 239] using radical scavengers such as glutathione (GSH) and (R)-α-lipoic acid. They are endogenous molecules capable of removing free radicals that accumulate in cells, thus maintaining ROS balance at physiological levels and preventing oxidative stress and cell damage. Furthermore, lipoic acid has been shown to mitigate Aβ aggregation and to protect from Aβ-mediated cytotoxicity in vitro. In light of these findings, Sozio and coworkers developed conjugates of memantine linked to GSH and LA through an amide bond and tested them as NMDAR antagonists, Aβ aggregation inhibitors, and radical scavengers [240].

6.4.1.6 Amantadine-Propargylamine Hybrids

Monoamine oxidase B (MAO-B) with monoamine oxidase A (MAO-A) belongs to flavoenzyme family bound to outer mitochondrial membrane. They exert a pivotal role in neurotransmission, regulating neurotransmitter metabolism and cellular redox balance. Particularly, MAO-B has been gaining particular interest in recent years due to its overexpression in neuroinflammatory condition and the consequential ROS overproduction and neurotransmitter impairment. Accordingly, development of MAO-B inhibitors seems to be an interesting strategy to deepen our knowledge on diseases characterized by neuroinflammatory conditions, like AD and Parkinson's disease (PD) [241].

6.4.1.7 Memantine-Polyamine Conjugates

Kumamoto and coworkers developed a series of memantine-polyamine conjugates showing very potent NMDAR antagonism, higher than memantine [242]. The rationale is based on the well-known potent blocking NMDAR activity of the polyamine backbone that the same authors have already conjugated with polycyclic scaffolds, developing compounds endowed with proved strong reversible and voltage-dependent NMDA blocking activity [243, 244]. These derivatives were further explored and optimized, merging polyamine moiety with memantine scaffold.

6.4.1.8 H_2S-Releasing Memantine Prodrug

In recent years, the possibility of linking moieties releasing endogenous gaseous signaling molecules, i.e., gasotransmitters, with a well-known pharmacophore, has assumed particular interest in multitarget drug design [245, 246]. Gasotransmitters, such as carbon monoxide (CO), nitric oxide (NO), and hydrogen sulfide (H_2S), are endogenously produced by enzymes and play important roles as antioxidants, neuromodulators, and anti-inflammatory agents [247]. Particularly, H_2S combines neuroprotection, anti-inflammatory, and antiapoptotic activities to NMDAR modulation ability both directly, with sulfhydration of critical cysteine residues, and indirectly, through regulation of intracellular calcium concentration. On this basis, the interesting biological outputs of a compound, an H_2S-releasing memantine derivative by Sestito and coworkers, can open a range of possibilities for new MTDL campaign [248]. The compound bears an isothiocyanate moiety as H_2S-donating moiety, replacing the characteristic free amino group of memantine, and it maintains memantine's NMDAR inhibition (K_i = 458 vs 329 nM) only in the presence of 4 mM *L*-cysteine, which can mediate H_2S release and memantine's unveiling. Anti-amyloidogenic properties of memantine were confirmed with this compound, which demonstrated to reduce $A\beta_{42}$ self-induced aggregation and to reduce Aβ-induced damage in rat microglia cells to roughly the same extent as its parent compound [248].

6.4.2 MTD in Parkinson's Disease (PD)

6.4.2.1 Safinamide as a MTD for PK

Safinamide, a broad-spectrum antiepileptic drug, was originally developed as anticonvulsant agent and approved in March 2017 as an adjunctive treatment for PD thanks to its particular multitarget profile [249]. In addition, evidence has been provided for safinamide in the treatment of motor symptoms in early PD patients [250]. It combines dopaminergic effects, including selective and reversible MAO-B and dopamine reuptake inhibition, which are largely responsible for its effects on motor symptoms, along with non-dopaminergic properties, i.e., blockade of voltage-dependent Na^+ and Ca^{2+} channels and consequent inhibition of glutamate release that is thought to confer neurorescue and neuroprotectant effects [249, 250].

6.4.2.2 Ladostigil as a MTD for PK

The cascade of neurotoxic events involved in neuronal degeneration suggests that it is naive to think mono-target drugs can induce disease modification by slowing the process of neurodegeneration in Parkinson's disease (PD). Employing molecular hybridization-based techniques with rasagiline (N-propargyl-1-R-aminoindan) as a pharmacophore, Youdim et al. [251] have developed a series of novel multitarget neuroprotective drugs, with ladostigil (N-propargyl-3R-aminoindan-5yl)-ethyl methylcarbamate) as one of them. Ladostigil possesses both cholinesterase-butyrylesterase (Ch-BuE) and brain-selective monoamine oxidase AB (MAO-AB) inhibitory activities. Ladostigil is a neuroprotective agent, currently undergoing clinical trials, that contains the characteristics of old drugs such as rasagiline (i.e., anti-monoamine oxidase (MAO)-B) and rivastigmine (i.e., anti-AChE). Ladostigil was studied in the treatment of AD, Lewy body dementia, and PD [59]. Moreover, this hybrid compound showed good reversible BuChE, AChE, and irreversible MAO-B inhibition properties and protected impairments of motor function and memory in mice [249]. Ladostigil (1 μM) also exhibited antioxidant properties via its direct scavenging activity on free radicals and also indirectly through the induction of the expression of cellular antioxidant enzymes [60]. It has been shown upon in vivo studies that ladostigil decreased the concentration of the holo-amyloid precursor protein (APP). In addition to this, ladostigil also increased the concentrations of protein kinase C (PKC)a, PKCe, and p-PKC in the mice hippocampus [27, 61].

6.4.2.3 Iron Chelator-Radical Scavenging Drugs (M30) as MTDs for PD

M30 was developed by combining the propargyl moiety of the anti-PD drug rasagiline with the skeleton of the neuroprotective iron chelator, VK-28. M30 possesses the vital pharmacophores (i.e., propargylamine moiety) from the Food and Drug Administration (FDA)-approved anti-PD rasagiline [27, 251]. The propargylamine moiety in the rasagiline's structure plays a crucial role in the disease-modifying potential of rasagiline since it can mediate the interaction of this compound with various neuroprotective/neurorescue pathways [252]. PBT2 is a metal-protein-attenuating compound (MPAC) of second generation, which is now under phase II clinical trials for the treatment of AD. M30 possesses a metal-protein-attenuating/ionophore moiety similar to that of PBT2 [63]. MPACs typically show intermediate reversible affinity toward metals and have the capacity to reach certain intracellular compartments to target the harmful upstream metal-protein reactions [64]. M30 displays a selective inhibition for MAO-A and MAO-B action in the brain, with a comparatively poor inhibition of these enzymes in the small intestine and liver in vivo [65]. M30 had a moderate affinity for metals (i.e., Zn, Fe, and Cu) like other MPACs, for example, clioquinol and PBT2. However, M30 showed a highly potent activity against the iron-stimulated mitochondrial membrane lipid peroxidation, which is comparable to the prototype iron chelator deferoxamine (DFO) [253]. The chronic treatment of aged mice with M30 (1 and 5 mg/kg; four times weekly for 6 months) in vivo exhibited a substantial positive effect on neuropsychiatry functions and age-related cognitive deficit [254]. Furthermore, in the treated mice, M30 considerably decreased the cerebral levels of Aβ and iron accumulation [255]. In this study, it was also revealed that the chronic administration of M30 reduces the cerebral pathology of Aβ and behavioral impairments in an APP/PS1 transgenic mouse model of AD [255]. M30 has both neuroprotective and neurorestorative activities for nigrostriatal dopamine neurons in post-lesion MPTP, lactacystin, and 6-hydroxydopamine animal models of PD. The neurorestorative activity has been identified as being related to the ability of the drug to activate hypoxia-inducible factor (HIF) by inhibiting prolyl-4-hydroxylase. M30 regulates cell cycle arrest and induces the neurotrophins brain-derived neurotrophic factor (BDNF), vascular endothelial growth factor (VEGF), erythropoietin (EPO), as well as glia-derived neurotrophic factor (GDNF). These unique multiple actions of M30 make it potentially useful as a disease-modifying drug for the treatment of PD [251]. Thus, M30 has been demonstrated to possess neuroprotective/neurorescue activities in vitro and in vivo against several insults applicable to various neurodegenerative diseases, such as Alzheimer's disease, Parkinson's disease, and amyotrophic lateral sclerosis [251].

6.4.3 MTD in Epilepsy and Seizures

6.4.3.1 Antiepileptic Drugs (AEDs) as MTDs for Epilepsy

Epilepsy is diverse, with many different seizure types and epilepsy syndromes. The complexity of epileptic pathogenesis may explain why the currently available drugs, developed according to the classic drug discovery paradigm of "one molecule, one target," have turned out to be effective only in a percentage of patient with epilepsy. The currently used antiepileptic drugs (AEDs) predominately target voltage-gated ion channels (Na^+, Ca^{2+}, and K^+), GABAA receptor, glutamate receptor, synaptic vesicle 2A (SV2A) protein, and carbonic anhydrase (CA) [102]. One group of AEDs acts on a single target while another group acts via multiple targets to control seizure episodes. AEDs which act via multiple mechanisms or polypharmacological mechanisms of action have appeared as broad-spectrum anticonvulsant agent, and therefore, they provide a better choice to clinicians to manage drug-resistant epilepsies and various other epileptic syndromes. For example, polypharmacological AEDs such as phenobarbitone, valproate, oxcarbazepine, felbamate, etc. are vital for managing epilepsy successfully, since decades [102].

6.4.3.2 Antiseizure Medications as MTDs for Seizures

The actions of most ASMs on molecular targets can be categorized into four broad groups [256]: **(a)** modulation of voltage-gated ion channels, including sodium, calcium, and potassium channels; **(b)** enhancement of GABA-mediated inhibition through effects on GABAA receptors, the GAT1 GABA transporter, GABA transaminase, or the GABA-synthesizing enzyme glutamate decarboxylase (GAD); **(c)** inhibition of synaptic excitation mediated by ionotropic glutamate receptors, including N-methyl-d-aspartate (NMDA) and α-amino-3-hydroxy-5-methyl-4-

isoxazole-propionate (AMPA) receptors; and **(d)** direct modulation of synaptic release through effects on components of the release machinery, including SV2A and the α2δ subunit of voltage-gated calcium channels. It is known that at least one-third of patients with epilepsy do not respond to treatment with current ASMs [257, 258]. Patients with drug-resistant epilepsy suffer from uncontrolled seizures, which may cause injuries, disability, and increased mortality. Although the MTD strategy has been proposed for epilepsy therapy >25 years ago, only one antiseizure medication (ASM), padsevonil, has been intentionally developed as a single molecular entity that could target two different mechanisms [256]. Overall, padsevonil is the first rationally designed ASM candidate that inhibits seizure activity by presynaptic modulation of all three SV2 isoforms and postsynaptic enhancement of GABA-mediated inhibition [256]. This novel drug exhibited promising effects in numerous preclinical models of difficult to treat seizures. In a randomized, double-blind, placebo-controlled phase IIa proof-of-concept trial in 55 adult patients with very frequent drug-resistant focal seizures, who had failed to respond to ≥4 ASMs, significant antiseizure efficacy of padsevonil was determined during add-on administration [259]. As predicted by the preclinical data, padsevonil was generally well tolerated at doses up to 400 mg b.i.d. However, in a subsequent larger randomized placebo-controlled phase IIb add-on trial, padsevonil had only a modest effect in treatment-resistant focal epilepsy and did not separate from placebo in its primary endpoints, which optimistically included ≥75% reduction in seizure frequency [256, 260, 261].

In contrast to padsevonil, the new ASM cenobamate, which was recently approved for adjunctive therapy of treatment-resistant focal epilepsy by the FDA, has demonstrated in two randomized controlled trials that it has a superior ability to render treatment-resistant focal epilepsy patients seizure-free versus any of the other novel ASMs (21% of treatment-resistant patients remained free of seizures during the 12-week maintenance phase) [256, 260, 262]. Yet, cenobamate was discovered purely through "standard" blind screening [256]. When put into further development, the MOA was completely unknown, but subsequently, evidence has supported a dual mechanism of GABA enhancement and preferential inhibition of the persistent sodium channel [256, 263]. Cenobamate was quite effective across various preclinical models of seizures or epilepsy, including models of drug-resistant focal seizures such as the 6 Hz mouse model and amygdala kindling. Interestingly, cenobamate was about 10 times more potent than padsevonil in the MES test, whereas the opposite was true for padsevonil vs cenobamate in the 6 Hz model at 44 mA [256].

Multitarget effects have also been described for numerous other ASMs (see next section) [256]. According to Löscher [256], one explanation for the failure of padsevonil is that most ASMs already act by more than one MOA, so the dual MOA drug padsevonil does not offer anything new. Furthermore, most patients with refractory epilepsy take two, three, or more ASMs with different MOAs, which render it unlikely that a novel and rationally designed multitarget ASM is more effective than already existing ASMs or their combinations unless the novel drug acts by unique mechanisms, such as presumably those of cenobamate. In fact, most ASMs act by more than one MOA and thus can be considered multitarget drugs. Furthermore, even many of those ASMs that are typically presented as single-target drugs presumably act by more than one MOA [264, 265]. Examples are the sodium channel modulator phenytoin, which also blocks high-voltage-activated Ca^{2+} channels, or the Kv7 (KCNQ) potassium channel opener retigabine, which also acts as an allosteric positive modulator (PAM) at GABAA receptors.

6.4.4 MTD in Schizophrenia

6.4.4.1 Antipsychotic Agents as MTDs for Schizophrenia

Nervous system application of polypharmacology has led to approval of four multitarget drugs [218]. Recently, an increased number of multitarget drugs have been developed to treat schizophrenia and major depressive disorders

[218–221]. The main strategy for treatment of schizophrenia is based on antagonizing dopamine D_2, serotonin 5-HT_{2A}, and α1-adrenergic receptors [266]. Incorporating effects at these broad targets has the objective of either improving antipsychotic efficacy or mitigating adverse effects. With these concepts in mind, starting from the arylpiperazine substructure as suitable scaffold for achieving a fine balancing of D_2, 5-HT_{1A}, and 5-HT_{2A} receptor activities, the prototype of the third generation of antipsychotics, aripiprazole, launched into the market in 2015, was the first designed serotonin-dopamine activity modulator (SDAM). However, it also displayed unwanted side effects, probably arising from a sustained interaction with postsynaptic D_2 receptors. Thus, brexpiprazole (an atypical antipsychotic) was licensed in October 2015 as a novel D_2 and 5-HT_{1A} partial agonist, with less intrinsic activity at D_2 receptors and more balanced activities at 5-HT_{2A}, 5-HT_{1A}, and α_{1B} receptor subtypes than aripiprazole [266].

Another excellent example is cariprazine (an atypical antipsychotic), approved by the FDA in September 2015 for the treatment of schizophrenia and bipolar disorders. Similarly to the abovementioned SDAM, cariprazine shares the arylpiperazine substructure linked to an alkyl-*N′*,*N′*-dimethyl urea, which determines a profile of partial agonism at D_2 and D_3 receptors [267]. This mechanism is innovative because many other antipsychotics are only D_2 and 5-HT_{2A} partial agonists.

Research findings suggested that other neurotransmission systems, including serotonin, glutamate, GABA, and acetylcholine, were implicated in the development of schizophrenia and that a more inclusive medication was needed. A new first-in-class antipsychotic that targets multiple neurotransmitter systems called lumateperone (ITI-007) was trialed and approved by the FDA in December 2019 for the treatment of schizophrenia in adults [192, 268, 269]. Lumateperone is a small molecule agent that shows improved safety and tolerance. It interacts with dopamine, serotonin, and glutamate in a complex, uniquely selective manner and is seen to improve negative and positive symptoms and social functioning [270]. Lumateperone was also found to reduce potential metabolic dysfunction, have lower rates of movement disorders, and have lower cardiovascular side effects such as a fast heart rate [192].

6.4.4.2 Psychoactive Drugs as MTDs for Schizophrenia

Benzodiazepines (BZ or BZD) are a class of gabaminergic psychoactive chemicals used in hypnotics, sedation, in the treatment of anxiety, and in other CNS disorders. These drugs include alprazolam (Xanax), diazepam (Valium), clonazepam (Klonopin), and others [271]. There are two distinct types of pharmacological binding sites for benzodiazepines in the brain (BZ1 and BZ2); these sites are on GABAA receptors and are classified as short, intermediate, or long acting [271]. From the thienobenzodiazepine class (TBZ), olanzapine (2-methyl-4-(4-methyl-l-piperazinyl)-10H-thieno[2,3-b][1, 5]benzodiazepine) (Zyprexa) was used as an example to demonstrate the antagonism of this class of compounds for multiple receptors including dopamine D1-D5, α-adrenoreceptor, histamine H1, muscarinic M1-M5, and 5-HT2A, 5-HT2B, 5-HT2C, 5-HT3, and 5-HT6 receptors [271]. Olanzapine is an atypical antipsychotic agent, structurally related to clozapine and extensively used for the treatment of schizophrenia, bipolar disorder-associated mania, and the behavioral symptoms of Alzheimer's disease. The functional blockade of these multiple receptors contributes to the wide range of its pharmacologic and therapeutic activities, having relatively few side effects when compared to other antipsychotics agents. Thienobenzodiazepines (such as olanzapine) are characterized as multi-receptor-targeted-acting agents [271].

AVN-101 is a novel drug candidate designed as a multitargeted therapy for the treatment of cognitive, neurodegenerative, and anxiety disorders, with an emphasis on dual 5-HT_6 and 5-HT_7 receptor antagonism [272]. As part of the phase I clinical studies, bio-distribution, bioavailability, in vivo efficacy, and safety profile of the multimodal drug candidate, AVN-101 was investigated [272]. The results showed that AVN-101 is a very potent 5-HT_7 receptor antagonist (K_i = 153 pM), with slightly lesser potency toward 5-HT_6, 5-HT2A, and 5HT-2C receptors (K_i = 1.2–2.0 nM). AVN-101 also exhibits a rather high affinity toward histamine H1 (K_i = 0.58 nM) and adrenergic α2A, α2B, and α2C (K_i = 0.41-3.6 nM) receptors. AVN-101 shows a good oral bioavailability and facilitated brain-blood barrier permeability, low toxicity, and reasonable efficacy in animal models of CNS diseases. The phase I clinical study indicates the AVN-101 to be well tolerated when taken orally at doses of up to 20 mg daily. It does not dramatically influence plasma and urine biochemistry, nor does it prolong QT ECG interval, thus indicating low safety concerns. AVN-101 exhibited significant anxiolytic and antidepressant capabilities, and due to its anxiolytic and antidepressive activities, there is a strong rational for it to be tested in schizophrenia and multiple sclerosis in future large-scale clinical trials [272].

References

1. Feigin VL, Vos T, Nichols E, et al. The global burden of neurological disorders: translating evidence into policy. Lancet Neurol. 2020;19(3):255–65.
2. Lehn A, Gelauff J, Hoeritzauer I, et al. Functional neurological disorders: mechanisms and treatment. J Neurol. 2016;263(3):611–20.
3. Pena SA, Iyengar R, Eshraghi RS, et al. Gene therapy for neurological disorders: challenges and recent advancements. J Drug Target. 2020;28(2):111–28.
4. Gilman C, McSweeney C, Mao Y. The applications of pharmacogenomics to neurological disorders. Curr Mol Med. 2014;14(7):880–90.
5. Hesdorffer DC. Comorbidity between neurological illness and psychiatric disorders. CNS Spectr. 2016;21(3):230–8.
6. Hellmann-Regen J, Piber D, Hinkelmann K, et al. Depressive syndromes in neurological disorders. Eur Arch Psychiatry Clin Neurosci. 2013;263(Suppl 2):S123–36.
7. Benedetti F, Bernasconi A, Pontiggia A. Depression and neurological disorders. Curr Opin Psychiatry. 2006;19(1):14–8.
8. Lopez JA, González HM, Léger GC. Alzheimer's disease. Handb Clin Neurol. 2019;167:231–55.
9. Serý O, Povová J, Míšek I, et al. Molecular mechanisms of neuropathological changes in Alzheimer's disease: a review. Folia Neuropathol. 2013;51(1):1–9.
10. Briggs R, Kennelly SP, O'Neill D. Drug treatments in Alzheimer's disease. Clin Med (Lond). 2016;16(3):247–53.
11. Mufson EJ, Counts SE, Perez SE, et al. Cholinergic system during the progression of Alzheimer's disease: therapeutic implications. Expert Rev Neurother. 2008;8:1703–18.
12. Manning FC. Tacrine therapy for the dementia of Alzheimer's disease. Am Fam Physician. 1994;50:819–26.
13. National Institute for Health and Care Excellence Donepezil, galantamine, rivastigmine and memantine for the treatment of Alzheimer's disease. London: NICE; 2011. Available online at www.nice.org.uk/guidance/ta217/chapter/3-The-technologies
14. Birks JS, Harvey R. Donepezil for dementia due to Alzheimer's disease. Cochrane Database Syst Rev. 2003;(3):CD001190.
15. Lipton SA. Paradigm shift in neuroprotection by NMDA receptor blockade: memantine and beyond. Nat Rev Drug Discov. 2006;5(2):160–70.
16. Robinson DM, Keating GM. Memantine: a review of its use in Alzheimer's disease. Drugs. 2006;66(11):1515–34.
17. Areosa SA, Sherriff F. Memantine for dementia. Cochrane Database Syst Rev. 2003;3:CD003154.
18. Atri A, Molinuevo JL, Lemming O, et al. Memantine in patients with Alzheimer's disease receiving donepezil: new analyses of efficacy and safety for combination therapy. Alzheimers Res Ther. 2013;5:6.
19. Cavazzoni P. FDA's decision to approve new treatment for Alzheimer's disease. The US Food and Drug Administration Website https://www.fda.gov/drugs/news-events-human-drugs/fdas-decision-approve-new-treatment-alzheimers-disease. Accessed on 17 July 2021.
20. Selkoe DJ, Hardy J. The amyloid hypothesis of Alzheimer's disease at 25 years. EMBO Mol Med. 2016;8(6):595–608.
21. Sevigny J, Chiao P, Bussière T, et al. The antibody aducanumab reduces Abeta plaques in Alzheimer's disease. Nature. 2016;537(7618):50–6.
22. Tolar M, Abushakra S, Hey JA, et al. Aducanumab, gantenerumab, BAN2401, and ALZ-801-the first wave of amyloid-targeting drugs for Alzheimer's disease with potential for near term approval. Alzheimers Res Ther. 2020;12(1):95.
23. Kandiah N, Ong PA, Yuda T, et al. Treatment of dementia and mild cognitive impairment with or without cerebrovascular disease: expert consensus on the use of Ginkgo biloba extract, EGb 761. CNS Neurosci Ther. 2019;25(2):288–98.
24. McKeage K, Lyseng-Williamson KA. Ginkgo Biloba extract EGb 761 in the symptomatic treatment of mild-to-moderate dementia: a profile of its use. Drugs Ther Perspect. 2018;34(8):358–66.
25. Vaz M, Silvestre S. Alzheimer's disease: recent treatment strategies. Eur J Pharmacol. 2020;887:173554.
26. Schmitt B, Bernhardt T, Moeller H-J, et al. Combination therapy in Alzheimer's disease:

a review of current evidence. CNS Drugs. 2004;18(13):827–44.
27. Kabir MT, Uddin MS, Mamun AA, et al. Combination drug therapy for the management of Alzheimer's disease. Int J Mol Sci. 2020;21(9):3272.
28. Patel L, Grossberg GT. Combination therapy for Alzheimer's disease. Drugs Aging. 2011;28:539–56.
29. Atri A, Shaughnessy LW, Locascio JJ, et al. Long-term course and effectiveness of combination therapy in Alzheimer disease. Alzheimer Dis Assoc Disord. 2008;22:209–21.
30. Choi SH, Park KW, Na DL, et al. EXPECT Study Group tolerability and efficacy of memantine add-on therapy to rivastigmine transdermal patches in mild to moderate Alzheimer's disease: a multicenter, randomized, open-label, parallel-group study. Curr Med Res Opin. 2011;27:1375–83.
31. Cummings JL, Schneider E, Tariot PN, et al. Memantine MEM-MD-02 Study Group. Behavioral effects of memantine in Alzheimer disease patients receiving donepezil treatment. Neurology. 2006;67:57–63.
32. Farlow MR, Alva G, Meng X, et al. A 25-week, open-label trial investigating rivastigmine transdermal patches with concomitant memantine in mild-to-moderate Alzheimer's disease: a post hoc analysis. Curr Med Res Opin. 2010;26:263–9.
33. Dantoine T, Auriacombe S, Sarazin M, et al. Rivastigmine monotherapy and combination therapy with memantine in patients with moderately severe Alzheimer's disease who failed to benefit from previous cholinesterase inhibitor treatment. Int J Clin Pract. 2005;60:110–8.
34. Feldman HH, Schmitt FA, Olin JT. Memantine MEM-MD-02 Study Group. Activities of daily living in moderate-to-severe Alzheimer disease: an analysis of the treatment effects of memantine in patients receiving stable donepezil treatment. Alzheimer Dis Assoc Disord. 2006;20:263–8.
35. Lopez OL, Becker JT, Wahed AS, et al. Long-term effects of the concomitant use of memantine with cholinesterase inhibition in Alzheimer disease. J Neurol Neurosurg Psychiatry. 2009;80:600–7.
36. Porsteinsson A, Grossberg G, Mintzer J, et al. Memantine MEM-MD-12 Study Group. Memantine treatment in patients with mild to moderate Alzheimer's disease already receiving a cholinesterase inhibitor: a randomized, double-blind, placebo-controlled trial. Curr Alzheimer Res. 2008;5:83–9.
37. Schmitt FA, van Dyck CH, Wichems CH, et al. Memantine MEM-MD-02 Study Group. Cognitive response to memantine in moderate to severe Alzheimer disease patients already receiving donepezil: an exploratory reanalysis. Alzheimer Dis Assoc Disord. 2006;20:255–62.
38. Riepe MW, Adler G, Ibach B, et al. Domain-specific improvement of cognition on memantine in patients with Alzheimer's disease treated with rivastigmine. Dement Geriatr Cogn Disord. 2007;23:301–6.
39. Tariot PN, Farlow MR, Grossberg GT, et al. Memantine Study Group. Memantine treatment in patients with moderate to severe Alzheimer disease already receiving donepezil: a randomized controlled trial. JAMA. 2004;291:317.
40. Weiner MW, Sadowsky C, Saxton J, et al. Magnetic resonance imaging and neuropsychological results from a trial of memantine in Alzheimer's disease. Alzheimer's Dement. 2011;7:425–35.
41. Wilkinson D. A review of the effects of memantine on clinical progression in Alzheimer's disease. Int J Geriatr Psychiatry. 2012;27:769–76.
42. Parsons CG, Danysz W, Dekundy A, et al. Memantine and cholinesterase inhibitors: complementary mechanisms in the treatment of Alzheimer's disease. Neurotox Res. 2013;24:358–69.
43. Posadas I, Lopez-Hernandez B, Cena V. Nicotinic receptors in neurodegeneration. Curr Neuropharmacol. 2013;11:298–314.
44. Takada-Takatori Y, Kume T, Sugimoto M, et al. Acetylcholinesterase inhibitors used in treatment of Alzheimer's disease prevent glutamate neurotoxicity via nicotinic acetylcholine receptors and phosphatidylinositol 3-kinase cascade. Neuropharmacology. 2006;51:474–86.
45. Simoni E, Daniele S, Bottegoni G, et al. Combining galantamine and memantine in multitargeted, new chemical entities potentially useful in Alzheimer's disease. J Med Chem. 2012;55:9708–21.
46. Koola MM. Galantamine-Memantine combination in the treatment of Alzheimer's disease and beyond. Psychiatry Res. 2020;293:113409.
47. Matsunaga S, Kishi T, Nomura I, et al. The efficacy and safety of memantine for the treatment of Alzheimer's disease. Expert Opin Drug Saf. 2018;17(10):1053–61.
48. Matsunaga S, Kishi T, Iwata N. Combination therapy with cholinesterase inhibitors and memantine for Alzheimer's disease: a systematic review and meta-analysis. Int J Neuropsychopharmacol. 2014;18(5):pyu115.
49. Lipton SA, Choi YB, Pan ZH, et al. A redox-based mechanism for the neuroprotective and neurodestructive effects of nitric oxide and related nitroso-compounds. Nature. 1993;364:626–32.
50. Zheng H, Fridkin M, Youdim M. From single target to multitarget/network therapeutics in Alzheimer's therapy. Pharmaceuticals. 2014;7:113–35.
51. Serrano-Pozo A, William CM, Ferrer I, et al. Beneficial effect of human anti-amyloid-β active immunization on neurite morphology and tau pathology. Brain. 2010;133:1312–27.
52. Dodel R, Rominger A, Bartenstein P, et al. Intravenous immunoglobulin for treatment of mild-to-moderate Alzheimer's disease: a phase 2, randomised, double-blind, placebo-controlled, dose-finding trial. Lancet Neurol. 2013;12:233–43.
53. Talantova M, Sanz-Blasco S, Zhang X, et al. Aβ induces astrocytic glutamate release, extrasynaptic

NMDA receptor activation, and synaptic loss. Proc Natl Acad Sci USA. 2013;110:E2518–27.
54. Prati F, Bergamini C, Fato R, et al. Novel 8-hydroxyquinoline derivatives as multitarget compounds for the treatment of Alzheimer's disease. ChemMedChem. 2016;11:1284–95.
55. Suh SW, Jensen KB, Jensen MS, et al. Histochemically-reactive zinc in amyloid plaques, angiopathy, and degenerating neurons of Alzheimer's diseased brains. Brain Res. 2000;852:274–8.
56. Curtain CC, Ali F, Volitakis I, et al. Alzheimer's disease Amyloid-β binds copper and zinc to generate an allosterically ordered membrane-penetrating structure containing superoxide dismutase-like subunits. J Biol Chem. 2001;276:20466–73.
57. Fernández-Bachiller MI, Peérez C, González-Munñoz GC, et al. Novel Tacrine–8-Hydroxyquinoline hybrids as multifunctional agents for the treatment of Alzheimer's disease, with neuroprotective, cholinergic, antioxidant, and copper-complexing properties. J. Med. Chem. 2010;53:4927–37.
58. Peña-Altamira E, Prati F, Massenzio F, et al. Changing paradigm to target microglia in neurodegenerative diseases: from anti-inflammatory strategy to active immunomodulation. Expert Opin Ther Targets. 2016;20:627–40.
59. Weinstock M, Bejar C, Wang R-H, et al. TV3326, a novel neuroprotective drug with cholinesterase and monoamine oxidase inhibitory activities for the treatment of Alzheimer's disease. In: Advances in Research on Neurodegeneration. Vienna: Springer; 2000. p. 157–69.
60. Bar-Am O, Weinreb O, Amit T, et al. The novel cholinesterase-monoamine oxidase inhibitor and antioxidant, ladostigil, confers neuroprotection in neuroblastoma cells and aged rats. J Mol Neurosci. 2009;37:135–45.
61. Bar-Am O, Yogev-Falach M, Amit T, et al. Regulation of protein kinase C by the anti-Parkinson drug, MAO-B inhibitor, rasagiline and its derivatives, in vivo. J Neurochem. 2004;89:1119–25.
62. Weinreb O, Amit T, Bar-Am O, et al. Rasagiline: a novel anti-Parkinsonian monoamine oxidase-B inhibitor with neuroprotective activity. Prog Neurobiol. 2010;92:330–44.
63. Faux NG, Ritchie CW, Gunn A, et al. PBT2 rapidly improves cognition in Alzheimer's disease: additional phase II analyses. J. Alzheimer's Dis. 2010;20:509–16.
64. Savelieff MG, Detoma AS, Derrick JS, et al. The ongoing search for small molecules to study metal-Associated amyloid-β species in alzheimer's disease. Acc Chem Res. 2014;47:2475–82.
65. Gal S, Zheng H, Fridkin M, et al. Novel multifunctional neuroprotective iron chelator-monoamine oxidase inhibitor drugs for neurodegenerative diseases. In vivo selective brain monoamine oxidase inhibition and prevention of MPTP-induced striatal dopamine depletion. J Neurochem. 2005;95:79–88.
66. Huat TJ, Camats-Perna J, Newcombe EA, et al. Metal toxicity links to Alzheimer's disease and Neuroinflammation. J Mol Biol. 2019;431:1843–68.
67. Lee HJ, Park MK, Seo YR. Pathogenic mechanisms of heavy metal induced-Alzheimer's disease. Toxicol Environ Health Sci. 2018;10:1–10.
68. Lannfelt L, Blennow K, Zetterberg H, et al. Safety, efficacy, and biomarker findings of PBT2 in targeting Aβ as a modifying therapy for Alzheimer's disease: a phase IIa, double-blind, randomised, placebo-controlled trial. Lancet Neurol. 2008;7:779–86.
69. Gannon M, Wang Q. Complex noradrenergic dysfunction in Alzheimer's disease: low norepinephrine input is not always to blame. Brain Res. 2019;1702:12–6.
70. Gannon M, Che P, Chen Y, et al. Noradrenergic dysfunction in Alzheimer's disease. Front Neurosci. 2015;9:220.
71. Mohs RC, Shiovitz TM, Tariot PN, et al. Atomoxetine augmentation of cholinesterase inhibitor therapy in patients with alzheimer disease: 6-month, randomized, double-blind, placebo-controlled, parallel-trial study. Am J Geriatr Psychiatry. 2009;17:752–9.
72. Kalia LV, Lang AE. Parkinson's disease. Lancet. 2015;386(9996):896–912.
73. Tysnes OB, Storstein A. Epidemiology of Parkinson's disease. J Neural Transm (Vienna). 2017;124(8):901–5.
74. Poewe W, Seppi K, Tanner CM, et al. Parkinson disease. Nat Rev Dis Primers. 2017;3:17013.
75. Axelsen TM, Woldbye DPD. Gene therapy for Parkinson's disease, an update. J Parkinsons Dis. 2018;8(2):195–215.
76. Elkouzi A, Vedam-Mai V, Eisinger RS, et al. Emerging therapies in Parkinson disease – repurposed drugs and new approaches. Nat Rev Neurol. 2019;15(4):204–23.
77. Greenland JC, Williams-Gray CH, Barker RA. The clinical heterogeneity of Parkinson's disease and its therapeutic implications. Eur J Neurosci. 2019;49(3):328–38.
78. Lenka A, Padmakumar C, Pal PK. Treatment of older Parkinson's disease. Int Rev Neurobiol. 2017;132:381–405.
79. The National Collaborating Centre for Chronic Conditions, editor. Symptomatic pharmacological therapy in Parkinson's disease. Parkinson's Disease. London: Royal College of Physicians; 2006. p. 59–100.
80. Armstrong MJ, Okun MS. Diagnosis and treatment of Parkinson disease: a review. JAMA. 2020;323(6):548–60.
81. Akhtar MJ, Yar MS, Grover G, et al. Neurological and psychiatric management using COMT inhibitors: a review. Bioorganic Chem. 2020;94:103418.
82. Goldenberg MM. Medical management of Parkinson's disease. P & T. 2008;33(10):590–606.
83. Crosby N, Deane KH, Clarke CE. Amantadine in Parkinson's disease. The. Cochrane Database Syst Rev. 2003;(1):CD003468.

84. The National Collaborating Centre for Chronic Conditions, editor. Non-motor features of Parkinson's disease. Parkinson's Disease. London: Royal College of Physicians; 2006. p. 113–33.
85. Elbers RG, Verhoef J, van Wegen EE, et al. Interventions for fatigue in Parkinson's disease. Cochrane Database Systemat Rev. 2015;(10):CD010925.
86. Oertel WH. Recent advances in treating Parkinson's disease. F1000Research. 2017;6:260.
87. Hauser RA, Hsu A, Kell S, et al. IPX066 ADVANCE-PD investigators. Extended-release carbidopa-levodopa (IPX066) compared with immediate-release carbidopa-levodopa in patients with Parkinson's disease and motor fluctuations: a phase 3 randomised, double-blind trial. Lancet Neurol. 2013;12(4):346–56.
88. Verhagen Metman L, Stover N, Chen C, et al. Gastroretentive carbidopa/levodopa, DM-1992, for the treatment of advanced Parkinson's disease. Mov Disord. 2015;30(9):1222–8.
89. Nyholm D, Lewander T, Gomes-Trolin C, et al. Pharmacokinetics of levodopa/carbidopa microtablets versus levodopa/benserazide and levodopa/carbidopa in healthy volunteers. Clin Neuropharmacol. 2012;35(3):111–7.
90. Da Prada M, Keller HH, Pieri L, et al. The pharmacology of Parkinson's disease: basic aspects and recent advances. Experientia. 1984;40(11):1165–72.
91. Barbeau A, Roy M. Six-year results of treatment with levodopa plus benzerazide in Parkinson's disease. Neurology. 1976;26(5):399–404.
92. Binde CD, Tvete IF, Gåsemyr J, et al. A multiple treatment comparison meta-analysis of monoamine oxidase type B inhibitors for Parkinson's disease. Br J Clin Pharmacol. 2018;84(9):1917–27.
93. Jiang DQ, Li MX, Jiang LL, et al. Comparison of selegiline and levodopa combination therapy versus levodopa monotherapy in the treatment of Parkinson's disease: a meta-analysis. Aging Clin Exp Res. 2020;32(5):769–79.
94. Jiang DQ, Wang HK, Wang Y, et al. Rasagiline combined with levodopa therapy versus levodopa monotherapy for patients with Parkinson's disease: a systematic review. Neurol Sci. 2020;41(1):101–9.
95. Rascol O, Brooks DJ, Melamed E, et al. LARGO Study Group. Rasagiline as an adjunct to levodopa in patients with Parkinson's disease and motor fluctuations (LARGO, Lasting effect in Adjunct therapy with Rasagiline Given Once daily, study): a randomised, double-blind, parallel-group trial. Lancet. 2005;365(9463):947–54.
96. Cattaneo C, Sardina M, Bonizzoni E. Safinamide as add-on therapy to levodopa in mid- to late-stage Parkinson's disease fluctuating patients: post hoc analyses of studies 016 and SETTLE. J Parkinsons Dis. 2016;6(1):165–73.
97. Hattori N, Tsuboi Y, Yamamoto A, et al. ME2125-3 Study Group. Efficacy and safety of safinamide as an add-on therapy to L-dopa for patients with Parkinson's disease: a randomized, double-blind, placebo-controlled, phase II/III study. Parkinsonism Relat Disord. 2020;75:17–23.
98. Talati R, Baker WL, Patel AA, et al. Adding a dopamine agonist to preexisting levodopa therapy vs. levodopa therapy alone in advanced Parkinson's disease: a meta analysis. Int J Clin Pract. 2009;63(4):613–23.
99. Jiang DQ, Zang QM, Jiang LL, Wang Y, Li MX, Qiao JY. Comparison of pramipexole and levodopa/benserazide combination therapy versus levodopa/benserazide monotherapy in the treatment of Parkinson's disease: a systematic review and meta-analysis. Naunyn Schmiedebergs Arch Pharmacol. 2021;394:1893–905.
100. Salat D, Tolosa E. Levodopa in the treatment of Parkinson's disease: current status and new developments. J Parkinsons Dis. 2013;3(3):255–69.
101. Seeberger LC, Hauser RA. Levodopa/carbidopa/entacapone in Parkinson's disease. Expert Rev Neurother. 2009;9(7):929–40.
102. Kumari S, Mishra C, Tiwari M. Polypharmacological drugs in the treatment of Epilepsy: the comprehensive review of marketed and new emerging molecules. Curr Pharm Des. 2016;22(21):3212–25.
103. Fisher R, van Emde BW, Blume W, et al. Epileptic seizures and epilepsy: definitions proposed by the International League Against Epilepsy (ILAE) and the International Bureau for Epilepsy (IBE). Epilepsia. 2005;46(4):470–2.
104. WHO. Epilepsy Fact sheet. Archived from the original on 11 March 2016. Retrieved 4 March 2016.
105. Hammer GD, McPhee SJ, editors. Pathophysiology of disease: an introduction to clinical medicine. 6th ed. New York: McGraw-Hill Medical; 2010. ISBN 978-0-07-162167-0
106. Goldberg EM, Coulter DA. Mechanisms of epileptogenesis: a convergence on neural circuit dysfunction. Nat Rev Neurosci. 2013;14(5):337–49.
107. Pandolfo M. Genetics of epilepsy. Semin Neurol. 2011;31(5):506–18.
108. Ghosh S, Sinha JK, Khan T, et al. Pharmacological and therapeutic approaches in the treatment of epilepsy. Biomedicines. 2021;9(5):470.
109. GBD 2013 Mortality Causes of Death Collaborators. Global, regional, and national age-sex specific all-cause and cause-specific mortality for 240 causes of death, 1990–2013: a systematic analysis for the Global Burden of Disease Study 2013. Lancet. 2015;385(9963):117–71.
110. Brodie MJ, Elder AT, Kwan P. Epilepsy in later life. Lancet Neurol. 2009;8(11):1019–30.
111. Holmes TR, Browne GL. Handbook of epilepsy. 4th ed. Philadelphia: Lippincott Williams & Wilkins; 2008. p. 7. ISBN 978-0-7817-7397-3
112. Wyllie's treatment of epilepsy: principles and practice (5th ed). Philadelphia: Wolters Kluwer/Lippincott Williams & Wilkins; 2010. ISBN 978-1-58255-937-7. Archived from the original on 24 June 2016.

113. Newton CR, Garcia HH. Epilepsy in poor regions of the world. Lancet. 2012;380(9848):1193–201.
114. Wilden JA, Cohen-Gadol AA. Evaluation of first nonfebrile seizures. Am Family Physician. 2012;86(4):334–40.
115. Berg AT. Risk of recurrence after a first unprovoked seizure. Epilepsia. 2008;49(Suppl1):13–8.
116. Devlin A, Odell M, Charlton J, et al. Epilepsy and driving: current status of research. Epilepsy Res. 2012;102(3):135–52.
117. National Clinical Guideline Centre. The Epilepsies: the diagnosis and management of the epilepsies in adults and children in primary and secondary care. National Institute for Health and Clinical Excellence; 2012. p. 21–8.
118. Wyllie E. Wyllie's treatment of Epilepsy: principles and practice. Lippincott Williams & Wilkins; 2012. p. 187. ISBN 978-1-4511-5348-4
119. National Clinical Guideline Centre. The Epilepsies: the diagnosis and management of the epilepsies in adults and children in primary and secondary care. National Institute for Health and Clinical Excellence; 2012. p. 57–83.
120. Kumar GP, Rambhau D, Apte SS. Development of NDDS of carbamazepine in epilepsy for medical discovery of formulation significance. J Med Discov. 2017;2(2):1–13.
121. Nevitt SJ, Marson AG, Tudur SC. Carbamazepine versus phenytoin monotherapy for epilepsy: an individual participant data review. Cochrane Database Syst Rev. 2019;7:CD001911.
122. Nevitt SJ, Marson AG, Weston J, et al. Sodium valproate versus phenytoin monotherapy for epilepsy: an individual participant data review. Cochrane Database Syst Rev. 2018;2018(8):CD001769.
123. Powell G, Saunders M, Rigby A, et al. Immediate-release versus controlled-release carbamazepine in the treatment of epilepsy. Cochrane Database Syst Rev. 2016;12:CD007124.
124. Mani J. Combination therapy in epilepsy: what, when, how and what not! J Assoc Physicians India. 2013;61(8 Suppl):40–4.
125. Leppik IE. Monotherapy and polypharmacy. Neurology. 2000;55(11 Suppl 3):S25–9.
126. Moeller JJ, Rahey SR, Sadler RM. Lamotrigine-valproic acid combination therapy for medically refractory epilepsy. Epilepsia. 2009;50(3):475–9.
127. Ferrie CD, Robinson RO, Knott C, Panayiotopoulos CP. Lamotrigine as an add-on drug in typical absence seizures. Acta Neurol Scand. 1995;91:200–2.
128. Panayiotopoulos CP, Ferrie CD, Knott C, et al. Interaction of lamotrigine with sodium valproate. Lancet. 1993;341:445.
129. Pisani F, DiPerri R, Perucca E, et al. Interaction of lamotrigine with sodium valproate. Lancet. 1993;341:1224.
130. Ferrie CD, Panayiotopoulos CP. Therapeutic interaction of lamotrigine and sodium valproate in intractable myoclonic epilepsy. Seizure. 1994;3:157–9.
131. Deckers CL, Czuczwar SJ, Hekster YA, et al. Selection of antiepileptic drug polytherapy based on mechanisms of action: the evidence reviewed. Epilepsia. 2000;41:1364–74.
132. Jonker DM, Voskuyl RA, Danhof M. Synergistic combinations of anticonvulsant agents: what is the evidence from animal experiments? Epilepsia. 2007;48:412–34.
133. Klitgaard H, Knudsen ML, Jackson HC. Synergism between drugs with diffrent mechanisms of action against audiogenic seizures in DBA/2 mice. Epilepsia. 1993;34(Suppl. 6):93–4.
134. Löscher W, Rundfeldt C, Honack D. Low doses of NMDA receptor antagonists synergistically increase the anticonvulsant effect of the AMPA receptor antagonist NBQX in the kindling model of epilepsy. Eur J Neurosci. 1993;5:1545–50.
135. Löscher W, Honack D. Over-additive anticonvulsant effect of memantine and NBQX in kindled rats. Eur J Pharmacol. 1994;259:R3–5.
136. Gasior M, Carter RB, Goldberg SR, et al. Anticonvulsant and behavioral effects of neuroactive steroids alone and in conjunction with diazepam. J Pharmacol Exp Ther. 1997;282:543–53.
137. Kaminski RM, Matagne A, Patsalos PN, et al. Benefit of combination therapy in epilepsy: a review of the preclinical evidence with levetiracetam. Epilepsia. 2009;50(3):387–97.
138. De Smedt T, Raedt R, Vonck K, et al. Levetiracetam: part II, the clinical profile of a novel anticonvulsant drug. CNS Drug Rev. 2007;13:57–8.
139. Compston A, Coles A. Multiple sclerosis. Lancet. 2008;372(9648):1502–17.
140. Compston A, Coles A. Multiple sclerosis. Lancet. 2002;359(9313):1221–31.
141. Murray ED, Buttner EA, Price BH. Depression and psychosis in neurological practice. In: Daroff R, Fenichel G, Jankovic J, Mazziotta J, editors. Bradley's neurology in clinical practice. 6th ed. Philadelphia: Elsevier/Saunders; 2012. ISBN 978-1-4377-0434-1.
142. Ascherio A, Munger KL. Environmental risk factors for multiple sclerosis. Part I: the role of infection. Ann Neurol. 2007;61(4):288–99.
143. Tsang BK, Macdonell R. Multiple sclerosis- diagnosis, management and prognosis. Aust Fam Physician. 2011;40(12):948–55.
144. Berer K, Krishnamoorthy G. Microbial view of central nervous system autoimmunity. FEBS Lett. 2014;588(22):4207–13.
145. GBD 2015 Disease and Injury Incidence and Prevalence Collaborators. Global, regional, and national incidence, prevalence, and years lived with disability for 310 diseases and injuries, 1990–2015: a systematic analysis for the Global Burden of Disease Study 2015. Lancet. 2016;388(10053):1545–602.
146. World Health Organization. Atlas: multiple sclerosis resources in the World 2008. Geneva: World Health Organization; 2008. p. 15–6.

147. GBD 2015 Mortality and Causes of Death Collaborators. Global, regional, and national life expectancy, all-cause mortality, and cause-specific mortality for 249 causes of death, 1980–2015: a systematic analysis for the Global Burden of Disease Study 2015. Lancet. 2016;388(10053):1459–544.
148. Milo R, Kahana E. Multiple sclerosis: geoepidemiology, genetics and the environment. Autoimmunity Rev. 2010;9(5):A387–94.
149. Clanet M. Jean-Martin Charcot. 1825 to 1893. Int MS J. 2008;15(2):59–61.
150. Cohen JA. Emerging therapies for relapsing multiple sclerosis. Arch Neurol. 2009;66(7):821–8.
151. Tavazzi E, Rovaris M, La Mantia L. Drug therapy for multiple sclerosis. CMAJ. 2014;186(11):833–40.
152. Huntley A. A review of the evidence for efficacy of complementary and alternative medicines in MS. Int MS J. 2006;13(1):5–12.
153. Weinshenker BG. Natural history of multiple sclerosis. Ann Neurol. 1994;36(Suppl):S6–11.
154. Miller AE. Multiple sclerosis: where will we be in 2020? The Mount Sinai J Med. 2011;78(2):268–79.
155. Jeffrey S. CONCERTO: a Third Phase 3 trial for Laquinimod in MS. Medscape Medical News. Archived from the original on 17 September 2012. Retrieved 21 May 2013.
156. He D, Han K, Gao X, et al. Laquinimod for multiple sclerosis. Cochrane Database Syst Rev. 2013;(8):CD010475.
157. Kieseier BC, Calabresi PA. PEGylation of interferon-β-1a: a promising strategy in multiple sclerosis. CNS Drugs. 2012;26(3):205–14.
158. Biogen Idec announces positive top-line results from Phase 3 study of Peginterferon Beta-1a in Multiple Sclerosis (Press release). Biogen Idec. 24 January 2013. Archived from the original on 4 October 2013. Retrieved 21 May 2013.
159. Gold SM, Voskuhl RR. Estrogen treatment in multiple sclerosis. J Neurological Sci. 2009;286(1–2):99–103.
160. Voskuhl RR, Wang H, Wu TC, et al. Estriol combined with glatiramer acetate for women with relapsing-remitting multiple sclerosis: a randomised, placebo-controlled, phase 2 trial. Lancet Neurol. 2016;15(1):35–46.
161. He D, Guo R, Zhang F, et al. Rituximab for relapsing-remitting multiple sclerosis. Cochrane Database Syst Rev. 2013;2013(12):CD009130.
162. Saidha S, Eckstein C, Calabresi PA. New and emerging disease modifying therapies for multiple sclerosis. Ann New York Acad Sci. 2012;1247(1):117–37.
163. Winslow R. After 40-year odyssey, first drug for aggressive MS wins FDA approval. STAT. Archived from the original on 1 April 2017.
164. Milo R, Panitch H. Combination therapy in multiple sclerosis. J Neuroimmunol. 2011;231(1–2):23–31.
165. Luessi F, Siffrin V, Zipp F. Neurodegeneration in multiple sclerosis: novel treatment strategies. Expert Rev Neurotherapeutics. 2012;12(9):1061–76.
166. Yang C, Hao Z, Zhang L, et al. Sodium channel blockers for neuroprotection in multiple sclerosis. Cochrane Database Syst Rev. 2015;2015(10):CD010422.
167. Fernández O. Combination therapy in multiple sclerosis. J Neurol Sci. 2007;259(1–2):95–103.
168. Conway D, Cohen JA. Combination therapy in multiple sclerosis. Lancet Neurol. 2010;9(3):299–308.
169. Stuart WH. Combination therapy for the treatment of multiple sclerosis: challenges and opportunities. Curr Med Res Opin. 2007;23(6):1199–208.
170. Jeffery DR. Use of combination therapy with immunomodulators and immunosuppressants in treating multiple sclerosis. Neurology. 2004;63(12 Suppl 6):S41–6.
171. Etemadifar M, Kazemi M, Chitsaz A, et al. Mycophenolate mofetil in combination with interferon beta-1a in the treatment of relapsing-remitting multiple sclerosis: a preliminary study J Res Med Sci 2011; 16(1): 1–5.
172. Remington GM, Treadaway K, Frohman T, et al. A one-year prospective, randomized, placebo-controlled, quadruple-blinded, phase II safety pilot trial of combination therapy with interferon beta-1a and mycophenolate mofetil in early relapsing-remitting multiple sclerosis (TIME MS). Ther Adv Neurol Disord. 2010;3(1):3–13.
173. Ravnborg M, Sørensen PS, Andersson M, et al. Methylprednisolone in combination with interferon beta-1a for relapsing-remitting multiple sclerosis (MECOMBIN study): a multicentre, double-blind, randomised, placebo-controlled, parallel-group trial Lancet Neurol 2010 l; 9(7): 672–680.
174. van Os J, Kapur S. Schizophrenia. Lancet. 2009;374(9690):635–45.
175. Insel TR. Rethinking schizophrenia. Nature. 2010;468(7321):187–93.
176. Elert E. Aetiology: searching for schizophrenia's roots. Nature. 2014;508(7494):S2–3.
177. Owen MJ, Sawa A, Mortensen PB. Schizophrenia. Lancet. 2016;388(10039):86–97.
178. Laursen TM, Nordentoft M, Mortensen PB. Excess early mortality in schizophrenia. Ann Rev Clin Psychol. 2014;10:425–48.
179. Javitt DC. Balancing therapeutic safety and efficacy to improve clinical and economic outcomes in schizophrenia: a clinical overview. Am J Managed Care. 2014;20(8 Suppl):S160–5.
180. James SL, Abate D. Global, regional, and national incidence, prevalence, and years lived with disability for 354 diseases and injuries for 195 countries and territories, 1990–2017: a systematic analysis for the Global Burden of Disease Study 2017. Lancet. 2018;392(10159):1789–858.
181. Schizophrenia Fact sheet. www.who.int. 4 October 2019. Retrieved 22 January 2020.
182. Saha S, Chant D, McGrath J. A systematic review of mortality in schizophrenia: is the differential mortality gap worsening over time? Arch General Psychiatry. 2007;4(10):1123–31.

183. van de Leemput J, Hess JL, Glatt SJ, et al. Genetics of schizophrenia: historical insights and prevailing evidence. Adv Genetics. 2016;96:99–141.
184. Siskind D, Siskind V, Kisely S. Clozapine response rates among people with treatment-resistant schizophrenia: data from a systematic review and meta-analysis. Can J Psychiatry. 2017;62(11):772–7.
185. Leucht S, Cipriani A, Spineli L, et al. Comparative efficacy and tolerability of 15 antipsychotic drugs in schizophrenia: a multiple-treatments meta-analysis. Lancet. 2013;382(9896):951–62.
186. Becker T, Kilian R. Psychiatric services for people with severe mental illness across western Europe: what can be generalized from current knowledge about differences in provision, costs and outcomes of mental health care? Acta Psychiatrica Scandinavica Supplementum. 2006;113(429):9–16.
187. Narayan KK, Kumar DS. Disability in a group of long-stay patients with schizophrenia: experience from a mental hospital. Indian J Psychological Med. 2012;34(1):70–5.
188. Schizophrenia – Treatment. nhs.uk. 23 October 2017. Retrieved 8 January 2020.
189. Ortiz-Orendain J, Covarrubias-Castillo SA, Vazquez-Alvarez AO, et al. Modafinil for people with schizophrenia or related disorders. Cochrane Database Syst Rev. 2019;12:CD008661.
190. Lally J, JH MC. Antipsychotic medication in schizophrenia: a review. Br Med Bull. 2015;114(1):169–79.
191. Harrow M, Jobe TH. Does long-term treatment of schizophrenia with antipsychotic medications facilitate recovery? Schizophrenia Bull. 2013;39(5):962–5.
192. Li P, Snyder GL, Vanover KE. Dopamine targeting drugs for the treatment of schizophrenia: past, present and future. Curr Topics Medicinal Chem. 2016;16(29):3385–403.
193. Lowe EJ, Ackman ML. Impact of tobacco smoking cessation on stable clozapine or olanzapine treatment. Ann Pharmacotherapy. 2010;44(4):727–32.
194. Wolff-Menzler C, Hasan A, Malchow B, et al. Combination therapy in the treatment of schizophrenia. Pharmacopsychiatry. 2010;43(4):122–9.
195. Baandrup L. Polypharmacy in schizophrenia. Basic Clin Pharmacol Toxicol. 2020;126(3):183–92.
196. Faden J, Kiryankova-Dalseth N, Barghini R, et al. Does antipsychotic combination therapy reduce the risk of hospitalization in schizophrenia? Expert Opin Pharmacother. 2021;22(5):635–46.
197. Lader M. Effectiveness of benzodiazepines: do they work or not? Expert Rev Neurotherapeutics. 2008;8(8):1189–91.
198. Gallego JA, Bonetti J, Zhang J, et al. Prevalence and correlates of antipsychotic polypharmacy: a systematic review and meta-regression of global and regional trends from the 1970s to 2009. Schizophr Res. 2012;138:18–28.
199. Taylor DM, Smith L. Augmentation of clozapine with a second antipsychotic–a meta-analysis of randomized, placebo-controlled studies. Acta Psychiatr Scand. 2009;119:419–25.
200. Baandrup L, Gasse C, Jensen VD, et al. Antipsychotic polypharmacy and risk of death from natural causes in patients with schizophrenia: a population-based nested case-control study. J Clin Psychiatry. 2010;71:103–8.
201. Lin CH, Wang FC, Lin SC, et al. Antipsychotic combination using low-dose antipsychotics is as efficacious and safe as, but cheaper, than optimal-dose monotherapy in the treatment of schizophrenia: a randomized, double-blind study. Int Clin Psychopharmacol. 2013;28(5):267–74.
202. Lin CH, Wang FC, Lin SC, et al. A randomized, double-blind, comparison of the efficacy and safety of low-dose olanzapine plus low-dose trifluoperazine versus full-dose olanzapine in the acute treatment of schizophrenia. Schizophr Res. 2017;185:80–7.
203. Tiihonen J, Suokas JT, Suvisaari JM, et al. Polypharmacy with antipsychotics, antidepressants, or benzodiazepines and mortality in schizophrenia. Arch Gen Psychiatry. 2012;69:476–83.
204. Tiihonen J, Mittendorfer-Rutz E, Torniainen M, et al. Mortality and cumulative exposure to antipsychotics, antidepressants, and benzodiazepines in patients with schizophrenia: an observational follow-up study. Am J Psychiatry. 2016;173:600–6.
205. Jennings L. Chapter 4: Antidepressants. In: Grossberg GT, Kinsella LJ, editors. Clinical psychopharmacology for neurologists: a practical guide. Springer; 2018. p. 45–71.
206. Correll CU, Rubio JM, Inczedy-Farkas G, et al. Efficacy of 42 pharmacologic cotreatment strategies added to antipsychotic monotherapy in schizophrenia: systematic overview and quality appraisal of the meta-analytic evidence. JAMA Psychiatry. 2017;74:675–84.
207. Huband N, Ferriter M, Nathan R, et al. Antiepileptics for aggression and associated impulsivity. Cochrane Database Syst Rev. 2010;2010:CD003499.
208. Stroup TS, Gerhard T, Crystal S, et al. Comparative effectiveness of adjunctive psychotropic medications in patients with schizophrenia. JAMA Psychiatry. 2019;76:508–15.
209. Calhoun A, King C, Khoury R, et al. An evaluation of memantine ER + donepezil for the treatment of Alzheimer's disease. Expert Opin Pharmacother. 2018;19(15):1711–7.
210. Deardorff WJ, Grossberg GT. A fixed-dose combination of memantine extended-release and donepezil in the treatment of moderate-to-severe Alzheimer's disease. Drug Des Devel Ther. 2016;10:3267–79.
211. Greig SL. Memantine ER/donepezil: a review in Alzheimer's disease. CNS Drugs. 2015;29(11):963–70.
212. Boinpally R, Chen L, Zukin SR, et al. A novel once-daily fixed-dose combination of memantine extended release and donepezil for the treatment of moderate to severe Alzheimer's disease: two phase

I studies in healthy volunteers. Clin Drug Investig. 2015;35(7):427–35.
213. Chen R, Chan PT, Chu H, et al. Treatment effects between monotherapy of donepezil versus combination with memantine for Alzheimer disease: a meta-analysis. PLoS One. 2017;12(8):e0183586.
214. Owen RT. Memantine and donepezil: a fixed drug combination for the treatment of moderate to severe Alzheimer's dementia. Drugs Today (Barc). 2016;52(4):239–48.
215. Shua-Haim J, Smith J, Picard F, et al. Steady-state pharmacokinetics of rivastigmine in patients with mild to moderate Alzheimer's disease not affected by co-administration of memantine: an open-label, crossover, single-centre study. Clin Drug Investig. 2008;28(6):361–74.
216. Stocchi F, Vacca L. A systematic review on the clinical experience with melevodopa/carbidopa fixed combination in patients with Parkinson disease. Minerva Med. 2019;110(6):575–58.
217. Sun L, McDonnell D, von Moltke L. Pharmacokinetics and short-term safety of ALKS 3831, a fixed-dose combination of Olanzapine and Samidorphan, in adult subjects with Schizophrenia. Clin Ther. 2018;40(11):1845–54.e2
218. Ramsay RR, Popovic-Nikolic MR, Nikolic K, et al. A perspective on multi-target drug discovery and design for complex diseases. Clin Transl Med. 2018;7:3.
219. Bannwart LM, Carter DS, Cai HY, et al. Novel 3,3-disubstituted pyrrolidines as selective triple serotonin/norepinephrine/dopamine reuptake inhibitors. Bioorg Med Chem Lett. 2008;18(23):6062–6.
220. Seeger TF, Seymour PA, Schmidt AW, et al. Ziprasidone (CP-88,059): a new antipsychotic with combined dopamine and serotonin receptor antagonist activity. J Pharmacol Exp Ther. 1995;275(1):101–13.
221. Felice D, Gardier AM, Sanchez C, et al. Innovative solutions to the development of novel antidepressants. In: Carvalho AF, Reus GZ, de Quevedo JL, editors. Frontiers in drug discovery: the search for antidepressants – an integrative view of drug discovery. Bentham Science Publishers, United Arab Emirates. 2017; V2: p. 1–40.
222. Marotta G, Basagni F, Rosini M, et al. Memantine derivatives as multitarget agents in Alzheimer's disease. Molecules. 2020;25(17):4005.
223. Gontijo VS, Viegas FPD, Ortiz CJC, et al. Molecular hybridization as a tool in the design of multi-target directed drug candidates for neurodegenerative diseases. Curr Neuropharmacol. 2020;18(5):348–407.
224. Spilovska K, Korabecny J, Nepovimova E, et al. Multitarget tacrine hybrids with neuroprotective properties to confront Alzheimer's disease. Curr Top Med Chem. 2017;17(9):1006–26.
225. Maramai S, Benchekroun M, Gabr MT, et al. Multitarget therapeutic strategies for Alzheimer's disease: review on emerging target combinations. Biomed Res Int. 2020;2020:5120230.
226. Ibrahim MM, Gabr MT. Multitarget therapeutic strategies for Alzheimer's disease. Neural Regen Res. 2019;14(3):437–40.
227. Benek O, Korabecny J, Soukup O. A perspective on multi-target drugs for Alzheimer's disease. Trends Pharmacol Sci. 2020;41(7):434–45.
228. Rosini M. Polypharmacology: the rise of multitarget drugs over combination therapies. Future Med Chem. 2014;6:485–7.
229. Romero A, Cacabelos R, Oset-Gasque MJ, et al. Novel tacrine-related drugs as potential candidates for the treatment of Alzheimer's Disease. Bioorg Med Chem Lett. 2013;23:1916–22.
230. Spilovska K, Korabecny J, Kral J, et al. 7-Methoxytacrine-adamantylamine heterodimers as cholinesterase inhibitors in Alzheimer's disease treatment–synthesis, biological evaluation and molecular modeling studies. Molecules. 2013;18:2397–418.
231. Spilovska K, Korabecny J, Horova A, et al. Design, synthesis and in vitro testing of 7-methoxytacrine-amantadine analogues: a novel cholinesterase inhibitors for the treatment of Alzheimer's disease. Med Chem Res. 2015;24:2645–55.
232. Giménez-Llort L, Ratia M, Pérez B, et al. Behavioural effects of novel multitarget anticholinesterasic derivatives in Alzheimer's disease. Behav Pharmacol. 2017;28(2 and 3-Spec Issue):124–31.
233. McGehee DS, Heath MJ, Gelber S, et al. Nicotine enhancement of fast excitatory synaptic transmission in CNS by presynaptic receptors. Science. 1995;269:1692–6.
234. Pieper AA, McKnight SL, Ready JM. P7C3 and an unbiased approach to drug discovery for neurodegenerative diseases. Chem Soc Rev. 2014;43:6716–26.
235. Bachurin SO, Shevtsova EF, Makhaeva GF, et al. Novel conjugates of aminoadamantanes with carbazole derivatives as potential multitarget agents for AD treatment. Sci Rep. 2017;7:45627.
236. Ballatore C, Brunden KR, Huryn DM, et al. Microtubule stabilizing agents as potential treatment for Alzheimer's disease and related neurodegenerative tauopathies. J Med Chem. 2012;55:8979–96.
237. Benchekroun M, Romero A, Egea J, et al. The antioxidant additive approach for Alzheimer's disease therapy: new Ferulic (Lipoic) acid plus Melatonin modified Tacrines as Cholinesterases inhibitors, direct antioxidants, and nuclear factor (Erythroid-Derived 2)-Like 2 activators. J Med Chem. 2016;59:9967–73.
238. Rosini M, Simoni E, Caporaso R, et al. Merging memantine and ferulic acid to probe connections between NMDA receptors, oxidative stress and amyloid-β peptide in Alzheimer's disease. Eur J Med Chem. 2019;180:111–20.
239. Jalili-Baleh L, Forootanfar H, Küçükkılınç TT, et al. Design, synthesis and evaluation of novel multi-target-directed ligands for treatment of Alzheimer's disease based on coumarin and lipoic acid scaffolds. Eur J Med Chem. 2018;152:600–14.

240. Sozio P, Cerasa LS, Laserra S, et al. Memantine-sulfur containing antioxidant conjugates as potential prodrugs to improve the treatment of Alzheimer's disease. Eur J Pharm Sci. 2013;49:187–98.
241. Edmonson DE, Binda C. Monoamine oxidases. In: Harris JR, Boekema EJ, editors. Membrane protein complexes: structure and function. Singapore: Springer; 2018. p. 117–40.
242. Kumamoto T, Nakajima M, Uga R, et al. Design, synthesis, and evaluation of polyamine-memantine hybrids as NMDA channel blockers. Bioorg Med Chem. 2018;26:603–8.
243. Jin L, Sugiyama H, Takigawa M, et al. Comparative studies of anthraquinone- and anthracene-tetraamines as blockers of N-methyl-D-aspartate receptors. J Pharmacol Exp Ther. 2007;320:47–55.
244. Takayama H, Yaegashi Y, Kitajima M, et al. Design, synthesis, and biological evaluation of tricyclic heterocycle-tetraamine conjugates as potent NMDA channel blockers. Bioorg Med Chem Lett. 2007;17:4729–32.
245. Sestito S, Nesi G, Pi R, et al. Hydrogen sulfide: a worthwhile tool in the design of new multitarget drugs. Front Chem. 2017;5:72.
246. Basagni F, Lanni C, Minarini A, et al. Lights and shadows of electrophile signaling: focus on the Nrf2-Keap1 pathway. Future Med Chem. 2019;11:707–21.
247. Shefa U, Yeo SG, Kim MS, et al. Role of gasotransmitters in oxidative stresses, neuroinflammation, and neuronal repair. Biomed Res Int. 2017;2017:1689341.
248. Sestito S, Daniele S, Pietrobono D, et al. Memantine prodrug as a new agent for Alzheimer's disease. Sci Rep. 2019;9:4612.
249. Marzo A, Dal Bo L, Monti NC, Crivelli F, Ismaili S, Caccia C, et al. Pharmacokinetics and pharmacodynamics of safinamide, a neuroprotectant with antiparkinsonian and anticonvulsant activity. Pharmacol Res. 2004;50(1):77–85.
250. Fabbri M, Rosa MM, Abreu D, et al. Clinical pharmacology review of safinamide for the treatment of Parkinson's disease. Neurodegener Dis Manag. 2015;5(6):481–96.
251. Youdim MB, Kupershmidt L, Amit T, et al. Promises of novel multi-target neuroprotective and neurorestorative drugs for Parkinson's disease. Parkinsonism Relat Disord. 2014 Jan;20(Suppl 1):S132–6.
252. Weinreb O, Amit T, Mandel S, et al. Novel therapeutic approach for neurodegenerative pathologies: multitarget iron-chelating drugs regulating hypoxia-inducible factor 1 signal transduction pathway. Neurodegener Dis. 2012;10(1–4):112–5.
253. Zheng H, Gal S, Weiner LM, et al. Novel multifunctional neuroprotective iron chelator-monoamine oxidase inhibitor drugs for neurodegenerative diseases: in vitro studies on antioxidant activity, prevention of lipid peroxide formation and monoamine oxidase inhibition. J Neurochem. 2005;95:68–78.
254. Kupershmidt L, Amit T, Bar-Am O, et al. Multi-target, neuroprotective and neurorestorative M30 improves cognitive impairment and reduces Alzheimer's-like neuropathology and age-related alterations in mice. Mol Neurobiol. 2012;46:217–20.
255. Kupershmidt L, Amit T, Bar-Am O, et al. Neuroprotection by the multitarget iron chelator M30 on age-related alterations in mice. Mech Ageing Dev. 2012;133:267–74.
256. Löscher W. Single-target versus multi-target drugs versus combinations of drugs with multiple targets: preclinical and clinical evidence for the treatment or prevention of epilepsy. Front Pharmacol. 2021;12:730257.
257. Chen Z, Brodie MJ, Liew D, et al. Treatment outcomes in patients with newly diagnosed epilepsy treated with established and new antiepileptic drugs: a 30-Year longitudinal cohort study. JAMA Neurol. 2018;75:279–86.
258. Devinsky O, Vezzani A, O'Brien TJ, et al. Epilepsy. Nat Rev Dis Primers. 2018;4:18024.
259. Muglia P, Hannestad J, Brandt C, et al. Padsevonil randomized phase IIa trial in treatment-resistant focal epilepsy: a translational approach. Brain Commun. 2020;2:fcaa183.
260. French JA. Cenobamate for focal seizures – a game changer? Nat Rev Neurol. 2020;1:133–4.
261. Werhahn K, Toledo M, Rademacher M, et al. Efficacy and safety of adjunctive padsevonil in adults with drug-resistant focal seizures: a double-blind, randomized, placebo-controlled dose-finding trial. 2020; AES Abstr.. Available at: https://cms.aesnet.org/abstractslisting/efficacy-and-safety-of-adjunctive-padsevonil-in-adults-with-drug-resistant-focal-seizures%2D%2Da-double-blind%2D%2Drandomized%2D%2Dplacebo-controlled-dose-finding-trial
262. French JA. Do you believe in magic (bullets)? Epilepsy Curr. 2020;20:S24–6.
263. Löscher W, Sills GJ, White HS. The ups and downs of alkyl-carbamates in epilepsy therapy: how does cenobamate differ? Epilepsia. 2021;62:596–614.
264. Rogawski MA, Löscher W, Rho JM. Mechanisms of action of antiseizure drugs and the ketogenic diet. Cold Spring Harb Perspect Med. 2016;6:a022780.
265. Sills GJ, Rogawski MA. Mechanisms of action of currently used antiseizure drugs. Neuropharmacology. 2020;168:107966.
266. Maeda K, Sugino H, Akazawa H, et al. Brexpiprazole I: in vitro and in vivo characterization of a novel serotonin-dopamine activity modulator. J Pharmacol Exp Ther. 2014;350(3):589–604.
267. Frankel JS, Schwartz TL. Brexpiprazole and cariprazine: distinguishing two new atypical antipsychotics from the original dopamine stabilizer aripiprazole. Ther Adv Psychopharm. 2017;7(1):29–41.
268. FDA Approves Caplyta (lumateperone) for the Treatment of Schizophrenia in Adults. drugs.com. 23 December 2019.

269. Blair HA. Lumateperone: first approval. Drugs. 2020;80(4):417–23.
270. Edinoff A, Wu N, deBoisblanc C, et al. Lumateperone for the Treatment of Schizophrenia. Psychopharmacol Bull. 2020;50(4):32–59.
271. Mendonça Júnior FJ, Scotti L, Ishiki H, et al. Benzo- and thienobenzo- diazepines: multi-target drugs for CNS disorders. Mini Rev Med Chem. 2015;15(8):630–4.
272. Ivachtchenko AV, Lavrovsky Y, Okun I. AVN-101: a multi-target drug candidate for the treatment of CNS disorders. J Alzheimers Dis. 2016;53(2):583–620.

7 Polypharmacology in Clinical Applications: Respiratory Polypharmacology

Abstract

The respiratory system is an open system that allows us to breathe, bring oxygen into our bodies (called inspiration or inhalation), and send carbon dioxide out (called expiration, or exhalation). It consists of the nose, mouth, throat, voice box, windpipe, and lungs. Because it is an open system, the respiratory system is highly prone to pathogens and other environmental insults and easy to get sick, like the digestive system, another open system. Respiratory diseases are leading causes of death and disability in the world. About 65 million people suffer from chronic obstructive pulmonary disease (COPD), and 3 million die from it each year, making it the third leading cause of death worldwide. The term lung or pulmonary disease refers to many disorders affecting the lungs, such as asthma; COPD; infections like influenza, pneumonia, and tuberculosis; lung cancer; and many other breathing problems. Some lung diseases can lead to respiratory failure. Corresponding to the multiplicity and diversity of respiratory disease, currently available drugs for the treatment of respiratory disease are also multitudinous. The recent application of polypharmacology to respiratory disease has further increased choices of therapeutic regimens and enhanced therapeutic efficacy as well. The present chapter first summarizes the basic knowledge of respiratory disease and the conventional drug therapy. In the next subsections, the CDT and FDC approaches for the treatment of COPD, pulmonary tuberculosis, and COVID-19 will be discussed, and multitarget tyrosine kinase inhibitors as MTD therapeutics for idiopathic pulmonary fibrosis and oligonucleotides as possible sequence-based MTD epi-drugs for respiratory disease will also be introduced.

Keywords

Respiratory polypharmacology · COPD · Tuberculosis · COVID-19 · Multitarget therapy

7.1 Respiratory Disease and Pharmacological Treatment

7.1.1 Basics of Respiratory Disease (RD)

Respiratory diseases are some of the most common medical conditions in the world [1–4]. Tens of millions of people have lung disease in the United States alone. Smoking, infections, and genes cause most lung diseases. The lungs are part of a complex system, expanding and relaxing thousands of times each day to bring in oxy-

Z. Wang, B. Yang, *Polypharmacology*, https://doi.org/10.1007/978-3-031-04998-9_7

gen and send out carbon dioxide. Lung disease can happen when there are problems in any part of this system.

There are three main types of lung disease [1–4], including the following:

7.1.1.1 Airway Diseases

The windpipe (trachea) branches into tubes called bronchi, which in turn become smaller tubes throughout the lungs. Airway conditions affect the tubes (or airways) that carry air into and out of the lungs. Diseases such as asthma, emphysema, and chronic bronchitis narrow or block these airways, making breathing through a straw.

Asthma. The airways are constantly inflamed and may spasm, causing wheezing and shortness of breath. Allergies, infections, or pollution can trigger asthma symptoms.

Chronic obstructive pulmonary disease (COPD). With this lung condition, patients cannot exhale the way as they usually would, which causes trouble breathing.

Chronic bronchitis. This form of COPD brings a long-term wet cough.

Emphysema. Lung damage allows air to be trapped in the lungs in this form of COPD. Trouble blowing air out is its hallmark.

Acute bronchitis. This sudden infection of the airways is usually caused by a virus.

Cystic fibrosis. With this condition, patients have trouble clearing mucus out of your bronchi. This leads to repeated lung infections.

7.1.1.2 Airway Diseases, Lung Tissue Diseases

The airways branch into tiny tubes (bronchioles) that end in clusters of air sacs called alveoli. These air sacs make up most of your lung tissue. The interstitium is the thin, delicate lining between your alveoli. Tiny blood vessels run through the interstitium and let gas transfer between the alveoli and the blood. Additionally, the lung is surrounded by the pleura, the thin lining that lines the inside of the chest wall. A tiny layer of fluid lets the pleura on the lung's surface slide along the chest wall with each breath. Moreover, the chest wall also plays an important role in breathing. Muscles connect ribs to each other, helping the chest expand, and the diaphragm descends with each breath, also causing chest expansion. Any part of the above-described lung tissues could get sick due to a variety of causes.

Lung Diseases Affecting the Alveoli

Lung diseases such as pulmonary fibrosis and sarcoidosis affect the structure of the lung tissue through inflammation or scarring. Lung tissue changes prevent the lungs from fully expanding, making it hard to take a deep breath to take in oxygen or expel carbon dioxide.

Lung Diseases Affecting the Structure of the Lungs

Pulmonary sarcoidosis. It is a disease characterized by the growth of tiny collections of inflammatory cells (granulomas) in any part of the lung tissue. It causes small lumps of inflammatory cells in the lungs. These lumps are called granulomas and can affect how the lungs work. The granulomas generally heal and disappear on their own. But, if they do not heal, the lung tissue can remain inflamed and become scarred and stiff. This is called pulmonary fibrosis. It changes the structure of the lungs and can affect breathing.

Pulmonary fibrosis. It is a lung disease that occurs when lung tissue becomes damaged and scarred. This thickened, stiff tissue makes it more difficult for your lungs to work properly. As pulmonary fibrosis worsens, patients become progressively more difficult to breath.

Lung Diseases Affecting the Alveoli

Pneumonia. It is an infection of the alveoli, usually by bacteria or viruses, including the coronavirus that causes COVID-19.

Tuberculosis (TB). It is pneumonia that slowly gets worse, caused by the bacteria *Mycobacterium tuberculosis*.

Emphysema. This happens when the fragile links between the alveoli are damaged. Smoking is the usual cause.

Pulmonary edema. Fluid leaks out of the small blood vessels of the lung into the air sacs and the area around them. One form is caused by heart failure and back pressure in the blood vessels of the lungs.

Lung cancer. It has many forms and may start in any part of the lungs. It most often happens in the main part of the lung, in or near the air sacs.

Acute respiratory distress syndrome (ARDS). This is a severe, sudden injury to the lungs from a serious illness. COVID-19 is one example. Many people who have ARDS need help breathing from a machine called a ventilator until their lungs recover.

Pneumoconiosis. This is a category of conditions caused by inhaling something that injures the lungs. Examples include black lung disease from coal dust and asbestosis from asbestos dust.

Lung Diseases Affecting the Interstitium

Interstitial lung disease (ILD). This is a group of lung conditions that includes sarcoidosis, idiopathic pulmonary fibrosis, and autoimmune disease.

Pneumonia and *pulmonary edema*. These conditions affect the interstitium.

Lung Diseases of the Pleura

Pleural effusion. Fluid collects in the space between the lung and the chest wall. It is usually caused by pneumonia or heart failure. Large pleural effusions can make it hard to breathe and may need to be drained.

Pneumothorax. Air may get into the space between the chest wall and the lung, collapsing the lung.

Mesothelioma. This is a rare form of cancer that forms on the pleura. Mesothelioma tends to happen several decades after patients come into contact with asbestos.

Lung Diseases Affecting the Chest Wall

Obesity hypoventilation syndrome. Extra weight on the chest and belly can make it hard for the chest to expand. This may cause serious breathing problems.

Neuromuscular disorders. One might have trouble breathing when the nerves that control the respiratory muscles do not work the way they should. Amyotrophic lateral sclerosis and myasthenia gravis are examples of neuromuscular lung disease.

7.1.1.3 Airway Diseases, Lung Circulation Diseases

These affect the blood vessels in the lungs, caused by inflammation, clotting, or scaring of the pulmonary blood vessels. This affects the ability of the lungs to take up oxygen or expel carbon dioxide, and this may also affect the functioning of the heart. The heart pumps blood into the lungs through the pulmonary arteries carrying low-oxygen blood from the veins. These blood vessels can have diseases, as well.

Pulmonary embolism (PE). A blood clot (usually in a deep leg vein, called deep vein thrombosis) breaks off, travels to your heart, and gets pumped into your lungs. The clot sticks in a pulmonary artery, often causing shortness of breath and low blood oxygen levels.

Pulmonary hypertension. Many conditions can cause high blood pressure in the pulmonary arteries. This can lead to shortness of breath and chest pain. If this condition is not obviously connected with a cause, it is considered idiopathic pulmonary arterial hypertension.

7.1.2 Drug Therapy of Pulmonary Disease

7.1.2.1 Airway Diseases, Categories of Therapeutic Agents for PD Treatment

Anti-inflammatory Agents

Certain types of pulmonary fibrosis (PF) respond well to treatment with steroids. Steroids, also known as glucocorticoids, are a class of medication that reduces inflammation (swelling). Corticosteroids can have harmful side effects, especially when used long term. You will need to be carefully monitored if you are on long-term steroids.

Other groups of medications prescribed to reduce pulmonary inflammation include azathioprine, cyclophosphamide and mycophenolate mofetil (CellCept®, Myfortic®), which work by suppressing the immune system.

Corticosteroid pills, also known as prednisone, are often used for short periods of time for COPD. Corticosteroid pills have more side effects than the inhaled corticosteroids that are in combination medicines.

Antibiotics

Bacterial infections in airways may cause COPD flare-ups. Antibiotics can help in these cases. However, antibiotics will not work if flare-up is caused by a viral infection. Prednisone is often prescribed along with antibiotics. Sometimes antibiotics are prescribed on a regular basis for those patients with frequent flare-ups.

Phosphodiesterase-4 Inhibitors

Phosphodiesterase-4 inhibitor (PDE-4) medications are a type of oral medication that decreases inflammation in the lungs. This pill is used for patients with a history of frequent flare-ups and chronic cough with mucus (phlegm). It can prevent flare-ups and should be used along with regular inhalers.

Mucolytics

Mucolytics help to remove secretions (unwanted contents like phlegm) from the lung by thinning mucus so it is easier to cough up. These medications are for patients with frequent flare-ups. Mucomyst (N-acetylcysteine—NAC) may also prevent flare-ups.

Anti-fibrotic Agents

Two drugs are FDA-approved for treatment of idiopathic pulmonary fibrosis (IPF), which is the most common form of pulmonary fibrosis (PF). These include nintedanib (Ofev®) and pirfenidone (Esbriet®). These medications are used to slow down the rate of fibrosis or scarring in the lungs for patients with mild, moderate, and severe IPF.

7.1.2.2 Airway Diseases, Major Respiratory Diseases and the Current Drug Therapies

Asthma, COPD, acute lower respiratory tract infections, TB, and lung cancer are among the most common pulmonary diseases, which cause severe illness and death worldwide [1–4].

Asthma

Asthma is a long-term inflammatory disease of the airways of the lungs [5]. It is characterized by variable and recurring symptoms, reversible airflow obstruction, and easily triggered bronchospasms [6, 7]. Asthma is thought to be caused by a combination of genetic and environmental factors [8]. Environmental factors include exposure to air pollution and allergens [5]. Other potential triggers include medications such as aspirin and β-blockers [5]. The classic hallmarks of this obstructive airway disease are airway inflammation, airway remodeling, mucus hypersecretion, and airway hyperresponsiveness (AHR) [9]. The hallmarks can be attributed to leukocyte infiltration in the lungs, including lymphocytes, eosinophils, and neutrophils, increased production of pro-inflammatory cytokines (IL-4, IL-5, IL-13, GM-CSF, and IL-17), and elevated lgE level [9, 10].

There are more than 300 million people who suffer from asthma, with an estimated death rate of approximately 250,000 patients annually [11, 12]. In 2007, US$56 billion in healthcare costs was incurred in the United States alone and about €18 billion in Europe [13]. Medications used to treat asthma are divided into two general classes: quick-relief medications used to treat acute symptoms and long-term control medications used to prevent further exacerbation [6], including inhaled corticosteroids (ICS) to suppress lung inflammation and β_2 agonists as bronchodilators [9, 14]. While such standard therapies are able to relieve asthma symptoms in most patients, there are approximately 10% of asthmatics who are diagnosed with severe asthma that are less responsive or resistant to ICS [15]. Alternative drug treatments such as leukotriene modifiers, mast cell stabilizers, and anti-lgE antibodies are also not capable of controlling asthma symptoms in severe asthmatics [9]. Antibiotics are generally not needed for sudden worsening of symptoms or for treating asthma at any time [16, 17]. Overall, there is an unmet medical need to discover a cure for asthma and treat severe asthmatics that are resistant to corticosteroid therapy.

Chronic Obstructive Pulmonary Disease (COPD)

Chronic obstructive pulmonary disease (COPD) is a type of progressive lung disease that is preventable and treatable. COPD is characterized by long-term respiratory symptoms and airflow limitation [18]. The main symptoms include shortness of breath and a cough, which may or may not produce mucus [19]. COPD progressively worsens with everyday activities such as walking or dressing becoming difficult [20]. The two most common conditions of COPD are emphysema and chronic bronchitis, and they have been the two classic COPD phenotypes [21]. Emphysema is defined as enlarged airspaces (alveoli) whose walls break down resulting in permanent damage to the lung tissue. Chronic bronchitis is defined as a productive cough that is present for at least 3 months each year for 2 years. Both of these conditions can exist without airflow limitation when they are not classed as COPD.

According to the World Health Organization, as of 2015, COPD affected about 174.5 million people (2.4% of the global population) [22]. It typically occurs in males and females over the age of 35–40 [22, 23]. In 2019 it caused 3.2 million deaths, 80% occurring in lower- and middle-income countries [22], up from 2.4 million deaths in 1990 [24, 25]. The number of deaths is projected to increase further because of continued exposure to risk factors and an aging population [18]. In the United States in 2010, the economic cost was put at US$32.1 billion and projected to rise to US$49 billion in 2020. In the United Kingdom, this cost is estimated at £3.8 billion annually.

The major risk factor for COPD is cigarette smoking, which contains massive amounts of reactive oxidants [26]. COPD is not curable, but the symptoms are treatable, and its progression can be delayed, particularly by stopping smoking [23, 27]. The major goals of management are to reduce exposure to risk factors including offering non-pharmacological treatments such as help with stopping smoking. Stopping smoking can reduce the rate of lung function decline and also reduce mortality from smoking-related diseases such as lung cancer and cardiovascular disease [23]. A number of medical treatments are used in the management of stable COPD and exacerbations. These include bronchodilators, corticosteroids, and antibiotics. In patients with a severe exacerbation, antibiotics improve outcomes [28]. A number of different antibiotics may be used including amoxicillin, doxycycline, and azithromycin; whether one is better than the others is unclear [29]. There is no clear evidence of improved outcomes for those with less severe cases [28]. The FDA recommends against the use of fluoroquinolones when other options are available due to higher risks of serious side effects [30]. In treating acute hypercapnic respiratory failure (acutely raised levels of carbon dioxide), bilevel positive airway pressure (BPAP) can decrease mortality and the need of intensive care [26]. In patients without acidosis from respiratory failure, home care may be able to help avoid some admissions [31]. In those with end-stage disease, palliative care is focused on relieving symptoms [32]. Morphine can improve exercise tolerance [27]. Noninvasive ventilation may be used to support breathing and also reduce daytime breathlessness [27, 33].

Pulmonary Fibrosis (PF)

Major pulmonary fibrosis diseases include cystic fibrosis and idiopathic pulmonary fibrosis (IPF). IPF is characterized by a chronic progressive decline in lung function, which is fatal at the end stage. The scarring of lung tissue causes worsening dyspnea (shortness of breath) in IPF. The prevalence of IPF was estimated to be 63 per 100,000 persons in the United States [34]. IPF is a restrictive airway disease and proposed to be the consequence of an aberrant tissue repair process, which results in inflammation and excessive collagen deposition in lung interstitium [35]. High levels of transforming growth factor (TGF)-β stimulate and differentiate airway fibroblasts into myofibroblasts, which are particularly active in extracellular matrix deposition, contributing to pulmonary fibrosis and airway remodeling [35].

Current treatments of pulmonary fibrosis aim at relieving symptoms and preventing acute exacerbations, by reducing mucus secretion and attenuating bacterial infection [36]. Current pharmacological interventions to IPF include pirfenidone (a small molecule that possesses anti-inflammatory, anti-oxidative, and anti-fibrotic effects) [37], *N*-acetyl cysteine (an antioxidant), and receptor tyrosine kinase inhibitor nintedanib [38]. Newer therapies can be designed to target essential growth factors and cytokines responsible for fibroblast proliferation, activation, transformation, and survival.

Tuberculosis (TB)

Tuberculosis (TB) is an infectious disease usually caused by *Mycobacterium tuberculosis* (MTB) bacteria [39]. TB generally affects the lungs but can also affect other parts of the body [39]. Most infections show no symptoms, in which case it is known as latent tuberculosis. About 10% of latent infections progress to active disease which, if left untreated, kills about half of those affected. TB is spread from one person to the next through the air when people who have active TB in their lungs cough, spit, speak, or sneeze [39, 40]. People with latent TB do not spread the disease [39]. Active infection occurs more often in people with HIV/AIDS and in those who smoke [39]. As of 2018, one quarter of the world's population was thought to have latent infection with TB [41]. New infections occur in about 1% of the population each year [42]. In 2020, an estimated 10 million people developed active TB, resulting in 1.5 million deaths and making it the second leading cause of death from an infectious disease after COVID-19 [43]. As of 2018, most TB cases occurred in the regions of South-East Asia (44%), Africa (24%), and the Western Pacific (18%).

Diagnosis of active TB is based on chest X-rays, as well as microscopic examination and culture of body fluids [44]. Diagnosis of latent TB relies on the tuberculin skin test (TST) or blood tests [44]. Treatment of TB uses antibiotics to kill the bacteria. Effective TB treatment is difficult, due to the unusual structure and chemical composition of the mycobacterial cell wall, which hinders the entry of drugs and makes many antibiotics ineffective [45]. Active TB is best treated with combinations of several antibiotics to reduce the risk of the bacteria developing antibiotic resistance [46]. The routine use of rifabutin instead of rifampicin in HIV-positive people with tuberculosis is of unclear benefit as of 2007 [47]. Latent TB is treated with either isoniazid or rifampin alone or a combination of isoniazid with either rifampicin or rifapentine [48–50]. The recommended treatment of new-onset pulmonary tuberculosis, as of 2010, is 6 months of a combination of antibiotics containing rifampicin, isoniazid, pyrazinamide, and ethambutol for the first 2 months and only rifampicin and isoniazid for the last 4 months.

Coronavirus Disease 2019 (COVID-19)

Coronavirus disease 2019 (COVID-19) is a highly contagious disease caused by severe acute respiratory syndrome coronavirus 2 (SARS-CoV-2), a novel coronavirus, that emerged in December 2019 and has led to a worldwide pandemic. COVID-19 transmits when people breathe in air contaminated by droplets and small airborne particles containing the virus [51, 52]. COVID-19 is a serious health threat and has caused millions of deaths around the world as well as lasting health problems in some who have survived the illness, which was first identified amid an outbreak of respiratory illness cases and is still evolving daily.

The SARS-CoV-2 virus can infect a wide range of cells and systems of the body, though COVID-19 is most known for affecting the upper respiratory tract (sinuses, nose, and throat) and the lower respiratory tract (windpipe and lungs) [53]. The lungs are the organs most affected by COVID-19 because the virus accesses host cells via the receptor for the enzyme angiotensin-converting enzyme 2 (ACE2), which is most abundant on the surface of type II alveolar cells of the lungs [54]. The virus uses a special surface glycoprotein called a "spike" to connect to the ACE2 receptor and enter the host cell [55]. The typical manifestations of COVID-19 include fever, sore throat, fatigue, cough, and dyspnea combined with recent exposure. Most of the patients with COVID-19 have mild or moderate disease; however, up to 5–10% present with severe and even life-threatening disease course. The mortality rates are approximately 2%.

The SARS-CoV-2 virus has demonstrated a high mutation rate, and current variants of concern include Alpha (B.1.1.7), Beta (B.1.351), Gamma (P.1), Delta (B.1.617.2), and Omicron (B.1.1.529). Evidence demonstrates that Alpha and Delta variants are at least 50% easier to spread than the original version of the virus identified at the beginning of the pandemic. As well, the Gamma, Beta, and Delta variants each have certain mutations that may have an impact on vaccine effectiveness.

Although most people with COVID-19 have mild to moderate symptoms, the disease can cause severe medical complications and lead to death in some people. Older adults or people with existing medical conditions (chronic diseases such as diabetes) are at a greater risk of becoming seriously ill with COVID-19. Complications may include pneumonia, acute respiratory distress syndrome (ARDS), multi-organ failure, septic shock, and death [56–59]. Cardiovascular complications may include heart failure, arrhythmias (including atrial fibrillation), heart inflammation, and thrombosis, particularly venous thromboembolism [60–65]. Approximately 20–30% of people who present with COVID-19 have elevated liver enzymes, reflecting liver injury [66, 67]. Neurologic manifestations include seizure, stroke, encephalitis, and Guillain-Barré syndrome (which includes loss of motor functions) [68, 69].

At present, there is no specific, effective treatment or cure for COVID-19 [70]. One year into the pandemic, highly effective vaccines have now been introduced and are beginning to slow the spread of SARS-CoV-2. COVID-19 drug development is the research process to develop preventative therapeutic prescription drugs that would alleviate the severity of COVID-19. From early 2020 through 2021, several hundred drug companies, biotechnology firms, university research groups, and health organizations were developing therapeutic candidates for COVID-19 disease in various stages of preclinical or clinical research (506 total candidates in April 2021), with 419 potential COVID-19 drugs in clinical trials, as of April 2021 [71]. As early as March 2020, the World Health Organization (WHO) [72], European Medicines Agency (EMA) [73], US Food and Drug Administration (FDA) [74], and the Chinese government and drug manufacturers [75, 76] were coordinating with academic and industry researchers to speed up the development of vaccines, antiviral drugs, and postinfection therapies [77–79]. The International Clinical Trials Registry Platform of the WHO recorded 536 clinical studies to develop postinfection therapies for COVID-19 infections [80, 81] with numerous established antiviral compounds for treating other infections under clinical research to be repurposed [77, 82–84].

Diverse categories of preclinical or early-stage clinical research for developing COVID-19 therapeutic candidates included [85] antibodies (81 candidates), antivirals (31 candidates), cell-based compounds (34 candidates), RNA-based compounds (6 candidates), scanning compounds to be repurposed (18 candidates), and various other therapy categories, such as anti-inflammatory, antimalarial, interferon, protein-based, antibiotics, and receptor-modulating compounds [85].

Several experimental treatments are being actively studied in clinical trials [70]. These include fluvoxamine, a cheap and widely available antidepressant [86], and the antivirals molnupiravir (developed by Merck) [87–89] and nirmatrelvir (developed by Pfizer) [90]. Others were thought to be promising early in the pandemic, such as hydroxychloroquine and lopinavir/ritonavir, but later research found them to be ineffective or even harmful [70, 91]. Currently, in the United States, two monoclonal antibody-based therapies are available for early use in cases with severe disease [92]. The antiviral remdesivir is available in the United States, Canada, Australia, and several other countries, with varying restrictions; however, it is not recommended for people needing mechanical ventilation and is discouraged altogether by the World Health Organization (WHO) [93] due to limited evidence of its efficacy [70]. In November 2021, the United Kingdom approved the use of molnupiravir as a COVID treatment for vulnerable patients recently diagnosed with the disease [94].

7.2 CDT in Respiratory Disease

7.2.1 CDT for the Treatment of Chronic Obstructive Pulmonary Disease (COPD)

The cornerstone of pharmacological therapy for COPD is bronchodilation, with a long-acting muscarinic antagonist (LAMA), a long-acting β_2 agonist (LABA), or a combination of the two [95–98]. Bronchodilator therapy has been shown to improve lung function, decrease the severity of symptoms, and reduce the risk of future exacerbations in COPD [99]. The efficacy of the LAMA umeclidinium (62.5 μg) was recently shown to be superior to that of the widely used tiotropium (18 μg) with a significant increase in trough forced expiratory volume in 1 s (FEV1) after 12 weeks of monotherapy [100, 101]. Multiple randomized controlled trials have demonstrated greater improvements in lung function and patient-reported outcomes, including exacerbations, with LAMA/LABA combinations compared with LAMA or LABA monotherapies in patients with stable COPD [102–108].

LAMA/LABA combinations can be administered as open dual combinations (using separate prescribed inhalers for each bronchodilator) and fixed-dose combinations (FDCs or FDC; use of a single inhaler delivering a fixed dose of each bronchodilator). However, in the majority of cases, FDC is used, which will be described in the next subsection.

7.2.2 CDT for the Treatment of Pulmonary Tuberculosis (PTB)

Rifampicin was approved by the Food and Drug Administration (FDA) in 1971 for the treatment of PTB [109]. Combination therapy had already been established as the standard of care, with the discovery that streptomycin monotherapy, first tested in 1946, resulted in short-lived improvements, with the frequent emergence of streptomycin-resistant bacteria [110]. Combination therapy of streptomycin together with *para*-aminosalicylic acid and, subsequently, with isoniazid in the early 1950s made for a considerably more robust therapy, although 18–24 months of treatment were required [111]. The inclusion of rifampicin resulted in fewer relapses, and it was found that therapy could be shortened to 9 months and with subsequent improvements to the current 6-month regimen [109].

As summarized by Rothstein, human studies of multiple combinations of drugs confirmed the importance of rifampicin [112]. Isoniazid was particularly active in the first week against metabolically active bacteria, whereas rifampicin contributed to bactericidal activity during the first week. Rifampicin continued to act in the coming months against the persistent, metabolically quiescent population of bacteria, often residing inside of host cells. More recently, pyrazinamide was found to contribute bactericidal activity against the persistent population [113]. These studies led to the combination therapy that is used to treat susceptible *M. tuberculosis* infections, that is, 2 months of isoniazid, rifampicin, pyrazinamide, and ethambutol, followed by 4 months of continuation therapy consisting of daily dosing with isoniazid and rifampicin [114].

Nausea is one common side effect of combination therapy, although generally it is not sufficient to stop therapy [109]. Drug-induced hepatitis is rare, occurring in 2.5% of patients, and is generally attributed to the isoniazid and/or pyrazinamide components. Flu-like syndrome is experienced by a small minority of patients exposed to rifampicin and, rarely among this minority, can include thrombocytopenia, hemolytic anemia, and acute renal failure, resulting in discontinuation of therapy [109].

Rifapentine, approved by the FDA in 1998 for PTB therapy, is a slightly more potent rifamycin derivative than is rifampicin and has a fivefold increased exposure as a result of a longer half-life [109, 110, 115, 116]. Preclinical studies suggested that if rifapentine replaced rifampicin,

PTB therapy might be shortened [113]. The results of a recent phase III clinical trial showed that the last 4 months of the 6-month therapy could be simplified with weekly dosing of rifapentine and moxifloxacin (a quinolone) but not shortened [114].

7.2.3 CDT for COVID-19

7.2.3.1 Novel Antibody Drugs

Convalescent Plasma

Passive immunization with convalescent plasma or hyperimmune serum has been proposed as a potential treatment for COVID-19 [117].

In the United States, the FDA has granted temporary authorization to convalescent plasma (plasma from the blood of people who have recovered from COVID-19, which thus contains antibodies against SARS-CoV-2) as an experimental treatment in cases where the person's life is seriously or immediately threatened [118]. However, convalescent plasma treatment has not undergone the randomized controlled or nonrandomized clinical studies needed to determine if it is safe and effective for treating people with COVID-19 [117, 119, 120].

Argentina, Brazil, Costa Rica, and Mexico have pursued development of antisera [121]. Brazil began development of an equine hyperimmune serum, obtained by inoculating horses with recombinant SARS-CoV-2 spike protein, in mid-2020. A consortium of Instituto Vital Brazil, UFRJ, the Oswaldo Cruz Foundation. and the D'Or Institute for Research and Education in Rio de Janeiro began preclinical trials in May 2020 [121], while Instituto Butantan in São Paulo completed animal testing in September [121]. In December 2020, Argentina granted emergency authorization to CoviFab, a locally developed formulation of equine hyperimmune serum, for use in cases of moderate to severe COVID-19, based on the initial results of a single phase II/III trial which suggested reductions in mortality, ICU admission, and mechanical ventilation requirements in patients who received the serum [122, 123]. This was harshly criticized by the Argentine Intensive Care Society, which stated that the trial failed to achieve its primary or secondary endpoints and did not demonstrate any statistically significant differences between the serum and placebo groups [123].

Casirivimab/Imdevimab Antibody Combination

Casirivimab/imdevimab, sold under the brand name REGEN-COV among others [124, 125], is a combination medicine used for the treatment and prevention of COVID-19 [125]. It consists of two human monoclonal antibodies, casirivimab and imdevimab that must be mixed together [124, 125]. The combination of two antibodies is intended to prevent mutational escape [126]. It is also available as a co-formulated product [127]. It was developed by the American biotechnology company Regeneron Pharmaceuticals [128, 129]. The most common side effects include allergic reactions, which include infusion-related reactions and injection site reactions [127].

This combination is approved under the brand name Ronapreve for medical use in Japan, the United Kingdom, the European Union, and Australia [125, 130–134]. In June 2021, Health Canada authorized the use of casirivimab and imdevimab to be administered together as a monoclonal antibody cocktail therapy. Similar to bamlanivimab, the drugs are used to treat mild to moderate COVID-19 in adults and children aged 12 and older who are at high risk for hospitalization or COVID-related complications. The WHO has backed the use of the antibodies in select high-risk COVID-19 patients.

Bamlanivimab/Etesevimab Antibody Combination

Bamlanivimab/etesevimab is a combination of two monoclonal antibodies, bamlanivimab and etesevimab, administered together via intravenous infusion as a treatment for COVID-19 [135–138]. Both types of antibody target the surface spike protein of SARS-CoV-2 [139, 140].

The bamlanivimab/etesevimab combination is authorized in the United States for the treatment of mild to moderate COVID-19 in people aged 12 years of age and older weighing at least 40 kg (88 lb) with positive results of direct SARS-

CoV-2 viral testing and who are at high risk for progression to severe COVID-19, including hospitalization or death [141, 142]. They are also authorized, when administered together, for use after exposure to the SARS-CoV-2 virus for post-exposure prophylaxis (prevention) for COVID-19 and are not authorized for pre-exposure prophylaxis to prevent COVID-19 before being exposed to the SARS-CoV-2 virus [141, 142].

Tixagevimab/Cilgavimab Antibody Combination

Tixagevimab/cilgavimab is a combination of two human monoclonal antibodies, tixagevimab (AZD8895) and cilgavimab (AZD1061) used to prevent COVID-19. It is being developed by British-Swedish multinational pharmaceutical and biotechnology company AstraZeneca [143, 144]. It is co-packaged and given as two separate consecutive intramuscular injections (one injection per monoclonal antibody, given in immediate succession) [145].

7.2.3.2 Novel Protease Inhibitor: PF-07321332/Nirmatrelvir/Paxlovid

PF-07321332 or nirmatrelvir, filed under the brand name Paxlovid [146], is an antiviral drug developed by Pfizer which acts as an orally active 3CL protease inhibitor. The combination of PF-07304814 with ritonavir (an antiretroviral medication used along with other medications to treat HIV/AIDS) is in phase III trials for the treatment of COVID-19 since September 2021 [147–149] After promising results preventing hospitalization and death if given within the first 3 days of symptoms, Pfizer submitted an application to the US Food and Drug Administration (FDA) for emergency use authorization for nirmatrelvir in combination with ritonavir in November 2021 [150]. In the drug combination, ritonavir serves to slow down metabolism of nirmatrelvir by cytochrome enzymes to maintain higher circulating concentrations of the main drug [151]. PF-07321332 was found to reduce the risk of hospitalization or death by 89% compared to placebo in nonhospitalized high-risk adults with COVID-19. In the overall study population through day 28, no deaths were reported in patients who received nirmatrelvir as compared to 10 deaths in patients who received placebo.

In April 2021, Pfizer began phase I trials [152] In September 2021, Pfizer began a phase II/III trial [153]. In November 2021, Pfizer announced 89% reduction in hospitalizations of high-risk patients studied when given within 3 days after symptom onset [154, 155]. On 14 December 2021, Pfizer announced that the combination of nirmatrelvir with ritonavir, when given within 3 days of symptom onset, reduced the risk of hospitalization or death by 89% compared with placebo in 2246 high-risk participants studied [149]. The drug, currently in phase III clinical trials in combination with ritonavir, relies on the electrophilic attack of a nitrile warhead to the catalytic cysteine of the protease. PF-07321332 was originally developed by Pfizer in 2002–2003 against SARS-CoV and repurposed for SARS-CoV-2 due to the aforementioned similarities between the two viruses' proteases [156, 157].

7.2.3.3 Novel Viral Replication Inhibitor: Molnupiravir

Molnupiravir, sold under the brand name Lagevrio among others, is an antiviral medication that inhibits the replication of certain RNA viruses and is used to treat COVID-19 patients [158]. It is a prodrug of the synthetic nucleoside derivative N4-hydroxycytidine and exerts its antiviral action through introduction of copying errors during viral RNA replication [159, 160]. Molnupiravir was originally developed to treat influenza at Emory University by the university's drug innovation company, Drug Innovation Ventures at Emory (DRIVE), but was reportedly abandoned for mutagenicity concerns [161, 162]. It was then acquired by Miami-based company Ridgeback Biotherapeutics, which later partnered with Merck & Co. to develop the drug further [163]. Molnupiravir was approved for medical use in the United Kingdom in November 2021 [158, 164–166].

Molnupiravir is a broad-spectrum antiviral [89, 90, 167]. It can act by causing G to A and C to U transition mutations in replicating coronaviruses. These increases in mutation frequencies can be linked to increases in antiviral effects. The study by Gordon et al. [167] generated the biochemical data in support of a mechanism of action of molnupiravir that is primarily based on RNA mutagenesis mediated via the template strand. Available evidence suggests that molnupiravir could become a paradigmatic example in the use of lethal mutagenesis as an antiviral strategy.

7.2.3.4 Drug Repositioning

Several existing antiviral medications, previously developed or used as treatments for severe acute respiratory syndrome (SARS), Middle East respiratory syndrome (MERS), HIV/AIDS, and malaria, are being researched as COVID-19 treatments by drug repositioning [84, 168], with some moving into clinical trials [82].

During the COVID-19 pandemic, drug repurposing is the clinical research process of rapidly screening and defining the safety and efficacy of existing drugs already approved for other diseases to be used for people with COVID-19 infection [82, 84, 169]. The WHO began in mid-March 2020 expediting international phase II–III trials on four promising treatment options—the SOLIDARITY trial [170–172]—with numerous other drugs having potential for repurposing in different disease treatment strategies, such as anti-inflammatory, corticosteroid, antibody, immune, and growth factor therapies, among others, being advanced into phase II or III trials during 2020 [82, 85, 169, 173].

The new anti-SARS-CoV-2 agents nirmatrelvir and molnupiravir can be considered repurposed drugs as they both were initially developed for different versions of virus: nirmatrelvir originally for SARS-CoV and now repurposed for SARS-CoV-2 and molnupiravir originally for influenza but abandoned and repurposed for SARS-CoV-2.

7.3 FDC in Respiratory Disease

7.3.1 FDC for the Treatment of COPD

COPD is characterized by persistent respiratory symptoms (e.g., dyspnea, cough, sputum production) and airflow limitation [174]. COPD is diagnosed primarily based on two main components, the forced expiratory volume in one second (FEV1), which is the greatest volume of air that can be breathed out in the first second of a breath, and the forced vital capacity (FVC), which is the greatest volume of air that can be breathed out in a single large breath [175]. Normally, 75–80% of the FVC comes out in the first second [175] and a FEV1/FVC ratio less than 70% in someone with symptoms of COPD defines a person as having the disease [176].

Forced expiratory volume (FEV) refers to the volume of air that an individual can exhale during a forced breath in *t* seconds. It is usually represented as FEV, followed by a subscript that indicates the number of seconds of the measurement's duration. For instance, forced expiratory volume in 1 second (FEV1) is the maximum amount of air that the subject can forcibly expel during the first second following maximal inhalation. The FEV1/FVC ratio, also called Tiffeneau-Pinelli index [174–176], is a calculated ratio used in the diagnosis of obstructive and restrictive lung disease. It represents the proportion of a person's vital capacity that they are able to expire in the first second of forced expiration (FEV1) to the full, forced vital capacity (FVC). The result of this ratio is expressed as FEV1%. Normal values are approximately 75%.

Several pharmacological agents are available for the treatment of COPD, including bronchodilators (e.g., β_2-adrenergic agonists/LABA, a long-acting muscarinic antagonist/LAMA, methylxanthines), inhaled corticosteroids (ICS), phosphodiesterase-4 inhibitors, and mucolytic agents. Maintenance bronchodilator therapy is key to the management of stable COPD, which aims to reduce symptoms and the frequency and severity of exacerbations, while improving health-related quality of life (HR-QoL) and exercise tolerance.

Oral inhalation is the preferred route of administration, with long-acting formulations preferred over short-acting agents. Combining two bronchodilators with different durations and mechanisms of action may increase the degree of bronchodilation and reduce the risk of adverse events compared with the individual components, providing a rationale for the development of fixed-dose single-pill combination (FDC) or FDC [177].

7.3.1.1 LAMA/LABA Dual Combination

Tiotropium/Olodaterol Combination

Tiotropium/olodaterol (Stiolto® Respimat®; Spiolto® Respimat®) is a fixed-dose combination of the long-acting muscarinic antagonist (LAMA) tiotropium bromide (hereafter referred to as tiotropium) and the long-acting β_2-adrenergic agonist (LABA) olodaterol, delivered via the Respimat® soft mist inhaler (SMI) [95, 178]. It is approved in several countries, including the United States [179], Japan [180], China [181], and those of the EU [182], for the long-term maintenance treatment of COPD. The pharmacological properties of tiotropium and olodaterol are well known and have been previously reviewed in detail [183–185].

Tiotropium is an LAMA with similar affinity for muscarinic receptors M_1 to M_5, a multitarget antagonist, displays kinetic subtype selectivity, binds competitively and reversibly to M_3 receptors on airway smooth muscle cells, and inhibits action of acetylcholine at muscarinic receptors, leading to smooth muscle relaxation.

Olodaterol is a potent and highly selective LABA, binds to and activates β_2 receptors on airway smooth muscle cells, stimulates intracellular adenyl cyclase, and enhances synthesis of cyclic-3′,5′ adenosine monophosphate, resulting in smooth muscle relaxation and bronchodilation.

Pharmacokinetics of inhaled tiotropium and olodaterol is FDC similar to those of each agent administered separately. Tiotropium/olodaterol, like other inhaled medicines, may cause potentially life-threatening paradoxical bronchospasm. Overall, tiotropium/olodaterol FDC is effective and generally well tolerated in adults with moderate to very severe COPD. Therefore, tiotropium/olodaterol is a useful option for the maintenance treatment of COPD [173, 186], with the convenience of once-daily administration via a single inhaler.

Umeclidinium/Vilanterol Combination

Feldman et al. [100] performed the first direct comparison of two once-daily fixed-dose LAMA/LABA combinations; superiority was found for the primary endpoint of trough FEV1 at week 8 with umeclidinium/vilanterol compared with tiotropium/olodaterol in patients with symptomatic COPD. Both treatments had similar safety profiles. These findings confirm the results of previous indirect LAMA/LABA comparisons and show that an efficacy gradient exists within the LAMA/LABA class. Similar results were also observed by other studies [177, 187].

On the other hand, umeclidinium/vilanterol and tiotropium/indacaterol, given once daily, provided similar improvements in lung function and patient-reported outcomes over 12 weeks in patients with COPD, with comparable tolerability and safety profiles [188]. A systematic review of direct (head-to-head) and indirect treatment comparisons revealed that currently available LAMA/LABA in the FDCs/FDC format have comparable efficacy and safety in patients with COPD and moderate to very severe airflow limitation [189].

Indacaterol/Glycopyrronium Combination

QVA149 (indacaterol/glycopyrronium) [Xoterna(®) Breezhaler(®), Ultibro(®) Breezhaler(®)] is an inhaled fixed-dose combination of indacaterol (an LABA) and glycopyrronium (an LAMA) that has been approved in Europe and Japan for the symptomatic control of COPD in adults [190]. In phase III studies, QVA149 significantly improved bronchodilation versus indacaterol, glycopyrronium, or tiotropium alone and the LABA/inhaled corticosteroid fixed-dose combination salmeterol/fluticasone.

These improvements in lung function, which were rapid in onset and maintained during long-term treatment, were generally associated with significant improvements in dyspnea, health status, COPD exacerbation risk, patient symptoms, and rescue medication use. The SHINE and ILLUMINATE studies in low (exacerbation)-risk patients with moderate to severe disease suggest that QVA149 may offer more symptomatic relief than tiotropium and salmeterol/fluticasone [190]. Similarly, the SPARK study in high (exacerbation)-risk patients with severe or very severe disease showed that QVA149 was more effective than glycopyrronium in preventing moderate to severe exacerbations and suggests that QVA149 may offer more symptomatic relief than LAMA monotherapy [190–192]. Another phase III study comparing QVA149 with salmeterol/fluticasone in high-risk patients with moderate to very severe disease (FLAME) is ongoing. QVA149 is generally well tolerated, with no new safety signals identified compared with its monocomponents. Furthermore, QVA149 achieved better bronchodilation than salmeterol 50 μg/fluticasone 500 μg twice daily [192]. Compared with tiotropium, a well-recognized treatment for COPD, the percentage of patients that exceed the minimal clinical important difference for dyspnea and health-related quality of life measurements was superior with QVA149. Other patient-oriented outcomes, such as daily symptoms, nighttime awakening, and use of rescue medication consistently, favored QVA149. Finally, QVA149 was significantly superior to LAMAs for reducing all types of exacerbation. Therefore, bronchodilators remain central to the symptomatic management of COPD. When dual bronchodilation is indicated, QVA149 offers the convenience of two bronchodilators in a single inhaler coupled with a simple, once-daily dosing regimen that may encourage better treatment adherence.

7.3.1.2 LAMA/ICS or LABA/ICS Dual Combination

Inhaled corticosteroid (ICS) plus long-acting β_2 agonist (LABA) combinations are commonly used in the treatment of patients with COPD. Currently, at least four fixed-dose LABA/ICS combinations are available, including formoterol/budesonide, formoterol/beclomethasone, vilanterol/fluticasone, and salmeterol/fluticasone [193].

Formoterol/Beclomethasone FDC/FDC

Extrafine formulation of the formoterol/beclomethasone combination has recently been approved for use in COPD. Small airway inflammation and remodeling are cardinal features of COPD; therefore, the ability of this extrafine formulation to reach the small, as well as the large, airways is likely to be therapeutically important by enabling treatment of inflammatory processes in the whole bronchial tree. The clinical development of extrafine formoterol/beclomethasone has demonstrated significant benefits over extrafine FF in terms of lung function improvement and reduction of the exacerbation rate, thus supporting the beneficial effect of an ICS combined to a LABA in COPD patients. Head-to-head comparison studies versus other LABA/ICS combinations have shown that the extrafine formulation enables the clinical benefits to be achieved with a lower dose of ICS. Extrafine formoterol/beclomethasone showed lung function and dyspnea improvements comparable to other LABA/ICS, and a significantly faster onset of action was observed when compared with a salmeterol-containing fixed-dose combination. Several other studies have also shown that the LABA/ICS FDC is more effective than the individual components alone for improving lung function and symptoms, reducing exacerbation rates, and improving general health status [194–199]. This is likely because β agonists and corticosteroids have different pharmacological targets and can exert synergistic interactions, e.g., ICS increase the β_2-adrenoreceptor density in the airways, which can counterbalance receptor downregulation and therefore prevent the development of LABA tolerance [200–202]; LABAs favor the nuclear translocation of glucocorticoid receptors, therefore enhancing their mechanism of action [203]. The clinical benefits of LABA/ICS combination treatment have also been demonstrated in several other trials [204–209].

Formoterol/Budesonide FDC/FDC

A study aimed to provide an in-depth review on studies that compared formoterol/budesonide with the other LABA/ICS combinations and assess the data on exacerbations, safety, and patient quality of life [210]. PubMed and Ovid databases were searched, and 14 studies were identified. Yet, there is general lack of prospective, randomized, controlled trials comparing LABA/ICS combinations in the treatment of COPD as only two such studies were identified. However, current evidence suggests that the effects of formoterol/budesonide FDC on reducing exacerbations and improving quality of life may be similar to, or more marked than, those of other LABA/ICS combinations in COPD and, compared with the other LABA/ICS combinations, formoterol/budesonide may be associated with a lower incidence of serious pneumonia events and oral candidiasis [210].

Collectively, commonly used dual combination agents used to treat COPD include the following:

- Combivent is the brand name for albuterol and ipratropium bromide (Atrovent®). In a Respimat® soft mist inhaler.
- Advair is the brand name for a combination of fluticasone propionate (Flovent®) and salmeterol inhalation powder (Serevent®). Advair comes in a Diskus® inhaler and an HFA metered dose inhaler.
- Symbicort is the brand name for a combination of formoterol (Foradil®) and budesonide (Pulmicort®). Symbicort comes in an HFA metered dose inhaler.
- Dulera is the brand name for a combination of formoterol and mometasone. Dulera comes in a metered dose inhaler.
- DuoNeb is the brand name for a combination of ipratropium bromide and albuterol sulfate. DuoNeb is supplied as a nebulizer solution.
- BREO is the brand name for a combination of fluticasone furoate and vilanterol in an ELLIPTA dry powder inhaler.
- Stiolto is a combination of tiotropium and olodaterol in a Respimat soft mist inhaler.
- Utibron is the brand name for glycopyrrolate and indacaterol in a Neohaler dry powder inhaler.
- Anoro is the brand name for a combination of umeclidinium and vilanterol in an ELLIPTA dry powder inhaler.
- Bevespi is the brand name for glycopyrrolate and formoterol fumarate in an Aersphere metered dose inhaler.

7.3.1.3 LAMA/LABA/ICS Triple Combination

Triple inhaled corticosteroid/long-acting muscarinic antagonist/long-acting β_2 agonist (LAMA/LABA/ICS) combination therapy is recommended for patients with chronic COPD who experience further exacerbations/symptoms on dual LAMA/LABA or LABA/ICS therapy. The relative efficacy of glycopyrronium/formoterol/budesonide metered dose inhaler (BGF MDI) in COPD was compared with other LAMA/LABA/ICS combinations and open combination therapies in a network meta-analysis (NMA). A systematic literature review was conducted to identify randomized controlled trials of at least 10-week duration, including at least one fixed-dose or open combination triple therapy arm, in patients with moderate to very severe COPD [211]. A three-level hierarchical Bayesian NMA model was used to determine the exacerbation rate per patient per year as well as the following outcomes at week 24: changes from baseline in pre-dose trough forced expiratory volume in 1 s (FEV1), post-dose peak FEV1, and St. George's Respiratory Questionnaire (SGRQ) total score, proportion of SGRQ responders, and Transition Dyspnea Index focal score. Change from baseline in rescue medication use over weeks 12–24 was also analyzed. Meta-regression and sensitivity analyses were used to assess heterogeneity across studies. Eighteen studies (n = 29,232 patients) contributed to the NMA. LABA/ICS dual combinations were combined as a single treatment group to create a connected network. Across all outcomes, there were no statistically significant differences between BGF MDI and other triple LAMA/LABA/ICS FDC (fluticasone furoate/umeclidinium/vilanterol and beclomethasone dipropionate/glycopyrronium/formoterol fumarate) and open combinations with data available within the network. The results suggested that BGF MDI has comparable efficacy to other LAMA/LABA/ICS FDCs and open triple combination therapies in reducing exacerbations and improving lung function and symptoms in patients with moderate to very severe COPD. Further research is warranted as additional evidence regarding triple therapies, especially fixed-dose combinations, becomes available.

Similar results and conclusions were obtained in another network meta-analysis (NMA) conducted to compare the relative efficacy, safety, and tolerability of BGF MDI with other fixed-dose and open combination triple therapies in COPD over 52 weeks [212].

KRONOS, a phase III, multicenter, randomized, double-blind study (NCT02497001) conducted in Canada, China, Japan, and the United States, assessed the efficacy and safety of budesonide/glycopyrrolate/formoterol fumarate metered dose inhaler (BGF MDI), a triple fixed-dose combination therapy, relative to dual therapies in patients with moderate to very severe COPD [213]. This study found that BGF MDI nominally significantly improved lung function and numerically improved symptoms *vs* glycopyrrolate/formoterol fumarate (GFF) MDI and budesonide/formoterol fumarate (BFF) MDI. BGF MDI nominally significantly reduced exacerbations *vs* GFF MDI in Japanese patients with COPD. Efficacy and safety findings were generally comparable to those in the global KRONOS population.

7.3.2 FDC for the Treatment of Pulmonary Tuberculosis (TB)

The currently preferred regimen for treatment of PTB is isoniazid, rifampin, pyrazinamide, and ethambutol, which has been used either as separate tablets (CDT) or as fixed-dose combination (FDC or FDC).

FDC/FDC drugs for PTB treatment have been advocated internationally to prevent the emergence of drug resistance attributable to inappropriate drug intake [214–216]. A systematic review of the literature was performed in accordance with Preferred Reporting Items for Systematic Reviews and Meta-Analyses guidelines to determine whether the four-drug FDC is safer or more effective than separate drugs for treatment of PTB [216]. In pooled results from 5 randomized controlled trials with 3502 patients across Africa, Asia, and Latin America, four-drug FDC therapy was no better than separate drugs therapy in terms of culture conversion after 2 and 6 months of treatment. There were no significant differences between the groups in overall incidence of adverse effects. However, the meta-analytic measure (log odds ratio) revealed that separate drugs treatment had a 1.65 [exp (0.5) = 1.65] increased chance of gastrointestinal adverse effects compared to four-drug FDC treatment. Thus, it appears that four-drug FDC therapy provides greater patient comfort by reducing the number of pills and the incidence of gastrointestinal adverse effects, as well as simplifying pharmaceutical management at all levels [216].

A randomized, open, multicenter, multinational study was conducted in 26 centers and included 1159 patients with smear-positive PTB [217]. Their data showed that the efficacy of the four-drug FDC regimen was non-inferior to that of the ST regimens, but patient acceptability significantly improved with the four-drug FDC.

A prospective, randomized controlled study was conducted using the directly observed treatment short-course strategy to compare the safety and efficacy of a FDC regimen and separate formulations for PTB treatment [218]. They observed transient higher serum bilirubin levels were noted for the FDC regimen compared with the separate formulations during treatment. However, no significant difference in safety or efficacy was found between the groups when the directly observed treatment short-course strategy was used.

7.4 MTD in Respiratory Disease

7.4.1 Multitarget Tyrosine Kinase Inhibitors as MTD Therapeutics for RD

Idiopathic pulmonary fibrosis (IPF) is a chronic, progressive interstitial lung disease caused by multiple causes, mostly in the elder and confined to the lung. The clinical manifestations are progressive respiratory aggravation and decreased lung function, which can lead to respiratory failure and even death [219]. IPF is a clinically heterogeneous disease but its prognosis is overall poor, with a median survival of 3–4 years [219, 220]. Currently, only pirfenidone and nintedanib were approved therapeutics for IPF, but they have limited efficacy [221], so it is urgently needed to identify new potential therapeutic agents for IPF patients.

Anlotinib (1-[[[4-(4-fluoro-2-methyl-1H--indol-5-yloxy)-6-methoxyquinolin-7-Yl]oxy] methyl]cyclopropanamine dihydrochloride; AL3818) is a novel multitarget small molecule tyrosine kinase inhibitor that has been approved for the treatment of advanced non-small-cell lung cancer (NSCLC) [222]. Its targets are VEGFR, FGFR, PDGFR, and c-Kit [222]. Ruan et al. [219] investigated the therapeutic effect of anlotinib on IPF and its possible mechanisms. Their results showed that anlotinib could attenuate bleomycin-induced pulmonary fibrosis through suppressing the TGF-β signaling pathway, inhibit EMT, and promote myofibroblasts apoptosis in vitro. It was found that anlotinib hydrochloride (AL3818) has the same targets as nintedanib, an approved drug for IPF through interfering with processes like fibroblast proliferation, differentiation, and laying down extracellular matrix [223, 224].

7.4.2 Natural Products as MTD Therapeutics for RD

Pulmonary rehabilitation mixture (PRM), also known as Fei-Fu-Kang (FFK), is a Chinese herbal medicine formula that is designed based on an empirical-based traditional Chinese medicine prescription. PRM has been used to treat patients with pulmonary fibrosis and lung cancer for several decades [225]. Zhao et al. [226] systematically evaluated the pharmacodynamic and pharmacokinetic performance of PRM. The pharmacodynamic results showed that PRM could improve the condition of $CoCl_2$-stimulated human type II alveolar epithelial cells, human pulmonary microvascular endothelial cells, human lung fibroblasts, and pulmonary fibrosis rats induced by bleomycin; PRM treatment reduced the expression of platelet-derived growth factor, fibroblast growth factor, Toll-like receptor 4, high-mobility group box protein 1, and hypoxia-inducible factor 1α. In the pharmacokinetic study, an accurate and sensitive ultrahigh-performance liquid chromatography tandem mass spectrometry method was developed and validated for the simultaneous determination of calycosin, calycosin-7-*O*-glucoside, formononetin, ononin, and mangiferin of PRM in the rat plasma for the first time. The method was then successfully applied to the comparative pharmacokinetic study of PRM in normal and pulmonary fibrosis rats. The five constituents could be absorbed in the blood after the oral administration of PRM and exhibited different pharmacokinetic behaviors in normal and pulmonary fibrosis rats.

Unfortunately, there has not been any clinical trials testing the efficacy and safety of PRM in patients with PD.

7.4.3 Oligonucleotides as Sequence-Based MTD Epi-drugs for RD

Chronic respiratory diseases such as asthma and COPD are disorders of the airways largely related to the presence of persistent inflammation. The approval of inhaled corticosteroids in the early 1970s pioneered a new age of therapy in treating chronic inflammatory airway diseases. This was the first time that an anti-inflammatory product was available to reduce the characteristic lung inflammation in airways and the associated obstruction, inflammation, and hyperresponsiveness. Since then, corticosteroids have been widely prescribed for the treatment of asthma, COPD, and IPF. However, the effectiveness of corticosteroids as an anti-inflammatory drug is limited by steroid resistance in severe asthma, the majority of COPD cases, and pulmonary fibrosis. There is an urgent medical need to develop target-specific drugs for the treatment of these respiratory conditions.

Oligonucleotide therapies are an emerging class which include the antisense, the RNAi (siRNA and miRNA), the immunomodulatory, the aptamer, and the decoy approaches. As these approaches are rather recent in the respiratory field, most are still early in development [227–230]. Nevertheless, with limitations of current small molecule therapies and the hurdles faced with biologics, the use of oligonucleotides is relevant, and the door is open to the development of this category of therapeutics.

Oligonucleotides comprise a diverse family of emerging drug candidates for the treatment of these diseases. There is a commonality in the chemistry of oligonucleotide drug molecules, but, importantly, individual candidates are distinct in their molecular targets and also possess different mechanisms of action. Oligonucleotides may be divided in two groups based on their main mechanism of action: (a) RNA targeting and (b) protein targeting [229]. Antisense oligonucleotides and siRNAs, both of which belong to the RNA-targeting group, and the immunostimulatory sequence-based drugs, which are members of the protein-targeting group, are progressing through clinical trials. In principle, the development of these agents is facilitated by a series of significant advantages, such as those associated with delivery to the lung via inhalation, and targeting to the site of action, resulting in reduced levels of systemic exposure. Although no oligonucleotide-based drug has been approved for the treatment of asthma or COPD, this class of compounds holds significant promise in the treatment of these two diseases. Several oligonucleotide drug candidates have been evaluated in the clinic with no significant adverse safety findings, while clinical proof of concept based on short-term inhalation protocols in challenge models in humans has also been established [229].

Inhaled oligonucleotide is an emerging therapeutic modality for various common respiratory diseases, including obstructive airway diseases like asthma and COPD and restrictive airway diseases like IPF [230]. The advantage of direct accessibility for oligonucleotide molecules to the lung target sites, bypassing systemic administration, makes this therapeutic approach promising with minimized potential systemic side effects. Oligonucleotide therapies, including antisense oligonucleotide (ASO), small interfering RNA (siRNA), and microRNA (miRNA), are now being evaluated both preclinically and clinically as potential therapeutics. The mechanisms of action of ASO and siRNA are highly target mRNA specific, ultimately leading to target protein knockdown. miRNA has both biomarker and therapeutic values, and its knockdown by a miRNA antagonist (antagomir) has a broader but potentially more nonspecific biological outcome.

Since the first report describing the use of ASO targeting Rous sarcoma virus 35S RNA as a potential oligonucleotide therapy, there has been an exponential increase in the discovery and development of oligonucleotides as a therapeutic modality [230, 231]. ASO, siRNA, and miRNA are the three major types of oligonucleotide therapeutics under development. There are a few advantages for targeting RNA over protein molecules. As a single copy of RNA could be a template for the synthesis of multiple copies of a protein, it would be more efficient to regulate the mRNA level rather than the protein level to block the protein function [232]. Another benefit of using oligonucleotide strategy comes from the complete decoding of human genome, which greatly shortens the time between identification of gene targets and the design and validation of oligonucleotide products. Finally, several different oligonucleotide sequences can be combined into one single drug product, which is especially important for diseases with multifactorial etiology. Aside from ASO, siRNA, and miRNA, another recently described RNA class, long noncoding RNA (lncRNA), has also been implicated in respiratory diseases including asthma, COPD, and fibrosis [230, 233–235]. A number of promising therapeutic strategies to regulate lncRNA are in the preclinical stage, even though big challenges and obstacles are expected [236]. One big challenge of lncRNAs for therapeutic development is their poor conservation between species. So even if one lncRNA can be demonstrated as a successful therapeutic target in preclinical animal models, it might not be translated into humans. Thus, it seems more important to investigate human lncRNAs directly for therapeutic purposes [236].

Chronic respiratory diseases are uniquely suitable for the inhalation route of administration. In the lungs, uptake of aerosolized oligonucleotides has been demonstrated in immune cells like alveolar macrophages, as well as in structural cells such as bronchial and alveolar epithelial cells [230, 237, 238]. Absorption of oligonucleotides by inhalation, occurring mainly within the respiratory tracts with minimum systemic absorption, is a proposed advantage over oral or systemic administration. However, two industry-originating reports revealed that the bulk dose of PS-modified and PS-LNA-modified oligonucleotides could accumulate systemically in a biologically active fashion in rodents subjected to intratracheal administration [230, 239, 240]. Nevertheless, the effective and uniform intracellular delivery of oligonucleotides to their site of action still remains a challenge. Oligonucleotide delivery strategies commonly used in in vitro cell culture models such as cationic lipids, polymers, and electroporation are not suitable for in vivo application due to large particle size, pharmacokinetics constraints, and toxicity [230, 241, 242]. To date, most inhaled oligonucleotide products that have demonstrated efficacy in humans through aerosolization of simple aqueous solutions do not contain any carriers [228, 230]. To improve in vivo delivery and efficacy, new strategies like ligand-oligonucleotide conjugates, antibody-targeted conjugates, and engineered viral vector-, lipid-, and polymer-based nanoparticles may be further evaluated and developed [230, 243, 244].

Several shortcomings arising from the intrinsic properties of oligonucleotides may challenge oligonucleotide therapies [230]. These include their stability in vivo, associated nonspecific immune responses, and off-target cytotoxicity. dsRNA can activate innate immune responses and increase interferon production through recognition of Toll-like receptors (TLR)3, TLR7, and retinoic acid-inducible gene I protein [230]. Specific chemical modification, such as 2′-OME modification, can largely block binding and activation of these receptors and reduce immune responses. In addition, partial sequence complementarity of mRNAs with oligonucleotides may lead to sequence-specific off-target effects, which can be minimized by chemical modification, alternative sequence selection, and the use of pools of oligonucleotides that dilute off-target effects by reducing the concentration of individual oligonucleotide [230, 245, 246]. Although the completed phase I and phase II studies with EPI-2010, TPI ASM8, and AIR645 have not shown any major side effects thus far, safety studies with longer duration will need to be monitored closely as asthma, COPD, and fibrosis are chronic disorders that require long-term drug management [230].

Although recently there have been a few monoclonal antibodies approved by the FDA for the treatment of asthma, namely, omalizumab, mepolizumab, and reslizumab, the only viable route of administration for these antibodies is via injection, which is considered invasive [230, 247–249]. Oligonucleotides, on the other hand, can be given locally into the airways in a less invasive way, such as through aerosolized inhalation. Although topically administrated dsRNA could trigger nonspecific innate immune responses in the airway, systemically delivered monoclonal antibodies may produce much more pronounced life-threatening adverse effect such as a hypersensitivity reaction [250]. Another advantage of oligonucleotide over monoclonal antibody therapy is that oligonucleotide can theoretically target any molecule as it can penetrate cells, whereas a monoclonal antibody can mainly target extracellular or cell surface molecules [230]. Nevertheless, monoclonal antibody and oligonucleotide strategies are two different treatment modalities with their own pros and cons.

In summary, oligonucleotide therapeutics is a promising drug modality and is especially ideal for respiratory diseases due to its convenient topical airway administration and minimum systemic toxicity [230]. Even though no clinical trials have been performed to assess the impacts of miRNA-targeted therapies in respiratory diseases, several human clinical trials that focus on miRNA as directed therapeutics in other human diseases are currently ongoing.

7.4.4 Drug Repositioning of MTDs for RD

Drug repositioning as a polypharmacology strategy of new drug discovery and development for the treatment of COVID-19 has been introduced above. Here the discussion is focused on drug repositioning of multitarget drugs.

Pinzi et al. [251] reported their research results of a computational drug repurposing campaign that aimed at identifying potential inhibitors of the main protease (M^{pro}) of the SARS-CoV-2. The performed in silico screening allowed the identification of 22 candidates with putative SARS-CoV-2 M^{pro} inhibitory activity. Interestingly, some of the identified compounds have recently entered clinical trials for COVID-19 treatment, albeit not being assayed for their SARS-CoV-2 antiviral activity. Some candidates present a polypharmacology profile that may be beneficial for COVID-19 treatment.

Another group created a quantitative structure-activity relationship (QSAR) model based on a machine learning strategy using hundreds of inhibitor molecules of the main protease (M^{pro}) of the SARS-CoV coronavirus [252]. The QSAR model was used for virtual screening of a large list of drugs from the DrugBank database. The best 20 candidates were then evaluated in silico against the M^{pro} of SARS-CoV-2 by using docking and molecular dynamics analyses. Docking was done by using the Gold software, and the free energies of binding were predicted with the MM-PBSA method as implemented in AMBER. Their results indicate that levothyroxine, amobarbital, and ABP-700 are the best potential inhibitors of the SARS-CoV-2 virus through their binding to the M^{pro} enzyme. Five other compounds showed also a negative but small free energy of binding: nikethamide, nifurtimox, rebimastat, apomine, and rebastinib.

In the study reported by Kumar et al. [253], the SARS-CoV-2 M^{pro} was targeted for the screening of FDA-approved antiviral drugs and carried out molecular docking-based virtual screening. Molecular dynamic simulation studies of the top three selected drugs were carried out to investigate their binding affinity and stability in the SARS-CoV-2 M^{pro} active site. The phylogenetic analysis was also performed to know the relatedness between the SARS-CoV-2 genomes isolated from different countries. The phylogenetic analysis of the SARS-CoV-2 genome revealed that the virus is closely related to the Bat-SL-CoV and does not exhibit any divergence at the genomic level. Molecular docking studies revealed that among the 77 drugs, screened top 10 drugs show good binding affinities, whereas the top 3 drugs, lopinavir-ritonavir, tipranavir, and raltegravir, were undergone for molecular dynamics simulation studies for their conformational stability in the active site of the SARS-CoV-2 M^{pro} protein. Thus, the top three inhibitors lopinavir-ritonavir, tipranavir, and raltegravir show the best molecular interaction with the main protease of SARS-CoV-2. However, the in vitro efficacy of the drug molecules screened in this study further needs to be corroborated by carrying out a biochemical and structural investigation [253].

On the other hand, Delre et al. [254] presented a virtual screening campaign to identify covalent and non-covalent inhibitors of the SARS-CoV-2 papain-like protease (PL^{pro}) showing potential multitarget activities (i.e., a desirable polypharmacology profile) for the COVID-19 treatment. A dataset including 688 phase III and 1702 phase IV clinical trial drugs was downloaded from ChEMBL (version 27.1) and docked to the recently released crystal structure of PL^{pro} in complex with a covalently bound peptide inhibitor [254]. The obtained results were analyzed by combining protein-ligand interaction fingerprint similarities, conventional docking scores, and

MM-GBSA-binding free energies and allowed the identification of some interesting candidates for further in vitro testing [254].

Ren et al. [255] implemented a multitargeted antiviral drug design strategy to discover highly potent SARS-CoV-2 inhibitors, which simultaneously act on the host ribosome, viral RNA as well as RNA-dependent RNA polymerases, and nucleocapsid protein of the virus, to impair viral translation, frameshifting, replication, and assembly. Driven by this strategy, they discovered three alkaloids, including lycorine, emetine, and cephaeline, which inhibit SARS-CoV-2 with EC_{50} values of low nanomolar levels potently. Their findings demonstrate the feasibility of this multitargeting drug design strategy and provide a rationale for designing more potent antivirus drugs [255].

In the absence of any effective treatment, various in silico approaches are being explored toward the therapeutic discovery against COVID-19 [256]. Targeting multiple key enzymes of SARS-CoV-2 with a single potential drug could be an important in silico strategy to tackle the therapeutic emergency. A number of Food and Drug Administration (FDA)-approved drugs entered into clinical stages were originated from multitarget approaches with an increased rate, 16–21% between 2015 and 2017. Kumar et al. [256] selected an FDA-approved library (Prestwick Chemical Library of 1520 compounds) and implemented in silico virtual screening against multiple protein targets of SARS-CoV-2 on the Glide module of Schrödinger software (release 2020–1). They analyzed the compounds for their docking scores and the top ranked against each targeted protein were further subjected to molecular dynamics (MD) simulations to assess the binding stability of ligand-protein complexes. A multitargeting approach was optimized that enabled the analysis of several compounds' binding efficiency with more than one protein target. It was demonstrated that diosmin showed the highest binding affinity toward multiple targets with binding free energy (kcal/mol) values of −63.39 (nsp3), −62.89 (nsp9), −31.23 (nsp12), and −65.58 (nsp15). These results suggests that diosmin possesses multitargeting capability, a potent inhibitor of various nonstructural proteins of SARS-CoV-2. The main objective of this study was to find out multitargeting inhibitors against large number of proteins of SARS-CoV-2. The four hit compounds, (a) hesperidin (nsp3), (b) diosmin (nsp9), (c) catenulin (nsp12), and (d) acarbose (nsp15), were selected for analysis of their interactions with other proteins (nsp3, nsp9, nsp12 and nsp15) individually to obtain an insight about their multitargeting capability. Hesperidin was found to show high binding affinity toward two more proteins, nsp9 and nsp15 other than nsp3. Likewise, diosmin targeted three more proteins, nsp3, nsp12, and nsp15 other than nsp9. Catenulin also targeted one additional protein, nsp15 other than nsp12. On the other hand, acarbose was found to be selective against nsp15 only. Overall, three compounds hesperidin, diosmin, and catenulin showed binding affinity toward multiple protein targets. The best multitargeting capacity was displayed by diosmin, which showed interaction with four proteins, whereas hesperidin (1) showed interactions with three targeted proteins [256].

Consistently, hesperidin and diosmin are the best binding molecules analyzed against different proteins of SARS-CoV-2 (TMPRSS, 3CL-pro, S2-RBD, and PD-ACE2) [257, 258]. These two compounds also possess chemopreventive properties against cancer, viral infection, and inflammatory symptoms that could further help in combating the comorbidities of COVID-19 [258]. Notably, these two compounds are under clinical trial phase I and explored as treatment option against COVID-19 [255]. Interestingly, nsp15 protein was found to be a common target for four compounds and displayed remarkable interactions with key residues, Gln^{245} and Glu^{340}, present on the active site of the protein. Therefore, nsp15 represents an attractive target for the development of therapeutics against SARS-CoV-2 [257, 258].

References

1. Barnes PJ. Pulmonary diseases and ageing. Subcell Biochem. 2019;91:45–74.
2. de Palo VA. Pulmonary disease: pneumonia, chronic obstructive pulmonary disease, asthma, and thromboembolic disease. J Am Podiatr Med Assoc. 2004;94(2):157–67.
3. Han MK, McLaughlin VV, Criner GJ, et al. Pulmonary diseases and the heart. Circulation. 2007;116(25):2992–3005.
4. Walker PA, White DA. Pulmonary disease. Med Clin North Am. 1996;80(6):1337–62.
5. Asthma Fact sheet №307. WHO. November 2013. Archived from the original on June 29, 2011. Retrieved 3 March 2016.
6. National Asthma Education and Prevention Program. Expert Panel Report 3: Guidelines for the Diagnosis and Management of Asthma. (PDF). National Heart Lung and Blood Institute; 2007. Archived from the original (PDF) on 2013-10-19. Retrieved 2005-08-31
7. Global Strategy for Asthma Management and Prevention. Global Initiative for Asthma. 2011. Archived from the original (PDF) on 2012-11-20.
8. Martinez FD. Genes, environments, development and asthma: a reappraisal. Eur Respir J. 2007;29(1):179–84.
9. Fanta CH. Asthma. N Engl J Med. 2009;360: 1002–14.
10. Fahy JV. Eosinophilic and neutrophilic inflammation in asthma: insights from clinical studies. Proc Am Thorac Soc. 2009;6:256–9.
11. Bush A, Zar HJ. WHO universal definition of severe asthma. Curr Opin Allergy Clin Immunol. 2011;11:115–21.
12. Martinez FD, Vercelli D. Asthma Lancet. 2013;382:1360–72.
13. Braman SS. The global burden of asthma. Chest. 2006;130:S 4–12.
14. Kirkland SW, Vandenberghe C, Voaklander B, et al. Combined inhaled beta-agonist and anticholinergic agents for emergency management in adults with asthma. Cochrane Database Syst Rev. 2017;2017(1):CD001284.
15. Ito K, Chung KF, Adcock IM. Update on glucocorticoid action and resistance. J Allergy Clin Immunol. 2006;117:522–43.
16. QRG 153. British guideline on the management of asthma (PDF). SIGN; 2016. Archived (PDF) from the original on 9 October 2016. Retrieved 6 October 2016
17. Normansell R, Sayer B, Waterson S, et al. Antibiotics for exacerbations of asthma. The. Cochrane Database Syst Rev. 2018;2018(6):CD002741.
18. COPD. Chapter 1: Definition and overview. In: Global strategy for prevention, diagnosis and management of COPD: 2021 report. Global Initiative for Chronic Obstructive Lung Disease; 2021. p. 4–8.
19. COPD. Chapter 2: Diagnosis and initial assessment. In: Global strategy for prevention, diagnosis and management of COPD: 2021 Report. Global Initiative for Chronic Obstructive Lung Disease; 2021. p. 20–7.
20. Chronic Obstructive Pulmonary Disease (COPD). www.who.int. Retrieved 1 July 2021.
21. Myc LA, Shim YM, Laubach VE, et al. Role of medical and molecular imaging in COPD. Clin Transl Med. 2019;8(1):12.
22. GBD 2015. Disease and Injury Incidence and Prevalence Collaborators (October 2016). Global, regional, and national incidence, prevalence, and years lived with disability for 310 diseases and injuries, 1990–2015: a systematic analysis for the Global Burden of Disease Study 2015. Lancet. 2016;388(10053):1545–602.
23. Chronic obstructive pulmonary disease. NICE. Retrieved 5 July 2021.
24. GBD 2015 Mortality and Causes of Death Collaborators. Global, regional, and national life expectancy, all-cause mortality, and cause-specific mortality for 249 causes of death, 1980–2015: a systematic analysis for the Global Burden of Disease Study 2015. Lancet. 2016; 388(10053): 1459–1544.
25. GBD 2013 Mortality and Causes of Death Collaborators. Global, regional, and national age-sex specific all-cause and cause-specific mortality for 240 causes of death, 1990–2013: a systematic analysis for the Global Burden of Disease Study 2013. Lancet. 2015; 385(9963): 117–171.
26. Decramer M, Janssens W, Miravitlles M. Chronic obstructive pulmonary disease. Lancet. 2012;379:1341–51.
27. COPD. Chapter 3: Evidence supporting prevention and maintenance therapy. In: Global strategy for prevention, diagnosis and management of COPD: 2021 report. Global Initiative for Chronic Obstructive Lung Disease; 2021. p. 60–5.
28. Vollenweider DJ, Frei A, Steurer-Stey CA, et al. Antibiotics for exacerbations of chronic obstructive pulmonary disease. Cochrane Database Syst Rev. 2018;2018(10):CD010257.
29. Mackay AJ, Hurst JR. COPD exacerbations: causes, prevention, and treatment. Med Clin North Am. 2012;96(4):789–809.
30. Fluoroquinolone Antibacterial Drugs: Drug Safety Communication – FDA Advises Restricting Use for Certain Uncomplicated Infections. FDA. 12 May 2016. Archived from the original on 16 May 2016. Retrieved 16 May 2016.
31. Vestbo J, Hurd SS, Agustí AG, et al. Global strategy for the diagnosis, management, and prevention of chronic obstructive pulmonary disease: GOLD executive summary. Am J Respir Crit Care Med. 2013;187(4):347–65.
32. Oliveira EP, Medeiros Junior P. Palliative care in pulmonary medicine. J Bras Pneumol. 2020;46(3):e20190280.

33. Wilson ME, Dobler CC, Morrow AS, et al. Association of home noninvasive positive pressure ventilation with clinical outcomes in chronic obstructive pulmonary disease: a systematic review and meta-analysis. JAMA. 2020;323(5):455–65.
34. Ley B, Collard HR. Epidemiology of idiopathic pulmonary fibrosis. Clin Epidemiol. 2013;5:483–92.
35. Harari S, Caminati A. IPF: new insight on pathogenesis and treatment. Allergy. 2010;65:537–53.
36. Loomis-King H, Flaherty KR, Moore BB. Pathogenesis, current treatments and future directions for idiopathic pulmonary fibrosis. Curr Opin Pharmacol. 2013;13:377–85.
37. Schaefer CJ, Ruhrmund DW, Pan L, Seiwert SD, Kossen K. Antifibrotic activities of pirfenidone in animal models. Eur Respir Rev. 2011;20:85–97.
38. Raghu G, Anstrom KJ, King TE, et al. Prednisone, azathioprine, and N-acetylcysteine for pulmonary fibrosis. N Engl J Med. 2012;366:1968–77.
39. Tuberculosis (TB). WHO. Archived from the original on 30 July 2020. Retrieved 8 May 2020.
40. Basic TB Facts. Centers for Disease Control and Prevention (CDC). 13 March 2012. Archived from the original on 6 February 2016. Retrieved 11 February 2016.
41. Tuberculosis (TB). World Health Organization (WHO). 16 February 2018. Archived from the original on 30 December 2013. Retrieved 15 September 2018.
42. Tuberculosis. World Health Organization (WHO). 2002. Archived from the original on 17 June 2013.
43. Tuberculosis (TB). WHO. Archived from the original on 30 July 2020. Retrieved 16 October 2021.
44. Konstantinos A. Testing for tuberculosis. Aust Prescr. 2010;33(1):12–8.
45. Brennan PJ, Nikaido H. The envelope of mycobacteria. Annu Rev Biochem. 1995;64:29–63.
46. Lawn SD, Zumla AI. Tuberculosis. Lancet. 2011;378(9785):57–72.
47. Davies G, Cerri S, Richeldi L. Rifabutin for treating pulmonary tuberculosis. Cochrane Database Syst Rev. 2007;2007(4):CD005159.
48. Latent tuberculosis infection. World Health Organization (WHO). 2018. p. 23. ISBN:978-92-4-155023-9. Archived from the original on 2 June 2021.
49. Borisov AS, Bamrah Morris S, Njie GJ, et al. Update of recommendations for use of once-weekly isoniazid-rifapentine regimen to treat latent mycobacterium tuberculosis infection. MMWR Morb Mortal Wkly Rep. 2018;67(25):723–6.
50. Sterling TR, Njie G, Zenner D, et al. Guidelines for the treatment of latent tuberculosis infection: recommendations from the National Tuberculosis Controllers Association and CDC, 2020. MMWR Recomm Rep. 2020;69(1):1–11.
51. Chams N, Chams S, Badran R, et al. COVID-19: a multidisciplinary review. Front Public Health. 2020;8:383.
52. Ahn DG, Shin HJ, Kim MH, et al. Current status of epidemiology, diagnosis, therapeutics, and vaccines for novel coronavirus disease 2019 (COVID-19). J Microbiol Biotechnol. 2020;30(3):313–24.
53. Harrison AG, Lin T, Wang P. Mechanisms of SARS-CoV-2 transmission and pathogenesis. Trends Immunol. 2020;41(12):1100–15.
54. Verdecchia P, Cavallini C, Spanevello A, et al. The pivotal link between ACE2 deficiency and SARS-CoV-2 infection. Eur J Intern Med. 2020;76:14–20.
55. Letko M, Marzi A, Munster V. Functional assessment of cell entry and receptor usage for SARS-CoV-2 and other lineage B betacoronaviruses. Nat Microbiol. 2020;5(4):562–9.
56. Hui DS, Azhar IE, Madani TA, et al. The continuing 2019-nCoV epidemic threat of novel coronaviruses to global health – the latest 2019 novel coronavirus outbreak in Wuhan, China. Int J Infect Dis. 2020;91:264–6.
57. Murthy S, Gomersall CD, Fowler RA. Care for critically ill patients with COVID-19. JAMA. 2020;323(15):1499–500.
58. Cascella M, Rajnik M, Cuomo A, et al. Features, evaluation and treatment coronavirus (COVID-19). Treasure Island: StatPearls. Retrieved 18 March 2020
59. Heymann DL, Shindo N, et al. WHO Scientific and Technical Advisory Group for Infectious Hazards. (February). COVID-19: what is next for public health? Lancet. 2020;395(10224):542–5.
60. Romiti GF, Corica B, Lip GY, et al. Prevalence and impact of atrial fibrillation in hospitalized patients with COVID-19: a systematic review and meta-analysis. J Clin Med. 2021;10(11):2490.
61. Wen W, Zhang H, Zhou M, et al. Arrhythmia in patients with severe coronavirus disease (COVID-19): a meta-analysis. Eur Rev Med Pharmacol Sci. 2020;24(21):11395–401.
62. Long B, Brady WJ, Koyfman A, et al. Cardiovascular complications in COVID-19. Am J Emerg Med. 2020;38(7):1504–7.
63. Puntmann VO, Carerj ML, Wieters I, et al. Outcomes of cardiovascular magnetic resonance imaging in patients recently recovered from coronavirus disease 2019 (COVID-19). JAMA Cardiol. 2020;5(11):1265–73.
64. Lindner D, Fitzek A, Bräuninger H, et al. Association of cardiac infection with SARS-CoV-2 in confirmed COVID-19 autopsy cases. JAMA Cardiol. 2020;5(11):1281–5.
65. Siripanthong B, Nazarian S, Muser D, et al. Recognizing COVID-19-related myocarditis: the possible pathophysiology and proposed guideline for diagnosis and management. Heart Rhythm. 2020;17(9):1463–71.
66. Xu L, Liu J, Lu M, et al. Liver injury during highly pathogenic human coronavirus infections. Liver Int. 2020;40(5):998–1004.
67. Sanders JM, Monogue ML, Jodlowski TZ, et al. Pharmacologic treatments for coronavi-

rus disease 2019 (COVID-19): a review. JAMA. 2020;323(18):1824–36.

68. Carod-Artal FJ. Neurological complications of coronavirus and COVID-19. Rev Neurol. 2020;70(9):311–22.
69. Toscano G, Palmerini F, Ravaglia S, et al. Guillain-Barré syndrome associated with SARS-CoV-2. N Engl J Med. 2020;382(26):2574–6.
70. Siemieniuk RA, Bartoszko JJ, Ge L, et al. Drug treatments for covid-19: living systematic review and network meta-analysis. BMJ. 2020;370:m2980.
71. COVID-19 vaccine and therapeutics tracker. BioRender. 5 April 2021. Retrieved 21 June 2021.
72. Kupferschmidt K, Cohen J. WHO launches global megatrial of the four most promising coronavirus treatments. Sci Magazine. 2020.
73. First regulatory workshop on COVID-19 facilitates global collaboration on vaccine development. European Medicines Agency. 18 March 2020. Retrieved 21 March 2020.
74. Coronavirus (COVID-19) Update: FDA Continues to Facilitate Development of Treatments (Press release). U.S. Food and Drug Administration (FDA). 19 March 2020. Retrieved 21 March 2020.
75. China approves first anti-viral drug against coronavirus Covid-19. Clinical Trials Arena. 18 February 2020. Retrieved 21 March 2020.
76. Chinese Vaccine Approved for Human Testing at Virus Epicenter. Bloomberg News. 19 March 2020. Retrieved 21 March 2020.
77. Dhama K, Sharun K, Tiwari R, et al. COVID-19, an emerging coronavirus infection: advances and prospects in designing and developing vaccines, immunotherapeutics, and therapeutics. Hum Vaccin Immunother. 2020;16(6):1232–8.
78. Zhang L, Liu Y. Potential interventions for novel coronavirus in China: a systematic review. J Med Virol. 2020;92(5):479–90.
79. Fox M (19 March 2020). Drug makers are racing to develop immune therapies for Covid-19. Will they be ready in time? Stat. Retrieved 21 March 2020.
80. COVID-19 Clinical Research Coalition. Global coalition to accelerate COVID-19 clinical research in resource-limited settings. Lancet. 2020;395(10233):1322–5.
81. Maguire BJ, Guérin PJ. A living systematic review protocol for COVID-19 clinical trial registrations. Wellcome Open Res. 2020;5:60.
82. Li G, De Clercq E. Therapeutic options for the 2019 novel coronavirus (2019-nCoV). Nat Rev Drug Discov. 2020;19(3):149–50.
83. Dong L, Hu S, Gao J. Discovering drugs to treat coronavirus disease 2019 (COVID-19). Drug Discov Therapeutics. 2020;14(1):58–60.
84. Harrison C. Coronavirus puts drug repurposing on the fast track. Nat Biotechnol. 2020;38(4):379–81.
85. COVID-19 vaccine and treatments tracker (Choose vaccines or treatments tab, apply filters to view select data). Milken Institute. 21 June 2021. Retrieved 11 June 2021.
86. Saima MS. Common antidepressant slashes risk of COVID death, study says. Nature. 2021; https://doi.org/10.1038/d41586-021-02988-4.
87. Pushkala A. Britain approves Merck's COVID-19 pill in world first. Reuters. Retrieved 8 November 2021.
88. Deena B. Pfizer says its antiviral pill slashes risk of severe COVID-19 by 89%. Reuters. Retrieved 8 November 2021.
89. Jayk Bernal A, Gomes da Silva MM, Musungaie DB, et al. MOVe-OUT Study Group. Molnupiravir for oral treatment of COVID-19 in nonhospitalized patients. N Engl J Med. 2021; https://doi.org/10.1056/NEJMoa2116044.
90. Mahase E. Covid-19: Molnupiravir reduces risk of hospital admission or death by 50% in patients at risk, MSD reports. BMJ. 2021;375:n2422.
91. Kim PS, Read SW, Fauci AS. Therapy for early COVID-19: a critical need. JAMA. AMA. 2020;324(21):2149–50.
92. COVID-19 Treatment Guidelines. www.nih.gov. National Institutes of Health. Retrieved 18 January 2021.
93. Hsu J. Covid-19: what now for remdesivir? BMJ. 2020;371:m4457.
94. Molnupiravir: First pill to treat Covid gets approval in UK Jim Reed, 4 November 2021. www.bbc.co.uk. Accessed 23 Nov 2021.
95. Blair HA. Tiotropium/olodaterol: a review in COPD. Drugs. 2019;79(9):997–1008.
96. Cazzola M, Page CP, Calzetta L, et al. Pharmacology and therapeutics of bronchodilators. Pharmacol Rev. 2012;64:450–504.
97. Tashkin DP, Cooper CB. The role of long-acting bronchodilators in the management of stable COPD. Chest. 2004;125:249–59.
98. Tashkin DP, Ferguson GT. Combination bronchodilator therapy in the management of chronic obstructive pulmonary disease. Respir Res. 2013;14:49.
99. Global Initiative for Chronic Obstructive Lung Disease. Global strategy for the diagnosis, management and prevention of chronic obstructive pulmonary disease. 2017. http://goldcopd.org/gold-2017-global-strategy-diagnosis-management-prevention-copd/. Accessed Feb 2021.
100. Feldman GJ, Sousa AR, Lipson DA, et al. Comparative efficacy of once-daily umeclidinium/vilanterol and tiotropium/olodaterol therapy in symptomatic chronic obstructive pulmonary disease: a randomized study. Adv Ther. 2017;34(11):2518–33.
101. Feldman G, Maltais F, Khindri S, et al. A randomized, blinded study to evaluate the efficacy and safety of umeclidinium 62.5 mcg compared with tiotropium 18 mcg in patients with COPD. Int J Chron Obstruct Pulmon Dis. 2016;11:719–30.
102. Bateman ED, Ferguson GT, Barnes N, et al. Dual bronchodilation with QVA149 versus single bronchodilator therapy: the SHINE study. Eur Respir J. 2013;42:1484–94.

103. Decramer M, Anzueto A, Kerwin E, et al. Efficacy and safety of umeclidinium plus vilanterol versus tiotropium, vilanterol, or umeclidinium monotherapies over 24 weeks in patients with chronic obstructive pulmonary disease: results from two multicentre, blinded, randomised controlled trials. Lancet Respir Med. 2014;2:472–86.
104. Maleki-Yazdi MR, Kaelin T, Richard N, et al. Efficacy and safety of umeclidinium/vilanterol 62.5/25 mcg and tiotropium 18 mcg in chronic obstructive pulmonary disease: results of a 24-week, randomized, controlled trial. Respir Med. 2014;108:1752–60.
105. Buhl R, Maltais F, Abrahams R, et al. Tiotropium and olodaterol fixed-dose combination versus monocomponents in COPD (GOLD 2–4). Eur Respir J. 2015;45:969–79.
106. ZuWallack R, Allen L, Hernandez G, et al. Efficacy and safety of combining olodaterol Respimat® and tiotropium HandiHaler® in patients with COPD: results of two randomized, double-blind, active-controlled studies. Int J Chron Obstruct Pulmon Dis. 2014;9:1133–44.
107. Singh D, Ferguson GT, Bolitschek J, et al. Tiotropium + olodaterol shows clinically meaningful improvements in quality of life. Respir Med. 2015;109:1312–9.
108. Wedzicha JA, Decramer M, Ficker JH, et al. Analysis of chronic obstructive pulmonary disease exacerbations with the dual bronchodilator QVA149 compared with glycopyrronium and tiotropium (SPARK): a randomised, double-blind, parallel-group study. Lancet Respir Med. 2013;1:199–209.
109. Rothstein DM. Rifamycins, alone and in combination. Cold Spring Harb Perspect Med. 2016;6(7):a027011.
110. Nuermberger EL, Spigelman MK, Yew WW. Current development and future prospects in chemotherapy of tuberculosis. Respirology. 2010;15:764–78.
111. Fox W, Ellard GA, Mitchison DA. Studies on the treatment of tuberculosis undertaken by the British Medical Research Council tuberculosis units, 1946–1986, with relevant subsequent publications. Int J Tuberc Lung Dis. 1999;3:S231–79.
112. Jindani A, Doré CJ, Mitchison DA. Bactericidal and sterilizing activities of antituberculosis drugs during the first 14 days. Am J Respir Crit Care Med. 2003;167:1348–54.
113. Mitchison D, Davies G. The chemotherapy of tuberculosis: past, present and future. Tubercul Lung Dis. 2012;16:724–32.
114. Jindani A, Harrison TS, Nunn AJ, et al. High-dose rifapentine with moxifloxacin for pulmonary tuberculosis. N Engl J Med. 2014;371:1599–608.
115. Mitnick CD, McGee B, Peloquin CA. Tuberculosis pharmacotherapy: strategies to optimize patient care. Expert Opin Pharmacother. 2009;10:381–401.
116. Burman WJ, Gallicano K, Peloquin C. Comparative pharmacokinetics and pharmacodynamics of the rifamycin antibacterials. Clin Pharmacokinet. 2001;40:327–41.
117. FDA now allows treatment of life-threatening COVID-19 cases using blood from patients who have recovered. TechCrunch. Retrieved 9 April 2020.
118. Blood plasma taken from covid-19 survivors might help patients fight off the disease. MIT Technology Review.
119. Trials of Plasma From Recovered Covid-19 Patients Have Begun. Wired.
120. Fioravanti C. Butantan develops serum against coronavirus. Revista Pesquisa Fapesp (in Portuguese). 2020. Retrieved 20 January 2021.
121. Monnerat A. Vital Brazil develops serum against COVID-19, but drug has yet to be tested in humans. O Estado de S. Paulo (in Portuguese). 2021. Retrieved 20 January 2021.
122. Argentina approves use of equine serum which reduces COVID-19 deaths by 45%. Deutsche Welle. 11 January 2021. Retrieved 20 January 2021.
123. Coronavirus: medics "strongly" reject use of equine serum in severe patients. Clarín (in Spanish). 17 January 2021. Retrieved 20 January 2021.
124. Casirivimab injection, solution, concentrate Imdevimab injection, solution, concentrate REGEN-COV – casirivimab and imdevimab kit. DailyMed. Retrieved 18 March 2021.
125. Ronapreve EPAR. European Medicines Agency. Text was copied from this source which is copyright European Medicines Agency. Reproduction is authorized provided the source is acknowledged. Retrieved 12 November 2021.
126. Baum A, Fulton BO, Wloga E, et al. Antibody cocktail to SARS-CoV-2 spike protein prevents rapid mutational escape seen with individual antibodies. Science. 2020;369(6506):1014–8.
127. Fact sheet for health care providers emergency use authorization (EUA) of casirivimab and imdevimab. U.S. Food and Drug Administration (FDA). June 2021.
128. Kelland K. Regeneron's antibody drug added to UK Recovery trial of COVID treatments. Reuters. Retrieved 14 September 2020.
129. Regeneron's COVID-19 response efforts. Regeneron Pharmaceuticals. Retrieved 14 September 2020.
130. Ronapreve. Therapeutic Goods Administration (TGA). 18 October 2021. Retrieved 22 October 2021.
131. Summary of Product Characteristics for Ronapreve. Medicines and Healthcare products Regulatory Agency (MHRA). 20 August 2021. Retrieved 29 August 2021.
132. Japan becomes first country to approve Ronapreve (casirivimab and imdevimab) for the treatment of mild to moderate COVID-19. Roche (Press release). 20 July 2021. Retrieved 29 August 2021.
133. Regulatory approval of Ronapreve. Medicines and Healthcare products Regulatory Agency (MHRA). 20 August 2021. Retrieved 29 August 2021.
134. First monoclonal antibody treatment for COVID-19 approved for use in the UK (Press release).

Medicines and Healthcare products Regulatory Agency (MHRA). 20 August 2021. Retrieved 29 August 2021.
135. FDA Authorizes Monoclonal Antibodies for Treatment of COVID-19. U.S. Food and Drug Administration (FDA) (Press release). 10 February 2021. This article incorporates text from this source, which is in the public domain. Retrieved 9 February 2021.
136. Lilly's bamlanivimab (LY-CoV555) administered with etesevimab (LY-CoV016) receives FDA emergency use authorization for COVID-19 (Press release). Eli Lilly and Company. 9 February 2021. Via PR Newswire. Retrieved 9 February 2021.
137. New data show treatment with Lilly's neutralizing antibodies bamlanivimab (LY-CoV555) and etesevimab (LY-CoV016) together reduced risk of COVID-19 hospitalizations and death by 70 percent (Press release). Eli Lilly and Company. 26 January 2021. Via PR Newswire. Retrieved 9 February 2021.
138. Hurt AC, Wheatley AK. Neutralizing antibody therapeutics for COVID-19. Viruses. 2021;13(4):628.
139. Etesevimab. IUPHAR/BPS Guide to Pharmacology. Retrieved 10 February 2021.
140. Lilly announces agreement with U.S. government to supply 300,000 vials of investigational neutralizing antibody bamlanivimab (LY-CoV555) in an effort to fight COVID-19. Eli Lilly and Company (Press release). 28 October 2020.
141. FDA authorizes bamlanivimab and etesevimab for COVID-19. U.S. Food and Drug Administration. 16 September 2021. Retrieved 16 September 2021.
142. Emergency Use Authorization 094. U.S. Food and Drug Administration (FDA). 16 September 2021.
143. Ray S. AstraZeneca's Covid-19 antibody therapy effective in preventing symptoms among high-risk groups, trial finds. Forbes. Archived from the original on 21 August 2021. Retrieved 21 August 2021.
144. Goriainoff AO. AstraZeneca says AZD7442 antibody phase 3 trial met primary endpoint in preventing Covid-19. MarketWatch. Archived from the original on 21 August 2021. Retrieved 21 August 2021.
145. Coronavirus (COVID-19) Update: FDA Authorizes New Long-Acting Monoclonal Antibodies for Pre-exposure Prevention of COVID-19 in Certain Individuals. U.S. Food and Drug Administration (FDA) (Press release). 8 December 2021. Retrieved 9 December 2021.
146. Vandyck K, Deval J. Considerations for the discovery and development of 3-chymotrypsin-like cysteine protease inhibitors targeting SARS-CoV-2 infection. Curr Opin Virol. 2021;49:36–40.
147. Şimşek-Yavuz S, Komsuoğlu Çelikyurt FI. Antiviral treatment of COVID-19: an update. Turk J Med Sci. 2021;
148. Ahmad B, Batool M, Ain QU, et al. Exploring the binding mechanism of PF-07321332 SARS-CoV-2 protease inhibitor through molecular dynamics and binding free energy simulations. Int J Mol Sci. 2021;22(17):9124.
149. Pfizer Announces Additional Phase 2/3 Study Results Confirming Robust Efficacy of Novel COVID-19 Oral Antiviral Treatment Candidate in Reducing Risk of Hospitalization or Death (Press release). 14 December 2021.
150. Anand K, Ziebuhr J, Wadhwani P, et al. Coronavirus main proteinase (3CLpro) structure: basis for design of anti-SARS drugs. Science. 2003;300(5626):1763–7.
151. Pfizer Covid pill can cut hospitalisations and deaths by nearly 90%. The Guardian. 5 November 2021. Retrieved 17 November 2021.
152. Nuki P. Pfizer is testing a pill that, if successful, could become first-ever home cure for COVID-19. National Post. Archived from the original on 27 April 2021.
153. Pfizer begins dosing in Phase II/III trial of antiviral drug for Covid-19. Clinical Trials Arena. 2 September 2021.
154. Pfizer's Novel COVID-19 Oral Antiviral Treatment Candidate Reduced Risk of Hospitalization or Death by 89% in Interim Analysis of Phase 2/3 EPIC-HR Study (Press release). Pfizer Inc. 5 November 2021.
155. Mahase E. Covid-19: Pfizer's paxlovid is 89% effective in patients at risk of serious illness, company reports. BMJ. 2021;375:n2713.
156. Macchiagodena M, Pagliai M, Procacci P. Characterization of the non-covalent interaction between the PF-07321332 inhibitor and the SARS-CoV-2 main protease. J Mol Graph Model. 2022;110:108042.
157. Pavan M, Bolcato G, Bassani D, et al. Supervised molecular dynamics (SuMD) insights into the mechanism of action of SARS-CoV-2 main protease inhibitor PF-07321332. J Enzyme Inhib Med Chem. 2021;36(1):1646–50.
158. Summary of Product Characteristics for Lagevrio. Medicines and Healthcare products Regulatory Agency (MHRA). 4 November 2021. Retrieved 4 November 2021.
159. Toots M, Yoon JJ, Cox RM, et al. Characterization of orally efficacious influenza drug with high resistance barrier in ferrets and human airway epithelia. Sci Transl Med. 2019;11(515):eaax5866.
160. Toots M, Yoon JJ, Hart M. Quantitative efficacy paradigms of the influenza clinical drug candidate EIDD-2801 in the ferret model. Transl Res. 2020;218:16–28.
161. Emails offer look into whistleblower charges of cronyism behind potential COVID-19 drug.
162. Cully M. A tale of two antiviral targets – and the COVID-19 drugs that bind them. Nat Rev Drug Discov. 2021;
163. Aleccia J. Daily pill to treat COVID could be just months away. ABC News. Kaiser Health News. Archived from the original on 29 September 2021. Retrieved 29 September 2021.

164. First oral antiviral for COVID-19, Lagevrio (molnupiravir), approved by MHRA (Press release). Medicines and Healthcare products Regulatory Agency (MHRA). 4 November 2021.
165. Merck and Ridgeback's Molnupiravir, an Oral COVID-19 Antiviral Medicine, Receives First Authorization in the World. Merck (Press release). 4 November 2021. Retrieved 4 November 2021.
166. U.K. Authorizes Merck's Antiviral Pill for Covid-19. The New York Times. 4 November 2021. Retrieved 27 November 2021.
167. Gordon CJ, Tchesnokov EP, Schinazi RF, et al. Molnupiravir promotes SARS-CoV-2 mutagenesis via the RNA template. J Biol Chem. 2021;297(1):100770.
168. Repurposed drug candidates. https://en.wikipedia.org/wiki/COVID-19_drug_development.
169. Kruse RL. Therapeutic strategies in an outbreak scenario to treat the novel coronavirus originating in Wuhan, China. F1000 Res. 2020;9:72.
170. UN health chief announces global 'solidarity trial' to jumpstart search for COVID-19 treatment. United Nations – News. World Health Organization. 18 March 2020. Retrieved 29 March 2020.
171. Kupferschmidt K, Cohen J. Race to find COVID-19 treatments accelerates. Science. 2020;367(6485):1412–3.
172. Kupferschmidt K, Cohen J. WHO launches global megatrial of the four most promising coronavirus treatments. Science Magazine. 2020.
173. COVID-19 drug development: Landscape analysis of therapeutics. United Nations, World Health Organization. 21 March 2020. Retrieved 29 March 2020.
174. Ponce MC, Sharma S. Pulmonary function tests. In: StatPearls [Internet]. Treasure Island: StatPearls Publishing; 2020.
175. Nathell L, Nathell M, Malmberg P, et al. COPD diagnosis related to different guidelines and spirometry techniques. Respir Res. 2007;8(1):89.
176. Qaseem A, Wilt TJ, Weinberger SE, et al. Diagnosis and management of stable chronic obstructive pulmonary disease: a clinical practice guideline update from the American College of Physicians, American College of Chest Physicians, American Thoracic Society, and European Respiratory Society. Ann Intern Med. 2011;155(3):179–91.
177. Kelly E. Umeclidinium bromide and vilanterol in combination for the treatment of chronic obstructive pulmonary disease. Expert Rev Clin Pharmacol. 2014;7(4):403–13.
178. Global Initiative for Chronic Obstructive Lung Disease. Global strategy for the diagnosis, management, and prevention of chronic obstructive pulmonary disease. 2019. http://goldcopd.org/gold--reports/. Accessed 7 May 2019.
179. Boehringer Ingelheim. Stiolto® Respimat® (tiotropium bromide and olodaterol): US prescribing information. 2018. http://www.fda.gov. Accessed 7 May 2019.
180. Pharmaceuticals and Medical Devices Agency. Spiolto Respimat 28 puffs/Spiolto Respimat 60 puffs: Japanese prescribing information. 2015. http://www.pmda.go.jp. Accessed 7 May 2019.
181. National Drug Administration Drug Evaluation Center. Tiotropium bromide olodaterol inhalation spray: Chinese prescribing information. 2018. http://202.96.26.102/. Accessed 7 May 2019.
182. European Medicines Agency. Spiolto® Respimat® (tiotropium bromide and olodaterol): summary of product characteristics. 2018. http://www.ema.europa.eu. Accessed 7 May 2019.
183. Deeks ED. Olodaterol: a review of its use in chronic obstructive pulmonary disease. Drugs. 2015;75(6):665–73.
184. Keating GM. Tiotropium Respimat® Soft Mist™ inhaler: a review of its use in chronic obstructive pulmonary disease. Drugs. 2014;74(15):1801–16.
185. Dhillon S. Tiotropium/olodaterol: a review in COPD. Drugs. 2016;76(1):135–46.
186. Rodrigo GJ, Price D, Anzueto A, et al. LABA/LAMA combinations versus LAMA monotherapy or LABA/ICS in COPD: a systematic review and meta-analysis. Int J Chron Obstruct Pulmon Dis. 2017;12:907–22.
187. Cazzola M, Segreti A, Matera MG. New developments in the combination treatment of COPD: focus on umeclidinium/vilanterol. Drug Des Devel Ther. 2013;7:1201–8.
188. Kalberg C, O'Dell D, Galkin D, et al. Dual bronchodilator therapy with umeclidinium/vilanterol versus tiotropium plus indacaterol in chronic obstructive pulmonary disease: a randomized controlled trial. Drugs R D. 2016;16(2):217–27.
189. Hurst JR, Gruffydd-Jones K, Biswas M, et al. Efficacy and safety of LAMA/LABA fixed-dose combination therapies in chronic obstructive pulmonary disease: a systematic review of direct and indirect treatment comparisons. Int J Chron Obstruct Pulmon Dis. 2020;15:1529–43.
190. Frampton JE. QVA149 (indacaterol/glycopyrronium fixed-dose combination): a review of its use in patients with chronic obstructive pulmonary disease. Drugs. 2014;74(4):465–88.
191. Banerji D, Fogel R, Patalano F. Indacaterol/glycopyrronium: a dual bronchodilator for COPD. Drug Discov Today. 2018;23(1):196–203.
192. Rossi A, Zanardi E, Poletti V, et al. Clinical role of dual bronchodilation with an indacaterol-glycopyrronium combination in the management of COPD: its impact on patient-related outcomes and quality of life. Int J Chron Obstruct Pulmon Dis. 2015;10:1383–92.
193. Singh D, Corradi M, Spinola M, et al. Extrafine beclometasone dipropionate/formoterol fumarate: a review of its effects in chronic obstructive pulmonary disease. NPJ Prim Care Respir Med. 2016;26:16030.
194. Calverley P, Pauwels R, Vestbo J, et al. Combined salmeterol and fluticasone in the treatment of chronic

obstructive pulmonary disease: a randomised controlled trial. Lancet. 2003;361:449–56.
195. Calverley P, Boonsawat W, Cseke Z, et al. Maintenance therapy with budesonide and formoterol in chronic obstructive pulmonary disease. Eur Respir J. 2003;22:912–9.
196. Ferguson GT, Anzueto A, Fei R, et al. Effect of fluticasone propionate/salmeterol (250/50 microg) or salmeterol (50 microg) on COPD exacerbations. Respir Med. 2008;102:1099–108.
197. Szafranski W, Cukier A, Ramirez A, et al. Efficacy and safety of budesonide-formoterol in the management of chronic obstructive pulmonary disease. Eur Respir J. 2003;21:74–81.
198. Dransfield MT, Bourbeau J, Jones PW, et al. Once-daily inhaled fluticasone furoate and vilanterol versus vilanterol only for prevention of exacerbations of COPD: two replicate double-blind, parallel-group, randomised controlled trials. Lancet Respir Med. 2013;1:210–23.
199. Rennard SI, Tashkin DP, McElhattan J, et al. Efficacy and tolerability of budesonide/formoterol in one hydrofluoroalkane pressurized metered-dose inhaler in patients with chronic obstructive pulmonary disease: results from a 1-year randomized controlled clinical trial. Drugs. 2009;69:549–65.
200. Boulet LP, Cartier A, Milot J, et al. Tolerance to the protective effects of salmeterol on methacholine-induced bronchoconstriction: influence of inhaled corticosteroids. Eur Respir J. 1998;11:1091–7.
201. Tan KS, McFarlane LC, Lipworth BJ. Concomitant administration of low-dose prednisolone protects against *in vivo* beta-2-adrenoceptor subsensitivity induced by regular formoterol. Chest. 1998;113:34–41.
202. Aziz I, Lipworth BJ. A bolus of inhaled budesonide rapidly reverses airway subsensitivity and b2-adrenoceptor down-regulation after regular inhaled formoterol. Chest. 1999;115:623–8.
203. Haque R, Hakim A, Moodley T, et al. Inhaled long-acting β_2 agonists enhance glucocorticoid receptor nuclear translocation and efficacy in sputum macrophages in COPD. J Allergy Clin Immunol. 2013;132:1166–73.
204. Calverley P, Anderson JA, Celli B, et al. Salmeterol and fluticasone propionate and survival in chronic obstructive pulmonary disease. N Engl J Med. 2007;356:775–89.
205. Celli BR, Thomas NE, Anderson JA, et al. Effect of pharmacotherapy on rate of decline of lung function in chronic obstructive pulmonary disease: results from the TORCH study. Am J Respir Crit Care Med. 2008;178:332–8.
206. Vestbo J, Anderson JA, Brook RD, et al. Fluticasone furoate and vilanterol and survival in chronic obstructive pulmonary disease with heightened cardiovascular risk (SUMMIT): a double-blind randomised controlled trial. Lancet. 2016;387:1817–26.
207. Dhillon S, Keating GM. Beclometasone dipropionate/formoterol: in an HFA-propelled pressurised metered-dose inhaler. Drugs. 2006;66:1475–83.
208. Hogg JC, Chu F, Utokaparch S, et al. The nature of small-airway obstruction in chronic obstructive pulmonary disease. N Engl J Med. 2004;350:2645–53.
209. Aggarwal B, Gogtay J. Use of pressurized metered dose inhalers in patients with chronic obstructive pulmonary disease: review of evidence. Expert Rev Respir Med. 2014;8:349–56.
210. Solidoro P, Patrucco F, Bagnasco D. Comparing a fixed combination of budesonide/formoterol with other inhaled corticosteroid plus long-acting beta-agonist combinations in patients with chronic obstructive pulmonary disease: a review. Expert Rev Respir Med. 2019;13(11):1087–94.
211. Ferguson GT, Darken P, Ballal S, et al. Efficacy of budesonide/glycopyrronium/formoterol fumarate metered dose inhaler (bgf mdi) versus other inhaled corticosteroid/long-acting muscarinic antagonist/long-acting β2-agonist (ICS/LAMA/LABA) triple combinations in COPD: a systematic literature review and network meta-analysis. Adv Ther. 2020;37(6):2956–75.
212. Bourdin A, Molinari N, Ferguson GT, et al. Efficacy and safety of budesonide/glycopyrronium/formoterol fumarate versus other triple combinations in COPD: a systematic literature review and network meta-analysis. Adv Ther. 2021;38(6):3089–112.
213. Ichinose M, Fukushima Y, Inoue Y, et al. Efficacy and safety of budesonide/glycopyrrolate/formoterol fumarate metered dose inhaler formulated using co-suspension delivery technology in Japanese patients with COPD: a subgroup analysis of the KRONOS Study. Int J Chron Obstruct Pulmon Dis. 2019;14:2979–91.
214. Lima GC, Silva EV, Magalhães PO, et al. Efficacy and safety of a four-drug fixed-dose combination regimen versus separate drugs for treatment of pulmonary tuberculosis: a systematic review and meta-analysis. Braz J Microbiol. 2017;48(2):198–207.
215. The Union Anti-tuberculosis Regimens of Chemotherapy Recommendations from the Committee on Treatment of the international union against tuberculosis and lung disease. Bull Int Union Tuberc Lung Dis. 1988;63(2):60–4.
216. Ferreira AC, Silva Junior JL, Conde MB, et al. Clinical treatment outcomes of tuberculosis treated with the basic regimen recommended by the Brazilian National Ministry of Health using fixed-dose combination tablets in the greater metropolitan area of Goiania, Brazil. J Bras Pneumol. 2013;39(1):76–83.
217. Bartacek A, Schütt D, Panosch B, et al. Comparison of a four-drug fixed-dose combination regimen with a single tablet regimen in smear-positive pulmonary tuberculosis. Int J Tuberc Lung Dis. 2009;13(6):760–6.
218. Wu JT, Chiu CT, Wei YF, et al. Comparison of the safety and efficacy of a fixed-dose combina-

tion regimen and separate formulations for pulmonary tuberculosis treatment. Clinics (Sao Paulo). 2015;70(6):429–34.
219. Ruan H, Lv Z, Liu S, et al. Anlotinib attenuated bleomycin-induced pulmonary fibrosis via the TGF-β1 signaling pathway. J Pharm Pharmacol. 2020;72(1):44–55.
220. Sauleda J, Núñez B, Sala E, et al. Idiopathic pulmonary fibrosis: epidemiology, natural history, phenotypes. Med Sci (Basel). 2018;6(4):110.
221. Raghu G, Rochwerg B, Zhang Y, et al. An official ATS/ERS/JRS/ALAT clinical practice guideline: treatment of idiopathic pulmonary fibrosis. An update of the 2011 clinical practice guideline. Am J Respir Crit Care Med. 2015;192:e3–19.
222. Syed YY. Anlotinib: first global approval. Drugs. 2018;78:1057–62.
223. Wollin L, Wex E, Pautsch A, et al. Mode of action of nintedanib in the treatment of idiopathic pulmonary fibrosis. Eur Respir J. 2015;45(5):1434–45.
224. Nintedanib for treating idiopathic pulmonary fibrosis. National Institute for Health and Care Excellence (NICE). 27 January 2016. TA379. Retrieved 7 August 2019.
225. Zhang L, Ji YX, Jiang WL, et al. Protective roles of pulmonary rehabilitation mixture in experimental pulmonary fibrosis in vitro and in vivo. Braz J Med Biol Res. 2015;48:545–52.
226. Zhao J, Ren Y, Qu Y, et al. Pharmacodynamic and pharmacokinetic assessment of pulmonary rehabilitation mixture for the treatment of pulmonary fibrosis. Sci Rep. 2017;7(1):3458.
227. Ferrari N, Seguin R, Renzi P. Oligonucleotides: a multi-targeted approach for the treatment of respiratory diseases. Future Med Chem. 2011;3(13):1647–62.
228. Séguin RM, Ferrari N. Emerging oligonucleotide therapies for asthma and chronic obstructive pulmonary disease. Expert Opin Investig Drugs. 2009;18(10):1505–17.
229. Parry-Billings M, Ferrari N, Seguin R. Oligonucleotides: new therapeutic approaches for asthma and chronic obstructive pulmonary disease. Curr Opin Investig Drugs. 2010;11(11):1276–85.
230. Liao W, Dong J, Peh HY, et al. Oligonucleotide therapy for obstructive and restrictive respiratory diseases. Molecules. 2017;22(1):139.
231. Stephenson ML, Zamecnik PC. Inhibition of Rous sarcoma viral RNA translation by a specific oligodeoxyribonucleotide. Proc Natl Acad Sci U S A. 1978;75:285–8.
232. Popescu FD. Antisense- and RNA interference-based therapeutic strategies in allergy. J Cell Mol Med. 2005;9:840–53.
233. Huang C, Yang Y, Liu L. Interaction of long noncoding RNAs and microRNAs in the pathogenesis of idiopathic pulmonary fibrosis. Physiol Genomics. 2015;47:463–9.
234. Bi H, Zhou J, Wu D, et al. Microarray analysis of long non-coding RNAs in COPD lung tissue. Inflamm Res. 2015;64:119–26.
235. Austin PJ, Tsitsiou E, Boardman C, et al. Transcriptional profiling identifies the long noncoding RNA plasmacytoma variant translocation (PVT1) as a novel regulator of the asthmatic phenotype in human airway smooth muscle. J Allergy Clin Immunol. 2017;139(3):7803–9.
236. Booton R, Lindsay MA. Emerging role of MicroRNAs and long noncoding RNAs in respiratory disease. Chest. 2014;146:193–204.
237. Zhang X, Shan P, Jiang D, et al. Small interfering RNA targeting heme oxygenase-1 enhances ischemia-reperfusion-induced lung apoptosis. J Biol Chem. 2004;279:10677–84.
238. Griesenbach U, Kitson C, Escudero Garcia S, et al. Inefficient cationic lipid-mediated siRNA and antisense oligonucleotide transfer to airway epithelial cells in vivo. Respir Res. 2006;7(1):26.
239. Nicklin PL, Bayley D, Giddings J, et al. Pulmonary bioavailability of a phosphorothioate oligonucleotide (CGP 64128A): comparison with other delivery routes. Pharm Res. 1998;15:583–91.
240. Moschos SA, Frick M, Taylor B, et al. Uptake, efficacy, and systemic distribution of naked, inhaled short interfering RNA (siRNA) and locked nucleic acid (LNA) antisense. Mol Ther. 2011;19:2163–8.
241. Juliano R, Alam MR, Dixit V, et al. Mechanisms and strategies for effective delivery of antisense and siRNA oligonucleotides. Nucleic Acids Res. 2008;36:4158–71.
242. Kuruba R, Wilson A, Gao X, et al. Targeted delivery of nucleic-acid-based therapeutics to the pulmonary circulation. AAPS J. 2009;11:23–30.
243. Juliano RL. The delivery of therapeutic oligonucleotides. Nucleic Acids Res. 2016;44:6518–48.
244. Silva AC, Lopes CM, Sousa Lobo JM, et al. Nucleic acids delivery systems: a challenge for pharmaceutical technologists. Curr Drug Metab. 2015;16:3–16.
245. Wittrup A, Lieberman J. Knocking down disease: a progress report on siRNA therapeutics. Nat Rev Genet. 2015;16:543–52.
246. Jackson AL, Linsley PS. Recognizing and avoiding siRNA off-target effects for target identification and therapeutic application. Nat Rev Drug Discov. 2010;9:57–67.
247. Pavord ID, Korn S, Howarth P, et al. Mepolizumab for severe eosinophilic asthma (DREAM): a multicentre, double-blind, placebo-controlled trial. Lancet. 2012;380:651–9.
248. Busse W, Corren J, Lanier BQ, et al. Omalizumab, anti-IgE recombinant humanized monoclonal antibody, for the treatment of severe allergic asthma. J Allergy Clin Immunol. 2001;108:184–90.
249. Spergel JM, Rothenberg ME, Collins MH, et al. Reslizumab in children and adolescents with eosinophilic esophagitis: results of a double-blind, randomized, placebo-controlled trial. J Allergy Clin Immunol. 2012;129:456–63.

250. Kang SP, Saif MW. Infusion-related and hypersensitivity reactions of monoclonal antibodies used to treat colorectal cancer – identification, prevention, and management. J Support Oncol. 2007;5:451–7.
251. Pinzi L, Tinivella A, Caporuscio F, et al. Drug repurposing and polypharmacology to fight SARS-CoV-2 through inhibition of the main protease. Front Pharmacol. 2021;12:636989.
252. Tejera E, Munteanu CR, López-Cortés A, et al. Drugs repurposing using QSAR, docking and molecular dynamics for possible inhibitors of the SARS-CoV-2 M^{pro} protease. Molecules. 2020;25(21):5172.
253. Kumar Y, Singh H, Patel CN. In silico prediction of potential inhibitors for the main protease of SARS-CoV-2 using molecular docking and dynamics simulation based drug-repurposing. J Infect Public Health. 2020;13(9):1210–23.
254. Delre P, Caporuscio F, Saviano M, et al. Repurposing known drugs as covalent and non-covalent inhibitors of the SARS-CoV-2 papain-like protease. Front Chem. 2020;8:594009.
255. Ren PX, Shang WJ, Yin WC, et al. A multi-targeting drug design strategy for identifying potent anti-SARS-CoV-2 inhibitors. Acta Pharmacol Sin. 2021;1–11
256. Kumar S, Sharma PP, Upadhyay C, et al. Multi-targeting approach for nsp3, nsp9, nsp12 and nsp15 proteins of SARS-CoV-2 by Diosmin as illustrated by molecular docking and molecular dynamics simulation methodologies. Methods. 2021;195:44–56.
257. Utomo RY, Putri DDP, Salsabila IA, et al. The chemopreventive potential of diosmin and hesperidin for COVID-19 and its comorbid diseases. Indones J Cancer Chemopreven. 2020;11(3):154–67.
258. Hesperidin and Diosmin for Treatment of COVID-19. https://clinicaltrials.gov/ct2/show/NCT04452799. 2020. Accessed 31 Jan 2021.

8 Polypharmacology in Clinical Applications: Gastrointestinal Polypharmacology

Abstract

The digestive system comprises the gastrointestinal track (GI track) and accessory organs/glands. Recent research discoveries prove that in addition to GI track itself, gut microbiota also plays crucial roles in regulating digestive function and other human physiological functionalities as well as pathological processes. We therefore proposed to include gut microbiota into the digestive system, making it now comprise three parts: GI track, digestive glands, and gut microbiota. Gastrointestinal diseases (GI diseases) refer to diseases involving the gastrointestinal tract from mouth through the esophagus, stomach, small intestine, large intestine, and rectum to anus, as well as the accessory organs of digestion, the liver, gallbladder, and pancreas. Gut microbiota are the microorganisms including bacteria, archaea, and microscopic eukaryotes that live in the digestive tracts. This chapter aims to introduce the basics of GI disease and highlight clinical applications of the polypharmacological approaches for the treatment of the GI disease.

Keywords

Gastrointestinal polypharmacology · Gastrointestinal diseases · Gut microbiota · Hepatitis C virus · Multitarget therapy

8.1 Gastrointestinal Disease

8.1.1 Basics of Gastrointestinal Disease (GI Disease)

The digestive system comprises the gastrointestinal track (GI track) and accessory organs/glands [1]. GI track, also called digestive tract or alimentary canal, consists of several organs of the digestive system including the oral cavity/mouth, pharynx, esophagus, stomach, small intestine, large intestine, rectum, and anus and forms the pathway by which food enters the body and solid wastes are expelled. The accessory organs are the teeth, tongue, and glandular organs such as salivary glands, liver, gallbladder, and pancreas. The function of the digestive system is to digest and absorb food and then excrete the waste products with highly cooperative activities of GI track and accessory glands. As an open system, the GI track is highly prone to pathogens and other pathological factors (even food); it is estimated that 80% human diseases belong to gastrointestinal diseases (GI diseases) [2]. The term GI diseases therefore covers a wide range of disorders, which can be either acute or chronic. Notably, recent research has let generation of a new concept of gut microbiota and highlighted its critical impact in shaping the digestive function and pathology in humans as well as in other animals. Gut microbiota are the microorganisms including bacteria,

Z. Wang, B. Yang, *Polypharmacology*, https://doi.org/10.1007/978-3-031-04998-9_8

archaea, and microscopic eukaryotes that live in the digestive tracts [3, 4]. Here we propose to include gut microbiota into the digestive system, making it now comprise three parts: GI track, digestive glands, and gut microbiota.

Diseases of the gastrointestinal system can be one of many causes, affecting anywhere from the mouth to the anal canal. Gastrointestinal diseases (GI diseases) refer to diseases involving the gastrointestinal tract from mouth through the esophagus, stomach, small intestine, large intestine, and rectum to anus, as well as the accessory organs of digestion, the liver, gallbladder, and pancreas.

8.1.1.1 Oral Disease

The oral cavity is part of the gastrointestinal system, and as such the presence of alterations in this district can be the first sign of both systemic and gastrointestinal diseases [5]. By far the most common oral conditions are plaque-induced diseases (e.g., gingivitis, periodontitis, dental caries). Oral symptoms can be similar to lesions occurring elsewhere in the digestive tract, with a pattern of swelling, inflammation, ulcers, and fissures. In addition, several GI diseases, especially those associated with malabsorption, can cause recurrent mouth ulcers, atrophic glossitis, and angular cheilitis (e.g., Crohn's disease is sometimes termed orofacial granulomatosis when it involves the mouth alone).

8.1.1.2 Esophageal Disease

Esophageal diseases include a spectrum of disorders affecting the esophagus. The most common condition of the esophagus in Western countries is gastroesophageal reflux disease [4], which in chronic forms is thought to result in changes to the epithelium of the esophagus, known as Barrett's esophagus [6].Acute disease might include infections such as esophagitis, trauma caused by the ingestion of corrosive substances, or rupture of veins such as esophageal varices, Boerhaave syndrome, or Mallory-Weiss tears. Chronic diseases might include congenital diseases, such as Zenker's diverticulum and esophageal webbing, and esophageal including the nutcracker esophagus, achalasia, diffuse esophageal spasm, and esophageal stricture [6].

8.1.1.3 Gastric Disease

Gastric diseases refer to diseases affecting the stomach. Inflammation of the stomach by infection from any cause is called gastritis and when including other parts of the gastrointestinal tract called gastroenteritis. When gastritis persists in a chronic state, it is associated with several diseases, including atrophic gastritis, pyloric stenosis, and gastric cancer. Another common condition is gastric ulceration, peptic ulcers. Ulceration erodes the gastric mucosa, which protects the tissue of the stomach from the stomach acids. Peptic ulcers are most commonly caused by a bacterial *Helicobacter pylori* infection [6].

8.1.1.4 Intestinal Disease

The small and large intestines may be affected by infectious, autoimmune, and physiological states. Inflammation of the intestines is called enterocolitis, which may lead to diarrhea. Acute conditions affecting the bowels include infectious diarrhea and mesenteric ischemia. Causes of constipation may include fecal impaction and bowel obstruction, which may in turn be caused by ileus, intussusception, and volvulus. Inflammatory bowel disease is a condition of unknown etiology, classified as either Crohn's disease or ulcerative colitis, that can affect the intestines and other parts of the gastrointestinal tract [6].

Small Intestine

The small intestine consists of the duodenum, jejunum and ileum. Inflammation of the small intestine is called enteritis, which if localized is called duodenitis, jejunitis, and ileitis, respectively. Peptic ulcers are also common in the duodenum [6]. Chronic diseases of malabsorption may affect the small intestine, including the autoimmune coeliac disease, infective tropical sprue, and congenital or surgical short bowel syndrome. Other rarer diseases affecting the small intestine include Curling's ulcer, blind loop syndrome, Milroy disease, and Whipple's disease.

Large Intestine

Diseases that affect the large intestine may affect it in whole or in part. Appendicitis is one such disease, caused by inflammation of the appendix. Generalized inflammation of the large intestine is referred to as colitis, which when caused by the bacteria *Clostridium difficile* is referred to as pseudomembranous colitis. Diverticulitis is a common cause of abdominal pain resulting from outpouchings that particularly affects the colon. Functional colonic diseases refer to disorders without a known cause, including irritable bowel syndrome and intestinal pseudo-obstruction. Constipation may result from lifestyle factors, impaction of a rigid stool in the rectum, or in neonates, Hirschsprung's disease [6].

Irritable Bowel Syndrome (IBS) Irritable bowel syndrome (IBS) [6, 7] is one of the most common functional GI disorders worldwide. Irritable bowel syndrome (also called spastic colon, irritable colon, IBS, or nervous stomach) is a functional condition where your colon muscle contracts more or less often than "normal." Certain foods, medicines, and emotional stress are some factors that can trigger IBS. In the management of IBS, various treatments have been used, such as fiber, interventions that modify the microbiota (e.g., probiotics, prebiotics, antibiotics), antispasmodics, antidiarrheals, antidepressants, psychological therapies, prosecretory agents, osmotic and stimulant laxatives, narcotic and nonnarcotic analgesics, and antibiotics. The choice of agents depends on the dominant symptom and the subtype of IBS.

Constipation Constipation [6, 8] is a digestive condition in which the person experiences hard, dry, and often painful bowel movements, occurring less frequently than normal (generally fewer than three bowel movements a week). Constipation is one of the most common digestive disorder symptoms and is estimated to affect approximately 2.5 million Americans. While there could be in some rare cases anatomical or inflammatory conditions causing it, constipation is typically caused by a low fiber diet, little or no physical activity, dehydration, certain meds including sedatives and some antidepressants, or anything that disrupts your normal diet/routine. All of this leads to a slower transit of the stools through the colon, so that they tend to sit in the rectum becoming harder and larger. When you're constipated, you tend to strain when passing stool, sometimes causing hemorrhoids and anal fissures.

Rectum and Anus

Diseases affecting the rectum and anus are extremely common, especially in older adults. Hemorrhoids, vascular outpouchings of skin, are very common, as is pruritus ani, referring to anal itchiness. Other conditions, such as anal cancer, may be associated with ulcerative colitis or with sexually transmitted infections such as HIV. Inflammation of the rectum is known as proctitis, one cause of which is radiation damage associated with radiotherapy to other sites such as the prostate. Pain on passing stool may result from anal abscesses, small inflamed nodules, anal fissures, and anal fistulas [6].

Rectal and anal disease may be asymptomatic or may present with pain when passing stools, fresh blood in stool, a feeling of incomplete emptying, or pencil-thin stools. In addition to regular tests, medical tests used to investigate the anus and rectum include the digital rectal exam and proctoscopy.

Hemorrhoids Hemorrhoids [6] are structural disease with dilated swollen veins in the anal canal. They are caused by chronic excess pressure from straining during a bowel movement, persistent diarrhea, or pregnancy. There are two types of hemorrhoids: internal and external.

8.1.1.5 Accessory Digestive Gland Disease

Hepatic Diseases

Hepatic diseases refer to those affecting the liver. Hepatitis refers to inflammation of liver tissue and may be acute or chronic. Infectious viral hepatitis, such as hepatitis A, B, and C, affect in excess of million people worldwide. Liver disease may also be a result of lifestyle factors, such as fatty liver and NASH. Alcoholic liver disease may also develop as a result of chronic alcohol use, which may also cause alcoholic hepatitis. Cirrhosis may develop as a result of chronic hepatic fibrosis in a chronically inflamed liver, such as one affected by alcohol or viral hepatitis [6].

Liver abscesses are often acute conditions, with common causes being pyogenic and amoebic. Acute liver disease rarely results in pain but may result in jaundice. Infectious liver disease may cause a fever.

Chronic liver disease, such as cirrhosis, may be a cause of liver failure, a state where the liver is unable to compensate for chronic damage and unable to meet the metabolic demands of the body. In the acute setting, this may be a cause of hepatic encephalopathy and hepatorenal syndrome. Other causes of chronic liver disease are genetic or autoimmune disease, such as hemochromatosis, Wilson's disease, autoimmune hepatitis, and primary biliary cirrhosis [6]. Chronic liver disease may result in a buildup of fluid in the abdomen, yellowing of the skin or eyes, easy bruising, immunosuppression, and feminization.

Pancreatic Disease

Pancreatic diseases that affect digestion refers to disorders affecting the exocrine pancreas, which is a part of the pancreas involved in digestion. One of the most common conditions of the exocrine pancreas is acute pancreatitis, which in the majority of cases relates to gallstones that have impacted in the pancreatic part of the biliary tree or due to acute or chronic hazardous alcohol use or as a side effect of ERCP. Other forms of pancreatitis include chronic and hereditary forms. Chronic pancreatitis may predispose to pancreatic cancer and is strongly linked to alcohol use [6]. In severe cases, pancreatitis may lead to rapid blood loss and systemic inflammatory response syndrome. When the pancreas is unable to secrete digestive enzymes, such as with a pancreatic cancer, occluding the pancreatic duct results in jaundice.

Gallbladder and Biliary Tract

Diseases of the hepatobiliary system affect the biliary tract (also known as the biliary tree), which secretes bile in order to aid digestion of fats. Diseases of the gallbladder and bile ducts are commonly diet-related and may include the formation of gallstones that impact in the gallbladder (cholecystolithiasis) or in the common bile duct (choledocholithiasis) [6]. Gallstones are a common cause of inflammation of the gallbladder, called cholecystitis. Inflammation of the biliary duct is called cholangitis, which may be associated with autoimmune disease, such as primary sclerosing cholangitis, or a result of bacterial infection, such as ascending cholangitis [6].

Collectively, among the various digestive conditions, the most common 10 are as follows:

- Gastroesophageal reflux disease (GERD)
- Peptic ulcer disease (PUD) and gastritis
- Stomach flu (e.g., diarrhea, vomiting, stomach pain, and cramps).
- Gluten sensitivity and celiac disease
- Inflammatory bowel disease (IBD; e.g., Crohn's disease and ulcerative colitis)
- Irritable bowel syndrome (IBS)
- Hemorrhoids
- Constipation
- Diverticulitis
- Gallstones

In terms of malignancy, gastrointestinal cancers are ranked within the top list in many countries. There were an estimated 4.8 million new cases of gastrointestinal (GI) cancers and 3.4 million related deaths, worldwide, in 2018. GI cancers account for 26% of the global cancer incidence and 35% of all cancer-related deaths. The most common types of gastrointestinal cancers are as follows:

- Esophageal cancer
- Gastric (stomach) cancer
- Colorectal cancer
- Pancreatic cancer
- Liver cancer

8.1.1.6 Gut Microbiota

Gut microbiota are the microorganisms including bacteria, archaea, and microscopic eukaryotes that live in the digestive tracts of humans [9–17] and other animals. Alternative terms include gut flora (an outdated term that technically refers to plants) and gut microbiome. The gastrointestinal metagenome (sometimes defined as the microbiome) is the aggregate of all the genomes of gut microbiota [10]. The gut is the main location of human microbiota [4]. The gut microbiota has broad impacts, including effects on colonization resistance to pathogens, maintaining the intestinal epithelium, metabolizing dietary and pharmaceutical compounds, controlling immune function, and even behavior through the gut-brain axis.

In humans, the gut microbiota has the largest numbers of bacteria and the greatest number of species compared to other areas of the body [11]. The microbial composition of the gut microbiota varies across the digestive tract. The colon contains the highest microbial density recorded in any habitat on Earth, representing between 300 and 1000 different species [12]. However, 99% of gut bacteria come from about 30 or 40 species [12]. Bacteria also make up to 60% of the dry mass of feces [12]. Over 99% of the bacteria in the gut are anaerobes, but in the cecum, aerobic bacteria reach high densities [12]. It is estimated that these gut flora have around a hundred times as many genes in total as there are in the human genome.

Without gut flora, the human body would be unable to utilize some of the undigested carbohydrates it consumes, because some types of gut flora have enzymes that human cells lack for breaking down certain polysaccharides [13]. Rodents raised in a sterile environment and lacking in gut flora need to eat 30% more calories just to remain the same weight as their normal counterparts [13]. Carbohydrates that humans cannot digest without bacterial help include certain starches, fiber, oligosaccharides, and sugars that the body failed to digest and absorb like lactose in the case of lactose intolerance and sugar alcohols, mucus produced by the gut, and proteins [11, 13]. Bacteria turn carbohydrates they ferment into short-chain fatty acids by a form of fermentation called saccharolytic fermentation [14]. Products include acetic acid, propionic acid, and butyric acid [14]. These materials can be used by host cells, providing a major source of energy and nutrients [14]. Gases (which are involved in signaling [15] and may cause flatulence) and organic acids, such as lactic acid, are also produced by fermentation [13]. Acetic acid is used by muscle, propionic acid facilitates liver production of ATP, and butyric acid provides energy to gut cells [14].

Gut flora also synthesize vitamins like biotin and folate and facilitate absorption of dietary minerals, including magnesium, calcium, and iron [12, 16]. Methanobrevibacter smithii is unique because it is not a species of bacteria, but rather a member of domain Archaea, and is the most abundant methane-producing archaeal species in the human gastrointestinal microbiota [17].

8.1.2 Pharmacological Therapy of GI Disease

Due to the multiplicity and diversity of GI disease, numerous medications for the conditions are in the market. Here we describe the commonly used therapeutic agents for GI disease based on the excellent review article by Litou et al. [18].

8.1.2.1 Agents Affecting Gastrointestinal Motility

Promotility Agents or Prokinetic Agents

Prokinetic agents promote gut wall contractions and increase their coordination, thus enhancing GI motility. However, they do not disrupt the normal physiological pattern of motility [18–20].

Anticholinergic Agents

Propantheline is an anticholinergic agent which reduces gastrointestinal motility and prolongs gastric emptying rate. It is usually used in combination with other medicines to treat stomach ulcers.

Laxatives

Laxatives promote defecation and are often used OTC for the treatment of constipation. They can be grouped in osmotic, stimulant, and bulk laxatives [18, 21]. Osmotic laxatives (indigestible disaccharides, sugar alcohols, synthetic macromolecules, saline laxatives) attract and retain water in the intestinal lumen by increasing the luminal osmotic pressure. Stimulant laxatives (such as bisacodyl, senna, and sodium picosulfate) act locally by increasing colonic motility and decreasing water absorption in the large intestine [22]. Bulk laxatives such as bran, ispaghula, and sterculia adsorb and retain luminal fluids and increase the fecal mass. For constipation linked with specific diseases, additional treatment options are available: linaclotide, an agonist of guanylate cyclase-C, stimulates fluid secretion, accelerates intestinal transit, and is used for constipation-predominant irritable bowel syndrome [23].

Antidiarrheal Agents

Antidiarrheal agents provide symptomatic relief of diarrhea by decreasing fluid loss, by slowing down the passage of the gastrointestinal contents through the digestive tract, by increasing fluid absorption, and/or by reducing intestinal secretions [18, 24]. They can be classified according to their mechanism of action. Opioids (such as loperamide, diphenoxylate, and codeine phosphate) inhibit intestinal transit by activating μ-opioid receptors. Adsorbents and bulking agents (kaolin, ispaghula, methylcellulose) adsorb water and increase the fecal mass, while the antisecretory action of racecadotril, an enkephalinase inhibitor, is linked to reducing chloride and fluid flux into the GI lumen.

8.1.2.2 Antiemetics

Antiemetics are classified according to their mechanism of action. There are five receptors that play a key role in the vomiting reflex; muscarinic, dopaminergic, histaminic, serotoninergic, and substance P/neurokinin receptors [18]. Antagonists to these receptors have been used to treat various GI conditions.

8.1.2.3 Gastric Acid-Reducing Agents and Antacids

Proton-pump inhibitors (PPIs), H_2-receptor antagonists (H_2RAs), and antacids are widely used in the treatment of various gastric acid-related disorders, such as peptic ulcers and gastroesophageal reflux disease [18]. Esomeprazole, a PPI, ranks among the top five most prescribed medications worldwide.

Proton-Pump Inhibitors

PPIs are a group of substituted benzimidazole sulfoxide drugs with strong inhibitory effects on gastric acid secretion from the parietal cells in the stomach. At present, six PPIs (dexlansoprazole, esomeprazole, lansoprazole, omeprazole, pantoprazole, and rabeprazole) are available on the market [18, 25]. PPIs are used in the treatment of acid-related disorders and for the prevention of gastrointestinal bleeding in patients receiving dual antiplatelet therapy of clopidogrel and aspirin. PPIs can affect the absorption of the coadministered drugs to a great extent, mainly due to the increase in gastric pH.

Histamine-2 (H_2) Blockers

The H_2RAs are another drug class used to treat gastric acid-related disorders. These compounds bind to histamine H_2 receptors on parietal cells and antagonize the action of histamine, which is the major transmitter for stimulation of acid secretion [18, 26]. H_2RAs mainly include cimetidine (Tagamet), ranitidine hydrochloride (Zantac), famotidine (Pepcid), and nizatidine (Axid).

Antacids

The term "antacids" describe a category of salts, formulated as the combination of polyvalent cations such as calcium, aluminum, or magnesium with a base, such as hydroxide, trisilicate, or carbonate [18]. Aluminum hydroxide alone, or in combination with magnesium hydroxide, is the main ingredient of many antacid products [18]. Owing to the introduction of PPIs and H_2RAs, which are more potent drugs and can be used for a wide variety of gastrointestinal disorders, antacids are now mainly marketed as OTC medications.

8.1.2.4 Antibiotics Used for Gastrointestinal Infections

Antibiotics aim to attack targets specific to bacterial organisms such as bacterial cell walls, bacterial cell membranes, and bacterial metabolism or replication, in order to avoid damage to human cells. However, antibiotics are not 100% selective for bacteria that are pathogenic for the host organism [18]. As a result, the GI microbiota is frequently disturbed after treatment with antibiotics [27, 28]. Depending on the antibiotic, 5–25% of patients treated experience diarrhea [29, 30].

8.1.2.5 Anti-inflammatory Drugs for Inflammatory Bowel Disease

Anti-inflammatory agents, such as aminosalicylates and corticosteroids, are the most commonly used drugs in IBD [18]. Treatment with aminosalicylates includes a range of prodrugs (sulfasalazine, olsalazine, balsalazide) or modified release formulations to deliver aminosalicylates to their target site in the intestine [18]. If remission cannot be achieved with aminosalicylates, the next treatment option consists of different corticosteroids ranging from locally acting drugs (budesonide) to systemic acting ones (hydrocortisone, prednisolone, dexamethasone).

8.1.2.6 Immunosuppressive Agents for Inflammatory Bowel Disease

Immunosuppressive agents are frequently used in gastroenterology for the treatment of IBD, autoimmune hepatitis, autoimmune pancreatitis, and sclerosing cholangitis and in the post-transplantation setting [18, 31]. In particular, in IBD, therapy with immunosuppressive agents has gained in importance over the last few years [18, 32]. Immunosuppressive agents can be classified in immunomodulators (e.g., thiopurines (6-mercaptopurine, azathioprine), methotrexate, tacrolimus, sirolimus, everolimus, cyclosporine A) and biologics (e.g., monoclonal antibodies, infliximab, adalimumab, vedolizumab, golimumab) [32].

8.1.2.7 Bile Acid Sequestrants

Bile acid sequestrants (BAS) such as cholestyramine, colesevelam, and colestipol are used for the treatment of primary hyperlipidemia, as monotherapy or in combination with statins or ezetimibe, and in the treatment of gastrointestinal diseases [18, 33]. Cholestyramine is indicated for diarrhea associated with Crohn's disease, ileal resection, vagotomy, diabetes, diabetic vagal neuropathy, and radiation [18]. While colesevelam is not licensed for the treatment of bile acid malabsorption, several clinical trials have demonstrated positive outcomes which have provoked its off-label use in this indication [18, 34, 35].

8.1.2.8 Gut Microbiota-Maintaining Supplements

Dietary Fibers

The use of dietary fibers in the treatment of various diseases, such as diabetes, hypercholesterolemia, obesity, chronic constipation, and gastrointestinal motility disorders, has increased over the last years [18]. However, there are few studies that have investigated the impact of concomitant use of dietary fibers with other drugs.

Dietary fiber (British spelling fibre) or roughage is the portion of plant-derived food that cannot be completely broken down by human digestive enzymes [9]. Dietary fibers are diverse in chemical composition and can be grouped generally by their solubility, viscosity, and fermentability, which affect how fibers are processed in the body. Dietary fiber has two main components: soluble fiber and insoluble fiber, which are components of plant foods, such as legumes, whole grains and cereals, vegetables, fruits, and nuts or seeds [10]. A diet high in regular fiber consumption is generally associated with supporting health and lowering the risk of several diseases [11].

Food sources of dietary fiber have traditionally been divided according to whether they provide soluble or insoluble fiber. Plant foods contain both types of fiber in varying amounts, according to the fiber characteristics of viscosity and fermentability [9, 12]. Advantages of consuming fiber depend upon which type of fiber is consumed and which benefits may result in the gastrointestinal system. Bulking fibers, such as cellulose, hemicellulose, and psyllium, absorb and hold water, promoting regularity [36]. Viscous fibers, such as beta-glucan and psyllium, thicken the fecal mass [36]. Fermentable fibers, such as resistant starch and inulin, feed the bacteria and microbiota of the large intestine and are metabolized to yield short-chain fatty acids, which have diverse roles in gastrointestinal health [11, 37].

Soluble fiber (fermentable fiber or prebiotic fiber) is generally fermented in the colon into gases and physiologically active by-products, such as short-chain fatty acids produced in the colon by gut bacteria. Examples are β-glucans (in oats, barley, and mushrooms) and raw guar gum. Psyllium, a soluble, viscous, nonfermented fiber, is a bulking fiber that retains water as it moves through the digestive system, easing defecation. Soluble fiber is generally viscous and delays gastric emptying which, in humans, can result in an extended feeling of fullness [38]. Inulin (in chicory root), wheat dextrin, oligosaccharides, and resistant starches [39] (in legumes and bananas) are soluble non-viscous fibers [2]. Regular intake of soluble fibers, such as β-glucans from oats or barley, has been established to lower blood levels of LDL cholesterol, a risk factor for cardiovascular diseases [13, 38, 39].

Insoluble fiber is inert to digestive enzymes in the upper gastrointestinal tract. Examples are wheat bran, cellulose, and lignin. Coarsely ground insoluble fiber triggers the secretion of mucus in the large intestine, providing bulking. Finely ground insoluble fiber does not have this effect and can actually have a constipating effect [38]. Some forms of insoluble fiber, such as resistant starches, can be fermented in the colon.

Dietary fiber consists of non-starch polysaccharides and other plant components such as cellulose, resistant starch, resistant dextrins, inulin, lignins, chitins (in fungi), pectins, β-glucans, and oligosaccharides [9, 10, 38].

Soluble fiber supplements may be beneficial for alleviating symptoms of IBS, such as diarrhea or constipation and abdominal discomfort [40]. Prebiotic soluble fiber products, like those containing inulin or oligosaccharides, may contribute to relief from inflammatory bowel disease as in Crohn's disease, ulcerative colitis, and *Clostridium difficile*, due in part to the short-chain fatty acids produced with subsequent anti--inflammatory actions upon the bowel. Fiber supplements may be effective in an overall dietary plan for managing irritable bowel syndrome by modification of food choices [41].

One insoluble fiber, resistant starch from high-amylose corn, has been used as a supplement and may contribute to improving insulin sensitivity and glycemic management [42, 43] as well as promoting regularity [44] and possibly relief of diarrhea [45–47]. One preliminary finding indicates that resistant corn starch may reduce symptoms of ulcerative colitis [48].

Fermentable fibers are consumed by the microbiota within the large intestines, mildly increasing fecal bulk and producing short-chain fatty acids as byproducts with wide-ranging physiological activities. Resistant starch, inulin, fructooligosaccharide, and galactooligosaccharide are dietary fibers which are fully fermented. These include insoluble as well as soluble fibers. This fermentation influences the expression of many genes within the large intestine, which affect digestive function and lipid and glucose metabolism, as well as the immune system, inflammation, and more.

Dietary fibers can change the nature of the contents of the gastrointestinal tract and can change how other nutrients and chemicals are absorbed through bulking and viscosity [10, 49]. Some types of soluble fibers bind to bile acids in the small intestine, making them less likely to reenter the body; this in turn lowers cholesterol levels in the blood from the actions of cytochrome P450-mediated oxidation of cholesterol.

The US FDA allows manufacturers of foods containing 1.7 g per serving of psyllium husk soluble fiber or 0.75 g of oat or barley soluble fiber as beta-glucans to claim that regular consumption may reduce the risk of heart disease [13, 49].

Prebiotics

Prebiotics are compounds in food that induce the growth or activity of beneficial microorganisms such as bacteria and fungi [50]. The most common example is in the gastrointestinal tract, where prebiotics can alter the composition of organisms in the gut microbiome. Dietary prebiotics are typically nondigestible fiber compounds that pass undigested through the upper part of the gastrointestinal tract and stimulate the growth or activity of advantageous bacteria that colonize the large bowel by acting as substrate for them [50]. They were first identified and named by Marcel Roberfroid in 1995 [50, 51]. As a functional food component, prebiotics, like probiotics, are a conceptual intermediary between foods and drugs. Depending on the jurisdiction, they typically receive an intermediate level of regulatory scrutiny, in particular of the health claims made concerning them for marketing purposes.

Compounds that can be classified as prebiotics must meet the following criteria [51, 53]: **(a)** nondigestible and resistant to breakdown by stomach acid and enzymes in the human gastrointestinal tract, **(b)** selectively fermented by intestinal microorganisms, and **(c)** selectively targeting and stimulating the growth and activity of beneficial bacteria.

Thus, consumption of prebiotics may facilitate the health of the host [54]. Based on the previous classifications, plant-derived carbohydrate compounds called oligosaccharides are the main source of prebiotics that have been identified [53, 55, 56]. Specifically, fructans and galactans are two oligosaccharide sources which have been found to stimulate the activity and growth of beneficial bacterial colonies in the gut [52, 54, 56]. Fructans are a category of carbohydrate consisting of fructooligosaccharides (FOS) and inulins, while galactans consist of galactooligosaccharides [52]. Other dietary fibers also fit the definition of prebiotics, such as resistant starch [57], pectin [58], β-glucans [59], and xylooligosaccharides [60].

The European Food Safety Authority (EFSA), the regulatory agency for product labeling, differentiates between "prebiotic" and "dietary fiber," stating that "a cause and effect relationship has not been established between the consumption of the food constituents which are the subject of the health claims and a beneficial physiological effect related to increasing numbers of gastrointestinal microbiota" [61]. Consequently, under EFSA rules, individual ingredients cannot be labeled as prebiotics, but only as dietary fiber and with no implication of health benefits [61].

Most prebiotic research has focused on the effects that prebiotics confer on *Bifidobacterium* and *Lactobacillus* [52, 53, 62]. These bacteria have been highlighted as key probiotics and beneficial gut bacteria as they may have several beneficial effects on the host in terms of improving digestion (including but not limited to enhancing mineral absorption) [63] and the effectiveness and intrinsic strength of the immune system [64]. Both *Bifidobacterium* and *Lactobacillus* have been shown to have differing prebiotic specificity and to selectively ferment prebiotic fiber based on the enzyme characteristic of the bacterial population [65]. Thus, lactobacilli prefer inulin and fructooligosaccharides, while bifidobacteria displays specificity for inulin, fructooligosaccharides, xylooligosaccharides, and galactooligosaccharides [16]. A product that stimulates bifidobacteria is described as a bifidogenic factor, a concept that overlaps, but is not identical with, being prebiotic [66]. Studies have also shown that prebiotics, besides stimulating the growth of beneficial gut bacteria, can also inhibit the growth of detrimental and potentially pathogenic microbes in the gut [53, 55] such as clostridia [53].

Fermentation is the main mechanism of action by which prebiotics are used by beneficial bacteria in the colon [53, 54, 56]. Both *Bifidobacterium* and *Lactobacillus* are bacterial populations which use saccharolytic metabolism to break down substrates [53]. The bifidobacterial genome contains many genes that encode for carbohydrate-modifying enzymes as well as genes that encode for carbohydrate uptake proteins [56]. The presence of these genes indicates that bifidobacteria contain specific metabolic pathways specialized for the fermentation and metabolism of plant-derived oligosaccharides, or prebiotics [56]. These pathways in bifidobacteria ultimately produce short-chain fatty acids [53, 54, 56], which have diverse physiological roles in body functions [52, 67].

Probiotics

As motioned above, the intestinal microflora plays a key role in physiological, metabolic, immunological and nutritional processes in the human body [18]. For this reason, there is currently great interest in influencing the composition of the microflora and its activity using probiotics for both the prevention and treatment of various diseases [68, 69]. According to the WHO, probiotics are "live microorganisms which, when administered in adequate amounts, confer a health benefit on the host" [70]. There are several clinical studies that have illustrated their beneficial effects on gastrointestinal disorders such as diarrhea and irritable bowel syndrome. The Gram-negative bacterium *Escherichia coli* Nissle 1917, for example, has been used since 1920 for the treatment or prevention of irritable bowel syndrome, chronic constipation, nonulcer dyspepsia, and other gastrointestinal disorders [71]. The mechanism of action of the probiotics is not yet fully understood. It seems that they may modulate the intestinal epithelial barrier and transport across it, noting that in inflammatory bowel diseases (IBDs), for example ulcerative colitis and Crohn's disease, the barrier properties of the epithelium are compromised due to secreted cytokines and/or medication [72].

Probiotics are live microorganisms promoted with claims that they provide health benefits when consumed, generally by improving or restoring the gut flora [73, 74]. Probiotics are considered generally safe to consume but may cause bacteria-host interactions and unwanted side effects in rare cases [75–77]. There is little evidence that probiotics bring the health benefits claimed for them [73].

The first discovered probiotic was a certain strain of bacillus in Bulgarian yogurt, called *Lactobacillus bulgaricus*. The discovery was made in 1905 by Bulgarian physician and microbiologist Stamen Grigorov. The modern-day theory is generally attributed to Russian Nobel laureate Élie Metchnikoff, who postulated around 1907 that yoghurt-consuming Bulgarian peasants lived longer [78].

Some probiotics are suggested as a possible treatment for various forms of gastroenteritis [79]. As a treatment for infectious diarrhea, probiotics are of no benefit to people who have the condition for more than 2 days, and there is no evidence they lessen the duration of diarrhea overall [80]. Probiotics are being studied for their potential to influence inflammatory bowel disease. Some evidence supports their use in conjunction with standard medications in treating ulcerative colitis, but no evidence shows their efficacy in treating Crohn's disease [81–83]. Probiotics are under study for their potential to affect irritable bowel syndrome, although uncertainty remains around which type of probiotic works best and around the size of possible effect [81, 84].

Supplements such as tablets, capsules, powders, and sachets containing the bacteria have been studied. However, probiotics taken orally can be destroyed by the acidic conditions of the stomach. As of 2010, a number of microencapsulation techniques were being developed to address this problem [85].

Multiple Probiotics Preliminary research is evaluating the potential physiological effects of multiple probiotic strains, as opposed to a single strain [86, 87]. As the human gut may contain several hundred microbial species, one theory indicates that this diverse environment may benefit from consuming multiple probiotic strains, an effect that remains scientifically unconfirmed.

Strains Only preliminary evidence exists for most probiotic health claims. Even for the most studied probiotic strains, few have been sufficiently developed in basic and clinical research to warrant approval for health claim status by a regulatory agency such as the FDA or EFSA, and as of 2010, no claims had been approved by those two agencies [88]. Some experts are skeptical about the efficacy of different probiotic strains and believe that not all subjects benefit from probiotics [88, 89].

Whereas the number of medications for GI disease is among the greatest for the illness of other systems, the application of polypharmacology to GI disease has been relatively rare.

8.2 CDT in GI Disease

Several different medical treatments for symptomatic esophageal reflux are available, besides the usual advice about weight loss, stopping smoking, and postural measures [90]. These include antacids, alginic acid preparations, H_2 receptor antagonists, and prokinetic agents. In view of the multiple etiological factors involved, it is not surprising that no single medical treatment has proved entirely satisfactory. H_2 receptor antagonists (H_2RAs) are known to heal esophagitis in about 60% of patients, rather less than the 80% success rate in healing peptic ulcer [90, 91]. The combination of treatments with complementary modes of action is therefore a logical step.

However, the combination of an H_2RA, cimetidine, and a prokinetic agent, metoclopramide, has been subjected to clinical trial [92] but produced little, if any, improvement over cimetidine alone and was associated with an increased incidence of adverse reactions. On the other hand, the pharmacological rationale for H_2 receptor antagonist and alginic acid combination has been confirmed [93].

8.2.1 H_2 Receptor Antagonist/Alginic Acid CDT for Esophageal Reflux Disease

Another randomized 12-week study comparing the efficacy of a chewable tablet containing cimetidine 200 mg and alginic acid 500 mg at a dosage of one tablet four times daily with the standard dosage of cimetidine (400 mg four times daily) in the management of esophageal reflux disease, however, observed no significant differences between the treatment groups in healing and improvement of the appearances of esophagitis [90]. Although two thirds of the patients receiving the combination tablet required only half the recommended daily dose of cimetidine, there was equivalent healing and improvement of endoscopic signs of esophagitis. Nonetheless, a superior control of the principal symptom heartburn was found. The study group therefore concluded that this combination tablet would be suitable for use in patients who have failed to respond to a simple alginate/antacid and in whom an H_2 receptor antagonist would be considered appropriate treatment.

8.2.2 H_2 Receptor Antagonist/Multiple Medications CDT for Esophageal Reflux Disease

Eriksen et al. [94] carried out a randomized double-blind study comparing the effects of cimetidine 400 mg four times daily, Gastrocote (alginic acid, aluminum hydroxide gel, magnesium trisilicate, sodium bicarbonate) two tablets four times daily, and a combination of both on ambulatory 24-h esophageal pH, endoscopic findings, and symptoms in patients with reflux esophagitis. The average number of days with heartburn per week was 5.0, 3.6, and 3.4 in the Gastrocote, cimetidine, and combined groups respectively. There was also a clear trend toward improvement of endoscopic esophagitis at 10 weeks in the combined therapy group when compared to single-agent treatment with cimetidine or Gastrocote although the difference was not significant due to inadequate numbers. This study indicates that combined cimetidine and Gastrocote is a more effective treatment for reflux esophagitis than single-agent therapy with either drug [94].

8.3 FDC in GI Disease

Application of the FDC approach or fixed-dose single-pill medication to GI disease has been extremely limited, if any, with regard to the conventional concept (or misunderstanding) of polypharmacology [95]. Nonetheless, as stated in Chap. 1 of this monograph, a single agent or combination drugs targeting multiple molecules for the treatment of multiple disease entities are also an important element of the modern polypharmacology. Up to date, the most frequent utilization of the FDC strategy is to combat hepatitis C virus (HCV) and the related condition particularly HIV.

HCV is spread mainly through contact with the blood of a person who has HCV. In the United States, HCV is spread mainly by sharing needles or other injection drug equipment (works) with someone who has HCV [96–99]. HCV can also be spread through sexual contact. Because both HIV and HCV can spread in blood, a major risk factor for both HIV and HCV infection is injection drug use. Sharing needles or other drug injection equipment increases the risk of contact with HIV- or HCV-infected blood. According to the Centers for Disease Control and Prevention (CDC), approximately 21% of people with HIV in the United States also have HCV [100, 101]. Infection with both HIV and HCV is called HIV/HCV coinfection.

In people with HIV/HCV coinfection, HIV may cause chronic HCV to advance faster. Approximately one-third of patients with chronic HCV infection progress to cirrhosis, at a median time of <20 years [96, 97]. The rate of progression increases with older age, alcoholism, male sex, and HIV infection [98–101]. A meta-analysis found that patients with HCV/HIV coinfection had a threefold greater risk of progression to cirrhosis or decompensated liver disease than patients with HCV mono-infection [100]. The treatment of HCV infection and HCV/HIV coinfection has been rapidly evolving. The subsections below are directed to these two particular conditions.

8.3.1 Daclatasvir/Sofosbuvir/Ribavirin FDC for the Treatment of HCV

Patients infected with genotype 3 hepatitis C virus (G3-HCV) that have compensated and decompensated liver cirrhosis are very difficult to treat. Although direct-acting antiviral agents (DAAs) have improved the sustained virological response (SVR) in all HCV genotypes, patients with G3-HCV-related cirrhosis, whether treatment-experienced or treatment-naïve, remain a therapeutic challenge. The suboptimal virological response in patients with G3-HCV is probably related to the high degree of viral steatosis and the relatively rapid progression to liver fibrosis [102–104]. Several cohorts and real-life studies have shown that SVR has improved with the combination of daclatasvir plus sofosbuvir, with or without ribavirin. Moreover, a 12-week or 16-week treatment duration was shown to be suboptimal in patients with cirrhosis [105]. A 24-week treatment duration is indicated in this population of patients with this antiviral regimen. The use of a full dose of ribavirin in addition to daclatasvir/sofosbuvir in patients with G3-HCV cirrhosis could lead to lower adherence and a higher rate of adverse events (AEs). In the present study, we explored SVR and the adherence to a 24-week treatment of daclatasvir/sofosbuvir plus a flat dose (800 mg) of ribavirin in a real, difficult-to-treat population of patients with G3-HCV and cirrhosis classified as Child-Pugh-Turcotte A or B (CPT-A or CPT-B). We compared SVR12 between patients treated with daclatasvir/sofosbuvir without ribavirin, daclatasvir/sofosbuvir plus flat dose ribavirin, and daclatasvir/sofosbuvir plus weight-based ribavirin.

Although current treatment regimens for different HCV genotypes are effective in most scenarios, SVR rates tend to be lower among patients with G3-HCV and advanced liver disease. G3 is commonly associated with steatosis, progression to cirrhosis, and the development of HCC [103, 104]. Overall, the available data from registry and real-world studies have suggested that patients with G3-HCV cirrhosis are more difficult to cure than patients with other HCV genotypes [105–107]. It is critical to improve the SVR in these patients, because achieving an undetectable viral load is associated with decreased hepatic morbidity and mortality. Previous antiviral DAA regimens with sofosbuvir/ribavirin showed low SVR12 achievement in patients with G3-HCV cirrhosis, whether they were treatment-naïve or treatment-experienced. Zeuzem et al. [108] demonstrated that patients treated with sofosbuvir/ribavirin weight-based doses for 24 weeks had a low SVR12 (62%). Similarly, Feld et al. [109] showed low SVR12 rates in treatment-naïve and treatment-experienced patients with cirrhosis that received the same antiviral regimen (44.8%). Paradoxically, treatments based on PegIFN/ribavirin achieved higher SVR12 rates but also higher AEs compared to treatments based on sofosbuvir ribavirin [110]. Association of sofosbuvir/ribavirin is not further recommended in cirrhotic G3 patients. Consequently, other antiviral combinations were attempted with sofosbuvir and daclatasvir. The Ally-3 trial analyzed a group of patients with G3-HCV cirrhosis treated with sofosbuvir/daclatasvir without ribavirin for 12 weeks; they found low SVR12 rates (58% in treatment-naïve and 69% in experienced groups) [107]. In the Ally-3+ study, Leroy et al. [105] analyzed how ribavirin added to a regimen of daclatasvir/sofosbuvir for a duration of 12 or 16 weeks increased SVR12 in patients with G3-HCV cirrhosis. They demonstrated higher SVR12 rates compared to those of the Ally-3 study, with SVR12 of 86% and 88% in patients treated for 12 and 16 weeks, respectively.

Although the small sample size of patients with G3-HCV cirrhosis prevented any solid conclusions regarding the role of ribavirin, those findings indicated that RBV provided important added value in this "difficult-to-treat population." However, it remains difficult to determine the optimal treatment duration for sofosbuvir/DCV with or without ribavirin for patients with G3-HCV. In addressing this question, different studies have shown that increasing the duration of therapy from 12 to 24 weeks was associated with a better SVR [110–113]. In a study by Herzer et al. [113], 126 patients with advanced liver fibrosis were analyzed; they received daclatasvir/sofosbuvir with or without a weight-based ribavirin dose for 24 weeks. They found SVR12 of 89% with daclatasvir/sofosbuvir (41/46 patients) and 88% (29/33) for DCV/sofosbuvir/ribavirin. Those authors concluded that response rates were similar in the two treatment groups, and they suggested that ribavirin did not confer an efficacy benefit to the daclatasvir/sofosbuvir regimen, even when treatment was extended to 24 weeks. Similarly, Hézode et al. [114] studied patients with compensated and decompensated cirrhosis treated for 24 weeks with daclatasvir/sofosbuvir or daclatasvir/sofosbuvir/ribavirin (weight-based). They showed SVR12 of 86% and 82%, respectively. In that study, it is important to consider that, at baseline, more than 50% of patients treated with daclatasvir/sofosbuvir/ribavirin had a Model for End-Stage Liver Disease category >10, a lower number of platelets, and lower albumin levels, compared to patients treated without ribavirin. These differences in patient characteristics indicated a more advanced liver disease in the group treated with ribavirin compared to those treated without ribavirin, which might explain the difference in SVR12. On the other hand, Alonso et al. [112] conducted a multicenter observational study with 94 patients with G3-HCV that had compensated and decompensated cirrhosis and were treated with daclatasvir/sofosbuvir ± ribavirin weight-based dose regimens for 24 weeks. They found SVR12 rates of 96% in patients treated with ribavirin and 90% in patients treated without ribavirin. Similarly, Cornberg et al. [110] showed that patients with G3-HCV cirrhosis that were either treatment-naïve or treatment-experienced achieved excellent SVR12 in 24-week treatments with a daclatasvir/sofosbuvir/ribavirin weight-based regimen (98%) compared to patients treated with daclatasvir/sofosbuvir (87%). These conflicting findings may result from different patient selection biases in clinical registered trials and real-world experiences.

8.3.2 Sofosbuvir/Velpatasvir/Ribavirin FDC for the Treatment of HCV

Other DAA regimens were also employed to treat patients with G3-HCV cirrhosis. Curry and Charlton [115] tested three different regimens sofosbuvir/velpatasvir for 12 weeks, sofosbuvir/velpatasvir/ribavirin for 12 weeks, and sofosbuvir/velpatasvir for 24 weeks, in patients with G3-HCV with decompensated cirrhosis. The SVR rates were, respectively, 50%, 85%, and 50%. Thus, in that case, ribavirin also played an important role in enhancing the SVR in this "difficult-to-treat population." In an interesting study by Pianko et al. [116], 52 patients with

G3-HCV cirrhosis were screened with a pretreatment NS5A resistance test. Patients were then treated with sofosbuvir/velpatasvir/ribavirin or sofosbuvir/velpatasvir without ribavirin for 12 weeks. They demonstrated a SRV12 of 95%, irrespective of the presence of ribavirin, in patients without NS5A RAVs. In contrast, a significant SVR reduction (respectively, of 75% vs 100%) was observed in patients with NS5A RAVs at baseline treated with regimen without ribavirin compared to patients treated with a regimen containing ribavirin, showing a penalizing effect of ribavirin-free regimes in those with NS5A RAVs at baseline [115, 116]. That study stressed the importance of ribavirin as additional antiviral component in patients with baseline RAVs. However, it is not possible to perform a NS5A RAV test routinely at baseline in all centers that provide DAA treatments. Therefore, it might be wise to include ribavirin in the DAA therapy whenever there is a suspicion that patients are at high risk of harboring RAVs (i.e., treatment-experienced patients). NS5A RAV screening was not performed at baseline, and due to the study design, we did not assign the administration of ribavirin based on the relative risk of RAVs. In particular, in the group of patients treated with daclatasvir/sofosbuvir without ribavirin, we found a high incidence of the Y93H mutation among patients that had relapsed. This mutation might explain the low efficiency of this antiviral regimen. Moreover, we speculate that suboptimal antiviral treatment might allow selective survival of RAVs and, thus, fail to provide antiviral activity. At present, European Association for the Study of the Liver guidelines recommend an association of glecaprevir/pibrentasvir for 12 or 16 weeks but only in compensated naïve or experienced G3 cirrhosis [117]. This regimen showed a SVR12 of 98% and 96%, respectively, in naïve and experienced compensated cirrhotic patients treated for 12 and 16 weeks [118]. But this DAA association is not indicated in decompensated cirrhosis or in cirrhotic with Model for End-Stage Liver Disease score >10. Similarly, sofosbuvir/velpatasvir did not show an optimal SVR12 in G3-naïve or G3-experienced cirrhotic patients as reported above.

Reig et al. [119] observed in patients treated with DAAs a high rate of HCC recurrence (27.6%) in patients with previous HCC complete radiological response; in our experience a lower recurrence rate of HCC at 6 months (7.7%) was disclosed. This result (9.2%) is similar to that reported by Cabibbo et al. [120]. In our study a correct evaluation by CT or MRI of all the patients previously treated for HCC to exclude posttreatment residual HCC was employed. As suggested from other authors, long-term observational studies will be needed to determine the effective risk for developing HCC recurrence or occurrence in patients treated with DAAs [121].

In conclusion, this study investigated the role of ribavirin in achieving SVR12 in a large multicenter cohort of 233 patients with G3-HCV cirrhosis, with and without compensation. Patients treated without ribavirin achieved a lower SVR12 compared to patients treated with either weight-based ribavirin or a flat ribavirin dose. Of interest, although high SVR12 rates were achieved with daclatasvir/sofosbuvir/ribavirin for 24 weeks, irrespective of how the RBV was dosed, patients treated with the daclatasvir/sofosbuvir/ribavirin flat dose (800 mg/day) had significantly fewer AEs than patients treated with the daclatasvir/sofosbuvir/ribavirin weight-based dose. Thus, ribavirin intolerance was lower with the flat dose than with the weight-based dose. We concluded that, in this "difficult-to-treat population," the antiviral regimen of daclatasvir/sofosbuvir/ribavirin flat dose (800 mg/day) was an effective regimen, in terms of high SVR12 rates and low AE incidence. In CPT-B patients, antiviral regimen of daclatasvir/sofosbuvir/ribavirin weight dose showed higher antiviral efficacy compared to other groups, but this regimen is burdened by more AEs.

Fixed-dose combination of sofosbuvir and ledipasvir has also been used for the treatment of chronic hepatitis C genotype 1 (G1-HCV) [122].

8.3.3 Daclatasvir/Sofosbuvir FDC for the Treatment of HCV-HIV Coinfection

The current recommendation for treating HCV in HIV patients includes the combination of sofosbuvir and daclatasvir. Daclatasvir should be used at different doses to compensate for interactions with antiretroviral therapy (ART). Up to three pills a day might be required which will significantly add to the pill burden of these patients. In this study, we have used a single-tablet approach to treating HCV-HIV coinfection.

A clinical study prospectively enrolled 233 patients coinfected with HIV and HCV from 10 centers [123]. The patients received a single once-daily fixed dose combination (FDC) pill containing 400 mg sofosbuvir and 30, 60, or 90 mg daclatasvir depending on the type of ART they were receiving for 12 or 24 weeks (ClinicalTrials.gov ID: NCT03369327). The results showed single-tablet combination of daclatasvir and sofosbuvir is an effective and safe treatment for patients coinfected with HIV and HCV. The combination works well in patients on ART in which dose adjustment is required. Patients with cirrhosis, previous treatment failure, and various genotypes respond identically. The expenses of genotyping can be saved.

The recent October 2014 approval of the fixed-dose combination of the NS5B polymerase inhibitor sofosbuvir and the NS5A inhibitor ledipasvir for the treatment of treatment-naive and treatment-experienced HCV genotype 1a/1b (HCV-1) has marked a new era of IFN, and ribavirin-free treatment for chronic HCV sofosbuvir/ledipasvir combination is approved for 12 weeks in treatment-naive patients with and without cirrhosis. For treatment-experienced patients, it is approved for 12 weeks in patients without cirrhosis but for 24 weeks in patients with cirrhosis. A shorter 8-week course of treatment can be considered for treatment-naive patients who have pretreatment HCV RNA of <6 million IU/ml and do not have cirrhosis.

The FDC with ledipasvir (90 mg) and sofosbuvir (400 mg) has a sustained virologic response of approximately 96% when given as a once-a-day pill for 3 months to both treatment-naive and treatment-experienced HCV-1 patients with the exception of prior null responders with cirrhosis. The latter group of patients also achieves high sustained virologic response of 95% but with therapy for 24 weeks. In addition, emerging data suggest that this FDC regimen may be effective in the treatment of HCV-1 coinfected patients with HIV, HCV-1, and HCV-4, patients with cirrhosis and hepatic decompensation, and those with post-liver transplant HCV recurrence.

8.4 MTD in GI Disease

The prokinetic cisapride, an important therapeutic option in functional gastrointestinal (GI) disorders, was withdrawn from the market 15 years ago due to rare severe side effects. Likewise in 2014, the use of metoclopramide (MCP) and domperidone in functional GI disorders (FGID) was restricted, consequently leaving a therapeutic gap in clinical practice. A systematic review revealed that the herbal medicinal product (HMP) STW5 presents a therapeutic option equivalent to MCP and cisapride. STW5 is the only HMP for which efficacy has been shown in randomized controlled clinical trials (RCTs) in functional dyspepsia and irritable bowel syndrome, based on its multitarget effect on numerous etiological factors. Due to an outstanding favorable safety profile, STW5 allows an effective and safe use in FGID without a limitation of the duration of the treatment [124, 125].

The herbal combination preparation STW5 (Iberogast®, Steigerwald Arzneimittelwerk GmbH, Darmstadt, Germany) has been comprehensively studied for the treatment of functional gastrointestinal disorders of the upper and lower abdomen. STW5 contains alcoholic extracts of *Iberis amara totalis recens*, *Angelicae radix*, *Cardui mariae fructus*, *Chelidonii herba*, *Liquiritiae radix*, *Matricariae flos*, *Melissae folium*, *Carvi fructus*, and *Menthae piperitae folium* and has been used for the therapy of functional gastrointestinal disorders for five decades. The extracts exert different proven pharmacological effects on different regions of the gastrointestinal tract and thus address the whole symptom complex of functional dyspepsia syndrome and IBS [126]. The tolerability of STW5 was favorable both in clinical studies and in post-marketing use.

Treatment of functional gastrointestinal disorders remains a major challenge as pathomechanisms are complex. Thus, a multitarget approach is likely to be more successful than a drug directed to an individual target [127]. STW5 as herbal combination preparation with individually active components [128] has not only shown its effectiveness in improving symptoms over time in noninterventional studies as well as in retrospective database surveillances. It also passed the higher hurdle of proving its efficacy in studies versus placebo and cisapride as an active control. The efficacy observed was clinically relevant, as indicated by the effect size versus placebo as well as by an overall similar and even numerically superior effect when compared to cisapride.

The prokinetic cisapride was a standard for the treatment of functional (upper) gastrointestinal disorders before its withdrawal from most markets for safety reasons, namely, a prolongation of the QT interval and resulting cardiac arrhythmia. Two other drugs of this class, domperidone and metoclopramide, have recently undergone labeling changes or are currently under review by the European Medicines Agency due to neurological and/or cardiovascular events [129, 130]. Other drugs for treating functional gastrointestinal disorders currently recommended in guidelines are proton-pump inhibitors, antidiarrhoics, anticholinergic spasmolytics, tricyclic antidepressants, and herbal drugs. All possess specific tolerability issues which have to be evaluated and considered when selecting a suitable therapy for patients. Proton-pump inhibitors tend to be over utilized. They have been associated with rebound acid hypersecretion, which may aggravate the symptoms of functional dyspepsia [131], whereas the compliance with spasmolytics, like mebeverine, and also with tricyclic antidepressants as drugs of the last resort for functional gastrointestinal disorders may be limited by their characteristic therapeutic profile. In addition to these safety issues, the drugs enumerated are rarely adequate for patients suffering from mixed complaints of the upper and lower gastrointestinal tract, as they address only one symptom at a time. The combination of drugs helpful for symptoms both of the upper and lower abdomen will generally further increase the side effects the patient has to tolerate.

In several countries, STW5 is approved as the only product both covering functional dyspepsia as well as IBS. It is furthermore included in the therapy guidelines for both upper and lower functional gastrointestinal disorders [132]. STW5 has been thoroughly evaluated for tolerability over approximately five decades. Preclinical and clinical studies have especially searched for signals of hepatotoxicity and abnormal liver function tests, as herbal drugs may possess hepatotoxic effects. They have shown that STW5 has no such effects. For celandine herbs, very rare and reversible dose-dependent cases of hepatic side effects had been reported. However, the daily doses for which these hepatic side effects of celandine herbs have been described are approximately 100–200 times higher than those applied with STW5. Furthermore, interventional and noninterventional clinical studies with STW5 as well as database surveillances specifically assessed liver function as a safety parameter. These studies under controlled conditions found no hepatotoxicity cases associated temporally or causally with STW5. The only adverse reactions considered attributable to STW5 are hypersensitivity reactions as described in the summary of product characteristics. They have been observed very rarely in predisposed patients and may take the form of pruritus, dyspnea, or skin reactions. Thus, STW5 shows the favorable tolerability profile important for a preparation used for functional gastrointestinal disorders, because the chronic nature of these diseases often requires long-term therapy.

References

1. Ogobuiro I, Gonzales J, Tuma F. Physiology, gastrointestinal. In: StatPearls [Internet]. Treasure Island (FL), StatPearls Publishing; 2021 Jan.

2. Arnold M, Abnet CC, Neale RE, et al. Global burden of 5 major types of gastrointestinal cancer. Gastroenterology. 2020;159(1):335–49.e15.
3. Gomaa EZ. Human gut microbiota/microbiome in health and diseases: a review. Antonie Van Leeuwenhoek. 2020;113(12):2019–40.
4. Adak A, Khan MR. An insight into gut microbiota and its functionalities. Cell Mol Life Sci. 2019;76(3):473–93.
5. Yamada T, Alpers DH, Kalloo AN, et al. Textbook of gastroenterology. 5th ed. Chichester: Blackwell Pub; 2009. p. 2774–84.
6. Ralston S, Penman I, Strachan M, et al., editors. Davidson's principles and practice of medicine. 23rd ed. Edinburgh: Churchill Livingstone/Elsevier; 2018. ISBN: 9780702070280
7. Chey WD, Kurlander J, Eswaran S. Irritable bowel syndrome: a clinical review. JAMA. 2015;313(9):949–58.
8. American Gastroenterological Association, Bharucha AE, Dorn SD, Lembo A, et al. American Gastroenterological Association medical position statement on constipation. Gastroenterology. 2013;144(1):211–7.
9. Moszak M, Szulińska M, Bogdański P. You are what you eat-the relationship between diet, microbiota, and metabolic disorders-a review. Nutrients. 2020;12(4):1096.
10. Saxena R, Sharma VK. A metagenomic insight into the human microbiome: its implications in health and disease. In: Kumar D, Antonarakis S, editors. Elsevier Science: Medical and Health Genomics; 2016. p. 117.
11. Quigley EM. Gut bacteria in health and disease. Gastroenterol Hepatol. 2013;9(9):560–9.
12. Guarner F, Malagelada J. Gut flora in health and disease. Lancet. 2003;361(9356):512–9.
13. Clarke G, Stilling RM, Kennedy PJ, et al. Minireview: Gut microbiota: The neglected endocrine organ. Mol Endocrinol. 2014;28(8):1221–38.
14. Gibson GR. Fibre and effects on probiotics (the prebiotic concept). Clin Nutr Suppl. 2004;1(2):25–31.
15. Hopper CP, De La Cruz LK, Lyles KV, et al. Role of carbon monoxide in host–gut microbiome communication. Chem Rev. 2020;120(24):13273–311.
16. O'Hara AM, Shanahan F. The gut flora as a forgotten organ. EMBO Rep. 2006;7(7):688–93.
17. Rajilić-Stojanović M, De Vos WM. The first 1000 cultured species of the human gastrointestinal microbiota. FEMS Microbiol Rev. 2014;38(5):996–1047.
18. Litou C, Effinger A, Kostewicz ES, et al. Effects of medicines used to treat gastrointestinal diseases on the pharmacokinetics of coadministered drugs: a PEARRL Review. J Pharm Pharmacol. 2019;71(4):643–73.
19. Longo WE, Vernava AM. Prokinetic agents for lower gastrointestinal motility disorders. Dis Colon Rectum. 1993;36:696–708.
20. Tonini M. Recent advances in the pharmacology of gastrointestinal prokinetics. Pharmacol Res. 1996;33:217–26.
21. Cash BD, Lacy BE. Systematic review: FDA-approved prescription medications for adults with constipation. Gastroenterol Hepatol (N Y). 2006;2:736–49.
22. Tack J, et al. Diagnosis and treatment of chronic constipation – a European perspective. Neurogastroenterol Motil. 2011;23:697–710.
23. Andresen V, et al. Effect of 5 days linaclotide on transit and bowel function in females with constipation-predominant irritable bowel syndrome. Gastroenterology. 2007;133:761–8.
24. Ippoliti C. Antidiarrheal agents for the management of treatment-related diarrhea in cancer patients. Am J Health Syst Pharm. 1998;55:1573–80.
25. Blanton WP, Wolfe MM. Proton pump inhibitors. In: Wolfe MM, Lowe RC, editors. Pocket Guide to Gastrointestinal Drugs. Chichester, UK: Wiley; 2014. p. 15–30.
26. Sugano K. Histamine H2-receptor antagonists. In: Wolf MM, Lowe RC, editors. Pocket Guide to Gastrointestinal Drugs. Chichester, UK: Wiley; 2014. p. 31–43.
27. Fröhlich EE, et al. Cognitive impairment by antibiotic-induced gut dysbiosis: analysis of gut microbiota–brain communication. Brain Behav Immun. 2016;56:140–55.
28. Sullivan A, et al. Effect of antimicrobial agents on the ecological balance of human microflora. Lancet Infect Dis. 2001;1:101–14.
29. Edlund C, Nord CE. Effect on the human normal microflora of oral antibiotics for treatment of urinary tract infections. J Antimicrob Chemother. 2000;46(Suppl 1):41–8. discussion 63–5
30. Beaugerie L, Petit J-C. Antibiotic-associated diarrhoea. Best Pract Res Clin Gastroenterol. 2004;18:337–52.
31. Orlicka K, et al. Prevention of infection caused by immunosuppressive drugs in gastroenterology. Ther Adv Chronic Dis. 2013;4:167–85.
32. Zenlea T, Peppercorn MA. Immunosuppressive therapies for inflammatory bowel disease. World J Gastroenterol. 2014;20:3146–52.
33. Scaldaferri F, et al. Use and indications of cholestyramine and bile acid sequestrants. Intern Emerg Med. 2013;8:205–10.
34. Wedlake L, et al. Effectiveness and tolerability of colesevelam hydrochloride for bile-acid malabsorption in patients with cancer: a retrospective chart review and patient questionnaire. Clin Ther. 2009;31:2549–58.
35. Odunsi-Shiyanbade ST, et al. Effects of chenodeoxycholate and a bile acid sequestrant, colesevelam, on intestinal transit and bowel function. Clin Gastroenterol Hepatol. 2010;8:159–165.e5.
36. Stephen AM, Cummings JH. The microbial contribution to human faecal mass. J Med Microbiol. 1980;13(1):45–56.

37. Sommer F, Bäckhed F. The gut microbiota—masters of host development and physiology. Nat Rev Microbiol. 2013;11(4):227–38.
38. Segata N, Boernigen D, Tickle TL, et al. Computational meta'omics for microbial community studies. Mol Syst Biol. 2013;9:666.
39. Faderl M, Noti M, Corazza N, et al. Keeping bugs in check: the mucus layer as a critical component in maintaining intestinal homeostasis. IUBMB Life. 2015;67(4):275–85.
40. Gibson GR. Fibre and effects on probiotics (the prebiotic concept). Clin Nutrition Supplements. 2004;1(2):25–31.
41. Yatsunenko T, Rey FE, Manary MJ, et al. Human gut microbiome viewed across age and geography. Nature. 2012;486(7402):222–7.
42. McNamara MP, Singleton JM, Cadney MD, et al. Early-life effects of juvenile Western diet and exercise on adult gut microbiome composition in mice. J Exp Biol. 2021;224(4):jeb.239699.
43. Yeh M-C, Glick-Bauer M. The health advantage of a vegan diet: Exploring the gut microbiota connection. Nutrients. 2014;6(11):4822–38.
44. David LA, Maurice CF, Carmody RN, et al. Diet rapidly and reproducibly alters the human gut microbiome. Nature. 2013;505(7484):559–63.
45. Jeffery I, O'Toole P. Diet-microbiota interactions and their implications for healthy living. Nutrients. 2013;5(1):234–52.
46. De Filippo C, Cavalieri D, Di Paola M, et al. Impact of diet in shaping gut microbiota revealed by a comparative study in children from Europe and rural Africa. Pro Nat Acad Sci. 2010;107(33):14691–6.
47. Jonkers DMAE. Microbial perturbations and modulation in conditions associated with malnutrition and malabsorption. Best Pract Res Clin Gastroenterol. 2016;30(2):161–72.
48. Million M, Diallo A, Raoult D. Gut microbiota and malnutrition. Microb Pathog. 2017;106:127–38.
49. Khanna S, Tosh PK. A clinician's primer on the role of the microbiome in human health and disease. Mayo Clin Proc. 2014;89(1):107–14.
50. Hutkins RW, Krumbeck JA, Bindels LB, et al. Prebiotics: why definitions matter. Curr Opin Biotechnol. 2016;37:1–7.
51. Gibson GR, Roberfroid MB. Dietary modulation of the human colonic microbiota: introducing the concept of prebiotics. J Nutr. 1995;125(6):1401–12.
52. Gibson GR, Hutkins R, Sanders ME, et al. Expert consensus document: The International Scientific Association for Probiotics and Prebiotics (ISAPP) consensus statement on the definition and scope of prebiotics. Nat Rev Gastroenterol Hepatol. 2017;14(8):491–502.
53. Slavin J. Fiber and prebiotics: mechanisms and health benefits. Nutrients. 2013;5(4):1417–35.
54. Lamsal BP. Production, health aspects and potential food uses of dairy prebiotic galactooligosaccharides. J Sci Food Agricul. 2012;92(10):2020–8.
55. Rajendran SR, Okolie CL, Udenigwe CC, et al. Structural features underlying prebiotic activity of conventional and potential prebiotic oligosaccharides in food and health. J Food Biochem. 2017;41(5):e12389.
56. Pokusaeva K, Fitzgerald GF, van Sinderen D. Carbohydrate metabolism in Bifidobacteria. Genes Nutrition. 2011;6(3):285–306.
57. Zaman SA, Sarbini SR. The potential of resistant starch as a prebiotic. Crit Rev Biotechnol. 2015;36(3):578–84.
58. Gómez B, Gullón B, Remoroza C, et al. Purification, characterization, and prebiotic properties of pectic oligosaccharides from orange peel wastes. J Agricul Food Chem. 2014;62(40):9769–82.
59. Arena MP, Caggianiello G, Fiocco D, et al. Barley β-glucans-containing food enhances probiotic performances of beneficial bacteria. Int J Mol Sci. 2014;15(2):3025–39.
60. Linares-Pasten JA, Aronsson A, Karlsson EN. Structural considerations on the use of endo-xylanases for the production of prebiotic Xylooligosaccharides from biomass. Curr Protein Peptide Sci. 2017;19(1):48–67.
61. Delcour JA, Aman P, Courtin CM, et al. Prebiotics, fermentable dietary fiber, and health claims. Advances in Nutrition. 2016;7(1):1–4.
62. Pandey KR, Naik SR, Vakil BV. Probiotics, prebiotics and synbiotics- a review. J Food Sci Technol. 2015;52(12):7577–87.
63. Coxam V. Current data with inulin-type fructans and calcium, targeting bone health in adults. J Nutrition. 2007;137(Suppl11):S2527–33.
64. Seifert S, Watzl B. Inulin and oligofructose: review of experimental data on immune modulation. J Nutrition. 2007;137(Suppl11):S2563–7.
65. Wilson B, Whelan K. Prebiotic inulin-type fructans and galacto-oligosaccharides: definition, specificity, function, and application in gastrointestinal disorders. J Gastroenterol Hepatol. 2017;32(Suppl1):64–8.
66. Prebiotics. Food-Info.net. Wageningen University.
67. Byrne CS, Chambers ES, Morrison DJ, et al. The role of short chain fatty acids in appetite regulation and energy homeostasis. Int J Obesity. 2015;39(9):1331–8.
68. Mack DR. Probiotics: mixed messages. Can Fam Physician. 2005;51(11):1455–7.
69. Gareau MG, et al. Probiotics and the gut microbiota in intestinal health and disease. Nat Rev Gastroenterol Hepatol. 2010;7:503–14.
70. Guidelines for the evaluation of probiotics in food report. Joint FAO/WHO Working Group report on drafting guidelines for the evaluation of probiotics in food. 2002. Available at: http://www.who.int/foodsafety/fs_management/en/probiotic_guidelines.pdf. Accessed 5 Sept 2021.
71. Westendorf AM, et al. Intestinal immunity of Escherichia coli NISSLE 1917: a safe carrier for therapeutic molecules. FEMS Immunol Med Microbiol. 2005;43:373–84.

72. Resta-Lenert SC, Barrett KE. Modulation of intestinal barrier properties by probiotics: role in reversing colitis. Ann N Y Acad Sci. 2009;1165:175–82.
73. Probiotics. National Health Service. 27 November 2018.
74. Probiotics: What you need to know. National Center for Complementary and Integrative Health, US National Institutes of Health. 1 August 2019. Retrieved 10 November 2021.
75. Doron S, Snydman DR. Risk and safety of probiotics. Clin Infect Dis. 2015;60(Suppl2):S129–34.
76. Singhi SC, Kumar S. Probiotics in critically ill children. F1000Res. 2016;5:407.
77. Durchschein F, Petritsch W, Hammer HF. Diet therapy for inflammatory bowel diseases: The established and the new. World J Gastroenterol. 2016;22(7):2179–94.
78. Brown AC, Valiere A. Probiotics and medical nutrition therapy. Nutr Clin Care. 2004;7(2):56–68.
79. King CK, Glass R, Bresee JS, et al. Managing acute gastroenteritis among children: oral rehydration, maintenance, and nutritional therapy. MMWR Recomm Rep. 2003;52(RR–16):1–16.
80. Collinson S, Deans A, Padua-Zamora A, et al. Probiotics for treating acute infectious diarrhea. Cochrane Database Syst Rev. 2020;12:CD003048.
81. Saez-Lara MJ, Gomez-Llorente C, Plaza-Diaz J, et al. The role of probiotic lactic acid bacteria and bifidobacteria in the prevention and treatment of inflammatory bowel disease and other related diseases: A systematic review of randomized human clinical trials. Biomed Res Int. 2015;2015:1–15.
82. Probiotics and prebiotics. World Gastroenterology Organisation Global Guidelines. World Gastroenterology Organisation. October 2011. Archived (PDF) from the original on 9 August 2016. Retrieved 1 November 2021.
83. Ghouri YA, Richards DM, Rahimi EF, et al. Systematic review of randomized controlled trials of probiotics, prebiotics, and synbiotics in inflammatory bowel disease. Clin Exp Gastroenterol. 2014;7:473–87.
84. Moayyedi P, Ford AC, Talley NJ, et al. The efficacy of probiotics in the treatment of irritable bowel syndrome: a systematic review. Gut. 2010;59(3):325–32.
85. Islam MA, Yun CH, Choi YJ, et al. Microencapsulation of live probiotic bacteria. J Microbiol Biotechnol. 2010;20(1367–1377):1367–77.
86. Timmerman HM, Koning CJ, Mulder L, et al. Monostrain, multistrain and multispecies probiotics – a comparison of functionality and efficacy. Int J Food Microbiol. 2004;96(3):219–33.
87. Williams EA, Stimpson J, Wang D, et al. Clinical trial: a multistrain probiotic preparation significantly reduces symptoms of irritable bowel syndrome in a double-blind placebo-controlled study. Aliment Pharmacol Ther. 2008;29(1):97–103.
88. Rijkers GT, de Vos WM, Brummer RJ, et al. Health benefits and health claims of probiotics: Bridging science and marketing. Brit J Nutrition. 2011;106(9):1291–6.
89. Bee P. Probiotics, not so friendly after all? The Times. London. 2008. Archived from the original on 11 May 2015. Retrieved 18 May 2021.
90. Cooperative Oesophageal Group. Combination of cimetidine and alginic acid: an improvement in the treatment of oesophageal reflux disease. Gut. 1991;32(7):819–22.
91. Tytgat GNJ. Medical therapy of reflux esophagitis. In: Siewart JR, Hoelscher AH (eds). Diseases of the esophagus. Berlin, Springer Verlag. 1988: pp1137–47.
92. Temple JG, Bradby GVH, O'Connor F, et al. Cimetidine and metoclopramide in oesophageal reflux disease. BMJ. 1983;286:1863–4.
93. Bennett JR, Buckton GK, Martin HD, et al. The effect of adding cimetidine to alginate–antacid in treating gastrooesophageal reflux. In: Siewert JR, Hoelscher AH, editors. Diseases of the esophagus. Berlin: Springer Verlag; 1988. p. 1111–5.
94. Eriksen CA, Cheadle WG, Cranford CA, et al. Combined cimetidine-alginate antacid therapy versus single agent treatment for reflux oesophagitis: results of prospective double-blind randomised clinical trial. Ann Chir Gynaecol. 1988;77(4):133–7.
95. Connell AM. Combination drugs in gastroenterology. Am J Gastroenterol. 1986;81(5):392–6.
96. Alter MJ, Margolis HS, Krawczynski K, et al. The natural history of community-acquired hepatitis C in the United States: the sentinel counties chronic non-A, non-B hepatitis study team. N Engl J Med. 1992;327(27):1899–905.
97. Thomas DL, Astemborski J, Rai RM, et al. The natural history of hepatitis C virus infection: host, viral, and environmental factors. JAMA. 2000;284(4):450–6.
98. Poynard T, Bedossa P, Opolon P. Natural history of liver fibrosis progression in patients with chronic hepatitis C. The OBSVIRC, METAVIR, CLINIVIR, and DOSVIRC groups. Lancet. 1997;349(9055):825–32.
99. Wiley TE, McCarthy M, Breidi L, Layden TJ. Impact of alcohol on the histological and clinical progression of hepatitis C infection. Hepatology. 1998;28(3):805–9.
100. Graham CS, Baden LR, Yu E, et al. Influence of human immunodeficiency virus infection on the course of hepatitis C virus infection: a meta-analysis. Clin Infect Dis. 2001;33(4):562–9.
101. Thein HH, Yi Q, Dore GJ, Krahn MD. Natural history of hepatitis C virus infection in HIV-infected individuals and the impact of HIV in the era of highly active antiretroviral therapy: a meta-analysis. AIDS. 2008;22(15):1979–91.
102. Pellicelli A, Messina V, Giannelli V, et al. High efficacy and safety of flat-dose ribavirin plus sofosbuvir/daclatasvir in genotype 3 cirrhotic patients. Gut Liver. 2020;14(3):357–67.

103. Kawaguchi Y, Mizuta T. Interaction between hepatitis C virus and metabolic factors. World J Gastroenterol. 2014;20:2888–901.
104. Kanwal F, Kramer JR, Ilyas J, et al. HCV genotype 3 is associated with an increased risk of cirrhosis and hepatocellular cancer in a national sample of U.S. Veterans with HCV. Hepatology. 2014, 60:98–105.
105. Leroy V, Angus P, Bronowicki JP, et al. Daclatasvir, sofosbuvir, and ribavirin for hepatitis C virus genotype 3 and advanced liver disease: a randomized phase III study (ALLY-3+). Hepatology. 2016;63:1430–41.
106. Lontok E, Mani N, Harrington PR, Miller V. Closing in on the target: sustained virologic response in hepatitis C virus genotype 1 infection response-guided therapy. Clin Infect Dis. 2013;56:1466–70.
107. Nelson DR, Cooper JN, Lalezari JP, et al. All-oral 12-week treatment with daclatasvir plus sofosbuvir in patients with hepatitis C virus genotype 3 infection: ALLY-3 phase III study. Hepatology. 2015;61:1127–35.
108. Zeuzem S, Dusheiko GM, Salupere R, et al. Sofosbuvir and ribavirin in HCV genotypes 2 and 3. N Engl J Med. 2014;370:1993–2001.
109. Feld JJ, Maan R, Zeuzem S, et al. Effectiveness and safety of sofosbuvir-based regimens for chronic HCV genotype 3 infection: results of the HCV-TARGET study. Clin Infect Dis. 2016;63:776–83.
110. Cornberg M, Petersen J, Schober A, et al. Real-world use, effectiveness and safety of anti-viral treatment in chronic hepatitis C genotype 3 infection. Aliment Pharmacol Ther. 2017;45:688–700.
111. European Association for Study of Liver. EASL clinical practice guidelines: management of hepatitis C virus infection. J Hepatol. 2014;60:392–420.
112. Alonso S, Riveiro-Barciela M, Fernandez I, et al. Effectiveness and safety of sofosbuvir-based regimens plus an NS5A inhibitor for patients with HCV genotype 3 infection and cirrhosis: results of a multicenter real-life cohort. J Viral Hepat. 2017;24:304–11.
113. Herzer K, Welzel TM, Spengler U, et al. Real-world experience with daclatasvir plus sofosbuvir ± ribavirin for post-liver transplant HCV recurrence and severe liver disease. Transpl Int. 2017;30:243–55.
114. Hézode C, Lebray P, De Ledinghen V, et al. Daclatasvir plus sofosbuvir, with or without ribavirin, for hepatitis C virus genotype 3 in a French early access programme. Liver Int. 2017;37:1314–24.
115. Curry MP, Charlton M. Sofosbuvir and velpatasvir for patients with HCV infection. N Engl J Med. 2016;374:1688.
116. Pianko S, Flamm SL, Shiffman ML, et al. Sofosbuvir plus velpatasvir combination therapy for treatment-experienced patients with genotype 1 or 3 hepatitis C virus infection: a randomized trial. Ann Intern Med. 2015;163:809–17.
117. European Association for the Study of the Liver. EASL recommendations on treatment of hepatitis C 2018. J Hepatol. 2018;69:461–511.
118. Wyles D, Poordad F, Wang S, et al. Glecaprevir/pibrentasvir for hepatitis C virus genotype 3 patients with cirrhosis and/or prior treatment experience: a partially randomized phase 3 clinical trial. Hepatology. 2018;67:514–23.
119. Reig M, Mariño Z, Perelló C, et al. Unexpected high rate of early tumor recurrence in patients with HCV-related HCC undergoing interferon-free therapy. J Hepatol. 2016;65:719–26.
120. Cabibbo G, Petta S, Calvaruso V, et al. Is early recurrence of hepatocellular carcinoma in HCV cirrhotic patients affected by treatment with direct-acting antivirals? A prospective multicentre study. Aliment Pharmacol Ther. 2017;46:688–95.
121. Chinchilla-López P, Qi X, Yoshida EM, et al. The direct-acting antivirals for hepatitis C virus and the risk for hepatocellular carcinoma. Ann Hepatol. 2017;16:328–30.
122. Kumari R, Nguyen MH. Fixed-dose combination of sofosbuvir and ledipasvir for the treatment of chronic hepatitis C genotype 1. Expert Opin Pharmacother. 2015;16(5):739–48.
123. Dehghan Manshadi SA, Merat S, et al. Single-pill sofosbuvir and daclatasvir for treating hepatis C in patients co-infected with human immunodeficiency virus. Int J Clin Pract. 2021;75(8):e14304.
124. Madisch A, Vinson BR, Abdel-Aziz H, et al. Modulation of gastrointestinal motility beyond metoclopramide and domperidone: pharmacological and clinical evidence for phytotherapy in functional gastrointestinal disorders. Wien Med Wochenschr. 2017;167(7–8):160–8.
125. Ottillinger B, Storr M, Malfertheiner P, et al. STW 5 (Iberogast®)—a safe and effective standard in the treatment of functional gastrointestinal disorders. Wien Med Wochenschr. 2013;163(3–4):65–72.
126. Vinson B. Development of Iberogast: clinical evidence for multicomponent herbal mixtures. In: Cooper R, Kronenberg F (eds). Botanical medicine: from bench to bedside. New Rochelle, Mary Ann Liebert. 2009; pp167–89.
127. Wagner H. Multitarget therapy: the future of treatment for more than just functional dyspepsia. Phytomedicine. 2006;13(Suppl. 5):122–99.
128. Wegener T, Wagner H. The active components and the pharmacological multi-target principle of STW 5 (Iberogast). Phytomedicine. 2006;13(Suppl. 5):20–35.
129. Bundesinstitut für Arzneimittel und Medizinprodukte. Abwehr von Gefahren durch Arzneimittel, Stufe II. Metoclopramid-

haltige Arzneimittel: Wirksamkeits- und Sicherheitsbedenken (hinsichtlich neurologischer und kardiovaskulärer Ereignisse). Bonn 2. Jan. 2012.

130. Arzneimittelkommission der deutschen Ärzteschaft UAW-News International. Ventrikuläre Arrhythmien und plötzlicher Herztod im Zusammenhang mit Domperidon. Deutsches Ärzteblatt. 2012;109(35/36):A1779–80.

131. Heidelbaugh JJ, Kim AH, Chang R. Overutilization of proton-pump inhibitors: what the clinician needs to know. Ther Adv Gastroenterol. 2012;5(4):219–32.

132. Malfertheiner P, Holtmann G, Peitz U. Guidelines of the German society of metabolic and digestive diseases for the therapy of dyspepsia. Z Gastroenterol. 2001;39(11):937–56.

9 Polypharmacology in Clinical Applications: Renal Polypharmacology

Abstract

The urinary system, also known as the renal system or urinary tract, consists of the kidneys, ureters, bladder, and the urethra. The purpose of the urinary system is to eliminate waste from the body, regulate blood volume and blood pressure, control levels of electrolytes and metabolites, and regulate blood pH. The urinary tract is the body's drainage system for the eventual removal of urine. Kidney is the major organ of the renal system, which is highly prone to various damaging factors. The most common form of kidney disease is chronic kidney disease (CKD), which is recognized as a leading public health problem worldwide. The global estimated prevalence of CKD is 13.4%, and patients with end-stage kidney disease or kidney failure needing renal replacement therapy are estimated between 4902 and 7083 million. Yet there is no cure for CKD and even no pharmaceutical compounds specific for the treatment of kidney diseases in general; currently available pharmacological treatment is essentially directed to the control of blood pressure but not to the kidney pathology per se. Clearly, there is an unmet need for therapeutics for kidney disease. This chapter gives a brief introduction to renal diseases and their treatment with particular emphasis on currently available combination drug treatment and single fixed-dose combination therapy in a hope to motivate the polypharmacology in kidney disease.

Keywords

Renal polypharmacology · Chronic kidney disease · Diabetic kidney disease · Hypertensive kidney disease · Multitarget therapy

9.1 Basics of Renal Diseases and Therapy

9.1.1 Renal Disease

Kidney disease, or renal disease, technically referred to as nephropathy, is damage to or disease of a kidney. Nephritis is an inflammatory kidney disease and has several types according to the location of the inflammation. Inflammation can be diagnosed by blood tests. Nephrosis is a noninflammatory kidney disease. Nephritis and nephrosis can give rise to nephritic syndrome and nephrotic syndrome, respectively. Kidney disease usually causes a loss of kidney function to some degree and can result in kidney failure, the complete loss of kidney function. Kidney failure is known as the end-stage kidney/renal disease (ESKD/ESRD), where dialysis or a kidney transplant is the only treatment option. Causes of kidney disease include deposition of the

Z. Wang, B. Yang, *Polypharmacology*, https://doi.org/10.1007/978-3-031-04998-9_9

immunoglobulin A antibodies in the glomerulus, administration of analgesics, xanthine oxidase deficiency, toxicity of chemotherapy agents, and long-term exposure to lead or its salts. Chronic conditions that can produce nephropathy include systemic lupus erythematosus, diabetes mellitus, and high BP (hypertension), which lead to diabetic nephropathy and hypertensive nephropathy, respectively.

9.1.2 Types of Renal Diseases

9.1.2.1 Chronic Kidney Disease (CKD)

Chronic kidney disease (CKD) is a condition characterized by a gradual loss of kidney function over time, which is defined as an estimated glomerular filtration rate (GFR) <60 mL/min/1.73 m^2. It is the most common form of renal disease and a major public health concern that affects approximately 47 million persons in the United States, or 14.8% of the US adult population [1, 2]. It is associated with significant healthcare costs, morbidity, and mortality [2, 3]. The presence of CKD increases the risk of hospitalization, cardiovascular events, and death [4, 5]. Recent data show that the prevalence of CKD has largely stabilized since 2004, possibly because of better awareness and treatment of obesity, hypertension, and diabetes mellitus [6]. A 2014 report showed that Medicare spending for patients with CKD was more than $52 billion, which represents 20% of all Medicare costs [7]. The per-person per-year Medicare expense for CKD rises with increasing disease severity, ranging from $1700 for stage 2 to $12,700 for stage 4, with costs rising exponentially in ESKD [3, 7]. Thus, early detection of CKD is critical to slow disease progression, prevent long-term morbidity and mortality, and decrease healthcare spending.

The two main causes of CKD are diabetes and high blood pressure (BP), both causing serious damage to the blood vessels in the kidneys, which are responsible for up to two-thirds of the cases. The kidney diseases induced by diabetes and hypertension are often called diabetic nephropathy and hypertensive kidney disease, respectively.

Diabetic Kidney Disease (DKD) or Diabetic Nephropathy (DN)

Diabetic nephropathy (DN) is a progressive kidney disease caused by angiopathy of the capillaries in the glomeruli [8, 9]. It is characterized by nephrotic syndrome and diffuse scarring of the glomeruli. It is particularly associated with poorly managed diabetes mellitus and is a primary reason for dialysis in many developed countries. It is classified as a small blood vessel complication of diabetes. DN is one of the main causes of ESKD. The pathogenesis of diabetic nephropathy is complex. DN is a major cause of CKD in diabetes; however, people with diabetes can also get CKD from other causes, including hypertensive nephrosclerosis or ischemic nephropathy from atherosclerotic changes to small or large renal arteries.

Hypertensive Kidney Disease (HKD) or Hypertensive Nephropathy (HN)

Hypertensive kidney disease (HKD) is considered one of the consequences of long-term and poorly controlled hypertension [10–12]. According to renal databases, it is a leading cause of end-stage renal failure (ESKF), second only to diabetic kidney disease. Yet, there is lack of specificity of both clinical and morphological presentations of hypertension-related kidney disease and low prevalence of HKD that is diagnosed based on kidney biopsy findings in registries despite that arterial hypertension plays a key role in the pathogenesis of chronic vascular disease and may lead to arterionephrosclerosis [11].

9.1.2.2 Kidney Failure

Kidney failure, also called ESRD, is the last stage of CKD; thus, CKD is also called chronic kidney failure [13]. Kidney function will eventually deteriorate to the point where the kidneys can no longer perform their job properly. In this case, a person would need to go on dialysis. Dialysis filters extra fluid and waste out of the blood. Dialysis can help treat kidney disease, but it can't cure it. A kidney transplant may be another treatment option depending on patients' circumstances. Diabetes is the most common cause of

ESRD. High BP is the second most common cause of ESRD.

9.1.2.3 Acute Renal Failure (ARF)

Acute renal failure (ARF), also called acute kidney injury, is marked by the sudden reduction in kidney function over 7 days [14]. Common causes of ARF include heart attack, illegal drug use and drug abuse, not enough blood flowing to the kidneys, urinary tract problems, etc. ARF is not always permanent, and the function of kidneys may go back to normal or almost normal with treatment.

9.1.2.4 Polycystic Kidney Disease (PKD)

Polycystic kidney disease (PKD), also known as polycystic kidney syndrome, is a genetic disorder in which the renal tubules become structurally abnormal, resulting in the development and growth of multiple cysts within the kidney [15]. These cysts may begin to develop in utero, in infancy, in childhood, or in adulthood. Cysts are nonfunctioning tubules filled with fluid pumped into them, which range in size from microscopic to enormous, crushing adjacent normal tubules and eventually rendering them nonfunctional as well, leading to kidney failure.

9.1.2.5 Drug-Induced Nephrotoxicity (DNT)

The kidney is an essential organ required by the body to perform several important functions including the maintenance of homeostasis; regulation of the extracellular environment, such as detoxification; and excretion of toxic metabolites and drugs [16]. Therefore, the kidney can be considered as a major target organ for exogenous toxicants. Drug-induced nephrotoxicity (DNT) can be defined as any renal injury caused directly or indirectly by medications, with acute renal failure, tubulopathies, and glomerulopathies as common clinical presentations [17, 18]. Some examples of drugs commonly associated with the acute reduction of glomerular filtration rate are anti-inflammatories; antibiotics, such as vancomycin and aminoglycosides; and chemotherapeutic agents, such as cisplatin and methotrexate. Approximately 20% of DNT is induced by drugs, but medication of the elderly increases the incidence of DNT up to 66% as the average life span increases.

9.1.2.6 Other Kidney Diseases

Kidney Stones

Kidney stones are another common kidney problem. They occur when minerals and other substances in the blood crystallize in the kidneys, forming solid masses (stones). Kidney stones usually come out of the body during urination.

Glomerulonephritis

Glomerulonephritis is an inflammation of the glomeruli. Glomeruli are extremely small structures inside the kidneys that filter the blood. Glomerulonephritis can be caused by infections, drugs, or congenital abnormalities.

Urinary Tract Infections

Urinary tract infections (UTIs) are bacterial infections of any part of the urinary system. Infections in the bladder and urethra are the most common. If left untreated, these infections can spread to the kidneys and cause kidney failure.

9.1.3 Pharmacological Therapy of Renal Diseases

There is no cure for CKD and even no pharmaceutical compounds specific for the treatment of kidney diseases. Current therapies aim to help control the signs and symptoms, reduce the risk of complications, and slow the progression of the disease. Some of the conditions that CKD causes that require treatment include high BP, anemia, phosphate and potassium accumulations, fluid retention, skin itching, etc. Severe kidney diseases require hemodialysis and kidney transplantation.

In 2021, the US Food and Drug Administration approved Farxiga (dapagliflozin) oral tablets to reduce the risk of kidney function decline, kidney failure, cardiovascular death, and hospitalization for heart failure in adults with CKD who are at risk of disease progression. Dapagliflozin is a highly potent, reversible, and selective sodium-glucose cotransporter-2 inhibitor indicated worldwide for the treatment of type 2 diabetes mellitus (T2DM) [19]. This compound reduces the risk of kidney failure and heart failure and all-cause mortality in patients with diabetic and non-diabetic CKD [20–22].

9.1.3.1 Blood Pressure Medicine

CKD is an increasingly prevalent condition globally and is strongly associated with incident cardiovascular disease (CVD). Hypertension is both a cause and effect of CKD and affects the vast majority of CKD patients. Control of hypertension is important in those with CKD as it leads to slowing of disease progression as well as reduced CVD risk. Non-pharmacological interventions are useful in reducing BP in CKD but are rarely sufficient to control BP adequately. Patients with CKD and hypertension will often require a combination of antihypertensive medications to achieve target BP. Certain pharmacological therapies provide additional BP-independent renoprotective and/or cardioprotective action. Managing hypertension in the context of hemodialysis and following kidney transplantation presents further challenges. Antihypertensives are among the most common medicine for CKD, including both hypertensive kidney disease and diabetic kidney disease [23–29]. However, existing guidelines do not offer a consensus on optimal BP targets.

9.1.3.2 Erythropoiesis-Stimulating Agents (ESAs)

With kidney disease, the kidneys do not make enough of a hormone called erythropoietin (EPO). This can cause a problem called anemia, manifested by reduced red blood cells. A medicine called an erythropoiesis-stimulating agent (ESA) can help prevent and treat anemia. Injections of ESAs are the most common treatment for CKD with anemia. ESAs mimic the protein erythropoietin and stimulate a person's bone marrow to make more red blood cells.

9.1.3.3 Iron Supplements

Another cause of anemia (not having enough red blood cells) is not having enough iron in your blood. If your doctor thinks you have anemia that is caused by not having enough iron, you may need to take an iron supplement. You can take iron pills or have iron put into your bloodstream using a shot or during your dialysis treatment.

9.1.3.4 Phosphate Binders

People with kidney disease may not be able to eliminate phosphate from their bodies properly. Treatment involves reducing the intake of nutritional phosphate intake. This usually means reducing consumption of dairy products, red meat, eggs, and fish.

9.1.3.5 Potassium Binders

Similarly, people with kidney disease may not be able to eliminate potassium from blood properly, which could make potassium accumulation in the body and cause serious problems with the heart and other muscles. In this case, a medicine called a potassium binder may be taken to help get rid of extra potassium from blood.

9.1.3.6 Calcium and Vitamin D

The kidneys activate the vitamin D obtained from the sun or food before the body can use it. People with kidney disease typically have low levels of vitamin D, which is essential for healthy bones. People with vitamin D deficiency may require supplementation.

9.2 CDT for Renal Disease

9.2.1 CDT in Chronic Kidney Disease (CKD)

The renin-angiotensin system (RAS) plays a fundamental and critical role in preserving the circulatory volume therefore preserving adequate organ perfusion through regulating vascular tone, vascular structure, and renal sodium handling, and yet, it may be injurious to the heart and blood vessels and may also allow, and sometimes hasten, kidney disease progression [30]. Thus, effective RAS inhibition may be a major pharmacologic necessity to control hypertension, to decrease cardiovascular complication, and to inhibit kidney disease progression [30]. Indeed, angiotensin-converting enzyme inhibitors (ACEIs), angiotensin II receptor blockers (ARBs), and direct renin inhibition with aliskiren (a direct renin inhibitor) have shown multiple benefits not only in lowering BP but also in heart failure, atherosclerosis, insulin resistance, and CKD of various etiologies including primary glomerulonephritis, vasculitis, diabetic nephropathy, nephrosclerosis, and others [30].

As summarized by Juncos and Caputo [30], two pharmacologic strategies have been proposed to overcome this drawback including higher than recommended doses of RAS inhibitors and the combination of two different RAS inhibitors [30]. For instance, reducing proteinuria in CKD is a vital objective, but it often cannot be achieved with standard recommended therapies. For this, intensive RAS inhibition with combination drugs has been recommended, not only for diabetic but also for nondiabetic nephropathies because this strategy could provide greater antihypertensive and renal and cardiovascular protective effects [30–32].

The antihypertensive regimen should be one that can both reduce BP sufficiently and reduce albuminuria [33]. Data from recent large hypertension clinical trials reveal that the typical antihypertensive regimen includes three or more different medications to achieve this level of BP reduction. The critical factor is simplification of the approach in patients who require multiple medications for control of their BP as well as control of their other cardiovascular comorbidities, such as elevated cholesterol, glucose, and anemia [33]. Large, randomized, clinical hypertension trials have shown that tighter BP control, compared with less tight BP control, can reduce progression of renal disease by 30–50% and cardiovascular disease by 40–70% [33]. Achieving BP levels of <130/80 mm Hg, as currently recommended for patients with diabetes or CKD, will often require three or more antihypertensive medications. Furthermore, reduction of BP should be accompanied by reductions in albuminuria and proteinuria to maximize potential benefits to the kidney [33]. Evidence from numerous randomized controlled trials mandate that agents that block the RAS should always be included in the antihypertensive regimens of patients with CKD, particularly with the excellent safety data with serum creatinine levels <3–4 mg/dL [33]. There have been more than 20 clinical trials on the efficacy of dual blockade of RAS with combinational regimens involving ACEIs and ARBs on CKD [34].

9.2.1.1 Dual RAS Inhibition: ACEI/ARB Combinations

Zoccali et al. [35] were among the first to investigate a potential effect of an AT1 receptor antagonist as add-on therapy in 11 patients treated with an ACE inhibitor. The patients had various renal diseases, including diabetic nephropathy. In this pilot study, losartan caused a 30% reduction of proteinuria in patients who were already on ACE inhibitors [35]. Although not significant, addition of losartan to ongoing ACE inhibitor therapy also caused a marked decrease in BP (from 105 ± 7 mm Hg to 98 ± 15 mm Hg mean arterial pressure). This report exemplifies the problems with which riddled almost all clinical studies investigating combination therapy in patients with CKD because BP was lower in the combination therapy arm [34]. Nevertheless, these smaller studies contributed important data in regard of side effects, investigated combination therapy of various ACE inhibitors with different AT1 receptor antagonists, and looked at patients with heterogeneous causes of CKD [34].

Candesartan (an angiotensin receptor blocker) was combined with lisinopril (an angiotensin-converting enzyme inhibitor) in the Candesartan and Lisinopril Microalbuminuria (CALM) trial in hypertensive patients with microalbuminuria [30, 36]. The authors reported a significant reduction in diastolic BP and in protein excretion rate with combined candesartan-lisinopril therapy compared with single treatment with either drug. There were no differences between groups in side effects.

Parving and colleagues found a significant fall in protein excretion rate in patients with diabetes type 1 (T1DM), receiving combination therapy with losartan and enalapril, compared with single drug treatment [37]. Similarly, in the Aliskiren in

Left Ventricular Hypertrophy (ALLAY) trial, combined therapy was superior to monotherapy [38].

However, three trials concluded that combination therapy was unsafe. These were the Ongoing Telmisartan Alone and in Combination with Ramipril Global Endpoint Trial (ONTARGET) [30, 39], the Aliskiren Trial in Type 2 Diabetes Using Cardio-Renal Endpoints (ALTITUDE) [30, 40], and more recently, the Veterans Affairs Combination Angiotensin Receptor Blocker and Angiotensin-converting Enzyme Inhibitor for Treatment of Diabetic Nephropathy study (VA NEPHRON D) [30, 41]. These apparent dissimilarities with previous reports could result from differences in inclusion and exclusion criteria [30]. Another cause might be that the use of antihypertensive agents, in particular RAS inhibitors, may be associated with decreased GFR due to a fall in renal perfusion pressure. Nonetheless, a fall in glomerular perfusion pressure not only is predictable, but it may also be desirable because it reveals a reduction in glomerular pressure. Indeed, RAS inhibitors have been shown to decrease GFR through a hemodynamic mechanism that is not overtly harmful. In fact, in CKD, elevations of serum-creatinine levels after RAS inhibition point to better outcomes [42, 43]. A post hoc analysis of the Reduction of Endpoints in Non-Insulin-Dependent Diabetes Mellitus with the Angiotensin II Antagonist Losartan (RENAAL) study showed that patients with greater falls in GFR had better long-term preservation of kidney function [44].

Nevertheless, the evidence indicates that intensive or dual blocking should be avoided in elderly patients with normal or borderline BP and at high cardiovascular risk [30]. When these treatments are indicated, BP, serum electrolytes, and renal function should be carefully monitored. In these settings, potassium-sparing drugs, such as spironolactone, eplerenone, triamterene, and amiloride, should be avoided [30].

9.2.1.2 RAS Inhibitor/ARB Combinations

A RAS-blocking agent can be combined with a diuretic, a calcium channel blocker (CCB), or a β-blocker or both to achieve optimal BP control as quickly as possible [45]. There is clinical evidence that earlier achievement of BP goals results in better long-term control rates [46, 47]. The pros and cons of using various drug classes with a RAS blocker depend on the comorbid conditions of the patient. In patients with CKD, one could consider using an ACE inhibitor and diuretic, ARB and diuretic, or ACE inhibitor and CCB. The decision about the diuretic component would depend on the level of GFR. Those patients with serum creatinine levels >2.0 mg/dL would likely derive greater benefit from a loop diuretic. A diuretic also would be appropriate for volume-overloaded patients. However, diuretic therapy has been shown to increase the risk of development of new-onset diabetes in patients at high risk [46–49] and for this reason should be used with caution in patients with elevated fasting glucose levels or other indicators of the metabolic syndrome [48].

A CCB would be another consideration in the patient with CKD. The combination of a CCB and an ACE inhibitor has been shown to produce additive BP-lowering effects as well as to reduce cardiovascular events in randomized clinical trials, including the Hypertension Optimal Treatment (HOT) trial [49–51]. The CCB class provides robust BP reduction regardless of age, gender, ethnicity, salt intake, and use of nonsteroidal anti-inflammatory drugs [52, 53]. Thus, an ACE inhibitor/CCB combination may have an important role in treating patients with diabetes or CKD; this low-dose combination has also shown efficacy in patients with isolated systolic hypertension or stage 2 hypertension, surpassing that shown with high-dose CCB monotherapy [47].

Collectively, data from randomized clinical trials and current guidelines [54, 55] indicate that there are significant renal and cardiovascular clinical benefits with agents that block the RAS in patients with concomitant hypertension and diabetes, all patients with diabetic nephropathy, and patients with hypertension and chronic renal insufficiency. These drugs have a well-established safety profile, even in patients with a serum creatinine level in the range of 3–4 mg/dL [43].

9.2.1.3 Mechanisms for the Efficacy of Combinational Therapy

Azizi et al. [34, 56] studied combination therapy in normotensive volunteers with mild sodium depletion and proposed the hypothesis that the combination abrogates the potentially adverse consequences of persistently high Ang II concentrations during ACE inhibition. ACE inhibition increased GFR and renal blood flow but had no significant effects on fractional sodium or water reabsorption. In contrast, AT1 receptor blockade shifted the curves for fractional sodium and water reabsorption to the left but had only small effects on renal blood flow and GFR [34, 57]. A combination of ACE inhibition and AT1 receptor blockade had an additive effect and shifted the pressure-natriuresis curves leftward more than did either drug alone [57].

9.2.2 CDT in Diabetic Kidney Disease (DKD)

Diabetic nephropathy (DN) is among the common and serious complications of diabetes and is also a major cause of ESKD. Early DN is also called diabetic microalbumin period, and the current treatment is only for a particular cause without multitarget therapeutic drugs. The main treatment is in the control of blood sugar on the basis of kidney protection and urine lowering protein. There are few effective methods of western medicine treatment, and most of them are accompanied by adverse reactions. But clinical studies have also suggested that the combination of ACEIs and ARBs offers greater benefits for patients with diabetic kidney disease (DKD) or diabetic nephropathy (DN) [9]. However, combination therapy with an ACE inhibitor and an ARB was found to be associated with an increased risk of adverse events among patients with diabetic nephropathy [58]. Better approaches are therefore needed for DKD.

It is commonly recognized that the comprehensive management of DKD depends on combined target therapies for hyperglycemia, hypertension, albuminuria, hyperlipidemia, etc. [59]. Sodium-glucose cotransporter-2 (SGLT2) inhibitors are the most recently developed oral hypoglycemic agents that act on renal proximal tubules to suppress glucose reabsorption and increase urinary glucose excretion. Besides improvements in glycemic control, they presented excellent performances in direct renoprotective effects and the cardiovascular safety by decreasing albuminuria and the independent cardiovascular risk factors such as body weight, BP, etc. Simultaneous use of SGLT2 inhibitors and RAS blockers is a novel strategy to slow the progression of DKD via reducing inflammatory and fibrotic markers induced by hyperglycemia more than either drug alone [59].

Monotherapy with metformin, insulin, or pioglitazone alone is poorly effective in maintaining long-term glycemic control and renoprotection in a majority of patients with DKD [59, 60]. Various pharmacological approaches may be added to them as dual therapies or combined together as triple therapies. The concept of combining SGLT2 inhibitors with ACEI/ARBs, DPP-4 inhibitors, and GLP1-RAs has got much attention [61].

9.2.2.1 SGLT2 Inhibitor/ACEI or ARB Combination

It has been reported that ACEIs/ARBs are associated with a lower incidence of the progression to end-stage renal disease in DKD patients and reduced the incidence of composite MACE outcome [59, 61]. SGLT2 inhibitors contribute to increased sodium levels delivered to the macula densa and secondary autoregulatory vasoconstriction of afferent glomerular arteriole to neutralize the vascular imbalance driven by local Ang II characterized by glomerular hypertension in T2DM patients [62]. In this way, SGLT2 inhibitors and ACEIs/ARBs likely generate synergistic effects on kidney.

Either SGLT2 inhibitors or ACEIs/ARBs can produce albuminuria- and BP-lowering effects in patients with DKD, the efficacy of monotherapy is from suboptimal to unsatisfactory. Several studies have demonstrated that the combination of SGLT2 inhibitors with ACEIs/ARBs satisfactorily afforded greater renoprotection than administration of either drug alone. These beneficial effects include a long-term control of hyperglycemia and BP, reduction of hyperfiltration and proteinuria, and attenuation of the development of renal injury in Dahl-STZ rat model and patients of DKD [59, 63–65].

In a phase III clinical trial, ACEIs/ARBs added on to dapagliflozin (a SGLT2 inhibitor) resulted in the albuminuria and SBP decreased by −33.2% compared with placebo, respectively, in patients with DKD. Addition of dapagliflozin produced greater decreases in albuminuria [66] and serum uric acid [59, 67–69].

9.2.2.2 SGLT2 Inhibitor/DPP-4 Inhibitor Combination

SGLT2 inhibitors could increase EGP (HGP) and plasma glucagon concentrations [59]. However, dipeptidyl peptidase 4 (DPP-4) inhibitors exert their hypoglycemic effects by preventing the degradation of endogenously released incretin hormones such as glucagon-like peptide-1 (GLP-1) and glucose-dependent insulinotropic polypeptide to enhance postprandial insulin secretion, suppress glucagon secretion, and decrease endogenous (hepatic) glucose [59, 70]. In addition, a study indicated that DPP-4 inhibitors could improve endothelial function and reduce renal and vascular oxidative stress, which was independent of albuminuria lowering or improvement in glucose control, in patients with T2DM and chronic kidney disease [71].

A phase III, randomized, double-blind, parallel-group study showed that the combination of SGLT2 inhibitors and DPP-4 inhibitors decreased systolic BP [72]. The SBP-lowering mechanisms of DPP-4 inhibitor may complement or augment SGLT2 inhibitor-induced plasma volume reduction and other mechanism. Of note, a higher proportion of subjects with microalbuminuria at baseline decreased to no albuminuria at the end of combination treatment [72]. Mean changes from baseline in UACR and eGFR were small and similar across treatment groups [73]. Furthermore, the combined therapy had CV safety and did not induce weight gain in several large prospective CV outcome studies, as well as potential renoprotection [70, 72]. Pharmacokinetics and pharmacodynamic (PK-PD) study investigated in healthy individuals found no drug-drug interaction in combination therapy of SGLT2 inhibitors and DPP-4 inhibitors [74].

9.2.2.3 SGLT2 Inhibitor/GLP1-RA Combination

GLP1-RAs act by inhibiting the master inflammatory mediator NF-κB, monocyte chemoattractant protein-1, intracellular cell adhesion molecule-1, and vascular cell adhesion molecule-1, which are associated with abnormalities in vascular function and progression of DKD [59, 75, 76]. Additionally, GLP1-RAs can also induce significant diuretic and natriuretic responses to improve renal sodium and water handling [77]. Several studies showed that combination therapy with SGLT2 inhibitors and GLP1-RAs produced more favorable effects on renoprotection than monotherapy [59, 78, 79].

9.2.2.4 SGLT2 Inhibitor/MRA Combination

It has been reported that aldosterone contributes to renal injury and mineralocorticoid receptor antagonists (MRAs, aldosterone receptor antagonists) have effect on renoprotection by reducing the concentration of aldosterone in DKD patients [59]. But MRAs are limited due to hyperkalemia. A new MRA compound was found to have better selective and greater affinity for the mineralocorticoid receptor than the older generation of MRAs like spironolactone eplerenone. Moreover, finerenone induces lower incidence of hyperkalemia as compared to spironolactone. Most notably, finerenone can efficiently reduce the concentration of aldosterone offering renoprotection [80]. Furthermore, a large randomized controlled trial (RCT) on the renoprotective effect of finerenone in DKD is currently ongoing (FIDELIO-DKD). SGLT2 inhibitor and MRA combination could potentially produce synergistic renoprotective effects to slow the progression of DKD.

9.2.3 CDT in Hypertensive Kidney Disease (HKD)

In the industrialized world, hypertension affects approximately 30% of the general population. Hypertensive kidney disease (HKD) is considered one of the consequences of long-term and poorly controlled hypertension [10, 11, 81]. According to renal databases, it is a leading cause of ESKF, second only to DKD. There is convincing evidence that malignant hypertension is associated with kidney injury that frequently evolves into progressive CKD [82]. On the other hand, numerous authors have previously suggested that nonmalignant hypertension may not be the cause of much of what is labeled hypertensive nephrosclerosis (and pathologically described as arterionephrosclerosis) or more provocatively have proposed that nonmalignant hypertension does not ever cause CKD. Proponents of the various versions of this skeptical position have included Weisstuch and Dworkin [83], Luft [84], Meyrier and Simon [85], Hsu [86], Skorecki and Wasser [87], and in particular Freedman and colleagues [88–91]. Hsu laid out several lines of evidence that nonmalignant hypertension does not cause CKD or at least is not a common cause of CKD [81, 86]. First, if hypertension causes CKD, then more effective treatment of hypertension would be expected to reduce the appearance of CKD. A meta-analysis by Hsu including 10 randomized controlled hypertension treatment trials published between 1966 and 1997 involving 25,000 patients and 114,000 treatment-years, which reflects a sufficiently long duration to see an effect of BP control on renal outcomes, demonstrated that lowering BP has no effect on the development of renal dysfunction [92]. More recently, the data from the African American Study of Kidney Disease and Hypertension (AASK) showed that even with excellent BP control and the use of a renin-angiotensin system inhibitor, progression of underlying CKD is relentless [93]. Second, the diagnosis of hypertension-attributed CKD is often erroneous [94], and even when great care has been taken in the clinical definition, as was the case in the AASK study, it is quite possible that many cases have underlying primary kidney disease due to genetic variants. Third, the apparent link between hypertension and CKD may be confounded by the effect of hypertension to accelerate underlying CKD of any origin, including primary glomerular disease and renal artery stenosis.

On the other hand, numerous studies have generated the evidence in support of the renoprotective effects of antihypertensives in patient with CKD, as already described above [35–55].

The controversy might be ascribed to the difficulty of dissecting the cause-effect relationship between hypertension and kidney disease since they are mutually causative and mutually reinforcing.

9.2.4 CDT in Polycystic Kidney Disease (PKD)

Polycystic kidney disease (PKD) is a group of monogenic disorders that result in renal cyst development and one of obstructive kidney diseases [15, 95]. Over the past 15 years, positional cloning approaches have identified more than 20 genes causing these disorders [96]. PKD is characterized by massive enlargement of fluid-filled renal tubular and/or collecting duct cysts [97]. Cystic kidneys are common causes of ESKD, both in children and in adults. There are two types of PKD: autosomal dominant PKD and autosomal recessive PKD. Autosomal dominant PKD causes cysts only in the kidneys. It is often called "adult PKD" because people with this type of PKD might not notice any symptoms until they are between 30 and 50 years old. Autosomal recessive PKD causes cysts to grow in both the kidneys and the liver [15, 95, 96].

The pharmacological treatment of PKD has been primarily focused on inhibition of cyst growth and control of BP [97–100]. Tolvaptan (Jynarque), a selective vasopressin V2-receptor antagonist (V2RI), was approved by the US Food and Drug Administration (FDA) in April 2018 to slow kidney function decline in adults at risk of rapidly progressing autosomal dominant polycystic kidney disease (ADPKD) [98, 99]. Approval was based on several phase III clinical trials. The drugs of choice for hypertension related to ADPKD are ACEIs and ARBs [100].

It has been documented that co-targeting two GPCRs, the V2R and the calcium sensing receptor (CSR), using a novel V2RI lixivaptan in combination with the calcimimetic R-568 (CSR activator), reduced cyst progression in two animal models of human PKD [101]. Lixivaptan is expected to have a safer liver profile compared to tolvaptan, the only drug approved to delay PKD progression, based on computational model results and initial clinical evidence. In animal models, the combined treatment appeared nominally more effective than the individual drugs used alone. These data point to an intriguing new application for two existing drugs in PKD treatment [101].

The V2RI tolvaptan delays the progression of ADPKD. However, some patients discontinue tolvaptan because of severe adverse aquaretic events. Uchiyama et al. [102] reported an open-label, randomized, controlled, counterbalanced, crossover trial investigating the effects of trichlormethiazide, a thiazide diuretic, in patients with ADPKD receiving tolvaptan who randomly received antihypertensive therapy with or without trichlormethiazide for 12 weeks. The primary and secondary outcomes included amount and osmolarity of 24-h urine and health-related quality-of-life (HRQOL) parameters assessed by the Kidney Disease Quality of Life-Short Form questionnaire, renal function slope, and plasma/urinary biomarkers associated with disease progression. The investigators observed a significant reduction in urine volume and a significant increase in urinary osmolarity in patients who received combinational treatment with tolvaptan/antihypertensive/trichlormethiazide. Addition of trichlormethiazide improved the HRQOL: effects of kidney disease, sleep, emotional role functioning, social functioning, and role/social component summary [102]. No significant differences were noted in renal function slope or plasma/urinary biomarkers between patients treated with and without trichlormethiazide. In patients with ADPKD treated with tolvaptan, trichlormethiazide may improve tolvaptan tolerability and HRQOL parameters [102].

In a study reported by Li et al. [103], the combination of curcumin and ginkgolide B was found to produce remarkable inhibition of cystogenesis with synergistic effect owing to their ability to regulate multiple signaling pathways, in vitro Madin-Darby canine kidney (MDCK) cyst model and an in vivo kidney-specific polycystin 1 transient receptor potential channel interacting (Pkd1) knockout mouse model. The combination of curcumin and ginkgolide was more effective compared with either treatment alone in inhibiting cystogenesis. It was proposed that curcumin mediates its anti-cyst effects by blocking the EGFR/ERK1/2, JNK, and PI3K/mTOR signaling pathway and ginkgolide via downregulating the EGFR/ERK1/2, JNK, and p38 pathway [103].

The current recommendation for ADPKD therapy is the combined pharmacological treatment with ACEIs or ARBs, statins (for patients with lipid disorders and cardiovascular disease), and tolvaptan (for patients with stage I–III CKD and rapidly progressive ADPKD) [104].

Both rapamycin and peroxisome proliferator-activated receptor-γ (PPARγ) have been reported to halt the progression of ADPKD owing to their antiproliferative effects. Liu et al. [105] investigated the effect of concomitant administration of rapamycin and the PPARγ agonist rosiglitazone on the proliferation of ADPKD cyst-lining epithelial cells and the progression of ADPKD in Han:SPRD rats. Their results clearly indicate that the effects of combination therapy were superior in inhibiting the proliferation of WT9-12 cells and delaying the progression of ADPKD to either drug alone [105].

It should be noted that most of the studies published to date on combination therapy were conducted in animal models and clinical trials are still lacking rendering an uncertainty of the approach to patients.

9.2.5 CDT in Drug-Induced Nephrotoxicity (DNT)

Drug-induced kidney injury or drug-induced nephrotoxicity (DNT) is a major side effect in clinical practice, frequently leading to acute renal failure (ARF) [106]. It accounts for more than 2–15% of cases of ARF in patients admitted to the hospital or in the intensive care unit, respectively. Many cases of DNT are caused by the use of antibiotics, particularly, the combination of different antibiotics, e.g., vancomycin/aminoglycoside combination and aminoglycoside and cephalosporin combination [107, 108].

Several natural antioxidants were found to be effective against DNT. Indeed, the protective effects of spirulina and pycnogenol alone or in combination on vancomycin-induced renal cortical oxidative stress were observed in an animal study [109]. Either monotherapy or combinational treatment mitigated the elevated serum creatinine, blood urea nitrogen, renal malondialdehyde, and immunoexpression of the proapoptotic protein (Bax), autophagic marker protein (LC3/B), and inducible nitric oxide synthase induced by vancomycin. Moreover, the antioxidants restored the reduced glutathione, glutathione peroxidase, superoxide dismutase, and immunoexpression of the antiapoptotic protein Bcl2. Notably, the combination therapy of spirulina and pycnogenol showed better protective effects than the corresponding monotherapy [109].

9.3 FDC Drugs for Renal Disease

In addition to the separate pill combinations as described above, fixed-dose single-pill combinations are also becoming more prevalent in the clinical treatment of kidney disease. The major reason and rationale for the improved efficacy of combination therapy (particularly fixed-dose single-pill combination or FDC) are that it brings high-risk hypertensive patients to appropriate BP goals, simplifying the complex medical regimens in these patients [110, 111].

9.3.1 FDC in Chronic Kidney Disease (CKD)

In light of the need to attain BP goals, combination therapy (especially fixed dose) is currently recommended. Among the various combinations available, calcium channel blockers (CCBs) and angiotensin-converting enzyme (ACE) inhibitors have been proven to be extremely effective, while also displaying good tolerability [112]. Individually, both the third-generation CCB lercanidipine and the ACE inhibitor enalapril are effective antihypertensive agents. In addition, both of these agents also show other beneficial effects when administered as monotherapy. Of particular importance is the fact that when lercanidipine plus enalapril is administered in combination, they show synergism, thus providing added efficacy with reduced side effects.

Several dual- and triple-drug, fixed-dose combinations with a RAS blocker, a CCB, and a diuretic (e.g., the combination of olmesartan (ARB)/amlodipine (CCB)/hydrochlorothiazide, a diuretic medication) have been developed and marketed, which verified the effectiveness of the FDC approach for the treatment of hypertension in patients with CKD [113]. Another study demonstrated that in patients with diabetes, CKD, or chronic CVD and short-term (12 weeks) and long-term treatment with fixed-dose combination of olmesartan/amlodipine/hydrochlorothiazide, the control of BP was better for participants receiving triple-combination treatment compared with the dual-combination treatments and was achieved in 55.0% of participants with CKD [114]. Moreover, at week 52, there was sustained BP lowering with the regimen, and overall, the triple combination was well tolerated [114].

Bakris et al. [115] assessed the effects of fixed-dose combinations between benazepril (an ACEI) and amlodipine (a CCB) and between benazepril and hydrochlorothiazide on progression of CKD in the Avoiding Cardiovascular Events through Combination Therapy in Patients Living with Systolic Hypertension (ACCOMPLISH) trial. ACCOMPLISH was a double-blind, randomized trial undertaken in five countries (USA, Sweden, Norway, Denmark, and Finland). 11,506 patients with hypertension who were at high risk for cardiovascular events are included in the study. The results showed that in patients with CKD, angioedema was more frequent in the benazepril plus amlodipine group than in the benazepril plus hydrochlorothiazide group. Initial antihypertensive treatment with benazepril plus amlodipine should be considered in preference to benazepril plus hydrochlorothiazide since it slows progression of nephropathy to a greater extent [115].

It has been reported that the ACEI/ARB combination was a better dual therapy than ACEI or ARB in combination with renin inhibitor or aldosterone receptor antagonist [116]. A meta-analysis of 72 randomized controlled trials with 10,296 patients demonstrated that dual RAS blockade therapy was superior to monotherapy in reducing the urine albumin excretion, urine protein excretion, and BP [116]. These beneficial effects were related to the decrease of glomerular filtration rate, the increase of serum potassium level, and higher rates of hyperkalemia and hypotension. Meanwhile, these effects did not lead to improvements in short-term or long-term outcomes, including doubling of serum creatinine, acute kidney injury, ESKD, mortality, and hospitalization. Compared with the single therapy, the ACEI/ARB combination was superior to the ACEI or ARB in combination with renin inhibitor or aldosterone receptor antagonist in decreasing urine albumin excretion, urine protein excretion, and BP, and the combination was not associated with a lower glomerular filtration rate [116].

9.3.2 FDC in Diabetic Kidney Disease (DKD)

T2DM is often accompanied by high BP and the clustering of several cardiovascular risk factors and is the most frequent cause of ESKD. The stages of development of overt nephropathy in T2DM patients range from an initial alteration in renal function with an increased GFR, followed by the development of microalbuminuria and macroalbuminuria or proteinuria, featuring an established diabetic nephropathy, which eventually progresses to ESKD. Early intervention is needed to prevent the development of diabetic nephropathy and requires effective control of the different risk factors and in particular high BP. In the initial stages of the disease, strict BP control is crucial to prevent the development of initial renal and vascular damage. Adequate BP control is particularly difficult in T2DM patients and in most cases requires the use of combination therapy [117]. Preterax, a fixed-dose combination of perindopril 2 mg and indapamide 0.625 mg, allows BP to be significantly reduced compared with conventional strategies; this combination can be uptitrated to BiPreterax when further BP control is needed. In the PREMIER study performed in T2DM over 12 months, the perindopril/indapamide combination brought about, in addition to excellent BP control, a significant reduction in urinary albumin excretion, compared with monotherapy with enalapril [117]. In more advanced degrees of renal damage, higher doses of the fixed combination have to be considered [117]. The pharmacological basis of the renoprotective effect of perindopril/indapamide is the demonstration that this combination prevented nephropathy as well as proteinuria in obese Zucker rats, independently of BP control. Strict BP control from the initial stages of nephropathy together with inhibition of the renin-angiotensin system is mandatory to prevent albuminuria. The fixed-dose combination of perindopril/indapamide can greatly help clinicians in achieving the above goals, using Preterax in the early and BiPreterax in the late stages of nephropathy. Renal protection in T2DM patients with combined ACEIs and ARBs for dual blockade of RAS has also been documented [118].

The presence of inadequately controlled hypertension in a diabetic patient with clinical signs of renal involvement portends a poor prognosis [119]. Intensive treatment requires finding a combination of medications which will reduce not only BP but also proteinuria. It is suggested that treatment should be started with an ACEI or an ARB often in a fixed combination with a low-dose thiazide diuretic. CCBs and β-blockers may be added if required as second- or third-line agents. In patients not responding to this combination, the dosages of the ACEI or ARB should be titrated upward in order to obtain the maximal therapeutic effect. However, if this is still insufficient, dual blockade of the RAS should be considered, and even an aldosterone receptor blocker may need to be added to the therapeutic regimen [119].

Schjoedt et al. [120] performed a combined analysis on three randomized, double-masked, crossover trials where 51 T1DM patients suffering from diabetic nephropathy received 8 weeks of dual RAS blockade using an ARB in combination with an ACEI and 8 weeks of monotherapy with the same ACEI. Albuminuria, 24 h BP, GFR, and plasma aldosterone were determined at the end of each treatment period. Compared to ACEI alone, dual RAS blockade lowered BP by 7/5 mm Hg from 137/76 mm Hg, decreased albuminuria by 37% and plasma aldosterone by 28%, and caused a reversible decline in GFR. There was a significant correlation between changes in 24-h diastolic BP and changes in albuminuria. Their data therefore indicate that dual RAS blockade offers renal protection in T1DM patients with diabetic nephropathy.

9.3.3 FDC in Hypertensive Kidney Disease (HKD)

Hypertension is a key pathogenic factor in the deterioration of renal function. Its presence accelerates the decline in eGFR and increases the onset of ESRD and the risk of CVD [121, 122]. Hypertension has been identified as an independent risk factor for ESRD in men [123] and women [124]; the risk of ESRD increases progressively with increasing severity of hypertension. The Multiple Risk Factor Intervention Trial (MRFIT) (N = 332,544) showed that, relative to men with systolic BP <120 mm Hg and diastolic BP <80 mm Hg, the risk for ESRD increased by 3-, 6-, 11-, and 22-fold in men with stage 1 (mild), stage 2 (moderate), stage 3 (severe), and stage 4 (very severe) hypertension, respectively [124].

Based on JNC 7 and the National Kidney Foundation recommendations for ACEI or ARB therapy as first-line antihypertensive treatment in patients with CKD, combination antihypertensive therapy in this population should include one of these RAAS inhibitors [121, 125, 126]. Clinical evidence supports the use of a RAAS inhibitor-based fixed-dose combination product as initial antihypertensive therapy. In the Strategies of Treatment in Hypertension: Evaluation (STRATHE) study, initial fixed low-dose RAAS inhibitor-based combination therapy was superior to either sequential monotherapy (atenolol replaced by losartan and then by amlodipine) or stepped-care treatment (valsartan 40 mg titrated to 80 mg, coadministered with hydrochlorothiazide if needed) in achieving the target BP [121, 127]. Fifty-six percent of patients in the low-dose combination group achieved their target BP without experiencing treatment-related adverse events, compared with 42% in the sequential monotherapy and the stepped-care groups. Furthermore, because hypertensive CKD patients are likely to require a combination of antihypertensive drugs, starting with fixed-dose combination therapy would appear logical [121].

In other studies, target BPs were achieved sooner (31) and by significantly more patients receiving initial fixed-dose RAAS inhibitor-based combination therapy than those receiving initial monotherapy [47, 127–130]. In a randomized, double-blind, placebo-controlled study (N = 214), hypertensive type 2 diabetic patients receiving fixed-dose combination therapy (benazepril/amlodipine) achieved the predetermined target BP (<130/85 mm Hg) in a mean of 5.3 weeks versus 6.4 weeks for (enalapril) monotherapy ($P = 0.001$) [131]. At 12 weeks, 70% of patients receiving fixed-dose combination therapy were at the currently recommended target BP (<130/80 mm Hg), compared with 31% receiving monotherapy [121, 131].

Rational drug combinations are characterized by synergistic or additive BP-lowering efficacy via complementary mechanisms (e.g., RAAS-inhibiting drugs plus a diuretic or a CCB) [47, 121, 128, 132, 133], and a mechanism of action of one component that offsets the other's adverse effects [47, 129, 133–135]. With respect to the latter, combining a RAAS inhibitor (ACEI or ARB) with a diuretic attenuates RAAS stimulation triggered by diuretic-induced salt excretion and reduced plasma volume [134]. Combining a RAAS inhibitor with a dihydropyridine CCB attenuates the reflex vasoconstriction and tachycardia resulting from increased sympathetic nervous system activity reflexive to CCB-induced systemic vasodilation [134]. Combining an ARB with a dihydropyridine CCB (amlodipine) can reduce peripheral edema [135].

Although the Kidney Disease Outcomes Quality Initiative working group recommends the combination of a RAAS inhibitor and a diuretic for most patients [121, 125], recent findings of the ACCOMPLISH (Avoiding Cardiovascular Events through Combination Therapy in Patients Living with Systolic Hypertension) trial suggest that RAAS inhibitor/CCB fixed-dose combination therapy may be a better treatment choice for patients at high risk for cardiovascular events, including patients with impaired renal function (18% of the study population had eGFR <60 mL/min/1.73 m^2; 6% had proteinuria or elevated serum creatinine) [136].

In the ACCOMPLISH trial, despite similar reductions in systolic and diastolic BP [121, 136], single-tablet, fixed-dose RAAS inhibitor/CCB (benazepril/amlodipine) combination therapy was superior to single-tablet, fixed-dose RAAS inhibitor/diuretic (benazepril/hydrochlorothiazide) combination therapy in reducing the composite risk of cardiovascular morbidity and mortality [121, 136]. Relative to RAAS inhibitor/diuretic therapy, RAAS inhibitor/CCB therapy was associated with a 20% lower risk of cardiovascular morbidity and mortality and a 22% lower risk of fatal and nonfatal myocardial infarction [121, 136]. After 6 months of single-tablet fixed-dose RAAS inhibitor/CCB or RAAS inhibitor/diuretic combination therapy, significant reductions from baseline in systolic BP were observed, and 40% of the patients with CKD were at the recommended target BP (<130/80 mm Hg) [121, 137]. This finding supports the initial use of fixed-dose combination therapy in hypertensive patients with CKD, even those with mild (stage 1) hypertension.

9.4 MTD Drugs for Renal Disease

There have been not any rationally designed therapeutic drugs for the treatment of kidney diseases. Nonetheless, some naturally occurring phytochemicals, especially curcumin, are contained in the rhizome of the plant *Curcuma longa* Linn [138]. Curcumin is chemically known as diferuloylmethane and well known for its multi-target properties. At the molecular level, this agent can elicit anti-inflammatory activity through the suppression of numerous cell signaling pathways including NF-κB, STAT3, Nrf2, ROS, and COX-2. Numerous studies have indicated that curcumin is a highly potent antimicrobial agent and has been shown to be active against various chronic diseases including various types of cancers, diabetes, obesity, cardiovascular, CKD, and pulmonary, neurological, and autoimmune diseases [139–145]. Furthermore, this compound has also been shown to be synergistic with other nutraceuticals such as resveratrol, piperine, catechins, quercetin, and genistein. To date, over 100 different clinical trials have been completed with curcumin, which clearly show its safety, tolerability, and effectiveness against various chronic diseases in humans.

For instance, Ali et al. [139] showed that curcumin was found to significantly abate adenine-induced toxic effects such as reduced creatinine clearance, elevated neutrophil gelatinase-associated lipocalin levels, and raised urinary N-acetyl-β-D-glucosaminidase activities. Curcumin markedly reduced renal morphological damage and histopathological markers of inflammation, fibrosis, and apoptosis. Curcumin further reduced adenine-induced hypertension; urinary albumin; the inflammatory cytokines IL-1β, IL-6, and TNF-α; cystatin C; and adiponectin. It restored plasma sclerostin concentrations and lowered oxidative stress in renal homogenates. In animals treated with the two higher curcumin concentrations, alone or in combination with adenine, an increased expression of the antioxidative transcription factor Nrf2 was found as well as upregulation of the activity of its direct target glutathione reductase and of an indirect target, the glutathione level.

CKD is essentially an inflammatory disease with various enzymes, transcription factors, and growth factors involved in modulating the production and action of inflammatory molecules [140]. Curcumin can blunt the generation and action of these inflammatory molecules and ameliorate CKD as well as associated inflammatory disorders. Recent studies have shown that increased intestinal permeability results in the leakage of pro-inflammatory molecules (cytokines and lipopolysaccharides) from the gut into the circulation in diseases such as CKD, diabetes, and atherosclerosis. This change in intestinal permeability is due to decreased expression of tight junction proteins and intestinal alkaline phosphatase (IAP). Curcumin increases the expression of IAP and tight junction proteins and corrects gut permeability. This action reduces the levels of circulatory inflammatory biomolecules. This effect of curcumin on intestine can explain why, despite poor bioavailability, curcumin has potential anti-inflammatory effects in vivo and beneficial effects on CKD [140].

In addition, curcumin was found to attenuate potassium oxonate-induced hyperuricemia and kidney inflammation in mice [141]; the degree of renal fibrosis [142]; the levels of blood glucose, serum creatinine, urea nitrogen and urine albumen, and attenuated renal histomorphological changes in DKD or DN rats [143]; the renal damage in adenine-induced chronic kidney disease [144]; and cisplatin-induced kidney damage by suppressing kidney inflammation and apoptosis in rats [145].

Xu et al. [146] reported a study aiming to search for drugs with multi-pathway effects on renoprotection using STITCH 3.1 [147], the world's largest database of well-defined and accepted chemical-protein associations. The Gene Prospector online tool identified 1841 nephropathy-related genes. The online functional annotation tool of DAVID identified 58 KEGG pathways related to nephropathy. The researchers further identified nine experimentally validated renoprotectants that exerted remarkable impact on KEGG pathways by targeting a limited number of proteins, based on databases [146]. They then selected curcumin as an illustrative compound to display the advantage of multi-pathway drugs on renoprotection. They compared curcumin with hemin, an agonist of heme oxygenase-1 (HO-1), which significantly affects only one KEGG pathway, porphyrin and chlorophyll metabolism. At the same concentration (10 μM), both curcumin and hemin equivalently mitigated oxidative stress in H_2O_2-treated glomerular mesangial cells. The benefit of using hemin was derived from its agonistic effect on HO-1, providing relief from oxidative stress. Selective inhibition of HO-1 completely blocked the action of hemin but not that of curcumin, suggesting simultaneous multi-pathway intervention by curcumin [146].

References

1. Gaitonde DY, Cook DL, Rivera IM. Chronic kidney disease: detection and evaluation. Am Fam Physician. 2017;96(12):776–83.
2. United States Renal Data System. Annual data report, Chapter 1: CKD in the general population. 2016. https://www.usrds.org/2016/view/v1_01.aspx. Accessed 20 Jan 2017.
3. Honeycutt AA, Segel JE, Zhuo X, et al. Medical costs of CKD in the Medicare population. J Am Soc Nephrol. 2013;24(9):1478–83.
4. Townsend RR. Stroke in chronic kidney disease: prevention and management. Clin J Am Soc Nephrol. 2008;3(suppl 1):S11–6.
5. Go AS, Chertow GM, Fan D, et al. Chronic kidney disease and the risks of death, cardiovascular events, and hospitalization. N Engl J Med. 2004;351(13):1296–305.
6. Murphy D, McCulloch CE, Lin F, et al. Centers for Disease Control and Prevention Chronic Kidney Disease Surveillance Team. Trends in prevalence of chronic kidney disease in the United States. Ann Intern Med. 2016;165(7):473–81.
7. United States Renal Data System. Annual data report. Vol 1, Chapter 6: Medicare expenditures for persons with CKD. 2016. https://www.usrds.org/2016/view/v1_06.aspx. Accessed 21 Jan 2017.
8. Bonner R, Albajrami O, Hudspeth J, et al. Diabetic kidney disease. Prim Care. 2020;47(4):645–59.
9. Stanton RC. Combination use of angiotensin converting enzyme inhibitors and angiotensin receptor blockers in diabetic kidney disease. Curr Diab Rep. 2013;13(4):567–73.
10. Bakris GL. Preventing hypertensive kidney disease: the critical role of combination therapy. Am J Hypertens. 2005;18(4 Pt 2):S 93–4.
11. Stompór T, Perkowska-Ptasińska A. Hypertensive kidney disease: a true epidemic or rare disease? Pol Arch Intern Med. 2020;130(2):130–9.
12. Barri YM. Hypertension and kidney disease: a deadly connection. Curr Hypertens Rep. 2008;10(1):39–45.
13. Elliott MJ, Tam-Tham H, Hemmelgarn BR. Alberta Kidney Disease Network. Age and treatment of kidney failure. Curr Opin Nephrol Hypertens. 2013;22(3):344–50.

14. Ferenbach DA, Bonventre JV. Acute kidney injury and chronic kidney disease: from the laboratory to the clinic. Nephrol Ther. 2016;121(Suppl 1):S41–8.
15. Bergmann C, Guay-Woodford LM, Harris PC, et al. Polycystic kidney disease. Nat Rev Dis Primers. 2018;4(1):50.
16. Ferguson MA, Vaidya VS, Bonventre JV. Biomarkers of nephrotoxic acute kidney injury. Toxicology. 2008;245:182–93.
17. Finn W, Porter G. Urinary biomarkers and nephrotoxicity. In: Clinical Nephrotoxins. 2nd ed. Massachusetts: Kluwer Academic Publishers; 2003. p. 621–55.
18. Galley HF. Can acute renal failure be prevented? J R Coll Surg Edinb. 2000;45:45–50.
19. Dhillon S. Dapagliflozin: a review in type 2 diabetes. Drugs. 20191;79(10):1135–46.
20. Wheeler DC, Stefánsson BV, Jongs N, DAPA-CKD Trial Committees and Investigators, et al. Effects of dapagliflozin on major adverse kidney and cardiovascular events in patients with diabetic and non-diabetic chronic kidney disease: a prespecified analysis from the DAPA-CKD trial. Lancet Diabetes Endocrinol. 2021;9(1):22–31.
21. Wheeler DC, Stefansson BV, Batiushin M, et al. The dapagliflozin and prevention of adverse outcomes in chronic kidney disease (DAPA-CKD) trial: baseline characteristics. Nephrol Dial Transplant. 2020;35(10):1700–11.
22. Heerspink HJL, Stefánsson BV, Correa-Rotter R, DAPA-CKD Trial Committees and Investigators, et al. Dapagliflozin in patients with chronic kidney disease. N Engl J Med. 2020;383(15):1436–46.
23. de la Sierra A, Salazar J. What is the role of direct renin inhibitors in the treatment of the hypertensive diabetic patient? Adv Ther. 2011;28(9):716–27.
24. Liu B, Wang Q, Wang Y, et al. C-STRIDE study group. Utilization of antihypertensive drugs among chronic kidney disease patients: results from the Chinese cohort study of chronic kidney disease (C-STRIDE). J Clin Hypertens (Greenwich). 2020;22(1):57–64.
25. Mennuni S, Rubattu S, Pierelli G, et al. Hypertension and kidneys: unraveling complex molecular mechanisms underlying hypertensive renal damage. J Hum Hypertens. 2014;28(2):74–9.
26. Pugh D, Gallacher PJ, Dhaun N. Management of hypertension in chronic kidney disease. Drugs. 2019;79(4):365–79.
27. Bonino B, Leoncini G, De Cosmo S, et al. Antihypertensive treatment in diabetic kidney disease: the need for a patient-centered approach. Medicina (Kaunas). 2019;55(7):382.
28. Leoncini G, Viazzi F, De Cosmo S, et al. Blood pressure reduction and RAAS inhibition in diabetic kidney disease: therapeutic potentials and limitations. J Nephrol. 2020;33(5):949–63.
29. Thomas MC, Atkins RC. Blood pressure lowering for the prevention and treatment of diabetic kidney disease. Drugs. 2006;66(17):2213–34.
30. Juncos LI, Caputo J. Combination therapy in chronic kidney disease? Ther Adv Cardiovasc Dis. 2015;9(4):147–52.
31. Dell'Italia L. Translational success stories: angiotensin receptor 1 antagonists in heart failure. Circ Res. 2011;109:437–42.
32. Schrader J, Lüders S, Kulschewski A, et al. Morbidity and mortality after stroke, eprosartan compared with nitrendipine for secondary prevention: principal results of a prospective randomized controlled study (MOSES). Stroke. 2005;36:1218–24.
33. Weir MR. The role of combination antihypertensive therapy in the prevention and treatment of chronic kidney disease. Am J Hypertens. 2005;18(4 Pt 2):S100–5.
34. Wolf G, Ritz E. Combination therapy with ACE inhibitors and angiotensin II receptor blockers to halt progression of chronic renal disease: pathophysiology and indications. Kidney Int. 2005;67(3):799–812.
35. Zoccali V, Valvo E, Russo D, et al. Antiproteinuric effect of losartan in patients with renal diseases [letter]. Nephrol Dial Transplant. 1997;12:234–5.
36. Mogensen C, Neldam S, Tikkanen I, et al. Randomised controlled trial of dual blockade of renin-angiotensin system in patients with hypertension, microalbuminuria, and non-insulin dependent diabetes: the candesartan and lisinopril microalbuminuria (CALM) study. Br Med J. 2000;321:1440–4.
37. Parving H, Persson F, Lewis J, AVOID Study Investigators, et al. Aliskiren combined with losartan in type 2 diabetes and nephropathy. N Engl J Med. 2008;358:2433–46.
38. Solomon S, Appelbaum E, Manning W, ALLAY trial investigators, et al. Effect of the direct renin inhibitor aliskiren, the angiotensin receptor blocker losartan, or both on left ventricular mass in patients with hypertension and left ventricular hypertrophy. Circulation. 2009;119:530–7.
39. Mann J, Schmieder R, McQueen M, ONTARGET Investigators, et al. Renal outcomes with telmisartan, ramipril, or both, in people at high vascular risk (the ONTARGET study): a multicentre, randomized, double-blind, controlled trial. Lancet. 2008;372:547–53.
40. Parving H, Brenner B, McMurray J, et al. Cardio renal end points in a trial of aliskiren for type 2 diabetes. N Engl J Med. 2012;367:2204–13.
41. Friend L, Emanuele N, Zhang J, et al. Combined angiotensin inhibition for the treatment of diabetic nephropathy. N Engl J Med. 2013;369:1892–903.
42. Apperloo A, de Zeeuw D, de Jong P. A short term antihypertensive treatment induced fall in glomerular filtration rate predicts long term stability of renal function. Kidney Int. 1997;51:793–7.
43. Bakris G, Weir M. Angiotensin-converting enzyme inhibitor-associated elevations in serum creatinine: is this a cause of concern? Arch Intern Med. 2000;160:685–93.

44. Holtkamp F, de Zeeuw D, Thomas M, et al. An acute fall in estimated glomerular filtration rate during treatment with losartan predicts a lower decrease in long-term renal function. Kidney Int. 2011;80:282–7.
45. Bakris GL, Williams M, Dworkin L, et al. National Kidney Foundation Hypertension and Diabetes Executive Committees Working Group. Preserving renal function in adults with hypertension and diabetes: a consensus approach. Am J Kidney Dis. 2000;36:646–61.
46. Bakris GL, Weir MR. Achieving goal blood pressure in patients with type 2 diabetes: conventional versus fixed-dose combination approaches. J Clin Hypertens (Greenwich). 2003;5:202–6.
47. Jamerson KA, Nwose O, Jean-Louis L, et al. Initial angiotensin-converting enzyme inhibitor/calcium channel blocker combination therapy achieves superior blood pressure control compared with calcium channel blocker monotherapy in patients with stage 2 hypertension. Am J Hypertens. 2004;17:495–501.
48. Verdecchia P, Reboldi G, Angeli F, et al. Adverse prognostic significance of new diabetes in treated hypertensive subjects. Hypertension. 2004;43(5):963–9.
49. Hansson L, Zanchetti A, Carruthers SG, et al. Effects of intensive blood-pressure lowering and low-dose aspirin in patients with hypertension: principal results of the Hypertension Optimal Treatment (HOT) randomised trial. HOT Study Group. Lancet. 1998;351(9118):1755–62.
50. Tuomilehto J, Rastenyte D, Birkenhäger WH, Systolic Hypertension in Europe Trial Investigators, et al. Effects of calcium-channel blockade in older patients with diabetes and systolic hypertension. N Engl J Med. 1999;340(9):677–84.
51. Brown MJ, Palmer CR, Castaigne A, et al. Morbidity and mortality in patients randomised to double-blind treatment with a long-acting calcium-channel blocker or diuretic in the International Nifedipine GITS study: intervention as a Goal in Hypertension Treatment (INSIGHT). Lancet. 2000;356(9227):366–72.
52. Weir MR, Chrysant SG, McCarron DA, et al. Influence of race and dietary salt on the antihypertensive efficacy of an angiotensin-converting enzyme inhibitor or a calcium channel antagonist in salt-sensitive hypertensives. Hypertension. 1998;31(5):1088–96.
53. Houston MC, Weir M, Gray J, et al. The effects of nonsteroidal anti-inflammatory drugs on blood pressures of patients with hypertension controlled by verapamil. Arch Intern Med. 1995;155(10):1049–54.
54. Americaen Diabetes Association. Prevention or delay of type 2 diabetes. Diabetes Care. 2004;27(Suppl 1):S47–54.
55. Chobanian AV, Bakris GL, Black HR, et al. The seventh report of the joint national committee on prevention, detection, evaluation, and treatment of high blood pressure: the JNC 7 report. JAMA. 2003;289(19):2560–72.
56. Azizi M, Chatellier G, Guyene TT, et al. Additive effects of combined angiotensin-converting enzyme inhibition and angiotensin II antagonism on blood pressure and renin release in sodium-depleted normotensives. Circulation. 1995;92:825–34.
57. Mervaala E, Dehmel B, Gross V, et al. Angiotensin-converting enzyme inhibition and AT1 receptor blockade modify the pressure-natriuresis relationship by additive mechanisms in rats with human renin and angiotensinogen genes. J Am Soc Nephrol. 1999;10:1669–80.
58. Fried LF, Emanuele N, Zhang JH, VA NEPHRON-D Investigators, et al. Combined angiotensin inhibition for the treatment of diabetic nephropathy. N Engl J Med. 2013;369(20):1892–903.
59. Zou H, Zhou B, Xu G. SGLT2 inhibitors: a novel choice for the combination therapy in diabetic kidney disease. Cardiovasc Diabetol. 2017;16(1):65.
60. Hershon KS. Options for empagliflozin in combination therapy in type 2 diabetes mellitus. Int J Gen Med. 2016;9:155–72.
61. de Boer IH, Kahn SE. SGLT2 inhibitors-sweet success for diabetic kidney disease? J Am Soc Nephrol. 2017;28:7–10.
62. Burrell LM, Johnston CI, Tikellis C, et al. ACE2, a new regulator of the renin–angiotensin system. Trends Endocrinol Metab. 2004;15:166–9.
63. Stanton RC. Sodium glucose transport 2 (SGLT2) inhibition decreases glomerular hyperfiltration: is there a role for SGLT2 inhibitors in diabetic kidney disease? Circulation. 2014;129:542–4.
64. Kojima N, Williams JM, Takahashi T, et al. Effects of a new SGLT2 inhibitor, luseogliflozin, on diabetic nephropathy in T2DN rats. J Pharmacol Exp Ther. 2013;345:464–72.
65. Kojima N, Williams JM, Slaughter TN, et al. Renoprotective effects of combined SGLT2 and ACE inhibitor therapy in diabetic Dahl S rats. Physiol Rep. 2015;3:E12436.
66. Heerspink HJ, Johnsson E, Gause-Nilsson I, et al. Dapagliflozin reduces albuminuria in patients with diabetes and hypertension receiving renin–angiotensin blockers. Diabetes Obes Metab. 2016;18:590–7.
67. Weber MA, Mansfield TA, Alessi F, et al. Effects of dapagliflozin on blood pressure in hypertensive diabetic patients on renin–angiotensin system blockade. Blood Press. 2016;25:93–103.
68. Weber MA, Mansfield TA, Cain VA, et al. Blood pressure and glycaemic effects of dapagliflozin versus placebo in patients with type 2 diabetes on combination antihypertensive therapy: a randomised, double-blind, placebo-controlled, phase 3 study. Lancet Diabetes Endocrinol. 2016;4:211–20.
69. Sha S, Polidori D, Heise T, et al. Effect of the sodium glucose co-transporter 2 inhibitor canagliflozin on plasma volume in patients with type 2 diabetes mellitus. Diabetes Obes Metab. 2014;16:1087–95.

70. Scheen AJ. Safety of dipeptidyl peptidase-4 inhibitors for treating type 2 diabetes. Expert Opin Drug Saf. 2015;14:505–24.
71. Sagara M, Suzuki K, Aoki C, et al. Impact of teneligliptin on oxidative stress and endothelial function in type 2 diabetes patients with chronic kidney disease: a case-control study. Cardiovasc Diabetol. 2016;15:76.
72. DeFronzo RA, Lewin A, Patel S, et al. Combination of empagliflozin and linagliptin as second-line therapy in subjects with type 2 diabetes inadequately controlled on metformin. Diabetes Care. 2015;38:384–93.
73. Hansen L, Iqbal N, Ekholm E, et al. Postprandial dynamics of plasma glucose, insulin, and glucagon in patients with type 2 diabetes treated with saxagliptin plus dapagliflozin add-on to metformin therapy. Endocr Pract. 2014;20:1187–97.
74. Brand T, Macha S, Mattheus M, et al. Pharmacokinetics of empagliflozin, a sodium glucose cotransporter-2 (SGLT-2) inhibitor, coadministered with sitagliptin in healthy volunteers. Adv Ther. 2012;29:889–99.
75. Chaudhuri A, Ghanim H, Vora M, et al. Exenatide exerts a potent antiinflammatory effect. J Clin Endocrinol Metab. 2012;97:198–207.
76. Goldberg RB. Cytokine and cytokine-like inflammation markers, endothelial dysfunction, and imbalanced coagulation in development of diabetes and its complications. J Clin Endocrinol Metab. 2009;94:3171–82.
77. Zhou X, Huang CH, Lao J, et al. Acute hemodynamic and renal effects of glucagon-like peptide 1 analog and dipeptidyl peptidase-4 inhibitor in rats. Cardiovasc Diabetol. 2015;14:29.
78. Lundkvist P, Sjostrom CD, Amini S, et al. Dapagliflozin once-daily and exenatide once-weekly dual therapy: a 24-week randomized, placebo-controlled, phase II study examining effects on body weight and prediabetes in obese adults without diabetes. Diabetes Obes Metab. 2017;19(4):9–60.
79. Gorgojo-Martinez JJ, Serrano-Moreno C, Sanz-Velasco A, et al. Real-world effectiveness and safety of dapagliflozin therapy added to a GLP1 receptor agonist in patients with type 2 diabetes. Nutr Metab Cardiovasc Dis. 2017;27:129–37.
80. Yang P, Huang T, Xu G. The novel mineralocorticoid receptor antagonist finerenone in diabetic kidney disease: progress and challenges. Metabolism. 2016;65:1342–9.
81. Kopp JB. Rethinking hypertensive kidney disease: arterionephrosclerosis as a genetic, metabolic, and inflammatory disorder. Curr Opin Nephrol Hypertens. 2013;22(3):266–72.
82. Lip GY, Beevers M, Beevers DG. Complications and survival of 315 patients with malignant-phase hypertension. J Hypertens. 1995;13:915–24.
83. Weisstuch JM, Dworkin LD. Does essential hypertension cause end-stage renal disease? Kidney international. Kidney Int Suppl. 1992;36:S33–7.
84. Luft FC. Hypertensive nephrosclerosis-a cause of end-stage renal disease? Nephrol Dial Transplant. 2000;15:1515–7.
85. Meyrier A, Simon P. Nephroangiosclerosis and hypertension: things are not as simple as you might think. Nephrol Dial Transplant. 1996;11:2116–20.
86. Hsu CY. Does nonmalignant hypertension cause renal insufficiency? Evidence-based perspective. Curr Opin Nephrol Hypertens. 2002;11:267–72.
87. Skorecki KL, Wasser WG. Hypertension-misattributed kidney disease in African Americans. Kidney Int. 2013;83:6–9.
88. Freedman BI, Hicks PJ, Bostrom MA, et al. Polymorphisms in the nonmuscle myosin heavy chain 9 gene (*MYH9*) are strongly associated with end-stage renal disease historically attributed to hypertension in African Americans. Kidney Int. 2009;75:736–45.
89. Freedman BI, Iskandar SS, Appel RG. The link between hypertension and nephrosclerosis. Am J Kidney Dis. 1995;25:207–21.
90. Freedman BI, Kopp JB, Winkler CA, et al. Polymorphisms in the nonmuscle myosin heavy chain 9 gene (*MYH9*) are associated with albuminuria in hypertensive African Americans: the HyperGEN study. Am J Nephrol. 2009;29:626–32.
91. Lipkowitz MS, Freedman BI, Langefeld CD, et al. Apolipoprotein L1 gene variants associate with hypertension-attributed nephropathy and the rate of kidney function decline in African Americans. Kidney Int. 2013;83:114–20.
92. Hsu CY. Does treatment of nonmalignant hypertension reduce the incidence of renal dysfunction? J Hum Hypertens. 2001;15:99–106.
93. Appel LJ, Wright JT Jr, Greene T, et al. Intensive blood-pressure control in hypertensive chronic kidney disease. N Engl J Med. 2010;363:918–29.
94. Schlessinger SD, Tankersley MR, Curtis JJ. Clinical documentation of end-stage renal disease due to hypertension. Am J Kidney Dis. 1994;23:655–60.
95. Kocyigit I, Eroglu E, Ecder T. Nebivolol can be used for combination therapy in patients with autosomal dominant polycystic kidney disease. Expert Opin Pharmacother. 2017;18(4):455.
96. Harris PC, Torres VE. Polycystic kidney disease. Annu Rev Med. 2009;60:321–37.
97. Sun Y, Zhou H, Yang B-X. Drug discovery for polycystic kidney disease. Acta Pharmacol Sin. 2011;32(6):805–16.
98. Shimoda N, Ikeda M, Yan T, et al. Long-term benefits of treatment with tolvaptan in patients with autosomal dominant polycystic kidney disease. J Nippon Med Sch. 2021; https://doi.org/10.1272/jnms.JNMS.2022_89-303.
99. Torres VE, Gansevoort RT, Perrone RD, et al. Tolvaptan in ADPKD patients with very low kidney function. Kidney Int Rep. 2021;6(8):2171–8.
100. Torres VE, Abebe KZ, Chapman AB, HALT-PKD Trial Investigators, et al. blockade in late autosomal

dominant polycystic kidney disease. N Engl J Med. 2014;371(24):2267–76.
101. Di Mise A, Wang X, Ye H, et al. Pre-clinical evaluation of dual targeting of the GPCRs CaSR and V2R as therapeutic strategy for autosomal dominant polycystic kidney disease. FASEB J. 2021;35(10):e21874.
102. Uchiyama K, Kitayama C, Yanai A, et al. The effect of trichlormethiazide in autosomal dominant polycystic kidney disease patients receiving tolvaptan: a randomized crossover controlled trial. Sci Rep. 2021;11(1):17666.
103. Li Y, Gao J, Yang X, et al. Combination of curcumin and ginkgolide B inhibits cystogenesis by regulating multiple signaling pathways. Mol Med Rep. 2021;23(3):195.
104. Rysz J, Gluba-Brzózka A, Franczyk B, et al. Combination drug versus monotherapy for the treatment of autosomal dominant polycystic kidney disease. Expert Opin Pharmacother. 2016;17(15):2049–56.
105. Liu C, Li H, Gao X, et al. Concomitant use of rapamycin and rosiglitazone delays the progression of polycystic kidney disease in Han:SPRD rats: a study of the mechanism of action. Am J Physiol Renal Physiol. 2018;314(5):F844–54.
106. Naughton CA. Drug-induced nephrotoxicity. Am Fam Physician. 2008;78(6):743–50.
107. Timpe EM. Nephrotoxicity with combination vancomycin-aminoglycoside therapy. J Pediatr Pharmacol Ther. 2005;10(3):174–82.
108. Rankin GO, Sutherland CH. Nephrotoxicity of aminoglycosides and cephalosporins in combination. Adverse Drug React Acute Poisoning Rev. 1989;8(2):73–88.
109. Bayomy NA, Abdelaziz EZ, Said MA, et al. Effect of pycnogenol and spirulina on vancomycin-induced renal cortical oxidative stress, apoptosis, and autophagy in adult male albino rat. Can J Physiol Pharmacol. 2016;94(8):838–48.
110. Deedwania P, Weber M, Reimitz PE, et al. Olmesartan-based monotherapy vs combination therapy in hypertension: a meta-analysis based on age and chronic kidney disease status. J Clin Hypertens (Greenwich). 2017;19(12):1309–18.
111. Hopkins KA, Bakris GL. Fixed-dose combination and chronic kidney disease progression: which is the best? Curr Opin Nephrol Hypertens. 2010;19(5):450–5.
112. Egan CG, Pontremoli R. Role of the fixed-dose combination lercanidipine-enalapril in renal protection. J Nephrol. 2011;24(4):428–37.
113. Chrysant SG. Effectiveness of the fixed-dose combination of olmesartan/amlodipine/hydrochlorothiazide for the treatment of hypertension in patients stratified by age, race and diabetes, CKD and chronic CVD. Expert Rev Cardiovasc Ther. 2013;11(9):1115–24.
114. Kereiakes DJ, Chrysant SG, Izzo JL Jr, et al. Olmesartan/amlodipine/hydrochlorothiazide in participants with hypertension and diabetes, chronic kidney disease, or chronic cardiovascular disease: a subanalysis of the multicenter, randomized, double-blind, parallel-group TRINITY study. Cardiovasc Diabetol. 2012;11:134.
115. Bakris GL, Sarafidis PA, Weir MR, ACCOMPLISH Trial Investigators, et al. Renal outcomes with different fixed-dose combination therapies in patients with hypertension at high risk for cardiovascular events (ACCOMPLISH): a prespecified secondary analysis of a randomised controlled trial. Lancet. 2010;375(9721):1173–81.
116. Zhao M, Qu H, Wang R, et al. Efficacy and safety of dual vs single renin-angiotensin-aldosterone system blockade in chronic kidney disease: an updated meta-analysis of randomized controlled trials. Medicine (Baltimore). 2021;100(35):e26544.
117. Ruilope LM, Segura J. Renal protection in diabetic patients: benefits of a first-line combination of perindopril-indapamide (Preterax). J Hypertens Suppl. 2006;24(3):S9–12.
118. Krairittichai U, Chaisuvannarat V. Effects of dual blockade of renin-angiotensin system in type 2 diabetes mellitus patients with diabetic nephropathy. J Med Assoc Thail. 2009;92(5):611–7.
119. Cooper M, Boner G. Dual blockade of the renin-angiotensin system in diabetic nephropathy. Diabet Med. 2004;21(Suppl 1):15–8.
120. Schjoedt KJ, Jacobsen P, Rossing K, et al. Dual blockade of the renin-angiotensin-aldosterone system in diabetic nephropathy: the role of aldosterone. Horm Metab Res. 2005;37(Suppl1):4–8.
121. Palmer BF, Fenves AZ. Optimizing BP control in patients with chronic kidney disease. Proc (Baylor Univ Med Cent). 2010;23(3):239–45.
122. Levey AS, Beto JA, Coronado BE, et al. National Kidney Foundation Task Force on Cardiovascular Disease Controlling the epidemic of cardiovascular disease in chronic renal disease: what do we know? What do we need to learn? Where do we go from here? Am J Kidney Dis. 1998;32(5):853–906.
123. Klag MJ, Whelton PK, Randall BL, et al. Blood pressure and end-stage renal disease in men. N Engl J Med. 1996;334(1):13–8.
124. Tozawa M, Iseki K, Iseki C, et al. Blood pressure predicts risk of developing end-stage renal disease in men and women. Hypertension. 2003;41(6):1341–5.
125. Kidney Disease Outcomes Quality Initiative (K/DOQI). K/DOQI clinical practice guidelines on hypertension and antihypertensive agents in chronic kidney disease: executive summary. Am J Kidney Dis. 2004;43(5 Suppl 1):S16–41.
126. Chobanian AV, Bakris GL, Black HR, et al. Joint National Committee on Prevention, Detection, Evaluation, and Treatment of High Blood Pressure. National Heart, Lung, and Blood Institute. National High Blood Pressure Education Program Coordinating Committee The seventh report of the Joint National Committee on Prevention, Detection, Evaluation, and Treatment of High

Blood Pressure: the JNC 7 report. Hypertension. 2003;42(6):1206–52.
127. Mourad JJ, Waeber B, Zannad F, Investigators of the STRATHE trial, et al. Comparison of different therapeutic strategies in hypertension: a low-dose combination of perindopril/indapamide versus a sequential monotherapy or a stepped-care approach. J Hypertens. 2004;22(12):2379–86.
128. Salerno CM, Demopoulos L, Mukherjee R, et al. Combination angiotensin receptor blocker/hydrochlorothiazide as initial therapy in the treatment of patients with severe hypertension. J Clin Hypertens (Greenwich). 2004;6(11):614–20.
129. Littlejohn TW 3rd, Majul CR, Olvera R, Study investigators, et al. Telmisartan plus amlodipine in patients with moderate or severe hypertension: results from a subgroup analysis of a randomized, placebo-controlled, parallel-group, 4 × 4 factorial study. Postgrad Med. 2009;121(2):5–14.
130. Lacourcière Y, Martin K. Comparison of a fixed-dose combination of 40 mg telmisartan plus 12.5 mg hydrochlorothiazide with 40 mg telmisartan in the control of mild to moderate hypertension. Am J Ther. 2002;9(2):111–7.
131. Bakris GL, Weir MR, Study of Hypertension and the Efficacy of Lotrel in Diabetes (SHIELD) Investigators. Achieving goal blood pressure in patients with type 2 diabetes: conventional versus fixed-dose combination approaches. J Clin Hypertens (Greenwich). 2003;5(3):202–20.
132. Chrysant SG, Melino M, Karki S, et al. The combination of olmesartan medoxomil and amlodipine besylate in controlling high blood pressure: COACH, a randomized, double-blind, placebo-controlled, 8-week factorial efficacy and safety study. Clin Ther. 2008;30(4):587–604.
133. Neutel JM, Franklin SS, Lapuerta P, et al. A comparison of the efficacy and safety of irbesartan/HCTZ combination therapy with irbesartan and HCTZ monotherapy in the treatment of moderate hypertension. J Hum Hypertens. 2008;22(4):266–74.
134. Reboldi G, Gentile G, Angeli F, et al. Choice of ACE inhibitor combinations in hypertensive patients with type 2 diabetes: update after recent clinical trials. Vasc Health Risk Manag. 2009;5(1):411–27.
135. Fogari R, Zoppi A, Derosa G, et al. Effect of valsartan addition to amlodipine on ankle oedema and subcutaneous tissue pressure in hypertensive patients. J Hum Hypertens. 2007;21(3):220–4.
136. Jamerson K, Weber MA, Bakris GL, ACCOMPLISH Trial Investigators, et al. Benazepril plus amlodipine or hydrochlorothiazide for hypertension in high-risk patients. N Engl J Med. 2008;359(23):2417–28.
137. Jamerson K, Bakris GL, Dahlöf B, ACCOMPLISH Investigators, et al. Exceptional early blood pressure control rates: the ACCOMPLISH trial. Blood Press. 2007;16(2):80–6.
138. Kunnumakkara AB, Bordoloi D, Padmavathi G, et al. Curcumin, the golden nutraceutical: multitargeting for multiple chronic diseases. Br J Pharmacol. 2017;174(11):1325–48.
139. Ali BH, Al-Salam S, Al Suleimani Y, et al. Curcumin ameliorates kidney function and oxidative stress in experimental chronic kidney disease. Basic Clin Pharmacol Toxicol. 2018;122(1):65–73.
140. Ghosh SS, Gehr TW, Ghosh S. Curcumin and chronic kidney disease (CKD): major mode of action through stimulating endogenous intestinal alkaline phosphatase. Molecules. 2014;19(12):20139–56.
141. Chen Y, Li C, Duan S, et al. Curcumin attenuates potassium oxonate-induced hyperuricemia and kidney inflammation in mice. Biomed Pharmacother. 2019;118:109195.
142. He X, Li G, Chen Y, et al. Pharmacokinetics and pharmacodynamics of the combination of rhein and curcumin in the treatment of chronic kidney disease in rats. Front Pharmacol. 2020;11:573118.
143. Tu Q, Li Y, Jin J, et al. Curcumin alleviates diabetic nephropathy via inhibiting podocyte mesenchymal transdifferentiation and inducing autophagy in rats and MPC5 cells. Pharm Biol. 2019;57(1):778–86.
144. Ghelani H, Razmovski-Naumovski V, Chang D, et al. Chronic treatment of curcumin improves hepatic lipid metabolism and alleviates the renal damage in adenine-induced chronic kidney disease in Sprague-Dawley rats. BMC Nephrol. 2019;20(1):431.
145. Soetikno V, Sari SDP, Ul Maknun L, et al. Pretreatment with curcumin ameliorates cisplatin-induced kidney damage by suppressing kidney inflammation and apoptosis in rats. Drug Res (Stuttg). 2019;69(2):75–82.
146. Xu J, Meng K, Zhang R, et al. The use of functional chemical-protein associations to identify multi-pathway renoprotectants. PLoS One. 2014;9(5):e97906.
147. Kuhn M, Szklarczyk D, Franceschini A, et al. STITCH 3: zooming in on protein-chemical interactions. Nucleic Acids Res. 2012;40:D876–88.

10 Polypharmacology in Clinical Applications: Anti-infection Polypharmacology

Abstract

Infectious diseases have for centuries ranked with wars and famine as major challenges to human progress and survival. They remain among the leading causes of death and disability worldwide. Against a constant background of established infections, epidemics of new and old infectious diseases periodically emerge, greatly magnifying the global burden of infections. According to the World Health Organization (WHO), about 15 million (>25%) of 57 million annual deaths worldwide are estimated to be related directly to infectious diseases. While "single-target drugs" (STDs) failed to provide optimal treatment of infectious diseases, polypharmacology is rising above the horizon for better cure. This chapter provides a brief introduction to the three main types of polypharmacology approaches (combination drug therapy [CDT], fixed-dose combination [FDC] and multitarget drug [MTD]) and their current status of application in the clinical setting as well as their potential to the discovery and development of new drugs for infectious diseases therapy.

Keywords

Anti-infection polypharmacology · Bacterial infection · Viral infection · Chagas disease · Multitarget therapy

10.1 Basics of Infectious Diseases

Infectious diseases are generally caused by pathogenic microorganisms, such as bacteria, viruses, parasites, or fungi [1]. These diseases can spread from the environment or from one person to another resulting in illness in communities, constituting a significant burden on public health and economic stability of societies all over the world. They have for centuries been among the leading causes of death and disability and presented growing challenges to health security and human progress. Mild infections may respond to rest and home remedies, while some life-threatening infections may need hospitalization. According to the WHO, about 15 million (>25%) of 57 million annual deaths worldwide are estimated to be related directly to infectious diseases; this figure does not include the additional millions of deaths that occur as a consequence of past infections (for example, streptococcal rheumatic heart disease), or because of complications associated with chronic infections, such as liver failure and hepatocellular carcinoma in people infected with hepatitis B or C viruses [1–4]. Moreover, the threat posed by infectious diseases is further deepened by the continued emergence of new, unrecognized, and old infectious disease epidemics of global impact. Over the past three and half decades, at least 30 new infectious agents affecting humans have emerged, most of which are zoonotic, and their

Z. Wang, B. Yang, *Polypharmacology*, https://doi.org/10.1007/978-3-031-04998-9_10

origins have been shown to correlate significantly with socioeconomic, environmental, and ecological factors.

Entry into host is mostly by routes such as the mouth, eyes, genital openings, nose, and the skin. Damage to tissues mainly results from the growth and metabolic processes of infectious agents, intracellular or within body fluids, with the production and release of toxins or enzymes that interfere with the normal functions of organs and/or systems [5–7]. These products may be distributed and cause damage to other organs and/or systems or function such that the pathogen consequently invades more organs and/or systems.

10.1.1 Emerging Infectious Diseases (EIDs)

Emerging infectious diseases (EIDs) are defined as infectious diseases that are newly recognized in a population or have existed but are rapidly increasing in incidence or geographic range. That is, EID falls under two major categories—newly emerging and reemerging infectious diseases. Newly emerging infections refer to diseases that have been discovered in the human host or a population for the first time; reemerging infectious diseases can be defined as infectious diseases that reappear, usually in more pathogenic form and in rapidly increasing incidence or new geographic locations after apparent control or eradication [8, 9]. Simply put, EID may be new infections resulting from changes or evolution of existing organisms, known infections spreading to new geographic areas or populations, previously unrecognized infections appearing in areas undergoing ecologic transformation, or old infections reemerging because of antimicrobial resistance in known agents or breakdowns in public health measures [2].

Emerging infections account for at least 15% of all human pathogens according to the 10th International *Conference on EID* [10]. A major concern is the synergistic communication between emerging diseases and other infectious and noninfectious conditions. Many emerging diseases are zoonotic or synoptic—an animal receptacle incubates the organism with random transmission into human populations. Likewise, EIDs may be foodborne, vector-borne, or airborne. Regardless, for an EID to become established, the infectious agent must be introduced into a vulnerable population, and the agent must have the ability to spread from human to human and cause disease [11]. EIDs have shaped the course of human history and have caused incalculable misery and death. In 1981, a new disease—acquired immune deficiency syndrome (AIDS)—was first recognized. As a global killer, AIDS now threatens to surpass the Black Death of the fourteenth century and the 1918–1920 influenza pandemic, each of which killed at least 50 million people [1, 12, 13]. Of the "newly emerging" and "re-emerging/resurging" diseases that have followed the appearance of AIDS, some have been minor curiosities, such as the 2003 cases of monkeypox imported into the United States [14], whereas others, such as severe acute respiratory syndrome (SARS), which emerged in the same year [15], have had a worldwide impact.

In contrast to other human diseases, infectious diseases may be unpredictable with the potential for global outbreaks. EIDs can mainly be divided into following two groups.

10.1.1.1 Zoonotic Diseases

Zoonotic diseases are those diseases transmitted from animals to humans through direct contact or through food, water, or the environment, contributing to 61% of infectious organisms affecting humans [2, 16–18]. Zoonotic diseases may be categorized by their ability to spread among humans through five stages, ranging from only spreading among animals (stage 1) to fully human pathogens (stage 5).

10.1.1.2 Vector-Borne Diseases

Vectors are blood-feeding insects and ticks capable of transmitting pathogens between hosts [19]. These diseases are major causes of mortality and morbidity globally. In the United States, the most common pathogens are transmitted by ticks and mosquitoes, including Lyme disease, Rocky Mountain spotted fever, West Nile, dengue, and Zika virus diseases. These diseases represent a growing public health problem for the United States and globally. Data are tracked by local and state health departments; however, national improvement in surveillance, diagnostics, reporting, and vector control has been noted as well as new vaccines have been developed [19]. Mosquito-transmitted EIDs can spread locally in the United States due to the presence of the specific vector. Likewise, global travel and immigration can bring these infections to the United States with potential transmission [20]. Four mosquito-borne viruses of concern are Zika virus, yellow fever, chikungunya virus, and dengue virus [17, 20–23].

10.1.2 Neglected Infectious Diseases (NIDs)

Neglected infectious diseases (NIDs), also known as neglected tropical diseases, are a group of infections that are prevalent in many of the tropical and subtropical developing countries where poverty is rampant [24–26]. Varied types of neglected tropical diseases, such as Chagas disease (CD), dengue, yellow fever, West Nile, Japanese encephalitis, and chikungunya, are also considered to be major global threats. Although such diseases emerge in tropical and subtropical regions, the risk of these infectious illnesses can be worldwide because of global economy and migration. NIDs claim the lives of more than half a million people every year and have rendered at least one billion people chronically infected [27]. According to the last meeting of the Strategic and Technical Advisory Group for NIDs, the World Health Organization (WHO) has in its portfolio a diverse group of heterogeneous diseases, such as: Chagas disease, Buruli ulcer, Rabies, Soil-transmitted helminthiases, Trachoma, Schistosomiasis, Snakebite envenoming, Scabies and other ectoparasites, Taeniasis/Cysticercosis, Dengue and Chikungunya, Dracunculiasis (guinea-worm disease), Leishmaniasis, Leprosy (Hansen's disease), Echinococcosis, Foodborne trematodiases, Human African trypanosomiasis (sleeping sickness), Onchocerciasis (river blindness), Lymphatic filariasis, Mycetoma, Yaws (Endemic treponematoses), Chromoblastomycosis, and other deep mycoses [26]. NTDs have been overlooked for decades but currently there is a growing awareness of their importance in the world's lower middle-income countries (LMICs) [28].

10.1.3 Treatment of Infectious Diseases

Emerging and neglected infections are some of the most lethal diseases with mortality rates ranging from 10% to 80%. At least 25% of about 60 million deaths that occur worldwide each year are estimated to be due to infectious diseases [29, 30]. Among different types of infectious diseases, human immunodeficiency virus (HIV), tuberculosis (TB), and malaria are known as the leading causes of deaths globally.

Over the past two decades, 52% (14 out of 27) of the FDA-approved antibiotics have demonstrated synergistic, multitarget mechanisms of action. Among these are three second-generation lipoglycopeptides, five new-generation quinolones, and six modernized β-lactams. The multitarget antibiotics help achieve the dual objectives of maximizing bactericidal or bacteriostatic efficacy and minimizing antibiotic resistance. The entrance of multitarget antibiotics into the FDA-approved regimens represents a milestone in the evolution of drug discovery as it has transcended from chemical library screening to rational drug design.

Researchers seeking anti-infective agents against pathogenic microorganisms have focused on genomics-based strategies to develop molecules with antibacterial or antiviral properties. A particular area of interest involves molecules that can combat neglected tropical diseases. The single-target strategy for new anti-infective drugs involves searching for single proteins that are essential for microorganism survival. The weakness in this approach is that a single amino acid mutation in the target protein is often enough to confer drug resistance. Indeed, researchers have retrospectively found that many effective antibiotics target multiple proteins simultaneously. Here too, multitarget strategies may offer an innovative approach to the discovery of novel and more efficacious drug candidates.

10.1.3.1 Antibiotics

Antibiotics are chemical compounds that kill bacteria or inhibit their growth. Antibiotics are usually reserved for bacterial infections because these types of drugs have no effect on illnesses caused by viruses. But sometimes it is difficult to tell which type of germ is at work. For example, pneumonia can be caused by a bacterium, a virus, a fungus, or a parasite. It must be noted that the overuse of antibiotics has resulted in several types of bacteria developing resistance to one or more varieties of antibiotics. This makes these bacteria much more difficult to treat.

10.1.3.2 Antivirals

Viruses are very tiny infectious organisms. They are even smaller than bacteria. On the most basic level, a virus is composed of a piece of genetic material that is surrounded by a protein shell. Some viruses may have an additional envelope or other features on their surface. Viruses are parasitic and require a host cell in which to carry out their life cycle. Once the virus has entered the host cell, it is able to use cellular components to reproduce. New viruses are released from the host cell, a process that will sometimes cause the host cell to die. Some examples of viral infections include: influenza (the flu), common cold, measles, rubella, chickenpox, norovirus, polio, herpes simplex virus (HSV), human papillomavirus (HPV), human immunodeficiency virus (HIV), viral hepatitis (including hepatitis A, B, C, D, and E), viral meningitis, rabies, Ebola, SARS coronavirus (SARS-CoV), Middle East respiratory syndrome coronavirus (MERS-CoV), and SARS-CoV-2.

Antibiotics are totally ineffective against viruses, and antivirus drugs have been developed to treat some, but not all, viruses. Examples include the viruses that cause HIV/AIDS, Herpes, Hepatitis B, Hepatitis C, influenza, and so on. For example, Veklury (Remdesivir) is an antiviral drug approved for use in adults and pediatric patients for the treatment of Coronavirus disease-2019 (COVID-19) requiring hospitalization caused by SARS-CoV-2 infection. Recently, Paxlovid™ (PF-07321332; ritonavir), a Pfizer's novel COVID-19 oral antiviral treatment candidate, has been officially announced to be nearly 90% effective at preventing severe disease symptoms or death when given to high-risk study participants in interim analysis of phase 2/3 EPIC-HR study (Evaluation of Protease Inhibition for Covid-19 in High-Risk Patients) [31]. The clinical trial results also show Paxlovid/ritonavir is highly effective against SARS-CoV-2 variants of concern (VOCs) as well as other types of coronaviruses [31].

10.1.3.3 Antifungals

Fungi are another diverse group of organisms that can include things like yeasts and molds. They can be found throughout the environment, including in the soil, indoors in moist areas like bathrooms, and on or in our bodies. Some examples of fungal infections include: vaginal yeast infections, ringworm, athlete's foot, thrush, aspergillosis, histoplasmosis, *Cryptococcus* infection, and fungal meningitis.

Topical antifungal medications can be used to treat skin or nail infections caused by fungi, for example, a topical antifungal cream may be prescribed for conditions like ringworm or athlete's foot. Some fungal infections, such as those affecting the lungs or the mucous membranes, can be treated with an oral antifungal. More-severe internal organ fungal infections, especially in people with weakened immune systems, may require intravenous antifungal medications.

10.1.3.4 Antiparasitics

Some diseases, including malaria, are caused by tiny parasites. Other examples of infections that are caused by parasites include: toxoplasmosis, trichomoniasis, giardiasis, tapeworm infection, roundworm infection, pubic and head lice, scabies, leishmaniasis, and river blindness. While there are drugs to treat these diseases, some varieties of parasites have developed resistance to the drugs.

10.2 CDT for Infectious Diseases

10.2.1 CDT in Bacterial Infection

Antimicrobial resistance has developed into a global healthcare crisis that has culminated in the emergence of multidrug-resistant bacteria that are no longer treatable with any common antibiotic [32–34]. Despite efforts to develop new antibiotics, antibacterial resistance still develops too fast for drug discovery to keep pace. Often, resistance against a new drug develops even before it reaches the market. This continued resistance crisis has demonstrated that resistance to antibiotics with single-protein targets develops too rapidly to be sustainable. Antibiotics that truly have one single-protein target, such as sulfonamides targeting an enzyme involved in folate synthesis or rifampicin targeting RNA polymerase, are famous for high resistance development and are therefore usually administered in combination regimes [35, 36]. Most successful long-established antibiotics target more than one molecule or possess targets that are encoded by multiple genes. This realization has motivated a change in antibiotic development toward drug candidates with multitarget drugs. Moreover, combination therapy approaches are being developed to break antibiotic resistance or to sensitize bacteria to antibiotic action.

A significant amount of research is being conducted to find combinations that are more effective than monotherapy or that are resistance-breaking. When two antibiotics are combined, they are usually either indifferent in their activity or more effective in killing bacteria due to additive effects. However, if the effects of the combination exceed the expected additive effects, the combination is called synergistic. This is, for example, the case for daptomycin and β-lactams [33, 37]. Synergistic drug combinations are highly sought after, since they allow one to lower drug doses and thus may reduce side effects. A different approach to combination strategies is using potentiators, sometimes also called antibiotic adjuvants. These are compounds that by themselves have little or no antibacterial activity but enhance the effects of antibiotics when administered in combination. This is, for example, the case for resistance-breaking compounds like β-lactamase inhibitors [38]. Such compounds typically target either acquired or intrinsic antibiotic resistance mechanisms. There have also been approaches to develop inhibitors of horizontal gene transfer to prevent the spread of resistance genes [39].

10.2.1.1 Synergistic Compound Combinations

Synergistic compound combinations are known against Gram-positive and Gram-negative bacteria and mycobacteria. As summarized by Gray and Wenzel in their excellent article [33], there are five widely accepted categories of possible synergistic mechanisms:

1. **Targeting the same molecule at multiple binding sites**

Combinational compounds targeting the same molecule at multiple binding sites: An example of this is the synergy of plectasin with teicoplanin and dalbavancin, which all bind the peptidoglycan precursor lipid II [33, 40]. However, this is not a universal phenomenon, and other antibiotics that target the same molecule might only have additive effects. This is, for example, the case for the penicillin-binding protein (PBP) inhibitors ampicillin and imipenem [41].

2. **Targeting independent targets in the same pathway**

Combinational compounds targeting independent targets in the same pathway: This is, for example, the case for D-cycloserine, an antituberculosis drug that inhibits an intracellular step of peptidoglycan synthesis [33, 42], and epigallocatechin gallate, a compound found in tea that is thought to disrupt cell wall peptidoglycan [43, 44]. Epigallocatechin gallate also acts synergistically with β-lactam antibiotics [43] and inhibits β-lactamase activity [45].

3. **Targeting a related process with different pathways**

Combinational compounds targeting a related process: These include the antibiotics that target cell wall synthesis and membrane-active compounds [33]; for example, β-lactams and daptomycin [41, 46, 47] or the cationic antimicrobial peptide LL-37 and the lipid II-binding antibiotic teicoplanin [48].

4. **Targeting unrelated pathways**

Combinational compounds targeting unrelated pathways: This is observed with penicillin and the ribosome inhibitor erythromycin. It has been suggested that the depletion of penicillin-binding proteins by translation inhibition might be the underlying mechanism [49].

5. **Modulating molecules to improve target accessibility**

Combinational compounds improving target accessibility: This is, for example, observed with the outer-membrane-permeabilizing peptide colistin in combination with drugs that have inner membrane-bound or cytosolic targets, such as the β-lactam meropenem, the ribosome inhibitor minocycline, and fosfomycin, which inhibits the first enzyme in the lipid II pathway [50]. This is an important anti-Gram-negative strategy.

10.2.1.2 Potentiative Compound Combinations

In addition to synergistic antibiotic combinations, potentiative combination approaches have recently attracted increased attention [33]. One of the advantages of potentiators is that they can potentially increase the activity of a whole array of existing antibiotics. Potentiators typically target intrinsic or acquired antibiotic resistance mechanisms to sensitize or resensitize bacteria to existing antibiotics [33].

The following mechanisms of actions of resistance-breaking and antibiotic-potentiating compounds are proposed by Gray and Wenzel [33]:

1. **Cell envelope permeabilizers**

The outer membrane is the main reason why most antibiotics are ineffective against Gram-negative bacteria, and the inactivation of this permeability barrier could make many of currently available antibiotic drugs for treating infections [33]. Two antibiotics in current clinical use are known to target the outer membrane of Gram-negative bacteria: colistin and polymyxin B. Both of them display synergy with antibiotics with intracellular targets, which can at least partially be attributed to increased outer membrane permeability [50–52]. Recent efforts have yielded lipophilic vancomycin analogues that were able to permeabilize both the inner and outer membranes of Gram-negative bacteria while retaining their ability to inhibit peptidoglycan activity [53]. Another very interesting example are short peptide sequences derived from the cell cycle proteins FtsA and MreB from *Escherichia coli*. These proteins bind to the cytoplasmic membrane with an amphipathic α-helix motif. Peptides derived from these membrane-anchoring sequences not only display potent activity against *E. coli* and other Gram-negative bacteria but also significantly increase their outer membrane permeability [54–56].

2. **Antibiotic sensitizers**

A different strategy to enhance the performance of existing antibiotic drugs is to sensitize bacteria to their action [33]. This can be achieved by inhibiting bacterial stress response systems that allow bacteria to adapt and survive antibiotic exposure. Many studies on bacterial stress responses have proposed this as an option for future drug development, yet this has almost never been pursued. One exception is the bacterial SOS response (the induction of multiple proteins that serve to promote the integrity of DNA) to DNA damage. DNA damage is sensed by the RecA protein, which induces autoproteolysis of the LexA repressor, leading to derepression of DNA repair genes. It has been shown that genetically impairing this stress response mechanism leads to reversion of quinolone resistance in *E.*

coli [57]. Few inhibitors of DNA repair have been verified so far. For example, *p*-coumaric acid was found to interfere with the DNA-binding ability of RecA in *Listeria monocytogenes* and increased the activity of ciprofloxacin against this pathogen [58]. Another study found that zinc inhibits the ability of RecA to bind to single-stranded DNA, suggesting a potential application for zinc ionophores [59].

3. **β-lactamase inhibitors**

Inhibition of β-lactamase activity is the first resistance-breaking strategy to be applied in the clinic [33, 60]. It is also the only combination of antibiotic and potentiator in current clinical use apart from synergistic combinations with the outer-membrane-permeabilizing peptides colistin and polymyxin B. The oldest combination of β-lactam antibiotic and β-lactamase inhibitor is amoxicillin–clavulanate, which was approved for clinical use in 1981. It is still in use today and remains the only one that is orally available [61]. Since then, several other β-lactamase inhibitors have been identified, and eight combinations are currently on the market: amoxicillin–clavulanate, ticarcillin–clavulanate, ampicillin–sulbactam, cefoperazone–sulbactam, piperacillin–tazobactam, ceftolozane–tazobactam, ceftazidime–avibactam, and Meropenem–vaborbactam [61, 62]. Several other combinations are in different stages of clinical development [62].

4. **Inhibitors of aminoglycoside-modifying enzymes**

Inhibitors of aminoglycoside-modifying enzymes have been identified by molecular docking studies with virtual compound libraries and by screening libraries for in vitro enzyme inhibition [63–65]. One of these has been confirmed to restore the activity of amikacin against a resistant *Acinetobacter baumannii* strain [63]. However, none of these compounds is close to clinical development at the present time.

5. **Efflux pump inhibitors**

The most important intrinsic resistance mechanism, next to the outer membrane permeability barrier, is constituted by antibiotic efflux pumps [33]. Particularly, Gram-negative bacteria possess different multidrug efflux pumps that can export a wide variety of antibiotics and antimicrobial molecules. The most prominent efflux pump type is the Gram-negative-specific resistance-nodulation-division (RND) superfamily, in particular the well-characterized AcrAB-TolC pump [66]. A wide variety of antibiotics can be the substrate of these export systems, and their overexpression leads to high-level drug resistance [66]. This makes efflux pumps one of the biggest challenges in overcoming antibiotic resistance and at the same time an attractive target for resistance-breaking antibiotic potentiators. Since the discovery of the very first efflux pump inhibitors reserpine and verapamil against the *Staphylococcus aureus* efflux pump NorA in 1991 [66], a multitude of inhibitors have been identified for both Gram-positive and Gram-negative efflux pumps. Many of them have been found to increase the activity of drugs, which are normally ineffective due to active export [66–70]. Despite these efforts, no efflux pump inhibitor has advanced to clinical development yet. Common limitations of these molecules appear to be a high toxicity for mammalian cells, low in vivo efficacy, and insufficient spectrum coverage [66, 71]. A very different strategy to target antibiotic resistance is inhibiting horizontal gene transfer [33]. Target mutations are only one way for bacteria to become antibiotic resistant. The much more threatening possibility is the spread of resistance genes that are encoded on plasmids or other mobile genetic elements within a bacterial population [33]. This is especially critical for otherwise slowly occurring resistance mechanisms like enzymatic resistance [72]. Importantly, the human (or animal) microbiota can act as a reservoir for resistance plasmids that may spread to pathogenic bacteria during infection. Thus, the administration of an inhibitor of horizontal gene transfer in combination with antibiotic treatment could decrease the incidence of resistance transfer from a reservoir to an infectious strain [73]. A small number of specific inhibitors have been found. For example, unsaturated fatty acid species from tropical fruits and their synthetic derivatives were able to inhibit the transfer of the most important resistance-related plasmid types in several Gram-negative pathogens [74, 75]. Secretion systems could be a promising target for such molecules [33]. Thus, the inhibition of type IV secretion, which is involved in conjugative DNA transfer, has been shown to be inhibited by small peptidomimetic compounds [76].

6. **Biofilm inhibitors**

One big problem in the clinic is the production of biofilms by bacteria like *S. aureus* and *Pseudomonas aeruginosa*, and biofilm inhibitors are potent potentiators for antibiotics used against such infections [33]. Strategies to prevent biofilm formation or disperse mature biofilms can commonly be divided into three categories: the inhibition of adhesion or extracellular matrix production, the inhibition of quorum sensing, and the dispersion of the extracellular matrix [33, 77]. Several antimicrobial peptides, for example, LL-37, prevent adhesion of bacterial cells by inhibiting the initiation of biofilm production [78]. The most prominent biofilm-dispersing agent is the glycoside hydrolase dispersin B, which directly targets the extracellular matrix [79]. Other small-molecule inhibitors have been identified for all these mechanisms [80–83]. Short glycans have been shown to disrupt *P. aeruginosa* biofilms and subsequently increase the efficacy of antibiotics against it [84]. The small peptide 1018 targets the stringent response (p)ppGpp signaling, which is involved in biofilm formation, and acts as a potent biofilm-dispersing agent not only for *P. aeruginosa* but also for other pathogens including *Klebsiella pneumoniae* and *S. aureus* [85]. Similarly, small-molecule quorum-sensing inhibitors have been shown to increase the efficacy of tobramycin (Tob) against *P. aeruginosa* in a foreign-body infection model in mice [86]. Micafungin, which inhibits the synthesis of the pseudomonal cell wall component 1,3-β-D-glucan, was demonstrated to prevent *P. aeruginosa* biofilm formation and improve the outcome of antibiotic therapy in *P. aeruginosa*-infected mice, particularly when combined with the fluoroquinolone levofloxacin [87].

Collectively, two-compound combination or CDT is the most common approach for anti-infection therapy [33]. Nonetheless, multi-compound combination is also a frequently employed strategy. For example, the tuberculosis treatment regimens often consist of a combination of rifampicin, isoniazid, ethambutol, and pyrazinamide [88]. A newly emerging approach is the combination of an antibiotic with more than one potentiator. To date, antibiotic combination therapy is rather a last resort for severe infections than a standard therapy option [33]. It is normally only employed for infections with multiresistant bacteria that cannot be treated with monotherapy, for latent infections like tuberculosis, or for antibiotics that have a high risk of encountering resistance when applied alone, such as trimethoprim–sulfonamide or β-lactam–β--lactamase combinations [33].

One important limitation is the development of clinically useful formulations and treatment regimes. Finding the right dosing, treatment intervals, and total treatment duration can already be challenging for single drugs and gets much more complicated with each additional drug added to the regime, since each individual drug has its own pharmacodynamic and pharmacokinetic properties and its own side effects. Matching these to elicit the desired outcome in patients can be a serious challenge [89, 90]. Another risk of CDT is antimicrobial resistance. While many empirical studies have found that drug combinations suppress resistance [90–98], other studies have warned that under certain conditions antibiotic combinations might even promote faster resistance development. This is a particular concern for synergistic drug combinations. While they allow one to lower the dose and thus reduce side effects [99, 100], they come with an inherent risk of treatment failure through resistance: If resistance to one drug occurs, synergistic effects are lost, resulting in the low-dose exposure to a single antibiotic [101–105].

In addition to bacterial infections, the CDT approach has also been successfully applied to antiretroviral therapy (ART) of a number of viral infectious diseases including HIV, and to anti-parasites. In the subsections to be presented below, brief introductions will be given to these infectious diseases.

10.2.2 CDT in Viral Infection

Acquired immunodeficiency syndrome (AIDS) is the disease caused by infection with the human immunodeficiency virus type 1 (HIV-1). According to the World Health Organization (WHO), more than 75 million people worldwide have been infected with human immunodeficiency virus (HIV), and there are now approximately 37 million individuals living with the infection [106–109]. Untreated HIV replication causes progressive $CD4^+$ T-cell loss and a wide range of immunological abnormalities, leading to an increased risk of infectious and oncological complications. HIV infection also contributes to cardiovascular disease, bone disease, renal and hepatic dysfunction, and several other common morbidities.

The current standard therapy for HIV-infected individual is daily administration of a cocktail of antiretroviral drugs that inhibit various stages of the viral life cycle, known as combination antiretroviral therapy (cART). As summarized by Chung et al. [109], treatments include drugs of six classes: (a) nucleoside/nucleotide reverse transcriptase (RT) inhibitors; (b) protease inhibitors; (c) non-nucleoside reverse transcriptase inhibitors (NNRTIs); (d) fusion inhibitors; (e) integrase inhibitors; and (f) CCR5 antagonists. Current best practice includes starting a cocktail of at least two or more drugs from each of at least three classes of drugs when patient $CD4^+$ T-cell levels fall below 350/mm^3. Recent evidence indicates that initiation of drugs before patients become symptomatic may provide a greater benefit [110]. Antiretroviral drugs are highly effective at inhibiting HIV replication, and, for individuals who can access and adhere to these drugs, cART leads to durable (and probably lifelong) suppression of viral replication. Viral suppression enables immune recovery and the near elimination of the risk for developing AIDS. The benefits of cART for HIV replication and transmission control have led to its universal recommendation. Despite effective treatment, HIV-infected individuals have a higher-than-expected risk of heart, bone, liver, kidney, and neurological diseases. When used optimally by an infected (or by an uninfected) person, antiretroviral drugs can virtually eliminate the risk of HIV transmission. While effective in reducing morbidity, improving the quality of life, and increasing median survival, cART is not curative. Patients with well-controlled viremia on cART (<50 infectious units/mm^3 of blood) are known to harbor a latent viral reservoir in resting $CD4^+$ T-cells. When cART is interrupted, latently infected cells produce infectious virus, which is rapidly followed by a loss of peripheral blood $CD4^+$ T-cells and progression toward immunodeficiency. Additionally, the prolonged use of cART is associated with other clinical sequelae including neural, renal, hepatic, and cardiovascular toxicities; diabetes; lipodystrophy; and other metabolic abnormalities. Moreover, the widespread use of cART and a lack of patient compliance with drug treatment schedules have resulted in the development of viral variants (escape mutants) that are drug resistant. Together, these limitations of cART emphasize the need for a more comprehensive approach to HIV therapy.

In addition, gene therapy approaches for the treatment of HIV/AIDS hold great promise for a functional cure of the disease [109]. Similar to the drug-based cART, anti-HIV genes can also be expressed in combination to reduce the frequency of viral escape. The clinical rHIV7-sh1-TAR-CCR5RZ lentiviral vector contains three anti-HIV small RNA with distinct mechanisms of action with synergy in inhibitory activity while prolonging antiviral protection. The former two RNA agents are driven by the RNA polymerase III (Pol III) human U6 promoter while the last is driven by the Pol III VA1 promoter. The ubiquitous Pol III promoters ensure strong small RNA expression in all hematopoietic lineages [111, 112]. The sh1 small RNA gene encodes a short hairpin RNA (shRNA) that mimics the natural pre-miRNA substrate in the endogenous RNA interference pathway in generation of small interfering RNA (siRNA) targeting an overlapping region shared between Tat and Rev. mRNA to mediate both transcripts for cleavage and protein knockdown. Tat and Rev. are early regulatory HIV-1 proteins that are essential for viral replication [113–120], with Tat essential for efficient viral transcription and Rev. essential for the export of intron-containing viral transcripts to cytoplasm for translation and packaging.

10.2.3 CDT in Chagas Disease (CD)

Chagas disease (CD), a tropical neglected illness, is caused by the parasite *Trypanosoma cruzi*, which is transmitted to animals and people by insect vectors that are found only in the Americas (mainly, in rural areas of Latin America where poverty is widespread). CD is affecting mainly populations of low socioeconomic status in Latin America. An estimated six to eight million people worldwide are infected with *Trypanosoma cruzi*, the etiological agent of CD. If untreated, infection is lifelong and can be life threatening. The impact of CD is not limited to only rural areas of Latin America in which vector-borne transmission (diseases transmitted by insects) occurs. Large-scale population movements from rural to urban areas of Latin America and to other regions of the world have increased the geographic distribution and changed the epidemiology of CD. Despite being one of the main global health problems, this disease continues without effective treatment during the chronic phase of the infection. The limitation of therapeutic strategies has been one of the biggest challenges on the fight against CD. Nifurtimox and benznidazole (BNZ), developed in the 1970s, are still the only commercial options with established efficacy on CD. However, the efficacy of these drugs has been proven only during early infection and the benefits in the chronic phase are questionable. Consequently, there is a growing need for new pharmacological alternatives, either by optimization of existing drugs or by the formulation of new compounds.

The concomitant or sequential use of two or more pharmaceutical compounds represents an alternative that is also being explored in CD treatment studies [121]. Theoretically, the combination of different compounds allows the reduction in the doses and/or treatment length, resulting in reduction in drug adverse side effects and costs. Synergistic treatment generally improves compounds' activity with distinct mechanisms of action, and additionally reduces the drug toxicity and the chance of developing resistance [121–123]. A successful association between itraconazole and BNZ was demonstrated in a murine model. The concomitant administration of both drugs allowed a decrease in BNZ dose by 25% and in time needed to curtail the parasitemia. In addition, mice treated with this drug combination showed lessened lesions in the cardiac tissue and fewer inflammatory cells associated with the chronic phase of the disease [124]. Itraconazole seems to alter the pharmacokinetic profile of BNZ, increasing its half-life and its volume distribution extent, allowing an increased accumulation of BNZ in the biological systems [124, 125].

Additionally, drug repositioning has been tested for its application to the treatment of CD. Several compounds of different pharmacological classes have already been tested against the *T. cruzi*. Among them, benidipine, a calcium channel blocker usually administered for the treatment of hypertension and angina, and the antibiotic clofazimine were tentatively used against the parasite since they act in inhibiting cruzipain, the main lysosomal cysteine protease with an important role in parasite infectivity. The inhibition of cruzipain has been much explored, because it is the most abundant protease of the parasite. Cruzipain has been shown to be crucial for all stages of the parasite life cycle and is involved in the parasite nutrition, invasion of mammalian cells, and evasion of the host immune response [126–129]. Thus, benidipine and clofazimine are considered promising candidates for the treatment of the disease in view of their capacity to reduce the parasitic burden in the blood and skeletal tissues of infected mice, additionally curtailing the inflammatory effects of the infection.

10.3 FDC for Infectious Diseases

10.3.1 FDC in Bacterial Infection

The single-pill fixed-dose combination (SP-FDC) or FDC strategy as anti-bacterial infection formulations is most highly recommended for the treatment of infectious diseases. Particularly, the efficacy of FDC drugs is well established in conditions such as tuberculosis (TB), malaria, and HIV treatment [130, 131]. Globally, the most sold antibiotic FDCs were amoxicillin/clavulanate and sulfamethoxazole/trimethoprim, which were sold in all countries and accounted for about 50% and 20% of the total antibiotic FDCs' consumption, respectively. This is consistent with these drugs being recommended for treatment of common infectious conditions (such as acute otitis media, community acquired pneumonia, and urinary tract infections) and with their inclusion on the erythroid myeloid lymphoid (EML) as Access antibiotics [132].

However, the FDC remained uncertain in terms of efficacy. A systematic review evaluated the risk of treatment failure or disease relapse, acquired drug resistance, bacterial conversion after 2 months of treatment, adverse events, adherence, and treatment satisfaction associated with treatment of active TB using FDC or separate drug formulations [133]. The authors searched four electronic databases for randomized controlled trials and cohort studies. Results from trials that directly compared FDC to separate drug formulations were pooled. Results from other studies were reported separately. Overall, 2450 citations, of which 15 were controlled trials, and 4 additional relevant studies were identified and included. In the 15 trials there were no differences in acquired drug resistance, bacterial conversion after 2 months of treatment, or adverse drug reactions with FDC or separate drug formulations. There was a trend toward higher risk of failure or relapse with FDC. Based on individual study results, only one of two trials that assessed treatment satisfaction, and none of five that assessed patient adherence, favored FDCs. Although FDC formulations simplify TB therapy, the current evidence does not indicate that these formulations improve treatment outcomes among patients with active TB.

Wang et al. [134] investigated the potential of new inhalable dry powder combinations consisting of a fixed dose of aztreonam (Azt) and tobramycin (Tob), using a spray drying process, against antibiotic-resistant Gram-negative respiratory pathogens. The interactions of Azt with Tob on resistant *Pseudomonas aeruginosa*, *Klebsiella pneumoniae*, and *Acinetobacter baumannii* were determined by calculating factional inhibitory concentration indices (FICI). A fixed concentration ratio of Azt and Tob that exhibited a synergistic antimicrobial effect was selected and formulated into inhalable dry powders by co-spray drying with and without L-leucine.

Chaudhary et al. [135] evaluated the efficacy and safety of fixed-dose combination of ceftriaxone-vancomycin in patients with various infections in 305 patients suffering from different bacterial infections (serious respiratory tract disease, bronchitis, gastrointestinal tract infections, urinary tract infection, cellulites, and meningitis). Patients were randomized into two groups depending on severity of illness. Group A (n = 106) and Group B (n = 199) patients were given intravenous dose of fixed-dose combination 1.5 g B.D. and 3.0 g B.D., respectively, for 3–10 days. Hemoglobin, total leukocyte count, erythrocyte sedimentation rate, urea, creatinine levels, and serum glutamyl oxaloacetic transaminase and glutamyl pyruvic transaminase activities were recorded before and on completion of the treatment. Most of patients (70%) were cured in seven or less than 7 days of treatment. Significant decrease in total leukocyte count and erythrocyte sedimentation rate levels was observed, indicating patients were cured from the infections. No significant alterations were observed in hemoglobin, serum urea, and creatinine levels, and in glutamyl oxaloacetic transaminase and glutamyl pyruvic transaminase activities on completion of treatment. The investigators concluded that the fixed-dose combination of ceftriaxone-vancomycin is effective in treating various bacterial infections without any toxic effect on liver and kidney [135].

10.3.2 FDC in Viral Infection

Single-pill fixed-dose combinations (SP-FDC or FDC) are a significant advance in antiretroviral treatment simplification, contributing to an increase in compliance with complex chronic therapies, thus improving patients' quality of life. Reducing the number of pills and daily doses is associated with higher adherence and better quality of life [136, 137]. As a fixed-dose combination tablet given once daily, efavirenz/emtricitabine/tenofovir alafenamide fumarate (EFV/FTC/TDF) was the first available single tablet regimen (STR) combining efficacy, tolerability, and convenience, with the simplest dosing schedule and smallest numbers of pills of any ART combination therapy [137]. The rilpivirine/emtricitabine/tenofovir (RPV/FTC/TDF) is a next-generation non-nucleoside reverse transcriptase inhibitor-based single tablet regimen (NNRTI-based STR), a once daily complete antiretroviral therapy (ART) regimen for the treatment of HIV-1 infection. Recently the combination of elvitegravir/cobicistat/emtricitabine/tenofovir alafenamide fumarate (EVG/COBI/FTC/TAF) was also approved by the European Commission and is the first integrase inhibitor-based STR. Receiving antiretroviral therapy as once daily STR is associated with both clinical and economic benefits, which confirms previous research [137].

The advent of oral direct-acting antiviral agents (DAAs) has dramatically improved the hepatitis C virus (HCV) treatment landscape in the last 4 years, providing cure rates over 95% with a shorter duration of treatment and a very good safety profile [138]. This has enabled access to treatment in nearly all HCV-infected patients. The launch of two pangenotypic fixed-dose combinations (FDCs) in 2017 made a new step forward in HCV treatment by slightly increasing efficacy and more importantly allowing the treatment of patients without HCV genotyping, and in some cases without fibrosis assessment. However, retreatment of the few DAA failure patients was still an issue for some HCV genotypes. The launch of the triple regimen FDC, sofosbuvir (SOF)/velpatasvir/voxilaprevir, solves this issue by providing a cure rate over 96% regardless of HCV genotype.

The current recommendation for treating hepatitis C virus (HCV) in HIV patients includes the combination of sofosbuvir (SOF) and daclatasvir (DCV) [139]. An FDC containing 400 mg SOF and 30, 60, or 90 mg DCV was proven an effective and safe treatment for patients coinfected with HIV and HCV.

10.3.3 FDC in Chagas Disease (CD)

Posaconazole (PCZ) and benznidazole (BNZ) are known to show synergetic effect in treating the acute and chronic phases of Chagas disease, a neglected parasitic disease [140]. However, as both compounds are poorly water soluble, the development of amorphous solid dispersions (ASDs) of a PCZ/BNZ fixed-dose combination in a water-soluble polymer becomes an attractive option to increase their apparent solubility and dissolution rate, potentially improving their oral bioavailability. The advantage of FDC with two synergistic antichagasic agents PCZ and BNZ together in amorphous enhanced dissolution is that it potentially improves their bioavailability and therapeutic activity in treating Chagas disease.

10.4 MTD for Infectious Diseases

10.4.1 MTD in Bacterial Infection

Multitarget antibiotics can be divided into multitarget compounds that target more than one isoenzyme or closely related proteins of the same pathway and multitarget compounds that bind to different molecules involved in separate cellular processes. Multifunctional compounds have a direct antibiotic target but an additional indirect antibacterial activity, for example, immunomodulatory properties.

10.4.1.1 Naturally Occurring Multitarget Antibiotics

According to Gray et al. [33], naturally occurring antibiotics often do not have strict single targets either. The most common class of natural antimicrobial substances are antimicrobial peptides, which occur in virtually all organisms. These molecules predominantly target complex bacterial structures, for example, the cell membrane, the cell wall, or both [141, 142]. Moreover, antibiotic producers often produce a mix of structurally related compounds. This is, for example, the case for gramicidin, tyrothricin, and polymyxins [143, 144]. Such small structural variations can result in slightly different target interactions [145–149] and may complicate resistance development by target alterations and antibiotic-cleaving or scavenging molecules. Naturally occurring mixes have been used in clinical settings with great success and in some cases are combined with each other or other antibiotics like neomycin [150, 151]. It is quite common in clinical practice to use antibiotic combinations for different reasons, as already described in a previous section above.

The most common antimicrobial target, whose inhibition results in a multitude of cellular effects, is the cytoplasmic membrane. It is the target of many natural antibiotics, most prominently host defense peptides [141]. The cytoplasmic membrane harbors essential cellular processes, such as the respiratory chain and cell wall synthesis, and many more important components, including nutrient uptake, secretion, and stress response systems. Thus, membrane-targeting antibiotics have the potential to (a) inhibit processes essential for survival, (b) reduce bacterial fitness, (c) impair virulence, and (d) interfere with stress adaptation. Depending on how an antibiotic interferes with the membrane, few or a whole array of membrane-bound processes can be affected [149]. Several membrane-active antibiotics are used in the clinic, although only a handful is applied systemically. For some of them, their multiple effects on bacterial cells have been studied.

10.4.1.2 Artificially Synthesized Multitarget Antibiotics

Attempts to fuse outer membrane-penetrating peptide sequences to the lantibiotic nisin, which by itself is not active against Gram-negative bacteria due to its inability to reach its inner membrane target, have resulted in hybrid molecules with significantly increased anti-Gram-negative activity [152, 153]. A similar approach is the hybridization of antibiotics with the aminoglycoside tobramycin [154]. Tobramycin is known as a ribosome inhibitor but at higher concentrations primarily attacks and permeabilizes the outer membrane [155, 156]. It has been conjugated with efflux pump inhibitors [157], quinolones [91, 158], and the chelator cyclam [159]. The resulting conjugates turned out to be potent antibiotic adjuvants and increased the activity of tetracyclines, fluoroquinolones, and β-lactams against Gram-negative bacteria by increasing uptake and limiting efflux. This is achieved by a multifaceted mechanism involving outer membrane permeabilization, impairment of efflux pumps, and depolarization of the inner membrane [91, 154, 157–160]. Similar results have been obtained for tobramycin dimers [161].

10.4.1.3 Multitargets in the Same Pathway

Most of the antibiotics described above target one specific cellular structure, which itself is involved in different processes. However, there are also several antibiotics that bind to two or more distinct molecules [33]. The most extreme examples of this are the β-lactam antibiotics, which typically target more than one penicillin-binding protein (PBP) [162]. For example, *Escherichia coli* has eight PBPs, six of which are inhibited by penicillin G [163–165]. Other examples include quinolone antibiotics, which target both topoisomerase II and IV involved in DNA supercoiling and resolving DNA concatemers, respectively [166], and the cell wall synthesis inhibitors fosfomycin and D-cycloserine. Gram-positive bacteria possess two copies of MurA, the first enzyme in the lipid II synthesis pathway, and fosfomycin inhibits both isoenzymes [167]. D-cycloserine competitively inhibits both alanine racemase and D-ala-D-ala ligase, which converts L-alanine to the D-alanine dipeptide that is part of the peptidoglycan interpeptide bridge [168]. Similarly, platencin, a fatty acid synthesis inhibitor that attracted significant attention upon its discovery but then encountered several hurdles in preclinical development [169], is a dual inhibitor of the FabF and FabH enzymes [170, 171].

10.4.1.4 Multitargets in Different Pathways

Several antibiotics have been found that inhibit two or more distinct targets that are not closely related components of the same pathway or process [33]. Not only is it more difficult to acquire target-based resistance in two different molecules, but also bacteria need to react with two different stress responses to mitigate antibiotic action, hampering their capabilities for efficient stress adaptation; thus, these antibiotics have an additional advantage [33].

There are several examples for such compounds in the clinic, and new molecules with multiple targets are regularly discovered. One example are the polymyxins, in particular polymyxin B and colistin, which are used as last resort antibiotics against multiresistant Gram-negative infections [172]. These polypeptides target both the outer and inner membranes of Gram-negative bacteria [173, 174]. Another example is clofazimine, which is used to treat leprosy [175]. This antibiotic binds to guanine bases in bacterial DNA but also increases phospholipase A1 activity, leading to toxic overproduction of lysophospholipids [176]. The atypical tetracycline chelocardin, which has recently attracted renewed interest in clinical development [177, 178], inhibits the bacterial ribosome and additionally depolarizes the cytoplasmic membrane [179]. Tetracycline itself, while not leading to depolarization, has recently been shown to disturb membrane organization in addition to translation inhibition [180]. The same was observed for anhydrotetracycline, suggesting that this dual activity could be a general activity of the tetracycline class [180]. It would be reasonable to assume that membrane activity may be a general consequence of translation inhibition, since ribosomes associate with the cell membrane to couple translation to protein secretion [181]. However, other ribosome inhibitors did not visibly disturb the cell membrane, and a ribosomal protein mutant that diminishes tetracycline binding to the ribosome showed the same tetracycline-induced membrane effects as the wild-type [180]. These findings suggest that the membrane activity of tetracycline is independent from translation inhibition and thus a separate secondary target. Similarly, tyrocidines, which primarily disrupt cytoplasmic membrane integrity, have long ago been proposed to bind DNA as secondary target. This notion, which was derived from test tube interactions of the peptides with isolated DNA [182], was only recently supported by in vivo experiments [149].

Another very prominent example are antibiotics with a dual mechanism on cell wall synthesis and the cell membrane. This is relatively common, since cell wall synthesis inhibitors often target membrane-bound steps of this pathway. Thus, both targets are in close proximity. The most prominent example of this is the lantibiotic nisin, which is not used clinically but is a common food preservative. Nisin has a two-step mechanism of action. It first establishes contact with lipid II, inhibiting peptidoglycan synthesis. It then inserts into the lipid bilayer to form a transmembrane pore [183–185]. More recently, it was found that nisin also binds to undecaprenyl phosphate, adding another structure to its list of molecular targets [186]. While nisin is able to bind to negatively charged lipids in the absence of lipid II or undecaprenyl phosphate [187], it needs the lipid-coupled cell wall precursor as a docking molecule to form a transmembrane pore. Other lantibiotics like gallidermin can impair both targets independently [188]. This property of lantibiotics also inspired the development of dual-function vancomycin derivatives. Telavancin and oritavancin carry a hydrocarbon tail, which allows them to insert into the cell membrane after binding lipid II. Thereby, they disrupt the permeability barrier, a feature that vancomycin is lacking and that restores activity against strains with high-level vancomycin resistance [189, 190]. Daptomycin could be included in this group as well, since it likewise inserts into the cell membrane and impairs cell wall synthesis [191]. However, at least to our current knowledge, it only binds a single molecular target, namely, phosphatidylglycerol-containing lipids [192–195].

10.4.1.5 Designer Dual-Target Compounds

Inspired by the clinical success of these multitarget antibiotics, approaches have been undertaken to develop dual-target molecules. These can either be stable hybrids combining two functions in one molecule or cleavable compounds, whereby the prodrug has a different target than the cleavage product [160, 196]. In the past, cleavable compounds have essentially behaved like simple prodrugs for the cleavage product rather than eliciting their intended dual effects. Hence, newer efforts rather focus on stable hybrids [196–198]. These approaches have been recently reviewed in detail [160, 199], but we want to give a brief overview of the molecules that are currently underway in clinical development.

Cadazolid is a stable hybrid of a quinolone and oxazolidinone and currently in phase III clinical trials for the treatment of *Clostridium difficile* infections [200, 201]. It primarily inhibits translation but also topoisomerase activity and shows low resistance development rates [202]. Cefilavancin is a stable hybrid molecule of vancomycin and a third-generation cephalosporin, which is currently undergoing phase III clinical trials for complicated skin and soft tissue infections [199]. This hybrid compound is active against both methicillin and vancomycin-resistant *Staphylococcus aureus*, yet it has not been verified to which extent it inhibits lipid II and PBPs [203, 204]. MCB-3681 is a stable fluoroquinolone–oxazolidinone hybrid similar to cadazolid [205]. It is active against strains that are resistant to both ciprofloxacin and linezolid [206]. MCB-3681 was scheduled to undergo phase II clinical trials since 2015 and was granted fast track status by the FDA in 2016 [205]. However, it was further developed into a prodrug molecule, MCB-3837. In 2018, its developing company Morphochem was acquired by Deinove, and the compound, now under the name DNV3837, finally went into phase II clinical trials for treatment of *C. difficile* infections in 2019 [207]. The study is planned to be completed in the summer of 2020 [208]. TNP-2092 is a stable rifamycin–quinolone hybrid currently in phase II trials for prosthetic joint infections. It is active against a range of quinolone and rifamycin-resistant strains [198]. TNP-2092 is also being evaluated against *Helicobacter pylori* infections, yet is at a preclinical stage [209]. TD-1607 is a stable glycopeptide–cephalosporin hybrid similar to cefilavancin. Two phase I clinical trials have been completed, yet results remain to be published [199, 210].

A different hybrid approach was followed by the company Visterra that developed an ultranarrow spectrum hybrid molecule against *P. aeruginosa* infections by coupling a specific antibody that targets cell surface glycan molecules with an antimicrobial peptide [211]. Here, the hybrid molecule does not possess two antibacterial targets, but the antibody strategy allows more targeted treatment by increasing the local peptide concentration at the cell surface and bringing the antibiotic group close to its target.

Another approach to new multifunctional antibiotic candidates against Gram-negative bacteria based on the polymyxin lead structure has recently been published [212]. These compounds are chimeric peptidomimetics combining features of polymyxin A with the outer membrane protein-targeting cyclic peptide murepavadin. They permeabilize the outer membrane by targeting lipopolysaccharides, inhibit the outer membrane protein complex Bam, which is involved in the folding and insertion of β-barrel proteins in the outer membrane, and permeabilize the inner membrane [212]. One of these compounds is currently undergoing preclinical toxicology studies [212].

It will be exciting to see whether these hybrid molecules will receive FDA approval and how they will ultimately perform in the clinic.

10.4.1.6 Multifunctional Compounds

The last category of multitarget molecules that we want to discuss are multifunctional compounds, which in addition to their antibacterial properties possess a second, unrelated activity that is beneficial for treatment. Most prominently, such compounds may have anti-inflammatory or immunomodulatory properties. This is well described for a range of antibiotics, and some are even used to treat conditions unrelated to infections due to these properties. This is, for example, the case for macrolides, which are used to treat diffuse panbronchiolitis [213]; dapsone, which has anti-inflammatory and immunomodulatory effects and is, for example, used against dermatitis herpetiformis [214]; or rifampicin, used against pruritus caused by primary biliary cholangitis [215]. For other indications, the dual effects of these compounds are crucial for their therapeutic efficacy. For example, clindamycin is used to treat acne and is known for both its direct antibacterial actien and its ability to decrease swelling and inflammation by modulating cytokine production [216, 217]. Acne is a very common indication for combining antibacterial and anti-inflammatory activities. Other antibiotics with this dual function are metronidazole, tetracycline, macrolides, and dapsone, which are all used to treat acne [218]. Clofazimine, which is used against leprosy, already has two distinct antibacterial targets, guanine bases and phospholipase A2 [176]. In addition to these activities, it also reduces neutrophil mobility, which is beneficial for treating patients with a type 2 lepra reaction characterized by strong inflammation [219, 220]. The range of antibiotics with immunomodulatory and anti-inflammatory effects and the detailed use of these compounds have been thoroughly reviewed before [218, 221–223] and should not be described here in more detail. However, it should be noted that daptomycin has been implicated to have immunomodulatory effects as well [224]. Daptomycin already has multiple effects on bacterial cells, and both tetracycline and clofazimine have two distinct antibacterial targets. This illustrates that multiple activities are quite common in clinically successful antibiotics.

Taken together, it is certain that all antibiotics that have been successfully employed for decades as monotherapeutics in the treatment of bacterial infections rely on mechanisms of bacterial growth inhibition that are by far more complex than inhibition of a single enzyme [225]. Such successful antibiotics have in common that they address several targets in parallel and/or that their targets are encoded by multiple genes. Such multiplicity of targets and of target genes has the advantage that the emergence of spontaneous target-related resistance is a comparatively slow process. Recently registered antibiotics and novel antibiotics in development are discussed in the light of this promising concept of antibacterial polypharmacology.

The current epidemic of antibiotic-resistant bacterial infections has fueled the demand for novel antibiotics exhibiting both antibacterial efficacy and antidrug resistance [34]. This need has not been fully satisfied by the conventional "one target–one molecule" approach. Consequently, there has been rising interest in the development of multitarget antibiotics. Over the past two decades, 52% (14 out of 27) of the FDA-approved antibiotics have demonstrated synergistic, multitarget mechanisms of action. Among these are three second-generation lipoglycopeptides, five new-generation quinolones, and six modernized β-lactams.

10.4.2 MTD Drugs in Viral Infection

10.4.2.1 MTD Drugs in Anti-HIV-1/HSV-2 Therapy

Balzarini et al. [226] reported an interesting and extraordinary study on the discovery of a compound designated 6-phosphonylmethoxyethoxy-2,4-diaminopyrimidine (PMEO-DAPym) as a multitargeted drug candidate with dual anti-HIV and anti-HSV activity. The contents below are essentially an introduction to their findings.

HIV infection is commonly associated with other sexually transmitted infections such as herpes simplex virus type 2 (HSV-2). Such infections with HSV-2 may facilitate the risk of HIV acquisition and often worsen the clinical course of the HIV disease [91, 129–231]. In fact, HIV-1 has been recovered frequently from genital herpes lesions in coinfected individuals [232]. Although HSV target cells in tissues are still poorly understood and it is not known whether macrophages are important targets for HSV in vivo, both HIV-1 and HSV-2 can infect macrophages. Cells of the monocyte/macrophage (M/M) lineage reside in genital mucosal tissues and are thought to be reservoirs of HIV-1 in the genital tract [233, 234]. Also, there is evidence that HSV infection can also stimulate macrophages in vitro and induce HIV-1 replication in these cells [235]. Thus, it would be beneficial if a microbicide has efficient suppressive activity against both HIV-1 and HSV-2.

Highly specific drugs, such as the acyclic nucleoside phosphonate (ANP) analogue 9-(2-phosphonylmethoxypropyl)adenine ((*R*)PMPA; tenofovir) against HIV [236] and the nucleoside analogue 9-(2-hydroxyethyloxymethyl)guanine (acyclovir; ACV) against HSV [237], have been developed. Unexpectedly, treatment of HIV-1/HSV-2-coinfected individuals with acyclovir diminishes both HSV-2 and HIV-1 load [238–241], while topically applied tenofovir diminishes transmission not only of HIV-1 [242] but also of HSV-2 [243]. Both drugs have been found to be directly active against these two viruses [244, 245], although suppression of HIV-1 by acyclovir and of HSV-2 by tenofovir is suboptimal. It would be advisable if antiviral agents can be designed and developed that concomitantly display pronounced inhibitory activity against both pathogens. Therefore, the different subclasses of acyclic nucleoside phosphonates were revisited because it has been previously shown that several members of the ANPs display potent activity against a variety of DNA viruses, including herpesviruses and hepatitis B virus, and retroviruses [246–248]. From a wide screen of hundreds of ANPs, a compound designated 6-phosphonylmethoxyethoxy-2,4-diaminopyrimidine (PMEO-DAPym) emerged as a novel prototype drug that concomitantly acts as an efficient inhibitor of both HIV-1 and HSV-2 replication, but that was also surprisingly endowed with a capacity to interact with HIV entry. This combination of unique properties in one single molecule makes it a promising new-generation multitargeted antiviral for dual-viral (HSV/HIV) infection therapy and HIV transmission prevention.

We demonstrated the potent anti-HIV and anti-HSV activity of PMEO-DAPym in a diverse set of clinically relevant in vitro, ex vivo, and in vivo systems including (a) $CD4^+$ T-lymphocyte (CEM) cell cultures, (b) embryonic lung (HEL) cell cultures, (c) organotypic epithelial raft cultures of primary human keratinocytes (PHKs), (d) primary human monocyte/macrophage (M/M) cell cultures, (e) human ex vivo lymphoid tissue, and (f) athymic nude mice. In all assay systems the drugs were administered prior to, or at the time of, virus infection, which is relevant from a microbicidal application viewpoint, for which the drugs are preferentially already present at the time of infection. However, it might well be possible that these drugs are also effective when exposed shortly after virus infection given the DNA polymerase/HIV reverse transcriptase as being one of the intracellular targets of PMEO-DAPym. In addition to our findings that PMEO-DAPym can efficiently suppress a wide variety of HSV clinical isolates including acyclovir-resistant virus strains, it has previously been demonstrated that PMEO-DAPym suppresses a wide variety of HIV-1 clinical strains belonging to different HIV-1 clades (i.e., A, B, C, A/E, G) [249]. This study also revealed that PMEO-DAPym showed a somewhat more favorable cross-resistance profile to various mutant HIV-1 isolates than adefovir and tenofovir. Likewise, we could demonstrate that the antivirally active PMEO-DAPym diphosphate metabolite potently inhibited mutant HIV-1 RT enzymes that harbor the tenofovir-characteristic K65R and K70R mutations (IC_{50} in the submicromolar range; data not shown), explaining the efficient suppression of a variety of (mutant) clinical virus isolates. Thus, PMEO-DAPym may have a rather high genetic barrier suppressing a wide variety of clinical HSV and HIV-1 clade isolates, including clinically relevant mutant viruses.

We investigated the underlying basis of the antiviral activity of PMEO-DAPym and found several different mechanisms that together explain the unique dual antiviral activity of this new compound: PMEO-DAPym efficiently suppressed HIV reverse transcriptase (RT)-catalyzed and HSV-1 DNA polymerase-catalyzed viral replication. Moreover, it also induced secretion of beta-chemokines, in particular MIP-1β, MIP-1α, and RANTES (Regulated upon Activation, Normal T Cell Expressed and Presumably Secreted), whose upregulation is associated with a pronounced downmodulation of the HIV-1 coreceptor CCR5, resulting in inhibition of HIV entry.

10.4.2.2 MTD Drugs in Anti-HIV-1/HCV-2 Therapy

Similarly, Wei et al. [250] reported an excellent in silico study aiming to develop multitarget compounds for the treatment of HIV/hepatitis C virus (HCV) coinfection, in which they constructed 60 classification models to predict compounds that are active against 11 HIV-1 targets and 4 HCV targets based on a multiple quantitative structure–activity relationship (multiple QSAR) method. Their findings are described below.

Hepatitis C virus (HCV) infection causes acute and chronic liver disease, including cirrhosis and hepatocellular carcinoma [251]. Unfortunately, since HIV-1 and HCV have similar routes of transmission, the risk of HIV/HCV coinfection is very high [252]. HIV-HCV coinfection is rapidly emerging as a major cause of morbidity and mortality throughout the world, due to the common rapid mutation characteristics of the two viruses as well as their similar complex influence on immunology system. According to the World Health Organization (WHO), approximately 2.3 million of the estimated 36.7 million people living with HIV globally had been infected with HCV as of 2015. It is necessary for these people to be diagnosed and provided with effective and reasonable treatment for both HIV and hepatitis as a priority [253]. However, in the treatment of HIV/HCV coinfection, special considerations must be made for potential drug–drug interactions and toxicities. The standard treatment for HIV-1 is highly active antiretroviral therapy (HAART), which improves the quality of life of AIDS patients by reducing viral load [254]. HAART utilizes a cocktail of drugs to inhibit multiple viral proteins involved in the viral life cycle, including reverse transcriptase, integrase, and protease [255]. Although it is effective, HAART requires the sequential administration of single-target drugs that may result in drug–drug interactions, poor treatment adherence, and the emergence of drug resistance. The current therapy for HIV/HCV coinfection includes regular injections of pegylated α-interferon (PEG-IFN) plus daily ribavirin (RBV) in combination with HAART [256]. As with the HAART treatment regimen, additive drug toxicity, high cost, and the long-term therapy cycle should be taken into account during HIV/HCV coinfection treatment. To avoid these undesirable factors, multitarget therapy using a single drug capable of simultaneously inhibiting two or more viral targets has been proposed, which can reduce the complexity of treatment too [254].

As a result of the development of computer-aided drug design (CADD) technology, several in silico methods have been exploited to predict chemical–protein interaction (CPI) pairs between drugs and target proteins [257–263]. These methods can be approximately classified into two categories: ligand-based [264–266] and structure-based [267, 268]. Conventional quantitative structure–activity relationship (QSAR) methods generally use a group of analogs against one target to elucidate the relationship between chemical structure and biological activity. In 2009, Vina et al. developed a multitarget quantitative structure–activity relationship (mt-QSAR) classification model to predict the probability of drug binding for more than 60 different receptors based on molecular connectivity and receptor sequences [269]. Since then, the mt-QSAR method has been widely applied to rational design of drugs for different biological targets.

In this study, a multiple QSAR method intended for constructing multiple binary classification models was applied to predict the chemical–protein interactions (CPIs) for 11 key targets related to HIV-1 and 4 targets related to HCV. First, by combining two machine learning algorithms (Naïve Bayes [NB] [270, 271] and the support vector machine [SVM] [272]) and two molecular fingerprinting methods [273] MELCOR accident consequence code system [MACCS] and extended connectivity fingerprints 6 [ECFP6] [274]), we developed multiple QSAR models for identifying inhibitors against 11 HIV-1-related and 4 HCV-related targets. Second, fivefold cross-validation and test set validation were used to evaluate the performance of all models. Third, to verify the application of this strategy, the multiple QSAR models that were developed were employed to predict the CPI for 27 approved HIV-1 drugs and 10 approved HCV drugs, as well as 9 compounds known to be active against one or more targets of HIV-1 or HCV. The predicted results showed 20 hits to be potential HIV/HCV coinfection multitarget inhibitors, which were further confirmed by the assessment of bioactivity, resulting in 7 compounds that actually interacted with HIV-1 and HCV targets simultaneously with diverse binding affinities. Our study indicates that machine-learning-based multiple QSAR approaches could potentially be applied to the discovery of HIV/HCV co-inhibitors and other multitarget drugs.

A similar study has also been presented by Liu et al. [275]. Indeed, rationally designed multitarget anti-HIV drugs has become a highly recommended and desirable strategy [276, 277].

10.4.2.3 MTD Strategy in Anti-SARS-CoV-2 Drug Discovery

Ren et al. [278] proposed and validated a multi-targeting drug design strategy for identifying potent anti-COVID-19/SARS-CoV-2 inhibitors. This MTD approach identified three alkaloid compounds—lycorine, emetine, and cephaeline—as potent inhibitors of SARS-CoV-2 by simultaneously acting on four targets of the virus.

The COVID-19 (Coronavirus disease-2019), caused by a novel coronavirus (SARS-CoV-2), has spread worldwide and remains out of control [279–282]. SARS-CoV-2 is an enveloped, positive-sensed, and single-stranded RNA beta coronavirus (CoV) [283, 284]. At present, there are seven human CoV strains identified, which includes: (a) four low-pathogenicity members: 229E (alpha coronavirus), NL63 (alpha coronavirus), OC43 (beta coronavirus), and HKU1 (beta coronavirus); and (b) three highly pathogenic members: MERS-CoV (the beta coronavirus that causes Middle East Respiratory Syndrome, abbreviated as MERS), SARS-CoV (the beta coronavirus that causes a severe acute respiratory syndrome, named as SARS), and SARS-CoV-2 [285]. SARS-CoV-2 encodes four structural proteins, including spike, envelope, membrane, and nucleocapsid protein (N protein), as well as 15 nonstructural proteins (abbreviated as Nsps) [279, 286]. While living cells are infected, the virus can directly access the ribosomes of the host cell to produce proteins for new viruses. Nsps are assembled into a multi-subunit polymerase complex to create an environment suitable for transcription and replication of the viral genome. The viral genomes then form mature virions with viral structural proteins [287]. It is worth noting that during the coronavirus replication cycle, these targets, including RNA-dependent RNA polymerase (RdRp), N protein, ribosome, and −1 programmed ribosomal frameshifting (−1 PRF) region, are involved in interacting with RNA, which herein provide a common feature for developing multitargeting drugs that could have maximal anti-viral efficacy.

Although ribosome is one necessary organelle in living cells for producing proteins for host cells, it can be regarded as one potential drug target because of its essential role in viral protein translation [288]. The human ribosome is known as the 80S ribosome, composed of a small 40S subunit and a large 60S subunit. There are four main functional sites embedded in the whole ribosome: (a) the mRNA-binding site; (b) the tRNA-binding sites that contain the A(aminoacyl)-site, P(peptidyl)-site, and E(exit)-site; (c) the peptidyl transfer center (PTC); and (d) the binding site for translational factors [289]. Previous studies showed that these functional sites were identified as druggable binding sites within ribosomes [290]. Besides, virus-infected cells produce more proteins than normal cells, which means that ribosome seems to be very important for replicating the virus. In total, ribosomes could be counted as a potential target for treating viral infection [291].

SARS-CoV-2 recruits 1 PRF—which involves a shift in the reading frame of the ribosome at a specific location in the RNA message, that is, five to seven nucleotides downstream of the "slippery" sequence where the reading-frame shift occurs, and where typically exists a pseudoknot—to control the expression of viral proteins, including RdRp [292]. Some works of literature showed that mutations or molecular binders of −1 PRF could significantly attenuate viral propagation [293, 294], demonstrating that the RNA genome of the virus could be a considerable druggable target.

RdRp, as a crucial drug target for anti-coronaviruses, plays an essential role in the process of replication and transcription of RNA viruses [295, 296]. The RdRp of SARS-CoV-2 is composed of nsp12, which is a catalytic subunit, and two accessory subunits, nsp7 and nsp8, which can stimulate nsp12 polymerase activity. The nsp12 subunit includes a nidovirus RdRp-associated nucleotidyltransferase (NiRAN) domain in the N-terminal portion, and an interface domain, which is responsible for connecting the C-terminal RdRp catalytic domain and NiRAN domain [297–299]. The core structure of RdRp presents a large and deep groove domain, resembling a human right hand, interconnected by "fingers," "palm," and "thumb" subdomains, which form the catalytic site of RNA synthesis. The catalytic site, comprising of the residues of SER 759, ASP 760, and ASP 761 (SDD) sulfadimidine, which lies in the motif C, is involved in primer binding for RdRp activity [298].

Being one of the multifunctional structural proteins in SARS-CoV-2, N protein plays a vital role in constructing helical ribonucleoproteins by interacting with the viral RNA genome and regulating viral RNA synthesis during the processes of replication and transcription [300]. N protein contains three highly conserved domains: (a) an N-terminal RNA-binding domain (NTD), which is primarily responsible for RNA-binding; (b) a C-terminal dimerization domain (CTD), which mediates N–N homotypic interactions; and (c) a poorly structured central Ser/Arg(SR)-rich linker that connects NTD and CTD and contains multiple putative sites of phosphorylation [301, 302]. Previous studies showed that N protein could be used as another potential target for developing small-molecular antivirus drugs via destroying the assembling process of virions [303].

In the work by Ren et al. [278], three alkaloid compounds—lycorine, emetine, and cephaeline—were identified to be potent inhibitors of SARS-CoV-2 by simultaneously working on these three potential targets. These three highly potent anti-SARS-CoV-2 inhibitors were identified by a multitarget-driven antivirus drug design strategy. It was found that these compounds may show the antivirus activities by simultaneously targeting the host ribosome, and viral RNA, RdRp, as well as N protein to interfere with the translating, propagating, replicating, and assembling process of the virus. The findings presented in this study show the applicability and feasibility of this novel multitarget-driven antivirus drug design strategy. Meanwhile, it indicated that more potent and safer drugs could be designed by improving their bindings with the targets of the virus but reducing their interactions with the host ribosome.

Lycorine, a natural alkaloid extracted from the amaryllidaceae plant *Lycoris radiata*, also a traditional Chinese medicine herb, is previously reported to have multiple biological properties, such as antiparasite, antibacterial, antitumor, anti-inflammatory, antiplasmodia, and antivirus [304]. Emetine, the principal alkaloid of ipecac root, is an antiprotozoal and emetine drug, which has also been demonstrated to have several biological functions, including contraceptive activities, and anticancer and antivirus functions [305]. During the preparation of this manuscript, the anti-SARS-CoV-2 activities of lycorine and emetine were reported, that is, lycorine has an anti-SARS-CoV-2 activities of $EC_{50} = 0.31\,\mu M$, and emetine of $EC_{50} = 0.46\,\mu M$ [306–308]. The reported anti-virus activity of lycorine is like ours ($EC_{50} = 0.439 \pm 0.122\,\mu M$), while emetine showed a much better inhibitory activity in our test ($EC_{50} = 0.00771 \pm 0.000117\,\mu M$). However, their molecular mechanisms of action are still not clear.

The broad-spectrum antiviral activities of emetine and lycorine have been reported [309–316]. It was reported that lycorine and emetine may be able to block the synthesis of the viral proteins of Zika virus, buffalopox virus (BPXV), or Newcastle disease virus (NDV) by inhibiting the function of the ribosomes of host cells [315, 317]. Moreover, the inhibition mechanisms of lycorine and emetine for eukaryotic ribosomes were further demonstrated respectively by complex structures released by X-ray diffraction and Cryo-EM analyses. According to their binding sites on the ribosome, it seems that lycorine inhibits protein synthesis by interfering with the peptide-bond formation, while emetine inhibits protein synthesis by blocking the E-site [290, 318], which is a site for deacylated tRNA molecules before they dissociate from the ribosome. By considering these findings, *human* ribosome structure complex with the three compounds were computationally modeled. Given the favorable molecular interactions formed between the ligands and ribosome as well as reports from other literature, the hindrance of the translation of viral proteins can be confirmed. However, this may lead to certain cytotoxicity.

Previous studies showed that lycorine and emetine could inhibit RNA polymerase of Zika virus [309, 317]. This study reported their potential inhibition ability against the RNA polymerase of SARS-CoV-2 and proposed its potential double-faced molecular mechanisms of suppressing replication and transcription, that is, emetine and cephaeline may block the RNA template-product duplex, or occupying the RNA growth site. On the other hand, our computational simulations indicate that these positively charged compounds could strongly bind to the negatively charged −1 RPF of SARS-CoV-2, contributing to impairing the propagation of the virus. Finally, we found that lycorine, emetine, and cephaeline may bind with N proteins by surface plasmon resonance (SPR) assay and molecular docking study, presenting a potential in disrupting the virus-assembly process. The cell-based viral infection assay was also performed to test the potent antiviral activities of lycorine, emetine, and cephaeline in Vero E6 cell. The effective concentration in the cell-based viral infection assay is smaller than the molecular level assay, which suggested the multiple-target activities of the three compounds.

There are other potential antiviral activities for lycorine, emetine, and cephaeline. Chloroquine was reported to inhibit the endocytic pathway of SARS-CoV-2 by altering the pH of endosomes due to its alkalinity [319–323]. Same as chloroquine, we speculate that the three alkaloids can also affect the virus with similar mechanisms, such as blocking the entry into the host cell by altering the pH of the endosomes.

Previous animal toxicological experiments exhibited low toxicity and mild side effects of lycorine [324]. In addition, it was reported that SARS-CoV-2 virus has an infect on the brain tissues [325]. Lycorine is able to cross the blood-brain barrier, which suggests that it has the potential of treating brain infection [326]. Although emetine has been known as a therapeutic drug since the seventeenth century, the sideways of myopathy and cardiomyopathy should be considered for further evaluation [327–329]. Altogether, lycorine, emetine, and cephaeline show certain anti-SARS-CoV-2 potential. More importantly, our multitargeted antivirus drug design strategy may provide a new avenue for rational drug design for SARS-CoV-2, or even other viruses.

References

1. Morens DM, Folkers GK, Fauci AS. The challenge of emerging and re-emerging infectious diseases. Nature. 2004;430(6996):242–9.
2. McArthur DB. Emerging infectious diseases. Nurs Clin North Am. 2019;54(2):297–311.
3. Esposito S. Infectious diseases: pathophysiology, diagnostics and prevention. Int J Mol Sci. 2016;17(9):1464.

4. Kumar H. Therapeutic approaches for genetic and infectious diseases. Int Rev Immunol. 2020;39(1):1–2.
5. National Institutes of Health (US). Understanding emerging and re-emerging infectious diseases. National Institutes of Health (US); 2007.
6. Tibayrenc M. Encyclopedia of infectious diseases: modern methodologies. Wiley; 2007.
7. Walker MJ, Barnett TC, McArthur JD, et al. Disease manifestations and pathogenic mechanisms of group A Streptococcus. Clin Microbiol Rev. 2014;27(2):264–301.
8. Racaniello VR. Emerging infectious diseases. J Clin Invest. 2004;113(6):796–8.
9. Barber NC, Stark LA. Online resources for understanding outbreaks and infectious diseases. CBE Life Sci Educ. 2015;14(1):fe1.
10. CDC International Conference on Emerging Infectious Diseases. 2018. https://www.cdc.gov/iceid/index.html. Available at: Accessed 3 Aug 2018.
11. Baylor College of Medicine Emerging infectious diseases. https://www.bcm.edu/departments/molecular-virology-and-microbiology/emerging-infections-and-biodefense/emerging-infectious-diseasesous-diseases. Available at: Accessed 3 Oct 2018.
12. Kohn GC. The Wordsworth encyclopedia of plague and pestilence. New York: Facts on File; 1995. p. 25–6.
13. Johnson NPAS, Mueller J. Updating the accounts: global mortality of the 1918–1920 "Spanish" influenza pandemic. Bull Hist Med. 2002;76:105–15.
14. Centers for Disease Control and Prevention. Multistate outbreak of monkeypox — Illinois, Indiana, and Wisconsin, 2003. MMWR52. 2003; 537–40.
15. Peiris JSM, et al. The severe acute respiratory syndrome. N Engl J Med. 2003;349:2431–41.
16. Ryu S, Kim BI, Lim J-S. One health perspectives on emerging public health threats. J Prev Med Public Health. 2017;50:411–4.
17. World Health Organization (WHO). Zoonoses. http://www.who.int/zoonoses. Available at: Accessed 3 Oct 2018.
18. Wolfe ND, Dunavan CP, Diamond J. Origins of major human infectious diseases. Nature. 2007;447:279–83.
19. Rosenberg R, Lindsey NP, Fischer M. Vital signs: trends in reported vectorborne disease cases- United States and territories, 2004–2016. MMWR Morb Mortal Wkly Rep. 2018;67(17):496–501.
20. Rathore MH, Runyon J, Haque TU. Emerging infectious diseases. Adv Pediatr Infect Dis. 2017;64:27–71.
21. CDC Yellow fever. 2018. https://www.cdc.gov/yellowfever/index.html. Available at: Accessed 3 Oct 2018.
22. Ramachandran VG, Das S, Roy P. Chikungunya: a reemerging infection spreading during 2010 dengue fever outbreak in National Capital Region of India. VirusDisease. 2016;27(2):183–1286.
23. Centers for Disease Control and Prevention Dengue. http://www.cdc.gov/dengue/index.html. Accessed 9 Oct 2018.
24. Martin-Plaza J, Chatelain E. Novel therapeutic approaches for neglected infectious diseases. J Biomol Screen. 2015;20(1):3–5.
25. Nii-Trebi NI. Emerging and neglected infectious diseases: insights, advances, and challenges. Biomed Res Int. 2017;2017:5245021.
26. World Health Organization. Neglected tropical diseases. https://www.who.int/neglected_diseases/diseases/en/. Accessed July 2020.
27. Bhutta ZA, Sommerfeld J, Lassi ZS, et al. Global burden, distribution, and interventions for infectious diseases of poverty. Infect Dis Poverty. 2014;3:21.
28. Feasey N, Wansbrough-Jones M, Mabey DC, et al. Neglected tropical diseases. Br Med Bull. 2010;93:179–200.
29. World Health Organization. WHO/The top 10 causes of death. http://www.who.int/mediacentre/factsheets/fs310/en/index2.html.
30. Fauci AS, Touchette NA, Folkers GK. Emerging infectious diseases: a 10-year perspective from the National Institute of Allergy and Infectious Diseases. Emerg Infect Dis. 2005;11(4):519–25.
31. Pfizer's novel COVID-19 oral antiviral treatment candidate reduced risk of hospitalization or death by 89% in interim analysis of phase 2/3 epic-hr study. Recent Press Releases of Pfizer's Homepage. https://www.pfizer.com/news/press-release/press-release-detail/pfizers-novel-covid-19-oral-antiviral-treatment-candidate. Accessed Nov 2021.
32. Sprenger M, Fukuda K. New mechanisms, new worries. Science. 2016;351:1263–4.
33. Gray DA, Wenzel M. Multitarget approaches against multiresistant superbugs. ACS Infect Dis. 2020;6(6):1346–65.
34. Wetzel C, Lonneman M, Wu C. Polypharmacological drug actions of recently FDA approved antibiotics. Eur J Med Chem. 2021;209:112931.
35. Chopra I. Bacterial RNA polymerase: a promising target for the discovery of new antimicrobial agents. Curr Opin Investig Drugs. 2007;8:600–7.
36. Sköld O. Sulfonamide resistance: mechanisms and trends. Drug Resist Updates. 2000;3:155–60.
37. Jorgensen SCJ, Zasowski EJ, Trinh TD, et al. Daptomycin plus beta-lactam combination therapy for methicillin-resistant staphylococcus aureus bloodstream infections: a retrospective, comparative cohort study. Clin Infect Dis. 2020;71(1):1–10.
38. Gonzalez-Bello C. Antibiotic adjuvants – a strategy to unlock bacterial resistance to antibiotics. Bioorg Med Chem Lett. 2017;27:4221–8.
39. Kwapong AA, Stapleton P, Gibbons S. Inhibiting plasmid mobility: the effect of isothiocyanates on bacterial conjugation. Int J Antimicrob Agents. 2019;53:629–36.

40. Breidenstein EBM, Courvalin P, Meziane-Cherif D. Antimicrobial activity of plectasin NZ2114 in combination with cell wall targeting antibiotics against Vana-type enterococcus faecalis. Microb Drug Resist. 2015;21:373–9.
41. Rand KH, Houck H. Daptomycin synergy with rifampicin and ampicillin against vancomycin-resistant enterococci. J Antimicrob Chemother. 2004;53:530–2.
42. Schneider T, Sahl HG. An oldie but a goodie – cell wall biosynthesis as antibiotic target pathway. Int J Med Microbiol. 2010;300:161–9.
43. Zhao WH, Hu ZQ, Okubo S, et al. Mechanism of synergy between epigallocatechin gallate and beta-lactams against methicillin-resistant staphylococcus aureus. Antimicrob Agents Chemother. 2001;45:1737–42.
44. Steinmann J, Buer J, Pietschmann T, et al. Anti-infective properties of epigallocatechin-3-gallate (EGCG), a component of green tea. Br J Pharmacol. 2013;168:1059–73.
45. Zhao W-H, Hu Z-Q, Hara Y, et al. Inhibition of penicillinase by epigallocatechin gallate resulting in restoration of antibacterial activity of penicillin against penicillinase-producing staphylococcus aureus. Antimicrob Agents Chemother. 2002;46:2266–8.
46. Berti AD, Theisen E, Sauer J-D, et al. Penicillin binding protein 1 is important in the compensatory response of staphylococcus aureus to daptomycin-induced membrane damage and is a potential target for beta-lactam-daptomycin synergy. Antimicrob Agents Chemother. 2016;60:451–8.
47. Smith JR, Barber KE, Raut A, et al. Beta-lactam combinations with daptomycin provide synergy against vancomycin-resistant enterococcus faecalis and enterococcus faecium. J Antimicrob Chemother. 2015;70:1738–43.
48. Koppen BC, Mulder PPG, de Boer L, et al. Synergistic microbicidal effect of cationic antimicrobial peptides and teicoplanin against planktonic and biofilm-encased staphylococcus aureus. Int J Antimicrob Agents. 2019;53:143–51.
49. Allen NE, Epp JK. Mechanism of penicillin-erythromycin synergy on antibiotic-resistant staphylococcus aureus. Antimicrob Agents Chemother. 1978;13:849–53.
50. Sertcelik A, Baran I, Akinci E, et al. Synergistic activities of colistin combinations with meropenem, sulbactam, minocycline, disodium fosfomycin, or vancomycin against different clones of carbapenem-resistant acinetobacter baumannii strains. Microb Drug Resist. 2020;26(5):429–33.
51. Zusman O, Avni T, Leibovici L, et al. Systematic review and meta-analysis of in vitro synergy of polymyxins and carbapenems. Antimicrob Agents Chemother. 2013;57:5104–11.
52. Lenhard JR, Nation RL, Tsuji BT. Synergistic combinations of polymyxins. Int J Antimicrob Agents. 2016;48:607–13.
53. Yarlagadda V, Manjunath GB, Sarkar P, et al. Glycopeptide antibiotic to overcome the intrinsic resistance of gram-negative bacteria. ACS Infect Dis. 2016;2:132–9.
54. Saikia K, Chaudhary N. Antimicrobial peptides from C-terminal amphipathic region of E. Coli FtsA. Biochim Biophys Acta Biomembr. 2018;1860:2506–14.
55. Saikia K, Chaudhary N. Interaction of MreB-derived antimicrobial peptides with membranes. Biochem Biophys Res Commun. 2018;498:58–63.
56. Saikia K, Sravani YD, Ramakrishnan V, et al. Highly potent antimicrobial peptides from N-terminal membrane-binding region of E. Coli MreB. Sci Rep. 2017;7:42994.
57. Recacha E, Machuca J, Diaz de Alba P, et al. Quinolone resistance reversion by targeting the SOS response. MBio. 2017;8:e00971–17.
58. Ojha D, Patil KN. P-coumaric acid inhibits the listeria monocytogenes RecA protein functions and SOS response: an antimicrobial target. Biochem Biophys Res Commun. 2019;517:655–61.
59. Crane JK, Cheema MB, Olyer MA, et al. Zinc blockade of SOS response inhibits horizontal transfer of antibiotic resistance genes in enteric bacteria. Front Cell Infect Microbiol. 2018;8:410.
60. Gonzalez-Bello C, Rodriguez D, Pernas M, et al. Beta-lactamase inhibitors to restore the efficacy of antibiotics against superbugs. J Med Chem. 2020;63:1859.
61. Drawz SM, Bonomo RA. Three decades of β-lactamase inhibitors. Clin Microbiol Rev. 2010;23:160–201.
62. Docquier J-D, Mangani S. An update on beta-lactamase inhibitor discovery and development. Drug Resist Updates. 2018;36:13–29.
63. Chiem K, Jani S, Fuentes B, et al. Identification of an inhibitor of the aminoglycoside 6′-n-acetyltransferase type ib [aac(6′)-ib] by glide molecular docking. MedChemComm. 2016;7:184–9.
64. Lin DL, Tran T, Adams C, et al. Inhibitors of the aminoglycoside 6′-N-acetyltransferase type Ib [AAC(6′)-Ib] identified by in silico molecular docking. Bioorg Med Chem Lett. 2013;23:5694–8.
65. Green KD, Chen W, Garneau-Tsodikova S. Identification and characterization of inhibitors of the aminoglycoside resistance acetyltransferase Eis from mycobacterium tuberculosis. ChemMedChem. 2012;7:73–7.
66. Li X-Z, Plesiat P, Nikaido H. The challenge of efflux-mediated antibiotic resistance in gram-negative bacteria. Clin Microbiol Rev. 2015;28:337–418.
67. Sharma A, Gupta VK, Pathania R. Efflux pump inhibitors for bacterial pathogens: from bench to bedside. Indian J Med Res. 2019;149:129–45.
68. Handzlik J, Matys A, Kieć-Kononowicz K. Recent advances in multi-drug resistance (MDR) efflux pump inhibitors of gram-positive bacteria S. Aureus. Antibiotics. 2013;2:28–45.

69. Spengler G, Kincses A, Gajdacs M, et al. New roads leading to old destinations: efflux pumps as targets to reverse multidrug resistance in bacteria. Molecules. 2017;22:468.
70. Song L, Wu X. Development of efflux pump inhibitors in antituberculosis therapy. Int J Antimicrob Agents. 2016;47:421–9.
71. Schillaci D, Spano V, Parrino B, et al. Pharmaceutical approaches to target antibiotic resistance mechanisms. J Med Chem. 2017;60:8268–97.
72. Soucy SM, Huang J, Gogarten JP. Horizontal gene transfer: building the web of life. Nat Rev Genet. 2015;16:472–82.
73. Graf FE, Palm M, Warringer J, et al. Inhibiting conjugation as a tool in the fight against antibiotic resistance. Drug Dev Res. 2019;80:19–23.
74. Fernandez-Lopez R, Machon C, Longshaw CM, et al. Unsaturated fatty acids are inhibitors of bacterial conjugation. Microbiology. 2005;151:3517–26.
75. Getino M, Sanabria-Rios DJ, Fernandez-Lopez R, et al. Synthetic fatty acids prevent plasmid-mediated horizontal gene transfer. MBio. 2015;6:e01032–510.
76. Shaffer CL, Good JAD, Kumar S, et al. Peptidomimetic small molecules disrupt type IV secretion system activity in diverse bacterial pathogens. MBio. 2016;7:e00221.
77. Grassi L, Maisetta G, Esin S, et al. Combination strategies to enhance the efficacy of antimicrobial peptides against bacterial biofilms. Front Microbiol. 2017;8:2409.
78. Roy R, Tiwari M, Donelli G, et al. Strategies for combating bacterial biofilms: a focus on anti-biofilm agents and their mechanisms of action. Virulence. 2018;9:522–54.
79. Ramasubbu N, Thomas LM, Ragunath C, et al. Structural analysis of dispersin b, a biofilm-releasing glycoside hydrolase from the periodontopathogen actinobacillus actinomycetemcomitans. J Mol Biol. 2005;349:475–86.
80. Chen X, Zhang L, Zhang M, et al. Quorum sensing inhibitors: a patent review (2014–2018). Expert Opin Ther Pat. 2018;28:849–65.
81. Xiang H, Cao F, Ming D, et al. Aloe-emodin inhibits staphylococcus aureus biofilms and extracellular protein production at the initial adhesion stage of biofilm development. Appl Microbiol Biotechnol. 2017;101:6671–81.
82. Wang J, Nong X-H, Amin M, et al. Hygrocin C from marine-derived streptomyces Sp. SCSGAA 0027 inhibits biofilm formation in bacillus amyloliquefaciens SCSGAB0082 isolated from South China Sea gorgonian. Appl Microbiol Biotechnol. 2018;102:1417–27.
83. Wunnoo S, Saising J, Voravuthikunchai SP. Rhodomyrtone inhibits lipase production, biofilm formation, and disorganizes established biofilm in propionibacterium acnes. Anaerobe. 2017;43:61–8.
84. Pritchard MF, Powell LC, Jack AA, et al. A low-molecular-weight alginate oligosaccharide disrupts pseudomonal microcolony formation and enhances antibiotic effectiveness. Antimicrob Agents Chemother. 2017;61:61.
85. Pletzer D, Wolfmeier H, Bains M, et al. Synthetic peptides to target stringent response-controlled virulence in a pseudomonas aeruginosa murine cutaneous infection model. Front Microbiol. 2017;8:1867.
86. Christensen LD, van Gennip M, Jakobsen TH, et al. Synergistic antibacterial efficacy of early combination treatment with tobramycin and quorum-sensing inhibitors against pseudomonas aeruginosa in an intraperitoneal foreign-body infection mouse model. J Antimicrob Chemother. 2012;67:1198–206.
87. Kissoyan KAB, Bazzi W, Hadi U, et al. The inhibition of pseudomonas aeruginosa biofilm formation by micafungin and the enhancement of antimicrobial agent effectiveness in BALB/c mice. Biofouling. 2016;32:779–86.
88. Worthington RJ, Melander C. Combination approaches to combat multidrug-resistant bacteria. Trends Biotechnol. 2013;31:177–84.
89. Tyers M, Wright GD. Drug combinations: a strategy to extend the life of antibiotics in the 21st century. Nat Rev Microbiol. 2019;17:141–55.
90. Drusano GL, Hope W, MacGowan A, et al. Suppression of emergence of resistance in pathogenic bacteria: keeping our powder dry, part 1. Antimicrob Agents Chemother. 2016;60:1194–201.
91. Gorityala BK, Guchhait G, Goswami S, et al. Hybrid antibiotic overcomes resistance in P. Aeruginosa by enhancing outer membrane penetration and reducing efflux. J Med Chem. 2016;59:8441–55.
92. Lee J, Patel G, Huprikar S, et al. Decreased susceptibility to polymyxin b during treatment for carbapenem-resistant klebsiella pneumoniae infection. J Clin Microbiol. 2009;47:1611–2.
93. Goss CH, Muhlebach MS. Review: staphylococcus aureus and MRSA in cystic fibrosis. J Cyst Fibros. 2011;10:298–306.
94. McCaughey G, Diamond P, Elborn JS, et al. Resistance development of cystic fibrosis respiratory pathogens when exposed to fosfomycin and tobramycin alone and in combination under aerobic and anaerobic conditions. PLoS One. 2013;8(7):e69763.
95. Monedero I, Caminero JA. Management of multidrug-resistant tuberculosis: an update. Ther Adv Respir Dis. 2010;4:117–27.
96. REX Consortium. Heterogeneity of selection and the evolution of resistance. Trends Ecol Evol. 2013;28:110–8.
97. Takesue Y, Nakajima K, Ichiki K, et al. Impact of a hospital-wide programme of heterogeneous antibiotic use on the development of antibiotic-resistant gram-negative bacteria. J Hosp Infect. 2010;75:28–32.
98. Raymond B. Five rules for resistance management in the antibiotic apocalypse, a road map for integrated microbial management. Evol Appl. 2019;12:1079–91.
99. Paul M, Lador A, Grozinsky-Glasberg S, et al. Beta lactam antibiotic monotherapy versus beta

lactam-aminoglycoside antibiotic combination therapy for sepsis. Cochrane Database Syst Rev. 2014;2014(1):CD003344.

100. Tamma PD, Cosgrove SE, Maragakis LL. Combination therapy for treatment of infections with gram-negative bacteria. Clin Microbiol Rev. 2012;25:450–70.
101. Lipsitch M, Levin BR. The population dynamics of antimicrobial chemotherapy. Antimicrob Agents Chemother. 1997;41:363–73.
102. Raymond B, Wright DJ, Crickmore N, et al. The impact of strain diversity and mixed infections on the evolution of resistance to bacillus thuringiensis. Proc R Soc. 2013;280(1769):20131497.
103. Pena-Miller R, Laehnemann D, Jansen G, et al. When the most potent combination of antibiotics selects for the greatest bacterial load: the smile-frown transition. PLoS Biol. 2013;11:e1001540.
104. MacLean RC, Hall AR, Perron GG, et al. The population genetics of antibiotic resistance: integrating molecular mechanisms and treatment contexts. Nat Rev Genet. 2010;11:405–14.
105. Hegreness M, Shoresh N, Damian D, et al. Accelerated evolution of resistance in multidrug environments. Proc Natl Acad Sci U S A. 2008;105:13977–81.
106. Ghosn J, Taiwo B, Seedat S, et al. HIV Lancet. 2018;392(10148):685–97.
107. Deeks SG, Overbaugh J, Phillips A, et al. HIV infection. Nat Rev Dis Primers. 2015;1:15035.
108. Atta MG, De Seigneux S, Lucas GM. Clinical pharmacology in HIV therapy. Clin J Am Soc Nephrol. 2019;14(3):435–44.
109. Chung J, DiGiusto DL, Rossi JJ. Combinatorial RNA-based gene therapy for the treatment of HIV/AIDS. Expert Opin Biol Ther. 2013;13(3):437–45.
110. Dybul M, Fauci AS, Bartlett JG, et al. Panel on clinical practices for the treatment of HIV. Guidelines for using antiretroviral agents among HIV-infected adults and adolescents. Recommendations of the Panel on Clinical Practices for Treatment of HIV. MMWR Recomm Rep. 2002;51(RR-7):1–55.
111. Deng H, Liu R, Ellmeier W, et al. Identification of a major co-receptor for primary isolates of HIV-1. Nature. 1996;381(6584):661–6.
112. Alkhatib G, Combadiere C, Broder CC, et al. CC CKR5: a RANTES, MIP-1alpha, MIP-1beta receptor as a fusion cofactor for macrophage-tropic HIV-1. Science. 1996;272(5270):1955–8.
113. Zhou C, Bahner IC, Larson GP, et al. Inhibition of HIV-1 in human T-lymphocytes by retrovirally transduced anti tat and rev hammerhead ribozymes. Gene. 1994;149(1):33–9.
114. Bauer G, Valdez P, Kearns K, et al. Inhibition of human immunodeficiency virus-1 (HIV-1) replication after transduction of granulocyte colony-stimulating factor-mobilized CD34+ cells from HIV-1-infected donors using retroviral vectors containing anti-HIV-1 genes. Blood. 1997;89(7):2259–67.
115. Michienzi A, Li S, Zaia JA, et al. A nucleolar TAR decoy inhibitor of HIV-1 replication. Proc Natl Acad Sci U S A. 2002;99(22):14047–52.
116. Lee NS, Dohjima T, Bauer G, et al. Expression of small interfering RNAs targeted against HIV-1 rev transcripts in human cells. Nat Biotechnol. 2002;20(5):500–5.
117. Dean M, Carrington M, Winkler C, et al. Genetic restriction of HIV-1 infection and progression to AIDS by a deletion allele of the CKR5 structural gene. Hemophilia Growth and Development Study, Multicenter AIDS Cohort Study, Multicenter Hemophilia Cohort Study, San Francisco City Cohort, ALIVE Study. Science. 1996;273(5283):1856–62.
118. Huang Y, Paxton WA, Wolinsky SM, et al. The role of a mutant CCR5 allele in HIV-1 transmission and disease progression. Nat Med. 1996;2(11):1240–3.
119. Michael NL, Louie LG, Sheppard HW. CCR5-delta 32 gene deletion in HIV-1 infected patients. Lancet. 1997;350(9079):741–2.
120. Li MJ, Kim J, Li S, et al. Long-term inhibition of HIV-1 infection in primary hematopoietic cells by lentiviral vector delivery of a triple combination of anti-HIV shRNA, anti-CCR5 ribozyme, and a nucleolar-localizing TAR decoy. Mol Ther. 2005;12(5):900–9.
121. Ribeiro V, Dias N, Paiva T, et al. Current trends in the pharmacological management of Chagas disease. Int J Parasitol Drugs Drug Resist. 2020;12:7–17.
122. Ferreira LG, Andricopulo AD. Drug repositioning approaches to parasitic diseases: a medicinal chemistry perspective. Drug Discov Today. 2016;21(10):1699–710.
123. Andrews KT, Fisher G, Skinner-Adams TS, et al. Drug repurposing and human parasitic protozoan diseases. Int J Parasitol Drugs Drug Resist. 2014;4(2):95–111.
124. Martins TAF, Diniz LF, Mazzeti AL, et al. Benznidazole/itraconazole combination treatment enhances anti-*Trypanosoma cruzi* activity in experimental Chagas disease. PLoS One. 2015;10:1–12.
125. Da Silva RM, Oliveira LT, Barcellos NMS, et al. Preclinical monitoring of drug association in experimental chemotherapy of Chagas' disease by a new HPLC-UV method. Antimicrob Agents Chemother. 2012;56(6):3444–8.
126. Haupt VJ, Uavalle JEA, Salentin S, et al. Computational drug repositioning by target hopping: a use case in chagas disease. Curr Pharm Des. 2016;22(21):3124–34.
127. Dias LC, Dessoy MA, Silva DJJ, et al. Chemotherapy of Chagas' Disease: State of the Art and Perspectives for the Development of New Drugs. 2009; pp 2444–2457.
128. Rogers KE, Keränen H, Durrant JD, et al. Novel cruzain inhibitors for the treatment of Chagas' disease. Chem Biol Drug Des. 2012;80(3):398–405.
129. Salas-Sarduy S, Landaburu LU, Karpiak J. Novel scaffolds for inhibition of Cruzipain identified from

high through put screening of anti kinetoplastid chemical boxes. Sci Rep. 2017;7(1):12073.
130. Bortone B, Jackson C, Hsia Y, et al. High global consumption of potentially inappropriate fixed dose combination antibiotics: analysis of data from 75 countries. PLoS One. 2021;16(1):e0241899.
131. Sayer B, Bortone B, Sharland M, et al. Fixed-dose combination antibiotics: the search for evidence using the example of ampicillin-cloxacillin. Br J Clin Pharmacol. 2021;87(7):2996–9.
132. World Health Organization. WHO model list of essential medicines 20th List. 2017.
133. Albanna AS, Smith BM, Cowan D, et al. Fixed-dose combination antituberculosis therapy: a systematic review and meta-analysis. Eur Respir J. 2013;42(3):721–32.
134. Wang J, Kutter JP, Mu H, et al. Synergistic antibacterial effect of inhaled aztreonam and tobramycin fixed dose combination to combat multidrug-resistant Gram-negative bacteria. Int J Pharm. 2020;590:119877.
135. Chaudhary M, Shrivastava SM, Sehgal R. Efficacy and safety study of fixed-dose combination of ceftriaxone-vancomycin injection in patients with various infections. Curr Drug Saf. 2008;3(1):82–5.
136. Gandhi M, Gandhi RT. Single-pill combination regimens for treatment of HIV-1 infection. N Engl J Med. 2014;371(3):248–59.
137. Aldir I, Horta A, Serrado M. Single-tablet regimens in HIV: does it really make a difference? Curr Med Res Opin. 2014;30(1):89–97.
138. Bourlière M, Pietri O, Castellani P, et al. Sofosbuvir, velpatasvir and voxilaprevir: a new triple combination for hepatitis C virus treatment. One pill fits all? Is it the end of the road? Ther Adv Gastroenterol. 2018;11:1756284818812358.
139. Dehghan Manshadi SA, Merat S, et al. Single-pill sofosbuvir and daclatasvir for treating hepatis C in patients co-infected with human immunodeficiency virus. Int J Clin Pract. 2021;75(8):e14304.
140. Figueirêdo CBM, Nadvorny D, Vieira ACQM, et al. Enhanced delivery of fixed-dose combination of synergistic antichagasic agents posaconazole-benznidazole based on amorphous solid dispersions. Eur J Pharm Sci. 2018;119:208–18.
141. Yeaman MR, Yount NY. Mechanisms of antimicrobial peptide action and resistance. Pharmacol Rev. 2003;55:27–55.
142. Brotz H, Josten M, Wiedemann I, et al. Role of lipid-bound peptidoglycan precursors in the formation of pores by nisin, epidermin and other lantibiotics. Mol Microbiol. 1998;30:317–27.
143. Velkov Y, Thompson P, Azad M, et al. History, chemistry and antibacterial spectrum. In: Li J, Nation R, Kaye K, editors. Polymyxin antibiotics: from laboratory bench to bedside, vol. 1145. 1st ed. Springer Nature; 2019. p. p17.
144. Tang X-J, Thibault P, Boyd RK. Characterisation of the tyrocidine and gramicidin fractions of the tyrothricin complex from bacillus brevis using liquid chromatography and mass spectrometry. Int J Mass Spectrom Ion Process. 1992;122:153–79.
145. Loll PJ, Upton EC, Nahoum V, et al. The high resolution structure of tyrocidine a reveals an amphipathic dimer. Biochim Biophys Acta Biomembr. 2014;1838:199–207.
146. Munyuki G, Jackson GE, Venter GA, et al. Beta-sheet structures and dimer models of the two major tyrocidines, antimicrobial peptides from Bacillus Aneurinolyticus. Biochemistry. 2013;52:7798–806.
147. Leussa AN-N, Rautenbach M. Detailed SAR and PCA of the tyrocidines and analogues towards leucocin a-sensitive and leucocin a-resistant listeria monocytogenes. Chem Biol Drug Des. 2014;84:543–57.
148. Spathelf BM, Rautenbach M. Anti-listerial activity and structure-activity relationships of the six major tyrocidines, cyclic decapeptides from Bacillus Aneurinolyticus. Bioorg Med Chem. 2009;17:5541–8.
149. Wenzel M, Rautenbach M, Vosloo JA, et al. The multifaceted antibacterial mechanisms of the pioneering peptide antibiotics tyrocidine and gramicidin S. MBio. 2018;9:e00802–18.
150. Palm J, Fuchs K, Stammer H, et al. Efficacy and safety of a triple active sore throat lozenge in the treatment of patients with acute pharyngitis: results of a multi-centre, randomised, placebo-controlled, double-blind, parallel-group trial (DoriPha). Int J Clin Pract. 2018;72:e13272.
151. Bosscha MI, van Dissel JT, Kuijper EJ, et al. The efficacy and safety of topical polymyxin b, neomycin and gramicidin for treatment of presumed bacterial corneal ulceration. Br J Ophthalmol. 2004;88:25–8.
152. Zhou L, van Heel AJ, Montalban-Lopez M, et al. Potentiating the activity of nisin against Escherichia Coli. Front Cell Dev Biol. 2016;4:7.
153. Li Q, Montalban-Lopez M, Kuipers OP. Increasing the antimicrobial activity of nisin-based lantibiotics against gram-negative pathogens. Appl Environ Microbiol. 2018;84:e00052-18.
154. Schweizer F. Enhancing uptake of antibiotics into gram-negative bacteria using nonribosome-targeting aminoglycoside-based adjuvants. Future Med Chem. 2019;11:1519–22.
155. Raulston JE, Montie TC. Early cell envelope alterations by tobramycin associated with its lethal action on Pseudomonas Aeruginosa. Microbiology. 1989;135:3023–34.
156. Bulitta JB, Ly NS, Landersdorfer CB, et al. Two mechanisms of killing of Pseudomonas Aeruginosa by tobramycin assessed at multiple inocula via mechanism-based modeling. Antimicrob Agents Chemother. 2015;59:2315–27.
157. Yang X, Goswami S, Gorityala BK, et al. A tobramycin vector enhances synergy and efficacy of efflux pump inhibitors against multidrug-resistant gram-negative bacteria. J Med Chem. 2017;60:3913–32.
158. Gorityala BK, Guchhait G, Fernando DM, et al. Adjuvants based on hybrid antibiotics overcome resistance in pseudomonas aeruginosa and enhance

fluoroquinolone efficacy. Angew Chem Int Ed. 2016;55:555–9.
159. Idowu T, Ammeter D, Arthur G, et al. Potentiation of beta-lactam antibiotics and beta-lactam/beta-lactamase inhibitor combinations against MDR and XDR pseudomonas aeruginosa using non-ribosomal tobramycin-cyclam conjugates. J Antimicrob Chemother. 2019;74:2640–8.
160. Domalaon R, Idowu T, Zhanel GG, et al. Antibiotic hybrids: the next generation of agents and adjuvants against gram-negative pathogens? Clin Microbiol Rev. 2018;31:e00077-17.
161. Idowu T, Zhanel GG, Schweizer F. A Dimer, but not monomer, of tobramycin potentiates ceftolozane against multidrug-resistant and extensively drug-resistant pseudomonas aeruginosa and delays resistance development. Antimicrob Agents Chemother. 2020;64:e02055–19.
162. Scheffers D-J, Pinho MG. Bacterial cell wall synthesis: new insights from localization studies. Microbiol Mol Biol Rev. 2005;69:585–607.
163. Spratt BG. Properties of the penicillin-binding proteins of Escherichia Coli K12. Eur J Biochem. 1977;72:341–52.
164. Denome SA, Elf PK, Henderson TA, et al. Escherichia Coli mutants lacking all possible combinations of eight penicillin binding proteins: viability, characteristics, and implications for peptidoglycan synthesis. J Bacteriol. 1999;181:3981–93.
165. Kocaoglu O, Carlson EE. Profiling of β-lactam selectivity for penicillin-binding proteins in Escherichia Coli Strain DC2. Antimicrob Agents Chemother. 2015;59:2785–90.
166. Drlica K, Malik M, Kerns RJ, et al. Quinolone-mediated bacterial death. Antimicrob Agents Chemother. 2008;52:385–92.
167. Du W, Brown JR, Sylvester DR, et al. Two active forms of UDP-N-acetylglucosamine enolpyruvyl transferase in gram-positive bacteria. J Bacteriol. 2000;182:4146–52.
168. Noda M, Kawahara Y, Ichikawa A, et al. Self-protection mechanism in D-cycloserine-producing streptomyces lavendulae. Gene cloning, characterization, and kinetics of its alanine racemase and D-alanyl-D-alanine ligase, which are target enzymes of D-cycloserine. J Biol Chem. 2004;79:46143–52.
169. Rudolf JD, Dong L-B, Shen B. Platensimycin and platencin: inspirations for chemistry, biology, enzymology, and medicine. Biochem Pharmacol. 2017;133:139–51.
170. Jayasuriya H, Herath KB, Zhang C, et al. Isolation and structure of platencin: a FabH and FabF dual inhibitor with potent broad-spectrum antibiotic activity. Angew Chem Int Ed. 2007;46:4684–8.
171. Wang J, Kodali S, Lee SH, et al. Discovery of platencin, a dual FabF and FabH inhibitor with in vivo antibiotic properties. Proc Natl Acad Sci U S A. 2007;104:7612–6.
172. Velkov T, Roberts KD, Nation RL, et al. Pharmacology of polymyxins: new insights into an "old" class of antibiotics. Future Microbiol. 2013;8:711–24.
173. Fu L, Wan M, Zhang S, et al. Polymyxin B loosens lipopolysaccharide bilayer but stiffens phospholipid bilayer. Biophys J. 2020;118:138–50.
174. Deris ZZ, Akter J, Sivanesan S, et al. A secondary mode of action of polymyxins against gram-negative bacteria involves the inhibition of NADH-quinone oxidoreductase activity. J Antibiot. 2014;67:147–51.
175. Stuart MC, Kouimtzi M, Hill S. WHO model formulary 2008. Geneva: WHO; 2009.
176. Arbiser JL, Moschella SL. Clofazimine: a review of its medical uses and mechanisms of action. J Am Acad Dermatol. 1995;32:241–7.
177. Lesnik U, Lukezic T, Podgorsek A, et al. Construction of a new class of tetracycline lead structures with potent antibacterial activity through biosynthetic engineering. Angew Chem Int Ed. 2015;54:3937–40.
178. Herrmann J, Lukezic T, Kling A, et al. Strategies for the discovery and development of new antibiotics from natural products: three case studies. Curr Top Microbiol Immunol. 2016;398:339–63.
179. Stepanek JJ, Lukezic T, Teichert I, et al. Dual mechanism of action of the atypical tetracycline chelocardin. Biochim Biophys Acta Proteins Proteomics. 2016;1864:645–54.
180. Wenzel M, Dekker MP, Wang B, et al. A flat embedding method for transmission electron microscopy reveals an unknown mechanism of tetracycline. Commun Biol. 2021;4(1):306.
181. Herskovits AA, Bibi E. Association of Escherichia Coli Ribosomes with the inner membrane requires the signal recognition particle receptor but is independent of the signal recognition particle. Proc Natl Acad Sci U S A. 2000;97:4621–6.
182. Ristow H, Schazschneider B, Vater J, et al. Some characteristics of the DNA-tyrocidine complex and a possible mechanism of the gramicidin action. Biochim Biophys Acta Nucleic Acids Protein Synth. 1975;414:1–8.
183. Ruhr E, Sahl HG. Mode of action of the peptide antibiotic nisin and influence on the membrane potential of whole cells and on cytoplasmic and artificial membrane vesicles. Antimicrob Agents Chemother. 1985;27:841–5.
184. Breukink E, Wiedemann I, van Kraaij C, et al. Use of the cell wall precursor lipid II by a pore-forming peptide antibiotic. Science. 1999;286:2361–4.
185. Hasper HE, de Kruijff B, Breukink E. Assembly and stability of Nisin-lipid II pores. Biochemistry. 2004;43:11567–75.
186. Hart P, Oppedijk SF, Breukink E, et al. New insights into Nisin's antibacterial mechanism revealed by binding studies with synthetic lipid II analogues. Biochemistry. 2016;55:232–7.
187. Bonev BB, Chan WC, Bycroft BW, et al. Interaction of the Lantibiotic Nisin with mixed lipid bilayers: a 31P and 2H NMR study. Biochemistry. 2000;39:11425–33.

188. Christ K, Al-Kaddah S, Wiedemann I, et al. Membrane lipids determine the antibiotic activity of the Lantibiotic Gallidermin. J Membr Biol. 2008;226:9–16.
189. Silver LL. Multi-targeting by monotherapeutic antibacterials. Nat Rev Drug Discov. 2007;6:41–55.
190. Higgins DL, Chang R, Debabov DV, et al. Telavancin, a multifunctional lipoglycopeptide, disrupts both cell wall synthesis and cell membrane integrity in methicillin-resistant Staphylococcus Aureus. Antimicrob Agents Chemother. 2005;49:1127–34.
191. Müller A, Wenzel M, Strahl H, et al. Daptomycin inhibits cell envelope synthesis by interfering with fluid membrane microdomains. Proc Natl Acad Sci U S A. 2016;113:E7077–86.
192. Muraih JK, Palmer M. Estimation of the subunit stoichiometry of the membrane-associated Daptomycin oligomer by FRET. Biochim Biophys Acta Biomembr. 2012;1818:1642–7.
193. Muraih JK, Pearson A, Silverman J, et al. Oligomerization of daptomycin on membranes. Biochim Biophys Acta Biomembr. 2011;1808:1154–60.
194. Muraih JK, Harris J, Taylor SD, et al. Characterization of daptomycin oligomerization with perylene excimer fluorescence: stoichiometric binding of phosphatidylglycerol triggers oligomer formation. Biochim Biophys Acta Biomembr. 2012;1818:673–8.
195. Kreutzberger MA, Pokorny A, Almeida PF. Daptomycin-phosphatidylglycerol domains in lipid membranes. Langmuir. 2017;33:13669–79.
196. Georgopapadakou NH, Bertasso A, Chan KK, et al. Mode of action of the dual-action cephalosporin Ro 23–9424. Antimicrob Agents Chemother. 1989;33:1067–71.
197. Stone GW, Zhang Q, Castillo R, et al. Mechanism of action of NB2001 and NB2030, Novel antibacterial agents activated by beta-lactamases. Antimicrob Agents Chemother. 2004;48:477–83.
198. Ma Z, Lynch AS. Development of a dual-acting antibacterial agent (TNP-2092) for the treatment of persistent bacterial infections. J Med Chem. 2016;59:6645–57.
199. Gupta V, Datta P. Next-generation strategy for treating drug resistant bacteria: antibiotic hybrids. Indian J Med Res. 2019;149:97–106.
200. Endres BT, Basseres E, Alam MJ, et al. Cadazolid for the treatment of clostridium difficile. Expert Opin Investig Drugs. 2017;26:509–14.
201. Locher HH, Seiler P, Chen X, et al. In vitro and in vivo antibacterial evaluation of cadazolid, a new antibiotic for treatment of clostridium difficile infections. Antimicrob Agents Chemother. 2014;58:892–900.
202. Locher HH, Caspers P, Bruyère T, et al. Investigations of the mode of action and resistance development of cadazolid, a new antibiotic for treatment of clostridium difficile infections. Antimicrob Agents Chemother. 2014;58:901–8.
203. Leuthner KD, Vidaillac C, Cheung CM, et al. In vitro activity of the new multivalent glycopeptide-cephalosporin antibiotic TD-1792 against vancomycin-nonsusceptible staphylococcus isolates. Antimicrob Agents Chemother. 2010;54:3799–803.
204. Blais J, Lewis SR, Krause KM, et al. Antistaphylococcal activity of TD-1792, a multivalent Glycopeptide-cephalosporin antibiotic. Antimicrob Agents Chemother. 2012;56:1584–7.
205. Webb E. FDA grants QIDP and Fast Track Designations to MCB3837, Morphochem's novel intravenous antibacterial to treat C. difficile infections. https://www.tvm-lifescience.com/fda-grants-qidp-fast-track-designations-mcb3837--morphochems-novel-intravenous-antibacterial-treat-c-difficile-infections/. Accessed 10 Mar 2020.
206. Pokrovskaya V, Baasov T. Dual-acting hybrid antibiotics: a promising strategy to combat bacterial resistance. Expert Opin Drug Discovery. 2010;5:883–902.
207. Deinove. DNV3837/DNV3681: First-in-class antibiotic candidate. https://www.deinove.com/en/antibiotics/portfolio/dnv3837/dnv3681. Accessed 10 Mar 2020.
208. Deinove. An Exploratory, Open-Label, Oligo-Center Study to Evaluate the Safety, Efficacy, and Pharmacokinetics of Intravenous DNV3837 in Subjects With Clostridium Difficile Infection. https://clinicaltrials.gov/ct2/show/NCT03988855?term=DNV3681&draw=2&rank=1. Accessed 10 Mar 2020.
209. Wang B, Zhao Q, Yin W, et al. In-vitro characterisation of a novel antimicrobial agent, TNP-2092, against helicobacter pylori clinical isolates. Swiss Med Wkly. 2018;148:w14630.
210. Parkes AL, Yule IA. Hybrid antibiotics – clinical progress and novel designs. Expert Opin Drug Discovery. 2016;11:665–80.
211. Burrows LL. The therapeutic pipeline for pseudomonas aeruginosa infections. ACS Infect Dis. 2018;4:1041–7.
212. Luther A, Urfer M, Zahn M, et al. Chimeric peptidomimetic antibiotics against gram-negative bacteria. Nature. 2019;576:452–8.
213. Keicho N, Kudoh S. Diffuse panbronchiolitis: role of macrolides in therapy. Am J Respir Med. 2002;1:119–1.
214. Zhu YI, Stiller MJ. Dapsone and sulfones in dermatology: overview and update. J Am Acad Dermatol. 2001;45:420–34.
215. Trivedi HD, Lizaola B, Tapper EB, et al. Management of pruritus in primary biliary cholangitis: a narrative review. Am J Med. 2017;130(744):e1–744.e7.
216. Del Rosso JQ, Schmidt NF. A review of the anti-inflammatory properties of clindamycin in the treatment of acne vulgaris. Cutis. 2010;85:15–24.
217. Nakano T, Hiramatsu K, Kishi K, et al. Clindamycin modulates inflammatory-cytokine induction in lipopolysaccharide-stimulated mouse peritoneal

macrophages. Antimicrob Agents Chemother. 2003;47:363–7.

218. Pradhan S, Madke B, Kabra P, et al. Anti-inflammatory and immunomodulatory effects of antibiotics and their use in dermatology. Indian J Dermatol. 2016;61:469–81.
219. van Rensburg CE, Gatner EM, Imkamp FM, et al. Effects of clofazimine alone or combined with dapsone on neutrophil and lymphocyte functions in normal individuals and patients with lepromatous leprosy. Antimicrob Agents Chemother. 1982;21:693–7.
220. Gatner EM, Anderson R, van Remsburg CE, et al. The in vitro and in vivo effects of clofazimine on the motility of neutrophils and transformation of lymphocytes from normal individuals. Lepr Rev. 1982;53:85–90.
221. Zimmermann P, Ziesenitz VC, Curtis N, et al. The immunomodulatory effects of macrolides—a systematic review of the underlying mechanisms. Front Immunol. 2018;9:302.
222. Pasquale TR, Tan JS. Nonantimicrobial effects of antibacterial agents. Clin Infect Dis. 2005;40:127–35.
223. Tauber SC, Nau R. Immunomodulatory properties of antibiotics. Curr Mol Pharmacol. 2008;1:68–79.
224. Ye Y, Xia Z, Zhang D, et al. Multifunctional pharmaceutical effects of the antibiotic daptomycin. Biomed Res Int. 2019;2019:8609218.
225. Brötz-Oesterhelt H, Brunner NA. How many modes of action should an antibiotic have? Curr Opin Pharmacol. 2008;8(5):564–73.
226. Balzarini J, Andrei G, Balestra E, et al. A multi-targeted drug candidate with dual anti-HIV and anti-HSV activity. PLoS Pathog. 2013;9(7):e1003456.
227. Blower S, Ma L. Calculating the contribution of herpes simplex virus type 2 epidemics to increasing HIV incidence: treatment implications. Clin Infect Dis. 2004;39(Suppl 5):S240–7.
228. Corey L. Synergistic copathogens–HIV-1 and HSV-2. N Engl J Med. 2007;356:854–6.
229. Buvé A. Can we reduce the spread of HIV infection by suppressing herpes simplex virus type 2 infection? F1000 Med Rep. 2010;2:41.
230. Freeman EE, Weiss HA, Glynn JR, et al. Herpes simplex virus 2 infection increases HIV acquisition in men and women: systematic review and meta-analysis of longitudinal studies. AIDS. 2006;20:73–83.
231. Martinelli E, Tharinger H, Frank I, et al. HSV-2 infection of dendritic cells amplifies a highly susceptible HIV-1 cell target. PLoS Pathog. 2011;7:e1002109.
232. Schacker T, Ryncarz AJ, Goddard J, et al. Frequent recovery of HIV-1 from genital herpes simplex virus lesions in HIV-1-infected men. JAMA. 1998;280:61–6.
233. Lehner T, Hussain L, Wilson J, et al. Mucosal transmission of HIV. Nature. 1991;353:709.
234. Spira AI, Marx PA, Patterson BK, et al. Cellular targets of infection and route of viral dissemination after an intravaginal inoculation of simian immunodeficiency virus into rhesus macaques. J Exp Med. 1996;183:215–25.
235. Moriuchi M, Moriuchi H, Williams R, et al. Herpes simplex virus infection induces replication of human immunodeficiency virus type 1. Virology. 2000;278:534–40.
236. Balzarini J, Holý A, Jindrich J, et al. Differential antiherpesvirus and antiretrovirus effects of the (S) and (R) enantiomers of acyclic nucleoside phosphonates: potent and selective in vitro and in vivo antiretrovirus activities of (R)-9-(2-phosphonomethoxypropyl)-2,6-diaminopurine. Antimicrob Agents Chemother. 1993;37:332–8.
237. Elion GB, Furman PA, Fyfe JA, et al. Selectivity of action of an antiherpetic agent, 9-(2-hydroxyethoxymethyl) guanine. Proc Natl Acad Sci U S A. 1977;74:5716–20.
238. Nagot N, Ouédraogo A, Foulongne V, et al. Reduction of HIV-1 RNA levels with therapy to suppress herpes simplex virus. N Engl J Med. 2007;356:790–9.
239. Delany S, Mlaba N, Clayton T, et al. Impact of aciclovir on genital and plasma HIV-1 RNA in HSV-2/HIV-1 co-infected women: a randomized placebo-controlled trial in South Africa. AIDS. 2009;23:461–9.
240. Ludema C, Cole SR, Poole C, et al. Meta-analysis of randomized trials on the association of prophylactic acyclovir and HIV-1 viral load in individuals coinfected with herpes simplex virus-2. AIDS. 2011;25:1265–9.
241. Mugwanya K, Baeten JM, Mugo NR, et al. High-dose valacyclovir HSV-2 suppression results in greater reduction in plasma HIV-1 levels compared with standard dose acyclovir among HIV-1/HSV-2 coinfected persons: a randomized, crossover trial. J Infect Dis. 2011;204:1912–7.
242. Abdool Karim Q, Abdool Karim SS, Frohlich JA, et al. Effectiveness and safety of tenofovir gel, an antiretroviral microbicide, for the prevention of HIV infection in women. Science. 2010;329:1168–74.
243. Cates W Jr. After CAPRISA 004: time to re-evaluate the HIV lexicon. Lancet. 2010;376:495–6.
244. Lisco A, Vanpouille C, Tchesnokov EP, et al. Acyclovir is activated into a HIV-1 reverse transcriptase inhibitor in herpesvirus-infected human tissues. Cell Host Microbe. 2008;4:260–70.
245. Andrei G, Lisco A, Vanpouille C, et al. Topical tenofovir, a microbicide effective against HIV, inhibits herpes simplex virus-2 replication. Cell Host Microbe. 2011;10:379–89.
246. De Clercq E, Sakuma T, Baba M, et al. Antiviral activity of phosphonylmethoxyalkyl derivatives of purine and pyrimidines. Antivir Res. 1987;8:261–72.
247. Pauwels R, Balzarini J, Schols D, et al. Phosphonylmethoxyethyl purine derivatives, a new class of anti-human immunodeficiency virus agents. Antimicrob Agents Chemother. 1988;32:1025–30.
248. Heijtink RA, De Wilde GA, Kruining J, et al. Inhibitory effect of 9-(2-phosphonylmethoxyethyl)-

adenine (PMEA) on human and duck hepatitis B virus infection. Antivir Res. 1993;21:141–53.
249. Balzarini J, Schols D, Van Laethem K, et al. Pronounced in vitro and in vivo antiretroviral activity of 5-substituted 2,4-diamino-6-[2-(phosphonomethoxy)ethoxy] pyrimidines. J Antimicrob Chemother. 2007;59:80–6.
250. Wei Y, Li W, Du T, et al. Targeting HIV/HCV coinfection using a machine learning-based multiple quantitative structure-activity relationships (multiple QSAR) method. Int J Mol Sci. 2019;20(14):3572.
251. Therese PJ, Manvar D, Kondepudi S, et al. Multiple e-pharmacophore modeling, 3D-QSAR, and high-throughput virtual screening of hepatitis C virus NS5B polymerase inhibitors. J Chem Inf Model. 2014;54:539–52.
252. Laguno M, Murillas J, Blanco JL, et al. Peginterferon ALFA-2b plus ribavirin compared with interferon alfa-2b plus ribavirin for treatment of HIV/HCV co-infected patients. AIDS. 2004;18:F27–36.
253. WHO Global Hepatitis Report. 2017. Available online: https://www.who.int/hepatitis/publications/global-hepatitis-report2017/en/. Accessed 24 Mar 2019.
254. Wang Z, Bennett EM, Wilson DJ, et al. Rationally designed dual inhibitors of HIV reverse transcriptase and integrase. J Med Chem. 2007;50:3416–9.
255. Cox BD, Prosser AR, Sun Y, et al. Pyrazolo-piperidines exhibit dual inhibition of CCR5/CXCR4 HIV entry and reverse transcriptase. ACS Med Chem Lett. 2015;6:753–7.
256. Operskalski EA, Kovacs A. HIV/HCV co-infection: pathogenesis, clinical complications, treatment, and new therapeutic technologies. Curr HIV/AIDS Rep. 2011;8:12–22.
257. Bajorath J. Computational analysis of ligand relationships within target families. Curr Opin Chem Biol. 2008;12:352–8.
258. Rognan D. Chemogenomic approaches to rational drug design. Br J Pharmacol. 2007;152:38–52.
259. Yang L, Wang KJ, Wang LS, et al. Chemical-protein interactome and its application in off-target identification. Interdiscip Sci Comput Life Sci. 2012;3:22–30.
260. Cheng F, Zhou Y, Li J, et al. Prediction of chemical-protein interactions: multitarget-QSAR versus computational chemogenomic methods. Mol BioSyst. 2012;8:2373–84.
261. Sawada R, Iwata H, Mizutani S, et al. Target-based drug repositioning using large-scale chemical-protein interactome data. J Chem Inf Model. 2015;55:2717–30.
262. Luo H, Chen J, Shi L, et al. DRAR-CPI: a server for identifying drug repositioning potential and adverse drug reactions via the chemical-protein interactome. Nucleic Acids Res. 2011;39:492–8.
263. Jenkins JL, Bender A, Davies JW. In silico target fishing: predicting biological targets from chemical structure. Drug Discov Today Technol. 2006;3:413–21.
264. Wang X, Shen Y, Wang S, et al. PharmMapper 2017 update: a web server for potential drug target identification with a comprehensive target pharmacophore database. Nucleic Acids Res. 2017;45:W356–60.
265. Keiser MJ, Roth BL, Armbruster BN, et al. Relating protein pharmacology by ligand chemistry. Nat Biotechnol. 2007;25:197–206.
266. Bauer RA, Bourne PE, Formella A, et al. Superimpose: a 3D structural superposition server. Nucleic Acids Res. 2008;36:W47–54.
267. Li H, Gao Z, Kang L, et al. TarFisDock: a web server for identifying drug targets with docking approach. Nucleic Acids Res. 2006;34:W219–24.
268. Chen YZ, Zhi DG. Ligand-protein inverse docking and its potential use in the computer search of protein targets of a small molecule. Proteins. 2001;43:217–26.
269. Viña D, Uriarte E, Orallo F, et al. Alignment-free prediction of a drug-target complex network based on parameters of drug connectivity and protein sequence of receptors. Mol Pharm. 2009;6:825–35.
270. Porwal A, Carranza EJM, Hale M. Bayesian network classifiers for mineral potential mapping. Comput Geosci. 2006;32:1–16.
271. Xia X, Maliski EG, Gallant P, et al. Classification of kinase inhibitors using a Bayesian model. J Med Chem. 2004;47:4463–70.
272. Jorissen RN, Gilson MK. Virtual screening of molecular databases using a support vector machine. J Chem Inf Model. 2005;45:549–61.
273. Durant JL, Leland BA, Henry DR, et al. Reoptimization of MDL keys for use in drug discovery. J Chem Inf Comput Sci. 2002;42:1273–80.
274. Rogers D, Hahn M. Extended-connectivity fingerprints. J Chem Inf Model. 2010;50:742–54.
275. Liu Q, Zhou H, Liu L, et al. Multi-target QSAR modelling in the analysis and design of HIV-HCV co-inhibitors: an in-silico study. BMC Bioinf. 2011;12:294.
276. Zhan P, Liu X. Rationally designed multitarget anti-HIV agents. Curr Med Chem. 2013;20(13):1743–58.
277. Andrae-Marobela K, Ghislain FW, Okatch H, et al. Polyphenols: a diverse class of multi-target anti-HIV-1 agents. Curr Drug Metab. 2013;14(4):392–413.
278. Ren PX, Shang WJ, Yin WC, et al. A multi-targeting drug design strategy for identifying potent anti-SARS-CoV-2 inhibitors. Acta Pharmacol Sin. 2021;43:1–11.
279. Zhu N, Zhang D, Wang W, et al. A novel coronavirus from patients with pneumonia in China, 2019. N Engl J Med. 2020;382:727–33.
280. Holshue ML, DeBolt C, Lindquist S, et al. First case of 2019 novel coronavirus in the United States. N Engl J Med. 2020;382:929–36.
281. Lillie PJ, Samson A, Li A, et al. Novel coronavirus disease (Covid-19): the first two patients in the UK with person to person transmission. J Infect. 2020;80:578–606.

282. Dong E, Du H, Gardner L. An interactive web-based dashboard to track COVID-19 in real time. Lancet Infect Dis. 2020;20:533–4.
283. Zhou P, Yang XL, Wang XG, et al. A pneumonia outbreak associated with a new coronavirus of probable bat origin. Nature. 2020;579:270–3.
284. Brian DA, Baric RS. Coronavirus genome structure and replication. In: Enjuanes L, editor. Coronavirus replication and reverse genetics. Berlin/Heidelberg: Springer Berlin Heidelberg; 2005. p. 1–30.
285. Su S, Wong G, Shi W, et al. Epidemiology, genetic recombination, and pathogenesis of coronaviruses. Trends Microbiol. 2016;24:490–502.
286. Kim D, Lee JY, Yang JS, et al. The architecture of SARS-CoV-2 transcriptome. Cell. 2020;181:914–21.e10.
287. Fehr AR, Perlman S. Coronaviruses: an overview of their replication and pathogenesis. In: Maier HJ, Bickerton E, Britton P, editors. Coronaviruses: methods and protocols. New York: Springer New York; 2015. p. 1–23.
288. Sonenberg N, Hinnebusch AG. Regulation of translation initiation in eukaryotes: mechanisms and biological targets. Cell. 2009;136:731–45.
289. Liljas A. Ribosomes. In: Maloy S, Hughes K, editors. Brenner's encyclopedia of genetics. 2nd ed. San Diego: Academic; 2013. p. 248–53.
290. Garreau de Loubresse N, et al. Structural basis for the inhibition of the eukaryotic ribosome. Nature. 2014;513:517–22.
291. Wu C-I, Wen H. Heightened protein-translation activities in mammalian cells and the disease/treatment implications. Natl Sci Rev. 2020;7:1851–5.
292. Neupane K, Munshi S, Zhao M, et al. Anti-frameshifting ligand active against SARS Coronavirus-2 is resistant to natural mutations of the frameshift-stimulatory pseudoknot. J Mol Biol. 2020;432:5843–7.
293. Haniff HS, Tong Y, Liu X, et al. Targeting the SARS-CoV-2 RNA genome with small molecule binders and ribonuclease targeting chimera (RIBOTAC) degraders. ACS Central Sci. 2020;6:1713–21.
294. Plant EP, Sims AC, Baric RS, et al. Altering SARS coronavirus frameshift efficiency affects genomic and subgenomic RNA production. Viruses. 2013;5:279–94.
295. Machitani M, Yasukawa M, Nakashima J, et al. RNA-dependent RNA polymerase, RdRP, a promising therapeutic target for cancer and potentially COVID-19. Cancer Sci. 2020;111:3976–84.
296. Allen CNS, Arjona SP, Santerre M, et al. Potential use of RNA-dependent RNA polymerase (RdRp) inhibitors against SARS-CoV2 infection. All Life. 2020;13:608–14.
297. Gao Y, Yan L, Huang Y, et al. Structure of the RNA-dependent RNA polymerase from COVID-19 virus. Science. 2020;368:779.
298. Yin W, Mao C, Luan X, et al. Structural basis for inhibition of the RNA-dependent RNA polymerase from SARS-CoV-2 by remdesivir. Science. 2020;368:1499–504.
299. Wang Q, Wu J, Wang H, et al. Structural basis for RNA replication by the SARS-CoV-2 polymerase. Cell. 2020;182:417–428.e13.
300. Cascarina SM, Ross ED. A proposed role for the SARS-CoV-2 nucleocapsid protein in the formation and regulation of biomolecular condensates. FASEB J. 2020;34:9832–42.
301. McBride R, van Zyl M, Fielding B. The coronavirus nucleocapsid is a multifunctional protein. Viruses. 2014;6:2991–3018.
302. Cong Y, Ulasli M, Schepers H, et al. Nucleocapsid protein recruitment to replication-transcription complexes plays a crucial role in coronaviral life cycle. J Virol. 2020;94:e01925-19.
303. Lin SY, Liu C-L, Chang Y-M, et al. Structural basis for the identification of the N-terminal domain of coronavirus nucleocapsid protein as an antiviral target. J Med Chem. 2014;57:2247–57.
304. Cao Z, Yang P, Zhou Q. Multiple biological functions and pharmacological effects of lycorine. Sci China Chem. 2013;56:1382–91.
305. Emmanuel S, Akinboye OB. Biological activities of emetine. Open Nat Prod J. 2011;4:8–15.
306. Choy KT, Wong AY-L, Kaewpreedee P, et al. Remdesivir, lopinavir, emetine, and homoharringtonine inhibit SARS-CoV-2 replication in vitro. Antivir Res. 2020;178:104786.
307. Ianevski A, Yao R, Fenstad MH, et al. Potential antiviral options against SARS-CoV-2 infection. Viruses. 2020;12:642.
308. Zhang YN, Zhang QY, Li X-D, et al. Gemcitabine, lycorine and oxysophoridine inhibit novel coronavirus (SARS-CoV-2) in cell culture. Emerg Microbes Infect. 2020;9:1170–3.
309. Yang S, Xu M, Lee EM, et al. Emetine inhibits Zika and Ebola virus infections through two molecular mechanisms: inhibiting viral replication and decreasing viral entry. Cell Discov. 2018;4:31.
310. Guo Y, Wang Y, Cao L, et al. A conserved inhibitory mechanism of a lycorine derivative against enterovirus and hepatitis C virus. Antimicrob Agents Chemother. 2016;60:913–24.
311. Khalifa SAM, Yosri N, El-Mallah MF, et al. Screening for natural and derived bio-active compounds in preclinical and clinical studies: one of the frontlines of fighting the coronaviruses pandemic. Phytomedicine. 2021;85:153311.
312. Liu J, Yang Y, Xu Y, et al. Lycorine reduces mortality of human enterovirus 71-infected mice by inhibiting virus replication. Virol J. 2011;8:483.
313. Zou G, Puig-Basagoiti F, Zhang B, et al. A single-amino acid substitution in West Nile virus 2K peptide between NS4A and NS4B confers resistance to lycorine, a flavivirus inhibitor. Virology. 2009;384:242–52.
314. Deng L, Dai P, Ciro A, et al. Identification of novel antipoxviral agents: mitoxantrone inhibits vaccinia virus replication by blocking virion assembly. J Virol. 2007;81:13392–402.
315. Khandelwal N, Chander Y, Rawat KD, et al. Emetine inhibits replication of RNA and DNA viruses with-

out generating drug-resistant virus variants. Antivir Res. 2017;144:196–204.
316. Andersen PI, Krpina K, Ianevski A, et al. Novel antiviral activities of obatoclax, emetine, niclosamide, brequinar, and homoharringtonine. Viruses. 2019;11:964.
317. Chen H, Lao Z, Xu J, et al. Antiviral activity of lycorine against Zika virus in vivo and in vitro. Virology. 2020;546:88–97.
318. Wong W, Bai X-C, Brown A, et al. Cryo-EM structure of the plasmodium falciparum 80S ribosome bound to the anti-protozoan drug emetine. Elife. 2014;3:e03080.
319. Quiros Roldan E, Biasiotto G, Magro P, et al. The possible mechanisms of action of 4-aminoquinolines (chloroquine/hydroxychloroquine) against Sars-Cov-2 infection (COVID-19): a role for iron homeostasis? Pharm Res. 2020;158:104904.
320. Satarker S, Ahuja T, Banerjee M, et al. Hydroxychloroquine in COVID-19: potential mechanism of action against SARS-CoV-2. Curr Pharm Rep. 2020;6:203–11.
321. Barrow E, Nicola AV, Liu J. Multiscale perspectives of virus entry via endocytosis. Virol J. 2013;10:177.
322. Sun Y, Tien P. From endocytosis to membrane fusion: emerging roles of dynamin in virus entry. Crit Rev Microbiol. 2013;39:166–79.
323. Oscanoa TJ, Romero-Ortuno R, Carvajal A, et al. A pharmacological perspective of chloroquine in SARS-CoV-2 infection: an old drug for the fight against a new coronavirus? Int J Antimicrob Agents. 2020;56:106078.
324. Vrijsen R, Vanden Berghe DA, Vlietinck AJ, et al. Lycorine: a eukaryotic termination inhibitor? J Biol Chem. 1986;261:505–7.
325. Bullen CK, Hogberg HT, Bahadirli-Talbott A, et al. Infectability of human BrainSphere neurons suggests neurotropism of SARS-CoV-2. Altex. 2020;37:665–71.
326. Lamoral-Theys D, et al. Lycorine, the main phenanthridine amaryllidaceae alkaloid, exhibits significant antitumor activity in cancer cells that display resistance to proapoptotic stimuli: an investigation of structure−activity relationship and mechanistic insight. J Med Chem. 2009;52:6244–56.
327. Hopf NJ, Goebel HH. Experimental emetine myopathy: enzyme histochemical, electron microscopic, and immunomorphological studies. Acta Neuropathol. 1993;85:414–8.
328. Pan SJ, Combs AB. Effects of pharmacological interventions on emetine cardiotoxicity in isolated perfused rat hearts. Toxicology. 1995;97:93–104.
329. Pan SJ, Combs AB. Emetine inhibits glycolysis in isolated, perfused rat hearts. Cardiovasc Toxicol. 2003;3:311–8.

11 Polypharmacology in Clinical Applications—Anti-inflammation Polypharmacology

Abstract

Inflammation, a vital defense mechanism for health, is defined as the immune system's response to harmful stimuli such as pathogens, damaged cells, toxic compounds, or irradiation. However, excessive inflammation could be rather detrimental and destructive; it has widely been recognized as a ubiquitous component of various chronic diseases such as cancer, asthma, and cardiovascular, neurodegenerative, and many autoimmune diseases, including rheumatoid arthritis, systemic lupus erythematosus, and chronic granulomatous disease. Recently, a link between inflammation and metabolic diseases, including obesity, nonalcoholic steatohepatitis, and fibrosis, has emerged. In this chapter, the basic concept, causes, and consequences of inflammation will be briefly introduced, and the focus will be placed on the applications of polypharmacological approaches to the treatment of inflammation and the associated pathological processes, and on the advancement of multitarget drug (MTD) discovery and development.

Keywords

Anti-inflammation polypharmacology · Chronic inflammation · COX-based MTDs · LOX-based MTDs · sEH-based MTDs

11.1 Anti-Inflammatory Medication

11.1.1 Basics of Inflammation

11.1.1.1 What Is Inflammation?

Inflammation refers to a biological response to stimuli interpreted by the body to have a potentially harmful effect [1]. Acute inflammation is the body's natural defense response intended to provide protection from injury and external pathogens [2–4]. However, evidence strongly suggests that uncontrolled chronic inflammation can cause abnormal inflammation, resulting in the immune system attacking the body's own cells or tissues leading to chronic pain, redness, swelling, stiffness, and damage to normal tissues [5, 6]. This is because the arterioles in the surrounding tissues widen, allowing increased blood flow to the area causing the redness. This reaction increases permeability of the arterioles causing the white blood cells, prostaglandins (PGs; chemical mediators), and blood proteins and other fluid to build up in the injured location.

Inflammatory diseases include a vast array of disorders and conditions that are characterized by inflammation. Examples include allergy, asthma, autoimmune diseases, coeliac disease, glomerulonephritis, hepatitis, inflammatory bowel disease, preperfusion injury, and transplant rejection.

Z. Wang, B. Yang, *Polypharmacology*, https://doi.org/10.1007/978-3-031-04998-9_11

11.1.1.2 Acute *vs.* Chronic Inflammation

Inflammation can be either short-lived (acute) or long-lasting (chronic). Acute inflammation goes away within hours or days.

Acute Inflammation Acute inflammation comes quickly and resolves in less than two weeks; it is the body's routine way of responding to harmful substances, repairing damage, and carrying away dead cells. Reactions are the redness, swelling, heat, and pain [7, 8].

Chronic Inflammation Chronic inflammation is slower and generally less severe but can last from months to years [7, 8]. It extends from the acute stage when the body can no longer remove the harmful substance or heal an injury, which causes the body to be in a state of inflammation for long periods of time. Reactions to this prolonged state is body pain, digestion problems, depression, anxiety, mood disorders, constant fatigue, insomnia, weight gain, and more frequent infections.

11.1.1.3 Inflammation *vs.* Infection

Inflammation is different from infection with distinct concepts and pathologies.

Inflammation is the normal tissue reaction following an injury from a traumatic agent or growth of organisms, basically an infection. It is an interior reaction from the body's response through the process of white blood cells. The purpose is to localize and eliminate whatever causes the injury and to remove damaged tissue components so the body can begin to heal.

Infection occurs via the invasion of an organism (virus, bacteria, fungi, or parasites) that enters the body and causes issues; thus, it is an exterior entry and growth of the organism. Specifically, infections are most commonly caused by microorganisms such as bacteria and viruses, but fungi (causing different types of mycoses) and macroparasites (ticks, worms) can also induce an infection.

Like inflammation, infection can be divided into two groups—acute infections and chronic infections—based on the time course of pathological progression.

Acute infections develop suddenly and last a few days or weeks and are generally caused by a virus, infection injury, accidents, or improper use of drugs or medications. In other words, infection is a contagion of the body tissues with disease-causing agents, which multiply in the body and cause the specific response of the host's immune system. These infectious agents and the toxins produced by them provoke an acute infection such as a cold, flu, bronchitis, pneumonia, broken bones, asthma attack, burns, respiratory infection, or strep throat, which often requires urgent short-term care.

Chronic infections, often caused by unhealthy behaviors like poor nutrition, lack of exercise, alcohol abuse, or excessive smoking, develop slowly and worsen over months to years. Chronic infection is a condition that progresses over time with possible warning signs. Such conditions are high blood pressure, diabetes, heart disease, obesity, high cholesterol, osteoporosis, arthritis, or Alzheimer's disease (AD). These conditions are often then maintained and not cured.

Symptoms of infection include fever or chills, body aches and pains, coughing or sneezing, digestive upset, and feeling tired or fatigued.

Infection can cause inflammation, where inflammation can be a result of an infection. While both an infection and inflammation are crucial to the body responding to injury, they should be resolved so they do not cause serious issues.

11.1.1.4 Inflammation and Pain

When inflammation happens, chemicals from white blood cells enter blood or tissues to protect body from invaders. This raises the blood flow to the area of injury or infection, meanwhile causing redness and warmth. Some of the chemicals cause fluid to leak into tissues, resulting in swelling. When this accumulation of fluid swelling happens, it compresses the nerves in the area—causing pain. Prostaglandins produced by the cells may cause irritation of the nerves and cause even more pain. At the site of injury, components transferred by the blood carry out the functions to protect the tissue by the white blood cells to eat the foreign bacteria, which begins the clotting process if bleeding is involved. This chemical process and increased blood supply cause the area to feel heated.

Higher numbers of white blood cells and the chemicals they make inside the joints cause irritation, swelling of the joint lining, and loss of cartilage (cushions at the end of bones) over time. Other painful conditions of the joints and musculoskeletal system that may not be related to inflammation include osteoarthritis (OA), fibromyalgia, muscular low back pain, and muscular neck pain.

On the other hand, an inflammatory disease might not cause pain because many organs do not have many pain-sensitive nerves.

Inflammatory conditions often end with the suffix "itis": periodontitis (bone and tissue), gingivitis (tissue), pulpitis (pulp), mucositis (mucous membranes), tonsillitis (tonsils), and stomatitis (mouth). Some types of arthritis are the result of inflammation, such as rheumatoid arthritis, psoriatic arthritis, and gouty arthritis.

Inflammation can affect internal organs as part of an autoimmune disorder. The symptoms depend on which organs are affected. For example: inflammation of your heart (myocarditis) may cause shortness of breath or fluid buildup; inflammation of the small tubes that take air to your lungs may cause shortness of breath; or inflammation of your kidneys (nephritis) may cause high blood pressure or kidney failure.

11.1.2 Causes and Consequences of Inflammation

11.1.2.1 Stimuli Causing Inflammation

- Microorganisms such as viruses, bacteria, and fungi
- Toxins
- Chemicals
- Allergens
- Impaired cell integrity (injury)
- Burning; frost
- Foreign body
- Radiation (including ultraviolet rays)

11.1.2.2 Symptoms of Inflammation

- Swelling
- Redness
- Warming at the affected place and/or fever
- Pain
- Disturbed function

11.1.2.3 Conditions Linked to Chronic Inflammation

- Cancer
- Heart disease
- Diabetes
- Asthma
- Alzheimer's disease

11.1.3 Treatment for Chronic Inflammation

The pain medications primarily comprise the following eight categories:

(a) *NSAIDS*: Nonsteroidal anti-inflammatory drugs (NSAIDs), such as aspirin, ibuprofen, or naproxen. These drugs inhibit the cells from producing prostaglandins, a group of hormones that will create the reactions that cause pain.
(b) *Corticosteroids* (such as prednisone): These inhibit prostaglandin formation by the cells and suppress the function of white blood cells.
(c) *Antihistamines*: Histamine is produced by the white blood cells and causes local inflammation; these medications block the white blood cells from producing and secreting histamines.
(d) *Antimalarial medications*: For example, hydroxychloroquine.
(e) *DMARDs*: Disease-modifying antirheumatic drugs, including azathioprine, cyclophosphamide, leflunomide, methotrexate, and sulfasalazine.
(f) *Biologic drugs*: For example, abatacept, adalimumab, certolizumab, etanercept, infliximab, golimumab, rituximab, and tocilizumab.

(g) *Drugs for dental inflammation*: Herpangina for coxsackie virus infection that causes painful, red ulcers inside the mouth; Tonsillitis for viral or bacterial infections causing swelling of the tonsils; Pulpitis for bacterial-related infection causing inflammation of the inner tooth; Mucositis for inflammation of the mucous lining of the inner mouth, commonly from cancer treatment; Stomatitis for an inflamed and sore mouth including the cheeks, tongue, and gums; Gingivitis for bacterial infection causing inflammation of the gums; Tooth abscess for a collection of pus caused by bacteria at the tip of the root; Periodontal abscess for a collection of pus caused by bacteria along the side of the tooth within the tissue; Periodontitis for a serious gum infection that damages not only the soft tissue but also the bone supporting the teeth.

Some of these analgesic agents or pain relievers are also used to treat conditions such as cancer or inflammatory bowel disease, or to prevent organ rejection after a transplant. But when "chemotherapy" types of medications (such as methotrexate or cyclophosphamide) are used to treat inflammatory diseases, they tend to have lower doses and less risk of side-effects than when they are prescribed for cancer treatment.

(h) *Hot and cold therapy*: Cold constricts the blood vessels, inhibiting inflammation by numbing the pain and cooling the area. Heat reduces inflammation and the build-up of fluid in tissues.

11.1.4 Inflammation and Polypharmacology

Arachidonic acid (ARA) undergoes enzyme-mediated oxidative metabolism, resulting in the formation of a number of biologically active metabolites. For over a century, these biochemical transformations have been the target of numerous pharmacological drugs for inflammation and pain. In particular, nonsteroidal anti-inflammatory drugs (NSAIDs) and cyclooxygenase-2 (COX-2) selective inhibitors (coxibs) are widely used in the treatment of inflammation and pain. Multitarget inhibition of proinflammatory mediators and metabolites represents a new and valuable option to enhance efficacy or reduce side-effects in the treatment of inflammation and pain; yet this field is a relatively new area of multitarget drug discovery. The past and current endeavors have been made to address different target combinations and to definitively assess the potential of multitarget approaches to the conditions. The targets validated to date can be grouped into the following classes.

11.1.4.1 Cyclooxygenases (COX-1/2)

Cyclooxygenases (COX-1/2) are well recognized as proinflammatory enzymes as well as vital enzymes in the metabolic pathway of arachidonic acid [9–13]. These enzymes are essential for the physiological production of eicosanoids, including leukotrienes (LTs), prostaglandins, and thromboxane (TXA), that primarily cause inflammation. Therefore, COX enzymes are the rate-limiting steps in the production of eicosanoids, and commonly used for screening and evaluating anti-inflammatory drugs.

COX-1

COX-1 is a fundamental enzyme that is localized mostly in the kidney, stomach, and platelets of the human body and helps to maintain the renal and gastric homeostasis [14]. COX-1 catalyzes the biosynthesis of eicosanoids propagating platelet aggregation, thromboxanes, and vasoconstriction [9, 10, 14]. Consequently, long-term use of selective COX-1 inhibitors such as aspirin in the treatment of some chronic inflammation may cause hemorrhage, kidney failure, and/or severe gastrointestinal (GI) side effects [15].

COX-2

COX-2 is an inducible enzyme that presents in low amounts in the human kidney, brain, and ovaries [9, 14]. Its expression is induced in the case of mitogenic stimulation, inflammation, or tissue damage [16]. Inhibition of COX-2 leads to reduced production of prostacyclin, a vasodilator, and anti-aggregatory prostanoid [17]. As a result, selective COX-2 inhibitors considerably minimize gastrointestinal side-effects but increase the likelihood of cardiovascular diseases such as congestive heart failure, systemic and pulmonary hypertension, and myocardial infarction [15, 18, 19].

11.1.4.2 Lipoxygenase (5-LOX)

5-Lipoxygenase (5-LOX), on the other hand, is one isozyme of LOX, which activates the second essential biosynthetic pathway to liberate eicosanoids [9, 20]. The 5-LOX pathway's end product is leukotriene B_4 (LTB_4), a mediator of several inflammatory and allergic diseases, including atherosclerosis, cancer, and cardiovascular diseases [20–23]. Thus, reducing the leukotrienes via inhibiting 5-LOX may help to reduce the potential risk of cardiovascular and gastrointestinal diseases caused by selective COX-2 and COX-1 inhibitors [24], respectively.

11.1.4.3 Soluble Epoxide Hydrolase (sEH)

The enzyme soluble epoxide hydrolase (sEH) plays a central role in metabolism of bioactive lipid-signaling molecules [25–28]. The substrate-specific hydrolase activity of sEH converts epoxyeicosatrienoic acids (EETs) to less bioactive dihydroxyeicosatrienoic acids (DHETs). EETs exhibit anti-inflammatory, analgesic, antihypertensive, cardio-protective, and organ-protective properties. Accordingly, sEH inhibition is a promising therapeutic strategy for addressing a variety of diseases. Soluble epoxy hydrolase is a prominent proinflammatory enzyme that is responsible for the infiltration of inflammatory mediators. Inhibition of sEH significantly reduces the conversion of EETs to corresponding DHETs, thereby highlighting sEH as an important target for addressing inflammation and carcinogenesis [26].

11.1.4.4 Microsomal Prostaglandin E Synthase-1 (mPGES-1)

Microsomal prostaglandin E synthase-1 (mPGES-1) [29, 30] or prostaglandin E synthase is an enzyme that in humans is encoded by the prostaglandin E synthase gene [31–33]. The protein encoded by this gene is a glutathione-dependent prostaglandin E synthase. The expression of this gene has been shown to be induced by various inflammatory stimuli and is primarily coupled to COX-2 expression [34, 35]. This synthase is a PGE_2-forming enzyme downstream of COX in the pathway, which has the potential to selectively affect PGE_2 but not other prostaglandins or thromboxane A2 (TXA_2) synthesis. Its expression can also be induced by tumor suppressor protein TP53, and may be involved in TP53-induced apoptosis. Knockout studies in mice suggest that this gene may contribute to the pathogenesis of collagen-induced arthritis and mediate acute pain during inflammatory responses [33]. Hence, mPGES-1 is one of the most attractive targets for polypharmacological treatment of inflammation.

11.1.4.5 Leukotriene A_4 Hydrolase (LTA_4H)

Leukotriene A_4 hydrolase (LTA_4H) is a proinflammatory enzyme that generates the inflammatory mediator leukotriene B_4 (LTB_4) [36–38]. As an epoxide hydrolase, LTA_4H acts to catalyze the conversion of LTA_4 to the proinflammatory mediator LTB_4, which mainly functions as a chemoattractant and an activator of inflammatory cells. As an aminopeptidase, LTA_4H may process peptides related to inflammation and host defense. Moreover, LTA_4H/LTB_4 is believed to play an important role in chronic inflammation-associated carcinogenesis.

11.1.4.6 Thromboxane 2 Synthase (TXA_2) Synthase

Eicosanoids, such as thromboxanes (TXAs), prostaglandins (PGs), and leukotrienes, are important mediators of inflammation [39–42]. PG and TX synthesis is controlled by the regulation of PG- and TX-synthesizing enzymes. TXA_2 synthase inhibitors and thromboxane prostanoid (TP) receptor antagonists have been developed as therapeutic drugs for asthma, an allergic pulmonary inflammatory disease [40].

11.1.4.7 11β-Hydroxysteroid Dehydrogenase Type 1 (11β-HSD1)

Cortisone, a glucocorticoid hormone, was first used to treat rheumatoid arthritis in humans in the late 1940s, for which Hench, Reichstein, and Kendall were awarded a Nobel Prize in 1950 and which led to the discovery of the anti-inflammatory effects of glucocorticoids. To be effective, the intrinsically inert cortisone must be converted to the active glucocorticoid, cortisol, by the intracellular action of 11-beta-hydroxysteroid dehydrogenase type 1 (11β-HSD1) [43–46]. Thus, 11β-HSD1 is considered an intracellular amplifier of glucocorticoid action. Recent work has highlighted an anti-inflammatory role for 11β-HSD1 within specific tissues, including in leukocytes. Acute inflammation is more severe with 11β-HSD1-deficiency or inhibition, yet in some inflammatory settings such as obesity or diabetes, 11β-HSD1-deficiency/inhibition is beneficial, reducing inflammation. Current evidence suggests both beneficial and detrimental effects may result from 11β-HSD1 inhibition in chronic inflammatory disease [45].

11.1.4.8 Fatty Acid Amide Hydrolase (FAAH)

The endogenous cannabinoid N-arachidonoyl ethanolamine (anandamide; AEA) produces most of its pharmacological effects by binding and activating CB_1 and CB_2 cannabinoid receptors within the CNS and periphery. However, the actions of AEA are short-lived because of its rapid catabolism by fatty acid amide hydrolase (FAAH) belonging to the amidase signature superfamily as a major endocannabinoid inactivating enzyme using an atypical catalytic mechanism involving hydrolysis of amide and occasionally ester bonds [47–50]. Indeed, FAAH knockout mice as well as animals treated with FAAH inhibitors are severely impaired in their ability to hydrolyze AEA as well as a variety of noncannabinoid lipid-signaling molecules and consequently possess greatly elevated levels of these endogenous ligands. FAAH-compromised animals reliably display antinociceptive and anti-inflammatory phenotypes with a similar efficacy as direct-acting cannabinoid receptor agonists. FAAH inhibitors are efficacious in experimental models of neuropathic pain, inflammation, and anxiety, among others.

11.1.4.9 CC Chemokine Ligand 5 (CCL5) and CC Chemokine Receptor 5 (CCR5)

Chemokines, a class of small secreted chemotactic cytokine, and their receptors mediate the inflammatory and immune responses during both infectious and noninfectious diseases [51–53]. The activation of CC chemokine receptor 5 (CCR5) by its ligand CCL5 is involved in leucocytes chemotaxis to sites of inflammation and plays an important role in the macrophages, T-cells, and monocytes recruitment. However, their continued activation and dysregulation are commonly associated with chronic inflammation. Frequently, affected sites are characterized by inflammatory cell infiltrates expressing CCR5 and high levels of CCL5. Recent studies have shown that the interaction between CCL5 and CCR5 is involved in various pathological processes including inflammation, chronic diseases, and cancers, as well as the infection of Coronavirus disease-2019 (COVID-19) [54, 55].

11.2 CDT Anti-Inflammatory Drugs

As reasoned by Hwang et al. [56], because selectively blocking one of the pathways in the arachidonic acid (ARA) cascade in most cases also causes the production of undesirable or unexpected metabolites, new approaches have been developed to combine therapies to modulate multiple pathway targets simultaneously [57–59]. However, administering two drugs in combination therapy raises safety concerns and presents unique challenges for optimizing dosage. For example, it is not safe to assume that two drugs with relatively high individual safety profiles will be safe when co-administered [60]. Co-administration or combination drug therapy, CDT treatment requires extensive investigations into optimal dose regimens including safety studies, a complex dosage ranging investigation, and drug–drug interaction analysis, all of which may significantly raise the practical cost and complexity of developing combination therapies [61]. During the drug development process, predicting pharmacodynamic and pharmacokinetic relationships is substantially less complex if multitarget inhibition is obtained from a single agent rather than from combination therapies (co-administration). In addition, improvement in patient compliance is expected with these multitarget inhibitors as fixed-dose drug combinations [62]. For these reasons, interest has grown in designed multiple ligands (also known as DMLs) [56, 57, 63] with the following aims: (a) to enhance drug efficacy, (b) to improve drug safety by acting specifically on multiple targets, (c) to amend patient compliance, and (d) to eliminate the challenges of formulation problems and drug–drug interactions in co-administrative therapies [56, 64].

11.3 FDC Anti-inflammatory Drugs

Like the case of CDT, there have not been any (fixed-dose combination) FDC drugs specifically designed and tested for anti-inflammation therapy. One study aimed to assess the effect of antidiabetic polypill containing metformin, aspirin, simvastatin, and an angiotensin-converting enzyme inhibitor on central pathology and cognitive impairment in a mixed model of Alzheimer's disease (AD) and type 2 diabetes (T2DM), which resulted in a reduction in spontaneous central bleeding and inflammation in diabetic mice [65]. The investigators believed that the anti-inflammatory property of the polypill contributes to the observed improvement in learning and memory processes in AD, T2DM, and AD-T2DM mice and proposed the polypill as an alternative or adjuvant, to slow down or delay the central complications associated with T2DM and AD.

11.4 MTD Anti-inflammatory Drugs

A number of compounds that possess co-inhibitory effects on varying combination of two anti-inflammatory targets among the validated ones as described above in Sect. 11.1.4, "Inflammation and Polypharmacology," have been tested in clinical trials, preclinical investigations, and laboratory studies. These studies have demonstrated encouraging and promising results for multitarget drugs in the treatment of inflammation and the associated pathological process and disease entities. Hwang et al. [56] provided an excellent summary on the rationally designed multitarget agents against inflammation and pain. Below is an introduction, based on the review article by Hwang et al. [56], of some examples of dual inhibitors of the proinflammatory pathways developed based on the MTD principle.

11.4.1 COX/5-LOX Dual Inhibitors as MTD Drugs for Inflammation

Nonsteroidal anti-inflammatory drugs (NSAIDs) are a mainstay in the treatment of inflammatory disease and are among the most widely used drugs worldwide. They are anti-inflammatory, anti-pyretic, and analgesic and are prescribed as first choice for the treatment of rheumatic disorders and, in general, inflammation [66]. But one of the main drawbacks of NSAIDs is the curtailed production of gastro-protective prostaglandins (PGs) being associated with the concurrent increased production of the gastro-damaging and broncho-constrictive leukotrienes (LTs) [66], leading to gastrointestinal ulcerogenic activity and broncho-spasm. Leukotrienes and cysteinyl-leukotrienes are moreover proinflammatory and increase microvascular permeability. One of the leukotrienes, LTB_4, is the most potent chemotactic agent and it induces chemotaxis of eosinophils, neutrophils, and monocytes in the inflamed tissue, as well as increases superoxide generation and pro-inflammatory cytokines production [67]. The mechanism of action of these drugs is attributed to the inhibition of cyclooxygenase (COX), and, consequently, the conversion of arachidonic acid into prostaglandins. It is hypothesized that the undesirable side-effects of NSAIDs are due to the inhibition of COX-1 (constitutive isoform), whereas the beneficial effects are related to the inhibition of COX-2 (inducible isoform). Arachidonic acid can also be converted to leukotrienes (LTs) by the action of 5-lipoxygenase (5-LOX). LTC_4, LTD_4, and LTE_4 are potent bronchoconstrictors, whereas LTB_4 is chemotactic for leukocytes and plays an important role in the development of gastrointestinal ulcers by contributing to the inflammatory process. Thus, developing dual-inhibitor compounds that will simultaneously inhibit COX and 5-LOX could enhance their individual anti-inflammatory effects and reduce the undesirable side-effects associated with NSAIDs, especially of the gastrointestinal tract. It is further advantageous for a drug to have both COX and 5-LOX inhibiting activities because prostaglandins enhance leukotriene-mediated inflammation. Various structural families of dual inhibitors have been designed and several compounds are currently undergoing clinical development. In the post-COX-2 selective inhibitors era, these dual-acting inhibitors may turn out to be promising new drugs to treat inflammatory diseases and possibly other diseases. Indeed, both COX-2 and 5-LOX are also involved in the development and progression of several types of cancer; in these conditions, selective inhibition of COX-2 alone may lead to a shunt of arachidonic acid metabolism toward the leukotriene pathway, and therefore the blockade of both COX-2 and 5-LOX may produce a better anticancer response. In addition, the dual inhibition of both COX and 5-LOX is neuroprotective by suppressing toxic actions of reactive microglia and macrophages that are increased in aging brain and in age-related degenerative conditions, particularly Alzheimer's and Parkinson's diseases. Finally, the blockade of 5-LOX does not impair the synthesis of lipoxins (LXs), which are mainly produced by further lipoxygenation of 15-HPETE, and which have potent anti-inflammatory properties and can be considered as stop-signal mediators. Leukocyte 15-LOX and platelet 12-LOX by intercellular mechanism via leukocyte/platelet cell–cell interaction convert 15-HPETE into lipoxins. Drugs able to inhibit both cyclooxygenases (COX-1 and COX-2) and 5-lipoxygenase (5-LOX; dual-acting anti-inflammatory drugs) have been designed in order to obtain compounds that retain the activity of classical nonsteroidal anti-inflammatory drugs (NSAIDs) while avoiding their main drawbacks [66].

Darbufelone and licofelone were first designed and clinically used as dual COX/5-LOX inhibitory drugs for anti-inflammatory therapy. Despite that due to the high toxicity and/or limited efficacy, they have never been able to be marketed [56, 68, 69], dual inhibition of COX and LOX by a single entity has been attracting attention in academia and industry. Many of the COX/LOX dual inhibitors developed have focused on inhibition of COX and specifically 5-LOX. They can be classified into three classes based on the type of 5-LOX inhibition: redox inhibitors, iron chelators, and nonredox/nonchelators.

The most promising COX/5-LOX inhibitor is ML3000 ([2,2-dimethyl-6-(4-chlorophenyl)-7-phenyl-2,3-dihydro-1H-pyrrolizine-5-yl]-acetic acid), now in phase III clinical trials. This new approach will certainly help to unravel the mechanisms at the root of the undesirable effects of NSAIDs and to develop safer NSAIDs [66].

11.4.1.1 Redox Inhibitors

Benoxaprofen, which is a radical scavenger, is the first commercially available COX-2/5-LOX inhibitor. However, it has been discontinued due to reports of causing photosensitivity, effects on nail and hair growth, and liver and kidney failure. These side-effects were not proposed to be associated with the inhibition of their target enzymes but remain significant obstacles to their clinical use [56, 70, 71].

Di-tert-Butylphenols

According to Hwang et al. [56], di-*tert*-butylphenol derivatives are the most extensively studied group of dual inhibitors of COX and 5-LOX enzymes. It was found that naturally occurring diarylheptanoids such as curcumin or yakuchinones exhibited COX and LOX inhibition in vitro [72], but these compounds failed to demonstrate inhibition of LTB_4 biosynthesis or anti-inflammatory activity in vivo [73]. Therefore, in an effort to improve both the in vitro and in vivo activities of these diarylheptanoids, various analogues containing similar structural 2,6-di-*tert*-butylphenols have been developed as dual inhibitors. Structure–activity relationship (SAR) studies have indicated that 2,6-di-*tert*-butyl-1-hydroxybenzene substituted in the fourth position is optimum for dual activity. The substituents are either five- or six-membered heterocycles or chains. The phenol moiety possesses the antioxidant and radical scavenging properties that are considered to be responsible for anti-inflammatory potency and low ulcerogenic potential. The therapeutic index of these compounds (ratio of anti-inflammatory efficacy to GI safety profile) has uniformly been shown to be superior to that of conventional NSAIDs [74]. For these reasons, a variety of COX/5-LOX dual inhibitors containing such a phenol have been developed including darbufelone (CI-1004) [75], S-2474 [76], and tebufelone [75]. Among them, S-2474 also displays cytokine-modulating properties and is currently being evaluated in clinical trials for arthritis [74, 76–80].

Pyrazoline Derivatives

BW-755C and phenidone are COX/5-LOX dual inhibitors functioning as one-electron reducing agents. These compounds were originally developed as antioxidant 5-LOX inhibitors and later demonstrated to possess inhibitory activity toward COX isozymes [70, 81, 82]. Phenidone significantly decreased eosinophils ($ED_{50} = 15$ mg/kg) in bronchoalveolar lavage (BAL) fluid of guinea pigs sensitized to ovalbumin and those challenged with aerosolized antigen, while indomethacin had no effect on eosinophil accumulation [83].

11.4.1.2 Iron Chelators

Hydroxamic Acids

Tepoxalin (Zubrin®) is a COX-2/5-LOX dual inhibitor that exhibited potent anti-inflammatory activity with excellent gastric protection. It is approved for veterinary use only (in canines) [84]. Oral administration in humans revealed that tepoxalin inhibits whole blood COX, LOX, and platelet function [85], but its further development has been stopped due to its hepatotoxicity [86].

Thiophene Derivatives

RWJ-63556 was developed as 5-LOX inhibitor [87] that was subsequently found to inhibit COX-2 selectively over COX-1 [88]. This compound displayed significant anti-inflammatory activity in carrageenan-induced inflammation in canines [87] and inhibited the development of neurogenic edema induced by saphenous nerve stimulation [89]. L-652,343 was demonstrated to inhibit COX and 5-LOX in vitro and was efficacious in acute and chronic inflammation models in vivo [90, 91]. However, L-652,343 failed to show inhibition of 5-LOX in vivo in human skin [92] or LTB_4 production ex vivo using stimulated human whole blood [93].

Pyridone Derivatives

Chowdhury et al. [94] developed a COX-2/5-LOX dual inhibitor by replacing the tolyl group in celecoxib with the *N*-difluoromethyl-1,2-dihydropyrid-2-one moiety, which exhibited good anti-inflammatory activity ($ED_{50} = 27.7$ mg/kg p.o.) and compared favorably with reference drugs such as celecoxib ($ED_{50} = 10.8$ mg/kg p.o.) and ibuprofen ($ED_{50} = 67.4$ mg/kg p.o.) [94]. The same strategy has also been applied to NSAIDs [95, 96].

11.4.1.3 Nonredox, Nonchelators

Due to the non-selective action of redox and chelating inhibitors that cause undesirable side-effects, there was an effort to design a new class of small molecule inhibitors without these actions [56]. Therefore, the following compounds are discovered or developed as new class of COX/LOX dual inhibitors.

DHDMBFs

7-*tert*-Butyl-2,3-dihydro-3,3-dimethylbenzofurans (DHDMBF) was originally found as a metabolite of tebufelone. Interestingly, this compound lacks the antioxidant phenolic moiety, but selectively inhibits both 5-LOX and COX-2 enzymes and displays anti-inflammatory activity equivalent to tebufelone in the rat carrageenan paw edema assay [74]. Among DHDMBF derivatives, PGV-20229 is of particular interest because it showed excellent gastric safety and was active in a rat arthritis model with an ED_{50} of 6.6 mg/kg [74]. The acute ulcerogenic dose (UD_{50}) was much higher, with no GI lesions observed at oral doses up to 1000 mg/kg. In addition, in a refed rat antral damage model, the UD_{50} for PGV-20229 was greater than naproxen and nabumetone and no GI damage was seen in 13-day studies in dogs and rats at oral doses up to 200 mg/kg/day. It has been demonstrated that a large variety of substituents (R) are compatible at the 5 position in DHDMBF retaining the anti-inflammatory and analgesic activities [79, 80].

(a) *Structural modification of NSAIDs*: In the attempt to find potent COX/LOX dual inhibitors, structural modifications were performed on conventional NSAIDs. Tenidap, a COX/5-LOX inhibitor and cytokine modulator, is more effective in the clinical treatment of rheumatic arthritis than diclofenac [97]. However, this compound has been withdrawn from the market due to liver and kidney toxicity attributed to its reactive oxidative metabolites with a thiophene moiety [98–100]. To avoid these toxic side-effects, a series of analogues have been developed replacing the thiophene ring in tenidap. These compounds showed comparable or stronger anti-inflammatory and analgesic activities, and better gastric tolerance in vivo than tenidap [101].

(b) *Structural modification of Coxibs*: The combination of a COX-2 pharmacophore and a 5-LOX pharmacophore led to potent selective COX-2/5-LOX inhibitors, for example, the combination of the pyrazole-based tricyclic moiety found in celecoxib (Pfizer) and the 4-(3-fluoro-5-oxy)phenyl-4-methoxy tetrahydropyran from the 5-LOX inhibitor ZD-2138 (Zeneca) [56, 102]. This dual inhibitor reduced ARA-induced ear edema by intravenous (i.v.) and oral administration. Several others in this class of COX-2/5-LOX dual inhibitors have also been developed using similar approach (e.g., a *p*-SO_2Me COX-2 pharmacophore [103] and a 3,4-diaryl-2(5*H*)furanone COX-2 pharmacophore as in Vioxx® [104]).

(c) *Pyrrolizine-based dual inhibitors*: Licofelone [105, 106] (ML3000, 2-(6-(4-chlorophenyl)-2,2-dimethyl-7-phenyl-2,3-dihydro-1*H*-pyrrazolin-5-yl) acetic acid) has shown anti-inflammatory, analgesic, and anti-asthmatic effects in several animal models [107]. Licofelone is one of the most promising COX/5-LOX inhibitors for human therapy and has successfully finished phase III clinical trial for the treatment of osteoarthritis (OA) [56, 108]. Notably, it possesses superior GI-protective effect by decreasing gastrotoxic LTs through 5-LOX inhibition [109]. In addition, it exhibits protective antithrombotic activity due to its inhibition of COX-1 [110]. This is an excellent example demonstrating that simultaneous inhibition of COX-1, COX-2, and 5-LOX can circumvent the side-effects associated with the individual inhibition of these enzymes [56]. Recent study [111] showed that licofelone might suppress PGE_2 formation by inhibiting mPGES-1 [112], and not through COX-2 inhibition.

11.4.2 COX/sEH Dual Inhibitors

Recent work evaluating the co-inhibition of COX-2 and sEH has led to the development of COX-2/sEH dual inhibitors [113]. The potencies of these compounds were evaluated using recombinant enzyme assays in vitro and the efficacy of the most promising candidate was evaluated using a lipopolysaccharide (LPS)-induced rat pain model in vivo. The COX-2/sEH dual inhibitor by subcutaneous (s.c.) injection exhibited anti-allodynic activity that is more efficacious than the same dose of either a celecoxib or an sEH inhibitor (*t*-AUCB) alone, or the co-administration of sEH inhibitor with celecoxib. This work demonstrated that the use of a dual inhibitor as a DML was more efficacious than combination therapy in a rat pain model [56, 113]. Use of COX-2/sEH dual inhibitors could be an alternative approach for treating inflammation and pain compared to well-studied COX/5-LOX dual inhibitors.

11.4.3 LOX/sEH Dual Inhibitors

5-LOX/sEH dual inhibitors have been recently discovered by dual-target virtual screening using an in silico approach [56, 114]. Among the 80 compounds selected by the virtual screening, 36 compounds have been tested with in vitro enzyme assays. One compound has inhibited both 5-LOX and sEH enzymes at a marginal concentration. Although no biology is associated with 5-LOX/sEH dual inhibitors, further development is warranted based on elevation of LOX products 5-HETE and 15-HETE with long-term sEH inhibition [56, 115].

11.4.4 Dual Inhibitors of Enzymes Downstream the COX or LOX Pathways as MTD Drugs for Inflammation

In addition to targeting 5-LOX and COX enzymes, a number of alternative multitarget approaches that focus on enzymes downstream of LOX and COX pathways have also been reported. The limitations of NSAIDs or coxibs' use are derived from on-target side-effects such as GI injury, renal irritation, and cardiovascular events, and off-target side-effects such as an increase in LT formation. Therefore, a new approach has been developed that targets enzymes downstream in the COX or LOX pathway to minimize potential on-target side-effects.

11.4.4.1 5-LOX and mPGES-1 Dual Inhibitors

Recent evidence demonstrated that microsomal prostaglandin E synthase-1 (mPGES-1) knockout mice did not display impaired cardiovascular function [116]. In addition, licofelone (Sect. 11.4.1.3.), the most promising COX/5-LOX dual inhibitor, is likely suppressing the formation of PGE_2 by inhibiting mPGES-1 and not through COX-2 inhibition [111]. Based on these findings several arylpyrrolizine derivatives have been developed as 5-LOX and mPGES-1 dual inhibitors [117]. In addition, pirinixic acid [118] and 2-mercaptohexanoic acid [119] derivatives have been reported as 5-LOX/mPGES-1 dual inhibitors. In comparison with indomethacin (5 mg/kg), YS121 (1.5 mg/kg) was equally efficient in reducing exudate formation and leukocyte infiltration with reduced pleural levels of PGE_2 and LTB_4 in the carrageenan-induced rat pleurisy model [120]. Recently, virtual screening using a comparative model of the human 5-LOX led to the discovery of new 5-LOX/mPGES-1 dual inhibitors possessing novel scaffolds [121].

Yarla et al. [122] used in silico molecular modeling to design an mPGES-1/5-LOX dual inhibitory and COX-1/2-sparing lead molecule licofelone analogue-9 (LFA-9) by modifying the pharmacophore of licofelone. They found that LFA-9 inhibited mPGES-1/5-LOX and their products PGE_2 and LTB_4, and spared COX-1/2 and its product PGI_2. LFA-9 bound strongly with human mPGES-1/5-LOX enzymes and induced changes in their secondary structure, thereby inhibiting their enzymatic activities. LFA-9 inhibited carrageenan-induced inflammation (70.4%) in rats and suppressed CD68 immune cell infiltration and tumor necrosis factor-a (TNF-α) expression. LFA-9 suppressed colon tumor stemness (60.2%) in vitro through inhibition of PGE_2 (82%) levels. Thus, targeting mPGES-1 and 5-LOX can be an ideal approach while sparing COX-1/2 activities for development of the next generation of anti-inflammatory drugs with better efficacy and safety.

11.4.4.2 COX-2 and LTA_4H Dual Inhibitors

In addition to targeting mPGES-1, downstream enzymes in the LOX or COX pathways such as LTA_4H and TXA_2 synthase are also being investigated as new approaches for developing dual inhibitors acting in the ARA cascade [56].

Recently, Chen et al. reported dual inhibitors that target COX and the downstream leukotriene metabolizing enzyme, LTA_4H, an epoxide hydrolase that catalyzes the conversion of LTA_4 into LTB_4 [123]. These dual inhibitors were developed by combining a COX-2 selective inhibitor nimesulide with an LTA_4H inhibitor, 1-(2-(4-phenoxy phenoxy) ethyl)pyrrolidine; however, no in vivo efficacy was reported [124].

11.4.4.3 5-LOX and TXA2 Synthase Dual Inhibitors

Hibi et al. [125] reported dual inhibitors such as E3040 targeting both 5-LOX and TXA2 synthase. E3040 inhibited the production of LTB_4 and TXB_2, but not PGE_2, and was efficacious at a dose of 100 mg/kg p.o. in the induced chronic colitis model compared to sulfasalazine (500 mg/kg). E6700, having better absorption compared to E3040, was under clinical trial for asthma treatment [126].

11.4.5 Dual Inhibitors Targeting Enzymes Downstream in the CYP Pathway and Enzymes Outside ARA Cascade

Extensive research on the benefits of therapeutic application of EETs has triggered development of dual inhibitors that include sEH inhibition. This approach is not limited to the enzymes in the arachidonic cascade [56]. Examples here outline the future direction of valuable dual inhibitors for inflammation and pain.

11.4.5.1 sEH/11β-HSD1 Dual Inhibitors

GlaxoSmithKline (GSK) reported dual inhibitors for sEH and 11β-hydroxysteroid dehydrogenase type 1 (11β-HSD1) enzymes. 11β-HSD1 converts the inactive glucocorticoids (e.g., cortisone) to the active glucocorticoids (e.g., cortisol) [46]. Cortisone injections have been a proven solution for short-term relief of pain associated with swelling from inflammation of a joint, tendon, or bursa. However, their use is restricted due to adverse side-effects. Thus, inhibition of sEH in concert with 11β-HSD1 is predicted to yield favorable results for the treatment of diseases mediated by the sEH or 11β-HSD1 [127].

11.4.5.2 Sorafenib

In a recent study that examined the structural similarities of the FDA-approved anticancer drug sorafenib (Nexavar®) and sEH inhibitors, Liu et al. observed that sorafenib possesses potent in vitro and in vivo sEH inhibitory activity (IC_{50} = 12 nM) in addition to multikinase inhibition [128]. Therefore, it has been suggested that sEH inhibition might ameliorate the toxicity of sorafenib during treatment through demonstrated anti-inflammatory and antihypertensive effects [56]. However, the contribution of sEH inhibition on the efficacy of sorafenib is yet to be fully described.

Inflammation and pain are extremely complex biological processes that involve a great number of lipid mediators providing a theoretically wide range of possible targets for pharmacological treatment. Of the current available drugs that inhibit a single branch of the ARA cascade, inhibition of the COX pathway with NSAIDs is by far the most commonly used approach in the treatment of inflammation and inflammatory pain. This drug class, as well as the second-generation coxibs, is not without serious GI and cardiovascular risks that stem from an imbalance in homeostatic eicosanoid metabolite levels. In addition, it is clear that inhibiting a single pathway can affect the metabolite output of the other pathways and crosstalk. This has led to a significant improvement in our understanding of the on-target adverse effect risks of COX inhibition [129]. However, these serious side-effects from inhibiting either of the COX enzymes remain and the risks are not dogmatically separated as were previously thought [130]. While attempts to mitigate side-effects based on this historical dogma drove the exploration of dual COX/LOX inhibition, little success has come from this approach in the past decades. Several of these compounds, especially COX/5-LOX inhibitors, have been extensively studied and exhibited significant increases in efficacy both in vitro and in vivo without the undesired on-target side-effects associated with single pathway inhibition. However, the off-target toxicity and adverse events limited many of these early dual inhibitors. Many of these initial dual inhibitors or multitarget drugs were not rationally designed; rather were subsequently found to have activity toward unintended targets. For example, compounds possessing antioxidant or iron-chelating properties originally developed as 5-LOX inhibitors also possessed COX inhibitory activity due to their nonselective inhibition, which eventually resulted in off-target adverse effects. Therefore, the need persists for safer and effective multitarget agents by rational design. Recent advances such as the recent X-ray crystal structure of human 5-LOX [131] and computer-aided drug design (CADD) will allow medicinal chemists to rationally design new therapies through virtual screening using molecular docking, ligand-based, and structure-based approaches. In addition, emerging roles of cytochrome P450 (CYP) metabolites in inflammation, pain, and other diseases will expand the scope of multitarget agents in ARA cascade. Dual inhibitors that target sEH and COX enzymes are an example of such a novel approach that holds great promise for treating pain and inflammation. This approach was fundamentally different from the previously developed COX/LOX dual inhibitors since the object was not to reduce undesirable metabolites, but to increase the lifetime of the CYP pathway metabolites (e.g., EETs) through sEH inhibition. The beneficial effects of the stabilized EETs allow dose-sparing of COX inhibitors, thereby potentially attenuating GI or prothrombic side-effects. Although sEH inhibitors share some of the same indications as coxibs, the synergy observed in lowering prostaglandins with combination therapy [132] and in antihyperalgesia with the dual inhibitor [113] merits inhibiting both target enzymes. Another promising approach is the development of dual inhibitors targeting enzymes downstream of the COX or LOX enzymes to prevent unavoidable on-target side-effects. While these new approaches are novel, recent results from use of these multitarget agents or dual inhibitors indicate they are a viable alternative pharmacological strategy for treatment of inflammation and pain.

11.4.6 FAAH/COX Dual Inhibitors as MTD Drugs for Inflammation

Fatty acid amide hydrolase (FAAH) is a homodimer of 64 kDa subunits and belongs to the amidase signature family, which is characterized by an unusual Ser241-Ser217-Lys142 catalytic triad [133]. The residues of the catalytic triad interact through a network of H-bonds that facilitates proton exchange, activating the nucleophilic residue Ser241 (via deprotonation) and the leaving group of the substrate (via protonation). Adjacent to the nucleophilic serine side chain is a circle of four amide N–H bonds that form the oxyanion hole of FAAH. In addition to the catalytic core, FAAH is characterized by a series of channels and cavities, which include the membrane access channel (MAC) connecting the active site to an opening located at the membrane anchoring face of the enzyme, the cytosolic port (CP) that allows exit of hydrophilic products from the active site, and the acyl chain-binding pocket (ABP)—a narrower, longer, and more hydrophobic channel that interacts with the acyl chain of the substrate during the catalytic reaction [134, 135]. On the other hand, COX-1 and COX-2 are homodimers of 70 and 72 kDa subunits, respectively. Dimerization is required for structural integrity and catalytic activity. Each subunit contains one cyclooxygenase and one peroxidase active sites, which are located on opposite sides of the catalytic domain, with the heme prosthetic group positioned at the base of the peroxidase site.

One strategy for the design of multi-target FAAH/COX inhibitors has been the identification of known single target inhibitors (e.g., COX) that also show weak activity at the other target (e.g.,

FAAH), and are amenable to structural modifications aimed at enhancing activities on that target [133]. Some NSAIDs directly bind to FAAH and inhibit its catalytic activity. In 1996, Paria et al. reported that indomethacin reduced anandamide hydrolysis in mouse uterus [136]. Subsequent in vitro studies showed that suprofen, ibuprofen, fenoprofen, naproxen, ketoprofen, diclofenac [137], ketorolac, flurbiprofen [138], and indomethacin [139] inhibit FAAH activity, albeit with weak potencies (median inhibitory concentrations, IC_{50}, in the low to high μM range) [140].

These observations stimulated interest in identifying NSAID-based compounds that also express FAAH-inhibitory activity [138]. In 2003, Cocco et al. described a series of heteroaromatic ibuprofen anilides, bearing substituted pyridine or pyrimidine groups, which showed improved analgesic activity and reduced GI side-effects relative to the parent acid [141]. This exploration identified a compound (ibu-am5) as the best analog. This compound produced a complete inhibition of acetic acid-induced writhing after oral administration in rats, which was associated with very low ulcerogenic effects compared to its parent molecule, ibuprofen [133].

FAAH degrades the endocannabinoid anandamide, which attenuates inflammation and promotes GI healing. Sasso et al. [47] described the first class of systemically active agents that simultaneously inhibit FAAH, Cox-1, and Cox-2 with high potency and selectivity. After oral administration in mice, ARN2508 engages its intended targets and exerts profound therapeutic effects on models of intestinal inflammation. Unlike NSAIDs, ARN2508 (2-[3-fluoro-4-[3-(hexylcarbamoyloxy)phenyl]phenyl]propanoic acid) causes no gastric damage and indeed protects the gastrointestine from NSAID-induced damage through a mechanism that requires FAAH inhibition.

Similarly, another study reported a mechanistic study that clarifies key aspects of the dual inhibition of FAAH and COX enzymes by a new multitarget-directed ligand named ARN2508 [142]. This potent dual inhibitor combines, in a single scaffold, the pharmacophoric elements often needed to block FAAH and COX, that is, a carbamate moiety and the 2-arylpropionic acid functionality, respectively. Molecular modeling and molecular dynamics simulations suggest that ARN2508 uses a noncovalent mechanism of inhibition to block COXs, while inhibiting FAAH via the acetylation of the catalytic Ser241, in line with previous experimental evidence for covalent FAAH inhibition. This study proposes the molecular basis for the dual FAAH/COX inhibition by this novel hybrid scaffold, stimulating further experimental studies and offering new insights for the rational design of novel anti-inflammatory agents that simultaneously act on FAAH and COX.

11.4.7 CCR5 Antagonist and mu Opioid Receptor (MOR) Agonist

Recent studies have demonstrated functional interactions between chemokine and opioid receptor-signaling pathways [143]. Reported heterodimerization of chemokine and opioid receptors led to development of bivalent compounds that bind both types of receptors, with the goal of targeting opioids to sites of inflammation. MCC22 is a novel bivalent compound containing a CCR5 antagonist and mu-opioid receptor (MOR) agonist pharmacophores linked through a 22-atom spacer. Dutta et al. [143] evaluated the efficacy of MCC22 in the K/B.g7 T-cell receptor transgenic mouse model of spontaneous inflammatory arthritis. Their results demonstrated that MCC22 elicited ~3000-fold more potent analgesia than morphine in this model. Daily treatment with MCC22 also led to a cumulative analgesic effect, reducing the daily baseline pain level. MCC22 produced no observable analgesic effect in nonarthritic control mice. Importantly, repeated administration of MCC22 did not induce pharmacologic tolerance, whereas a similar regimen of morphine did. Both grip strength and nesting behaviors improved among arthritic mice treated with MCC22. Ankle thickness and arthritis scores were not affected by MCC22. The analgesic effect of MCC22 was abolished in K/B.g7 mice genetically lacking CCR5, demonstrating the receptor specificity of the antagonist pharmacophore. The investigators concluded that MCC22 is a novel bivalent ligand that targets CCR5 and MOR and provides highly potent analgesia and improved functional outcomes in a model of inflammatory arthritis, without inducing typical opioid tolerance. These findings suggest that MCC22 or similar compounds could be used to treat the pain associated with inflammatory arthritis and related conditions, while minimizing the risks typically associated with chronic opioid use [143].

11.4.8 Benzimidazoles as a Scaffold of MTD Drugs for Inflammation

Benzimidazole, a heterocyclic aromatic organic compound, consists of the fusion of benzene and imidazole. Privileged benzimidazole scaffold being historically versatile structural motif could offer a viable starting point in the search for novel multitarget ligands against multifactorial inflammation process since, when appropriately substituted, it can selectively modulate diverse receptors, pathways, and enzymes associated with the pathogenesis of inflammation [143–145]. Recently, benzimidazole scaffold has emerged as a pharmacophore of choice for designing analgesic and anti-inflammatory agents active on different clinically approved targets. A series of benzimidazole scaffold-based hybrid molecules were designed by integrating benzimidazoles (containing pharmacophoric elements for COX and LOX inhibitors) with phthalimide subunit of thalidomide (pharmacophore element for TNF-α inhibitor) under one construct via molecular hybridization strategy [146]. The designed molecules were synthesized and evaluated for their inhibitory activity against COX (COX-1, COX-2) and LOX (5-LOX, 15-LOX) enzymes as well as TNF-α inhibitory effect. The results revealed that compounds (3a-l) obtained showed inhibition in submicromolar range against COX and LOX targets whereas milder inhibitory activity was obtained against lipopolysaccharide (LPS)-induced TNF-α secretion by murine macrophage-like cells (RAW264.7). Within this class of compounds, 3j emerged as having alluring multiple inhibitory effects on set of COX-1/2 and 5-/15-LOX enzymes in conjunction with a good anti-inflammatory and analgesic activities. Additionally, compound 3j showed gastrointestinal safety with reduced lipid peroxidation. Docking results of compound 3j with COX-2 and 5-LOX were also consistent with the in vivo anti-inflammatory results.

Benzimidazole derivatives exert anti-inflammatory effects mainly by interacting with transient receptor potential vanilloid-1 (TRPV1), cannabinoid receptors, bradykinin receptors, specific cytokines, 5-lipoxygenase activating protein (FLAP), and cyclooxygenase [147]. Literature on structure–activity relationship (SAR) and investigations of benzimidazoles highlight that the substituent's tendency and position on the benzimidazole ring significantly contribute to the anti-inflammatory activity. Reported SAR analyses indicate that substitution at the N1, C2, C5, and C6 positions of the benzimidazole scaffold greatly influences the anti-inflammatory activity. For example, benzimidazole substituted with anacardic acid on C2 inhibits COX-2, and 5-carboxamide or sulfamoyl or sulfonyl benzimidazole antagonizes the cannabinoid receptor, whereas the C2 diarylamine and C3 carboxamide substitutions of the benzimidazole scaffold result in antagonism of the bradykinin receptor. Various heterocyclic substitutions at N1 of benzimidazole account for the effective anti-inflammatory effects of various agents.

(a) Cyclooxygenase inhibitors: Paramashivappa et al. [148], in the year 2003, prepared 2-[[2-alkoxy-6-pentadecyl-phenyl)methyl]thio]-1*H*-benzimidazoles by incorporating benzimidazole with anacardic acid and evaluated its COX-2-inhibitory activity in humans. A compound with R = H and R1 = methoxy moiety showed potent anti-inflammatory activity, with 384-fold selectivity for COX-2 over COX-1 inhibition. Meanwhile, the other compound with R = methyl and R1 = H moiety showed 470-fold selectivity, comparable to the 375-fold and

200-fold selectivity of celecoxib and rofecoxib, respectively.

(b) Cannabinoid receptor agonist: Cannabinoid receptor agonists are effective for the treatment of inflammation. These effects are mediated via two subtypes of cannabinoid receptors: CB1, located centrally and peripherally, and CB2, on immune cells or peripheral tissues. Scientists from AstraZeneca reported cannabinoid receptor agonist, CB2 agonist, and CB1/CB2 dual-agonist activity for benzimidazole derivatives as selective for the management of pain and inflammation. A compound having a carboxamide substitution at C5 of benzimidazole was a highly selective CB2 agonist, with 970-fold selectivity over CB1 receptors [147, 149]. It also showed Ki values of 3170 and 3.3 nM for the human CB1 and CB2 receptors, respectively. The replacement of the carboxamide group in the previous compound with the sulfamoyl group produced another comparable cannabinoid receptor agonist.

(c) Bradykinin receptor antagonists: Several acute and chronic inflammatory pathways resulting in pain, edema, and vasodilation are aggravated by kinins, bradykinin, and kallidin [147, 150]. In most cell types, the bradykinin B2 receptor is expressed under normal conditions, whereas the bradykinin B1 receptor is expressed in infections, inflammatory diseases, and traumatic tissue injury [151]. Guo et al, in 2008, introduced a benzimidazole nucleus into a benzodiazepine derivative with potent bradykinin B1 receptor antagonistic activity, by replacing the phenethyl benzodiazepine moiety, and reported anti-inflammatory activity. Their study showed a potent antagonistic effect at bradykinin B1 receptors with better bioavailability. The combination of the β-alanine linker and 2-imidazoline-5-aminopyridine group at the C2 of benzimidazole resulted in excellent potency, with IC_{50} = 2 nM. They also revealed that the 2-carboxamide group on benzimidazole was necessary for the activity. Still, the activity was reduced when the ethyl linker between the two carboxamide groups was replaced with a methyl or any other longer alkyl group [152]. They concluded that the hydrogen bond acceptor in these compounds seemed to enhance the effect.

(d) Anticytokines: Anticytokine therapies are used in the treatment of chronic inflammatory diseases, particularly autoimmune diseases such as rheumatoid arthritis [147]. A number of *N*-acridin-9-yl-4-benzimidazo-2-ylbenzamides were synthesized and screened for anti-inflammatory and CDK-inhibitory activities. The study revealed that the substitution at C5 of benzimidazole plays a crucial role in anti-inflammatory and CDK-inhibitory activities. Pronounced activity against CDK1 and CDK5 was shown by a compound bearing a nitro group at C5. By contrast, an amino or methyl group at C5 led to the complete loss of the anti-inflammatory and CDK-inhibitory activity [150].

(e) FLAP inhibitors: Leukotrienes are responsible for initiating and amplifying the inflammatory response by regulating the recruitment and activation of leukocytes in inflamed tissues, and they are lipid mediators [147]. The inhibition of leukotriene biosynthesis may also be achieved by targeting FLAP. Banoglu et al. [153] examined benzimidazoles as 5-lipoxygenase-activating protein (FLAP) inhibitors. They performed virtual screening using combined ligand- and structure-based pharmacophore modeling to find compounds targeting FLAP. Their study led to the identification of 1-(2-chlorobenzyl)-2-(1-(4-isobutylphenyl)ethyl)-1*H*-benzimidazole, which inhibited leukotriene biosynthesis with IC_{50} = 0.31 mM. In order to optimize the above compound, the authors synthesized a few potent benzimidazole derivatives, with IC_{50} = 0.12–0.19 mM for intact neutrophils. They also stated that no substitution at C5 of benzimidazole, a methyl linker between the N of benzimidazole and the aryl or heteroaryl groups, favored the inhibition of leukotriene biosynthesis

(f) TRPV-1 antagonists: TRPV-1 is a member of the ion channels that allow the transient influx of Ca^{2+} ions when activated and predominantly expressed in peripheral sensory neurons involved in nociception and neurogenic inflammation. Ognyanov et al. [154] evaluated the TRPV-1-antagonising property of 2-(4-pyridin-2-ylpiperazin-1-yl)-1*H*-benzo-[d]imidazoles, which blocked capsaicin-induced flinch in rats in a dose-dependent manner. One of the compounds, (*R*,*S*)-1-(5-chloro-6- ((*R*)-3-methyl-4-(6-(trifluoromethyl)-4-(3,4,5-trifluorophenyl)-1*H*-benzo[d]imidazol-2-yl)piperazin-1-yl) pyridin-3-yl)ethane-1,2-diol, was the most effective orally bioavailable TRPV-1 antagonist [154]. Conclusively, the simultaneous introduction of bulky lipophilic groups at position 4 of the benzimidazole, a methyl group at R4, hydrophilic groups at R6, and CF3 or Cl at R7 was favorable for potent TRPV-1-antagonising activity. Additionally, a patent from Amgen has also demarcated the TRPV-1-antagonising action of a piperazine-linked benzimidazole derivative, WO04035549 [155].

Available data suggest that for developing benzimidazoles as anti-inflammatory agents, substitutions at N1, C2, C5, and C6 were preferable [147]. Specifically, an aryl/heteroaryl substitution at C5 or C6 enhances the anti-inflammatory activity, while being unsubstituted at C4 and C7 is better for anti-inflammatory activity [147]. However, an amine group at C2 consistently enhanced the anti-inflammatory activity, while a methyl or ethyl linker between the N1 of benzimidazole and the substituent groups was essential for potent anti-inflammatory activity [147].

References

1. Inflammation. National Institute of Environmental Health Sciences. March 16, 2021. https://www.niehs.nih.gov/health/topics/conditions/inflammation/index.cfm
2. Fullerton JN, Gilroy DW. Resolution of inflammation: a new therapeutic frontier. Nat Rev Drug Discov. 2016;15:551–67.
3. Recchiuti A, Serhan CN. Pro-resolving lipid mediators (SPMs) and their actions in regulating miRNA in novel resolution circuits in inflammation. Front Immunol. 2012;3:298.
4. Serhan CN, Chiang N, Van Dyke TE. Resolving inflammation: dual anti-inflammatory and pro-resolution lipid mediators. Nat Rev Immunol. 2008;8:349–61.
5. Serhan CN, Savill J. Resolution of inflammation: the beginning programs the end. Nat Immunol. 2005;6:1191–7.
6. Dr Chris. Inflammation—Causes, Symptoms, Process, Treatment. Health Hype. https://www.healthhype.com/inflammation-causes-symptoms-process-treatment.html
7. Acute and chronic inflammation. CBHS Health. March 20, 2020. https://www.cbhs.com.au/mind-and-body/blog/acute-and-chronic-inflammation
8. Chronic vs. acute medical conditions: What's the Difference? August 11, 2016. https://www.ncoa.org/article/chronic-versus-acute-disease
9. Nguyen HT, Vu TY, Chandi V, et al. Dual COX and 5-LOX inhibition by clerodane diterpenes from seeds of Polyalthia longifolia (Sonn.) Thwaites. Sci Rep. 2020;10(1):15965.
10. Vane JR, Bakhle YS, Botting RM. Cyclooxygenases 1 and 2. Annu Rev Pharmacol Toxicol. 1998;38:97–120.
11. Turini ME, DuBois RN. Cyclooxygenase-2: a therapeutic target. Annu Rev Med. 2002;53:35–57.
12. Vane JR, Botting RM. New insights into the mode of action of anti-inflammatory drugs. Inflamm Res. 1995;44(1):1–10.
13. Kutil Z, Temml V, Maghradze D, et al. Impact of wines and wine constituents on cyclooxygenase-1, cyclooxygenase-2, and 5-lipoxygenase catalytic activity. Mediat Inflamm. 2014;2014:1–8.
14. Pairet M, Engelhardt G. Distinct isoforms (COX-1 and COX-2) of cyclooxygenase: possible physiological and therapeutic implications. Fund Clin Pharmacol. 1996;10:1–15.
15. Harirforoosh S, Asghar W, Jamali F. Adverse effects of nonsteroidal antiinflammatory drugs: an update of gastrointestinal, cardiovascular and renal complications. J Pharm Pharm Sci. 2014;16:821.
16. de Gaetano G, Donati MB, Cerletti C. Prevention of thrombosis and vascular inflammation: benefits and limitations of selective or combined COX-1, COX-2 and 5-LOX inhibitors. Trends Pharmacol Sci. 2003;24:245–52.
17. Catella-Lawson F, et al. Effects of specific inhibition of cyclooxygenase-2 on sodium balance, hemodynamics, and vasoactive eicosanoids. J Pharmacol Exp Ther. 1999;289:735–41.
18. McGettigan P, Henry D. Cardiovascular risk and inhibition of cyclooxygenase: a systematic review of the observational studies of selective and nonselective inhibitors of cyclooxygenase 2. JAMA J Am Med Assoc. 2006;296:1633–44.

19. Warner TD, Mitchell JA. COX-2 selectivity alone does not define the cardiovascular risks associated with non-steroidal anti-inflammatory drugs. Lancet. 2008;371:270–3.
20. Kaur G, Silakari O. Multiple target-centric strategy to tame inflammation. Future Med Chem. 2017;9:1361–76.
21. Dwyer JH, et al. Arachidonate 5-lipoxygenase promoter genotype, dietary arachidonic acid, and atherosclerosis. N Engl J Med. 2004;350:29–37.
22. Hyde CAC, Missailidis S. Inhibition of arachidonic acid metabolism and its implication on cell proliferation and tumour-angiogenesis. Int Immunopharmacol. 2009;9:701–15.
23. Koukoulitsa C, Hadjipavlou-Litina D, Geromichalos G, et al. Inhibitory effect on soybean lipoxygenase and docking studies of some secondary metabolites, isolated from *Origanum vulgare* L. ssp. hirtum. J Enzyme Inhib Med Chem. 2007;22:99–104.
24. Martel-Pelletier J, Lajeunesse D, Reboul P, et al. Therapeutic role of dual inhibitors of 5-LOX and COX, selective and non-selective non-steroidal anti-inflammatory drugs. Ann Rheum Dis. 2003;62:501–9.
25. Das Mahapatra A, Choubey R, Datta B. Small molecule soluble epoxide hydrolase inhibitors in multitarget and combination therapies for inflammation and cancer. Molecules. 2020;25(23):5488.
26. Zhang W, Liao J, Li H, et al. Reduction of inflammatory bowel disease-induced tumor development in IL-10 knockout mice with soluble epoxide hydrolase gene deficiency. Mol Carcinog. 2013;52(9):726–38.
27. Pillarisetti S, Khanna I. Targeting soluble epoxide hydrolase for inflammation and pain – an overview of pharmacology and the inhibitors. Inflamm Allergy Drug Targets. 2012;11(2):143–58.
28. Deng Y, Theken KN, Lee CR. Cytochrome P450 epoxygenases, soluble epoxide hydrolase, and the regulation of cardiovascular inflammation. J Mol Cell Cardiol. 2010;48(2):331–41.
29. Jakobsson PJ, Thorén S, Morgenstern R, et al. Identification of human prostaglandin E synthase: a microsomal, glutathione-dependent, inducible enzyme, constituting a potential novel drug target. Proc Natl Acad Sci USA. 1999;96(13):7220–5.
30. Forsberg L, Leeb L, Thorén S, et al. Human glutathione dependent prostaglandin E synthase: gene structure and regulation. FEBS Lett. 2000;471(1):78–82.
31. Han R, Smith TJ. Cytoplasmic prostaglandin E2 synthase is dominantly expressed in cultured KAT-50 thyrocytes, cells that express constitutive prostaglandin-endoperoxide H synthase-2. Basis for low protaglandin E2 production. J Biol Chem. 2002;277(39):36897–903.
32. Giannoulias D, Alfaidy N, Holloway AC, et al. Expression of prostaglandin I_2 synthase, but not prostaglandin E synthase, changes in myometrium of women at term pregnancy. J Clin Endocrinol Metab. 2002;87(11):5274–82.
33. Ouellet M, Falgueyret JP, Ear PH, et al. Purification and characterization of recombinant microsomal prostaglandin E synthase-1. Protein Expr Purif. 2003;26(3):489–95.
34. Murakami M, Nakashima K, Kamei D, et al. Cellular prostaglandin E2 production by membrane-bound prostaglandin E synthase-2 via both cyclooxygenases-1 and -2. J Biol Chem. 2003;278:37937–47.
35. Murakami M, Naraba H, Tanioka T, et al. Regulation of prostaglandin E2 biosynthesis by inducible membrane-associated prostaglandin E2 synthase that acts in concert with cyclooxygenase-2. J Biol Chem. 2000;275:32783–92.
36. Snelgrove RJ, Jackson PL, Hardison MT, et al. A critical role for LTA4H in limiting chronic pulmonary neutrophilic inflammation. Science. 2010;330(6000):90–4.
37. Haeggström JZ. Structure, function, and regulation of leukotriene A4 hydrolase. Am J Respir Crit Care Med. 2000;161(2 Pt 2):S25–31.
38. Funk CD. Prostaglandins and leukotrienes: advances in eicosanoid biology. Science. 2001;294(5548):1871–5.
39. Dogné JM, de Leval X, Benoit P, et al. Thromboxane A2 inhibition: therapeutic potential in bronchial asthma. Am J Respir Med. 2002;1(1):11–7.
40. Ristimäki A. Cyclooxygenase 2: from inflammation to carcinogenesis. Novartis Found Symp. 2004;256:215–21.
41. Hayashi M, Koya T, Kawakami H, et al. A prostacyclin agonist with thromboxane inhibitory activity for airway allergic inflammation in mice. Clin Exp Allergy. 2010;40(2):317–26.
42. Liu T, Laidlaw TM, Feng C, et al. Prostaglandin E2 deficiency uncovers a dominant role for thromboxane A2 in house dust mite-induced allergic pulmonary inflammation. Proc Natl Acad Sci U S A. 2012;109(31):12692–7.
43. Chapman KE, Seckl JR. 11beta-HSD1, inflammation, metabolic disease and age-related cognitive (dys)function. Neurochem Res. 2008;33(4):624–36.
44. Chapman KE, Coutinho AE, Gray M, et al. The role and regulation of 11beta-hydroxysteroid dehydrogenase type 1 in the inflammatory response. Mol Cell Endocrinol. 2009;301(1-2):123–31.
45. Chapman KE, Coutinho AE, Zhang Z, et al. Changing glucocorticoid action: 11β-hydroxysteroid dehydrogenase type 1 in acute and chronic inflammation. J Steroid Biochem Mol Biol. 2013;137:82–92.
46. Draper N, Stewart PM. 11beta-hydroxysteroid dehydrogenase and the pre-receptor regulation of corticosteroid hormone action. J Endocrinol. 2005;186:251–71.
47. Sasso O, Migliore M, Habrant D, et al. Multitarget fatty acid amide hydrolase/cyclooxygenase blockade suppresses intestinal inflammation and protects against nonsteroidal anti-inflammatory drug-dependent gastrointestinal damage. FASEB J. 2015;29(6):2616–27.

48. Schlosburg JE, Kinsey SG, Lichtman AH. Targeting fatty acid amide hydrolase (FAAH) to treat pain and inflammation. AAPS J. 2009;11(1):39–44.
49. Ligresti A, Silvestri C, Vitale RM, et al. FAAH-catalyzed C-C bond cleavage of a new multitarget analgesic drug. ACS Chem Neurosci. 2019;10(1):424–37.
50. Lodola A, Castelli R, Mor M, et al. Fatty acid amide hydrolase inhibitors: a patent review (2009–2014). Expert Opin Ther Pat. 2015;25(11):1247–66.
51. Marques RE, Guabiraba R, Russo RC, et al. Targeting CCL5 in inflammation. Expert Opin Ther Targets. 2013;17(12):1439–60.
52. Aldinucci D, Colombatti A. The inflammatory chemokine CCL5 and cancer progression. Mediators Inflamm. 2014;2014:292376.
53. Ness TL, Kunkel SL, Hogaboam CM. CCR5 antagonists: the answer to inflammatory disease? Expert Opin Ther Pat. 2006;16(8):1051–65.
54. Zeng Z, Lan T, Wei Y, et al. CCL5/CCR5 axis in human diseases and related treatments. 2021 Aug 26.
55. de Oliveira CE, Oda JM, Losi Guembarovski R, et al. CC chemokine receptor 5: the interface of host immunity and cancer. Dis Markers. 2014;2014:126954.
56. Hwang SH, Wecksler AT, Wagner K, et al. Rationally designed multitarget agents against inflammation and pain. Curr Med Chem. 2013;20(13):1783–99.
57. Morphy R, Kay C, Rankovic Z. From magic bullets to designed multiple ligands. Drug Discov Today. 2004;9:641–51.
58. Millan MJ. Multi-target strategies for the improved treatment of depressive states: conceptual foundations and neuronal substrates, drug discovery and therapeutic application. Pharmacol Therapeut. 2006;110:135–370.
59. Keith CT, Borisy AA, Stockwell BR. Multicomponent therapeutics for networked systems. Nat Rev Drug Discov. 2005;4:71–8.
60. Sreedhar D, Subramanian G, Udupa N. Combination drugs: Are they rational? Curr SciIndia. 2006;91:406.
61. Frantz S. The trouble with making combination drugs. Nat Rev Drug Discov. 2006;5:881–2.
62. Bergmann JF. Review: fixed-dose drug combinations improve medication compliance compared with free-drug regimens. Evid Based Med. 2008;13:18.
63. Morphy R, Rankovic Z. Designing multiple ligands – medicinal chemistry strategies and challenges. Curr Pharm Des. 2009;15:587–600.
64. Morphy R, Rankovic Z. Designed multiple ligands. An emerging drug discovery paradigm. J Med Chem. 2005;48:6523–43.
65. Infante-Garcia C, Ramos-Rodriguez JJ, Hierro-Bujalance C, et al. Antidiabetic polypill improves central pathology and cognitive impairment in a mixed model of Alzheimer's disease and type 2 diabetes. Mol Neurobiol. 2018;55(7):6130–44.
66. Fiorucci S, Meli R, Bucci M, et al. Dual inhibitors of cyclooxygenase and 5-lipoxygenase. A new avenue in anti-inflammatory therapy? Biochem Pharmacol. 2001;62(11):1433–8.
67. Leone S, Ottani A, Bertolini A. Dual acting anti-inflammatory drugs. Curr Top Med Chem. 2007;7(3):265–75.
68. Huang Y, Li J, Zhang B, et al. Design, synthesis, biological evaluation and docking study of novel indole-2-amide as anti-inflammatory agents with dual inhibition of COX and 5-LOX. Eur J Med Chem. 2019;180:41–50.
69. Tries S, Neupert W, Laufer S. The mechanism of action of the new anti-inflammatory compound ML3000: inhibition of 5-LOX and COX-1/2. Inflamm Res. 2002;51:135–43.
70. Leval X, Julemont F, Delarge J, et al. New trends in dual 5-LOX/COX inhibition. Curr Med Chem. 2002;9:941–62.
71. Brahams D. Late benoxaprofen claims. Lancet. 1991;337:483–4.
72. Flynn DL, Rafferty MF, Boctor AM. Inhibition of 5-hydroxyeicosatetraenoic acid (5-HETE) formation in intact human neutrophils by naturally-occurring diarylheptanoids: inhibitory activities of curcuminoids and yakuchinones. Prostaglandins Leukot Med. 1986;22:357–60.
73. Flynn DL, Belliotti TR, Boctor AM, et al. Styrylpyrazoles styrylisoxazoles, styrylisothiazoles. Novel 5-lipoxygenase and cyclooxygenase inhibitors. J Med Chem 1991; 34: 518–525.
74. Janusz JM, Young PA, Ridgeway JM, et al. New cyclooxygenase-2/5-lipoxygenase inhibitors1. 7-tert-buty1-2,3-dihydro-3,3-dimethylbenzofuran derivatives as gastrointestinal safe antiinflammatory and analgesic agents: discovery and variation of the 5-keto substituent. J Med Chem. 1998;41:1112–23.
75. Weisman SM, Doyle MJ, Wehmeyer KR, et al. Effects of tebufelone (NE-11740), a new anti-inflammatory drug, on arachidonic acid metabolism. Agents Actions. 1994;41:156–63.
76. Unangst PC, Shrum GP, Connor DT, et al. Novel 1,2,4-oxadiazoles and 1,2,4-thiadiazoles as dual 5-lipoxygenase and cyclooxygenase inhibitors. J Med Chem. 1992;35:3691–8.
77. Unangst PC, Connor DT, Cetenko WA, et al. Synthesis and biological evaluation of 5-[[3,5-bis(1,1-dimethylethyl)-4-hydroxyphenyl] methylene]oxazoles, -thiazoles, and -imidazoles: novel dual 5-lipoxygenase and cyclooxygenase inhibitors with antiinflammatory activity. J Med Chem. 1994;37:322–8.
78. Inagaki M, Tsuri T, Jyoyama H, et al. Novel anti-arthritic agents with 1,2-isothiazolidine-1,1-dioxide (gamma-sultam) skeleton: cytokine suppressive dual inhibitors of cyclooxygenase-2 and 5-lipoxygenase. J Med Chem. 2000;43:2040–8.
79. Janusz JM, Young PA, Ridgeway JM, et al. New cyclooxygenase-2/5-lipoxygenase inhibitors. 3. 7-tert-butyl-2, 3-dihydro-3,3-dimethylbenzofuran

derivatives as gastrointestinal safe antiinflammatory and analgesic agents: variations at the 5 position. J Med Chem. 1998;41:3515–29.
80. Janusz JM, Young PA, Scherz MW, et al. New cyclooxygenase-2/5-lipoxygenase inhibitors. 2. 7-tert-butyl-2,3-dihydro-3,3-dimethyl benzofuran derivatives as gastrointestinal safe antiinflammatory and analgesic agents: variations of the dihydrobenzofuran ring. J Med Chem. 1998;41:1124–37.
81. Stanton BJ, Coupar IM. The effect of BW 755C and nordihydroguaiaretic acid in the rat isolated perfused mesenteric vasculature. Prostaglandins Leukot Med. 1986;25:199–207.
82. Radmark O, Malmsten C, Samuelsson B. The inhibitory effects of BW 755C on arachidonic acid metabolism in human polymorphonuclear leukocytes. FEBS Lett. 1980;110:213–5.
83. Gulbenkian AR, Fernandez X, Kreutner W, et al. Anaphylactic challenge causes eosinophil accumulation in bronchoalveolar lavage fluid of guinea pigs. Modulation by betamethasone, phenidone, indomethacin, WEB 2086; and a novel antiallergy agent, SCH 37224. Am Rev Respir Dis. 1990;142:680–5.
84. Argentieri DC, Ritchie DM, Ferro MP, et al. Tepoxalin: a dual cyclooxygenase/5-lipoxygenase inhibitor of arachidonic acid metabolism with potent anti-inflammatory activity and a favorable gastrointestinal profile. J Pharmacol Exp Ther. 1994;271:1399–408.
85. Waldman SA, Vitow C, Osborne B, et al. Pharmacokinetics and pharmacodynamics of tepoxalin after single oral dose administration to healthy volunteers. J Clin Pharmacol. 1996;36:462–8.
86. Knight EV, Kimball JP, Keenan CM, et al. Preclinical toxicity evaluation of tepoxalin, a dual inhibitor of cyclooxygenase and 5-lipoxygenase, in Sprague-Dawley rats and beagle dogs. Fundam Appl Toxicol. 1996;33:38–48.
87. Beers SA, Malloy EA, Wu W, et al. *N*-(5-substituted) thiophene-2-alkylsulfonamides as potent inhibitors of 5-lipoxygenase. Bioorg Med Chem. 1997;5:779–86.
88. Kirchner T, Argentieri DC, Barbone AG, et al. Evaluation of the antiinflammatory activity of a dual cyclooxygenase-2 selective/5-lipoxygenase inhibitor, RWJ 63556; in a canine model of inflammation. J Pharmacol Exp Ther. 1997;282:1094–101.
89. Le Filliatre G, Sayah S, Latournerie V, et al. Cyclo-oxygenase and lipoxygenase pathways in mast cell dependent-neurogenic inflammation induced by electrical stimulation of the rat saphenous nerve. Br J Pharmacol. 2001;132:1581–9.
90. Bailey PJ, Dallob AL, Allison DL, et al. Pharmacology of the dual inhibitor of cyclo-oxygenase and 5-lipoxygenase 3-hydroxy-5-trifluoromethyl-*N*-(2-(2-thienyl)-2-phe nyl-ethenyl)-benzo (b)thiophene-2-carboxamide. Arzneimittel forschung. 1988;38:372–8.
91. Tischler A, Bailey P, Dallob A, et al. L-652,343: a novel dual 5-lipoxygenase/cyclooxygenase inhibitor. Adv Prostaglandin Thromboxane Leukot Res. 1986;16:63–6.
92. Barr RM, Black AK, Dowd PM, et al. The *in vitro* 5-lipoxygenase and cyclooxygenase inhibitor L-652,343 does not inhibit 5-lipoxygenase *in vivo* in human skin. Br J Clin Pharmacol. 1988;25:23–6.
93. Gresele P, Arnout J, Deckmyn H, et al. L-652,343, a novel dual cyclo/lipoxygenase inhibitor, inhibits LTB4-production by stimulated human polymorphonuclear cells but not by stimulated human whole blood. Biochem Pharmacol. 1987;36:3529–31.
94. Chowdhury MA, Abdellatif KR, Dong Y, et al. Synthesis of celecoxib analogues possessing a N-difluoromethyl-1,2-dihydropyrid-2-one 5-lipoxygenase pharma cophore: biological evaluation as dual inhibitors of cyclooxy genases and 5-lipoxygenase with anti-inflammatory activity. J Med Chem. 2009;52:1525–9.
95. Chowdhury MA, Abdellatif KR, Dong Y, et al. Synthesis and biological evaluation of salicylic acid and N-acetyl-2-carboxy benzenesulfonamide regioisomers possessing a N-difluoromethyl-1,2-dihydropyrid-2-one pharmacophore: dual inhibitors of cyclo oxygenases and 5-lipoxygenase with anti-inflammatory activity. Bioorg Med Chem Lett. 2009;19:6855–61.
96. Yu G, Praveen Rao PN, Chowdhury MA, et al. Synthesis and biological evaluation of N-difluoromethyl-1,2-dihydropyrid-2-one acetic acid regioisomers: dual inhibitors of cyclooxygenases and 5-lipoxygenase. Bioorg Med Chem Lett. 2010;20:2168–73.
97. Wylie G, Appelboom T, Bolten W, et al. A comparative study of tenidap, a cytokine-modulating anti-rheumatic drug, and diclofenac in rheumatoid arthritis: a 24-week analysis of a 1-year clinical trial. Br J Rheumatol. 1995;34:554–63.
98. Aleo MD, Wang T, Giebisch G, et al. Model development and analysis of tenidap-induced proteinuria in the rat. J Pharmacol Exp Ther. 1996;279:1318–26.
99. Nelson SD. Structure toxicity relationships–how useful are they in predicting toxicities of new drugs? Adv Exp Med Biol. 2001;500:33–43.
100. Fouda HG, Avery MJ, Dalvie D, et al. Disposition and metabolism of tenidap in the rat. Drug Metab Dispos. 1997;25:140–8.
101. Lai Y, Ma L, Huang W, et al. Synthesis and biological evaluation of 3-[4-(amino/methyl sulfonyl)phenyl]methylene-indolin-2-one derivatives as novel COX-1/2 and 5-LOX inhibitors. Bioorg Med Chem Lett. 2010;20:7349–53.
102. Barbey S, Goossens L, Taverne T, et al. Synthesis and activity of a new methoxytetrahydropyran derivative as dual cyclooxygenase-2/5-lipoxygenase inhibitor. Bioorg Med Chem Lett. 2002;12:779–82.
103. Rao PN, Chen QH, Knaus EE. Synthesis and structure-activity relationship studies of 1,3-diarylprop-2-yn-1-ones: dual inhibitors of

cyclooxygenases and lipoxygenases. J Med Chem. 2006;49:1668–83.
104. Chen QH, Rao PN, Knaus EE. Synthesis and biological evaluation of a novel class of rofecoxib analogues as dual inhibitors of cyclooxygenases (COXs) and lipoxygenases (LOXs). Bioorg Med Chem. 2006;14:7898–909.
105. Laufer SA, Augustin J, Dannhardt G, et al. (6,7-Diaryldihydropyrrolizin-5-yl)Acetic acids, a novel class of potent dual inhibitors of both cyclooxygenase and 5-lipoxygenase. J Med Chem. 1994;37:1894–7.
106. Cossy J, Belotti D. Synthesis of ML-3000, an Inhibitor of Cyclooxygenase and 5-Lipoxygenase. J Org Chem. 1997;62:7900–1.
107. Rotondo S, Dell'Elba G, Krauze-Brzosko K, et al. Licofelone, a dual lipoxygenase-cyclooxygenase inhibitor, downregulates polymorphonuclear leukocyte and platelet function. Eur J Pharmacol. 2002;453:131–9.
108. Kulkarni SK, Singh VP. Licofelone–a novel analgesic and antiinflammatory agent. Curr Top Med Chem. 2007;7:251–63.
109. Alvaro-Gracia JM. Licofelone–clinical update on a novel LOX/COX inhibitor for the treatment of osteoarthritis. Rheumatology. 2004;43(Suppl 1):i21–5.
110. Vidal C, Gomez-Hernandez A, Sanchez-Galan E, et al. Licofelone, a balanced inhibitor of cyclooxygenase and 5-lipoxygenase, reduces inflammation in a rabbit model of atherosclerosis. J Pharmacol Exp Ther. 2007;320:108–16.
111. Koeberle A, Siemoneit U, Buhring U, et al. Licofelone suppresses prostaglandin E2 formation by interference with the inducible microsomal prostaglandin E2 synthase-1. J Pharmacol Exp Ther. 2008;326:975–82.
112. Samuelsson B, Morgenstern R, Jakobsson PJ. Membrane prostaglandin E synthase-1: a novel therapeutic target. Pharmacol Rev. 2007;59:207–24.
113. Hwang SH, Wagner KM, Morisseau C, et al. Synthesis and structure-activity relationship studies of urea-containing pyrazoles as dual inhibitors of cyclooxygenase-2 and soluble epoxide hydrolase. J Med Chem. 2011;54:3037–50.
114. Moser D, Wisniewska JM, Hahn S, et al. Dual-target virtual screening by pharmacophore elucidation and molecular shape filtering. ACS Med Chem Lett. 2012;3:155–8.
115. Jung O, Jansen F, Mieth A, et al. Inhibition of the soluble epoxide hydrolase promotes albuminuria in mice with progressive renal disease. PLoS One. 2010;5:e11979.
116. Cheng Y, Wang M, Yu Y, et al. Cyclooxygenases, microsomal prostaglandin E synthase-1, and cardiovascular function. J Clin Invest. 2006;116:1391–9.
117. Liedtke AJ, Keck PR, Lehmann F, et al. Arylpyrrolizines as inhibitors of microsomal prostaglandin E2 synthase-1 (mPGES-1) or as dual inhibitors of mPGES-1 and 5-lipoxygenase (5-LOX). J Med Chem. 2009;52:4968–72.
118. Koeberle A, Zettl H, Greiner C, et al. Pirinixic acid derivatives as novel dual inhibitors of microsomal prostaglandin E2 synthase-1 and 5-lipoxygenase. J Med Chem. 2008;51:8068–76.
119. Greiner C, Zettl H, Koeberle A, et al. Identification of 2-mercaptohexanoic acids as dual inhibitors of 5-lipoxygenase and microsomal prostaglandin E2 synthase-1. Bioorg Med Chem. 2011;19:3394–401.
120. Koeberle A, Rossi A, Zettl H, et al. The molecular pharmacology and *in vivo* activity of 2-(4-chloro-6-(2,3-dimethylphenylamino)pyrimidin-2-ylthio)octanoic acid (YS121), a dual inhibitor of microsomal prostaglandin E2 synthase-1 and 5-lipoxygenase. J Pharmacol Exp Ther. 2010;332:840–8.
121. Wu Y, He C, Gao Y, et al. Dynamic modeling of human 5-lipoxygenase-inhibitor interactions helps to discover novel inhibitors. J Med Chem. 2012;55:2597–605.
122. Yarla NS, Pathuri G, Gali H, et al. Discovery and development of a novel mPGES-1/5-lox dual inhibitor LFA-9 for prevention and treatment of chronic inflammatory diseases. J Inflamm Res. 2020;13:1261–78.
123. McGee JE, Fitzpatrick FA. Erythrocyte-neutrophil interactions: formation of leukotriene B4 by transcellular biosynthesis. Proc Natl Acad Sci USA. 1986;83:1349–53.
124. Chen Z, Wu Y, Liu Y, et al. Discovery of dual target inhibitors against cyclooxygenases and leukotriene A4 hydrolyase. J Med Chem. 2011;54:3650–60.
125. Hibi S, Okamoto Y, Tagami K, et al. Novel dual inhibitors of 5-lipoxygenase and thromboxane A2 synthetase: synthesis and structure-activity relationships of 3-pyridylmethyl-substituted 2-amino-6-hydroxybenzothiazole derivatives. J Med Chem. 1994;37:3062–70.
126. Hibi S, Okamoto Y, Tagami K, et al. Structure-activity relationships of (E)-3-(1,4-benzoquinonyl)-2-[(3-pyridyl)-alkyl]-2-propenoic acid derivatives that inhibit both 5-lipoxygenase and thromboxane A2 synthetase. J Med Chem. 1996;39:3148–57.
127. Marino JPBCA, Eidam P, Mcatee JJ. sEH and 11β-HSD1 dual inhibitors. 2010/011917 A1. WO. 2008 Jul 25.
128. Liu JY, Park SH, Morisseau C, et al. Sorafenib has soluble epoxide hydrolase inhibitory activity, which contributes to its effect profile *in vivo*. Mol Cancer Ther. 2009;8:2193–203.
129. Warner TD, Mitchell JA. COX-2 selectivity alone does not define the cardiovascular risks associated with non-steroidal antiinflammatory drugs. Lancet. 2008;371:270–3.
130. Marnett LJ. The COXIB experience: a look in the rearview mirror. Annu Rev Pharmacol Toxicol. 2009;49:265–90.
131. Gilbert NC, Bartlett SG, Waight MT, et al. The structure of human 5-lipoxygenase. Science. 2011;331:217–9.
132. Schmelzer KR, Inceoglu B, Kubala L, et al. Enhancement of antinociception by coadministra-

tion of nonsteroidal anti-inflammatory drugs and soluble epoxide hydrolase inhibitors. Proc Natl Acad Sci USA. 2006;103:13646–51.
133. Scarpelli R, Sasso O, Piomelli D. A double whammy: Targeting both fatty acid amide hydrolase (FAAH) and cyclooxygenase (COX) to treat pain and inflammation. ChemMedChem. 2016;11(12):1242–51.
134. Seierstad M, Breitenbucher JG. Discovery and development of fatty acid amide hydrolase (FAAH) inhibitors. J Med Chem. 2008;51(23):7327–43.
135. Bracey MH, Hanson MA, Masuda KR, et al. Structural adaptations in a membrane enzyme that terminates endocannabinoid signaling. Science. 2002;298(5599):1793–6.
136. Paria BC, Deutsch DD, Dey SK. The uterus is a potential site for anandamide synthesis and hydrolysis: differential profiles of anandamide synthase and hydrolase activities in the mouse uterus during the periimplantation period. Mol Reprod Dev. 1996;45(2):183–92.
137. Fowler CJ, Tiger G, Stenstrom A. Ibuprofen inhibits rat brain deamidation of anandamide at pharmacologically relevant concentrations. Mode of inhibition and structure-activity relationship. J Pharmacol Exp Ther. 1997;283(2):729–34.
138. Fowler CJ, Janson U, Johnson RM, et al. Inhibition of anandamide hydrolysis by the enantiomers of ibuprofen, ketorolac, and flurbiprofen. Arch Biochem Biophys. 1999;362(2):191–6.
139. Fowler CJ, Borjesson M, Tiger G. Differences in the pharmacological properties of rat and chicken brain fatty acid amidohydrolase. Br J Pharmacol. 2000;131(3):498–504.
140. Tiger G, Stenstrom A, Fowler CJ. Pharmacological properties of rat brain fatty acid amidohydrolase in different subcellular fractions using palmitoylethanolamide as substrate. Biochem Pharmacol. 2000;59(6):647–53.
141. Cocco MT, Congiu C, Onnis V, et al. Synthesis of ibuprofen heterocyclic amides and investigation of their analgesic and toxicological properties. Eur J Med Chem. 2003;38(5):513–8.
142. Palermo G, Favia AD, Convertino M, et al. The molecular basis for dual fatty acid amide hydrolase (FAAH)/cyclooxygenase (COX) inhibition. ChemMedChem. 2016;11(12):1252–8.
143. Dutta R, Lunzer MM, Auger JL, et al. A bivalent compound targeting CCR5 and the mu opioid receptor treats inflammatory arthritis pain in mice without inducing pharmacologic tolerance. Arthritis Res Ther. 2018;20(1):154.
144. Gaba M, Singh S, Mohan C. Benzimidazole: an emerging scaffold for analgesic and anti-inflammatory agents. Eur J Med Chem. 2014;76:494–505.
145. Kaur G, Kaur M, Silakari O. Benzimidazoles: an ideal privileged drug scaffold for the design of multitargeted anti-inflammatory ligands. Mini Rev Med Chem. 2014;14(9):747–67.
146. Kaur G, Silakari O. Benzimidazole scaffold based hybrid molecules for various inflammatory targets: synthesis and evaluation. Bioorg Chem. 2018;80:24–35.
147. Veerasamy R, Roy A, Karunakaran R, et al. Structure-activity relationship analysis of benzimidazoles as emerging anti-inflammatory agents: an overview. Pharmaceuticals (Basel). 2021;14(7):663.
148. Paramashivappa R, Phani Kumar P, Subba Rao PV, et al. Design, synthesis and biological evaluation of benzimidazole/benzothiazole and benzoxazole derivatives as cyclooxygenase inhibitors. Bioorg Med Chem Lett. 2003;13(4):657–60.
149. Cheng Y, Hitchcock SA. Targeting cannabinoid agonists for inflammatory and neuropathic pain. Expert Opin Investig Drugs. 2007;16(7):951–65.
150. Sondhi SM, Singh N, Kumar A, et al. Synthesis, anti-inflammatory, analgesic and kinase (CDK-1, CDK-5 and GSK-3) inhibition activity evaluation of benzimidazole/benzoxazole derivatives and some Schiff's bases. Bioorg Med Chem. 2006;14(11):3758–65.
151. Mader M, de Dios A, Shih C, et al. Imidazolyl benzimidazoles and imidazo[4,5-b]pyridines as potent p38alpha MAP kinase inhibitors with excellent in vivo antiinflammatory properties. Bioorg Med Chem Lett. 2008;18(1):179–83.
152. Guo Q, Chandrasekhar J, Ihle D, et al. 1-Benzylbenzimidazoles: the discovery of a novel series of bradykinin B_1 receptor antagonists. Bioorg Med Chem Lett. 2008;18(18):5027–31.
153. Banoglu E, Çalişkan B, Luderer S, et al. Identification of novel benzimidazole derivatives as inhibitors of leukotriene biosynthesis by virtual screening targeting 5-lipoxygenase-activating protein (FLAP). Bioorg Med Chem. 2012;20(12):3728–41.
154. Ognyanov VI, Balan C, Bannon AW, et al. Design of potent, orally available antagonists of the transient receptor potential vanilloid 1. Structure-activity relationships of 2-piperazin-1-yl-1H-benzimidazoles. J Med Chem. 2006;49(12):3719–42.
155. Rami HK, Gunthorpe MJ. The therapeutic potential of TRPV1 (VR1) antagonists: clinical answers await. Drug Discov Today Ther Strateg. 2004;1:97–104.

12 Polypharmacology in Drug Design and Discovery—Basis for Rational Design of Multitarget Drugs

Abstract

Chapter 3 through Chap. 11 introduces the current status of clinical applications of polypharmacology to disease therapy. This chapter will describe the applications of polypharmacology to drug discovery, mainly from the theoretical account. Specifically, it will explain the theoretical basis for the rational design of new drugs of the "single agent on multiple targets for single disease" category, or the rational design of MTD drugs. Seven essential strategies for MTD drug discovery, including sequence-based epigenetic drugs, sequence-based small-molecule drugs, function-based epigenetic drugs, function-based small-molecule drugs, function-based macromolecule drugs, naturally occurring MTD drugs, and repurposed MTD drugs, will be described in detail with real examples. The contents of this chapter lay the foundation for better understanding of Section II chapters on the methodologies of MTD polypharmacology-based drug discovery and development. Several new concepts on the rational design and discovery of MTDs are proposed by us and presented in this chapter.

Keywords

Drug design/discovery · Epigenetic polypharmacology · Sequence-based MTDs · Medicinal chemistry-based MTDs · Biopharmaceutical-based MTDs

12.1 Sequenced-Based Strategy for Designing Epigenetic MTDs (Epi-MTDs)

The sequence-based MTD approach implies the discovery and application of therapeutic drugs that act by sequence-complementary mechanisms to base pair the target DNAs, RNAs, or peptides/proteins. It offers remarkable opportunities for the easy and precise design of MTD drugs able to act on multiple targets (multitargeticity) with preselected specificities (polyspecificity), manifesting the "selective non-selectivity" and representing the core and essence of polypharmacology. This approach is virtually the only one that can afford exact polyspecificity or preselected specificities.

The majority of human diseases are influenced by genetic factors; yet some of disorders are determined by single gene defects [1]. For monogenic diseases, one therapeutic avenue, gene therapy which works by substituting the disease-associated gene with the "healthy" version of the gene or gene products has been a focus of drug discovery for the past decades. The very first approved human gene therapeutic was approved in 1990 to treat adenosine deaminase deficiency [2]. The advantage of gene therapy is that it offers successful cure of monogenetic diseases, as demonstrated in patients with severe combined immunodeficiency (SCID) [3]. Currently, gene therapies for more complex disorders like polygenic neurodegenerative diseases [4] and cancers are in active development [5].

Z. Wang, B. Yang, *Polypharmacology*, https://doi.org/10.1007/978-3-031-04998-9_12

However, for most of polygenic diseases, the strength of gene therapy targeting single genes may not match its ambitions; instead, targeting multiple genes or gene products is a better, more rational approach. While gene therapy and epigenetic therapy both belong to sequence-based therapeutic strategy, the latter offers a broader range of multitargeting for polygenic and/or multifactorial diseases. Herein, the following subsections will focus on epigenetic therapy as a representative of sequence-based MTD approach.

12.1.1 Epigenetic Polypharmacology

As a branch of Polypharmacology, Epigenetic Polypharmacology is the converge of Epigenetics and Polypharmacology. Multitarget epigenetic drugs (epi-MTDs) are the main theme of Epigenetic Polypharmacology that is a new frontier for epi-drug discovery [6].

Epi-MTD include two major classes:

(a) Sequence-based epi-MTDs, including DNA/RNA sequence-based and amino acid sequence-based MTD drugs.
(b) Function-based epi-MTDs, mainly include "small-molecule drugs" and "macromolecular drugs." Before jumping into these topics, we need to first educate ourselves a bit about Epigenetics and Epigenetic Polypharmacology.

12.1.1.1 Epigenetics

Genetic variation and epigenetic modification of DNA, in both protein-coding regions and in the non-coding intron region of genes, are both critically involved in the pathogenesis of human diseases because they can result in phenotypic variations. It is now clear that the growth, development, and evolution information acquired by the offspring of organisms from the parental genome do not only depend on the gene sequence but also on the changes in the process of gene expression. While genetic variations refer to the differences in DNA among individuals or populations caused by gene mutation and genetic recombination as the traditional genetic basis for inheritance, epigenetic modifications create the heritable and reversible changes in gene expression and the associated phenotypes without affecting the underlying DNA sequence of the genes. The term "epigenetics" has also been used to describe processes which have not been demonstrated to be heritable, such as some forms of histone modification; there are therefore attempts to redefine "epigenetics" in broader terms that would avoid the constraints of requiring heritability [7], though such a definition is not universally accepted and are still subject to debate [8]. The changes in epigenetics can be stably transmitted during growth and development and have reversible potential. Throughout the course of life, genetics provides information on the synthesis of various proteins including epigenetic regulators, while epigenetics provides instructions on how to apply genetic information.

Epigenetic regulation of gene expression can be roughly divided into two routes: regulation during gene transcription (transcriptional regulation) and regulation after gene transcription (post-transcriptional regulation). The former mainly studies the causes of changes in offspring gene expression caused by environmental factors acting in parents on the chromatin remodeling including DNA methylation and histone modifications, and gene silencing and RNA editing. And the latter primarily involves the regulatory mechanisms of RNA, including ncRNAs, asRNA, riboswitch, and RNA methylation. Epigenetics is a natural and necessary addition to the central dogma of molecular biology that the flow of genetic information follows a simple route from DNA to RNA to protein. It tells us that the genetic information can be controlled by genomic and mRNA modifications for more precise and adaptive manifestation and the changes in gene expression caused by these modifications are heritable [9].

The epigenetic alterations are governed by at least three main mechanisms:

(a) Covalent modifications to the cytosine residues of DNA.
(b) Histone covalent chemical modifications such as acetylation, methylation, and phosphorylation are also known on the whole as chromatin remodeling. Proteins taking part in chromatin remodeling complexes can add,

remove, or read such epigenetic modifications, and are classified as "writers," "erasers," or "readers," respectively. These proteins work together to determine either transcription or repression of specific genes; thereby the expression of genes responsible for cellular functions and the dysfunction of these proteins can therefore cause many human diseases [10]. These epigenetic changes may last through cell divisions for the duration of the cell's life, and may also last for multiple generations, even though they do not involve changes in the underlying DNA sequence of the organism.

(c) Gene regulation by noncoding RNAs (ncRNAs).

Apparently, epigenetic control of transcription is regulated by enzymes that mediate covalent modifications at gene-regulatory regions and histone proteins around which chromosomal DNA is wound. Many of the enzymes that mediate chromatin epigenetic reactions are deregulated in diseases such as cancer [11]. Moreover, gene expression is also under the tight control by epigenetic mechanisms at the post-transcriptional level.

12.1.1.2 Epigenetic Drugs (Epi-Drugs)

Recent studies have shown that epigenetics plays a central role in many types of diseases, including cardiovascular diseases, neurological diseases, metabolic disorders, inflammatory diseases and cancer, and epigenetic events that are involved throughout the entire human lifecycle, from embryogenesis to adulthood [12]. Because of the reversible nature of epigenetic events, researchers postulate that inhibition of epigenetic therapy could be of valuable strategy for the treatment of a wide spectrum of human diseases. Epigenetic therapy refers the use of drugs or other epigenome-influencing techniques to treat medical conditions. The modulation of epigenetic processes is presently considered an innovative and promising therapeutic and drug discovery strategy, and the goal of the MTD epi-drug approach consists in the development of compounds able to simultaneously and (often) reversibly bind different specific epi-targets [13–16]. Epigenetic drug discovery is a rapidly advancing field, and drugs targeting epigenetic enzymes are in the clinic for the treatment of hematological cancers. Approved and experimental epigenetic drugs are undergoing clinical trials in combination with other therapeutic agents via fused or linked pharmacophores in order to benefit from synergistic effects of polypharmacology [17]. In fact, in some neurological diseases and cancers, epigenetic drugs are already in use [18, 19]. For example, nucleoside-like compounds are the oldest form of methylation inhibitors as epi-drugs, and several of these epi-drugs have been FDA approved for the treatment of certain cancers, such as 5-azacytidine [20], zebularine [21], and MG98 (antisense oligonucleotides to inhibit methylation) [22]. Many other epi-drugs including multitarget epi-drugs or MTD epi-drugs have entered advanced stage of clinical trials [6, 17].

12.1.1.3 Multitarget Epigenetic Drugs (Epi-MTDs)

Sequence-based MTD epigenetic drugs refer to the compounds that are designed according to the unique nucleotide or amino acid sequences to modify the multiple target genes or gene products at fault or the endogenous molecules that act by regulating multiple disease-related genes and/or gene products through sequence-based complementary mechanisms. Existing pharmaceutical agents to treat diseases through epigenetic mechanisms are nonspecific. Epigenetic drugs, such as the DNA methyltransferase (DNMT) inhibitor 5-aza-2′-deoxycytidine, and histone deacetylase (HDAC) inhibitors, such as sodium valproate, depsipeptide, tetrapeptide suberoylanilide hydroxamic acid (SAHA), and CI-994 (N-acetyldinaline) [12], have all been used to treat various types of cancer; however, their modes of action alter chromatin structure throughout the genome. Histone- or DNA-modifying enzymes attached to gene-specific zinc finger proteins may in the future lead to targeted treatments that will specifically bind to targeted epimutation sites [23].

Non-coding RNAs (ncRNAs) provide the best of all opportunities and resources for sequence-based MTD epi-drug discovery and rational design, not only because each individual ncRNA in general acts on multiple disease-related molecules simultaneously in a specific signaling pathway or in different pathways within a signaling network through different mechanisms but also each individual ncRNA may be expressed in a tissue-specific manner. ncRNAs are an emerging superfamily of transcripts which are coded by the genome but essentially lack protein/peptide-coding potential and capacity [24, 25]. Out of 3 billion base pairs in human genome only ~2% code for proteins; and out of 180,000 transcripts in human cells, about 20,000 code for protein, and remaining 160,000 are non-coding transcripts. Though not translated, ncRNAs are crucial players in a variety of cellular and physiological functions through affecting normal gene expression, thereby controlling disease progression [26]. ncRNAs include microRNAs (miRNAs), long noncoding RNAs (lncRNAs), circular RNAs (circRNAs), endogenous antisense RNA (asRNAs), PIWI-interacting RNAs (piRNAs), small nucleolar RNAs (snoRNAs), and extracellular RNAs [27, 28]. While ncRNAs lack potential to encode proteins, they can affect the expression of other genes through a variety of mechanisms. As a general rule, ncRNAs exert their function through sequence-based mechanisms (i.e., classical base pairing) or forming complex secondary and tertiary structures, including bulges, hairpins, stem-loops, duplex-, triplex-, and quadruplex-motifs, which allow interactions with both proteins and other nucleic acids [27, 28]. The ability of ncRNAs to control gene expression makes them a new class of targets for lead identification and development. However, we are only beginning to understand the nature and extent of the involvement of ncRNAs in disease, and uncertainty about the biological functionalities, pathophysiological roles, and mechanisms of action makes ncRNA-based drug discovery challenging.

From the therapeutic point of view, RNA/DNA sequence-based epi-MTD discovery can be viewed as oligonucleotide- or polynucleotide-based approach as endogenous or synthetic oligonucleotide or polynucleotide fragments function by the sequence complementary mechanism based on Watson–Crick base pairing to the functional domain and/or unique motif of its target genes and an individual oligonucleotide or polynucleotide fragment may target multiple genes.

Collectively, the sequence-based epi-MTD strategy bears some clear strengths.

(a) It offers the opportunity for the maximum precision of desired specificities towards multiple target genes or gene products essentially using their sequences as the templates to yield the property of specific non-specificity or polyspecificity.
(b) It is relatively easy and efficient to design a multitarget nucleotide fragment or peptide and to hybrid multiple nucleotide fragments or peptides by incorporating linkers into neighboring fragments.
(c) It is relatively easy and straightforward to experimentally and preclinically validate the effectiveness of sequence-based epi-MTDs and identify the side effects in cells and animal models.
(d) Sequence-based epi-MTDs could be conjugated with certain "guided missile" molecules (e.g., antibodies) or moieties (e.g., cholesterol) to improve, at least to a certain extent, the penetration into targeted cells and the association with target molecules in the cell membrane or inside the cells.
(e) Sequence-based epi-MTDs could be rapidly engineered to fit the conditions of individual patients based on their pharmacogenomic characteristics thereby being a patient-orientated precision medicine approach.

Nonetheless, the sequence-based epi-MTD strategy also has some limitations.

(a) It is difficult to achieve the industrial scale of large quantity of sequence-based epi-MTDs.
(b) Delivery of sequence-based epi-MTDs into living cells, particularly the target cells within the diseased tissues, could be rather difficult to achieve in spite of the possibility of chemical conjugation or bioconjugation as mentioned above.

(c) Industrial production of sequence-based epi-MTDs could be of very high costs.

12.1.2 RNA Sequence-Based Strategy for Designing Epi-MTDs

All functional RNAs, including both coding RNAs (mRNA) and noncoding RNAs (ncRNAs; regulatory RNAs), can be the targets for RNA sequence-based MTD epi-drug discovery. Currently, many pharmaceutical and biotech companies are developing RNA-based therapeutics to specifically regulate disease-causing genes and their variants. A handful of RNA-based products has successfully been approved for use in the clinic, with many more in varying stages of the drug development pipeline, and the majority of these products are synthetic ncRNAs [27–30].

Both mRNA and ncRNAs are encoded in the genome and have essential functions in the regulation of gene expression. Owing to the increasing accuracy and sensitivity of sequencing technologies, our understanding of the specificity of the expression of these RNAs has been rapidly expanding and deepening. Many mRNAs and ncRNAs are differentially expressed in disease states and are thought to contribute to disease development and progression. Therefore, they have great potential as diagnostic markers, therapeutic targets, or even drugs [30]. Many functions of RNA, not only as a messenger of genetic information, but as essential and specific regulators of numerous steps of gene expression, are becoming increasingly apparent. Especially, our understanding of ncRNA-based regulatory mechanisms created opportunities to develop potential therapeutics that harness the endogenous regulatory machinery and is designed accurately and specifically to regulate the target of choice by complementary base-pairing. Many ncRNA-based mechanisms of gene regulation depend on small, relatively stable, RNA molecules (e.g., miRNA). As these molecules can also be isolated from endosomes and microvesicles [31], they have great potential for therapeutic use or as biomarkers for disease.

Sequencing technologies have also drastically improved the identification of new drug targets. However, a new tool, Clustered Regularly Interspersed Short Palindromic Repeats (CRISPR), is also promising to accelerate drug target identification [32]. The CRISPR complex is composed of a nuclease (e.g., CRISPR-associated protein 9 (Cas9)) and a guide RNA (gRNA) which specifically binds to and irreversibly cleaves complementary DNA sequences [33]. Synthesizing specific gRNA will direct the complex to cleave desired regions of the genome and is a suitable tool for accurate target identification and replicating disease mutations. This tool could, therefore, rapidly identify new therapeutic targets [34]. Additionally, CRISPR's ability to manipulate DNA could be directly used as a therapeutic and has shown potential in combating HIV and other infections [35]. However, CRISPR-based therapies are still in their infancy.

RNA sequence-based epi-MTDs can be roughly divided into two major categories: endogenously existing or native RNAs and rationally designed or artificial RNA fragments with each the two including several subclasses of theoretically druggable ncRNAs or artificial RNAS fragments.

12.1.2.1 Endogenous RNAs as Epi-MTDs

Endogenously existing/native RNAs are those being found in human cells and are known to be deregulated in a variety of pathological states. The deregulation is reflected by either abnormal upregulation or downregulation or both of ncRNAs and the pathological alterations of downstream coding genes (mRNAs), leading to either abnormal gain-of-function or loss-of-function phenotypes. As therapeutic strategies, we can either use endogenous ncRNAs for replacement therapy if their abnormal downregulation is disease-related or artificial RNA fragments as inhibitors to knockdown the disease-causing or disease-related ncRNAs. For multifactorial diseases related to epigenetic anomalies, multiple gene targeting is necessary to overcome escape mutants or isoform redundancy, either by applying multiple native RNAs or a multiplex native RNA or by introducing multiple RNA inhibitors or a complex artificial RNA inhibitor to knockdown multiple RNA targets. The following subsections will focus on miRNA, lncRNA, circRNA, and asRNA sequence-based epi-MTDs.

miRNAs Sequence-Based epi-MTDs—miRNA Replacement Therapy

MicroRNAs (miRNAs or miRs) are members of ncRNAs that range in size from 17 to 25 nucleotides [36–41]. miRNAs are endogenous to the human genome and are derived from regions of RNA transcripts called pre-miRNAs that possess short hairpins or stem-loop structures with the "stem" forming a double-stranded RNA (dsRNA) region, similar to but distinct from linear longer dsRNA for siRNAs generation [41]. Plant miRNAs usually have near-perfect pairing with their mRNA targets, which induce gene repression through cleavage of the target transcripts. In animal cells, however, miRNAs recognize their target genes (mRNAs) and imperfectly bind to the targets, with a minimum binding requirement of nucleotides 2–8 of the miRNA known as the seed sequence [26, 40, 41]. A miRNA is a double-stranded fragment with the strand carrying seed site as the "guide strand" and the complementary strand as "passenger strand" [37]. The guide strand of a miRNA first binds to an Ago protein to form RISC that then guides the miRNA-Ago complex to bind to complementary sites in target transcripts [36]. Once bound, the miRNA regulatory complex represses translation of the target [41]. In essence, miRNAs act by antisense mechanism to base-pair mRNAs forming miRNA:mRNA duplex so as to induce repression of gene expression at the post-transcriptional level. The partial complementarity of miRNAs to their target mRNAs is usually not sufficient to induce cleavage of the target mRNAs as plant miRNAs and siRNAs do. If siRNAs are viewed as STD agents, then miRNAs are equivalent to MTD drugs.

In addition to direct binding to target mRNAs, miRNAs can also directly interact with proteins. A recent study demonstrated that a cardiac enriched miR-1 can physically bind with cardiac membrane proteins, including an inward rectifier K^+ channel α-subunit Kir2.1 [42], which was observed in mouse, guinea pig, canine, and human cardiomyocytes. Through this interaction, miR-1 quickly and significantly suppressed I_{K1} at sub-pM concentration, which is close to endogenous miRNA expression level. Acute presence of miR-1 depolarized resting membrane potential and prolonged action potential duration in cardiomyocytes. In this study, three miR-1-binding residues on the C-terminus of Kir2.1 were identified. Mechanistically, miR1 binds to the pore-facing G-loop of Kir2.1 through the core sequence AAGAAG, which is outside its miRNA interference seed region. This biophysical modulation is involved in the dysregulation of gain-of-function Kir2.1–M301K mutation in short QT or atrial fibrillation. Moreover, the study revealed that an arrhythmia-associated human single nucleotide polymorphism of miR-1 (hSNP14A/G) specifically disrupts the biophysical modulation while retaining the miRNA interference function. The most remarkable finding is that miR1 but not hSNP14A/G relieved the hyperpolarized resting membrane potential in miR1-deficient cardiomyocytes, improved the conduction velocity, and eliminated the high inducibility of arrhythmia in miR1-deficient hearts *ex vivo* [42]. It is likely that such a protein-binding mechanism is common to miRNAs and therefore might be of therapeutic implications.

In general, miRNAs have the following unique characteristics:

(a) One striking characteristic of miRNAs is the ability of an individual miRNA sequence to regulate multiple genes that have corresponding seed match sites, often in the same biological process [36–41, 43]. There are an average of 300 conserved targets predicted for each evolutionarily conserved miRNA [40, 41], and miRNAs can potentially regulate the expression of many genes in a tissue- or cell-specific manner [44]. miRNAs, therefore, have the potential to regulate an entire signaling pathway known to be misregulated in a pathogenic state, and artificial introduction or depletion of a single miRNA can directly alter the abundance of hundreds of proteins [45, 46]. On the other hand, each single mRNA may bind to a variety of miRNAs, either simultaneously or in context-dependent fashion [37]. With nearly 60% of human protein-coding genes having conserved seed match sites in their 3′ UTRs [40, 41], miRNAs are able to form regulatory networks that modulate the expression of many genes in concert [46–48]. These facts imply that each miRNA can be viewed as a multi-target "drug" or a MTD epi-drug. The multi-target property of miRNAs is ascribed to the nature of their binding with partial base pairing, and creates a vast and intricate regulatory network, making them very suitable therapeutic candidates [49].

(b) According to Alles et al [50] and Plotnikova et al [51], human genome encodes approximately 2300–2600 mature miRNAs (miRBase v22), each with a distinct expression pattern and a specific set of target miRNAs. Therefore, almost every biological process is controlled by miRNA-mediated mechanisms of regulation [46].

(c) Target recognition for metazoan miRNAs is based on partial complementarity between the miRNA and the 3′ untranslated region (UTR) of each target mRNA, although targeting is also observed in open reading frames (ORFs) at lower frequency [36, 52–54]. In particular, perfect or near-perfect Watson–Crick base pairing occurs between the miRNA seed region (nucleotides 2–8) and the 3′ UTRs of multiple target transcripts, while complementarity is incomplete across the remaining miRNA sequence. Pairing between the seed region and 3′ UTR is generally necessary, and in some cases sufficient, for repression [52–55], and conservation of seed region base pairing is a key predictor of miRNA targeting [53, 54]. Further evidence for the importance of the seed region for target recognition was provided by the crystal structure of the miRNA effector protein Argonaute 2, which revealed that the miRNA seed region is held in an A-form helix suitable for pairing with the mRNA [56]. Beyond the seed region, 3′ UTR context, secondary structure and accessibility have been suggested as additional contributors to targeting [48, 57–60], while base pairing at the miRNA 3′ end can increase site efficacy [59] or compensate for imperfect seed matches [52].

(d) Many miRNAs are epigenetically regulated. About 50% of miRNA genes are associated with CpG islands [61], which may be repressed by epigenetic methylation. Transcription from methylated CpG islands is strongly and heritably repressed [62]. Other miRNAs are epigenetically regulated by either histone modifications or by combined DNA methylation and histone modification [61].

(e) Notably, the target regions for particular miRNAs commonly cluster together, thus resulting in the cooperative and coordinated repression of multiple mRNA transcripts [59]. Moreover, miRNAs can exert strong biological effects with significant phenotypic manifestations [63, 64]. Numerous studies have confirmed that miRNAs have an inherent and critical role in development and cell differentiation, proliferation, and survival [65]. Gene regulation by miRNA mechanisms is essential in cell maintenance, homeostasis, and responding to dynamic environments. Deregulation of miRNA expression and function due to genomic alterations or changes to miRNA biogenesis, such as mutations, deletions, amplifications, transcriptional changes, and enzymatic differences, are important factors in pathogenesis and progression of many human diseases, such as cardiovascular disease, cancer, and neurological disorders [41, 66–73].

(f) Furthermore, as changes in miRNA expression can be indicative of mechanisms of disease development, progression and tissue of disease origin, quantifying miRNA changes in a patient could provide crucial insight required for diagnosing and tailoring treatment regimens to the individual [74, 75]. miRNAs are ubiquitously expressed, can be extracted from body fluids, and quantification techniques are fast and robust. The advances in RNA chemical modifications and delivery systems have led to successful ASO and siRNA therapeutics and have also been applied to the development of miRNA drugs [49, 76]. These attributes of miRNAs further increase their potential as essential therapeutic tools as we move towards precision medicine [77, 78].

As individual miRNAs can be either over or under expressed in different disease states, miRNA therapeutics aim to either boost a particular miRNA expression (miRNA gain-of-function) or reduce the level of a specific miRNA (miRNA loss-of-function). That is, miRNAs could be artificially engineered or manipulated in both of their biosynthesis, expression, stability, and sequence to alter their functional states thereby the expression of their target protein-coding genes. We proposed the "MicroRNA-Interference" concept (miRNAi) [37, 38]. miRNAi is different from RNA interference (RNAi) in that it artificially alters gene expression through manipulating miRNAs and can either generate "gain-of-function" or "loss-of-function" changes of miRNAs and their target genes, whereas the only target of RNAi is mRNAs and that the only outcome of RNAi is silencing of mRNAs (loss-of-function). For example, in cancer, the majority of miRNAs are expressed at a lower level when compared to healthy tissue. Therefore, there is great investment in developing miRNAs as antitumor drugs as replacement therapy. Many miRNAs regulate the cell cycle and are considered to have tumor-suppressing functions. Replacement of these miRNAs may complement or augment current cancer treatments. For example, normal miRNA regulation is critical to controlling the expression of oncogenes, for example, let-7 regulates Ras, Myc, and HMGA-2 [79, 80]. p53 expression is promoted by the miR-34 family, which is commonly deleted from cancers. miR-15 and 16 are frequently deleted in B-cell lymphocytic leukemia and prostate cancer. However, other miRNAs are encoded at fragile sites in the genome where chromosomes can break, contributing to cancer development [81]. Furthermore, when genomic instability is a component of disease development, mutations accumulate in the genome. If a mutation occurred genomic sequences encoding miRNAs or their associated regulatory sites, this can have a drastic effect on the function of the miRNA, either eliminating its ability to bind to its targets, redirecting the miRNA to bind to and repress a new set of transcripts, or losing its tissue-specific expression pattern. In order to increase the level of a miRNA whose level has been lowered by the development or progression of disease [82, 83], a number of approaches can be used.

(a) miRNA mimics can be used. miRNA mimics are synthetically produced RNA molecules that have the same sequence as the under expressed endogenous miRNA [37, 84]. These mimics can re-establish normal expression and function of the miRNA of interest.
(b) Alternatively, short hairpin RNAs (shRNAs) or pre-miRNA mimics can also be used as a miRNA replacement tool. These synthetic RNA molecules are processed by the endogenous processing machinery of the miRNA biogenesis pathway into mature miRNAs [81, 85, 86]. As they are processed at a rate acceptable to the cell, they minimize toxicity, while still effectively regulating their targets [87, 88].
(c) Several miRNAs can be polycistronically encoded to be under the control of a single promoter, producing different functional regulatory complexes with low cytotoxicity [88–90], targeting a wide range of targets in one treatment. This can be viewed as a multiple multitarget miRNA drug.
(d) A study reports a flexible multiplexed RNAi and transgene platform using endogenous intronic primary miRNAs (pri-miRNAs) as a scaffold located in the green fluorescent protein (GFP) as a model for any functional transgene [91]. The multiplexed intronic miRNA/GFP transgene platform was designed to co-express multiple small RNAs within the polycistronic cluster from a Pol II promoter at more moderate levels to reduce potential vector toxicity.

Yet, despite being lauded as a potential drug target or therapeutic nearly since their discovery in 1993 and many studies showing great therapeutic promise [29, 42, 92–98], some obstacles remain in this field and a miRNA-based drug is yet to reach the clinic. It should be noted that the ubiquitous nature of miRNA binding, to perhaps as yet undiscovered targets, may have unanticipated and adverse off-target effects [99–101]. Furthermore, miRNA:mRNA interactions are incompletely understood as a well characterized miRNA:mRNA association displays poor seed region binding [102], while high seed sequence complementarity does not guarantee physiologically relevant regulation of the mRNA target [103]. Therefore, there is currently a relatively small number of experimentally validated miRNA:mRNA interactions and a need for more accurate bioinformatic prediction methodologies [104]. Many bioinformatic tools currently exist to make these predictions. However, as miRNA regulatory networks are highly complex, the accuracy rate of these predictions is low (approximately 26%) [105]. It is, therefore, critical to validate the predicted miRNA targets in vitro and in vivo, to assess their suitability as a drug target or therapeutic. Many high-throughput techniques have been developed to facilitate in vitro validation. Identifying and validating the suitability of a miRNA as a therapeutic or target are still, therefore, major challenges in the field. Moreover, miRNA replacement therapy must deliver the drug at effective, but safe levels. Overexpression of the drug can cause hepatotoxicity, organ failure and death [106]. As the drug depends on endogenous machinery to facilitate the action, and processing if applicable, of the drug, this limits the amount of the drug a cell can tolerate [106–110]. The potential toxicity caused by miRNA replacement therapies may be mitigated by using local delivery methods.

miRNA mimic drugs are in development by miRagen Therapeutics Inc. Remlarsen (MRG-201), delivered by intradermal injection, is designed to mimic miR-29b, a negative regulator of extracellular matrix deposition, in order to decrease fibrous scar formation. miR-29b levels are decreased in fibrotic diseases and it is hoped that this drug will help patients with a predisposition to forming keloid scars. Remlarsen also has potential uses for pathological cutaneous fibrosis and idiopathic pulmonary fibrosis [111].

EnGeneIC Limited has successfully completed phase 1 clinical trials of a miRNA mimic drug. This drug is delivered as a TargomiR, encapsulated in a bacterial minicell. This particular TargomiR, MesomiR 1, mimics miR-16, which is targeted to epidermal growth factor (EGFR)-expressing cancer cells via an EGFR-bispecific antibody and shows success treating malignant pleural mesothelioma and non-small-cell lung cancer [112]. Phase 2 clinical trials for mesomiR 1 are planned for the near future [113].

Unfortunately, severe reactions to some miRNA mimic drugs have been recorded in clinical trials, which is of great concern to the field. Mirna Therapeutics Inc. (Synlogic) developed a miRNA mimic for miR-34 (MRX34). Mir-34 functions as a tumor suppressor and is downregulated in a broad range of cancers [114–116]. MRX34 was delivered as a double-stranded RNA encapsulated in a liposome nanoparticle. While the drug showed promise in several preclinical studies in different cancers including renal cell carcinoma and hepatocellular carcinoma [117], clinical phase 1 trials were halted when immune-related serious adverse events resulting in death were reported. Consequently, all further development of the drug was halted.

lncRNA Sequence-Based epi-MTDs—lncRNA Replacement Therapy

In addition to miRNAs, long non-coding RNAs or lncRNAs (ncRNAs that are >200 nt in length) also play key roles in regulating gene expression, growth, differentiation, and development through a wide variety of mechanisms [24–29, 118–121]. It is now well-recognized that more than 75% of the human genome is functional and encodes large numbers of ncRNAs [119]. Based on the ENCODE project, it is estimated that the human genome encodes more than 28,000 distinct long noncoding RNAs (lncRNAs) many of which are still being discovered and are yet to be annotated [122, 123]. While understanding the functions of so many lncRNAs and their detailed characterization are challenging tasks, analysis of transcriptome profiles using next-generation sequencing in the last few years has revealed that thousands of lncRNAs are aberrantly expressed or mutated in various cancers [120].

(1) lncRNAs are diverse transcription products emanating from thousands of loci in mammalian genomes and can be classified into following subclasses based on their genomic localization, modes of action, and function [119].
   (a) Intronic lncRNAs originate from the introns of protein coding genes in either direction and terminate without overlapping exon.
   (b) Intergenic lncRNAs (also termed large intervening noncoding RNAs or lincRNAs) locate in the interval regions between two protein-coding genes and have their own promoter with separate transcriptional units from protein-coding gene.
   (c) Enhancer lncRNAs (elncRNAs) originate from the promoter enhancer regions.
   (d) Bidirectional lncRNAs are localized within the vicinity of a coding transcript of the opposite strand. These lncRNAs are transcripts that initiate in a divergent fashion from the promoter of a protein-coding gene; the precise distance cutoff that constitutes bidirectionality is not defined but is generally within a few hundred base pairs.
   (e) Sense-overlapping lncRNAs overlap with one or more introns and exons of different protein coding genes in the sense strand of the DNA.
   (f) Antisense-transcripts originate from the antisense-strands of the DNA or are located on the same strand, of the nearest protein-coding genes, and overlap at least one protein-coding exon. They may or may not be complementary to protein coding sequences in the sense-strand [120, 121].
(2) Functionally, lncRNAs can be classified as signaling, decoy, guide, and scaffold lncRNAs [120, 121].
   (a) Signaling lncRNAs are associated with specific signaling pathways and their expression indicates an active signaling event, irrespective of their roles (direct/indirect) in the signaling process [121].
   (b) Decoy lncRNAs act like molecular sinks for transcription factors and repressors. They interact with and titrate away transcription factors from binding to the target gene promoters facilitating gene activation or silencing [122]. Examples of decoy lncRNAs include GAS5 (growth arrest specific 5), TERRA (telomeric repeat-containing RNAs), and others [122].
   (c) Guide lncRNAs bind to the regulatory or enzymatically active protein complexes and direct them to specific target gene promoters or genomic loci regulating downstream signaling events and gene expressions. Examples of guide lncRNAs include AIR, CCND1 (cyclin D promoter associated lncRNA), lincRNA-p21, and others [121, 122].
   (d) Scaffold lncRNAs act as a central platform to which various protein complexes tether and get directed to specific genomic location or target gene promoter regulating gene expression and chromosomal dynamics. Examples of scaffold lncRNAs are HOTAIR, TERC, and others.
(3) Accumulating evidence indicates that lncRNAs act as molecular switches in cellular differentiation, movement, apoptosis, and in the reprogramming of cell states by altering gene expression patterns. They act through a wide spectrum of mechanisms with the following four modes as the predominant mechanisms of action [121, 124–127].
   (a) LncRNA can fold into a tertiary structure and provide a scaffold for the formation of a quaternary structure for proteins and other regulatory RNA. For

example, lncRNA participates in gene silencing by regulating polycomb repressive complex 2 activity, which has histone methyltransferase activity and primarily methylate histone H3 on lysine 27 (H3K27me3) a mark of transcriptionally silent chromatin.

(b) LncRNAs may directly bind DNA sequences and form RNA-DNA triplex complex. This can trigger the expression of genes by providing a more open chromatin domain for mRNA transcription. LncRNAs can also silence the gene expression by occupying the transcription factor binding sites. In the nucleus, lncRNAs also control the epigenetic state of particular genes, involved in transcriptional regulation and splicing. It is well established that biochemically-purified chromatin contains a significant amount of RNA, which may influence chromatin structure and gene regulation and/or DNA methylation.

(c) LncRNAs regulate the gene expression at post-transcriptional level in the cytoplasm by affecting the stability of mRNAs, altering the translation efficacy of target mRNAs and may function as a precursor for miRNAs or compete for miRNA-mediated inhibition of mRNA, leading to increased mRNA expression. The lncRNA-miRNA-mRNA regulatory signaling pathways interplay to form a complex network for maintaining homeostatic and adaptive gene expression [128].

(d) Another important regulatory mechanism of lncRNAs is the transcriptional and post-transcriptional regulation mediated by RNA-binding proteins (RBPs) to impact, either positively, or negatively, the function of bound protein and gene expression [129–132]. In fact, most of cellular functions of lncRNAs require the interaction with one or more RNA-binding proteins (RBPs), histone-modifying complexes DNA binding proteins (including transcription factors), and even with RNA polymerase II [131]. For example, we found that a lncRNA ZFAS1 (lncR-ZFAS1) causes intracellular Ca2+ overload and contractile dysfunction in myocardial infarction by acting as a SERCA2a inhibitor through direct binding to this critical intracellular Ca2+ handling protein [129].

LncRNAs play versatile roles in many aspects of gene regulation, including transcription, mRNA splicing, translation, epigenetic silencing, genomic imprinting, X-chromosome inactivation, and the processing of small ncRNAs [133, 134]. Validated functions of lncRNAs suggest that they are master regulators of gene expression and often exert their influences via epigenetic mechanisms by modulating chromatin structure and miRNA and protein functions.

Dysregulated or mutated lncRNAs play a critical role in the etiology and pathogenesis of many diseases. Many different types of cancers and syndromes are associated with mutations in lncRNA genes [134–139]. Genome-wide array studies show differential lncRNA expression patterns in comparisons between normal and diseased cells [138, 139]. For example, lncR-HOTAIR is associated with multiple types of cancer [118, 140, 141] where it interacts with PRC2 and LSD1 to repress target gene transcription. LncR-ANRIL binds to chromobox 7 (CBX7), which is a component of the PRC1 complex, to induce gene silencing, and both ANRIL and CBX7 are upregulated in prostate cancer [142]. LincR-p21 is a p53 repressor and is associated with the development and progression of prostate cancer, chronic lymphocytic leukemia, atherosclerosis, and rheumatoid arthritis [143, 144]. LncR-ZFAS1 is a proapoptotic lncRNA in various types of cells through various mechanisms of action, by causing cytosolic Ca^{2+} overload to trigger the mitochondrial death pathway in cardiomyocytes [129, 145], by activating the PI3K/AKT pathway via upregulating miR-421 in hippocampal neurons [146], by promoting small nucleolar RNA-mediated 2′-O-methylation via NOP58 recruitment in colorectal cancer cells

[147]. Among the lncRNAs identified to be associated with cardiovascular disease (CVD), myocardial infarction-associated transcript (lncR-MIAT) was found to confer risk of myocardial infarction (MI) upon its upregulation by the single nucleotide polymorphism (SNP) and to predict left ventricular dysfunction in acute MI (AMI) [148]. Another study supported this finding by demonstrating that the expression level of MIAT is a significant univariate predictor of left ventricular dysfunction in patients with AMI [149]. The role of lncR-MIAT as a pro-hypertrophic lncRNA in the heart was attributed to its sponging effect on miR-150 [150]. Our research group reported that lncR-MIAT acts as a competing endogenous RNA (ceRNA) for miR-24 to promote cardiac fibrosis in post-infarct myocardium [151]. lncR-MIAT was also found to be involved in the pathogenesis of diabetic cardiomyopathy by upregulating DAPK2 expression via sponging miR-22-3p [152]. The involvement of lncR-MIAT in the pathological angiogenesis through regulating retinal microvascular function by absorbing miR-150-5p in the setting of diabetes mellitus (DM) was also demonstrated [153]. The findings from these example studies indicate that an individual lncRNA is a multitarget regulator of genes (e.g., lncR-MIAT can bind to several miRNAs) that are associated with certain closely related pathological processes.

(4) The following characteristics of lncRNAs afford some advantages for epi-MTD discovery.
   (a) An individual lncRNA may act on multiple mediators (i.e., targeting several mRNAs or miRNAs) of a specific signaling pathway or of multiple signaling pathways. This property provides an opportunity for epi-MTD discovery.
   (b) Since lncRNAs normally regulate relatively small sets of genes, often with related functions, the ability to target lncRNAs pharmacologically should result in improved specificity and lowered incidence of side effects. Gene clusters often contain genes with similar functions (i.e., they function in a shared pathway), and these gene clusters are typically regulated by one or few specific lncRNAs.
   (c) lncRNAs are generally expressed in a spatiotemporal fashion with certain tissue specificity at certain timepoints under certain conditions.
   (d) The expression profile of a given lncRNA is different between normal and diseased tissues and among different pathological entities. These above two properties allow for disease- and target tissue-specific treatment of certain diseases.
   (e) lncRNAs are generally expressed at low abundance, which may permit the use of lower concentrations/doses of mimics or their inhibitors for replacement or knockdown to achieve therapeutic effectiveness. Thus, this property may help to reduce the likelihood of adverse events or to alleviate some toxicities.

(5) Owing to the above-described unique features, lncRNAs constitute attractive and specific molecular targets for the development of multitarget drugs of either gain-of-function or loss-of-function. Notwithstanding the opportunities and advantages afforded by lncRNAs, challenges co-exist for epi-drug discovery using the lncRNA sequence-based MTD strategy due to the following issues.
   (a) lncRNAs are normally very long in their sequences, ranging from 200 nt to over 10 kb [154, 155] and being typically 1000–10,000 nt, to be taken up by cells [156]. For example, lncR-MIAT is a ~10-kb long polyadenylated transcript containing seven exons and can be alternatively spliced to produce more than 10 isoforms [157]. This property creates a barrier for lncRNA replacement therapy—the difficulty to introduce full-length lncRNAs into cells.
   (b) lncRNA sequences are in general poorly conserved among the homologs across species, which not only creates difficulties of pursuing basic research to eluci-

date the functionalities of lncRNAs and the associated mechanisms of action, but also forms barriers to validating preclinical animal studies and extrapolating the results to humans.

To resolve these issues, we have developed a strategy along with a specific protocol that could be applied to effective lncRNA replacement therapy in the preclinical setting [129]. The overall strategy is to identify the "functional domain (FD)" of a given lncRNA of interest. The basis for this strategy is hypothesis that lncRNAs may work as flexible modular scaffolds with discrete domains for RNA–protein interactions and sequences for selecting RNA and DNA targets [158]. In general, the "functional domain" is a short stretch of sequence region of a given lncRNA, which confers the ability of this lncRNA to bind a protein or a miRNA to elicit its cellular function. Moreover, such a functional domain is likely conserved across species. For example, in investigating the function and mechanism of lncR-ZFAS1, we identified a region of high-degree conservation across varying species, including man and mouse, by sequence alignment, that we termed functional domain. We then performed theoretical analysis using the computational docking software Hex 8.050,51 to see whether this functional domain could bind SERCA2a. The docking results revealed that the nucleotides U19, G20, G23, G25, and G26 encompassed by *the functional domain* are likely the core motif for specific binding to of lncR-ZFAS1 SERCA amino acids Ala^{241}, Glu^{243}, Glu^{689}, Ser^{693}, Arg^{667}, Arg^{671}, and Arg^{672}. We then synthesized an oligonucleotide fragment corresponding to the conserved region/functional domain. We transfected this fragment into cardiomyocytes and verified its ability to bind SERCA2a using RNA-binding protein immunoprecipitation (RIP) analysis and affect Ca^{2+} handling. We found that application of the functional domain reproduced the effects of full-length lncR-*ZFAS1* on the expression of SERCA2a, a critical regulatory protein of Ca^{2+}-handling, and intracellular Ca^{2+} status with high fidelity. The negative control oligonucleotides fragment did not produce any effects. Next, we mutated the core motif of the functional domain by nucleotide substitution to disrupt the binding site for SERCA and examined the effects of this construct on SERCA2a function. The computational docking results confirmed the loss of the original binding sites for SERCA in the mutated functional domain, and the functional study also confirmed the loss-of-function mutation on SERCA2a expression and intracellular Ca^{2+} handling. We then subsequently constructed an oligonucleotide fragment antisense to the functional domain and verified the abrogation of the repressive effect of full-length lncR-ZFAS1 on SERCA2a expression and enhancement of Ca^{2+} overload [129]. Such an ASO fragment bears a significant therapeutic potential as by binding to the functional domain of lncR-ZFAS1 it can specifically block the interaction between lncR-ZFAS1 and SERCA2a protein by masking the binding site without affecting other possible cellular functions of this lncRNA.

An important note about lncRNAs is the ability of a single lncRNA to bind multiple proteins. Still taking lncR-ZFAS as an example, in addition to binding SERCA2a in cardiomyocytes to induce apoptosis, this lncRNA has also been shown to directly interact with cyclin-dependent kinase 1 (CDK1) to suppress apoptosis in colorectal cancer cells [159] and with zinc finger E-box-binding homeobox 2 (ZEB2) to enhance the growth and metastatic capacity of osteosarcoma cells *in vitro* and *in vivo* [160]. These findings suggest a possibility of developing a combination therapy (the CDT strategy) with a mixture of multiple lncR-ZFAS1 fragments targeting their respective binding proteins or of a MTD epi-drug with a hybrid molecule containing multiple fragments targeting all lncR-ZFAS1–interacting proteins. Such a strategy could in theory be applied to any other lncRNAs.

Alternatively, lncRNA–miRNA interactions can also be used as a MTD epi-drug strategy based on the fact that an individual lncRNA can simultaneously bind and sequestrate multiple miRNAs to counter the repressive effects of these miRNAs. Each of miRNA-binding domains in a lncRNA is expected not to exceed 25 nt in length, and such a short fragment can be used as a substitute of a full-length lncRNA for a target miRNA. Pooled miRNA-binding domains from a given lncRNA can therefore be used as a CDT regimen, and hybrid chain with linkers connecting multiple miRNA-binding domains of a given lncRNA can be considered a MTD epi-drug.

Collectively, to resolve poor-homology and lengthy-sequence problems of lncRNAs for replacement therapy, identifying functional domains (protein-binding and miRNA-binding regions) is a feasible approach. At least three prerequisites must be met by a druggable functional domain. First, a functional domain must be conserved across species. Second, it should be long enough to retain the functional effects of its full-length parent and short enough to be easily delivered into cells. Third, it should be unique in sequence to be able to produce selective effects towards target molecules.

Unfortunately, the nature of the domains of lncRNAs remains elusive. Computational methods for systematically identifying lncRNA secondary structure modules are still unsatisfactory and deletion analysis to assign biochemical activities or biological function to specific structures is restricted to very few examples [161, 162]. As stated by Zucchelli et al [163], the ability to identify lncRNA domains would represent a major breakthrough because it would enable us to predict a priori the functions of the many thousands of annotated lncRNAs, and such knowledge may also lead to the construction of synthetic RNAs to be used for manipulating gene expression *in vivo*.

Alternatively, we can use SINEUPs molecules to enhance protein synthesis of their target mRNA. SINEUPs are short (< 500 bp) interspersed nuclear elements (SINEs) or non-coding repetitive sequences that can transpose into new parts of the genome by means of an RNA intermediate and reverse transcriptase. They are a new class of natural antisense or synthetic lncRNAs that stimulate translation of sense mRNAs. SINEUPs are known to enhance protein synthesis of their target mRNA and function post-transcriptionally [163, 164]. SINEUPs contain two functional domains: a Binding Domain (BD), which defines specificity (via base-pairing) to the targeted mRNA, and an Effector Domain (ED) that acts as an enhancer of translation. The BD consists of 72 nucleotides, which include sequences that are complimentary to the 5′-UTR, the translation initiation codon and initial codons. The ED contains repetitive SINEB2 sequences. The SINEUP lncRNA base-pairs with the mRNA, and the SINEB2 element facilitates association of the mRNA with polysomes, thus increasing the rate of translational initiation [164]. The use of synthetic SINEUPs to increase protein synthesis has been demonstrated both by *in vitro* and *in vivo* studies [165, 166]. Notably, it is possible to elevate the level of proteins encoded by the targeted mRNAs of interest acting at post-transcriptional level by swapping the BD with an appropriate sequence present in SINEUPs. This notion was validated with a synthetic SINEUP targeting GFP which strongly increased GFP protein levels in HEK cells, while it had no effects on GFP mRNA [165]. Synthetic SINEUPs can thus be targeted to endogenous mRNA of interest that do not present a natural SINEUP in their genomic locus opening up their use for virtually every protein coding transcript of mammalian cells [163]. According to Zucchelli et al [163], while natural SINEUPs suggest that embedded TE may represent functional domains in lncRNAs, synthetic SINEUPs have two major advantages over competing technologies. First, synthetic SINEUPs modulate translation of target mRNAs without introducing stable genomic changes into target cells. Second, their induction of selected proteins is typically in a more physiological range (twofold) than most conventional gene replacement strategies. These features render SINEUPs a potentially interesting platform for a number of applications.

circRNA Sequence-Based epi-MTDs—circRNA Replacement Therapy

Beyond traditional ncRNAs, circular RNAs (circRNAs) are also emerging as a novel class of endogenous ncRNAs that form covalently-closed continuous loops instead of traditional linear forms, thereby lacking the 5′ or 3′ ends and polyadenylation. As endogenous biomolecules, circRNAs are conserved across species and are found to be associated with a variety of important biological processes and human diseases including cancer. Generally, circRNAs are classified as exonic, intronic, and retained-intronic circRNAs. They may be derived from exons, introns, untranslated regions, antisense transcripts, and intergenic regions. circRNA biogenesis has been explained by various models, incorporating a range of spliceosomes and RNA binding proteins. The most accepted model suggests that circRNA biogenesis involves joining of a 5′ splice site and a 3′ splice site as the result of back splicing [167–169]. In mammalian genome over 20% of the expressed genes in cells and tissues can produce these transcripts.

circRNAs have the following characteristics:

(a) The mechanism of action of circRNAs is relatively simple; they appear to function as miRNA sponges to counteract miRNA-mediated repression of mRNA forming the circRNA-miRNA-mRNA regulatory networks or as protein inhibitors by sequestration forming the circRNA-protein complexes [167, 170]. Two studies in 2013 not only identified thousands of circRNAs in different mammalian systems, but also showed that two circRNAs, *CDR1as* (also known as *ciRS-7*) and circSry, can bind to and likely modulate activities of specific miRNAs [171, 172]. Some circRNAs possess many miRNA-binding sites. circRNAs are involved in the regulation of mRNA splicing, transcription, and gene expression [172, 173].
(b) circRNAs are generally expressed at low levels and often demonstrate tissue-specific and cell-specific expression patterns in eukaryotes [173–175]. Like miRNAs and lncRNAs, circRNAs are expressed in a tissue- and developmental stage-specific manner in humans, mice, and flies [175–177]. Genes hosting circRNAs are highly enriched for brain-related genes [175, 177, 178]. This CNS-specific enrichment seems to be a general feature of circRNAs in all the studied species [175, 177–179]. Notably, the expression of circRNAs within the brain is highly specific and, in many cases, independent of the expression of the linear isoform generated from the hosting gene [175, 177–179]. A large fraction of the genes hosting circRNAs encode for synaptic proteins and many brain circRNAs are strongly enriched in synaptosomes.
(c) Because of their unique circular structure lacking 5′ and 3′ends, circRNAs are resistant to nucleases and are stable with a relatively long half-life [178]. They strongly accumulate with age in the brains of flies and mice [176–178], leading to the suggestion that circRNAs may be involved in age-related brain diseases [180].
(d) There is a strong anti-correlation between the amount of circRNAs and cell replication rate, which is manifested in brain [181]. However, cardiomyocytes usually do not divide, but circRNAs do not accumulate in the heart as the animal ages [176].
(e) circRNAs are predominantly cytoplasmic [171, 182, 183] and may exist in tissues, serum, and urine and display aberrant expression in various disease entities and possess strong promise toward development of novel biomarkers and therapeutics.

In terms of the cellular function of circRNAs, the most well-known aspect is the gene regulation through miRNA-sponging or ceRNA action. However, the circRNA-miRNA sponge hypothesis must be seen critically. The abundance of specific circRNAs (besides MBL in flies or *CDR1as* in humans) is generally low. This matter contradicts possible sponge theories [184, 185], since the miRNA might be more abundant than the total RNA in which the miRNA can bind and titrate, and there are more miRNA-binding sites available than miRNA molecules.

Several groups reported that circRNAs are translated [186–188]. While these studies do not provide much insight into the functions of the circRNA-encoded protein products, a recent report showed a potential role of a circRNA-encoded peptide in cancer [189].

Probably a more important function of circRNAs is their protein-binding ability to form circRNA-protein complexes and regulate the function of targeted proteins. By binding to RNA binding proteins (RBPs), circRNAs can act as decoys or transporters for these factors. In some of the cases, there might be a direct or indirect crosstalk between the circRNA and the protein product of the host gene [178, 190–192].

circRNAs have been implicated in diseases such as cancer [193–195], neurological disorders [196], cardiovascular diseases [197], metabolic diseases including diabetes and obesity [198, 199], aging, and aging-related diseases [200]. However, the functions of most circRNAs remain largely unexplored due to the challenges at multiple levels to understanding of the regulation of circRNAs because of their circular conformation and sequence overlap with linear mRNA counterparts. As far as the potential as therapeutics or therapeutic targets concerned, the capacity of circRNAs binding to miRNAs and proteins may be the most attractive feature and feasible approach among all, just like lncRNAs. Acting as competing endogenous RNA (ceRNA), or miRNA sponge, circRNAs can bind miRNAs through a miRNA response element (MRE) and negatively regulate their activity [171]. Since the length of exon circRNAs ranges from 100 nt to >4 kb [201, 202], the potential number of MREs could be large and variable. Thus, a given circRNA may likely regulate more than one miRNA. For example, circR-ITCH acts as a miRNA sponge by binding miR-17, miR-214, and miR-7 in esophageal squamous cell carcinoma [203]. This property forms the basis for the potential of utilizing circRNAs for MTD epi-drug discovery.

Similarly, the ability of individual circRNAs to bind multiple proteins lays the foundation for the development of a distinct subclass of circRNA sequence-based MTD epi-drugs. For instance, circR-Foxo3 can inhibit cell proliferation and cell cycle progression by simultaneously binding to CDK2 and p21 proteins to form a complex to repress CDK2–cyclin A/E complex formation resulting in block of cell cycle progression [204].

asRNA Sequence-Based epi-MTDs—asRNA Replacement Therapy

Messenger RNA (mRNA) is a single-stranded molecule that is used as the template for protein translation, and the level of mRNA present in cells is a critical determinant of the level of protein thereby of the cellular functions. It is possible for mRNA to form duplexes, similar to DNA, with the other sequence of RNA complementary to the mRNA strand, and this complementary sequence is called endogenous antisense RNA (asRNA).

Ionically, whilst asRNAs have recently been well recognized and appreciated as unique but diverse new players in the regulatory network of gene expression, there has not been a consensus and conformity to the definition of this class of RNAs. Often, the term "asRNAs" or "natural antisense transcripts (NATs)" is coined for the single-stranded ncRNA molecules that can bind and inhibit specific mRNA translation to protein [205–209]. Yet, with the recent discovery of several new classes of non-coding functional RNAs and their mechanisms of action, elucidation of new functions of mRNAs, and emergence of new concepts on these RNA species, the abovementioned definition of asRNAs has become inadequate and imprecise to cover the large variety of existing asRNA species. First, new studies have shown that some protein-coding mRNAs can also act as asRNAs. Second, asRNAs, while primarily inhibit gene expression, can also activate or enhance gene expression. And finally, while the majority of asRNAs are antisense to mRNAs, others are antisense to ncRNAs. Therefore, in a broad sense, any RNA species that contains complementary sequence to and can hybridize with another RNA to ultimately alter the level and function of mRNAs can be considered an asRNA. Or more specifically, asRNAs can be defined as single-stranded RNA molecules that can bind other RNA molecules (coding or non-coding) through full or partial base-paring mechanism so as to regulate the activities of the targeted RNA molecules that ultimately alter the gene expression. According to this scenario, small interference RNAs (siRNAs), miRNAs, lncRNAs, circRNAs, and competing endogenous RNAs (ceRNAs) can all be considered asRNAs. asRNAs are not a uniform group of regulatory RNAs but instead comprise a diverse array of transcripts with some common features and similar mechanisms and mode of actions. Like, miRNAs, lncRNAs and circRNAs, asRNAs are native or endogenous multitarget gene regulators for each asRNAs may concurrently regulate the expression of more than one gene. Furthermore, miRNAs, lncRNAs, circRNAs, and even some mRNAs that possess antisense properties can be considered asRNAs because they all act by "antisense" complementarity to their target RNAs (mRNAs and/or ncRNAs).

(1) *Natural antisense transcripts (NATs)*: Many genomic loci contain transcription units on both strands; therefore, two oppositely oriented transcripts can overlap, with one of the double strands being NATs [205, 208–210]. Thus, NATs are transcribed RNA products from the DNA strand complementary to a region harboring a sense transcript of either protein-coding or noncoding genes [205–207]. NATs can be produced against both protein-coding and non-coding genes and some NATs are protein-coding genes, while others are lncRNAs [207]. There are two types of NATs in eukaryotic genomes: cis-encoded antisense NATs, which are transcribed from the opposite strand of the same genomic locus as the sense RNA and have a long (or perfect) overlap with the sense transcripts and trans-encoded antisense NATs, which are transcribed from a different genomic locus of the sense RNA and have a short (or imperfect) overlap with the sense transcripts [205]. They can also arise from intronic regions of coding genes or near transcriptional start sites of neighboring genes.

NATs can function within the nucleus or in the cytoplasm and in a cis or trans manner [207] to regulate the expression of their cognate genes. Nuclear NATs involves epigenetic changes proximal to a target gene, whereas cytoplasmic NATs act through RNA–RNA interactions. Cis-NATs are usually related in a one-to-one fashion to their sense transcripts originating from the same locus, primarily due to the full and perfect sequence complementarity, as part of self-regulatory circuits that allow genes to regulate their own expression. In contrast, a single trans-NAT may target several sense transcripts presumably owing to the partial imperfect complementarity mechanism similar to miRNAs [205, 208–210]. Trans-acting NATs can regulate the expression and/or function of genes from other genomic loci. Cis-regulation of NATs can lead to activation or silencing of their sense partners via transcriptional activation, transcriptional interference, RNA silencing, mRNA stabilization [211, 212], alternative splicing, RNA editing, translational interference, genomic imprinting, and X-inactivation. In general, there were two types of gene regulation by NATs: (a) discordant regulation, where the NATs inhibit the sense RNAs or proteins; and (b) concordant regulation, where NATs increase the level of the sense RNAs or the corresponding proteins.

NATs are highly prevalent in a wide range of animal species—approximately 70% of human protein-encoding genes have an associated NAT. A study showed that at least 21,000 human and 16,000 mouse genes are involved in NATs-sense pairing [213], and an accumulating body of evidence suggests that they play pivotal roles in a wide range of human diseases including cancers of various types.

(a) *Non-protein-coding NATs*: The vast majority of NATs are not translated into proteins and therefore belong to the category of ncRNAs. They include lncRNAs antisense to mRNAs and antisense to lncRNAs, endogenous siRNAs, miRNAs, and non-coding ceRNAs.
(b) *Protein-coding NATs: mRNAs as NATs*: A prominent and striking finding in the asRNA research is that mRNAs can also be asRNAs; a small proportion of NATs belong to the category of protein-coding genes. For example, a recent study revealed that DHPS, the gene encoding deoxyhypusine synthase, can also serve as a NAT of WDR83 (WD repeat domain 83, a member of the WD-40 protein family), concordantly regulating the expression of WDR83 mRNA and protein

[212]. Conversely, WDR83 can also regulates DHPS by antisense pairing in a concordant manner. These two distinct mRNAs are capable of forming an RNA duplex at overlapping 3′ untranslated regions (3′UTR) to enhance their mutual stability.

(2) *Antisense lncRNA (aslncRNA) or lncRNA antisense to mRNA*: LncRNAs, the mRNA-like transcripts ranging from 200 nt to 100 kb in length featured by lack of protein-coding activity, are newly recognized regulatory molecules of gene expression that are implicated in a variety of fundamental biological processes and pathophysiological events. In particular, the demonstration of their roles in human cancers has garnered tremendous research interest [135, 214]. LncRNAs comprise a diverse group of transcripts including large intervening noncoding RNAs (lincRNAs), NATs and intronic lncRNAs. There have been numerous cases of lncRNAs acting as NATs or aslncRNAs to mRNAs to alter gene expression with some of them implicated in tumorigenesis.
(3) *Long non-coding antisense RNA (lncasRNA) or the antisense transcript of lncRNA*: Different from aslncRNAs, lncasRNAs refer to the antisense transcripts of lncRNAs; that is, they are complementary to lncRNAs but not to mRNAs. An example of lncasRNAs is 91H, an antisense to lncRNA H19 [215]. The H19/IGFf2 locus belongs to a large, imprinted domain located on human chromosome 11p15.5. The H19 gene is expressed from the maternal allele, whereas IGF2 is paternally expressed. 91H was found to a nuclear and short-lived RNA corresponding to a short-lived and 120-kb-long transcript, overlapping the H19 imprinted gene. 91H is not imprinted in mouse but is expressed predominantly from the maternal allele in both mice and humans within the H19 gene region. Intriguingly, in humans, 91H, rather than affecting H19 expression, regulates IGF2 expression in trans manner [215].
(4) *miRNAs*: By the mechanism of action, all miRNAs are antisense to mRNAs.
(5) *Competing endogenous RNA (ceRNA)*: On one hand, miRNAs can act as asRNAs to mRNAs; on the other hand, they also can be the targets for antisense binding and action by other asRNAs. ceRNAs are such a miRNA-regulating category of asRNAs. In 2009, Seitz proposed that many natural miRNA target sites could act as miRNA decoys or sponge, and changes in the expression of some miRNA targets would alter the availability of a miRNA thereby affecting the activity of its other targets (43). In 2010, Pandolfi et al found that some protein-coding genes and their pseudogenes contain the same conservative miRNA binding sites in their 3′UTRs, and that they can regulate their respective expression levels by competing for miRNA binding (44). Based on these findings, one year later in 2011, Pandolfi et al proposed the concept of competing endogenous RNA (ceRNA) (45). Central to this hypothesis is the presence of the so-called miRNA responsive elements (MREs) in other RNAs (i.e., ceRNAs) and miRNAs as "messengers" for crosstalk among ceRNAs acting through sequence complementarity. The ceRNAs with specific MREs can communicate with others via the "miRNA messenger." Any RNA transcript harboring one or more MREs may function as a ceRNA and de-repress the activity of other ceRNAs with similar MREs by competing for the same miRNAs in a common pool of available miRNAs. Each individual miRNA has multiple RNA targets and a single miRNA binding to MREs in one target ceRNA would relieve the repressive activity of that miRNA on other target genes. This mode of interactions can be viewed as a mechanism for balanced asRNA regulation of gene expression under physiological conditions. ceRNAs have been implicated in numerous biological processes including the development of cancer when the balanced interactions on MREs are shifted towards the direction in favor of tumorigenesis. ceRNAs typically include

mRNAs, pseudogenes, lncRNAs, and circRNAs.

The unique and common feature of action of asRNAs is sequence complementarity as the prerequisite for gene regulation; yet, the subsequent steps, leading to alterations of gene expression involve a number of distinct mechanisms [208, 209, 216]. These mechanisms are primarily rooted in modifications of mRNA transcription (transcriptional interference and attenuation), mRNA destabilization or stabilization (mRNA cleavage through RNAi, miRNA degradation by exonucleases, and miRNA availability by ceRNA sponge), translational block, epigenetic regulation. Though not completely deciphered and experimentally validated, there is little doubt that these multiple mechanisms are operating in general, and there might be other yet-to-be-identified mechanisms too. It should be noted that asRNAs can also act through non-antisense mechanisms to alter gene expression. Collectively, each individual asRNA possesses the potential of acting simultaneously on multiple target RNAs/DNAs (many binding sites for a particular RNA or DNA species) and diverse target RNAs/DNAs (different species of RNAs/DNAs, mRNA, lncRNA, miRNAs, etc.).

(a) *RNA cleavage through RNAi*: Double-stranded RNA corresponding to a particular gene is a powerful suppressant of that gene through the RNAi pathway. In fact, the suppressive effect of an asRNA largely depends on its ability to form dsRNA. Normally, RNA molecules are present in the cytoplasm of a cell as single-stranded transcripts. If the cell finds molecules of dsRNA, for example, asRNA:mRNA duplex, it uses Dicer to cleave them into siRNA fragments containing ~21 base pairs. This is because the presence of dsRNA signals is perceived as an abnormality by the cells. This siRNA then binds to RISC, as already mentioned above. The RISC is subsequently activated when the siRNA unwinds and binds to the corresponding mRNA using the antisense strand. The RISC contains an enzyme to cleave the bound mRNA and therefore cause gene silencing.

(b) *Translation block*: asRNAs can compete with the ribosomes to bind 5′ end of mRNA, hence physically hindering the translation initiation. Or if asRNAs bind to coding regions of mRNAs, the ribosome cannot gain access to the nucleotides and translation can be interrupted. It has been observed that if an asRNA is not complementary to the 5′ end of mRNA, the extent of translational block is significantly reduced [217]. On the other hand, once the ribosome binds to mRNA, mRNA:asRNA duplex formation is greatly reduced and so is the translational block. Since the formation of mRNA:asRNA duplex is the critical step for translational block, the degree of homology and homology in certain specific regions is important.

(c) *RNA masking*: The formation of sense:antisense duplex can provide a physical barrier against post-transcriptional interactions by "masking" the regions for normal splicing to influence RNA stability [210, 218]. Such a masking property has been widely utilized in miRNA interference (miRNAi) to manipulate the function of miRNAs thereby the expression of miRNA target genes [37, 38, 219–229].

(d) *Transcriptional interference and attenuation*: asRNAs can alter transcription of the genes encoded on the opposite strand in two different ways: transcription interference or transcription attenuation. Transcription interference occurs when transcription from one promoter is suppressed by a second promoter present in cis. Transcriptional attenuation is a regulatory mechanism that causes premature termination of transcription under certain conditions, thereby preventing the expression of the mRNA required for expression of the corresponding gene products.

(e) *Epigenetic regulation*: Epigenetic trait is defined as "stably heritable phenotype resulting from changes in a chromosome without altering the underlying DNA sequence". DNA methylation and histone modifications are the two most important epigenetic regulations in gene expression regulation. DNA methylation is a process by which methyl groups are added to DNA. When located in a gene promoter, DNA methylation typically acts to repress gene transcription. Aberrant DNA methylation patterns (hypermethylation and hypomethylation) have been associated with a large number of human malignancies. Histone modifications refer to covalent post-translational modifications to histone proteins which includes methylation, phosphorylation, acetylation, ubiquitylation, and sumoylation. Histone modifications can impact gene expression by altering chromatin structure or recruiting histone modifiers. asRNAs can silence gene expression by promoter DNA methylation dependent and independent mechanisms. For instance, they can interact with key epigenetic regulators such as histone methyltransferase PRC2 and histone demethylase LSD1. Conversely, the promoter regions of asRNAs can also be epigenetically modified to alter the oncogenic potential and status of cancer development.

(f) *mRNA stabilization*: The formation of asRNA:mRNA duplex, especially within their 3′UTRs can protect mRNA from being attacked by exosomes to reduce mRNA degradation at the post-transcriptional level. In addition, it also can mask the miRNA-binding sites in mRNA, as well as in asRNA, thereby preventing the repressive effect of miRNA.

Given the wide spectrum of function and mechanism of asRNAs and the "one asRNA, multiple and diverse targets" properties, asRNAs open up new horizons with rich resources for rational design of sequence-based epi-MTDs in new drug discovery and development.

Ribozymes as epi-MTDs

Ribozymes (ribonucleic acid enzymes) are RNA molecules that have the ability to catalyze specific biochemical reactions, including RNA splicing in gene expression, similar to the action of protein enzymes. They are hammerhead-shaped, approximately 30-nucleotide long, self-cleaving, and nuclease-resistant catalytic RNA oligonucleotides that can bind to and attack the 2′-OH that is 5′- to the scissile bond (a covalent bond, which can be broken by enzymes) in specific RNA targets, resulting in destabilization of the phosphodiester backbone of the targeted RNA molecule [230–232]. Once cleaved, the ribozymes dissociate from the products and may cleave other target RNAs. The cleavage is highly sequence-specific and is sensitive to single nucleotide mismatches; hence, toxicity due to off-target effects can be minimal [233]. The most common activities of natural or in vitro-evolved ribozymes are the cleavage or ligation of RNA and DNA and peptide bond formation. Within the ribosome, ribozymes function as part of the large subunit ribosomal RNA to link amino acids during protein synthesis. They also participate in a variety of RNA processing reactions, including RNA splicing, viral replication, and transfer RNA biosynthesis. Examples of ribozymes include the hammerhead ribozyme, the VS ribozyme, Leadzyme, and the hairpin ribozyme.

Stability of short ribozyme sequences in the presence of endogenous ribonucleases poses a problem, but this has been addressed by the chemical modification of the phosphate and sugar moieties. Incorporation of four phosphorothioate linkages at the 5′- end of the ribozymes and 2′-O-Me nucleotides in place of the 2′- hydroxyl group has been shown to stabilize ribozymes without altering their catalytic activity [234, 235]. Targeted delivery of ribozymes to specific cellular compartments where lncRNAs are located using liposomes or peptide-based delivery systems provides an alternative to RNAi [236]. Synthetic ribozymes against VEGF mRNA administered *in vivo* decreased growth and metastasis of solid tumors, and the same methodology can be adapted to target specific lncRNAs [231, 237]. Thus, ribozymes are an exciting future therapeutic tool that can be used to treat various lncRNA-associated diseases.

Ribozymes have been proposed and developed for the treatment of disease through gene therapy [238]. A type of synthetic ribozyme directed against HIV RNA called gene shears has been developed and has entered clinical testing for HIV infection [239, 240]. Similarly, ribozymes have been designed to target the hepatitis C virus RNA, SARS coronavirus (SARS-CoV) [241], Adenovirus [241] and influenza A and B virus RNA [241–244]. The ribozyme is able to cleave the conserved regions of the virus's genome which has been shown to reduce the virus in mammalian cell culture [245]. Despite these efforts by researchers, these projects have remained in the preclinical stage.

One major challenge of using RNA based enzymes as a therapeutic is the short half-life of the catalytic RNA molecules in the body. To combat this, the 2′ position on the ribose is modified to improve RNA stability. One area of ribozyme gene therapy has been the inhibition of RNA-based viruses.

12.1.2.2 Rationally Designed RNA Fragments as Epi-MTDs

While the above subsection introduces endogenous RNA species for sequence-based epi-MTD discovery, this one is to describe artificial RNAs or RNA fragments (oligonucleotides) and DNA fragments or ODNs (oligodeoxynucleotides) as novel strategies for rational design of sequence-based MTD epi-drugs. Artificial constructs that are druggable mainly include artificial miRNAs, anti-miRNA antisense inhibitor, antisense ODNs, antisense RNAs, and siRNAs.

Oligonucleotide therapy or sequence-based epi-MTDs has great potential for the treatment of human diseases, with FDA approval of antisense drugs like Fomivirsen [246] for the treatment of cytomegalovirus retinitis; and Kynamro for the treatment of homozygous familial hypercholesterolemia [247]; and with ongoing clinical trials, in using small interfering RNA (siRNA) [248] and splice switching oligonucleotides [249]. The entry of oligonucleotide into cells depends on the structure and concentration of the oligonucleotide and the cell types. Adsorptive endocytosis and fluid phase pinocytosis are believed to be the major mechanisms of oligonucleotide internalization. At a relatively low concentration, the internalization of oligonucleotide occurs via interaction with membrane-bound receptors in a process called adsorptive endocytosis, whereas, at a relatively high concentration, pinocytosis plays a major role in oligonucleotide internalization due to the saturation of membrane-bound receptors [250]. However, effective delivery of the oligonucleotides to their intracellular sites of action still remains a major challenge. Current approaches to facilitate oligonucleotide delivery include lipid- and polymer-based nanoparticles and ligand-oligonucleotide conjugates such as lipid, peptide, antibody, and aptamer conjugates, and the like [251]. While the majority of oligonucleotide therapies are given systemically, the respiratory track provides an ideal site for topical delivery for respiratory diseases. The airways are uniquely lined with pulmonary surfactants, which are primarily composed of zwitterionic lipids. These surfactant lipids possess cationic properties at the pH of the respiratory tract. When anionic oligonucleotides are inhaled, they tend to be adsorbed by the surfactants, resulting in reformulated particles that have been hypothesized to be efficiently taken up into the cells [252].

Since the first report describing the use of ASO targeting Rous sarcoma virus 35S RNA as a potential oligonucleotide therapy, there has been an exponential increase in the discovery and development of oligonucleotides as a therapeutic modality [253]. ASO, siRNA, and miRNA are the three major types of oligonucleotide therapeutics under development. There are a few advantages for targeting RNA over protein molecules. As a single copy of RNA could be a template for the synthesis of multiple copies of a protein, it would be more efficient to regulate the mRNA level rather than the protein level to block the protein function [254]. Another benefit of using oligonucleotide strategy comes from the complete decoding of human genome, which greatly shortens the time between identification of gene targets and the design and validation of oligonucleotide products. Finally, several different oligonucleotide sequences can be combined into one single drug product, which is especially important for diseases with multifactorial etiology.

Multitarget Artificial miRNAs (MT-amiRNAs)

RNA interference (RNAi) technology has been used for almost two decades to study gene functions and in therapeutic approaches. It uses cellular machinery and small, designed RNAs in the form of synthetic small interfering RNAs (siRNAs) or vector-based short hairpin RNAs (shRNAs), and artificial miRNAs (amiRNAs) to inhibit a gene of interest. Artificial miRNAs, known also as miRNA mimics, shRNA-miRs, or pri-miRNA-like shRNAs have the most complex structures and undergo two-step processing in cells to form mature siRNAs, which are RNAi effectors. Unlike short interfering RNAs (siRNAs), which are commonly designed to repress a single mRNA target through perfect base pairing, amiRNAs are constructed for concurrent repression of multiple mRNA targets through imperfect complementarity in a similar manner to endogenous miRNAs using endogenous miRNAs as scaffolds [89, 255–259]. amiRNAs are generally constructed by replacing the mature miRNA sequence in the pre-miRNA stem-loop with a sequence targeting a gene of interest. amiRNAs have mostly been applied to gene knockdown in plants; they have been examined to a lesser extent in animal cells [247]. amiRNAs are embedded in sequences driven by RNA Polymerase II (Pol II) promoters, whereas shRNA expression is typically driven by constitutive Pol III promoters, such as H1 or U6 [137, 258]. amiRNA sequences embedded in natural miRNA backbones have proven to be useful tools for RNAi. amiRNAs have reduced off-target and toxic effects compared to other RNAi-based methods such as short-hairpin RNAs (shRNAs) or siRNAs involving the use of vector-based expression. amiRNAs are often less effective for knockdown, however, compared to their shRNA counterparts [257].

Functional backbones for amiRNA expression have been made from a number of naturally-occurring miRNAs [89, 90, 258–260], and the two most commonly used amiRNA scaffolds are derived from either miR-30 or the synthetic inhibitory BIC/miR-155 RNA (SIBR) [261, 262], especially since commercial vectors have been developed using these backbones (Open Biosystems Expression Arrest™, GE Dharmacon and Block-iT Pol II miR RNAi™, Life Technologies, respectively).

According to Fowler et al [257], amiRNAs have the following strengths over shRNAs/siRNAs.

(a) Potent knockdown using shRNAs often comes at a cost, however, as at least three problems frequently arise from shRNA overexpression. First, high levels of shRNAs can cause toxicity due to oversaturation of the endogenous miRNA pathway [263, 264]. Oversaturation may even result in lethality during *in vivo* studies [265]. amiRNAs are typically expressed at much lower levels and do not saturate endogenous RNAi pathways [87, 266]. As such, amiRNAs are suitable for studies where shRNAs were toxic [108, 267]. Second, shRNA sequences are usually designed with perfectly matched guide and passenger strands. In contrast, amiRNAs are often designed with central mismatches between the guide and passenger strands, which may reduce off-target effects by decreasing unwanted passenger strand incorporation into RISC [266]. Third, shRNAs often induce an immune response which may compound or mask RNAi-specific effects. Use of amiRNAs may circumvent this problem by avoiding immune activation [109, 266].

(b) amiRNAs allow greater vector design flexibility and diversity of application compared to shRNAs. For example, amiRNAs can be co-expressed with transgenes from a single cistron driven by Poll II promoters, such as cell-specific or conditional promoters [268, 269]. amiRNAs can also be placed in an intron so that miRNA processing does not interfere with transgene expression [269, 270]. Delivered as viral vectors and expressed under tissue-specific polymerase II (Pol II) promoters, amiRNAs provide long-lasting silencing and expression in selected tissues.

(c) More importantly, amiRNAs can easily be chained in tandem, either to increase knockdown efficiency or to target multiple genes, without the need of dedicated promoters for each hairpin sequence [259, 261, 271, 272]. Chained amiRNAs are particularly powerful tools for systems requiring multiple-target knockdown, such as studying functionally redundant genes and attractive approaches for sequence-based MTD epidrug discovery. Targeted knockdown of multiple genes can be used to overcome functional redundancy, which is presently hard, if not impossible, to achieve in humans with technologies such as conditional multigene knockout. Thus, combinatorial amiRNA holds great promise for gene therapies of those requiring targeting rapidly evolving pathogens [273, 274] or human diseases caused by multiple factors, including cancer [275, 276].

(d) Another important characteristic of amiRNAs is that the dynamics of maturation and steady-state level of siRNA-like stretch incorporated into a pri-miRNA scaffold in the cell resemble those of endogenous miRNAs. This property ensures that amiRNAs are safer than other forms of RNAi.

amiRNAs are composed of a target-specific siRNA insert and a scaffold based on a natural primary miRNA (pri-miRNA). siRNAs serve as a guide to search for complementary sequences in transcripts, whereas pri-miRNA scaffolds ensure proper processing and transport. Rational design of amiRNAs need to follow several criteria [277–281].

(a) First, specific nucleotides include a U or A (preference for U) at guide position 1 relative to the 5′ microprocessor cleavage site, U/A preferences in positions 2–7, 10–14, and 17, and G/C preferences in positions 19–21.

(b) Second, design sequences with an optimal G/C content between 36.4% and 45.5% when possible. Guide strands are designed to be 2 nucleotides longer than the passenger strand as in the original report [259].

(c) Mismatches were positioned when it is possible to make three general structural categories: a "loop" mismatch where 3–5 adjacent nucleotides of the guide strand were not base paired to the target strand; a "3bp-spaced" mismatch where 2 single guide strand nucleotide mismatches were separated by 3 guide/passenger base pairs and a "4bp-spaced" mismatch where 2 single guide strand nucleotide mismatches were separated by 4 guide/passenger base pairs. For all amiRNA sequences, mismatches ranged between guide positions 11 and 17. Predicted RNA secondary structures were generated using Mfold [282].

amiRNAs are believed to be useful therapeutic tools for a broad spectrum of human diseases, including neurodegenerative diseases, cancers, and viral infections. Recent reports on the role of sequence and structure in pri-miRNA processing may contribute to the improvement of the amiRNA tools. The success of a recently initiated clinical trial for Huntington's disease could pave the way for other amiRNA-based therapies, if proven effective and safe [283]. While miRNAs are naturally occurring small RNA repressors of gene expression, artificial short interfering RNAs (siRNAs) have been developed as powerful tools for experimental gene repression. In contrast to the partial complementarity observed with miRNA target interactions, siRNAs are designed to be perfectly complementary to a single mRNA, enabling the repression of individual genes [284]. An approach for the experimental repression of multiple genes in concert has been reported but not extensively developed [285]. Since miRNAs have a well-studied, naturally evolved seed-based mechanism for targeting multiple genes, they are appealing as the basis for developing a multi-target RNA interference (RNAi) system.

An approach to multiple amiRNA expression was established by Stegmeier et al [286] in their report that a mouse miR30-based shRNA expression cassette can be driven by Pol II promoters and provide higher knockdown efficiency than those driven by the Pol III U6 promoter. The combination of Pol II promoters and miRNA-based design offers some significant advantages. First, the use of tissue-specific or inducible Pol II promoters allows more effective control of the timing and the level of miRNA expression [287]. Second, the Pol II promoter supports expression of both miRNA cassettes and reporter genes, such as GFP, from the same transcript, thus allowing easier tracking of miRNA expression. Third, using artificial miRNA cassettes modeled after endogenous miRNAs seems to avoid the induction of cellular immune responses and apoptosis [88, 108, 109].

Different strategies have been investigated to express multiple artificial miRNAs from the same lentiviral vector. Zhou et al [288] reported that two tandem copies of the miR30-based cassette can be expressed in a single transcript driven by a Pol II promoter [288, 289]. Subsequently, Sun et al showed that a single Pol II promoter can drive three artificial miR30 cassettes to express siRNAs all targeting GFP, resulting in further knockdown of the GFP intensity in the cells [270]. A similar miR30-based approach was utilized by Zhu et al to knockdown multiple genes [290]. In addition to miR30-based designs, mouse miR155-based design has also been used to knockdown multiple genes [259]. Some miRNAs are present in the genome as a cluster, such as the miR17–92 cluster with six pre-miRNAs encoded in a ~1 kb pri-miRNA. To explore the miRNA polycistrons for artificial miRNA expression [90], four different anti-HIV artificial miRNAs under the control of the CMV promoter were expressed simultaneously and shown to inhibit virus production in transduced cells.

Another group developed a novel chicken miR126-based miRNA expression cassette using lentiviral vectors capable of expressing multiple anti-influenza miRNAs [261]. The miR126-based cassette can express one, two or three miRNAs from a single cassette in the context of a lentiviral vector. These researchers show that each of the miRNAs expressed from the same lentiviral vector is capable of potent inhibition of reporter genes in both transient transfection and stable integration assays. Cells transduced with lentivirus expressing two or three, but not one, anti-influenza miRNAs also inhibit influenza virus production. Although the expression cassette is based on chicken miR126, the resulting lentiviral vectors also effectively inhibit reporter gene expression in human, dog, mouse and monkey cells. These results demonstrate the versatility of the miR126-based miRNA expression cassette for potent and effective silencing of target genes [261].

Multitarget Anti-miRNA Antisense Inhibitor (MT-AMO)

One of the indispensable approaches in miRNA research as well as in miRNA therapy is inhibition or loss-of-function of miRNAs. With respect to miRNA interference (miRNAi) [37], endogenous miRNA and amiRNA belong to miRNA "gain-of-function" and target mRNA "loss-of-function" approach, whereas anti-miRNA antisense inhibitor (AMO or antimiR) is considered miRNA "loss-of-function" and target mRNA "gain-of-function" strategy. The base-pair interaction between miRNAs and mRNAs is essential for the function of miRNAs; therefore, a logical approach of silencing miRNAs is to use an oligonucleotide fragment (ON) that is antisense to the miRNA. These anti-miRNA oligonucleotides specifically and stoichiometrically bind, and efficiently and irreversibly silence their target miRNAs and derepress the target mRNAs of these targeted miRNAs. AMOs are synthetic ON fragments that can bind to and sequester the endogenous miRNA, preventing it from binding to and repressing the expression of its mRNA target. This AMO approach has been used in numerous studies to identify many cellular functions of miRNAs [37, 39, 291–295] and has been considered a plausible strategy for miRNA therapy of human disease. It is the most straightforward and apparently most effective strategy tested so far to block the function of miRNAs in RISC. It mediates potent and miRNA-specific inhibition of miRNA function, providing a powerful "loss-of-function" strategy for interfering miRNA expression. When conjugated with cholesterol moiety, AMO is also called antagomiR. AntagomiRs are used to suppress the function of a specific miRNA that is overexpressed and has a function in disease development/progression [82, 83]. The AMO approach has been used in numerous studies to identify many cellular functions of miRNAs and has been considered a plausible strategy for miRNA therapy of human disease.

AMOs usually contain chemical modifications to the backbone and sugar rings of the nucleotides increasing their resistance to nuclease degradation, its binding to plasma proteins to maintain stable serum concentrations, its complementarity to target RNAs and its ability to activate the innate immune system. The AMOs can be designed to contain high-affinity miRNA binding sites, to which the endogenous miRNA will preferably bind, and, therefore, sequesters the miRNA machinery away from the physiologically relevant targets. Since individual miRNAs are endogenous multitarget regulators of gene expression, inhibition of a single miRNA by a miRNA-specific AMO could be considered.

To date, no miRNA-based therapies have broken through the translation barrier and been FDA approved as yet. However, several potential therapies have reached phase I and phase II clinical trials, with many more in clinical development. Many biotech companies are exclusively working on miRNA-based drugs such as Miragen, Synlogic, and Regulus Therapeutics.

Miravirsen (Roche/Santaris Pharma) is a phosphorothioate-modified LNA antagomiR, which targets miR-122 to repress hepatitis C viral (HCV) infections [296]. This drug entered phase 2 clinical trials in 2017. The miRNA regulatory mechanisms are vulnerable to attack from invading pathogens. Some viruses express miRNAs, which enter the endogenous silencing complex, to repress mRNAs that interfere with their replication [297]. Others, such as HCV, hijack specific cellular miRNAs and use it to its own advantage, rendering it essential for its own life cycle. HCV requires the action of miR-122, a liver-specific expressing miRNA [296, 298, 299], using it as a transcription factor to increase expression of the HCV genome, a mechanism essential for viral replication [300], making this an ideal candidate to inhibit the HCV infection. Miravirsen readily accumulates in the liver and as this is the site of action of the drug, does not require a specific delivery strategy. The outcome of the phase 2 clinical trials is positive, as patients who received miravirsen had undetectable HCV RNA levels [296], indicating that the replication of the virus was successfully repressed.

Another miR-122 targeting drug was developed by Regulus Therapeutics, RG-101. RG-101 is an N-acetyl-D-galactosamine-conjugated RNA antagomiR that targets miR-122 in HCV infected hepatocytes [301]. Like miravirsen, this drug showed great promise in lowering levels of HCV RNA to undetectable levels. However, due to adverse effects experienced by patients in clinical trials, further development of this drug, and other antagomiRs developed by the company, have been halted.

miRagen Therapeutics Inc. have a suite of miRNA antagomiRs in different drug development stages, with one—MRG-106—showing promise treating some cancers. MRG-106, a LNA antagomiR that targets miR-155, is in phase 1 (NCT02580552) and phase 2 clinical trials (NCT03713320). miR-155 regulates differentiation and proliferation of blood and lymphoid cells and is a suitable target for treating certain kinds of lymphoma and leukemia [302]. MRG-107 also targets miR-155 but has promise in treating patients with amyotrophic lateral sclerosis (ALS). However, this drug is not yet in clinical trials.

However, it has become clear that many pathological conditions are associated with multiple miRNAs and a given gene may be regulated by multiple miRNAs. These properties of miRNA regulation may well create some uncertainties of outcomes by applying the AMO technology to silence miRNAs since knocking down a single miRNA may not be sufficient, and certainly not optimal, to achieve the expected interference of cellular process and gene expression which are regulated by multiple miRNAs. These facts also prompted us to raise several pertinent questions.

If targeting a single miRNA is adequate for tackling a pertinent pathological condition? If simultaneously targeting multiple miRNAs relevant to a particular condition offers an improved approach than targeting a single miRNA using the regular AMO techniques? How can we concomitantly silence multiple miRNAs to achieve an interference of a cellular function? To answer these questions, our group innovated the AMO approach by engineering multiple AMO units into a single sequence that are designed to simultaneously knockdown multiple miRNAs of different seed sites, which we named "the multiple-target AMO or MT-AMO" [292]. We validated the technique with two separate MT-AMOs: antimiR-21/antimiR-155/antimiR-17-5p and antimiR-1/antimiR-133. We first verified the ability of the MT-AMOs to antagonize the repressive effects of their target miRNAs using luciferase reporter activity assays and to specifically knock down the levels of their target miRNAs using real-time RT-PCR methods. We then used the MT-AMO approach to identify several tumor suppressors-TGFBI, APC and BCL2L11 as the target genes for oncogenic miR-21, miR-155 and miR-17-5p, respectively, and two cardiac ion channel genes HCN2 (encoding a subunit of cardiac pacemaker channel) and CACNA1C (encoding the alpha-subunit of cardiac L-type Ca^{2+} channel) for the muscle-specific miR-1 and miR-133. We further demonstrated that the MT-AMO targeting miR-21, miR-155 and miR-17-5p produced a greater inhibitory effect on cancer cell growth, compared with the regular single-target AMOs. Similarly, the MT-AMO for miR-1 and miR-133 knocks down both miR-1 and miR-133. Therefore, MT-AMO as an example of sequence-based MTD drugs may be used for its broad applications to the treatment of diseases involving multiple miRNAs and multiple genes [292–295].

The beauty of MT-AMO "single-agent, multiple-targets" strategy is that it can be either homogeneous carrying multiple identical antisense units targeting a specific miRNA to enhance the antagonizing effect or heterogeneous with multiple distinct antisense units targeting different miRNAs to expand targeting capacity. Considering that a given mRNA is regulated by multiple miRNAs and concordantly targeting multiple miRNAs might be required to achieve optimal interference of that gene, the MT-AMO approach may be superior to the regular AMO technique. Just taking cancer therapy as an example, an MT-AMO can be designed to target the miRNAs that are involved in tumor cell proliferation, cell death, metastasis, and vascular angiogenesis. Targeting any one of these aspects may not be optimal for controlling tumorigenesis and progression. Interfering concomitantly with all these processes should expectedly yield a better anticancer outcome. Our findings therefore suggest the MT-AMO approach as a useful tool in miRNA target finding and the potential of this novel strategy as a miRNA therapy for cancer and cardiac disorders.

Indeed, the MT-AMO approach has been widely used in basic research and translational studies since the development of this technology [292]. For example, an MT-AMO for miR-221, 21, and 106a, known to be oncomiRs (oncogenic miRNAs), were reported to effectively antagonize proliferation and migration of gastric cancer cells promoted by these three oncomiRs [303]. Another research team reported the development and validation of a short AMO multimers via phosphodiester linkers that are stable in plasma, but cleaved inside cells, releasing the active AMO monomers [304]. Compared to single miRNA-target AMOs, these multimers and MT-AMOs enhanced plasma protein binding and biodistribution to liver and in vivo efficacy against single or multiple targets with a single construct. In vivo, MT-AMOs synthesized in both RNase H-activating and steric-blocking oligonucleotide designs induced 4-5-fold higher potency and approximately twofold greater efficacy. An MT-AMO carrying the antisense sequences targeting miR-106b, miR-27a and miR-30d (MT-AMO-106b/27a/30d) was examined for its efficacy in knocking down the three miRNAs and regulating glucose metabolism by targeting the GLUT4 signaling pathway [305]. MT-AMO-106b/27a/30d was also proved to offer more potent effects than each of the singular AMOs.

MT-AMO has been used to inhibit oncogenic miRNAs [306].

There are essentially two different designs of single-stranded MT-AMOs that act by different mechanisms [307–309].

(a) First, "gapmer" (short DNA antisense oligonucleotide structures with RNA-like segments on both sides of the sequence) AMOs contain a DNA center that activates RNase H to cleave the target RNA: this core generally is flanked by 5′ and 3′ wings of nucleotides with ribose modifications such as 2′-methoxyethyl (MOE) or LNA which do not support RNase H cleavage, but which confer an increased affinity for hybridization.

(b) Second, "shortmer" (oligonucleotides or oligomers: short DNA or RNA molecules) and "mixmer" (an oligomer consisting of alternating short stretches of LNA and DNA) AMOs act by steric blocking and contain ribose modifications distributed in such a manner that the AMO does not activate RNase H. Historically, the standard AMO gapmers were ~20 nucleotides in length which was derived based on *in vitro* transfection experiments in which AMOs of at least 20 bases in length were found to be more effective than shorter AMOs. More recently, several groups have reported that the incorporation of LNA or cET makes it possible to use shorter gapmer AMOs of less than 15 nucleotides in length [310–313]: steric blocking shortmer ASOs can be as short as 8 nucleotides long [314].

One of the major factors affecting AMO pharmacokinetics and *in vivo* efficacy is their plasma protein binding. "Standard" 20mer phosphorothioate (PS) backbone AMO (PS-AMO) typically are ~95% protein bound in plasma (i.e., ~5% unbound) [315]. High plasma protein binding of AMO reduces renal excretion of the unbound AMO, the major route of *in vivo* elimination. This high plasma protein binding of PS-ASO is low affinity, and therefore facilitates uptake into peripheral tissues via higher affinity receptors [316]. A study tested the properties of the MT-AMO design with various AMOs, starting with a previously reported 8mer shortmer blocking miR-122 [317]. This 8mer was synthesized as a monomer, and as a homodimer or homotrimer linked via nuclease-sensitive phosphodiester DNA. The *in vivo* efficacy of these AMOs was tested by measuring the derepression of expression of the miR-122-regulated genes BCKDK and AldoA1 in the livers of normal mice dosed subcutaneously. The 8mer demonstrated a dose-dependent induction of BCKDK expression up to approximately fourfold at the highest dose of 10 mg/kg. A homodimer of this same oligo showed increased *in vivo* efficacy, reaching the same 4-fold induction of expression of the BCKDK target at a dose of just 3 mg/kg, and a homotrimer of the steric blocker had an even greater activity, reaching an approximately sevenfold induction of the target at the 3 mg/kg dose. Thus, the MT-AMO design in this case provides a greater than threefold improvement in *in vivo* potency compared to a traditional monomeric shortmer, with the added potential benefit of an increased peak induction of BCKDK expression to almost twice that achieved with the monomer.

Yet, we consider the work on MT-AMO published to date is still preliminary; to completely validate the MT-AMO technology as a gene therapy strategy, many important issues are to be unresolved. Future studies are warranted to test more MT-AMOs for the advantages of this approach and to get insight into more detailed mechanisms for the actions of MT-AMOs. Dose–response actions were not evaluated in this study due to the difficulties of optimizing the ratio of construct/lipofectamine to ensure efficient transfection. The optimal combination of targets for an MT-AMO remains unknown. In our study [292], we tested "two-in-one" and "three-in-one" MT-AMOs. In theory, "*N*-in-one" MT-AMOs (*N* could be any number of AMO units) can be designed to include more relevant AMOs. More rigorous studies are warranted to define the optimal combination of length of MT AMOs to optimize desired effectiveness. This work does not allow us to draw any conclusions as to what the optimal organization is for multiple AMO units (the order of the units) to be placed in a single MT-AMO molecule.

Multiple Seed-Sequence Anti-miRNA Antisense Inhibitor

Ebert et al [283] invented a miRNA inhibitor AMO containing multiple, tandem binding sites to a specific miRNA of interest, which they termed "miRNA sponges." When vectors encoding this sponge are transiently transfected into cultured cells, sponges derepress the targeted miRNA strongly. Notably, the sponge specifically inhibits miRNAs with the same seed sequence (5′end 2–8 nucleotides), such that a single sponge can be used to block an entire miRNA seed family. Such a mode of action can be viewed as a special example of MTD as one AMO can target multiple miRNAs of an identical seed site.

Synthetic Antisense Oligonucleotide (ASO)

Antisense oligonucleotides (ASOs) are single-stranded, highly-modified, synthetic RNA or DNA sequences designed to selectively bind via complementary base-pairing to RNA which encodes the gene of interest, and have been tested in a number of disorders [318]. They are short (12–20-mer) nucleic acid sequences that bind, via Watson–Crick hybridization, to their complementary RNA substrates. Upon hybridization to their target mRNA or ncRNAs, ASOs inactivate these RNAs via two predominant termination mechanisms:

(a) Occupancy only, in which the binding of the ASO leads to translational arrest by preventing targeted RNA from attaching to the ribosome, inhibition of splicing, or induction of alternatively spliced variants [319–322].
(b) Occupancy-induced destabilization, in which AOSs can recruit RNase H, a ubiquitously expressed cellular endonuclease, to the target transcript promoting the degradation of the targeted RNA or modulating RNA function without degradation [319–322]. ASOs employing the RNase H pathway are widely used in the clinical setting, and this destabilization mechanism is utilized by all of the ASOs.

In addition, when binding to complementary sequences in pre-mRNAs, ASOs can alter the recruitment of splicing factors to the molecule regulating splicing events.

ASOs are highly selective therapeutics that have demonstrated broad in vitro and in vivo activities and produced clinical benefit based upon a concept originally proposed in 1978 [323]. This first original work was soon followed by the founding of commercial companies focused on developing antisense therapeutics in the late 1980s [321]. Since the late 1980s, there has been steady progress in improving the technology platform, taking advantage of advances in oligonucleotide chemistry and formulations as well as increased understanding of the distribution and safety of ASOs. There are several approved ASO drugs and a broad pipeline in development. In addition, advances in understanding human disease, including the genetic basis for most monogenic diseases and the availability of the full human genome sequence, have created numerous therapeutic applications for the technology.

In addition, the beauty or further advantages of ASO technology is the possibility and feasibility of designing ASO heteromultimers and multi-targeting oligonucleotides (MTOs) to effectively inhibit multiple targets with a single MTO molecule. In NHBE cells, PDE4B/4D and 7A AONs dose-dependently and specifically inhibited expression of their respective target mRNA. When used in combination, PDE4B/4D and 7A AONs significantly abrogated the cytokine-induced secretion of IL-8 and MCP-1 to near baseline levels [324].

To investigate this type of MTO, Subramanian et al [304] selected a published phosphorothioate backbone 13mer LNA targeting apolipoprotein B (ApoB) [312] and synthesized heterodimer MTOs linking this 13mer to a newly developed 14mer LNA gapmer targeting apolipoprotein C3 (ApoC3) [311, 325]. To determine whether the cleavable phosphodiester linker was required for the improved *in vivo* activity, the investigators also synthesized a phosphorothioate-linked MTO. The MTO-mediated target knockdown was consistently superior to monomeric ASOs administered either individually or in combination. MTOs can be designed to target any RNA, including for example, mRNA, miRNA, and/or lncRNA. Compared to current short ASO designs, MTO should enable reduced dose levels and frequency of administration, which could have potential for broad experimental and human therapeutic applications.

Synthetic Small Interfering RNAs (siRNAs)
Small interfering RNA (siRNA), also named short interfering RNA (siRNA), is a small (21–23 nucleotide long), double-stranded complex with 3′-dinucleotide overhangs produced by cleavage of double-stranded RNAs by the enzyme Dicer, which triggers the RNAi pathway. The RNAi pathway evolved as a natural cellular defense mechanism against RNA viruses, which identifies pathogenic double-stranded RNA molecules and targets them for cleavage [326]. One strand of the siRNA duplex is loaded onto an Argonaute 2 (Ago2) protein, forming a protein-RNA complex, the RNA-induced Silencing Complex (RISC). Once RISC binds to the target mRNA at specific, perfectly complementary, binding sites, Ago2 cleaves the mRNA, leading to the recruitment of ribonucleases and degradation of the transcript and, therefore, inhibiting its translation [327]. siRNA binding to its target is highly selective and potent and can discriminate between sequences that differ by a single nucleotide [328]. This specificity in binding makes siRNA suitable therapeutic tools.

However, in humans, and other mammals, the RNAi pathway does not naturally occur. Instead, endogenously expressed human Ago proteins bind to small RNA molecules encoded in the human genome, called miRNAs, forming RISC, which has essential functions in gene regulation. However, as the RNAi and miRNA pathway share common proteins (such as Ago2) the proteins of the miRNA pathway can be used by synthetic siRNAs to bind to and cleave their targets. Introduced siRNA drugs compete with endogenous miRNAs for Ago2 binding, a limitation of all siRNA drugs. Therefore, the concentration of the RNA drug must be carefully optimized, as excess may saturate the RISC machinery, inhibiting its normal regulatory functions [265, 329, 330]. Despite this, the RNAi pathway can successfully be induced in humans and used to therapeutically to lower gene expression by adding synthetic siRNA to cleave its complementary target. Approximately twenty siRNA-based therapeutics have reached the clinical trial stages of drug development [331]. siRNA-based drugs are not limited to treating infections. Many in the drug development pipeline target disease-causing genes and gene variants involved in a wide variety of diseases including cancers, inflammatory disorders, and neuropathies [332].

When using synthetic, therapeutic siRNA to inhibit the replication of an invading pathogen, there is a chance that the targeted pathogen becomes resistant to treatment or escape mutants form. Targeting more than one gene of the pathogen or using different siRNA duplexes that target the same transcript can help reduce this risk. However, this increases the possibility of off-target effects. Short hairpin RNAs (shRNAs), RNA molecules that are processed by endogenous machinery to form siRNA duplexes, that encode two or more siRNAs have shown promise in delivering a multiplexed approach to treatment [332, 333]. siRNA drugs designed to disrupt mRNAs involved in cancer are currently in clinical trials [334]. Similar siRNA strategies can be applied to target lncRNAs. For instance, a siRNA directed against MALAT-1 lncRNA in prostate cancer cells resulted in downregulation of MALAT-1, inhibited cell growth, invasion, migration, and induced cell cycle arrest [335].

In general, synthetic RNA-based oligonucleotides are often chemically modified to contain 2′OMe, 2′-fluoro, 2′-*O*-methoxyethyl (2′MOE) or 2′4′-methylene (locked nucleic acids (LNAs)) ribose sugars [336–338]. These modified RNA-based oligos have an increased affinity to bind to and inhibit their target (e.g., target miRNA) in vivo, are more resistant to degradation and have increased bioavailability [339]. Furthermore, several animal studies have concluded that modified RNA-based oligonucleotides are specific, stable and non-toxic when delivered intravenously [340, 341]. Some RNA-based drugs are modified with a 3′ cholesterol [342]. Cholesterol is an essential component of animal cell membranes, is an efficient transmembrane transporter and is not toxic. The addition of a cholesterol moiety to siRNA will aid association of the siRNA to lipoprotein particles, lipoprotein receptors and transmembrane proteins. Different cholesterol conjugates can target the siRNA to specific tissues [343]. For instance, siRNA conjugated to a cholesterol that associates with high density lipoprotein (HDL) will preferentially target the liver, kidney, and the intestines. Low density lipoprotein (LDL) associated cholesterol will target the liver and is dependent on the liver cells expressing LDL receptors. Lipid-conjugated siRNAs are efficiently delivered to cancer cells [344] and have been found to efficiently decrease HIV-1 replication in infected cells [345]. Some RNA-based drugs in development are "self-delivering," which means that the therapeutic siRNA duplex requires no additional delivery vehicles for targeted distribution in the body. A self-delivering siRNA designed to treat malignant melanoma, conjugated to a 3′ cholesterol moiety via a tetraethylene glycol (TEG) linker, targets T cells [346].

However, many RNA-based drugs require specific nanoparticle delivery systems. For cell-specific delivery of RNA-based therapeutics, one strategy being investigated is attaching an aptamer to the drug. An aptamer is a nucleic acid sequence with a unique 3D structure that has a high affinity to a chosen cell target. Aptamer-siRNA molecules have been generated as one covalently bound polymer which aids receptor-targeting and siRNA delivery. This delivery method was tested for targeted delivery using an aptamer designed with high affinity for prostate-specific membrane antigen (PSMA) expressed on the surface of prostate cancer cells. The aptamer-siRNA molecule was internalized into the cell, allowing the RNAi directed silencing of the siRNA target [347]. Furthermore, in vivo tests revealed that the addition of a 20 kDa polyethylene glycol (PEG) moiety to the 5′ end to strand of an siRNA duplex increased circulation and bioavailability of the aptamer-siRNA molecule [348]. Targeting aptamers have also been designed to target HIV-1 specific proteins to selectively deliver siRNAs into HIV-infected cells, allowing for a more potent inhibition of HIV target transcripts [349]. Efficient delivery of siRNAs using nanoparticles and lipid-encapsulation will increase uptake and pharmacokinetic duration of drug delivery [350].

Antibodies are also used to direct siRNAs to specific cells, by binding specific surface receptors to ensure directed delivery of the therapeutic. These antibodies were modified to allow for a charge interaction with siRNA at their C-terminus, forming an antibody-siRNA delivery complex that specifically bound to the T cell surface protein CD7 to suppress HIV infection [351]. To enhance the targeting antibody, cell-penetrating peptides, less than 30 aa, can also be linked to aid movement of drug into specific tissues [352].

12.1.3 DNA Sequence-Based Strategy for Designing Epi-MTDs

There are currently two major strategies for DNA sequence-based epi-MTD discovery.

12.1.3.1 Multitarget Antisense Oligodeoxynucleotides (MT-asODN)

Synthetic antisense oligodeoxynucleotides (asODNs) are single-stranded DNA fragments or polydeoxynucleotides of around 20~30nt in length, which can be designed as single-target ODNs (ST-asODNs) or multitarget ODNs (MT-asODNs) [353–357]. asODNs are different from asRNAs at least in two aspects, though they both are inhibitors of gene expression via the specificity of Watson-Crick base pair hybridization. First, asODNs are artificially designed and synthetic DNA fragments, while asRNAs are either endogenous or synthetic RNA fragments. Second, asODNs are single-stranded constructs whereas asRNAs could either be single-stranded (e.g., ASOs, mRNAs, lncRNAs) or double-stranded (e.g., miRNAs, siRNAs) molecules or exist in double-stranded form but function in single-stranded form (e.g., miRNAs, siRNAs).

Normal DNA or phosphodiester oligodeoxynucleotides cannot be used as drugs because they are nuclease sensitive. However, oligodeoxynucleotides can become nuclease resistant after phosphorothioate modification with a sulfur atom substituted for one of the phosphate oxygen atoms at a nonbridging position [353–357]. asODNs technology is based on a simple and rational principle of Watson-Crick complementary base pairing of a short oligonucleotide with the targeted mRNA to downregulate the disease-causing gene product. The principle of asODN technology is the sequence-specific binding to target mRNA, resulting in the prevention of gene translation.

asODNs as therapeutic agents were proposed as far back as in the 1970s when the antisense strategy was initially developed [358, 359]. Nonetheless, it has taken almost a quarter of a century for this potential to be realized. asODNs have great potential for the treatment of many severe and debilitating diseases. The use of antisense oligonucleotides as therapeutic agents has generated considerable enthusiasm in the research and medical community, and concerted efforts from both industry and academia have made significant progress in turning these nucleic acid drugs into therapeutic. While the first-generation phosphorothioate antisense oligonucleotides are in clinical trials, a number of factors, including sequence motifs that could lead to unwanted mechanisms of action and side effects, have been identified [360]. The severity of the side effects of first-generation antisense oligonucleotides is mostly dependent on the presence of certain sequence motifs, such as CpG dinucleotides. A number of second-generation chemical modifications have been proposed to overcome the limitations of the first-generation antisense oligonucleotides. The safety and efficacy of several second-generation mixed-backbone antisense oligonucleotides are being evaluated in clinical trials. The immune stimulation affects observed with CpG-containing antisense oligonucleotides are being exploited as a novel therapeutic modality, with several CpG oligonucleotides being evaluated in clinical trials. A number of medicinal chemistry studies performed to date suggest that the immunomodulatory activity of CpG oligonucleotides can be fine-tuned by site-specific incorporation of chemical modifications in order to design disease-specific oligonucleotide therapeutics [360].

The last few years have seen a rapid increase in the number of asODN molecules progressing past Phase I, II and III clinical trials [359]. The first therapeutic asODN was approved in 1998 by the US FDA for clinical use. Twenty years later, in 2018, the first therapeutic asRNA (ASO) was approved [361]. Yet, despite the success of one product and several other ongoing clinical trials, challenges still exist in their stability, cellular uptake, disposition, site-specific delivery and therapeutic efficacy [355].

A study tested the hypothesis that a multitargeted asODNs approach is better than a single-targeted one in a rat model of asthma [362]. The investigators simultaneously delivered asODNs targeting the chemokine receptor CCR3 and the common beta chain subunit of the receptors for IL-3, IL-5, and GM-CSF at the time of ovalbumin challenge in sensitized rats. Fewer eosinophils were detected in bronchoalveolar lavage (BAL) of rats treated with both asODNs as compared to each asODN alone. Moreover, airway responsiveness to LTD4 was significantly decreased at lower doses in the two asODN-treated groups compared to a single asODN. Administration of individual asODNs was efficacious in inhibiting airway hyperresponsiveness, whereas concordant delivery of the two asODNs (targeting CCR3 and common beta chain) reduced the influx not only of eosinophils but also lymphocytes and macrophages in the lungs of rats as compared to the unmodified ODNs. This CDT asODN approach is more effective in preventing eosinophilia and airway hyperresponsiveness than inhibiting either pathway alone [362].

The first multitarget asODN based on MTD strategy is exemplified by a study demonstrating the effectiveness of a single asODN molecule incorporating the antisense sequence stretches for targeting three miRNAs (miR-221, miR-21, and miR-106a) in suppressing proliferation and migration of gastric cancer cells induced by these oncomiRs [303]. Such a multitarget asODN is superior to the single-target asODN for each of the three miRNAs and the combination of the three single-target asODNs.

Most recently, Zhang et al [363] developed a new drug delivery system that targets biofilms. They designed a multitarget asODN sequence based on conserved regions of the VicK protein-binding, which was carried by framework nucleic acid (FNAs) with tetrahedral conformations (FNAs) into bacterial biofilms, for inhibiting biofilm formation and virulence. Biofilm formation is responsible for numerous chronic infections and represents a serious health challenge. Bacteria and the extracellular polysaccharides (EPS) cause biofilms to become adherent, toxic, resistant to antibiotics, and ultimately difficult to remove. Inhibition of EPS synthesis can prevent the formation of bacterial biofilms, reduce their robustness, and promote removal [363–365]. VicK proteins bind specifically to the *gtfBCD*, *gbpB*, and *ftf* promoter regions [366] with a consensus binding sequence of TGTWAHNNNNNTGTWAH (where W is A or T and H is A, T, or C) [367]. Based on these findings, Zhang et al [354] designed an asODN with sequence DTWACANNNNNDTAACA (where D is A, G, or T and W is A or T) to target the conserved sequences of the gtfBCD, gbpB, and ftf promoter regions [368] and inhibit the *Streptococcus mutans* biofilm, which is related to the occurrence of various oral diseases. Once delivered to bacterial cells, the asODN significantly decreased EPS synthesis and biofilm thickness. The investigators attributed the superior efficacy over other approaches to that it simultaneously reduces the expression of all targeted genes (*gtfBCD*, *gbpB*, *ftf*) [363].

12.1.3.2 Complex Decoy Oligodeoxynucleotides (cdODN)

As another DNA sequence-based approach for epi-drug discovery, the decoy oligodeoxynucleotides (dODN) serves as a decoy sequence for a target transcription factor (TF), then inhibiting its binding to the authentic sequence at the promoter, and consequently hinders the gene expression [369–372]. A decoy is referred to something, or someone, that is intended to attract attention or deceive others by leading them into a trap or getting them away from a particular place. Transfer of double-stranded dODN corresponding to the cis-sequence at the promoter region of target gene of interest will bind to sequence-specific DNA-binding proteins (specifically TFs) resulting in attenuation of the authentic cis-trans interaction, leading to removal of TFs from the endogenous cis-elements with subsequent modulation of gene expression. In other words, a dODN carries a DNA recognition motif for the binding of target TF in order to trap it and block its effects. There are two major distinctions between dODN and asODN: **(a)** dODN is a double-stranded DNA fragment of around 15–22 bp long, whereas asODN is a single-stranded DNA fragment of approximately 20–30 nt in length and **(b)** dODN targets proteins to interfere transcription, whereas asODN targets mRNAs to block translation.

TFs are known to bind to cis elements in a cooperative manner, where one molecule of TF binds weakly but multiple molecules of the same TF engage in protein–protein interactions that increase each of their bindings to the cis element. To facilitate TF binding to a dODN, one can elevate molar concentration of dODNs, but this may well elicit toxicity. Otherwise, one can engineer multiple consensus sites into one dODN so that one molecule of dODN could provide a number of binding sites for the target TF, even at lower concentrations. For the sake of clarity, we call the originally defined dODN simplex decoy ODN (sdODN), because it generally contains only one binding site for a TF, and we call the dODN incorporating multiple binding sites for multiple TFs complex decoy ODN (cdODN).

The first experimental validation of sequence-based epi-MTD approach was reported on the application of cdODN in 2006 by our group [372]. In this study, we developed an innovative dODN technology that we term complex decoy oligodeoxynucleotide (cdODN) in which multiple cis elements are engineered into single dODNs attacking multiple target transcription factors. Specifically, we designed dODNs targeting NF-κB, E2F, and Stat3 separately and a cdODN targeting NF-κB, E2F, and Stat3 concomitantly. We evaluated effects of the cdODN on expression of cancer-related genes, viability of human cancer cell lines, and in vivo tumor growth in nude mice. The cdODN targeting all NF-κB, E2F, and Stat3 together demonstrated markedly enhanced efficacy and increased potency by approximately two-orders of magnitude compared with each of the dODNs or the combination of all three dODNs. The cdODN also showed earlier onset and longer-lasting action. Most strikingly, the cdODN has the ability to attack multiple molecules critical to cancer progression via multiple mechanisms, leading to elimination of regression. The findings suggest that the cdODN knocked down expression of the genes regulated by the target transcription factors. The cdODN strategy offers resourceful combinations of varying cis elements for concomitantly targeting multiple molecules in cancer biological processes and opens the door to "one-drug, multiple-target" therapy for a broad range of human cancers. In this sense, a cdODN is a MTD agent. The advantages of cdODN as a MTD strategy are likely to be ascribed to simultaneous interference of expression of multiple genes controlled by the target TFs. The beauty of this cdODN one-drug, multiple-target strategy is that it can be either homogeneous (carrying multiple consensus sites corresponding to a specific TF to enhance the efficacy and potency of desired effects) or heterogeneous.

This MTD strategy was subsequently employed by many other researchers in their studies [373–378]. For example, a study investigated the effects of a cdODN for combination targeting of Sox2/Oct4/c-Myc TFs that confer the stemness of cancer stem cells (CSC) and found that this cdODN efficiently eliminated CSC-like phenotypes and minimized the tumorigenic capability of the side proportion cells and the resistance to chemotherapy in three pancreatic cancer cell lines h-PCCLs: PANC-1, SW1990, and BxPc-3 and a xenograft model in nude mice [373]. Another group tested the effects of different simplex and complex dODNs against Nanog, OCT-4, Sox2, and Klf4, the master regulators in the maintenance of stemness in both cancer cells and CSCs, and the results showed that cdODN was superior to sdODNs in decreasing the expression of downstream genes [374]. Swindell et al examined the efficacy of a cdODN that concomitantly target psoriasis-activated TFs (i.e., FOXM1, ISGF3, IRF1, and NF- Swindell B) [375]. Johari and Zargan investigated the anti-stemness properties of a cdODN targeting Sox2 and Oct4 in mouse embryonic stem cells (ESCs). The results revealed that the synthesized cdODN significantly repressed the stemness properties of ESCs compared to controls through decreasing cell viability, arresting cell cycle in G0 /G1 phases, inducing apoptosis, and modulating differentiation in ESCs [376]. Another study which is also first-authored by Johari demonstrated that the same cdODN (targeting Sox2 and Oct4) could increase the sensitivity of cancer cells toward irradiation, which has a potential to eliminate the cancerous cells from tumors and support cancer treatment [377]. The efficacy of Oct4-Sox2 cdODN in inhibiting the cell growth and triggering differentiation of HT29-ShE cells, human colon cancer stem-like cells, by reversing epithelial to mesenchymal transition (EMT) [378].

The potential of dODNs as therapeutics have been well recognized [379–381]. It is known that TFs make up 6% of the human genome, ranking second because of their abundance, and have recently been considered a new class of candidate targets for drug discovery [382]. This offers resourceful combinations of varying cis elements for concomitantly targeting multiple molecules using cdODNs as a MTD epi-drug strategy, particularly biological processes. On the other hand, the dODN technology using TFs as molecular targets is emerging as a powerful strategy for gene therapy of a broad range of human diseases [379–381, 383, 384]. In theory, the cdODN one-drug, multiple-target strategy mimics the well-known drug cocktail therapy. Nevertheless, this MTD strategy is devoid of the weaknesses of the drug cocktail therapy, involving complicated treatment regimens, undesirable drug–drug interactions, and increased side effects. It noteworthy that the cdODN strategy has the potential to target specific types of cancer. For example, a cdODN can be designed to treat breast cancer in particular by targeting SNAIL (a zincfinger transcription factor) [385], the estrogen receptor responsive element [386], Brn-3b [387], SLUG (a zinc-finger transcription factor of the SNAIL family) [388], and Ets-binding sites. The estrogen receptor responsive element in the form of decoy has been shown to be effective in suppressing breast cancer cell growth. Brn-3b is a repressor of BRCA1 and SLUG is a repressor of BRCA2 (down-regulation and/or mutations of BRCA1/2 have been shown to be critical for breast cancer development). Moreover, the cdODN strategy can also be applied to other disorders in addition to cancer. For instance, tumor necrosis factor-α, GATA-4, FOG-2, and Janus tyrosine kinase-signal transducer and activator of transcription could be a reasonable combination for a cdODN aiming to treat heart failure by reducing apoptosis [389, 390]. A cdODN targeting Irx5, Irx3, and Etv1 may be applied to reduce regional heterogeneity of cardiac repolarization to minimize arrhythmogenesis, because these TFs have been shown to be expressed in transmural gradients across the ventricular wall [391, 392] and to be responsible for the transmural difference of a K^+ channel [392]. Therefore, the cdODN technology opens the door to one-drug, multiple-target intervention, providing promising prototypes of gene therapeutic agents for a wide range of human diseases.

Nevertheless, to validate cdODNs as a strategy for DNA sequence-based MTD epi-drug discovery, many important issues remain yet to be resolved. The optimal combination of targets for a cdODN remains unknown. In our study [372], we tested "three-in-one" cdODNs. In theory, "N-in-one" cdODNs (N could be any number of cis-acting elements) can be designed to include more relevant target TFs; however, larger cdODNs may hinder their penetration into the cells and nuclei and compromise the effectiveness. More rigorous studies are warranted to define the optimal combination of length and accessibility of cdODNs to optimize desired effectiveness. Our work does not either allow us to draw any conclusions as to what the optimal organization is for multiple cis elements to be integrated in a single cdODN molecule. Nonetheless, the study lays the groundwork for future exploitation on these subjects. Efficient delivery of dODNs into a cell is another challenge to using dODNs as therapeutic agents, as in other nucleotide-based technologies such as small interfering RNA, antisense, ribozyme, aptamers. Still another difficulty is to maintain an effective concentration of dODN within a cell for a sufficient period of time. At present, investigation on modifications of dODNs to enhance efficiency of transfection and to strengthen the stability within a cell so as to prolong the duration of actions is an active field of research. Constructing cdODN into virus vectors, such as adenovirus, lentivirus, might be a reasonable choice to offset, at least partially, the weakness of the nucleotide technologies.

12.1.4 Protein/Peptide Sequence-Based Strategy for Designing Epi-MTDs

Protein sequence-based drugs refer to agents that act by targeting DNA–protein interaction, RNA–protein interaction, protein–ligand interaction, or protein–effector interaction, based on the amino acid sequence of DNA-binding region, RNA-binding region or ligand-binding region in proteins. Specifically, protein sequence-based drugs primarily include decoy oligodeoxynucleotides (dODNs), inhibitors of RNA-binding protein (RBP), and decoy peptides (DPs) or peptide decoys (PDs).

12.1.4.1 Decoy Oligodeoxynucleotides (dODNs)

Decoy ODNs carry a DNA recognition motif for the binding of transcription factors in order to trap them and block their effects. The dODN approach has been introduced above as a DNA sequence-based MTD epi-drug discovery strategy for dODNs are double-stranded DNA fragments encompassing the binding sites for certain transcription factors [369–384]. On the other hand, however, dODNs can also be viewed as amino acid sequence-based approach since their sequences are designed based on amino acid–nucleotide interactions.

12.1.4.2 RNA-Binding Protein (RBP)/ Ribonucleoprotein (RNP)

Once an mRNA is transcribed, it is subjected to various processing, mainly controlled by RNA-binding proteins (RBPs) [393–400]. RBPs coordinately bind to each mRNA and form ribonucleoprotein (RNP) complexes; thus, a RNP is a complex of RNA and RNA-binding protein [393]. The composition of the RNP complexes is dynamic as RBPs can bind to various classes of RNAs such as ribosomal RNAs, mRNAs, snRNA, snoRNA, tRNAs, and non-coding RNAs. Approximately half of the RBPs manifest their function by binding to mRNAs and exert their distinct roles in regulating mRNA fate [393, 401]. Given the important role of RBPs in various cellular processes, such as transcription, splicing, mRNA stability, mRNA transport and translation, it comes as no surprise that alterations in RBP expression or RBP mutations or RBP function can lead to various diseases, including cancer [402–404].

RBPs were initially identified due to their capability to bind various types of RNAs through an RNA-binding domain (RBD) that forms steady secondary and tertiary structures. The classical RBDs include the K-homology domain (KH), RNA recognition motif (RRM), Zinc finger domain (ZNF), Pumilio homology domain (PUM), double-stranded RNA binding domain (dsRBD), and others [405]. Due to the recently developed high-throughput approaches, the number of RBPs has been largely increased to 1542 [406–411], comprising about ~7.5% of the protein coding genes. Interestingly, among them, only a quarter of RBPs contain classical RNA-binding domains (RBDs) and the rest of them contain non-canonical RBDs that have previously uncharacterized motifs [393, 406]. These non-canonical RBPs are unique in that they have intrinsically disordered regions that adapt their structure upon contacting to RNAs and subsequently mediate cell cycle regulation, metabolism, and signal transduction.

According to Mohibi et al [393], altered RNA metabolism due to an RBP malfunction can lead to genome-wide changes in the transcriptome and proteome of the cells and subsequently, affect cell growth, proliferation, invasion and death. Thus, it is not a surprise that altered expression of RBPs is a common phenomenon during development and progression of cancers. Thus, targeting RBPs for cancer therapy has gained more attention in the last decade. Recent years have seen increase in studies targeting the previously "undruggable" RBPs as it has become increasingly possible to target these molecules using diverse approaches.

On the other hand, over 2000 RNPs can be found in the RCSB Protein Data Bank (PDB) [412]. Furthermore, the Protein-RNA Interface Data Base (PRIDB) possesses a collection of information on RNA-protein interfaces based on data drawn from the PDB [413]. "RNP" can also refer to ribonucleoprotein particles. Ribonucleoprotein particles are distinct intracellular foci for post-transcriptional regulation. These particles play an important role in influenza A virus replication [398]. Thus, both RBP and RNP databases provide us rich resources for design and development of therapeutics targeting RBP/RNP.

Currently, there are an array of strategies for interfering RBP/RNP, which have been applied to basic research and bear the potential for protein/peptide sequence-based epi-drug discovery.

Antisense Oligonucleotides (ASOs) Against RBPs

To date, six ASO drugs have been approved for the treatment of diseases such as viral infections, hyperlipidemias, and neurological diseases [414], and several ASOs targeting RBPs have been designed for the treatment of neurodegenerative disorders [415–419]. ASOs against RBPs are being tested clinically and preclinically. For example, ASO against eIF4E [LY2275796] has been developed and has completed a phase I clinical trial [420]. Phase II clinical trials are now under way combining ISIS EIF4E Rx with carboplatin and paclitaxel for non-small cell lung cancer (NCT01234038) or with docetaxel and prednisone for castration-resistant prostate cancer (NCT01234025). Similarly, another eIF4E ASO, ISIS 183750, showed significant tumor suppression in a mouse model and has also been tested in phase I/II clinical trials in patients with advanced colorectal cancer [421]. Additionally, ASO against Musashi was recently developed and found to effectively block tumor growth in preclinical models of pancreatic cancer [422].

Silencing RBP by siRNAs

A liposome nanoparticle siRNA delivery system was developed by using transferrin as a targeting ligand and showed to reduce the tumor burden in mouse models of lung cancer [423]. This therapy was also able to reduce the metastasis in A549 based metastasis mouse model compared to control siRNA-based nanoparticles. siRNA targeting the RNA-binding protein Hu antigen R (HuR), also known as a universally expressed RBP, has been shown to impair malignant characteristics of pancreatic ductal adenocarcinoma cells [424]. Moreover, HuR silencing sensitizes triple negative breast cancer cells to radiotherapy due to induction of oxidative stress and DNA damage [425]. Furthermore, conjugation of siRNAs against HuR with Cy3-labeled folic acid-coated DNA dendrimer nanocarrier has shown dramatic anti-oncogenic activity in ovarian cancer [426]. In addition to HuR, siRNAs against eIF4E have also been tested for their efficacy in multiple tumors [427–430].

Targeting lncRNA:Protein Interaction

Many lncRNAs can interact with RBPs. As already described above, our research team found that a lncRNA ZFAS1 (lncR-ZFAS1) triggers intracellular Ca^{2+} overload, contractile dysfunction and the mitochondrial apoptotic cell death pathway in cardiomyocytes of myocardial infarction by acting as a SERCA2a inhibitor through direct binding to this critical intracellular Ca^{2+} handling protein [129, 145]. We also identified and characterized the "functional domain (FD)" of lncR-ZFAS1 that is capable of reproducing the effects of full-length lncR-ZFAS1: binding to SERCA2a protein and causing apoptosis and contractile impairment (see Sect. 12.1.2 for more details). Notably, the functional domain is well conserved across species and an ASO fragment antisense to the functional domain related to SERCA2a in the lncR-ZFAS1 sequence can specifically mask the binding of SERCA2a to endogenous lncR-ZFAS1 and disrupt the interaction so as to mitigate the detrimental effect of this lncRNA on SERCA2a leaving the other cellular functions of this same lncRNA unaffected [129]. This approach could well be applied to rescuing the cardiomyocyte apoptosis and contractile dysfunction in myocardial infarction.

In another study [431], we investigated the role and mechanism of lncRNA named myocardial infarction associated transcript (lncR-MIAT) in regulating cell homeostasis in the heart. lncR-MIAT was identified as a risk factor of myocardial infarction (MI) upon its upregulation by the single nucleotide polymorphism (SNP) and to predict left ventricular dysfunction in acute MI (AMI) [148]. One of the problems of infarcted myocardium is reduced supply of and relatively increased demand for energy. Mitochondria are cellular organelles serving as the powerhouse of cells, which through oxidative phosphorylation produce more than 95% of the energy in the form of ATP for cardiac cells under normal physiological conditions. Mitochondrial dysfunction is a hallmark of multiple cardiovascular disorders, particularly of ischemic heart disease [432]. Mitochondrial function is critically and delicately determined by its membrane integrity for maintaining the appropriate membrane potential ($\Delta\Psi_m$). Normal $\Delta\Psi_m$ is required for many proteins that are anchored in the mitochondrial membrane to function normally, of which the translocator protein (TSPO) that spans the inner and outer mitochondrial membranes is highly expressed in the cardiac cells, has been found to play a crucial role in regulating mitochondrial function through modulating the mitochondrial membrane potential [433]. We found that lncR-MIAT interacted directly with TSPO to induce opening of mitochondrial permeability transition pore (mPTP), result in a loss of mitochondrial membrane potential with declined $\Delta\Psi_m$ which in turned triggered cytochrome c release into cyto-sol, leading to apoptosis. We then identified the functional domain that is responsible for lncR-MIAT:TSPO protein interaction or binding and is highly conserved across various mammalian species. An artificially synthesized nucleotide fragment encompassing this functional domain can bind TSPO like full-length lncR-MIAT and reproduce the effects of the full-length lncR-MIAT on TSPO, mitochondria, apoptosis, and cardiac function. Most importantly, an antisense oligonucleotide fragment to the functional domain of lncR-MIAT can efficiently disrupt lncR-MIAT:TSPO binding and abrogate the deleterious influence of lncR-MIAT on mitochondrial function in MI.

Our studies proofed that both lncR-ZFAS1 and lncR-MIAT are detrimental lncRNAs in the setting of myocardial infarction through RNA:protein interactions and antisense fragments targeting their respective functional domains are able to block the interactions and reverse the deleterious effects of these lncRNAs. These findings prompted us to propose a rational design of a single nucleotide fragment embedding two antisense sequences to the functional domains of lncR-ZFAS1 and lncR-MIAT with a by linker in between. Such a fragment is expected to destruct both lncR-ZFAS1:SERCA2a and lncR-MIAT:TSPO. This approach might be generalized to other RNA:RBP interactions and is currently under investigations.

circRNAs as RBP Decoys

circRNAs have recently been demonstrated to regulate a wide spectrum of biological functions and pathophysiological processes. It is recognized that one circRNA may contain several sites for a single or multiple RBPs and thus regulate the function of RBPs by acting as an RBP sponge [193]. This phenomenon is observed in the binding of circRNAs derived from the Drosophila muscleblind (mbl) locus to MBL protein itself, a RBP and regulatory factor required for the differentiation of photoreceptor cells and muscle Z-bands. The binding of circRNA inhibits MBL function by preventing its binding to its mRNA targets [434]. Another example of circRNAs acting as an RBP sponge comes from the binding of circRNA derived from PABPN1 gene to HuR. HuR has been shown to bind PABPN1 mRNA and increase its translation; however, binding of circR-PABPN1 to HuR prevents its binding to PABPN1 mRNA, resulting in decreased PABPN1 translation [435] More importantly, sequestering of HuR by circR-PABPN1 may also affect other cancer relevant HuR targets such p53, c-myc and bcl-2. In addition to this, circRNAs are extremely stable and have a long half-life in cells. These properties of circRNAs make them a promising therapeutic strategy for targeting RBP functions in cancer. Natural or recombinant circRNAs capable of sequestering oncogenic RBPs can thus be used to treat cancers overexpressing the particular RBP. Although there have been no preclinical reports that have utilized circRNAs sequestering RBPs as therapeutic molecules for cancer treatment, this is likely going to change in the near future.

Aptamers

Aptamers represent another strategy used for targeting RBPs for disease therapy, albeit less frequently. Aptamers can be either nucleic acid- or peptide-based scaffolds and are selected on the basis of binding to a specific target [436]. These molecules fold into sequence specific three-dimensional structures that can recognize their unique target. They act like antibodies while also possessing several desirable properties such as smaller size, better stability, and no immunogenicity [437].

Nucleic acid aptamers consist of short, single-stranded DNA or RNA. Nucleic acid aptamers are selected using the procedure SELEX (systemic evolution of ligands by exponential enrichment), whereas peptide aptamers are selected using yeast two hybrid technique, phage display, mRNA display or other techniques [436, 438–440].

In recent years, peptide aptamers have emerged as novel molecular tools that are useful for both basic and applied aspects of molecular medicine. Peptide aptamers are small combinatorial proteins that are selected to bind to specific sites on their target molecules [440–444]. Peptide aptamers consist of short, 5–20 amino acid residues long sequences, typically embedded as a loop within a stable protein scaffold [440]. The short peptide region is responsible for binding with its target protein and the scaffold protein helps to enhance the binding affinity and specificity through restriction on the conformation of the binding peptide [442]. This unique structural feature allows peptide aptamers to bind with their target proteins with strong affinity and high specificity. The peptide aptamer approach employs high-throughput selection to identify members of a randomized peptide library displayed from a scaffold protein by virtue of their interaction with a target molecule. Peptide aptamers are combinatorial protein reagents that bind to target proteins with a high specificity and a strong affinity. By so doing, they can modulate the function of their cognate targets [443]. Due to their ability to specifically bind to and inactivate a given target protein at the intracellular level, they provide a new experimental strategy for functional protein analyses, both in vitro and in vivo [444]. In addition, by using peptide aptamers as "pertubagens," they can be employed for genetic analyses, in order to identify biochemical pathways, and their components, that are associated with the induction of distinct cellular phenotypes. Furthermore, peptide aptamers may be developed into diagnostic tools for the detection of a given target protein or for the generation of high-throughput protein arrays. Finally, the peptide aptamer technology has direct therapeutic implications. Because peptide aptamers introduce perturbations that are similar to those caused by therapeutic molecules, they can be used to identify and/or validate therapeutic targets at the intracellular level with a higher confidence level than is typically provided by methods that act upon protein expression levels. Moreover, the peptide aptamer molecules themselves should possess therapeutic potential, both as lead structures for drug design and as a basis for the development of protein drugs.

The first and only aptamer approved by FDA to date is the anti-VEGF aptamer pegaptanib (Macugen®; Pfizer/Eyetech) for the treatment of macular degeneration in 2004. Although initially designed for use in cancer treatment, it did not show efficacy in pre-clinical studies and hence has not been tested in a clinical trial for cancer. While several aptamers are in the development stages as anticancer agents, only two therapeutic aptamers have made it to clinical trials for the treatment of cancer [445, 446]. One of them is the aptamer AS1411 (previously called AGRO100) against the RBP nucleolin and it was the first aptamer in clinical trial for cancer treatment. Nucleolin regulates several essential cellular processes including regulation of RNA polymerase I transcription, maturation, and proper folding of the pre-ribosomal RNA, mRNA translation, mRNA stability and is overexpressed in cancers [447]. AS1411 is a 26-nucleotide DNA-based aptamer that forms stable G-quadruplex structures which are resistant to nucleases. AS1411 binds to the external domain of nucleolin and has shown growth inhibition in vitro and in vivo xenograft models of breast, lung, and renal cancers [448].

Peptides

Therapeutic peptides are receiving increasing attention for biomedical applications. Peptides present promising opportunities in targeted drug delivery due to their high specificity, selectivity, small dimension, ease of modification, and high biocompatibility. During the last decade, the potential of peptides as therapeutics or targeting ligands has boomed in nanotechnology and cancer research. There are several types of peptides being developed for cancer treatment [449] and some of these peptides are designed to target RBPs. For example, Yo and colleagues developed several 4EBP-based peptides that bound eIF4E and subsequently, disrupted the interaction of eIF4E to eIF4G, leading to inhibited cap-dependent translation [450]. Notably, when fused with gonadotropin-releasing hormone agonist, these peptides exerted potent antitumor effects in a mouse xenograft model of epithelial ovarian cancer without eliciting significant toxicity. Recently, Lucchesi et al [451] developed a small peptide that can inhibit tumor growth in vitro and in vivo, based on their previous finding that RBM38 inhibits p53 mRNA translation by preventing the binding of eIF4E to the p53 mRNA

[452]. The designed synthetic peptides correspond to the binding interface between RBM38 and eIF4E and are effective in relieving RBM38-mediated repression of p53 translation.

12.1.4.3 Decoy Peptides (DPs) and Cell-Penetrating Peptides (CPPs)

One approach to the rational design of protein interaction inhibitors is to use decoy peptides or peptide decoys. Decoy peptides are relatively short synthetic amino acid sequences or segments of protein primary sequence, which are derived from natural protein interfaces that mediate functional protein interactions by mimicking their corresponding endogenous proteins for binding target molecules but without eliciting the functionalities [453–457]. Thus, decoy peptides often retain the ability of the full-length prototype to bind the docking site of the folded protein and thereby block the signal transduction. Decoy peptides can be either soluble proteins, namely binding proteins, or inactive cell surface receptors. For example, decoy receptor proteins are the same as their respective natural receptor proteins but that additionally include a molecular trap that binds to, and identifies, certain ligands (e.g., inflammatory cytokines) with high affinity and specificity. That is, they mimic the binding structure of the receptor protein, bind to the docking site of the target ligand, and prevent the interaction of the ligand and receptor. However, these decoys are structurally inert and cannot initiate signaling pathways [453–457]. In recent years much research has been published describing how peptide decoys can offer new therapeutic approaches for the treatment of a wide range of illnesses.

Decoy Peptides in Drug Discovery for Immune Function

Toll-like receptors (TLRs) play crucial roles in the innate immune system by recognizing pathogen-associated molecular patterns derived from various microbes. TLRs signal through the recruitment of specific adaptor molecules, leading to activation of the transcription factors NF-κB and IRFs, which dictate the outcome of innate immune responses. Toll/IL-1R (TIR) domain-containing adapter-inducing IFN-β (TRIF)-related adapter molecule (TRAM) serves as a bridging adapter that enables recruitment of TRIF to activate TLR4 and thereby mediates the induction of TRIF-dependent cytokines [453–457]. A library of cell-permeating decoy peptides derived from TRAM TIR domain has been screened for the ability of individual peptides to inhibit TLR4 signaling in primary murine macrophages. Peptides derived from TRAM TIR BB loop (TM4) and C helix (TM6) inhibited the lipopolysaccharide (LPS)-induced activation of MyD88-dependent and TRIF-dependent cytokines, as well as MAPK activation. TM4 and TM6 did not block macrophage activation induced by TLR2, TLR9, or retinoic acid-inducible gene 1-like receptor agonists. Both TM4 and TM6 blocked coimmunoprecipitation of TRAM and TLR4 ectopically expressed in HEK293T cells. Both peptides also blocked the LPS-induced recruitment of MyD88 to TLR4 in primary murine macrophages. In vivo examination of TRAM-derived peptides demonstrated that all peptides that were inhibitory in vitro profoundly suppressed systemic inflammatory response elicited in mice by a sublethal LPS dose, and protected mice against a lethal LPS challenge. This research identifies novel TLR inhibitors effective in vitro and in vivo and validates the approach taken in this study as a rational way for development of signaling inhibitors and lead therapeutics.

Decoy Peptides in Anticancer Drug Discovery

Recent research has demonstrated the successful use of peptide decoys in assisting the development of potential new therapies for cancer [458–461]. A study outlines how peptide decoy technology has advanced and established itself as a novel therapeutic approach for breast cancer, the most common cancer among women. Absence of hormone receptors (estrogen and progesterone) and lack of overexpression of Human Epidermal Growth Factor 2 (HER2) make triple-negative breast cancer (TNBC) an aggressive subtype of breast cancer that is resistant to conventional therapies. Peptide decoys have emerged as a novel therapeutic approach for the treatment of TNBC. Peptide decoys can suppress oncogenic signaling pathways through a high-affinity interaction with specific ligands (such as inflammatory cytokines) that are involved in the pathogenesis of breast cancer. Previous studies have shown the effectiveness of peptide decoys in targeting other forms of cancer [461], with some studies suggesting that they could potentially be used in the future to develop vaccines and cancer immunotherapies.

Decoy Peptides in Drug Discovery for Cardiovascular Disease

Cardiac sarcoplasmic reticulum Ca^{2+}-ATPase (SERCA2a) plays a crucial role in Ca^{2+} handling in cardiomyocytes [129, 462]. Phospholamban (PLB) is an endogenous inhibitor of SERCA2a, and its inhibitory activity is enhanced via dephosphorylation by protein phosphatase 1 (PP1). Therefore, the inhibition of PP1-mediated dephosphorylation of PLB might be an efficient strategy for the restoration of reduced SERCA2a activity in failing hearts. Oh and colleagues developed decoy peptides that mimic phosphorylated PLB and thus competitively inhibit the PP1-mediated dephosphorylation of endogenous PLB [463]. The phosphorylation sites Ser16 and Thr17 are located within the flexible loop region (amino acids 14–22) of PLB. A synthesized a 9-mer peptide derived from this region (ΨPLB-wt) and two pseudo-phosphorylated peptides where Ser16 was replaced with Glu (ΨPLB-SE) or Thr17 was replaced with Glu (ΨPLB-TE). These peptides were coupled to the cell-permeable peptide TAT to facilitate cellular uptake. Treatment of adult rat cardiomyocytes with ΨPLB-SE or ΨPLB-TE, but not with ΨPLB-wt, significantly elevated the phosphorylation levels of PLB at Ser16 and Thr17. This increased phosphorylation of PLB correlated with an increase in contractile parameters in vitro. Furthermore, the perfusion of isolated rat hearts with ΨPLB-SE or ΨPLB-TE, but not with ΨPLB-wt, significantly improved left ventricular developed pressure that had been previously impaired by ischemia. These data indicate that ΨPLB-SE and ΨPLB-TE efficiently prevented dephosphorylation of PLB by serving as decoys for PP1, which provides an effective modality to regulate SERCA2a activity in failing hearts.

Decoy peptide (R10PIFLKRMPSI19P) has been documented to elicit beneficial role in ameliorating the end-stage organ damage related disorders [463–465]. These studies with decoy peptide including the "handle" sequence have focused on the association of the (pro)renin receptor and prorenin in the pathogenesis of hypertension, diabetes and chronic kidney disease. The rationale behind this decoy peptide is the discovery of the (pro)renin receptor that uncovered a novel function of renin/prorenin as the receptor ligands in addition to the enzyme and its precursor. The bindings of renin and prorenin to the (pro)renin receptor trigger two major pathways: the angiotensin II-dependent pathway as a result of the enzymatic activation of renin/prorenin and the angiotensin II-independent intracellular pathway involving hypertrophic, hyperplastic, and profibrotic signals [464, 465]. A specific blocker of the receptor or a decoy receptor was discovered through identification of the amino acid sequence of prorenin prosegment that binds to the receptor and leads to non-proteolytic conversion of prorenin to its active form. A peptide containing this sequence was found to block the binding of prorenin to its receptor. Its infusion in animal models of diabetes and low-renin hypertension significantly inhibited the development and progression of nephropathy, but (pro)renin receptor blockade had no benefit in the clipped kidney of 2K1C rats or rat models of high-renin hypertension. Since renin is still active without a (pro)renin receptor, (pro)renin-receptor blockade elicits a maximum benefit under low-renin conditions. Thus, (pro)renin-receptor blockade can be a useful therapy for chronic kidney disease with low renin levels in the plasma.

Decoy Peptides in Drug Discovery for COVID-19

The spike protein of SARS-CoV-2 recognizes angiotensin-converting enzyme 2 (ACE2) receptors on the host cell membrane to initiate entry. Soluble decoy receptors, in which the ACE2 ectodomain is engineered to block the spike protein with high affinity, potently neutralize infection and, because of close similarity with the natural receptor, hold out the promise of being broadly active against virus variants without opportunity for escape [466, 467]. A study described an engineered decoy receptor, sACE22v2.4 that tightly binds spike protein of SARS-associated viruses from humans and bats [466]. Saturation mutagenesis of the receptor-binding domain followed by in vitro selection, with wild-type ACE2 and the engineered decoy competing for binding sites, failed to find spike protein mutants that discriminate in favor of the wild-type receptor. Another report summarizes the efforts to engineer higher affinity variants of soluble ACE2 that rival the potency of affinity-matured antibodies [467]. Strategies have also been used to increase the valency of ACE2 decoys for avid spike interactions and to improve pharmacokinetics via IgG fusions. The intrinsic catalytic activity of ACE2 for the turnover of the vasoconstrictor angiotensin II may directly address COVID-19 symptoms and protect against lung and cardiovascular injury, conferring dual mechanisms of action unachievable by monoclonal antibodies. Soluble ACE2 decoy peptides therefore have the potential to be next-generation therapeutics for addressing the immediate needs of the current pandemic and possible future outbreaks.

Decoy Peptides in Drug Discovery for Hemophilia A

Other recent studies have suggested novel approaches to treat hemophilia A by neutralizing pathologic antibodies with peptide decoys [468, 469]. Hemophilia is an inherited disorder that can be fatal. Its symptoms are a result of factor VIII (FVIII) deficiency. It has been observed that patients with hemophilia develop anti-FVIII antibodies that inhibit FVIII procoagulant activity. Research has shown that peptide decoys can be used to neutralize the activity of FVIII inhibitors, allowing for an increase in FVIII activation and reducing symptoms of the illness. While research is still in its early stages, it is thought that peptide decoys could be developed to create new therapeutic approaches for neutralizing pathologic antibodies.

Decoy Peptides in Drug Discovery for Pain Relief and Gout

Research has also shown that peptide decoys can be used in deepening our knowledge of pain modulation and may even open the door to new pharmacological targets to reduce pain [455–471]. Mounting evidence has implicated a role for the interaction between the signaling adapter protein known as Toll-like receptor (TLR) and Toll/Interleukin-1 Receptor (TIR) in the modulation of pain by the sensory neurons. It has been observed that after ligand interaction with TLRs, intracellular signaling is activated by TIR, which also initiates the activation of signaling adapter proteins. While TLR and TIR have long been known to be vital to TLR signaling cascades, the exact function of these receptors in pain modulation had remained elusive. Recent research using peptide decoys has addressed this gap in knowledge. Scientists have used modified TLR TIR decoy peptides to discover that these domains affect LPS-induced intracellular calcium flux and excitation in sensory neurons, leading to induced tactile allodynia (pain sensitization). Through the use of peptide decoys, scientists were able to determine that the TLR TIR domains may serve as new pharmacological targets for pain relief.

There are several approaches to the intracellular delivery of biologically active macromolecules including decoy peptides. One such approach uses cell-permeable peptides (CPPs). The CPPs are a class of short peptides, often derived from a natural protein that can efficiently translocate across cell membranes and carry the "cargo" macromolecules inside cells [472, 473]. The CPPs do not require a protein carrier to translocate and therefore are effective in a large variety of cell types. The discovery of first cell-permeable proteins (and short peptides within them responsible for cell permeation) have laid a foundation for a new approach to the targeted delivery of diverse biologically active molecules inside cells [474, 475]. Approximately 1000 CPPs have been reported to date [476]. The length of the CPPs varies from 5 to 30

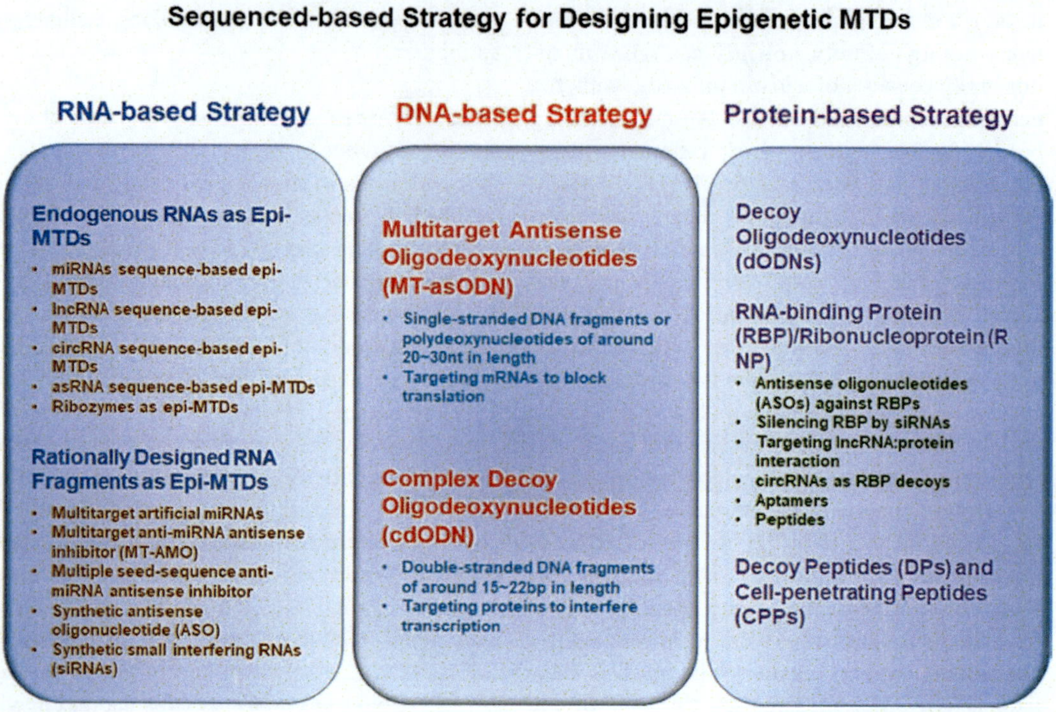

Fig. 12.1 A diagram providing an overview of various sequence-based strategies for designing multitarget epigenetic drugs (epi-MTDs)

amino acids. The CPPs have been used for biological delivery of diverse substances including peptides, nucleic acids, proteins, drugs, and imaging agents, and also large multi-molecular structures, such as liposomes or nanoparticles [472, 477, 478].

One particular CPP application is the intracellular delivery of decoy peptides [474, 479, 480]. This application uses the ability of many decoy peptides to bind and obstruct the protein docking sites. The described approach has been successfully used for specific inhibition of diverse signaling pathways, including the TLR signaling. This review summarizes available literature on the TLR-targeted cell-permeable decoy peptides (CPDPs) and assesses these molecules as potential therapeutics and tools to study TLR physiology.

The sequence-based strategies for epi-MTD design are summarized in Fig. 12.1 as a conspectus.

12.2 Sequenced-Based Strategy for Designing Gene Therapy MTDs

Gene therapy is a medical field which focuses on the genetic modification of cells to produce a therapeutic effect or the treatment of disease by repairing or reconstructing defective genes through transferring genetic material into cells of the patients [481, 482] . The first attempt at modifying human DNA was performed in 1980 by Martin Cline, but the first successful nuclear gene transfer in humans, approved by the National Institutes of Health, was performed in May 1989 [483]. The first therapeutic use of gene transfer as well as the first direct insertion of human DNA into the nuclear genome was performed by French Anderson in a trial starting in September 1990. The concept of gene therapy is to fix a genetic problem at its source and the objective for human gene therapy is to express DNA or RNA constructs (genetic materials) at a desired site in vivo for long enough and to reach sufficient levels to produce a therapeutic effect [482, 484]. If, for instance, in an (usually recessively) inherited disease an error or a mutation in a certain gene results in the production of a dysfunctional protein, gene therapy could be used to deliver a copy of this gene that does not contain the deleterious mutation, and thereby produces a functional protein. This strategy is referred to as gene replacement therapy and is employed to treat inherited retinal diseases. Over two dozen of gene therapeutics have been approved for clinical use by the drug regulatory agencies from different countries [485]. While the concept of gene replacement therapy is mostly suitable for recessive diseases, novel strategies have been suggested that are capable of also treating conditions with a dominant pattern of inheritance. For example, the introduction of CRISPR gene editing has opened new doors for its application and utilization in gene therapy, as instead of pure replacement of a gene, it enables correction of the particular genetic defect [486, 487]. Solutions to medical hurdles, such as the eradication of latent human immunodeficiency virus (HIV) reservoirs and correction of the mutation that causes sickle cell disease, may be available as a therapeutic option in the future [488].

As alternatives, therapeutic polynucleotides or oligos that generally carry a limited number (between 1 and 5) of modifying bases within a segment of DNA that is otherwise homologous to the target sequences have been used for gene therapy. The polynucleotide-mediated homologous exchange process relies on the endogenous cellular machinery to affect an exchange between the exogenous nucleotide sequences and the endogenous target sequences. This enzymatically driven "genetic surgery" maintains the integrity of the gene on one hand and minimizes permanent regional chromatin remodeling that could change gene expression patterns [489]. The cellular machinery involved in polynucleotide homologous exchanges is not well defined for human cells, but it is likely that DNA repair (particularly nucleotide excision repair, NER) and replication genes and pathways play a primary role the process of sequence-specific modification. However, the actual mechanisms that underlie /polynucleotide-mediated modification require further elucidation and are the subject of numerous studies. In general, therapeutic polynucleotides mainly include the following forms.

12.2.1 Single-Stranded Oligodeoxynucleotides (ssODNs)

In the middle of the nineties, the field of targeted gene alteration (TGA) emerged as a possible method to correct diseases caused by single-base mutations [490, 491]. Initially, the approach focused on stimulating the endogenous gene repair mechanisms using various single- or double-stranded oligonucleotides. These are complementary to part of the targeted gene except for one mismatched base specifically located at the site of the endogenous mutation. Upon cellular introduction these molecules will interact with the targeted gene sequence by different mechanisms. The mismatch is then recognized by components of the gene repair pathways, which subsequently can be stimulated to correct the mismatch by the use of the introduced targeting molecule [492–495]. Using TGA, mutated genes can be targeted and corrected without interfering with the endogenous promoter as well as enhancer/silencer elements and reading frames [496]. Such an impact has otherwise been seen with certain aspects of gene therapy introducing a complete gene sequence including all its associated elements [497, 498]. Several methods have been developed in order to optimize and effectively implement the TGA strategy *in vitro* as well as *in vivo*. These methods all constitute different structures of targeting molecules, pathways of integration and gene repair pathways stimulated, resulting in variable success rates [493, 499–501].

Single-stranded oligodeoxyribonucleotides (ssODNs) have been used for TGA. ssODNs are <200 bp in length and their common structure is simple comprising a single-stranded DNA sequence complementary to the target site except for a single mismatched nucleotide located centrally in the molecule [476]. Phosphorothioate-conjugates as well as 2′-O-methylated uracil bases can be used to create modified ssODNs which exhibit high levels of stability through resistance to endogenous RNase H activity [502, 503]. The invasion mechanism of these ssODNs is still unclear. However, several experimental results point to the involvement of DNA replication in the incorporation process with replication forks destabilizing the genomic nucleosome structure [504, 505]. Upon invasion, a three-stranded heteroduplex is formed between the ssODN and the double-stranded target site [492, 502].

The specific repair mechanism underlying ssODN-mediated TGA is still disputed. However, a general consensus on the suppressive role of the DNA mismatch repair (MMR) pathway has been established with several groups reporting a correction efficiency increase in Msh2-deficient cells [506–508]. Likewise, cells lacking the mismatch repair endonuclease Pms2 also showed a higher level of ssODN-mediated TGA [509]. The cellular introduction of ssODNs leads to an increase in the amount of genomic double strand breaks (DSBs) [508, 510]. This indicates a genotoxic effect of ssODNs but more notably that homology directed repair (HDR) could be involved in the TGA mechanism, despite the fact that ssODNs are complementary and not homologous to their target strands.

For multitarget drug discovery, there is a possibility of incorporating multiple ssODN units into one polynucleotides fragment for multiple TGAs. This notion is yet to be examined.

12.2.2 Chimeric RNA/DNA Oligonucleotides (RDOs)

Chimeric RNA/DNA oligonucleotides (RDOs) are another type of oligonucleotides which have been investigated for TGA. Compared to ssODNs, the RDO structure is more complex with a hairpin structure comprising a DNA strand, homologous to the targeted strand, pairing with RNA-nucleotides flanking the mismatched base [476]. The all-DNA strand of the RDO has been shown to be the only active player in the TGA process [511, 512]. To avoid degradation of the RNA-moieties by cellular nucleases these nucleotides are usually modified by 2′-O-methylation of the sugar units [512]. It is believed that upon target invasion a heteroduplex is formed causing cellular recognition of the newly formed mismatch and leading to nucleotide correction using the all-DNA RDO-strand as template [492]. However, RDOs are rarely used in gene correction studies today, primarily due to a lack of reproducibility of correction efficiencies [491, 492, 503, 512].

12.2.3 Small DNA Fragments (SDFs)/ Small Fragment Homologous Replacement (SFHR)

Small DNA-fragments (SDFs), also known as small homologous DNA fragments, can be used for TGA. The fragments usually comprise 400–1000 bp and are homologous to their DNA target sequence being able to concurrently modify up to 4 sequential base pairs in vitro as well as in vivo [513, 514]. SDFs induce genetic modification by means of a homology-based mechanism known as small fragment homologous replacement (SFHR) [514, 515]. The details of the SFHR mechanism are still unknown [515]. However, homologous pairing is believed to cause the endogenous DNA target sequence to be replaced by the exogenous SDF after the introduction of this fragment into the cell nucleus [516]. This replacement causes a genetic modification of the targeted mismatch. Surprisingly, the HDR repair pathway does not seem to be directly involved in the SFHR-mechanism [517, 518].

SDFs can be created as either double-stranded or single-stranded DNA molecules (dsSDFs or ssSDFs)—the latter by heat-denaturation of the double-stranded molecule [508]. Studies conducted using mammalian cells indicate no difference in correction efficiency between ss- and ds-SDFs [518, 519]. However, a study carried out using *E. Coli* indicates a higher efficiency using ssSDFs compared to dsSDFs [496]. This may be due to circumvention of an SDF unpairing process, which in this study is suggested to be the rate-limiting step of the bacterial SFHR process [496]. Like several other TGA techniques such as ssODNs, SDFs have shown relatively high correction efficiencies within episomal target genes in vitro as well as in vivo [493, 514, 520]. SDF-mediated episomal gene repair has been reported in mouse embryonic stem cells and in human hematopoietic stem/progenitor cells [496, 514, 515].

An advantage of SDF-mediated gene modification is the reproducibility of results [513, 520]. Small fragment homologous replacement (SFHR) modification using exogenous wild-type DNA fragments (SDFs) is distinct from ssODN-mediated modification in that the SDF polynucleotides used in SFHR (generally between 200 and 2000 bp/nt in length) are larger than the ssODNs and are either individual ssDNA, complementary ssDNA, or double-stranded DNA. SFHR effectively relies on **(a)** an SDF, carrying a single or multiple base alterations, finding its sequence homolog (genomic or episomal), and **(b)** the cellular enzymatic pathways that facilitate homologous exchange between target sequences and the SDF. Although there has been speculation about the mechanism(s) that underlie SFHR [514].

SFHR has been shown to occur in a variety of cell types both *in vitro* and *in vivo*, targeting a number of different genes related to inherited diseases. Previous studies have demonstrated SFHR-mediated modification of genes associated with sickle cell disease [496] and β-thalassemia [520], cystic fibrosis (CFTR) [514], Duchene's muscular dystrophy (dystrophin) [521], severe combined immune deficiency (DNA-dependent kinase catalytic subunit) [513], α1 antitrypsin deficiency (α1 antitrypsin) [522], spinal muscular atrophy (survival motor neuron-1) [523], and Lesch-Nyhan syndrome (hypoxanthine-guanine phosphoribosyl transferase, HPRT) [515].

Although most corrections are single base substitutions, studies with the CFTR concentrated on the most common mutation, a 3-bp deletion in human CFTR exon 10 resulting in a phenylalanine deletion at codon 508 (ΔF508) [524]. Approximately 1% of immortalized CF airway epithelial cells homozygous for the ΔF508 mutation have been functionally and/or genotypically corrected following transfection with an SDF that reinserts the 3 deleted bases [525–527], indicating a multitarget mechanism of the SDF strategy.

12.2.4 Triplex-Forming Oligodeoxynucleotides (TFOs)

Triplex-forming oligonucleotides (TFOs) have emerged as powerful tools for site-specific gene modification. Their sequence specificity, binding affinity, and ability to provoke repair and recombination make them promising reagents for altering gene expression [528–530]. TFOs and related molecules to direct site-specific DNA damage, inducing mutagenesis or sensitizing a site to recombination. TFOs were initially used as positioning devices for nonspecific mutagens but were later discovered to have mutagenic properties of their own in cells with functional nucleotide excision repair (NER) and transcription-coupled repair (TCR) pathways. In subsequent studies triplex formation was able to induce both intramolecular and intermolecular homologous recombination, revealing its potential application for gene therapy.

TFOs are short oligonucleotides (10–50 bp) consisting of RNA, DNA or synthetic derivatives (described later), which bind to the major groove of duplex DNA [531]. Hereby, the TFO functions as a 3rd strand in a DNA-TFO-DNA triplex [531, 532]. The specific binding is limited to homopurine tracts of the target sequence because the triplex is based on Hoogsteen bonds which are dependent on the available H-bond existing in purines [532, 533]. Once bound to the targeted DNA, electrostatic repulsions originating between the TFO and DNA duplex are believed to trigger an, as yet, unknown series of DNA repair pathways [531, 534]. These triple helices, originally described >50 years ago [535], were formed at polypurine or polypyrimidine regions of DNA bound via Hoogsteen hydrogen bonds to TFOs [536]. Because of their high binding affinity, TFOs have been used to manipulate genes and gene function, for example, for sequence-specific induced mutagenesis in plasmids [537] or mice [538] or for enhanced intrachromosomal recombination [539, 540]. TFOs have also been used successfully for gene targeting by tethering an SDF to a TFO directed to the proximity of the sequence to be modified. These studies showed correction of an episomal reporter gene that had been transfected into mammalian cells [541, 542]. In addition, TFO-mediated correction of adenosine deaminase-deficient human lymphocytes and p53 mutant human glioblastoma cells has been demonstrated [543]. The NER pathway has been shown important for this repair process [537, 544].

As sequences rich in polypurine and polypyrimidine often occur in the human genome [545], this sequence specificity also limits the ability of TFOs to target any DNA lesion. One approach to overcome this limitation has been to develop TFOs that have modified nucleotides such as locked nucleic acid or peptide nucleic acid residues to increase the affinity of the TFO for specific target sequences [546].

12.2.5 Triplex-Forming Peptide Nucleic Acids (PNAs)

In addition to TFOs, peptide nucleic acids (PNAs) are single-stranded triplex-forming molecules exhibiting target sequence complementarity [531, 547], which provide a functional alternative to TFOs. PNAs are 12–18 nucleotides with a DNA backbone completely substituted by uncharged N-(2-aminoethyl)-glycine polyamides [548]. This modification highly increases the stability of the molecule through nuclease and protease resistance [548]. Furthermore, it enables a stable complex formation with the target DNA because of no electrostatic repulsions between the molecules [528]. This stability can be further enhanced by PNA-conjugation of the DNA intercalator molecule, 9-aminoacridine [547]. PNAs exist as three different variants: PNA oligomers, bis-PNAs and pseudo-complementary PNAs (pcPNAs) [549, 550].

(a) PNA oligomers can engage in either DNA-PNA-DNA triplexes like TFOs or in a PNA-DNA-PNA triplex invasion complex with the second DNA strand displaced as a P-loop [551]. Both of these complexes depend, at least partly, on Hoogsteen bonds causing a similar restriction to homopurine tracts as seen with TFOs.

(b) Likewise, bis-PNAs (2 PNA oligomers connected by a linker) induce PNA-DNA-PNA triplex invasion complexes [548]. These molecules have been shown to successfully correct a β-globin splice site mutation in primary hematopoietic progenitor cells [547]. However, target restriction to homopurine tracts is considered to be a major drawback of the triplexing method. Thus, double-duplex forming pcPNAs are the primary molecules used in PNA-mediated TGA today.

(c) In pcPNAs, A and T nucleobases of the backbone are replaced with pseudo-complementary 2,6-diaminopurine (D) and 2-thiouracil (U_s) bases, respectively [552]. This incorporation sterically inhibits the otherwise stable PNA-PNA duplex formation and results in a double duplex invasion complex with the targeted DNA [524]. This type of invasion is solely dependent on Watson-Crick base pairing exempting pcPNAs from the homopurine target restriction [531]. Using N-(aminoethyl)-D-lysine entities the pcPNA backbone can be positively charged resulting in stable DNA duplex invasion complexes because of the electrostatic attraction between pcPNA and target [552]. The induced polarity furthermore enables invasion of G-C rich target sequences, which has otherwise been complicated by the lack of pseudo-complementary G-C nucleobases [552]. The modification has resulted in episomal correction frequencies of 0.65% [533]. However, the target sequence is still required to contain ≥50% A-T's in order to avoid PNA-PNA duplex formation [531]. Histone deacetylase (HDAC) inhibitor treatment following S-phase synchronization has furthermore led to chromosomal correction efficiencies of 0.78% indicating a role for DNA replication in the mechanism of pcPNA-mediated TGA [533]. The uncertainties concerning the TFO-mediated repair mechanism apply for PNA-based technology as well, with the mechanism employed by these techniques believed to be similar, if not identical [531]. Since this mechanism has yet to be elucidated the use of pcPNAs for TGA is still not fully exploited.

12.2.6 Adeno-Associated Virus Vectors

Targeted gene alteration using vectors based on adeno-associated viruses (AAVs) has been studied for more than a decade. AAVs are icosahedral viruses consisting of a 4.7 kb single-stranded genome encoding rep- and cap-genes important for viral replication and capsid formation, respectively [553]. These genes are flanked by two inverted terminal repeats (ITRs, 145nt each), which are cis-acting elements necessary for viral transduction and functionality in TGA. The ITRs are the only original viral elements present in recombinant AAV vectors (rAAV), where rep- and cap-genes have been replaced by the homologous target-specific DNA before cellular introduction [493]. For production of the viral vectors the rep- and cap-genes are provided in trans.

After entry of the vector into the cell, target-specific homologous DNA is believed to activate and recruit HR-dependent repair factors, such as members of the MRN complex, a protein complex consisting of Mre11, Rad50 and Nbs1, as well as Rad51 and Rad54 [554]. As seen with SDFs [504], the introduction of double-strand breaks (DSBs) as well as single-strand breaks (SSBs) following the transduction process has demonstrated a significant increase in AAV-mediated correction efficiency reaching levels as high as 65% [555]. This increase supports the involvement of homology directed repair (HDR) and non-homologous end joining (NHEJ) in AAV-induced genetic correction. Furthermore, S-phase dependency seems important with the S/G_2-arresting drug phleomycin leading to a ten-fold increase in the chromosomal correction efficiency of AAVs [493]. A direct correlation between intracellular AAV copy numbers and gene targeting frequency has been confirmed [555]. An advantage of AAV-based TGA is the success with which mesenchymal, hematopoietic and embryonic stem cells as well as induced pluripotent stem cells have been genetically targeted—with correction efficiencies ranging from 0.07–1% [556–559]. However, despite most groups only reaching stem cell efficiencies around 0.01–0.1%, the potential use of this technique to modify stem cells is revolutionary [493, 555, 557]. Based on high fidelity gene targeting, lack of pathogenicity and efficient in vivo delivery, AAV-mediated TGA shows great promise for the future.

12.3 Sequenced-Based Strategy for Designing Small-Molecule MTDs (SM-MTDs)

12.3.1 Small Molecules (SM) and Small-Molecule Drugs (SMDs)

A small molecule is a low molecular weight (< 900 daltons) organic compound that may regulate a biological process, with a size on the order of 1 nm. Many drugs belong to small molecules [560, 561]. Larger structures such as nucleic acids and proteins, and many polysaccharides are not small molecules, although their constituent monomers (ribo- or deoxyribonucleotides, amino acids, and monosaccharides, respectively) are often considered small molecules. Once inside the cells, small molecules can affect other molecules, such as proteins [562]. Small molecules have a wide spectrum of biological functions or applications, serving as cell signaling molecules, drugs in medicine, pesticides in farming, and in many other roles.

Pharmacology usually restricts the term "small molecule drugs" to molecules that bind specific biological macromolecules and act as an effector, altering the activity or function of the target. The upper molecular-weight limit for a small molecule is approximately 900 daltons, which allows for the possibility to rapidly diffuse across cell membranes so that it can reach intracellular sites of action [560, 561]. This molecular weight cutoff is also a necessary but insufficient condition for oral bioavailability as it allows for transcellular transport through intestinal epithelial cells. In addition to intestinal permeability, the molecule must also possess a reasonably rapid rate of dissolution into water and adequate water solubility and moderate to low first pass metabolism [563, 564]. A somewhat lower molecular weight cutoff of 500 daltons (as part of the "rule of five") has been recommended for oral small molecule drug candidates based on the observation that clinical attrition rates are significantly reduced if the molecular weight is kept below this limit [564, 565].

Small-molecule drugs (SMDs) have some unique characteristics and distinct advantages as therapeutics:

(a) SMDs are relatively simple chemical compounds and can be manufactured by chemical synthesis. These compounds are typically comprised of 20 to 100 atoms and have a molecular mass of less than 1000 g/mol or 1 kilodalton [kDa] (0.1–1 kDA)
(b) Most SMDs can be administered orally and can pass through cell membranes to reach intracellular targets and enter cells easily because they have low molecular weights. With the exception of therapeutic antibodies, many proteins are degraded if administered orally and most often cannot cross cell membranes. The oral bioavailability of small-molecule drugs is the most obvious advantage over biologics in drug development, whereas biologics generally require injection or another parenteral administration [566].
(c) SMDs can be designed to engage biological targets by various modes of action and their distribution can further be tailored, for example to allow for systemic exposure with or without brain penetration.
(d) SMDs can be natural (such as secondary metabolites) or artificial (such as antiviral drugs); they may have a beneficial effect against a disease (such as drugs) or may be detrimental (such as teratogens and carcinogens).
(e) SMDs include biological molecules such as fatty acids, glucose, amino acids, and cholesterol and secondary metabolites such as lipids, glycosides, alkaloids, and natural phenols. Small molecules do not include larger molecules such as polysaccharides, proteins, and nucleic acids and can be involved in biological reactions as a product or substrate [566].
(f) SMDs are easy to synthesize by chemical reactions and are cheaper than biologics. The small-molecule drugs are developed to follow Lipinski's rule of five to be made bioavailable to the patient and be cleared from the body after its action [564]. The Lipinski's rule of five-ADME governs that SMD has properties to be adsorbed by the human body, be easily distributed inside the human body, metabolized to an active drug, and then later excreted out form the system.

SMDs have dominated the pharmaceutical industry since the beginning of modern medicine, making up 90% of pharmaceutical drugs, although intertest in biologics is rapidly increasing. Small-molecule compounds and proteins/antibodies remain as the major forms of medications for medical use and the preferred modalities in drug development, acting mainly on protein targets such as enzymes, receptors, ion channels, transporters, and kinases [567–569]. Recent studies, however, revealed that small molecules are capable of interacting with and modulating RNA and DNA and these findings along with the development of related technologies open new avenues to target challenging disease pathways that have previously been considered undruggable. DNA- and RNA-modulating small molecules also provide new opportunities for epi-drug and gene therapeutics discovery based on the MTD strategy.

12.3.2 Small-Molecule Modulators of RNAs as Epigenetic SM-MTDs

The vast majority of SMDs in clinical use target proteins. Successful protein-targeted drugs include small molecules that bind proteins expressed by humans or those expressed by bacteria, viruses and other infectious organisms [570, 571]. According to Warner et al [570], if SMDs could be extended to modulate RNA, the landscape of targetable macromolecules would be expanded by more than an order of magnitude. For example, only roughly 1.5% of the human genome encodes proteins [572, 573]. Within the roughly 20,000 expressed human proteins, 10–15% of these are thought to be disease-related [574–576], meaning that disrupting or altering their activity is likely to have a therapeutically useful consequence. Within the group of disease-related proteins, many are termed "undruggable," meaning that these molecules lack distinctive cleft-like motifs into which small molecules can bind with high specificity and affinity [576]. The net implication of this analysis is that de novo human-designed small-molecule drugs target a sliver (the 3.5% of drugged proteins) of a sliver (1.5% of the human genome that encodes proteins) of the human genome. Currently approved protein-targeted drugs interact therapeutically with fewer than 700 gene products [577], meaning that only 0.05% of the human genome has been drugged.

12.3.2.1 Advantages of Targeting RNAs for Disease Therapy

RNA functions as a central conduit of information transfer in all biological systems and pathophysiological processes, and in principle, there are numerous opportunities are open to the discovery and development of SMDs targeting RNA [578, 579]. Mechanistically, targeting mRNA can lead to modulation of protein levels thereby functions, by either upregulating or downregulating translation efficiency or by altering mRNA abundance or stability. Moreover, by targeting mRNAs, activities of proteins that are very difficult to drug or undruggable might be modulated before or during their biogenesis. In addition, a large fraction of the human genome (~70%) is transcribed into ncRNAs [580]. For example, miRNAs are now well-validated therapeutic targets [27, 581]. It was estimated that human genome can produce more than 15,000 lncRNAs [582], and a subset of these may eventually prove to be good drug targets. If a fraction of the tens of thousands of mRNAs and ncRNAs were targetable, the extent of the druggable human genome could increase considerably.

12.3.2.2 Structural Requirements of RNAs for Small-Molecule Targeting

Although disease-related mRNAs and ncRNAs could be targeted by natural/endogenous or artificial/synthetic RNAs DNAs such as ASOs, asRNAs, siRNAs and miRNAs, conventional small-molecule compounds remain the preferred entities in drug discovery and development [571]. With the understanding of mechanistic actions of many natural antibiotic drugs, including aminoglycosides, tetracyclines, macrolides, and oxazolidinones, on rRNAs in the inhibition of microbial protein synthesis [583, 584], there is growing interest in discovering and developing RNA-targeting SMDs [570, 585, 586]. The notion of developing small-molecule drugs targeting RNAs is also supported by the fact that riboswitches can be selectively bound by small-molecule ligands in the control of gene expression [587].

Similar to protein targets, macromolecule RNAs are folded into highly structured entities for their interactions with small molecules or proteins [410, 571, 588]. Through complementary base pairings and other forms of physicochemical interactions, RNAs are folded into secondary (e.g., helices or stems, loops, and bulges), tertiary (e.g., junctions, pseudoknot, and motifs), and quaternary (e.g., complexes) structures [589–591]; thus, small-molecule compounds have the potential to directly interact with unique higher-order structures that are conferred by the primary sequences. Specific RNA structural elements or motifs, such as bulges, loops, junctions, pseudoknots, and complexes, may be recognized and bound by particular small molecules beyond the attenuation by RNAs with complementary sequences. Highly structured RNAs tend to contain pockets permitting specific and high-affinity binding by small molecules with particular functional groups and electrostatic surfaces, similar to those of protein targets. The interactions between small-molecule and disease-related RNA or RNA motif, therefore, could lead to the inhibition or activation of RNA functions or change of gene expression and cellular processes toward the control of disease [571].

Different from proteins as drug targets, which comprise a total of 22 proteinogenic amino acids, RNAs consist of only four primary nucleotide monomeric units (A, U, G, and C) as building blocks [571]. In comparison with 22 proteinogenic amino acids, showing a combination of physicochemical properties (e.g., basic and acidic, positively and negatively charged, hydrophilic and hydrophobic amino acids), the four monomeric nucleotides seem less variable. However, with each containing a nucleobase, ribose, and phosphate group, nucleotide monomers are much more complex than individual amino acids. As each phosphate unit carries one negatively charged residue, RNAs are highly charged macromolecules. Under low-salt conditions, many functional RNA molecules may not fold. RNAs are generally more hydrophilic than a typical protein [571, 587, 592]. In addition, a RNA molecule is surrounded by an ion atmosphere under physiologic conditions, and the folding and structure of electrostatic RNA, especially the sites for ligand binding, are relatively complicated and dynamic, and they are very different from those found in proteins [571, 586, 593, 594].

Warner et al [570] compiled and analyzed a collection of instructional molecules that target RNA and are active in a cell culture or animal model or in humans, focusing on molecules with reasonable drug-like properties, high potencies or both. They grouped the compounds first by the type of RNA structure they target: tertiary structures involving multiple closely packed helices, irregular and usually bulge-containing secondary structures or triplet repeats. They described that the four named molecules—linezolid, ribocil, branaplam, and SMA-C5—representing the current gold standards in biologically effective small molecules that target RNA target a complex RNA structural motif with each of them having a high quantitative estimate of drug-likeness (QED) score. Nonetheless, only after the functional screen was completed was it discovered that these molecules function by interaction with RNA. Thus, none of these examples provide clear guidance on how to intentionally target RNA with small molecules. The authors proposed RNA-targeted SMDs should look like conventional protein-targeted drugs; good RNA targets should have high information content; relationship between RNA information content and quality of the ligand pocket.

The SMDs discovered to date via strategies intentionally designed to target RNA generally suffer from trade-offs in drug-likeness and potency due to low information content and low pocket quality. They believed that a wide variety of screening strategies — including phenotypic assays, high-throughput screening, fragment-based screening, small-molecule microarrays, other biophysical partitioning approaches, and structure-inspired and computationally assisted design—will eventually prove useful. In this light, they proposed the following guidelines for future efforts directed at RNA-targeted SMD discovery.

(a) Focus on RNA motifs of sufficient complexity that they will be unique, able to form high-quality pockets and capable of binding small-molecule ligands with high affinity. Good RNA targets should have 30 bits and a pocket. Such motifs are likely to be common in large RNAs.
(b) Be cautious of early attempts to define specialized privileged chemical scaffolds for targeting RNA. These might exist, but a lot more information is likely to be required before promising scaffolds can be defined. Instead, emphasize rules derived from decades of development of protein-targeted drugs.
(c) De-emphasize "lessons" from the ribosome. Due to their uniquely high concentration in cells, ribosomal RNAs are a special case. The only engagement expected for a ligand

with micromolar affinity is a target that is present at micromolar concentration. Drugs targeting ribosomal RNAs thus achieve a unique kind of specificity by binding relatively weakly to the ribosome.

(d) Be cautious in interpreting the results of studies focused on compounds that are highly basic, intercalating, strongly stacking and/or highly hydrophobic. These kinds of compounds may bind RNA with high affinity but are unlikely to show strong discrimination between target and off-target sites and are likely to be difficult to improve in a medicinal chemistry campaign.

(e) Target RNA–protein interactions or RNA elements stabilized by protein binding. Recognition of a target-specific RNA element in conjunction with a general protein component of cellular metabolism or regulatory network might be a highly productive strategy for achieving both specificity (via a selective RNA-targeted component) and affinity (owing to the high information content of a protein interface).

(f) Use methods that enable rapid identification of RNA motifs with high-information-content structures and high-quality pockets. A clear advantage of RNA over proteins is that tools, such as quantitative chemical probing approaches, already exist that make it possible to identify high-information-content and structurally well-determined (low-entropy) target sites, to examine the influence of the cellular environment on a specific RNA structure and to ensure that simplified screening constructs recapitulate the native RNA motif.

(g) Develop additional tools and new refinements of current tools, including advanced structural biology and modelling strategies for RNAs with complex structures. These technologies can then be leveraged to identify and rapidly characterize the subset of complex RNA structures with high-quality ligand-binding pockets and to rapidly assess the impact of ligand binding on the functions of these sites.

(h) Implement reproducible strategies for connecting ligand binding to complex targets with clear therapeutic mechanisms. There is modest-to-good evidence that disrupting or stabilizing RNA structures near translation initiation sites can modulate gene expression in therapeutically useful ways. There is some evidence that the binding of small molecules to mRNAs can modulate the efficiency of translation. Both disrupting and stabilizing RNA–protein interactions are therapeutically promising. Covalent drugs are gaining acceptance in the protein-targeting field; analogously, ligands that can form covalent linkages or induce cleavage at an RNA target site also hold promise. (For more detailed information, see the articles by Warner et al., Nat Rev Drug Discov. 2018; 17(8): 547–58 and Yu et al., Pharmacol Rev. 2020; 72(4): 862–98).

12.3.2.3 Classes of RNA Targets

Ribosomal RNAs (rRNA)

Many natural, semisynthetic, or synthetic antibiotics (e.g., aminoglycosides, tetracyclines, macrolides, oxazolidinones, and phenicols) can directly bind to ribosomal RNAs to achieve the inhibition of bacterial infections. Discovery of the mechanistic actions of natural antibiotics on bacterial ribosome is a major driver for the discovery and development of RNA-targeted drugs [571]. Bacterial ribosome (70S) catalyzing protein synthesis comprises two subunits (50S and 30S) that are assembled from three rRNA chains (16S, 23S, and 5S rRNAs) and around 55 ribosomal proteins (e.g., L16) [584, 595, 596]. After the binding of 70S ribosome with the initiator tRNA and start codon of the mRNA positioned at the peptidyl-tRNA site (P-site), protein synthesis is initiated. Going through the cycle of accommodating aminoacyl-tRNA at the A-site, forming peptide bond between the A- and P-site tRNAs, translocating the peptidyl-tRNA from the A- to the P-site, and releasing the deacylated tRNAs from the exit site (E-site) that are stringently controlled by corresponding factors, a nascent polypeptide is synthesized. The termination of translation involves the hydrolysis of mature polypeptide from tRNA at the P-site, and the 70S ribosome is recycled after the release of deacylated tRNA. The rRNA-assembled ribosome is highly structured for its interactions with individual components required for protein synthesis, in which specific clefts and pockets can be directly blocked by small molecules, leading to the inhibition of translation.

Many aminoglycoside antibiotics, such as paromomycin and streptomycin, directly bind to 16S rRNA in the decoding region A-site on the bacterial 30S subunit to induce misreading of the genetic code and inhibit protein synthesis [596–599]. The drug–target interactions include the involvement of crucial nucleotides, such as A1408 in 16S rRNA, which when mutated (e.g., A1408G) or methylated at the N1 position (A1408m1A), confers resistance to aminoglycosides [600, 601]. The unfavorable base pairing between C1409 and A1491 in human rRNA versus perfect U1409-A1491 pairing in *Leishmania donovai* may contribute to the selective inhibition of pathogenic organisms by paromomycin instead of human ribosome [602]. The human ribosome as a cancer target for antibiotics has also been recognized [603]. (For more detail, see Yu, et al. Pharmacol Rev. 2020; 72(4): 862–98).

Viral RNA Motifs (VRMs)

Viral genomes contain certain well conserved RNA elements that are key to gene regulation and viral replication. The highly structured viral RNA motifs have emerged as potential targets for the development of small-molecule antiviral drugs [586, 604]. The HIV transactivation response (TAR) element located within the first 59 nt of the viral genome is one of the most studied RNA elements [605]. The TAR RNA element, folded into a hairpin structure [606], binds to the Tat protein to form a Tat/TAR complex that stimulates transcription and HIV-1 replication [607, 608]. Through viral screening of TAR dynamic structures against >50,000 compounds and fluorescence-based experimental assays, one study identified and validated six small molecules that were able to bind to TAR (Kd = 55 nM to 22 μM) and inhibit its interaction with Tat (Ki = 710 nM to 169 μM) [609]. Another well studied viral RNA target is the internal ribosome entry site (IRES) element located in the 5′UTR of the hepatitis C virus (HCV) viral genome [586, 604, 610]. Although the HCV IRES arises as a promising target for the treatment of HCV infections, no small-molecule inhibitors have entered clinical trials yet.

Additionally, other viral RNA motifs are also the potential targets for small molecules. For example, the SARS-CoV genome consists of an atypical three-stemmed RNA pseudoknot that stimulates-1 programmed ribosomal frameshifting to initiate translation of viral proteins essential for replication [611, 612]. Through virtual screening of about 80,000 compounds against a computationally constructed, three-dimensional SARS-CoV pseudoknot structure and experimental studies on a number of high-ranked compounds, a 1,4-diazepine analog, 2-[[4-(2-methylthiazol-4-ylmethyl)-(1,4) diazepane-1-carbonyl]amino]benzoic acid ethyl ester (MTDB), was identified to inhibit the −1 ribosomal frameshifting of SARS-CoV with an IC_{50} of 0.45 μM [613]. Indeed, MTDB directly interacts with SARS-CoV pseudoknot to block the formation of alternate conformers, as revealed by single-molecule force spectroscopy studies [614]. These findings support the concept of developing small-molecule inhibitors against SARS-CoV RNA pseudoknot for the treatment of pandemic severe acute respiratory syndrome, including the ongoing SARS-CoV-2, in which the pseudoknot is well conserved. (For more detail, see Yu, et al. Pharmacol Rev. 2020; 72(4): 862–98).

Riboswitches

Riboswitches are a diverse group of natural RNA "receptors" commonly found in bacteria that directly control gene expression critical for survival [587, 615, 616]. Riboswitches are usually located in the 5′UTR of particular mRNAs, and each consists of a ligand-sensing domain (or aptamer domain) and a regulatory domain (or expression platform). The specific binding of small-molecule ligand to an "aptamer" domain triggers conformational changes and stimulates the regulatory domain to govern target gene expression. Most bacterial riboswitches modulate gene expression through transcriptional or translational regulation, although some, including eukaryotic riboswitches, may provoke alternative splicing, and others can induce self-cleavage to cause mRNA degradation. In particular, various endogenous metabolite ligands, such as amino acids, vitamins, and flavin mononucleotide (FMN), present within cells above threshold concentrations can directly and specifically bind to the aptamer domains of riboswitches [617–619]. Entrapment of a ligand can stabilize the incorporation of the switching sequence into the aptamer domain, and thus the expression platform is induced to fold into a specific structure to modulate the expression of genes controlling the biosynthesis and transport of metabolites [616, 620].

Riboswitches have emerged as potential therapeutic targets because they are highly structured to form unique ligand binding pockets and function as gene regulatory factors, similar to the structural and functional selectivity of proteins. Many riboswitches are unique in bacteria, which upon binding to small-molecule ligands, turn off or on the transcriptional or translational expression of genes essential for pathogen survival or virulence [621]. Therefore, small molecules may be discovered and developed to mimic or block the actions of natural ligands to interact with riboswitches with reasonable affinity and specificity for the control of infections. Although challenges remain in the discovery of high-affinity small molecules to target bacterial riboswitches, recent efforts have led to the discovery of a few inhibitors with improved drug-like properties. Roseoflavin, a natural analog of riboflavin and FMN, was revealed to directly bind to the aptamer domain of bacterial FMN riboswitch with a Kd value of ~100 nM to exhibit antimicrobial activity, in comparison with that of FMN (5 nM) and riboflavin (3 μM) [618, 622]. Through phenotypic screening, a structurally distinct compound, ribocil, was discovered to mimic FMN actions to suppress FMN riboswitch–mediated gene expression and inhibit bacterial growth via direct binding to the aptamer domain (Kd = 16 nM) [623]. Although these compounds are still under preclinical investigation, these findings shall enlighten the development of a novel class of riboswitch-targeted antibiotics. (For more detail, see Yu, et al. Pharmacol Rev. 2020; 72(4): 862–98).

Precursor Messenger RNAs (pre-mRNAs)
Orally bioavailable small molecules have been reported to directly interact with pre-mRNAs to modulate splicing [571]. This strategy seems more feasible for the treatment of genetic disorders, especially those diseases with verified RNA targets. For instance, the neuromuscular disorder spinal muscular atrophy (SMA) is amenable to *SMN2* genetic variations that alter pre-mRNA splicing and thus deficiency of functional SMN protein and targeting SMN2 pre-mRNA splicing as an effective therapeutic strategy is proven by the approval of the ASO drug nusinersen [624, 625]. A pyrimidin-4-one analog, SMN-C3, was first identified as an effective SMN2 splicing modifier through chemical screening and optimization [626]. Being orally bioavailable, SMN-C3 was able to increase SMN protein levels in severe SMA disease Δ7 mouse models, improve motor function, and protect the neuromuscular circuit. Additional chemical modifications and preclinical efficacy and safety studies [627] led to clinical development of RG7916 (risdiplam). As risdiplam was well tolerated among healthy volunteers and exhibited effectiveness in shifting SMN2 splicing [628], phase II/III studies (NCT02913482) are underway to evaluate the safety and efficacy of risdiplam in patients with SMA. Branaplam (LMI070 and NVS-SM1) represents another class of small-molecule pyridazine analogs identified to modulate SMN2 pre-mRNA splicing for the treatment of SMA caused by deficiency of SMN protein [629, 630]. Branaplam was found to bind to the pre-mRNA of SMN2 and stabilize the transient dsRNA structure formed by the SMN2 pre-mRNA and U1 small nuclear RNA protein complex, a key component of the spliceosome, which enhances SMN2 pre-mRNA splicing and increases the expression of full-length SMN mRNA and functional SMN protein. With a favorable PK profile, branaplam was effective at improving survival of

SMA Δ7 mice in a dose-dependent manner. Branaplam has entered a first-in-human study on the safety, tolerability, PK, pharmacological actions, and efficacy of oral branaplam in infants with type 1 SMA (NCT02268552). (For more detail, see Yu, et al. Pharmacol Rev. 2020; 72(4): 862–98).

Primary and Precursor MicroRNAs (Pri-miRNAs and Pre-miRNAs)

The biogenesis of oncogenic miRNAs has been considered a target for SMDs, which can be achieved through the interference with Drosha-mediated processing of pri-miRNA within nucleus or Dicer-controlled cleavage of pre-miRNA in cytoplasm because both precursors are folded into hairpin structures available for ligand binding [571]. Computational approach has been used to design and identify lead small molecules based on the predicted human miRNA hairpin precursor structures [631]. Among 27 lead compounds, a benzimidazole analog was shown to selectively inhibit the processing of pri-miR-96 into oncogenic miR-96 and thus alter miR-96 target gene expression and induce apoptosis in cancer cells. The focus on the Drosha processing site of oncogenic pri-miR-96 and optimization of benzimidazole led to the discovery of a dimeric benzimidazole and bisbenzimide compound, targaprimir-96, which showed a favorable PK profile and was effective at releasing tumor burden in a triple-negative breast cancer xenograft mouse model [632]. In addition, the conjugate of targaprimir-96 with bleomycin A5 resulted in over 100-fold higher selectivity toward pri-miR-96 than DNA [633]. Very recently, another dimeric benzimidazole and bisbenzimide analog, targaprimir (TGP)-515, was identified to target pri-miR-515, leading to an induction of human epidermal growth factor receptor 2 expression levels in breast cancer cells and enhancement of the efficacy of human epidermal growth factor receptor 2 antibody therapeutics [634].

Likewise, a bisbenzimide (Hoechst 33342) analog called targarpremir-210, or TGP-210, was identified to bind to pre-miR-210, leading to the inhibition of Dicer-mediated processing to mature miR-210 in human carcinoma cells and the outgrowth of xenograft tumors in mice [635]. The attachment of a nuclease (RNase L) recruitment module onto targarpremir-210 offered a conjugate, TGP-210-RL, which recruited RNase L onto pre-miR-210 to induce the degradation of pre-miR-210 [636]. Interestingly, this TGP-210-RL conjugate exhibited high binding selectivity to the pre-miR-210 while lacking interactions with DNA as compared with TGP-210.

A small-molecule microarray-based approach was also developed for the identification of binders of RNA motifs [637]. After screening against a library of 727 pharmacologically active compounds being used in the clinic or human trials, a number of DNA-intercalating agents (and topoisomerases inhibitors), such as doxorubicin, ellipticine, and mitoxantrone, as well as other entities, were revealed to bind to RNAs. Further computational and experimental studies demonstrated that mitoxantrone was able to directly bind to pre-miR-21 and subsequently inhibit Dicer-mediated biogenesis of oncogenic miR-21. Although these pharmacological agents commonly exhibit multiple levels of mechanisms of action and druggability of these pri- and pre-miRNA inhibitors awaits further investigations, these findings support the development of small molecules to target oncogenic ncRNAs as anticancer drugs. (For more detail, see Yu, et al. Pharmacol Rev. 2020; 72(4): 862–98).

lncRNAs

SMDs can alter gene expression by perturbing the interactions of lncRNAs with chromatin modifying proteins. lncRNAs are folded into secondary structures that often change conformation upon interaction with ribonucleoproteins [133, 162]. PAN RNAs produced by KSHV and MALAT-1 both have an ENE in their structure, which stabilizes these lncRNAs and retains them in the nucleus. ENEs are activated when the 3′-poly (A) tail hybridizes to the U-rich internal loop in the ENE to form a triple helical structure. This triple helix confers stability upon the lncRNA from rapid deadenylation-dependent decay [133, 638]. Future small molecule inhibitors that can disrupt stable lncRNA secondary structure or inhibit lncRNA association with accessory proteins or target genes may provide other avenues to target dysregulated lncRNAs. Fatemi et al. used Amplified Luminescent Proximity Homogeneous Assay (ALPHA screen assay) to analyze the RNA–protein interactions and identified small molecules using high-throughput compound screening methods [631]. They reported the specific and quantifiable binding of brain-derived neurotrophic factor antisense (BDNF-AS) lncRNA to protein EZH2 (component of PCR2) and also identified a small-molecule inhibitor Ellipticine that upregulated its downstream target genes [133, 639, 640]. In-depth studies involving lncRNA secondary structures are necessary in the future to allow the possibility of identifying new and efficient small-molecule inhibitors that can specifically bind to the target lncRNAs.

Extensive secondary structures and long lncRNA size may hinder the design of effective small interfering RNAs (siRNAs) and SMDs. Toxicity might be observed with siRNA or antisense oligonucleotide (ASO)-mediated knockdown strategies designed to disrupt lncRNA functions. Single-stranded ASOs are highly unstable in cells and subject to nucleases. Other difficulties encountered with ASOs include low target affinity and low potency, which may require the use of higher concentrations that in turn could lead to off-target effects. Chemical modifications to ASOs, such as phosphorothioate modifications, heterocyclic modifications, 2′-O-methyl modifications, and 5′-, 3′-end-locked nucleic acid (LNA) modifications, are known to increase affinity and cellular uptake and decrease toxicity [133, 641].

Other RNA Targets

Some small molecules were also found to act on other types of disease-related RNAs or pathogenic RNA motifs. Many of them are for RNA repeats, such as the expanded GGGGCC repeat [r(GGGGCC)^{exp}, or r(G4C2)^{exp}, or G-quadruplex RNA], which causes frontotemporal dementia and amyotrophic lateral sclerosis [571, 642, 643]; the CGG repeat (rCGG^{exp}), which causes fragile X–associated tremor/ataxia syndrome [644–647]; and the CUG repeat r(CUG)^{exp} [648–650], which is associated with myotonic dystrophy type 1. Interestingly, there are several G-quadruplex motifs located on the 5′UTR of oncogenic KRAS transcript that could be targeted by 4,11-bis(2-aminoethylamino)anthra[2,3-b]furan-5,10-dione and 4,11-bis(2-aminoethylamino)anthra[2,3-b]thiophene-5,10-dione for the control of cancer cell proliferation and colony formation [650]. Another polyaromatic molecule, namely RGB-1, was found to stabilize RNA G-quadruplex, leading to the inhibition of RNA translation and suppression of proto-oncogene neuroblastoma RAS viral oncogene homolog expression in breast cancer cells [651]. In addition, a benzopyrrole analog named synucleozid was identified very recently as a binder to the structured iron-responsive element located on the 5′UTR of α-synuclein mRNA, thus reducing α-synuclein protein levels in human cells [652], which may provide insight into developing small-molecule remedies for α-synuclein–caused Parkinson disease and other α-synucleinopathies. (For more detail, see Yu, et al. Pharmacol Rev. 2020; 72(4): 862–98).

12.3.2.4 Challenges in the Discovery of RNA-Targeted Small-Molecule Drugs

Strong proof of principle for RNA-targeted SMDs has been provided by experiments and clinical trials using antisense oligonucleotides (several of which have gained regulatory approval), as well as preclinical studies with synthetic RNAs that redirect the cellular RNA interference (RNAi) machinery or that activate CRISPR-based systems [27, 590, 653–655]. Depending on the approach employed, it is possible to inhibit or to upregulate expression [652, 654] of particular mRNAs and to inhibit ncRNA function by targeting specific RNA sequences. However, these nucleic-acid-based approaches involve large, often highly charged molecules and present delivery challenges. Given the choice between an oligonucleotide-based therapy and a (currently hypothetical) small-molecule-based therapy that targets RNA, the small-molecule therapy would be preferred in many cases.

A number of small molecules that alter RNA function have been identified, providing encouraging evidence that RNA-targeted SMDs could be developed. Some of these studies are based on molecules identified in natural systems. For example, many RNA riboswitch regulatory elements that bind diverse small-molecule metabolites and regulate gene expression have been characterized [587], and small-molecule antibiotics produced by numerous bacterial species bind ribosomal RNA and interfere with translation [595]. Furthermore, a growing number of synthetic monovalent and multivalent molecules that bind to secondary structure elements or repeat-containing RNAs have been shown to modulate biological function [656]. While this progress is helping to address skepticism that RNA could be a viable small-molecule drug target, several key challenges remain to be addressed.

As stressed by Warner and colleagues [570], three key components are necessary to enable effective, repeatable and scalable RNA-targeted drug discovery: **(a)** a therapeutically compelling RNA target, **(b)** a screening approach that will identify drug-like lead molecules with appropriate pharmacological properties and **(c)** the identification of RNA motifs with sufficient information content such that high-specificity and high-potency binding to a high-quality pocket is achieved.

Difficulties in Identifying and Validating Druggable RNA Targets

There are a limited number of verified druggable RNA targets, and validation of a new therapeutic target requires tedious and extensive investigations. Two critical questions should be addressed for the identification of a druggable RNA target to control chosen disease [571]. The first question is whether the RNA target can be accessed by SMDs. Beyond the intrinsic properties of small molecules, the accessibility of target RNA is dictated by RNA properties such as the abundance of target RNA in diseased cells, the availability of highly structured and functional sites for small-molecule binding, and so on. rRNAs accounting for the majority of cellular RNAs (e.g., >80%) are folded to form functional binding pockets and clefts that are accessible by small molecules to intervene specifically, leading to the disruption of protein synthesis and thus bacterial growth. As oncogenic miRNAs are commonly overexpressed in tumor tissues and carcinoma cells, the unique hairpin structures of miRNA precursors amenable to Drosha or Dicer processing make them promising targets for the development of small-molecule inhibitors. Rather, it is necessary to verify selective binding of small molecules with the projected RNA target among complex cellular substances, such as other forms of RNAs, as well as DNAs and large quantities of proteins.

The second question is whether interactions between small molecules and RNA targets in cells and body can lead to an effective control of disease [571]. Prior knowledge of RNA functions related to the pathogenesis and progression of a chosen disease shall be helpful, which could have already involved extensive epidemiologic, genetic, and biochemical studies. Starting from phenotype, specific RNA targets might be pinpointed. On the other hand, the appearance or apparent management of disease with the gain or loss of RNA coding gene or function will strongly support the role of target RNA. As such, pharmacological perturbation of RNA target with small molecules becomes possible, which awaits critical and comprehensive studies using phenotypic assays and animal disease models before entering clinical investigations.

Specificity in Targeting RNAs

Similar to the selective actions of a drug on corresponding protein target, new RNA-targeted SMDs should bind to RNAs with reasonable specificity to elicit on-target therapeutic efficacy and avoid off-target adverse effects. This is actually a common principle for drug development, but it poses a big challenge for the discovery and development of novel RNA-targeted small-molecule modalities. As different RNA targets possess their intrinsic characteristics, the requirements for and strategies to achieve "selective" targeting may vary [585]. Although structural studies have clearly elucidated the binding of small-molecule antibiotics with rRNAs, it is inevitable that their interactions are based on the participation of essential proteins to assemble ribosomes [584, 595].

Moreover, the specificity of RNA-targeted small molecules could be further confounded by the fact that RNAs and DNAs are both negatively charged nucleic acids and have common structural features, as well as the observation that the majority of small-molecule binders identified thus far for RNAs are either fluorescent dye derivatives, polyaromatic molecules, or highly basic or polar molecules. Many compounds themselves (e.g., doxorubicin and mitoxantrone) are well known as multifunctional DNA-intercalating agents [657]. Some approaches have developed and successfully improved the selectivity toward target RNA motifs over DNA [633, 636]; however, the resulting conjugate becomes a much bigger molecule, with a molecular mass over 1500 Da, whose druggability warrants further and more extensive investigations. In any cases, further structural modification and optimization of lead RNA binders are necessary to improve the selectivity toward RNA target for safe and effective control or eradication of particular diseases [658].

Difficulties in Creating RNA-Targeted SMDs in a Repeatable and Scalable Way

We currently lack knowledge and technologies to create drug-like, RNA-targeted SMDs in a repeatable and scalable way. There is currently only a single human-designed and approved drug class that functions by binding RNA alone — the linezolid antibiotics, linezolid and tedizolid. According to Warner et al [570], successful targeting of RNA in therapeutically useful ways with small molecules requires three distinct components:

(a) Identification of RNAs and RNA motifs with disease-related function.
(b) Use of screening approaches and libraries likely to identify drug-like molecules with appropriate pharmacological properties.
(c) Identification of and focus on RNA motifs with sufficient structural sophistication that make it likely that high affinity and specificity in small-molecule binding can be achieved.

12.3.3 Small-Molecule Modulators of DNAs as Gene-Therapy SM-MTDs

DNA as carrier of genetic information is a major target for a wide range of drugs including SMDs, and drug–DNA interactions have been extensively and intensively studied in the recent years. SMDs may specifically or non-specifically interact with DNA and affect its functions [659–662]. These DNA-targeted SMDs include "small-molecule anti-genomic therapeutics (or SMATs)" that are considered new, broad-spectrum drugs that unify antibacterial, antiviral, and antimalarial activities into a single therapeutic that offers substantial cost benefits and logistic advantages for physicians and the military. Interaction between SMDs and DNA are primarily of two types: covalent interactions and noncovalent interactions.

12.3.3.1 Covalent Interactions between SMDs and DNA

Covalent binding in DNA is irreversible and invariably leads to complete inhibition of DNA processes and subsequent cell death [659, 660]. Cis-platin (cisdiamminedichloroplatinum) is a typical example of covalent binder used as an anticancer drug and makes an intra/interstrand cross-link via the chloro groups with the nitrogens on the DNA bases. Noncovalent binding is reversible and is typically preferred over covalent adduct formation keeping the drug metabolism and toxic side effects in mind. However, the high binding strength of covalent binders is a major advantage. Proteins are large molecules and bind quite strongly to the DNA, with binding constants in the nanomolar range. It has been difficult to achieve similar specificity and affinity using small noncovalent binders and remains a major challenge to the design of drugs for DNA.

12.3.3.2 Noncovalent Interactions between SMDs and DNA

A noncovalent interaction differs from a covalent bond in that it does not involve the sharing of electrons, but rather involves more dispersed variations of electromagnetic interactions between molecules or within a molecule [662]. Noncovalent interactions are critical in maintaining the three-dimensional structure of large molecules, such as proteins and nucleic acids. In addition, they are also involved in many biological processes in which large molecules bind specifically but transiently to one another. These interactions also heavily influence drug design, crystallinity and design of materials, particularly for self-assembly, and, in general, the synthesis of many organic molecules. Over the last four decades, intense research has focused on the effects of SMDs that noncovalently bind to nucleic acids. These interactions have been shown to disrupt replication and/or transcription culminating in cellular death. Accordingly, DNA binding compounds have potential applications as anticancer and antiviral agents. There are three major modes of noncovalent interactions: groove binding, intercalative binding, and electrostatic interactions.

Minor Groove Binders

Minor groove binding drugs are usually crescent shaped, which complements the shape of the groove and facilitates binding by promoting van der Waals interactions [663–665]. Additionally, these drugs can form hydrogen bonds to bases, typically to N3 of adenine and O_2 of thymine. Most minor groove binding drugs bind to A/T rich sequences. This preference in addition to the designed propensity for the electronegative pockets of AT sequences is probably due to better van der Waals contacts between the ligand and groove walls in this region, since A/T regions are narrower than G/C groove regions and also because of the steric hindrance in the latter, presented by the C2 amino group of the guanine base. However, a few synthetic polyamides like lexitropsins and imidazole-pyrrole polyamides have been designed which have specificity for G-C and C-G regions in the grooves.

For example, certain bis-distamycins and related lexitropsins show activities against human immunodeficiency virus (HIV)-1 and HIV-2 at low nanomolar concentrations [665, 666]. 4′,6-diamidino-2-phenylindole (DAPI) binds strongly to AT-containing polymers and is located in the minor groove of DNA. DAPI intercalates in DNA sequences that do not contain at least three consecutive AT bp. Berenil can also exhibit intercalative, as well as minor groove binding, properties depending on sequence. Furan-containing analogues of Berenil play an important role in their activities against Pneumocystis carinii and Cryptosporidium parvuam infections in vivo. Pt(II)--Berenil conjugates show a good activity profile against HL60 and U-937 human leukemic cells. Pt-pentamidine shows higher antiproliferative activity against small cell lung, non-small cell lung, and melanoma cancer cell lines compared with many other tumor cell lines. trans-Butenamidine shows good anti-P. carinii activity in rats. Pentamidine is used against P. carinii pneumonia in individuals infected with HIV who are at high risk from this infection. A comparison of the cytotoxic potencies of adozelesin, bizelesin, carzelesin, cisplatin, and doxorubicin indicates that adozelesin is a potent analog of CC-1065. Naturally occurring pyrrolo[2,1-c][1,4]benzodiazepines such as anthramycin have a 2- to 3-bp sequence specificity, but a synthetic PBD dimer spans 6 bp, actively recognizing a central 5′-GATC sequence. The crosslinking efficiency of PBD dimers is much greater than that of other major groove crosslinkers, such as cisplatin and melphalan. Neothramycin is used clinically for the treatment of superficial carcinoma of the bladder.

Intercalators

Intercalators contain planar heterocyclic groups which stack between adjacent DNA base pairs.

The complex, among other factors, is thought to be stabilized by π-π stacking interactions between the drug and DNA bases [667]. Intercalators introduce strong structural perturbations in DNA.

A study revealed that compared with the minor groove binding of the drug to [d(GCGATCGC)]2, previously studied by NMR spectroscopy, the interaction of DAPI with [d(CGACGTCG)]2 appears markedly different and gives results typical of a binding mechanism by intercalation [668]. However, another study investigated the mechanism of imidazoacridinone binding to DNA for a series of 13 analogs that differ in their cytotoxic potency [669]. These results show that the ability to intercalate into DNA is a prominent attribute of IA, although factors other than intercalative binding seem to be required for the biological activities of IA drugs.

Electrostatic Interactions

Electrostatic interaction (van der Waals interaction) refers to the attractive or repulsive interaction between objects having electric charges. There are three different types of electrostatic interactions namely, dipole–dipole interactions, London dispersion forces, and hydrogen bonds. For SMD–DNA interactions, electrostatic interactions play an important role in defining the binding affinity thereby the efficacy of SMDs in modifying DNA structure and function [670, 671]. For instance, at higher ratios of mitoxantrone, a promising antitumor drug having considerably reduced cardiotoxicity as compared to anthracyclines, over DNA, electrostatic interactions are dominant [672].

12.3.3.3 Forces Involved in DNA-Drug Recognition

Understanding the forces involved in the binding of proteins or SMDs to DNA is of prime importance due to two major reasons. First, the design of sequence specific drugs having requisite affinity for DNA requires a knowledge how the structure of the drug is related to the specificity/affinity of binding and what structural modifications could result in a drug with desired qualities. Second, identifying the forces/energetics involved in such processes is fundamental to unraveling the mystery of molecular recognition in general and DNA binding in particular. Some of the forces that are known to contribute to biomolecular recognition and also to DNA-drug binding are direct electrostatic interactions, direct van der Waals/packing interactions, complex hydration/dehydration contributions composed of hydrophobic component, solvation electrostatics, solvation van der Waals, ion effects and entropy terms.

Consider DNA-drug binding in an aqueous environment. DNA is polyanionic in nature and the drug molecule is also often charged. The associated counterions lie near the charged groups and are also partially solvated. When binding occurs, it results in a displacement of solvent from the binding site on both the DNA and drug. Also, since there would be partial compensation of charges as the DNA and drug are oppositely charged, some counterions would be released into the bulk solvent and are solvated fully. Also, the binding process would be associated with some structural deformation/adaptation of the DNA as well as the drug molecule in order to accommodate each other. All these events are associated with some energetic gains/losses, the comprehensive estimation of which is a major challenge. We are attempting to understand the energetics of DNA–drug interaction by theoretically estimating the above contributions employing classical and statistical mechanical methods. Developing a theoretical protocol for detailed quantitative analysis of DNA-ligand binding in solution is a daunting task due to some major challenges. Simulations of DNA with solvent and the attendant counterion atmosphere require careful consideration to ensure system stability. Also, evolving a computationally efficient technique using statistical mechanical principles for quantitative estimates of binding free energies in large biomolecular systems is an equally challenging task. Our study is aimed at providing such a theoretical protocol for complementing experimental techniques and facilitating a minute study of the structure-energy relationships in DNA-drug complexes.

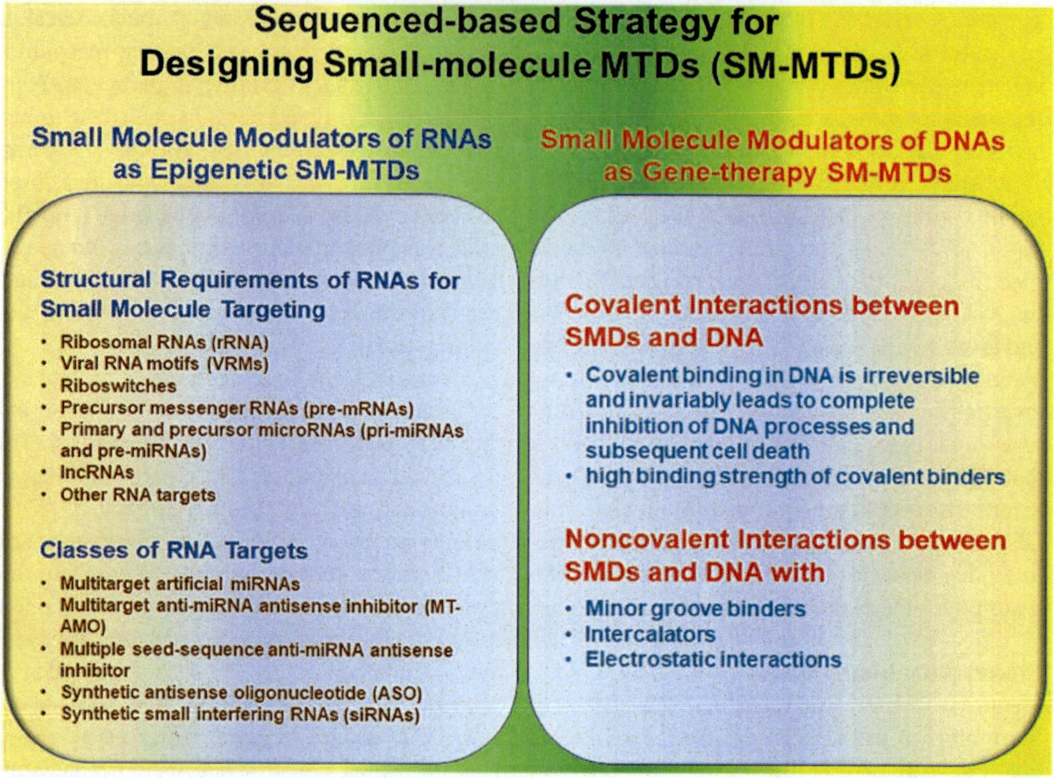

Fig. 12.2 A summary of various approaches for sequenced-based design of small-molecule multitarget drugs (SM-MTDs)

Structural and conformational changes in the DNA and drug on binding in solution are associated with enthalpic and entropic contributions to the binding free energy, which can be theoretically estimated from ensembles of structures generated via simulations. The only drawback of this approach is the long time taken for the simulations. The other terms, namely, electrostatics, van der Waals, hydrophobic component, rotational and translational entropy can be estimated from single structures. The web tool, PreDDICTA, estimates the components of DNA-drug binding free energy which can be calculated from a single structure, and correlates it with experimental binding free energy and ΔTm, thus providing a swift method for the evaluation of potential lead candidates for researchers pursuing structure-based drug design for DNA (Fig. 12.2).

12.4 Function-Based Strategy for Designing Epi-MTDs

Tomaselli and colleagues published an insightful review article on epigenetic polypharmacology with a focus on the strategies of the discovery of SMDs that act on epigenetic targets through interfering the key enzymes responsible for epigenetic regulation of gene expression [6]. It is clear that in addition to the sequence-based therapeutic strategies, the function-based strategies are equally important for SMAT epi-drug discovery and therapy. While the sequence-based approaches primarily rely on the sequence information of epigenetic targets, the function-based ones mainly take advantages of the functional roles of epigenetic targets. Additionally, the sequence-based approaches mostly deal with nucleotides and peptides, and the function-based strategies work primarily with SMDs.

12.4.1 Function-Based Epigenetic Targets

As already introduced at the beginning of this chapter, epigenetics is the study of heritable changes in gene expression (active versus inactive genes) that do not involve changes to the underlying DNA sequence—a change in phenotype without a change in genotype. Epigenetic change can have more damaging effects that can result in diseases like cancer. At least three systems including DNA methylation, histone modification and ncRNA-associated gene silencing are currently considered to initiate and sustain epigenetic change, which all involve in chromatin remodeling [6, 7, 673, 674]. New and ongoing research is continuously uncovering the role of epigenetics in a variety of human disorders and fatal diseases [673]. The renewed interest in epigenetics has led to new findings about the relationship between epigenetic changes and a host of disorders including various cancers, mental retardation associated disorders, immune disorders, neuropsychiatric disorders, and pediatric disorders.

Chromatin is the complex of DNA and proteins that are packed within the nucleus of mammalian cells. To form chromatin, DNA is tightly condensed by being wrapped around nuclear proteins called histones. This repeating DNA-histone complex, which consists of 146 base pairs of double-stranded DNA wrapped around eight histone proteins, is called a nucleosome. The repeating nucleosomal units are commonly referred to appear as "beads on a string" when seen under a microscope. Chromatin remodeling is the dynamic modification of chromatin architecture to allow access of condensed genomic DNA to the regulatory transcription machinery proteins, and thereby control gene expression [675–677]. Such remodeling involves a dynamic interchange between open (transcriptionally active) and closed (transcriptionally repressed or silenced) chromatin structures and is controlled by enzymatic modification of the DNA and/or histone proteins, such as DNA methylation and histone acetylation, respectively. In general, the more condensed the chromatin, the harder it is for transcription factors and other DNA binding proteins to access DNA and perform their duties. Chromatin remodeling is highly implicated in epigenetics. Epigenetic modifications to histone proteins such as methylation/demethylation and acetylation/deacetylation can alter the structure of chromatin resulting in transcriptional activation or repression. In addition, RNA methylation plays an important role in epigenetic regulation of gene expression at the post-transcriptional level. Methylation represents the most common and important chemical modifications in biological systems, which is catalyzed by enzymes and can be classified into three primary categories: DNA methylation, RNA methylation and protein methylation that essentially represent transcriptional, post-transcriptional and post-translational regulation mechanisms, respectively.

12.4.1.1 DNA Methylation Modifications as Epigenetic Targets

Currently, DNA methylation is one of the most broadly studied and well-characterized epigenetic modifications dating back to studies done by Griffith and Mahler in 1969 which suggested that DNA methylation may be important in long term memory function [674].

DNA Methylation

DNA methylation is an epigenetic mechanism that occurs by the addition of a methyl (CH3) group to DNA, thereby often modifying the function of the genes and affecting gene expression. The most widely characterized DNA methylation process is the covalent addition of the methyl group at the 5-carbon of the cytosine ring resulting in 5-methylcytosine (5-mC), also informally known as the "fifth base" of DNA. These methyl groups project into the major groove of DNA and inhibit transcription. In human DNA, 5-methylcytosine is found in approximately 1.5% of genomic DNA.1 In somatic cells, 5-mC occurs almost exclusively in the context of paired symmetrical methylation of a CpG site, in which a cytosine nucleotide is located next to a guanidine nucleotide.

The addition of methyl groups is controlled at several different levels in cells and is carried out by a family of enzymes called DNA methyltransferases (DNMTs) that catalyze the transfer of a methyl unit from *S*-adenosylmethionine (SAM), the methyl donor co-substrate, to the C5 of cytosine residues at CpG dinucleotide sequences within DNA. Three DNMTs (DNMT1, DNMT3a and DNMT3b) are required for establishment and maintenance of DNA methylation patterns. Two additional enzymes (DNMT2 and DNMT3L) may also have more specialized but related functions. DNMT1 is the most abundant and is able to catalyze the methylation of hemi-methylated DNA maintaining the methylation status during cell replication, while DNMT3a and 3b seem to mediate establishment of new or *de novo* DNA methylation patterns [675–677]. 5-mC recruits proteins containing methyl binding domains (MBDs), leading to a global silencing of gene expression [678]. DNMTs are therefore considered natural epigenetic targets for modulating DNA methylation thereby gene expression and DNMT inhibitors (DNMTi) bear tremendous potential as epi-drugs.

DNA Demethylation

DNA demethylation is the removal of a methyl group from DNA. This mechanism is equally important as and coupled with DNA methylation [679]. The demethylation process is necessary for epigenetic reprogramming of genes and is also directly involved in many important disease mechanisms such as tumor progression [680]. Demethylation of DNA can either be passive or active, or a combination of both. Passive DNA demethylation usually takes place on newly synthesized DNA strands via DNMT1 during replication rounds. Active DNA demethylation mainly occurs by the removal of 5-methylcytosine via the sequential modification of cytosine bases that have been converted by ten-eleven translocation (TET) enzyme-mediated oxidation [681–684]. The TET family of 5-mC hydroxylases includes TET1, TET2 and TET3. These proteins may promote DNA demethylation by binding to CpG rich regions to prevent unwanted DNA methyltransferase activity, and by converting 5-mC to 5-hmC, 5-hmC to 5-fC (5-formylcytosine), and 5-fC to 5-caC (5-carboxylcytosine) through hydroxylase activity. The TET proteins have been shown to function in transcriptional activation and repression (TET1), tumor suppression (TET2), and DNA methylation reprogramming processes (TET3).

12.4.1.2 Histone Modifications as Epigenetic Targets

A histone modification is a covalent post-translational modification (PTM) to histone proteins which includes methylation, phosphorylation, acetylation, ubiquitylation, and sumoylation [685]. The PTMs made to histones can impact gene expression by altering chromatin structure or recruiting histone modifiers. Histone proteins act to package DNA, which wraps around the eight histones, into chromosomes. Histone modifications have diverse biological functions such as transcriptional activation/inactivation, chromosome packaging, and DNA damage/repair. In most species, histone H3 is primarily acetylated at lysines 9, 14, 18, 23, and 56, methylated at arginine 2 and lysines 4, 9, 27, 36, and 79, and phosphorylated at ser10, ser28, Thr3, and Thr11. Histone H4 is primarily acetylated at lysines 5, 8, 12 and 16, methylated at arginine 3 and lysine 20, and phosphorylated at serine 1. Thus, quantitative detection of various histone modifications would provide useful information for a better understanding of epigenetic regulation of cellular processes and the development of histone modifying enzyme-targeted drugs. Among histone modifications, (de)acetylation and (de) methylation of lysine residues are the most studied.

Histone Acetylation/Deacetylation

Histone acetylation occurs by the enzymatic addition of an acetyl group (COCH3) from acetyl coenzyme A. The process of histone acetylation is tightly involved in the regulation of many cellular processes including chromatin dynamics and transcription, gene silencing, cell cycle progression, apoptosis, differentiation, DNA replication, DNA repair, nuclear import, and neuronal repression [686]. The modifying enzymes involved in histone acetylation are called histone acetyltransferases (HATs) and they play a critical role in controlling histone H3 and H4 acetylation [687–689]. More than 20 HATs have been identified which can be classified into five families: GNAT1, MYST, TAFII250, P300/CBP, and nuclear receptor coactivators such as ACTR. Histone H3 acetylation may be increased by inhibition of histone deacetylases (HDACs) and decreased by HAT inhibition.

HDACs catalyze the hydrolytic removal of acetyl groups from histone lysine residues [690–692]. An imbalance in the equilibrium of histone acetylation has been associated with tumorigenesis and cancer progression. Detecting whether histone H3 is acetylated at its lysine residues would provide useful information for further characterization of acetylation patterns or sites, thereby leading to a better understanding of epigenetic regulation of gene activation as well as the development of HAT-targeted drugs. Similar to HATs, HDACs play a critical role in various cellular processes involving histone H3 and H4. So far, at least 4 classes of HDACs have been identified. Class I HDACs include 1, 2, 3, and 8. Class II HDACs are comprised of 4, 5, 6, 7, 9, and 10. Class III enzymes, known as sirtuins, require NAD+ cofactors and include SIRTs 1-7. The Class IV enzyme, which contains only HDAC11, has features of both Class I and II. HDAC inhibition displays significant effects on apoptosis, cell cycle arrest, and differentiation in cancer cells. HDAC inhibitors are currently being developed as anticancer agents.

In addition, acetylation modifications are mediated through histone acetyltransferases (HAT) and deacetylases (HDAC), and the balance of these enzymes regulates neuronal homeostasis, maintaining the pre-existing acetyl marks responsible for the global chromatin structure, as well as regulating specific dynamic acetyl marks that respond to changes and facilitate neurons to encode and strengthen long-term events in the brain circuitry (e.g., memory formation) [688]. Recent studies have implicated a fundamental role of reversible protein acetylation in the regulation of CVDs such as hypertension, pulmonary hypertension, diabetic cardiomyopathy, coronary artery disease, arrhythmia, and heart failure [689].

In many pathological entities, HDAC inhibitors (HDACi) are epi-drugs able to reactivate signaling pathways silenced by deacetylation, in cancer as well as in non-cancer diseases [6]. To date, 11 zinc-dependent human HDAC isoforms have been classified into three classes (I, II and IV), differing in structure, enzymatic function, sub-cellular localization, and tissue expression. In addition to these "classical" HDACs, the mammalian genome encodes another group of deacetylases, known as sirtuins (class III HDACs), which hydrolyze acyl-lysines through a different mechanism involving nicotinamide adenine dinucleotide (NAD^+) as co-substrate. Importantly, the HATs/HDACs have also thousands of non-histone protein substrates. A pharmacophoric model for HDACi has been described including:

(a) A cap group binding at the enzyme surface and partially exposed to the solvent.
(b) A spacer mimicking the lysine side-chain of the substrate.
(c) A moiety able to complex the zinc ion, crucial for the HDAC catalytic action.

Histone Methylation/demethylation

Methylation modifications at these sites usually affect the structural properties of chromatin, which in turn regulates gene transcription and affects various physiological processes of the organism. As a relatively stable type of post-translational modification, histones have methylation modifications on several lysine and arginine sites, but the main sites occur at the N-terminal lysine and arginine of H3 and H4 acid residues. A considerable number of methyltransferases and demethylases have been identified with the two cooperating to maintain the balance between methylation and demethylation of amino acids.

The precise temporal and spatial coordination of histone lysine methylation dynamics across the epigenome regulates virtually all DNAtemplated processes [684]. A large number of histone lysine methyltransferase (KMT) enzymes catalyze the various lysine methylation events decorating the core histone proteins. Mutations, genetic translocations and altered gene expression involving these KMTs are frequently observed in cancer, developmental disorders and other pathologies [693–695].

Histone methylation is defined as the transfer of one, two, or three methyl groups from S-adenosyl-L-methionine to lysine or arginine residues of histone proteins by histone methyltransferases (HMTs) [696–698]. HMTs control or regulate DNA methylation through chromatin-dependent transcriptional repression or activation. In the cell nucleus, when histone methylation occurs, specific genes within the DNA complexed with the histone may be activated or silenced. Several different histone methyltransferases exist that are specific for the lysine or arginine residue which they modify. Human histone methyltransferases (HMTs)constitute a large family that includes histone lysine methyltransferases (HKMTs) and histone/protein arginine methyltransferases (PRMTs). On histone H3 for example, SET1, SET7/9, Ash1, ALL-1, MLL, ALR, Trx, and SMYD3 are histone methyltransferases that catalyze methylation of histone H3 at lysine 4 (H3-K4) in mammalian cells. ESET, G9a, SUV39-h1, SUV39-h2, SETDB1, Dim-5, and Eu-HMTase are histone methyltransferases that catalyze methylation of histone H3 at lysine 9 (H3-K9) in mammalian cells. G9a and polycomb group enzymes such as EZH2 are histone methyltransferases that catalyze methylation of histone H3 at lysine 27 (H3-K27) in mammalian cells. Both H3-K9 and H3-K27 methylation mediates heterochromatin formation and also participates in silencing gene expression at euchromatic sites. Increased global H3-K27 methylation is also found to be involved in some pathological processes such as cancer progression. Therapeutic compounds targeting specific KMTs are currently being tested in the clinic, although overall drug discovery in the field is relatively underdeveloped [6, 699, 700].

On the other hand, arginine methylation of histones H3 and H4 promotes transcriptional activation and is mediated by a family of protein arginine methyltransferases (PRMTs) [701–703]. There are 9 types of PRMTs found in humans but only 7 members are reported to methylate histones. They can mediate mono or dimethylation of arginine residues. Based on the position of the methyl group addition, PRMTs can be classified into type I (CARM1, PRMT1, PRMT2, PRMT3, PRMT6, and PRMT8) and type II (PRMT5 and PRMT7). Type II PRMTs are found to be strongly implicated in diseases like cancer. For example, PRMT5 plays a role in the repression of certain tumor suppressor genes such as RB tumor suppressors, while PRMT7 overexpression is observed in breast cancer. Detection of activity and inhibition of type II PRMTs as well as other HMTs would be important in elucidating mechanisms of epigenetic regulation of gene activation and silencing, as well as benefiting cancer diagnostics and therapeutics. Yet, although small molecules specific to PRMTs with high potency in vitro have been identified [703], development of small molecules that can regulate the activity of PRMTs in vivo is urgently required for therapeutic purposes.

Intriguingly, methylation of DNA and of histone 3 at Lys 9 (H3K9) are highly correlated with gene silencing in eukaryotes from fungi to humans. Both of these epigenetic marks need to be established at specific regions of the genome and then maintained at these sites through cell division. The combinatorial pattern of DNA and histone modifications constitutes an epigenetic "code" that shapes gene-expression patterns by enabling or restricting the transcriptional potential of genomic domains. DNA methylation is associated with histone modifications, particularly the absence of histone H3 lysine 4 methylation (H3K4me0) and the presence of H3K9 methylation. The inhibition of DNA and histone methyltransferases (DNMTs and HMTs) is a promising approach for the therapeutic treatment of numerous diseases, including cancer [696, 697].

In addition to histones, many proteins can also be methylated. The non-histone proteins with methylation and demethylation modifications identified thus far include components involved in the transcriptional regulation of cell signaling pathways, such as receptor kinases and effector proteins, effector protein activators and repressors, transcription factors, and tumor suppressor p53.

Histone demethylation is the removal of methyl groups in modified histone proteins via histone demethylases [704]. These demethylases have been found to have potential oncogenic functions and involvement in other pathological processes. The discovery of histone demethylases demonstrates that histone methylation is not a permanent modification but rather a more dynamic process. Two major families of demethylases have been discovered: Lysine specific demethylase 1 (LSD1) and Jumonji domain containing (JmjC domain) histone demethylases (JMJD2, JMJD3/UTX and JARIDs) [705–708]. The specific amino acid residue and degree of methylation determines the demethylation enzyme. For example, on histone H3, mono- and di-methylated lysine 4 are demethylated by LSD1 (BHC110, KDM1) and tri-methylated lysine 4 by JARID (1A-1D); di- and tri-methylated lysine 27 are demethylated by JMJD3 and UTX (KDM6A) and mono- and di-methylated lysine 9 are demethylated by JMJD1 and tri-methylated lysine 9 is demethylated by JMJD2. Inhibition of histone demethylases may lead to histone re-methylation at specific residues important for chromatin dynamics and gene expression. Furthermore, detection of the activity and inhibition of these enzymes would be important in elucidating mechanisms of epigenetic regulation of gene activation and silencing and may benefit cancer diagnostics and therapeutics.

12.4.1.3 RNA Methylation Modifications as Epigenetic Targets

More than 170 modifications have been identified in native cellular RNAs including mRNAs, tRNAs, rRNAs, small nuclear RNA (snRNAs), and small nucleolar RNAs (snoRNAs) [709–713]. As one of the most common modifications, methylation occurs on either nitrogen or oxygen atoms at the post-transcriptional level with S-adenosylmethionine serving as the methyl donor for m6A formation. The most prevalent methylated nucleoside in eukaryotic mRNA is N6-methyl adenosine (m6A), which accounts for more than 80% of all RNA base methylations and exists in various species [709, 710, 714–719]. Around 7000 mRNAs have been found to possess m6A sites in both human and mouse cells [275, 720].

The biological function of m6A modification is dynamically and reversibly mediated by methyltransferases (writers), demethylases (erasers) and m6A binding proteins (readers) [721–724]. The methyltransferase complex is responsible for the catalyzation of m6A modification and is typically made up of methyltransferase-like 3 and 14 (METTL3 and METTL14, respectively) and Wilms tumor 1-associated protein (WTAP). METTL3 plays a central role in m^6A methylation by forming a dimer complex with METTL14 in a ratio of 1:1, which is mainly located in the nuclear spot area[21]. WTAP itself does not have methylation activity, but it can interact with METTL3:METTL14 dimer to regulate the recruitment of m^6A methyltransferase complex to mRNA targets, which in turn affects methylation efficiency. Erasers remove methylation by fat mass and obesity-associated protein and ALKB homolog 5. Readers play a role through the recognition of m6A-modified targeted RNA. The YT521-B homology domain family, heterogeneous nuclear ribonucleoprotein and insulin-like growth factor 2 mRNA-binding protein serve as m6A readers. The m6A methylation on transcripts plays a pivotal role in the regulation of downstream molecular events and biological functions, such as RNA splicing, transport, stability and translatability at the post-transcriptional level. The m6A modification is catalyzed by an unidentified methyltransferase complex containing at least one subunit METTL3. m6A modification on mRNAs mainly occurs in the exonic regions and 3′-untranslated region (3′-UTR) as revealed by high-throughput m6A-seq [722–724]. One significant advance in m6A research is the recent discovery of the first two m6A RNA demethylases fat mass and obesity-associated (FTO) gene and ALKB homolog 5 (ALKBH5), which catalyze m6A demethylation in an α-ketoglutarate (α-KG)- and Fe^{2+}-dependent manner [709]. These findings support the view that m6A methylation is a dynamic reversible RNA modification, or more specifically, m6A methylation catalyzed by METTL3 and demethylation catalyzed by RNA demethylases FTO and ALKBH5 are in constant balance [720].

Recent studies in model organisms demonstrate that METTL3, FTO and ALKBH5 play important roles in many biological processes, ranging from development and metabolism to fertility. Moreover, perturbation of activities of these enzymes leads to the disturbed expression of thousands of genes at the cellular level, implicating a regulatory role of m6A in RNA metabolism. And has been implicated in cancer [722, 723], drug resistance, virus replication and the pluripotency of embryonic stem cells. Recently, a number of studies have identified aberrant m6A methylation in cardiovascular diseases (CVDs), including cardiac hypertrophy, heart failure, arterial aneurysm, vascular calcification, and pulmonary hypertension [724]. METTL3 is believed to be an oncogene to promote the initiation and development of a variety of cancers, including hematopoietic malignancies and solid tumors, through depositing m6A modification on critical transcripts [725]. While METTL3 exhibits its oncogenic functions in most cancer types, it has also been reported as a tumor suppressor in some cases [726]. Given the vital roles of DNA and histone methylations in epigenetic regulation of basic life processes in mammals, the dynamic and reversible chemical m6A modification on RNA may also serve as a novel epigenetic marker of profound biological significances.

In addition to m6A methyltransferase complex (MTC) that comprises of a METTL3-METTL14 heterodimer core and other binding partners, such as WTAP, ZC3H13, VIRMA, and RBM15/15B [727–730]. In addition to the MTC, other m6A writers have also been identified in recent years, including METTL16, METTL5, and ZCCHC4, which are responsible for the deposition of m6A into structured RNAs, such as U6 snRNA, 28S rRNA, and 18S rRNA, and in some cases, the introns of mRNA [731–738]. These other enzymes could well be therapeutic targets too, for the diseases associated with their deregulation and dysfunction.

12.4.2 Chromatin-Targeted SMDs as Epi-MTDs

All three families of epigenetic proteins-readers, writers, and erasers-are druggable targets that can be addressed through SMDs. To date, nearly 30 chromatin-targeted drugs targeting epigenetic enzymes as well as analogues of epigenetic modifications have been approved since early 2000 for clinical use in the treatment of various diseases associated with epigenetic deregulation (e.g., to treat hematological malignancies). In addition, a wide range of epigenetic-based drugs are undergoing clinical trials. These agents were primarily developed based on the STD strategy (single agent on single target) for specific targeting of chromatin remodeling. These drugs were mainly designed based on functional characteristics of enzymes involved in chromatin remodeling. While these STD-based epi-drugs are effective to a certain extent, they all suffer from certain undesirable to unacceptable issues. The CDT strategy (multiple agents on multiple targets) helps relieve the problems but does not lead to an ultimate solution. The MTD strategy is expected to yield optimal therapeutic epi-drugs.

12.4.2.1 Epi-STDs in Clinical Use

The first epigenetic modulators reaching the FDA approval for the treatment of hematological malignancies (myelodysplastic syndrome) were the two DNMTi, azacytidine and decitabine, in 2004 and 2006, respectively [6]. However, despite their high efficacy in inhibiting DNA methylation, these drugs suffer from low selectivity, poor bioavailability, chemical instability, and toxic side-effects.

As for histone deacetylation, Vorinostat (SAHA, suberoylanilide hydroxamic acid) was the first pan-HDACi (non-isoform selective HDAC inhibitor) approved by FDA in 2006 for the treatment of refractory cutaneous T-cell lymphoma (CTCL).

Romidepsin (FK-228) is a natural prodrug approved by FDA in 2009 for the treatment of refractory CTCL, carrying a disulfide bridge that is reduced *in vivo* to release the thiol zinc binding moiety.

Belinostat and panobinostat are two other hydroxamate-containing pan-HDACi approved by FDA, the first in 2014 for the treatment of refractory peripheral T-cell lymphoma, the latter in 2015 for the treatment of refractory or relapsed multiple myeloma.

Several agents including entinostat and mocetinostat are two benzamide-containing, class I-selective HDACi currently in clinical trials for the treatment of numerous solid tumors. Abexinostat, pracinostat, quisinostat, resminostat, givinostat, and rocilinostat are examples of hydroxamates pan-HDACi (with the exception of rocilinostat, quite selective for HDAC6) in clinical trials for the treatment of several hematological and solid tumors. Among them, pracinostat, resminostat, and givinostat granted the status of orphan drugs for AML, hepatocellular carcinoma, and Duchenne muscular dystrophy, respectively [6, 739–741]. Nicotinamide is the only sirtuin inhibitor currently used in clinics for the treatment of solid tumors.

According to Tomaselli and colleagues [6], the main reasons for which a single epi-target approach can be ineffective is the emergence of pharmacological resistance. Phenotypic aberrations at the base of neoplastic diseases are the result of a complicated signaling network system that evolves and changes over time to evade drug-induced events such as cell death, growth inhibition, DNA repair, and metabolic alterations. The goal of drug discovery in the epi-drug research is no longer simply the design of highly potent and selective molecules acting against a known epi-target but is moving towards the identification of network-active compounds with multi-targeting properties. This latter strategy could be obtained by using the multitarget single drug approach [6, 15].

12.4.2.2 Epi-CDTs in Clinical Use

Epigenetic modulators, such as pan-HDACi or DNMTi, showed efficacy in the treatment of different kinds of tumors such as ovarian cancer [742], CTCL [743], breast cancer [744], and myelodysplastic syndromes [745]. To date, the combination therapy is more extensively explored in the clinic than the use of multi-targeting single drugs. The first clinical success of the CDT approach dates back to 1980 and consists in the treatment of childhood acute lymphoblastic leukemia, based on the co-administration of methotrexate, vincristine, prednisone and 6-mercaptopurine [746]. The combination of HDACi with cytostatic agents (such as doxorubicin and paclitaxel) has been widely evaluated in both preclinical and clinical contexts. The most commonly evaluated epigenetic combinations are those based on HDACi/DNMTi, since DNMTi incorporation into DNA is restricted to cell cycling cells [747]. For example, combination therapy based on decitabine and vorinostat has been recently evaluated against pancreatic cancer by Han et al [748], demonstrating improved effects *in vitro* when compared to the single drug application. Treatment with vorinostat or decitabine alone induced pancreatic cancer cells cycle arrest, as well as inhibition of migration and proliferation. Combination led to stronger effect compared to each epigenetic agent alone, resulting more effective in proliferation and migration inhibition, G2 cell cycle arrest, apoptosis induction, and in enhancing tumor suppressor genes (*P16* and *TP53*) expression levels.

Previous studies conducted by Meeran et al [749] revealed that a diet containing compounds with HDAC and DNMT inhibitory activities altered the epigenome, resulting in the modulation of various aberrant epigenetics patterns in tumor-related genes. Therefore, Paul et al [750] decided to evaluate the effects deriving from the combination of genistein and sulforaphane. Genistein/sulforaphane combination decreased the HDAC2 and HDAC3 expression both at protein and mRNA levels, downregulating KLF4, essential in stem cells formation, as well as *h*TERT that is known to be activated when KLF4 binds its promoter region. The combination resulted more effective than the single agents alone in decreasing cell viability, lowering colony formation and inducing apoptosis in breast cancer cell lines (MCF-7 and MDA-MB-231). The combination of platinum-based compounds/HDACi against non-small cell lung carcinoma (NSCLC) founded its rationale mainly in two findings: **(a)** HDACs play a main role in resistance and adaptation to genotoxic treatment in different ways [751–753]; **(b)** platinum chemotherapy is the first choice for the treatment of such tumors [754–756].

As pointed out by Tomaselli et al [6], one of the central aspects of multi-medication therapy (or the CDT strategy) consists in the possibility to choose different dosages, establishing a prolonged administration of drugs and increasing the number and type of potential therapeutic schedules, paving the way to a personalized treatment. The CDT approach allows to modulate the dose of each single drug within the cocktail (if necessary), according to the patient's needs [15, 757–759]. Thus, the CDT could increase or maintain the desired therapeutic efficacy potentially requiring a lower dose of each drug, thus minimizing drug resistance and adverse reactions compared to single-drug treatment [760]. Despite the MTD overcame some of the typical limitations of the CDT and FDC ("Single Pill with Multiple Agents on Multiple Targets for Single Disease" or polypill) strategies, with in principle better results at both PK and PD level (e.g., better patient compliance, no solubility/biodistribution issues, greater efficacy, and lower toxicity), the CDT still proves easier to apply and customize, due to the virtually infinite drug combination and dosage possibilities [15, 759]. Yet, despite the success of CDT at pre-clinical and clinical stage, the potential formulation problems, drug–drug interactions and the consequence toxicity led researchers to hypothesize that the MTD approach could be, even if not always, a true promise of epigenetics-based therapy.

12.4.2.3 Epi-MTDs in Development

An epi-MTD could specifically act on one epi-target and another target not related to epigenetics, or on more epigenetic targets. Moreover, the active components of a MTD can interact with molecular factors belonging to the same or different cellular pathways, in order to increase the therapeutic efficacy especially in diseases characterized by a frequent insurgence of acquired resistance during the therapy.

Most of the MTD developed so far are hybrid molecules characterized by the scaffold of HDACi linked to another drug able to hit a related target involved in cancer. Pharmacophore portions of HDACi can be linked to other epi-drugs as well as to conventional cytotoxic agents (or to other drugs) to generate HDAC-based multi-target agents. The preference of HDACi's structural elements among various epi-drugs to design multi-targeting agents is due to the following reasons:

(a) In the 2000–2010 years, there was a boom of studies on HDACi by medicinal chemists belonging to both industrial and academic institutions, and many HDACi characterized by a wide structural variety have been reported.
(b) The cap group that characterizes all HDACi lays out of the catalytic tunnel of the enzymes and is tolerant of a high degree of structural variation without compromising the enzyme inhibition activity [761, 762]. This allows the introduction in this portion of a specific warhead able to interact with another target different from HDACs [15–17].

The currently in-clinic-use or under-test epi-MTDs are nearly exclusively the results of conjugation of two active substances in which only one portion has an epigenetic target or both fragments interact with two (or more) epigenetic targets. These include dual HDAC/kinase inhibitors, dual LSD1/kinase inhibitors, dual BET/kinase inhibitors, and dual inhibitors hitting two or more epigenetic targets.

For example, LSD1 and JmjC KDM4A/C enzymes play a key role in human prostate cancer because they are co-expressed and co-localized with androgen receptor. Rotili et al [763] designed and synthesized several LSD1/JmjC hybrid molecules by coupling the skeleton of the well-known LSD1 inhibitor tranylcypromine, to the carboxylate group of either the 4-carboxy-4′-carbomethoxy-2,2′-bipyridine or the 5-carboxy-8-hydroxyquinoline (5-carboxy-

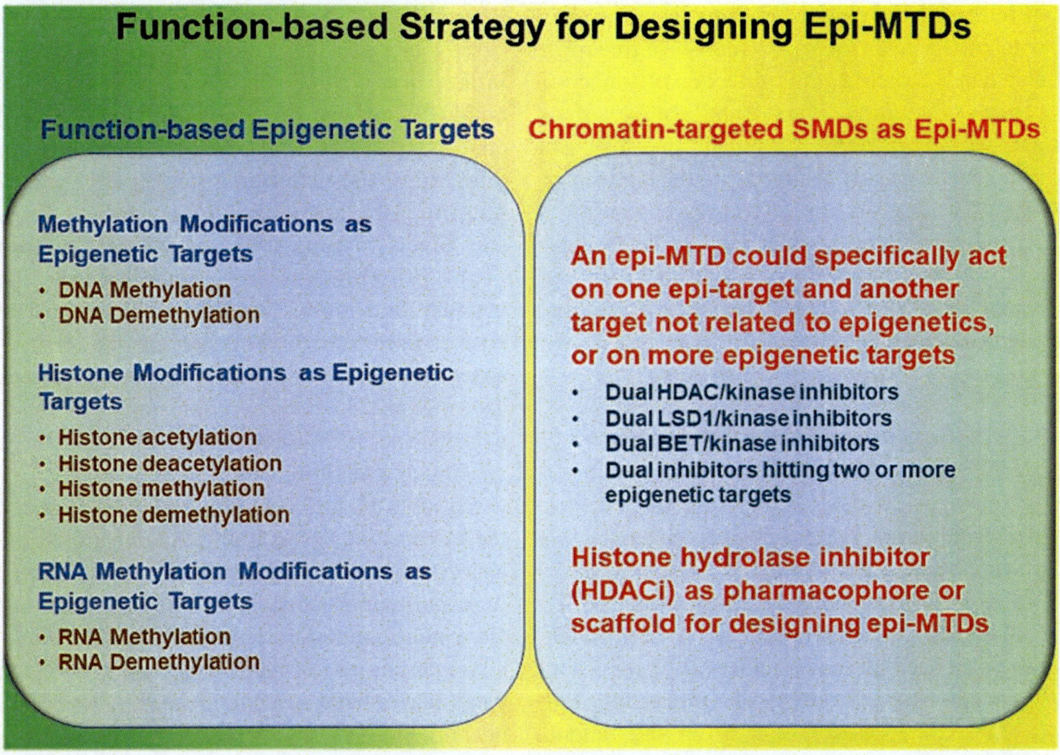

Fig. 12.3 A summary of various function-based strategy for the design of multitarget epigenetic drugs (epi-MTDs)

8-HQ), two 2-oxoglutarate competitive scaffolds developed as JmjC inhibitors. They found that pan-histone demethylase inhibitors simultaneously targeting Jumonji C and lysine-specific demethylases display high anticancer activities [763].

Hepatocellular, pulmonary, prostatic carcinoma and leukemia are associated with a global histone deacetylation state and with the double methylation of H3K9 [764, 765]. From these evidences, it emerged the need to design inhibitors capable to interact with both G9a and HDAC enzymes. Merging the quinazoline scaffold typical of BIX01294 with polymethylene hydroxamate chains allowed the identification of dual G9a/HDAC inhibitors. Among them, one showed low toxicity, good physico-chemical properties, and antiproliferative activity in the micromolar range in HeLa and K562 cell lines [766, 767] (Fig. 12.3).

12.5 Medicinal Chemistry-Based Strategy for Designing SM-MTDs

Multitarget single drugs could either be the natural structures that possess two or more functional activities highly desirable for a disease, the core structure-based, chemically modified compounds, or the result of the conjugation of two or more active molecules that individually show a known activity against specific targets. This subsection will focus on the multitarget single small-molecule drugs (SM-MTDs) leaving the multitarget single macromolecule drugs (MM-MTDs) to the following subsection.

The medicinal chemistry approaches to designed multi-target ligands can be classified in three types: linked, fused, and merged pharmacophores.

Linking of two (or more) pharmacophores by a flexible linker often leads to compounds exhibiting sufficient *in vitro* potency toward all targets of interest, however the larger molecular weight combined with an increased number of free rotatable bonds have a negative effect on pharmacokinetics. Therefore, the identification of a pharmacophore element to enable the fusion of two ligands usually leads to smaller molecules, while the identification of a merged pharmacophore for both targets can be regarded as the most promising way to develop a multi-target ligand exhibiting a promising pharmacodynamics and pharmacokinetic profile.

12.5.1 Multitarget Small-Molecule Drugs (SM-MTDs)

The definition of SMDs has been discussed in a previous Sect. 12.3 entitled "Sequenced-based Strategy for Designing Small-Molecule MTDs (SM-MTDs)," and a few examples of SM-MTDs have also been presented under the Sect. 12.4 of "Function-based Strategy for Designing Epi-MTDs". Here, we will focus on the rational design of multitarget small-molecule drugs (SM-MTDs).

Multitarget pharmacology of small-molecule cancer drugs significantly contributes to their mechanism of action, side effects, and emergence of drug resistance and opens ways to repurpose, combine, or customize drug therapy. In most cases, the set of targets affected at therapeutic concentrations is not fully characterized and/or the interaction efficacy values are not accurately quantified.

Discovery of new SM-MTDs relies on the network maps of the complex interactions of multiple drugs and multiple targets. Building such network maps however is not an easy task but is full of challenges. One approach to achieve this goal is to take full advantage of relevant databases and platforms. Currently, there are several the drug-target databases that can be utilized for mining and designing SM-MTDs after being transformed into networks of drugs and their targets, and the most frequently referred databases are DrugBank [768], Therapeutic Target Database (TTD) [769], ChEMBL [770], PubChem [771], BindingDB [772], and SuperTarget [773]. However, none of these databases alone provides complete and quantitative information. Several attempts have been made to build network maps of drugs and targets by: (**a**) connecting drugs with shared targets to form drug maps; (**b**) connecting targets with shared drugs to form target maps; (**c**) connecting drugs and targets to form drug-target bipartite maps [774–778]. However, these maps are both too complex and highly variable because the data sources used to build the maps may not be consistent, the list of drug–target interactions keeps growing, and the considerations for quantitative contribution or threshold for each edge may be missing or oversimplified.

To solve the problem, Shi et al [779] collected the information about multiple targets for each cancer drug and the "therapeutic" concentrations in experimental models or binding activities from multiple sources and analyzed the multitarget properties of cancer drugs. They then used these data to build two comprehensive proximity network pharmacology maps of anticancer drugs and their targets, respectively. They found that the majority of the small-molecule anticancer drugs have substantial multitargets. In addition, most of the small-molecule anticancer drugs can simultaneously act on macromolecular targets of different classes and types. These authors integrated the network maps of the small-molecule anticancer drugs and their targets along with their polypharmacological activity data into a freely available database, CancerDrugMap (http://ruben.ucsd.edu/dnet/maps/drugnet.html). This database may help drug repurposing on one hand and SM-MTDsdiscovery/designing on the other hand.

p53-Murine double minute 2 (MDM2) interaction and histone deacetylases (HDACs) are important targets in antitumor drug development. He et al identified a small-molecule anticancer drug with MDM2/HDACs dual inhibitor activities, the synergistic effects between MDM2 and HDACs [780]. They confirmed the potent and effective inhibition of MDM2 and HDACs and antitumor efficacy of this compound in the A549 xenograft model after oral administration.

12.5.2 Privileged Scaffold-Based SM-MTDs

"Privileged scaffold (PS)" or "privileged structure (PS)" refers to effective templates or core structures for drug discovery and design or construct high-affinity ligands that can bind more than one receptor. This term was first coined in 1988 by Evans et al [781], and ever since then, the PS-based strategy for drug discovery and design has evolved from a stand-alone technology to an integral component of various lead generation platforms [782–784]. On some occasions, the terms like "molecular framework," "chemotype," and "molecular fragment," are used as synonyms for privileged scaffold or privileged structure. The major feature and advantage of PSs is their multitargeting property because of their potential to bind to a diverse range of biologically important proteins with high affinities, which forms the basis for the discovery and design of novel bioactive compounds or SM-MTDs.

In medicinal chemistry, the choice of relevant "drug-likeness" scaffold is a starting point for the design of the structure dedicated to specific molecular targets. A PS is usually identified based on simple structural subunits present in the molecules of several drugs, with distinctive therapeutic uses, or affinities to several different receptors. Many well-known PSs have been identified and applied to drug discovery and development. Some examples are described below.

12.5.2.1 Chalcone

Chalcones, known as α, β-unsaturated ketones belonging to a class of small drug/druglike molecules with different and multitarget biological activities, are widely used as PSs for the design of antiinfective, antimalarial, anti-inflammatory, and anticancer agents [785–789]. Chalcones are a group of plant-derived polyphenolic compounds belonging to the flavonoids family that possess a wide variety of cytoprotective and modulatory functions, which may have therapeutic potential for multiple diseases [790–792]. The physicochemical characteristics and multitargeting properties define the extent of their biological activity and polypharmacological applications. Indeed, several natural and synthetic chalcones and their derivatives have demonstrated their potential as MTD anti-inflammatory and anticancer SMDs [790, 791, 793–795].

Liargkova et al [785] designed and synthesized potential a series of new small multi-target agents with lipoxygenase (LOX), acetyl cholinesterase (AChE) and lipid peroxidation inhibitory activities, as well as antioxidant activity based on 2-/4- hydroxy-chalcones and the bis-etherified bis-chalcone skeleton. Functional evaluation revealed that bis-chalcones c11 and c12 produced potent anti-LOX, anti-AChE, and anti-lipid peroxidation activities. Compounds c2 and c4 display additional protective actions against Alzheimer's disease (AD). Moreover, predicted results indicate that the majority of the compounds with the exception of c11 (144 Å) can cross the blood brain barrier and act in CNS. The investigators reached a conclusion that the presence of a double enone group supports better biological activities.

In addition, numerous published reviews underline the medicinal significance of the enone moiety of chalcones [786, 788, 791]. This conjugation of the double bond with the carbonyl moiety seems to be responsible for the biological activities of chalcones, as removal of the enone group renders them inactive. While a number of synthetic processes have been reported for the synthesis of chalcones, the general and more widely used synthetic route involves Claisen–Schmidt condensation under acidic or basic homogeneous conditions [789, 790, 794, 796, 797]. Changes in their structure have offered a high degree of diversity which has been extremely useful and highly advantageous for medicinal chemistry purposes aiming to develop new therapeutic agents with improved pharmacokinetic properties, as well as a better therapeutic profile [785]. Thus, chalcones represent a key structural scaffold for many synthetic and natural products.

12.5.2.2 Chromanone

Chromanone, also known as chroman-4-one or benzo-dihydropyran or benzopyran, is another important and interesting heterobicyclic compound and acts as a building block in medicinal chemistry for isolation, designing and synthesis of novel lead compounds [798, 799]. Structurally, chromanone consists of two rings in which 2, 3-dihydro-γ-pyranone fused with an aromatic benzene nucleus and derivatization at 2, 3 and 4-positions of chromanone skeleton yields more effective families of flavonoids like 3-benzylidene-chromanones, spirochromanone, hydrazones, oximes, flavanones, homisoflavonones and isoflavanones. The absence of a double bond in chromanone between C-2 and C-3 shows a minor difference from chromone but exhibits significant variations in biological activities. existing in natural compounds as polyphenols and as synthetic compounds like Taxifolin.

Natural and synthetic chromanone analogs can elicit various biological activities such as anticancer, tumor necrosis factor-α (TNF-α) inhibitors, antivascular, antidiabetic, antioxidant, antimicrobial, antifungal, antiviral, antileishmanial, insecticidal, spasmolytic, analgesic, anti-inflammatory, anticoagulant, estrogenic inhibitor, anti-acetylcholinesterase (AChE) inhibitor, antihuman immunodeficiency virus (HIV), anticonvulsant, antidepressants, anticoronal and antitubercular activity [798, 799]. Due to of these activities, several analogues of chromane are also available in market like tocopherols (vitamin E), taxifolin (antidiabetic), tetrazole (antidiabetic), troglitazone (antidiabetic), ormeloxifene (anticancer), and nebivolol (beta-blocker) [800]. Moreover, several isolated flavanones, flavonols, homoisoflavonoids like naringenin, naringin, myricetin, dihydroquercetin, and kaempferol are also under clinical studies [801]. Therefore, among the diverse array of chromane, chroman-4-one/4-chromanone are of considerable interest to researchers due to potency of this clinically useful pharmacophore in the treatment of cancer and several other diseases [802]. A large number of research has been carried out on modified naturally occurring chromanone scaffolds and/or synthesized new analogues. For example, several chromanone derivatives include DSP-1053 (novel serotonin reuptake inhibitor and fast antidepressant), calanolide A (anti-HIV), Silibinin and chrysin(anticancer), taxifolin (antidiabetic), tetrazole (antidiabetic), troglitazone (antidiabetic), ormeloxifene (anticancer), and nebivolol (β-blocker) [798, 803, 804].

12.5.2.3 Coumarin

Coumarins are a group of plant-derived polyphenolic compounds belonging to the benzopyrones family. They possess some peculiar physicochemical features and the versatile and easy synthetic transformation into a large variety of functionalized coumarins [805]. Coumarin was first isolated in 1820 by Vogel [806] from tonka beans (*Dipteryx odoranta* Wild; *Fabaceae* family) called also *Coumarou*, a vernacular French name. Since then, isolation, structural characterization, synthesis, and biological activity of thousands of natural coumarins from plants, bacteria, fungi [807–809], and chemical synthesis [810], have been reported.

Coumarin is characterized by a 2*H*-chromen-2-one (1,2-benzopyrone, or 2*H*-1-benzopyran-2-one) oxa-heterocycle with its structural skeleton being present in many biologically active agents. A huge number of coumarin derivatives have been designed, synthesized, and tested to address many pharmacological targets in a selective way, for example, selective enzyme inhibitors, and more recently, a number of selected targets (multitarget ligands) involved in multifactorial diseases, such as Alzheimer's and Parkinson's diseases due to their inherent inhibition of monoamine oxidase (MAO) and cholinesterase enzymes [811, 812]. Coumarin and some of its derivatives have become therapeutic drugs, as the anticoagulants warfarin, acenocoumarin, and phenprocoumon, all acting as vitamin K antagonists, the choleretics armillarisin A and hymecromone (umbelliferone), and the antibiotic novobiocin, which is a potent inhibitor of bacterial DNA gyrase (GyrB) [805]. In addition, industrial interest on coumarins has led to the development of cosmetics, agrochemicals, polymeric [6], and optoelectronic materials [813, 814].

12.5.2.4 Benzisoxazole

Heterocyclic compounds are an exceedingly important class of compounds and they have attracted more attention for diverse biological studies. All available natural and synthetic heterocyclic compounds can participate in biochemical reactions in the human body [815, 816]. The engineering and rationale behind drug design are closely related to the strategic incorporation of heterocyclic-like fragments with specific physicochemical properties. The potency and selectivity through bioisosteric replacements, lipophilicity, polarity, and aqueous solubility can ultimately be fine-tuned to the point of altering and conditioning the possible mechanisms of action of pharmaceutical drugs in an attempt to obtain molecularly targeted agents [815, 816].

Numerous benzisoxazole moieties have been found to possess a broad spectrum of pharmacological activities, which encouraged research activities in these areas. The presence of an amide bond, benzisoxazoles, chromans and fluorine atom substituents can alter the chemical properties, disposition, and biological activities of drugs [817]. Many fluorinated compounds, 1,2-benzisoxazole derivatives and various amides are currently used in the treatment of diseases [818]. These include antidepressants, anti-inflammatory agents, antimalarial drugs, antipsychotics, antiviral agents, steroids, and anesthetics. The 1,2-benzisoxazole ring containing drugs zonisamide and risperidone have been used as anticonvulsants [819]. Fluorine substitution can also have a profound effect on drug disposition, in terms of distribution, drug clearance, route(s), and extent of drug metabolism [820].

Benzisoxazole scaffolds are well represented in pharmacology, having shown significant importance in anti-HIV [821], antimicrobial [813–825], antipsychotic [826, 827], anti-inflammatory [824, 828, 829], antioxidation [824, 830], anticancer [830–833], anticonvulsant [834], and antidiabetic [835, 836] research. Moreover, 1,2-benzisoxazole derivatives, which include zonisamide, risperidone, paliperidone, and iloperidone, are all FDA approved and currently in use for epilepsy, mood disorders, and schizophrenia, respectively [837–841]. A most recent study demonstrated that a naturally occurring 1,2-benzisoxazole with minimum inhibitory concentrations as low as 6.25 μg/mL against clinical strains of multi-drug resistant A. baumannii [842]. This molecule represents a new chemotype for antibacterial agents against A. baumannii and is easily accessed in two steps via de novo synthesis. In vitro testing of structural analogs suggests that the natural compound may already be optimized for activity against this pathogen.

12.5.2.5 Stilbene

The stilbene scaffold as a privileged structure is a basic element for a number of biologically active natural and synthetic compounds [843]. Stilbenes are exemplified by resveratrol, combretastatin A-4, pterostilbene, and piceatannol that have great potential in therapeutic and preventive applications.

Resveratrol (3,5,4′-trihydroxy-trans-stilbene), present in grapes and other food products, plays a role in the prevention of several human pathological processes and has been suggested as an anticancer agent [844, 845]. In addition, it has many known and potential therapeutic applications in human general homeostasis; it mediates a great number of biological responses relevant for human health such as cardiovascular protective [846, 847], neuroprotective [848], antioxidant [849], and antimicrobial activities.

Combretastatin A-4 (CA-4) is one of the most potent antimitotic and antiangiogenic agents of natural origin from the bark of South African bush willow Combretum caffrum [850]. CA-4 also has significant antitumor activity in a wide variety of preclinical tumor models. CA-4 shows strong cytotoxicity against a variety of cancer cells, including multi-drug resistant cancer cell lines. In addition, CA-4 disodium phosphate (CA4DP), a water-soluble prodrug of CA-4, shows potent antivascular and antitumor effects in a wide variety of preclinical tumor models.

Pterostilbene (trans-3,5-dimethoxy-4-hydroxystilbene) is a natural dietary compound and the primary antioxidant component of blueberries [851]. It is closely related to resveratrol, sharing the same unique therapeutic potential as anti-inflammatory, antineoplastic and antioxidant agent. Multiple studies have demonstrated the antioxidant activity of pterostilbene in both in vitro and in vivo models illustrating both preventative and therapeutic benefits. The antioxidant activity of pterostilbene has been implicated in anticarcinogenesis, modulation of neurological disease, anti-inflammation, attenuation of vascular disease, and amelioration of diabetes [851]. Studies also suggest that pterostilbene exhibits the hallmark characteristics of an effective anticancer agent based on its antineoplastic properties in several common malignancies. In vitro models have shown that pterostilbene inhibits cancer growth through alteration of the cell cycle, induction of apoptosis, and inhibition of metastasis. In vivo, pterostilbene inhibits tumorigenesis and metastasis with negligible toxicity. Pterostilbene has also been shown to be effective as an inducer of antioxidant capacity in multiple cancer cell lines that may facilitate its function as an anticarcinogenic compound. Additionally, preliminary studies show that pterostilbene exhibits much greater bioavailability compared with other stilbene compounds [852].

Piceatannol is a naturally occurring polyphenolic stilbene found in various fruits and vegetables and has been reported to exhibit anticancer and anti-inflammatory properties [853, 854]. In addition, recently reported beneficial effects of piceatannol on hypercholesterolemia, atherosclerosis, and angiogenesis underscore its therapeutic potential in cardiovascular disease. The natural stilbene known as "piceatannol" for its anticancer properties. Numerous preclinical studies have reported the potential of Piceatannol to prevent or impede the growth of various cancers originating from different organs such as brain, breast, cervical, colon, liver, lung, prostate, and skin.

12.5.3 Molecular Hybrid SM-MTDs

In addition to privileged scaffold, molecular hybrid is another widely used strategy for generating multitarget compounds by taking advantage of structural commonalities to combine or overlap two distinct frameworks into a single chemical entity [759, 855, 856]. It is a relatively new concept and technology in drug design and development based on the combination of pharmacophoric moieties of different bioactive substances to produce a new hybrid compound with improved affinity and efficacy, when compared to the parent drugs [857]. Hybrid drugs can provide combination therapies in a single multi-functional agent; that is, the SM-MTDs.

In other words, SM-MTDs routinely result from the integration of pharmacophores of selective molecules, either already known drugs or drug candidates, starting from compounds with the desired activity towards the targets of interest. To date, the majority of SM-MTDs are hybrid compounds [6, 15, 17]. Intuitively, pharmacophores featuring similar scaffolds, usually ring systems, can be fused or merged depending on the overlapping degree between the starting frameworks. Alternatively, if pharmacophores present different structural elements required for the interaction with each target, they can be conjugated with cleavable or non-cleavable linkers, although this strategy often leads to molecules with poor drug-likeness properties [858].

Molecular hybridization in drug discovery can be achieved in three different approaches [6, 15, 17]: cleavable linker, non-cleavable linker (or stable linker) and molecular merge, which primarily serve to connecting two or more different pharmacophores into one to form a single compound.

12.5.3.1 Cleavable Linker

The "Cleavable Linker" approach connects two or more molecules with a linker that is cleaved in physiological conditions (mutual pro-drugs or codrugs) [859–862]. "Codrugs" or "mutual prodrugs" are single molecules obtained by the conjugation of two therapeutic compounds with synergistic activity, via a cleavable linker [860, 861]. When both pharmacophores are connected by a linker that is labile or can be easily cleaved in vivo, they are called dual or mutual pro-drugs since each of them uses another pharmacologically active compound instead of some inert molecule as carrier. Being released simultaneously inside the cancer cells, they might act synergistically and affect distinctive cellular targets, in contrast to the simultaneous administration of two individual synergistic agents, which are usually transported to the site of action with different efficiencies.

A marketed example is the antibacterial sultamicillin, an ampicillin/sulbactam codrug. A cleavable linker is stable in the systemic blood circulation for a long period of time but is readily cleaved within the target cells. The fact that the two starting compounds, following metabolic transformation, have the potential to be released in the same target cells and at the same time is a peculiar feature of codrugs with respect to combinations (two single compounds, each one with an individual pharmacokinetic profile) [861, 862]. On the other hand, it should be mentioned that the codrug approach is applicable only to starting compounds carrying functional groups suitable for conjugation [860].

12.5.3.2 Non-Cleavable Linker

A non-cleavable linker conjugates two or more molecules, allowing each molecular moiety to interact with its specific target without interfering with the interactions established by the other portion without obvious drug release mechanism [863]. One of the advantages of non-cleavable linkers over cleavable linkers is their increased plasma stability, which can improve the therapeutic index. Studies have shown that non-cleavable linked SMDs generally perform better than their cleavable counterparts in vivo. Moreover, non-cleavable linkers can potentially provide a larger therapeutic window compared to cleavable linkers. Finally, it is expected to reduce off-target toxicity compared to cleavable linker conjugates because non-cleavable compounds can provide greater stability and tolerance.

12.5.3.3 Molecular Merge

The use of linkers often leads to compounds with unfavorable biopharmaceutic or pharmacokinetic profile (e.g., compounds that violate more than two of the Lipinski's rules) [864]. Although the use of cleavable linkers might be advantageous, it also limits some of the merits of the multi-target approach in comparison with combination therapies (simplified pharmacokinetics, reduced chance of drug interactions). Moreover, the linker-based approaches could lead to poor ligand efficiency metrics [864, 865], which refer to the binding efficiency per atom. It might be speculated that, since only a part of the molecule can interact with each of the proposed targets, the other part can become an obstacle for the binding event, reducing the binding efficiency because of enthalpic and/or entropic reasons. Therefore, the overlapping or merging approach (searching partially or highly integrated pharmacophores in a small molecule) seems more attractive from a biopharmaceutical viewpoint. Additionally, the linking approaches produce molecules with poor pharmacokinetic properties and high molecular weight. Therefore, smaller hybrid molecules might prove beneficial as they possess better pharmacokinetic properties. Two or more different molecules/compounds, or more specifically two or more pharmacophores that are similar and share common substructures, usually hydrophobic or basic ring systems, can be synthetically fused or merged into one by adding a bond between two atoms, one from each molecule [864, 866]. Therefore. merged molecules, in contrast to pro-drugs, contain two (or more than two) domains with different biological functions and dual or multiple activities that ideally act as distinct pharmacophores, although not necessarily on the same biological target [864, 866]. Thus, multitarget ligands usually consist of the combination of pharmacophores of selective ligands (either already known drugs or candidates) [864, 867]. Alternatively, if pharmacophores are dissimilar, they can be joined as conjugates with cleavable or non-cleavable linkers as described above, although these latter strategies often lead to structures of high molecular weight and lipophilicity [759]. The drawback of molecular merge is that this method may not align properly the two parts of the new molecule.

Taking the drug treatment of Alzheimer's disease (AD) as an example, it is known that targeting glutamatergic and cholinergic systems concurrently is the present standard of care for individuals with AD. Memantine (N-methyl-D-aspartate receptor (NMDAR) antagonist) constitutes the most recent treatment option which was approved more than 10 years ago [868]. On the other hand, four out of five of the standard treatments including memantine (NMDAR antagonist), rivastigmine, galantamine, and donepezil (cholinesterase inhibitors (ChEIs)) are licensed in the European Union [869–871]. The fifth treatment option is the combination of memantine and donepezil and this combination therapy (CDT; i.e., Namzaric®) was approved in 2014 to treat individuals with moderate to severe AD, who are stabilized on donepezil and memantine therapy [872]. It involves the combination of two proven therapeutic agents (i.e., donepezil and memantine) in a fixed-dose combination product, providing the most effective way to start CDT therapy in individuals with AD. Memoquin is a novel drug, which has been developed as a potential anti-AD candidate because of its MTD design approaches [873]. Moreover, SAMTs are formulated by the molecular hybridization of various pharmacophore subunits, from recognized biologically active molecules, which work as diverse ligands and which affect diverse biological targets [874]. Since AD is a multifactorial disorder, the combination of therapeutic agents may thus prove more effective as compared to single-agent therapy.

It has been suggested by biochemical evidence that there is a dysfunction in the activity of glutamatergic neurons in AD. Furthermore, their mutual dysfunction along with the cholinergic system is vital in Alzheimer's pathology [875, 876]. Therefore, CDT or SAMTs that target glutamatergic and cholinergic systems concurrently are considered as the present standard of care for individuals with AD [877]. It is known that NMDAR is accountable for learning and memory. Furthermore, glutamate and glycine are the agonists that need to simultaneously bind with the NMDAR for activating this receptor. It has been shown clinically that NMDAR overactivation (i.e., through an excess of glutamate) can cause an uncontrolled influx of Ca^{2+} into the neuron, which can eventually lead to excitotoxicity. On the other hand, along with its inhibitory effects on AChE, galantamine exhibits neuroprotective activities against glutamate toxicity through the activation of nicotinic ACh receptors [878]. Interestingly, smokers are less prone to Parkinson's disease (PD) because of the induction of the nicotinic acetylcholine receptor (nAChR) in these individuals, which thus confers this neuroprotective effect [879]. Therefore, memantine and galantamine can act together in the same excitotoxic cascade to deliver a synergistic neuroprotective action [880]. The study by Lopes et al. [872] reinforced this hypothesis by demonstrating that galantamine's neuroprotective action against NMDA-induced neurotoxicity in primary cortical neurons primarily involves nicotinic receptors. Furthermore, they also mentioned in their study that full neuroprotection might be achieved by combining subactive concentrations of memantine and galantamine. Additionally, their study also provides a new design strategy in the development of a MTD drug with maximized efficiency [880]. A new series of hybrid compounds were designed by following this strategy that combined the pharmacological effects of memantine and galantamine [881]. On the other hand, following a dual-binding strategy, new hybrid compounds were designed by linking two drug molecules via a variable-length polymethylene linker. In this series, compounds that carry a hexamethylene spacer are considered to display the most auspicious properties. A 6-methylene spacer offered the optimal distance to allow a concurrent contact with both the catalytic active site and the peripheral anionic binding site on AChE [882]. A second generation of memantine analogs was developed to improve the efficacy of memantine and to provide it with a substantial disease-modifying effect [883]. An important pharmacophore (nitrooxy moiety, –ONO2) from nitroglycerin was linked with memantine to design these memantine analogs [884]. Only the nitromemantine-treated group exhibited considerably better functions on the location-novelty recognition test. A phase II RCT revealed that nitromemantine was able to reverse the loss of brain connections and bring the number of synapses all the way back to normal within a few months of treatment in mouse AD models [885].

In addition, designing a natural product hybrid bearing triple antiplatelet profile with cyclooxygenase-1 (COX-1) inhibiting, a P2Y12 receptor antagonizing, and a protease-activated receptor-1 (PAR-1) antagonizing efficacies by the molecular merging technology has also been reported [886]. Several alkylphosphocholines-gefitinib conjugates as MTD anticancer agents have also been reported [887].

Collectively, the generation of SM-MTDs is typically driven by the nature of the targets, the availability of starting frameworks and the chemical tractability. Beyond that, the essential requirement of the SM-MTDs is that each framework retains the ability to interact with its specific target [859]. This undoubtedly challenging task involves considering structure–activity relationships that govern the interaction of the starting molecules with their specific targets and becomes most challenging when these targets are only slightly related or unrelated (i.e., when they belong to different protein families) [859].

12.5.4 Fragment-Based SM-MTDs

Fragments are small organic molecules which are small in size and low in molecular weight [888, 889]. It is based on identifying small chemical fragments, which may bind only weakly to the biological target, and then growing them or combining them to produce a lead with a higher affinity. Fragment-based drug discovery (FBDD) is a new approach, increasingly used in the pharmaceutical industry, for reducing attrition and providing leads for previously intractable biological targets by screening fragment libraries with a few thousand compounds with molecular weights of around 200 Da. FBDD identifies low-molecular-weight ligands (~150 Da) that bind to biologically important macromolecules. The three-dimensional experimental binding mode of these fragments is determined using X-ray crystallography or NMR spectroscopy and is used to facilitate their optimization into potent molecules with drug-like properties. Compared with high-throughput-screening, the fragment approach requires fewer compounds to be screened, and, despite the lower initial potency of the screening hits, offers more efficient and fruitful optimization campaigns. The FBDD strategy could also help design MTDs for multiple diseases.

FBDD usually generates a compound starting from a chemical fragment with a low binding affinity to the target, low complexity in chemical structures and low molecular weight (less than 300 Da) [890–892]. These starting hits are usually identified from a compound library using sensitive biophysical methods. The identified hit is then grown into drug-like molecules through different strategies. Although FBDD cannot replace high-throughput screening (HTS) campaigns in drug discovery, it has some attractive advantages such as saving experimental cost, offering diverse hits, and exhibiting multiple ways to develop novel compounds [890, 893]. These advantages have encouraged researchers to adopt this method to develop inhibitors of different types of targets. With the development of new approaches in screening and progress made in structural biology, FBDD has been readily carried on and playing important roles in target-based drug discovery [890, 894, 895]. Several drugs such as vemurafenib-an inhibitor of oncogenic B-RAF kinase activity derived from fragment-based approach have been approved by FDA [890, 896]. With more and more compounds derived from FBDD entering different stages of clinical studies, this method has been highly recognized in drug discovery. To carry out a fragment screening experiment, following procedures are usually required, namely selecting a compound library, setting up a method for hits identification, determining structures of fragment-target complexes, developing an assay for analyzing structure-activity relationship (SAR) and designing a strategy to grow the fragment into a potent inhibitor. In this review, fragment library, methods utilized in fragment screening, strategies applied in fragment optimization and targets that have been studied using FBDD are summarized. With more and more compounds developed through this method, FBDD will be playing essential roles in target-based drug discovery

[890, 897, 898]. FADD roughly includes the following procedures.

12.5.4.1 Identification of Druggable Targets for Fragments

The FBDD strategy relies on the identification and validation of druggable targets and is mainly applicable to target-based drug discovery. Druggability of a target is predicted with computer-assisted methodologies and validated by identifying a pocket favoring binding to SMDs [890, 899, 900]. In most cases, druggable targets are of great interests as the probability to develop small molecule drugs is very high [901]. Draggability is utilized to evaluate whether a small-molecule drug can be developed to affect the biological function of a protein [902]. Several methods are applied to determine the druggability of targets [903], which is important in drug discovery. A druggable target usually contains a hydrophobic pocket favoring its binding to hydrophobic compounds. However, undruggable targets do not have a pocket or the pocket is highly hydrophilic and shallow. Many undruggable targets are important for cancer development. These targets include protein–protein interactions (PPIs), transcription factors, phosphatases, Ras proteins, and others [901, 904]. These undruggable targets are usually not considered in small-molecule drug discovery due to challenges in hit identification and lead optimization. (For detail, see the excellent review article by Li Q. Front Mol Biosci. 2020; 7: 180; Ref. [881]).

12.5.4.2 Primary Screening of Fragments

Binding affinities between fragments and their targets are normally very low in micromolar to millimolar range. At present, a number of approaches such as differential scanning fluorimetry (DSF), isothermal titration calorimetry (ITC), nuclear magnetic resonance (NMR), surface plasmon resonance (SPR), and X-ray crystallography are being widely used in FBDD [890]. (For detail, see the excellent review article by Li Q. Front Mol Biosci. 2020; 7: 180; Ref. [881]).

12.5.4.3 Compound Optimization

As fragments usually bind weakly to targets and exhibit no potent inhibitory effect on the activity of the targets, further chemical modification of the hits is required in hit-to-lead step. Therefore, for the hits to be developed into leads, their target binding affinities must be strengthened to enhance the potency of activities against the targets [890, 903, 905, 906]. The three most frequently utilized strategies for this purpose are fragment growing, fragment hopping/molecular merging, and fragment linking. (For detail, see the excellent review article by Li Q. Front Mol Biosci. 2020; 7: 180; Ref. [881]).

Most previous multitarget-optimization studies have relied on a rational combination of the pharmacophores. This approach is in many cases highly efficient, yet, it requires deep understanding of the pharmacophores and the underlying structure–activity relationship for the targets of interest, which is not always available. Furthermore, linking, merging, or fusing pharmacophores mostly yields previously described chemical scaffolds or combinations thereof, whereas a systematic fragment-based approach is unbiased by previously described ligands [907–909]. FADD for MTD can be based either on the properties of multitargeticity and polyspecificity or drug promiscuity or on single-target compounds for molecular hybridization.

Howard et al [910] described the identification of a clinical candidate via structure-based optimization of a ligand efficient pyrazole-benzimidazole fragment. Aurora kinases play a key role in the regulation of mitosis and in recent years have become attractive targets for the treatment of cancer. X-ray crystallographic structures were generated using a novel soakable form of Aurora A and were used to drive the optimization toward potent (IC_{50} approximately 3 nM) dual Aurora A/Aurora B inhibitors. These compounds inhibited growth and survival of HCT116 cells and produced the polyploid cellular phenotype typically associated with Aurora B kinase inhibition. Optimization of cellular activity and physicochemical properties ultimately led to the identification of compound 16 (AT9283), a multitargeted kinase inhibitor. In addition to Aurora A and Aurora B, another compound was also found to inhibit a number of other kinases including JAK2 and Abl (T315I). This compound demonstrated in vivo efficacy in mouse xenograft models and is currently under evaluation in phase I clinical trials.

A research team developed the first fragment-based approach by exploring quantitative-structure-activity relationships (QSAR) for in silico design of multitarget (mt) inhibitors against five proteins associated with AD [911]. Their approach was focused on the construction of an multitarget QSAR discriminant model using a large and heterogeneous database of compounds and substructural descriptors, which permitted the simultaneous classification and prediction of inhibitors against five proteins associated with AD. The model correctly classified more than 90% of active and inactive compounds in both, training and prediction series. Similar approach has been applied extensively for MTD drug discovery against various diseases.

Another group introduced the first multitasking computational model devoted to the in-silico fragment-based design of new molecules with high inhibitory activity against 19 different proteins involved in breast cancer [912]. The multitasking computational model was created from a dataset formed by 24,285 cases, and it exhibited accuracy around 93% in both training and prediction (test) sets. Several molecular fragments were extracted from the molecules present in the dataset, and their quantitative contributions to the inhibitory activities against all the proteins under study were calculated. The combined use of the fragment contributions and the physicochemical interpretations of the different molecular descriptors in the multitasking computational model allowed the design of eight new molecular entities not reported in our

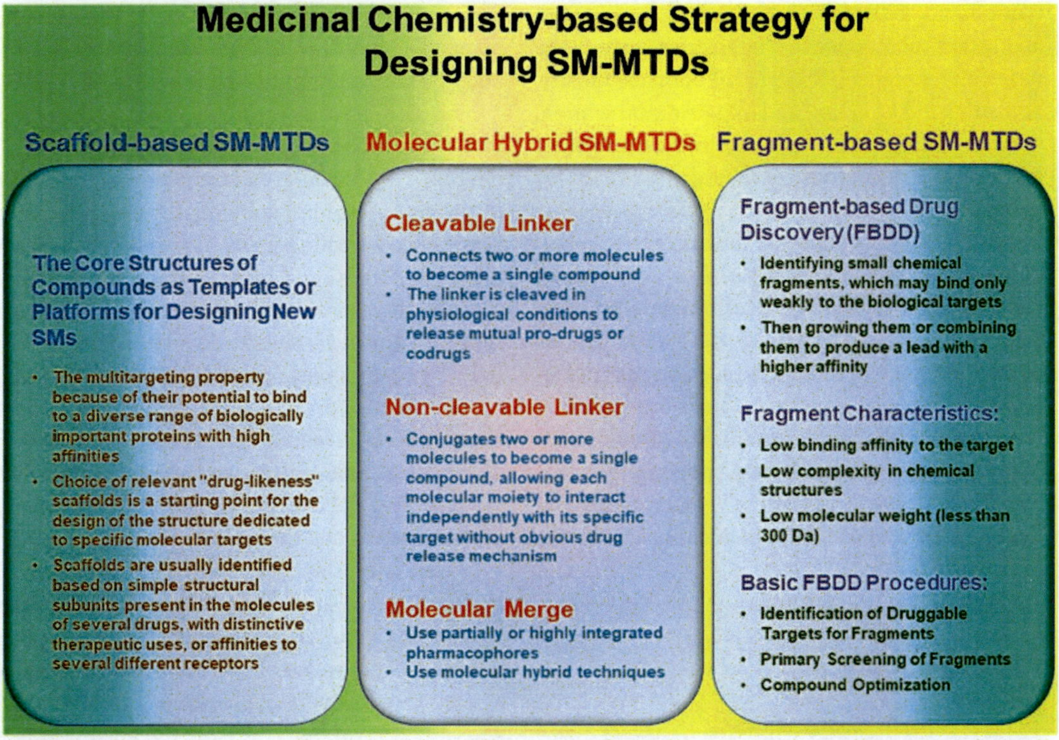

Fig. 12.4 A summary of medicinal chemistry-based strategies for designing small-molecule multitarget drugs (SM-MTDs)

dataset. These molecules were predicted as potent multitarget inhibitors against all the proteins, and they exhibited a desirable drug-likeness according to the Lipinski's rule of five and its variants (Fig. 12.4).

12.6 Biopharmaceutical-Based Strategy for Designing Macromolecular MTDs (MM-MTDs)

There has not been a commonly accepted definition of macromolecular drugs. Biologically, macromolecules refer to the four major classes of high-molecular weight molecules including carbohydrates, lipids, proteins, and nucleic acids that constitute the functional and structural substances of the cells of living organisms. Pharmacologically, macromolecules are described as macromolecular drugs or biotech drugs >1000g/mol (peptides, proteins, and antibodies) and macromolecular therapeutics (polymeric nanomedicines, polymer-drug conjugates, and other compounds characterized by large molecular weight) [913–915]. Pharmaceutically, according to U.S. Food and Drug Administration, macromolecules can be classified as vaccines, blood and blood products, allergen extracts for diagnosis and treatment (such as allergy vaccine injections), human cells and tissues for transplantation (such as tendons, ligaments and bones), gene therapy preparations, cell therapy preparations, and reagents for detecting infectious agents. Herein, we intend to focus on pharmacological consent of macromolecular drugs, which can be classified into the following three categories: antibody and biologics, polymeric nanomedicines, and drug-free macromolecular therapeutics (DFMT) in connection with the MTD strategy.

12.6.1 Antibody

12.6.1.1 Antibody Mixture (AbMix)

From 2015 to 2017, biotech drugs (proteins, peptides, and monoclonal antibodies) represent 31% of the novel new molecular entities (NMEs), nearly approaching the number of single-target drugs (34%) [916]. With the enormous success of recombinant monoclonal antibodies (rMAbs) as human therapeutics, there are increasing efforts underway to explore NMEs that mimic rMAbs to replicate this huge success. In addition to naked intact rMAbs, antibody drug conjugates (ADCs), Fab and F(ab')2 fragments, and also Fc fusion proteins have been developed and/or marketed as human therapeutics to treat different human diseases, including life-threatening diseases such as cancer. Several hundreds more intact rMAbs, ADCs, Fab, F(ab')2 fragments, and Fc fusion proteins are currently undergoing human clinical trials. In addition to these molecules, new type of antibody fragments such as single-chain Fvs (scFvs), VH, scFv-Fc, scFv-CH, scFAb, scFv-zipper, diabodies, bispecific antibodies, and similar types of constructs are also being investigated to be developed as human monotherapeutics. Further, there are quite a few current examples of combinations of biologics being developed. For example, currently, several biopharmaceutical companies are developing combinations of antibody mixtures as human therapeutics [916–918].

For the past several decades, intravenous immunoglobulins have successfully been used in treating various autoimmune diseases. Intravenous immunoglobulins (IVIGs) have long been used in treating many autoimmune diseases [918–920]. IVIG is a highly heterogeneous mixture of polyclonal antibodies derived from the serum of multiple healthy human donors [919]. IVIG preparations, which have been very successful on the market, are being used to treat many human diseases throughout the world. In this context, several biotechnology companies are exploring the use of combinations of antibody mixtures as human therapeutics.

Similar to the CDT approach with SMDs, mixtures of rMAbs are now being developed as human therapeutics [918, 921]. These antibody mixtures contain two or more antibodies directed against either a single antigen or different antigens [922]. Antibody mixtures against a single antigen may recognize different binding epitopes and may exhibit synergies in neutralizing and/or cell killing, whereas antibody mixtures against different antigens is perhaps more similar to IVIG antibodies, with the exception that the recombinant antibody mixtures contain well-defined and well-characterized antibodies that target well-defined antigens. There are at least eight different examples of antibody mixtures that are currently in clinical development [923]. These antibody mixtures may provide advantages over antibody monotherapy because the individual antibody components present in the mixture may complement each other to target the diseases [924]. For example, Xoma is developing an antibody mixture containing three antibodies to target and neutralize the toxin from botulism poisoning [921, 922, 924]. Xoma 3AB, which is a lyophilized formulation of three different antibodies in equimolar proportions directed against three distinct domains on the BoNT/A neurotoxin protein, is an example of an antibody mixture against a single antigen [921, 922, 924]. Antibody mixtures using a combination of highly successful marketed rMAbs or developing new combinations of novel rMAbs may prove to be a useful alternative to monotherapy to treat complex human diseases.

In Addition, for several human therapies, therapeutic rMAbs can also be used in combination with SMDs to treat patients. One classic example of this paradigm is the treatment of rheumatoid arthritis patients with the small-molecule, methotrexate, and an anti-TNF-α antibody such as infliximab, or adalimumab, or golimumab [918].

12.6.1.2 Antibody–Antibody Conjugates (AACs) or Chimeric Antibodies

Over the past few years, the interest in developing bispecific and multispecific antibodies as human therapeutics has significantly increased [925]. Bispecific antibodies, which can be generated by a variety of different approaches [925], may be intact rMAbs that contain two or more variable domains with specific affinity to bind different antigens. One example of a bispecific antibody format is the DuoBody developed by Genmab, which is formed by combining two antibodies, each separately mutated in the CH3 domain, resulting in a more favorable recombination over formation of the parental antibodies when reduced and re-oxidized [925]. Another example of a bispecific antibody format is the dual-variable domain-immunoglobulin (DVD-Ig) [919]. Unlike DuoBody, in DVD-Ig, each heavy and light chains contain two variable fused domains and hence each IgG molecule of DVD-Ig carries four variable-binding domains. Currently, there is a rapidly increasing interest by many companies to develop bispecific and/or multispecific antibodies as human therapeutics because of their significant therapeutic potential [919, 925]. At least seven new bispecific antibodies are in early stages of clinical development, with many more in preclinical studies.

The term bispecific antibody (bsAb) is used to describe a large family of molecules designed to recognize two different epitopes or antigens [926]. BsAbs come in many formats, ranging from relatively small proteins, merely consisting of two linked antigen-binding fragments, to large immunoglobulin G (IgG)-like molecules with additional domains attached. Hence, these are also called genetically engineered antibodies (GEAs) or chimeric antibodies [927]. An attractive bsAb feature is their potential for novel functionalities—that is, activities that do not exist in mixtures of the parental or reference antibodies. In these so-called obligate bsAbs, the physical linkage of the two binding specificities creates a dependency that can be temporal, with binding events occurring sequentially, or spatial, with binding events occurring simultaneously, such as in linking an effector to a target cell. To date, more than 20 different commercialized technology platforms are available for bsAb creation and development, 2 bsAbs are marketed, and over 85 are in clinical development [926]. One of the major advantages of expressing genetically engineered antibodies is that one is not limited to using antibodies as they occur in nature [927, 928]. In particular, non-immunoglobulin sequences can be joined to antibody sequences creating multi-functional chimeric antibodies. Creation of a family of multi-functional chimeric antibodies with a growth factor joined to a combining specificity may be useful in targeting therapy to malignant cells and delivering drugs into specific locales in the human body. Presence of the growth factor may facilitate transcytosis of chimeric antibody across the blood-brain barrier using growth factor receptors. These novel chimeric antibodies constitute a new family of immunotherapeutic molecules for cancer therapy.

12.6.1.3 Antibody–Drug Conjugates (ADCs)

Antibody–drug conjugates or ADCs are a class of biopharmaceutical drugs designed as a targeted therapy for treating cancer [929–931]. Unlike chemotherapy, ADCs are intended to target and kill tumor cells while sparing healthy cells. ADCs are complex molecules composed of an antibody linked to a biologically active cytotoxic (anticancer) payload or drug [932]. ADCs combine the targeting capabilities of monoclonal antibodies with the cancer-killing ability of cytotoxic drugs. They can be designed to discriminate between healthy and diseased tissue [933, 934]. As of 2019, some 56 pharmaceutical companies were developing ADCs.

As for the mechanism concerned, the ADC approach comprises a monoclonal antibody (mAb) conjugated to the cytotoxic payload via a chemical linker that directed toward a target antigen expressed on the cancer cell surface, reducing systemic exposure and therefore toxicity. Antibodies attach themselves to the antigens on the surface of cancerous cells. The biochemical reaction between the antibody and the target protein (antigen) triggers a signal in the tumor cell, which then absorbs or internalizes the antibody together with the linked cytotoxin. After the ADC is internalized, the cytotoxin kills the cancer [935]. This targeting limits side effects and gives a wider therapeutic window than other chemotherapeutic agents. Therefore, one of the most important aspects of ADC development for cancer is the identification of the unique antigenic target of the mAb component. There are 328 unique antigens used in antibody-based therapy as a target [930]. The success of ADCs is dependent on four factors--target antigen, antibody, linker, and payload. A stable link between the antibody and cytotoxic (anticancer) agent is a crucial aspect of an ADC [936].

According to Khongorzul et al [930], the selected antigen needs to fulfill several requirements. First, the target antigen needs to have high expression in the tumor and no or low expression in the healthy cell [937]. For example, the HER2 receptor is almost 100-fold highly expressed in the tumor cell compared with the healthy cell [938]. Second, the target antigen should be displayed on the surface of the tumor cell to be available to the circulated mAb [939]. Third, the target antigen should possess internalization properties as it will facilitate the ADC to transport into the cell, which will in turn enhance the efficacy of the cytotoxic agent [940], although some studies have demonstrated that noninternalized ADC product directed against components of the tumor microenvironment can efficiently detach their drug in the extracellular space and arbitrate a potent therapeutic activity in some cases and that ADCs often induce a strong "bystander effect" [941].

A stable ADC linker ensures that less of the cytotoxic payload falls off before reaching a tumor cell, improving safety, and limiting dosages. Linkers are based on chemical motifs including disulfides, hydrazones or peptides (cleavable), or thioethers (noncleavable). Cleavable and noncleavable linkers were proved to be safe in preclinical and clinical trials. Brentuximab vedotin includes an enzyme-sensitive cleavable linker that delivers the antimicrotubule agent monomethyl auristatin E or MMAE, a synthetic antineoplastic agent, to human-specific CD30-positive malignant cells. MMAE inhibits cell division by blocking the polymerization of tubulin. Because of its high toxicity MMAE cannot be used as a single-agent chemotherapeutic drug. However, MMAE linked to an anti-CD30 monoclonal antibody (cAC10, a cell membrane protein of the tumor necrosis factor or TNF receptor) was stable in extracellular fluid. It is cleavable by cathepsin and safe for therapy. Trastuzumab emtansine is a combination of the microtubule-formation inhibitor mertansine (DM-1) and antibody trastuzumab that employs a stable, non-cleavable linker. The availability of better and more stable linkers has changed the function of the chemical bond. The type of linker, cleavable or noncleavable, lends specific properties to the cytotoxic drug. For example, a non-cleavable linker keeps the drug within the cell. As a result, the entire antibody, linker and cytotoxic (anticancer) agent enter the targeted cancer cell where the antibody is degraded into an amino acid. The resulting complex—amino acid, linker, and cytotoxic agent—is considered to be the active drug. In contrast, cleavable linkers are detached by enzymes in the cancer cell. The cytotoxic payload can then escape from the targeted cell and, in a process called "bystander killing," attack neighboring cells [942]. Another type of cleavable linker, currently in development, adds an extra molecule between the cytotoxin and the cleavage site. This allows researchers to create ADCs with more flexibility without changing cleavage kinetics. Researchers are developing a new method of peptide cleavage based on Edman degradation, a method of sequencing amino acids in a peptide [943]. Also under development are site-specific conjugation (TDCs) [944] and novel conjugation techniques [945, 946] to further improve stability and therapeutic index, α emitting immunoconjugates [947], antibody-conjugated nanoparticles [948] and antibody-oligonucleotide conjugates [949].

ADCs harness the highly specific targeting capabilities of an antibody to deliver a cytotoxic payload to specific cell types. They have garnered widespread interest in drug discovery, particularly in oncology, as discrimination between healthy and malignant tissues or cells can be achieved. Nine ADCs have received approval from the US Food and Drug Administration and more than 80 others are currently undergoing clinical investigations for a range of solid tumors and hematological malignancies. Whilst early generation ADCs, that employ cysteine or lysine residues as conjugation sites, were primarily synthesized as heterogeneous mixtures, these were found to have sub-optimal pharmacokinetics, stability, tolerability and/or efficacy. Efforts have now shifted towards generating homogeneous constructs with precise drug loading and predetermined, controlled sites of attachment. Homogeneous ADCs have repeatedly demonstrated superior overall pharmacological profiles compared to their heterogeneous counterparts. A wide range of methods have been developed in the pursuit of homogeneity, comprising chemical or enzymatic methods or a combination thereof to afford precise modification of specific amino acid or sugar residues [950–954].

12.6.1.4 Antibody–siRNA Conjugates (ARCs)

Antibody–siRNA conjugates (ARCs) have been successfully used for targeted delivery of siRNA to specific types of cells expressing receptor-antigens; however, the effectiveness of ARCs varies significantly. For example, it has been shown that an ARC with an antibody against the insulin receptor suppresses expression of the target gene by 90% in HEK293 cells at a concentration of 115 nM [955, 956]. Another ARC with an antibody to the Lewis-Y protein inhibited the expression of the target gene by 60% at a concentration of 300 nM only when the cells were treated with chloroquine, an agent that inhibits endosome maturation [955, 957]. However, the noncovalent siRNA complex with the same antibody, formed by electrostatic interaction of oligo-arginine and siRNA, showed 60% biological activity (300 nM) in the absence of chloroquine. It is likely that the efficiency of endosomal escape mediated by chloroquine or oligo-arginine is an important factor for the biological activity of ARCs. The biological activity of both covalent and noncovalent siRNA complexes and antibodies has been shown *in vivo* in a number of studies [955, 958, 959]. However, a systemic comparison of the effectiveness of the biological activity of ARCs differing in target antigens was carried out only in one study [955, 959], which showed that along with the level of expression of the receptor-antigen, the type of intracellular transport of the receptor influences the interfering activity of the ARC. However, no direct correlation was found between the type of penetration of the antibody complex with the receptor-antigen and the biological activity of the ARC. Such a factor as the presence of a cleavable bond between the siRNA and antibody did not affect the interfering activity of the ARC. Since in this study [959], the efficiency of the binding of antibodies to the corresponding receptors was not compared, it is not possible to evaluate the efficiency of their dissociation and the degree of endosomal escape of the conjugates. However, it is likely that this is a significant factor in determining the biological activity of ARCs.

Conjugation of siRNAs with antibodies to deliver siRNA to target cells has several advantages compared with the conjugation of siRNAs with other molecules, such as high ligand binding efficiency ($K_d < 10^{-9\text{-}10}$) and prolonged presence in blood due to high molecular weight (~150 kDa) [960]. However, the immune response and low efficiency of the endosomal yield are the main disadvantage of this approach. Thus, further optimization, including the use of humanized antibodies or fragments, endosomolytic agents, and fully modified siRNAs, is required for effective use of ARCs in the clinic.

12.6.2 Biologics

A biologic drug (biologics) is a product that is produced from living organisms or contain components of living organisms and are sometimes referred to as biologic response modifiers because they change the manner of operation of natural biologic intracellular and cellular actions [961–963]. Biologic drugs include a wide variety of products derived from human, animal, or microorganisms by using biotechnology. Types of biologic drugs include vaccines, blood, blood components, cells, allergens, genes, tissues, and recombinant proteins. Biologic products may contain proteins that control the action of other proteins and cellular processes, genes that control production of vital proteins, modified human hormones, or cells that produce substances that suppress or activate components of the immune system.

Biologic drugs are used for the treatment of numerous diseases and conditions and are the most advanced therapies available. Some biologic drugs are used for the treatment of Crohn's disease, ulcerative colitis, rheumatoid arthritis, and other autoimmune diseases. Available biologics have revolutionized cancer treatment, delayed or reversed the course of immune related conditions, changed the lives of people with rare diseases, and have offered hope for many patients who previously had no effective treatment options for their condition. Examples of medical uses for biologic drugs include various cancers, rheumatoid arthritis (RA), ankylosing spondylitis, Crohn's disease, ulcerative colitis, psoriasis, diabetes, diabetic retinopathy, multiple sclerosis, heart attack, systemic lupus erythematous, age-related macular degeneration, prevention of pneumonia, anemia, chronic migraine, hepatitis B, hemophilia, respiratory syncytial virus prophylaxis, HPV prevention, varicella virus vaccine, growth hormone deficiency, osteoporosis, allergic asthma, chronic idiopathic urticaria, infertility, and cystic fibrosis.

TNF blockers are the most common type of modern biologic. TNF blockers include but are not necessarily limited to certolizumab (Cimzia), etanercept (Enbrel), adalimumab (Humira), infliximab (Remicade), golimumab (Simponi). Examples of other biologics (and their mode of action) include tocilizumab (Actemra), targets the interleukin-6 pathway; anakinra (Kineret), an interleukin-1 receptor antagonist; rituximab (Rituxan), a CD20-directed cytolytic antibody; tofacitinib (Xeljanz), first in a class of treatments RA called Janus kinase (JAK) inhibitors.

It should be noted that many biologics themselves are multifunctional or multitarget molecules in human cells, which is consistent with the principles of MTD drugs.

12.6.3 Polymeric Nanomedicines

Macromolecular therapeutics, also referred to as polymeric nanomedicines, are a diverse group of drugs characterized by their large molecular weight, including polymer-drug conjugates, polymeric micelles, and polymer-modified liposomes. The advantages of macromolecular therapeutics when compared to low-molecular-weight compounds are reviewed elsewhere [914, 915, 964]. In particular, water-soluble polymeric drugs (MW > 40 kDa) attain prolonged plasma half-lives and achieve tumoritropic accumulation due to the enhanced permeability and retention (EPR) effect [965, 966]. Conventional polymeric nanomedicines utilize polymers as delivery vehicles to carry anticancer therapeutic agents. Many of these approaches are under clinical development [967–970]. Increasingly the role of nanomedicine is not only to deliver a given drug to diseased tissues efficiently but also to trigger or improve therapeutic effects through innate biological responses [971, 972]. The design of macromolecular therapeutics has extended towards a unique paradigm where biomimetic strategies are employed to incite or control specific cellular activities [973, 974]. For instance, receptor coupling (or clustering) can be used to sensitize diseased tissues to a therapeutic agent [975, 976].

Polymeric nanoconjugates are characterized by multiple covalently attached functional moieties that can target cellular and molecular markers of diseased tissues specifically to achieve maximum treatment efficacy [977]. An outstanding feature of polymeric nanoconjugates is that they can carry a variety of covalently bound drugs that may act simultaneously on several targets (e.g., mRNA and/or protein). The benefit of such drug carriers is that they deliver more than one prodrug by single conjugate molecule, increasing the probability of coordinate and synergistic action and thus providing an efficient inhibition of multiple aberrant tumor pathways. Owing to the specific targeting of tumor tissue/cells, nanoconjugates provide therapeutically efficacious drug concentrations at the site of treatment and minimal side effects on healthy tissue. Nanoconjugate delivery systems are significantly different from nonconjugated nanodelivery vehicles, for example, micelles and liposomes, which also combine drugs, targeting and/or other functional moieties but do not form a covalently linked chemical entity. Biodegradability and lack of immunogenicity are other criteria important for treatment that can be met by nanoconjugates. Biodegradability without accumulation of potentially harmful metabolic byproducts in nontargeted tissues is an important feature that enables repeated treatments. Biodegradable moieties can be obtained from natural sources or can be synthesized chemically. Biodegradability involves nanoconjugate metabolism and eventually decomposition to water and carbon dioxide. Absence or low levels of immunogenicity can be met by choosing nanoconjugate platforms and functional moieties that do not elicit an immune response [977].

An important idea behind the nanoconjugate concept is multitargeting, and the multitargeting nanostructures form a single chemical entity with one or more targeting moieties, one or more prodrugs and other auxiliary molecular groups, releasing free drug within the specifically targeted tissue. For example, nanoconjugates bear groups that target cancer markers of tumor cell-surface antigens, which provide specific homing to tumor tissue. In addition, "passive" targeting of tumor tissue is achieved owing to the enhanced permeability and retention effect of nanoconjugates [978]; the efficiency increasing as a function of particle size of less than 200 nm. Delivery molecules carrying and releasing at once several drugs that are active exclusively against tumor-specific molecules may provide enhanced drug specificity and efficacy [977]. Targeting of several different markers at the same time has a potentially more powerful additive/synergistic impact. Simultaneous delivery of several drugs targeting different tumor markers on one carrier molecule (MTD) has a better chance for a potent combined drug effect than a mixture of nanoconjugates carrying each drug separately (CDT or FDC).

12.6.4 Drug-Free Macromolecular Therapeutics (DFMT)

Drug-free macromolecular therapeutics (DFMT) is a novel paradigm in the nanomedicine research area [915, 979–981]. The seminal work from Kopeček's laboratory et al [973] on N-(2-hydroxypropyl)methacrylamide (HPMA) copolymers containing biorecognition motifs (peptides or oligonucleotides) for crosslinking cell surface non-internalizing receptors of malignant cells and inducing their apoptosis, without containing any low molecular weight drug, led to the definition of a special group of DFMT [982, 983]. The basic idea is to induce apoptosis by crosslinking of cell-surface non-internalizing receptors mediated by the bio-recognition of high-fidelity natural binding motifs, such as antiparallel coiled-coil peptides or complementary oligonucleotides. An important feature of these designs is the absence of low-molecular-weight cytotoxic compounds (thus named "drug-free"). The therapeutic efficacy of DFMTs is based on the tailored multiple presentations of biologically active vectors, that is, peptides, oligopeptides, or oligosaccharides [982, 984]. Thus, they enable, for example, to directly induce the apoptosis of malignant cells by the crosslinking of surface slowly internalizing receptors, or to deplete the efficacy of tumor-associated proteins. The precise bio-recognition of natural binding motifs by multiple vectors on the polymer construct remains the crucial part in the designing of these drug-free nanomedicines.

12.7 Natural Products for MTD Discovery

Natural products (NPs) are chemical compounds or substances produced by living organisms, animals, plants, etc., and some of them are the active components of traditional and modern medicines. As the structural diversity of NPs helps in the chemical synthesis, and synthetic analogs can be prepared with improved potency and safety, these are often used as the starting points of drug discovery, thus being a source of useful biological activity for the development of new drugs. Yet, the once-popular approach of using natural products as a prime mining source for medicinal chemistry and drug discovery has waned considerably in the past two decades due to the advent of high-throughput screening of small-molecule mega libraries. Nevertheless, the recent revitalization of interest in NPs and their derivatives owing to the growing appreciation of polypharmacology as the next drug-discovery paradigm has made a quick return [985–987]. NPs and their unique polypharmacology as the best examples of multitarget compounds offer significant advantages for finding novel therapeutics particularly for the treatment of complex and multifactorial diseases. Recent advances and developments of systems pharmacology and computational (in silico) approaches provide powerful tools for exploring the polypharmacological profiles of natural products. It was reported that >60% of the current anticancer drugs have been derived from various NPs [986, 988]. Compared with the traditionally synthetic small-molecule drugs, NPs possess enormous chemical diversities and extensive bioactive spaces. The unprecedented chemical scaffolds of NPs provide incredible opportunities for overcoming cross-resistance to many of the known cancer drugs, serving as great inspiration for the next generation of cancer therapeutics [980, 981]. NPs constitute a huge source of MTDs, NP-MTDs.

As documented by Fang et al [984], NPs commonly exhibit a higher polypharmacologial profiles than synthetic/traditional small-molecule compounds because of their chemical diversities [989]. Our recent study has suggested a significant polypharmacological profile (a stronger degree in the drug–target network) of NPs compared with US Food and Drug Administration (FDA)-approved nonnatural product drugs based on data from DrugBank database. Specifically, the average degree (connectivity) of a natural product is 6.03 in the drug–target interaction (DTI) network, which is stronger than 2.22 of nonnatural product drugs (P-value <2.2× 10−16, Wilcoxon test) [989, 990].

Consistent with this view, Gu et al [990] collected 19,7201 NPs and used principal component analysis to explore the chemical space. They found a large portion of overlap between NPs and FDA-approved drugs in the chemical space, indicating a large quantity of potential lead compounds. They then explored the network properties of natural product-target networks and found that polypharmacology is enriched to those compounds with large degree and high betweenness centrality. With high throughput virtual screening, the investigators were able to dock these NPs to 332 target proteins of FDA-approved drugs. These results allowed them to predict the druggability of the NPs based on a docking score-weighted prediction model.

12.7.1 Reconstructing Drug–Target Network of Natural Products

Reconstruction of high-quality drug–target interaction (DTI) network is a key step for identifying polypharmacology of NPs [991, 992]. To this end, Fang et al [984] surveyed two types of chemoinformatics and bioinformatics resources: general drug target databases and natural products-specific drug target databases.

12.7.1.1 General Drug Target Databases

Several commonly used databases provide large-scale experimental drug–protein binding affinity data of natural products, such as ChEMBL [993], BindingDB [772], PubChem [994], DrugBank [995], Therapeutics Target Database (TTD) [996], and DGIdb [997]. Currently, ChEMBL (v22.1) contains 14 371 197 bioactivity records, covering 2 036 512 compounds against 11 224 targets (accessed in December 2016) [993]. BindingDB contains 1 279 670 experimentally measured binding affinity data for 6612 proteins and 565 136 molecules (accessed in December 2016) [772].

DrugBank (v5.0) is the most commonly used database, containing 8250 drug entries including 2016 FDA-approved small-molecule drugs and over 6000 experimental drugs (accessed in December 2016) [995]. The TTD (v2016) includes 2589 targets (397 approved targets, 723 clinical trial targets and 1469 research targets) and 31,614 drugs (2071 approved drugs, 9528 drugs under clinical trials and 17,803 experimental drugs, accessed in December 2016) [996]. DGIdb is a Web resource that consolidates different data sources with drug–gene interactions and gene druggability. DGIdb 2.0 includes 26 298 unique drug–gene interactions connecting 7569 drugs and 7 524 unique genes [997].

Some other common databases, such as STITCH [998] and SuperPred [999], include both experimental binding affinity data and computationally predicted data. For example, STITCH is a resource that contains both experimentally validated and computationally predicted interactions between chemicals and proteins with several different level evidences: experimental data, manually curated literature data with text-mining information and computationally inferred data [998]. SuperPred provides experimentally reported DTIs and predicted DTIs derived from molecular similarity approach, covering 665,000 DTIs connecting 341,000 compounds and 1800 targets [999].

12.7.1.2 Natural Products-Specific Drug Target Databases

Several natural product-specific databases provide useful information to reconstruct drug–target network of natural products, such as Herb Ingredients' Targets (HIT) [1000], Traditional Chinese Medicine Integrated Database (TCMID) [1001] and Traditional Chinese Medicine Systems Pharmacology database (TCMSP) [1002]. HIT contains drug–target information between 586 active NPs derived from >1300 reputable Chinese herbs and 1301 known protein targets by manually curating data from 3250 literatures [1000]. TCMID provides information on different types of traditional Chinese medicine (TCM) information: formulae, herbs, herbal ingredients, and targets [1001]. Currently, TCMID contains 47,000 prescriptions, 8159 herbs, 43,413 ingredients, 3854 diseases and 17,521 related targets.

TCMSP includes natural products, predicted targets and drug–target networks, and associated drug–target-disease networks, as well as the predicted pharmacokinetic properties for natural products. In November 2016, TCMSP contains 499 Chinese herbs registered in the Chinese pharmacopoeia with 29 384 ingredients, 3311 targets and 837 associated diseases [1002]. In addition, naturally occurring Plant-based Anti-cancer Compound-activity-Target database (NPACT) [1003] and anticancer herbs database of systems pharmacology (CancerHSP) [1004] contain target information for anticancer natural products. Altogether, these databases have provided useful resources to reconstruct the comprehensive drug–target networks for various natural products.

12.7.2 Systems Biology Resources

Systems biology resources play essential roles to build network-based or systems biology-based models for identification of drug targets of natural products. Here, we primarily focus on two types of systems biology resources: chemogenomics resources [1005] and PPI databases.

12.7.2.1 Chemogenomics Resources

Nowadays, several chemogenomics resources are reported via integrating multiple scale biological data sets for serving system biology-based drug discovery paradigm, such as connectivity map (CMap) [1006], Library of Integrated Cellular Signatures (LINCS) [1007], open TG-GATEs [1008] and DrugMatrix [1009]. CMap v2.0 includes >7000 gene expression profiles representing 1309 compounds tested in four different human cancer cell lines [1010]. Currently, a limited number of compounds and drug-induced gene expression profiles tested in four cancer types in CMap v2.0 limit the cancer-type-specific analysis. The LINCS [1007] is a National Institutes of Health (NIH)-funded program for the generation of large-scale drug-induced expression data across multiple cell types and perturbations (http://www.lincscloud.org/). As of December of 2016, LINCS contains >1 million genome-wide expression profiles representing >10 000 drugs and drug-like compounds tested in ~80 different cancer cell lines. LINCS attempts to create a network-based understanding of biology by cataloging changes in gene expression and other cellular phenotypes perturbed by various drugs or compounds compared with previous CMap [1010, 1011].

12.7.2.2 PPI Databases

Several well-known PPI databases, such as BioGRID [1012], HPRD [1013], Interactome3D [1014], STRING [1015], and MINT [1016], have been commonly used for PPI network analysis, providing useful information in the characterization of disease-related proteins and drug targets [1017–1020]. BioGRID is an interactive repository with data compiled from the primary biomedical literatures. As of December of 2016, the BioGRID (version 3.4.143) contains 1 410 457 protein and genetic interactions (including 369 198 interactions in *Homo sapiens*) in 61 model organisms by extracting data from 57 763 publications [1012]. HPRD contains annotation data pertaining to human proteins based on experimental evidence from the literatures. The HPRD release v9 covers 30 047 protein entries, 41 327 PPIs and 453 521 PubMed literatures [1013]. Interactome3D is a Web service for the structural annotation of PPI networks. Through the integration of interaction data from 10 different databases, Interactome3D provides structural details at atomic resolution for >12 000 PPIs in eight model organisms [1014]. STRING collects and integrates the known and predicted PPIs for a large number of organisms. STRING (v10.5) contains 932 553 897 interactions with quantitative confidence scores in 2031 organisms (accessed in December of 2016) [1015]. MINT is a repository of experimentally validated PPIs with special emphasis on mammalian interaction. As of December of 2016, MINT includes 125,464 interactions from 611 organisms curated from 5941 publications [1016].

12.7.3 Computational Approaches for Predicting New Targets of Natural Products

Given the tremendous growth of compound bioactivity databases, the application of computational approaches or tools to identify new DTIs for NPs has been gaining importance recently. The details about *in silico* target prediction can be found in several recent reviews [1021–1024]. In this review, we mainly focused on five types of computational approaches: target-based, ligand-based, chemogenomics-based, network-based, and omics-based systems biology approaches for identifying new targets of natural products.

12.7.3.1 Target-Based Approaches

Target-based approach, also known as protein structure-based approach, predicts target proteins of natural product, which relies on knowledge of the three-dimensional (3D) structure of the biological target [1025]. Nowadays, the rapid advances in the number of available structural data of macromolecules (such as radiography, nuclear magnetic resonance spectroscopy, and cryo-electron microscope) have significantly accelerated the development of target-based approaches [1026]. Molecular docking is the most common target-based approach. Various molecular docking approaches for drug target identification are well summarized in several recent reviews [1027–1029]. Those approaches can also be applied to identify potential targets of NPs in two aspects when 3D structure of target protein is available, that is, molecular docking-based virtual screening and inverse docking. Molecular docking-based virtual screening docks multiple ligands for a given target, while inverse docking docks multiple targets for a given ligand.

Here, we briefly summarized several recently successful examples through using molecular docking-based virtual screening approaches for discovery of bioactive NPs toward known targets in cancer therapies [1030–1036]. Signal transducer and activator of transcription 3 (STAT3) is an important target in a variety of solid and hematological tumors, such as leukemia and lymphomas. In 2014, Liu and his colleagues [1030] applied molecular docking to identify benzofuran derivative 1 as an inhibitor of STAT3 transcriptional activity from a natural product database. The mixed-lineage leukemia (MLL) proteins are histone methyltransferases that act as epigenetic transcriptional regulators. The menin–MLL interaction represents an attractive therapeutic target for cancers with MLL rearrangements, such as liver carcinomas [1034]. Zhong et al [1031] identified cytisine alkaloid 1 as an inhibitor of the menin–MLL interaction by molecular docking-based virtual screening. Singh and his coworkers [1032] reported that honokiol inhibited the growth of head and neck squamous cell carcinoma (HNSCC) by targeting epidermal growth factor receptor (EGFR) using molecular docking approach.

Typically, inverse docking approaches search a given molecule against a panel of proteins where their 3D structures are available, such as INVDOCK [1037] and IDTarget [1038]. Recently, Lauro et al [1039, 1040] proposed an application of inverse docking for screening on 10 phenolic natural compounds tested on a panel of 163 cancer-associated proteins. They identified that xanthohumol and isoxanthohumol inhibited phosphoinositide-dependent kinase 1 (PDK1) and PKC protein kinases during *in vitro* assays [1039, 1040].

Although target-based approaches have been successfully applied in the identification of new targets for natural products, several potential limitations must be recognized. First, only 15–20% proteins at the human proteome have been reported to have the availability of 3D structures [1041, 1042]. Second, high false-positive rate of molecular docking scoring functions must be recognized [1043]. Furthermore, high chemical diversity and complexity of NPs with multiple chiral atoms further limit the accuracy of current docking scores.

12.7.3.2 Ligand-Based Approaches

Ligand-based approach is commonly used when the 3D structure of a target protein is absent. This approach predicts targets of natural product, which relies on knowledge of other similar molecules that bind to the biological target of interest [1044]. Ligand-based approaches can be divided broadly into similarity-based and machine learning (ML)-based approaches [1045]. Similarity-based approach assumes that structurally similar compounds exhibit similar biological activity more often than not, via using two-dimensional molecular fingerprint [1046], 3D pharmacophore models [1047] or 3D molecular shape queries [1048], to find potential targets. Similarity-based approach is a rapid method to search all possible targets. However, it only compares molecules with each other as a whole, while it does not weight the features of the chemical structure according to their influence on the biologic activity of interest. ML-based approach solves this problem via correlating relationships between the features of molecular and target of interest, and it yields high robustness and high accuracy. Numerous methods that use ML algorithms for target prediction or hits identification toward specific targets have been published [1049–1052]. Generally, chemical structures and biological activities are integrated with different ML algorithms to train predictive models [1053], such as support vector machine (SVM), naive Bayesian and artificial neural network.

Sprague et al developed predictive classifiers to discriminate chemopreventive agents from inactive samples using random forest (RF) and SVM algorithms. They identified 18 NPs with potential *in vitro* anticancer activities via searching a natural product library containing nearly 23,000 compounds [1054]. A recent study by Bilsland et al reported a ML-based *in silico* screen to identify potential senescence agonists for cancer therapy. They identified a selective G1-Phase benzimidazolone inhibitor (CB-20903630), which induced senescence-associated β-galactosidase activity via a neural network classifier [1055].

Compared with molecular docking, ML approaches reveal some distinct advantages, including highly predictive accuracy and low computational cost. However, several limitations must be recognized for target prediction of natural products. First, the high quality of negative samples validated by experiments is limited. Most of ML models are built based on the unknown DTIs or the randomly selected DTIs as negative samples, which limits the accuracy of ML models. To address this problem, Liu et al proposed a new approach for searching highly credible negative samples. Their method assumed that the proteins that were dissimilar to each target of a given compound were not much likely to be targeted by the compound and vice versa [1056]. Second, the predictive accuracy of the method depends on large and diverse chemical space in the training set and test set. For the targets that only have few ligands or no ligand (such as orphan receptors), this approach does not work. Third, it is difficult to select proper molecular descriptors to represent the features of natural product, as there are multiple chiral atoms for natural products. Fourth, most ML models are like "black box" models owing to the lack of the detailed biological or molecular mechanisms of the predictions. Finally, conventional ML models only consider a series of ligands against a single target or phenotype. It is a challenge to predict all thousands of drug targets for a compound with a single classifier. Chemogenomics-based approaches that integrate both drug and target information can effectively address the above-mentioned limitations [1057].

12.7.3.3 Chemogenomics-Based Approaches

To overcome challenges of traditional ML methods, chemogenomics-based approaches have recently gained much attention to predict multiple drugs against multiple targets. A key issue in chemogenomics is to infer unknown DTIs by integrating both drug- and target-based descriptors, such as target protein sequence information, drug chemical structures and the available known drug–target network topology information simultaneously [1057–1059].

Yamanishi et al [1060] reported a pivotal study to predict DTIs with chemogenomic approach that integrates the sequence similarities between pairwise proteins and the chemical structural similarities between pairwise compounds for predictions of DTIs. High performance was yielded when evaluating their models using known DTIs covering four types of targets: enzymes, ion channels, G-protein-coupled receptors (GPCRs) and nuclear receptors. Belakley et al [1061] proposed a supervised inference method, named the bipartite local model (BLM). BLM predicted target proteins of a given molecule using the structural similarity of compounds, genomic similarity and network topological information of known interactions between drugs and targets. Recently, Keum et al [1062] adopted this method to construct classification models with good performance for six classes of DTIs in humans including GPCRs, ion channel, enzyme, receptors, transporters, and other proteins. Yu et al [1063] reported a chemogenomics approach that efficiently incorporated the chemical, genomic and pharmacological information using the largest available data set from the DrugBank database. In the follow-up analyses, they successfully identified unknown DTIs for ingredients of traditional Chinese herbs via their chemogenomics models [1064].

Overall, one merit of chemogenomic approaches or models is that it can simultaneously search huge numbers of drug candidates and candidate targets at the same time. However, there are two shortcomings have to be recognized. First, the performance of chemogenomic models depends heavily on the relevant descriptors used for model building. The suitable feature selection approaches are recommended to use for selecting the biological or pharmacological relevant descriptors. Hundreds of descriptors used in one model often generate a high risk of "overfitting" [1065]. The best way to avoid "overfitting" risk is that all computational models are suggested to be built based on more relevant features and have to be passed the testing by an external validation set that are not used to develop the model. Second, compared with conventional ML models, chemogenomic models contain target descriptors, while the increasing of target descriptor variables increases the risk of "over fitting" and computational complexity. A recent study examined two methods: **(a)** multitarget-quantitative structure–activity relationships (mt-QSARs) and **(b)** computational chemogenomics, for prediction of DTIs. They found that the performance of the chemogenomic models was worse than that of mt-QSARs [1066]. Several possible explanations are suggested: **(a)** traditional protein sequence-based descriptors cannot describe the natural interactions between drugs and targets, and **(b)** a highly imbalanced number of positive samples versus negative samples often occurred owing to lack of literature-reported negative biological or pharmacological data.

12.7.3.4 Network-Based Approaches

Network-based approaches for prediction of DTIs have shown a great promise in rational drug discovery. In a bipartite drug–target network, nodes denote drugs and targets, and edges denote their interactions, such as the binding affinity between drugs and targets via half maximal inhibitory concentration (IC_{50}) or inhibition constant/potency (Ki) value <10 μM. Recently, graph-based network approaches have demonstrated potential for identifying novel targets for known drugs and drug repositioning. Cheng et al [113] proposed three supervised inference models to predict DTIs, namely, drug-based similarity inference (DBSI), target-based similarity inference (TBSI) and network-based inference (NBI). DBSI and TBSI rely on chemical structural similarity and target sequence similarity, respectively, whereas NBI is only based on drug–target bipartite network topology similarity. Among three algorithms, NBI performed the best on four benchmark data sets covering GPCRs, enzymes, ion channels and nuclear receptors. Moreover, five old drugs were identified via NBI and then experimentally validated to have potential binding effects on ERs or dipeptidyl peptidase IV *in vitro* [775].

As an improvement of NBI, Cheng et al [1067] further developed two weighted NBI methods, edge-weighted NBI (EWNBI) and node-weighted NBI (NWNBI), by assigning the weighted values to edges and nodes in the drug–target network, respectively. They found that the weighted NBI methods marginally outperformed the unweighted ones [1067]. However, both of NBI or the weighted NBI cannot be applied to the new drugs without any known target information in the training set. Recently, they further developed a novel algorithm, namely, substructure-drug-target network-based inference (SDTNBI), to prioritize potential targets for new chemical entities [1068]. SDTNBI imports chemical substructures to bridge the gap between new chemical entities and known DTI network. Systematic evaluation based on cross-validation and external validation showed high accuracy and robustness of SDTNBI. Finally, a global network was constructed, and thousands of new potential DTIs were computationally identified via SDTNBI [115]. However, there are still several potential limitations in SDTNBI yet. For instance, whether introducing tunable parameters to SDTNBI could improve the method performance remains unclear. Hub nodes that have high degree in the drug–target network often generate high risk of false-positive rate for network-based prediction, limiting the application of two previous unweighted network approaches (NBI and SDTNBI). The same group further proposed an improvement on SDTNBI [1069], named balanced substructure-drug-target network-based inference (bSDTNBI), to identify the MOA for both old drugs and new chemical entities. Finally, they demonstrated that bSDTNBI showed potential application for identifying the specifically antagonistic or agonistic ERα ligands, and dual-effect ERα ligands with both antagonistic and agonistic activities for the development of potential therapies to breast cancer or osteoporosis. While these network-based approaches have made significant advances, several improvements could be further achieved in the future. For example, current network-based approaches such as bSDTNBI and NBI cannot predict potential DTIs for those target proteins that are absent in the known substructure-drug-target network. In addition, the interaction types (e.g., antagonistic and agonistic effects) and binding affinities of drug–target pairs have not been considered yet. Integrating human PPI network and gene regulatory network [1069–1072] (e.g., microRNA network) with the aforementioned network approaches would provide potential solution to identify DTIs for many novel target proteins at the human proteome level.

12.7.3.5 Omics-Based Systems Biology Approaches

Owing to recent advances of "omics" technologies, massive omics data were generated, including transcriptomics, proteomics, and metabolomics. These omics resources provide huge opportunities for the development of systems biology models for identifying new targets for NPs [1073, 1074]. Here, we briefly describe three omics-based systems biology approaches, including transcriptomics, proteomics, and metabolomics, for prediction of DTIs for NPsas below.

Transcriptomics

Genome-wide transcriptomics data from the cultured human cells treated with bioactive small molecules have provide a unique resource for drug repurposing and identifying MOA for various drugs, including NPs [1075, 1076]. Qu et al [1077] integrated drug-induced transcriptomics resources and computational approaches for target identification, lead compound identification and drug repurposing studies. Hieronymus and his colleagues [1078] introduced a chemical genomic approach for discovery and target prediction of modulators of cancer phenotypes, as exemplified by androgen acceptor-mediated signaling. Among the hits that emerged, two natural products, celastrol and gedunin, were identified to have potential to treat prostate cancer as heat shock protein 90 (HSP90) inhibitors [1078]. With a similar approach, Wei *et al.* identified drugs to overcome resistance to glucocorticoid treatment based on a gene expression signature of glucocorticoid sensitivity/resistance in acute lymphoblastic leukemia via downregulation of myeloid cell leukemia 1 (MCL1) [1079]. Compared with previous "black box" ML-based models and chemogenomics models, transcriptomics-based systems biology models provide novel biological or molecular mechanisms for characterizing their specific anticancer effects.

Proteomics

In addition to transcriptomics, proteomics, representing the large-scale analysis of proteins, has also driven much of systems biology studies over the past two decades [1080]. For example, mass spectrometry (MS)-based proteomics refers to the application of MS to the identification, quantification, and characterization of the proteins [1081]. Savitski et al [1082] proposed an approach named thermal proteome profiling (TPP) to monitor the changes in protein thermal stability across the proteome using quantitative MS. They used the histone deacetylase inhibitor panobinostat to show the whole workflow and demonstrated that this methodology had the potential to identify direct and indirect drug targets in living cells [1083]. In subsequent work, they further extended TPP to profile membrane–protein targets of small molecules, either in cell extracts or for *in situ* studies on intact cells [1084]. Sacco et al [1085] systematically investigated signaling network perturbation after metformin treatment in the proteome and phosphoproteome of breast cancer cells via combining high-resolution MS and network analysis. They examined perturbations of 7875 proteins and 15,813 phosphosites after metformin treatment, and further found that metformin inhibited the p70S6K-rpS6 in a PP2A-phosphatase dependent manner. This study demonstrated that deep proteomics and network analysis provided novel strategy for exploring both detailed and global mechanisms involving to the anticancer effects of metformin [1085]. Compared with previous transcriptomics-based approaches, proteomics-based approaches provide more useful strategies for directly exploring both detailed and global molecular mechanisms of NPs at the human proteome level, not transcriptomics level.

Metabolomics

Metabolomics monitors the biochemical changes of many small molecules and their metabolites at various systems biology levels (e.g., cells, tissues, organs) via integrating high-throughput assays and bioinformatics analyses [1086, 1087]. It provides a unique strategy to facilitate the identification of new drug targets of NPs or their metabolites [1088–1093]. Birkenstock et al [1088] used metabolic profiling to demonstrate that the antimicrobial agent triphenylbismuth dichloride (TPBC) blocked bacterial pyruvate catabolism in *Staphylococcus aureus* by quantifying the extracellular metabolites of bacterial cultures and helped to identify pyruvate dehydrogenase as a target for TPBC. Zhang et al [1089] used cellular metabolic profiling to quantitate six isoprenoid precursor metabolites. They demonstrated that fosmidomycin treatment blocked isoprenoid metabolism in two microbes (*Escherichia coli* and *Plasmodium falciparum*), suggesting that methylerythritol phosphate cytidylyltransferase (IspD) is a potential target for fosmidomycin. Bayet-Robert et al [1090] used metabolomics to elucidate the MOA of three marine NPs (MNPs) against biochemical disorders induced in human MCF7 breast cancer cells. Metabolite profiling revealed that ascididemin involved enzyme blockade in citrate metabolism and kahalalide F increased lipid membrane, suggesting new metabolic targets for these MNPs [1092].

Owing to the dramatically increasing of multi-scale omics data, omics-based systems biology approaches offer powerful tools for drug target identifications of NPs in drug discovery and development. In the future, new computational systems biology approaches to analyze and integrate new multi-scale omics data are urgently needed in this field.

12.7.3.6 Performance Evaluation of Computational Models

Several indicators are commonly used to evaluate the performance and robustness of computational models, including accuracy, precision, and recall. The details of these measurements are discussed in a previous study [775]. Specifically, a receiver operating characteristic (ROC) curve is often drawn via measuring a series of true-positive rate and false-positive rate under different conditions. In addition, a precision–recall curve is often used via measuring an array of precision and recall values under different cutoff. Finally, the areas under these ROC curves or precision–recall curves will be yielded to evaluate the performance of various computational models.

The N-fold cross-validation is often used for the evaluation of the performance of computational models, such as fivefold cross-validation and leave-one-out cross-validation. There are two differential cross-validation protocols for currently computational approaches or models: samples-based cross-validation versus links-based cross-validation. For example, during leave-one-out cross-validation in general ML-based models, each drug or target (not a drug–target link) in the entire drug–target network is extracted to be the test set in turn. The remnant data set containing N-1 drugs or N-1 targets will be used as training set for building the models. For a N-fold cross-validation (e.g., five-fold cross-validation) in network-based link predictions [775], the entire links in drug–target network are randomly divided into five parts. One part is used as the test set in turn, which is not included in training set. The remnant network is used as the training set to build network-based model. However, N-fold cross-validations cannot measure the real generalization ability for most computational models [1094]. It is crucial that all computational models are recommended to pass the external validation (model performance is evaluated by an external validation set that are not used to develop the model). Finally, a short list of predicted candidates with high-confidence score can be selected for further experimental validation (*in vitro* assays, *in vivo* animal models or patient electronic medical records) to test the performance of computational models in practice.

12.7.4 Macromolecular Targets of Complex Natural Products

A study presented the development and experimental validation of a computational method for the discovery of such targets [1095]. The technique does not require three-dimensional target models and may be applied to structurally complex natural products. The algorithm dissects the NPs into fragments and infers potential pharmacological targets by comparing the fragments to synthetic reference drugs with known targets. We demonstrate that this approach results in confident predictions. In a prospective validation, we show that fragments of the potent antitumor agent archazolid A, a macrolide from the myxobacterium Archangium gephyra, contain relevant information regarding its polypharmacology. Biochemical and biophysical evaluation confirmed the predictions. The results obtained corroborate the practical applicability of the computational approach to natural product "de-orphaning."

12.8 Drug Repurposing for MTD Discovery

"Drug repurposing," also known as "drug repositioning," "drug reprofiling," "drug redirecting," "drug re-tasking," and "drug rediscovery," refers to the research and development to identify novel indications of existing drugs outside the scope of the original medical indication [1096–1098]. As an alternative strategy in drug development for both common and rare human diseases, drug repurposing represents an attractive opportunity for rare diseases. This is because that more than 7000 rare diseases have been identified, only about 5% benefit from a licensed treatment and the majority of these diseases is life threatening. It is an "old wine in new bottles" drug discovery strategy, which is generally believed to be efficient, economical, and low-risk.

12.8.1 Advantages and Challenges of Drug Repurposing in MTD Discovery

12.8.1.1 Strengths of Drug Repurposing Strategy in MTD Discovery

The whole process of the traditional model of drug discovery and development implies a high investment by the pharmaceutical industry, which ultimately impacts on high drug prices. Given the high attrition rates, substantial costs and slow pace of new drug discovery and development, repurposing of "old" drugs to treat both common and rare diseases is increasingly becoming an attractive proposition and a universal strategy owing to the following strengths.

(a) Drug repurposing involves the use of de-risked compounds with known safety profiles, which helps limit potential biosafety risks and minimize toxic effects in patient consumers [1099, 1100].
(b) Unlike the traditional approach, drug repurposing enters both the preclinical evaluation and clinical trials of the compound of interest faster with reduced number of required

clinical trial steps. This could not only reduce budgeting research but also reduce the development costs and shorter development timelines to reach market. Lower costs could potentially lower the drug price and relive the economic burden of patients and would also help meet the needs of patients to access useful and novel treatments. It is estimated that drug repurposing saves up to 5–7 years in average drug development time [1096].

(c) Existing pharmaceutical supply chains for existing compounds could facilitate "formulation and distribution" of the drug.

(d) Known possibility of combining with other drugs could allow more effective treatment.

(e) Drug repurposing could facilitate the discovery of :new mechanisms of action for old drugs and new classes of medicines."

(f) Drug repurposing has the potential to "rescue" the abandon

(g) In recent years, the need for new approaches to drug research and development, combined with the advent of big data repositories and associated analytical methods, has generated interest in developing systematic approaches to drug repurposing. Various data-driven and experimental approaches have been suggested for the identification of repurposable drug candidates. Particularly, a variety of innovative computational methods to enable systematic repurposing screens, experimental as well as through in silico approaches, have emerged [1101–1103].

12.8.1.2 Challenges of Drug Repurposing Strategy in MTD Discovery

However, drug repositioning has a number of downsides too.

(a) The dosage required for the treatment of a novel disease usually differs from that of its original target disease, and if this happens, the discovery team will have to begin from Phase I clinical trials, which effectively strips drug repositioning of its advantages of over de novo drug discovery [1104].

(b) For new formulation or technology, an almost complete clinical program may be required in order to characterize the new technology/formulation, beyond the efficacy for the new indication [1100, 1104]. This can lead to lengthy development timelines and high costs and create another barrier to bringing the repurposed medicine to the clinic.

(c) Like new drug development, drug repurposing also relies on the generation of new intellectual property (IP). Such IP might arise from the repurposing concept itself, a new formulation or any of a number of other avenues (e.g., dose level, drug combination and route of administration) that might improve the drug in tailoring it specifically to a new patient population. Nevertheless, for compounds for which substantial data exist, new IP may be challenging and may constrain the portfolio of compounds that can be pursued by pharmaceutical companies. Patent right issues can be very complicated for drug repurposing due to the lack of experts in the legal area of drug repositioning, the disclosure of repositioning online or via publications, and the extent of the novelty of the new drug purpose [1104].

12.8.2 Basis for Drug Repurposing

The conceptual basis of drug repurposing is drug promiscuity, or in other words, polypharmacology of existing drugs is the foundation for drug repurposing. Drug database is the gold mine or rich resources for drug repurposing, and the accessibility to the databases and the expertise in using these data are the technological requirements for drug repurposing. Additionally, the principles and some key element in the drug development procedures for new chemical entity (NCE) are also strictly applied to drug repurposing.

12.8.2.1 Drug Promiscuity: The Theoretical Basis

A pre-requisite for drug repurposing is drug promiscuity. The "repurposibility" of a drug depends on the extent of its pharmacological promiscuity or a drug's ability to bind to multiple molecular targets and appropriateness of promiscuity or a drug's ability to bind to multiple desirable targets as well.

Quantifying the degree of drug promiscuity is the first step towards drug repurposing. The extent of promiscuity of a drug was found to be correlated to structural similarity as well as binding site similarity of protein targets [1105–1107]. According to a systematic study of drug promiscuity [1105], 71% of the drugs have at least two targets with similar binding sites. Another study presented a systematic analysis of public biological screening data and revealed more than 1000 compounds that were active against 10 or more targets from different classes [1106]. These properties are supported by the finding that there is significant extent of structural homology in many protein families, which confers ligands promiscuity. Promiscuous compounds might be active against closely related target proteins from the same family or against distantly related or unrelated targets. Currently, the molecular basis of multitarget activity is not well understood, although insights from X-ray structures of complexes involving promiscuous compounds are beginning to emerge [1107]. In theory, a higher degree of promiscuity would provide more opportunities and broader options for drug repurposing; however, this property could also compromise the "repurposibility" with a higher potential of off-target effects.

For drug repurposing, to quantify the extent of promiscuity is not enough, and equally important is to identify appropriateness of promiscuity. In this sense, drug promiscuity can be either bad or good/undesirable or desirable, often judged based on the intentional application to specific disease entities of interest. Bad promiscuity causes off-target activity, whereas good promiscuity is characterized by desired multitargeticity and polyspecificity. It is necessary to distinguish "undesirable promiscuity" resulting from chemical liabilities, nonspecific binding, or assay artifacts, from "desirable promiscuity" representing true multitarget activities.

Another issue is that the exploration of multitarget activity for both the degree and appropriateness is complicated by the fact that apparent compound promiscuity may often be due to undesired experimental artifacts. In particular, "false positive" promiscuity may result from a tendency of compounds to aggregate or react under assay conditions or engage in other nonspecific interactions with different targets [1108–1111]. Therefore, in the study of multitarget activity, care must be taken to distinguish "true" promiscuity from assay interference effects and other potential artifacts. This also affects computational analysis where data confidence criteria must be carefully considered and potentially liable compounds be identified [1112, 1113].

Fortunately, the emerging "big data" era in medicinal chemistry is already yielding unprecedentedly large amounts of compounds and activity data, which form a sound basis for careful promiscuity assessment, even if significant numbers of potentially questionable compounds and activity data are disregarded [1113].

12.8.2.2 Drug Data: The Technological Basis

Drugs constituting potential candidates for repurposing can be divided into three categories [1100]:

(a) Generic drugs: drugs approved that are no longer protected by patents.
(b) Failed drugs: compounds that have been through some stage of clinical development but are currently not on the market.
(c) Patented drugs: either approved or in the late stage of clinical development.

Generic drugs constitute, by and large, a common resource for all major players in drug repurposing, that is, academia, biotech and pharmaceutical companies. These drugs have been commercially available for a significant amount of time, have well-known and largely documented safety profiles and are readily accessible for studying preclinically as well as for clinical development. However, with regard to drugs falling into the two other categories, that is, failed drugs or those still under patent, the pharmaceutical companies hold an advantage in access to drug-related data, for their respective compounds. Thus, in many cases, access to data is biased toward the pharmaceutical companies, rendering them a key partner in an effort to systematically identify new indications for existing molecules [1100].

12.8.2.3 Clinical Development Principles

Clinical development aims to "translate" findings in fundamental research into medical practice and meaningful health outcomes [1114]. Clinical development principles for drug repurposing should be adopted from those applied for new chemical entity (NCE) drug development, in which the pharmaceutical companies are well versed [1100]. There may be elements in the process that can be shortened because of existing knowledge. However, the trajectory of clinical development is generally similar between drug discovery and drug repurposing. Two major apparent categories in drug development can be linked to failure in drug approval, namely, safety and efficacy [1100, 1115]. The first relates to the toxicological and pharmacokinetic profiles of the NCE, which are mainly addressed in Phases I and IIa clinical trials, following many preclinical evaluations. The second refers to target engagement and biological activities that are subject to therapeutic interference and is linked to the clinical efficacy of the NCE under investigation.

The clinical testing of a repurposed drug necessitates the integrated understanding of the fundamental pharmacokinetic/pharmacodynamic principles of exposure at the site of action, target binding and expression of functional pharmacological activity (termed together as the "three pillars of survival") [1116]. Early stages of development (pre-proof of concept) determine the likelihood of either newly or repurposed candidate compound to continue in Phase II trials and improve the chance of progression to Phase III and ultimate approval success. For drug repurposing programs, some elements may have been elucidated during the development or approval of another indication providing some measure of early risk reduction in terms of both cost and time. On one end of the drug repurposing spectrum is an approved drug being examined at similar or lower doses than the maximum dose already approved by regulatory agencies to target the same molecular pathway/mechanism in a different patient population [1100, 1117]. The large body of clinical data and the experience accumulated in Phase III (efficacy) and Phase IV (post-marketing) trials for the drug in question offer a good understanding of its profile in terms of adverse events, long-term and chronic toxicity and on- and off-label effects limiting the need for Phase I and Phase IIa/IIb clinical trials, resulting in shorter development timelines.

On the other end of the spectrum are approved drugs being evaluated for novel therapeutic indications through the discovery of biologically and clinically relevant affinities for new targets, which play a determinant role in those indications [1100, 1118]. In this case, identifying the "right compound" during repurposing efforts is just the beginning and caution should be applied against over-relying on prior clinical experience; all elements—far beyond the "three pillars"—should be thoroughly re-explored with the following objective: the right dose, reaching the right compartment and to engage the right target, for the right time, in the right patient population, measured for the right duration, with the right tools/methodology. Consequently, under these conditions, a full clinical development program similar to that of an NCE might ensue without necessarily saving time or cost [1100].

12.8.3 Common Approaches for Drug Repurposing

Often considered as a serendipitous approach, where repurposable drugs are discovered by chance, drug repurposing has heavily benefited from advances in human genomics and network biology. It is now possible to identify serious repurposing candidates by finding genes involved in a specific disease and checking if they interact, in the cell, with other genes which are targets of known drugs. It was shown that drugs against targets supported by human genetics are twice as likely to succeed than overall drugs in the pharmaceutical pipeline. Drug repurposing can be a time and cost-effective strategy for treating dreadful diseases such as cancer and is applied as a means of solution-finding to combat the COVID-19 pandemic. An efficient drug repurposing pipeline requires the combination of access to molecular data, appropriate analytical expertise to enable robust insights, expertise, and experimental setup for validation and clinical development know-how. According to Cha et al [1100], systematic drug repurposing approaches can be largely divided into those based on experimental screening approaches/functional approaches, *in silico* approaches/computational approaches and combinatory, or mixed approaches that employ existing data to identify potential new drug–disease associations.

12.8.3.1 Experimental Screening Approaches

Experimental screening approaches can be used as a source of hits for both drug discovery and drug repurposing, with notable differences in their application and outcomes. Such approaches include target screening approaches, cell assay approaches, animal model approaches, and clinical approaches, whereas computational approaches are, for instance, structure-, network- or text-mining-based [1100, 1119, 1120]. Searches in drug discovery programs are typically done for *de novo* candidate hits, fueled by an HTS campaign, which requires highly specialized screening facilities and compound libraries containing several million compounds. Repurposing programs focus on advanced known molecules either approved or failed with some knowledge of their safety or mechanism of action available, led by in-depth screening, and with smaller size compound libraries. Typical approved compound libraries containing 500–2000 compounds and a similar number of existing but unapproved compounds are thought to be available. Some may contain annotation and/or data on safety and mechanism of action. Compound libraries represent a scale that is manageable in many academic drug discovery laboratories, opening up opportunities for additional players and approaches to identification of candidate hits for a clinical development program.

Another key difference between drug discovery and repurposing is the fate of the hits. For a repurposing screen, a compelling hit is a drug molecule that is a candidate to be advanced into development. This is quite distinct from an HTS hit in drug discovery that becomes a starting point for a newly found medicinal chemistry program which is then evolved iteratively. Lastly, for drug discovery screens, an efficient assay must be simple and fast in order to manage the number of potential compounds for review, whereas for repurposing screens, the limited scale allows a broader range of complexity of assay types.

Two types of screens, commonly used and adaptable for HTS, are cell-free or cell-based, target-focused screens and phenotypical screens [1100]. The first is based on a specific compound activity in relation to a specific mechanism. The second is based on cell behavior (e.g., growth and death) with or without the compound addition. Phenotypical screens have historically been more successful than target screening in generating new drugs [1121, 1122]. A potentially powerful variant of phenotypical assays is the use of screens that report signals relevant to disease biology. However, the use of *in vitro* assays is challenged by its validity in presenting the human disease, and an ongoing debate around the use of more representative cell-based models such as human-induced pluripotent stem cell-derived tissues for the case of degenerative diseases is likely to continue [1123]. Screening with multicellular model organisms [e.g., worms (*Caenorhabditis* elegans), fruit flies (*Drosophila* melanogaster) and larval zebrafish (*Danio rerio*)] introduced with human genes and/or disease-causing mutations often provides a more compelling approach for phenotypical screening [1124]. While there are obviously limitations to the extent to which the human disease phenotype is fully represented in these models, they clearly have the potential to contribute to the identification and prioritization of active molecules, early in a repurposing effort.

12.8.3.2 *In silico* Repurposing Approaches

The browsing functionality can be used as a global entry point to collect further information with regard to small molecules (~1 million), side-effects (~110 000), or drug–target interactions (~3 million). The drug-repositioning tab for small molecules will also suggest possible drug-repositioning opportunities to the user by using structural similarity measurements for small molecules using two different approaches [1125]. *In silico* repurposing approaches apply sophisticated analytical methods to existing data identifying new potential associations between drug and disease. Approaches can be broadly divided into two categories [1100]:

(a) Molecular approaches, which are based on understanding of drug activity and disease pathophysiology and are often powered by large-scale molecular data (i.e., "omic data"), such as genomic, transcriptomic or proteomic data, as well as data on drug targets and chemical structure.
(b) Real-world data (RWD) approaches, focusing on identification of unknown, and at times unexpected, relationships between drugs and diseases or their symptoms, based on RWD—data on individuals' health, habits, and behavior that are captured without intervention from the environment or biases introduced through data collection methodologies.

Molecular Approaches

Understanding the mechanism of action of a drug and matching it with a disease indication different from that for which it was originally approved or developed are at the heart of molecular approaches to drug repurposing. Access to increasing amounts of data largely attributed to the rapid advance of molecular biology methodologies and omic data enables the application of systematic analyses to drug and disease mechanism of action [1100, 1126, 1127].

Of the various types of omic data available, transcriptomics and genomics are the two data types most widely used to support drug repurposing, due to the combination of availability of datasets on drugs and diseases and the robustness and reproducibility of the data [1006]. Transcriptomics involves measuring the expression levels of thousands of genes, often by quantifying RNA using RNA-Seq or gene expression microarrays. One approach of applying transcriptomics to drug repurposing is based on the concept that reversal of gene expression signatures may translate into clinical benefit [1006, 1128]. In an early example of this approach, published along with the report of one of the first large-scale datasets of drug-induced gene expression profiles (the Broad Institute's Connectivity Map), sirolimus, an mTOR inhibitor, was shown to reverse the signature of dexamethasone resistance in acute lymphoblastic leukemia [1129]. The *in-silico* findings were subsequently confirmed *in vitro* through a dramatic reduction in viability of cells exposed to sirolimus in combination with dexamethasone, compared with dexamethasone alone. Furthermore, the sirolimus–dexamethasone combination provided durable remission in xenografted acute lymphoblastic leukemia [1129], supporting the transcriptomic-based repurposing approach. Clinical validation for the concept of using sirolimus or related drugs to overcome resistance to glucocorticoids is, however, still pending [1130].

The utility of gene expression data to support disease indications has been further demonstrated in a recent publication [1131] which found that pridopidine upregulates expression of genes associated with various neuroprotective pathways downregulated in Huntington's disease. In addition to transcriptomic approaches, genomics (mutations and genetic variants associated with disease) is also commonly utilized as means to identify repurposing candidate, for example, drugs targeting genes associated with genome-wide association study hits in a given disease [1125]. Consistently, it was demonstrated that drugs with genetic support have substantially higher likelihood of success in clinical trials [1132]. Another recent publication demonstrated methods by which the combination of transcriptomic and genomic data can be used to identify potential repurposing candidates for Alzheimer's disease [1133].

In recent years, a rich set of resources has become available for repurposing activities, falling into three broad categories. Drug-focused databases provide information on the gene expression profiles, known targets, clinical status, and other aspects of various drugs. Disease-focused databases provide detailed information on the gene expression profiles and proteomic, genetic, and epigenetic characteristics of diseases. Finally, various tools to link drugs with diseases have been made available. Each of these databases and tools has strengths and limitations: listing here does not imply endorsement, and appropriate caution and quality control are advised.

Real-World Data (RWD) Approaches

RWD, non-interventional data on individuals' activities and health, are characterized by large, complex, intricately structured datasets, often containing several years of data on millions of patients. Data sources for RWD can stem either from observational, simple trials (i.e., pragmatic trials) or from registries, administrative data, health surveys, electronic medical records (EMRs), medical chart reviews, or adverse event reporting and even social media. This type of data can address a wide range of challenges across drug development and has been mainly used to support health economics research. However, RWD constitute a fertile, and largely untapped, ground for generating and validating drug repurposing candidates. For example, Xu et al [1134] used EMR data to demonstrate that metformin, a common oral medication for type 2 diabetes, was associated with decreased mortality after a cancer diagnosis compared with diabetic cancer patients on metformin, and compared with non-diabetic cancer patients not on metformin. The ability to utilize RWD, often stemming from disparate sources each with its own annotation system and limited accompanying clinical phenotypical data, is highly dependent on the advent of sophisticated analytical methods such as artificial intelligence and deep learning, and their application to healthcare. The appropriate demonstration of utility in combining and integrating multiple data types with varying levels of ascertainment and metadata is critical for the growing need to enable the extraction of additional insights from RWD and other big data modalities.

12.8.3.3 Combinatory Approaches

Each of the methodologies described above addresses the question of pairing diseases to potentially beneficial drugs from a different perspective. HTS approaches aim to affect a specific phenotype or target; molecular/omic approaches are based on the mechanistic principle that efficacy relies on the ability of the drug to alter disease-related pathways, and RWD assess unexpected occurrences and biases that may reflect novel utility for a given drug. Ultimately, robust repurposing candidates with high likelihood of success will stem from a systematic and rigorous combination of various data types, approaches, and analytics. One such approach is based on the notion that shared drug side effects may imply similar molecular mechanisms. The former may then be utilized to identify drugs similar to those already known to treat a disease by combination of RWD [e.g., the Food and Drug Administration (FDA) adverse events reporting system] and databases on drug side effects from randomized-controlled clinical trials [1135, 1136]. Additional methodologies combining multiple approaches have been demonstrated, for example, through the use of phenotypical, chemical, and genomic similarity among drugs [1137, 1138]. Structural similarity can be used to infer new targets for existing therapies and combined with both phenotypical and gene expression-based signatures to infer drugs with similar functions to existing therapies for a disease [1139].

A repurposing approach involving combination of different methodologies and datasets was recently reported by Paik et al [1140], who generated disease and drug pair similarity scores independently in genomics data and in EMR-extracted lab test data, based on the underlying assumption that similar drugs can be used to treat similar diseases. The result was identification of terbutaline sulfate, a β_2-adrenoceptor agonist widely used for the treatment of asthma, as a candidate for the treatment of amyotrophic lateral sclerosis (ALS), based on similarity between terbutaline sulfate and ursodeoxycholic acid, on the one hand, and Kawasaki syndrome and ALS, on the other. The potential therapeutic benefit of terbutaline sulfate for ALS was then demonstrated and validated via prevention of defects in axons and neuromuscular junction degeneration in a zebrafish model of ALS, but to date, no clinical trial has demonstrated a similar benefit in humans.

Currently, a variety of biological databases such as Drugbank [768] and ChEMBL [770] exist, which can provide the information which is necessary to build a potential drug repositioning pipeline. Unfortunately, the scope of existing databases that focus directly on drug repurposing is limited, as they either serve as a pure compilation of successfully repositioned drugs [1141, 1142] or completely lack interactive features [1143, 1144].

Promiscuous 2.0 provides a powerful yet easy to use resource, which enables experts and non-experts alike to create complex interaction networks in an intuitive and interactive way, and potentially infer new indications for existing compounds [1119]. Moreover, it is possible to search for existing drugs or even submit new molecular structures and be presented with suggested application areas or, vice versa, get potential drug candidates for disease indications of interest; a completely unique feature across all published databases.

Another database on drug promiscuity, DrugCentral [1145], is a public resource (http://drugcentral.org) that serves the scientific community by providing up-to-date drug information, as described in previous papers. DrugCentral integrates a broad spectrum of drug resources related to chemical structures, biological activities, regulatory data, pharmacology, and drug formulations [1146]. Since 2018, DrugCentral has continuously strengthened its role as a key resource for the worldwide scientific community being additionally cross-referenced by several resources, such as UniProt [1147], ChEBI [1148], Hetionet [1149], GUILDify [1150], UniChem [1151], and Guide to Pharmacology [1152]. DrugCentral served as primary resource for RepoDB, a drug repurposing database [1153], a time-resolved computational drug repurposing algorithm [1154], and an adverse drug event network for computational toxicology predictions [1155]. First introduced and published in the 2017 NAR database issue [1146], DrugCentral reconciles the basic scientist's understanding of the "drug" concept (active pharmaceutical ingredient) with the view of the patient and healthcare practitioner (pharmaceutical formulation). Since its initial launch, the two DrugCentral papers [1146, 1156] were cited more than 160 times cf. Google Scholar, and the website is accessed on average by ~8000 visitors monthly, with a monthly average of ~20 000 page views and over 20 000 full database downloads per year (as of 15 September 2020). Throughout regulatory and scientific documents, several terms are often used interchangeably: drug substance, new chemical (or molecular) entity, and active (pharmaceutical) ingredient. While these terms have precise contextual meaning, in this paper preference is given to the term "drug" as synonymous with these three concepts. The term "formulation" is used when discussing pharmaceutical products. The current update adds newly approved drugs by the US Food and Drug Administration (FDA, https://www.fda.gov/home) and the European Medicines Agency (EMA, https://www.ema.europa.eu/en) up to 31 March 2020. Drugs approved by Japan Pharmaceuticals and Medical Devices Agency (PMDA, https://www.pmda.go.jp/english/index.html) were also monitored up to the latest information available, that is, November 2019. In addition, for numerous drugs present in DrugCentral since 2018, regulatory agency information was added according to their approval status. Sudden outbreaks can rapidly impact global health, as evidenced by the COVID-19 pandemic, caused by severe acute respiratory syndrome coronavirus 2 (SARS-CoV-2). This pandemic has accelerated the need to rely on computational platforms [1157] capable of identifying and advancing novel therapeutics for clinical evaluation. In this regard, the current DrugCentral update enables computational and medicinal chemists with **(a)** drug repositioning categories, that is, an in-depth classification of drugs based on current market status and intellectual property rights in the United States [1158], to prioritize new therapeutic uses for "old drugs"; and **(b)** a suite of machine learning models that predict anti-SARS-CoV-2 activities,

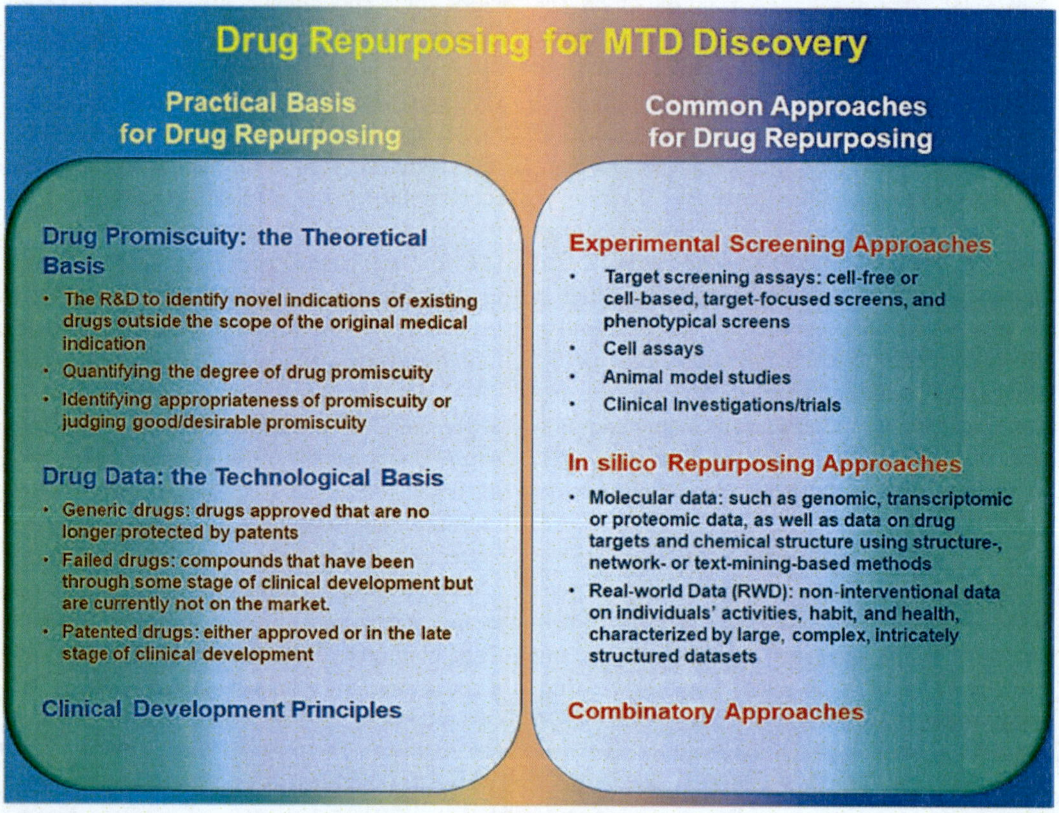

Fig. 12.5 A summary of drug repurposing for multitarget drug (MTD) discovery

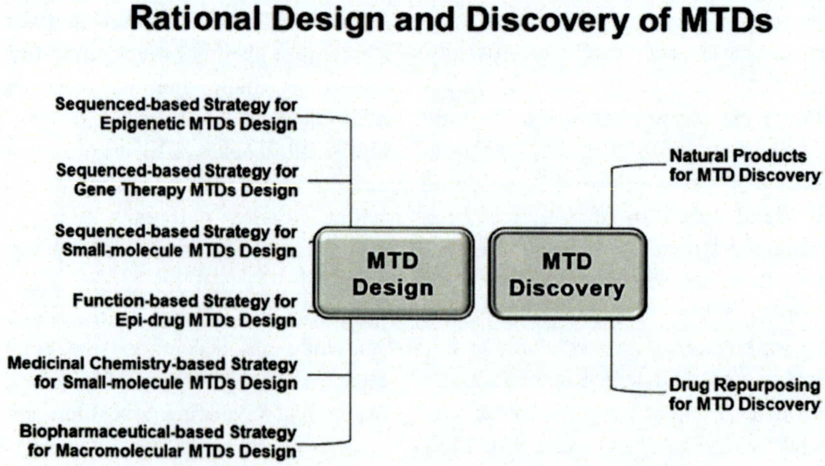

Fig. 12.6 A summary of eight major strategies for rational design and discovery of MTDs

REDIAL-2020 [1159], to prioritize compounds against COVID-19.

Collectively, the design of this multitarget drugs must focus on the selection of suitable targets, which have to be well characterized and preferably implicated in different pathways of the disease, and on the optimization of the relative potency of the compound towards each receptor. Screening and knowledge-based approaches represent the classical way to design multi-target ligands; nevertheless fragment-based and in silico screening have been proposed as a potentially attractive route for the multi-target drug design process. The knowledge-based approach also known as "framework combination" relies upon structure–activity relationship (SAR) knowledge of every single target involved, in which the pharmacophores are combined in different ways, where they can be connected by a linker, or they can be overlapping in the same compound or highly integrated in a multi-target drug [759]. Another way is to screen sets of compounds against several targets. A modern approach is the high-throughput screening (HTS), which allows large, diverse compound sets to be screened against several targets of interest, in parallel. Finally, *in silico* methods gained an increasingly popularity as multi-target drug design tools. A promising tool is the fragment-based drug design (FBDD) approach where fragments, even when binding weakly to the biological target, are identified. These fragments are then expanded or linked together to produce drug leads with a high affinity. Other well-known *in silico* methods, such as molecular docking, pharmacophore analysis, quantitative structure activity relationship (QSAR), machine learning, and their various combinations, have been extensively used (Figs. 12.5 and 12.6).

References

1. SoRelle R. Who Owns you DNA? Who Will own it? Circulation. 2000;101:e67–8.
2. Anderson WF. Human gene therapy. Nature. 1998;392:25–30.
3. Otsu M, Candotti F. Gene therapy in infants with severe combined immunodeficiency. BioDrugs. 2002;16:229–39.
4. Baekelandt V, De Strooper B, Nuttin B, et al. Gene therapeutic strategies for neurodegenerative diseases. Curr Opin Mol Ther. 2000;2:540–54.
5. Galanis E, Russell S. Cancer gene therapy clinical trials: Lessons for the future. Br J Cancer. 2001;85:1432–6.
6. Tomaselli D, Lucidi A, Rotili D, et al. Epigenetic polypharmacology: A new frontier for epi-drug discovery. Med Res Rev. 2020;40(1):190–244.
7. Bird A. Perceptions of epigenetics. Nature. 2007;447(7143):396–8.
8. Ledford H. Language: Disputed definitions. Nature. 2008;455(7216):1023–8.
9. Chandler VL. Paramutation: from maize to mice. Cell. 2007;128(4):641–5.
10. Dupont C, Armant DR, Brenner CA. Epigenetics: definition, mechanisms and clinical perspective. Seminars Reprod Med. 2009;27(5):351–7.
11. Franci G, Miceli M, Altucci L. Targeting epigenetic networks with polypharmacology: A new avenue to tackle cancer. Epigenomics. 2010;2(6):731–42.
12. Ptak C, Petronis A. Epigenetics and complex disease: from etiology to new therapeutics. Annu Rev Pharmacol Toxicol. 2008;48:257–76.
13. Ganesan A, Arimondo PB, Rots MG, et al. The timeline of epigenetic drug discovery: From reality to dreams. Clin Epigenetics. 2019;11(1):174.
14. Heerboth S, Lapinska K, Snyder N, et al. Use of epigenetic drugs in disease: an overview. Genet Epigenet. 2014;6:9–19.
15. Benedetti R, Conte M, Iside C, et al. Epigenetic-based therapy: From single- to multi-target approaches. Int J Biochem Cell Biol. 2015;69:121–31.
16. Ganesan A. Multitarget Drugs: An epigenetic epiphany. ChemMedChem. 2016;11(12):1227–41.
17. de Lera AR, Ganesan A. Epigenetic polypharmacology: from combination therapy to multitargeted drugs. Clin Epigenetics. 2016;8:105.
18. Mann BS, Johnson JR, Cohen MH, et al. Approval summary: vorinostat for treatment of advanced primary cutaneous T-cell lymphoma. Oncologist. 2007;12:1247–52.
19. Mikaelsson MA, Miller CA. The path to epigenetic treatment of memory disorders. Neurobiol Learn Mem. 2011;96:13–8.
20. Kaminskas E, Farrell AT, Wang Y-C, et al. FDA drug approval summary: azacitidine (5-azacytidine, Vidaza™) for injectable suspension. Oncologist. 2005;10:176–82.
21. Cheng JC, Matsen CB, Gonzales FA, et al. Inhibition of DNA methylation and reactivation of silenced genes by zebularine. J Natl Cancer Inst. 2003;95:399–409.

22. Amato RJ. Inhibition of DNA methylation by antisense oligonucleotide MG98 as cancer therapy. Clin Genitourin Cancer. 2007;5:422–6.
23. Jamieson AC, Miller JC, Pabo CO. Drug discovery with engineered zinc-finger proteins. Nat Rev Drug Discov. 2003;2(5):361–8.
24. Bhan A, Soleimani M, Mandal SS. Long noncoding RNA and cancer: A new paradigm. Cancer Res. 2017;77(15):3965–81.
25. Khalil AM, Guttman M, Huarte M, et al. Many human large intergenic noncoding RNAs associate with chromatin-modifying complexes and affect gene expression. Proc Natl Acad Sci. 2009;106:11667–72.
26. Guttman M, Amit I, Garber M, et al. Chromatin signature reveals over a thousand highly conserved large non-coding RNAs in mammals. Nature. 2009;458:223–7.
27. Matsui M, Corey DR. Non-coding RNAs as drug targets. Nat Rev Drug Discov. 2017;16(3):167–79.
28. Esteller M. Non-coding RNAs in human disease. Nat Rev Genet. 2011;12(12):861–74.
29. Rupaimoole R, Slack FJ. MicroRNA therapeutics: Towards a new era for the management of cancer and other diseases. Nat Rev Drug Discov. 2017;16:203–22.
30. Harries LW. RNA biology provides new therapeutic targets for human disease. Front Genet. 2019;10:1–12.
31. Rani A, O'Shea A, Ianov L, et al. miRNA in circulating microvesicles as biomarkers for age-related cognitive decline. Front Aging Neurosci. 2017;9:1–10.
32. le Sage C, Lawo S, Cross BCS. CRISPR: A screener's guide. SLAS Discov. 2020;25(3):233–40.
33. Jansen R, Van Embden JDA, Gaastra W, et al. Identification of genes that are associated with DNA repeats in prokaryotes. Mol Microbiol. 2002;43:1565–75.
34. Kurata M, Yamamoto K, Moriarity BS, et al. CRISPR/Cas9 library screening for drug target discovery. J Hum Genet. 2018;63:179–86.
35. Herrera-Carrillo E, Gao Z, Berkhout B. CRISPR therapy towards an HIV cure. Brief Funct Genomics. 2020;19(3):201–8.
36. Bartel DP. MicroRNAs: Target recognition and regulatory functions. Cell. 2009;136:215–33.
37. Wang Z. MicroRNA-Interference Technologies. Springer-Verlag, Heidelberg, Germany; New York, USA. 2009; ISBN-13: 978-3-642-00488-9.
38. Wang Z, Luo X, Lu Y, et al. miRNAs at the heart of the matter. J Mol Med. 2008;86:771–3.
39. Wang Z, Yang B. MicroRNA Expression Detection Methods. Springer-Verlag, Heidelberg, Germany; New York, USA. 2010; ISBN-13: 978-3-642-04927-9.
40. Friedman RC, Farh KK-H, Burge CB, et al. Most mammalian mRNAs are conserved targets of microRNAs. Genome Res. 2009;19:92–105.
41. Bartel DP. MicroRNAs: Genomics, biogenesis, mechanism, and function. Cell. 2004;116:281–97.
42. Yang D, Wan X, Dennis AT, et al. MicroRNA biophysically modulates cardiac action potential by direct binding to ion channel. Circulation. 2021;143(16):1597–613.
43. Bonneau E, Neveu B, Kostantin E, et al. How close are miRNAs from clinical practice? A perspective on the diagnostic and therapeutic market. Ejifcc. 2019;30:114–27.
44. Lim LP, Lau NC, Garrett-engele P, et al. Microarray analysis shows that some microRNAs downregulate large numbers of target mRNAs. Nature. 2005;292:288–92.
45. Baek D, Villen J, Shin C, et al. The impact of microRNAs on protein output. Nature. 2008;455:64–71.
46. Selbach M, Schwanhausser B, Thierfelder N, et al. Widespread changes in protein synthesis induced by microRNAs. Nature. 2008;455:58–63.
47. Farh KK, Grimson A, Jan C, et al. The widespread impact of mammalian microRNAs on mRNA repression and evolution. Science. 2005;310:1817–21.
48. Satoh J, Tabunoki H. Comprehensive analysis of human microRNA target networks. BioData Mining. 2011;4:17.
49. Li Z, Rana TM. Therapeutic targeting of microRNAs: Current status and future challenges. Nat Rev Drug Discov. 2014;13:622–38.
50. Alles J, Fehlmann T, Fischer U, et al. An estimate of the total number of true human miRNAs. Nucleic Acids Res. 2019;47(7):3353–64.
51. Plotnikova O, Baranova A, Skoblov M. Comprehensive analysis of human microRNA-mRNA interactome. Front Genet. 2019;10:933.
52. Brennecke J, Stark A, Russell RB, et al. Principles of microRNA-target recognition. PLoS Biol. 2005;3:e85.
53. Lewis BP, Burge CB, Bartel DP. Conserved seed pairing, often flanked by adenosines, indicates that thousands of human genes are microRNA targets. Cell. 2005;120:15–20.
54. Lewis BP, Shih IH, Jones-Rhoades MW, et al. Prediction of mammalian microRNA targets. Cell. 2003;115:787–98.
55. Doench JG, Sharp PA. Specificity of microRNA target selection in translational repression. Genes Dev. 2004;18:504–11.
56. Schirle NT, MacRae IJ. The crystal structure of human Argonaute2. Science. 2012;336:1037–40.
57. Didiano D, Hobert O. Perfect seed pairing is not a generally reliable predictor for miRNA-target interactions. Nat Struct Mol Biol. 2006;13:849–51.
58. Hausser J, Landthaler M, Jaskiewicz L, et al. Relative contribution of sequence and structure features to the mRNA binding of Argonaute/EIF2C-miRNA complexes and the degradation of miRNA targets. Genome Res. 2009;19:2009–20.
59. Grimson A, Farh KK, Johnston WK, et al. MicroRNA targeting specificity in mammals:

determinants beyond seed pairing. Mol Cell. 2007;27:91–105.
60. Kertesz M, Iovino N, Unnerstall U, et al. The role of site accessibility in microRNA target recognition. Nat Genet. 2007;39:1278–84.
61. Saini HK, Griffiths-Jones S, Enright AJ. Genomic analysis of human microRNA transcripts. Proc Natl Acad Sci U S A. 2007;104(45):17719–24.
62. Glaich O, Parikh S, Bell RE, et al. DNA methylation directs microRNA biogenesis in mammalian cells. Nat Commun. 2019;10(1):5657.
63. Ivey KN, Muth A, Arnold J, et al. MicroRNA regulation of cell lineages in mouse and human embryonic stem cells. Cell Stem Cell. 2008;2:219–29.
64. Anokye-Danso F, Trivedi CM, Juhr D, et al. Highly efficient miRNA-mediated reprogramming of mouse and human somatic cells to pluripotency. Cell Stem Cell. 2011;8:376–88.
65. Chi SW, Hannon GJ, Darnell RB. An alternative mode of microRNA target recognition. Nat. Struct. Mol. Biol. 2012;19:321–7.
66. Rupaimoole R, Calin GA, Lopez-Berestein G, et al. MiRNA deregulation in cancer cells and the tumor microenvironment. Cancer Discov. 2016;6:235–46.
67. Ha M, Kim VN. Regulation of microRNA biogenesis. Nat Rev Mol Cell Biol. 2014;15:509–24.
68. Lin S, Gregory RI. MicroRNA biogenesis pathways in cancer. Nat Rev Cancer. 2015;15:321–33.
69. Wang Z. MicroRNAs and Cardiovascular Disease. Bentham Science Publishers, Potomac. 2010; eISBN: 978-1-60805-184-7.
70. Yang B, Lin H, Xiao J, et al. The muscle-specific microRNA miR-1 regulates cardiac arrhythmogenic potential by targeting GJA1 and KCNJ2. Nat Med. 2007;13:486–91.
71. Lu Y, Zhang Y, Wang N, et al. Control of experimental atrial fibrillation by microRNA-328. Circulation. 2010;122:2378–87.
72. Xiao J, Lin H, Luo X, et al. miRNA-605 joins the p53 network to form a p53:miRNA-605:Mdm2 positive feedback loop in response to cellular stress. EMBO J. 2011;30:524–32.
73. Luo X, Pan Z, Shan H, et al. MicroRNA-26 governs profibrillatory inward-rectifier potassium current changes in atrial fibrillation. J Clin Invest. 2013;123:1939–51.
74. Zhang S, Yue W, Xie Y, et al. The four-microRNA signature identified by bioinformatics analysis predicts the prognosis of nasopharyngeal carcinoma patients. Oncol Rep. 2019;1767–80
75. Andersen GB, Tost J. Circulating miRNAs as biomarker in cancer. In: Schaffner F, Merlin JL, von Bubnoff N (eds). Tumor liquid biopsies. Volume 215. Springer, Cham. 2020. Recent Results in Cancer Research.
76. Rooij E, Kauppinen S. Development of micro RNA therapeutics is coming of age. EMBO Mol Med. 2014;6:851–64.
77. Cavalcante P, Mizrachi T, Barzago C, et al. MicroRNA signature associated with treatment response in myasthenia gravis: A further step towards precision medicine. Pharmacol Res. 2019;148:104388.
78. Wang Y, Ru J, Jin T, et al. An approach to identify individual functional single nucleotide polymorphisms and isoform microRNAs. Biomed Res Int. 2019;2019:6193673.
79. Johnson CD, Esquela-Kerscher A, Stefani G, et al. The let-7 microRNA represses cell proliferation pathways in human cells. Cancer Res. 2007;67:7713–22.
80. Yu F, Yao H, Zhu P, et al. let-7 Regulates self renewal and tumorigenicity of breast cancer cells. Cell. 2007;131:1109–23.
81. Bonci D, Coppola V, Musumeci M, et al. The miR-15a-miR-16-1 cluster controls prostate cancer by targeting multiple oncogenic activities. Nat Med. 2008;14:1271–7.
82. Metias SM, Lianidou E, Yousef GM. MicroRNAs in clinical oncology: At the crossroads between promises and problems. J Clin Pathol. 2009;62:771–6.
83. Farooqi AA, Fayyaz S, Shatynska-Mytsyk I, et al. Is miR-34a a well-equipped swordsman to conquer temple of molecular oncology? Chem Biol Drug Des. 2016;87:321–34.
84. Wang Z. The guideline of the design and validation of miRNA mimics. Methods Mol Biol. 2011;676:211–24.
85. Kumar MS, Erkeland SJ, Pester RE, et al. Suppression of non-small cell lung tumor development by the let-7 microRNA family. Proc Natl Acad Sci USA. 2008;105:3903–8.
86. Esquela-Kerscher A, Trang P, Wiggins JF, et al. The let-7 microRNA reduces tumor growth in mouse models of lung cancer. Cell Cycle. 2008;7:759–64.
87. Boudreau RL, Monteys AM, Davidson BL. Minimizing variables among hairpin-based RNAi vectors reveals the potency of shRNAs. RNA. 2008;14:1834–44.
88. Bauer M, Kinkl N, Meixner A, et al. Prevention of interferon-stimulated gene expression using microRNA-designed hairpins. Gene Ther. 2009;16:142–7.
89. Aagaard LA, Zhang J, von Eije KJ, et al. Engineering and optimization of the miR-106b cluster for ectopic expression of multiplexed anti-HIV RNAs. Gene Ther. 2008;15:1536–49.
90. Liu YP, Haasnoot J, ter Brake O, et al. Inhibition of HIV-1 by multiple siRNAs expressed from a single microRNA polycistron. Nucleic Acids Res. 2008;36:2811–24.
91. Seyhan AA. A multiplexed miRNA and transgene expression platform for simultaneous repression and expression of protein coding sequences. Mol Biosyst. 2016;12(1):295–312.
92. Rooij E, Kauppinen S. Development of microRNA therapeutics is coming of age. EMBO Mol Med. 2014;6:851–64.
93. Mollaei H, Safaralizadeh R, Rostami Z. MicroRNA replacement therapy in cancer. J Cell Physiol. 2019;234(8):12369–84.

94. Simonson B, Das S. MicroRNA therapeutics: The next magic bullet? Mini Rev Med Chem. 2015;15(6):467–74.
95. Fiedler J, Batkai S, Thum T. MicroRNA-based therapy in cardiology. Herz. 2014;39(2):194–200.
96. Mellis D, Caporali A. MicroRNA-based therapeutics in cardiovascular disease: screening and delivery to the target. Biochem Soc Trans. 2018;46(1):11–21.
97. Bader AG, Brown D, Winkler M. The promise of microRNA replacement therapy. Cancer Res. 2010;70(18):7027–30.
98. Greenberg DS, Soreq H. MicroRNA therapeutics in neurological disease. Curr Pharm Des. 2014;20(38):6022–7.
99. Singh S, Narang AS, Mahato RI. Subcellular fate and off-target effects of siRNA, shRNA, and miRNA. Pharm Res. 2011;28:2996–3015.
100. Chen Y, Zhao H, Tan Z, et al. Bottleneck limitations for microRNA-based therapeutics from bench to the bedside. Pharmazie. 2015;70:147–54.
101. Jackson AL, Burchard J, Schelter J, et al. Widespread siRNA "off-target" transcript silencing mediated by seed region sequence complementarity. RNA. 2006;12:1179–87.
102. Lal A, Navarro F, Maher CA, et al. miR-24 inhibits cell proliferation by targeting E2F2, MYC, and other cell-cycle genes via binding to "seedless" 3′UTR microRNA recognition elements. Mol Cell. 2009;35:610–25.
103. Seitz H. Redefining microRNA targets. Curr Biol. 2009;19:870–3.
104. Lee YJ, Kim V, Muth DC, et al. Validated microRNA target databases: An evaluation. Drug Dev Res. 2015;76:389–96.
105. Leclercq M, Diallo AB, Blanchette M. Prediction of human miRNA target genes using computationally reconstructed ancestral mammalian sequences. Nucleic Acids Res. 2017;45:556–66.
106. Grimm D, Wang L, Lee JS, et al. Argonaute proteins are key determinants of RNAi efficacy, toxicity, and persistence in the adult mouse liver. J Clin Investig. 2010;20:3106–19.
107. Diederichs S, Jung S, Rothenberg SM, et al. Coexpression of Argonaute-2 enhances RNA interference toward perfect match binding sites. Proc Natl Acad Sci USA. 2008;105:9284–9.
108. McBride JL, Boudreau RL, Harper SQ, et al. Artificial miRNAs mitigate shRNA-mediated toxicity in the brain: Implications for the therapeutic development of RNAi. Proc Natl Acad Sci USA. 2008;105:5868–73.
109. Boudreau RL, Martins I, Davidson BL. Artificial MicroRNAs as siRNA shuttles: Improved safety as compared to shRNAs in vitro and In vivo. Mol Ther. 2009;17:169–75.
110. Beer S, Bellovin DI, Lee JS, et al. Low-level shRNA cytotoxicity can contribute to MYC-induced hepatocellular carcinoma in adult mice. Mol Ther. 2010;18:161–70.
111. Gallant-Behm CL, Piper J, Lynch JM, et al. A microRNA-29 mimic (Remlarsen) represses extracellular matrix expression and fibroplasia in the skin. J Investig Dermatol. 2019;139:1073–81.
112. Reid G, Pel ME, Kirschner MB, et al. Restoring expression of miR-16: A novel approach to therapy for malignant pleural mesothelioma. Ann Oncol. 2013;24:3128–35.
113. van Zandwijk N, Pavlakis N, Kao SC, et al. Safety and activity of microRNA-loaded minicells in patients with recurrent malignant pleural mesothelioma: A first-in-man, phase 1, open-label, dose-escalation study. Lancet Oncol. 2017;18:1386–96.
114. Bouchie A. First microRNA mimic enters clinic. Nat Biotechnol. 2013;31:577.
115. Adams BD, Parsons C, Slack FJ. The tumor-suppressive and potential therapeutic functions of miR-34a in epithelial carcinomas. Expert Opin Ther Targets. 2016;20:737–53.
116. Misso G, Di Martino MT, De Rosa G, et al. Mir-34: A new weapon against cancer? Mol Ther Nucleic Acids. 2014;3:e195.
117. Ling H, Girnita L, Buda O, et al. Non-coding RNAs: The cancer genome dark matter that matters! Clin Chem Lab Med. 2017;55:705–14.
118. Bhan A, Mandal SS. LncRNA HOTAIR: A master regulator of chromatin dynamics and cancer. Biochim Biophys Acta. 2015;1856:151–64.
119. Ma L, Bajic VB, Zhang Z. On the classification of long non-coding RNAs. RNA Biol. 2013;10:925–33.
120. Bhan A, Mandal SS. Long noncoding RNAs: emerging stars in gene regulation, epigenetics and human disease. ChemMedChem. 2014;9:1932–56.
121. Wang Kevin C, Chang HY. Molecular mechanisms of long noncoding RNAs. Molecular cell. 2011;43:904–14.
122. Sanfilippo PG, Hewitt AW. Translating the ENCyclopedia of DNA Elements Project findings to the clinic: ENCODE's implications for eye disease. Clin Exp Ophthalmol. 2014;42:78–83.
123. Tragante V, Moore JH, Asselbergs FW. The ENCODE project and perspectives on pathways. Genet Epidemiol. 2014;38:275–80.
124. Statello L, Guo CJ, Chen LL, et al. Gene regulation by long non-coding RNAs and its biological functions Nat Rev Mol Cell Biol. 2021; 22(2): 96–118.
125. Dykes IM, Emanueli C. Transcriptional and post-transcriptional gene regulation by long non-coding RNA. Genom Proteom Bioinform. 2017;15(3):177–86.
126. Gil N, Ulitsky I. Regulation of gene expression by cis-acting long non-coding RNAs Nat Rev Genet. 2020; 21(2): 102–117.
127. Zhang Y, Du W, Yang B. Long non-coding RNAs as new regulators of cardiac electrophysiology and arrhythmias: Molecular mechanisms, therapeutic implications and challenges. Pharmacol Ther. 2019;203:107389.

128. Kim TK, Shiekhattar R. Diverse regulatory interactions of long noncoding RNAs. Curr Opin Genet Dev. 2016;36:73–82.
129. Zhang Y, Jiao L, Sun L, et al. LncRNA ZFAS1 as a SERCA2a inhibitor to cause intracellular Ca2+ overload and contractile dysfunction in a mouse model of myocardial infarction. Circ Res. 2018;122(10):1354–68.
130. Ferrè F, Colantoni A, Helmer-Citterich M. Revealing protein-lncRNA interaction. Brief Bioinform. 2016;17(1):106–16.
131. Long Y, Wang X, Youmans DT, et al. How do lncRNAs regulate transcription? Sci Adv. 2017;3(9):eaao2110.
132. Jonas K, Calin GA, Pichler M. RNA-binding proteins as important regulators of long non-coding RNAs in cancer. Int J Mol Sci. 2020;21(8):2969.
133. Prabhakar B, Zhong XB, Rasmussen TP. Exploiting long noncoding RNAs as pharmacological targets to modulate epigenetic diseases. Yale J Biol Med. 2017;90(1):73–86.
134. Wapinski O, Chang HY. Long noncoding RNAs and human disease. Trends Cell Biol. 2011;21(6):354–61.
135. Schmitz SU, Grote P, Herrmann BG. Mechanisms of long noncoding RNA function in development and disease. Cell Mol Life Sci. 2016;73(13):2491–509.
136. Uchida S, Dimmeler S. Long noncoding RNAs in cardiovascular diseases. Circ Res. 2015;116(4):737–50.
137. Yu B, Wang S. Angio-LncRs: LncRNAs that regulate angiogenesis and vascular disease. Theranostics. 2018;8(13):3654–75.
138. Chi Y, Wang D, Wang J, et al. Long non-coding RNA in the pathogenesis of cancers. Cells. 2019;8(9):1015.
139. Song H, Sun W, Ye G, et al. Long non-coding RNA expression profile in human gastric cancer and its clinical significances. J Transl Med. 2013;11(225):5876.
140. Hajjari M, Salavaty A. HOTAIR: An oncogenic long non-coding RNA in different cancers. Cancer Biol Med. 2015;12(1):1–9.
141. Gupta RA, Shah N, Wang KC, et al. Long noncoding RNA HOTAIR reprograms chromatin state to promote cancer metastasis. Nature. 2010;464(7291):1071–6.
142. Yap KL, Li S, Munoz-Cabello AM, et al. Molecular interplay of the noncoding RNA ANRIL and methylated histone H3 lysine 27 by polycomb CBX7 in transcriptional silencing of INK4a. Mol Cell. 2010;38(5):662–74.
143. Huarte M, Guttman M, Feldser D, et al. A large intergenic noncoding RNA induced by p53 mediates global gene repression in the p53 response. Cell. 2010;142(3):409–19.
144. Tang SS, Zheng BY, Xiong XD. LincRNA-p21: Implications in human diseases. Int J Mol Sci. 2015;16(8):18732–40.
145. Jiao L, Li M, Shao Y, Zhang Y, et al. lncRNA-ZFAS1 induces mitochondria-mediated apoptosis by causing cytosolic Ca2+ overload in myocardial infarction mice model. Cell Death Dis. 2019;10(12):942.
146. Hu F, Shao L, Zhang J, et al. Knockdown of ZFAS1 inhibits hippocampal neurons apoptosis and autophagy by activating the PI3K/AKT pathway via upregulating miR-421 in epilepsy. Neurochem Res. 2020;45(10):2433–41.
147. Wu H, Qin W, Lu S, et al. Long noncoding RNA ZFAS1 promoting small nucleolar RNA-mediated 2′-O-methylation via NOP58 recruitment in colorectal cancer. Mol Cancer. 2020;19(1):95.
148. Ishii N, Ozaki K, Sato H, et al. Identification of a novel non-coding RNA, MIAT, that confers risk of myocardial infarction. J Hum Genet. 2006;51(12):1087–99.
149. Vausort M, Wagner DR, Devaux Y. Long noncoding RNAs in patients with acute myocardial infarction. Circ Res. 2014;115:668–77.
150. Zhu XH, Yuan YX, Rao SL, et al. LncRNA MIAT enhances cardiac hypertrophy partly through sponging miR-150. Eur Rev Med Pharmacol Sci. 2016;20:3653–60.
151. Qu X, Du Y, Shu Y, et al. MIAT is a pro-fibrotic long non-coding RNA governing cardiac fibrosis in postinfarct myocardium. Sci Rep. 2017;7:42657.
152. Zhou X, Zhang W, Jin M, et al. lncRNA MIAT functions as a competing endogenous RNA to upregulate DAPK2 by sponging miR-22-3p in diabetic cardiomyopathy. Cell Death Dis. 2017;8:e2929.
153. Yan B, Yao J, Liu JY, et al. LncRNA-MIAT regulates microvascular dysfunction by functioning as a competing endogenous RNA. Circ Res. 2015;116:1143–56.
154. Sleutels F, Zwart R, Barlow DP. The non-coding Air RNA is required for silencing autosomal imprinted genes. Nature. 2002;415:810–3.
155. Pandey RR, Mondal T, Mohammad F, et al. Kcnq1ot1 antisense noncoding RNA mediates lineage-specific transcriptional silencing through chromatin-level regulation. Mol Cell. 2008;32:232–46.
156. Novikova IV, Hennelly SP, Sanbonmatsu KY. Sizing up long non-coding RNAs. Do lncRNAs have secondary and tertiary structure? Bioarchitecture. 2012;2(6):189–99.
157. Sone M, Hayashi T, Tarui H, et al. The mRNA-like noncoding RNA Gomafu constitutes a novel nuclear domain in a subset of neurons. J Cell Sci. 2017;120:2498–506.
158. Guttman M, Rinn JL. Modular regulatory principles of large non-coding RNAs. Nature. 2012;482(7385):339–46.
159. Thorenoor N, Faltejskova-Vychytilova P, Hombach S, et al. Long non-coding RNA ZFAS1 interacts with CDK1 and is involved in p53-dependent cell cycle control and apoptosis in colorectal cancer. Oncotarget. 2016;7(1):622–37.
160. Liu G, Wang L, Han H, et al. LncRNA ZFAS1 promotes growth and metastasis by regulating BMI1 and ZEB2 in osteosarcoma. Am J Cancer Res. 2017;7(7):1450–62.

161. Quinn JJ, Ilik IA, Qu K, et al. Revealing long noncoding RNA architecture and functions using domain-specific chromatin isolation by RNA purification. Nat Biotechnol. 2014;32(9):933–40.
162. Mercer TR, Mattick JS. Structure and function of long noncoding RNAs in epigenetic regulation. Nat Struct Mol Biol. 2013;20(3):300–7.
163. Zucchelli S, Cotella D, Takahashi H, et al. SINEUPs: A new class of natural and synthetic antisense long non-coding RNAs that activate translation. RNA Biol. 2015;12(8):771–9.
164. Zucchelli S, Fasolo F, Russo R, et al. SINEUPs are modular antisense long non-coding RNAs that increase synthesis of target proteins in cells. Front Cell Neurosci. 2015;9:174.
165. Carrieri C, Cimatti L, Biagioli M, et al. Long non-coding antisense RNA controls Uchl1 translation through an embedded SINEB2 repeat. Nature. 2012;491(7424):454–7.
166. Indrieri A, Grimaldi C, Zucchelli S, et al. Synthetic long non-coding RNAs (SINEUPs) rescue defective gene expression in vivo. Sci Rep. 2016;6:27315.
167. Kristensen LS, Andersen MS, Stagsted LVW, et al. The biogenesis, biology and characterization of circular RNAs. Nat Rev Genet. 2019;20(11):675–91.
168. Vicens Q, Westhof E. Biogenesis of circular RNAs. Cell. 2014;159:13–4.
169. Zhang XO, Dong R, Zhang Y, et al. Diverse alternative back-splicing and alternative splicing landscape of circular RNAs. Genome Res. 2016;26:1277–87.
170. Panda AC. Circular RNAs act as miRNA sponges. Adv Exp Med Biol. 2018;1087:67–79.
171. Memczak S, Jens M, Elefsinioti A, et al. Circular RNAs are a large class of animal RNAs with regulatory potency. Nature. 2013;495:333–8.
172. Hansen TB, Jensen TI, Clausen BH, et al. Natural RNA circles function as efficient microRNA sponges. Nature. 2013;495:384–8.
173. Salzman J. Circular RNA expression: Its potential regulation and function. Trends Genet. 2016;32:309–16.
174. Jeck WR, Sharpless NE. Detecting and characterizing circular RNAs. Nat Biotechnol. 2014;32:453–61.
175. Rybak-Wolf A, Stottmeister C, Glazar P, et al. Circular RNAs in the mammalian brain are highly abundant, conserved, and dynamically expressed. Mol Cell. 2015;58:870–85.
176. Gruner H, Cortés-López M, Cooper DA, et al. CircRNA accumulation in the aging mouse brain. Sci Rep. 2016;6:38907.
177. Westholm JO, Miura P, Olson S, et al. Genome-wide analysis of Drosophila circular RNAs reveals their structural and sequence properties and age-dependent neural accumulation. Cell Rep. 2014;9:1966–80.
178. Patop IL, Wüst S, Kadener S. Past, present, and future of circRNAs. EMBO J. 2019;38(16):e100836.
179. Guo JU, Agarwal V, Guo H, et al. Expanded identification and characterization of mammalian circular RNAs. Genome Biol. 2014;15:409.
180. Hanan M, Soreq H, Kadener S. CircRNAs in the brain. RNA Biol. 2017;14:1028–34.
181. Bachmayr-Heyda A, Reiner AT, Auer K, et al. Correlation of circular RNA abundance with proliferation–exemplified with colorectal and ovarian cancer, idiopathic lung fibrosis, and normal human tissues. Sci Rep. 2015;5:8057.
182. Werfel S, Nothjunge S, Schwarzmayr T, et al. Characterization of circular RNAs in human, mouse and rat hearts. J Mol Cell Cardiol. 2016;98:103–7.
183. Salzman J, Gawad C, Wang PL, et al. Circular RNAs are the predominant transcript isoform from hundreds of human genes in diverse cell types. PLoS One. 2012;7:e30733.
184. Denzler R, Agarwal V, Stefano J, et al. Assessing the ceRNA hypothesis with quantitative measurements of miRNA and target abundance. Mol Cell. 2014;54:766–76.
185. Jens M, Rajewsky N, et al. Competition between target sites of regulators shapes post-transcriptional gene regulation. Nat Rev Genet. 2015;16:113–26.
186. Legnini I, Di Timoteo G, Rossi F, et al. Circ-ZNF609 is a circular RNA that can be translated and functions in myogenesis. Mol Cell. 2017;66:22–37.
187. Yang Y, Fan X, Mao M, et al. Extensive translation of circular RNAs driven by N6-methyladenosine. Cell Res. 2017;27:626–41.
188. Pamudurti NR, Bartok O, Jens M, et al. Translation of circRNAs. Mol Cell. 2017;66:9–21.
189. Yang Y, Gao X, Zhang M, et al. Novel role of FBXW7 circular RNA in repressing glioma tumorigenesis. J Natl Cancer Inst. 2018;110:304–15.
190. Holdt LM, Stahringer A, Sass K, et al. Circular non-coding RNA ANRIL modulates ribosomal RNA maturation and atherosclerosis in humans. Nat Commun. 2016;7:12429.
191. Abdelmohsen K, Panda AC, Munk R, et al. Identification of HuR target circular RNAs uncovers suppression of PABPN1 translation by CircPABPN1. RNA Biol. 2017;14:361–9.
192. Du WW, Yang W, Liu E, et al. Foxo3 circular RNA retards cell cycle progression via forming ternary complexes with p21 and CDK2. Nucleic Acids Res. 2016;6:2846–58.
193. Kristensen LS, Hansen TB, Venø MT, et al. Circular RNAs in cancer: opportunities and challenges in the field. Oncogene. 2018;37(5):555–65.
194. Wang Y, Mo Y, Gong Z, et al. Circular RNAs in human cancer. Mol Cancer. 2017;16:25.
195. Ng WL, Mohd Mohidin TB, Shukla K. Functional role of circular RNAs in cancer development and progression. RNA Biol. 2018;15(8):995–1005.
196. Floris G, Zhang L, Follesa P, et al. Regulatory role of circular RNAs and neurological disorders. Mol Neurobiol. 2017;54(7):5156–65.
197. Altesha MA, Ni T, Khan A, et al. Circular RNA in cardiovascular disease. J Cell Physiol. 2019;234(5):5588–600.

198. Wang T, Pan W, Hu J, et al. Circular RNAs in metabolic diseases. Adv Exp Med Biol. 2018;1087:275–85.
199. Zeng Y, Zheng Z, Liu F, et al. Circular RNAs in metabolism and metabolic disorders. Obes Rev. 2021;22(7):e13220.
200. Yang D, Yang K, Yang M. Circular RNA in aging and age-related diseases. Adv Exp Med Biol. 2018;1086:17–35.
201. Verduci L, Strano S, Yarden Y, et al. The circRNA-microRNA code: emerging implications for cancer diagnosis and treatment. Mol Oncol. 2019;13(4):669–80.
202. Lasda E, Parker R. Circular RNAs: diversity of form and function. RNA. 2014;20:1829–42.
203. Li F, Zhang L, Li W, et al. Circular RNA ITCH has inhibitory effect on ESCC by suppressing the Wnt/β--catenin pathway. Oncotarget. 2015;6:6001–13.
204. Du WW, Yang W, Liu E, et al. Foxo3 circular RNA retards cell cycle progression via forming ternary complexes with p21 and CDK2. Nucleic Acids Res. 2016;44:2846–58.
205. Wang Z. Antisense RNA and cancer. In: Cancer and noncoding RNAs by Jaybokas Chakrabarti & Sanga Mitra (eds), Translational epigenetics series. Elsevier AP Academic Press, London. 2017; ISBN: 978-0-12-811022-5.
206. Katayama S, Tomaru Y, Kasukawa T, et al. Antisense transcript in the mammalian transcriptome. Science. 2005;309:1564–6.
207. Pelechano V, Steinmetz LM. Gene regulation by antisense transcription. Nat Rev Genet. 2013;14(12):880–93.
208. Faghihi MA, Wahlestedt C. Regulatory roles of natural antisense transcripts. Nat Rev Mol Cell Biol. 2009;10(9):637–43.
209. Piatek MJ, Henderson V, Zynad HS, et al. Natural antisense transcription from a comparative perspective. Genomics. 2016;108(2):56–63.
210. Wight M, Werner A. The functions of natural antisense transcripts. Essays Biochem. 2013;54:91–101.
211. Mahmoudi S, Henriksson S, Corcoran M, et al. Wrap53, a natural p53 antisense transcript required for p53 induction upon DNA damage. Mol Cell. 2009;33:462–71.
212. Su WY, Li JT, Cui Y, et al. Bidirectional regulation between WDR83 and its natural antisense transcript DHPS in gastric cancer. Cell Res. 2012;22(9):1374–89.
213. Galante PA, Vidal DO, de Souza JE, et al. Sense-antisense pairs in mammals: functional and evolutionary considerations. Genome Biol. 2007;8(3):R40.
214. Bartonicek N, Maag JL, Dinger ME. Long noncoding RNAs in cancer: mechanisms of action and technological advancements. Mol Cancer. 2016;15(1):43.
215. Berteaux N, Aptel N, Cathala G, et al. A novel H19 antisense RNA overexpressed in breast cancer contributes to paternal IGF2 expression. Mol Cell Biol. 2008;28(22):6731–45.
216. Ling MHT, Ban Y, Wen H, et al. Conserved expression of natural antisense transcripts in mammals. BMC Genomics. 2013;14:243.
217. Pestka S, Daugherty BL, Jung V, et al. Anti-mRNA: specific inhibition of translation of single mRNA molecules. Proc Natl Acad Sci U S A. 1984;81(23):7525–8.
218. Munroe SH. Antisense RNA inhibits splicing of pre-mRNA in vitro. EMBO J. 1988;7:2523–32.
219. Wang Z. The principle of miRNA-Masking antisense oligonucleotides technology. Methods Mol Biol. 2011;676:43–50.
220. Wang Z. The principle of miRNA-Masking antisense oligonucleotides technology. In MicroRNA and cancer: Methods and protocols by Wei Wu (ed), pp 43–50. Springer-Verlag/Humana Press, New York. 2011; ISBN: 978-1-60761-862-1.
221. Murakami K, Miyagishi M. Tiny masking locked nucleic acids effectively bind to mRNA and inhibit binding of microRNAs in relation to thermodynamic stability. Biomed Rep. 2014;2(4):509–12.
222. Ors-Kumoglu G, Gulce-Iz S, Biray-Avci C. Therapeutic microRNAs in human cancer. Cytotechnology. 2019;71(1):411–25.
223. Gambari R, Brognara E, Spandidos DA, et al. Targeting oncomiRNAs and mimicking tumor suppressor miRNAs: nuew trends in the development of miRNA therapeutic strategies in oncology. Int J Oncol. 2016;49(1):5–32.
224. Zhang S, Yan ML, Yang L, et al. MicroRNA-153 impairs hippocampal synaptic vesicle trafficking via downregulation of synapsin I in rats following chronic cerebral hypoperfusion. Exp Neurol. 2020;332:113389.
225. Wang N, Sun LY, Zhang SC, et al. MicroRNA-23a participates in estrogen deficiency induced gap junction remodeling of rats by targeting GJA1. Int J Biol Sci. 2015;11(4):390–403.
226. Duan MJ, Yan ML, Wang Q, et al. Overexpression of miR-1 in the heart attenuates hippocampal synaptic vesicle exocytosis by the posttranscriptional regulation of SNAP-25 through the transportation of exosomes. Cell Commun Signal. 2018;16(1):91.
227. Zhang T, Hu Y, Ju J, et al. Downregulation of miR-522 suppresses proliferation and metastasis of non-small cell lung cancer cells by directly targeting DENN/MADD domain containing 2D. Sci Rep. 2016;6:19346.
228. Hong H, Tao T, Chen S, et al. MicroRNA-143 promotes cardiac ischemia-mediated mitochondrial impairment by the inhibition of protein kinase Cepsilon. Basic Res Cardiol. 2017;112(6):60.
229. Qiu Y, Cheng R, Liang C, et al. MicroRNA-20b promotes cardiac hypertrophy by the inhibition of mitofusin 2-mediated inter-organelle Ca2+ cross-talk. Mol Ther Nucleic Acids. 2020;19:1343–56.

230. Haseloff J, Gerlach WL. Simple RNA enzymes with new and highly specific endoribonuclease activities. Biotechnology. 1992;24:264–9.
231. Scott WG, Murray JB, Arnold JR, et al. Capturing the structure of a catalytic RNA intermediate: The hammerhead ribozyme. Science. 1996;274(5295):2065–9.
232. Citti L, Rainaldi G. Synthetic hammerhead ribozymes as therapeutic tools to control disease genes. Curr Gene Ther. 2005;5(1):11–24.
233. Phylactou LA, Tsipouras P, Kilpatrick MW. Hammerhead ribozymes targeted to the FBN1 mRNA can discriminate a single base mismatch between ribozyme and target. Biochem Biophys Res Commun. 1998;249(3):804–10.
234. Beigelman L, McSwiggen JA, Draper KG, et al. Chemical modification of hammerhead ribozymes. Catalytic activity and nuclease resistance. J Biol Chem. 1995;270(43):25702–8.
235. Heidenreich O, Benseler F, Fahrenholz A, et al. High activity and stability of hammerhead ribozymes containing 2′-modified pyrimidine nucleosides and phosphorothioates. J Biol Chem. 1994;269(3):2131–8.
236. Serganov A, Patel DJ. Ribozymes, riboswitches and beyond: Regulation of gene expression without proteins. Nat Rev Genet. 2007;8(10):776–90.
237. Pavco PA, Bouhana KS, Gallegos AM, et al. Antitumor and antimetastatic activity of ribozymes targeting the messenger RNA of vascular endothelial growth factor receptors. Clin Cancer Res. 2000;6(5):2094–103.
238. Johnston WK, Unrau PJ, Lawrence MS, et al. RNA-catalyzed RNA polymerization: accurate and general RNA-templated primer extension. Science. 292(5520): 1319–25.
239. de Feyter R, Li P. Technology evaluation: HIV ribozyme gene therapy, Gene Shears Pty Ltd. Curr Opin Mol Therap. 2000;2(3):332–5.
240. Khan AU. Ribozyme: a clinical tool Clinica Chimica Acta/Intern J Clin Chem. 2006; 367(1–2): 20–27.
241. Asha K, Kumar P, Sanicas M, et al. Advancements in nucleic acid based therapeutics against respiratory viral infections. J Clin Med. 2018;8(1):6.
242. Khanna M, Saxena L, Rajput R, et al. Gene silencing: a therapeutic approach to combat influenza virus infections. Future Microbiol. 2015;10(1):131–40.
243. Kumar B, Khanna M, Kumar P, et al. Nucleic acid-mediated cleavage of M1 gene of influenza A virus is significantly augmented by antisense molecules targeted to hybridize close to the cleavage site. Mol Biotechnol. 2012;51(1):27–36.
244. Kumar B, Asha K, Khanna M, et al. The emerging influenza virus threat: status and new prospects for its therapy and control. Arch Virol. 2018;163(4):831–44.
245. Lieber A, He CY, Polyak SJ, et al. Elimination of hepatitis C virus RNA in infected human hepatocytes by adenovirus-mediated expression of ribozymes. J Virol. 1996;70(12):8782–91.
246. Marwick C. First "antisense" drug will treat CMV retinitis. JAMA. 1998; 280 (10): 871.
247. Hair P, Cameron F, McKeage K. Mipomersen sodium: First global approval. Drugs. 2013;73:487–93.
248. Suhr OB, Coelho T, Buades J, et al. Efficacy and safety of patisiran for familial amyloidotic polyneuropathy: A phase II multi-dose study. Orphanet J Rare Dis. 2015;10:109.
249. Mendell JR, Rodino-Klapac LR, Sahenk Z, et al. Eteplirsen for the treatment of Duchenne muscular dystrophy. Ann Neurol. 2013;74:637–47.
250. Dias N, Stein CA. Antisense oligonucleotides: Basic concepts and mechanisms. Mol Cancer Ther. 2002;5:347–55.
251. Juliano RL. The delivery of therapeutic oligonucleotides. Nucleic Acids Res. 2016;44:6518–48.
252. Tanaka M, Nyce JW. Respirable antisense oligonucleotides: A new drug class for respiratory disease. Respir Res. 2001;2:5–9.
253. Stephenson ML, Zamecnik PC. Inhibition of Rous sarcoma viral RNA translation by a specific oligodeoxyribonucleotide. Proc Natl Acad Sci USA. 1978;75:285–8.
254. Popescu FD. Antisense- and RNA interference-based therapeutic strategies in allergy. J Cell Mol Med. 2005;9:840–53.
255. Arroyo JD, Gallichotte EN, Tewari M. Systematic design and functional analysis of artificial microRNAs. Nucleic Acids Res. 2014;42(9):6064–77.
256. Calloni R, Bonatto D. Scaffolds for artificial miRNA expression in animal cells. Hum Gene Ther Methods. 2015;26(5):162–74.
257. Fowler DK, Williams C, Gerritsen AT, et al. Improved knockdown from artificial microRNAs in an enhanced miR-155 backbone: a designer's guide to potent multi-target RNAi. Nucleic Acids Res. 2016;44(5):e48.
258. Lebbink RJ, Lowe M, Chan T, et al. Polymerase II promoter strength determines efficacy of microRNA adapted shRNAs. PLoS One. 2011;6:e26213.
259. Chung KH, Hart CC, Al-Bassam S, et al. Polycistronic RNA polymerase II expression vectors for RNA interference based on BIC/miR-155. Nucleic Acids Res. 2006;34:e53.
260. Yang JS, Maurin T, Robine N, et al. Conserved vertebrate mir-451 provides a platform for Dicer-independent, Ago2-mediated microRNA biogenesis. Proc Natl Acad Sci USA. 2010;107:15163–8.
261. Chen SC, Stern P, Guo Z, et al. Expression of multiple artificial microRNAs from a chicken miRNA126-based lentiviral vector. PLoS One. 2011;6:e22437.
262. Zeng Y, Wagner EJ, Cullen BR. Both natural and designed micro RNAs can inhibit the expression of cognate mRNAs when expressed in human cells. Mol Cell. 2002;9:1327–33.
263. Grimm D, Wang L, Lee JS, et al. Argonaute proteins are key determinants of RNAi efficacy, toxicity, and persistence in the adult mouse liver. J Clin Invest. 2010;120:3106–19.
264. Snove O Jr, Rossi JJ. Toxicity in mice expressing short hairpin RNAs gives new insight into RNAi. Genome Biol. 2006;7:231.

265. Grimm D, Streetz KL, Jopling CL, et al. Fatality in mice due to oversaturation of cellular microRNA/short hairpin RNA pathways. Nature. 2006;441:537–41.
266. Amendola M, Passerini L, Pucci F, et al. Regulated and multiple miRNA and siRNA delivery into primary cells by a lentiviral platform. Mol Ther. 2009;17:1039–52.
267. Giering JC, Grimm D, Storm TA, et al. Expression of shRNA from a tissue-specific pol II promoter is an effective and safe RNAi therapeutic. Mol Ther. 2008;16:1630–6.
268. Shin KJ, Wall EA, Zavzavadjian JR, et al. A single lentiviral vector platform for microRNA-based conditional RNA interference and coordinated transgene expression. Proc Natl Acad Sci USA. 2006;103:13759–64.
269. Stegmeier F, Hu G, Rickles RJ, et al. A lentiviral microRNA-based system for single-copy polymerase II-regulated RNA interference in mammalian cells. Proc Natl Acad Sci USA. 2005;102:13212–7.
270. Du G, Yonekubo J, Zeng Y, et al. Design of expression vectors for RNA interference based on miRNAs and RNA splicing. FEBS J. 2006;273:5421–7.
271. Hu T, Fu Q, Chen P, et al. Construction of an artificial MicroRNA expression vector for simultaneous inhibition of multiple genes in mammalian cells. Int J Mol Sci. 2009;10:2158–68.
272. Sun D, Melegari M, Sridhar S, et al. Multi-miRNA hairpin method that improves gene knockdown efficiency and provides linked multi-gene knockdown. Biotechniques. 2006;41:59–63.
273. Grimm D, Kay MA. Combinatorial RNAi: a winning strategy for the race against evolving targets. Mol Ther. 2007;15:878–88.
274. Herrera-Carrillo E, Berkhout B. The impact of HIV-1 genetic diversity on the efficacy of a combinatorial RNAi-based gene therapy. Gene Ther. 2015;22:485–95.
275. Aagaard L, Rossi JJ. RNAi therapeutics: principles, prospects and challenges. Adv Drug Deliv Rev. 2007;59:75–86.
276. Wang SL, Yao HH, Qin ZH. Strategies for short hairpin RNA delivery in cancer gene therapy. Expert Opin Biol Ther. 2009;9:1357–68.
277. Amarzguioui M, Prydz H. An algorithm for selection of functional siRNA sequences. Biochem Biophys Res Commun. 2004;316:1050–8.
278. Fellmann C, Zuber J, McJunkin K, et al. Functional identification of optimized RNAi triggers using a massively parallel sensor assay. Mol Cell. 2011;41:733–46.
279. Pei Y, Tuschl T. On the art of identifying effective and specific siRNAs. Nat Methods. 2006;3:670–6.
280. Reynolds A, Leake D, Boese Q, et al. Rational siRNA design for RNA interference. Nat Biotechnol. 2004;22:326–30.
281. Ui-Tei K, Naito Y, Takahashi F, et al. Guidelines for the selection of highly effective siRNA sequences for mammalian and chick RNA interference. Nucleic Acids Res. 2004;32:936–48.
282. Zuker M. Mfold web server for nucleic acid folding and hybridization prediction. Nucleic Acids Res. 2003;31:3406–15.
283. Kotowska-Zimmer A, Pewinska M, Olejniczak M. Artificial miRNAs as therapeutic tools: Challenges and opportunities. Wiley Interdiscip Rev RNA. 2021;12(4):e1640.
284. Elbashir SM, Harborth J, Lendeckel W, et al. Duplexes of 21-nucleotide RNAs mediate RNA interference in cultured mammalian cells. Nature. 2001;411:494–8.
285. Passioura T, Gozar MM, Goodchild A, et al. Interfering ribonucleic acids that suppress expression of multiple unrelated genes. BMC Biotechnol. 2009;9:57.
286. Stegmeier F, Hu G, Rickles RJ, Hannon GJ, Elledge SJ. A lentiviral microRNA-based system for single-copy polymerase II-regulated RNA interference in mammalian cells. Proc Natl Acad Sci U S A. 2005;102:13212–7.
287. Wang J, Theunissen TW, Orkin SH. Site-directed, virus-free, and inducible RNAi in embryonic stem cells. Proc Natl Acad Sci U S A. 2007;104:20850–5.
288. Zhou HX, Xia XG, Xu ZS. An RNA polymerase II construct synthesizes short-hairpin RNA with a quantitative indicator and mediates highly efficient RNAi. Nucleic Acids Research. 2005;33:e62.
289. Xia XG, Zhou HX, Xu ZS. Multiple shRNAs expressed by an inducible pol II promoter can knock down the expression of multiple target genes. Biotechniques. 2006;41:64–8.
290. Zhu XC, Santa LA, Chang MS, et al. A versatile approach to multiple gene RNA interference using microRNA-based short hairpin RNAs. BMC Molecular Biology. 2007;8:98.
291. Ebert MS, Neilson JR, Sharp PA. MicroRNA sponges: competitive inhibitors of small RNAs in mammalian cells. Nat Methods. 2007;4(9):721–6.
292. Lu Y, Xiao J, Lin H, et al. A single anti-microRNA antisense oligodeoxyribonucleotide (AMO) targeting multiple microRNAs offers an improved approach for microRNA interference. Nucleic Acids Res. 2009;37:e24–33.
293. Wang Z. New concepts of miRNAi technologies. In: MicroRNA interference technologies by Wang Z. Heidelberg: Springer-Verlag; 2009.
294. Wang Z. The concept of multiple-target anti-miRNA antisense oligonucleotides technology. Methods Mol Biol. 2011;676:51–7.
295. Wang Z. The concept of multiple-target anti-miRNA antisense oligonucleotides technology. In: Wu W, editor. MicroRNA and cancer. Totowa: Humana Press; 2011.
296. Lindow M, Kauppinen S. Discovering the first microRNA-targeted drug. J Cell Biol. 2012;199:407–12.

297. Umbach JL, Cullen BR. The role of RNAi and microRNAs in animal virus replication and antiviral immunity. Genes Dev. 2009;23:1151–64.
298. Gebert LFR, Rebhan MAE, Crivelli SEM, et al. Miravirsen (SPC3649) can inhibit the biogenesis of miR-122. Nucleic Acids Res. 2014;42:609–21.
299. Jopling C. Liver-specific microRNA-122. RNA Biol. 2012;9:137–42.
300. Jopling CL, Schütz S, Sarnow P. Position-Dependent Function for a Tandem MicroRNA miR-122-Binding Site Located in the Hepatitis C Virus RNA Genome. Cell Host Microbe. 2008;4:77–85.
301. Baek J, Kang S, Min H. MicroRNA-targeting therapeutics for hepatitis C. Arch Pharm Res. 2014;37:299–305.
302. Seto AG, Beatty X, Lynch JM, et al. Cobomarsen, an oligonucleotide inhibitor of miR-155, co-ordinately regulates multiple survival pathways to reduce cellular proliferation and survival in cutaneous T-cell lymphoma. Br J Haematol. 2018;183:428–44.
303. Xu L, Dai WQ, Xu XF, et al. Effects of multiple-target anti-microRNA antisense oligodeoxyribonucleotides on proliferation and migration of gastric cancer cells. Asian Pac J Cancer Prev. 2012;13(7):3203–7.
304. Subramanian RR, Wysk MA, Ogilvie KM, et al. Enhancing antisense efficacy with multimers and multi-targeting oligonucleotides (MTOs) using cleavable linkers. Nucleic Acids Res. 2015;43(19):9123–32.
305. Zhou T, Meng X, Che H, et al. Regulation of insulin resistance by multiple miRNAs via targeting the GLUT4 signaling pathway. Cell Physiol Biochem. 2016;38(5):2063–78.
306. Zaman MS, Maher DM, Khan S, et al. Current status and implications of microRNAs in ovarian cancer diagnosis and therapy. J Ovarian Res. 2012;5(1):44.
307. Kolse R, Krainer AR, Altman S. RNA therapeutics: beyond RNA interference and antisense oligonucleotides. Nat Rev Drug Discov. 2012;11:125–40.
308. Lennox KA, Behlke MA. Chemical modification and design of anti-miRNA oligonucleotides. Gene Therapy. 2011;18:1111–20.
309. Kurreck J, Wyszko E, Gillen C, et al. Design of antisense oligonucleotides stabilized by locked nucleic acids. Nucleic Acids Res. 2002;30:1911–8.
310. Seth PP, Siwkowski A, Allerson CR, et al. Short antisense oligonucleotides with novel 2′-4′ conformationaly restricted nucleoside analogues show improved potency without increased toxicity in animals. J Med Chem. 2009;52:10–3.
311. Stanton R, Sciabola S, Salatto C, et al. Chemical modification study of antisense gapmers. Nucleic Acid Ther. 2012;22:344–59.
312. Straarup EM, Fisker N, Hedtjärn M, et al. Short locked nucleic acid antisense oligonucleotides potently reduce apolipoprotein B mRNA and serum cholesterol in mice and non-human primates. Nucleic Acids Res. 2010;38:7100–11.
313. Murray S, Ittig D, Koller E, et al. TricycloDNA-modified oligo-2′-deoxyribonucleotides reduce scavenger receptor B1 mRNA in hepatic and extrahepatic tissues–a comparative study of oligonucleotide length, design and chemistry. Nucleic Acids Res. 2012;40:6135–43.
314. Obad S, dos Santos CO, Petri A, et al. Silencing of microRNA families by seed-targeting tiny LNAs. Nature Genet. 2011;43:371–8.
315. Watanabe TA, Geary RS, Levin AA. Plasma protein binding of an antisense oligonucleotide targeting human ICAM-1 (ISIS 2302). Oligonucleotides. 2006;16:169–80.
316. Koller E, Vincent TM, Chappell A, et al. Mechanisms of single-stranded phosphorothioate modified antisense oligonucleotide accumulation in hepatocytes. Nucleic Acids Res. 2011;39:4795–807.
317. Elmen J, Lindow M, Schütz S, et al. LNA-mediated microRNA silencing in non-human primates. Nature. 2008;452:896–9.
318. Rossor AM, Reilly MM, Sleigh JN. Antisense oligonucleotides and other genetic therapies made simple. Pract Neurol. 2018;18:126–31.
319. Crooke ST, editor. Antisense drug technology: principles, strategies, and applications. 1. New York: Marcel Dekker; 2001.
320. Bennett CF, Baker BF, Pham N, et al. Pharmacology of antisense drugs. Annu Rev Pharmacol Toxicol. 2017;57:81–105.
321. Bennett CF. Therapeutic antisense oligonucleotides are coming of age. Annu Rev Med. 2019;70:307–21.
322. Lee RG, Crosby J, Baker BF, et al. Antisense technology: an emerging platform for cardiovascular disease therapeutics. J Cardiovasc Transl Res. 2013;6(6):969–80.
323. Stephenson ML, Zamecnik PC. Inhibition of Rous sarcoma viral RNA translation by a specific oligodeoxyribonucleotide. Proc Nat Acad Sci U S A. 1978;75(1):285–8.
324. Fortin M, D'Anjou H, Higgins ME, et al. A multitarget antisense approach against PDE4 and PDE7 reduces smoke-induced lung inflammation in mice. Respir Res. 2009;10(1):39.
325. Burdick AD, Sciabola S, Mantena SR, et al. Sequence motifs associated with hepatotoxicity of locked nucleic acid–modified antisense oligonucleotides. Nucleic Acids Res. 2014;42:4882–91.
326. Schuster S, Miesen P, van Rij RP. Antiviral RNAi in insects and mammals: Parallels and differences. Viruses. 2019;11:448.
327. Sharp PA. RNAi and double-strand RNA. Genes Dev. 1999:139–41.
328. Schwarz DS, Ding H, Kennington L, et al. Designing siRNA that distinguish between genes that differ by a single nucleotide. PLoS Genet. 2006;2:1307–18.
329. Jackson AL, Linsley PS. Recognizing and avoiding siRNA off-target effects for target identification and therapeutic application. Nat Rev Drug Discov. 2010;9:57–67.

330. Grimm D. The dose can make the poison: Lessons learned from adverse in vivo toxicities caused by RNAi overexpression. Silence. 2011;2:8.
331. Chakraborty C, Sharma AR, Sharma G, et al. Therapeutic miRNA and siRNA: Moving from bench to clinic as next generation medicine. Mol Ther Nucl Acid. 2017;8:132–43.
332. Liu YP, Haasnoot J, Berkhout B. Design of extended short hairpin RNAs for HIV-1 inhibition. Nucl Acid Res. 2007;35:5683–93.
333. Liu YP, von Eije KJ, Schopman NCT, et al. Combinatorial RNAi against HIV-1 using extended short hairpin RNAs. Mol Ther. 2009;17:1712–23.
334. Lam JK, Chow MY, Zhang Y, et al. siRNA versus miRNA as therapeutics for gene silencing. Mol Ther Nucleic Acids. 2015;4:e252.
335. Ren S, Liu Y, Xu W, et al. Long noncoding RNA MALAT-1 is a new potential therapeutic target for castration resistant prostate cancer. J Urol. 2013;190(6):2278–87.
336. Davis S, Lollo B, Freier S, et al. Improved targeting of miRNA with antisense oligonucleotides. Nucleic Acids Res. 2006;34:2294–304.
337. Esau C, Davis S, Murray SF, et al. miR-122 regulation of lipid metabolism revealed by in vivo antisense targeting. Cell Metab. 2006;3:87–98.
338. Ørom UA, Kauppinen S, Lund AH. LNA-modified oligonucleotides mediate specific inhibition of microRNA function. Gene. 2006;372:137–41.
339. Lennox KA, Behlke MA. A direct comparison of anti-microRNA oligonucleotide potency. Pharm Res. 2010;27:1788–99.
340. Elmén J, Lindow M, Schütz S, et al. LNA-mediated microRNA silencing in non-human primates. Nature. 2008;452:896–9.
341. Lanford RE, Hildebrandt-Eriksen ES, Petri A, et al. Therapeutic silencing of microRNA-122 in primates with chronic hepatitis C virus infection. Science. 2010;327:198–201.
342. Soutschek J, Akinc A, Bramlage B, et al. Therapeutic silencing of an endogenous gene by systemic administration of modified siRNAs. Nature. 2004;432:173–8.
343. Wolfrum C, Shi S, Jayaprakash KN, et al. Mechanisms and optimization of in vivo delivery of lipophilic siRNAs. Nat Biotechnol. 2007;25:1149–57.
344. Petrova NS, Chernikov IV, Meschaninova MI, et al. Carrier-free cellular uptake and the gene-silencing activity of the lipophilic siRNAs is strongly affected by the length of the linker between siRNA and lipophilic group. Nucleic Acids Res. 2012;40:2330–44.
345. Letsinger RL, Zhang G, Sun DK, et al. Cholesteryl-conjugated oligonucleotides: Synthesis, properties, and activity as inhibitors of replication of human immunodeficiency virus in cell culture. Proc Natl Acad Sci USA. 1989;86:6553–6.
346. Ligtenberg MA, Pico de Coaña Y, Shmushkovich T, et al. Self-Delivering RNAi Targeting PD-1 Improves Tumor-Specific T Cell Functionality for Adoptive Cell Therapy of Malignant Melanoma. Mol Ther. 2018;26:1482–93.
347. McNamara JO, Andrechek ER, Wang Y, et al. Cell type-specific delivery of siRNAs with aptamer-siRNA chimeras. Nat. Biotechnol. 2006;24:1005–15.
348. Dassie JP, Liu XY, Thomas GS, et al. Systemic administration of optimized aptamer-siRNA chimeras promotes regression of PSMA-expressing tumors. Nat Biotechnol. 2009;27:839–46.
349. Zhou J, Swiderski P, Li H, et al. Selection, characterization and application of new RNA HIV gp 120 aptamers for facile delivery of Dicer substrate siRNAs into HIV infected cells. Nucleic Acids Res. 2009;37:3094–109.
350. Morrissey DV, Lockridge JA, Shaw L, et al. Potent and persistent in vivo anti-HBV activity of chemically modified siRNAs. Nat Biotechnol. 2005;23(8):1002–7.
351. Kumar P, Ban HS, Kim SS, et al. T Cell-Specific siRNA Delivery Suppresses HIV-1 Infection in Humanized Mice. Cell. 2008;134:577–86.
352. Song E, Zhu P, Lee SK, et al. Antibody mediated in vivo delivery of small interfering RNAs via cell-surface receptors. Nat Biotechnol. 2005;23:709–17.
353. Wagner RW. Gene inhibition using antisense oligodeoxynucleotides. Nature. 1994;372(6504):3339–5.
354. Stein CA, Narayanan R. Antisense oligodeoxynucleotides. Curr Opin Oncol. 1994;6(6):587–94.
355. Mahato RI, Cheng K, Guntaka RV. Modulation of gene expression by antisense and antigene oligodeoxynucleotides and small interfering RNA. Expert Opin Drug Deliv. 2005;2(1):3–28.
356. Ghosh MK, Cohen JS. Oligodeoxynucleotides as antisense inhibitors of gene expression. Prog Nucleic Acid Res Mol Biol. 1992;42:79–126.
357. Feng J, Wible B, Li GR, et al. Antisense oligodeoxynucleotides directed against Kv1.5 mRNA specifically inhibit ultrarapid delayed rectifier K^+ current in cultured adult human atrial myocytes. Circ Res. 1997;80(4):572–9.
358. Aboul-Fadl T. Antisense oligonucleotides: the state of the art. Curr Med Chem. 2005;12(19):2193–214.
359. Pirollo KF, Rait A, Sleer LS, Chang EH. Antisense therapeutics: from theory to clinical practice. Pharmacol Ther. 2003;99(1):55–77.
360. Agrawal S, Kandimalla ER. Antisense and/or immunostimulatory oligonucleotide therapeutics. Curr Cancer Drug Targets. 2001;1(3):197–209.
361. Bajan S, Hutvagner G. RNA-based therapeutics: From antisense oligonucleotides to miRNAs. Cells. 2020;9(1):137.
362. Allakhverdi Z, Allam M, Guimond A, et al. Multitargeted approach using antisense oligonucleotides for the treatment of asthma. Ann N Y Acad Sci. 2006;1082:62–73.
363. Zhang Y, Xie X, Ma W, et al. Multi-targeted antisense oligonucleotide delivery by a framework

nucleic acid for inhibiting biofilm formation and virulence. Nanomicro Lett. 2020;12(1):74.
364. Lewis K. Persister cells, dormancy and infectious disease. Nat Rev Microbiol. 2007;5(1):48–56.
365. Hall-Stoodley L, Costerton JW, et al. Bacterial biofilms: from the natural environment to infectious diseases. Nat Rev Microbiol. 2004;2(2):95–108.
366. Senadheera DB, Cordova M, Ayala EA, et al. Regulation of bacteriocin production and cell death by the VicRK signaling system in Streptococcus mutans. J Bacteriol. 2012;194(6):1307–16.
367. Dubrac S, Msadek T. Identification of genes controlled by the essential YycG/YycF two-component system of Staphylococcus aureus. J Bacteriol. 2004;186(4):1175–81.
368. Senadheera MD, Guggenheim B, Spatafora GA, et al. A VicRK signal transduction system in *Streptococcus mutans* affects gtfBCD, gbpB, and ftf expression, biofilm formation, and genetic competence development. J Bacteriol. 2005;187(12):4064–76.
369. Bielinska A, Shivdasani RA, Zhang L, et al. Regulation of gene expression with double-stranded phosphorothioate oligonucleotides. Science. 1990;250:997–1000.
370. Morishita R, Gibbons GH, Horiuchi M, et al. A gene therapy strategy using a transcription factor decoy of the E2F binding site inhibits smooth muscle proliferation in vivo. Proc Natl Acad Sci USA. 1995;92:5855–9.
371. Morishita R, Sugimoto T, Aoki M, et al. In vivo transfection of cis element "decoy" against nuclear factor-kappaB binding site prevents myocardial infarction. Nat Med. 1997;13:894–9.
372. Gao H, Xiao J, Yang B, et al. A single decoy oligodeoxynucleotides targeting multiple oncoproteins produces strong anti-cancer effects. Mol Pharmacol. 2006;70:1621–9.
373. Wang X, Liu Q, Hou B, et al. Concomitant targeting of multiple key transcription factors effectively disrupts cancer stem cells enriched in side population of human pancreatic cancer cells. PLoS One. 2013;8(9):e73942.
374. Rad SM, Bamdad T, Sadeghizadeh M, et al. Transcription factor decoy against stem cells master regulators, Nanog and Oct-4: a possible approach for differentiation therapy. Tumour Biol. 2015;36(4):2621–9.
375. Swindell WR, Sarkar MK, Stuart PE, et al. Psoriasis drug development and GWAS interpretation through in silico analysis of transcription factor binding sites. Clin Transl Med. 2015;4:13.
376. Johari B, Zargan J. Simultaneous targeted inhibition of Sox2-Oct4 transcription factors using decoy oligodeoxynucleotides to repress stemness properties in mouse embryonic stem cells. Cell Biol Int. 2017;41(12):1335–44.
377. Johari B, Rezaeejam H, Moradi M, et al. Increasing the colon cancer cells sensitivity toward radiation therapy via application of Oct4-Sox2 complex decoy oligodeoxynucleotides. Mol Biol Rep. 2020;47(9):6793–805.
378. Bigdelou Z, Mortazavi Y, Saltanatpour Z, et al. Role of Oct4-Sox2 complex decoy oligodeoxynucleotides strategy on reverse epithelial to mesenchymal transition (EMT) induction in HT29-ShE encompassing enriched cancer stem-like cells. Mol Biol Rep. 2020;47(3):1859–69.
379. Tehran MM, Rezaei S, Jalili A, et al. Decoy oligodeoxynucleotide technology: an emerging paradigm for breast cancer treatment. Drug Discov Today. 2020;25(1):195–200.
380. Osako MK, Nakagami H, Morishita R. Modification of decoy oligodeoxynucleotides to achieve the stability and therapeutic efficacy. Curr Top Med Chem. 2012;12(15):1603–7.
381. Morishita R, Aoki M, Kaneda Y. Decoy oligodeoxynucleotides as novel cardiovascular drugs for cardiovascular disease. Ann N Y Acad Sci. 2001;947:294–301.
382. Roth M. Transcription factors: Are they a real target for future therapeutic strategies? Pharmacologyonline. 2005;1:45–66.
383. Morishita R, Aoki M, KanedaY. Decoy oligodeoxynucleotides as novel cardiovascular drugs for cardiovascular disease. Ann NY Acad Sci. 2001;947:294–301.
384. Mann MJ, Dzau VJ. Therapeutic applications of transcription factor decoy oligonucleotides. J Clin Invest. 2000;106:1071–5.
385. Martin TA, Goyal A, Watkins G, et al. Expression of the transcription factors snail, slug, and twist and their clinical significance in human breast cancer. Ann Surg Oncol. 2005;12:488–96.
386. Wang LH, Yang XY, Zhang X, et al. The cis decoy against the estrogen response element suppresses breast cancer cells via target disrupting c-fos not mitogen-activated protein kinase activity. Cancer Res. 2003;63:2046–51.
387. Budhram-Mahadeo V, Ndisang D, Ward T, et al. The Brn-3b POU family transcription factor represses expression of the BRCA-1 antioncogene in breast cancer cells. Oncogene. 1999;18:6684–91.
388. Tripathi MK. Regulation of BRCA2 gene expression by the SLUG repressor protein in human breast cells. J Biol Chem. 2005;280:17163–71.
389. Suzuki YJ, Evans T. Regulation of cardiac myocyte apoptosis by the GATA-4 transcription factor. Life Sci. 2004;74:1829–38.
390. Kassiri Z, Oudit GY, Sanchez O, et al. Combination of tumor necrosis factor-alpha ablation and matrix metalloproteinase inhibition prevents heart failure after pressure overload in tissue inhibitor of metalloproteinase-3 knock-out mice. Circ Res. 2005;97:380–90.
391. Rosati B, Grau F, McKinnon D. Regional variation in mRNA transcript abundance within the ventricular wall. J Mol Cell Cardiol. 2006;40:295–302.
392. Costantini DL, Arruda EP, Agarwal P, et al. The homeodomain transcription factor Irx5 establishes

the mouse cardiac ventricular repolarization gradient. Cell. 2006;23:347–58.
393. Mohibi S, Chen X, Zhang J. Cancer the 'RBP'eutics-RNA-binding proteins as therapeutic targets for cancer. Pharmacol Ther. 2019;203:107390.
394. DeJong ES, Luy B, Marino JP. RNA and RNA-protein complexes as targets for therapeutic intervention. Curr Top Med Chem. 2002;2(3):289–302.
395. Cheng MS, Su MX, Wang MX, et al. Probes and drugs that interfere with protein translation via targeting to the RNAs or RNA-protein interactions. Methods. 2019;167:124–33.
396. Baker JD, Uhrich RL, Strovas TJ, et al. Targeting pathological tau by Small molecule inhibition of the Poly(A):MSUT2 RNA-protein interaction. ACS Chem Neurosci. 2020;11(15):2277–85.
397. Hermann T. Strategies for the design of drugs targeting RNA and RNA-protein complexes. Angew Chem Int Ed Engl. 2000;39(11):1890–904.
398. Baudin F, Bach C, Cusack S, et al. Structure of influenza virus RNP. I. Influenza virus nucleoprotein melts secondary structure in panhandle RNA and exposes the bases to the solvent. EMBO J. 1994;13(13):3158–65.
399. Lukong KE, Chang K-W, Khandjian EW, et al. RNA-binding proteins in human genetic disease. Trends Genet. 2008;24(8):416–25.
400. Keene JD. RNA regulons: coordination of post-transcriptional events. Nat Rev Genet. 2007;8(7):533–43.
401. Zhang J, Chen X. Posttranscriptional regulation of p53 and its targets by RNA-binding proteins. Curr Mol Med. 2008;8(8):845–9.
402. Kechavarzi B, Janga SC. Dissecting the expression landscape of RNA-binding proteins in human cancers. Genome Biol. 2014;15(1):R14.
403. Neelamraju Y, Gonzalez-Perez A, Bhat-Nakshatri P, et al. Mutational landscape of RNA-binding proteins in human cancers. RNA Biol. 2018;15(1):115–29.
404. Pereira B, Billaud M, Almeida R. RNA-binding proteins in cancer: Old players and new actors. Trends Cancer. 2017;3(7):506–28.
405. Lunde BM, Moore C, Varani G. RNA-binding proteins: modular design for efficient function. Nat Rev Mol Cell Biol. 2007;8(6):479–90.
406. Baltz AG, Munschauer M, Schwanhäusser B, et al. The mRNA-bound proteome and its global occupancy profile on protein-coding transcripts. Mol Cell. 2012;46(5):674–90.
407. Castello A, Fischer B, Eichelbaum K, et al. Insights into RNA biology from an atlas of mammalian mRNA-binding proteins. Cell. 2012;149(6):1393–406.
408. Gerstberger S, Hafner M, Tuschl T. A census of human RNA-binding proteins. Nat Rev Genet. 2014;15(12):829–45.
409. Wang ZL, Li B, Luo YX, et al. Comprehensive genomic characterization of RNA-binding proteins across human cancers. Cell Rep. 2018;22(1):286–98.
410. Hentze MW, Castello A, Schwarzl T, et al. A brave new world of RNA-binding proteins. Nat Rev Mol Cell Biol. 2018;19(5):327–41.
411. Moore S, Järvelin AI, Davis I, et al. Expanding horizons: new roles for non-canonical RNA-binding proteins in cancer. Curr Opin Genet Dev. 2018;48:112–20.
412. Bank, RCSB Protein data. RCSB Protein Data Bank-RCSB PDB. Retrieved 2021-08-28.
413. Lewis BA, Walia RR, Terribilini M, et al. PRIDB: a protein–RNA interface database. Nucleic Acids Res. 2016;39:D277–82.
414. Stein CA, Castanotto D. FDA-approved oligonucleotide therapies in 2017. Mol Ther. 2017;25(5):1069–75.
415. Hua Y, Sahashi K, Hung G, et al. Antisense correction of SMN2 splicing in the CNS rescues necrosis in a type III SMA mouse model. Genes Dev. 2010;24(15):1634–44.
416. Rigo F, Hua Y, Krainer AR, et al. Antisense-based therapy for the treatment of spinal muscular atrophy. J Cell Biol. 2012;199(1):21–5.
417. Chiriboga CA, Swoboda KJ, Darras BT, et al. Results from a phase 1 study of nusinersen (ISIS-SMN(Rx)) in children with spinal muscular atrophy. Neurology. 2016;86(10):890–7.
418. Mercuri E, Darras BT, Chiriboga CA, et al. CHERISH Study Group. Nusinersen versus sham control in later-onset spinal muscular atrophy. N Engl J Med. 2018;378(7):625–35.
419. Nussbacher JK, Tabet R, Yeo GW, et al. Disruption of RNA metabolism in neurological diseases and emerging therapeutic interventions. Neuron. 2019;102(2):294–320.
420. Hong DS, Kurzrock R, Oh Y, et al. A phase 1 dose escalation, pharmacokinetic, and pharmacodynamic evaluation of eIF-4E antisense oligonucleotide LY2275796 in patients with advanced cancer. Clin Cancer Res. 2011;17(20):6582–91.
421. Duffy AG, Makarova-Rusher OV, Ulahannan SV, et al. Modulation of tumor eIF4E by antisense inhibition: A phase I/II translational clinical trial of ISIS 183750-an antisense oligonucleotide against eIF4E-in combination with irinotecan in solid tumors and irinotecan-refractory colorectal cancer. Int J Cancer. 2016;139(7):1648–57.
422. Fox RG, Lytle NK, Jaquish DV, et al. Image-based detection and targeting of therapy resistance in pancreatic adenocarcinoma. Nature. 2016;534(7607):407–11.
423. Muralidharan R, Babu A, Amreddy N, et al. Tumor-targeted nanoparticle delivery of HuR siRNA inhibits lung tumor growth in vitro and in vivo by disrupting the oncogenic activity of the RNA-binding protein HuR. Mol Can Therapeut. 2017;16(8):1470–86.
424. Jimbo M, Blanco FF, Huang YH, et al. Targeting the mRNA-binding protein HuR impairs malignant characteristics of pancreatic ductal adenocarcinoma cells. Oncotarget. 2015;6(29):27312–31.

425. Mehta M, Basalingappa K, Griffith JN, et al. HuR silencing elicits oxidative stress and DNA damage and sensitizes human triple-negative breast cancer cells to radiotherapy. Oncotarget. 2016;7(40):64820–35.
426. Huang YH, Peng W, Furuuchi N, et al. Delivery of therapeutics targeting the mRNA-binding protein HuR using 3DNA nanocarriers suppresses ovarian tumor growth. Cancer Res. 2016;76(6):1549–59.
427. Dong K, Wang R, Wang X, et al. Tumor-specific RNAi targeting eIF4E suppresses tumor growth, induces apoptosis and enhances cisplatin cytotoxicity in human breast carcinoma cells. Breast Cancer Res Treat. 2009;113(3):443–56.
428. Chen H, Liu J, Wang H, et al. Inhibition of RNA-binding protein musashi-1 suppresses malignant properties and reverses paclitaxel resistance in ovarian carcinoma. J Cancer. 2019;10(6):1580–92.
429. Sheng W, Dong M, Chen C, et al. Musashi2 promotes the development and progression of pancreatic cancer by down-regulating Numb protein. Oncotarget. 2017;8(9):14359–73.
430. Sureban SM, May R, George RJ, et al. Knockdown of RNA binding protein musashi-1 leads to tumor regression in vivo. Gastroenterology. 2008;134(5):1448–58.
431. Bai X, Yang C, Jiao L, et al. LncRNA MIAT impairs cardiac contractile function by acting on mitochondrial translocator protein TSPO in a mouse model of myocardial infarction. Signal Transduct Target Ther. 2021;6(1):172.
432. Mercer JR. Mitochondrial bioenergetics and therapeutic intervention in cardiovascular disease. Pharmacol Ther. 2014;141:13–20.
433. Morin D, Musman J, Pons S, et al. Mitochondrial translocator protein (TSPO): From physiology to cardioprotection. Biochem Pharmacol. 2016;105:1–13.
434. Ashwal-Fluss R, Meyer M, Pamudurti NR, et al. Circ RNA Biogenesis competes with Pre-mRNA splicing. Molecular Cell. 2014;56(1):55–66.
435. Abdelmohsen K, Panda AC, Munk R, et al. Identification of HuR target circular RNAs uncovers suppression of PABPN1 translation by CircPABPN1. RNA Biology. 2017;14(3):361–9.
436. Mascini M, Palchetti I, Tombelli S. Nucleic acid and peptide aptamers: Fundamentals and bioanalytical aspects. Angewandte Chemie International Edition. 2012;51(6):1316–32.
437. Zhou J, Rossi J. Aptamers as targeted therapeutics: Current potential and challenges. Nat Rev Drug Discov. 2017;16(3):181–202.
438. Nimjee SM, White RR, Becker RC, et al. Aptamers as therapeutics. Annu Rev Pharmacol Toxicol. 2017;57:61–79.
439. Wu YX, Kwon YJ. Aptamers: The "evolution" of SELEX. Methods. 2016;106:21–8.
440. Reverdatto S, Burz DS, Shekhtman A. Peptide aptamers: development and applications. Curr Top Med Chem. 2015;15(12):1082–101.
441. New RRC, Bui TTT, Bogus M. Binding interactions of peptide aptamers. Molecules. 2020;25(24):6055.
442. Li J, Tan S, Chen X, et al. Peptide aptamers with biological and therapeutic applications. Curr Med Chem. 2011;18(27):4215–22.
443. Baines IC, Colas P. Peptide aptamers as guides for small-molecule drug discovery. Drug Discov Today. 2006;11(7-8):334–41.
444. Hoppe-Seyler F, Crnkovic-Mertens I, Tomai E, et al. Peptide aptamers: specific inhibitors of protein function. Curr Mol Med. 2004;4(5):529–38.
445. Crawford M, Woodman R, Ko FP. Peptide aptamers: tools for biology and drug discovery. Brief Funct Genomic Proteomic. 2003;2(1):72–9.
446. Morita Y, Leslie M, Kameyama H, et al. Aptamer therapeutics in cancer: Current and future. Cancers (Basel). 2018;10(3):80.
447. Berger CM, Gaume X, Bouvet P. The roles of nucleolin subcellular localization in cancer. Biochimie. 2015;113:78–85.
448. Ireson CR, Kelland LR. Discovery and development of anticancer aptamers. Mol Can Therapeut. 2006;5(12):2957–62.
449. Marqus S, Pirogova E, Piva TJ. Evaluation of the use of therapeutic peptides for cancer treatment. J Biomed Sci. 2017;24(1):21.
450. Song YK, Guo H, Barengo N, et al. Inhibition of ovarian cancer growth by a tumor-targeting peptide that binds eukaryotic translation initiation factor 4E. Clin Can Res. 2009;15(13):4336–47.
451. Lucchesi CA, Zhang J, Ma B, et al. Disruption of the RBM38-eIF4E complex with a synthetic peptide PEP8 increases p53 expression. Can Res. 2019;79(4):807–18.
452. Zhang J, Cho SJ, Shu L, et al. Translational repression of p53 by RNPC1, a p53 target overexpressed in lymphomas. Genes Dev. 2011;25(14):1528–43.
453. Ebner S, Trieb M, Schönfeld M, et al. Decoy peptides derived from the extracellular domain of toll-like receptor 2 (TLR2) show anti-inflammatory properties. Bioorg Med Chem. 2018;26(16):4615–23.
454. Piao W, Vogel SN, Toshchakov VY. Inhibition of TLR4 signaling by TRAM-derived decoy peptides in vitro and in vivo. J Immunol. 2013;190(5):2263–72.
455. Allette Y, Kim Y, Randolph A, et al. Decoy peptide targeted to the Toll-IL-1R domain inhibits LPS and TLR4-active metabolite morphine-3 glucuronide sensitization of sensory neurons. Sci Rep.;7(1):3741.
456. Mantovani A, Locati M, Vecchi A, et al. Decoy receptors: a strategy to regulate inflammatory cytokines and chemokines. Trends Immunol. 2001;22(6):328–36.
457. Husain M, Becker EJ Jr, Bone NB, et al. NOX2 decoy peptides disrupt trauma-mediated neutrophil immunosuppression and protect against lethal peritonitis. Redox Biol. 2020;36:101651.
458. Mahjoubin-Tehran M, Rezaei S, Jalili A, et al. Peptide decoys: a new technology offering thera-

peutic opportunities for breast cancer. Drug Discov Today. 2020;25(3):593–8.
459. Wu FY, Ou ZL, Feng LY, et al. Chemokine decoy receptor d6 plays a negative role in human breast cancer. Mol Cancer Res. 2008;6(8):1276–88.
460. Kwon YJ, Leibovitch BA, Bansal N, et al. Targeted interference of SIN3A-TGIF1 function by SID decoy treatment inhibits Wnt signaling and invasion in triple negative breast cancer cells. Oncotarget. 2016;8(51):88421–36.
461. Zhang Y, Li D, Zhao X, et al. Decoy receptor 3 suppresses FasL-induced apoptosis via ERK1/2 activation in pancreatic cancer cells. Biochem Biophys Res Commun. 2015;463(4):1144–51.
462. Oh JG, Kim J, Jang SP, et al. Decoy peptides targeted to protein phosphatase 1 inhibit dephosphorylation of phospholamban in cardiomyocytes. J Mol Cell Cardiol. 2013;56:63–71.
463. Nabi AN, Biswas KB, Arai Y, et al. Functional characterization of the decoy peptide, [R10P] IFLKRMPSI[19P]. Front Biosci (Elite Ed). 2010;2:1211–7.
464. Mahjoubin-Tehran M, Rezaei S, Atkin SL, et al. Decoys as potential therapeutic tools for diabetes. Drug Discov Today. 2021;26(7):1669–79.
465. Ichihara A, Sakoda M, Kurauchi-Mito A, et al. Drug discovery for overcoming chronic kidney disease (CKD): new therapy for CKD by a (pro)renin-receptor-blocking decoy peptide. J Pharmacol Sci. 2009;109(1):20–3.
466. Chan KK, Tan TJC, Narayanan KK, et al. An engineered decoy receptor for SARS-CoV-2 broadly binds protein S sequence variants. Sci Adv. 2021;7(8):eabf1738.
467. Jing W, Procko E. ACE2-based decoy receptors for SARS coronavirus 2. Proteins. 2021;89(9):1065–78.
468. Villard S, Piquer D, Raut S, et al. Low molecular weight peptides restore the procoagulant activity of factor VIII in the presence of the potent inhibitor antibody ESH8. J Biol Chem. 2002;277(30):27232–9.
469. Ananyeva NM, Lacroix-Desmazes S, Hauser CA, et al. Inhibitors in hemophilia A: mechanisms of inhibition, management and perspectives. Blood Coagul Fibrinolysis. 2004;15(2):109–24.
470. Schlesinger N. Anti-interleukin-1 therapy in the management of gout. Curr Rheumatol Rep. 2014;16(2):398.
471. Woods AS, Kaminski R, Oz M, et al. Decoy peptides that bind dynorphin noncovalently prevent NMDA receptor-mediated neurotoxicity. J Proteome Res. 2006;5(4):1017–23.
472. Bechara C, Sagan S. Cell-penetrating peptides: 20 years later, where do we stand? FEBS letters. 2013;587:1693–702.
473. Guo Z, Peng H, Kang J, et al. Cell-penetrating peptides: Possible transduction mechanisms and therapeutic applications. Biomedical reports. 2016;4:528–34.
474. Toshchakov VY, Javmen A. Targeting the TLR signalosome with TIR domain-derived cell-permeable decoy peptides: the current state and perspectives. Innate Immun. 2020;26(1):35–47.
475. Copolovici DM, Langel K, Eriste E, et al. Cell-penetrating peptides: design, synthesis, and applications. ACS Nano. 2014;8:1972–94.
476. Kauffman WB, Fuselier T, He J, et al. Mechanism matters: A taxonomy of cell penetrating peptides. Trends Biochem Sci. 2015;40:749–64.
477. Heitz F, Morris MC, Divita G. Twenty years of cell-penetrating peptides: from molecular mechanisms to therapeutics. Br J Pharmacol. 2009;157:195–206.
478. Milletti F. Cell-penetrating peptides: classes, origin, and current landscape. Drug discovery today. 2012;17:850–60.
479. Järver P, Langel K, El-Andaloussi S, et al. Applications of cell-penetrating peptides in regulation of gene expression. Biochem Soc Trans. 2007;35(Pt 4):770–4.
480. Toshchakov VY, Vogel SN. Cell-penetrating TIR BB loop decoy peptides a novel class of TLR signaling inhibitors and a tool to study topology of TIR-TIR interactions. Expert Opinion Biol Therapy. 2007;7:1035–50.
481. Wirth T, Parker N, Ylä-Herttuala S. History of gene therapy. Gene. 2013;525(2):162–9.
482. Brody H. Gene therapy. Nature. 2018;564(7735):S5.
483. Ylä-Herttuala S. The pharmacology of gene therapy. Mol Ther. 2017;25(8):1731–2.
484. Tang R, Xu Z. Gene therapy: a double-edged sword with great powers. Mol Cell Biochem. 2020;474(1-2):73–81.
485. Mullard A. Gene therapy boom continues. Nat Rev Drug Discov. 2019;18(10):737.
486. Bak RO, Gomez-Ospina N, Porteus MH. Gene editing on center stage. Trends in Genetics. 2018;34(8):600–11.
487. Memi F, Ntokou A, Papangeli I. CRISPR/Cas9 gene-editing: Research technologies, clinical applications and ethical considerations. Semin Perinatol. 2018;42(8):487–500.
488. Sanches-da-Silva GN, Medeiros LF, Lima FM. The potential use of the CRISPR-Cas system for HIV-1 gene therapy. Intl J Genom. 2019;2019:8458263.
489. Cavazzana-Calvo M, Thrasher A, Mavilio F. The future of gene therapy. Nature. 2004;427:779–81.
490. Kotani H, Kmiec EB. A role for RNA synthesis in homologous pairing events. Mol Cell Biol. 1994;14:6097–106.
491. Yoon K, Cole-Strauss A, Kmiec EB. Targeted gene correction of episomal DNA in mammalian cells mediated by a chimeric RNA/DNA oligonucleotide. Proc Natl Acad Sci USA. 1996;93:2071–6.
492. Andersen MS, Sorensen CB, Bolund L, et al. Mechanisms underlying targeted gene correction using chimeric RNA/DNA and single-stranded DNA oligonucleotides. J Mol Med. 2002;80:770–81.
493. Leclerc X, Danos O, Scherman D, et al. A comparison of synthetic oligodeoxynucleotides, DNA fragments

and AAV-1 for targeted episomal and chromosomal gene repair. BMC Biotechnol. 2009;9:35.
494. Liu X, Yan Z, Luo M, et al. Targeted correction of single-base-pair mutations with adeno-associated virus vectors under nonselective conditions. J Virol. 2004;78:4165–75.
495. Urnov FD, Miller JC, Lee YL, et al. Highly efficient endogenous human gene correction using designed zinc-finger nucleases. Nature. 2005;435:646–51.
496. Gruenert DC, Bruscia E, Novelli G, et al. Sequence-specific modification of genomic DNA by small DNA fragments. J Clin Invest. 2003;112:637–41.
497. Nickerson HD, Colledge WH. A comparison of gene repair strategies in cell culture using a lacZ reporter system. Gene Ther. 2003;10:1584–91.
498. Bertoni C, Morris GE, Rando TA. Strand bias in oligonucleotide-mediated dystrophin gene editing. Hum Mol Genet. 2005;14:221–33.
499. Katada H, Komiyama M. Artificial restriction DNA cutters as new tools for gene manipulation. Chembiochem. 2009;10:1279–88.
500. Miller DG, Wang PR, Petek LM, et al. Gene targeting in vivo by adeno-associated virus vectors. Nat Biotechnol. 2006;24:1022–6.
501. Olsen PA, Solhaug A, Booth JA, et al. Cellular responses to targeted genomic sequence modification using single-stranded oligonucleotides and zinc-finger nucleases. DNA Repair (Amst). 2009;8:298–308.
502. Engstrom JU, Suzuki T, Kmiec EB. Regulation of targeted gene repair by intrinsic cellular processes. Bioessays. 2009;31:159–68.
503. Igoucheva O, Peritz AE, Levy D, et al. A sequence-specific gene correction by an RNA-DNA oligonucleotide in mammalian cells characterized by transfection and nuclear extract using a lacZ shuttle system. Gene Ther. 1999;6:1960–71.
504. Radecke S, Radecke F, Peter I, et al. Physical incorporation of a single-stranded oligodeoxynucleotide during targeted repair of a human chromosomal locus. J Gene Med. 2006;8:217–28.
505. Parekh-Olmedo H, Ferrara L, Brachman E, et al. Gene therapy progress and prospects: targeted gene repair. Gene Ther. 2005;12:639–46.
506. Maguire KK, Kmiec EB. Multiple roles for MSH2 in the repair of a deletion mutation directed by modified single-stranded oligonucleotides. Gene. 2007;386:107–14.
507. Pierce EA, Liu Q, Igoucheva O, et al. Oligonucleotide-directed single-base DNA alterations in mouse embryonic stem cells. Gene Ther. 2003;10:24–33.
508. Dekker M, Brouwers C, te Riele H. Targeted gene modification in mismatch-repair-deficient embryonic stem cells by single-stranded DNA oligonucleotides. Nucleic Acids Res. 2003;31:e27.
509. Parekh-Olmedo H, Kmiec EB. Progress and prospects: targeted gene alteration (TGA). Gene Ther. 2007;14:1675–80.
510. Bonner M, Kmiec EB. DNA breakage associated with targeted gene alteration directed by DNA oligonucleotides. Mutat Res. 2009;669:85–94.
511. Gamper HB Jr, Cole-Strauss A, Metz R, et al. A plausible mechanism for gene correction by chimeric oligonucleotides. Biochemistry. 2000;39:5808–16.
512. Cole-Strauss A, Gamper H, Holloman WK, et al. Targeted gene repair directed by the chimeric RNA/DNA oligonucleotide in a mammalian cell-free extract. Nucleic Acids Res. 1999;27:1323–30.
513. Zayed H, McIvor RS, Wiest DL, et al. In vitro functional correction of the mutation responsible for murine severe combined immune deficiency by small fragment homologous replacement. Hum Gene Ther. 2006;17:158–66.
514. Sangiuolo F, Scaldaferri ML, Filareto A, et al. Cftr gene targeting in mouse embryonic stem cells mediated by Small Fragment Homologous Replacement (SFHR). Front Biosci. 2008;13:2989–99.
515. Bedayat B, Abdolmohamadi A, Ye L, et al. Sequence-specific correction of genomic hypoxanthine-guanine phosphoribosyl transferase mutations in lymphoblasts by small fragment homologous replacement. Oligonucleotides. 2010;20:7–16.
516. Gatz SA, Wiesmuller L. p53 in recombination and repair. Cell Death Differ. 2006;13:1003–16.
517. Tsuchiya H, Harashima H, Kamiya H. Increased SFHR gene correction efficiency with sense single-stranded DNA. J Gene Med. 2005;7:486–93.
518. Goncz KK, Prokopishyn NL, Abdolmohammadi A, et al. Small fragment homologous replacement-mediated modification of genomic beta-globin sequences in human hematopoietic stem/progenitor cells. Oligonucleotides. 2006;16:213–24.
519. Gruenert DC. Gene correction with small DNA fragments. Curr Res Mol Ther. 1998;1:607–13.
520. Colosimo A, Guida V, Antonucci I, et al. Sequence-specific modification of a beta-thalassemia locus by small DNA fragments in human erythroid progenitor cells. Haematologica. 2007;92:129–30.
521. Todaro M, Quigley A, Kita M, et al. Effective detection of corrected dystrophin loci in mdx mouse myogenic precursors. Hum Mutat. 2007;28:816–23.
522. Mcnab GL, Ahmad A, Mistry D, et al. Modification of gene expression and increase in alpha1-antitrypsin (alpha1-AT) secretion after homologous recombination in alpha1-AT-deficient monocytes. Hum Gene Ther. 2007;18:1171–7.
523. Sangiuolo F, Filareto A, Spitalieri P, et al. In vitro restoration of functional SMN protein in human trophoblast cells affected by spinal muscular atrophy by small fragment homologous replacement. Hum Gene Ther. 2005;16:869–80.
524. Consortium TCFGA. World-wide survey of ΔF508 mutation—report from Cystic Fibrosis Genetic Analysis Consortium. Am J Hum Genet. 1990;47:354–7.
525. Bruscia E, Sangiuolo F, Sinibaldi P, et al. Isolation of CF cell lines corrected at DeltaF508-CFTR

locus by SFHR-mediated targeting. Gene Ther. 2002;9:683–5.
526. Kunzelmann K, Legendre JY, Knoell DL, et al. Gene targeting of CFTR DNA in CF epithelial cells. Gene Ther. 1996;3:859–67.
527. Sangiuolo F, Bruscia E, Serafino A, et al. In vitro correction of cystic fibrosis epithelial cell lines by small fragment homologous replacement (SFHR) technique. BMC Med Genet. 2002;3:8.
528. Knauert MP, Glazer PM. Triplex forming oligonucleotides: sequence-specific tools for gene targeting. Hum Mol Genet. 2001;10(20):2243–51.
529. Kuan JY, Glazer PM. Targeted gene modification using triplex-forming oligonucleotides. Methods Mol Biol. 2004;262:173–94.
530. Hélène C. The anti-gene strategy: control of gene expression by triplex-forming-oligonucleotides. Anticancer Drug Des. 1991;6(6):569–84.
531. Hansen ME, Bentin T, Nielsen PE. High-affinity triplex targeting of double stranded DNA using chemically modified peptide nucleic acid oligomers. Nucleic Acids Res. 2009;37:4498–507.
532. Chin JY, Glazer PM. Repair of DNA lesions associated with triplex-forming oligonucleotides. Mol Carcinog. 2009;48:389–99.
533. Lonkar P, Kim KH, Kuan JY, et al. Targeted correction of a thalassemia-associated beta-globin mutation induced by pseudo-complementary peptide nucleic acids. Nucleic Acids Res. 2009;37:3635–44.
534. Wang G, Seidman MM, Glazer PM. Mutagenesis in mammalian cells induced by triple helix formation and transcription-coupled repair. Science. 1996;271:802–5.
535. Felsenfeld G, Rich A. Studies on the formation of two- and three-stranded polyribonucleotides. Biochim Biophys Acta. 1957;26:457–68.
536. Kallenbach NR, Daniel WE Jr, Kaminker MA. Nuclear magnetic resonance study of hydrogen-bonded ring protons in oligonucleotide helices involving classical and nonclassical base pairs. Biochemistry. 1976;15:1218–24.
537. Wang G, Glazer PM. Altered repair of targeted psoralen photo-adducts in the context of an oligonucleotide-mediated triple helix. J Biol Chem. 1995;270:22595–601.
538. Vasquez KM, Dagle JM, Weeks DL, et al. Chromosome targeting at short polypurine sites by cationic triplex-forming oligonucleotides. J Biol Chem. 2001;276:38536–41.
539. Luo Z, Macris MA, Faruqi AF, et al. High-frequency intrachromosomal gene conversion induced by triplex-forming oligonucleotides microinjected into mouse cells. Proc Natl Acad Sci U S A. 2000;97:9003–8.
540. Datta HJ, Glazer PM. Intracellular generation of single-stranded DNA for chromosomal triplex formation and induced recombination. Nucleic Acids Res. 2001;29:5140–7.
541. Chan PP, Lin M, Faruqi AF, et al. Targeted correction of an episomal gene in mammalian cells by a short DNA fragment tethered to a triplex-forming oligonucleotide. J Biol Chem. 1999;274:11541–8.
542. Maurisse R, Feugeas JP, Biet E, et al. A new method (GOREC) for directed mutagenesis and gene repair by homologous recombination. Gene Ther. 2002;9:703–7.
543. Culver KW, Hsieh WT, Huyen Y, et al. Correction of chromosomal point mutations in human cells with bifunctional oligonucleotides. Nat Biotechnol. 1999;17:989–93.
544. Datta HJ, Chan PP, Vasquez KM, et al. Triplex-induced recombination in human cell-free extracts. Dependence on XPA and HsRad51. J Biol Chem. 2001;276:18018–23.
545. Goni JR, De La Cruz X, Orozco M. Triplex-forming oligonucleotide target sequences in the human genome. Nucleic Acids Res. 2004;32:354–60.
546. Simon P, Cannata F, Concordet JP, et al. Targeting DNA with triplex-forming oligonucleotides to modify gene sequence. Biochimie. 2008;90:1109–16.
547. Chin JY, Kuan JY, Lonkar PS, et al. Correction of a splice-site mutation in the beta-globin gene stimulated by triplex-forming peptide nucleic acids. Proc Natl Acad Sci USA. 2008;105:13514–9.
548. Nielsen PE. PNA Technology. Mol Biotechnol. 2004;26:233–48.
549. Bentin T, Nielsen PE. Superior duplex DNA strand invasion by acridine conjugated peptide nucleic acids. J Am Chem Soc. 2003;125:6378–9.
550. Kim KH, Nielsen PE, Glazer PM. Site-directed gene mutation at mixed sequence targets by psoralen-conjugated pseudo-complementary peptide nucleic acids. Nucleic Acids Res. 2007;35:7604–13.
551. Nielsen PE, Egholm M, Buchardt O. Evidence for (PNA)2/DNA triplex structure upon binding of PNA to dsDNA by strand displacement. J Mol Recognit. 1994;7:165–70.
552. Yamamoto Y, Yoshida J, Tedeschi T, et al. Highly efficient strand invasion by peptide nucleic acid bearing optically pure lysine residues in its backbone. Nucleic Acids Symp Ser (Oxf). 2006; pp109–10.
553. Coura RS, Nardi NB. The state of the art of adeno-associated virus-based vectors in gene therapy. Virol J. 2007;4:99.
554. Vasileva A, Linden RM, Jessberger R. Homologous recombination is required for AAV-mediated gene targeting. Nucleic Acids Res. 2006;34:3345–60.
555. Gellhaus K, Cornu TI, Heilbronn R, et al. Fate of recombinant adeno-associated viral vector genomes during DNA double-strand break-induced gene targeting in human cells. Hum Gene Ther. 2010;21:543–53.
556. Chamberlain JR, Deyle DR, Schwarze U, et al. Gene targeting of mutant COL1A2 alleles in mesenchymal stem cells from individuals with osteogenesis imperfecta. Mol Ther. 2008;16:187–93.
557. Khan IF, Hirata RK, Wang PR, et al. Engineering of human pluripotent stem cells by AAV-mediated gene targeting. Mol Ther. 2010;18:1192–9.
558. Paiboonsukwong K, Ohbayashi F, Shiiba H, et al. Correction of mutant Fanconi anemia gene by

homologous recombination in human hematopoietic cells using adeno-associated virus vector. J Gene Med. 2009;11:1012–9.
559. Mitsui K, Suzuki K, Aizawa E, et al. Gene targeting in human pluripotent stem cells with adeno-associated virus vectors. Biochem Biophys Res Commun. 2009;388:711–7.
560. Macielag MJ. Chemical properties of antibacterials and their uniqueness. In: Dougherty TJ, Pucci MJ (eds). Antibiotic Discovery and Development. 2012; pp. 801–2.
561. Ngo HX, Garneau-Tsodikova S. What are the drugs of the future? MedChemComm. 2018;9(5):757–8.
562. Arkin MR, Wells JA. Small-molecule inhibitors of protein-protein interactions: progressing towards the dream. Nat Rev Drug Discov. 2004;3(4):301–17.
563. Veber DF, Johnson SR, Cheng HY, et al. Molecular properties that influence the oral bioavailability of drug candidates. J Med Chem. 2002;45(12):2615–23.
564. Lipinski CA. Lead-and drug-like compounds: the rule-of-five revolution. Drug Discov Today: Technol. 2004;1(4):337–41.
565. Leeson PD, Springthorpe B. The influence of drug-like concepts on decision-making in medicinal chemistry. Nat Rev Drug Discov. 2007;6(11):881–90.
566. Samanen J. Chapter 5.2 How do SMDs differ from biomolecular drugs? In Ganellin CR, Jefferis R, Roberts SM (eds.). Introduction to Biological and Small Molecule Drug Research and Development: theory and case studies (Kindle ed.). New York: Academic Press. 2013; pp161–203.
567. Santos R, Ursu O, Gaulton A, et al. A comprehensive map of molecular drug targets. Nat Rev Drug Discov. 2017;16(1):19–34.
568. Usmani SS, Bedi G, Samuel JS, et al. THPdb: Database of FDA-approved peptide and protein therapeutics. PLoS One. 2017;12(7):e0181748.
569. Yin W, Rogge M. Targeting RNA: A transformative therapeutic strategy. Clin Transl Sci. 2019;12(2):98–112.
570. Warner KD, Hajdin CE, Weeks KM. Principles for targeting RNA with drug-like small molecules. Nat Rev Drug Discov. 2018;17(8):547–58.
571. Yu AM, Choi YH, Tu MJ. RNA drugs and RNA targets for small molecules: principles, progress, and challenges. Pharmacol Rev. 2020;72(4):862–98.
572. Clamp M. Distinguishing protein-coding and noncoding genes in the human genome. Proc Natl Acad Sci USA. 2007;104:19428–33.
573. Ezkurdia I. Multiple evidence strands suggest that there may be as few as 19,000 human protein-coding genes. Hum Mol Genet. 2014;23:5866–78.
574. Hopkins AL, Groom CR. The druggable genome. Nat Rev Drug Discov. 2002;1:727–30.
575. Overington JP, Al-Lazikani B, Hopkins AL. How many drug targets are there? Nat Rev Drug Discov. 2006;5:993–6.
576. Dixon SJ, Stockwell BR. Identifying druggable disease-modifying gene products. Curr Opin Chem Biol. 2009;13:549–55.
577. Santos R. A comprehensive map of molecular drug targets. Nat Rev Drug Discov. 2017;16:19–34.
578. Sharp PA. The centrality of RNA. Cell. 2009;136:577–80.
579. Cech TR, Steitz JA. The noncoding RNA revolution-trashing old rules to forge new ones. Cell. 2014;157:77–94.
580. Djebali S. Landscape of transcription in human cells. Nature. 2012;489:101–8.
581. Adams BD, Parsons C, Walker L, et al. Targeting noncoding RNAs in disease. J Clin Invest. 2017;127:761–71.
582. Harrow J. GENCODE: the reference human genome annotation for The ENCODE Project. Genome Res. 2012;22:1760–74.
583. Wilson DN. The A-Z of bacterial translation inhibitors. Crit Rev Biochem Mol Biol. 2009;44:393–33.
584. Lin J, Zhou D, Steitz TA, et al. Ribosome-targeting antibiotics: modes of action, mechanisms of resistance, and implications for drug design. Annu Rev Biochem. 2018;87:451–78.
585. Donlic A, Hargrove AE. Targeting RNA in mammalian systems with small molecules. Wiley Interdiscip Rev RNA. 2018;9:e1477.
586. Hermann T. Small molecules targeting viral RNA. Wiley Interdiscip Rev RNA. 2016;7:726–43.
587. McCown PJ, Corbino KA, Stav S, et al. Riboswitch diversity and distribution. RNA. 2017;23:995–1011.
588. Cruz JA, Westhof E. The dynamic landscapes of RNA architecture. Cell. 2009;136:604–9.
589. Butcher SE, Pyle AM. The molecular interactions that stabilize RNA tertiary structure: RNA motifs, patterns, and networks. Acc Chem Res. 2011;44:1302–11.
590. Jones CP, Ferré-D'Amaré AR. RNA quaternary structure and global symmetry. Trends Biochem Sci. 2015;40:211–20.
591. Schlick T. Adventures with RNA graphs. Methods. 2018;143:16–33.
592. Mustoe AM, Busan S, Rice GM, et al. Pervasive regulatory functions of mRNA structure revealed by high-resolution SHAPE probing. Cell. 2018;173:181–95.
593. Morgan BS, Forte JE, Culver RN, et al. Discovery of key physicochemical, structural, and spatial properties of RNA-targeted bioactive ligands. Angew Chem Int Ed Engl. 2017;56:13498–502.
594. Morgan BS, Forte JE, Hargrove AE. Insights into the development of chemical probes for RNA. Nucleic Acids Res. 2018;46:8025–37.
595. Wilson DN. Ribosome-targeting antibiotics and mechanisms of bacterial resistance. Nat Rev Microbiol. 2014;12:35–48.
596. Steitz TA. A structural understanding of the dynamic ribosome machine. Nat Rev Mol Cell Biol. 2008;9:242–53.

597. Lynch SR, Gonzalez RL, Puglisi JD. Comparison of X-ray crystal structure of the 30S subunit-antibiotic complex with NMR structure of decoding site oligonucleotide-paromomycin complex. Structure. 2003;11:43–53.
598. Fourmy D, Recht MI, Blanchard SC, et al. Structure of the A site of Escherichia coli 16S ribosomal RNA complexed with an aminoglycoside antibiotic. Science. 1996;274:1367–71.
599. Demirci H, Murphy F IV, Murphy E, et al. A structural basis for streptomycin-induced misreading of the genetic code. Nat Commun. 2013;4:1355.
600. Kanazawa H, Baba F, Koganei M, et al. A structural basis for the antibiotic resistance conferred by an N1-methylation of A1408 in 16S rRNA. Nucleic Acids Res. 2017;45:12529–35.
601. Kondo J. A structural basis for the antibiotic resistance conferred by an A1408G mutation in 16S rRNA and for the antiprotozoal activity of aminoglycosides. Angew Chem Int Ed Engl. 2012;51:465–8.
602. Zhang X, Lai M, Chang W, et al. Structures and stabilization of kinetoplastid-specific split rRNAs revealed by comparing leishmanial and human ribosomes. Nat Commun. 2016;7:13223.
603. Myasnikov AG, Kundhavai Natchiar S, et al. Structure-function insights reveal the human ribosome as a cancer target for antibiotics. Nat Commun. 2016;7:12856.
604. Di Giorgio A, Duca M. Synthetic small-molecule RNA ligands: future prospects as therapeutic agents. MedChemComm. 2019;10:1242–55.
605. Stevens M, De Clercq E, Balzarini J. The regulation of HIV-1 transcription: molecular targets for chemotherapeutic intervention. Med Res Rev. 2006;26:595–625.
606. Kulinski T, Olejniczak M, Huthoff H, et al. The apical loop of the HIV-1 TAR RNA hairpin is stabilized by a cross-loop base pair. J Biol Chem. 2003;278:38892–901.
607. Le Grice SF. Targeting the HIV RNA genome: high-hanging fruit only needs a longer ladder. Curr Top Microbiol Immunol. 2015;389:147–69.
608. Connelly CM, Moon MH, Schneekloth JS., Jr. The emerging role of RNA as a therapeutic target for small molecules. Cell Chem Biol. 2016; 23:1077–1090.
609. Stelzer AC, Frank AT, Kratz JD, et al. Discovery of selective bioactive small molecules by targeting an RNA dynamic ensemble. Nat Chem Biol. 2011;7:553–9.
610. Dibrov SM, Parsons J, Carnevali M, et al. Hepatitis C virus translation inhibitors targeting the internal ribosomal entry site. J Med Chem. 2014;57:1694–707.
611. Plant EP, Pérez-Alvarado GC, Jacobs JL, et al. A three-stemmed mRNA pseudoknot in the SARS coronavirus frameshift signal. PLoS Biol. 2005;3:e172.
612. Su MC, Chang CT, Chu CH, et al. An atypical RNA pseudoknot stimulator and an upstream attenuation signal for -1 ribosomal frameshifting of SARS coronavirus. Nucleic Acids Res. 2005;33:4265–75.
613. Park SJ, Kim YG, Park HJ. Identification of RNA pseudoknot-binding ligand that inhibits the -1 ribosomal frameshifting of SARS-coronavirus by structure-based virtual screening. J Am Chem Soc. 2011;133:10094–100.
614. Ritchie DB, Soong J, Sikkema WK, et al. Anti-frameshifting ligand reduces the conformational plasticity of the SARS virus pseudoknot. J Am Chem Soc. 2005;136:2196–9.
615. Tucker BJ, Breaker RR. Riboswitches as versatile gene control elements. Curr Opin Struct Biol. 2005;15:342–8.
616. Hallberg ZF, Su Y, Kitto RZ, et al. Engineering and in vivo applications of riboswitches. Annu Rev Biochem. 2017;86:515–39.
617. Chauvier A, Picard-Jean F, Berger-Dancause JC, et al. Transcriptional pausing at the translation start site operates as a critical checkpoint for riboswitch regulation. Nat Commun. 2017;8:13892.
618. Lee ER, Blount KF, Breaker RR. Roseoflavin is a natural antibacterial compound that binds to FMN riboswitches and regulates gene expression. RNA Biol. 2009;6:187–94.
619. Winkler WC, Cohen-Chalamish S, Breaker RR. An mRNA structure that controls gene expression by binding FMN. Proc Natl Acad Sci USA. 2002;99:15908–13.
620. Serganov A, Nudler E. A decade of riboswitches. Cell. 2013;152:17–24.
621. Zhang R, Ou HY, Zhang CT. DEG: a database of essential genes. Nucleic Acids Res. 2004;32:D271–2.
622. Serganov A, Huang L, Patel DJ. Coenzyme recognition and gene regulation by a flavin mononucleotide riboswitch. Nature. 2009;458:233–7.
623. Howe JA, Wang H, Fischmann TO, et al. Selective small-molecule inhibition of an RNA structural element. Nature. 2015;526:672–7.
624. Ottesen EW. ISS-N1 makes the first FDA-approved drug for spinal muscular atrophy. Transl Neurosci. 2017;8:1–6.
625. Aartsma-Rus A, Krieg AM. FDA approves eteplirsen for Duchenne muscular dystrophy: the next chapter in the eteplirsen saga. Nucleic Acid Ther. 2017;27:1–3.
626. Naryshkin NA, Weetall M, Dakka A, et al. Motor neuron disease. SMN2 splicing modifiers improve motor function and longevity in mice with spinal muscular atrophy. Science. 2014;345:688–93.
627. Ratni H, Ebeling M, Baird J, et al. Discovery of risdiplam, a selective survival of motor neuron-2 (SMN2) gene splicing modifier for the treatment of spinal muscular atrophy (SMA). J Med Chem. 2018;61:6501–17.
628. Sturm S, Günther A, Jaber B, et al. A phase 1 healthy male volunteer single escalating dose study of the pharmacokinetics and pharmacodynamics of risdiplam (RG7916, RO7034067), a SMN2 splicing modifier. Br J Clin Pharmacol. 2019;85:181–93.

629. Palacino J, Swalley SE, Song C, et al. SMN2 splice modulators enhance U1-pre-mRNA association and rescue SMA mice. Nat Chem Biol. 2015;11:511–7.
630. Cheung AK, Hurley B, Kerrigan R, et al. Discovery of small molecule splicing modulators of survival motor neuron-2 (SMN2) for the treatment of spinal muscular atrophy (SMA). J Med Chem. 2018;61:11021–36.
631. Velagapudi SP, Gallo SM, Disney MD. Sequence-based design of bioactive small molecules that target precursor microRNAs. Nat Chem Biol. 2014;10:291–7.
632. Velagapudi SP, Cameron MD, Haga CL, et al. Design of a small molecule against an oncogenic noncoding RNA. Proc Natl Acad Sci USA. 2016;113:5898–903.
633. Li Y, Disney MD. Precise small molecule degradation of a noncoding RNA identifies cellular binding sites and modulates an oncogenic phenotype. ACS Chem Biol. 2018;13:3065–71.
634. Costales MG, Hoch DG, Abegg D, Childs-Disney JL, Velagapudi SP, Adibekian A, Disney MD. (2019a) A designed small molecule inhibitor of a non-coding RNA sensitizes HER2 negative cancers to herceptin. J Am Chem Soc. 2019;141:2960–74.
635. Costales MG, Haga CL, Velagapudi SP, et al. Small molecule inhibition of microRNA-210 reprograms an oncogenic hypoxic circuit. J Am Chem Soc. 2017;139:3446–55.
636. Costales MG, Suresh B, Vishnu K, et al. Targeted degradation of a hypoxia-associated non-coding RNA enhances the selectivity of a small molecule interacting with RNA. Cell Chem Biol. 2019;26:1180–6.
637. Velagapudi SP, Costales MG, Vummidi BR, et al. Approved anti-cancer drugs target oncogenic noncoding RNAs. Cell Chem Biol. 2018;25:1086–94. e7.
638. Brown JA, Bulkley D, Wang J, et al. Structural insights into the stabilization of MALAT1 noncoding RNA by a bipartite triple helix. Nat Struct Mol Biol. 2014;21(7):633–40.
639. Pedram Fatemi R, Salah-Uddin S, Modarresi F, et al. Screening for small-molecule modulators of long noncoding RNA-protein interactions using AlphaScreen. J Biomol Screen. 2015;20(9):1132–41.
640. Fatemi RP, Velmeshev D, Faghihi MA. De-repressing LncRNA-targeted genes to upregulate gene expression: Focus on small molecule therapeutics. Mol Ther Nucleic Acids. 2014;3:e196.
641. Bennett CF, Swayze EE. RNA targeting therapeutics: Molecular mechanisms of antisense oligonucleotides as a therapeutic platform. Annu Rev Pharmacol Toxicol. 2010;50(1):259–93.
642. Simone R, Balendra R, Moens TG, et al. G-quadruplex-binding small molecules ameliorate C9orf72 FTD/ALS pathology in vitro and in vivo. EMBO Mol Med. 2018;10:22–31.
643. Su Z, Zhang Y, Gendron TF, et al. Discovery of a biomarker and lead small molecules to target r(GGGGCC)-associated defects in c9FTD/ALS. Neuron. 2014;83:1043–50.
644. Disney MD, Liu B, Yang WY, et al. A small molecule that targets r(CGG)(exp) and improves defects in fragile X-associated tremor ataxia syndrome. ACS Chem Biol. 2012;7:1711–8.
645. Qurashi A, Liu H, Ray L, et al. Chemical screen reveals small molecules suppressing fragile X premutation rCGG repeat-mediated neurodegeneration in Drosophila. Hum Mol Genet. 2012;21:2068–75.
646. Green KM, Sheth UJ, Flores BN, et al. High-throughput screening yields several small-molecule inhibitors of repeat-associated non-AUG translation. J Biol Chem. 2019;294:18624–38.
647. Parkesh R, Childs-Disney JL, Nakamori M, et al. Design of a bioactive small molecule that targets the myotonic dystrophy type 1 RNA via an RNA motif-ligand database and chemical similarity searching. J Am Chem Soc. 2012;134:4731–42.
648. Angelbello AJ, Rzuczek SG, Mckee KK, et al. Precise small-molecule cleavage of an r(CUG) repeat expansion in a myotonic dystrophy mouse model. Proc Natl Acad Sci USA. 2019;116:7799–804.
649. Rzuczek SG, Colgan LA, Nakai Y, et al. Precise small-molecule recognition of a toxic CUG RNA repeat expansion. Nat Chem Biol. 2017;13:188–93.
650. Miglietta G, Cogoi S, Marinello J, et al. RNA G-quadruplexes in Kirsten Ras (KRAS) oncogene as targets for small molecules inhibiting translation. J Med Chem. 2017;60:9448–61.
651. Katsuda Y, Sato S, Asano L, et al. A small molecule that represses translation of G-quadruplex-containing mRNA. J Am Chem Soc. 2016;138:9037–40.
652. Zhang P, Park HJ, Zhang J, et al. Translation of the intrinsically disordered protein α-synuclein is inhibited by a small molecule targeting its structured mRNA. Proc Natl Acad Sci USA. 2020;117:1457–67.
653. Liang X-H. Translation efficiency of mRNAs is increased by antisense oligonucleotides targeting upstream open reading frames. Nat Biotechnol. 2016;34:875–80.
654. Liang X-H. Antisense oligonucleotides targeting translation inhibitory elements in 5′ UTRs can selectively increase protein levels. Nucleic Acids Res. 2017;45:9528–46.
655. Fellmann C, Gowen BG, Lin P-C, et al. Cornerstones of CRISPR-Cas in drug discovery and therapy. Nat Rev Drug Discov. 2017;16:89–100.
656. Childs-Disney JL, Disney MD. Approaches to validate and manipulate RNA targets with small molecules in cells. Annu Rev Pharmacol Toxicol. 2016;56:123–40.
657. Wakelin LP. Polyfunctional DNA intercalating agents. Med Res Rev. 1986;6:275–340.
658. Costales MG, Childs-Disney JL, Haniff HS, et al. How we think about targeting RNA with small molecules. J Med Chem. 2020;63(17):8880–900.
659. Chaires JB. Drug-DNA interactions. Curr Opin Struc Biol. 1998;8:314–20.

660. Graves DE. Targeting DNA through-covalent interactions of reversible binding drugs. Methods Enzymol. 2001;340:377–95.
661. Rehman SU, Sarwar T, Husain MA, et al. Studying non-covalent drug-DNA interactions. Arch Biochem Biophys. 2015;576:49–60.
662. Strekowski L, Wilson B. Noncovalent interactions with DNA: an overview. Mutat Res. 2007;623(1-2):3–13.
663. Turner PR, Denny WA. The genome as a drug target: sequence specific minor groove binding ligands. Curr Drug Targ. 2000;1:1–14.
664. Geierstanger BH, Wemmer DE. Complexes of the minor groove of DNA. Annu Rev Biophys Biomol Struct. 1995;24:463–93.
665. Reddy BS, Sondhi SM, Lown JW. Synthetic DNA minor groove-binding drugs. Pharmacol Ther. 1999;84:1–111.
666. Dervan PB, Edelson BS. Recognition of the DNA minor groove by pyrrole-imidazole polyamides. Curr Opin Struct Biol. 2003;13:284–99.
667. Neidle S. Structural aspects of drug-DNA complexes: molecular modelling of intercalative interactions. Drugs Exp Clin Res. 1986;12(6-7):455–62.
668. Trotta E, D'Ambrosio E, Ravagnan G, et al. Evidence for DAPI intercalation in CG sites of DNA oligomer [d(CGACGTCG)]2: a 1H NMR study. Nucleic Acids Res. 1995;23(8):1333–40.
669. Dziegielewski J, Slusarski B, Konitz A, et al. Intercalation of imidazoacridinones to DNA and its relevance to cytotoxic and antitumor activity. Biochem Pharmacol. 2002;63(9):1653–62.
670. Misra VK, Honig B. On the magnitude of the electrostatic contribution to ligand-DNA interactions. Proc Nat Acad Sci USA. 1995;92:4691–5.
671. Manning GS. The molecular theory of polyelectrolyte solutions with applications to the electrostatic properties of polynucleotides. Q Rev Biophys. 1978;11:179–246.
672. Dogra S, Awasthi P, Tripathi S, et al. NMR-based structure of anticancer drug mitoxantrone stacked with terminal base pair of DNA hexamer sequence d-(ATCGAT)2. J Biomol Struct Dyn. 2014;32(7):1164–83.
673. Egger G, Liang G, Aparicio A, et al. Epigenetics in human disease and prospects for epigenetic therapy. Nature. 2004;429:457–63.
674. Holliday R. Epigenetics: A historical overview. Epigenetics. 2006;1(2):76–80.
675. Lorch Y, Maier-Davis B, Kornberg RD. Mechanism of chromatin remodeling. PNAS. 2010;107(8):3458–62.
676. Lorch Y, Kornberg RD. Chromatin-remodeling for transcription. Q Rev Biophys. 2017;50:e5.
677. Kaur J, Daoud A, Eblen ST. Targeting chromatin remodeling for cancer therapy. Curr Mol Pharmacol. 2019;12(3):215–29.
678. Lyko F. The DNA methyltransferase family: a versatile toolkit for epigenetic regulation. Nat Rev Genet. 2018;19(2):81–92.
679. Wu X, Zhang Y. TET-mediated active DNA demethylation: mechanism, function and beyond. Nat Rev Genet. 2017;18(9):517–34.
680. Mirfattah B, Herring J, Tang H, et al. Probes and targets of DNA methylation and demethylation in drug development. Curr Top Med Chem. 2017;17(15):1727–40.
681. Kohli RM, Zhang Y. TET enzymes, TDG and the dynamics of DNA demethylation. Nature. 2013;502(7472):472–9.
682. Pastor WA, Aravind L, Rao A. TETonic shift: biological roles of TET proteins in DNA demethylation and transcription. Nat Rev Mol Cell Biol. 2013;14(6):341–56.
683. Lu X, Zhao BS, He C. TET family proteins: oxidation activity, interacting molecules, and functions in diseases. Chem Rev. 2015;115(6):2225–39.
684. Bannister AJ, Zegerman P, Partridge JF, et al. Selective recognition of methylated lysine 9 on histone H3 by the HP1 chromo domain. Nature. 2001;410(6824):120–4.
685. Strahl BD, Allis CD. The language of covalent histone modifications. Nature. 2000;403(6765):41–5.
686. Grunstein M. Histone acetylation in chromatin structure and transcription. Nature. 1997;389(6649):349–52.
687. Roth SY, Denu JM, Allis CD. Histone acetyltransferases. Annu Rev Biochem. 2001;70:81–120.
688. Schneider A, Chatterjee S, Bousiges O, et al. Acetyltransferases (HATs) as targets for neurological therapeutics. Neurotherapeutics. 2013;10(4):568–88.
689. Wang Y, Miao X, Liu Y, et al. Dysregulation of histone acetyltransferases and deacetylases in cardiovascular diseases. Oxid Med Cell Longev. 2014;2014:641979.
690. Falkenberg KJ, Johnstone RW. Histone deacetylases and their inhibitors in cancer, neurological diseases and immune disorders. Nat Rev Drug Discov. 2014;13(9):673–91.
691. Ropero S, Esteller M. The role of histone deacetylases (HDACs) in human cancer. Mol Oncol. 2007;1(1):19–25.
692. Benedetti R, Conte M, Altucci L. Targeting Histone deacetylases in diseases: Where are we? Antioxid Redox Signal. 2015;23(1):99–126.
693. Husmann D, Gozani O. Histone lysine methyltransferases in biology and disease. Nat Struct Mol Biol. 2019;26(10):880–9.
694. McCabe MT, Mohammad HP, Barbash O, et al. Targeting Histone methylation in cancer. Cancer J. 2017;23(5):292–301.
695. Yi X, Jiang X, Li X, et al. Histone lysine methylation and congenital heart disease: From bench to bedside. Int J Mol Med. 2017;40(4):953–64.
696. Hashimoto H, Vertino PM, Cheng X. Molecular coupling of DNA methylation and histone methylation. Epigenomics. 2010;2(5):657–6.

697. Du J, Johnson LM, Jacobsen SE, et al. DNA methylation pathways and their crosstalk with histone methylation. Nat Rev Mol Cell Biol. 2015;16(9):519–32.
698. Michalak EM, Burr ML, Bannister AJ, et al. The roles of DNA, RNA and histone methylation in ageing and cancer. Nat Rev Mol Cell Biol. 2019;20(10):573–89.
699. Richart L, Margueron R. Drugging histone methyltransferases in cancer. Curr Opin Chem Biol. 2020;56:51–62.
700. Liu Q, Wang MW. Histone lysine methyltransferases as anti-cancer targets for drug discovery. Acta Pharmacol Sin. 2016;37(10):1273–80.
701. Blanc RS, Richard S. Arginine methylation: The coming of age. Mol Cell. 2017;65(1):8–24.
702. Guccione E, Richard S. The regulation, functions and clinical relevance of arginine methylation. Nat Rev Mol Cell Biol. 2019;20(10):642–57.
703. Cha B, Jho EH. Protein arginine methyltransferases (PRMTs) as therapeutic targets. Expert Opin Ther Targets. 2012;16(7):651–64.
704. Zhang J, Jing L, Li M, et al. Regulation of histone arginine methylation/demethylation by methylase and demethylase. Mol Med Rep. 2019;19(5):3963–71.
705. Hashizume R, Andor N, Ihara Y, et al. Pharmacologic inhibition of histone demethylation as a therapy for pediatric brainstem glioma. Nat Med. 2014;20(12):1394–6.
706. Berry WL, Janknecht R. KDM4/JMJD2 histone demethylases: epigenetic regulators in cancer cells. Cancer Res. 2013;73(10):2936–42.
707. Maiques-Diaz A, Somervaille TC. LSD1: biologic roles and therapeutic targeting. Epigenomics. 2016;8(8):1103–16.
708. Harmeyer KM, Facompre ND, Herlyn M, et al. JARID1 histone demethylases: Emerging targets in cancer. Trends Cancer. 2017;3(10):713–25.
709. Niu Y, Zhao X, Wu YS, et al. N6-methyl-adenosine (m6A) in RNA: an old modification with a novel epigenetic function. Genomics Proteom Bioinform. 2013;11(1):8–17.
710. Frye M, Harada BT, Behm M, et al. RNA modifications modulate gene expression during development. Science. 2018;361(6409):1346–9.
711. Cantara WA, Crain PF, Rozenski J, et al. The RNA modification database, RNAMDB: 2011 update. Nucleic Acids Res. 2011;39:D195–201.
712. Czerwoniec A, Dunin-Horkawicz S, Purta E, et al. MODOMICS: a database of RNA modification pathways 2008 update. Nucleic Acids Res. 2009;37:D118–21.
713. Narayan P, Rottman FM. Methylation of mRNA. Adv Enzymol Relat Areas Mol Biol. 1992;65:255–85.
714. Adams JM, Cory S. Modified nucleosides and bizarre 5′-termini in mouse myeloma mRNA. Nature. 1975;255:28–33.
715. Desrosiers RC, Friderici KH, Rottman FM. Characterization of Novikoff hepatoma mRNA methylation and heterogeneity in the methylated 5′ terminus. Biochemistry. 1975;14:4367–74.
716. Canaani D, Kahana C, Lavi S, et al. Identification and mapping of N6-methyladenosine containing sequences in simian virus 40 RNA. Nucleic Acids Res. 1979;6:2879–99.
717. Chen-Kiang S, Nevins JR, Darnell JE Jr. N-6-methyl-adenosine in adenovirus type 2 nuclear RNA is conserved in the formation of messenger RNA. J Mol Biol. 1979;135:733–52.
718. Traube FR, Carell T. The chemistries and consequences of DNA and RNA methylation and demethylation. RNA Biol. 2017;14(9):1099–107.
719. Chen X, Sun YZ, Liu H, et al. RNA methylation and diseases: experimental results, databases, Web servers and computational models. Brief Bioinform. 2019;20(3):896–917.
720. Meyer KD, Saletore Y, Zumbo P, et al. Comprehensive analysis of mRNA methylation reveals enrichment in 3′ UTRs and near stop codons. Cell. 2012;149:1635–46.
721. Dominissini D, Moshitch-Moshkovitz S, Schwartz S, et al. Topology of the human and mouse m6A RNA methylomes revealed by m6A-seq. Nature. 2012;485:201–6.
722. Chen XY, Zhang J, Zhu JS. The role of m6A RNA methylation in human cancer. Mol Cancer. 2019;18(1):103.
723. Zhao W, Qi X, Liu L, et al. Epigenetic regulation of m6A modifications in human cancer. Mol Ther Nucleic Acids. 2020;19:405–12.
724. Qin Y, Li L, Luo E, et al. Role of m6A RNA methylation in cardiovascular disease. Int J Mol Med. 2020;46(6):1958–72.
725. Zeng C, Huang W, Li Y, et al. Roles of METTL3 in cancer: mechanisms and therapeutic targeting. J Hematol Oncol. 2020;13(1):117.
726. Ma S, Chen C, Ji X, et al. The interplay between m6A RNA methylation and noncoding RNA in cancer. J Hematol Oncol. 2019;12(1):121.
727. Yue Y, Liu J, Cui X, et al. VIRMA mediates preferential m6a mRNA methylation in 3′UTR and near stop codon and associates with alternative polyadenylation. Cell Discov. 2018;4:10.
728. Wen J, Lv R, Ma H, et al. Zc3h13 regulates nuclear RNA m6A methylation and mouse embryonic stem cell self-renewal. Mol Cell. 2018;69(6):1028–38.
729. Patil DP, Chen CK, Pickering BF, et al. m6A RNA methylation promotes XIST-mediated transcriptional repression. Nature. 2016;537(7620):369–73.
730. Liu J, Yue Y, Han D, et al. A METTL3-METTL14 complex mediates mammalian nuclear RNA N6-adenosine methylation. Nat Chem Biol. 2014;10(2):93–5.
731. Mendel M, Chen KM, Homolka D, et al. Methylation of structured RNA by the m6A writer METTL16 is essential for mouse embryonic development. Mol Cell. 2018;71(6):986–1000.

732. Warda AS, Kretschmer J, Hackert P, et al. Human METTL16 is a N6-methyladenosine (m6A) methyltransferase that targets pre-mRNAs and various noncoding RNAs. EMBO Rep. 2017;18(11):2004–14.
733. Pendleton KE, Chen B, Liu K, et al. The U6 snRNA m6A methyltransferase METTL16 regulates SAM synthetase intron retention. Cell. 2017;169(5):824–35.
734. Ignatova VV, Stolz P, Kaiser S, et al. The rRNA m6A methyltransferase METTL5 is involved in pluripotency and developmental programs. Genes Dev. 2020;34(9-10):715–29.
735. van Tran N, Ernst FGM, Hawley BR, et al. The human 18S rRNA m6A methyltransferase METTL5 is stabilized by TRMT112. Nucleic Acids Res. 2019;47(15):7719–33.
736. Pinto R, Vagbo CB, Jakobsson ME, et al. The human methyltransferase ZCCHC4 catalyses N6-methyladenosine modification of 28S ribosomal RNA. Nucleic Acids Res. 2020;48(2):830–46.
737. Ren W, Lu J, Huang M, et al. Structure and regulation of ZCCHC4 in m6A-methylation of 28S rRNA. Nat Commun. 2019;10(1):5042.
738. Ma H, Wang X, Cai J, et al. N6-Methyladenosine methyltransferase ZCCHC4 mediates ribosomal RNA methylation. Nat Chem Biol. 2019;15(1):88–94.
739. Blanquart C, Linot C, Cartron PF, et al. Epigenetic metalloenzymes. Curr Med Chem. 2019;26(15):2748–85.
740. Zwergel C, Valente S, Jacob C, et al. Emerging approaches for histone deacetylase inhibitor drug discovery. Expert Opin Drug Discov. 2015;10(6):599–613.
741. Faria Freitas M, Cuendet M, Bertrand P. HDAC inhibitors: A 2013–2017 patent survey. Expert Opin Ther Pat. 2018;1–17
742. Yang XF, Zhao ZJ, Liu JJ, et al. SAHA and/or MG132 reverse the aggressive phenotypes of glioma cells: An in vitro and vivo study. Oncotarget. 2017;8(2):3156–69.
743. Glaser KB. HDAC inhibitors: clinical update and mechanism-based potential. Biochem Pharmacol. 2007;74(5):659–71.
744. Munster PN, Thurn KT, Thomas S, et al. A phase II study of the histone deacetylase inhibitor vorinostat combined with tamoxifen for the treatment of patients with hormone therapy-resistant breast cancer. Br J Cancer. 2011;104(12):1828–35.
745. Gore SD. Combination therapy with DNA methyltransferase inhibitors in hematologic malignancies. Nat Clin Pract Oncol. 2005; 2 Suppl1: S30–S35.
746. Berry DH, Fernbach DJ, Herson J, et al. Comparison of prednisolone, vincristine, methotrexate and 6-mercaptopurine vs. 6-mercaptopurine and prednisone maintenance therapy in childhood acute leukemia: A Southwest Oncology Group Study. Cancer. 1980;46(5):1098–103.
747. Hollenbach PW, Nguyen AN, Brady H, et al. A comparison of azacitidine and decitabine activities in acute myeloid leukemia cell lines. PLoS One. 2010;5(2):e9001.
748. Han T, Zhuo M, Hu H, et al. Synergistic effects of the combination of 5-Aza-CdR and suberoylanilide hydroxamic acid on the anticancer property of pancreatic cancer. Oncol Rep. 2018;39(1):264–70.
749. Meeran SM, Patel SN, Tollefsbol TO. Sulforaphane causes epigenetic repression of hTERT expression in human breast cancer cell lines. PLoS One. 2010;5(7):e11457.
750. Paul B, Li Y, Tollefsbol TO. The Effects of combinatorial genistein and sulforaphane in breast tumor inhibition: Role in epigenetic regulation. Int J Mol Sci. 2018;19(6):1754.
751. Easwaran H, Tsai HC, Baylin SB. Cancer epigenetics: tumor heterogeneity, plasticity of stem-like states, and drug resistance. Mol Cell. 2014;54(5):716–27.
752. Miller KM, Tjeertes JV, Coates J, et al. Human HDAC1 and HDAC2 function in the DNA-damage response to promote DNA nonhomologous end-joining. Nat Struct Mol Biol. 2010;17(9):1144–51.
753. Lee JH, Choy ML, Ngo L, et al. Histone deacetylase inhibitor induces DNA damage, which normal but not transformed cells can repair. Proc Natl Acad Sci U S A. 2010;107(33):14639–44.
754. Schiller JH, Harrington D, Belani CP, et al. Comparison of four chemotherapy regimens for advanced non-small-cell lung cancer. N Engl J Med. 2002;346(2):92–8.
755. Group NM-AC. Chemotherapy in addition to supportive care improves survival in advanced non-small-cell lung cancer: a systematic review and meta-analysis of individual patient data from 16 randomized controlled trials. J Clin Oncol. 2008;26(28):4617–25.
756. D'Addario G, Pintilie M, Leighl NB, et al. Platinum-based versus non-platinum-based chemotherapy in advanced non-small-cell lung cancer: a meta-analysis of the published literature. J Clin Oncol. 2005;23(13):2926–36.
757. Mroz EA, Rocco JW. The challenges of tumor genetic diversity. Cancer. 2017;123(6):917–27.
758. Dumbrava EI, Meric-Bernstam F. Personalized cancer therapy-leveraging a knowledge base for clinical decision-making. Cold Spring Harb Mol Case Stud. 2018;4(2):a001578.
759. Morphy R, Rankovic Z. Designed multiple ligands. An emerging drug discovery paradigm. J Med Chem. 2005;48(21):6523–43.
760. Chou TC. Theoretical basis, experimental design, and computerized simulation of synergism and antagonism in drug combination studies. Pharmacol Rev. 2006;58(3):621–81.
761. Wang DF, Helquist P, Wiech NL, et al. Toward selective histone deacetylase inhibitor design: homology modeling, docking studies, and molecular dynamics simulations of human class I histone deacetylases. J Med Chem. 2005;48(22):6936–47.

762. Butler KV, Kozikowski AP. Chemical origins of isoform selectivity in histone deacetylase inhibitors. Curr Pharm Des. 2008;14(6):505–28.
763. Rotili D, Tomassi S, Conte M, et al. Pan-histone demethylase inhibitors simultaneously targeting Jumonji C and lysine-specific demethylases display high anticancer activities. J Med Chem. 2014;57(1):42–55.
764. Dokmanovic M, Clarke C, Marks PA. Histone deacetylase inhibitors: overview and perspectives. Mol Cancer Res. 2007;5(10):981–9.
765. Inoue S, Riley J, Gant TW, et al. Apoptosis induced by histone deacetylase inhibitors in leukemic cells is mediated by Bim and Noxa. Leukemia. 2007;21(8):1773–82.
766. Zang L, Kondengaden SM, Zhang Q, et al. Structure based design, synthesis and activity studies of small hybrid molecules as HDAC and G9a dual inhibitors. Oncotarget. 2017;8(38):63187–207.
767. Kargbo RB. Histone deacetylase inhibitors as treatment for targeting multiple components in cancer therapy. ACS Med Chem Lett. 2018;9(3):167–8.
768. Wishart DS, Feunang YD, Guo AC, et al. DrugBank 5.0: A major update to the DrugBank database for 2018. Nucleic Acids Res. 2018;46:D1074–82.
769. Li YH, Yu CY, Li XX, et al. Therapeutic target database update 2018: Enriched resource for facilitating bench-to-clinic research of targeted therapeutics. Nucleic Acids Res. 2018;46:D1121–7.
770. Gaulton A, Hersey A, Nowotka M, et al. The ChEMBL database in 2017. Nucleic Acids Res. 2017;45:D945–54.
771. Kim S, Thiessen PA, Bolton EE, et al. PubChem substance and compound databases. Nucleic Acids Res. 2016;44:D1202–13.
772. Gilson MK, Liu T, Baitaluk M, et al. BindingDB in 2015: A public database for medicinal chemistry, computational chemistry and systems pharmacology. Nucleic Acids Res. 2016;44:D1045–53.
773. Günther S, Kuhn M, Dunkel M, et al. SuperTarget and matador: Resources for exploring drug-target relationships. Nucleic Acids Res. 2008;36:D919–22.
774. Yıldırım MA, Goh K-I, Cusick ME, et al. Drug—target network. Nat Biotechnol. 2007;25:1119–26.
775. Cheng F, Liu C, Jiang J, et al. Prediction of drug-target interactions and drug repositioning via network-based inference. PLoS Comput Biol. 2012;8:e1002503.
776. Lu J-J, Pan W, Hu Y-J, et al. Multi-target drugs: The trend of drug research and development. PLoS One. 2012;7:e40262.
777. Santos R, Ursu O, Gaulton A, et al. A comprehensive map of molecular drug targets. Nat Rev Drug Discovery. 2016;16:19–34.
778. Hert KMJ, Irwin JJ, et al. Quantifying the Relationships among Drug Classes. J Chem Inf Model. 2008;48:755–65.
779. Shi D, Khan F, Abagyan R. Extended multitarget pharmacology of anticancer drugs. J Chem Inf Model. 2019;59(6):3006–17.
780. He S, Dong G, Wu S, et al. Small molecules simultaneously inhibiting p53-murine double minute 2 (MDM2) interaction and histone deacetylases (HDACs): Discovery of novel multitargeting antitumor agents. J Med Chem. 2018;61(16):7245–60.
781. Evans BE, Rittle KE, Bock MG, et al. Methods for drug discovery: development of potent, selective, orally effective cholecystokinin antagonists. J Med Chem. 1988;31(12):2235–46.
782. Zhao H, Dietrich J. Privileged scaffolds in lead generation. Expert Opin Drug Discov. 2015;10(7):781–90.
783. Silva CFM, Pinto DCGA, Silva AMS. Chromones: privileged scaffolds for the production of multi-target-directed-ligand agents for the treatment of Alzheimer's disease. Expert Opin Drug Discov. 2018;13(12):1141–51.
784. Faheem, Karan Kumar B, Venkata Gowri Chandra Sekhar K, et al. 1,2,3,4-Tetrahydroisoquinoline (THIQ) as privileged scaffold for anticancer de novo drug design Expert Opin Drug Discov. 2021: 16(10): 1119–1147.
785. Liargkova T, Eleftheriadis N, Dekker F, et al. Small multitarget molecules incorporating the enone moiety. Molecules. 2019;24(1):199.
786. Gupta D, Jaina DK, Trivedi P. Recent advances in chalcones as antiinfective agents. Int J Chem Sci. 2010;8:649–54.
787. Sinha S, Medhi B, Sehga R. Chalcones as an emerging lead molecule for antimalarial therapy: A review. J Mod Med Chem. 2013;1:64–77.
788. Yadav VR, Prasad S, Sung B, et al. The role of chalcones in suppression of NF-κB-mediated inflammation and cancer. Int Immunopharmacol. 2011;11:295–309.
789. Nowakowska Z. A review of anti-infective and anti-inflammatory chalcones. Eur J Med Chem. 2007;42:125–37.
790. Katsori AM, Hadjipavlou-Litina D. Recent progress in therapeutic applications of chalcones. Expert Opin Ther Pat. 2011;21:1575–96.
791. Rahman MA. Chalcone: A valuable insight into the recent advances and potential pharmacological activities. Chem Sci J. 2011;29:1–16.
792. Di Carlo G, Mascolo N, Izzo AA, et al. Flavonoids: Old and new aspects of a class of natural therapeutic drugs. Life Sci. 1999;65:337–53.
793. Nasir S, Bukhari A, Jasamai M, et al. Synthesis and biological evaluation of chalcone derivatives (mini review). Med Chem. 2012;12:1394–403.
794. Eddarir S, Cotelle N, Bakkour Y, et al. An efficient synthesis of chalcones based on the Suzuki reaction. Tetrahedron Lett. 2003;44:5359–63.
795. Dhar DN. Chemistry of chalcones and related compounds. John Wiley & Sons, Inc.; New York, NY, USA: 1981; pp213.
796. Guida A, Lhouty MH, Tichit D, et al. Hydrotalcites as base catalysts. Kinetics of Claisen-Schmidt condensation, intramolecular condensation of acetonylacetone and synthesis of chalcone. Appl Catal A. 1997;164:251–64.

797. Romanelli G, Pasquale G, Sathicq A, et al. Synthesis of chalcones catalyzed by aminopropylated silica sol–gel under solvent-free conditions. J Mol Catal A Chem. 2011;340:24–32.
798. Kamboj S, Singh R. Chromanone-A prerogative therapeutic scaffold: An overview. Arab J Sci Eng. 2021: 1–37.
799. Emami S, Ghanbarimasir Z. Recent advances of chroman-4-one derivatives: synthetic approaches and bioactivities. Eur J Med Chem. 2015;93:539–63.
800. Panche AN, Diwan AD, Chandra R. Flavonoids: an overview. J Nutr Sci. 2016;5(47):1–15.
801. Christophe Carola H, Ralf Rosskopf M. Use of chroman-4-one derivatives. United States Patent application publication. US 2010/0028278 A1,1, 1–24 (2010).
802. Cotelle N. Role of flavonoids in oxidative stress. Curr Top Med Chem. 2001;I(2001):569–90.
803. Park JH, Lee SU, Kim SH, et al. Chromone and chromanone derivatives as strand transfer inhibitors of HIV-1 integrase. Arch Pharm Res. 2008;31(1):1–5.
804. Jalili-Baleh L, Babaei E, Abdpour S, et al. A review on flavonoid-based scaffolds as multi-target-directed ligands (MTDLs) for Alzheimer's disease. Eur J Med Chem. 2018;152:570–89.
805. Stefanachi A, Leonetti F, Pisani L. Coumarin: A natural, privileged and versatile scaffold for bioactive compounds. Molecules. 2018;23(2):250.
806. Vogel A. Darstellung von Benzoesaure aus der Tonka-Boline und aus den Meliloten-oder Steinklee-Blumen. Ann Phys. 1820;64:161–6.
807. Dean FM. Naturally occurring coumarins. Fortschr Chem Org Naturst IX. 1952;9:225–91.
808. Murray RDH, Mendez J, Brown SA. The Natural coumarins: Occurrence, chemistry and biochemistry. New York, NY, USA: John Wiley & Sons; 1982.
809. Murray RDH. Naturally occurring plant coumarins. Fortschr Chem Org Naturst. 1991;58:84–322.
810. O'Kennedy R. Thornes RD (eds). Biology, applications, and mode of action. John Wiley & Sons, New York, NY, USA: Coumarins; 1997.
811. Mzezewa SC, Omoruyi SI, Zondagh LS, et al. Design, synthesis, and evaluation of 3,7-substituted coumarin derivatives as multifunctional Alzheimer's disease agents. J Enzyme Inhib Med Chem. 2021;36(1):1607–21.
812. Jameel E, Umar T, Kumar J, et al. Coumarin: A privileged scaffold for the design and development of antineurodegenerative agents. Chem Biol Drug Des. 2016;87(1):21–38.
813. Detsi A, Kontogiorgis C, Hadjipavlou-Litina D. Coumarin derivatives: An updated patent review (2015–2016). Expert Opin Ther Pat. 2017;27:1201–27.
814. Barot KP, Jain SV, Kremer L, et al. Recent advances and therapeutic journey of coumarins: Current status and perspectives. Med Chem Res. 2015;24:2771–98.
815. Uto Y. 1, 2-Benzisoxazole: A privileged structure with a potential for polypharmacology. Curr Pharm Des. 2016;22(21):3201–11.
816. Rakesh KP, Shantharam CS, Sridhara MB. Benzisoxazole: a privileged scaffold for medicinal chemistry. Medchemcomm. 2017;8(11):2023–39.
817. Kirk KL, Filler R. In Biomedical frontiers of fluorine chemistry, symposium series. American Chemical Society, Washington, DC. 1996; vol. 639: pp 1–24.
818. Gelders YG, Heylen SLE, Vander BG, et al. Pilot clinical investigation of risperidone in the treatment of psychotic patients. J Pharmacopsychiatry. 1990;23:206–11.
819. Dollery C. Therapeutic drugs. Edinburgh, UK: Churchill Livingstone; 1999.
820. Park BK, Kitteringham NR. Effects of fluorine substitution on drug metabolism: pharmacological and toxicological implications. Drug Metab Rev. 1994;26:605–43.
821. Deng BL, Cullen MD, Zhou Z, et al. Synthesis and anti-HIV activity of new alkenyldiarylmethane (ADAM) non-nucleoside reverse transcriptase inhibitors (NNRTIs) incorporating benzoxazolone and benzisoxazole rings. Bioorg Med Chem. 2006;14:2366–74.
822. Chaker A, Najahi E, Chatriant O, et al. New 3-subs tituted-2,1-benzisoxazoles: synthesis and antimicrobial activities. Arab J Chem. 2017;10:S2464–70.
823. Sivala MR, Chintha V, Potla KM, et al. In silico docking studies and synthesis of new phosphoramidate derivatives of 6-fluoro-3-(piperidin-4-yl) benzo[d]isoxazole as potential antimicrobial agents. J Recept Signal Transduct. 2020;40:1–7.
824. Shivaprasad CM, Jagadish S, Swaroop TR, et al. Synthesis of new benzisoxazole derivatives and their antimicrobial, antioxidant and anti-inflammatory activities. Eur J Chem. 2014;5:91–5.
825. Wierenga W, Evans BR, Zurenko GE. Benzisoxazolones: antimicrobial and antileukemic activity. J Med Chem. 1984;27:1212–5.
826. Hrib NJ, Jurcak JG, Burgher KL, et al. Benzisoxazole- and benzisothiazole-3-carboxamides as potential atypical antipsychotic agents. J Med Chem. 1994;37:2308–14.
827. Davis L, Effland RC, Klein JT, et al. 3-substituted-1,2-benzisoxazoles: Novel antipsychotic agents. Drug Des Disco. 1992;8:225–40.
828. Saunders JC, Williamson WRN. Potential antiinflammatory compounds. 2. Acidic antiinflammatory 1,2-benzisoxazoles. J Med Chem. 1979;22:1554–8.
829. Khedekar P, Bahekar R, Chopadec R, et al. Synthesis and anti-inflammatory activity of alkyl/arylidene-2-aminobenzothiazoles and 1-benzothi azol-2-yl-3-chloro-4-substituted-azetidin-2-ones. Arzneimittelforschung. 2011;53:640–7.
830. Anand M, Selvaraj V, Alagar M. Synthesis, characterization and evaluation of antioxidant and anticancer activities of novel benzisoxazole-substituted-allyl derivatives. Korean J Chem Eng. 2014;31:659–63.
831. Chandra S, Mahadimane V. Effect of novel benzisoxazole derivatives against ehrlich ascites carcinoma cells in swiss albino mice: cytotoxic

and haematalogical studies. Int J Pharma Sci Res. 2015;6:3606–11.
832. Jain M, Kwon CH. 1,2-Benzisoxazole phosphorodiamidates as novel anticancer prodrugs requiring bioreductive activation. J Med Chem. 2003;46:5428–36.
833. Ashwini N, Garg M, Mohan CD, et al. Synthesis of 1,2-benzisoxazole tethered 1,2,3-triazoles that exhibit anticancer activity in acute myeloid leukemia cell lines by inhibiting histone deacetylases, and inducing p21 and tubulin acetylation. Bioorg Med Chem. 2015;23:6157–65.
834. Aiello S, Wells G, Stone EL, et al. Synthesis and biological properties of benzothiazole, benzoxazole, and chromen-4-one analogues of the potent antitumor agent 2-(3,4-dimethoxyphenyl)-5- fluorobenzothiazole (PMX 610, NSC 721648). J Med Chem. 2008;51:5135–9.
835. Rakesh KP, Shantharam CS, Sridhara MB, et al. Benzisoxazole: a privileged scaffold for medicinal chemistry. Med Chem Commun. 2017;8:2023–39.
836. Purohit SS, VeerapurV P. Benzisoxazole containing thiazolidinediones as peroxisome proliferator activated receptor-γ agonists: design, molecular docking, synthesis & antidiabetic studies. Sch Acad J Pharm. 2014;3:26–37.
837. Sills GJ. Zonisamide. In: Enna SJ, Bylund DB, editors. xPharm: the comprehensive pharmacology reference. University of Liverpool, Liverpool, United Kingdom, Elsevier. 2009; p1–6.
838. Vasquez AR, Bobo WV. Mood stabilizers: risperidone for treating bipolar disorders in adults. In: Riederer P, Laux G, Mulsant B, Le W, Nagatsu T, editors. NeuroPsychopharmacotherapy. Cham: Springer; 2020; pp1–32.
839. Green B. Paliperidone: a clinical review. Curr Drug Ther. 2009;4:7–11.
840. Weiden P. Iloperidone for the treatment of schizophrenia: an updated clinical review. Clin Schizophr Relat Psychoses. 2012;6:34–44.
841. Uto Y. 1,2-Benzisoxazole compounds: a patent review (2009 - 2014). Expert Opin Ther Pat. 2015; 25(6): 643–662.
842. Deering RW, Whalen KE, Alvarez I, et al. Identification of a bacteria-produced benzisoxazole with antibiotic activity against multi-drug resistant Acinetobacter baumannii. J Antibiot (Tokyo). 2021;74(6):370–80.
843. Giacomini E, Rupiani S, Guidotti L, et al. The use of stilbene scaffold in medicinal chemistry and multi- target drug design. Curr Med Chem. 2016;23(23):2439–89.
844. De Filippis B, Ammazzalorso A, Fantacuzzi M, et al. Anticancer activity of stilbene-based derivatives. ChemMedChem. 2017;12(8):558–70.
845. Rauf A, Imran M, Butt MS, et al. Resveratrol as an anti-cancer agent: A review. Crit Rev Food Sci Nutr. 2018;58(9):1428–47.
846. Dyck GJB, Raj P, Zieroth S, et al. The effects of resveratrol in patients with cardiovascular disease and heart failure: A narrative review. Int J Mol Sci. 2019;20(4):904.
847. Bonnefont-Rousselot D. Resveratrol and cardiovascular diseases. Nutrients. 2016;8(5):250.
848. Komorowska J, Wątroba M, Szukiewicz D. Review of beneficial effects of resveratrol in neurodegenerative diseases such as Alzheimer's disease. Adv Med Sci. 2020;65(2):415–23.
849. De Filippis B, Ammazzalorso A, Amoroso R, et al. Stilbene derivatives as new perspective in antifungal medicinal chemistry. Drug Dev Res. 2019;80(3):285–93.
850. Chaudhary A, Pandeya SN, Kumar P, et al. Combretastatin a-4 analogs as anticancer agents. Mini Rev Med Chem. 2007;7(12):1186–205.
851. McCormack D, McFadden D. A review of pterostilbene antioxidant activity and disease modification. Oxid Med Cell Longev. 2013;2013:575482.
852. McCormack D, McFadden D. Pterostilbene and cancer: current review. J Surg Res. 2012;173(2):e53–61.
853. Kershaw J, Kim KH. The therapeutic potential of piceatannol, a Natural stilbene, in metabolic diseases: A review. J Med Food. 2017;20(5):427–38.
854. Banik K, Ranaware AM, Harsha C, et al. Piceatannol: A natural stilbene for the prevention and treatment of cancer. Pharmacol Res. 2020;153:104635.
855. Morphy R, Rankovic Z. Design of multitarget ligands. In: Rankovic Z, Morphy R, editors. Lead generation approaches in drug discovery. Hoboken: Wiley; 2010. p. 141–64.
856. Morphy R, Rankovic Z. Designing multiple ligands—medicinal chemistry strategies and challenges. Curr Pharm Des. 2009;15(6):587–600.
857. Bérubé G. An overview of molecular hybrids in drug discovery. Expert Opin Drug Discov. 2016;11(3):281–305.
858. Morphy R, Rankovic Z. The physicochemical challenges of designing multiple ligands. J Med Chem. 2006;49(16):4961–70.
859. Decker M. Design of hybrid molecules for drug development. Amsterdam: Elsevier; 2017.
860. Das N, Dhanawat M, Dash B, et al. Codrug: an efficient approach for drug optimization. Eur J Pharm Sci. 2010;41(5):571–88.
861. Vu CB, Bemis JE, Benson E, et al. Synthesis and characterization of fatty acid conjugates of niacin and salicylic acid. J Med Chem. 2016;5(3):1217–31.
862. Fornasari E, Di Stefano A, Cacciatore I. Direct- and spacer-coupled codrug strategies for the treatment of Alzheimer's disease. Austin Alzheimers J Parkinsons Dis. 2014;1(2):9.
863. Poudel YB, Chowdari NS, Cheng H, et al. Chemical modification of linkers provides stable linker-payloads for the generation of antibody-drug conjugates. ACS Med Chem Lett. 2020;11(11):2190–4.
864. Talevi A. Multi-target pharmacology: possibilities and limitations of the "skeleton key approach" from a medicinal chemist perspective. Front Pharmacol. 2015;6:205.

865. Hopkins AL, Keserü GM, Leeson PD, et al. The role of ligand efficiency metrics in drug discovery. Nat Rev Drug Discov. 2014;13:105–21.
866. Meunier B. Hybrid molecules with a dual mode of action: dream or reality? Acc Chem Res. 2008;41(1):69–77.
867. Zimmermann GR, Lehar J, Keith CT. Multi-target therapeutics: when the whole is greater than the sum of the parts. Drug Discovery Today. 2007;12(1-2):34–42.
868. Kabir MT, Uddin MS, Mamun AA, et al. Combination drug therapy for the management of Alzheimer's disease. Int J Mol Sci. 2020;21(9):3272.
869. Cummings JL, Morstorf T, Zhong K. Alzheimer's disease drug-development pipeline: Few candidates, frequent failures. Alzheimers Res Ther. 2014;6:37.
870. Kabir MT, Abu Sufian M, Uddin MS, et al. NMDA receptor antagonists: Repositioning of memantine as multitargeting agent for Alzheimer's therapy. Curr Pharm Des. 2019;25:3506–18.
871. Uddin MS, Kabir MT, Tewari D, et al. Emerging signal regulating potential of small molecule biflavonoids to combat neuropathological insults of Alzheimer's disease. Sci Total Environ. 2020;700:1–11.
872. Kabir MT, Uddin MS, Begum MM, et al. Cholinesterase inhibitors for Alzheimer's disease: Multitargeting strategy based on anti-Alzheimer's drugs repositioning. Curr Pharm Des. 2019;25:3519–35.
873. Deardorff WJ, Grossberg GT. A fixed-dose combination of memantine extended-release and donepezil in the treatment of moderate-to-severe Alzheimer's disease. Drug Des Devel Ther. 2016;10:3267–79.
874. Agatonovic-Kustrin S, Kettle C, Morton DW. A molecular approach in drug development for Alzheimer's disease. Biomed Pharmacother. 2018;106:553–65.
875. Capurro V, Busquet P, Lopes JP, et al. Pharmacological characterization of memoquin, a multi-target compound for the treatment of Alzheimer's disease. PLoS ONE. 2013;8:e56870.
876. Parsons CG, Danysz W, Dekundy A, et al. Memantine and cholinesterase inhibitors: Complementary mechanisms in the treatment of Alzheimer's disease. Neurotox Res. 2013;24:358–69.
877. Patel L, Grossberg GT. Combination therapy for Alzheimer's disease. Drugs Aging. 2011;28:539–46.
878. Posadas I, Lopez-Hernandez B, Cena V. Nicotinic receptors in neurodegeneration. Curr Neuropharmacol. 2013;11:298–314.
879. Jun Kawamata J, Syuuichirou SS, Shun SS. α7 nicotinic acetylcholine receptor mediated neuroprotection in Parkinson's disease. Curr Drug Targets. 2012;13:623–30.
880. Takada-Takatori Y, Kume T, Sugimoto M, et al. Acetylcholinesterase inhibitors used in treatment of Alzheimer's disease prevent glutamate neurotoxicity via nicotinic acetylcholine receptors and phosphatidylinositol 3-kinase cascade. Neuropharmacology. 2006;51:474–86.
881. Lopes JP, Tarozzo G, Reggiani A, et al. Galantamine potentiates the neuroprotective effect of memantine against NMDA-induced excitotoxicity. Brain Behav. 2013;3:67–74.
882. Rosini M, Simoni E, Bartolini M, et al. Inhibition of acetylcholinesterase, β-amyloid aggregation, and NMDA receptors in Alzheimer's disease: A promising direction for the multi-target-directed ligands gold rush. J Med Chem. 2008;51:4381–4.
883. Simoni E, Daniele S, Bottegoni G, et al. Combining galantamine and memantine in multitargeted, new chemical entities potentially useful in Alzheimer's disease. J Med Chem. 2012;55:9708–21.
884. Lipton SA. Paradigm shift in neuroprotection by NMDA receptor blockade: Memantine and beyond. Nat Rev Drug Discov. 2006;5:160–70.
885. Lipton SA, Choi YB, Pan ZH, et al. A redox-based mechanism for the neuroprotective and neurodestructive effects of nitric oxide and related nitrosocompounds. Nature. 1993;364:626–32.
886. Tsiailanis A, Tsoumani M, Stylos EK, et al. Designing natural product hybrids bearing triple antiplatelet profile and evaluating their human plasma stability. Methods Mol Biol. 2018;1824:371–85.
887. Alam MM, Hassan AHE, Kwon YH, et al. Design, synthesis and evaluation of alkylphosphocholine-gefitinib conjugates as multitarget anticancer agents. Arch Pharm Res. 2018;41(1):35–45.
888. Price AJ, Howard S, Cons BD. Fragment-based drug discovery and its application to challenging drug targets. Essays in Biochem. 2017;61(5):475–84.
889. Tounge BA, Parker MH. Designing a diverse high-quality library for crystallography-based FBDD screening. Fragment-based drug design—Tools, practical approaches, and examples. Methods in Enzymol. 2011;493:3–20.
890. Li Q. Application of fragment-based drug discovery to versatile targets. Front Mol Biosci. 2020;7:180.
891. Doak BC, Norton RS, Scanlon MJ. The ways and means of fragment-based drug design. Pharmacol Therapeut. 2016;167:28–37.
892. Murray CW, Rees DC. The rise of fragment-based drug discovery. Nat Chem. 2009;1:187–92.
893. Erlanson DA, Fesik SW, Hubbard RE, et al. Twenty years on: the impact of fragments on drug discovery. Nat Rev Drug Discov. 2016;15:605–19.
894. Bollag G, Hirth P, Tsai J, et al. Clinical efficacy of a RAF inhibitor needs broad target blockade in BRAF-mutant melanoma. Nature. 2010;467:596.
895. Bollag G, Tsai J, Zhang J, et al. Vemurafenib: the first drug approved for BRAF-mutant cancer. Nat Rev Drug Discovery. 2012;11:873.
896. Erlanson DA. Introduction to fragment-based drug discovery. In Fragment-Based Drug Discovery and X-Ray Crystallography. Davies TG, Hyvönen M (eds). Springer-Verlag, Berlin. 2012; pp1–32.

897. Jacquemard C, Kellenberger E. A bright future for fragment-based drug discovery: What does it hold? Exp Opin Drug Discov. 2019;14:413–6.
898. Whittaker M, Law RJ, Ichihara O, et al. Fragments: past, present and future. Drug Discov Today. 2010;7:e163–71.
899. Aretz J, Kondoh Y, Honda K, et al. Chemical fragment arrays for rapid druggability assessment. Chem Commun. 2016;52:9067–70.
900. Gee CT, Arntson KE, Urick AK, et al. Protein-observed 19F-NMR for fragment screening, affinity quantification and druggability assessment. Nat Protoc. 2016;11:1414.
901. Dang CV, Reddy EP, Shokat KM, et al. Drugging the 'undruggable' cancer targets. Nat Rev Cancer. 2017;17:502–8.
902. Owens J. Determining druggability. Nat Rev Drug Discov. 2007;6:187.
903. Cheng AC, Coleman RG, Smyth KT, et al. Structure-based maximal affinity model predicts small-molecule druggability. Nat Biotechnol. 2007;25:71–5.
904. Arkin MR, Tang Y, Wells JA. Small-molecule inhibitors of protein-protein interactions: progressing toward the reality. Chem Biol. 2014;21:1102–14.
905. Erlanson DA. Fragment-based lead discovery: a chemical update. Curr Opin Biotechnol. 2006;17:643–52.
906. Lamoree B, Hubbard RE. Current perspectives in fragment-based lead discovery (FBLD). Essays Biochem. 2017;61:453–64.
907. Brunst S, Kramer JS, Kilu W, et al. Systematic assessment of fragment identification for multitarget drug design. ChemMedChem. 2021;16(7):1088–92.
908. Morphy R, Rankovic Z. Fragments, network biology and designing multiple ligands. Drug Discovery Today. 2007;12:156–60.
909. Bottegoni G, Favia AD, Recanatini M, et al. The role of fragment-based and computational methods in polypharmacology. Drug Discovery Today. 2012;17:23–34.
910. Howard S, Berdini V, Boulstridge JA, et al. Fragment-based discovery of the pyrazol-4-yl urea (AT9283), a multitargeted kinase inhibitor with potent aurora kinase activity. J Med Chem. 2009;52(2):379–88.
911. Speck-Planche A, Kleandrova VV, Luan F, et al. Multi-target inhibitors for proteins associated with Alzheimer: in silico discovery using fragment-based descriptors. Curr Alzheimer Res. 2013;10(2):117–24.
912. Speck-Planche A, Cordeiro MNDS. Fragment-based in silico modeling of multi-target inhibitors against breast cancer-related proteins. Mol Divers. 2017;21(3):511–23.
913. Moroz E, Matoori S, Leroux JC. Oral delivery of macromolecular drugs: Where we are after almost 100 years of attempts. Adv Drug Deliv Rev. 2016;101:108–21.
914. Yang J, Kopeček J. Macromolecular therapeutics. J Control Release. 2014; 0: 288–303.
915. Rütter M, Milošević N, David A. Say no to drugs: Bioactive macromolecular therapeutics without conventional drugs. J Control Release. 2021;330:1191–207.
916. Ramsay RR, Popovic-Nikolic MR, Nikolic K, et al. A perspective on multi-target drug discovery and design for complex diseases. Clin Transl Med. 2018;7:3.
917. Dimitrov DS. Therapeutic proteins. Methods Mol Biol. 2012;899:1–26.
918. Raju TS, Strohl WR. Potential therapeutic roles for antibody mixtures. Expert Opin Biol Ther. 2013;13(10):1347–52.
919. Jakob CG, Edalji R, Judge RA, et al. Structure reveals function of the dual variable domain immunoglobulin (DVD-Ig™) molecule. MAbs. 2013;5:358–63.
920. Günther G, Dreger B. Post-marketing observational study on 5% intravenous immunoglobulin therapy (Alphaglobin®/Flebogamma®) in secondary immunodeficiency with recurrent serious bacterial infections. Microbiol Immunol. 2013;57(7):527–35.
921. Meng Q, Garcia-Rodriguez C, Manzanarez G, et al. Engineered domain-based assays to identify individual antibodies in oligoclonal combinations targeting the same protein. Anal Biochem. 2012;430:141–50.
922. Haurum JS. Recombinant polyclonal antibodies: the next generation of antibody therapeutics? Drug Discov Today. 2006;11:655–60.
923. Strohl WR, Strohl LM. Therapeutic antibody engineering: current and future advances driving the strongest growth area in the pharma industry. Woodhead Publishing Series in Biomedicine No. 11 Woodhead Publishing; Cambridge. 2012.
924. Skartved NJ, Jacobsen HJ, Pedersen MW, et al. Preclinical pharmacokinetics and safety of Sym004: a synergistic antibody mixture directed against epidermal growth factor receptor. Clin Cancer Res. 2011;17:5962–72.
925. Labrijn AF, Meesters JI, de Goeij BE, et al. Efficient generation of stable bispecific IgG1 by controlled Fab-arm exchange. Proc Natl Acad Sci USA. 2013;110:5145–50.
926. Labrijn AF, Janmaat ML, Reichert JM, et al. Bispecific antibodies: a mechanistic review of the pipeline. Nat Rev Drug Discov. 2019;18(8):585–608.
927. Shin SU. Chimeric antibody: potential applications for drug delivery and immunotherapy. Biotherapy. 1991;3(1):43–53.
928. Wright A, Shin SU, Morrison SL. Genetically engineered antibodies: progress and prospects. Crit Rev Immunol. 1992;12(3-4):125–68.
929. Panowski S, Bhakta S, Raab H, et al. Site-specific antibody drug conjugates for cancer therapy. MAbs. 2014;6(1):34–45.

930. Khongorzul P, Ling CJ, Khan FU, et al. Antibody-drug conjugates: A comprehensive review. Mol Cancer Res. 2020;18(1):3–19.
931. Hamilton GS. Antibody-drug conjugates for cancer therapy: The technological and regulatory challenges of developing drug-biologic hybrids. Biologicals. 2015;43(5):318–32.
932. Fitzpatrick-Dimond PF. Antibody-drug conjugates stage a comeback. GEN: Genetic Engineering and Biotechnology News. 2010.
933. DiJoseph JF, Armellino DC, Boghaert ER, et al. Antibody-targeted chemotherapy with CMC-544: a CD22-targeted immunoconjugate of calicheamicin for the treatment of B-lymphoid malignancies. Blood. 2004;103(5):1807–14.
934. Mullard A. Maturing antibody-drug conjugate pipeline hits 30. Nat Rev Drug Discov. 2013;12(5):329–32.
935. Chari RV, Martell BA, Gross JL, et al. Immunoconjugates containing novel maytansinoids: promising anticancer drugs. Can Res. 1992;52(1):127–31.
936. Beck A, Goetsch L, Dumontet C, et al. Strategies and challenges for the next generation of antibody-drug conjugates. Nat Rev Drug Discov. 2017;16(5):315–37.
937. Damelin M, Zhong W, Myers J, et al. Evolving strategies for target selection for antibody-drug conjugates. Pharm Res. 2015;32:3494–507.
938. Diamantis N, Banerji U. Antibody-drug conjugates–an emerging class of cancer treatment. Br J Can. 2016;114:362–7.
939. Tipton TR, Roghanian A, Oldham RJ, et al. Antigenic modulation limits the effector cell mechanisms employed by type I anti-CD20 monoclonal antibodies. Blood. 2015;125:1901–9.
940. Donaghy H. Effects of antibody, drug and linker on the preclinical and clinical toxicities of antibody-drug conjugates. mAbs. 2016;8:659–71.
941. Casi G, Neri D. Noninternalizing targeted cytotoxics for cancer therapy. Mol Pharm. 2015;12:1880–4.
942. Kovtun YV, Goldmacher VS. Cell killing by antibody-drug conjugates. Can Lett. 2007;255(2):232–40.
943. Bąchor R, Kluczyk A, Stefanowicz P, et al. New method of peptide cleavage based on Edman degradation. Mol Divers. 2013;17(3):605–11.
944. Axup JY, Bajjuri KM, Ritland M, et al. (October 2012). Synthesis of site-specific antibody-drug conjugates using unnatural amino acids. Proc Nat Acad Sci U S A. 109 (40): 16101–16106.
945. Lyon RP, Setter JR, Bovee TD, et al. Self-hydrolyzing maleimides improve the stability and pharmacological properties of antibody-drug conjugates. Nat Biotechnol. 2014;32(10):1059–62.
946. Kolodych S, Koniev O, Baatarkhuu Z, et al. CBTF: new amine-to-thiol coupling reagent for preparation of antibody conjugates with increased plasma stability. Bioconjugate Chem. 2015;26(2):197–200.
947. Wulbrand C, Seidl C, Gaertner FC, et al. Alpha-particle emitting 213Bi-anti-EGFR immunoconjugates eradicate tumor cells independent of oxygenation. PLOS ONE. 2013;8(5):e64730.
948. Cardoso MM, Peça IN, Roque AC. Antibody-conjugated nanoparticles for therapeutic applications. Curr Medicinal Chem. 2012;19(19):3103–27.
949. Dovgan I, Koniev O, Kolodych S, et al. Antibody-oligonucleotide conjugates as therapeutic, imaging, and detection agents. Bioconjugate Chem. 2019;30(10):2483–501.
950. Walsh SJ, Bargh JD, Dannheim FM, et al. Site-selective modification strategies in antibody-drug conjugates. Chem Soc Rev. 2021;50(2):1305–53.
951. Axup JY, Bajjuri KM, Ritland M, et al. Synthesis of site-specific antibody-drug conjugates using unnatural amino acids. Proc Natl Acad Sci U S A. 2012;109(40):16101–6.
952. Schumacher D, Hackenberger CP, Leonhardt H, et al. Current status: Site-specific antibody drug conjugates. J Clin Immunol. 2016;36(Suppl 1):100–7.
953. Sochaj AM, Świderska KW, Otlewski J. Current methods for the synthesis of homogeneous antibody-drug conjugates. Biotechnol Adv. 2015;33(6 Pt 1):775–84.
954. Hussain AF, Grimm A, Sheng W, et al. Toward homogenous antibody drug conjugates using enzyme-based conjugation approaches. Pharmaceuticals (Basel). 2021;14(4):343.
955. Chernikov IV, Vlassov VV, Chernolovskaya EL. Current development of siRNA bioconjugates: From research to the clinic. Front Pharmacol. 2019;10:444.
956. Xia CF, Boado RJ, Pardridge WM. Antibody-mediated targeting of siRNA via the human insulin receptor using avidin-biotin technology. Mol Pharm. 2009;6:747–51.
957. Ma Y, Kowolik CM, Swiderski PM, et al. Humanized Lewis-Y specific antibody-based delivery of STAT3 siRNA. ACS Chem Biol. 2011;6:962–70.
958. Sugo T, Terada M, Oikawa T, et al. Development of antibody-siRNA conjugate targeted to cardiac and skeletal muscles. J Control Release. 2016;237:1–13.
959. Ibtehaj N, Huda R. High-dose BAFF receptor specific mAb-siRNA conjugate generates Fas-expressing B cells in lymph nodes and high-affinity serum autoantibody in a myasthenia mouse model. Clin Immunol. 2017;176:122–30.
960. Cuellar TL, Barnes D, Nelson C, et al. Systematic evaluation of antibody-mediated siRNA delivery using an industrial platform of THIOMAB-siRNA conjugates. Nucleic Acids Res. 2014;43:1189–203.
961. Inumaru S. Introduction to advanced biologics. Vet Immunol Immunopathol. 2012;148(1-2):126–8.
962. Lacaná E, Amur S, Mummanneni P, et al. The emerging role of pharmacogenomics in biologics. Clin Pharmacol Ther. 2007;82(4):466–71.

963. O'Neill ID, Scully C. Biologics in oral medicine: principles of use and practical considerations. Oral Dis. 2012;18(6):525–36.
964. Greish K, Fang J, Inutsuka T, et al. Macromolecular therapeutics. Clin Pharmacokinet. 2003;42:1089–105.
965. Matsumura Y, Maeda H. A new concept for macromolecular therapeutics in cancer chemotherapy: mechanism of tumoritropic accumulation of proteins and the antitumor agent SMANCS. Cancer Res. 1986;46:6387–92.
966. Kopeček J. Polymer-drug conjugates: origins, progress to date and future directions. Adv Drug Deliv Rev. 2013;65:49–59.
967. Seymour LW, et al. Phase II studies of polymer-doxorubicin (PK1, FCE28068) in the treatment of breast, lung and colorectal cancer. Int J Oncol. 2009;34:1629–36.
968. Chipman SD, Oldham FB, Pezzoni G, Singer JW. Biological and clinical characterization of paclitaxel poliglumex (PPX, CT-2103), a macromolecular polymer-drug conjugate. Int J Nanomedicine. 2006;1:375–83.
969. Weiss GJ, et al. First-in-human phase 1/2a trial of CRLX101, a cyclodextrin-containing polymer-camptothecin nanopharmaceutical in patients with advanced solid tumor malignancies. Invest New Drugs. 2013;31:986–1000.
970. Yurkovetskiy AV, Fram RJ. XMT-1001, a novel polymeric camptothecin pro-drug in clinical development for patients with advanced cancer. Adv Drug Deliv Rev. 2009;61:1193–202.
971. Couvreur P, Vauthier C. Nanotechnology: Intelligent design to treat complex disease. Pharm Res. 2006;23:1417–50.
972. Vicent MJ, Ringsdorf H, Duncan R. Polymer therapeutics: clinical applications and challenges for development. Adv Drug Deliv Rev. 2009;61:1117–20.
973. Destouches D, Page N, Hamma-Kourbali Y, et al. A simple approach to cancer therapy afforded by multivalent pseudopeptides that target cell-surface nucleoproteins. Cancer Res. 2011;71:3296–305.
974. Zhang Z, et al. DNA-scaffolded multivalent ligands to modulate cell function. Chembiochem. 2014;15:1268–73.
975. Kiessling LL, Gestwicki JE, Strong LE. Synthetic multivalent ligands in the exploration of cell-surface interactions. Curr Opin Chem Biol. 2000;4:696–703.
976. Stephens B, Handel TM. Chemokine receptor oligomerization and allostery. Prog Mol Biol Transl Sci. 2013;115:375–420.
977. Ljubimova JY, Fujita M, Ljubimov AV, et al. Poly(malic acid) nanoconjugates containing various antibodies and oligonucleotides for multitargeting drug delivery. Nanomedicine (Lond). 2008;3(2):247–65.
978. Maeda H, Wu J, Sawa T, et al. Tumor vascular permeability and the EPR effect in macromolecular therapeutics: a review. J Control Release. 2000;65:271–84.
979. Li L, Yang J, Wang J, et al. Drug-free macromolecular therapeutics exhibit amplified apoptosis in G2/M phase arrested cells J Drug Target. 2019; 27(5-6): 566–72.
980. Chu TW, Kopeček J. drug-free macromolecular therapeutics--a new paradigm in polymeric nanomedicines. Biomater Sci. 2015; 3(7): 908–922.
981. Tavares MR, Pechar M, Chytil P, et al. Polymer-based drug-free therapeutics for anticancer, anti-inflammatory, and antibacterial treatment. Macromol Biosci. 2021;21(8):e2100135.
982. Wu K, Liu J, Johnson RN, et al. Drug-free macromolecular therapeutics: induction of apoptosis by coiled-coil-mediated cross-linking of antigens on the cell surface. Angew Chem Int Ed. 2010;49:1451–5.
983. Ho TT, Tran QT, Chai CL. The polypharmacology of natural products. Future Med Chem. 2018;10(11):1361–8.
984. Fang J, Liu C, Wang Q, et al. In silico polypharmacology of natural products. Brief Bioinform. 2018;19(6):1153–71.
985. Kibble M, Saarinen N, Tang J, et al. Network pharmacology applications to map the unexplored target space and therapeutic potential of natural products. Nat Prod Rep. 2015;32(8):1249–66.
986. Butler MS, Robertson AA, Cooper MA. Natural product and natural product derived drugs in clinical trials. Nat Prod Rep. 2014;31:1612–61.
987. Shen B. A new golden age of natural products drug discovery. Cell. 2015;163:1297–300.
988. Lee KW, Bode AM, Dong Z. Molecular targets of phytochemicals for cancer prevention. Nat Rev Cancer. 2011;11:211–8.
989. Fang J, Cai C, Wang Q, et al. Systems pharmacology-based discovery of natural products for precision oncology through targeting cancer mutated genes. CPT Pharmacometrics Syst Pharmacol. 2017;6:177–87.
990. Gu J, Gui Y, Chen L, et al. Use of natural products as chemical library for drug discovery and network pharmacology. PLoS One. 2013;8(4):e62839.
991. Yildirim MA, Goh KI, Cusick ME, et al. Drug-target network. Nat Biotechnol. 2007;25:1119–26.
992. Barneh F, Jafari M, Mirzaie M. Updates on drug-target network; facilitating polypharmacology and data integration by growth of DrugBank database. Brief Bioinform. 2016;17:1070–80.
993. Bento AP, Gaulton A, Hersey A, et al. The ChEMBL bioactivity database: an update. Nucleic Acids Res. 2014;42:D1083–90.
994. Wang Y, Bolton E, Dracheva S, et al. An overview of the PubChem BioAssay resource. Nucleic Acids Res. 2010;38:D255–66.
995. Law V, Knox C, Djoumbou Y, et al. DrugBank 4.0: shedding new light on drug metabolism. Nucleic Acids Res. 2014;42:D1091–7.
996. Yang H, Qin C, Li YH, et al. Therapeutic target database update 2016: enriched resource for bench to

clinical drug target and targeted pathway information. Nucleic Acids Res. 2016;44:D1069–74.
997. Wagner AH, Coffman AC, Ainscough BJ, et al. DGIdb 2.0: mining clinically relevant drug-gene interactions. Nucleic Acids Res. 2016;44:D1036–44.
998. Kuhn M, Szklarczyk D, Pletscher-Frankild S, et al. STITCH 4: integration of protein-chemical interactions with user data. Nucleic Acids Res. 2014;42:D401–7.
999. Nickel J, Gohlke BO, Erehman J, et al. SuperPred: update on drug classification and target prediction. Nucleic Acids Res. 2014;42:W26–31.
1000. Ye H, Ye L, Kang H, et al. HIT: linking herbal active ingredients to targets. Nucleic Acids Res. 2011;39:D1055–9.
1001. Xue R, Fang Z, Zhang M, et al. TCMID: Traditional Chinese Medicine integrative database for herb molecular mechanism analysis. Nucleic Acids Res. 2013;41:D1089–95.
1002. Ru J, Li P, Wang J, et al. TCMSP: a database of systems pharmacology for drug discovery from herbal medicines. J Cheminform. 2014;6:13.
1003. Mangal M, Sagar P, Singh H, et al. NPACT: naturally occurring plant-based anti-cancer compound-activity-target database. Nucleic Acids Res. 2013;41:D1124–9.
1004. Tao W, Li B, Gao S, et al. CancerHSP: anticancer herbs database of systems pharmacology. Sci Rep. 2015;5:11481.
1005. Bredel M, Jacoby E. Chemogenomics: an emerging strategy for rapid target and drug discovery. Nat Rev Genet. 2004;5:262–75.
1006. Lamb J, Crawford ED, Peck D, et al. The Connectivity Map: using gene-expression signatures to connect small molecules, genes, and disease. Science. 2006;313:1929–35.
1007. Duan Q, Flynn C, Niepel M, et al. LINCS Canvas Browser: interactive web app to query, browse and interrogate LINCS L1000 gene expression signatures. Nucleic Acids Res. 2014;42:W449–60.
1008. Igarashi Y, Nakatsu N, Yamashita T, et al. Open TG-GATEs: a large-scale toxicogenomics database. Nucleic Acids Res. 2015;43:D921–7.
1009. Ganter B, Snyder RD, Halbert DN, et al. Toxicogenomics in drug discovery and development: mechanistic analysis of compound/class-dependent effects using the DrugMatrix database. Pharmacogenomics. 2006;7:1025–44.
1010. Lamb J. The connectivity map: a new tool for biomedical research. Nat Rev Can. 2007;7:54–60.
1011. Adams JU. Genetics: big hopes for big data. Nature. 2015;527:S108–9.
1012. Chatr-Aryamontri A, Breitkreutz BJ, Oughtred R, et al. The BioGRID interaction database: 2015 update. Nucleic Acids Res. 2015;43:D470–8.
1013. Keshava Prasad TS, Goel R, Kandasamy K, et al. Human protein reference database–2009 update. Nucleic Acids Res. 2009;37:D767–72.
1014. Mosca R, Ceol A, Aloy P. Interactome3D: adding structural details to protein networks. Nat Methods. 2013;10:47–53.
1015. Szklarczyk D, Morris JH, Cook H, et al. The STRING database in 2017: quality-controlled protein-protein association networks, made broadly accessible. Nucleic Acids Res. 2017;45:D362–8.
1016. Licata L, Briganti L, Peluso D, et al. MINT, themolecular interaction database: 2012 update. Nucleic Acids Res. 2012;40:D857–61.
1017. Cheng F, Jia P, Wang Q, et al. Quantitative network mapping of the human kinome interactome reveals new clues for rational kinase inhibitor discovery and individualized cancer therapy. Oncotarget. 2014;5:3697–710.
1018. Cheng F, Murray JL, Zhao J, et al. Systems biology-based investigation of cellular antiviral drug targets identified by gene-trap insertional mutagenesis. PLoS Comput Biol. 2016;12:e1005074.
1019. Cheng F, Liu C, Shen B, et al. Investigating cellular network heterogeneity and modularity in cancer: a network entropy and unbalanced motif approach. BMC Syst Biol. 2016;10(Suppl 3):65.
1020. Cheng F, Jia P, Wang Q, et al. Studying tumorigenesis through network evolution and somatic mutational perturbations in the cancer interactome. Mol Biol Evol. 2014;31:2156–69.
1021. Chen X, Yan CC, Zhang X, et al. Drug-target interaction prediction: databases, web servers and computational models. Brief Bioinform. 2016;17:696–712.
1022. Koutsoukas A, Simms B, Kirchmair J, et al. From in silico target prediction to multi-target drug design: current databases, methods and applications. J Proteomics. 2011;74:2554–74.
1023. Yue R, Shan L, Yang X, et al. Approaches to target profiling of natural products. Curr Med Chem. 2012;19:3841–55.
1024. Jenkins JL, Bender A, Davies JW. In silico target fishing: predicting biological targets from chemical structure. Drug Discov Today Technol. 2007;3:413–21.
1025. Harren J, Andrew RL. Structure-based Drug Discovery. Dordrecht: Netherlands, Springer; 2007.
1026. Taboureau O, Baell JB, Fernandez-Recio J, et al. Established and emerging trends in computational drug discovery in the structural genomics era. Chem Biol. 2012;19:29–41.
1027. Luo H, Mattes W, Mendrick DL, et al. Molecular docking for identification of potential targets for drug repurposing. Curr Top Med Chem. 2016;16:3636–45.
1028. Ye H, Wei J, Tang K, et al. Drug repositioning through network pharmacology. Curr Top Med Chem. 2016;16:3646–56.
1029. Sakkiah S, Ng HW, Tong W, et al. Structures of androgen receptor bound with ligands: advancing understanding of biological functions and drug discovery. Expert Opin Ther Targets. 2016;20:1267–82.

1030. Liu LJ, Leung KH, Chan DS, et al. Identification of a natural product-like STAT3 dimerization inhibitor by structure-based virtual screening. Cell Death Dis. 2014;5:e1293.
1031. Zhong HJ, Lee BR, Boyle JW, et al. Structure-based screening and optimization of cytisine derivatives as inhibitors of the menin-MLL interaction. Chem Commun. 2016;52:5788–91.
1032. Singh T, Gupta NA, Xu S, et al. Honokiol inhibits the growth of head and neck squamous cell carcinoma by targeting epidermal growth factor receptor. Oncotarget. 2015;6:21268–82.
1033. Zhong HJ, Ma VP, Cheng Z, et al. Discovery of a natural product inhibitor targeting protein neddylation by structure-based virtual screening. Biochimie. 2012;94:2457–60.
1034. Lee HM, Chan DS, Yang F, et al. Identification of natural product fonsecin B as a stabilizing ligand of c-myc G-quadruplex DNA by high-throughput virtual screening. Chem Commun. 2010;46:4680–2.
1035. Ma D-L, Chan DS-H, Leung C-H. Molecular docking for virtual screening of natural product databases. ChemSci. 2011;2:1656–65.
1036. Cierpicki T, Grembecka J. Challenges and opportunities in targeting the menin-MLL interaction. Future Med Chem. 2014;6:447–62.
1037. Chen YZ, Zhi DG. Ligand-protein inverse docking and its potential use in the computer search of protein targets of a small molecule. Proteins. 2001;43:217–26.
1038. Wang JC, Chu PY, Chen CM, et al. idTarget: a web server for identifying protein targets of small chemical molecules with robust scoring functions and a divide-and-conquer docking approach. Nucleic Acids Res. 2012;40:W393–9.
1039. Lauro G, Romano A, Riccio R, et al. Inverse virtual screening of antitumor targets: pilot study on a small database of natural bioactive compounds. J Nat Prod. 2011;74:1401–7.
1040. Lauro G, Masullo M, Piacente S, et al. Inverse virtual screening allows the discovery of the biological activity of natural compounds. Bioorg Med Chem. 2012;20:3596–602.
1041. Vuong H, Cheng F, Lin CC, et al. Functional consequences of somatic mutations in cancer using protein pocket-based prioritization approach. Genome Med. 2014;6:81.
1042. Zhao J, Cheng F, Wang Y, et al. Systematic prioritization of druggable mutations in approximately 5000 genomes across 16 cancer types using a structural genomics-based approach. Mol Cell Proteomics. 2016;15:642–56.
1043. Lu W, Cheng F, Jiang J, et al. FXR antagonism of NSAIDs contributes to drug-induced liver injury identified by systems pharmacology approach. Sci Rep. 2015;5:8114.
1044. Sliwoski G, Kothiwale S, Meiler J, et al. Computational methods in drug discovery. Pharmacol Rev. 2014;66:334–95.
1045. Lavecchia A. Machine-learning approaches in drug discovery: methods and applications. Drug Discov Today. 2015;20:318–31.
1046. Yan X, Liao C, Liu Z, et al. Chemical structure similarity search for ligand-based virtual screening: methods and computational resources. Curr Drug Targets. 2016;17:1580–5.
1047. Liu X, Ouyang S, Yu B, et al. PharmMapper server: a web server for potential drug target identification using pharmacophore mapping approach. Nucleic Acids Res. 2010;38:W609–14.
1048. Yan X, Li J, Liu Z, et al. Enhancing molecular shape comparison by weighted Gaussian functions. J Chem Inf Model. 2013;53:1967–78.
1049. Fang J, Yang R, Gao L, et al. Predictions of BuChE inhibitors using support vector machine and naive Bayesian classification techniques in drug discovery. J Chem Inf Model. 2013;53:3009–20.
1050. Fang J, Yang R, Gao L, et al. Consensus models for CDK5 inhibitors in silico and their application to inhibitor discovery. Mol Divers. 2015;19:149–62.
1051. Fang J, Li Y, Liu R, et al. Discovery of multitarget-directed ligands against Alzheimer's disease through systematic prediction of chemical-protein interactions. J Chem Inf Model. 2015;55:149–64.
1052. Fang J, Pang X, Wu P, et al. Discovery of neuroprotective compounds by machine learning approaches. RSC Adv. 2016;6:9857.
1053. Cheng F, Li W, Liu G, et al. In silico ADMET prediction: recent advances, current challenges and future trends. Curr Top Med Chem. 2013;13:1273–89.
1054. Sprague B, Shi Q, Kim MT, et al. Design, synthesis and experimental validation of novel potential chemopreventive agents using random forest and support vector machine binary classifiers. J Comput Aided Mol Des. 2014;28:631–46.
1055. Bilsland AE, Pugliese A, Liu Y, et al. Identification of a selective G1-phase benzimidazolone inhibitor by a senescence-targeted virtual screen using artificial neural networks. Neoplasia. 2015;17:704–15.
1056. Liu H, Sun J, Guan J, et al. Improving compound-protein interaction prediction by building up highly credible negative samples. Bioinformatics. 2015;31:i221–9.
1057. Mousavian Z, Masoudi-Nejad A. Drug-target interaction prediction via chemogenomic space: learning-based methods. Expert Opin Drug Metab Toxicol. 2014;10:1273–87.
1058. Yamanishi Y. Chemogenomic approaches to infer drug-target interaction networks. Methods Mol Biol. 2013;939:97–113.
1059. Zhao S, Li S. Network-based relating pharmacological and genomic spaces for drug target identification. PLoS One. 2010;5:e11764.
1060. Yamanishi Y, Araki M, Gutteridge A, et al. Prediction of drug target interaction networks from the integration of chemical and genomic spaces. Bioinformatics. 2008;24:i232–40.

1061. Bleakley K, Yamanishi Y. Supervised prediction of drug target interactions using bipartite local models. Bioinformatics. 2009;25:2397–403.
1062. Keum J, Yoo S, Lee D, et al. Prediction of compound-target interactions of natural products using large-scale drug and protein information. BMC Bioinformatics. 2016;17(Suppl 6):219.
1063. Yu H, Chen J, Xu X, et al. A systematic prediction of multiple drug-target interactions from chemical, genomic, and pharmacological data. PLoS One. 2012;7:e37608.
1064. Huang C, Zheng C, Li Y, et al. Systems pharmacology in drug discovery and therapeutic insight for herbal medicines. Brief Bioinform. 2014;15:710–33.
1065. Sawada R, Kotera M, Yamanishi Y. Benchmarking a wide range of chemical descriptors for drug-target interaction prediction using a chemogenomic approach. Mol Inform. 2014;33:719–31.
1066. Cheng F, Zhou Y, Li J, et al. Prediction of chemical-protein interactions: multitarget-QSAR versus computational chemogenomic methods. Mol Biosyst. 2012;8:2373–84.
1067. Cheng F, Zhou Y, Li W, et al. Prediction of chemical-protein interactions network with weighted network-based inference method. PLoS One. 2012;7:e41064.
1068. Wu Z, Cheng F, Li J, et al. SDTNBI: an integrated network and chemoinformatics tool for systematic prediction of drug-target interactions and drug repositioning. Brief Bioinform. 2017;18:333–47.
1069. Wu Z, Lu W, Wu D, et al. In silico prediction of chemical mechanism of action via an improved network-based inference method. Br J Pharmacol. 2016;173:3372–85.
1070. Cheng F, LiW ZY, et al. Prediction of human genes and diseases targeted by xenobiotics using predictive toxicogenomic derived models (PTDMs). Mol Biosyst. 2013;9:1316–25.
1071. Li J, Wu Z, Cheng F, et al. Computational prediction of microRNA networks incorporating environmental toxicity and disease etiology. Sci Rep. 2014;4:5576.
1072. Li J, Lei K, Wu Z, et al. Network-based identification of microRNAs as potential pharmacogenomic biomarkers for anticancer drugs. Oncotarget. 2016;7:45584–96.
1073. Kitano H. Systems biology: a brief overview. Science. 2002;295:1662–4.
1074. Kitano H. Computational systems biology. Nature. 2002;420:206–10.
1075. Berg EL. Systems biology in drug discovery and development. Drug Discov Today. 2014;19:113–25.
1076. Cheng F, Murray JL, Rubin DH. Drug repurposing: new treatments for Zika virus infection? Trends Mol Med. 2016;22:919–21.
1077. Qu XA, Rajpal DK. Applications of connectivity map in drug discovery and development. Drug Discov Today. 2012;17:1289–98.
1078. Hieronymus H, Lamb J, Ross KN, et al. Gene expression signature-based chemical genomic prediction identifies a novel class of HSP90 pathway modulators. Cancer Cell. 2006;10:321–30.
1079. Wei G, Twomey D, Lamb J, et al. Gene expression-based chemical genomics identifies rapamycin as a modulator of MCL1 and glucocorticoid resistance. Cancer Cell. 2006;10:331–42.
1080. Langley SR, Dwyer J, Drozdov I, et al. Proteomics: from single molecules to biological pathways. Cardiovasc Res. 2013;97:612–22.
1081. Bensimon A, Heck AJ, Aebersold R. Mass spectrometry-based proteomics and network biology. Annu Rev Biochem. 2012;81:379–405.
1082. Savitski MM, Reinhard FB, Franken H, et al. Tracking cancer drugs in living cells by thermal profiling of the proteome. Science. 2014;346:1255784.
1083. Franken H, Mathieson T, Childs D, et al. Thermal proteome profiling for unbiased identification of direct and indirect drug targets using multiplexed quantitative mass spectrometry. Nat Protoc. 2015;10:1567–93.
1084. Reinhard FB, Eberhard D. Thermal proteome profiling monitors ligand interactions with cellular membrane proteins. Nat Methods. 2015;12:1129–31.
1085. Sacco F, Silvestri A, Posca D, et al. Deep proteomics of breast cancer cells reveals that metformin rewires signaling networks away from a pro-growth state. Cell Syst. 2016;2:159–71.
1086. Harvey AL, Edrada-Ebel R, Quinn RJ. The re-emergence of natural products for drug discovery in the genomics era. Nat Rev Drug Discov. 2015;14:111–29.
1087. Kaddurah-Daouk R, Kristal BS, Weinshilboum RM. Metabolomics: a global biochemical approach to drug response and disease. Annu Rev Pharmacol Toxicol. 2008;48:653–83.
1088. Birkenstock T, Liebeke M, Winstel V, et al. Exometabolome analysis identifies pyruvate dehydrogenase as a target for the antibiotic triphenylbismuthdichloride in multiresistant bacterial pathogens. J Biol Chem. 2012;287:2887–95.
1089. Zhang B, Watts KM, Hodge D, et al. A second target of the antimalarial and antibacterial agent fosmidomycin revealed by cellular metabolic profiling. Biochemistry. 2011;50:3570–7.
1090. Bayet-Robert M, Lim S, Barthomeuf C, et al. Biochemical disorders induced by cytotoxic marine natural products in breast cancer cells as revealed by proton NMR spectroscopy-based metabolomics. Biochem Pharmacol. 2010;80:1170–9.
1091. Pulido MR, Garcia-Quintanilla M, Gil-Marques ML, et al. Identifying targets for antibiotic development using omics technologies. Drug Discov Today. 2016;21:465–72.
1092. Zhao Y, Hu Q, Cheng F, et al. SoNar, a highly responsive NADþ/NADH sensor, allows high-

throughput metabolic screening of anti-tumor agents. Cell Metab. 2015;21:777–89.
1093. Kim HK, Wilson EG, Choi YH, et al. Metabolomics: a tool for anticancer lead-finding from natural products. Planta Med. 2010;76:1094–102.
1094. Collins GS, de Groot JA, Dutton S, et al. External validation of multivariable prediction models: a systematic review of methodological conduct and reporting. BMC Med Res Methodol. 2014;14:40.
1095. Reker D, Perna AM, Rodrigues T, et al. Revealing the macromolecular targets of complex natural products. Nat Chem. 2014;6(12):1072–8.
1096. Ashburn TT, Thor KB. Drug repositioning: identifying and developing new uses for existing drugs. Nat Rev Drug Discov. 2004;3(8):673–83.
1097. Pushpakom S, Iorio F, Eyers PA, et al. Drug repurposing: progress, challenges and recommendations. Nat Rev Drug Discov. 2019;18(1):41–58.
1098. Schcolnik-Cabrera A, Juárez-López D, Duenas-Gonzalez A. Perspectives on drug repurposing. Curr Med Chem. 2021;28(11):2085–99.
1099. Sleigh SH, Barton CL. Repurposing strategies for therapeutics. Pharmaceutical Med. 2012;24(3):151–9.
1100. Cha Y, Erez T, Reynolds IJ, et al. Drug repurposing from the perspective of pharmaceutical companies. Br J Pharmacol. 2018;175(2):168–80.
1101. Sam E, Athri P. Web-based drug repurposing tools: a survey. Brief Bioinform. 2019;20(1):299–316.
1102. Masoudi-Sobhanzadeh Y, Omidi Y, Amanlou M, et al. Drug databases and their contributions to drug repurposing Genomics. 2020; 112(2): 1087–1095.
1103. Karaman B, Sippl W. Computational drug repurposing: Current trends. Curr Med Chem. 2019;26(28):5389–409.
1104. Oprea TI, Bauman JE, Bologa CG, et al. Drug repurposing from an academic perspective. Drug Discov Today. 2011;8(3-4):61–9.
1105. Haupt VJ, Daminelli S, Schroeder M. Drug promiscuity in PDB: Protein binding site similarity is key. PLoS One. 2013;8(6):e65894.
1106. Feldmann C, Miljković F, Yonchev D, et al. Identifying promiscuous compounds with activity against different target classes. Molecules. 2019;24(22):4185.
1107. Gupta MN, Alam A, Hasnain SE. Protein promiscuity in drug discovery, drug-repurposing and antibiotic resistance. Biochimie. 2020;175:50–7.
1108. Gilberg E, Gütschow M, Bajorath J. Promiscuous ligands from experimentally determined structures, binding conformations, and protein family-dependent interaction hotspots. ACS Omega. 2019;4(1):1729–37.
1109. Irwin JJ, Duan D, Torosyan H, et al. An aggregation advisor for ligand discovery. J Med Chem. 2015;58:7076–87.
1110. Baell JB, Walters MA. Chemistry: Chemical con artists foil drug discovery. Nature. 2014;513:481–3.
1111. Aldrich C, Bertozzi C, Georg GI, et al. The ecstasy and agony of assay interference compounds. J Chem Inf Model. 2017;57:387–90.
1112. Stumpfe D, Tinivella A, Rastelli G, et al. Promiscuity of inhibitors of human protein kinases at varying data confidence levels and test frequencies. RSC Adv. 2017;7:41265–71.
1113. Hu Y, Bajorath J. Entering the 'big data' era in medicinal chemistry: Molecular promiscuity analysis revisited. Future Sci OA. 2017;3:FSO179.
1114. Woolf SH. The meaning of translational research and why it matters. JAMA. 2008;299:211–3.
1115. Schuster D, Laggner C, Langer T. Why drugs fail—a study on side effects in new chemical entities. Curr Pharm Des. 2005;11:3545–59.
1116. Morgan P, Van Der Graaf PH, Arrowsmith J, et al. Can the flow of medicines be improved? Fundamental pharmacokinetic and pharmacological principles toward improving phase II survival. Drug Discov Today. 2012;17:419–24.
1117. Oprea TI, Bauman JE, Bologa CG, et al. Drug repurposing from an academic perspective. Drug Discov Today Ther Strateg. 2011;8:61–9.
1118. Oprea TI, Mestres J. Drug repurposing: far beyond new targets for old drugs. AAPS J. 2012;14:759–63.
1119. Gallo K, Goede A, Eckert A, et al. PROMISCUOUS 2.0: a resource for drug-repositioning. Nucleic Acids Res. 2021;49(D1):D1373–80.
1120. Xue H, Li J, Xie H, et al. Review of drug repositioning approaches and resources. Int J Biol Sci. 2018;14:1232–44.
1121. Swinney DC, Anthony J. How were new medicines discovered? Nat Rev Drug Discov. 2011;10:507–19.
1122. Eder J, Sedrani R, Wiesmann C. The discovery of first-in class drugs: origins and evolution. Nat Rev Drug Discov. 2014;13:577–87.
1123. Horvath P, Aulner N, Bickle M, et al. Screening out irrelevant cell-based models of disease. Nat Rev Drug Discov. 2016;15:715–69.
1124. Pandey UB, Nichols CD. Human disease models in Drosophila melanogaster and the role of the fly in therapeutic drug discovery. Pharmacol Rev. 2011;63:411–36.
1125. Sanseau P, Agarwal P, Barnes MR, et al. Use of genome-wide association studies. Nat Biotechnol. 2012;30:317–20.
1126. Rastegar-Mojarad M, Zhan Y, Kolesar JM, et al. Opportunities for drug repositioning from phenome-wide association studies. Nat Biotechnol. 2015;33:342–5.
1127. Iorio F, Rittman T, Ge H, et al. Transcriptional data: a new gateway to drug repositioning? Drug Discov Today. 2013;18:350–7.
1128. Sirota M, Dudley JT, Kim J, et al. Discovery and preclinical validation of drug indications using compendia of public gene expression data. Sci Transl Med. 2011;3:96ra77.

1129. Teachey DT, Sheen C, Hall J, et al. mTOR inhibitors are synergistic with methotrexate: an effective combination to treat acute lymphoblastic leukemia. Blood. 2008;112:2020–3.
1130. Fransecky L, Mochmann LH, Baldus CD. Outlook on PI3K/AKT/mTOR inhibition in acute leukemia. Mol Cell Ther. 2015;3:2–19.
1131. Geva M, Kusko R, Soares H, et al. Pridopidine activates neuroprotective pathways impaired in Huntington disease. Hum Mol Genet. 2016;25(18):3975–87.
1132. Nelson MR, Tipney H, Painter JL, et al. The support of human genetic evidence for approved drug indications. Nat Genet. 2015;47:856–60.
1133. Fowler KD, Funt JM, Artyomov MN, et al. Leveraging existing data sets to generate new insights into Alzheimer's disease biology in specific patient subsets. Sci Rep. 2015;5:14324–32.
1134. Xu H, Aldrich MC, Chen Q, et al. Validating drug repurposing signals using electronic health records: a case study of metformin associated with reduced cancer mortality. J Am Med Inform Assoc. 2014;22:179–91.
1135. Yang L, Agarwal P. Systematic drug repositioning based on clinical side-effects. PLoS One. 2011;6:e28025.
1136. Kuhn M. The SIDER database of drugs and side effects. Nucleic Acids Res. 2016;44:D1075–9.
1137. Gottlieb A, Stein GY, Ruppin ER, et al. PREDICT: a method for inferring novel drug indications with application to personalized medicine. Mol Syst Biol. 2011;7:496–505.
1138. Yu L, Ma X, Zhang L, et al. Prediction of new drug indications based on clinical data and network modularity. Sci Rep. 2016;6:32530.
1139. Tan F, Yang R, Xu X, et al. Drug repositioning by applying 'expression profiles' generated by integrating chemical structure similarity and gene semantic similarity. Mol Biosyst. 2014;10:1126–38.
1140. Paik H, Chung AY, Park HC, et al. Repurpose terbutaline sulfate for amyotrophic lateral sclerosis using electronic medical records. Sci Rep. 2015;5:8580–8.
1141. Shameer K, Glicksberg BS, Hodos R, et al. Systematic analyses of drugs and disease indications in RepurposeDB reveal pharmacological, biological and epidemiological factors influencing drug repositioning. Brief Bioinform. 2018;19:656–78.
1142. Brown AS, Patel CJ. A standard database for drug repositioning. Sci Data. 2017;4:170029.
1143. Zhao C, Dai X, Li Y, et al. EK-DRD: a comprehensive database for drug repositioning inspired by experimental knowledge. J Chem Inf Model. 2019;59:3619–24.
1144. Wu H, Huang J, Zhong Y, et al. DrugSig: A resource for computational drug repositioning utilizing gene expression signatures. PLoS One. 2017;12:e0177743.
1145. Avram S, Bologa CG, Holmes J, et al. DrugCentral 2021 supports drug discovery and repositioning. Nucleic Acids Res. 2021;49(D1):D1160–9.
1146. Ursu O, Holmes J, Knockel J, et al. DrugCentral: Online drug compendium. Nucleic Acids Res. 2017;45:D932–9.
1147. UniProtConsortium UniProt. a worldwide hub of protein knowledge. Nucleic Acids Res. 2019;47:D506–15.
1148. Hastings J, Owen G, Dekker A, et al. ChEBI in 2016: Improved services and an expanding collection of metabolites. Nucleic Acids Res. 2016;44:D1214–9.
1149. Himmelstein DS, Lizee A, Hessler C, et al. Systematic integration of biomedical knowledge prioritizes drugs for repurposing. Elife. 2017;6:e26726.
1150. Aguirre-Plans J, Piñero J, Sanz F, et al. GUILDify v2.0: A tool to identify molecular networks underlying human diseases, their comorbidities and their druggable targets. J Mol Biol. 2019;431:2477–84.
1151. Chambers J, Davies M, Gaulton A, et al. UniChem: a unified chemical structure cross-referencing and identifier tracking system. J Cheminform. 2013;5:3.
1152. Alexander SPH, Kelly E, Mathie A, et al. The concise guide to pharmacology 2019/20: Introduction and other protein targets. Br J Pharmacol. 2019;176:S1–20.
1153. Brown AS, Patel CJ. A standard database for drug repositioning. Scientific Data. 2017;4:170029.
1154. Mayers M, Li TS, Queralt-Rosinach N, et al. Time-resolved evaluation of compound repositioning predictions on a text-mined knowledge network. BMC Bioinformatics. 2019;20:653.
1155. Wu Q, Taboureau O, Audouze K. Development of an adverse drug event network to predict drug toxicity. Curr Res Toxicol. 2020;1:48–55.
1156. Ursu O, Holmes J, Bologa CG, et al. DrugCentral 2018: an update. Nucleic Acids Res. 2019;47:D963–70.
1157. Levin JM, Oprea TI, Davidovich S, et al. Artificial intelligence, drug repurposing and peer review. Nat Biotechnol. 2020;38:1127–31.
1158. Avram S, Curpan R, Halip L, et al. Off-patent drug repositioning. J Chem Inf Model. 2020;60(12):5746–53.
1159. KC G, Bocci G, Verma S, et al. REDIAL-2020: A suite of machine learning models to estimate anti-SARS-CoV-2 activities. ChemRxiv. 2020.

13 Polypharmacology in Old Drug Rediscovery: Drug Repurposing

Abstract

The concept of drug repurposing has been introduced in this chapter (Sect. 13.8) as a new strategy for MTD drug discovery based on "promising promiscuity." Here in this chapter, a different prospect of drug repurposing will be discussed by highlighting the latest applications of drug repurposing in various diseases in the clinical settings exemplified by some representative repurposed drugs for comparison of old and new therapeutic uses, with the assessment of their novel mechanisms of action and pharmacological effects in the management of various disorders; specifically, the applications of repurposed drugs to the treatment of cancer, cardiovascular disease, metabolic disorders, Alzheimer's disease, COVID-19, and rare diseases. Finally, the drug repurposing from the perspective of pharmaceutical industries will be discussed.

Keywords

Drug repurposing · Cancer · Cardiovascular disease · Rare and orphan diseases · COVID-19

13.1 Drug Repurposing Approaches and Resources

Drug repurposing is viewed as a revolutionary breakthrough in drug development that focuses on rediscovering new uses for old therapeutic agents or on the use of a drug in an indication other than the one for which it was initially marketed, including approved, discontinued, shelved, and experimental drugs [1–4]. Drug repurposing can be defined more precisely as the process of exploring new indications for an already approved drug while drug repurposing includes overall redevelopment approaches grounded in the identical chemical structure of the active drug moiety as in the original product [1]. The theoretical basis and practical guideline for drug repurposing is drug promiscuity (see Chaps. 13 and 15) or more accurately polypharmacology (see Chaps. 1 and 2). The repurposing approach, as an emerging tool for drug reuse, recycling, and discovery, accelerates tremendously the drug development process, curtails the cost and risk inherent to drug development. The strategy focuses on the polypharmacology of drugs to unlock novel opportunities for logically designing more efficient therapeutic agents for unmet medical needs. Various restrictions and challenges that repurposed drugs come across during their development and regulatory phases are also discussed.

According to the analysis reported by Xue et al. [5], traditional drug development strategies usually involve five stages: lead discovery and preclinical validation, safety review, clinical studies, FDA review, and FDA post-market safety monitoring [6, 7]. For drug repurposing, however, only four steps are obligatory, which include compound identification, compound acquisition, development, and FDA post-market safety monitoring [5, 8]. The time for a compound to get into market is tremendously shortened for repurposed

Z. Wang, B. Yang, *Polypharmacology*, https://doi.org/10.1007/978-3-031-04998-9_13

drug and the failure rate is also substantially reduced, relative to a new drug; it takes 1–2 years to identify new drug targets and 8 years to develop a repurposed drug on average, compared to 10–15 years to develop a new drug with a success rate of only 2%. The research and development investment required for drug repurposing is also reduced relative to traditional strategies; it costs only $1.6 billion to develop a repurposed drug, while the cost of the traditional strategy is $12 billion [9]. Repurposing drugs is a generally lower-risk process because repositioned drugs have passed all clinical tests in Phase I, Phase II, and Phase III, their safety has been confirmed.

Most of the successful drug repurposing cases have been found by serendipity or through exploitation of the original action mechanism of a drug for new indications (so-called on target repurposing) [2]. Multitarget agents are natural candidates for more innovative, so-called off-target drug repurposing [2]. Computational approaches to drug repurposing have so far focused on so-called retrospective drug repurposing: screening known drug collections/libraries to find novel indications for already known therapeutic agents [2]. Prospective drug repurposing, in contrast, would explore drug repurposing possibilities much earlier in the drug discovery process. There are generally three different approaches for drug repurposing: computational approaches, biological experimental approaches, and mixed approaches [2, 5, 10–12]. In a comprehensive review article, Xu et al. [5] summarized computational approaches and 76 related resources for drug repurposing, highlighted the important findings obtained using these approaches, and discussed the challenges and opportunities in drug repurposing from multiple perspectives, including technology, commercial models, patents, and investment. The approaches commonly used for drug repurposing have been described in Chap. 13, and such contents will therefore not be repeated in this chapter but merely outlined for the sake of refreshing mind and of the completeness and integrity of the chapter content.

13.1.1 Computational/In Silico Approaches

With the fast development of biology microarray techniques, various drug and disease knowledge databases such as DrugBank [13], ChemBank [14], OMIM [15], KEGG [16], and PubMed [17] have appeared, and massive genomic databases such as MIPS [18], PDB [19], GEO [20], and GenBank [21] have been built. These databases tremendously promote the development of computational approaches that have some advantages over biological experimental approaches in terms of lower costs and fewer barriers [22].

Most existing computational approaches are based on the gene expression response of cell lines after treatment or merging several types of information about disease–drug relationships [23] that can be divided into different types from different viewpoints [24–27]. For instance, some researchers grouped drug repurposing methods according to the biological networks used [24], and others divided drug repurposing methods into two types: data-driven and hypothesis-driven [26]. To emphasize the core methodologies of drug repurposing approaches, Xu et al. [5] assign them into three categories: network-based approaches [28–30], text-mining approaches [31–38], and semantic approaches [39–41]. Computational drug-repurposing methods can also be classified into target-based, knowledge-based, signature-based, pathway- or network-based, and targeted-mechanism-based methods [42]. By combining these different views, we tentatively classified the computational approaches into the following methodological categories.

13.1.1.1 Knowledge-Based Methods

Knowledge-based drug-repurposing methods refer to the applications of bioinformatics or cheminformatics relevant to drugs, drug–target networks [42–47], chemical structures of targets and drugs [47], clinical trial information (adverse effects) [48, 49], FDA approval labels [50], signaling or metabolic pathways [51], and so forth, to drug repurposing. Knowledge-based methods incorporate known information into predicting unknown mechanisms, such as unknown targets for drugs, unknown drug–drug similarities, and new biomarkers for diseases. The advantage of knowledge-based methods is that they include a large amount of known information into the drug-repurposing process to improve its prediction accuracy [52].

13.1.1.2 Signature-Based Methods

Signature-based drug-repurposing methods make use of gene signatures derived from disease-omics data with or without treatments [53–60] to discover unknown off-targets or unknown disease mechanisms. As the advancement of microarray and next-generation sequencing techniques speed up the generation of vast volumes of genomics data pertinent to drug repurposing, gene signatures can be used to discover unknown mechanisms. One can easily access such genomics data in publicly available databases, such as NCBI-GEO (http://www.ncbi.nlm.nih.gov/geo/), SRA (http://www.ncbi.nlm.nih.gov/Traces/sra/), CMAP [55], and CCLE [61]. The advantage of signature-based methods is that they are useful to uncover unknown mechanisms of action of molecules and drugs [42]. Compared with knowledge-based methods, signature-based methods involve more molecular-level mechanisms, such as the significantly changed genes, by using computational approaches [42].

13.1.1.3 Targeted Mechanism-Based Methods

Targeted mechanism-based drug-repurposing methods integrate treatment-omics data, available signaling pathway information, and protein interaction networks to delineate the unknown mechanisms of action of drugs [62–64]. The era of precision medicine motivates such drug-repurposing studies. For instance, drug resistance remains an unresolved issue in cancer therapy. Although patients respond well to a drug initially, they often acquire resistance to that drug after a few months of treatment. This indicates that deriving a successful drug treatment needs additional information about the mechanisms of action of drugs to find better drug targets. Systems biology approaches are promising in addressing this challenge. The advantage of these methods is that their goals are not only to discover the mechanisms related to diseases or drugs, but also to identify those directly related to treatments of drugs to specific diseases. Owing to the difficulties in deriving effective computational models, there are only a few studies on these targeted mechanism-based methods [62–64] that developed elegant computational models to predict the drug effects and related targeted pathways.

13.1.1.4 Network-Based Approaches

Pathway- or network-based drug-repurposing methods utilize disease-omics data, available signaling or metabolic pathways and protein interaction networks to reconstruct disease-specific pathways that provide the key targets for repurposed drugs [65–67]. The advantage of these methods is that they are helpful in narrowing general signaling networks from a large number of proteins down to a specific network with a few proteins (or targets). A recent study of drug repurposing addressed distinct signaling mechanisms of metastatic subtypes of breast cancer [67]. Neither knowledge-based nor signature-based methods can address these repurposing results because the subtype signaling mechanisms are hard to elucidate from existing breast cancer pathways or the gene signatures [42].

Network-based computational biology, with the emphasis on biomolecular interactions and omics-data integration, has not only promoted and accelerated the drug development based on traditional approaches but also opened a new arena for MTD drug discovery through drug repurposing. Rational large-scale drug repurposing is driven by the accumulation of high-throughput genome-wide data. These data can be utilized to capture the interplay among targets and biological molecules, uncover the resulting network structures, and bridge molecular profiles and phenotypes. Network-based computational approaches primarily include two methods [5]: network-based cluster approaches [27–30] and network-based propagation approaches [28, 29, 68–70].

Network-Based Cluster Approaches (NBC)
The use of network-based cluster (NBC) approaches for uncover drug–disease relationships or drug–target relationships is based on the fact that biologic entities (disease, drug, protein, etc.) in the same module of biological networks share similar characteristics. These approaches aim to find several modules (also known as subnetworks, groups, or cliques) using cluster algorithms according to the topology structures of networks. These modules include various relationships such as drug–disease, drug–drug, or drug–target relationships. The most common network-based cluster approaches include DBSCAN [71], CLIQUE [72], STING [73], OPTICS [74], and ClusterONE [5, 30, 75, 76].

Network-Based Propagation Approaches (NBP)
Network-based propagation (NBP) approaches involve the workflow with prior information propagating from the source node to all network nodes and some subnetwork nodes. According to the different propagation ways, these approaches can be divided into two types: local approaches and global approaches. Several studies have proven that these methods perform well in finding disease–targets, disease–genes, and disease–drug relationships [28]. Local propagation approaches only take the limited information of the network into account and may fail to make correct predictions in some cases [77]. By contrast, global approaches containing information from the entire network perform better than local approaches [68].

Network-based approaches are vital for drug repurposing and the networks employed for such a purpose can be divided into two classes: homogeneous and heterogeneous. The homogeneous networks mainly involve protein–protein interaction (PPI) network analysis [78–80], which are used for identifying drug targets involved in multiple pathways. The limitation of PPI networks is that they only include protein information without considering other necessary information. With the advent of the era of big data, the accumulation of various medical data (such as drugs, diseases, and targets data, among others) has made it possible to construct complex heterogeneous networks. Heterogeneous networks, which integrate multiple sources of information including genomes, proteomes and metabolic pathways [27], typically contain two-layer (i.e., disease–drug) or three–layer (i.e., disease–drug–gene) relationships.

Overall, NBC algorithms are frequently used to find interesting modules, and NBP algorithms are often used to infer new relationships between biological entities. NBC approaches are more general because most network-based cluster algorithms can be employed for detecting biological modules. NBP approaches are easier to implement and can provide more accurate predictions.

13.1.1.5 Text Mining-Based Approaches

Rapid advance of biomedical research results in rapid accumulation of an immense amount of medical and biological publications containing ever-updating novel information and knowledge on biological entity relationships, which can be used for drug repurposing. However, efficiently extracting information useful for drug purposing from the literature is a challenge. Text mining (TM) techniques were born to address this problem and have been increasingly developed to mine new knowledge from scientific literature and identify connections between biological concepts or biological entities.

"Text mining," also referred to as "text data mining," is the process of deriving high-quality information from text in a similar manner to text analytics. It involves "the discovery by computer of new, previously unknown information, by automatically extracting information from different written resources" [81]. Written resources may include websites, books, emails, reviews, and articles. High-quality information is typically obtained by devising patterns and trends by means such as statistical pattern learning. According to Hotho et al. [82] we can differ three different perspectives of text mining: information extraction, data mining, and a Knowledge Discovery in Databases (KDD) process. Text mining usually involves the process of structuring the input text (usually parsing, along with the addition of some derived linguistic features and the removal of others, and subsequent insertion into a database), deriving patterns within the structured data, and finally evaluation and interpretation of the output. "High quality" in text mining usually refers to some combination of relevance, novelty, and interest. Typical text mining tasks include text categorization, text clustering, concept/entity extraction, production of granular taxonomies, sentiment analysis, document summarization, and entity relation modeling (i.e., learning relations between named entities).

Text analysis involves information retrieval, lexical analysis to study word frequency distributions, pattern recognition, tagging/annotation, information extraction, data mining techniques including link and association analysis, visualization, and predictive analytics. The overarching goal is, essentially, to turn text into data for analysis, via application of natural language processing (NLP), different types of algorithms and analytical methods. An important phase of this process is the interpretation of the gathered information. A typical application is to scan a set of documents written in a natural language and either model the document set for predictive classification purposes or populate a database or search index with the information extracted. The document is the basic element while starting with text mining. Here, a document is defined as a unit of textual data, which normally exists in many types of collections [83].

For drug repurposing, the main pipeline of biological text mining includes four phrases: information retrieval (IR), biological name entity recognition (BNER), biological information extraction (BIE), and biological knowledge discovery (BKD) [5]. In the IR stage, relevant documents are extracted from the literature. These relevant documents need to be filtered because there are some useless concepts in documents. In the BNER step, valuable biological concepts are identified with controlled vocabularies. In the BIE and BKD processes, useful information is extracted to uncover the knowledge about biological concepts and build a knowledge graph. Meanwhile, potential associations between knowledge of different sets, such as drug–disease and drug–target relationships, can also be detected.

The origin of text mining methods in the medical field is the Swanson "ABC" model, which states that if concept A is connected with concept B, and concept B is involved with concept C, then concept A may have a novel connection with concept C [5, 84]. Based on the "ABC" model, various text mining methods have been proposed to find potential disease–drug relationships in the literature. A number of studies have been devoted to applying text mining techniques in drug repurposing [5]. Li et al. [31] developed an approach to building disease-specific drug–protein connectivity maps combining network mining and text mining. In the proposed method, they first extracted disease–protein relationships from molecular interaction networks using network mining. Then, they searched for drug terms indirectly associated with certain diseases such as Alzheimer's disease (AD) in PubMed abstracts through text mining techniques. Finally, drugs and proteins could be linked through drug–disease or disease–protein relationships [5, 31]. In this way, they found that diltiazem and quinidine, as anti-hypertensive and antiarrhythmic drugs, have the potential to cure Alzheimer's disease, which has been confirmed by clinical evidence. Gramatica et al. [85] proposed an approach to building sentence graph networks using text mining techniques. The proposed network can be used to discover relationships between any drug and any disease. These relationships are specific paths among the biomedical entities in the graph network. A novel disease target of sarcoidosis was identified using this approach. The investigators extracted drug–gene and gene–disease relationships from medical abstracts to obtain drug–disease relationships using a rank score. To evaluate the performance of the proposed approach, the obtained drug–disease relationships were validated in the Comparative Toxicogenomics Database (CTD). Experimental results indicated that the discovered relationships confirmed in CTD were highly confident. Jang et al. [86] developed an approach to building dependency graph networks through extracting sentences with genes, drugs, and phenotypes from biomedical literature. They calculated the possibility that a drug treats a phenotype based on drug-phenotype associations. The authors compared the predicted drug-phenotype associations using known drug-phenotype associations in databases and proved the good performance of their method. Kuusisto et al. [32] proposed another text mining method named KinderMiner to identify potential indications of some old drugs. The method is based on co-occurrence statistics between drugs and diseases in the literature. As a result, new indications of some drugs such as Zestoretic, Zebeta, and Tiazac were found. Zhang et al. [33] reported an algorithm to prioritize anti-AD (Alzheimer's disease) targets. The authors extracted 224 genetic variations, 14 epigenetic modifications, 98 proteins, and 86 metabolites associated with AD using text mining and integrated these interactions to construct a weighted sum model to prioritize potential anti-AD drug targets.

With the development of natural language processing (NLP) techniques, increasing numbers of text mining tools have been developed and used to discover repurposable drugs [87, 88]. These tools can be divided into two categories: static tools and dynamic tools. Text mining tools reduce the time complexity of drug repurposing and assist researchers in verifying their experimental results by returning massive amounts of biological entity relationships.

13.1.1.6 Semantics-Based Approaches

Semantics-based approaches are widely used in information retrieval, image retrieval, and other fields [5]. Recently, these methods have been applied to drug repurposing. The protocol mainly includes three steps. First, biological entity relationships are extracted from prior information in massive medical databases to build the semantic network. Then, semantics networks based on existing ontology networks are constructed by adding the prior information obtained in the previous step. Finally, mining algorithms are designed to predict novel relationships in the semantic network.

Based on a hypothesis in which similar drugs are correlated with similar targets and similar targets are connected to similar drugs, Palma et al. [39] proposed an unsupervised algorithm to predict drug–target relationships. The authors constructed a semantic network including drug–drug, target–target, and drug–target relationships. The proposed approach, which combines semantics link prediction methods and edge partition methods, was evaluated on a network. Due to substantial semantics knowledge being used, the proposed method made accurate predictions about drug–target relationships. Mullell et al. [40] presented a semantics data-driven algorithm for drug repurposing. The authors used a Bayesian statistics approach to rank drug–disease relationships according to prior knowledge. Then, they integrated ranked relationships with other biological entity associations to construct a semantical drug discovery network. To infer drug–disease relationships, the author applied an algorithm for detecting semantic subgraphs. As a result, nitrendipine, a potent blocker of the calcium channel used to treat hypokalemic periodic paralysis, was found.

Chen et al. [41] built a semantic linked network consisting of over 290,000 nodes and 720,000 edges with multisource data including drugs, targets, proteins, and disease pathways. Then, the authors applied a statistical model to predict drug–target relationships. Consequently, the proposed model identified some drug–targets pairs and drugs for repurposing. For example, barbiturate, a drug used for treating migraines, was predicted for use in curing insomnia with literature support. Zhu et al. [89] proposed an automatic reasoning approach for heterogeneous semantics networks. Biological entities (such as drugs) are converted to labels in a semantics network. Then, disease–drug relationships are obtained from automatic reasoning techniques. As a demonstration, the authors reported that tamoxifen, a drug used for treating breast cancer, can treat ovarian cancer, which was confirmed by the literature [90].

Semantics-based approaches take full advantage of semantics information included in massive amounts of literature. Therefore, the precision of predicting biological entity relationships was improved. However, there is a still challenge in constructing a semantic network by integrating multisource data. It is urgent to construct semantic networks that contain fruitful medical data.

13.1.2 Biological Experimental Screening Approaches

Pharmaceutical companies rely on traditional drug discovery methods to seek repurposing opportunities. Among the 75 agents (50 small molecules and 25 biologics) approved between 1999 and 2008, 28 first-in-class small molecules were discovered by phenotypic drug screening and 17 were identified by target-based methods [91, 92], accounting for more than 50% of the FDA-approved small molecules and biologics. Phenotypic drug-screening approaches discover drug candidates from libraries serendipitously. Alternatively, target-based methods improve the repurposing process by including known target information into drug-repurposing studies.

13.1.2.1 Target-Based Methods

Target-based drug-repurposing methods comprise in vitro and in vivo high-throughput and/or high-content screening (HTS/HCS) of drugs for a protein or a biomarker of interest [42, 93–95] and in silico screening of drugs or compounds from drug libraries [93, 96], such as ligand-based screening or docking [97, 98]. Compared with blinded methods, targeted-based methods significantly improve the likelihood of drug discovery because most targets link directly with the disease mechanisms. Integration of target information into the drug repurposing process ensures a higher possibility of finding useful drugs compared with traditional blinded methods. The advantage of target-based methods, such as docking, is that these methods enable researchers to screen nearly all drugs or compounds with known chemical structure information (e.g., SMILES) within a few days.

13.1.2.2 Phenotypic Screening Methods

Another viable method for drug repurposing is in vitro screening of known drugs or drug-like molecules, initially in disease-relevant phenotypic assays, to identify and validate candidates for repurposing. This approach can use large compound libraries or can focus on subsets of known drugs or drug-like molecules [99–101]. In vitro, ex vivo, or phenotypic assays can be defined as modeling some aspect of disease biology in cells or tissues derived from an experimental species or humans, and that provide a reason to believe that activity has translatable relevance [99]. In other words, these are not recombinant cell lines with an overexpressed target of the type that would be used to screen or rank sets of compounds for activity during lead generation campaigns.

Phenotypic assays have the following benefits when used to identify repurposing activity: Direct knowledge is gained in relation to a potential new disease setting; multiple compounds with different modes of action can be tested and, depending on the throughput of the assay, over a full concentration–effect range; it allows for serendipitous, or hypothesis-free, assessment of compounds; and it provides data-driven choices for subsequent evaluation in more complex phenotypic or in vivo tests [99].

It should be noted that a wide range of cell-based assay types, typically in 96-well formats, could differ markedly in their outcome and throughput when compared to, for example, biochemical or in vivo assays. Various whole–organism phenotypic assays are being increasingly used both for drug screening and for repurposing [102, 103]. However, plate-based screening technologies have been applied to both larval zebrafish (*Danio rerio*) and the nematode worm *Caenorhabditis elegans* (*C. elegans*), and are being extensively used in drug discovery, with this use now extending to drug repurposing [99]. The drugs identified by phenotypic screenings are best described by the use of the theraTRACE in vivo screening platform for finding and developing new indications for discontinued clinical compounds. Clinical proof-of-concept for the platform is exemplified by MLR-1023, a repositioned compound that has recently shown significant clinical efficacy in Type 2 diabetes patients.

13.1.3 Mixed or Combinatory Approaches

The literature indicates a significant trend toward the use of computational- or informatics-based methods for generating initial repurposing hypotheses, followed by focused assessment of biological activity in phenotypic assays. This is one way of combining computational and biological screening approaches for drug repurposing.

Another viable method for combinatory drug repurposing is in vitro screening of known drugs or drug-like molecules, initially in disease-relevant phenotypic assays, to identify and validate candidates for repurposing. This approach can use large compound libraries or can focus on subsets of known drugs or drug-like molecules to generate and validate repurposing candidates in disease-related in vitro and phenotypic assays. Coupling high-throughput technologies with genome-wide essentiality screens, network analysis, genome-scale metabolic modeling, and machine learning techniques enables the proposal of new drug–target signatures and uncovers unanticipated modes of action for available drugs [99, 104] (Fig. 13.1).

13.1.4 Resources for Drug Repurposing

Along with the development of biological technology and the accumulation of various omics data (genomics, proteomics, or metabolomics), more databases from chemical, medical, pharmacological, and biological fields have been established. Xu et al. [5] described 80 widely used databases or resources that can be used for drug repurposing, as summarized below.

Pharmacological databases [13, 105–110] are crucial resources for drug repurposing. These databases collect not only drug property data but also data on interactions between drugs and other biological entities. Pharmacological data lays the foundation of various computational approaches. For instance, DrugBank [13] is a unique bioinformatics and cheminformatics resource that combines detailed drug data with comprehensive drug target information. Many computational approaches, especially network-based approaches [30, 75, 76], were designed based on this database, which achieved excellent results.

Proteomics databases [18, 19, 80, 111, 112] are another type of the data resource for drug repurposing. Most important protein–protein interaction (PPI) networks from proteomics databases are the basis of network-based drug repurposing approaches. In addition, proteomics databases are also important resources for building heterogeneous networks such as drug–protein–disease networks. One of the famous proteomics databases is MIPS [18], which includes manual high-quality PPI data from scientific literature. MIPS can provide PPI information for some network approaches [112, 113]. Moreover, it is a good resource for evaluating the experimental results of some computational approaches.

Chemical properties of drugs provide important information for designing chemical-based approaches. Publicly available databases [14, 113–118] of chemical structures contain massive amounts of useful information such as 2D topological fingerprints and 3D conformations. Chemical information on drugs is usually employed for predicting novel drug structures to find new indications for drugs having similar structures. PubChem [119] is a famous database for chemical molecules structures. The database contains a massive amount of 2D data that can be used to measure the similarity of drugs and to construct chemical networks.

With rapid advance of the field of drug repurposing, enormous amounts of medical and biological literature that contain novel biological entity relations have been published, which has

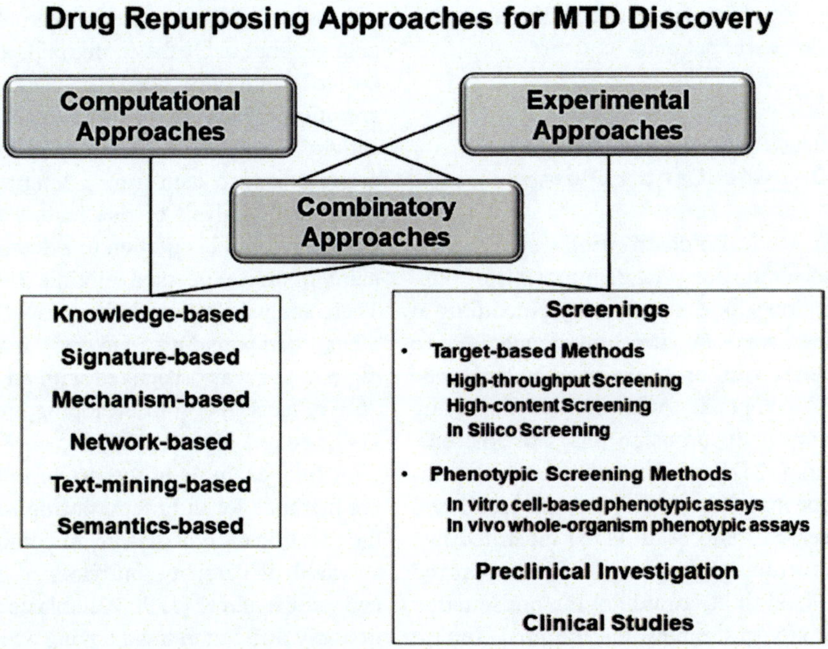

Fig. 13.1 Diagram illustrating drug repurposing approaches for MTD discovery

prompted and accelerated the medical literature databases building and building-up [15, 17, 120]. Based on these valuable medical textual data, text mining approaches were proposed for drug repurposing. PubMed [17] is the most widely used literature database. It comprises more than 27 million citations for biomedical literature from MEDLINE, life science journals, and online books. Citations may include links to full-text content from PubMed Central and publisher websites. Many text mining tools and search engines have been built using these databases [5].

The specific applications of various databases to new drug discovery and drug repurposing will be described in more detail in the relevant chapters in Section II of this book.

13.2 Drug Repurposing in Oncology

The worldwide increase in the occurrence of cancer associated with the limitations of immunotherapy and the emergence of resistance have impaired the prognosis of cancer patients, which leads to the search for alternative treatment methods. Serafin et al. [121] presented 63 studies in which 52 non-anticancer drugs with anticancer activity against a number of malignancies are described. Some of these drugs have already been the target of phase III studies, such as the Add-Aspirin trial for nonmetastatic solid tumors, as well as 9 other drugs (prepatent, artesunate, auranofin, captopril, celecoxib, disulfiram, itraconazole, ritonavir, and sertraline) in the CUSP9* clinical trial for the treatment of recurrent glioblastoma. Others have already been successful in repurposing such as thalidomide, zoledronic acid, celecoxib, methotrexate, and gemcitabine.

The group of compounds that could be considered as promising candidates for repurposing in oncology includes the central nervous system drugs, especially selected antidepressant and antipsychotic agents [122, 123]]. Antoszczak et al. [123] provided an overview of some antidepressants (citalopram, fluoxetine, paroxetine, sertraline) and antipsychotics (chlorpromazine, pimozide, thioridazine, trifluoperazine) that have the potential to be repurposed as novel chemotherapeutics in cancer treatment, as they have been found to exhibit preventive and/or therapeutic action in cancer patients.

13.2.1 Repurposing Non-anticancer Drugs for Cancer Therapy

Wilkinson and Pritchard [99] summarized in vitro screening for drug repurposing of non-anticancer drugs for cancer therapy. According to their review, many in vitro tumor models have been screened with drug-like or FDA-approved libraries of compounds with the accompanying identification of repurposing hits. For instance, Jachan et al. [124] have shown that tricyclic antidepressants may have utility in non-small cell lung cancer; Pessetto et al. [125] identified two compounds from a screen of 796 FDA-approved drugs for their ability to inhibit the proliferation of metastatic gastrointestinal stromal tumor; while a number of compounds from an FDA clinical library were identified as sensitizers of tumor necrosis factor-α (TNF-α)-related apoptosis-inducing ligand (TRAIL), indicating potential as chemotherapeutic agents [126].

Using a library of 1600 annotated clinical compounds (Pharmakon 1600 from MicroSource Inc.) initially tested at 10 μM, Nygren et al. [127] identified mebendazole, an anthelminthic drug used to treat gut worm infections, as having cytotoxic activity in the RKO and HCT 116 colon cancer cell lines. The same group also reported a case study in a patient with refractory metastatic colon cancer whose metastases were significantly reduced by twice-daily dosing with the standard anthelminthic dose of mebendazole.

To identify novel potential therapies for prostate cancer, Platz et al. [128] screened the Johns Hopkins Drug Library (JHDL) of 3187 compounds, of which 1811 are FDA-approved drugs, against two prostate cancer cell lines, LNCaP and PC3, which are androgen insensitive and androgen sensitive, respectively. In their primary screen, Platz et al. measured the ability of the JHDL compounds (5 μM) to inhibit overall DNA synthesis and cell proliferation using a 96-well plate-based assay [^{3}H]-thymidine incorporation. Actives were followed up against a larger panel of prostate cell lines, and concentration-response data generated. Of these, digoxin, a cardiac glycoside commonly used for heart failure and arrhythmia, was found to have antiproliferative activity. To validate the potential for digoxin in prostate cancer, data from a longitudinal epidemiological analysis of risk factors for cancer in men were used in addition to a detailed questionnaire of those deemed eligible for this study. Their analysis, taking into account a variety of other compounding factors, indicated that digoxin use was associated with an approximate 25% reduced risk for developing prostate cancer in this cohort of men [128].

A final example of the use of in vitro screening methods for drug repurposing in oncology is the identification of statins, in combination with standard of care, as inhibitors of glioblastoma cell proliferation [129]. Glioblastomas are particularly difficult to treat, having a high incidence of mortality [130], and are also considered as orphan diseases with relatively little associated drug development [131]. Importantly, an additional constraint for a repurposed drug would be the need for compounds to cross the blood–brain barrier, increasing the challenge involved. Jiang et al. [129] screened a collection of 446 compounds available from the NIH Clinical Collection library in a U87 glioblastoma cell line, monitoring effects on cell proliferation. The most potent compounds, determined by ranking IC_{50} values, were subsequently retested against three additional glioma cell lines and also against a number of stem cell lines derived from patient material. Alongside a number of known antineoplastic agents, two statins, pitavastatin and cerivastatin, were found to be active. When subsequently screened in combination with low doses of chemotherapeutic agents such as irinotecan, pitavastatin significantly increased their effectiveness. The authors suggested that use of pitavastatin in combination with irinotecan may be useful for patients who develop side effects to chemotherapy such as diarrhea or neutropenia, allowing a reduction in the dose of chemotherapeutic while maintaining efficacy.

Perhaps the best-known example is the use of thalidomide and its structural derivate lenalidomide (Revlimid) for the treatment of multiple myeloma [132]. The recognition that thalidomide was able to exert immunomodulatory effects came via its use as a sedative in leprosy patients suffering from erythema nodosum leprosum (ENL). ENL produces painful skin lesions and other systemic effects, and at the time it was intractable. However, patients treated with thalidomide showed significant reduction in the number of lesions [133]. Thalidomide was later shown to inhibit angiogenesis, via suppression of TNF-α [134], and ultimately to be effective in multiple myeloma [135]. Another example of clinical data informing repurposing in the oncology field is the potential for use of metformin, a widely used glucose-lowering drug [136, 137]. Metformin targets activated protein kinase (AMP), which is an important regulator of tumor suppressor genes such as LKB1. In a variety of cell models, and in vivo, metformin is able to reduce cell proliferation [138] as well as inhibit cancer stem cell production [139, 140]. Finally, the antiretroviral drug nelfinavir, which is used in the treatment of HIV and which is an inhibitor of the PI3K-Akt-mTOR pathway, is the subject of a number of additional clinical oncology studies [141].

13.2.2 Repurposing of Approved Cardiovascular Drugs for Cancer Therapy

In theory, any class of non-anticancer drugs in the market has the potential to be repurposed for cancer therapy, and this notion has indeed been proven by numerous investigational, preclinical, and clinical studies including those described above. One of the hottest categories of approved drugs for anticancer repurposing is the agents for cardiovascular diseases [142–144]. Numerous experimental and epidemiological studies have reported promising anticancer activities for cardiovascular drugs, and a number of excellent review articles provide comprehensive overview of such a topic [142–144].

Despite that existing data in this area are still somewhat controversial and no final conclusion has yet been reached and no final recommendations can be made, several classes of cardiovascular drugs such as β-blockers, angiotensin converting enzyme inhibitors (ACE-Is), angiotensin II receptor blockers (ARBs), statins, aspirin, cardiac glycosides, and low-molecular-weight heparins (LMWHs) have demonstrated promising results in randomized controlled trials (RCTs). Research on the concept of biological overlap between cardiovascular and oncological diseases is gaining momentum. In fact, in both conditions, the malfunction of common regulatory mechanisms, such as the renin–angiotensin system (RAS), sympathetic nervous system (SNS), coagulation cascade, sodium–potassium ATPases, and mevalonate pathway, occurs [144]. Thus, targeting these mechanisms with well-known cardiology drugs could be a novel, promising adjuvant strategy in cancer management. Ishida et al. [142] provided analytical discussions on the idea of repurposing cardiovascular drugs in oncology based on available preclinical and clinical data.

13.2.2.1 Blood Thinner Aspirin

Aspirin is commonly used for the treatment and prevention of atherosclerotic diseases, and pharmacological targets of aspirin are two isoforms of cyclooxygenase (COX) enzyme, COX-1 and COX-2 [145]. COX-1 is a constitutive enzyme expressed in most mammalian tissues and produces thromboxane A2 (TXA2) in platelets, which promotes platelets aggregation and adherence of platelets to tumor cells and thus prevents immune cells from recognizing and eliminating them, resulting in increased distant metastasis [142]. On the other hand, COX-2 is a rapidly inducible enzyme during inflammation and dominantly produces prostaglandin E2 (PGE2) in tumor cells compared with COX-1, and PGE2 is thought to play an important role in accelerating cell proliferation and tumor growth [142]. Aspirin administered at low doses (50–100 mg daily) and high doses (>325 mg daily) selectively blocks COX-1 and COX-2 in an irreversible manner, respectively. Since the antitumor effect of aspirin was first reported in tumor-bearing mice in 1972 [146, 147], a number of subsequent experimental studies have supported this evidence [142, 148]. As most clinical trials have shown a significant reduction in cancer risk and cancer-associated death in patients taking aspirin at a low dose, but not a high dose [142, 149, 150], one important mechanism of tumor suppression by aspirin is proposed as inhibition of COX-1. With this activity, low-dose aspirin could prevent platelets from binding to tumor cells, resulting in suppression of distant metastasis and improved survival [142]. In addition, PGE2 was upregulated in colon cancer [151] and administration of PGE2 enhanced tumor growth and angiogenesis [152]. As PGE2 was significantly suppressed in human colons when aspirin was administered even at a low dose (81 mg daily) [153], suppression of PGE2 might be also important in antitumor activity of aspirin [142].

In clinical fields, a case-control study first demonstrated that aspirin use was associated with reduced risk of colorectal cancer (CRC) [142, 154]. Since then, a number of observational studies have shown that regular aspirin use significantly reduced risk of several cancers including CRC [155], esophageal cancer [156], gastric cancer [156], breast cancer [156], and prostate cancer [157–159]. In addition, Rothwell et al. [160] reported that regular aspirin use reduced not only risk of distant metastasis, but also cancer-related death [150]. Regarding the dose and the duration of aspirin, a meta-analysis of the five RCTs showed that aspirin at low dose (75–300 mg daily) reduced the 20-year incidence and mortality of CRC and that the effects of aspirin increased with the duration of the treatment [149]. Thus, aspirin could be effective for the prevention and/or the treatment of cancers [142]. However, currently aspirin should be administered only for patients with cardiovascular diseases, not for the prevention of cancer [142].

13.2.2.2 β-Blockers

Previous experimental studies have demonstrated that chronic stress, depression, and social isolation are associated with tumor progression [142, 161–164]. As catecholamines such as norepinephrine and epinephrine are elevated under chronic stress and their effects are mainly mediated through β-adrenoreceptors, activation of β-adrenoreceptors by catecholamines is believed to play an important role in tumor progression. Indeed, the presence of β-adrenoreceptors has been shown in the cell lines of breast cancer [165], pancreatic cancer [166], nasopharyngeal cancer [167], and ovarian cancer [168], and catecholamines significantly increased cell proliferation as well as cell migration in human cancer cell lines [169, 170]. Furthermore, in a mouse model of ovarian cancer, β-adrenergic stimulation not only increased angiogenesis and tumor invasion through the cyclic adenosine monophosphate (cAMP)-protein kinase A (PKA) pathway [168], but also prevented the cancer cells from apoptosis by activating focal adhesion kinase (FAK) [171]. Additionally, a recent study showed that β-adrenergic activation increased distant metastasis by 30 times through M2 macrophage infiltration in a mouse model of breast cancer [142, 172].

On the other hand, β-blockers mainly block β-adrenoreceptors, and have been investigated as treatment for malignancies as well as cardiovascular diseases [142]. Generally prescribed β-blockers are classified into three types depending on selectivity of receptor subtypes, β1 selective β-blockers such as bisoprolol, non-selective β-blockers (NSBB) such as propranolol and nadolol, and α- and β-blockers such as carvedilol, which block β1, all types of, and α- and β-adrenoreceptors, respectively [142]. All of these β-blockers are widely used for the treatment of heart failure, hypertension, ischemic heart disease, and arrhythmias in the cardiovascular field. A number of experimental studies have also shown the antitumor effects of β-blockers. In human cancer cell lines, catecholamine-induced proliferation and migration were inhibited by NSBB [169, 170], and enhanced invasiveness [173] and activated FAK by catecholamine were completely blocked by propranolol in a mouse model of ovarian cancer [171]. Moreover, in a mouse model of breast cancer, propranolol counteracted catecholamine-induced metastasis to distant tissues through M2 macrophage infiltration [172], which indicates the importance of β-adrenergic signaling in M2 macrophage. Interestingly, a recent publication demonstrated that bisoprolol improved cardiac function and survival in a dose-dependent manner in rats with cancer cachexia, suggesting the favorable effects of β-blockers in the terminal stage of cancer [142, 174].

In clinical settings, several epidemiological studies have examined the potential effect of β-blockers on the incidence and the outcome of cancer [142]. The results have been inconsistent [175–180], but some of them demonstrated that the use of β-blockers was associated with improved overall survival in patients with certain types of cancer such as breast cancer [175], ovarian cancer [176], and non-small cell lung carcinoma [177]. In addition, a recent meta-analysis of 12 clinical studies has shown that β-blocker usage was associated with significantly improved overall survival [181]. β-Blockers appeared to have a greater effect in patients with early-stage cancer or cancer treated with primary surgery than those with late-stage cancer or cancer treated without primary surgery [181]. Thus, experimental and clinical evidences have indicated the efficacy of β-blockers for the treatment of cancer. Although there are no clinical trials to clarify it prospectively, to confirm these promising effects of β-blockers on cancer, two RCTs are ongoing to investigate the preventive role of β-blockers in patients with breast cancer (NCT 00502684) and CRC (NCT 00888797) undergoing surgery with curative intent [142].

13.2.2.3 Antihypertensive ACE Inhibitors and ARBs

Angiotensin II promotes vasoconstriction and sodium reabsorption via angiotensin II type 1 (AT1) receptors in the renin–angiotensin system (RAS), resulting in increased blood pressure. Both of ACE inhibitors (ACEIs) and ARBs, agents blocking RAS, are widely used in the cardiovascular fields for the treatment of heart failure, hypertension, and old myocardial infarction [142]. In addition to the systemic RAS, recently much attention has been paid to the existence of the local RAS in tumor cells. Indeed, AT1 receptors have been identified in various types of human cancer, such as renal cell carcinoma [182], laryngeal carcinoma [183], pancreatic cancer [184], ovarian cancer [185], breast cancer [186], melanoma [187], and intrahepatic cholangiocarcinoma [188]. Moreover, immunohistochemical analysis showed co-localization of AT1 receptor and vascular endothelial growth factor (VEGF), a major angiogenic protein, in pancreatic cancer cells [189], and expression of AT1 receptor was detected more frequently in high-grade invasive ovarian cancer than in benign ovarian tumor and was positively correlated with expression of VEGF [185]. Administration of angiotensin II upregulated VEGF [185, 186, 189] and increased tumor angiogenesis [185], tumor growth [188], and tumor invasiveness [185]. Thus, these findings suggested that the local RAS controlled VEGF and tumor progression, and then blockade of the local RAS has been noted as a promising strategy for the treatment of cancer. Indeed, ACEIs suppressed VEGF expression, VEGF-induced angiogenesis and tumor growth [190, 191] and ARBs also showed similar effects in certain cancer cell lines and animal cancer models [184–186, 188, 189].

In 1998, an observational study first demonstrated that hypertensive patients taking ACEIs had a reduced cancer risk compared with patients in the control group [192]. However, following clinical studies failed to show the favorable effects of ACEIs on cancer risk or outcome [193–197]. For example, Lindholm et al. also investigated a protective role of ACEIs for cancer in elderly patients with hypertension. In this study, ACEI usage was not associated with decreased risk of new cancer occurrence [195]. Furthermore, in 2010 a meta-analysis of five RCTs revealed that the use of ARBs increased cancer risk [198]. However, subsequent several meta-analyses showed no significant association between the use of ARBs and new cancer risk [199, 200]. Judging from the results of clinical studies, there is no reason to administer ACEIs or ARBs for the prevention and/or treatment of cancer despite the favorable findings in experimental research, and further long-term prospective trials are needed [142].

13.2.2.4 Cardiac Glycosides

Cardiac glycosides were originally derived from the foxglove. Two types of cardiac glycosides, digoxin and digitoxin, have been currently prescribed to treat heart failure or to reduce heart rate in the cardiovascular field [142], whereas the relationship between cardiac glycosides and cancer have been noted since Shiratori et al. [201] reported the antiproliferative effect of cardiac glycosides in cancer cells in 1967. Sodium- and potassium-activated adenosine triphosphatase (Na^+-K^+-ATPase), the primary target of cardiac glycosides, exports three sodium ions in exchange for two potassium ions using ATP as an energy source, thereby maintaining the cell membrane potential. Indeed, increased expression of Na^+-K^+-ATPase was found in gastric [202] and bladder cancer cells [203], and elevated Na^+-K^+-ATPase activity was observed in highly invasive human renal carcinoma cells [204]. Cardiac glycosides bind to Na^+-K^+-ATPase and disrupt its abilities in tumor cells. Namely, cardiac glycosides decreased the membrane potential and increase intracellular Na^+, which caused the induction of apoptosis in human cancer cells [205], and also increased intracellular Ca^{2+} in human prostate adenocarcinoma cells, resulting in activation of calcineurin and transcriptional upregulation of Fas ligand, which can induce apoptosis [206]. Some other mechanisms of cardiac glycosides such as suppression nuclear factor-κB [207] and inhibition of DNA topoisomerase II [208] were also shown to induce apoptosis in tumor cells. The complex mechanism in cardiac glycoside-induced apoptosis has been already well documented [209–211].

In clinical settings, there were several studies evaluating the effects of cardiac glycosides on the development and progression of cancer [142]. Early observational studies showed that the use of cardiac glycosides reduced the rate of recurrence and the malignant grade of breast cancer [212, 213] and a subsequent 20-year follow-up revealed that patients with cardiac glycosides had a significantly lower mortality rate (6%) than those without cardiac glycosides (34%) [142, 214]. These findings seemed to indicate the antitumor effect of cardiac glycosides, but the results from recent reports have been inconsistent with them. In 2001, a large cohort study (n = 9271) revealed that patients taking cardiac glycosides had a higher incidence of cancer compared with controls [215]. Moreover, the use of cardiac glycosides was associated with an increased risk of invasive breast cancer among postmenopausal women [216] and the incidence of estrogen receptor-positive breast cancer was significantly higher than that of estrogen receptor-negative breast cancer in women taking cardiac glycosides [217]. Cardiac glycosides also increased the risk of corpus uteri cancer, whereas did not affect the incidence of ovary cancer and cervix cancer [218] and reduced the risk of prostate cancer [128]. These findings suggest that the estrogenic effect of cardiac glycosides might play an important role in inhibiting cancer, but further experimental studies and RCTs setting cancer-related events as primary endpoints should be performed to confirm this hypothesis.

13.2.2.5 Lipid-Lowering Drugs Statins

Statins, competitive inhibitors of 3-hydroxy-3-methylglutaryl-coenzyme A reductase, block the formation of mevalonate in the mevalonate pathway to synthesize cholesterol in liver and are commonly administered for the treatment of hyperlipidemia and the secondary prevention of atherosclerotic diseases such as myocardial infarction and ischemic stroke [142]. Geranylgeranyl pyrophosphate (GGPP) and farnesyl pyrophosphate (FPP) are downstream products of the mevalonate pathway and used as substrates in protein prenylation, which is essential for localization of proteins in cell membranes [219]. As GGPP and FPP are also inhibited by statins, it has been proposed that statins could cause apoptosis in tumor cells. Indeed, experimental studies have shown that statins suppressed proliferation in multiple human cancer cell lines through this inhibitory activity [220–222]. Despite this promising antitumor effect of statins, the results of clinical studies are controversial. Initial clinical trials showed that statins were associated with a significant reduction in overall cancer incidence [223, 224]. Some subsequent meta-analyses have demonstrated that statins significantly reduced the risk of prostate [225], esophageal [226], and gastric cancer [227], while other recent meta-analyses failed to show benefits of statins in cancer-associated mortality [228], and in the risk of lung [229] and skin cancer [230]. On the other hand, statins enhanced the efficacy of treatment in acute myeloid leukemia [231] and hepatocellular carcinoma [232], when combined with chemotherapeutic agents. Further clinical investigation should be performed, since in most of clinical studies to investigate the role of statins, the primary endpoint was not set on cancer-related events and the observation period was not long enough to observe the development and/or prevention of cancer [142].

13.2.2.6 Antiarrhythmic Drug Lidocaine

Lidocaine, one of the most used local anesthetics and class Ib antiarrhythmic agent as a sodium channel blocker, in clinical settings, has been found to show multi-activities, including potential in cancer treatment [233]. Growing evidence shows that lidocaine may not only work as a chemosensitizer that sensitizes other conventional chemotherapeutics to certain resistant cancer cells, but also could suppress cancer cells growth by single use at different doses or concentrations. Lidocaine could suppress cancer cell growth in vitro and in vivo via multiple mechanisms, such as regulating epigenetic changes and promoting pro-apoptosis pathways, as well as regulating ABC transporters, metastasis, and angiogenesis [233, 234]. In addition to acting as a chemosensitizer in the treatment of cancer, lidocaine also exerts inhibitory effects toward various cancer cells and in tumor xenograft models by single use at higher concentrations [233]. For example, lung cancer, categorized into two main subtypes, small-cell lung cancer (SCLC) and NSCLC, is the primary cause of cancer-related death worldwide. Lidocaine has been found to kill lung cancer cells in vitro and in vivo [233]. Beck-Schimmer's group confirmed that certain amide-type local anesthetics, including lidocaine, could inhibit cancer cells by suppressing tumor necrosis factor alfa (TNFα)-induced Src activation, a mechanism related to its anti-inflammatory effects [235] (Piegeler et al. 2015). Further study showed that lidocaine (1 nM to 100 μM, at clinically relevant concentrations) remarkably suppressed the activation/phosphorylation of AKT, focal adhesion kinase (FAK), and caveolin-1 induced by TNFα in lung adenocarcinoma NCI-H838 cells. Matrix-metalloproteinase 9 (MMP-9) is an enzyme that catalyzes the degradation of the extracellular matrix and basal lamina to allow migrating cancer cells to escape the primary tumor site, or to locate to a new spot to form the satellite lesions [236, 237], which is important to form new solid tumors. Lidocaine significantly suppressed the MMP-9 secretion as confirmed by the immunosorbent assay, and it completely inhibited the cancer cell invasion, indicating its potential in preventing cancer invasion and metastasis [235]. Another local anesthetic ropivacaine exhibited similar efficacy but at lower concentrations compared to lidocaine [235], which necessitates a broad screening of these type anesthetics.

To reveal more details on the mechanisms of the anticancer effects of lidocaine in lung cancer, Zhang et al. checked the related proteins by real-time quantitative PCR (qPCR) and Western blot assay in the samples of lung cancer patients. They confirmed that Golgi phosphor protein 2 (GOLPH2), a Golgi phosphoprotein that had been described as a potential serum marker of hepatocellular carcinoma, prostate cancer, and renal cell cancer [238, 239], was overexpressed in the patients with higher grade malignancy. Lidocaine (0.5, 2, and 8 mM, which were determined by MTT and BrdU assays) was found to suppress the expression of GOLPH2 in NSCLC A549 cells, leading to cell cycle arrest at the G1 phase and cell proliferation inhibition [240]. In addition, lidocaine showed activity in suppressing postoperative atrial fibrillation (POAF), which occurs frequently after lung cancer surgery [241], indicating its broad potent applications in lung cancer treatment.

In addition, lidocaine also exhibits activities in suppressing breast cancer growth and cancer cell metastasis [233, 242–244], liver cancer [233, 245], gastric cancer [233, 246], colorectal cancer [233, 247], glioma [233, 248], etc.

13.2.3 Repurposing Antidiabetic Metformin for Cancer Therapy

Metformin, a biguanide, is the first-line treatment for type 2 diabetes mellitus as an oral glucose-lowering agent. It decreases hepatic gluconeogenesis and improves insulin sensitivity by increasing peripheral glucose uptake and use, lowering both basal and postprandial plasma glucose [249, 250]. It was first approved in the United Kingdom in 1958, followed by Canada in 1972, and then the United States in 1994 [251]. The most common side effects that occur in >5% of patients include diarrhea, nausea/vomiting, flatulence, indigestion, abdominal discomfort, lack of energy, and headache [250]. Lactic acidosis is a rare but potentially fatal side effect that can be avoided by checking and monitoring renal function. This tolerable and safe profile, in addition to its low cost as a generic drug, has made metformin an excellent candidate for repurposing. The ease of access to metformin has led to a multitude of preclinical and clinical investigations, as a single agent and in combination with other agents, for both cancer prevention and treatment, according to Web of Science (http://wok.mimas.ac.uk).

Metformin has been associated with a decreased risk of breast, colon, liver, pancreas, prostate, endometrium, and lung cancer across meta-analyses [249]. Besides time bias, individual organ-specific confounders are also important. For lung cancer, overall and time-unbiased analyses point to a protective effect of metformin, whereas adjustment for smoking, the leading cause of lung cancer, resulted in loss of significance. A recent meta-analysis confirmed the role of metformin in colorectal adenoma, a precursor to colorectal cancer [252]. However, they also noted publication bias against negative studies. Metformin use has previously been shown to be associated with a reduced risk of colorectal cancer [253–255].

Compared with cancer incidence, the effect of metformin on cancer mortality is informed by fewer studies and, thus, site-specific analyses of the effects of metformin on overall mortality have not been performed in older meta-analyses [256]. However, more recent meta-analyses have focused on the impact of metformin on site-specific cancer mortality and overall survival [257–267]. These analyses have shown marked reductions in cancer-specific mortality for colon, lung and early-stage prostate cancer, and improvements in overall survival for breast, colon, gynecological (endometrial and ovarian), liver, lung, prostate, and pancreatic cancer.

To date, there are 17 clinical trials that have examined the effect of metformin on cancer-related biomarkers and reported results [249, 268]. The studies were conducted in various disease settings, spanning at risk participants, individuals with preneoplastic lesions, individuals who were awaiting surgical excision, and those receiving metformin in an adjuvant setting. An analysis of clinical trials registered on ClinicalTrials.gov (clinicaltrials.gov/) revealed an additional 23 trials examining the effect of metformin in participants at risk of cancer, 30 presurgical studies and 30 studies in the adjuvant setting [249]. This is a significant increase from a similar analysis of ongoing studies conducted in 2011, in which there were three studies examining the effect of metformin in participants at risk for cancer, nine presurgical studies and seven studies in the adjuvant setting. The ongoing studies examine similar endpoints, including pharmacodynamic markers, such as tumor/tissue levels of pS6K, or other measures of AMPK signaling. Many of the studies examine surrogate endpoints for cancer, such as atypia or breast density for breast cancer, or change in prostate specific antigen (PSA) for prostate cancer. As has been reported in many studies, accrual to clinical trials is challenging with many pilot studies closing because of poor accrual.

13.3 Drug Repurposing in Cardiometabolic Disorders (CMDs)

An estimated 47 million people in the United States are living with cardiometabolic disorders (CMDs), putting them at an increased risk of developing heart disease or type 2 diabetes. CMDs represent a cluster of diseases, including cardiovascular diseases (CVDs), metabolic syndrome (MS) and diabetes mellitus (DM), consequent to a cluster of interrelated risk factors, primarily hypertension, elevated fasting blood sugar, dyslipidemia, abdominal obesity, and elevated triglycerides [269]. CMDs remain the principal cause of death in both developed and developing countries, accounting for nearly 32% of all deaths worldwide per year [270]. CMDs are now recognized as a disease entity by the World Health Organization and the American Society of Endocrinology [271]. People with CMS are two times more likely to die from coronary heart disease and three times more likely to have a heart attack or stroke than those who do not have the syndrome, and CVDs remain the leading cause of death and disability worldwide. In addition, dyslipidemia, angina, arrhythmia, cardiac failure, myocardial infarction (MI), and diabetes mellitus represent the leading killer with an estimated 19 million people died from CMDs in 2012. By 2030 more than 23 million people will die annually from CVDs [270]. There is therefore a great need to implement the pharmacological armamentarium, considering also the need to balance the therapeutic and the side effects.

Nonetheless, existing drugs are not efficient enough to reduce the disease burden as well as mortality. Therefore, there is an urgent demand for new drugs in this area to reduce the mortality and control the associated disability. New drug discovery (NDD) in CMDs has become more challenging for the last couple of decades due to increased expenses and decreased success rate [272, 273]. In such a scenario, drug repurposing in the CMDs appears promising for introducing existing drugs for new therapeutic indication. Repurposing is quite an old strategy dating back to 1960s and mainly followed by serendipitous observations during clinical use of drugs [272]. With rapid advancing of polypharmacology-based approaches in drug discovery, drug repurposing has become an important alternative to NDD for the development of CVD and CMD drugs for which safety profile is an exceptionally strict requirement. A major advantage of repurposing is that the safety profile of the drug is well established thus reducing the chances of failure due to adverse toxic effects.

Though only a few compounds have been approved for new indications in the field of metabolic disorders, there are a number of substances which have the potential to become reprofiled in a new indication. Finsterer and Frank [274] categorized repurposed drugs for metabolic disorders into three groups. Group A contains both the original and repurposed indications concerning metabolic disorders. Group B comprises drugs that were originally approved for non-metabolic disorders but show beneficial effects for metabolic disorders after repurposing. Group C comprises drugs that were originally approved for metabolic disorders and are effective for non-metabolic disorders in their repurposed indication. Repurposed drugs in the field of metabolic disorders of group A include tetrahydrobiopterin that was originally indicated for phenylketonuria and now also approved for tetrahydrobiopterin deficiency, coenzyme-Q that was originally approved for primary coenzyme-Q deficiency and repurposed for statin-myopathy, and colesevelam that was originally approved for reducing elevated low-density lipoprotein (LDL)-cholesterol (LDL-C) and now being approved for type-2 diabetes. An example of group C is phenylbutyrate that was originally approved for urea-cycle disorders and meanwhile gained approval for progressive familial intrahepatic cholestasis type 2 due to mutations in the ABCB11 gene. Still additional compounds used to treat metabolic (non-metabolic) disorders show promising effects in non-metabolic (metabolic disorders) after repurposing in cell and tissue models. Future investigations will need to identify which candidate drugs may leave the pipeline status to acquire approval for new indications [274].

Zhang et al. [275] analyzed data from genome wide association studies (GWAS), proteomics and metabolomics studies and revealed a total of 992 proteins as potential antidiabetic targets in human. They then retrieved the information on the drugs that target these 992 proteins from the Therapeutic Target Database (TTD) and identified 108 of these proteins as drug targets with drug projects information. Research and preclinical drug targets were excluded and 35 of the 108 proteins were selected as druggable proteins. Among them, five proteins were known targets for treating diabetes. Based on the pathogenesis knowledge gathered from the OMIM and PubMed databases, 12 protein targets of 58 drugs were found to have a new indication for treating diabetes. CMap (connectivity map) was used to compare the gene expression patterns of cells treated by these 58 drugs and that of cells treated by known antidiabetic drugs or diabetes risk causing compounds. As a result, 9 drugs were found to have the potential to treat diabetes. Among the 9 drugs, 4 drugs (diflunisal, nabumetone, niflumic acid, and valdecoxib) targeting COX2 (prostaglandin G/H synthase 2) were repurposed for treating type 1 diabetes, and 2 drugs (phenoxybenzamine and idazoxan) targeting ADRA2A (α_{2A} adrenergic receptor) had a new indication for treating type 2 diabetes. These findings indicated that "omics" data mining-based drug repurposing is a potentially powerful tool to discover novel antidiabetic indications from marketed drugs and clinical candidates.

Two representative compounds undergoing drug repurposing for CMDs are described below.

13.3.1 Colchicine

Atherosclerosis is a slowly progressive disease leading to ischemic heart disease that is responsible for >15% of all global deaths, being among the top 3 most important noncommunicable diseases in almost every developed country. Management of atherosclerosis has improved significantly during the last 50 years. Controlling lifestyle, dyslipidemia, and thrombosis has reduced associated mortality rates in the last 2 decades by 25–50%. However, even with these treatment strategies, any given individual with stable coronary artery disease (SCAD) may have a 5-year risk of at least 15% for cardiovascular death or any major adverse cardiovascular event. This residual risk reflects a mechanism in the biology of atherothrombosis that is incompletely addressed by lipid lowering and antithrombotic therapy, which is modulation of the inflammatory response that drives the atherosclerotic process. Cholesterol crystal-induced inflammation leading to plaque destabilization is believed to be an important factor contributing to plaque instability and might well be amenable to treatment with anti-inflammatory drugs [276]. The ability of novel anti-inflammatory drugs to modify this inflammatory response has been the subject of recent research, yielding varying effects.

The decades-old drug colchicine has emerged with promising results in the treatment of SCAD through drug repurposing. Colchicine is an inexpensive drug that is readily available throughout the world, with proven safety and efficacy for several inflammatory diseases. It has anti-inflammatory properties with the potential to address both the direct and indirect inflammatory mechanisms in the plaque. Fiolet et al. [276] performed a literature search in MEDLINE (PubMed), EMBASE, and the Cochrane Central Register of Controlled Trials, as well as in the clinical trial registries, to identify finished and ongoing clinical studies on colchicine in SCAD. Preclinical findings of colchicine in SCAD have shown protective effects on surrogate outcomes, such as myocardial infarction size and postangioplasty restenosis. Retrospective cohort studies in patients with gout report a lower incidence of combined cardiovascular outcomes in those treated with colchicine. Thus far, one prospective, randomized clinical trial has provided evidence on a possible protective effect of colchicine in stable coronary artery disease. Meta-analysis of trials of colchicine in multiple CVD revealed a decrease in myocardial infarction with varying levels of evidence. Currently, 5 major clinical trials involving >10,000 patients are recruiting patients, all focusing on major cardiovascular outcomes.

Preliminary evidence demonstrating that adding 0.5 mg of colchicine per day to statin and antiplatelet therapy reduced the risk of acute coronary events in patients with stable coronary artery disease has raised the hope that it may prove effective for the long-term secondary prevention of cardiovascular disease [277, 278]. This was supported by the LoDoCo (Low Dose Colchicine) pilot trial showing that the anti-inflammatory drug colchicine 0.5 mg once daily appears safe and effective for secondary prevention of cardiovascular disease [279]. Colchicine's low cost and long-term safety suggest that if its efficacy can be confirmed in a rigorous trial, repurposing it for secondary prevention of CVD would have the potential to impact the global burden of CVD. In a randomized, controlled, double-blind trial, we assigned patients with chronic coronary disease to receive 0.5 mg of colchicine once daily or matching placebo [280]. The primary end point was a composite of cardiovascular death, spontaneous (nonprocedural) myocardial infarction, ischemic stroke, or ischemia-driven coronary revascularization. The key secondary end point was a composite of cardiovascular death, spontaneous myocardial infarction, or ischemic stroke. A total of 5522 patients underwent randomization; 2762 were assigned to the colchicine group and 2760 to the placebo group [280]. The median duration of follow-up was 28.6 months. A primary end-point event occurred in 187 patients (6.8%) in the colchicine group and in 264 patients (9.6%) in the placebo group. A key secondary end-point event occurred in 115 patients (4.2%) in the colchicine group and in 157 patients (5.7%) in the placebo group. The incidence rates of spontaneous myocardial infarction or ischemia-driven coronary revascularization (composite end point), cardiovascular death or spontaneous myocardial infarction (composite end point), ischemia-driven coronary revascularization, and spontaneous myocardial infarction were also significantly lower with colchicine than with placebo. However, the incidence of death from noncardiovascular causes was higher in the colchicine group than in the placebo group. The conclusion is that the risk of cardiovascular events is significantly lower among those who received 0.5 mg of colchicine once daily than among those who received placebo.

13.3.2 Metformin

Although introduced for use as a diabetic medication in 1957, metformin remains the cornerstone of diabetic drug management in patients with type 2 diabetes mellitus (T2D). Its widespread use has largely been underpinned by the United Kingdom Prospective Diabetes Study that reported lower cardiovascular mortality and morbidity in patients treated with metformin in comparison with alternative glucose-lowering drugs, despite similar glycemic control. A recent meta-analysis suggests that the cardiovascular effects of metformin could be smaller than that reported by United Kingdom Prospective Diabetes Study; however, this should be interpreted with caution because there have only been a small number of randomized controlled trials [281]. Although patients who have CVDs with T2D comorbidity are likely to benefit most from metformin, indications of cardiovascular benefit over other diabetes treatments have driven interest in repurposing metformin to treat CVDs, irrespective of diabetes status.

Identified before the era of target-driven drug discovery programs, metformin's cellular mechanism is poorly established. The most likely cellular effect underlying antihyperglycemic responses is the inhibition of mitochondrial enzymes, including complex I. Mitochondrial suppression by metformin activates AMP-activated protein kinase; however, AMP-activated protein kinase-independent targets also contribute to effects on glycemia, and, in addition, benefits of the drug in CVD may also be exerted by mechanisms distinct from its metabolic actions. Metformin has been widely used as an antidiabetic drug to treat patients with DM. Its cardioprotective role is being increasingly realized beyond its glucose lowering effect, although there may be some overlap of these two properties in the systemic cardioprotective effects.

In a recent work, for example, investigators have been studying the anti-inflammatory effects of the drug through suppression of the NF-κB (nuclear factor κB) inflammatory signaling pathway [282]. These effects can be traced back to mitochondrial inhibition, consistent with other evidence of mitochondrial regulation of inflammation [283]. Rena and Lang [283] found that metformin suppressed plasma cytokines in patients without diabetes mellitus who had heart failure, including the aging-associated cytokine CCL11 (C-C motif chemokine ligand 11). Blockade of CCL11 can suppress aspects of age-related cellular dysfunction, and further investigation is required to determine whether the observed effects of metformin on mammalian longevity, where suppression of NF-κB is also observed, can be attributed, at least in part, to cardioprotective benefits arising from suppression of this and other cytokines.

Intriguingly, existing nonsteroidal anti-inflammatory drugs and anti-tumor necrosis factor drugs have shown limited utility in CVD treatment [283], despite that there is still a strong belief that anti-inflammatory strategies are worth pursuing, key exemplars of which are the recent CIRT (Cardiovascular Inflammation Reduction Trial; ClinicalTrials.gov. Unique identifier: NCT01594333) and CANTOS (Cardiovascular Risk Reduction Study [Reduction in Recurrent Major VC Disease Events]; ClinicalTrials.gov. Unique identifier: NCT01327846) trials [284]. If inflammation is to be targeted successfully in CVD, it is likely that other agents, with different anti-inflammatory mechanisms, will need to be studied. Control of immunity through changes in metabolism-immunometabolism [285], such as is exhibited by metformin, is becoming recognized as an additional potential site of therapy even in nondiabetic age-related disease.

Despite encouraging data in T2D, recent findings on CVD benefit from prospective studies in groups without T2D have so far been mixed [283]. In subjects without diabetes mellitus who have cardiac syndrome X with normal coronary arteriography, but two consecutive positive exercise tolerance tests, an 8-week period of metformin treatment improved maximal ST-segment depression, Duke score, and chest pain incidence in comparison with placebo [286]. The proof-of-concept TAYSIDE trial (Metformin in Insulin Resistant Left Ventricular Dysfunction; ClinicalTrials.gov. Unique identifier: NCT00473876) in patients without diabetes mellitus who have heart failure and insulin resistance found a beneficial effect of metformin on its secondary outcome of V_E/V_{CO_2} slope, a prognostic measure of exercise capacity. However, the CAMERA trial (Carotid Atherosclerosis: Metformin for Insulin Resistance Study; ClinicalTrials.gov. Unique identifier: NCT00723307) on patients with insulin resistance but without diabetes mellitus who had coronary artery disease already taking statins, showed no effect of metformin on carotid artery intima-media thickness. In addition, the GIPS-III trial (Metformin to Reduce Heart Failure After Myocardial Infarction; ClinicalTrials.gov. Unique identifier: NCT01217307) on metformin in patients who have ST-segment–elevation myocardial infarction without diabetes mellitus showed little or no effect of metformin. In contrast, the recent REMOVAL trial (Reducing with Metformin Vascular Adverse Lesions of Type 1 Diabetes; ClinicalTrials.gov. Unique identifier: NCT01483560) in type 1 diabetes mellitus did show an effect on the prespecified tertiary end point of maximal carotid artery intima-media thickness, despite no sustained effect on glycemia, possibly consistent with a cardiovascular effect of the drug that is not attributable to changes in glycemia. Further studies with metformin are ongoing. Recent evidence from the EMPA-REG OUTCOME trial (BI 10773 [Empagliflozin] Cardiovascular Outcome Event Trial in Type 2 Diabetes Mellitus Patients; ClinicalTrials.gov. Unique identifier: NCT01131676) showed that the sodium-glucose cotransporter 2 inhibitor, empagliflozin, another T2D drug, lowered CVD mortality in patients with T2D with prevalent atherosclerotic cardiovascular disease. This may largely be independent of effects on glycemia, similar to the multifactorial picture emerging with metformin. The present clinical equipoise for metformin in CVD without T2D has led to a call for cardiovascular end point trials of metformin in the absence of diabetes mellitus [281].

Taken together, a variety of evidence suggests that metformin has potential as a treatment for CVD irrespective of diabetes status, but prospective randomized trials to repurpose the drug in individuals without diabetes mellitus have so far been largely neutral on cardiovascular-related outcomes. Future randomized controlled trials, incorporating better stratification/targeting, using, for example, novel preclinical markers of diabetes-independent inflammation, might establish the utility of metformin and therefore help to prioritize metformin repurposing over other emerging CVD treatments.

Nonetheless, the anti-atherosclerotic effects of metformin, independent of its glycemic control remain unclear and are an area of active research [287]. Evidence backing the pleiotropic effects of metformin in reducing CVD-related events necessitates further exploration with regard to the mechanism of drug action in various tissue compartments. The recent evidence pointing toward its therapeutic benefit in heart failure with persevered ejection fraction in clinical and preclinical studies underlines the need to explore the breadth of the pleiotropic systemic effects of metformin [288]. Recent study also reported that metformin offers therapeutic benefit during heart failure with preserved ejection fraction by lowering titin-based passive stiffness in mice model [289], is associated with improved survival and decreased incidence of adverse cardiac events in peripheral arterial disease patients [290] and with a lower below-the-knee arterial calcification score [291]. Moreover, metformin is able to prevent cardiac dysfunction in a murine model of adult congenital heart disease [288]. A proper understanding of these pleotropic effects will allow us to tailor the dose of the drug and remain abreast about its potential side effects in patients receiving them. Apart from their use in DM patients, its role in primary prevention of cardiovascular outcomes in "at-risk" population, akin to statins, requires further exploration. Prospective population studies and randomized clinical trials need to be conducted to identify will allow us to identify a subset of patients who may benefit from early administration of metformin. This will further aid in disease surveillance and intervention through enhancement of primary and secondary prevention of CVD.

13.4 Drug Repurposing in Alzheimer's Disease

There is much in vitro and in vivo evidence that drugs designed for use in a wide range of non-central nerve system (non-CNS) disorders have activity in CNS models. However, many of these CNS-focused repurposing examples have primarily arisen from specific hypotheses or serendipity, such as the use of the antiviral drug adamantine for early-onset Parkinson's disease, rather than screening efforts [99, 292].

In a recent wide-ranging review, Corbett et al. [293] highlighted the potential for drug repurposing into Alzheimer's disease. Using literature analysis, they evaluated the preclinical and, when available, clinical evidence and suggested a number of compounds and associated mechanisms that might be worthy of future study. Of particular note was the effect of calcium channel blockers (CCBs). In one clinical trial, nitrendipine, a widely used antihypertensive and angina drug, appeared to reduce the risk of developing dementia in a cohort of subjects older than 60 years by up to 55% [294]. Preclinical studies have indicated that CCBs are able to reduce the production and oligomerization of Aβ as well as improve cell survival. They went on to highlight an in vitro study whereby nine CCBs prevented Aβ production in a cell model, although the concentrations required were higher than those used therapeutically for CVD. In addition to the activity of CCBs, angiotensin receptor inhibitors, glucagon-like peptide 1 (GLP1) analogues, tetracycline antibiotics, and retinoids were also implicated.

Alzheimer's disease (AD) is the most prevalent form of dementia and represents one of the highest unmet requirements in medicine today. There is shortage of novel molecules entering into market because of poor pharmacokinetic properties and safety issues. Drug repurposing offers an innovative approach to drug discovery with great potential in the field of AD and dementia therapeutics [295]. In AD, existing therapeutic agents only provide symptomatic benefits and do not play a role in disease modification, therefore, an alternative strategy of repurposing can be used to inhibit neurodegeneracy process and other pathological complications. In a study focusing on Alzheimer's, Hayes et al. [296] screened a panel of 89 FDA-approved oncology drugs in a Chinese hamster ovary (CHO) cell line expressing the amyloid precursor protein 751WT. Aβ production was reduced in response to bischloroethylnitrosourea (BCNU; Carmustine; BiCNU), a licensed chemotherapeutic used for the treatment of a number of tumor types, including Hodgkin and non-Hodgkin lymphomas. Importantly, the effect of BCNU was also translated in vivo; in an APdE9 mouse model, in which the mice are predisposed to the generation of amyloid plaques, total amyloid plaque burden was significantly reduced after chronic (60-day) administration in comparison to vehicle-treated littermates [99, 296]. In a different neurodegenerative disorder, Parkinson's disease, the Michael J. Fox Foundation (https://www.michaeljfox.org) has brought considerable focus to the potential of drug repurposing, with research grants being awarded to a number of programs, such as for the study of amitriptyline, a tricyclic antidepressant, and of GLP1 agonists and DPP4 inhibitors, which are used to treat type 2 diabetes [297].

Drug repurposing in CNS disorders has proven particularly intractable, such as for ischemic stroke [298–300]. Hess and Fagan [301], for example, have shown that the broad-spectrum antibiotic minocycline is an effective adjunct to the use of tissue plasminogen activator (tPA) in a rat model of ischemic stroke, and this observation has been clinically tested with benefit being observed in combination with aspirin in male patients [302]. However, in a number of clinical trials, the blood pressure-lowering effects of angiotensin-II antagonists, such as candesartan, have not proven beneficial [303], indicating that, at least for stroke, the source of the drug is no predictor of success in humans.

Clinical trials have failed to produce any new AD drugs for Food and Drug Administration (FDA) approval since 2003 [304, 305]. Bauzon et al. assessed repurposed AD therapies represented in Phase I, Phase II, and Phase III of the current AD pipeline as registered on ClinicalTrials.gov as of February 27, 2020 [305]. They identified 53 clinical trials involving 58 FDA-approved agents. Seventy-eight percent of the agents in trials had putative disease-modifying mechanisms of action. Of the repurposed drugs in the pipeline 20% are hematologic–oncologic agents, 18% are drugs derived from cardiovascular indications, 14% are agents with psychiatric uses, 12% are drug used to treat diabetes, 10% are neurologic agents, and the remaining 26% of drugs fall under other conditions. Intellectual property strategies utilized in these programs included using the same drug but altering doses, routes of administration, or formulations. Most repurposing trials were supported by Academic Medical Centers and were not funded through the biopharmaceutical industry. They compared the results to a European trial registry and found results similar to those derived from ClinicalTrials.gov. They therefore concluded that drug repurposing is a common approach to AD drug development and represents 39% of trials in the current AD pipeline. Therapies from many disease areas provide agents potentially useful in AD. Most of the repurposed agents are generic and a variety of intellectual property strategies have been adopted to enhance their economic value.

A computational method based on ligand–protein interaction was adopted in a study reported by Kumar et al. to explore potential antipsychotic drugs for the treatment of AD [306]. The screening of approximately 150 antipsychotic drugs was performed on five major protein targets (AChE, BuChE, BACE 1, MAO, and NMDA) by molecular docking. For each protein target, the best drug was identified on the basis of dock score and glide energy. The top hits were then compared with the already known inhibitor of the respective proteins. Some of the drugs showed relatively better docking score and binding energies as compared to the already known inhibitors of the respective targets. Molecular descriptors like molecular weight, number of hydrogen bond donors, acceptors, predicted octanol/water partition coefficient and percentage human oral absorption were also analyzed to determine the in silico pharmacokinetic (ADME) properties of these drugs and all were found in the acceptable range and follows Lipinski's rule. Their results led to unravel the potential of leading antipsychotic drugs such as pimozide, bromperidol, melperone, anisoperidone, benperidol, and anisopirol against multiple targets associated with AD. Benperidol was found to be the best candidate drug interacting with different target proteins involved in AD.

At present, about 30% of clinical trials for AD in progress around the world use the drug repurposing method and hold potential in halting the current deadlock in treatment options [300]. By using drugs approved for other indications, these clinical trials target dysregulated pathways in AD with different or a combination of modes of action, including anti-amyloid, cardiovascular, anti-tau, anti-inflammatory, immunomodulatory, metabolic, neuroprotective, and neurotransmission-based approaches. For instance, antidiabetic drugs, such as insulin, metformin, liraglutide, and dapagliflozin, and cardiovascular drugs, such as cilostazol, candesartan, telmisartan, prazosin, and dabigatran, could serendipitously provide previously unearthed benefits in AD. This is in line with recent thinking, which views AD as a complex multifactorial disorder, not dominated by one dominant biological factor, such as amyloid-β, and likely a confluence of many pathobiological mechanisms, including vascular dysregulation.

13.5 Drug Repurposing for Rare and Orphan Diseases

A rare disease is any disease that affects a small percentage of the population. Orphan diseases, including rare diseases, are neglected conditions whose treatments are often not considered profitable due to their cost to develop and limited patient population, except by the government granting economically advantageous conditions to creating and selling such treatments [307–309]. Orphan drugs are ones so created or sold. There is no single, widely accepted definition for rare diseases. Some definitions rely solely on the number of people living with a disease, and other definitions include other factors, such as the existence of adequate treatments or the severity of the disease. Diseases that are statistically rare, and no single cut-off number has been agreed upon for which a disease is considered rare [307–309]. A disease may be considered rare in one part of the world, or in a particular group of people, but be common in another. In the United States, the Rare Diseases Act of 2002 defines rare disease strictly according to prevalence, specifically "any disease or condition that affects fewer than 200,000 people in the United States" or about 1 in 1500 people. This definition is essentially the same as that of the Orphan Drug Act of 1983, a federal law that was written to encourage research into rare diseases and possible cures. However, the European Commission on Public Health defines rare diseases as "life-threatening or chronically debilitating diseases which are of such low prevalence that special combined efforts are needed to address them." The term low prevalence is later defined as generally meaning fewer than 1 in 2000 people. The US organization Global Genes has estimated that more than 300 million people worldwide are living with one of the approximately 7000 diseases they define as "rare" in the United States [310].

Most rare diseases are genetic and thus are present throughout the person's entire life, even if symptoms do not immediately appear. Many of these diseases are caused by infrequent genetic disorders and loss-of-function mutations in specific proteins that may not be amenable to small-molecule therapies and appear early in life, and about 30% of children with rare diseases will die before reaching their fifth birthday [311]. With only four diagnosed patients in 27 years, ribose-5-phosphate isomerase deficiency is considered the rarest known genetic disease [312]. Most rare diseases still lack approved treatments despite major advances in research providing the tools to understand their molecular basis, as well as legislation providing regulatory and economic incentives to catalyze the development of specific therapies [313]. While more than 7000 rare diseases have been identified, only about 5% benefit from a licensed treatment. As the majority of these diseases are life-threatening, these facts underscore the need for new drugs. Addressing this translational gap is a multifaceted challenge, for which a key aspect is the selection of the optimal therapeutic modality for translating advances in rare disease knowledge into potential medicines, known as orphan drugs [314].

It is logical, therefore, that drug repurposing can be a more attractive approach than de novo compound discovery [315]. In addition, it may also be particularly fruitful to combine computational approaches in orphan diseases as a means of identifying repurposing opportunities and also for target validation [316]. The literature reports a number of approaches aimed at generating frameworks to identify both druggable targets and repositionable drugs for rare diseases [317–320]. Over 200 orphan drugs are registered by the European Medicines Agency (EMA) so far, for an estimated total number of between 5000 and 8000 rare diseases [321]. The discovery and development of innovative "new molecular entities" is the least common way. More often, existing drugs are repurposed, which involves lower R&D costs while companies may still benefit from intellectual property (IP) rights.

As an example, the antifungal ciclopirox was repurposed for the treatment of congenital erythropoietic porphyria (CEP) [322]. There are currently no approved therapies for CEP, a multisystem disease resulting from deficient activity of the hem biosynthetic pathway enzyme uroporphyrinogen III synthase (UROIIIS) [323]. Through a screen of 2500 chemical fragments, followed by structural comparison of the most validated fragments against a library of 1800 FDA-approved drugs, Urquiza et al. identify the antifungal ciclopirox as being able to bind to and stabilize UROIIIS, thereby restoring activity of the enzyme [322, 323]. In a genetic mouse model of CEP, oral ciclopirox administered for up to 45 days alleviated clinical symptoms of the disease, without signs of toxicity.

Govindaraj et al. employed structure-based matching of drug-binding pockets for drug repurposing [324]. In order to find new targets for known drugs ultimately leading to drug repositioning, they recently developed eMatchSite, a new computer program to compare drug-binding sites. In this study, eMatchSite is combined with virtual screening to systematically explore opportunities to reposition known drugs to proteins associated with rare diseases. The effectiveness of this integrated approach is demonstrated for a kinase inhibitor, which is a confirmed candidate for repositioning to synapsin Ia. The resulting dataset comprises 31,142 putative drug–target complexes linked to 980 orphan diseases. The modeling accuracy is evaluated against the structural data recently released for tyrosine–protein kinase HCK. To illustrate how potential therapeutics for rare diseases can be identified, the investigators discuss a possibility to repurpose a steroidal aromatase inhibitor to treat Niemann–Pick disease type C. Overall, the exhaustive exploration of the drug repositioning space exposes new opportunities to combat orphan diseases with existing drugs (DrugBank/Orphanet repositioning data are freely available to research community at https://osf.io/qdjup/).

The International Rare Diseases Research Consortium (IRDiRC) set up the multi-stakeholder Data Mining and Repurposing (DMR) Task Force to examine the potential of applying biomedical data mining strategies to identify new opportunities to use existing pharmaceutical compounds in new ways and to accelerate the pace of drug development for rare disease patients [325]. In reviewing past successes of data mining for drug repurposing, and planning for future biomedical research capacity, the DMR Task Force identified four strategic infrastructure investment areas to focus on in order to accelerate rare disease research productivity and drug development: (a) improving the capture and sharing of self-reported patient data, (b) better integration of existing research data, (c) increasing experimental testing capacity, and (d) sharing of rare disease research and development expertise. Additionally, the DMR Task Force also recommended a number of strategies to increase data mining and repurposing opportunities for rare diseases research as well as the development of individualized and precision medicine strategies.

13.6 Drug Repurposing for Neglected Tropical Diseases (NTDs)

Diseases prioritized for action by the World Health Organization (WHO) include tuberculosis, malaria, dengue fever (viral), buruli ulcer (mycobacterial), and a range of diseases that, unlike the infectious diseases of developed countries, are caused by eukaryotic parasites—unicellular kinetoplastid protozoa or multicellular helminths (flukes, tapeworms, and nematodes). WHO estimated that up to 1 billion of the world's population are affected by one or more NTDs, and the resulting sickness and long-term disability impede economic productivity, perpetuating the cycle of deprivation [99, 326]. There remains a great need for more efficacious and safer treatments for several of the NTDs. These diseases are somewhat "neglected" because those who need them are unable to pay the economic cost of treatment, and there is no prospect of recovering research and development (R&D) costs. Funds for R&D into NTDs come from philanthropic organizations, research councils, and governments. Pharmaceutical companies have been generous in providing in-kind help, and especially in sharing compounds via, for example, the Drugs for Neglected Diseases Initiative.

Although there have been some notable large-scale primary screening efforts (e.g., 6 million compounds screened by Medicines for Malaria Venture, leading to an open-access "box" of leads) [327] the cost of de novo drug discovery is such that, given the absence of commercial returns, it is logical that nongovernmental organizations and academics have been very actively attempting to repurpose existing mature compounds for NTDs. The repurposing approach has many advantages (covered in detail elsewhere in this article) and also some specific challenges when applied to the NTD field. First of all is throughput: High-throughput, biologically relevant screening assays do not exist for all NTDs. The smaller number of mature compounds available for screening, normally up to a few hundreds, is a good match to the rather low throughput of some biologically relevant phenotypic screens (e.g., for parasitic nematodes). Second are drug targets in NTDs: Because causative agents of many NTDs are eukaryotic, there is a realistic possibility that drug target classes in the pathogens may overlap with the targets of the diseases in Pharma pipelines. Some important pathogen targets are known empirically, and efforts to prioritize possible targets computationally continue [45, 328] (also see the TDR Targets database: www.tdrtargets.org). It has proven productive to screen panels of mature compounds that are active on related mammalian targets. Good examples of this approach include inhibitors of cysteine proteases, HMGCoA reductase, and a variety of kinases. Cysteine proteases have important physiological roles in eukaryotic pathogens (mammals would have serine proteases instead in some of the biological contexts). Inhibitors of mammalian cathepsins are potent against *Trypanosoma cruzi* (Chagas disease), *Plasmodium* spp. (malaria) [329], and hookworm [330]. The parasite enzymes have been cloned and expressed, and parallel biochemical and phenotypic screening is useful to provide correlative evidence of mode of action and to compare potencies against the pathogen target with data on the human target. HMGCoA reductase inhibitors have been shown to have activity against schistosoma parasites (www.wipo.int/research/en/collaboration), and investigation of their potential continues [331, 332]. Kinase inhibitors, many of which have been developed in the pharmaceutical industry as oncology treatments, are promising targets in parasitic protozoa, which have related kinases [333], for example kinetoplastid protozoa [334].

An alternative approach is serendipitous screening of panels of mature compounds in whole-pathogen assays; for example, Gloeckner et al. [335] used the Johns Hopkins Clinical Collection to identify Closantel, a veterinary antihelmintic, as active on *Onchocerca volvulus*, the nematode that causes river blindness; whereas Abdulla et al. [336] have identified compounds that have hit- and lead-like properties for the treatment of schistosomiasis. This approach has the potential to reveal new mechanisms and/or identify candidates directly, without initially depending on knowledge of parasite target biology (which is less well investigated than mammalian target biology). Compound panels can be selected on the basis of compound quality and availability and/or may be enriched for classes of targets known to exist in pathogens. The evolutionary distance of NTD parasites from humans limits the confidence with which target validity can be deduced from activity data, and it ought not to be assumed that a compound that produces a phenotypic effect in a cell assay is acting via the target for which the compound was originally discovered. If mode of action is considered important, then the target in the pathogen will need to be identified experimentally.

The compounds being tested will be pharmacologically active in humans. Equal or greater potency in pathogens will be a fortunate accident, except for the most highly conserved targets. High sequence or active site conservation may be associated with essentiality fundamental cellular functions. For example, human cyclin-dependent kinase can regulate the yeast cell cycle, and elucidating this mechanism of cell cycle control depended on discoveries in yeasts, sea urchins, and human cells [337]. Of course, cell cycle inhibition as a side effect in humans would raise safety concerns. Therefore, we could expect that further chemical optimization would be unavoidable when repurposing such compounds for pathogens. Nevertheless, the prospects for direct repurposing are more hopeful than they might appear. Drug candidates intended to treat chronic human conditions could, depending on primary pharmacology, be sufficiently tolerable to treat infections. Other compounds that would not be desirable for chronic human administration (e.g., those with cytochrome p450 activity) could be tolerable for short courses of treatment and/or for use in populations that, unlike the aging inhabitants of the most affluent countries, are unlikely to be receiving multiple medications. For example, benznidazole is used successfully to treat Chagas disease in spite of its suboptimal properties [338]. Systematic screening of publicly available drugs and candidate drugs recently yielded promising leads for dysentery caused by the protozoa *Entamoeba* and *Giardia* [339]. One lead was the antirheumatic compound aurofin, which has been used as a human treatment for more than 20 years [340]. Similarly, clinical experience with inhibitors of cathepsin K showed that such compounds are well tolerated when administered chronically [341], so there is good reason to be optimistic that antiparasitic cysteine–protease inhibitors will also prove to be well-tolerated treatments. Fortuitously, the ADME properties might favor activity in the pathogen. Pathogens might be less able than mammals to transport particular compounds out of their cells—ivermectin is a classic example of a compound that would be unsafe in mammals were it not an excellent substrate for the multidrug transporter P-glycoprotein [342, 343]. Such helpful quirks of comparative biochemistry cannot, of course, be relied on. The projects that delivered the drug candidates will have generated large numbers of related compounds, and screening these should at least provide some understanding of comparative structure–activity relationships of the pathogen target versus the human target. Divergence between pathogen and human target active-site sequence might make it possible to design parasite-specific compounds even when the side effects of the parent candidate drug might not be acceptable. An example is the redesign of epidermal growth factor receptor compounds to give selectivity for *Trypanosoma brucei* kinases [344, 345]. Aurora kinase inhibition is currently being pursued as a cancer treatment; however, intrinsic mechanistic side effects limit the dose and duration of treatment of inhibitors of the human enzymes. Related kinases are important for mitosis in the malaria parasite *Plasmodium* [346]; selectivity for the parasite over host cells will be essential. This should be achievable through a combination of analogue screening and rational design [346].

13.7 Drug Repurposing in Coronavirus Disease 2019 (COVID-19)

The outbreak of severe acute respiratory syndrome coronavirus 2 (SARS-CoV-2) has led to a worldwide pandemic with high morbidity and mortality since the end of 2019. Thus far, there have been no effective drugs or other regimen for curing of the condition. Development of new drugs is obviously unrealistic given the research-intensive, time-consuming, and costly process with a high rate of failure. Drug repurposing has therefore been naturally taken as an alternative, or even the main strategy, for anti-coronavirus disease (COVID-19) drug discovery [347–359]. In fact, many categories of existing drugs have been investigated on their potential for anti-COVID-19 therapy, including antimalarial, antiviral drugs, antibiotics, glucocorticoids, monoclonal antibodies, anthelmintics, anti-protozoal, anti-HIVs, anti-influenza, anti-hepacivirus, antineoplastics, immunoglobulins, interferons, among others. In vitro, in vivo, or preliminary trials of these drugs in the treatment of COVID-19 have been encouraging, leading to new research projects and trials to find effective and safe drugs. The findings, although preliminary, have established a starting point in the search for an urgent solution [348]. Benavides-Cordoba [348] presented a selection of drugs, of different classes and with potential activity against COVID-19, whose clinical trials are ongoing; and as proofs of concept, double blind, add-on event-driven, would allow proposing research that generates results in less time and preserving quality criteria for drug development and approval by regulatory agencies. Repurposing existing drugs such as hydroxychloroquine, remdesivir, ivermectin, and baricitinib has shown reasonable potential for COVID-19 treatment [349].

This subsection aims to discuss the drug repurposing strategies in case of emergent event like the outbreak of COVID-19 pandemic and some specific examples of promising compounds with the potential of anti-COVID-19 therapy based on drug repurposing.

13.7.1 The Drug Repurposing Approaches in Emergent Situations: COVID-19

13.7.1.1 Transcriptome-Based Drug Repurposing for COVID-19

Using bronchoalveolar lavage fluid transcriptome data of COVID-19 patients, Jia et al. [350] found that the endocytosis and lysosome pathways are highly involved in the condition and the regulation of genes involved in neutrophil degranulation was disrupted. They performed cogena co-expressed gene-set enrichment analysis for drug repurposing and identified two antiviral drugs (saquinavir and ribavirin) and several other candidate drugs (such as dinoprost, dipivefrine, dexamethasone, and (−)-isoprenaline). Notably, the two antiviral drugs have also previously been identified using molecular docking methods, and ribavirin is a recommended drug in the diagnosis and treatment protocol for COVID pneumonia (trial version 5–7) published by the National Health Commission of the P. R. China [350]. Their study demonstrated the value of the cogena-based drug repurposing method for emerging infectious diseases and provides potential drugs for the prevention and treatment of COVID-19.

Chen et al. [351] identified a set of repurposable and investigational drugs as potential therapeutics against COVID-19. These drugs were deduced from the gene expression signature of SARS-CoV-2-infected A549 cells screened against Connectivity Map and prioritized by network proximity analysis with respect to disease modules in the viral–host interactome. The group also identified immunomodulating compounds aiming at suppressing hyperinflammatory responses in severe COVID-19 patients, based on the transcriptome of ACE2-overexpressing A549 cells. Experiments with Vero-E6 cells infected by SARS-CoV-2, as well as independent syncytia formation assays for probing ACE2/SARS-CoV-2 spike protein-mediated cell fusion using HEK293T and Calu-3 cells, showed that several predicted compounds had inhibitory activities. Among them, salmeterol, rottlerin, and mTOR inhibitors exhibited antiviral activities in Vero-E6 cells; imipramine, linsitinib, hexylresorcinol, ezetimibe, and brompheniramine impaired viral entry. These novel findings provide new paths for broadening the repertoire of compounds pursued as therapeutics against COVID-19.

13.7.1.2 Structure-Based Drug Repurposing for COVID-19

SARS-CoV-2 relies on non-structural proteins such as Nsp13 helicase and nsp14 which are the key components of the replication–transcription complex (RTC) to complete its infectious life cycle [352]. Targeting these essential viral proteins with small molecules is considered one of the strategies for anti-COVID-19 therapy. The lack of experimental structures of these proteins deters the process of structure-based identification of their specific inhibitors. In a study reported by Ayman et al. [352], the in silico models of SARS-CoV-2 nsp13 helicase and nsp14 protein were elucidated using a comparative homology modeling approach. These in silico model structures were validated using various parameters such as Ramachandran plot, Verify 3D score, ERRAT score, knowledge-based energy, and Z-score. The in silico models were further used for virtual screening of the Food and Drug Administration (FDA) approved antiviral drugs. Simeprevir (SMV), paritaprevir (PTV), and grazoprevir (GZR) were the common leads identified which show higher binding affinity to both nsp13 helicase and nsp14 as compared to the control inhibitors, and therefore, they might be potential dual-target inhibitors. The leads also establish a network of hydrogen bonds and hydrophobic interactions with the key residues lining the active site pockets.

13.7.1.3 Systematic and Integrative Drug Repurposing for COVID-19

Isgrò et al. [353] described their systematic and integrative antiviral drug repurposing effort aimed at identifying, among the drugs already authorized for clinical use, some active inhibitors of the SARS-CoV-2 main protease. They revealed that ethacrynic acid, a powerful diuretic, is an effective inhibitor of SARS-CoV-2 main protease. Even with all the necessary cautions, given the particular nature of this drug, these data can be the starting point for the development of an effective therapeutic strategy against SARS-CoV-2.

13.7.1.4 Network Pharmacology-Based Drug Repurposing for COVID-19

A recent study used an integrated network-based pharmacologic and transcriptomic approach to screen drug candidates novel for COVID-19 treatment [354]. Network-based proximity scores were calculated to identify the drug–disease pharmacological effect between drug–target relationship modules and COVID-19 related genes. Gene-set enrichment analysis (GSEA) was then performed to determine whether drug candidates influence the expression of COVID-19 related genes and examine the sensitivity of the repurposing drug treatment to peripheral immune cell types. The complementary exposure model was employed to recommend potential synergistic drug combinations. The analyses identified for the first time 18 individual drug candidates including nicardipine, orantinib, tipifarnib, and promethazine, which have the potential as repurposed drugs for the treatments of COVID-19. Additionally, 30 synergistic drug pairs were ultimately recommended, including fostamatinib plus tretinoin and orantinib plus valproic acid. Differential expression genes of most repurposing drugs were enriched significantly in B cells.

Another recent study analyzed the genome sequence of SARS-CoV-2 and confirmed SARS as the closest virus by genome similarity, followed by MERS and other human coronaviruses [355]. Using text mining and database searches, the investigators obtained 34 COVID-19-related genes to seed the construction of a molecular network where our module detection and drug prioritization algorithms identified 24 disease-related human pathways, 5 modules, and 78 drugs for repurposing. They then re-prioritized 30 potentially repurposable drugs against COVID-19 (including pseudoephedrine, andrographolide, chloroquine, abacavir, and thalidomide), based on clinical knowledge.

A research group also utilized a network-based drug repurposing strategy to rapidly identify potential drugs targeting SARS-CoV-2 [356]. To this end, seven potential drugs were proposed for COVID-19 treatment using protein–protein interaction (PPI) network analysis. First, 524 proteins in humans that have interaction with the SARS-CoV-2 virus were collected, and then the PPI network was reconstructed for these collected proteins. Next, the target miRNAs of the mentioned module genes were separately obtained from the miRWalk 2.0 database because of the important role of miRNAs in biological processes and were reported as an important clue for future analysis. Finally, the list of the drugs targeting module genes was obtained from the DGIDb database, and the drug–gene network was separately reconstructed for the obtained protein modules. Based on the network analysis of the PPI network, seven clusters of proteins were specified as the complexes of proteins which are more associated with the SARS-CoV-2 virus. Moreover, seven therapeutic candidate drugs were identified to control gene regulation in COVID-19. PACLITAXEL, as the most potent therapeutic candidate drug and previously mentioned as a therapy for COVID-19, had four gene targets in two different modules. The other six candidate drugs, namely, Bortezomib, Carboplatin, Crizotinib, Cytarabine, Daunorubicin, and Vorinostat, some of which were previously discovered to be efficient against COVID-19, had three gene targets in different modules. Eventually, Carboplatin, Crizotinib, and Cytarabine drugs were found as novel potential drugs to be investigated as a therapy for COVID-19 [356].

13.7.2 Examples of Repurposed Anticancer Drugs for COVID-19

In their review article, El Bairi et al. [359] discussed some repurposed anticancer drugs for the treatment of COVID-19, which are under investigation in clinical trials or proposed for the clinical testing.

1. *Ruxolitinib.*

Ruxolitinib is a potent orally bioavailable Janus kinase (JAK) 1 and JAK2 inhibitor indicated for various myeloproliferative malignancies including myelofibrosis and polycythemia vera [360, 361]. These neoplasms, alongside with the acute myeloid leukemia, are known by their aberrant activation of the JAK-STAT signaling [362, 363]. This pathway plays a pivotal role in cell proliferation and survival and seems to be targetable for a spectrum of conditions beyond malignancies.

Regarding COVID-19, ruxolitinib was recently repurposed to quell the immune-hyperactivation, thus dampening the cytokine storm and improving ARDS. An artificial intelligence-driven study has identified ruxolitinib as a promising drug for COVID-19 along with other molecules capable to disrupt the JAK-STAT signaling, used for myelofibrosis and rheumatoid arthritis, such as baricitinib and fedratinib [364]. The clinical trials using ruxolitinib for COVID-19 are ongoing. In particular, ruxolitinib is being investigated against placebo for COVID-19 patients with ARDS in two randomized phase III trials (NCT04362137 and NCT04377620). The RUXCOVID study (n = 402, NCT04362137) is a phase III, multicentered, randomized (2:1) and placebo-controlled trial, which evaluates the safety and efficacy of ruxolitinib added to best supportive care in patients with cytokine storm. The RUXCOVID-DEVENT (NCT04377620) is a phase III, placebo-controlled trial, which is currently evaluating ruxolitinib for the treatment of COVID-19 patients with ARDS requiring mechanical ventilation. Both the phase III trials are designed for clinically critical patients, aiming to tackle the pathological immune response and control the hyperinflammatory syndrome in more severe patients. In addition, another ongoing phase III trial (NCT04424056) combines multiple immune suppressors including anakinra (an anti-IL-1 inhibitor), tocilizumab, in association with ruxolitinib for more severe COVID-19 cases, including ARDS, associated with possible cytokine storm and other immune dysregulations.

2. *Carmofur.*

Carmofur (1-hexylcarbamoyl-5-fluorouracil) is an old pyrimidine analogue and a lipophilic masked compound of fluorouracil. Its mechanism of action is independent of 5-fluorouracil-antitumour activity, based on the inhibition of the lysosomal acid ceramidase [365, 366]. The upregulation of acid ceramidase is linked to survival and growth of cancer cells as well as drug resistance [367]. This enzyme is an emerging target for several malignancies [368], such as prostate cancer [369] and glioblastoma [370] as well as pediatric brain tumors [371]. In addition, carmofur exerts its antiproliferative and antimetastatic effects through the blockade of the Wnt/β-catenin signaling pathway [372]. This molecule has been previously tested in clinical trials for the treatment of colon cancer as single or in combination with other chemotherapeutics [373] and in early breast cancer [374]. Recently, carmofur has been found to block SARS-CoV-2 replication by inhibiting the viral protease through the covalent binding between the carbonyl reactive group and the catalytic Cys145. In another molecular modeling report, Gao et al. [375] indicated that carmofur strongly bound to SARS-CoV-2 M^{pro}. This suggests that repurposing low-molecular weight drugs such as carmofur against the main protease of SARS-CoV-2 based on in silico studies may show a therapeutic potential. Importantly, the identification of the specific mechanisms of viral blockade provided the paradigm to design analogues of carmofur, with improved efficacy for COVID-19 and less neurotoxicity. This last aspect is particularly relevant because of its association with drug-related leukoencephalopathy.

3. *Bevacizumab.*

Bevacizumab is a monoclonal antibody with antiangiogenic properties, capable to target the vascular endothelial growth factor [376]. This biological agent has been successfully implemented in the clinic management of various cancer types, including advanced ovarian cancer, renal cell carcinoma as well as colorectal cancer in combination with chemotherapy [376, 377]. The key mechanism of bevacizumab is related to the disruption of the malignant neo-angiogenesis, ultimately reducing the abnormal vascular permeability and nutrients supply to cancer cells. Based on this mechanism [378, 379], the use of bevacizumab has been proposed for the control of the abnormal angiogenesis and the related vascular permeability undermining the accumulation of intra-alveolar fibrin deposition, resulting in ARDS, and occurring during COVID-19 disease. A recent study found that new vessel growth induced by intussusceptive angiogenesis in the lungs of COVID-19 patients is approximately three times higher than in the lungs of patients with influenza [380]. Therefore, bevacizumab is being investigated also in this field (NCT04344782, NCT04305106, and NCT04275414). More in detail, the BEST-CP clinical trial (NCT04275414) is a Chinese pilot study that is currently recruiting patients with COVID-19 to be treated with intravenous bevacizumab with the aim to control the pulmonary edema. Another multicentered and randomized trial in China is investigating the safety and efficacy of bevacizumab (7.5 mg/kg) in COVID-19 patients with severe pneumonia (NCT04305106). Additionally, the French CORIMMUNO-BEVA phase II trial (NCT04344782) is randomizing COVID-19 patients to receive bevacizumab versus best of the standard supportive care. Based on the proposed mechanism of action, the use of bevacizumab is now included in clinical trials for more severe patients, presenting with the clinical features of ARDS.

4. *Carfilzomib.*

Carfilzomib is an approved antineoplastic drug for the treatment of relapsed or refractory multiple myeloma patients who have received ≤3 previous lines of therapies [381–383]. It has been investigated in several clinical trials for other malignancies such as Hodgkin lymphoma (NCT02867618) and non-Hodgkin lymphoma like in diffuse large B-cell lymphoma (NCT02073097) and mantle cell lymphoma (NCT03891355). In detail, carfilzomib is a second-generation antiproteasome with higher maximal percentage of proteasome inhibition at the maximum tolerated dose than bortezomib [384]. This mechanism leads to the irreversible inhibition of the 26S proteasome, which plays a key role in the hallmarks of myeloma, by binding to its chymotrypsin-like site of the proteolytic core [385, 386]. The multiple targets of proteasome targeting encompass the NF-κB pathway, ER stress, cell-cycle arrest, downregulation of growth factor receptors and adhesion molecules expression, angiogenesis inhibition as well as immunogenic cell death [384]. Carfilzomib was recently repurposed for COVID-19 based on virtual docking screening, a structure-based virtual screening method for molecules capable of binding specific proteins in key binding sites or domains [387]. Among several other promising drugs, including eravacycline, elbasvir, lopinavir, and valrubicin; carfilzomib was excellent in terms of binding free energy, which is an important parameter for a rational drug design targeting the SARS-CoV-2 main protease, anticipating a potent antiviral activity in silico [387].

5. *Toremifene.*

Toremifene is a selective modulator of estrogen receptors used in the endocrine therapy of breast cancer since 1997 [388]. Toremifene represents an alternative drug to tamoxifen, showing a similar efficacy in patients with advanced breast cancer [389]. Additionally, this antiestrogen drug was shown to destabilize the envelope glycoprotein of Ebola virus [390], leading to the inhibition of the fusion between viral and endosome membranes during intracellular trafficking [390]. The subsequent in vivo studies confirmed the anti-Ebola activity of toremifene regardless of its estrogen-modulating activity, by inhibiting virus entry and internalization [391].

Finally, a recent in silico study identified toremifene as a potential therapeutic approach against SARS-CoV-2, particularly when combined to emodin [392]. Emodin is a natural-derived compound with anti-coronavirus pharmacological properties, belonging to the traditional Chinese medicine *Lianhua Qingwen* used for the prevention and treatment of viral influenza [393, 394]. The authors suggested that toremifene interacts with several key host coronavirus proteins including ribosomal proteins like RPL19, HNRNPA1, NPM1, EIF3I, EIF3F, and EIF3E [392]. However, the exact mechanism of its activity is still unclear. Based on the action on viral entry, the use of this agent can be proposed for prophylaxis of the infection and in patients with milder disease, in the earlier stages of infection.

6. *Zotatifin.*

Zotatifin is a potent and selective small molecule inhibiting the eukaryotic initiation factor 4A (eIF4A) [395]. eIF4A is an RNA helicase activated by the B-cell receptor to stimulate several oncogenes linked to tumor growth, cell survival, and metastasis [395, 396]. The inhibition of this target by zotatifin and other synthetic molecules demonstrated promising antineoplastic preclinical efficacy [395, 397, 398]. For example, zotatifin is being studied in a phase I/II dose-escalation and cohort expansion clinical trial as monotherapy for pancreatic cancer and other malignancies (NCT04092673). Promisingly, a recent study published in *Nature* suggests a potential of zotatifin repurposing for COVID-19 [399]. Based on affinity-purification mass spectrometry, the authors revealed 332 SARS-CoV-2-human protein–protein interactions after cloning and expressing the key SARS-CoV-2 proteins in human cells [399]. Importantly, 66 targetable human proteins were identified to be pharmacologically druggable by 69 compounds including 29 FDA-approved drugs [399]. Several of these compounds exhibited antiviral activities against mRNA translation and Sigma1 and Sigma2 virus receptors expressed on host cells [399]. In this study, zotatifin emerged as an effective drug in reducing viral infectivity by inhibiting protein biogenesis through eIF4A blockade along with other compounds such as ternatin-4 [399, 400]. The proposed disruption of the translation of the Sigma proteins coupled to the direct inhibitory activity against them seems a promising strategy of drug targeting against SARS-CoV-2 infection and propagation, possibly indicated in the earlier stages of the infection.

7. *Plitidepsin.*

Plitidepsin (also known as dehydrodidemnin B) is a natural compound discovered in a Caribbean marine organism (*Aplidium albicans*) [401]. Its antitumor effects encompass several cancer targets such as angiogenesis and cell survival [402, 403]. Additionally, plitidepsin may also enhance radiotherapy response by a bystander effect [404]. The activation of c-Jun *N*-terminal kinase is a key mediator of plitidepsin sensitivity along with the eukaryotic elongation factor 1 alfa 2 that were recently suggested as potential predictive biomarkers of drug response [405, 406]. So far, the drug has been investigated in several clinical trials for multiple myeloma, lymphoma, liposarcoma, and prostate cancer [401]. Only one phase III randomized and controlled trial has evaluated the clinical activity of plitidepsin combined to dexamethasone in relapsed/refractory multiple myeloma (NCT01102426/ADMYRE), showing mild improvements in survival outcomes [407]. The development of this drug has been halted or terminated because of the arrival of competitive targeted agents as well as the poor enrolment in clinical trials.

On the other hand, plitidepsin has also antiviral properties, showing promising findings in the COVID-19 treatment on in vitro models (Press release 1, [408, 409]). According to these models, plitidepsin blocks the eEF1A protein in human cells that is required for SARS-CoV-2 infection (Press release 2, [408, 409]). Therefore, the drug is being tested in a phase I Spanish trial (NCT04382066/APLICOV) with the aim to assess the safety and early efficacy of three doses of plitidepsin for three consecutive days in patients requiring hospitalization for COVID-19. Based on the activity on a key mechanism of the viral pathogenesis, the use of this drug can be considered either in the earlier stages of infection, in a multidrug regimen and for patients with more advanced disease, together with proper immunomodulators.

8. *Dactinomycin.*

Historically, the anticancer drug dactinomycin was discovered in *Streptomyces parvulus* bacterium by Selman Waksman and colleagues in the 1940s [410] and launched after its initial FDA-approval in 1964. This intravenously administered cytotoxic antibiotic acts by intercalation to DNA and interruption of RNA transcription [411, 412]. Recently, dactinomycin was shown to induce immunogenic cell death, suggesting a direct antineoplastic activity enhanced by immunomodulatory properties [413]. In fact, this drug is used in combination with other cytotoxic drugs in the management of a broad array of solid malignancies, such as gestational trophoblastic disease, Wilms' tumor and Ewing sarcoma [414–417].

The use of dactinomycin against coronaviruses was firstly described in 1979 by Kennedy et al. [418] in the context of alfa-coronavirus 229E, involved in a spectrum of respiratory diseases from upper to lower airways infections. Later on, Lewis et al. [419] showed a potent inhibition of the feline enteric coronavirus strain 79–1683 in whole feline embryo cells. Based on these and subsequent evidence, dactinomycin was widely used as an in vitro inhibitor of viral cellular transcription in infected cells [420]. Recently, a combination of dactinomycin and sirolimus showed a synergistic action against SARS-CoV-2-host proteins [392]. In fact, the association may inhibit DNA topoisomerase required for RNA synthesis as well as mammalian target of rapamycin (mTOR) signaling in human coronavirus infected cells [392]. Notably, the mTOR pathway is a target of sirolimus that has also shown a promising activity against MERS-CoV in addition to its known anticancer, immunosuppressive and antifungal properties [421]. However, these results for SARS-CoV-2 are limited by their in silico nature and warrant more clinical experiences beyond speculations. In addition, the use of this antibiotic is associated with an adverse safety profile when used at therapeutic doses as an antineoplastic agent that may be less suitable for severely ill patients with an acute infectious disease. In particular, the emetogenesis and myelosuppression can deserve a special consideration, and antiviral doses may be acceptable only if lower, as expecting a better safety profile.

9. *Valrubicin.*

Valrubicin is a semisynthetic derivative of the anthracycline doxorubicin, used as intravesical topic treatment for the management of refractory non-muscle invasive bladder cancer [422]. The mechanism of action of valrubicin affects DNA synthesis and metabolism by inhibiting the incorporation of nucleosides and topoisomerase II [422, 423]. Based on computational hierarchical virtual screening, recently Wang showed that valrubicin alongside with other drugs, such as lopinavir and carfilzomib inhibits the main protease of SARS-CoV-2 [387]. Moreover, another report based on a machine-learning simulation has recently shown that valrubicin is a potential COVID-19 protease inhibitor [424]. However, no considerations on the effective doses have been modeled: the clinical use as an agent for systemic administration is not validated, being approved only as a topic solution, and any speculation for the design of clinical trials should account for the adverse safety profile of this chemotherapeutic.

10. *Leronlimab.*

Leronlimab (PRO 140) is a humanized monoclonal antibody targeting the C–C chemokine receptor type 5 (CCR5) on T lymphocytes [425]. The role of CCR5 in the immunopathology of disease has been extensively studied since its discovery as a co-receptor for human immunodeficiency virus entry in 1996 [426] (NCT03902522 and NCT02859961). In addition, CCR5 is also expressed on a broad array of cancers and it is linked to various cancer hallmarks, particularly "avoiding immune destruction" [425]. It is under investigation in trials for few cancers, such as metastatic colon and breast cancers (NCT03838367), for which several other inhibitors of CCR5 are being developed (reviewed elsewhere: [425]. Previously, Chen et al. [427] demonstrated that mRNA transcripts of CCR5 and other chemokine receptors were found upregulated in animals infected by SARS-CoV. Additionally, enhanced production of these mediators was associated with recruitment of inflammatory cells and T lymphocytes influx into the lung parenchyma [427].

PRO 140 has shown a promising antiviral potency with a high barrier to resistance and encouraging short-term safety and tolerability [428]. A recently published *Preprint* suggested that leronlimab disrupts the CCL5/RANTES-CCR5 axis and restores immune homeostasis in COVID-19 patients [429]. The authors observed a marked CCR5 receptor occupancy on macrophages and T lymphocytes, rapid drop of circulating IL-6, CD4/CD8 ratio restoration and a significant decrease of patients' viremia [429]. These encouraging findings boosted the rapid entrance of leronlimab into two clinical trials for mild to moderate and severe COVID-19 infection (NCT04343651 and NCT04347239). The largest trial using this drug is a randomized, double-blind and placebo-controlled, two-arm phase IIb/III trial investigating weekly leronlimab at a dose of 700 mg given subcutaneously (NCT04347239). A multicentered enrolment will include 390 patients with severe or critical disease with a mortality rate after 4 weeks of treatment as a primary end-point. Likewise, the other study is a phase II ($n = 75$) with a similar design that will enroll patients with mild to moderate symptoms (NCT04343651). Patients will be randomized to receive leronlimab (700 mg) or placebo weekly in a 2:1 ratio with changes in total symptom score based on fever, myalgia, dyspnea, and cough at day 14 as a primary end-point. Notably, the other outcome measures in this clinical trial include serum changes of cytokines and chemokines from baseline, CCR5 occupancy levels on macrophages and T^{reg} cells and CD3+, CD4+, and CD8+ T cell count from baseline. Based on the specific action in-between the viral pathogenesis and the immune-dysregulation, the use of this agent has been proposed for patients with mild clinical COVID-19 to prevent severe outcomes and reduce the related mortality.

11. *Antineoplastic Immune-checkpoint Inhibitors.*

Based on the hypothesis suggesting the upregulation of immune-checkpoints such as PD-1/PD-L1, cytotoxic T lymphocyte antigen-4 (CTLA-4), lymphocyte activation-gene-3 (LAG-3), and T cell membrane protein-3 (TIM-3) pathways in excessive inflammatory response during sepsis [430], these types of drugs were repurposed beyond their use for the treatment of cancers [430, 431]. Importantly, a recent meta-analysis including four studies showed PD-1 blockade improves patient outcome in sepsis [432].

Camrelizumab is a humanized IgG4-kappa antiprogrammed cell death 1 (PD-1) monoclonal antibody used in cancer immunotherapy [433]. The blockade of PD-1 binding to its ligand PD-L1 enables reactivation of cell-mediated immune response against cancer cells [433]. Camrelizumab was tested in phases II/III clinical trials for advanced or metastatic esophageal squamous cell carcinoma [434, 435], osteosarcoma [436], relapsed or refractory classical Hodgkin lymphoma [437, 438] and advanced hepatocellular carcinoma [439]. It received its first approval in China for treating RRHCL patients who received at least two lines of chemotherapy [433]. Additionally, it is currently being studied in two ongoing phase III trials for local advanced nasopharyngeal carcinoma (NCT03427827) and gastric and gastroesophageal junction adenocarcinoma (NCT04208347). The use of anti-PD-1 inhibitor camrelizumab in COVID-19 patients with pneumonia is being evaluated in a Chinese clinical trial (ChiCTR2000029806) in comparison with the standard supportive care and the immunomodulatory agent thymosin.

Nivolumab and pembrolizumab are other immune-checkpoint inhibitors (ICIs) that were successfully introduced into the management of various solid cancers, particularly for melanoma [440, 441]. These ICIs are both humanized and fully human anti-PD-1 IgG4 monoclonal antibodies. Their marked clinical activity depends on PD-L1 expression in some tumors, as well as on emerging predictive biomarkers, such as tumor-mutational burden and microbiota that seem to mediate drug response [442–444].

Nivolumab is currently used in a multicentered phase II trial comparing its efficacy against tocilizumab, GNS561 (a chloroquine analogue), and standard supportive care in advanced or metastatic cancer patients with COVID-19 (NCT04333914). A total of 273 participants will be randomized to receive one of three treatments based on two cohorts. Patients that are asymptomatic or with mild disease will be prospectively assigned to the group comparing nivolumab versus GNS561 versus standard of care in a 1:1:1 ratio. Similarly, in patients with moderate to severe symptoms, nivolumab will be compared to GNS561 and tocilizumab. Survival after 4 weeks of randomized treatment is the primary end-point in this trial. Moreover, another ongoing French phase II trial is investigating the safety and efficacy of nivolumab in COVID-19 patients requiring hospitalization (NCT04343144). Patients will be randomly allocated to receive nivolumab or standard of care. One pilot trial is also testing nivolumab in the earlier setting of COVID-19, in patients presenting with no or mild symptoms and not hospitalized—to examine the role of this ICI as an immune-enhancer, to accelerate the clearance of SARS-CoV-2 (NCT04356508).

Pembrolizumab was also repurposed for COVID-19. The COPERNICO is a multicentered two-arm phase II trial (NCT04335305) that is currently enrolling patients with COVID-19 associated with pneumonia in Spain and who are non-responsive to frontline therapy. Twenty-four participants will be prospectively randomized to receive intravenous tocilizumab plus pembrolizumab or standard treatments (at physician's choice). The use of ICIs for COVID-19 is essentially intended as an immunomodulator, based on the proposed role of PD-L1 axis in the pathogenesis of ARDS and the expedited clearance of SARS-CoV-2 with anti-PD-L1 therapies [445]. Accordingly, the studies ongoing are testing these ICIs both for more severe cases, intending to attenuate the exuberant cytokine production [446], and for mild presentations, to prevent the onset of the hyper-immune pathological response and then dampen the viral replication [445]. Moreover, a recent report showed that PD-1 checkpoint inhibition in lung cancer patients who were diagnosed with COVID-19 does not appear to affect the prognostic outcomes [447]. Therefore, this provides a promising rationale for immunotherapy, which seems to have no unfavorable effects in this setting.

12. *Kinase Inhibitors.*

The blockade of kinase cascade is one of the successful targeted strategies in treating cancer in the modern era of oncology. Several kinase inhibitors including palbociclib, crizotinib, dabrafenib, and imatinib were introduced in practice for various cancers [448]. The rationale for using kinase inhibitors in COVID-19 is principally based on their ability to inhibit host intracellular-viral events particularly during trafficking induced during the viral life cycle [449]. Viruses commonly use a number of host kinases during their invasion and thus, they are considered potential targets for kinase inhibitors repurposing as broad-spectrum antivirals [450, 451]. Several previous reports demonstrated potent antiviral properties for kinase inhibitors in various virus-induced diseases, including SARS-CoV infections [421, 452, 453].

Imatinib is a marketed tyrosine kinase inhibitor for chronic myeloid leukemia and gastrointestinal stromal tumors in the first-line setting [454, 455]. The BCR-ABL1 oncoprotein with kinase activity was found to mediate virus entry in several reports encompassing hepatitis C and Ebola viruses [456, 457]. In coronaviruses, recent studies indicated that imatinib blocks cell–cell and cell–virus membrane fusions [458, 459]. Additionally, in in vivo and ex vivo mouse models, the use of imatinib protected the lungs from ischemia/reperfusion injury [460, 461], attenuated inflammation and vascular leak [462] and prevented lung injury and death [463].

Clinically, the IMAGE-19 trial is a single-arm phase II study in France that will enroll 90 elderly COVID-19 patients (NCT04357613). Early imatinib therapy at a dose of 800 mg daily for 2 weeks will be evaluated to prevent the progression to severe disease in patients >70 years with COVID-19. The efficacy of imatinib at a dose of 100 mg is also being examined in combination with hydroxychloroquine in the Spanish COVID-19HUF phase II trial (NCT04346147). Participants will be randomized to receive either oral lopinavir/ritonavir (protease inhibitors), imatinib or baricitinib (JAK inhibitor), all combined to hydroxychloroquine. The trial completion date is estimated in September 2020.

The COVERAGE study is a large French multiarm phase III trial (NCT04356495) that will investigate the efficacy of imatinib and several other experimental treatments. A total of 1057 symptomatic COVID-19 patients will be randomly allocated to receive imatinib, or telmisartan, or favipiravir, or hydroxychloroquine as compared to a control group with the objective to reduce the risk of hospitalization or death in elderly patients. Another phase III trial using imatinib versus placebo for hospitalized COVID-19 patients is currently recruiting in the United States (NCT04394416). Given the acceptable safety profile of imatinib and the several available generics, this seems to be an effective strategy for clinical repurposing. The drug repurposing of imatinib in COVID-19 has been embedded for the early phases of the disease, commonly referred to as the "viral response" phase [464]. Imatinib seems to tackle the mechanisms of viral entry and cell–cell spread, functioning as an antiviral agent. In this initial stage, the drug development is intended to reduce the duration of the symptoms, minimize the contagiousness among people and prevent the progression to severe COVID-19.

Duvelisib is an orally bioavailable phosphatidylinositol 3-kinase (PI3K) selective inhibitor that is being studied for hematologic malignancies [465]. The PI3K–AKT–mTOR signaling is frequently dysregulated in cancer as well as during hyperinflammation and its therapeutic targeting is highly represented in ongoing clinical trials. Duvelisib was investigated in the DUO phase III trial and compared to ofatumumab for refractory and relapsed chronic lymphocytic leukemia and small lymphocytic lymphoma [466]. It was approved by the FDA in 2018 for treating adult patients with relapsed or refractory chronic lymphocytic leukemia or small lymphocytic leukemia after at least two prior therapies [467]. There is currently one pilot phase II trial examining the efficacy of duvelisib twice daily at a dose of 25 mg for 10 days in COVID-19 patients as monotherapy (NCT04372602). This single-arm study will test the hypothesis that PI3K blockade could hamper the immune system hyperactivation and therefore reduce lung inflammation and interfere with the viral cycle. It is estimated to enroll 25 participants, with overall survival as a primary end-point. Notably, the PI3K inhibitors including the drug duvelisib can cause lung inflammation and increase the risk for infections, as consistently demonstrated for multiple compounds of this pharmacological class, and a special caution is warranted during the clinical trials using this class of molecules.

The aberrant activation of the Bruton tyrosine kinase has a key role in the tumorigenesis of B-cell lymphoma [468]. Its abrogation seems to protect lungs from acute immune-induced or lethal influenza-induced injuries [469, 470] as well as sepsis-induced injury [471]. Zanubrutinib is an orally administered irreversible inhibitor of the Bruton tyrosine kinase [472–474]. This drug showed a durable and high overall response rate in patients with relapsed/refractory Mantle cell lymphoma [475]. Recently, zanubrutinib received accelerated FDA approval for treating adult patients with this rare and aggressive cancer who have received at least one line of therapy [474]. For COVID-19, a phase II trial in the United States is ongoing, aiming to dampen the disease-related immune dysregulation and hyperinflammation (NCT04382586). This study is expected to randomize 52 participants to receive zanubrutinib combined to supportive care versus placebo and to test the hypothesis of increasing respiratory failure-free survival rates in patients with pulmonary distress. Ibrutinib is another inhibitor of Bruton kinase inhibitor, approved to treat indolent B-cell malignancies and chronic graft-versus-host disease, now being repurposed for COVID-19 [476]. An earlier experience with this drug in COVID-19 patients with cancer showed marked improvements and resolution or near resolution of COVID-19 symptoms during the follow-up [476]. In the same way, the iNSPIRE (NCT04375397) is a phase II that will enroll 46 COVID-19 patients to randomly receive ibrutinib versus placebo along with standard supportive care. The trialists expect that ibrutinib will reduce respiratory failure and their findings are expected to be released in September 2020.

Opaganib (ABC294640) is a first-in-class selective inhibitor of sphingosine kinase-2 (sk2). This key enzyme in lipid signaling promotes tumor growth, cell proliferation and cytokine production during inflammation [437, 438, 477]. It was previously noticed that sk2 is the host factor that impacts the viral replication process of the chikungunya virus [478] and maintains viral latency and survival in Kaposi's sarcoma-associated herpes virus [479]. This drug has an orphan drug designation for cholangiocarcinoma [480] and is being tested in clinical trials for patients with metastatic castration-resistant prostate cancer (NCT04207255) and advanced cholangiocarcinoma (NCT03377179). It is also under development for COVID-19-associated pneumonia in a phase II study (NCT04414618). This is a randomized, double-blind, placebo-controlled trial in 40 adult patients with positive pneumonia, using daily opaganib for 2 weeks.

Saracatinib (AZD0530) is a tyrosine kinase inhibitor of the c-SRC family of proto-oncogenes. This pathway mediates signal transduction for several cell functions such as proliferation, cell-to-cell transmission, epithelial-to-mesenchymal transition, vascular permeability, invasion and metastasis [481, 482]. Moreover, it interferes with additional cell signaling such as PI3K/mTOR and MAPK [483]. Saracatinib selectively abrogates this enzyme and was investigated in several clinical trials for recurrent osteosarcoma [484], platinum-resistant ovarian cancer [485], unresectable thymic malignancy [486] and cancer-induced bone pain [487]. SRC kinases have also a role in endocytic trafficking [488] and there is a growing body of evidence indicating the inhibitory properties of viral replication of dengue viruses and MERS-CoV by the anticancer agent saracatinib [489–491]. This makes this drug a possible candidate for further investigations on other members of the Coronaviridae family. In a recently published *Preprint*, molecular modeling showed that saracatinib binds to the nucleocapsid protein sites of SARS-CoV-2 along with other previously discussed drugs such as imatinib and camostat [492]. To the best of our knowledge, saracatinib has not entered into any COVID-19 clinical trial. The clinical use of molecular blockers of PI3K–AKT–mTOR, Bruton tyrosine kinase, sk2 and c-SRC is anticipated to exert immune-modulating rather than antiviral effect. To date, the drug development of this group of tyrosine kinase inhibitors is included in the protocols for hospitalized patients with severe COVID-19, to rescue the immune-dysregulation and improve the outcome.

13. *Selinexor.*

Selinexor is a first-in-class orally bioavailable inhibitor of Exportin-1 (XPO1) [493]. XPO1 is an export receptor for various proteins and RNA molecules with oncogenic properties and is frequently altered in cancer [494]. Moreover, it is implicated in several cancer pathways including cell growth and survival, differentiation, and drug resistance [495]. XPO1 suppression by selinexor in human clinical trials demonstrated promising results for patients with heavily treated relapsed or refractory multiple myeloma [496–498]. The drug received first an accelerated approval by the FDA in the United States of America (USA) in 2019 in combination with dexamethasone [497]. The antiviral activity of selinexor is based on its potential to interfere with host targets including XPO1 that interacts with SARS-CoV-2 proteins [399, 499]. In fact, blockade of this nucleocytoplasmic transporter can sequester viral materials in the host cell nucleus and therefore reduce its replication mechanisms and viral load [499]. Moreover, selinexor has also selective and potent anti-inflammatory effects on sepsis pathophysiology [500]. Based on these concepts, two phase II randomized and controlled trials of low dose selinexor were initiated for COVID-19 patients with moderate to severe disease (NCT04355676 and NCT04349098). The largest trial that is currently recruiting patients is single-blind and placebo-controlled and is expected to enroll 230 hospitalized participants with severe COVID-19 (NCT04349098).

13.7.3 Repurposing Colchicine for Anti-COVID-19 Therapy

Colchicine, made from preparations of the meadow saffron *Colchicum autumnale*, has been used for over 2000 years as a poison and as a remedy for gout flares [501]. Colchicine for oral use (capsule/tablet/liquid) is currently FDA approved for the prevention and treatment of gout flares in adults with gout and Familial Mediterranean fever (FMF) (https://www.fda.gov/drugs/postmarket-drug-safety-information-patients-and-providers/colchicine-marketed-colcrys-information). Off-label uses for colchicine are many and include acute calcium pyrophosphate (CPP) arthritis (pseudogout), sarcoid and psoriatic arthritis, Behcet's disease, and pericarditis, and recently, studies have shown colchicine's efficacy in preventing major cardiovascular adverse events among patients who suffered a recent myocardial infarction [502].

Recently, the potential of colchicine as a repurposed anti-COVID-19 drug has aroused worldwide attention. A meta-analysis screened a total of 112 articles on the subject, of which 106 were excluded since they did not meet inclusion criteria [503]. A total of six studies, four published after peer-review [504–507] and two preprints [508, 509] including 881 patients with confirmed COVID-19 were evaluated, 406 of whom were treated with colchicine in addition to standard-of-care. The findings of the present meta-analysis suggest a definite signal of benefit of mortality with the addition of colchicine in patients with COVID-19. The small number of studies and the relatively small absolute number of patients included certainly warrant caution as to any definitive conclusion, but this signal cannot be ignored, considering the scarcity of effective treatments for COVID-19. The need for randomized controlled trials (RCTs) involving adequately numbered populations has been well stressed during this pandemic. On the other hand, a year after the first COVID-19 cases, current standards-of-care is not based upon such evidence. Study limitations include the lack of patient-level data which did not allow to assess or control for possible differences in baseline or procedural variables.

Currently, ten colchicine clinical trials are in progress for the treatment of SARS-CoV-2 infection, which are listed in clinicaltrials.gov [510].

Colchicine's potential for COVID-19 treatment is speculated to be mainly mediated by its inhibitory effect on the activation, destabilization, and degradation of inflammasomes, while a potential antiviral effect could be exerted through microtubule polymerization inhibition [510, 511]. Colchicine's mechanism of action is ascribed to the formation of tubulin-colchicine complex and colchicine interfere with several inflammatory pathways, including inhibition of neutrophil chemotaxis, adhesion, and mobilization; disruption of superoxide production, inflammasome inhibition, and tumor necrosis factor reduction.

13.8 Drug Repurposing from the Perspective of Pharmaceutical Industries

In recent years, the need for new approaches to drug research and development, combined with the advent of big data repositories and associated analytical methods, has generated interest in developing systematic approaches to drug repurposing [512]. Identification of novel indications for existing compounds through drug repurposing holds the potential to complement traditional drug discovery by mitigating the high monetary- and time-related costs and risks associated with the latter [512]. With a failure rate of ~45% associated with safety or toxicity issues [3], mitigating the safety risk, in addition to the saving of up to 5–7 years in average drug development time, confers an attractive prospect for drug developers and patients alike. The latter stand to gain significant advantage in quicker access to drugs that not only would otherwise not be available but also have well-documented safety profiles [3]. A variety of innovative computational methods to enable systematic repurposing screens, experimental as well as through in silico approaches, have emerged. An efficient drug repurposing pipeline requires the combination of access to molecular data, appropriate analytical expertise to enable robust insights, expertise and experimental setup for validation and clinical development know-how [512].

In this review on the approach of pharmaceutical companies to repurposing, we will describe the methodologies, tools and data types that support the new field of systematic, high-throughput, drug repurposing. Covered methodologies span high-throughput screening (HTS) platforms and in silico approaches including omics-based repurposing and the use of real-world data (RWD) to support utility of approved drugs in new indications.

As detailed by Cha et al. [512], there primarily three key players in the field of drug repurposing, including academia and research institutes, pharmaceutical companies and repurposing technology companies. While all share elements of a common approach to drug development through identifying new indications for existing drugs, the emphasis, field of expertise or capabilities, as well as business models and incentives, differ greatly. Academia can be viewed as divided between the approach of developing a molecule or indication and that of focusing on methodologies such as in silico or HTS. It is less constrained by the need of an economical or commercial success but is dependent on scientific breakthroughs that attract talent, government funding, and partnerships with for-profit institutions.

Repurposing technology companies, on the other hand, are bound by the choice of their business model, which may vary depending on the company's vision and capability [512]. These models include consulting services, service provider such as compound evaluation, database and/or platform for drug libraries, screening platform, and drug pipeline. Often, a mixed model is used, which is likely to increase both the development outcome of repurposing and the business success. Repurposing technology companies often utilize advanced technologies, expertise, focused research and agility but may lack access to the resources needed to rapidly and successfully pursue preclinical and clinical development of a drug in a regulatory-oriented environment. On the other hand, larger pharmaceutical companies may be more focused on LCM activities of a specific product or molecule, which is often done as part of the late-stage development of the product or post-marketing. For these reasons, partnerships between smaller repurposing technology companies and larger pharmaceutical companies, when effectively executed, represent an attractive means to combine advanced repurposing capabilities with deep expertise in drug development. Such an example is the case of Adamas Pharmaceuticals, a small biotech company that developed a new drug combination for the innovative marketed memantine, which was combined with donepezil for the treatment of Alzheimer's disease [513]. Forrest Labs, the original proprietor of memantine, bought the new therapeutic entity for exclusivity rights [512].

Drug repurposing activities involve the identification of robust candidates with promising new indications and the subsequent clinical development of these drugs in their new indications, as far as regulatory submission and approval [512]. The academic contribution to the development of analytical methods and approaches such as those described above to enable systematic analysis of data to generate drug repurposing insights is critical, as is the concentrated focus of repurposing-focused biotech companies. The core expertise of pharmaceutical companies, including that of clinical development, together with their overall remit to bring effective drugs to patients, positions them well to pursue a systematic approach to drug repurposing, while relying on partnerships and collaborations to increase likelihood of success in candidate identification. Within this overall scheme, pharmaceutical companies have several advantages, in particular with regard to data access and clinical development, as well as some specific constraints, inherent in their need for an underlying business model.

While some pharmaceutical companies now consider exploring repurposing alternatives for drugs in the pipeline, the approach could be taken much further, by designing multi-purpose drugs to treat different conditions; prominently, frequently co-morbid disorders (e.g., diabetes and cardiac disease; anxiety and peptic ulcer disease, epilepsy, and depression) or, alternatively, underlying pathologies plus disease symptoms. The case of amiodarone and related compounds and Chagas disease can be illustrative. Chagas disease is a tropical parasitic disease historically endemic to Latin America. The late phase of the disease is characterized by life-threatening heart disorder in around one third of the patients. Amiodarone is a class III antiarrhythmic agent that shares many characteristics of other electrophysiological anti-arrhythmic drugs, including inhibition of sodium and potassium channels and L-type calcium channels. Interestingly, some studies showed that patients with Chagasic cardiomyopathy treated with amiodarone had a more rapid recovery when compared with other patients treated with class I and class IV antiarrhythmics [2]. This fact suggested that other mode of action could be in play. It was later demonstrated that amiodarone was able to act directly on the parasite survival, affecting the growth of *T. cruzi* extracellular epimastigotes and *T. cruzi* amastigotes (that is, amiodarone could act on the underlying pathology). The mechanism of action of the drug was elucidated, showing that this drug directly disrupts the intracellular calcium regulation of the parasite [3]. Similar results were later observed with dronedarone [2, 514, 515]. Still, this example is another case of retrospective drug repurposing, since the new medical use emerged from clinical observations. A future challenge is to define whether this kind of indication expansion oriented to the treatment of co-morbid conditions could be anticipated through rational approaches at early stages of the drug development process, thus helping to provide evidence on possible advantages of new treatments compared to the existent ones, and additional criteria to decide which drug candidates should be prioritized to clinical trials and to conveniently choose the clinical endpoints of the trial that will be used to test superiority or non-superiority of the treatments under comparison. Computational network-based approximations could prove valuable to unveil hidden connections between diseases and assist these types of initiatives.

13.8.1 Drug Data Access

As already mentioned at the beginning of this chapter, drugs constituting potential candidates for repurposing can be divided into three categories: (i) generic drugs, that is, drugs approved that are no longer protected by patents; (ii) failed drugs—those that have been through some stage of clinical development but are currently not on the market; and (iii) patented drugs, either approved or in the late stage of clinical development [512].

Generic drugs constitute, by and large, a common resource for all major players in drug repurposing, that is, academia, biotech and pharmaceutical companies. These drugs have been commercially available for a significant amount of time, have well-known and largely documented safety profiles and are readily accessible for studying preclinically as well as for clinical development. However, with regard to drugs falling into the two other categories, that is, failed drugs or those still under patent, the pharmaceutical companies hold an advantage in access to drug-related data, for their respective compounds. Thus, in many cases, access to data is biased toward the pharmaceutical companies, rendering them a key partner in an effort to systematically identify new indications for existing molecules [512].

13.8.2 Clinical Development Principles for Drug Repurposing

Clinical development applies findings from basic science to enhance human health and well-being. In a medical research context, it aims to 'translate' findings in fundamental research into medical practice and meaningful health outcomes [512, 516]. These efforts often continue past the approval of the drug and extend our knowledge of the compound and its use. Clinical development principles for drug repurposing should be adopted from those applied for new chemical entity (NCE) drug development, in which the pharmaceutical companies are well versed. There may be elements in the process that can be shortened because of existing knowledge. However, the trajectory of clinical development is generally similar between drug discovery and repurposing efforts. Two major apparent categories in drug development can be linked to failure in drug approval, namely, safety and efficacy [517]. The first relates to the toxicological and pharmacokinetic profiles of the NCE, which are mainly addressed in Phases I and IIa clinical trials, following many preclinical evaluations. The second refers to target engagement and biological activities that are subject to therapeutic interference and is linked to the clinical efficacy of the NCE under investigation [512, 517].

The clinical testing of a repurposed drug necessitates the integrated understanding of the fundamental pharmacokinetic/pharmacodynamic principles of exposure at the site of action, target binding and expression of functional pharmacological activity (termed together as the 'three pillars of survival') [518]. Early stages of development (pre-proof of concept) determine the likelihood of either newly or repurposed candidate compound to continue in Phase II trials and improve the chance of progression to Phase III and ultimate approval success. For drug repurposing programs, some elements may have been elucidated during the development or approval of another indication providing some measure of early risk reduction in terms of both cost and time. On one end of the drug repurposing spectrum is an approved drug being examined at similar or lower doses than the maximum dose already approved by regulatory agencies to target the same molecular pathway/mechanism in a different patient population [519]. The large body of clinical data and the experience accumulated in Phase III (efficacy) and Phase IV (post-marketing) trials for the drug in question offer a good understanding of its profile in terms of adverse events, long-term and chronic toxicity and on- and off-label effects limiting the need for Phase I and Phase IIa/IIb clinical trials, resulting in shorter development timelines [512, 519].

On the other end of the spectrum are approved drugs being evaluated for novel therapeutic indications through the discovery of biologically and clinically relevant affinities for new targets, which play a determinant role in those indications [520]. In this case, identifying the 'right compound' during repurposing efforts is just the beginning and caution should be applied against over-relying on prior clinical experience; all elements—far beyond the "three pillars"—should be thoroughly re-explored with the following objective: the right dose, reaching the right compartment and to engage the right target, for the right time, in the right patient population, measured for the right duration, with the right tools/methodology. Consequently, under these conditions, a full clinical development program similar to that of an NCE might ensue without necessarily saving time or cost [512].

13.8.3 Challenges and Opportunities in Drug Repurposing for Pharmaceutical Companies

Drug repurposing by pharmaceutical companies faces many and varied challenges. In addition to the scientific challenges of identifying favorable and robust candidate compounds, there is a need to establish business models to support bringing existing molecules as therapies for new indications. Despite the potentially abbreviated clinical development path for repurposed drugs, there remains a significant commitment in the need to demonstrate the efficacy of the molecules in new indications, and pharmaceutical companies are faced with a challenge in trying to recoup the investment needed to bring a repurposed product to market [512]. Existing incentives to support investment in the development of repurposing drugs include 3 years of market exclusivity in the USA and 10 years in Europe [512]. In order to take advantage of this exclusivity, products rely on the generation of new intellectual property (IP). Such IP might arise from the repurposing concept itself, a new formulation or any of a number of other avenues (e.g., dose level, drug combination and route of administration) that might improve the drug in tailoring it specifically to a new patient population. Nevertheless, and in particular for compounds for which substantial data exist, new IP may be challenging and may constrain the portfolio of compounds that can be pursued by pharmaceutical companies. Additionally, the inventions underlying the new IP often create an additional tier of risk associated with the product's development process [512]. As opposed to the abbreviated regulatory pathway that repurposed drugs are expected to follow, such as the FDA's 505(b)(2), for new formulation or technology, an almost complete clinical program may be required in order to characterize the new technology/formulation, beyond the efficacy for the new indication. This can lead to lengthy development timelines and high costs and create another barrier to bringing the repurposed medicine to the clinic [512]. Giovannoni et al. [521] proposes several policy-oriented approaches, beyond extended exclusivity, for incentivizing drug repurposing in the pharmaceutical companies, including the development of more flexible regulatory paradigms to enable earlier approvals followed by post-marketing studies. The FDA's Qualified Infectious Disease Product designation, which provides expedited approval and additional years of marketing exclusivity for the approval of new antibiotics [522], may serve as a template for such new policies for drug repurposing. Importantly, the business model limitations described above often do not apply to academics and non-profit patient advocacy organizations, and novel ways of engaging in multi-partner collaborations may help support drug repurposing activities by combining the needs, expertise, and demands of the various entities. For instance, a variety of non-profit organizations provide funding for repurposing activities, including Cures Within Reach, the Michael J. Fox Foundation and the Alzheimer's Drug Discovery Foundation [523].

Traditional drug development strategies are costly, failure-prone, and expensive ventures. Therefore, drug repurposing has recently drawn attention and brings drugs out faster for clinical use. However, drug repurposing is a complex process involving multiple factors such as technology, commercial models, patents, investment and market demands [2]. Although many medical databases have been established, selecting the appropriate approach to make full use of massive amounts of medical data is still a challenge. It is urgent to develop new approaches for drug repurposing. The intellectual property (IP) issue is another highlighted issue to be solved. For repurposing drugs, IP protection is limited [519]. For example, some novel drug-target-disease associations found by repurposing researchers were confirmed by publications or online databases; however, it is difficult to seek IP protection for such associations because of the law. The IP issue prevents some repositioned drugs from entering the market. Moreover, some repurposing projects are forced to be abandoned, which is a waste of time and money [524]. It is necessary to develop a new commercial model because the traditional commercial model is a serial model and causes overlapping investment issues.

Opportunities come with challenges. The first example of drug repurposing was an accidental discovery in the 1920s. After about a century of development, more approaches have been proposed for accelerating the process of drug repurposing. For this reason, drug repurposing has acquired great achievements. Massive machine learning algorithms were introduced to improve the performance of drug repurposing in this scenario. In addition to computational approaches, experimental approaches that give direct evidence of links between drugs and diseases were developed, such as target screening approaches [91, 524–526], cell assay approaches [527–530], animal model approaches [531–534], and clinical approaches [89]. These approaches are reliable and credible. In recent years, increasing numbers of researchers have combined computational approaches and experimental approaches to find new indications for drugs, called mixed approaches [78, 535, 536], wherein the result of computational methods was validated by biological experiments and clinical tests. Mixed approaches offer opportunities for developing repositioned drug effectively and rapidly.

Generating secondary patents provides an opportunity for researchers to find new indications for existing drugs [2, 512]. With the IP problem solved, many repurposing projects have been conducted smoothly with low cost, which has aroused concern from many countries. With regard to the commercial model, parallel strategies bring significant improvement in the efficiency of drug repurposing. For example, multiple tests or studies are conducted for a candidate drug, which reduces the time cost for drug repurposing.

From the market perspective, a large number of diseases require new drugs to be treated, which brings potential economic benefits. Taking rare diseases as an example, there are over 7000 rare diseases that need to be studied. However, only 5% of them are being researched [113]. Rare diseases are a large potential market to explore.

References

1. Roy S, Dhaneshwar S, Bhasin B. Drug repurposing: an emerging tool for drug reuse, recycling and discovery. Curr Drug Res Rev. 2021;
2. Talevi A. Multi-target pharmacology: possibilities and limitations of the "skeleton key approach" from a medicinal chemist perspective. Front Pharmacol. 2015;6:205.
3. Ashburn TT, Thor KB. Drug repositioning: identifying and developing new uses for existing drugs. Nat Rev Drug Discov. 2004;3(8):673–83.
4. Novac N. Challenges and opportunities of drug repositioning. Trends Pharmacol Sci. 2013;34:267–72.
5. Xue H, Li J, Xie H, et al. Review of drug repositioning approaches and resources. Int J Biol Sci. 2018;14(10):1232–44.
6. Drug development process. https://www.fda.gov/Drugs/default.htm.
7. Drug approval process. https://www.fda.gov/Drugs/DevelopmentApprovalProcess/default.htm.
8. Alaimo S, Pulvirenti A. Network-based drug repositioning: approaches, resources, and research directions. Methods Mol Biol. 2019;1903:97–113.
9. Deotarse PP, Jain AS, Baile MB, et al. Drug repositioning: a review. Int J Pharma Res Rev. 2015;4:51–8.
10. Pushpakom S, Iorio F, Eyers PA, et al. Drug repurposing: progress, challenges and recommendations. Nat Rev Drug Discov. 2019;18(1):41–58.
11. Schcolnik-Cabrera A, Juárez-López D, Duenas-Gonzalez A. Perspectives on drug repurposing. Curr Med Chem. 2021;28(11):2085–99.
12. Sleigh SH, Barton CL. Repurposing strategies for therapeutics. Pharmaceut Med. 2012;24(3):151–9.
13. Wishart DS, Knox C, Guo AC, et al. DrugBank: a comprehensive resource for in silico drug discovery and exploration. Nucleic Acids Res. 2006;34:668–72.
14. Seiler KP, George GA, Happ MP, et al. ChemBank: a small-molecule screening and cheminformatics resource database. Nucleic Acids Res. 2008;36:351–9.
15. Hamosh A, Scott AF, Amberger JS, et al. Online Mendelian Inheritance in Man (OMIM), a knowledgebase of human genes and genetic disorders. Nucleic Acids Res. 2005;33:D514–7.
16. Ogata H, Goto S, Sato K, et al. KEGG: Kyoto encyclopedia of genes and genomes. Nucleic Acids Res. 1999;27:29–34.
17. PubMed: US National Library of Medicine National Institutes of Health. https://www.ncbi.nlm.nih.gov/pubmed/
18. Mewes HW, Hani J, Pfeiffer F, et al. MIPS: a database for protein sequences and complete genomes. Nucleic Acids Res. 1998;26:33–7.
19. Bernstein FC, Koetzle TF, Williams GJ, et al. The Protein Data Bank: a computer-based archival file for macromolecular structures. J Mol Biol. 1977;112:535–42.

20. Barrett T, Wilhite SE, Ledoux P, et al. NCBI GEO: archive for functional genomics data sets-update. Nucleic Acids Res. 2013;41:D991–5.
21. Benson DA, Cavanaugh M, Clark K, et al. GenBank. Nucleic Acids Res. 2012;41:36–42.
22. Oprea TI, Overington JP. Computational and practical aspects of drug repositioning. Assay Drug Dev Technol. 2015;13:299–306.
23. Napolitano F, Zhao Y, Moreira VM, et al. Drug repositioning: a machine-learning approach through data integration. J Cheminform. 2013;5:30.
24. Lotfi Shahreza M, Ghadiri N, Mousavi SR, et al. A review of network-based approaches to drug repositioning. Brief Bioinform. 2018;19(5):878–92.
25. Zou J, Zheng MW, Li G, et al. Advanced systems biology methods in drug discovery and translational biomedicine. Biomed Res Int. 2013;2013:742835.
26. Gönen M. Predicting drug-target interactions from chemical and genomic kernels using Bayesian matrix factorization. Bioinformatics. 2012;28:2304–10.
27. Wu Z, Wang Y, Chen L. Network-based drug repositioning. Mol BioSyst. 2013;9(6):1268–81.
28. Emig D, Ivliev A, Pustovalova O, et al. Drug target prediction and repositioning using an integrated network-based approach. PLoS One. 2013;8(4):e60618.
29. Vanunu O, Magger O, Ruppin E, et al. Associating genes and protein complexes with disease via network propagation. PLoS Comput Biol. 2010;6(1):e1000641.
30. Wu C, Gudivada RC, Aronow BJ, et al. Computational drug repositioning through heterogeneous network clustering. BMC Syst Biol. 2013;7(Suppl 5):S6.
31. Li J, Zhu X, Chen JY. Building disease-specific drug-protein connectivity maps from molecular interaction networks and PubMed abstracts. PLoS Comput Biol. 2009;5:e1000450.
32. Kuusisto F, Steill J, Kuang Z, et al. A simple text mining approach for ranking pairwise associations in biomedical applications. AMIA Jt Summits Transl Sci Proc. 2017;2017:166–74.
33. Zhang M, Schmitt-ulms G, Sato C, et al. Drug repositioning for Alzheimer's disease based on systematic 'omics' data mining. PLoS One. 2016;11(12):e0168812.
34. Papanikolaou N, Pavlopoulos GA, Theodosiou T, et al. DrugQuest – a text mining workflow for drug association discovery. BMC Bioinformatics. 2016;17(Suppl 5):182.
35. Cheng D, Knox C, Young N, et al. PolySearch: a web-based text mining system for extracting relationships between human diseases, genes, mutations, drugs and metabolites. Nucleic Acids Res. 2008;36:W399–405.
36. Perovšek M, Kranjc J, Erjavec T, et al. TextFlows: a visual programming platform for text mining and natural language processing. Sci Comput Program. 2016;121:128–52.
37. Moreno I, Moreda P, Romá-Ferri MT. MaNER: a MedicAl named entity recogniser. Appl Nat Lang Inform Syst. 2015:418–23.
38. Lee S, Kim D, Lee K, et al. BEST: next-generation biomedical entity search tool for knowledge discovery from biomedical literature. PLoS One. 2016;11:e0164680.
39. Palma G, Vidal M-E, Raschid L. Drug-target interaction prediction using semantic similarity and edge partitioning. ISWC. 2014;1:131–46.
40. Mullen J, Cockell SJ, Woollard P, et al. An integrated data driven approach to drug repositioning using gene-disease associations. PLoS One. 2016;11:e0155811.
41. Chen B, Ding Y, Wild DJ. Assessing drug target association using semantic linked data. PLoS Comput Biol. 2012;8:e1002574.
42. Jin G, Wong ST. Toward better drug repositioning: prioritizing and integrating existing methods into efficient pipelines. Drug Discov Today. 2014;19(5):637–44.
43. Yildirim MA, et al. Drug-target network. Nat Biotechnol. 2007;25:1119–26.
44. Zhao S, Li S. Network-based relating pharmacological and genomic spaces for drug target identification. PLoS One. 2010;5:e11764.
45. Cheng F, Liu C, Jiang J, et al. Prediction of drug-target interactions and drug repositioning via network-based inference. PLoS Comput Biol. 2012;8:e1002503.
46. Alaimo S, Pulvirenti A, Giugno R, et al. Drug-target interaction prediction through domain-tuned network-based inference. Bioinformatics. 2013;29(16):2004–8.
47. Kinnings SL, Liu N, Buchmeier N, et al. Drug discovery using chemical systems biology: repositioning the safe medicine Comtan to treat multi-drug and extensively drug resistant tuberculosis. PLoS Comput Biol. 2009;5:e1000423.
48. Yang L, Agarwal P. Systematic drug repositioning based on clinical side-effects. PLoS One. 2011;6:e28025.
49. Campillos M, Kuhn M, Gavin A-C, et al. Drug target identification using side-effect similarity. Science. 2008;321:263–6.
50. Bisgin H, Liu Z, Kelly R, et al. Investigating drug repositioning opportunities in FDA drug labels through topic modeling. BMC Bioinformatics. 2012;13(Suppl 15):S6.
51. An SM, Ding QP, Li L-S, et al. Stem cell signaling as a target for novel drug discovery: recent in the WNT and Hedgehog pathways. Acta Pharmacol Sin. 2013;34:777–83.
52. Blatt J, Corey SJ. Drug repurposing in pediatrics and pediatric hematology oncology. Drug Discov Today. 2013;18:4–10.
53. Haeberle H, Dudley JT, Liu JT, et al. Identification of cell surface targets through meta-analysis of microarray data. Neoplasia. 2012;14:666–9.

54. Sanseau P, Agarwal P, Barnes MR, et al. Use of genome-wide association studies for drug repositioning. Nat Biotechnol. 2012;30:317–20.
55. Lamb J, Crawford ED, Peck D, et al. The connectivity map: using gene-expression signatures to connect small molecules, genes, and disease. Science. 2006;313:1929–35.
56. Lamb J. The connectivity map: using gene-expression profiling to identify new therapeutics and potential adverse drug effects. Chem Res Toxicol. 2007;20:2018.
57. Qu XYA, Rajpal DK. Applications of connectivity map in drug discovery and development. Drug Discov Today. 2012;17:1289–98.
58. Dudley JT, Sirota M, Shenoy M, et al. Computational repositioning of the anticonvulsant topiramate for inflammatory bowel disease. Sci Transl Med. 2011;3:96ra76.
59. Lussier YA, Chen JL. The emergence of genome-based drug repositioning. Sci Transl Med. 2011;3:96ps35.
60. Sirota M, Dudley JT, Kim J, et al. Discovery and preclinical validation of drug indications using compendia of public gene expression data. Sci Transl Med. 2011;3:96ra77.
61. Barretina J, Caponigro G, Stransky N, et al. The cancer cell line encyclopedia enables predictive modelling of anticancer drug sensitivity. Nature. 2012;483:603–7.
62. Jin GX, Fu C, Zhao H, et al. A novel method of transcriptional response analysis to facilitate drug repositioning for cancer therapy. Cancer Res. 2012;72:33–44.
63. Jin GX, Zhao H, Zhou X, et al. An enhanced Petrinet model to predict synergistic effects of pairwise drug combinations from gene microarray data. Bioinformatics. 2011;27:I310–6.
64. Iskar M, Zeller G, Blattmann P, et al. Characterization of drug-induced transcriptional modules: towards drug repositioning and functional understanding. Mol Syst Biol. 2013;9:662.
65. Ideker T, Thorsson V, Ranish JA, et al. Integrated genomic and proteomic analyses of a systematically perturbed metabolic network. Science. 2001;292:929–34.
66. Segal E, Shapira M, Regev A, et al. Module networks: identifying regulatory modules and their condition-specific regulators from gene expression data. Nat Genet. 2003;34:166–76.
67. Hong Z, Jin G, Cui K, et al. Novel modeling of cancer cell signaling pathways enables systematic drug repositioning for distinct breast cancer metastases. Cancer Res. 2013;73(20):6149–63.
68. Köhler S, Bauer S, Horn D, et al. Walking the interactome for prioritization of candidate disease genes. Am J Hum Genet. 2008;82(4):949–58.
69. Martínez V, Navarro C, Cano C, et al. DrugNet: network-based drug-disease prioritization by integrating heterogeneous data. Artif Intell Med. 2015;63(1):41–9.
70. Sun Y, Sheng Z, Ma C, et al. Combining genomic and network characteristics for extended capability in predicting synergistic drugs for cancer. Nat Commun. 2015;6:8481.
71. Sander JR, Ester M, Kriegel H-P, et al. Density-based clustering in spatial databases: the algorithm GDBSCAN and its applications. Data Min Knowl Disc. 1998;2:169–94.
72. Agrawal R, Gehrke J, Gunopulos, et al. Automatic subspace clustering of high dimensional data for data mining applications. ACM. 1998;27:94–105.
73. Wang W, Yang J, Muntz R. STING: a statistical information grid approach to spatial data mining. VLDB. 1997;1:186–95.
74. Ankerst M, Breunig MM, Kriegel H, et al. OPTICS: ordering points to identify the clustering structure. ACM. 1999;28:49–60.
75. Subelj L, Bajec M. Unfolding communities in large complex networks: combining defensive and offensive label propagation for core extraction. Phys Rev E Stat Nonlinear Soft Matter Phys. 2011;83:036103.
76. Luo H, Wang J, Li M, et al. Drug repositioning based on comprehensive similarity measures and Bi-Random walk algorithm. Bioinformatics. 2016;32:2664–71.
77. Mei J-P, Kwoh C-K, Yang P, et al. Drug-target interaction prediction by learning from local information and neighbors. Bioinformatics. 2012;29:238–45.
78. Ai N, Wood RD, Welsh WJ. Identification of nitazoxanide as a group imetabotropic glutamate receptor negative modulator for the treatment of neuropathic pain: an in silico drug repositioning study. Pharm Res. 2015;32:2798–807.
79. King AD, Przulj N, Jurisica I. Protein complex prediction via cost-based clustering. Bioinformatics. 2004;20:3013–20.
80. Macropol K, Can T, Singh AK. RRW: repeated random walks on genome-scale protein networks for local cluster discovery. BMC Bioinformatics. 2009;10:283.
81. Hearst MA. Untangling text data mining. In Proceedings of the 37th annual meeting of the association for computational linguistics on computational linguistics: association for computational linguistics. 1999; 1:3–10.
82. Hotho A, Nürnberger A, Paaß G. A brief survey of text mining. In Ldv Forum. 2005;20(1):19–62.
83. Feldman R, Sanger J. The text mining handbook. New York: Cambridge University Press; 2007.
84. Weeber M, Vos R, Klein H. Using concepts in literature-based discovery: simulating Swanson's Raynaud-fish oil and migraine-magnesium discoveries. J Am Soc Inf Sci Technol. 2001;52:972–6.
85. Gramatica R, Di Matteo T, Giorgetti S, et al. Graph theory enables drug repurposing-how a mathematical model can drive the discovery of hidden mechanisms of action. PLoS One. 2014;9:e84912.
86. Jang G, Lee T, Lee BM, et al. Literature-based prediction of novel drug indications consider-

ing relationships between entities. Mol BioSyst. 2017;13:1399–405.

87. Fleuren WW, Alkema W. Application of text mining in the biomedical domain. Methods. 2015;74:97–106.
88. Krallinger M, Erhardt RA-A, Valencia A. Text-mining approaches in molecular biology and biomedicine. Drug Discov Today. 2005;10:439–45.
89. Zhu Q, Tao C, Shen F, et al. Exploring the pharmacogenomics knowledge base (PharmGKB) for repositioning breast cancer drugs by leveraging Web ontology language (OWL) and cheminformatics approaches. Pac Symp Biocomput. 2014;1:172–82.
90. Lee JY, Shin JY, Kim HS, et al. Effect of combined treatment with progesterone and tamoxifen on the growth and apoptosis of human ovarian cancer cells. Oncol Rep. 2012;27:87–93.
91. Swinney DC, Anthony J. How were new medicines discovered? Nat Rev Drug Discov. 2011;10:507–19.
92. Hurle MR, Yang L, Xie Q, et al. Computational drug repositioning: from data to therapeutics. Clin Pharmacol Ther. 2013;93:335–41.
93. Swamidass SJ. Mining small-molecule screens to repurpose drugs. Brief Bioinform. 2011;12:327–35.
94. Crisman TJ, Parker CN, Jenkins JL, et al. Understanding false positives in reporter gene assays: in silico chemogenomics approaches to prioritize cell-based HTS data. J Chem Inf Model. 2007;47:1319–27.
95. Feng BY, Simeonov A, Jadhav A, et al. A high-throughput screen for aggregation-based inhibition in a large compound library. J Med Chem. 2007;50:2385–90.
96. Ekins S, Mestres J, Testa B, et al. In silico pharmacology for drug discovery: methods for virtual ligand screening and profiling. Br J Pharmacol. 2007;152:9–20.
97. Doman TN, McGovern SL, Witherbee BJ, et al. Molecular docking and high-throughput screening for novel inhibitors of protein tyrosine phosphatase-1B. J Med Chem. 2002;45:2213–21.
98. Kolb P, Ferreira RS, Irwin JJ, et al. Docking and chemoinformatic screens for new ligands and targets. Curr Opin Biotechnol. 2009;20:429–36.
99. Wilkinson GF, Pritchard K. In vitro screening for drug repositioning. J Biomol Screen. 2015;20(2):167–79.
100. Ciallella JR, Reaume AG. In vivo phenotypic screening: clinical proof of concept for a drug repositioning approach. Drug Discov Today Technol. 2017;23:45–52.
101. Lipinski CA, Reaume AG. High throughput in vivo phenotypic screening for drug repurposing: discovery of MLR-1023 a novel insulin sensitizer and novel Lyn kinase activator with clinical proof of concept. Bioorg Med Chem. 2020;28(9):115425.
102. Reaume A. Drug repurposing through nonhypothesis driven phenotypic screening. Drug Disc Today. 2011;8:85–8.
103. Saporito MS, Lipinski C, Reaume A. Phenotypic in vivo screening to identify new, unpredicted indications for existing drugs and drug candidates. Hoboken: Wiley; 2012.
104. Mohammadi E, Benfeitas R, Turkez H, et al. Applications of genome-wide screening and systems biology approaches in drug repositioning. Cancers (Basel). 2020;12(9):2694.
105. Pacini C, Iorio F, Gonçalves E, et al. DvD: an R/Cytoscape pipeline for drug repurposing using public repositories of gene expression data. Bioinformatics. 2013;29(1):132–4.
106. Liu Y, Hu B, Fu C, Chen X. DCDB: drug combination database. Bioinformatics. 2009;26:587–8.
107. Ekins S, Bunin BA. The Collaborative Drug Discovery (CDD) database. Methods Mol Biol. 2013;993:139–54.
108. Fu C, Jin G, Gao J, et al. DrugMap Central: an on-line query and visualization tool to facilitate drug repositioning studies. Bioinformatics. 2013;29:1834–6.
109. Barabási AL, Gulbahce N, Loscalzo J. Network medicine: a network-based approach to human disease. Nat Rev Genet. 2011;12:56–68.
110. Thorn CF, Klein TE, Altman RB. PharmGKB: the pharmacogenomics knowledge base. Methods Mol Biol. 2013;1015:311–20.
111. Goel R, Muthusamy B, Pandey A, et al. Human protein reference database and human proteinpedia as discovery resources for molecular biotechnology. Mol Biotechnol. 2011;48:87–95.
112. Hodis E, Prilusky J, Martz E, et al. Proteopedia – a scientific 'wiki' bridging the rift between three-dimensional structure and function of biomacromolecules. Genome Biol. 2008;9:R121.
113. Sardana D, Zhu C, Zhang M, et al. Drug repositioning for orphan diseases. Brief Bioinformatics. 2011;12:346–56.
114. Gaulton A, Bellis LJ, Bento AP, et al. ChEMBL: a large-scale bioactivity database for drug discovery. Nucleic Acids Res. 2012;40:D1100–7.
115. Chen B, Dong X, Jiao D, et al. Chem2Bio2RDF: a semantic framework for linking and data mining chemogenomic and systems chemical biology data. BMC Bioinformatics. 2010;11:255.
116. Chen J, Swamidass SJ, Dou Y, et al. ChemDB: a public database of small molecules and related chemoinformatics resources. Bioinformatics. 2005;21:4133–9.
117. Wang Y, Xiao J, Suzek TO, et al. PubChem: a public information system for analyzing bioactivities of small molecules. Nucleic Acids Res. 2009;37:623–33.
118. Ma DL, Chan DS, Leung CH. Drug repositioning by structure-based virtual screening. Chem Soc Rev. 2013;42(5):2130–41.
119. Bolton EE, et al. PubChem: integrated platform of small molecules and biological activities. Annu Rep Comput Chem. 2008;4:217–41.
120. Szklarczyk D, Franceschini A, Kuhn M, et al. The STRING database in 2011: functional interaction networks of proteins, globally integrated and scored. Nucleic Acids Res. 2011;39:D561–8.

121. Serafin MB, Bottega A, da Rosa TF, et al. Drug repurposing in oncology. Am J Ther. 2021;28(1):e111–7.
122. Antoszczak M, Markowska A, Markowska J, et al. Old wine in new bottles: drug repurposing in oncology. Eur J Pharmacol. 2020;866:172784.
123. Antoszczak M, Markowska A, Markowska J, et al. Antidepressants and antipsychotic agents as repurposable oncological drug candidates. Curr Med Chem. 2021;28(11):2137–74.
124. Jahchan NS, Dudley JT, Mazur PK, et al. A drug repositioning approach identifies tricyclic antidepressants as inhibitors of small cell lung cancer and other neuroendocrine tumors. Cancer Discov. 2013;3:1364–77.
125. Pessetto ZY, Weir SJ, Sethi G, et al. Drug repurposing for gastrointestinal stromal tumor. Mol Cancer Ther. 2013;12:1299–309.
126. Taylor DJ. Parallel screening of FDA-approved antineoplastic drugs for identifying sensitizers of TRAIL-induced apoptosis in cancer cells. BMC Cancer. 2011;11:470.
127. Nygren P, Fryknäs M, Agerup B, et al. Repositioning of the antihelmintic drug mebendazole for the treatment for colon cancer. J Cancer Res Clin Oncol. 2013;139:2133–40.
128. Platz EA, Yegnasubramanian S, Liu JO, et al. A novel two-stage, transdisciplinary study identifies digoxin as a possible drug for prostate cancer treatment. Cancer Discov. 2011;1:68–77.
129. Jiang P, Mukthavavam R, Chao Y, et al. Novel anti-glioblastoma agents and therapeutic combinations identified from a collection of FDA approved drugs. J Transl Med. 2014;12:13.
130. Wen PY, Kesari S. Malignant gliomas in adults. N Engl J Med. 2008;359:492–507.
131. Sharma A, Jacob A, Tandon M, et al. Orphan drug: development trends and strategies. J Pharm Bioallied Sci. 2010;2:290–9.
132. Greenstone G. The revival of thalidomide: from tragedy to therapy. Br Columb Med J. 2011;53:230–3.
133. Iyer CGS, Languillon J, Ramanujam K, et al. WHO co-ordinated short-term doubleblind trial with thalidomide in the treatment of acute lepra reactions in male lepromatous patients. Bull WHO. 1971;45:719–32.
134. D'Amato R, Loughnan MS, Flynn E, et al. Thalidomide is an inhibitor of angiogenesis. Proc Natl Acad Sci U S A. 1994;91:4082–5.
135. Singhal S, Mehta J, Desikan R, et al. Antitumor activity of thalidomide in refractory multiple myeloma. N Engl J Med. 1999;341:1565–71.
136. Evans JMM, Donnelly LA, Emslie-Smith AM, et al. Metformin and the reduced risk of cancer in diabetic patients. BMJ. 2005;330:1304–5.
137. Micic D, Cvijovic G, Trajkovic V, et al. Metformin: its emerging role in oncology. Hormones (Athens). 2011;10:5–15.
138. Liu B, Fan Z, Edgerton SM, et al. Metformin induces unique biological and molecular responses in triple negative breast cancer cells. Cell Cycle. 2009;8:2031–40.
139. Buzzai M, Jones RG, Amaravadi RK, et al. Systemic treatment with the antidiabetic drug metformin selectively impairs p53-deficient tumor cell growth. Cancer Res. 2007;67:6745–52.
140. Hirsch HA, Iliopoulos D, Tsichlis PN, et al. Metformin selectively targets cancer stem cells, and acts together with chemotherapy to block tumor growth and prolong remission. Cancer Res. 2009;69:7507–11.
141. National Cancer Institute. A phase I trial of nelfinavir (Viracept) in adults with solid tumors. ClinicalTrialsgov, NCT01445106.
142. Ishida J, Konishi M, Ebner N, et al. Repurposing of approved cardiovascular drugs. J Transl Med. 2016;14:269.
143. Gelosa P, Castiglioni L, Camera M, et al. Repurposing of drugs approved for cardiovascular diseases: opportunity or mirage? Biochem Pharmacol. 2020;177:113895.
144. Regulska K, Regulski M, Karolak B, et al. Can cardiovascular drugs support cancer treatment? The rationale for drug repurposing. Drug Discov Today. 2019;24(4):1059–65.
145. Smith WL, Garavito RM, DeWitt DL. Prostaglandin endoperoxide H synthases (cyclooxygenases)-1 and -2. J Biol Chem. 1996;271:33157–60.
146. Gasic GJ, Gasic TB, Murphy S. Anti-metastatic effect of aspirin. Lancet. 1972;2:932–3.
147. Kolenich JJ, Mansour EG, Flynn A. Haematological effects of aspirin. Lancet. 1972;2:714.
148. Thun MJ, Henley SJ, Patrono C. Nonsteroidal anti-inflammatory drugs as anticancer agents: mechanistic, pharmacologic, and clinical issues. J Natl Cancer Inst. 2002;94:252–66.
149. Rothwell PM, Wilson M, Elwin CE, et al. Long-term effect of aspirin on colorectal cancer incidence and mortality: 20-year follow-up of five randomised trials. Lancet. 2010;376:1741–50.
150. Rothwell PM, Fowkes FG, Belch JF, et al. Effect of daily aspirin on long-term risk of death due to cancer: analysis of individual patient data from randomised trials. Lancet. 2011;377:31–41.
151. Pugh S, Thomas GA. Patients with adenomatous polyps and carcinomas have increased colonic mucosal prostaglandin E2. Gut. 1994;35:675–8.
152. Shao J, Jung C, Liu C, et al. Prostaglandin E2 stimulates the beta-catenin/T cell factor-dependent transcription in colon cancer. J Biol Chem. 2005;280:26565–72.
153. Ruffin MT, Krishnan K, Rock CL, et al. Suppression of human colorectal mucosal prostaglandins: determining the lowest effective aspirin dose. J Natl Cancer Inst. 1997;89:1152–60.
154. Kune GA, Kune S, Watson LF. Colorectal cancer risk, chronic illnesses, operations, and medications: case control results from the Melbourne Colorectal Cancer Study. Cancer Res. 1988;48:4399–404.

155. Flossmann E, Rothwell PM. Effect of aspirin on long-term risk of colorectal cancer: consistent evidence from randomized and observational studies. Lancet. 2007;369:1603–13.
156. González-Pérez A, Rodríguez LAG, López-Ridaura R. Effects of non-steroidal anti-inflammatory drugs on cancer sites other than the colon and rectum: a meta-analysis. BMC Cancer. 2003;3:28.
157. Elwood PC, Gallagher AM, Duthie GG, et al. Aspirin, salicylates, and cancer. Lancet. 2009;373:1301–9.
158. Algra AM, Rothwell PM. Effects of regular aspirin on long-term cancer incidence and metastasis: a systematic comparison of evidence from observational studies versus randomised trials. Lancet Oncol. 2012;13:518–27.
159. Bosetti C, Rosato V, Gallus S, et al. Aspirin and cancer risk: a quantitative review to 2011. Ann Oncol. 2012;23:1403–15.
160. Rothwell PM, Wilson M, Price JF, et al. Effect of daily aspirin on risk of cancer metastasis: a study of incident cancers during randomised controlled trials. Lancet. 2012;379:1591–601.
161. Bukberg J, Penman D, Holland JC. Depression in hospitalized cancer patients. Psychosom Med. 1984;46:199–212.
162. Spiegel D, Giese-Davis J. Depression and cancer: mechanisms and disease progression. Biol Psychiatry. 2003;54:269–82.
163. Spiegel D. Health caring. Psychosocial support for patients with cancer. Cancer. 1994;74:1453–7.
164. Chida Y, Hamer M, Wardle J, et al. Do stress-related psychosocial factors contribute to cancer incidence and survival? Nat Clin Pract Oncol. 2008;5:466–75.
165. Vandewalle B, Revillion F, Lefebvre J. Functional beta-adrenergic receptors in breast cancer cells. J Cancer Res Clin Oncol. 1990;116:303–6.
166. Weddle DL, Tithoff P, Williams M, et al. Beta-adrenergic growth regulation of human cancer cell lines derived from pancreatic ductal carcinomas. Carcinogenesis. 2001;22:473–9.
167. Yang EV, Sood AK, Chen M, et al. Norepinephrine up-regulates the expression of vascular endothelial growth factor, matrix metalloproteinase (MMP)-2, and MMP-9 in nasopharyngeal carcinoma tumor cells. Cancer Res. 2006;66:10357–64.
168. Thaker PH, Han LY, Kamat AA, et al. Chronic stress promotes tumor growth and angiogenesis in a mouse model of ovarian carcinoma. Nat Med. 2006;12:939–44.
169. Wu WK, Wong HP, Luo SW, et al. 4-(Methylnitrosamino)-1-(3-pyridyl)-1-butanone from cigarette smoke stimulates colon cancer growth via β-adrenoceptors. Cancer Res. 2005;65:5272–7.
170. Masur K, Niggemann B, Zanker KS, et al. Norepinephrine-induced migration of SW 480 colon carcinoma cells is inhibited by β-blockers. Cancer Res. 2001;61:2866–9.
171. Sood AK, Armaiz-Pena GN, Halder J, et al. Adrenergic modulation of focal adhesion kinase protects human ovarian cancer cells from anoikis. J Clin Invest. 2010;120:1515–23.
172. Sloan EK, Priceman SJ, Cox BF, et al. The sympathetic nervous system induces a metastatic switch in primary breast cancer. Cancer Res. 2010;70:7042–52.
173. Sood AK, Bhatty R, Kamat AA, et al. Stress hormone-mediated invasion of ovarian cancer cells. Clin Cancer Res. 2006;12:369–75.
174. Springer J, Tschirner A, Haghikia A, et al. Prevention of liver cancer cachexia-induced cardiac wasting and heart failure. Eur Heart J. 2014;35:932–41.
175. Barron TI, Connolly RM, Sharp L, et al. Beta blockers and breast cancer mortality: a population- based study. J Clin Oncol. 2011;29:2635–44.
176. Diaz ES, Karlan BY, Li AJ. Impact of beta blockers on epithelial ovarian cancer survival. Gynecol Oncol. 2012;127:375–8.
177. Wang HM, Liao ZX, Komaki R, et al. Improved survival outcomes with the incidental use of beta-blockers among patients with non-small-cell lung cancer treated with definitive radiation therapy. Ann Oncol. 2013;24:1312–9.
178. Grytli HH, Fagerland MW, Fosså SD, et al. Association between use of β-blockers and prostate cancer–specific survival: a cohort study of 3561 prostate cancer patients with high-risk or metastatic disease. Eur Urol. 2014;65:635–41.
179. Ganz PA, Habel LA, Weltzien EK, et al. Examining the influence of beta blockers and ACE inhibitors on the risk for breast cancer recurrence: results from the LACE cohort. Breast Cancer Res Treat. 2011;129:549–56.
180. Lemeshow S, Sørensen HT, Phillips G, et al. β-Blockers and survival among Danish patients with malignant melanoma: a population-based cohort study. Cancer Epidemiol Biomark Prev. 2011;20:2273–9.
181. Choi CH, Song T, Kim TH, et al. Meta-analysis of the effects of beta blocker on survival time in cancer patients. J Cancer Res Clin Oncol. 2014;140:1179–88.
182. Goldfarb DA, Diz DI, Tubbs RR, et al. Angiotensin II receptor subtypes in the human renal cortex and renal cell carcinoma. J Urol. 1994;151:208–13.
183. Marsigliante S, Resta L, Muscella A, et al. AT1 angiotensin II receptor subtype in the human larynx and squamous laryngeal carcinoma. Cancer Lett. 1996;110:19–27.
184. Fujimoto Y, Sasaki T, Tsuchida A, et al. Angiotensin II type 1 receptor expression in human pancreatic cancer and growth inhibition by angiotensin II type 1 receptor antagonist. FEBS Lett. 2001;495:197–200.
185. Suganuma T, Ino K, Shibata K, et al. Functional expression of the angiotensin II type 1 receptor in human ovarian carcinoma cells and its blockade therapy resulting in suppression of tumor invasion, angiogenesis, and peritoneal dissemination. Clin Cancer Res. 2005;11:2686–94.

186. Herr D, Rodewald M, Fraser HM, et al. Potential role of renin-angiotensin-system for tumor angiogenesis in receptor negative breast cancer. Gynecol Oncol. 2008;109:418–25.
187. Otake AH, Mattar AL, Freitas HC, et al. Inhibition of angiotensin II receptor 1 limits tumor-associated angiogenesis and attenuates growth of murine melanoma. Cancer Chemother Pharmacol. 2010;66:79–87.
188. Okamoto K, Tajima H, Ohta T, et al. Angiotensin II induces tumor progression and fibrosis in intrahepatic cholangiocarcinoma through an interaction with hepatic stellate cells. Int J Oncol. 2010;37:1251–9.
189. Anandanadesan R, Gong Q, Chipitsyna G, et al. Angiotensin II induces vascular endothelial growth factor in pancreatic cancer cells through an angiotensin II type 1 receptor and ERK1/2 signaling. J Gastrointest Surg. 2008;12:57–66.
190. Neo JH, Malcontenti-Wilson C, Muralidharan V, et al. Effect of ACE inhibitors and angiotensin II receptor antagonists in a mouse model of colorectal cancer liver metastases. J Gastroenterol Hepatol. 2007;22:577–84.
191. Yasumatsu R, Nakashima T, Masuda M, et al. Effects of the angiotensin-I converting enzyme inhibitor perindopril on tumor growth and angiogenesis in head and neck squamous cell carcinoma cells. J Cancer Res Clin Oncol. 2004;130:567–73.
192. Lever AF, Hole DJ, Gillis CR, et al. Do inhibitors of angiotensin-I-converting enzyme protect against risk of cancer? Lancet. 1998;352:179–84.
193. Meier CR, Derby LE, Jick SS, et al. Angiotensin-converting enzyme inhibitors, calcium channel blockers, and breast cancer. Arch Intern Med. 2000;160:349–53.
194. Friis S, Sorensen HT, Mellemkjaer L, et al. Angiotensin-converting enzyme inhibitors and the risk of cancer: a population-based cohort study in Denmark. Cancer. 2001;92:2462–70.
195. Lindholm LH, Anderson H, Ekbom T, et al. Relation between drug treatment and cancer in hypertensives in the Swedish trial in old patients with hypertension 2: a 5-year, prospective, randomised, controlled trial. Lancet. 2001;358:539–44.
196. Sjoberg T, Garcia Rodriguez LA, Lindblad M. Angiotensin-converting enzyme inhibitors and risk of esophageal and gastric cancer: a nested case-control study. Clin Gastroenterol Hepatol. 2007;5(1160–1166):e1161.
197. Ronquist G, Rodriguez LA, Ruigomez A, et al. Association between captopril, other antihypertensive drugs and risk of prostate cancer. Prostate. 2004;58:50–6.
198. Sipahi I, Debanne SM, Rowland DY, et al. Angiotensin-receptor blockade and risk of cancer: meta-analysis of randomised controlled trials. Lancet Oncol. 2010;11:627–36.
199. Pasternak B, Svanstrom H, Callreus T, et al. Use of angiotensin receptor blockers and the risk of cancer. Circulation. 2011;123:1729–36.
200. ARB Trialists Collaboration Effects of telmisartan, irbesartan, valsartan, candesartan, and losartan on cancers in 15 trials enrolling 138,769 individuals. J Hypertens. 2011;29:623–35.
201. Shiratori O. Growth inhibitory effect of cardiac glycosides and aglycones on neoplastic cells: in vitro and in vivo studies. Gann. 1967;58:521–8.
202. Avila J, Lecuona E, Morales M, et al. Opposite expression pattern of the human Na, K-ATPase β1 isoform in stomach and colon adenocarcinomasa. Ann N Y Acad Sci. 1997;834:653–5.
203. Espineda C, Seligson DB, James Ball W, et al. Analysis of the Na, K-ATPase α- and β-subunit expression profiles of bladder cancer using tissue microarrays. Cancer. 2003;97:1859–68.
204. Rajasekaran SA, Ball WJ, Bander NH, et al. Reduced expression of beta-subunit of Na, K-ATPase in human clear-cell renal cell carcinoma. J Urol. 1999;162:574–80.
205. Kawazoe N, Aiuchi T, Masuda Y, et al. Induction of apoptosis by bufalin in human tumor cells is associated with a change of intracellular concentration of Na^+ ions. J Biochem. 1999;126:278–86.
206. Raghavendra PB, Sreenivasan Y, Ramesh GT, et al. Cardiac glycoside induces cell death via FasL by activating calcineurin and NFAT, but apoptosis initially proceeds through activation of caspases. Apoptosis. 2007;12:307–18.
207. Manna SK, Sah NK, Newman RA, et al. Oleandrin suppresses activation of nuclear transcription factor-kappaB, activator protein-1, and c-Jun NH2-terminal kinase. Cancer Res. 2000;60:3838–47.
208. Bielawski K, Winnicka K, Bielawska A. Inhibition of DNA topoisomerases I and II, and growth inhibition of breast cancer MCF-7 cells by ouabain, digoxin and proscillaridin A. Biol Pharm Bull. 2006;29:1493–7.
209. Chen J-Q, Contreras RG, Wang R, et al. Sodium/potasium ATPase (Na^+, K^+-ATPase) and ouabain/related cardiac glycosides: a new paradigm for development of anti-breast cancer drugs? Breast Cancer Res Treat. 2006;96:1–15.
210. Nesher M, Shpolansky U, Rosen H, et al. The digitalis-like steroid hormones: new mechanisms of action and biological significance. Life Sci. 2007;80:2093–107.
211. Schoner W, Scheiner-Bobis G. Endogenous and exogenous cardiac glycosides: their roles in hypertension, salt metabolism, and cell growth. Am J Physiol-Cell Physiol. 2007;293:C509–36.
212. Stenkvist B, Bengtsson E, Eriksson O, et al. Cardiac glycosides and breast cancer. Lancet. 1979;1:563.
213. Stenkvist B, Bengtsson E, Dahlqvist B, et al. Cardiac glycosides and breast cancer, revisited. N Engl J Med. 1982;306:484.
214. Stenkvist B. Is digitalis a therapy for breast carcinoma? Oncol Rep. 1999;6:493–6.

215. Haux J, Klepp O, Spigset O, et al. Digitoxin medication and cancer; case control and internal dose-response studies. BMC Cancer. 2001;1:11.
216. Ahern TP, Lash TL, Sorensen HT, et al. Digoxin treatment is associated with an increased incidence of breast cancer: a population-based case-control study. Breast Cancer Res. 2008;10:R102.
217. Biggar RJ, Wohlfahrt J, Oudin A, et al. Digoxin use and the risk of breast cancer in women. J Clin Oncol. 2011;29:2165–70.
218. Biggar RJ, Wohlfahrt J, Melbye M. Digoxin use and the risk of cancers of the corpus uteri, ovary and cervix. Int J Cancer. 2012;131:716–21.
219. Demierre M-F, Higgins PDR, Gruber SB, et al. Statins and cancer prevention. Nat Rev Cancer. 2005;5:930–42.
220. Ogunwobi OO, Beales IL. Statins inhibit proliferation and induce apoptosis in Barrett's esophageal adenocarcinoma cells. Am J Gastroenterol. 2008;103:825–37.
221. Kang S, Kim ES, Moon A. Simvastatin and lovastatin inhibit breast cell invasion induced by H-Ras. Oncol Rep. 2009;21:1317–22.
222. Denoyelle C, Vasse M, Korner M, et al. Cerivastatin, an inhibitor of HMG-CoA reductase, inhibits the signaling pathways involved in the invasiveness and metastatic properties of highly invasive breast cancer cell lines: an in vitro study. Carcinogenesis. 2001;22:1139–48.
223. Graaf MR, Beiderbeck AB, Egberts AC, et al. The risk of cancer in users of statins. J Clin Oncol. 2004;22:2388–94.
224. Blais L, Desgagné A, LeLorier J. 3-Hydroxy-3-methylglutaryl coenzyme A reductase inhibitors and the risk of cancer: a nested case-control study. Arch Intern Med. 2000;160:2363–8.
225. Bansal D, Undela K, D'Cruz S, et al. Statin use and risk of prostate cancer: a meta-analysis of observational studies. PLoS One. 2012;7:e46691.
226. Singh S, Singh AG, Singh PP, et al. Statins are associated with reduced risk of esophageal cancer, particularly in patients with Barrett's esophagus: a systematic review and meta-analysis. Clin Gastroenterol Hepatol. 2013;11:620–9.
227. Wu X-D, Zeng K, Xue F-Q, et al. Statins are associated with reduced risk of gastric cancer: a meta-analysis. Eur J Clin Pharmacol. 2013;69:1855–60.
228. Emberson JR, Kearney PM, Blackwell L, et al. Lack of effect of lowering LDL cholesterol on cancer: meta-analysis of individual data from 175,000 people in 27 randomised trials of statin therapy. PLoS One. 2012;7:e29849.
229. Tan M, Song X, Zhang G, et al. Statins and the risk of lung cancer: a meta-analysis. PLoS One. 2013;8:e57349.
230. Li X, Wu X, Chen Q. Statin use is not associated with reduced risk of skin cancer: a meta-analysis. Br J Cancer. 2014;110:802–7.
231. Kornblau SM, Banker DE, Stirewalt D, et al. Blockade of adaptive defensive changes in cholesterol uptake and synthesis in AML by the addition of pravastatin to idarubicin + high-dose Ara-C: a phase 1 study. Blood. 2007;109:2999–3006.
232. Graf H, Jüngst C, Straub G, et al. Chemoembolization combined with pravastatin improves survival in patients with hepatocellular carcinoma. Digestion. 2008;78:34–8.
233. Zhou D, Wang L, Cui Q, et al. Repurposing lidocaine as an anticancer drug: the role beyond anesthesia. Front Cell Dev Biol. 2020;8:565.
234. Grandhi RK, Perona B. Mechanisms of action by which local anesthetics reduce cancer recurrence: a systematic review. Pain Med. 2020;21(2):401–14.
235. Piegeler T, Schläpfer M, Dull RO, et al. Clinically relevant concentrations of lidocaine and ropivacaine inhibit TNFα-induced invasion of lung adenocarcinoma cells in vitro by blocking the activation of Akt and focal adhesion kinase. Br J Anaesth. 2015;115(5):784–91.
236. El-Badrawy MK, Yousef AM, Shaalan D, et al. Matrix metalloproteinase-9 expression in lung cancer patients and its relation to serum mmp-9 activity, pathologic type, and prognosis. J Bronchol Interv Pulmonol. 2014;21(4):327–34.
237. Lee CY, Shim HS, Lee S, et al. Prognostic effect of matrix metalloproteinase-9 in patients with resected non-small cell lung cancer. J Cardiothorac Surg. 2015;10:44.
238. Duan J, Li X, Huang S, et al. GOLPH2, a gene downstream of ras signaling, promotes the progression of pancreatic ductal adenocarcinoma. Mol Med Rep. 2018;17(3):4187–94.
239. Li H, Yang LL, Xiao Y, et al. Overexpression of Golgi phosphoprotein 2 is associated with poor prognosis in oral squamous cell carcinoma. Am J Clin Pathol. 2018;150(1):74–83.
240. Zhang L, Hu R, Cheng Y, et al. Lidocaine inhibits the proliferation of lung cancer by regulating the expression of GOLT1A. Cell Prolif. 2017;50(5):e12364.
241. Akkuş M, Öner E. Can local infiltration of lidocaine reduce the postoperative atrial fibrillation rate in patients undergoing lobectomy for lung cancer? Acta Chir Belg. 2020;120(4):265–70.
242. D'Agostino G, Saporito A, Cecchinato V, et al. Lidocaine inhibits cytoskeletal remodelling and human breast cancer cell migration. Br J Anaesth. 2018;121(4):962–8.
243. Freeman J, Crowley PD, Foley AG, et al. Effect of perioperative lidocaine, propofol and steroids on pulmonary metastasis in a murine model of breast cancer surgery. Cancers (Basel). 2019;11(5):613.
244. Chamaraux-Tran TN, Mathelin C, Aprahamian M, et al. Antitumor effects of lidocaine on human breast cancer cells: an in vitro and in vivo experimental trial. Anticancer Res. 2018;38(1):95–105.
245. Xing W, Chen DT, Pan JH, et al. Lidocaine induces apoptosis and suppresses tumor growth in human hepatocellular carcinoma cells in vitro and in a xenograft model in vivo. Anesthesiology. 2017;126(5):868–81.

246. Kaneishi K, Kawabata M. Continuous subcutaneous infusion of lidocaine for persistent hiccup in advanced cancer. Palliat Med. 2013;27(3):284–5.
247. Tat T, Jurj A, Selicean C, et al. Antiproliferative effects of propofol and lidocaine on the colon adenocarcinoma microenvironment. J BUON. 2019;24(1):106–15.
248. Leng T, Lin S, Xiong Z, et al. Lidocaine suppresses glioma cell proliferation by inhibiting TRPM7 channels. Int J Physiol Pathophysiol Pharmacol. 2017;9(2):8–15.
249. Heckman-Stoddard BM, DeCensi A, Sahasrabuddhe VV, et al. Repurposing metformin for the prevention of cancer and cancer recurrence. Diabetologia. 2017;60(9):1639–47.
250. Zhou M, Xia L, Wang J. Metformin transport by a newly cloned proton-stimulated organic cation transporter (plasma membrane monoamine transporter) expressed in human intestine. Drug Metab Dispos. 2007;35:1956–62.
251. Bailey C. Metformin: historical overview. Diabetologia. 2017;60(9):1566–76.
252. Jung YS, Park CH, Eun CS, Park DI, Han DS. Metformin use and the risk of colorectal adenoma: a systematic review and meta-analysis. J Gastroenterol Hepatol. 2017;32:957–65.
253. He XK, Su TT, Si JM, Sun LM. Metformin is associated with slightly reduced risk of colorectal cancer and moderate survival benefits in diabetes mellitus: a meta-analysis. Medicine. 2016;95:e2749.
254. Nie Z, Zhu H, Gu M. Reduced colorectal cancer incidence in type 2 diabetic patients treated with metformin: a meta-analysis. Pharm Biol. 2016;54:2636–42.
255. Rokkas T, Portincasa P. Colon neoplasia in patients with type 2 diabetes on metformin: a meta-analysis. Eur J Intern Med. 2016;33:60–6.
256. Gandini S, Puntoni M, Heckman-Stoddard BM, et al. Metformin and cancer risk and mortality: a systematic review and meta-analysis taking into account biases and confounders. Cancer Prev Res. 2014;7:867–85.
257. Lega IC, Shah PS, Margel D, et al. The effect of metformin on mortality following cancer among patients with diabetes. Cancer Epidemiol Biomark Prev. 2014;23:1974–84.
258. Zhang ZJ, Li S. The prognostic value of metformin for cancer patients with concurrent diabetes: a systematic review and meta-analysis. Diabetes Obes Metab. 2014;16:707–10.
259. Yang T, Yang Y, Liu S. Association between metformin therapy and breast cancer incidence and mortality: evidence from a meta-analysis. J Breast Cancer. 2015;18:264–70.
260. Perez-Lopez FR, Pasupuleti V, Gianuzzi X, et al. Systematic review and meta-analysis of the effect of metformin treatment on overall mortality rates in women with endometrial cancer and type 2 diabetes mellitus. Maturitas. 2017;101:6–11.
261. Ma SJ, Zheng YX, Zhou PC, et al. Metformin use improves survival of diabetic liver cancer patients: systematic review and meta-analysis. Oncotarget. 2016;7:66202–11.
262. Wan G, Yu X, Chen P, et al. Metformin therapy associated with survival benefit in lung cancer patients with diabetes. Oncotarget. 2016;7:35437–45.
263. Coyle C, Cafferty FH, Vale C, et al. Metformin as an adjuvant treatment for cancer: a systematic review and meta-analysis. Ann Oncol. 2016;27:2184–95.
264. Stopsack KH, Ziehr DR, Rider JR, et al. Metformin and prostate cancer mortality: a meta-analysis. Cancer Causes Control. 2016;27:105–13.
265. Zhou PT, Li B, Liu FR, et al. Metformin is associated with survival benefit in pancreatic cancer patients with diabetes: a systematic review and meta-analysis. Oncotarget. 2017;8:25242–50.
266. Tian RH, Zhang YG, Wu Z, et al. Effects of metformin on survival outcomes of lung cancer patients with type 2 diabetes mellitus: a meta-analysis. Clin Transl Oncol. 2016;18:641–9.
267. Raval AD, Thakker D, Vyas A, et al. Impact of metformin on clinical outcomes among men with prostate cancer: a systematic review and meta-analysis. Prostate Cancer Prostatic Dis. 2015;18:110–21.
268. Heckman-Stoddard BM, Gandini S, Puntoni M, Dunn BK, et al. Repurposing old drugs to chemoprevention: the case of metformin. Semin Oncol. 2016;43:123–33.
269. Katare PB, Banerjee SK. Repurposing of drugs in cardiometabolic disorders: importance and current scenario. Curr Top Med Chem. 2016;16(19):2189–200.
270. Kelli HM, Kassas I, Lattouf OM. Cardiometabolic syndrome: a global epidemic. J Diabetes Metab. 2015;6:1–14.
271. Srivastava AK. Challenges in the treatment of cardiometabolic syndrome. Indian J Pharmacol. 2012;44(2):155–6.
272. Castro JP, El-Atat FA, MacFarlane SI, et al. Cardiometabolic syndrome: pathophysiology and treatment. Curr Hypertens Rep. 2003;5:393–401.
273. Gelosa P, Castiglioni L, Camera M, et al. Drug repurposing in cardiovascular diseases: opportunity or hopeless dream? Biochem Pharmacol. 2020;177:113894.
274. Finsterer J, Frank M. Repurposed drugs in metabolic disorders. Curr Top Med Chem. 2013;13(18):2386–94.
275. Zhang M, Luo H, Xi Z, et al. Drug repurposing for diabetes based on 'omics' data mining. PLoS One. 2015;10(5):e0126082.
276. Fiolet ATL, Nidorf SM, Mosterd A, et al. Colchicine in stable coronary artery disease. Clin Ther. 2019;41(1):30–40.
277. Nidorf SM, Eikelboom JW, Budgeon CA, et al. Low-dose colchicine for secondary prevention of cardiovascular disease. J Am Coll Cardiol. 2013;61(4):404–10.
278. Nidorf SM, Eikelboom JW, Thompson PL. Colchicine for secondary prevention of cardiovascular disease. Curr Atheroscler Rep. 2014;16(3):391.

279. Nidorf SM, Fiolet ATL, Eikelboom JW, et al. LoDoCo2 investigators. The effect of low-dose colchicine in patients with stable coronary artery disease: the LoDoCo2 trial rationale, design, and baseline characteristics. Am Heart J. 2019;218:46–56.
280. Nidorf SM, Fiolet ATL, Mosterd A, et al. LoDoCo2 trial investigators. Colchicine in patients with chronic coronary disease. N Engl J Med. 2020;383(19):1838–47.
281. Griffin SJ, Leaver JK, Irving GJ. Impact of metformin on cardiovascular disease: a meta-analysis of randomised trials among people with type 2 diabetes. Diabetologia. 2017;60:1620–9.
282. Cameron AR, Morrison VL, Levin D, et al. Anti-inflammatory effects of metformin irrespective of diabetes status. Circ Res. 2016;119:652–65.
283. Rena G, Lang CC. Repurposing metformin for cardiovascular disease. Circulation. 2018;137:422–4.
284. Murphy E, Ardehali H, Balaban RS, et al. American Heart Association Council on Basic Cardiovascular Sciences, Council on Clinical Cardiology, and Council on Functional Genomics and Translational Biology. Mitochondrial function, biology, and role in disease: a scientific statement from the American Heart Association. Circ Res. 2016;118:1960–91.
285. O'Neill LA, Kishton RJ, Rathmell J. A guide to immunometabolism for immunologists. Nat Rev Immunol. 2016;16:553–65.
286. Jadhav S, Ferrell W, Greer IA, et al. Effects of metformin on microvascular function and exercise tolerance in women with angina and normal coronary arteries: a randomized, double-blind, placebo-controlled study. J Am Coll Cardiol. 2006;48:956–63.
287. Luo F, Das A, Chen J, et al. Metformin in patients with and without diabetes: a paradigm shift in cardiovascular disease management. Cardiovasc Diabetol. 2019;18(1):54.
288. Wilmanns JC, Pandey R, Hon O, et al. Metformin intervention prevents cardiac dysfunction in a murine model of adult congenital heart disease. Mol Metab. 2019;20:102–14.
289. Slater RE, Strom JG, Methawasin M, et al. Metformin improves diastolic function in an HFpEF-like mouse model by increasing titin compliance. J Gen Physiol. 2019;151(1):42–52.
290. Khan SZ, Rivero M, Nader ND, et al. Metformin is associated with improved survival and decreased cardiac events with no impact on patency and limb salvage after revascularization for peripheral arterial disease. Ann Vasc Surg. 2019;55:63–77.
291. Mary A, Hartemann A, Liabeuf S, et al. Association between metformin use and below-the-knee arterial calcification score in type 2 diabetic patients. Cardiovasc Diabetol. 2017;16(1):24.
292. Hubsher G, Haider M, Okun MS. Amantadine: the journey from fighting flu to treating Parkinson disease. Neurology. 2012;3:1096–9.
293. Corbett A, Pickett J, Burns A, et al. Drug repositioning for Alzheimer's disease. Nat Rev Drug Discov. 2012;11:833–46.
294. Forette F, Seux ML, Staessen JA, et al. The prevention of dementia with antihypertensive treatment. Arch Intern Med. 2002;162:2046–52.
295. Corbett A, Williams G, Ballard C. Drug repurposing in Alzheimer's disease. Front Biosci (Schol Ed). 2015;7:184–8.
296. Hayes CD, Dey D, Palavicini JP, et al. Striking reduction of amyloid plaque burden in an Alzheimer's mouse model after chronic administration of carmustine. BMC Med. 2013;26:81.
297. Michael J. Fox foundation. MJFF grants catalog 2013. https://www.michaeljfox.org/files/071414. Accessed Oct 23 2014.
298. Fagan SC. Drug repurposing for drug development in stroke. Pharmacotherapy. 2010;30:S51–4.
299. Singh RK. Recent trends in the management of Alzheimer's disease: current therapeutic options and drug repurposing approaches. Curr Neuropharmacol. 2020;18(9):868–82.
300. Ihara M, Saito S. Drug repurposing for Alzheimer's disease: finding hidden clues in old drugs. J Alzheimers Dis. 2020;74(4):1013–28.
301. Hess DC, Fagan S. Repurposing an old drug to improve the safety and use of tissue plasminogen activator for acute ischemic stroke: minocycline. Pharmacotherapy. 2010;30:S55.
302. Amiri-Nikpour MR, Nazarbaghi S, Hamdi-Holasou M, et al. An open-label evaluator-blinded clinical study of minocycline neuroprotection in ischemic stroke: gender-dependent effect. Acta Neurol Scand. 2015;131(1):45–50.
303. Hornslien AG, Sandset EC, Bath PM, et al. Effects of candesartan in acute stroke on cognitive function and quality of life: results from the Scandinavian Candesartan Acute Stroke Trial. Stroke. 2013;44:2022–4.
304. Shoaib M, Kamal MA, Rizvi SMD. Repurposed drugs as potential therapeutic candidates for the management of Alzheimer's disease. Curr Drug Metab. 2017;18(9):842–52.
305. Bauzon J, Lee G, Cummings J. Repurposed agents in the Alzheimer's disease drug development pipeline. Alzheimers Res Ther. 2020;12(1):98.
306. Kumar S, Chowdhury S, Kumar S. In silico repurposing of antipsychotic drugs for Alzheimer's disease. BMC Neurosci. 2017;18(1):76.
307. Ferreira CR. The burden of rare diseases. Am J Med Genet A. 2019;179(6):885–92.
308. Tambuyzer E, Vandendriessche B, Austin CP, et al. Therapies for rare diseases: therapeutic modalities, progress and challenges ahead. Nat Rev Drug Discov. 2020;19(2):93–111.
309. Micallef J, Blin O. Orphan drug designation in Europe: a booster for the research and development of drugs in rare diseases. Therapie. 2020;75(2):133–9.
310. "RARE List". Global genes. https://globalgenes.org/rare-list/. Retrieved 15 October 2021.
311. Wright CF, FitzPatrick DR, Firth HV. Paediatric genomics: diagnosing rare disease in children. Nat Rev Genet. 2018;19(5):253–68.

312. Wamelink MM, Grüning NM, Jansen EE, et al. The difference between rare and exceptionally rare: molecular characterization of ribose 5-phosphate isomerase deficiency. J Mol Med. 2010;88(9):931–9.
313. Hendrickx K, Dooms M. Orphan drugs, compounded medication and pharmaceutical commons. Front Pharmacol. 2021;12:738458.
314. Roessler HI, Knoers NVAM, van Haelst MM, et al. Drug repurposing for rare diseases. Trends Pharmacol Sci. 2021;42(4):255–67.
315. Sardana D, Zhu C, Zhang M. Drug repositioning for orphan diseases. Brief Bioinform. 2011;12:346–56.
316. Xu K, Coté TR. Database identifies FDA-approved drugs with potential to be repurposed for treatment of orphan diseases. Brief Bioinform. 2011;12:341–5.
317. Van Herle K, Behne JM, Van Herle A, et al. Integrative continuum: accelerating therapeutic advances in rare autoimmune diseases. Ann Rev Pharmacol Toxicol. 2012;52:523–47.
318. Beaulieu CL, Samuels ME, Ekins S, et al. A generalizable pre-clinical research approach for orphan disease therapy. Orphanet J Rare Dis. 2012;15:39.
319. Scherman D, Fetro C. Drug repurposing for rare diseases: knowledge-based success stories. Therapie. 2020;75(2):161–7.
320. Fetro C, Scherman D. Drug repurposing in rare diseases: myths and reality. Therapie. 2020;75(2):157–60.
321. European Commission. Public Health – Union Register of Medicinal Products. Available at: https://ec.europa.eu/health/documents/community-register/html/index_en.htm. Accessed Oct 2021.
322. Urquiza P, Laín A, Sanz-Parra A, et al. Repurposing ciclopirox as a pharmacological chaperone in a model of congenital erythropoietic porphyria. Sci Transl Med. 2018;10:eaat7467.
323. Crunkhorn S. Rare diseases: repurposing ciclopirox. Nat Rev Drug Discov. 2018;17(11):788.
324. Govindaraj RG, Naderi M, Singha M, et al. Large-scale computational drug repositioning to find treatments for rare diseases. NPJ Syst Biol Appl. 2018;4:13.
325. Southall NT, Natarajan M, Lau LPL, et al. IRDiRC data mining and repurposing task force. The use or generation of biomedical data and existing medicines to discover and establish new treatments for patients with rare diseases – recommendations of the IRDiRC Data Mining and Repurposing Task Force. Orphanet J Rare Dis. 2019;14(1):225.
326. World Health Organization. Working to overcome the burden of neglected tropical diseases. Geneva: WHO Press; 2010.
327. Spangenberg T, Burrows JN, Kowalczyk P, et al. The open access malaria box: a drug discovery catalyst for neglected diseases. PLoS One. 2013;8:e62906.
328. Li YY, An J, Jones SJ. A computational approach to finding novel targets for existing drugs. PLoS Comput Biol. 2011;7:e1002139.
329. Rosenthal PJ, Sijwali PS, Singh A, et al. Cysteine proteases of malaria parasites: targets for chemotherapy. Curr Pharm Des. 2002;8:1659–72.
330. Vermeire JJ, Lantz LD, Caffrey CR. Cure of hookworm infection with a cysteine protease inhibitor. PLoS Negl Trop Dis. 2012;6:e1680.
331. Chen GZ, Foster L, Bennett JL. Antischistosomal action of mevinolin: evidence that 3-hydroxy-methylglutaryl-coenzyme a reductase activity in schistosoma mansoni is vital for parasite survival. Naunyn Schmiedeberg's Arch Pharmacol. 1990;342:477–82.
332. Vandewaa EA, Mills G, Chen GZ, et al. Physiological role of HMG-CoA reductase in regulating egg production by schistosoma mansoni. Am J Physiol – Regul Int Comp Physiol. 1989;257:R618–25.
333. Doerig C, Meijer L, Mottram JC. Protein kinases as drug targets in parasitic protozoa. Trends Parasitol. 2002;18:366–71.
334. Urbaniak MD, Mathieson T, Bantscheff M, et al. Chemical proteomic analysis reveals the drugability of the kinome of trypanosoma brucei. ACS Chem Biol. 2012;7:1858–65.
335. Gloeckner C, Garner AL, Mersha F, et al. Repositioning of an existing drug for the neglected tropical disease onchocerciasis. Proc Natl Acad Sci U S A. 2010;107:3424–9.
336. Abdulla M-H, Ruelas DS, Wolff B, et al. Drug discovery for schistosomiasis: hit and lead compounds identified in a library of known drugs by medium-throughput phenotypic screening. PLoS Negl Trop Dis. 2009;3:e478.
337. Nurse P. A long twentieth century of the cell cycle, and beyond. Cell. 2000;100:71–8.
338. Viotti R, Vigliano C, Lococo B, et al. Side effects of benznidazole as a treatment in chronic Chagas disease: fears and realities. Exp Rev Anti-infect Ther. 2009;7:157–63.
339. Debnath A, Ndao M, Reed SL. Reprofiled drug targets ancient protozoans. Gut Microbes. 2013;4:66–71.
340. Tejman-Yarden N, Miyamoto Y, Leitsch D, et al. A reprofiled drug, auranofin, is effective against metronidazole-resistant giardia lamblia. Antimicrob Agents Chemother. 2013;57:2029–35.
341. Bromme D, Lacaille F. Cathepsin K inhibitors for osteoporosis and potential off-target effects. Exp Opin Invest Drugs. 2009;18:585–600.
342. Dent JA, Davis MW, Avery L. Avr-15 encodes a chloride channel subunit that mediates inhibitory glutamatergic neurotransmission and ivermectin sensitivity in Caenorhabditis elegans. EMBO J. 1997;16:5867–79.
343. Geyer J, Gavrilova O, Petzinger E. Brain penetration of ivermectin and selamectin in MDR1a,b P-glycoprotein- and BCRP-deficient knockout mice. J Vet Pharmacol Ther. 2009;32:87–96.
344. Patel G, Karver CE, Behera R, et al. Kinase scaffold repurposing for neglected disease drug discovery: discovery of an efficacious, lapatanib-derived lead compound for trypanosomiasis. J Med Chem. 2013;56:3820–32.
345. Ochiana SO, Pandarinath V, Wang Z, et al. The human aurora kinase inhibitor danusertib is a

lead compound for anti-trypanosomal drug discovery via target repurposing. Eur J Med Chem. 2013;62:777–84.
346. Carvalho TG, Doerig C, Reininger L. Nima- and aurora-related kinases of malaria parasites. Biochim Biophys Acta. 2013;1834:1336–45.
347. Serafin MB, Bottega A, Foletto VS, et al. Drug repurposing is an alternative for the treatment of coronavirus COVID-19. Int J Antimicrob Agents. 2020;55(6):105969.
348. Benavides-Cordoba V. Drug repurposing for COVID-19. Colomb Med (Cali). 2020;51(2):e4279.
349. Low ZY, Farouk IA, Lal SK. Drug repurposing: new approaches and future prospects for life-debilitating diseases and the COVID-19 pandemic outbreak. Viruses. 2020;12(9):1058.
350. Jia Z, Song X, Shi J, et al. Transcriptome-based drug repurposing for coronavirus disease 2019 (COVID-19). Pathog Dis. 2020;78(4):ftaa036.
351. Chen F, Shi Q, Pei F, et al. A systems-level study reveals host-targeted repurposable drugs against SARS-CoV-2 infection. Mol Syst Biol. 2021;17(8):e10239.
352. Ayman F, Ping W, Mahmoud A, et al. Identification of FDA approved drugs targeting COVID-19 virus by structure-based drug repurposing. 2020;21:100860.
353. Isgrò C, Sardanelli AM, Palese LL. Systematic search for SARS-CoV-2 main protease inhibitors for drug repurposing: ethacrynic acid as a potential drug. Viruses. 2021;13(1):106.
354. Liu DY, Liu JC, Liang S, et al. Drug repurposing for COVID-19 treatment by integrating network pharmacology and transcriptomics. Pharmaceutics. 2021;13(4):545.
355. Li X, Yu J, Zhang Z, et al. Network bioinformatics analysis provides insight into drug repurposing for COVID-19. Med Drug Discov. 2021;10:100090.
356. Adhami M, Sadeghi B, Rezapour A, et al. Repurposing novel therapeutic candidate drugs for coronavirus disease-19 based on protein-protein interaction network analysis. BMC Biotechnol. 2021;21(1):22.
357. Rameshrad M, Ghafoori M, Mohammadpour AH, et al. A comprehensive review on drug repurposing against coronavirus disease 2019 (COVID19). Naunyn Schmiedeberg's Arch Pharmacol. 2020;393(7):1137–52.
358. Luo L, Qiu Q, Huang F, et al. Drug repurposing against coronavirus disease 2019 (COVID-19): a review. J Pharm Anal. 2021.
359. El Bairi K, Trapani D, Petrillo A, et al. Repurposing anticancer drugs for the management of COVID-19. Eur J Cancer. 2020;141:40–61.
360. Ajayi S, Becker H, Reinhardt H. Ruxolitinib. Recent results. Cancer Res. 2018;212:119–32.
361. Vannucchi AM, Kiladjian JJ, Griesshammer M, et al. Ruxolitinib versus standard therapy for the treatment of polycythemia vera. N Engl J Med. 2015;372(5):426–35.
362. Verstovsek S, Mesa RA, Gotlib J, et al. A double-blind, placebo-controlled trial of ruxolitinib for myelofibrosis. N Engl J Med. 2012;366(9):799–807.
363. Venugopal S, Bar-Natan M, Mascarenhas JO. JAKs to STATs: a tantalizing therapeutic target in acute myeloid leukemia. Blood Rev. 2020;40:100634.
364. Stebbing J, Phelan A, Griffin I, et al. COVID-19: combining antiviral and anti-inflammatory treatments. Lancet Infect Dis. 2020;20(4):400–2.
365. Dementiev A, Joachimiak A, Nguyen H, et al. Molecular mechanism of inhibition of acid ceramidase by carmofur. J Med Chem. 2019;62(2):987–92.
366. Realini N, Solorzano C, Pagliuca C, et al. Discovery of highly potent acid ceramidase inhibitors with in vitro tumor chemosensitizing activity. Sci Rep. 2013;3:1035.
367. Canals D, Perry DM, Jenkins RW, et al. Drug targeting of sphingolipid metabolism: sphingomyelinases and ceramidases. Br J Pharmacol. 2011;163(4):694–712.
368. Govindarajah N, Clifford R, Bowden D, et al. Sphingolipids and acid ceramidase as therapeutic targets in cancer therapy. Crit Rev Oncol Hematol. 2019;138:104–11.
369. Liu X, Cheng JC, Turner LS, et al. Acid ceramidase upregulation in prostate cancer: role in tumor development and implications for therapy. Expert Opin Ther Targets. 2009;13(12):1449–58.
370. Nguyen HS, Awad AJ, Shabani S, et al. Molecular targeting of acid ceramidase in glioblastoma: a review of its role, potential treatment, and challenges. Pharmaceutics. 2018;10(2):45.
371. Doan NB, Nguyen HS, Montoure A, et al. Acid ceramidase is a novel drug target for pediatric brain tumors. Oncotarget. 2017;8(15):24753–61.
372. Liu P, Ma S, Liu H, et al. HCFU inhibits cervical cancer cells growth and metastasis by inactivating Wnt/β-catenin pathway. J Cell Biochem. 2017. https://doi.org/10.1002/jcb.26570.
373. Sakamoto J, Hamada C, Rahman M, et al. An individual patient data meta-analysis of adjuvant therapy with carmofur in patients with curatively resected colon cancer. Jpn J Clin Oncol. 2005;35(9):536–44.
374. Morimoto K, Koh M. Postoperative adjuvant use of carmofur for early breast cancer. Osaka City Med J. 2003;49(2):77–83.
375. Gao J, Zhang L, Liu X, et al. Repurposing low-molecular-weight drugs against the main protease of severe acute respiratory syndrome coronavirus 2. J Phys Chem Lett. 2020;11:7267–72.
376. Garcia J, Hurwitz HI, Sandler AB, et al. Bevacizumab (Avastin®) in cancer treatment: a review of 15 years of clinical experience and future outlook. Cancer Treat Rev. 2020;86:102017.
377. El Bairi K, Amrani M, Kandhro AH, et al. Prediction of therapy response in ovarian cancer: where are we now? Crit Rev Clin Lab Sci. 2017;54(4):233–66.
378. Bates DO. Vascular endothelial growth factors and vascular permeability. Cardiovasc Res. 2010;87(2):262–71.

379. Weis SM, Cheresh DA. Pathophysiological consequences of VEGF-induced vascular permeability. Nature. 2005;437(7058):497–504.
380. Ackermann M, Verleden SE, Kuehnel M, et al. Pulmonary vascular endothelialitis, thrombosis, and angiogenesis in covid-19. N Engl J Med. 2020;383(2):120–8.
381. Sheng Z, Li G, Li B, et al. Carfilzomib-containing combinations as frontline therapy for multiple myeloma: a meta-analysis of 13 trials. Eur J Haematol. 2017;98(6):601–7.
382. Stewart AK, Rajkumar SV, Dimopoulos MA, et al. Carfilzomib, lenalidomide, and dexamethasone for relapsed multiple myeloma. N Engl J Med. 2015;372(2):142–52.
383. Xu W, Sun X, Wang B, et al. Pooled analysis of the reports of carfilzomib/ixazomib combinations for relapsed/refractory multiple myeloma. Ann Hematol. 2018;97(2):299–307.
384. Landgren O, Sonneveld P, Jakubowiak A, et al. Carfilzomib with immunomodulatory drugs for the treatment of newly diagnosed multiple myeloma. Leukemia. 2019;33(9):2127–43.
385. Gandolfi S, Laubach JP, Hideshima T, et al. The proteasome and proteasome inhibitors in multiple myeloma. Cancer Metastasis Rev. 2017;36(4):561–84.
386. Ziogas DC, Terpos E, Kastritis E, et al. An overview of the role of carfilzomib in the treatment of multiple myeloma. Expert Opin Pharmacother. 2017;18(17):1883–97.
387. Wang J. Fast identification of possible drug treatment of coronavirus disease-19 (COVID-19) through computational drug repurposing study. J Chem Inf Model. 2020;60(6):3277–86.
388. Vogel CL, Johnston MA, Capers C, et al. Toremifene for breast cancer: a review of 20 years of data. Clin Breast Cancer. 2014;14(1):1–9.
389. Mao C, Yang ZY, He BF, et al. Toremifene versus tamoxifen for advanced breast cancer. Cochrane Database Syst Rev. 2012;2012(7):CD008926.
390. Zhao Y, Ren J, Harlos K, et al. Toremifene interacts with and destabilizes the Ebola virus glycoprotein. Nature. 2016;535(7610):169–72.
391. Johansen LM, Brannan JM, Delos SE, et al. FDA-approved selective estrogen receptor modulators inhibit Ebola virus infection. Sci Transl Med. 2013;5(190):190ra79.
392. Zhou Y, Hou Y, Shen J, et al. Network-based drug repurposing for novel coronavirus 2019-nCoV/SARS-CoV-2. Cell Discov. 2020;6:14.
393. Ho TY, Wu SL, Chen JC, et al. Emodin blocks the SARS coronavirus spike protein and angiotensin-converting enzyme 2 interaction. Antivir Res. 2007;74(2):92–101.
394. Schwarz S, Wang K, Yu W, et al. Emodin inhibits current through SARS-associated coronavirus 3a protein. Antivir Res. 2011;90(1):64–9.
395. Thompson PA. Preclinical evaluation of eFT226, a novel, potent and selective eIF4A inhibitor with anti-tumor activity in B-cell malignancies. Blood. 2017;130(Suppl 1):1530.
396. Heerma van Voss MR, van Diest PJ, Raman V. Targeting RNA helicases in cancer: the translation trap. Biochim Biophys Acta Rev Cancer. 2017;1868(2):510–20.
397. Chan K, Robert F, Oertlin C, et al. eIF4A supports an oncogenic translation program in pancreatic ductal adenocarcinoma. Nat Commun. 2019;10(1):5151.
398. Müller D, Shin S, Goullet de Rugy T, et al. eIF4A inhibition circumvents uncontrolled DNA replication mediated by 4E-BP1 loss in pancreatic cancer. JCI. Insight. 2019;4(21):e121951.
399. Gordon DE, Jang GM, Bouhaddou M, et al. A SARS-CoV-2 protein interaction map reveals targets for drug repurposing. Nature. 2020;583(7816):459–68.
400. Pessoa C, Silveira ER, Lemos TL, et al. Antiproliferative effects of compounds derived from plants of Northeast Brazil. Phytother Res. 2000;14(3):187–91.
401. El Bairi K, Amrani M, Afqir S, et al. Starvation tactics using natural compounds for advanced cancers: pharmacodynamics, clinical efficacy, and predictive biomarkers. Cancer Med. 2018;7(6):2221–46.
402. Alonso-Álvarez S, Pardal E, Sánchez-Nieto D, et al. Plitidepsin: design, development, and potential place in therapy. Drug Des Devel Ther. 2017;11:253–26.
403. Gomes NGM, Valentão P, Andrade PB, et al. Plitidepsin to treat multiple myeloma. Drugs Today. 2020;56(5):337–47.
404. Rockwell S, Liu Y. Aplidin as a potential adjunct to radiation therapy: in vitro studies. Int J Radiat Biol. 2010;86(1):63–70.
405. El Bairi K, Atanasov AG, Amrani M, et al. The arrival of predictive biomarkers for monitoring therapy response to natural compounds in cancer drug discovery. Biomed Pharmacother. 2019;109:2492–8.
406. Muñoz-Alonso MJ, Álvarez E, Guillén-Navarro MJ, et al. c-Jun N-terminal kinase phosphorylation is a biomarker of plitidepsin activity. Mar Drugs. 2013;11(5):1677–92.
407. Spicka I, Ocio EM, Oakervee HE, et al. Randomized phase III study (ADMYRE) of plitidepsin in combination with dexamethasone vs. dexamethasone alone in patients with relapsed/refractory multiple myeloma. Ann Hematol. 2019;98(9):2139–50.
408. PharmaMar has submitted a phase II clinical trial of Aplidin® (plitidepsin) for the treatment of COVID-19 to the Spanish Medicines Agency. PharmaMar. 2020. http://pharmamar.com/wp--content/uploads/2020/04/PR_clinical_trial_plitidepsin_covid19_DEF.pdf.
409. PharmaMar reports positive results for Aplidin® against coronavirus HCoV-229E. PharmaMar. 2020. http://pharmamar.com/wp-content/uploads/2020/03/PR_Results_Aplidin_coronavirus.pdf.
410. Waksman SA, Woodruff HB. Bacteriostatic and bactericidal substances produced by a soil actinomyces. Proc Soc Exp Biol Med. 1940;45:609.

411. National Institute of Diabetes and Digestive and Kidney Diseases. Cytotoxic antibiotics. LiverTox: clinical and research information on drug-induced liver injury. Bethesda: National Institute of Diabetes and Digestive and Kidney Diseases; 2012.
412. Hollstein U. Actinomycin. Chemistry and mechanism of action. Chem Rev. 1974;74:625–52.
413. Humeau J, Sauvat A, Cerrato G, et al. Inhibition of transcription by dactinomycin reveals a new characteristic of immunogenic cell stress. EMBO Mol Med. 2020;12(5):e11622.
414. Alazzam M, Tidy J, Osborne R, et al. Chemotherapy for resistant or recurrent gestational trophoblastic neoplasia. Cochrane Database Syst Rev. 2016;2016(1):CD008891.
415. Gaspar N, Hawkins DS, Dirksen U, et al. Ewing sarcoma: current management and future approaches through collaboration. J Clin Oncol. 2015;33(27):3036–46.
416. Li J, Li S, Yu H, et al. The efficacy and safety of first-line single-agent chemotherapy regimens in low-risk gestational trophoblastic neoplasia: a network meta-analysis. Gynecol Oncol. 2018;148(2):247–53.
417. van den Heuvel-Eibrink MM, Hol JA, Pritchard-Jones K, et al. Position paper: rationale for the treatment of Wilms tumour in the UMBRELLA SIOP-RTSG 2016 protocol. Nat Rev Urol. 2017;14(12):743–52.
418. Kennedy DA, Johnson-Lussenburg CM. Inhibition of coronavirus 229E replication by actinomycin D. J Virol. 1979;29(1):401–4.
419. Lewis EL, Harbour DA, Beringer JE, et al. Differential in vitro inhibition of feline enteric coronavirus and feline infectious peritonitis virus by actinomycin D. J Gen Virol. 1992;73(Pt 12):3285–8.
420. van Hemert MJ, van den Worm SH, Knoops K, et al. SARS-coronavirus replication/transcription complexes are membrane-protected and need a host factor for activity in vitro. PLoS Pathog. 2008;4(5):e1000054.
421. Kindrachuk J, Ork B, Hart BJ, et al. Antiviral potential of ERK/MAPK and PI3K/AKT/mTOR signaling modulation for Middle East respiratory syndrome coronavirus infection as identified by temporal kinome analysis. Antimicrob Agents Chemother. 2015;59(2):1088–99.
422. Sharma P, Zargar-Shoshtari K, Sexton WJ, et al. Valrubicin in refractory non-muscle invasive bladder cancer. Expert Rev Anticancer Ther. 2015;15(12):1379–87.
423. Barlow LJ, Benson MC. Experience with newer intravesical chemotherapy for high-risk non-muscle-invasive bladder cancer. Curr Urol Rep. 2013;14(2):65–70.
424. Khandelwal R, Nayarisseri A, Madhavi M, et al. Shape-based machine learning models for the potential novel COVID-19 protease inhibitors assisted by molecular dynamics simulation. Curr Top Med Chem. 2020;20(24):2146–67.
425. Jiao X, Nawab O, Patel T, et al. Recent advances targeting CCR5 for cancer and its role in immuno-oncology. Cancer Res. 2019;79(19):4801–7.
426. Brelot A, Chakrabarti LA. CCR5 revisited: how mechanisms of HIV entry govern AIDS pathogenesis. J Mol Biol. 2018;430(17):2557–89.
427. Chen J, Lau YF, Lamirande EW, et al. Cellular immune responses to severe acute respiratory syndrome coronavirus (SARS-CoV) infection in senescent BALB/c mice: CD4+ T cells are important in control of SARS-CoV infection. J Virol. 2010;84(3):1289–301.
428. Thompson MA. The return of PRO 140, a CCR5-directed mAb. Curr Opin HIV AIDS. 2018;13(4):346–53.
429. Patterson B, Seetthamraju H, Dhody K. ResearchSquare; 2020. Disruption of the CCL5/RANTES-CCR5 pathway restores immune homeostasis and reduces plasma viral load in critical COVID-19.
430. Patil NK, Guo Y, Luan L, et al. Targeting immune cell checkpoints during sepsis. Int J Mol Sci. 2017;18(11):2413.
431. Qin W, Hu L, Zhang X, et al. The diverse function of PD-1/PD-L pathway beyond cancer. Front Immunol. 2019;10:2298.
432. Zhang Q, Qi Z, Bo-Liu, et al. Programmed cell death-1/programmed death-ligand 1 blockade improves survival of animals with sepsis: a systematic review and meta-analysis. Biomed Res Int. 2018;2018:1969474.
433. Markham A, Keam SJ. Camrelizumab: first global approval. Drugs. 2019;79(12):1355–61.
434. Huang C, Wang Y, Li X, et al. Clinical features of patients infected with 2019 novel coronavirus in Wuhan. China Lancet. 2020;395(10223):497–506.
435. Huang J, Xu J, Chen Y, et al. Camrelizumab versus investigator's choice of chemotherapy as second-line therapy for advanced or metastatic oesophageal squamous cell carcinoma (ESCORT): a multicentre, randomised, open-label, phase 3 study. Lancet Oncol. 2020;21(6):832–42.
436. Xie L, Xu J, Sun X, et al. Apatinib plus camrelizumab (anti-PD1 therapy, SHR-1210) for advanced osteosarcoma (APFAO) progressing after chemotherapy: a single-arm, open-label, phase 2 trial. J Immunother Canc. 2020;8(1):e000798.
437. Song K, Dai L, Long X, et al. Sphingosine kinase 2 inhibitor ABC294640 displays anti-epithelial ovarian cancer activities in vitro and in vivo. OncoTargets Ther. 2019;12:4437–49.
438. Song Y, Wu J, Chen X, et al. A single-arm, multicenter, phase II study of camrelizumab in relapsed or refractory classical Hodgkin lymphoma. Clin Cancer Res. 2019;25(24):7363–9.
439. Qin C, Zhou L, Hu Z, et al. Dysregulation of immune response in patients with COVID-19 in Wuhan. China Clin Infect Dis. 2020;71(15):762–8.
440. Luke JJ, Flaherty KT, Ribas A, et al. Targeted agents and immunotherapies: optimizing outcomes in melanoma. Nat Rev Clin Oncol. 2017;14(8):463–82.
441. Waldman AD, Fritz JM, Lenardo MJ. A guide to cancer immunotherapy: from T cell basic sci-

ence to clinical practice. Nat Rev Immunol. 2020;20(11):651–68.
442. El Bairi K, Jabi R, Trapani D, et al. Can the microbiota predict response to systemic cancer therapy, surgical outcomes, and survival? The answer is in the gut. Expet Rev Clin Pharmacol. 2020;13(4):403–21.
443. Havel JJ, Chowell D, Chan TA. The evolving landscape of biomarkers for checkpoint inhibitor immunotherapy. Nat Rev Cancer. 2019;19(3):133–50.
444. Roy S, Trinchieri G. Microbiota: a key orchestrator of cancer therapy. Nat Rev Cancer. 2017;17(5):271–85.
445. Schönrich G, Raftery MJ. The PD-1/PD-L1 axis and virus infections: a delicate balance. Front Cell Infect Microbiol. 2019;9:207.
446. Shah KP, Song H, Ye F, et al. Demographic factors associated with toxicity in patients treated with anti-programmed cell death-1 therapy. Cancer Immunol. 2019;8(7):851–5.
447. Luo J, Rizvi H, Egger JV, et al. Impact of PD-1 blockade on severity of COVID-19 in patients with lung cancers. Cancer Discov. 2020;10(8):1121–8.
448. Gharwan H, Groninger H. Kinase inhibitors and monoclonal antibodies in oncology: clinical implications. Nat Rev Clin Oncol. 2016;13(4):209–27.
449. Pu SY, Xiao F, Schor S, et al. Feasibility and biological rationale of repurposing sunitinib and erlotinib for dengue treatment. Antivir Res. 2018;1(55):67–75.
450. García-Serradilla M, Risco C, Pacheco B. Drug repurposing for new, efficient, broad spectrum antivirals. Virus Res. 2019;264:22–31.
451. Schor S, Einav S. Repurposing of kinase inhibitors as broad-spectrum antiviral drugs. DNA Cell Biol. 2018;37(2):63–9.
452. Neveu G, Barouch-Bentov R, Ziv-Av A, et al. Identification and targeting of an interaction between a tyrosine motif within hepatitis C virus core protein and AP2M1 essential for viral assembly. PLoS Pathog. 2012;8(8):e1002845.
453. Neveu G, Ziv-Av A, Barouch-Bentov R, et al. AP-2-associated protein kinase 1 and cyclin G-associated kinase regulate hepatitis C virus entry and are potential drug targets. J Virol. 2015;89(8):4387–404.
454. Farag S, Smith MJ, Fotiadis N, et al. Revolutions in treatment options in gastrointestinal stromal tumours (GISTs): the latest updates. Curr Treat Options in Oncol. 2020;21(7):55.
455. Hochhaus A, Baccarani M, Silver RT, et al. European LeukemiaNet 2020 recommendations for treating chronic myeloid leukemia. Leukemia. 2020;34(4):966–84.
456. García M, Cooper A, Shi W, et al. Productive replication of Ebola virus is regulated by the c-Abl1 tyrosine kinase. Sci Transl Med. 2012;4(123):123ra24.
457. Min S, Lim YS, Shin D, et al. Abl tyrosine kinase regulates hepatitis C virus entry. Front Microbiol. 2017;8:1129.
458. Coleman CM, Sisk JM, Mingo RM, et al. Abelson kinase inhibitors are potent inhibitors of severe acute respiratory syndrome coronavirus and Middle East respiratory syndrome coronavirus fusion. J Virol. 2016;90(19):8924–33.
459. Sisk JM, Frieman MB, Machamer CE. Coronavirus S protein-induced fusion is blocked prior to hemifusion by Abl kinase inhibitors. J Gen Virol. 2018;99(5):619–30.
460. Magruder JT, Grimm JC, Crawford TC. Imatinib is protective against ischemia-reperfusion injury in an ex vivo rabbit model of lung injury. Ann Thorac Surg. 2018;105(3):950–6.
461. Tanaka S, Chen-Yoshikawa TF, Kajiwara M, et al. Protective effects of imatinib on ischemia/reperfusion injury in rat lung. Ann Thorac Surg. 2016;102(5):1717–24.
462. Rizzo AN, Sammani S, Esquinca AE, et al. Imatinib attenuates inflammation and vascular leak in a clinically relevant two-hit model of acute lung injury. Am J Physiol Lung Cell Mol Physiol. 2015;309(11):L1294–304.
463. Stephens RS, Johnston L, Servinsky L, et al. The tyrosine kinase inhibitor imatinib prevents lung injury and death after intravenous LPS in mice. Phys Rep. 2015;3(11):e12589.
464. Siddiqi HK, Mehra MR. COVID-19 illness in native and immunosuppressed states: a clinical-therapeutic staging proposal. J Heart Lung Transplant. 2020;39(5):405–7.
465. Janku F, Yap TA, Meric-Bernstam F. Targeting the PI3K pathway in cancer: are we making headway? Nat Rev Clin Oncol. 2018;15(5):273–91.
466. Flinn IW, Hillmen P, Montillo M, et al. The phase 3 DUO trial: duvelisib vs ofatumumab in relapsed and refractory CLL/SLL. Blood. 2018;132(23):2446–55.
467. https://www.fda.gov/drugs/resources-information-approved-drugs/duvelisib-copiktra-verastem-inc-adult-patients-relapsed-or-refractory-chronic-lymphocytic-leukemia.
468. Guo Y, Liu Y, Hu N, et al. Discovery of zanubrutinib (BGB-3111), a novel, potent, and selective covalent inhibitor of Bruton's tyrosine kinase. J Med Chem. 2019;62(17):7923–40.
469. Florence JM, Krupa A, Booshehri LM, et al. Inhibiting Bruton's tyrosine kinase rescues mice from lethal influenza-induced acute lung injury. Am J Physiol Lung Cell Mol Physiol. 2018;315(1):L52–8.
470. Krupa A, Fol M, Rahman M, et al. Silencing Bruton's tyrosine kinase in alveolar neutrophils protects mice from LPS/immune complex-induced acute lung injury. Am J Physiol Lung Cell Mol Physiol. 2014;307(6):L435–48.
471. Zhou P, Ma B, Xu S, et al. Knockdown of Burton's tyrosine kinase confers potent protection against sepsis-induced acute lung injury. Cell Biochem Biophys. 2014;70(2):1265–75.
472. Zhang X, Tan Y, Ling Y, et al. Viral and host factors related to the clinical outcome of COVID-19. Nature. 2020;583(7816):437–40.
473. Zhang C, Tian B. Nonclinical safety assessment of zanubrutinib: a novel irreversible BTK inhibitor. Int J Toxicol. 2020;39(3):232–40.

474. Syed YY. Zanubrutinib: first approval. Drugs. 2020;80(1):91–7.
475. Song Y, Zhou KS, Zou D, et al. Treatment of patients with relapsed or refractory mantle cell lymphoma with zanubrutinib, a selective inhibitor of Bruton's tyrosine kinase. Clin Cancer Res. 2020;26(16):4216–24.
476. Treon SP, Castillo JJ, Skarbnik AP, et al. The BTK inhibitor ibrutinib may protect against pulmonary injury in COVID-19-infected patients. Blood. 2020;135(21):1912–5.
477. Song DD, Zhou JH, Sheng R. Regulation and function of sphingosine kinase 2 in diseases. Histol Histopathol. 2018;33(5):433–45.
478. Reid SP, Tritsch SR, Kota K, et al. Sphingosine kinase 2 is a chikungunya virus host factor co-localized with the viral replication complex. Emerg Microb Infect. 2015;4(10):e61.
479. Dai L, Plaisance-Bonstaff K, Voelkel-Johnson C, et al. Sphingosine kinase-2 maintains viral latency and survival for KSHV-infected endothelial cells. PLoS One. 2014;9(7):e102314.
480. https://www.redhillbio.com/RedHill/Templates/showpage.asp?DBID=1&LNGID=1&TMID=178&FID=1365&PID=0&IID=5106Accessed. 06-06-2020.
481. Choi YR, Kim JB, Kang SJ, et al. The dual role of c-src in cell-to-cell transmission of α-synuclein. EMBO Rep. 2020;21(7):e48950.
482. Patel A, Sabbineni H, Clarke A, et al. Novel roles of Src in cancer cell epithelial-to-mesenchymal transition, vascular permeability, microinvasion and metastasis. Life Sci. 2016;157:52–61.
483. Rho O, Kim DJ, Kiguchi K, et al. Growth factor signaling pathways as targets for prevention of epithelial carcinogenesis. Mol Carcinog. 2011;50(4):264–79.
484. Baird K, Glod J, Steinberg SM, et al. Results of a randomized, double-blinded, placebo-controlled, phase 2.5 study of saracatinib (AZD0530), in patients with recurrent osteosarcoma localized to the lung. Sarcoma. 2020;2020:7935475.
485. McNeish IA, Ledermann JA, Webber L, et al. A randomised, placebo-controlled trial of weekly paclitaxel and saracatinib (AZD0530) in platinum-resistant ovarian, fallopian tube or primary peritoneal cancer. Ann Oncol. 2014;25(10):1988–95.
486. Gubens MA, Burns M, Perkins SM, et al. A phase II study of saracatinib (AZD0530), a Src inhibitor, administered orally daily to patients with advanced thymic malignancies. Lung Cancer. 2015;89(1):57–60.
487. Danson S, Mulvey MR, Turner L, et al. An exploratory randomized-controlled trial of the efficacy of the Src-kinase inhibitor saracatinib as a novel analgesic for cancer-induced bone pain. J Bone Oncol. 2019;19:100261.
488. Reinecke J, Caplan S. Endocytosis and the Src family of non-receptor tyrosine kinases. Biomol Concepts. 2014;5(2):143–55.
489. de Wispelaere M, LaCroix AJ, Yang PL. The small molecules AZD0530 and dasatinib inhibit dengue virus RNA replication via Fyn kinase. J Virol. 2013;87(13):7367–81.
490. Shin JS, Jung E, Kim M, et al. Saracatinib inhibits Middle East respiratory syndrome-coronavirus replication in vitro. Viruses. 2018;10(6):283.
491. Nabavi SF, Habtemariam S, Clementi E, et al. Lessons learned from SARS-CoV and MERS-CoV: FDA-approved Abelson tyrosine-protein kinase 2 inhibitors may help us combat SARS-CoV-2. Arch Med Sci. 2020;16(3):519–21.
492. Tatar G, Turhan K. Investigation of N terminal domain of SARS CoV 2 nucleocapsid protein with antiviral compounds based on molecular modeling approach. Sci Open; 2020.
493. Syed YY. Selinexor: first global approval. Drugs. 2019;79(13):1485–94.
494. Azizian NG, Li Y. XPO1-dependent nuclear export as a target for cancer therapy. J Hematol Oncol. 2020;13(1):61.
495. Nachmias B, Schimmer AD. Targeting nuclear import and export in hematological malignancies. Leukemia. 2020;34(11):2875–86.
496. Chari A, Vogl DT, Gavriatopoulou M, et al. Oral selinexor-dexamethasone for triple-class refractory multiple myeloma. N Engl J Med. 2019;381(8):727–38.
497. Richter J, Madduri D, Richard S, et al. Selinexor in relapsed/refractory multiple myeloma. Ther Adv Hematol. 2020;11:2040620720930629.
498. Vogl DT, Dingli D, Cornell RF, et al. Selective inhibition of nuclear export with oral selinexor for treatment of relapsed or refractory multiple myeloma. J Clin Oncol. 2018;36(9):859–66.
499. Uddin MH, Zonder JA, Azmi AS. Exportin 1 inhibition as antiviral therapy. Drug Discov Today. 2020;25(10):1775–81.
500. Wu M, Gui H, Feng Z, et al. KPT-330, a potent and selective CRM1 inhibitor, exhibits anti-inflammation effects and protection against sepsis. Biochem Biophys Res Commun. 2018;503(3):1773–9.
501. Weede RP. Poison in the pot: the legacy of lead. Southern Illinois University Press: Carbondale and Edwardsville; 1984. p. 83.
502. Tardif JC, Kouz S, Waters DD, et al. Efficacy and safety of low-dose colchicine after myocardial infarction. N Engl J Med. 2019;381(26):2497–505.
503. Vrachatis DA, Giannopoulos GV, Giotaki SG, et al. Impact of colchicine on mortality in patients with COVID-19: a meta-analysis. Hell J Cardiol. 2021;S1109-9666(20):30285–2.
504. Deftereos SG, Giannopoulos G, Vrachatis DA, et al. Effect of colchicine vs standard care on cardiac and inflammatory biomarkers and clinical outcomes in patients hospitalized with coronavirus disease 2019: the GRECCO-19 randomized clinical trial. JAMA Netw Open. 2020;3(6):e2013136.
505. Scarsi M, Piantoni S, Colombo E, et al. Association between treatment with colchicine and improved

survival in a single-centre cohort of adult hospitalised patients with COVID-19 pneumonia and acute respiratory distress syndrome. Ann Rheum Dis. 2020;79(10):1286–9.
506. Brunetti L, Diawara O, Tsai A, et al. Colchicine to weather the cytokine storm in hospitalized patients with COVID-19. J Clin Med. 2020;9(9):2961.
507. Sandhu T, Tieng A, Chilimuri S, et al. A case control study to evaluate the impact of colchicine on patients admitted to the hospital with moderate to severe COVID-19 infection. Can J Infect Dis Med Microbiol. 2020;2020:8865954.
508. Lopes MIF, Bonjorno LP, Giannini MC, et al. Beneficial effects of colchicine for moderate to severe COVID-19: an interim analysis of a randomized, double-blinded, placebo controlled clinical trial. medRxiv. 2020.
509. Alejandro Pinzón M, Arango MDC, Felipe Betancur J, et al. Clinical Outcome of Patients with COVID-19 Pneumonia Treated with Corticosteroids and Colchicine in Colombia; 2020.
510. Schlesinger N, Firestein BL, Brunetti L. Colchicine in COVID-19: an old drug, new use. Curr Pharmacol Rep. 2020;6(4):137–45.
511. Deftereos SG, Siasos G, Giannopoulos G, et al. The GReek study in the effects of colchicine in COvid-19 complications prevention (GRECCO-19 study): rationale and study design. Hell J Cardiol. 2020;61(1):42–5.
512. Cha Y, Erez T, Reynolds IJ, et al. Drug repurposing from the perspective of pharmaceutical companies. Br J Pharmacol. 2018;175(2):168–80.
513. Howard R, McShane R, Lindesay J, et al. Donepezil and memantine for moderate-to-severe Alzheimer's disease. N Engl J Med. 2012;366:893–903.
514. Benaim G, Sanders JM, García-Marchan Y, et al. Amiodarone has intrinsic anti-*Trypanosoma cruzi* activity and acts synergistically with posaconazole. J Med Chem. 2006;49:892–9.
515. Benaim G, Hernández-Rodriguez V, Mujica-Gonzalez S, et al. *In vitro* anti-*Trypanosoma cruzi* activity of dronedarone, a novel amiodarone derivative with an improved safety profile. Antimicrob Agents Chemother. 2012;56:3720–5.
516. Woolf SH. The meaning of translational research and why it matters. JAMA. 2008;299:211–3.
517. Schuster D, Laggner C, Langer T. Why drugs fail – a study on side effects in new chemical entities. Curr Pharm Des. 2005;11:3545–59.
518. Morgan P, Van Der Graaf PH, Arrowsmith J, et al. Can the flow of medicines be improved? Fundamental pharmacokinetic and pharmacological principles toward improving phase II survival. Drug Discov Today. 2012;17:419–24.
519. Oprea TI, Bauman JE, Bologa CG, et al. Drug repurposing from an academic perspective. Drug Discov Today Ther Strateg. 2011;8:61–9.
520. Oprea TI, Mestres J. Drug repurposing: far beyond new targets for old drugs. AAPS J. 2012;14:759–63.
521. Giovannoni G, Baker D, Schmierer K. The problem with repurposing: is there really an alternative to Big Pharma for developing new drugs for multiple sclerosis? Mult Scler Relat Disord. 2015;4:3–5.
522. Woodcock J. Three encouraging steps toward new antibiotics. 2014. Available at: https://blogs.fda.gov/fdavoice/index.php/2014/09/three-encouraging-steps-towards-new-antibiotics. Accessed Oct 19 2021.
523. Shineman DW, Alam J, Anderson M, et al. Overcoming obstacles to repurposing for neurodegenerative disease. Ann Clin Transl Neurol. 2014;1:512–8.
524. Roin BN. Unpatentable drugs and the standards of patentability. Tex L Rev. 2008;87:503.
525. Fields JD, Bhardwaj A. Non-peptide arginine-vasopressin antagonists (vaptans) for the treatment of hyponatremia in neurocritical care: a new alternative? Neurocrit Care. 2009;11:1–4.
526. Kuter DJ. New thrombopoietic growth factors. Blood. 2007;109:4607–16.
527. Selby P, Buick RN, Tannock I. A critical appraisal of the "human tumor stem-cell assay". N Engl J Med. 1983;308:129–34.
528. Shah ET, Upadhyaya A, Philp LK, et al. Repositioning "old" drugs for new causes: identifying new inhibitors of prostate cancer cell migration and invasion. Clin Exp Metastasis. 2016;33:385–99.
529. Eriksson A, Österroos A, Hassan S, et al. Drug screen in patient cells suggests quinacrine to be repositioned for treatment of acute myeloid leukemia. Blood Cancer J. 2015;5:e307.
530. Santánna R, Gallego P, Robinson LZ, et al. Repositioning tolcapone as a potent inhibitor of transthyretin amyloidogenesis and associated cellular toxicity. Nat Commun. 2016;7:10787.
531. Camus S, Quevedo C, Menéndez S, et al. Identification of phosphorylase kinase as a novel therapeutic target through high-throughput screening for anti-angiogenesis compounds in zebrafish. Oncogene. 2012;31:4333–42.
532. Ridges S, Heaton WL, Joshi D, et al. Zebrafish screen identifies novel compound with selective toxicity against leukemia. Blood. 2012;119:5621–31.
533. Tomlinson ML, Hendry AE, Wheeler GN. Chemical genetics and drug discovery in Xenopus. Methods Mol Biol. 2012;917:155–66.
534. Tat J, Liu M, Wen XY. Zebrafish cancer and metastasis models for in vivo drug discovery. Drug Discov Today Technol. 2013;10:e83–9.
535. Amelio I, Gostev M, Knight RA, et al. DRUGSURV: a resource for repositioning of approved and experimental drugs in oncology based on patient survival information. Cell Death Dis. 2014;5:e1051.
536. Lee H, Kang S, Kim W. Drug repositioning for cancer therapy based on large-scale drug-induced transcriptional signatures. PLoS One. 2016;11:e0150460.

14 Polypharmacology in Predicting Drug Toxicity: Drug Promiscuity

Abstract

Drug promiscuity refers to multitarget activity of a drug and can elicit two distinct or opposing actions: adverse side effects and improved therapeutic efficacy. On one hand, drug promiscuity is the source and mechanism of off-target effects; on the other hand, it forms the basis for polypharmacology-based drug repurposing, thereby a source of drug rediscovery. These opposing effects reflect two sides of a coin: positive/good/desirable drug promiscuity and negative/bad/undesirable drug promiscuity. In Chap. 13, the "good" side of drug promiscuity for drug repurposing and how the "good" side should be used for drug rediscovery have been discussed. The topic in this chapter focuses on the "bad" side of drug promiscuity, specifically the application of polypharmacology principles to predicting the drug toxicity induced by drug promiscuity, which is a critical issue in drug development.

Keywords

Drug promiscuity · Prediction · Drug toxicity · Drug promiscuity cliff · Big data

14.1 The Connotations of Drug Promiscuity

14.1.1 The "Good" Side or "Angel Face" of Drug Promiscuity

Drug promiscuity or pharmacological promiscuity results from specific interactions of a small molecule with multiple biological targets to exhibit distinct pharmacological effects and is an emerging concept in medicinal chemistry and drug discovery as it is the root of polypharmacology [1–3]. The majority of marketed drugs and bioactive compounds, if not all, have been found to interact with more than one target. The ability of a small-molecule drug (SMD) to interact with multiple target molecules (proteins or noncoding RNAs) constitutes the molecular basis of polypharmacology being the major determinant of the efficacy of promiscuous drugs. From this point of view, the phenomenon of drug promiscuity may facilitate the "design in" of multitarget ligands for the treatment of complicated diseases by providing a new paradigm of drug discovery and drug repositioning. Elucidating and understanding drug promiscuity can boost up MTD

Z. Wang, B. Yang, *Polypharmacology*, https://doi.org/10.1007/978-3-031-04998-9_14

drug discovery on one hand and help predict the potential adverse effects of drugs in the early stage of drug development and in clinical use. Therefore, one of the primary goals in drug design is to scrutinize and manipulate the "merits and faults" of promiscuity.

14.1.2 The "Bad" Side or "Devil Face" of Drug Promiscuity

Pharmacological promiscuity is undesired in typical drug discovery projects, which focus on the "one drug–one target" paradigm because it is the major source of off-target activity, which can lead to adverse drug reactions or can obscure pharmacodynamic effects [1–3]. Additionally, drug promiscuity is a major safety concern that needs to be detected as early as possible in the drug discovery process. Therefore, advanced lead compounds, pharmacological tool compounds, and drug candidates are usually screened against panels of safety-relevant targets to detect unwanted pharmacological activities.

Drug-related adverse events affect approximately 2 million patients in the United States each year, resulting in about 100,000 deaths [4]. For example, highly publicized cases of severe adverse reactions recently resulted in a US Food and Drug Administration advisory panel suggesting that the popular pain relievers Percocet and Vicodin be banned [5]. The anti-obesity drug fenfluramine/phentermine, also known as fen-phen, may cause heart disease and hypertension [6].

During the pre-marketing stages, many clinical trials fail due to lack of drug efficacy and safety concerns [7]. Furthermore, even if a drug reaches the market, post-marketing studies can characterize some critical adverse drug reactions (ADR) not observed during the clinical studies. This is the case for some drugs described to cause teratogenicity and cardiovascular toxicity, which have been withdrawn from the market [8]. Recently, some studies have reported liver and respiratory failure risk associated with the antioxidant drug Limbrel [9]. Globally, lack of efficacy (therapeutic effect) and toxicity (adverse effect) have been identified as two major reasons for the drug attrition in late development stages and drug safety [10]. According to the definition of WHO, adverse effects are divided into three groups: adverse drug event (ADE), adverse drug reaction (ADR), and side effect (SE). ADR and SE are unexpected effects that occurred during the use of drug at normal dose. The causality of both, ADR and SE, are related to the drug pharmacological properties. The main distinction between ADR and SE is that ADR (e.g., myocardial infarction) is always harmful for the patient, whereas SE (e.g., weight increase) may be beneficial for some patients [11]. For example, the drug mirtazapine, used in anorexic patients, may be administered to some patients to solve weight problems, as one of the SEs causes weight gain [12]. The third group of events belonging to adverse effects is the ADE (e.g., intentional product misuse). Adverse drug event corresponds to an event that occurs during the drug treatment, which is not necessarily caused by the drug itself. ADEs include ADR, SE, as well as other events (e.g., product issues, medical errors, social factors) [13]. Adverse drug reactions (ADRs) are one of the leading causes of morbidity and mortality in healthcare. Understanding which drug targets are linked to ADRs can lead to the development of safer medicines.

Some adverse events are predictable consequences of the known mechanism of a drug, but others are not predicted and seem to result from "off-target" pathways. Adverse side effects and lack of efficacy are the two most important reasons a drug will fail clinical trials, each accounting for around 30% of failures [14]. When developing novel chemical entities (NCEs) for a therapeutic application, knowledge of binding partners and affected biological pathways is useful for predicting both efficacy and side effects. The development of tools that can predict adverse events and system-wide effects might thus reduce the attrition rate. Such tools will most certainly include emerging information about protein–protein interactions, signaling pathways, and pathways of drug action and metabolism.

14.2 Predicting Drug Promiscuity and Associated Off-Target Effects

No matter it shows up with its "angel" face or "devil" face, promiscuity is common and inevitable to all compounds in clinical use or in developmental stage. Taking full advantage of the "angel" nature and minimizing "evil deeds" of the "devil" nature of drug promiscuity is one of the tasks of polypharmacology. In this sense, repositioning drugs deals with the "angel" nature, whereas elucidating off-target side effects copes with the "devil" nature. Predicting promiscuity not only helps avoiding and minimizing side effects of therapeutic drugs but also aids the development of new compounds for therapeutic purpose. Therefore, predicting drug promiscuity has recently become a hot research field attracting intensive efforts from both scientific community and pharmaceutical enterprises.

In theory, functional evaluation is the most direct and conclusive way of elucidating and clarifying drug promiscuity. Two-year rodent bioassay is the current gold standard for determining the carcinogenicity of a NCE [15]. Nonetheless, the laboratory assays are generally costly and time-consuming and labor-intensive. Moreover, some assays are also of doubtful utility; for example, only around 15% of gene knockouts in the standard pharmaceutical model organisms show any fitness defect [14]. Furthermore, apparent promiscuity can also be caused by experimental artifacts leading to false-positive activity annotations such as assay interference effects. Such unwanted effects include a tendency of liable compounds to aggregate, form nonspecific interactions with target proteins, or react in various ways under assay conditions [16–18]. Additionally, when comparing the promiscuity of multiple compounds, apparent differences might be influenced by limited assay overlap or significantly different test frequencies. For example, if two close structural analogs are considered and one has been tested in 100 assays against 90 targets and the other in 10 assays against 2 targets, observed differences in promiscuity might be largely due to the much higher test frequency and target coverage of one of the analogs. Compound inactivity and test frequencies are typically not reported in the medicinal chemistry literature and are thus not available in ChEMBL [19], the major public repository of compounds from medicinal chemistry.

One of the issues relating the challenges in predicting drug promiscuity is the current incomplete understanding of molecular determinants of promiscuity. It is largely unclear at present if specific structural characteristics might trigger defined multitarget activity of compounds. Different binding mechanisms of promiscuous compounds are just beginning to be elucidated on the basis of X-ray structures of relevant complexes [20]. Some studies have suggested that promiscuous compounds might often interact with similar binding sites in proteins [21]. These also include correlation analysis of binding-site similarity and compound promiscuity [22, 23]. However, promiscuous binding events are not restricted to similar binding sites. Other studies have identified a large number of promiscuous compounds interacting with proteins from different families and functional classes [24]. The multiclass compounds displayed a variety of binding modes with often only little shape similarity [24]. Hence, compound promiscuity cannot be rationalized by assuming the presence of canonical binding modes.

Currently, there are four major approaches for predicting drug promiscuity: **(a)** "pattern" databases of tissue or organ response to drugs, compiled from high-throughput experiments; **(b)** structure–activity relationship (SAR) computational models based on compound structure; **(c)** "systems biology" databases of metabolic pathways, genes, and regulatory networks [25]; and **(d)** chemical systems biology networks.

14.2.1 Using High-Throughput Screening Data for Drug Promiscuity

Databases of tissue or organ response to drugs were mostly compiled from high-throughput screening (HTS) experiments. HTS and virtual screening are widely used to discover novel leads for drug design. HTS searches large libraries of chemical compounds for those that can modulate the activity of a particular biological target; it is the dominant technique used in early-stage drug discovery. An issue associated with HTS is the prevalence of nonspecific or "promiscuous" inhibitors. These molecules have peculiar properties, act on unrelated targets, and can dominate the results from screening campaigns, which provides an excellent opportunity for exploiting drug promiscuity. Today, with the acquisition of knowledge in drug-induced toxicities and ADRs through HTS studies (LINCS, ToxCast) [26–28] and databases (SIDER, DrugCentral, ChemProt) [27, 29, 30], development of innovative computational methods is feasible to decipher and predict biological targets associated with toxicity and ADE [27, 31–34].

To identify determinants of pharmacological promiscuity, Peters et al. [35] compared the panel screening outcomes of 213 recent Roche compounds with their molecular properties. In their study, pronounced promiscuity was not observed below a threshold Clog P value of 2. For basic compounds, the propensity for weak off-target activity was found to increase with calculated basicities, whereas the potential for strong off-target activity depends more qualitatively on the presence of a positive charge at physiological pH. Compounds originating from projects with an aminergic receptor or transporter as a therapeutic target are particularly prone to promiscuity; the promiscuity of such compounds is mainly caused by their activity at other aminergic targets in the screening panel. The analysis of large HTS datasets reveals that the majority of promiscuous compounds are characterized by recognizable molecular properties and structural motifs, the most important one being a basic center with a $pK_{aB}>6$ [36]. These compounds interact with a small set of targets such as aminergic GPCRs; some of these targets attract surprisingly high hit rates.

Another study used HTS data to discriminate compounds with single-target effects from those with side effects and explored the use of multiple assays of a chemical library and a statistic based on entropy to identify lead compound classes that have patterns of assay activity resulting primarily from small-molecule action on a single target [37]. This statistic, called the coincidence score, discriminates with 88% accuracy compound classes known to act primarily on a single target from compound classes with significant side effects on nonhomologous targets. Furthermore, a significant number of the compound classes predicted to have primarily single-target effects contain known bioactive compounds.

McGovern et al. [38] investigated 45 diverse screening hits, of which 15 had been previously reported as inhibitors of various receptors, including β-lactamase, malarial protease, dihydrofolate reductase, HIV Tar RNA, thymidylate synthase, kinesin, insulin receptor, tyrosine kinases, farnesyltransferase, gyrase, prions, triosephosphate isomerase, nitric oxide synthase, phosphoinositide 3-kinase, and integrase; 30 were from an in-house screening library of a major pharmaceutical company. In addition to their original targets, 35 of these 45 compounds were shown to inhibit several unrelated model enzymes. These 35 screening hits included compounds, such as fullerenes, dyes, and quercetin, that have repeatedly shown activity against diverse targets. They found that nonspecific inhibitors formed particles of 30–400 nm diameter, but a well-behaved competitive inhibitor did not. Consistently, the particle size and IC_{50} values of the promiscuous inhibitors varied monotonically with ionic strength; a competitive inhibitor was unaffected by changes in ionic strength. Notably, aggregate formation appears to explain the activity of many nonspecific inhibitors and may account for the activity of many promiscuous screening hits. The investigators believed that molecules acting via this mechanism may be widespread in drug discovery screening databases. The "aggregate" hypothesis was supported by the study reported by Feng et al. [39] who employed a single mechanism to explain the effects of many promiscuous inhibitors: some organic molecules form large colloid-like aggregates that sequester and thereby inhibit enzymes.

Recently, Mateen and colleagues [40] reported a printable hydrogel-based drug-screening platform capable of non-ambiguously differentiating true enzyme inhibitors from promiscuous aggregating inhibitors, critical for accelerating the drug discovery process and identifying off-target drug effects. The printed hydrogels can both immobilize and support the activity of entrapped enzymes against drying or treatment with a protease or chemical denaturant. Furthermore, the printed hydrogel can be applied in a high-throughput microarray-based screening platform (consistent with current practice) to rapidly (<25 min) and inexpensively identify only clinically promising lead compounds with true inhibitory potential as well as to accurately quantify the dose–response relationships of those inhibitors, all while using 95% less sample than required for a solution assay.

14.2.2 Computational Approaches for Drug Promiscuity

14.2.2.1 Quantitative Structure–Activity Relationships

With the development of sophisticated computational methods such as deep learning, a new paradigm is emerging to predict chemical toxicity [25, 41]. Quantitative structure–activity relationships (QSARs) are a computational modeling method for revealing relationships between structural properties of chemical compounds and biological activities, which are widely used for drug safety prediction [25, 42]. QSARs are regression or classification models used in the chemical and biological sciences and engineering. Like other regression models, QSAR regression models relate a set of "predictor" variables (X) to the potency of the response variable (Y), while classification QSAR models relate the predictor variables to a categorical value of the response variable. In QSAR modeling, the predictors consist of physicochemical properties or theoretical molecular descriptors [43] of chemicals; the QSAR response variable could be a biological activity of the chemicals. QSAR models first summarize a supposed relationship between chemical structures and biological activity in a dataset of chemicals. Second, QSAR models predict the activities of new chemicals [43].

QSARs are also known as "quantitative structure–property relationships (QSPRs)" when a chemical property is modeled as the response variable [44]. Different properties or behaviors of chemical molecules have been investigated in the field of QSPR. Some examples are quantitative structure–reactivity relationships (QSRRs), quantitative structure–chromatography relationships (QSCRs), and quantitative structure–toxicity relationships (QSTRs), quantitative structure–electrochemistry relationships (QSERs), and quantitative structure–biodegradability relationships (QSBRs) [45]. Principal steps of QSARs/QSPRs include **(a)** selection of dataset and extraction of structural/empirical descriptors, **(b)** variable selection, **(c)** model construction, and **(d)** validation evaluation [45]. The biological activity of molecules is usually measured in assays to establish the level of inhibition of particular signal transduction or metabolic pathways. Drug discovery often involves the use of QSAR to identify chemical structures that could have good inhibitory effects on specific targets and have low toxicity (nonspecific activity). Of special interest is the prediction of partition coefficient log P, which is an important measure used in identifying "druglikeness" according to Lipinski's rule of five. In other words, QSARs attempt to relate a numerical description of molecular structure/properties to known biological activity. Large numbers of readily computable descriptors are available that in combination with sophisticated techniques improve the initial linear regression analysis methods used in deriving QSAR equations. In general, QSAR equations relate one objective (such as activity) with a number of descriptors.

QSAR modeling is essential for drug discovery [46, 47]. While many quantitative structure–activity relationship analyses involve the interactions of a family of molecules with an enzyme or receptor-binding site, QSARs can also be used to study the interactions between the structural domains of proteins. Protein–protein interactions can be quantitatively analyzed for structural variations resulted from site-directed mutagenesis [48]. It is part of the machine learning method to reduce the risk for a SAR paradox, especially taking into account that only a finite amount of data is available (see also MVUE).

Several studies have used such computational approach to identify the correlation between cardiac adverse effects and drug properties [49], to detect drug-hepatotoxicity associations [50], to predict SEs using combined canonical correlation analysis with network-based diffusion methods [51], and to identify drugs that may cause ADRs [52]. Nevertheless, most of these studies do not explore potential mechanism of actions linking a drug to an ADE. A study described a method for mining and modeling binding data obtained from a large panel of targets (in vitro safety pharmacology) to distinguish differences between promiscuous and selective compounds [53]. Two naïve Bayes models for promiscuity and selectivity were generated and validated on a test set as well as publicly available drug databases. The model shows a higher score (lower promiscuity) for marketed drugs than for compounds in early development or compounds that failed during clinical development. Such models can be used in triaging high-throughput screening data or for lead optimization.

14.2.2.2 Matched Molecular Pair Analysis for Drug Promiscuity

Utilizing the matched molecular pair formalism, Dimova et al. [54] analyzed patterns of compound promiscuity in a publicly available small-molecule microarray dataset. On the basis of their analysis, the authors introduced "promiscuity cliffs" (to be explained in a following subsection) as pairs of structural analogs with single-site substitutions that lead to large-magnitude differences in apparent compound promiscuity involving between 50 and 97 unrelated targets. They found no substructures or substructure transformations that are generally responsible for introducing promiscuity. However, within a given structural context, small chemical replacements were found to lead to dramatic promiscuity effects. The author proposed that promiscuity is not an inherent feature of molecular scaffolds but can be induced by small chemical substitutions [54].

14.2.2.3 Chemical Structure Analysis for Drug Promiscuity

Preclinical Safety Pharmacology (PSP) attempts to anticipate adverse drug reactions (ADRs) during early phases of drug discovery by testing compounds in simple, in vitro binding assays (i.e., preclinical profiling). The selection of PSP targets is based largely on circumstantial evidence of their contribution to known clinical ADRs, inferred from findings in clinical trials, animal experiments, and molecular studies going back more than 40 years. In this work, we explore PSP chemical space and its relevance for the prediction of adverse drug reactions [55]. Firstly, in silico (computational) Bayesian models for 70 PSP-related targets were built, which are able to detect 93% of the ligands binding at IC_{50}< or = 10 mM at an overall correct classification rate of about 94%. Secondly, employing the World Drug Index (WDI), a model for adverse drug reactions was built directly based on normalized side effect annotations in the WDI, which does not require any underlying functional knowledge. This is, to our knowledge, the first attempt to predict adverse drug reactions across hundreds of categories from chemical structure alone. On average, 90% of the adverse drug reactions observed with known, clinically used compounds were detected, an overall correct classification rate of 92%. Drugs withdrawn from the market (rapacuronium, suprofen) were tested in the model and their predicted ADRs align well with known ADRs. The analysis was repeated for acetylsalicylic acid and benperidol which are still on the market. Importantly, features of the models are interpretable and back-projectable to chemical structure, raising the possibility of rationally engineering out adverse effects. By combining PSP and ADR models, new hypotheses linking targets and adverse effects can be proposed, and examples for the opioid mu and the muscarinic M_2 receptors, as well as for cyclooxygenase-1, are presented.

The research-based pharmaceutical industry has increasingly employed modern medicinal chemistry methods, including molecular modeling. This field has progressed together with advances in biomolecular spectroscopic techniques such as X-ray crystallography and nuclear magnetic resonance (NMR), which have enabled striking progress in molecular and structural biology [24, 56, 57]. Structure-based drug design (SDBB) methods (using three-dimensional structural information gathered from biological targets) are a prominent component of modern medicinal chemistry. In most cases, the protein structure used in the structure-based design process has been determined by X-ray crystallography. Molecular docking, structure-based virtual screening (SBVS), and ligand-based virtual screening (LBVS) are among the most frequently used SDBB strategies due to their wide range of applications in the analysis of molecular recognition events. Along with the prediction of the binding mode, SBVS provides a ranking of the docked molecules for the sole criterion for selecting promising molecules, or it can be combined with other evaluation methods [56, 57].

In general, X-ray crystallography is the preferred technique for elucidating the three-dimensional(3D)structureoftarget–small-molecule ligand complexes. There are, however, several cases where NMR is indispensable. One obvious case is systems where crystallization fails due to the dynamic nature of the target. Other more classical cases involve targets that can be crystallized but where soaking or crystallization fails because the binding site is not accessible. Experienced scientists at Creative Biolabs are glad to apply our special expertise to offer NMR fragment-based immune checkpoint targeted small-molecule drug discovery [24, 56, 57].

14.2.2.4 Structural Similarity Analysis for Drug Promiscuity

Hu and Bajorath reported a study of promiscuity by focusing on structurally similar promiscuous compounds, comparing their activity profiles, and determining multitarget activity patterns [58]. Their analysis revealed that most structurally similar promiscuous compounds have identical or similar activity profiles but also detected different types of structure–activity pattern relationships. In addition, the propensity of activity cliff formation among promiscuous compounds was determined. Activity cliffs, defined as pairs or groups of structurally similar compounds that are active against the same target but have large differences in potency which capture chemical modifications that strongly influence biological activity, were formed at a high rate among promiscuous compounds indicating that many structurally similar promiscuous compounds have greatly varying activity against one or more targets they share. Their findings suggest that compound promiscuity and target selectivity are not mutually exclusive and further refine current views of promiscuity [58].

14.2.2.5 Chemical–Chemical and Protein–Chemical Interactions for Drug Promiscuity

As described above, the side effects of drugs can be predicted based on drug-induced perturbation of gene expression profiles or biological pathways uncovered by HT microarray analysis [6, 59] or on chemical structures [55, 60, 61]. Although most of the methods can only provide whether the query drug has some side effects, they cannot determine which side effects are most likely to happen or even the order information of the side effects [62]. Chen et al. [62] proposed a novel computational method to predict the side effects of drugs based on chemical–chemical interaction and protein–chemical interaction. Compared to most of the previous studies, our method can provide the order information of the side effects, that is, prioritizing the side effects from the most likely one to the least likely one. These researchers utilized the compound databases KEGG (Kyoto Encyclopedia of Genes and Genomes) [63] and STITCH (Search Tool for Interactions of Chemicals) [64]. KEGG provides the information of chemical substances and reactions, while STITCH provides the interaction information of chemicals and proteins, which allow for acquisition of the properties of many compounds and their other information. For those compounds not being covered by these databases, their properties can be inferred from the property-known compounds stored in the databases [65–68]. Likewise, the drugs' side effects can also be inferred as elaborated below.

Moreover, it was evidenced that interactive proteins are more likely to share common biological functions [69–72] and that interactive compounds are also more likely to share common biological functions [65, 68]. Since the side effects are part of biological functions of drugs, it would be feasible to use the chemical–chemical interactions to identify the drugs' side effects. Unfortunately, some of the query drugs cannot be predicted for their side effects by this way because their interactive counterparts do not have any information of the side effects. To overcome such difficulty, we proposed to utilize the information of indirect interactions, including both the chemical–chemical interaction and the protein–chemical interaction, to identify the drugs' side effects of which the direct chemical–chemical interaction data are not available. To evaluate the method, a benchmark dataset retrieved from SIDER [73] was constructed, which consisted of 835 drug compounds, and it was divided into one training dataset and one test dataset. By a jackknife test on the training dataset, the first-order prediction accuracy was 86.30%, while it was 89.16% on the test dataset. To confirm the effectiveness of the method, another method based on chemical structure similarity obtained by SMILES string [74] was also conducted on the training and test datasets. Encouraged by the good performance of the method and superiority to the method based on chemical structure similarity, we hope that the proposed method can become a useful tool to predict drugs' side effects and screen out drugs with undesired side effects.

14.2.2.6 Adverse Drug Event Network

Recently, network-based approaches have given new insights of drug action. Chen et al. performed an ADR–protein network and identified 41 network modules related to specific ADRs [75]. Campillos et al. constructed a drug–target–SE network and demonstrated that drugs with similar side effect profile have a similar protein target profile [76]. Oprea et al. included some tissue information on a similar network and reported that a drug is more likely to cause SE in the organ/tissue where it is more likely to accumulate [77]. Scheiber et al. have developed a network-based model using the structural information of the drugs to predict ADEs [60]. In this study, the authors show that compounds with similar chemical structures share ADE profile.

The objective of the present study was to develop an ADE–ADE network model based on drug–ADE information, in order to predict uncharacterized drug–ADE linkage, and to assess ADEs that are more often associated with each other. Such network may also suggest ADE complexes occurring when a new drug–ADE is observed. To perform this model, we took advantage of a protein–protein association network-based approach developed internally, previously applied, and experimentally validated [32]. Such models allowed to predict linkage between chemicals and proteins and diseases, therefore reflecting the complexity of a biological system [32, 33]. A pull-down scoring system was implemented to automatically identify novel drug–ADE associations within the developed ADE network. As an example, three case studies (i.e., safinamide, sonidegib, rufinamide) were used, and findings were validated through literature supports.

According to Wu et al. [25], the construction of an adverse drug event network (ADEN) generally involves the following steps:

Data Compilation

The researchers downloaded ADE–drug associations and their corresponding likelihood-ratio test (LRT) from the DrugCentral database and extracted a total of 487,698 associations between 9998 unique ADE and 1524 unique drug [25]. The maximum number of ADEs assigned to a single drug "methotrexate" was 3598, and the maximum number of drugs associated with one ADE "vomiting" was 1044. To facilitate further analysis, they set the ADE–drug associations with LRT at a ratio$\geq$1, which resulted in a reduction of the association number to 90,827 between 930 unique drugs and 6221 unique ADEs. The average number of ADEs for a drug was found to be 7. The maximum number of ADE assigned to "methotrexate" was 1575, and the maximum number of drugs associated with one ADE "nausea" was 409.

Generating the ADE–ADE Network

Using the selected drug–ADE associations, we developed a network model, using the previously established PPAN procedure [25, 32, 33]. The protein–protein association network (PPAN) was based under the assumption that nodes (ADE) are connected to each other if they share at least one common interacting partner (e.g., a drug). Therefore, ADEs are associated together in the drug space. Based on the distribution of the number of overlapping drugs between each pair of ADE, only ADE–ADE associations having at least three shared drugs were kept for further analysis [25, 32]. The completed ADE network contains 1,515,860 ADE–ADE unique associations between 3721 ADEs. The ADE–ADE association with the higher number of overlapping drugs was "nausea–vomiting" with 363 drugs.

System Organ Classes (SOC) were then integrated in the ADE–ADE network to facilitate further visualization, and exploration of targeted systems by drugs. In the constructed ADE network, each ADE was classified into the 27 System Organ Classes using the Preferred Term (PT). Taking all the information into account, the maximum number of ADEs assigned to one SOC "nervous system disorders" was found to be 442 [25].

Case Studies Using the ADE–ADE Network

To validate the ADE–ADE network, Wu et al. [25] further assessed the uncharacterized ADEs of three drugs approved by the FDA including safinamide, sonidegib, and rufinamide. First, the known ADEs for each of these three drugs were individually screened on the constructed ADE model to identify potential ADE associations. Then, to verify the accuracy of the predictions, they manually evaluated the results using a two-way procedure: **(a)** finding some literature-based evidence supporting the potential linkages between the most significant predicted ADEs and **(b)** comparing the predicted ADEs with toxicity information reported in FDA documents.

Overall, their approach demonstrated that construction of a network-based model using known information of drug–ADE associations provides a global mapping of ADE relationships in the drug space, without considering the structural information of the drugs and a predictive power of such approach to decipher uncharacterized ADEs for a specific drug. Integration of the SOC information into the ADE network uncovered that drugs do not affect only the targeted systems, but multiple systems leading to various pathway perturbations that may cause ADEs.

According to Wu et al. [25], the major limitation of this approach is that the ADE predictions are limited to the drugs included in the ADE model. Hence, drug effects that may be reported in the literature or in FDA reports (such as hematuria for rufinamide, or more complex effect as retinal degeneration and loss of photoreceptor cells reported for safinamide), but not in the DrugCentral database, cannot be predicted by the model. Such gaps could be paved when more data with LRT information will be available from diverse data sources.

14.2.2.7 Scaffold Analysis of SAR Profiles [Hu Y 2011]

The scaffold concept is widely applied in chemoinformatics and medicinal chemistry to organize bioactive compounds according to common core structures or associate compound classes with specific biological activities, and it is a central concept in medicinal chemistry. A variety of scaffold analyses have been carried out to derive statistics for scaffold distributions, generate structural organization schemes, or identify scaffolds that preferentially occur in given compound activity classes. Hu and Bajorath [78] extended scaffold analysis by identifying scaffolds that display defined profiles of structure–activity relationships (SARs) consisting of multiple properties. A structural relationship-based scaffold network was designed as the basic data structure underlying their analysis. From network representations of scaffolds extracted from compounds active against 32 different target families, scaffolds with different SAR profiles were extracted on the basis of decision trees that capture structural and functional characteristics of scaffolds in different ways. More than 600 scaffolds and 100 scaffold clusters were assigned to ten SAR profiles. These scaffold sets represent different activity and target selectivity profiles.

The same authors have mined publicly available compound and bioactivity data for promiscuous chemotypes [79]. They constructed a hierarchy of active compounds, atomic property-based scaffolds, and unique molecular topologies and analyzed the activity annotations using this framework. Starting from ~35,000 compounds active against human targets with at least 1 μM potency, 33 chemotypes with distinct topology were identified that represent molecules active against at least three different target families. They then utilized network representations to study scaffold–target family relationships and activity profiles of scaffolds corresponding to promiscuous chemotypes. A subset of promiscuous chemotypes displayed a significant enrichment in drugs over bioactive compounds. A total of 190 drugs were identified that had on average only two known target annotations but belonged to the seven most promiscuous chemotypes that were active against 8–15 target families.

Another group described an algorithm for identifying likely promiscuous compounds via associated scaffolds which combines general and domain-specific features to assist and facilitate prediction of ADRs and accelerate drug discovery informatics, called Badapple: bioassay-data associative promiscuity pattern learning engine [80]. Results are described from an analysis using data from MLP assays via the BioAssay Research Database (BARD) http://bard.nih.gov. Specific examples are analyzed in the context of medicinal chemistry, to illustrate associations with mechanisms of promiscuity. Badapple is a method for rapidly identifying likely promiscuous compounds via associated scaffolds. Badapple generates a score associated with a pragmatic, empirical definition of promiscuity, with the overall goal to identify "false trails" and streamline workflows. Unlike methods reliant on expert curation of chemical substructure patterns, Badapple is fully evidence-driven, automated, self-improving via integration of additional data, and focused on scaffolds. Badapple is robust with respect to noise and errors, and skeptical of scanty evidence.

Baell et al. [81] presented a method that is fully automatic, fully empirical, and focused on scaffolds. That is, the promiscuous scaffolds are perceived by the algorithm, and somewhat counterintuitively, by virtue of being scaffolds, are generally as meaningful to chemists, or more so, than substructure patterns. This method requires no code revision to accommodate new data, new assay methods, or new compound classes. The generality of its inferences is limited solely by the breadth and accuracy of the data from which those inferences are derived. Although library design and hit selection are separate tasks, informatically their effects are merged to determine which compounds are carried forward. Filtering and ranking for compound promiscuity and other aspects of suitability in powerful and flexible ways is relevant in a variety of scenarios. The literature does reflect increasing interest in informatics-based approaches to analysis of promiscuity, polypharmacology, and non-selectivity assessment [79, 81, 82].

14.2.3 Systems Biology Approaches for Drug Promiscuity

Systems biology means an approach to biology that looks at networks of molecular interactions (including gene products, endogenous small molecules, and therapeutic drugs) and processes these using qualitative graphical models or quantitative mathematical modeling [83]. Examples of implementations of quantitative methods include flux balance analysis [84], differential equations [85], and Petri nets [86]. Implementations of qualitative methods include Cytoscape [87], a graphical network representation, and Genoscape [88], a network-based knowledge integration extension tool. When the principles of systems biology are extended to medications, we get a network of interactions between drugs and the naturally occurring metabolic and signaling networks. These drugs may connect otherwise disconnected and independent subnetworks, and this may cause both expected and unexpected effects. Pharmacological systems biology must combine the biological and chemical characteristics of small and large molecules to develop an understanding of drug action. These protein–drug joint networks provide two opportunities. First, they can provide more detailed descriptions (even signatures) of drug effects, and second, they can provide a framework for the design of novel therapeutic strategies [15].

The information obtained from analysis of protein three-dimensional structure often kick-starts the modern drug design process, providing information for designing SMDs that are complementary to the shape of an investigated protein. Thus, the basic idea is to take advantage of complementarity built into designed SMDs to inactivate the protein function and stop downstream cellular processes. Subsequent design revisions are typically aimed at optimizing the binding, reducing molecular weight, or increasing lipophilicity, which all improve the molecule's drug-like properties. With the assistance of computational techniques, a number of drugs have been refined with this mindset, and a number of reviews have discussed the impact made by computational drug design based on network or systems biology level of [89–92]. The target and ligand promiscuities can be regarded as a very reasonable mechanism for subtle control of complicated biological systems because the existence of multiple combinations of the limited number of endogenous ligands and target proteins is a potential reason for explaining the diversity of input patterns into the downstream signaling pathways, suggesting that understanding and controlling of the promiscuities is a critical issue for drug design. A systematic way to detect pairs of proteins that could share a ligand is to compare binding sites in their 3D structures [93–96]. Using these methods, today, it is possible to identify alternative drug targets [97], predict molecular functions [98], and uncover links between remote proteins [99].

Duran-Frigola et al. [93] applied binding-site similarity analysis in several systems biology scenarios through exhaustively comparing pockets across a large fraction of the human proteome, finding connections between close and distant proteins belonging to families. They then applied systems biology techniques and pinpointed those cases that could have an impact on protein–protein interaction networks (PPIs) related to several therapeutic areas and tumor types [100] and to a genome-scale metabolic model (GSMM) of cancer cell lines [101]. Specifically, these investigators performed a pairwise comparison of 90,000 putative binding pockets detected in 3700 proteins and found that 23,000 pairs of proteins have at least one similar cavity that could, in principle, accommodate similar ligands [93]. By inspecting these pairs, they demonstrated how the detection of similar binding sites expands the space of opportunities for the rational design of drug polypharmacology [93]. Finally, they illustrated how to leverage these opportunities in protein–protein interaction networks related to several therapeutic classes and tumor types, and in a genome-scale metabolic model of leukemia [93].

14.2.4 Chemical Systems Biology for Drug Promiscuity

The ability to predict and even design the effects of new drugs is critical for the future pharmaceutical industry. By integrating biological and chemical knowledge, the pharmacological effects of drugs can be more completely understood and used to create predictive models. Recent work has focused on relating drugs to targets by chemical similarity, target structural similarity, and even side effect similarity. In each case, the results have illustrated the power of thinking about drug responses in the context of a network of interactions, and from a systems perspective. Chemical approaches, both synthetic and analytical, provide researchers with powerful new technologies to perturb, dissect, and even reconstruct complex biological systems. Chemical biology and systems biology have grown and evolved in parallel during the past decade, but the mindsets of the two disciplines remain quite different [102, 103]. As the inevitable intersections between the disciplines become more frequent, chemical biology has an opportunity to assimilate the most powerful ideas from systems biology. Although much is known about the molecular components of cellular signaling pathways, very little is known about how these multicomponent biochemical machineries process complex extracellular signals to generate a consolidated cellular response. On one hand, chemical biologists are exploring biological phenomena of greater complexity, and on the other hand, systems biologists are increasingly adopting chemical tools. Both fields have begun to investigate, often with different techniques and perspectives, the potentially pleiotropic biological effects of small molecules. Understanding the components and dynamics of cellular networks is essential for a global view of cellular function. Russell and Aloy [104] discuss interaction networks and the ways in which these maps are currently being exploited for applications from biomedicine to synthetic biology. For a few well-studied prokaryotic systems, computational modeling of large collections of genomics data has been used to successfully predict the dynamic response of the organism to genetic or environmental perturbations. Bonneau [105] describes computational methods for deciphering regulatory networks and discusses the outlook for extending these approaches to metabolomic and proteomic data and across a wider range of species. The importance of integrating multiple levels of omics data is highlighted by the significant differences observed between quantitative proteomic and RNA interference data in *Drosophila melanogaster* [106]. Proteomic approaches are contributing to a better understanding of apoptosis with two reports doubling known caspase substrates [107]. Chemical tools, with their potential for rapid and dose-dependent action, are proving to be important for systems-wide approaches to understanding biology. Simon and Cravatt [108] highlight chemical methods, including global approaches for profiling posttranslational modifications and enzyme activity, that are particularly promising chemical parts in the systems biology toolbox. Zamir and Bastiaens [109] describe how chemical perturbations can be combined with quantitative detection and a recently developed mathematical approach to gain new insights into the dynamics of intracellular signaling pathways. Lehár et al. [110] describe how combinations of chemical–genetic or chemical–chemical perturbations can advance chemical genetics from investigating protein function to probing pathway structure. “Systems chemistry” is also contributing to our understanding of molecular interaction networks at the chemical level [111]. A systems-level understanding of biology can provide new approaches for targeting disease and for understanding drug action. For instance, metabolomics approaches have recently been used to identify palmitoleate as a new adipose lipid hormone with an important role in regulating systemic metabolism and, in a separate study, to implicate fatty acid synthesis as a potentially broad-spectrum antiviral target [106]. Hopkins [112] puts forth the case for targeting perturbed disease networks rather than disease-causing genes and describes the new chemical challenges that emerge in seeking to design small molecules that act on multiple targets. In a step toward rational polypharmacology, Apsel et al. [113] design compounds that selectively target Src, a tyrosine kinase, and PI3K, a lipid kinase, and then investigate the biological effects of these dual inhibitors [114]. Finally, Peterson [83] proposes that chemical biologists could benefit from embracing a systems biology perspective and moving away from a one molecule–one target mindset toward an “integrationist” view of small-molecule action.

Predicting the behavior of living organisms is an enormous challenge given their vast complexity [110]. Efforts to model biological systems require large datasets generated by physical binding experiments and perturbation studies. Genetic perturbations have proven important and are greatly facilitated by the advent of comprehensive mutant libraries in model organisms. Small-molecule chemical perturbagens provide a complementary approach, especially for systems that lack mutant libraries, and can easily probe the function of essential genes. Though single chemical or genetic perturbations provide crucial information associating individual components (e.g., genes, proteins, or small molecules) with pathways or phenotypes, functional relationships between pathways and modules of components are most effectively obtained from combined perturbation experiments.

Yabuuchi et al. [115] proposed a drug design concept, "Chemical Genomics-Based Drug Design" (CGBDD) for incorporating system-level multi-interaction networks connecting chemistry and biology. The set of all possible small organic molecules, referred to as chemical space, has been estimated to consist of more than 10^{60} compounds [115, 116]. Chemical space is as vast as the diversity of biological systems, and the vastness of the two domains creates difficulty in comprehensive understanding of the interface between chemical space and biological systems [115, 117, 118]. Recently, chemical genomics has emerged as a promising area of research applicable to exploration of novel bioactive molecules, and researchers are currently striving toward the identification of all possible ligands for all target protein families [119]. Large-scale datasets of compound–protein interactions (CPIs) are being collected, and chemical genomics studies have shown that patterns of protein–ligand interactions are too diverse to be understood as simple one-to-one events. For example, multiple structurally different compounds have been shown to bind the same protein or express similar biological activities [120, 121]. In other cases, one drug has been shown to affect multiple targets from different protein families [122, 123]. Therefore, an integrative understanding of multiple interactions among chemical and biological components beyond a one-compound/one-target simplification could open up new opportunities in drug development, but the need to develop appropriate data mining methods for characterizing and visualizing the full complexity of interactions between chemical space and biological systems is urgent [124]. Recently, mining of multiple CPI datasets has been used to identify new protein targets for known drugs and thereby predict unreported polypharmacology [125]. However, this approach only identifies additional targets for known drugs. No existing screening approach has so far succeeded in identifying novel bioactive compounds using multiple interactions among compounds and target proteins, and the potential application of analysis of multiple CPIs to identify novel bioactive molecules remains unknown.

HTS and computational screening have greatly aided in the identification of early lead compounds for drug discovery. However, the large numbers of assays required for HTS to identify drugs that target multiple proteins render this process very costly and time-consuming. Therefore, interest in using in silico strategies for screening has been increasing. The most common computational approaches, ligand-based virtual screening (LBVS) and structure-based virtual screening (SBVS) [126–128], have been used for practical drug development. Unfortunately, these methods have important limitations. LBVS aims to identify molecules that are very similar to known active molecules and generally has difficulty identifying compounds with novel structural scaffolds that differ from reference molecules. Attempts to scaffold-hop using LBVS are prone to identification of increased numbers of false positives [120]. Therefore, the primary objective of virtual screening, reduction of the number of candidate compounds to be assayed, remains unachievable using this method. The other popular strategy, SBVS, is constrained by the number of three-dimensional crystallographic structures available and, more importantly, by the difficulty of accurately simulating molecular docking processes for targets, including membrane-spanning G-protein-coupled receptors (GPCRs). To circumvent these limitations, we have shown that a new computational screening strategy, chemical genomics-based virtual screening (CGBVS), has the potential to identify novel, scaffold-hopping compounds and assess their polypharmacology by using a machine learning method to recognize conserved molecular patterns in comprehensive CPI datasets.

The intersection of systems biology and chemical biology opens new avenues of research. In particular, there are opportunities to combine data from genomics, three-dimensional structure, large chemical screens, protein–protein interactions, protein–drug-binding interactions, and cellular imaging and localization to assemble a high-fidelity model of how and where small molecules interact with cellular components. A harbinger of the opportunities that exist is the work by Apsel et al. [113], who have integrated chemical biology and systems biology techniques to design drugs that act as dual inhibitors of two families of oncogenes.

In a recent study, Philip Bourne and colleagues [129] have used a chemical systems biology approach to explain the serious side effects of a drug that was being trialed for prevention of cardiovascular disease, which is an excellent example of the use of networks combining proteins and drugs. They investigated the reasons for the serious side effects of torcetrapib, an inhibitor of cholesteryl ester transfer protein (CETP) that was in clinical trials as a preventive treatment for cardiovascular disease. The aim of torcetrapib was to raise the levels of the desirable high-density lipoprotein cholesterol (HDL-C), but torcetrapib turned out to have the side effect of raising blood pressure, with potentially fatal effects in high-risk patients, and was withdrawn from development in 2006. The investigators generated off-target binding networks by comparing the structure of ligand-binding sites in all known protein structures. The proteins identified as having similar binding domains were ranked by a normalized docking score and clustered by their structural and functional characteristics into a gene network that includes metabolic and regulation pathways. Using this analysis, the authors identified possible off-targets for torcetrapib even though the binding site of CETP itself is not fully described. Perhaps most interestingly, they incorporated biological pathways into their off-target networks and found a potential explanation for the poorly understood effects of torcetrapib on blood pressure. By combining a simple gene regulation model with the predicted binding affinities to activators and inhibitors of the renin–angiotensin–aldosterone system (RAAS), they showed that torcetrapib caused more severe effects since it has a higher affinity for more RAAS activators. To validate this approach, they compared the off-target networks for drugs with different side effect profiles, and show that the networks are quite different and consistent with the different effects of the drugs on blood pressure [129]. Their method can, however, only use proteins with known structures—a small fraction of the human proteome. As a result, pharmacologists may become fans of high-throughput structural biology!

An alternative approach to discovering off-target effects relies on identifying common chemical features among drugs with the same set of adverse reactions [60]. This approach links chemical substructures to specific toxicities and can be used to determine the potential side effects of a drug with a novel chemical structure. An implementation of this technique is described by Scheiber et al. [60] to relate chemical substructures to side effects and by Campillos et al. [76] to combine drug chemical similarity to side effect similarity to predict shared drug targets. Recent work by Shoichet and colleagues [130] in this field uses the similarity ensemble approach with a Bayesian method to build chemoinformatics networks based on chemical similarities between drugs, instead of on structural or sequence similarities between drug targets. Comparisons between the ligand-based network of Hert et al. [130] and the target-based network of Xie et al. [129] might provide interesting insights. If the networks' information content is complementary, as opposed to redundant, then a method that utilized both networks may outperform either one alone.

Other investigators have taken a complementary approach. Instead of looking for common chemical substructures, they focus on common adverse reactions. Scheiber et al. [4] have incorporated data from a variety of databases and identified drugs with shared toxicities. They then apply an understanding of the molecular pathways underlying these toxicities to predict drug targets. In this way, they can develop data-driven hypotheses about the mechanisms of a particular side effect. This approach is particularly useful when chemicals with very different structures (not likely to be recognized using measures of chemical similarity) interact with the same biological pathway. The toxicities are effectively used as a proxy for the biological pathways that the drug is involved with.

The success of network-based methods relies heavily on the development and curation of high-quality biological and pharmacological databases. The new high-throughput technologies have provided a huge amount of data on protein–protein and gene–gene interaction networks. The meta-database pathguide.org [131] currently lists more than 70 such databases that are freely available. However, as Blow points out in a recent review [132], no one database is complete, and combining datasets will yield more information. The study by Xie et al. [129], for example, incorporates data from eight different sources. The availability of these databases will fuel the next generation of chemical systems biology tools and lead to major advances in drug discovery and repositioning. Databases that attempt to integrate these different sources of data are becoming available. One such, STITCH, tries to consolidate knowledge about interactions between proteins and small molecules [64]. Although undoubtedly useful, these huge databases do raise the issue of false discovery. Incorporating domain knowledge to rank genes by their propensity to cause a modulated drug response may be one way of addressing this issue [133].

14.3 Structure–Promiscuity Relationships of Therapeutic Compounds

14.3.1 The Concept of Drug Promiscuity Cliff

Currently available volumes of compounds and biological activity data enable large-scale analyses of compound promiscuity (multitarget activity). To aid in the exploration of structure–promiscuity relationships, promiscuity cliffs (PCs) were introduced by Bajorath [54, 134–138]. In analogy to activity cliffs (defined as pairs or groups of structurally similar compounds that are active against the same target but have large differences in potency), PCs were defined as pairs of structurally analogous compounds with large differences in the number of targets they are active against. The PC concept aids in promiscuity analysis for better understanding of the molecular basis of multi-target activities and structure–promiscuity relationships (SPRs) and reveal small chemical modifications that lead to large apparent differences in promiscuity between structurally analogous compounds. PCs can be identified by applying the matched molecular pair formalism and are thus confined to changes at a single substitution site. PCs have been identified among compounds with activity against many therapeutic targets [134] including protein kinases [136, 137]. A large-scale analysis of currently available kinase inhibitors covering more than 80% of the human kinome yielded nearly 16,000 PCs [136, 137].

The formation of PCs can be visualized in network representations where compounds are nodes and edges pairwise PC relationships between nodes [136, 137]. In such networks, PCs form clusters of varying size and complexity. PC clusters are disjoint subgraphs in a PC network. These clusters are rich in structure–promiscuity relationship information, but difficult to analyze. Therefore, as an extension of the PC concept, the PC pathway (PCP) data structure was introduced [136, 137]. PCPs are formed in PC clusters and consist of sequences of PCs with overlapping compounds. PCP compounds have alternating high and low promiscuity (or are highly promiscuous and nonpromiscuous). They can be systematically extracted from PC clusters using a computational search method [136]. Nearly 16,000 PCs formed by inhibitors of human kinases were organized in more than 600 separate network clusters [137], and from these clusters, 8900 PCPs were isolated [136]. PCPs often contain so-called promiscuity hubs (PHs). Following network terminology, hubs are densely connected nodes. PHs are highly promiscuous PCP compounds that form large numbers of PCs with weakly or nonpromiscuous structural analogs outside the PCP [136]. PHs also occur in network regions outside PCPs.

PCs, PCPs, and PHs provide a wealth of hypotheses for structural determinants of promiscuity and also for additional targets of weakly or nonpromiscuous compounds. For example, structural analogs of a PH might not have been tested against many confirmed PH targets and thus additional targets might be inferred for individual analogs. Taken together, PCs, PCPs, and PHs provide valuable information for medicinal chemistry or chemical biology projects. This data note details an open-access deposition of PCs identified across the human kinome [137], PCPs extracted from their network clusters [136], and PHs formed by individual kinase inhibitors including clinical compounds. These data are made freely available in an organized and easily accessible form.

14.3.2 Identification of Promiscuity Cliff by Systematic Analysis

Selectivity of a compound for a given target over other—related or unrelated—target(s) usually results from potency differences [135, 139]. This means that a compound would be selective for a target that it binds with high potency over another for which it has only low potency. Intuitively, one might expect that promiscuity and selectivity might go hand in hand, i.e., that most promiscuous compounds would be selective for a primary (high potency) target over other(s), but this is not the case. Only less than 20% of promiscuous compounds for which high-confidence activity data are available bind their targets with significant differences in potency [135, 140, 141]. Many promiscuous compounds are active against multiple targets in the nanomolar range [140], as further discussed below.

For a systematic assessment of selectivity, we define for a compound that is active against a pair of targets T1 and T2 a selectivity ratio SR=potency(T1)−potency(T2) on the basis of pK_i or pIC_{50} values. Accordingly, an SR value of 2 means that the compound is 100-fold more potent against T1 than T2; that is, it is selective for T1 over T2. If the SR value is −2, the compound is 100-fold more potent against T2 than T1 (and hence selective for T2). Albeit simple, SR values for individual compounds and pairs of targets are very useful for defining selectivity cliffs, as discussed in the following.

14.3.2.1 From Activity and Selectivity Cliffs to Promiscuity Cliffs

As stated by Dimova and Bajorath [135], the classical definition of an activity cliff (AC) is a pair of structurally similar compounds having a large difference in potency [142]. This AC definition has been further evolved for medicinal chemistry applications by considering alternative similarity and potency difference criteria and extending the initial focus on compound pairs to groups of structural analogs and AC clusters [143, 144]. ACs account for strong discontinuity of structure–activity relationships (SARs) and are therefore of interest in compound optimization. As an extension of the AC concept, selectivity cliffs (SCs) were defined as pairs of structurally analogous compounds with a large difference in selectivity against a target pair [135, 139, 141]. SR values were used as a measure of selectivity. An at least 100-fold difference in selectivity ($|\Delta SR| \geq 2$) was set as a SC criterion. The idea underlying the introduction of SCs was the exploration of confined structural changes in closely related compounds that influence or determine target selectivity. In SC networks in which nodes represent compounds and edges pairwise SCs, selectivity patterns emerge that complemented and further refined SAR analysis [139]. SCs can be formed by two compounds with inverse selectivity or by a nonselective and a highly selective compound. However, more than 90% of nearly 10,000 SCs that were identified by compound data mining belonged to the latter category [141]

Going beyond the target pair-based framework of SCs, another extension of the AC concept was introduced to account for unexpected differences in multitarget activity of closely related compounds. Accordingly, promiscuity cliffs (PCs) were defined as pairs or groups of structural analogs with large differences in promiscuity [54, 134–138]. For PC analysis, the promiscuity degree (PD) of a compound was introduced as the number of unique targets it is active against [54]. Thus, a prespecified |ΔPD| threshold serves as a PC criterion.

On a methodological note, after evaluating a variety of similarity criteria for all types of small-molecule-based "cliffs" that the researchers have been investigating, Bajorath and colleagues [135, 145] focused on a substructure-based similarity assessment applying the matched molecular pair (MMP) formalism [146]. An MMP is defined as a pair of compounds that are only distinguished by a structural modification at a single site [146], the exchange of a pair of substructures, termed a (chemical) transformation [147]. For cliff identification, they designed transformation size-restricted MMPs (tsr-MMPs) to limit structural changes encoded by MMPs to small chemical modifications typically observed in analog series [148]. The formation of a tsr-MMP is used as a similarity criterion for cliff formation. This also applies to PCs, as discussed in the following.

14.3.2.2 Systematic Analysis of Promiscuity Cliffs

Systematic searches for PCs were carried out by considering compounds with single-target activity as well as compounds that were consistently inactive in assays, depending on data availability [145]. Under these conditions, a PC might be formed by a promiscuous compound and another one with multi- or single-target (PD = 1) activity or an inactive compound (PD = 0). Activity data mining had revealed earlier that the majority of promiscuous compounds were active in the submicromolar range against two to five targets from the same family, having similar potency against these targets [140]. This was a rather unexpected characteristic of promiscuous compounds. A systematic search for PCs was thought to uncover most discontinuous SPRs among compounds from different sources. The frequency at which PCs might be formed was unknown but anticipated to be low, given the observations above.

PCs were systematically identified in compounds from medicinal chemistry available in ChEMBL [19] and from biological screening available in PubChem [149]. Database compounds that might potentially cause activity artifacts in assays [16, 150] were excluded from the analysis. For the ChEMBL (release 22) search [134], compounds were preselected for which high-confidence activity data were available. In addition, screening compounds incorporated from PubChem into ChEMBL were eliminated. As a PC criterion, a |ΔPD| ≥10 threshold was applied, together with the formation of a tsr-MMP by candidate compounds. PD differences are arbitrarily set and can hence be adjusted for a given application. ChEMBL only contains active compounds. Accordingly, the lowest possible PD combination for a PC was a compound with single-target activity (PD=1) and a structural analog with activity against at least 11 targets (PD=11), respectively. The predefined PD difference of 10 represented a stringent criterion in light of the predominantly observed activity characteristic of promiscuous compounds [140], as discussed above. Applying these criteria, a total of 784 PCs were detected that involved 763 compounds including 77 highly promiscuous compounds with PD≥11 [134]. Forty-two of these compounds were active against targets from different families. Hence, PCs were formed by about 10% of highly promiscuous compounds and 90% of weakly promiscuous or single-target analogs. More than 70% of PCs from ChEMBL combined a highly promiscuous compound with an analog having single-target activity [134]. Given their composition, the majority of PCs from ChEMBL suggested additional targets for structural analogs of highly promiscuous compounds [145]. It is conceivable that analogs of promiscuous compounds might be active against additional targets, thus suggesting experimentally testable hypotheses.

To further extend PC analysis to screening data, compounds tested in both primary (single compound dose) and confirmatory (dose–response assays) assays were collected, and their assay frequency and activity/inactivity counts were determined. A pool of more than 430,000 extensively tested compounds was obtained, with a median test frequency of 437 assays per compound. Then, PCs were identified separately for compounds from primary and confirmatory assays, applying a |ΔPD| ≥20 threshold as a PC criterion [151]. Compounds tested in primary and confirmatory assays yielded 2070 PCs (with 2158 compounds) and 282 PCs (318 compounds), respectively, which included 1024 (primary assays) and 82 (confirmatory) compounds that were active against 20 or more targets, mostly from different families. PCs were prioritized for which participating analogs were tested in comparably large numbers of assays with at least 85% assay overlap. From compounds tested in at least 400 primary or 100 confirmatory assays, 713 and 142 PCs were isolated, respectively [151]. Thus, the use of screening data made it possible to counteract data sparseness and take inactivity records into account. For screening data, the proportion of highly promiscuous compounds involved in PCs was greater than for ChEMBL, with 47.5% for primary and 22.0% for confirmatory assays. Nonetheless, many PCs were also centered on individual highly promiscuous compounds forming multiple PCs [151]. An exemplary cluster from the PubChem PC network was centered on a few highly promiscuous compounds and contained PC pathways. One of these pathways comprised a sequence of pairwise overlapping PCs formed by compounds that were consistently inactive in many assays or highly promiscuous.

PCs were designed to aid in compound promiscuity (multitarget activity) analysis and capture most interesting (discontinuous) SPRs [151, 152]. They can be rationalized as an extension of ACs, similar to SCs. PCs can be used as an easy-to-generate and intuitive data format for large-scale analysis of compound structures and activity data. A systematic search in ChEMBL and PubChem has identified many interesting and in part puzzling SPRs, which are subject to further experimental controls. However, analysis of PCs should also help to better understand structural determinants of promiscuity, which represent a prime motivation for exploring PCs in a systematic manner. As a computational tool, PCs are of interest for practical applications because they encode an abundance of experimental target hypotheses for promiscuous compounds and analogs across many different families. PC analysis of screening data was focused on extensively assayed compounds, thereby directly addressing the issues of test frequency and data sparseness. Given the prevalence of comparable potency values for different targets among many promiscuous compounds, the information content of PCs might not substantially increase by comparing potency values of participating analogs.

14.3.3 Identification of Promiscuity Cliff by Machine Learning

The fact that similar compounds may have very different properties has a large impact in several areas of chemistry. In drug discovery, almost every medicinal chemist working on lead optimization has faced unexpected large "jumps" in activity due to small changes in structure, that is, activity cliffs (ACs) [148]. According to Hu et al. [148, 153], ACs are generally understood as pairs or groups of similar compounds with large differences in potency. The study of ACs is of high interest for the characterization of activity landscapes of compound datasets and the identification of SAR determinants. ACs can be divided into five different classes dependent on whether the participating compounds are only distinguished by chirality, topology, R-group sets, core structures (scaffolds), or core structures and R-group topology. All cliff types are frequently detected in the ChEMBL database. R-group cliffs occur with higher propensity than other cliff types, as one might expect. However, many scaffold and R-group cliffs are not identified on the basis of whole-molecule similarity calculations, although they are often chemically intuitive. This makes the activity cliff classification attractive for medicinal chemistry analysis, independent of similarity calculations. Assignment of activity cliffs on the basis of well-defined structural criteria complements and further extends current approaches to identify and represent cliffs.

On the other hand, Blaschke et al. [145] reported a systemic study which extracted different structural analogs with varying promiscuity using the matched molecular pair (MMP) formalism from public biological screening and medicinal chemistry data [145]. After eliminating all compounds with potential false-positive activity annotations from the analysis, the investigators predicted promiscuity at the level of compound pairs representing PCs (formed by analogs with large promiscuity differences) and corresponding non-PC MMPs (analog pairs without significant promiscuity differences). They then generated different machine learning models to address this prediction task by comparing the results from single compound predictions. PCs encoding promiscuity differences were found to contain more structure–promiscuity relationship information than sets of individual promiscuous compounds. Their approach can be summarized as the following steps.

14.3.3.1 Compound Pair-Based Promiscuity Predictions

To further investigate compound promiscuity with the aid of machine learning, Blaschke et al. [145] devised an analysis involving five different approaches and two distinct data sources including a large set of extensively assayed screening compounds with experimentally confirmed (non-)promiscuity and another manually curated set of human kinase inhibitors from the medicinal chemistry literature. The two intrinsically different datasets were intentionally selected to enable independent assessments of promiscuity via machine learning in the presence or absence of experimental test frequency information. This also assigned different levels of confidence to PCs and non-PC MMPs, directly addressing the data incompleteness issue. The primary task of their analysis was promiscuity prediction at the level of compound pairs [154].

To this end, they systematically calculated MMPs for both datasets and PCs as well as non-PC MMPs. In addition, training and test sets for single compound promiscuity predictions were also extracted from both data sources. On the basis of PubChem data, a total of 170,606 compounds formed 362,362 PCs with a ΔPD of 2 or larger. Among these, there were 5750 PC with $\Delta PD \geq 10$. For the formation of kinase inhibitor PCs, which exclusively consisted of active compounds, the $\Delta PD \geq 10$ criterion was consistently applied. This also ensured that kinase inhibitor PCs exclusively consisted of compounds with large differences in the number of kinase targets they were active against.

14.3.3.2 Selection of Training and Test Set

Machine learning models were trained on the basis of carefully assembled balanced training and test sets to distinguish between PCs and non-PC MMPs [145]. To aid in the selection of unique pairs, each of the six sets of PCs described above was organized in a PC network. In addition, non-PC MMPs were obtained by randomly sampling an MMP network built from all nonpromiscuous compounds.

In addition, pair-based promiscuity difference predictions with promiscuity predictions of individual compounds taken from PCs were compared, which served as a reference for PC predictions. Therefore, promiscuous and nonpromiscuous compounds were taken from PCs and non-PC MMPs, respectively, and the same number of the training and test instances was selected in each case [145].

14.3.3.3 Molecular Representation for Machine Learning

Each compound was encoded using the ECFP4 format. For each training and test PCs, the ΔPD was assigned in a direction-dependent manner such that the promiscuous compound was presented preceding the nonpromiscuous analog. This ordering provided the basis for the generation of the MMPFP. Here, separate fingerprint components representing the common core and distinguishing substituents were combined. MMPFP similarity was calculated as the product of pairwise Tanimoto similarity comparisons of the three components. This similarity assessment was inspired by the design of MMP-based kernel functions that were successfully used for the prediction of ACs [154]. For prospective applications predicting PCs among new MMPs, ordering is not essential for test instances.

14.3.3.4 Prediction of Promiscuity Cliffs

For systematically distinguishing between PCs and non-PC MMPs, 1-NN, k-NN, SVM, DF, and DNN classification models were generated. The predictive performance of each model was assessed using ACC, MCC, F1, and ROC AUC values. The models were used to predict PCs with $\Delta PD \geq 10$ from the screening compound and kinase inhibitor datasets. These PCs encoded largest differences in analog promiscuity compared to non-PC MMPs. For pairs of screening compounds, all models were found to be predictive with an accuracy of at least 70%. The accuracy of SVM and RF approached 80%, indicating a clear tendency to distinguish between substitutions in the context of specific core structures leading to a large change in the promiscuity of structural analogs. Given the consistently high MCC (max. 0.56), F1 (0.77), and ROC AUC (0.86) values, there was no significant difference in prediction accuracy between the SVM, RF, and DNN machine learning models.

For kinase inhibitors, the models displayed similar trends. Although global model performance decreased relative to screening compounds, the machine learning models were comparably predictive, with max. ACC and F1 values of 0.71 and max. MCC value of 0.42. For all machine learning models, MCC values were ~0.10 lower than for screening compounds. The overall lower predictive performance on the kinase inhibitor set might be due to two factors including the general structural similarity of many (ATP site-directed) kinase inhibitors and the uncertainties associated with unknown test frequencies, potentially resulting in underestimated PDs for a number of inhibitors, due to data incompleteness.

While these models were also predictive, their accuracy was generally lower than for pair-based predictions. These findings indicated that PCs and non-PC MMPs captured more structure–promiscuity relationship information than individual compounds. For single compound and PC predictions, results for PubChem compounds were slightly or moderately superior to kinase inhibitors. This was likely due to the fact that most kinase inhibitors target the ATP binding site and are often structurally similar. PubChem targets are diverse, and the screening hits cover a variety of chemical classes.

The difference in predictive performance between simple nearest neighbor classifiers and machine learning models was only small across all prediction tasks, consistent with observations made when first attempting to predict individual promiscuous compounds [152]. Notably, for individual kinase inhibitors, prediction accuracy of the 1-NN and *k*-NN classifiers exceeded the accuracy achieved with the complex DNN. Hence, predictions of individual promiscuous compounds were mostly determined by nearest neighbor effects. On the other hand, for pair-based predictions, the accuracy of machine learning models was overall higher than for the *k*-NN classifiers, indicating that nearest neighbor effects alone could not fully explain predictive performance and reinforcing the notion of higher promiscuity-relevant information content of analog pairs compared to single compounds.

14.3.3.5 Predicting Promiscuity Cliffs with Varying ΔPDs

For screening compounds, the authors also investigated the influence of different ΔPD threshold values [145, 152]. In these calculations, PC training and test sets were varied but the same set of non-PC MMPs was used. The performance of machine learning models was generally very similar for PC predictions and higher than of NN classifiers. Equivalent observations were made for PCs at all ΔPD thresholds. However, the accuracy of PC predictions generally decreased with decreasing ΔPD threshold values. The same trend was observed for all models. For the NN classifiers, prediction accuracy for PCs with ΔPD of 2–3 was very close to random classification. Thus, PCs with decreasing ΔPD values between structural analogs presented increasingly difficult prediction tasks.

14.3.3.6 Structure–Promiscuity Relationships

The findings discussed above raised the question whether one might be able to identify structural features determining accurate PC predictions. The influence of individual fingerprint features on the predictions of screening compound PCs with ΔPD≥10 was assessed using a surrogate model approach based on linear SVM and feature elimination. The use of linear SVM models permitted direct comparison of the importance of different features at the cost of lower predictive performance compared to nonlinear SVM models (using the MMP kernel). To these ends, linear SVM models were iteratively built using reduced feature sets relative to the previous step. In each iteration, the top 100 features with largest positive or negative weights in SVM predictions were identified and removed. This process was repeated for 40 iterations. Training data for predicting PCs with ΔPD≥10 yielded a total of 51,146 distinct fingerprint features. Of these, 5178 were positive and 5432 negative. The relatively small size of subsets of positive or negative features compared to all features was attributable to inherent feature redundancy of fingerprints capturing layered atom environments. On the other hand, feature redundancy enabled the iterative generation of models with reduced feature sets retaining predictive power (i.e., eliminated features could be replaced with others).

14.4 Big Data in Molecular Promiscuity Analysis

14.4.1 The Concept of Big Data

The term "big data" refers to datasets, structured or unstructured, that multiply quickly and are so large and multifaceted that they are impossible to treat using personal computers and traditional computational approaches [155]. Datasets with big data require advanced tools such as heterogeneous and cloud computing [156] that have capabilities beyond those of conventional data processing and handling techniques as well as dynamic data curation and sharing using algorithms such as those used to handle data streams [157, 158]. These advanced techniques allow for rapid identification of target entities in these massive datasets in ways that manual data compilation and curation could never efficiently match, which has radical implications for the improvement of traditional computational toxicology modeling techniques like read-across [159, 160].

Thus, big data is a field that treats ways to analyze, systematically extract information from, or otherwise deal with datasets that are too large or complex to be dealt with by traditional data processing application software. Data with many fields (columns) offer greater statistical power, while data with higher complexity (more attributes or columns) may lead to a higher false discovery rate. Big data analysis challenges include capturing data, data storage, data analysis, search, sharing, transfer, visualization, querying, updating, information privacy, and data source. Big data was originally associated with three key concepts: volume, variety, and velocity. The analysis of big data presents challenges in sampling, and thus previously allowing for only observations and sampling. Therefore, big data often includes data with sizes that exceed the capacity of traditional software to process within an acceptable time and value.

Volume The amount of data matters. With big data, you will have to process high volumes of low-density, unstructured data. This can be data of unknown value, such as Twitter data feeds, clickstreams on a web page or a mobile app, or sensor-enabled equipment. For some organizations, this might be tens of terabytes of data. For others, it may be hundreds of petabytes.

Velocity Velocity is the fast rate at which data is received and (perhaps) acted on. Normally, the highest velocity of data streams directly into memory versus being written to disk. Some Internet-enabled smart products operate in real time or near real time and will require real-time evaluation and action.

Variety Variety refers to the many types of data that are available. Traditional data types were structured and fit neatly in a relational database. With the rise of big data, data comes in new unstructured data types. Unstructured and semistructured data types, such as text, audio, and video, require additional preprocessing to derive meaning and support metadata.

14.4.2 Big Data in Medicinal Chemistry

In science, big data has been on the agenda for at least a decade, in some fields more so than in others. In particle physics, colliders have produced massive amounts of data since the last decade [161, 162]. In biology, the big data wave hits more recently, due to the advent of efficient genomics technologies providing high throughput [161, 163]. Consequently, genomic sequencing data began to grow exponentially in 2008 [163]. Thus, biology big data is no news, and building and evolving computational infrastructures for handling and analyzing big data, especially genomic sequencing data, still is a major topic in bioinformatics. In physics and biology, big data is measured in "petabytes" (1 petabyte = 10^{15} bytes), and cloud computing has become an indispensable part of big data storage and analysis. Big data originating from biology has also entered drug discovery [161, 164, 165], where it is complemented with, for example, data from HTS, array experiments, imaging, pharmacology, and clinical investigations. These developments considerably challenge drug discovery environments and result in the need to explore new training concepts for discovery scientists, further emphasize interdisciplinary approaches, and add data science to the spectrum of drug discovery-relevant disciplines [165].

Compared to the situation in biology, big data is still in its infancy in medicinal chemistry, which is another pillar of drug discovery. Medicinal chemistry is experiencing the advent of the big data era, which biology already entered more than a decade ago, due to the availability of high-throughput genomics technologies. In medicinal chemistry, which is an integral part of drug discovery and traditionally a conservative scientific discipline, big data primarily comprise rapidly increasing numbers of compounds and volumes of associated activity data. Herein, we have discussed important big data criteria including data heterogeneity, complexity, and confidence that—in addition to mere data volumes—play a decisive role in utilizing big data and learning from it. The computational analysis of promiscuity of drugs and other bioactive compounds as the molecular basis of polypharmacology was selected as an instructive example of how big data influences and challenges research in medicinal chemistry. Clearly, the variety of data associated with cells and organisms is principally much larger than of data associated with chemical compounds—and so are the ensuing data volumes that can be generated. However, although big data trends are only beginning to emerge in medicinal chemistry [166, 167], it is evident that this field will also be increasingly influenced by big data issues. For example, proprietary medicinal chemistry projects in the pharmaceutical industry will inevitably need to take compound activity data into consideration that is rapidly accumulating in the public domain [166]. Merging internal and external data and viewing chemical optimization of compound series in an overarching context represents a departure from the long-established operating culture of medicinal chemistry and presents new challenges to practicing chemists. However, the opportunities provided by extracting knowledge from rapidly growing amounts of compounds and publicly available activity data cannot be disregarded. Big data issues in medicinal chemistry primarily focus on experimental data, which typically require computational analysis—but not on computationally generated data.

Big data in medicinal chemistry has also been associated with specific attributes. While stressing complexity, heterogeneity, and confidence as additional big data criteria of particular relevance for medicinal chemistry [165], for reasons detailed below, Lusher et al. [167] have promoted "5Vs" for medicinal chemistry by adding veracity and value to the original "3Vs" (volume, velocity, variety).

Veracity the "truth" of data. It can be called into question at varying confidence levels (and is often—but not always—also related to variability).

Value the value of data. It refers to the relevance medicinal chemistry.

In addition, data complexity, heterogeneity, and confidence are related to some of the "Vs" but are distinct from them.

Structure–activity relationships (SARs) can be extracted on a large scale from compounds active against current pharmaceutical targets and used to complement chemical optimization efforts. It is also possible to systematically explore multitarget activities of small molecules on the basis of available activity data. Such multitarget activities provide the foundation of polypharmacology [1, 14, 123, 168–170], an emerging concept in drug discovery, according to which many bioactive compounds and drugs elicit their physiological and therapeutic effects through interactions with multiple—rather than single—targets. The ability of small molecules to specifically interact with multiple targets is also rationalized as promiscuity [140], which needs to be clearly distinguished from nonspecific interactions and assay artifacts caused by aggregating molecules [38, 150] or (reactive) interference compounds [16, 81].

Another point of consideration is that promiscuity analysis typically does not take relative compound potencies into account. Even weak activities should be considered, which could give rise to side effects. Therefore, promiscuity must be differentiated from selectivity. A promiscuous compound might not be equally potent against all of its targets. Rather, it might display higher potency against one or more targets over others. Accordingly, a promiscuous compound might or might not display target selectivity. Interestingly, however, the majority of promiscuous compounds we have identified through data mining were found to display comparably high potency against their targets [171], whereas promiscuous compounds that were selective for a primary target over others (on the basis of high vs low potency values) were less frequently observed [171].

Furthermore, in the context of promiscuity analysis, the terms "frequent hitter" [172] and "privileged structure" [173], which are commonly used in medicinal chemistry, might also be considered. Frequent hitters refer to compounds that are often active in biological screens, which might either be due to promiscuity or undesired artifacts. Privileged structures refer to common core structures of compounds that are primarily active against a given target family over others, but not specific for individual members of the family, which would result in intrafamily promiscuity [161, 166].

Exploring compound promiscuity as the molecular basis of polypharmacology represents an interesting application for emerging big data in medicinal chemistry. This is the case because views concerning the potential magnitude of promiscuity and polypharmacology widely differ in the field and are often articulated on the basis of intuition or subjective expectations [170], without carefully considering available activity data [174]. With the advent of big data in medicinal chemistry, opportunities increase to arrive at quantitative estimates of promiscuity that are backed up by statistically significant data volumes.

While the practice of medicinal chemistry is just beginning to experience big data phenomena, it is evident that big data will play an increasingly important role going forward. More awareness of big data issues and potential caveats will still need to be raised to positively impact the field. For example, albeit rather obvious, opportunities provided by complementing in-house chemical optimization efforts with public domain SAR data are not yet widely appreciated, although they should be routinely considered. Incorporating external data—and knowledge derived from such data—into discovery projects and decision-making challenges the practice of medicinal chemistry. Big data characteristics further complicate matters. Promiscuity analysis serves as a good example. Currently, multitarget activities of small molecules and their potential utility or drawbacks for drug discovery are still only little understood. In particular, judging about the magnitude of promiscuity among drugs and candidate compounds is often much more subject to speculation and subjective views than rigorous scientific assessment. Albeit difficult to analyze, big data provides an attractive opportunity for a more detailed scientific assessment of molecular promiscuity in drug discovery.

14.4.3 Big Data in Chemical Toxicology

Recent HTS programs and their associated data sharing efforts have revolutionized the landscape in many health fields, highlighted by the Big Data to Knowledge (BD2K) initiative by the National Institutes of Health (NIH), which emphasizes the usefulness of big data in biomedical research and critical need to capitalize on the amount of data available in the health field [175, 176]. A significant HTS effort in toxicology is the US Environmental Protection Agency (US EPA) research program called Toxicity Forecaster (ToxCast), which employed in vitro HTS tests and toxicogenomics techniques to quickly evaluate the toxicity of compounds and prioritize compounds for experimental testing [28, 177, 178]. Phase I of this project evaluated 300 unique compounds, mostly of agricultural interest (i.e., pesticides), using about 500 HTS assays [177]. Phase II evaluated an additional 767 compounds, including some failed pharmaceutical compounds, using about 700 HTS assays [178]. Recently, the ToxCast initiative advanced to the Tox21 collaboration between the US EPA Office of Research and Development/National Center for Computational Toxicology (NCCT), the NIH/National Institute of Environmental Health Sciences (NIEHS)/National Toxicology Program (NTP), and the NIH/National Chemical Genomics Center (NCGC), now part of the National Center for Advancing Translational Sciences (NCATS) [179–182]. Phase I of Tox21 used 75 HTS assays, which were selected and refined from ToxCast assays, to screen an initial set of about 2800 compounds [179]. Phase II began in 2010 to screen a more extensive set of approximately 10,000 environmental compounds [179, 181, 182]. As of 2018, the Tox21 program generated over 120 million data points for approximately 8500 chemicals [180].

Publicly available databases store much of the data obtained from the toxicology community, including data from HTS programs such as the ToxCast and Tox21 programs [28, 177, 178, 183]. Among them, Aggregated Computational Toxicology Resource (ACToR) [184, 185], Registration, Evaluation, Authorisation and Restriction of Chemicals (REACH) [186–189], RepDose [190], Safety Evaluation Ultimately Replacing Animal Testing (SEURAT) [191], and Toxicology Data Network (ToxNET) [192] were specifically developed to share toxicity data. Chemical Effects in Biological Systems (CEBS) [193], ChEMBL [194], Connectivity Map [195, 196], Comparative Toxicogenomics Database (CTD) [197], DrugMatrix [198], Gene Expression Omnibus (GEO) [199, 200], and PubChem share general biological data, including toxicity data. Most of these data portals are being updated frequently, and the total number of available data is increasing quickly with the abovementioned HTS programs. In 2013, the total toxicity data pool contained over 70 million compounds and around 1 million assays [201]. From 2008 to 2018, the number of compounds in PubChem increased over threefold from 25.6 million [202] to 96.5 million [203]. Similarly, the number of bioassay records increased from approximately 1500 [202] to over 1 million [203]. In vitro assays often investigate mechanistically relevant information on toxicants. However, one or few assays cannot represent the complexity of whole organisms, and the results obtained from in vitro and in vivo tests always have obscure relationships, making in vitro–in vivo extrapolation (IVIVE) challenging [204].

Mechanism-driven modeling, initiated and advanced by the concept of the adverse outcome pathway (AOP), allows for mechanistic extrapolation of toxicity evaluations of new compounds, filling a critical need of applying alternatives for regulatory toxicology studies [205]. An AOP starts with a molecular feature (e.g., a chemical fragment), which indicates potential interactions with biomolecules such as receptors. This molecular initiating event (MIE) triggers a cascade of measurable key events at the cellular level, and leads to tissue, organ, and eventually in vivo organism level adverse outcomes [205]. The identification and organization of MIEs and key events in a pathway that leads to an adverse outcome define the associated toxicity mechanisms of interest for risk assessments. Mechanism-based assay outcomes can be used within this pathway to systematically assess whether a compound is likely to induce the target adverse outcome [206]. Currently, AOPs are being developed for various types of toxicities such as acute inhalation toxicity [207], neurotoxicity [208–210], skin sensitization [211–213], estrogen receptor bindings [214, 215], forestomach tumors not induced by genotoxic events [216], and drug-induced cholestatic liver injury [217].

Computational techniques and models based on chemical structures, such as structural alerts, also could advance traditional QSAR studies by predicting toxicity mechanisms for large datasets [206, 218]. For example, recently, there have been reports of modeling studies to predict MIEs relevant to hepatic steatosis [219, 220]. Mellor et al. evaluated binding interactions of 12,713 compounds in the ChEMBL database with nuclear receptor structure files in Protein Data Bank to develop a basic alert-based workflow to identify compounds that may bind to nuclear receptors and induce hepatic steatosis [219]. Another study by Gadaleta et al. involved the development of QSAR models to predict compound activity in ToxCast assays that are relevant to MIEs that lead to hepatic steatosis [220, 221].

References

1. Hopkins AL. Drug discovery: predicting promiscuity. Nature. 2009;462(7270):167–8.
2. Feldmann C, Miljković F, Yonchev D, et al. Identifying promiscuous compounds with activity against different target classes. Molecules. 2019;24(22):4185.
3. Gupta MN, Alam A, Hasnain SE. Protein promiscuity in drug discovery, drug-repurposing and antibiotic resistance. Biochimie. 2020;175:50–7.
4. Scheiber J, Chen B, Milik M, et al. Gaining insight into off-target mediated effects of drug candidates with a comprehensive systems chemical biology analysis. J Chem Inf Model. 2009;49:308–17.
5. Fosnocht D, Taylor JR, Caravati EM. Emergency department patient knowledge concerning acetaminophen (paracetamol) in over-the-counter and prescription analgesics. Emerg Med J. 2008;25:213–6.
6. Huang T, Cui W, Hu L, et al. Prediction of pharmacological and xenobiotic responses to drugs based on time course gene expression profiles. PLoS ONE. 2009;4(12):e8126.
7. Kola I, Landis J. Can the pharmaceutical industry reduce attrition rates? Nat Rev Drug Discov. 2004;3:711–6.
8. FitzGerald GA. Coxibs and cardiovascular disease. N Engl J Med. 2004;351:1709–11.
9. O. of R. Affairs, Primus Announces a Voluntary Nationwide Recall of All Lots Within Expiry of Prescription Medical Food Limbrel® Due to Rare But Serious and Reversible Adverse Events While Seeking FDA's Cooperation to Restore Access for Patients with Medical Necessity, U.S. Food and Drug Administration. 2019. http://www.fda.gov/safety/recalls-market-withdrawals-safety-alerts/primus-announces-voluntary-nationwide-recall-all-lots-within-expiry-prescription-medical-food. Accessed 10 Mar 2020.
10. Waring MJ, Arrowsmith J, Leach AR, et al. An analysis of the attrition of drug candidates from four major pharmaceutical companies. Nat Rev Drug Discov. 2015;14:475–86.
11. Thomas D, Clinical pharmacy education, practice and research: clinical pharmacy, drug information, pharmacovigilance, pharmacoeconomics and clinical research. Elsevier; 2018.
12. Hrdlicka M, Beranova I, Zamecnikova R, et al. Mirtazapine in the treatment of adolescent anorexia nervosa. Eur Child Adolesc Psychiatry. 2008;17:187–9.
13. Schatz SN, Weber RJ. Adverse drug reactions. In: Pharmacotherapy Self-Assessment Program (PSAP); 2015. p. 5–21.
14. Hopkins AL. Network pharmacology: the next paradigm in drug discovery. Nat Chem Biol. 2008;4:682–90.
15. Plant N. Can systems toxicology identify common biomarkers of non-genotoxic carcinogenesis? Toxicology. 2008;254:164–9.
16. Baell JB, Walters MA. Chemistry: chemical con artists foil drug discovery. Nature. 2014;513:481–3.
17. Aldrich C, Bertozzi C, Georg GI, et al. The ecstasy and agony of assay interference compounds. J Chem Inf Model. 2017;57:387–90.
18. Irwin JJ, Duan D, Torosyan H, et al. An aggregation advisor for ligand discovery. J Med Chem. 2015;58:7076–87.
19. Gaulton A, Hersey A, Nowotka M, et al. The ChEMBL database in 2017. Nucleic Acids Res. 2017;45(D1):D945–54.

20. Gilberg E, Gütschow M, Bajorath J. Promiscuous ligands from experimentally determined structures, binding conformations, and protein family-dependent interaction hotspots. ACS Omega. 2019;4(1):1729–37.
21. Sturm N, Desaphy J, Quinn RJ, et al. Structural insights into the molecular basis of the ligand promiscuity. J Chem Inf Model. 2012;52:2410–21.
22. Haupt VJ, Daminelli S, Schroeder M. Drug promiscuity in PDB: protein binding site similarity is key. PLoS One. 2013;8:e65894.
23. Pinzi L, Caporuscio F, Rastelli G. Selection of protein conformations for structure-based polypharmacology studies. Drug Discov Today. 2018;23:1889–96.
24. Feldmann C, Bajorath J. X-ray structure-based chemoinformatic analysis identifies promiscuous ligands binding to proteins from different classes with varying shapes. Int J Mol Sci. 2020;21(11):3782.
25. Bugrim A, Nikolskaya T, Nikolsky Y. Early prediction of drug metabolism and toxicity: systems biology approach and modeling. Drug Discov Today. 2004;9(3):127–35.
26. Wu Q, Taboureau O, Audouze K. Development of an adverse drug event network to predict drug toxicity. Curr Res Toxicol. 2020;1:48–55.
27. Stathias V, Turner J, Koleti A, et al. LINCS Data Portal 2.0: next generation access point for perturbation-response signatures. Nucleic Acids Res. 2020;48:D431–9.
28. Dix DJ, Houck KA, Martin MT, et al. The ToxCast program for prioritizing toxicity testing of environmental chemicals. Toxicol Sci. 2007;95:5–12.
29. Kuhn M, Letunic I, Jensen LJ, et al. The SIDER database of drugs and side effects. Nucleic Acids Res. 2016;44:D1075–9.
30. Taboureau O, Nielsen SK, Audouze K, et al. ChemProt: a disease chemical biology database. Nucleic Acids Res. 2011;39:D367–72.
31. Ciallella HL, Zhu H. Advancing computational toxicology in the big data era by artificial intelligence: data-driven and mechanism-driven modeling for chemical toxicity. Chem Res Toxicol. 2019;32:536–47.
32. Audouze K, Juncker AS, Roque FJ, et al. Deciphering diseases and biological targets for environmental chemicals using toxicogenomics networks. PLoS Comput Biol. 2010;6(5):e1000788.
33. Taboureau O, Audouze K. Human environmental disease network: a computational model to assess toxicology of contaminants. ALTEX. 2017;34:289–300.
34. Hodos RA, Kidd BA, Khader S, et al. Computational approaches to drug repurposing and pharmacology. Wiley Interdiscip Rev Syst Biol Med. 2016;8:186–210.
35. Peters JU, Schnider P, Mattei P, et al. Pharmacological promiscuity: dependence on compound properties and target specificity in a set of recent Roche compounds. ChemMedChem. 2009;4(4):680–6.
36. Peters JU, Hert J, Bissantz C, et al. Can we discover pharmacological promiscuity early in the drug discovery process? Drug Discov Today. 2012;17(7-8):325–35.
37. Klekota J, Brauner E, Roth FP, et al. Using high-throughput screening data to discriminate compounds with single-target effects from those with side effects. J Chem Inf Model. 2006;46(4):1549–62.
38. McGovern SL, Caselli E, Grigorieff N, et al. A common mechanism underlying promiscuous inhibitors from virtual and high-throughput screening. J Med Chem. 2002;45(8):1712–22.
39. Feng BY, Shelat A, Doman TN, et al. High-throughput assays for promiscuous inhibitors. Nat Chem Biol. 2005;1(3):146–8.
40. Uesawa Y. Quantitative structure–activity relationship analysis using deep learning based on a novel molecular image input technique. Bioorg Med Chem Lett. 2018;28:3400–3.
41. Mateen R, Ali MM, Hoare T. A printable hydro-gel microarray for drug screening avoids false positives associated with promiscuous aggregating inhibitors. Nat Commun. 2018;9(1):602.
42. Bloomingdale P, Housand C, Apgar JF, et al. Quantitative systems toxicology. Curr Opin Toxicol. 2017;4:79–87.
43. Ghasemi P-S, Mehri P-G. Neural network and deep-learning algorithms used in QSAR studies: merits and drawbacks. Drug Discov Today. 2018;23(10):1784–90.
44. Nantasenamat C, Isarankura-Na-Ayudhya C, Prachayasittikul V. Advances in computational methods to predict the biological activity of compounds. Expert Opin Drug Discov. 2010;5(7):633–54.
45. Yousefinejad S, Hemmateenejad B. Chemometrics tools in QSAR/QSPR studies: a historical perspective. Chemom Intell Lab Syst. 2015;149(B):177–204.
46. Perkins R, Fang H, Tong W, et al. Quantitative structure-activity relationship methods: perspectives on drug discovery and toxicology. Environ Toxicol Chem. 2003;22(8):1666–79.
47. Kwon S, Bae H, Jo J, et al. Comprehensive ensemble in QSAR prediction for drug discovery. BMC Bioinform. 2019;20(1):521.
48. Freyhult EK, Andersson K, Gustafsson MG. Structural modeling extends QSAR analysis of antibody-lysozyme interactions to 3D-QSAR. Biophysical J. 2003;84(4):2264–72.
49. Mauri A, Consonni V, Todeschini R. Molecular descriptors. In: Handbook of computational chemistry. Springer; 2017. p. 2065–93.
50. Zhu X, Kruhlak NL. Construction and analysis of a human hepatotoxicity database suitable for QSAR modeling using post-market safety data. Toxicology. 2014;321:62–72.
51. Atias N, Sharan R. An algorithmic framework for predicting side effects of drugs. J Comput Biol. 2011;18:207–18.
52. Hammann F, Gutmann H, Vogt N, et al. Prediction of adverse drug reactions using decision tree modeling. Clin Pharmacol Ther. 2010;88:52–9.

53. Azzaoui K, Hamon J, Faller B, et al. Modeling promiscuity based on in vitro safety pharmacology profiling data. ChemMedChem. 2007;2(6):874–80.
54. Dimova D, Hu Y, Bajorath J. Matched molecular pair analysis of small molecule microarray data identifies promiscuity cliffs and reveals molecular origins of extreme compound promiscuity. J Med Chem. 2012;55(22):10220–8.
55. Bender A, Scheiber J, Glick M, et al. Analysis of pharmacology data and the prediction of adverse drug reactions and off-target effects from chemical structure. ChemMedChem. 2007;2(6):861–73.
56. Ferreira LG, Dos Santos RN, Oliva G, et al. Molecular docking and structure-based drug design strategies. Molecules. 2015;20(7):13384–421.
57. Batool M, Ahmad B, Choi S. A Structure-based drug discovery paradigm. Int J Mol Sci. 2019;20(11):2783.
58. Hu Y, Bajorath J. Activity profile relationships between structurally similar promiscuous compounds. Eur J Med Chem. 2013;69:393–8.
59. Fukuzaki M, Seki M, Kashima H, et al. Side effect prediction using cooperative pathways. In: Proceedings of the IEEE international conference on bioinformatics and biomedicine (BIBM '09); 2009. p. 42–147.
60. Scheiber J, Jenkins JL, Sukuru SCK, et al. Mapping adverse drug reactions in chemical space. J Med Chem. 2009;52(9):3103–7.
61. Pauwels E, Stoven V, Yamanishi Y. Predicting drug side-effect profiles: a chemical fragment-based approach. BMC Bioinformatics. 2011;12:169.
62. Chen L, Huang T, Zhang J, et al. Predicting drugs side effects based on chemical-chemical interactions and protein-chemical interactions. Biomed Res Int. 2013;2013:485034.
63. Kanehisa M, Goto S. KEGG: Kyoto encyclopedia of genes and genomes. Nucleic Acids Res. 2000;28(1):27–30.
64. Kuhn M, von Mering C, Campillos M, et al. STITCH: interaction networks of chemicals and proteins. Nucleic Acids Res. 2008;36(1):D684–8.
65. Hu L, Chen C, Huang T, et al. Predicting biological functions of compounds based on chemical-chemical interactions. PLoS ONE. 2011;6(12):e29491.
66. Yamanishi Y, Araki M, Gutteridge A, et al. Prediction of drug-target interaction networks from the integration of chemical and genomic spaces. Bioinformatics. 2008;24(13):i232–40.
67. Chen L, He Z, Huang T, et al. Using compound similarity and functional domain composition for prediction of drug-target interaction networks. Medicinal Chem. 2010;6(6):388–95.
68. Chen L, Zeng W, Cai Y, et al. Predicting anatomical therapeutic chemical (ATC) classification of drugs by integrating chemical-chemical interactions and similarities. PLoS ONE. 2012;7(4):e35254.
69. Sharan R, Ulitsky I, Shamir R. Network-based prediction of protein function. Mol Syst Biol. 2007;3:88.
70. Bogdanov P, Singh AK. Molecular function prediction using neighborhood features. IEEE/ACM Trans Comput Biol Bioinform. 2010;7(2):208–17.
71. Kourmpetis YAI, van Dijk ADJ, Bink MCAM, et al. Bayesian Markov random field analysis for protein function prediction based on network data. PLoS ONE. 2010;5(2):e9293.
72. Hu L, Huang T, Shi X, et al. Predicting functions of proteins in mouse based on weighted protein-protein interaction network and protein hybrid properties. PLoS ONE. 2011;6(1):e14556.
73. Kuhn M, Campillos M, Letunic I, et al. A side effect resource to capture phenotypic effects of drugs. Mol Syst Biol. 2010;6:343.
74. Weininger DSMILES. a chemical language and information system. 1. Introduction to methodology and encoding rules. J Chem Inf Computer Sci. 1988;28:31–6.
75. Chen X, Liu X, Jia X, et al. Network characteristic analysis of ADR-related proteins and identification of ADR-ADR associations. Sci Rep. 2013;3:1–7.
76. Campillos M, Kuhn M, Gavin A-C, et al. Drug target identification using side-effect similarity. Science. 2008;321:263–6.
77. Oprea TI, Nielsen SK, Ursu O, et al. Associating drugs, targets and clinical outcomes into an integrated network affords a new platform for computer-aided drug repurposing. Mol Inform. 2011;30:100–11.
78. Hu Y, Bajorath J. Target family-directed exploration of scaffolds with different SAR profiles. J Chem Inf Model. 2011;51(12):3138–48.
79. Hu Y, Bajorath J. Polypharmacology directed compound data mining: identification of promiscuous chemotypes with different activity profiles and comparison to approved drugs. J Chem Inf Model. 2010;50(12):2112–8.
80. Yang JJ, Ursu O, Lipinski CA, et al. Badapple: promiscuity patterns from noisy evidence. J Cheminform. 2016;8:29.
81. Baell JB, Holloway GA. New substructure filters for removal of pan assay interference compounds (PAINS) from screening libraries and for their exclusion in bioassays. J Med Chem. 2010;53:2719–40.
82. Yang Y, Chen H, Nilsson I, et al. Investigation of the relationship between topology and selectivity for druglike molecules. J Med Chem. 2010;53:7709–14.
83. Peterson RT. Chemical biology and the limits of reductionism. Nat Chem Biol. 2008;4:635–8.
84. Lee JM, Gianchandani EP, Papin JA. Flux balance analysis in the era of metabolomics. Brief Bioinformatics. 2006;7:140–50.
85. McAdams HH, Shapiro L. Circuit simulation of genetic networks. Science. 1995;269:650–6.
86. Peleg M, Rubin D, Altman RB. Using Petri Net tools to study properties and dynamics of biological systems. J Am Med Inform Assoc. 2005;12:181–99.
87. Shannon P, Markiel A, Ozier O, et al. Cytoscape: a software environment for integrated models of biomolecular interaction networks. Genome Res. 2003;13:2498–504.

88. Clément-Ziza M, Malabat C, Weber C, et al. Genoscape: a cytoscape plug-in to automate the retrieval and integration of gene expression data and molecular networks. Bioinformatics. 2009;25(19):2617–8.
89. Brown JB, Okuno Y. Systems biology and systems chemistry: new directions for drug discovery. Chem Biol. 2012;19(1):23–8.
90. Schneider G, Fechner U. Computer-based de novo design of drug-like molecules. Nat Rev Drug Discov. 2005;4:649–63.
91. Jacoby E. Computational chemogenomics. Comput Mol Sci. 2011;1:57–67.
92. Maggiora GM. The reductionist paradox: Are the laws of chemistry and physics sufficient for the discovery of new drugs? J Comput Aided Mol Des. 2011;25:699–708.
93. Duran-Frigola M, Siragusa L, Ruppin E, et al. Detecting similar binding pockets to enable systems polypharmacology. PLoS Comput Biol. 2017;13(6):e1005522.
94. Siragusa L, Cross S, Baroni M, et al. BioGPS: navigating biological space to predict polypharmacology, off-targeting, and selectivity. Proteins. 2015;83(3):517–32.
95. Siragusa L, Luciani R, Borsari C, et al. Comparing drug images and repurposing drugs with BioGPS and FLAPdock: the thymidylate synthase case. ChemMedChem. 2016;11(15):1653–66.
96. Konc J, Janezic D. Binding site comparison for function prediction and pharmaceutical discovery. Curr Opin Struct Biol. 2014;25:34–9.
97. Xie L, Xie L, Bourne PE. Structure-based systems biology for analyzing off-target binding. Curr Opin Struct Biol. 2011;21(2):189–99.
98. Wong MT, Choi SB, Kuan CS, et al. Structural modeling and biochemical characterization of recombinant KPN_02809, a zinc-dependent metalloprotease from Klebsiella pneumoniae MGH 78578. Int J Mol Sci. 2012;13(1):901–17.
99. Lin H, Sassano MF, Roth BL, et al. A pharmacological organization of G protein-coupled receptors. Nat Methods. 2013;10(2):140–6.
100. Rubio-Perez C, Tamborero D, Schroeder MP, et al. In silico prescription of anticancer drugs to cohorts of 28 tumor types reveals targeting opportunities. Cancer Cell. 2015;27(3):382–96.
101. Yizhak K, Gaude E, Le Devedec S, et al. Phenotype-based cell-specific metabolic modeling reveals metabolic liabilities of cancer. Elife. 2014;3:e03641.
102. Tatonetti NP, Liu T, Altman RB. Predicting drug side-effects by chemical systems biology. Genome Biol. 2009;10(9):238.
103. Networking chemical biology. Nat Chem Biol. 2008;4:633.
104. Russell RB, Aloy P. Targeting and tinkering with interaction networks. Nat Chem Biol. 2008;4:666–73.
105. Bonneau R. Learning biological networks: from modules to dynamics. Nat Chem Biol. 2008;4:658–64.
106. Research highlights. Nat Chem Biol. 2008;4:657.
107. Enoksson M, Salvesen GS. Proteolytic needles in the cellular haystack. Nat Chem Biol. 2008;4:651–2.
108. Simon GM, Cravatt BF. Challenges for the 'chemical-systems' biologist. Nat Chem Biol. 2008;4:639–42.
109. Zamir E, Bastiaens PIH. Reverse engineering intracellular biochemical networks. Nat Chem Biol. 2008;4:643–7.
110. Lehár J, Stockwell BR, Giaever G, et al. Combination chemical genetics. Nat Chem Biol. 2008;4:674–81.
111. Seelig B. An autocatalytic network for ribozyme self-construction. Nat Chem Biol. 2008;4:654–5.
112. Cipriano A, Sbardella G, Ciulli A. Targeting epigenetic reader domains by chemical biology. Curr Opin Chem Biol. 2020;57:82–94.
113. Apsel B, Blair JA, Gonzalez B, et al. Targeted polypharmacology: discovery of dual inhibitors of tyrosine and phosphoinositide kinases. Nat Chem Biol. 2008;4:691–9.
114. Bilanges B, Torbett N, Vanhaesebroeck B. Killing two kinase families with one stone. Nat Chem Biol. 2008;4:648–9.
115. Yabuuchi H, Niijima S, Takematsu H, et al. Analysis of multiple compound-protein interactions reveals novel bioactive molecules. Mol Syst Biol. 2011;7:472.
116. Dobson CM. Chemical space and biology. Nature. 2004;432:824–8.
117. Lipinski C, Hopkins A. Navigating chemical space for biology and medicine. Nature. 2004;432:855–61.
118. Renner S, van Otterlo WA, Dominguez Seoane M, et al. Bioactivity-guided mapping and navigation of chemical space. Nat Chem Biol. 2009;5:585–92.
119. Wang Y, Xiao J, Suzek TO, et al. PubChem: a public information system for analyzing bioactivities of small molecules. Nucleic Acids Res. 2009;37:W623–33.
120. Eckert H, Bajorath J. Molecular similarity analysis in virtual screening: foundations, limitations and novel approaches. Drug Discov Today. 2007;12:225–3.
121. Young DW, Bender A, Hoyt J, et al. Integrating high-content screening and ligand-target prediction to identify mechanism of action. Nat Chem Biol. 2008;4:59–68.
122. MacDonald ML, Lamerdin J, Owens S, et al. Identifying off-target effects and hidden phenotypes of drugs in human cells. Nat Chem Biol. 2006;2:329–33.
123. Paolini GV, Shapland RH, van Hoorn WP, et al. Global mapping of pharmacological space. Nat Biotechnol. 2006;24:805–15.
124. Oprea TI, Tropsha A, Faulon JL, et al. Systems chemical biology. Nat Chem Biol. 2007;3:447–50.
125. Keiser MJ, Setola V, Irwin JJ, et al. Predicting new molecular targets for known drugs. Nature. 2009;462:175–81.
126. Oprea TI, Matter H. Integrating virtual screening in lead discovery. Curr Opin Chem Biol. 2004;8:349–58.

127. Muegge I, Oloff S. Advances in virtual screening. Drug Discov Today Technol. 2006;3:405–11.
128. McInnes C. Virtual screening strategies in drug discovery. Curr Opin Chem Biol. 2007;11:494–502.
129. Xie L, Li J, Xie L, Bourne PE. Drug discovery using chemical systems biology: identification of the protein-ligand binding network to explain the side effects of CETP inhibitors. PLoS Comput Biol. 2009;5:e1000387.
130. Hert J, Keiser MJ, Irwin JJ, et al. Quantifying the relationships among drug classes. J Chem Inf Model. 2008;48:755–65.
131. Pathguide: the pathway resource list http://www.pathguide.org
132. Blow N. Systems biology: untangling the protein web. Nature. 2009;460:415–8.
133. Hansen N, Brunak S, Altman R. Generating genome-scale candidate gene lists for pharmacogenomics. Clin Pharmacol Ther. 2009;86:183–9.
134. Dimova D, Gilberg E, Bajorath J. Identification and analysis of promiscuity cliffs formed by bioactive compounds and experimental implications. RSC Adv. 2017;7(1):58–66.
135. Dimova D, Bajorath J. Rationalizing promiscuity cliffs. ChemMedChem. 2018;13(6):490–4.
136. Miljković F, Vogt M, Bajorath J. Systematic computational identification of promiscuity cliff pathways formed by inhibitors of the human kinome. J Comput Aided Mol Des. 2019;33(6):559–72.
137. Miljković F, Bajorath J. Data structures for compound promiscuity analysis: promiscuity cliffs, pathways and promiscuity hubs formed by inhibitors of the human kinome. Future Sci OA. 2019;5(7):FSO404.
138. Hu H, Bajorath J. Exploring structure-promiscuity relationships using dual-site promiscuity cliffs and corresponding single-site analogs. Bioorg Med Chem. 2020;28(1):115238.
139. Peltason L, Hu Y, Bajorath J. From structure-activity to structure-selectivity relationships: quantitative assessment, selectivity cliffs, and key compounds. ChemMedChem. 2009;4(11):1864–73.
140. Hu Y, Bajorath J. Compound promiscuity: what can we learn from current data? Drug Discov Today. 2013;18(13–14):644–50.
141. Hu Y, Bajorath J. Systematic assessment of molecular selectivity at the level of targets, bioactive compounds, and structural analogues. ChemMedChem. 2016;11(12):1362–70.
142. Maggiora GM. On outliers and activity cliffs--why QSAR often disappoints. J Chem Inf Model. 2006;46(4):1535.
143. Stumpfe D, Bajorath J. Exploring activity cliffs in medicinal chemistry. J Med Chem. 2012;55(7):2932–42.
144. Stumpfe D, Hu Y, Dimova D, et al. Recent progress in understanding activity cliffs and their utility in medicinal chemistry. J Med Chem. 2014;57(1):18–28.
145. Blaschke T, Feldmann C, Bajorath J. Prediction of promiscuity cliffs using machine learning. Mol Inform. 2021;40(1):e2000196.
146. Kenny PW, Sadowski J. Chemoinformatics in drug discovery. Weinheim: Wiley-VCH; 2004. p. 271–85.
147. Hussain J, Rea C. Computationally efficient algorithm to identify matched molecular pairs (MMPs) in large data sets. J Chem Inf Model. 2010;50(3):339–48.
148. Hu Y, Hu Y, Vogt M, et al. MMP-Cliffs: systematic identification of activity cliffs on the basis of matched molecular pairs. J Chem Inf Model. 2012;52(5):1138–45.
149. Wang Y, Suzek T, Zhang J, et al. PubChem BioAssay: 2014 update. Nucleic Acids Res. 2014;42(Database issue):D1075–82.
150. Shoichet BK. Screening in a spirit haunted world. Drug Discov Today. 2006;11(13-14):607–15.
151. Hu Y, Jasial S, Gilberg E. Structure-promiscuity relationship puzzles-extensively assayed analogs with large differences in target annotations. AAPS J. 2017;19(3):856–64.
152. Blaschke T, Miljković F, Bajorath J. Prediction of different classes of promiscuous and nonpromiscuous compounds using machine learning and nearest neighbor analysis. ACS Omega. 2019;4:6883–90.
153. Hu Y, Bajorath J. Extending the activity cliff concept: structural categorization of activity cliffs and systematic identification of different types of cliffs in the ChEMBL database. J Chem Inf Model. 2012;52(7):1806–11.
154. Heikamp K, Hu X, Yan A, et al. Chem Inf Model. 2012;52:2354–65.
155. Gandomi A, Haider M. Beyond the hype: big data concepts, methods, and analytics. Int J Inf Manage. 2015;35(2):137–44.
156. Schadt EE, Linderman MD, Sorenson J, et al. Cloud and heterogeneous computing solutions exist today for the emerging big data problems in biology. Nat Rev Genet. 2011;12:224.
157. Liu W, Schmidt B, Voss G, et al. Streaming algorithms for biological sequence alignment on GPUs. IEEE Trans Parallel Distrib Syst. 2007;18(9):1270–81.
158. Charikar M, O'Callaghan L, Panigrahy R. Better streaming algorithms for clustering problems. Proc thirty-fifth ACM Symp Theory Comput – STOC. 2003;03:30–9.
159. Zhu H, Bouhifd M, Donley E, et al. Supporting read-across using biological data. ALTEX. 2016;33(2):167–82.
160. Hartung T. Making big sense from big data in toxicology by read-across. ALTEX. 2016;33(2):83–93.
161. Hu Y, Bajorath J. Entering the 'big data' era in medicinal chemistry: molecular promiscuity analysis revisited. Future Sci OA. 2017;3:FSO179.
162. Lynch C. Big data: how do your data grow? Nature. 2008;455(7209):28–9.
163. Marx V. The big challenges of big data. Nature. 2013;498(7453):255–60.

164. Al-Lazikani B, Workman P. Minimizing bias in target selection by exploiting multidisciplinary Big Data and the protein interactome. Future Med Chem. 2016;8(14):1711–6.
165. Bajorath J, Jenkins J, Overington J, et al. Drug discovery and development in the era of big data. Future Med Chem. 2016;8(15):1807–13.
166. Hu Y, Bajorath J. Learning from 'big data': compounds and targets. Drug Discov Today. 2014;19(4):357–60.
167. Lusher SJ, McGuire R, van Schaik RC, et al. Data-driven medicinal chemistry in the era of big data. Drug Discov Today. 2014;19(7):859–68.
168. Boran AD, Iyengar R. Systems approaches to polypharmacology and drug discovery. Curr Opin Drug Discov Devel. 2010;13(3):297–309.
169. Jalencas X, Mestres J. On the origins of drug polypharmacology. Med Chem Commun. 2013;4(1):80–7.
170. Anighoro A, Bajorath J, Rastelli G. Polypharmacology: challenges and opportunities in drug discovery: miniperspective. J Med Chem. 2014;57(19):7874–87.
171. Hu Y, Bajorath J. Promiscuity profiles of bioactive compounds: potency range and difference distributions and the relation to target numbers and families. Med Chem Commun. 2013;4:1196–201.
172. Schneider G, Neidhart W, Giller T, et al. "Scaffold-hopping" by topological pharmacophore search: a contribution to virtual screening. Angew Chem Int Ed Engl. 1999;38(19):2894–6.
173. Müller G. Medicinal chemistry of target family-directed masterkeys. Drug Discov Today. 2003;8(15):681–91.
174. Hu Y, Bajorath J. How promiscuous are pharmaceutically relevant compounds? A data-driven assessment. AAPS J. 2013;15(1):104–11.
175. National Institutes of Health. Big Data to Knowledge; National Institutes of Health. 2018. https://commonfund.nih.gov/bd2k. Accessed 10 Nov 2018.
176. Margolis R, Derr L, Dunn M, et al. The National Institutes of Health's Big Data to Knowledge (BD2K) initiative: capitalizing on biomedical big data. J Am Med Informatics Assoc. 2014;21(6):957–8.
177. Judson RS, Houck KA, Kavlock RJ, et al. In Vitro screening of environmental chemicals for targeted testing prioritization: the ToxCast project. Environ Health Perspect. 2010;118(4):485–92.
178. Kavlock R, Chandler K, Houck K, et al. Update on EPA's ToxCast program: providing high throughput decision support tools for chemical risk management. Chem Res Toxicol. 2012;25(7):1287–302.
179. Attene-Ramos MS, Miller N, Huang R, et al. The Tox21 robotic platform for the assessment of environ-mental chemicals – from vision to reality. Drug Discov Today. 2013;18(15–16):716–23.
180. Thomas RS, Paules RS, Simeonov A, et al. The US federal Tox21 program: a strategic and operational plan for continued leadership. ALTEX. 2018;35(2):163–8.
181. Shukla SJ, Huang R, Austin CP, et al. The future of toxicity testing: a focus on in vitro methods using a quantitative high-throughput screening platform. Drug Discov Today. 2010;15(23–24):997–1007.
182. Hsu C-W, Huang R, Attene-Ramos MS, et al. Advances in high-throughput screening technology for toxicology. Int J Risk Assess Manage. 2017;20:109.
183. Zhao L, Zhu H. Big data in computational toxicology: challenges and opportunities. In: Ekins S, editor. Computational toxicology. Hoboken: Wiley; 2018. p. 291–312.
184. Judson RS, Martin MT, Egeghy P, et al. Aggregating data for computational toxicology applications: the U.S. Environmental Protection Agency (EPA) Aggregated Computational Toxicology Resource (ACToR) system. Int J Mol Sci. 2012;13(2):1805–31.
185. Judson R, Richard A, Dix D, et al. ACToR – aggregated computational toxicology resource. Toxicol Appl Pharmacol. 2008;233(1):7–13.
186. Luechtefeld T, Maertens A, Russo DP, et al. Analysis of publically available skin sensitization data from REACH registrations 2008–2014. ALTEX. 2016;33(2):135–48.
187. Luechtefeld T, Maertens A, Russo DP, et al. Analysis of public oral toxicity data from REACH registrations 2008–2014. ALTEX. 2016;33(2):111–22.
188. Luechtefeld T, Maertens A, Russo DP, et al. Global analysis of publicly available safety data for 9,801 substances registered under REACH from 2008–2014. ALTEX. 2016;33(2):95–109.
189. Luechtefeld T, Maertens A, Russo DP, et al. Analysis of draize eye irritation testing and its prediction by mining publicly available 2008–2014 REACH data. ALTEX. 2016;33(2):123–34.
190. Bitsch A, Jacobi S, Melber C, et al. REPDOSE: a database on repeated dose toxicity studies of commercial chemicals – a multifunctional tool. Regul Toxicol Pharmacol. 2006;46(3):202–10.
191. Vinken M, Pauwels M, Ates G, et al. Screening of repeated dose toxicity data present in SCC(NF)P/SCCS safety evaluations of cosmetic ingredients. Arch Toxicol. 2012;86(3):405–12.
192. Fonger GC, Stroup D, Thomas PL, et al. TOXNET: a computerized collection of toxicological and environ-mental health information. Toxicol Ind Health. 2000;16(1):4–6.
193. Lea IA, Gong H, Paleja A, et al. CEBS: a comprehensive annotated database of toxicological data. Nucleic Acids Res. 2017;45(D1):D964–71.
194. Gaulton A, Bellis LJ, Bento AP, et al. ChEMBL: a large-scale bioactivity database for drug discovery. Nucleic Acids Res. 2012;40(D1):1100–7.
195. Subramanian A, Narayan R, Corsello SM, et al. A next generation connectivity map: L1000 platform and the first 1,000,000 profiles. Cell. 2017;171(6):1437–52.
196. Lamb J, Crawford ED, Peck D, et al. The connectivity map: using gene-expression signatures to con-

nect small molecules, genes, and disease. Science. 2006;313(5795):1929–35.
197. Davis AP, Grondin CJ, Johnson RJ, et al. The comparative toxicogenomics database: update 2017. Nucleic Acids Res. 2017;45(D1):D972–8.
198. Ganter B, Tugendreich S, Pearson CI, et al. Development of a large-scale chemogenomics database to improve drug candidate selection and to understand mechanisms of chemical toxicity and action. J Biotechnol. 2005;119(3):219–44.
199. Barrett T, Wilhite SE, Ledoux P, et al. NCBI GEO: archive for functional genomics data sets – update. Nucleic Acids Res. 2012;41(D1):D991–5.
200. Edgar R, Domrachev M, Lash AE. Gene expression omnibus: NCBI gene expression and hybridization array data repository. Nucleic Acids Res. 2002;30(1):207–10.
201. Sayers EW, Barrett T, Benson DA, et al. Database resources of the National Center for Biotechnology Information. Nucleic Acids Res. 2009;37(D1):D5–15.
202. Sayers EW, Barrett T, Benson DA, et al. Database resources of the National Center for Biotechnology Information. Nucleic Acids Res. 2010;38(D1):D5–16.
203. Sayers EW, Agarwala R, Bolton EE, et al. Database resources of the National Center for Biotechnology Information. Nucleic Acids Res. 2019;47(D1):D23–8.
204. Yoon M, Campbell JL, Andersen ME, et al. Quantitative in vitro to in vivo extrapolation of cell-based toxicity assay results. Crit Rev Toxicol. 2012;42(8):633–52.
205. Ankley GT, Bennett RS, Erickson RJ, et al. Adverse outcome pathways: a conceptual framework to support ecotoxicology research and risk assessment. Environ Toxicol Chem. 2010;29(3):730–41.
206. Patlewicz G, Simon TW, Rowlands JC, et al. Proposing a scientific confidence framework to help support the application of adverse outcome pathways for regulatory purposes. Regul Toxicol Pharmacol. 2015;71(3):463–77.
207. Clippinger AJ, Allen D, Behrsing H, et al. Pathway-based predictive approaches for non-animal assessment of acute inhalation toxicity. Toxicol In Vitro. 2018;52:131–45.
208. Bal-Price A, Lein PJ, Keil KP, et al. Developing and applying the adverse outcome pathway concept for understanding and predicting neurotoxicity. NeuroToxicology. 2017;59:240–55.
209. Bal-Price A, Meek ME. Adverse outcome pathways: application to enhance mechanistic understanding of neurotoxicity. Pharmacol Ther. 2017;179:84–95.
210. Sachana M, Rolaki A, Bal-Price A. Development of the adverse outcome pathway (AOP): Chronic binding of antagonist to N-methyl-D-aspartate receptors (NMDARs) during brain development induces impairment of learning and memory abilities of children. Toxicol Appl Pharmacol. 2018;354:153–75.
211. Maxwell G, MacKay C, Cubberley R, et al. Applying the skin sensitization adverse outcome pathway (AOP) to quantitative risk assessment. Toxicol In Vitro. 2014;28(1):8–12.
212. Patlewicz G, Kuseva C, Kesova A, et al. Towards AOP application – Implementation of an Integrated Approach to Testing and Assessment (IATA) into a pipeline tool for skin sensitization. Regul Toxicol Pharmacol. 2014;69(3):529–45.
213. Organization for Economic Co-Operation and Development. The adverse outcome pathway for skin sensitisation initiated by covalent binding to proteins part 1: scientific evidence. Ser Test Assess 2012; 168.
214. Browne P, Noyes PD, Casey WM, et al. Application of adverse outcome pathways to U.S. EPA's endocrine disruptor screening program. Environ Health Perspect. 2017;125(9):096001.
215. Benigni R, Battistelli CL, Bossa C, et al. Endocrine disruptors: data-based survey of in vivo tests, predictive models and the adverse outcome pathway. Regul Toxicol Pharmacol. 2017;86:18–24.
216. Proctor DM, Suh M, Chappell G, et al. An adverse outcome pathway (AOP) for forestomach tumors induced by non-genotoxic initiating events. Regul Toxicol Pharmacol. 2018;96:30–40.
217. Vinken M, Landesmann B, Goumenou M, et al. Development of an adverse outcome pathway from drug-mediated bile salt export pump inhibition to cholestatic liver injury. Toxicol Sci. 2013;136(1):97–106.
218. Xia M, Huang R, Shi Q, et al. Comprehensive analyses and prioritization of tox21 10k chemicals affecting mitochondrial function by in-depth mechanistic studies. Environ Health Perspect. 2018;126(7):077010.
219. Mellor CL, Steinmetz FP, Cronin MTD. Using molecular initiating events to develop a structural alert based screening workflow for nuclear receptor ligands associated with hepatic steatosis. Chem Res Toxicol. 2016;29(2):203–12.
220. Gadaleta D, Manganelli S, Roncaglioni A, et al. QSAR modeling of ToxCast assays relevant to the molecular initiating events of AOPs leading to hepatic steatosis. J Chem Inf Model. 2018;58(8):1501–17.
221. Frid AA, Matthews EJ. Prediction of drug-related cardiac adverse effects in humans-B: use of QSAR programs for early detection of drug-induced cardiac toxicities. Regul Toxicol Pharmacol. 2010;56:276–89.

15 Polypharmacology and Natural Products

Abstract

Natural products (NPs) play an important role in drug discovery. No less than 50% of the Food and Drug Administration (FDA)-approved drugs were NPs or NPs derivatives. More importantly, NPs have become an indispensable component of polypharmacology and a constant source of bioactive compounds or a golden mine of multitarget drugs (MTD drugs) because of their multitargeting property: recent studies have unequivocally demonstrated that FDA-approved, clinically tested, and experimentally investigational NPs commonly act on multiple molecular targets (e.g., proteins, DNAs, RNAs). With diverse chemical structures, NPs and their derivatives contribute greatly to the landscape of new chemical entities (NCEs) for de novo drug discovery and development, as well as drug repositioning, in the past, at present, and in the future. This chapter will first explain the relationship between NPs and polypharmacology, which will be followed by an introduction to the relationships between systems pharmacology/network pharmacology and NPs polypharmacology. Then, the current applications of NPs as MTDs to cancer therapy will be summarized. Finally, resources and tools for exploiting NPs polypharmacology will be discussed.

Keywords

Natural products (NPs) · Systems pharmacology · Network polypharmacology · New chemical entities (NCEs) · Anticancer therapy

15.1 Natural Products and Polypharmacology

Natural products (NPs) play an important role in drug discovery [1–5]. No less than 50% of the Food and Drug Administration (FDA)-approved drugs were NPs or NPs derivatives [6, 7]. More importantly, NPs have become an indispensable component of polypharmacology and a constant source of bioactive compounds or a golden mine of multitarget drugs (MTD drugs) because of their multitargeting property [8–13]: recent studies have unequivocally demonstrated that FDA-approved, clinically tested, and experimentally investigational NPs commonly act on multiple molecular targets (e.g., proteins, DNAs, RNAs) [1, 4, 8]. With diverse chemical structures, NPs and their derivatives contribute greatly to the landscape of new chemical entities (NCEs) for de novo drug discovery and development, as well as drug repositioning, in the past, at present, and in the future [14–17].

Z. Wang, B. Yang, *Polypharmacology*, https://doi.org/10.1007/978-3-031-04998-9_15

Gu et al. [10] reported biological activities and virtual screening results on 197,201 NPs structures. Through principal component analysis, they found that there was a large portion of overlap between NPs and FDA-approved drugs in the chemical space, indicating that NPs offer a large quantity of potential lead compounds. The network properties of NP–target networks revealed that polypharmacology is greatly enriched in those compounds with a large degree and high betweenness of centrality. High-throughput virtual screening (HTVS) unraveled that all 19,7201 NPs are docked to 332 target proteins of FDA-approved drugs. These investigators predicted the NPs with greatest potential for drug discovery and their indications based on a docking score-weighted prediction model. Their analyses suggest that NPs have vast chemical diversity, have good drug-like properties, and can interact with multiple cellular target proteins.

Moreover, NPs have special selectivity to cellular targets [18, 19]. Biologically active NPs can provide selective ligands for disease-related targets [18, 20], and influence the disease-related pathways and eventually shift the biological network from disease status to the healthy status.

As pointed out by Mayr et al. [21], NPs are remarkable in many regards, particularly for being the main source of drugs in the past and, today, by serving as a source for innovative leads [22]. NPs bear privileged structural features that were "shaped" by evolution, yielding compounds that can serve as promising starting points for drug development [3, 23]. Further, NPs often show polypharmacological properties, interacting with more than one target [24]. NPs are more likely to act as true polypharmacological agents rather than unspecific binders—a property instead associated with synthetic compounds [25, 26].

In the next two subsections, approaches used for discovery of NPs as multitarget drugs and some specific examples will be introduced.

15.2 Systems Pharmacology and Network Pharmacology in NPs Polypharmacology

As described in Chap. 1 (Basics of Polypharmacology), systems pharmacology and network pharmacology with the holistic and integrative view and approaches are the core of polypharmacology.

Notwithstanding the recent advances in omics, the data collection of NPs is largely incomplete. Firstly, the inventory of NPs remains incomplete and new chemical structures are being discovered [18]. Secondly, researchers explored only a small part of biological functions of NPs. Thirdly, there were mistakes and errors in existing data. Many chemical structures of NPs are questionable. Data of biological activity obtained from different laboratories for one compound would vary greatly. While no adequate data is available, a good and useful complement is virtual screening results. Last but not least, more research methods both experimental and computational to afford more overall and more accurate data are needed urgently. We are extending the computational targets to all proteins if it has protein–ligand complex structure.

With the development of large-scale network analysis, researchers have recently begun to explore the action mechanism of bioactive compounds in the context of biological networks, e.g., drug–target network (DTN) [12, 19, 20], protein–protein interaction network [27], metabolic network [28, 29], and disease pathway [30]. However, most studies focused on few molecules. NPs possess vast chemical diversity and so have enormous potential to find various different kinds of bioactive molecules [31]. Researchers have done statistics and analysis for natural products in several aspects, such as chemical diversity [31–34], property distribution [35], molecular scaffold [36–38], chemical space [39, 40], and comparison between NPs and other compound collections [38, 41, 42].

In 2002, Gu et al. [10] established a 3D structure database of components from Chinese traditional medicinal herbs (CHDD) [43]. One year later, they constructed the Universal Natural Products Database (UNPD) to facilitate HTVS for NPs, and the database comprised nearly 200,000 NPs from plants, animals, and microorganisms. UNPD is the largest noncommercial and freely available database for NPs (http://pkuxxj.pku.edu.cn/UNPD). However, less than 2% known NPs has been reported with their biological activities to date, and there is a need for elucidating and clarifying a large quantity of binding data between NPs and target proteins. Their analysis using AutoDock4 revealed that all NPs collected in the UNPD can be docked to 332 target proteins with their protein–ligand complex structures provided in the RCSB protein data bank of FDA-approved drugs and screened according to docking score [10].

UNPD contained more than 65 million of docked conformations of NPs and FDA-approved drugs. Although the potential binding of NPs in cavities may be different from the binding site of drugs, most proteins had limited binding sites. In most cases, the binding sites of NPs and drugs were identical. Generally, the hit rate of HTVS is about 35% [44]. In the study by Gu et al. [10], the number of NPs with docking score was greater than 7, and the score of original ligands of complex structure of the target protein was 62,918, accounting for 32% of total compound. Their results also revealed that each of NPs can act on an average of 2.14 target proteins on one hand, and each target protein contains an average of 25 hits (NPs) on the other hand. After extracting the target-related diseases from the Therapeutic Targets Database [45], the investigators constructed a docking score-weighted prediction model to predict the possibility of an NP to treat diseases [10]. In this way, they identified UNPD194973 and UNPD43323 with great potential as drugs for antibacterial and anticancer therapy, respectively.

Given the huge number of NPs collected in the UNPD, it is believed that traditional approaches may not be applicable to or practical for de novo drug discovery and/or drug repositioning from NPs, and systems and network pharmacology might be a better alternative for the purpose. Indeed, search for pharmacological targets for new or known NPs using modern computer-aided methods is a current endeavor. Yet, reliable pipelines for the large-scale target prediction of natural products are still rare. To this end, Fang et al. [46] conducted systematic study to identify drug–target interactions (DTI) and constructed a DTN for NPs at the human proteome and established a systems pharmacology platform for in silico identification of drug targets and anticancer indications of NPs [9, 47, 48]. They also reconstructed a global drug–target network of natural products and built the predictive network models via their previous network-based approach, named bSDTNBI [9, 46]. They further developed the statistical network models for identification of new anticancer indications of natural products through integration of both experimentally validated and computationally predicted DTIs with the known cancer proteins. In this way, they computationally identified multiple anticancer indications for several typical NPs with new mechanism of actions (MOAs) across 13 cancer types [46]. Specifically, they predicted 42,225 new DTIs between 2388 unique NPs and 680 targets using top 20 as a cutoff. Their DTI bipartite network contains 124 DTIs (76 known DTI pairs and 48 predicted DTI pairs) connecting 3 natural products and 80 targets (48 cancer proteins and 32 non-cancer proteins). We systematically searched previously reported literatures in PubMed for the 48 predicted DTI pairs to kaempferol, resveratrol, and genistein. Among the 48 predicted DTI pairs, 19 (19/48, 39.6% success rate) had been experimentally validated in previously published data, suggesting a reasonable accuracy of our network model. The remaining 29 predicted DTIs are needed to be validated by experimental assays in the future. To examine how the network predictive model helps to identify new drug targets of natural products, they selected three typical FDA-approved or investigational natural products (kaempferol, resveratrol, and genistein) as example drugs.

One advantage of the DTIs strategy is that it does not rely on either 3D structures of target proteins or negative samples [49–57]. For example, Cheng et al. [49] proposed several network-based approaches, such as network-based inference (NBI) and the edge-weighted/node-weighted NBI [50], for prediction of DTIs and drug repositioning. However, neither NBI nor the weighted NBIs can predict targets of new chemical entities (NCEs) whose targets are absent in the known drug–target network. To overcome this drawback, they further proposed a novel network-based approach, named substructure–drug–target network-based inference (SDTNBI) [53], to identify potential targets for NCEs via integrating molecular fingerprints of drugs into drug–target network. As an improvement of SDTNBI, they further developed a balanced substructure–drug–target network-based inference (bSDTNBI) for identification of targets for both old drugs and NCEs by introducing three tunable parameters to balance the weights of each nodes and edges in networks. Via bSDTNBI, they identified several novel antagonistic or agonistic estrogen receptor alpha (ERα) ligands, and dual-effect ERα ligands by in vitro assays for the development of potential therapies in breast cancer or osteoporosis [54]. Therefore, systematic identification of new target spaces of natural products via network-based approaches would provide unexpected opportunities for drug discovery and development by exploiting the pharmaceutical wealth of natural products.

The integration of chemoinformatic and bioinformatic tools into drug discovery in the early 1990s and the recent advances in big data handling have leveraged access to a myriad of massive public datasets [21, 58, 59] and powerful tools, e.g., HTVS [60]. In the past decade, the concept of drug repurposing has emerged as an attractive strategy to rededicate approved drugs or partially developed compounds to new molecular targets [61, 62]. This development is, next to the intention to reduce R&D costs, also owed to advances in computational chemistry [21, 63]. Many such tools have been made public in the past decade [64–66], aiming to boost both drug repurposing efforts and drug discovery as a whole.

Based on this concept, tools or workflows that allow for the accurate prediction of new molecular targets for natural products are of great interest. We here propose a workflow to specifically search for yet unreported protein targets of known compounds, using a combination of in silico and in vitro methods. The compounds of interest are virtually screened against our in-house resources, representing a panel of 39 drug targets expressed as 387 pharmacophore models. Additionally, the compounds are subjected to three independent open-access target prediction servers. The results generated by these four diverse methods are combined, and those with the highest consensus (protein targets predicted by several methods independently of each other) are selected for in vitro evaluation. Moreover, already known protein targets are excluded from further investigations by checking their appearance in PubChem, a pertinent open knowledge base for bioactivities [67].

15.3 NPs as MTDs in Cancer Therapy

According to the work presented by Fang et al. [9], systematic identification of polypharmacology of NPs provides exceptional opportunities for the development of targeted cancer therapies (e.g., drug repurposing) in two conceptual aspects: (i) NPs exert anticancer activities via directly targeting various oncogenic pathways by their polypharmacology, and (ii) NPs act as immunomodulatory or inflammatory agents by modulating the tumor microenvironments (TMEs) [68]. For example, immune cells in the TME, such as regulatory T cells or B cells, and tumor-associated macrophages (TAMs), contribute to tumor progression and evasion of immune surveillance by expressing several immunosuppressive factors, such as interleukin-6, interleukin-10, and transforming growth factor-β [69]. Many natural products targeting immune system-mediated proteins as well as immunomodulators are in preclinical or clinical studies, suggesting a promising strategy for prevention and treatment of cancer [70, 71].

Below are some examples of anticancer NPs identified by systems pharmacology approaches and verified by experimental and clinical studies.

15.3.1 Flavonoids

15.3.1.1 Kaempferol

Kaempferol is a natural flavonol mainly found in plants and plant-derived foods such as apples [21, 72]. It possesses anti-breast cancer [73] and anti-ovarian cancer activities [74]. Kaempferol interacts with 39 known (20) and predicted (19) targets (including 23 cancer proteins as well as 16 non-cancer proteins). Oncoproteins MAPK10 and CBR1 were predicted as molecular targets of kaempferol. Experimental studies have also demonstrated kaempferol as a weak inhibitor of MAPK10 (19.1 μM) [75] as well as a strong inhibitor of CBR1 (0.13 μM) [21, 76].

15.3.1.2 Curcumin

Curcumin possesses various anticancer effects in preclinical studies [9]. Currently, 59 clinical trials (https://clinicaltrials.gov) on curcumin for its potential to treat or prevent various cancer types are undergoing or completed, such as breast cancer (NCT01042938), pancreatic cancer (NCT00094445), and prostate cancer (NCT01917890) [67]. These clinical trials have demonstrated the safety profile and potential efficacy of curcumin in several cancer types. Various preclinical studies have suggested that curcumin targeted several key cancer pathways: cell proliferation, signaling pathways, transcription factors, and tumor angiogenesis and metastasis [77–79]. For example, curcumin targets several specific cancer signaling pathways, such as tumor protein p53, Ras proteins, phosphatidylinositol 3-kinase/protein kinase B/mammalian target of rapamycin (PI3K/AKT/mTOR), and Wnt/beta-catenin [77]. In addition, curcumin also targets several cancer-related transcription factors, such as nuclear factor-kappa B (NF-κB), activator protein 1 (AP-1), and signal transducer and activator of transcription (STAT) families [77]. Finally, curcumin also inhibits several key targets involved in tumor angiogenesis and metastasis, including vascular endothelial growth factor, matrix metalloproteinases, tyrosine–protein phosphatase non-receptor, and inflammatory cytokine CXCL1 [78].

Recent studies have identified that curcumin exerted immunomodulatory effects on several cell types and organs in the immune systems, including T cells, B cells, macrophages, natural killer (NK) cells, and dendritic cells [80–82]. For instance, curcumin has been reported to suppress inflammatory responses by inhibiting nitric oxide production, cyclooxygenase-2 (COX-2), NF-κB, inducible nitric oxide synthase (iNOS), lipoxygenase in NK cells, interferon-γ, and tumor necrosis factor-α-activated macrophages [80, 81]. In addition, curcumin micelle has been reported to improve therapeutic efficacy of cancer vaccine and the cytotoxic T-cell responses [82].

15.3.1.3 Quercetin

Quercetin is another bioactive flavonol enriched in plant foods such as onions, apples, and berries [9, 83]. Several clinical trials for quercetin to treat or prevent several cancer types are undergoing or completed, including prostate cancer (NCT01912820), renal cell carcinoma (NCT02446795), and advanced pancreatic cancer (NCT01879878). Previous in vitro and in vivo studies have showed that quercetin effectively exerted anticancer activities via targeting multiple cancer pathways: apoptosis, cell cycle, detoxification, antioxidant replication, and angiogenesis [84, 85]. For example, quercetin inhibits the invasiveness in cancer cells via downregulating STAT3, NF-κB, and mitogen-activated protein kinase (MAPK) AKT pathways [86]. The immunomodulatory effects of quercetin are also recognized recently in several immune cell types [87–90]. Quercetin modulates the level of interleukin-1β and TNF-α released by peripheral blood mononuclear cells and reduces the matrix metalloproteinases 9/metalloproteinase inhibitor 1 (MMP-9/TIMP-1) ratio by reducing MMP-9 production [87]. Moreover, quercetin inhibits COX-2 and iNOS via suppressing AP-1, NF-kB, and STAT-1 signaling in cytokine-induced or lipopolysaccharide-induced human umbilical vein endothelial cells and macrophages [88].

15.3.2 Isoflavonoid

15.3.2.1 Genistein

Genistein, an isoflavonoid derived from soy products, has been reported to show the chemo-preventive properties against several cancer types [9, 91], including breast cancer (NCT00244933), bladder cancer (NCT00118040), and prostate cancer (NCT01325311). Genistein was predicted to bind with 15 targets potentially. For example, nuclear factor-kappa B (NF-*k*B), playing crucial roles in tumor cell proliferation, survival, and angiogenesis [92], was predicted to be interacted by genistein. A recent study has suggested that genistein inhibited the proliferation of human multiple myeloma (MM) cells via suppressing the expression of NF-*k*B [93].

15.3.2.2 Naringenin (DB03467)

Naringenin (DB03467), a flavanone, is mainly found in grapefruit, oranges, and tomatoes. It has been reported to possess various anticancer activities in breast, colon, pancreatic, lung, and prostate cancers [21, 94]. Naringenin is predicted to have potential indications for six cancer types: bladder, lung, uterine, colon, prostate, and breast. Among of them, breast, prostate, and colon are the three new anticancer indications with non-significance based on ExpNet only. Previous studies have suggested that naringenin showed strong anticancer activities in three cancer types [95–98]. For example, a previous study showed that naringenin prolonged the survival in breast cancer mice model [96]. In addition, naringenin was reported to induce apoptotic cell death via regulating activation of PI3K/AKT and MAPK signaling pathways in prostate cancer cells [98].

15.3.3 Non-flavonoid Polyphenol

15.3.3.1 Resveratrol

Resveratrol, a non-flavonoid polyphenol derived from the skin of grapes, exerts a wide range of anticancer effects [9, 21, 99, 100]. Currently, approximate 20 clinical trials (http://clinicaltrials.gov/) are being conducted to treat various cancers by resveratrol, such as colon cancer (NCT00256334) and liver cancer (NCT02261844). The exact MOAs of anticancer activities by resveratrol are still unclear. Resveratrol interacts with 23 cancer proteins and 21 non-cancer proteins, consisting of 30 known and 14 predicted ones. Interestingly, 15-lipoxygenase (ALOX15) and estrogen receptor-β (ERβ encoded by ESR2) are the two predicted cancer proteins for resveratrol by our predictive network model. Recent studies have showed that resveratrol acted as a competitive weak inhibitor of ALOX15 with an IC_{50} value of 25 μM [101] and inhibited ERβ to suppress cancer cell growth [102], indicating potential anticancer mechanism of resveratrol. A previous study has suggested that natural products often interacted with human proteins with moderate or weak binding affinities at micromolar level [103]. A recent analysis on the binding affinities of natural products also suggests that therapeutic efficacy is not necessarily related to high binding affinity [104].

Anticancer mechanisms of resveratrol are involved in inflammatory pathway (e.g., down-regulating reactive oxygen species, COX-2 and cytokines), transcription factors (e.g., downregulating NF-κB and AP-1), cell cycle (e.g., down-regulating cyclins and cyclin-dependent kinases), apoptosis (e.g., upregulating p53 and BAX, while downregulating BCL2 and survivin), hormone receptors [e.g., modulating estrogen receptors (ERs)], angiogenesis and metastasis, and immune surveillance [99]. A resveratrol analog (HS-1793) was reported to reverse the properties of TAMs via increasing the level of INF-γ production [105]. Resveratrol was recently reported to show potent immune enhancement activity in immuno-suppressive mice via activating NF-*κ*B [106].

15.3.4 Other NPs

15.3.4.1 Disulfiram (DB00822)

Disulfiram is an FDA-approved carbamate derivative for the treatment of chronic alcoholism, which has showed potential in treating several cancer types [21, 107]. It was predicted to have potential indications for eight cancer types. Five of them are the new anticancer indications using Exp&ComNet, including breast, colon, lung, thyroid, and uterine. The prediction by Fang et al. [9] is consistent with the findings by several previous in vitro or in vivo studies of disulfiram in cancer: inducing cell apoptosis in breast cancer cells through suppressing the activity of proteasome [108], and eliciting apoptosis and excluding cancer stemlike cells via the suppression of HER2/Akt signaling pathway in HER2-positive breast cancer [109, 110].

15.3.4.2 Metformin (DB00331)

Metformin (DB00331) is a biguanide (mainly found in *Galega officinalis*) oral agent for treating type 2 diabetes. Nowadays, several clinical trials of metformin are being conducted to treat several cancer types, such as breast (NCT01266486), lung (NCT02109549), and ovarian (NCT01579812) [21]. In this work, none of these cancer types was predicted using ExpNet only. Among the top five predicted targets for metformin, three targets were cancer proteins (TSHR, BLM, and ALDH1A1). Via Exp&ComNet, metformin was predicted to have potential in treating two cancer indications: breast and ovarian. Previous in vitro and clinical studies have suggested that metformin showed potential in treating breast cancer [111]. In addition, metformin exerted anticancer effect in ovarian cancer cells via targeting AMPK/GSK3β signaling axis [112].

15.3.4.3 Epigallocatechin Gallate (EGCG)

EGCG, a bioactive compound in green tea, shows potential in vitro and in vivo anticancer effects in various cancer types, including breast, colorectal, gastric, and prostate cancers [9, 113]. Several clinical studies of EGCG are being investigated in bladder cancer (NCT00666562), breast cancer (NCT00949923), and colorectal cancer (NCT01606124). The potential mechanism of actions (MOAs) of anticancer effects for EGCG include antioxidant activities, induction of apoptosis, cell cycle arrest, modulation of intracellular signaling cascades, inhibition of epigenetic modifications, dihydrofolate reductase, telomerase, and angiogenesis [114]. In addition, EGCG also enhances anticancer efficacy by inducing antitumor immunity. For example, EGCG was reported to inhibit an immunomodulatory protein indoleamine 2,3-dioxygenase via blocking of gamma-interferon-induced janus kinase (JAK) protein kinase C (PKC) delta-STAT1 signaling in human oral cancer cells [115] and regulate CD3-mediated T-cell receptor signaling in leukemia via inhibition of an immunomodulatory target tyrosine–protein kinase ZAP-70 [116].

15.3.4.4 Berberine

Berberine, a quaternary ammonium protoberberine alkaloid, is widely distributed in several medicinal herbs such as *Coptis chinensis*. A recent study has suggested that berberine exerted extensive anticancer effects in vitro and in vivo [9, 117]. Currently, there are 37 completed and ongoing registered clinical studies (http://clinicaltrials.gov/) regarding berberine to treat or prevent different diseases and conditions including cancer. Several recent preclinical studies have showed that berberine exerted anticancer effects in various cancer cell lines (e.g., hepatoma, colorectal cancer, and gastric carcinoma) [118, 119]. The possible anticancer mechanisms include cell cycle arrest induction, inhibition of invasion, apoptosis induction, and inhibition of proliferation [120, 121]. The antitumor effect of berberine could be strengthened by melatonin in lung cancer cells via regulating several specific signaling pathways (e.g., NF-κB/COX-2 and AKT/extracellular regulated protein kinases (ERK)) [122]. In addition, berberine has been further reported to exert anti-inflammatory or immunomodulatory effects in several emerging studies. For example, berberine was reported to inhibit migration and invasion of human SCC-4 tongue squamous carcinoma cells through cancer-related inflammation pathways (e.g., mediating IKK and NF-κB) [123, 124].

15.4 Resources and Tools for Systems/Network Pharmacology on NPs

Nowadays, there enormous number of resources and tools available for de novo drug discovery and drug repositioning from NPs based on systems/network pharmacology principles. Here, the descriptions on this topic will be mainly based on the excellent review articles presented by Fang et al. [9] and Mayr et al. [21].

15.4.1 Reconstructing Drug–Target Network of Natural Products

Reconstruction of high-quality DTI network is a key step for identifying polypharmacology of natural products [9, 13, 125]. To facilitate the identification of drug targets for natural products, one can use two types of chemoinformatics and bioinformatics resources: general drug–target databases and natural products-specific drug–target databases [9].

15.4.1.1 General Drug–Target Databases

Several commonly used databases provide large-scale experimental drug–protein binding affinity data of natural products [9], such as ChEMBL [126], BindingDB [129], PubChem [128], DrugBank [129], Therapeutic Target Database (TTD) [130], and DGIdb [131]. Currently, ChEMBL (v22.1) contains 14,371,197 bioactivity records, covering 203,512 compounds against 11,224 targets (accessed in December 2016) [126]. BindingDB contains 1,279,670 experimentally measured binding affinity data for 6612 proteins and 565,136 molecules (accessed in December 2016) [127]. DrugBank (v5.0) is the most commonly used database, containing 8250 drug entries including 2016 FDA-approved small-molecule drugs and over 6000 experimental drugs [129]. The TTD (v2016) includes 2589 targets (397 approved targets, 723 clinical trial targets, and 1469 research targets) and 31,614 drugs (2071 approved drugs, 9528 drugs under clinical trials, and 17,803 experimental drugs) [130]. DGIdb is a web resource that consolidates different data sources with drug–gene interactions and gene druggability. DGIdb 2.0 includes 26,298 unique drug–gene interactions connecting 7569 drugs and 7524 unique genes [131].

Some other common databases, such as STITCH [132] and SuperPred [133], include both experimental binding affinity data and computationally predicted data. For example, STITCH is a resource that contains both experimentally validated and computationally predicted interactions between chemicals and proteins with several lines of evidence: experimental data, manually curated literature data with text-mining information, and computationally inferred data [134]. SuperPred provides experimentally reported DTIs and predicted DTIs derived from molecular similarity approach, covering 665,000 DTIs connecting 341,000 compounds and 1800 targets [133].

15.4.1.2 Natural Product-Specific Resources

Several natural product-specific databases provide useful information to reconstruct drug–target network of natural products, such as Herb Ingredients' Targets (HIT) [9138], Traditional Chinese Medicine Integrated Database (TCMID) [135], and Traditional Chinese Medicine Systems Pharmacology database (TCMSP) [136]. HIT contains drug–target information between 586 active natural products derived from >1300 reputable Chinese herbs and 1301 known protein targets by manually curating data from 3250 literatures [134]. TCMID provides information on different types of traditional Chinese medicine (TCM) information: formulas, herbs, herbal ingredients, and targets [135]. Currently, TCMID contains 47,000 prescriptions, 8159 herbs, 43,413 ingredients, 3854 diseases, and 17,521 related targets. TCMSP includes natural products, predicted targets and drug–target networks, and associated drug–target–disease networks, as well as the predicted pharmacokinetic properties for natural products. In November 2016, TCMSP contains 499 Chinese herbs registered in the Chinese pharmacopoeia with 29,384 ingredients, 3311 targets, and 837 associated diseases [136].

In addition, Naturally occurring Plant-based Anti-cancer Compound-activity-Target (NPACT) database [137] and anticancer herbs database of systems pharmacology (CancerHSP) [138] contain target information for anticancer natural products. Altogether, these databases have provided useful resources to reconstruct the comprehensive drug–target networks for various natural products.

15.4.2 Systems Biology Resources

Systems biology resources play essential roles to build network-based or systems biology-based models for identification of drug targets of natural products. Two major types of systems biology resources are chemogenomics resources [139] and PPI databases.

15.4.2.1 Chemogenomics Resources

Several chemogenomics resources can be used for integrating multiple scale biological datasets for serving systems biology-based drug discovery paradigm, such as connectivity map (CMap) [9, 140], Library of Integrated Cellular Signatures (LINCS) [141], open TG-GATEs [142], and DrugMatrix [143]. CMap v2.0 includes >7000 gene expression profiles representing 1309 compounds tested in four different human cancer cell lines [144]. Currently, a limited number of compounds and drug-induced gene expression profiles tested in four cancer types in CMap v2.0 limit the cancer-type-specific analysis. The LINCS [141] is a National Institutes of Health (NIH)-funded program for the generation of large-scale drug-induced expression data across multiple cell types and perturbations (http://www.lincscloud.org/). As of December of 2016, LINCS contains >1 million genome-wide expression profiles representing >10,000 drugs and drug-like compounds tested in ~80 different cancer cell lines. LINCS attempts to create a network-based understanding of biology by cataloging changes in gene expression and other cellular phenotypes perturbed by various drugs or compounds compared with previous CMap [145].

15.4.2.2 PPI Databases

Several well-known PPI databases [9], such as BioGRID [146], HPRD [147], Interactome3D [148], STRING [149], and MINT [150], have been commonly used for PPI network analysis, providing useful information in the characterization of disease-related proteins and drug targets [151–157]. BioGRID is an interactive repository with data compiled from the primary biomedical literatures. As of December of 2016, the BioGRID (version 3.4.143) contains 1,410,457 protein and genetic interactions (including 369,198 interactions in *Homo sapiens*) in 61 model organisms by extracting data from 57,763 publications [146]. HPRD contains annotation data pertaining to human proteins based on experimental evidence from the literatures. The HPRD release v9 covers 30,047 protein entries, 41,327 PPIs, and 453,521 PubMed literatures [147]. Interactome3D is a web service for the structural annotation of PPI networks. Through the integration of interaction data from ten different databases, Interactome3D provides structural details at atomic resolution for >12,000 PPIs in eight model organisms [148]. STRING collects and integrates the known and predicted PPIs for a large number of organisms. STRING (v10.5) contains 932,553,897 interactions with quantitative confidence scores in 2031 organisms (accessed in December of 2016) [149]. MINT is a repository of experimentally validated PPIs with special emphasis on mammalian interaction. As of December of 2016, MINT includes 125,464 interactions from 611 organisms curated from 5941 publications [150].

15.4.2.3 Cancer Genomics Resources

Recent advances in next-generation sequencing technologies and computational tools have generated massive cancer genomics data, such as The Cancer Genome Atlas (TCGA) and the International Cancer Genome Consortium (ICGC) [9, 158]. These global and local cancer genome projects have provided somatic mutation landscapes for better understanding of cancer biology and improving cancer diagnosis and therapy [159, 160]. Currently, TCGA and ICGC contain somatic mutations of coding regions in 412,000 samples. Since 2006, TCGA has led to comprehensive, multidimensional maps of genomic profiles in 33 types of cancer, and generated 2.5 petabytes of data describing tumor tissue and matched normal tissues from >11,000 patients (http://cancergenome.nih.gov) [161]. ICGC was launched in 2010 with an aim to catalog the genomic, epigenomic, and transcriptomic profiles for >25,000 cancer genomes across 50 different tumor types or subtypes [162]. As of August of 2016 (data release 22), ICGC comprised data from over 16,000 cancer donors spanning 70 projects and 21 tumor sites (https://dcc.icgc.org). The cBioPortal was developed by the Memorial Sloan Kettering Cancer Center to systematically visualize and analyze the cancer genome landscape. As of December of 2016, cBioPortal curated genomic data for >26,000 samples from 147 cancer studies (http://www.cbioportal.org) [163]. In addition, the Catalogue of Somatic Mutations in Cancer (COSMIC) is the most comprehensive database of somatic mutations in cancer. As of December of 2016 (version v79), COSMIC contains 4,175,878 coding point mutations, 18,165 fusions, 29,112 whole genomes, and 2,113,866 copy number variants by curating cancer genomics data from 23,870 papers (http://cancer.sanger.ac.uk/cosmic) [164, 165].

15.4.3 Computational Approaches for Predicting New Targets of NPs

Given the tremendous growth of compound bioactivity databases, the application of computational approaches or tools to identify new DTIs for natural products has been gaining importance recently. The details about in silico target prediction can be found in several recent reviews [166–169]. Below, focus will be placed on five types of computational approaches [9]: (a) target-based, (b) ligand-based, (c) chemogenomics-based, (d) network-based, and (e) omics-based systems biology approaches for identifying new targets of natural products.

15.4.3.1 Target-Based Approaches

Target-based approach, also known as protein structure-based approach, predicts target proteins of natural product, which relies on knowledge of the three-dimensional (3D) structure of the biological target [9, 170]. Nowadays, the rapid advances in the number of available structural data of macromolecules (such as radiography, nuclear magnetic resonance spectroscopy, and cryo-electron microscope) have significantly accelerated the development of target-based approaches [171]. Molecular docking is the most common target-based approach. Various molecular docking approaches for drug–target identification are well summarized in several recent reviews [47, 48, 172]. Those approaches can also be applied to identify potential targets of natural products in two aspects when 3D structure of target protein is available, i.e., molecular docking-based virtual screening and inverse docking. Molecular docking-based virtual screening docks multiple ligands for a given target, while inverse docking docks multiple targets for a given ligand.

Several recently successful examples through using molecular docking-based virtual screening approaches for discovery of bioactive natural products toward known targets in cancer therapies are summarized below [9, 173–178]. Signal transducer and activator of transcription 3 (STAT3) is an important target in a variety of solid and hematological tumors, such as leukemia and lymphomas. In 2014, Liu and his colleagues [173] applied molecular docking to identify benzofuran derivative 1 as an inhibitor of STAT3 transcriptional activity from a natural product database. The mixed-lineage leukemia (MLL) proteins are histone methyltransferases that act as epigenetic transcriptional regulators. The menin–MLL interaction represents an attractive therapeutic target for cancers with MLL rearrangements, such as liver carcinomas [179]. Zhong et al. [174] identified cytisine alkaloid 1 as an inhibitor of the menin–MLL interaction by molecular docking-based virtual screening. Singh and his coworkers [175] reported that honokiol inhibited the growth of head and neck squamous cell carcinoma (HNSCC) by targeting epidermal growth factor receptor (EGFR) using molecular docking approach.

Typically, inverse docking approaches search a given molecule against a panel of proteins where their 3D structures are available, such as INVDOCK [180] and idTarget [181]. Recently, Lauro et al. [182, 183] proposed an application of inverse docking for screening on 10 phenolic natural compounds tested on a panel of 163 cancer-associated proteins. They identified that xanthohumol and isoxanthohumol inhibited phosphoinositide-dependent kinase 1 (PDK1) and PKC protein kinases during in vitro assays [182, 183].

Although target-based approaches have been successfully applied in the identification of new targets for natural products, several potential limitations have to be recognized. First, only 15–20% proteins at the human proteome have been reported to have the availability of 3D structures [184, 185]. Second, high false-positive rate of molecular docking scoring functions must be recognized [57]. Furthermore, high chemical diversity and complexity of natural products with multiple chiral atoms further limit the accuracy of current docking scores.

15.4.3.2 Ligand-Based Approaches

Ligand-based approach is commonly used when the 3D structure of a target protein is absent. This approach predicts targets of natural product, which relies on knowledge of other similar molecules that bind to the biological target of interest [9, 60]. Ligand-based approaches can be divided broadly into similarity-based and machine learning (ML)-based approaches [186]. Similarity-based approach assumes that structurally similar compounds exhibit similar biological activity more often than not, via using two-dimensional molecular fingerprint [187], 3D pharmacophore models [188], or 3D molecular shape queries [189], to find potential targets. Similarity-based approach is a rapid method to search all possible targets. However, it only compares molecules with each other as a whole, while it does not weight the features of the chemical structure according to their influence on the biologic activity of interest. ML-based approach solves this problem via correlating relationships between the features of molecular and target of interest, and it yields high robustness and high accuracy. Numerous methods that use ML algorithms for target prediction or hits identification toward specific targets have been published [190–193]. Generally, chemical structures and biological activities are integrated with different ML algorithms to train predictive models [194], such as support vector machine (SVM), naive Bayesian, and artificial neural network.

Sprague et al. [195] developed predictive classifiers to discriminate chemopreventive agents from inactive samples using random forest (RF) and SVM algorithms. They identified 18 natural products with potential in vitro anticancer activities via searching a natural product library containing nearly 23,000 compounds. A recent study by Bilsland et al. [196] reported a ML-based in silico screen to identify potential senescence agonists for cancer therapy. They identified a selective G1-Phase benzimidazolone inhibitor (CB-20903630), which induced senescence-associated β-galactosidase activity via a neural network classifier.

Compared with molecular docking, ML approaches reveal some distinct advantages, including highly predictive accuracy and low computational cost. However, several limitations must be recognized for target prediction of natural products. First, the high quality of negative samples validated by experiments is limited. Most of ML models are built based on the unknown DTIs or the randomly selected DTIs as negative samples, which limits the accuracy of ML models. To address this problem, Liu et al. proposed a new approach for searching highly credible negative samples. Their method assumed that the proteins that were dissimilar to each target of a given compound were not much likely to be targeted by the compound and vice versa [198]. Second, the predictive accuracy of the method depends on large and diverse chemical space in the training set and test set. For the targets that only have few ligands or no ligand (such as orphan receptors), this approach does not work. Third, it is difficult to select proper molecular descriptors to represent the features of natural product, as there are multiple chiral atoms for natural products. Fourth, most ML models are like "black box" models owing to the lack of the detailed biological or molecular mechanisms of the predictions. Finally, conventional ML models only consider a series of ligands against a single target or phenotype. It is a challenge to predict all thousands of drug targets for a compound with a single classifier. Chemogenomics-based approaches that integrate both drug and target information can effectively address the abovementioned limitations [198].

15.4.3.3 Chemogenomics-Based Approaches

To overcome challenges of traditional ML methods, chemogenomics-based approaches have recently gained much attention to predict multiple drugs against multiple targets [9]. A key issue in chemogenomics is to infer unknown DTIs by integrating both drug- and target-based descriptors, such as target protein sequence information, drug chemical structures, and the available known drug–target network topology information simultaneously [198–200].

Yamanishi et al. [201] reported a pivotal study to predict DTIs with chemogenomic approach that integrates the sequence similarities between pairwise proteins and the chemical structural similarities between pairwise compounds for predictions of DTIs. High performance was yielded when evaluating their models using known DTIs covering four types of targets: enzymes, ion channels, G-protein-coupled receptors (GPCRs), and nuclear receptors. Bleakley et al. [202] proposed a supervised inference method, named the bipartite local model (BLM). BLM predicted target proteins of a given molecule using the structural similarity of compounds, genomic similarity and network topological information of known interactions between drugs and targets. Recently, Keum et al. [203] adopted this method to construct classification models with good performance for six classes of DTIs in humans including GPCRs, ion channel, enzyme, receptors, transporters and other proteins. Yu et al. [204] reported a chemogenomic approach that efficiently incorporated the chemical, genomic and pharmacological information using the largest available dataset from the DrugBank database. In the follow-up analyses, they successfully identified unknown DTIs for ingredients of traditional Chinese herbs via their chemogenomic models [205].

Overall, one merit of chemogenomic approaches or models is that it can simultaneously search huge numbers of drug candidates and candidate targets at the same time. However, two shortcomings have to be recognized. First, the performance of chemogenomic models depends heavily on the relevant descriptors used for model building. The suitable feature selection approaches are recommended to use for selecting the biological or pharmacological relevant descriptors. Hundreds of descriptors used in one model often generate a high risk of "over-fitting" [206]. The best way to avoid "over-fitting" risk is that all computational models are suggested to be built based on more relevant features and have to be passed the testing by an external validation set that are not used to develop the model. Second, compared with conventional ML models, chemogenomic models contain target descriptors, while the increasing of target descriptor variables increases the risk of "over-fitting" and computational complexity. A recent study examined two methods: multitarget-quantitative structure–activity relationships (mt-QSARs) and computational chemogenomics, for prediction of DTIs. They found that the performance of the chemogenomic models was worse than that of mt-QSARs [207]. Several possible explanations are suggested: (a) traditional protein sequence-based descriptors cannot describe the natural interactions between drugs and targets, and (b) a highly imbalanced number of positive samples versus negative samples often occurred owing to lack of literature-reported negative biological or pharmacological data.

15.4.3.4 Network-Based Approaches

Network-based approaches for prediction of DTIs have demonstrated their great value in rational drug discovery [9]. In a bipartite drug–target network, nodes denote drugs and targets, and edges denote their interactions, such as the binding affinity between drugs and targets via half maximal inhibitory concentration (IC50) or inhibition constant/potency (Ki) value <10 μM. Recently, graph-based network approaches have demonstrated potential for identifying novel targets for known drugs and drug repositioning. Cheng et al. [49] proposed three supervised inference models to predict DTIs, namely, drug-based similarity inference (DBSI), target-based similarity inference (TBSI), and network-based inference (NBI). DBSI and TBSI rely on chemical structural similarity and target sequence similarity, respectively, whereas NBI is only based on drug–target bipartite network topology similarity. Among three algorithms, NBI performed the best on four benchmark datasets covering GPCRs, enzymes, ion channels, and nuclear receptors. Moreover, five old drugs were identified via NBI and then experimentally validated to have potential binding effects on ERs or dipeptidyl peptidase IV in vitro [49].

As an improvement of NBI, Cheng et al. further developed two weighted NBI methods, edge-weighted NBI (EWNBI) and node-weighted NBI (NWNBI), by assigning the weighted values to edges and nodes in the drug–target network, respectively. They found that the weighted NBI methods marginally outperformed the unweighted ones [50]. However, both NBI and the weighted NBI cannot be applied to the new drugs without any known target information in the training set. Recently, they further developed a novel algorithm, namely, substructure–drug–target network-based inference (SDTNBI), to prioritize potential targets for new chemical entities [53]. SDTNBI imports chemical substructures to bridge the gap between new chemical entities and known DTI network. Systematic evaluation based on cross-validation and external validation showed high accuracy and robustness of SDTNBI. Finally, a global network was constructed, and thousands of new potential DTIs were computationally identified via SDTNBI [53]. However, there are still several potential limitations in SDTNBI yet. For instance, whether introducing tunable parameters to SDTNBI could improve the method performance remains unclear. Hub nodes that have high degree in the drug–target network often generate high risk of false-positive rate for network-based prediction, limiting the application of two previous unweighted network approaches (NBI and SDTNBI). The same group further proposed an improvement on SDTNBI [54], named balanced substructure–drug–target network-based inference (bSDTNBI), to identify the MOA for both old drugs and new chemical entities. Finally, they demonstrated that bSDTNBI showed potential application for identifying the specifically antagonistic or agonistic ERα ligands, and dual-effect ERα ligands with both antagonistic and agonistic activities for the development of potential therapies to breast cancer or osteoporosis. While these network-based approaches have made significant advances, several improvements could be further achieved in the future. For example, current network-based approaches such as bSDTNBI and NBI cannot predict potential DTIs for those target proteins that are absent in the known substructure–drug–target network. In addition, the interaction types (e.g., antagonistic and agonistic effects) and binding affinities of drug–target pairs have not been considered yet. Integrating human PPI network and gene regulatory network [51, 52, 208] (e.g., microRNA network) with the network approaches would provide potential solution to identify DTIs for many novel target proteins at the human proteome level.

15.4.3.5 Omics-Based Systems Biology Approaches

Owing to recent advances of "omics" technologies, massive omics data were generated, including transcriptomics, proteomics, and metabolomics. These omics resources provide huge opportunities for the development of systems biology models for identifying new targets for natural products [209, 210]. Here, we briefly describe three omics-based systems biology approaches, including transcriptomics, proteomics, and metabolomics, for prediction of DTIs for natural products as below.

Transcriptomics

Genome-wide transcriptomics data from the cultured human cells treated with bioactive small molecules have provided a unique resource for drug repurposing and identifying MOA for various drugs, including natural products [9, 211, 212]. Qu et al. [213] integrated drug-induced transcriptomics resources and computational approaches for target identification, lead compound identification, and drug repurposing studies. Hieronymus and his colleagues [214] introduced a chemical genomic approach for discovery and target prediction of modulators of cancer phenotypes, as exemplified by androgen acceptor-mediated signaling. Among the hits that emerged, two natural products, celastrol and gedunin, were identified to have potential to treat prostate cancer as heat shock protein 90 (HSP90) inhibitors [214]. With a similar approach, Wei et al. [215] identified drugs to overcome resistance to glucocorticoid treatment based on a gene expression signature of glucocorticoid sensitivity/resistance in acute lymphoblastic leukemia via downregulation of myeloid cell leukemia 1 (MCL1). Compared with previous "black box" ML-based models and chemogenomic models, transcriptomics-based systems biology models provide novel biological or molecular mechanisms for characterizing their specific anticancer effects.

Proteomics

In addition to transcriptomics, proteomics, representing the large-scale analysis of proteins, has also driven much of systems biology studies over the past two decades [9, 216]. For example, mass spectrometry (MS)-based proteomics refers to the application of MS to the identification, quantification, and characterization of the proteins [217]. Savitski et al. [218] proposed an approach named thermal proteome profiling (TPP) to monitor the changes in protein thermal stability across the proteome using quantitative MS. They used the histone deacetylase inhibitor panobinostat to show the whole workflow and demonstrated that this methodology had the potential to identify direct and indirect drug targets in living cells [219]. In subsequent work, they further extended TPP to profile membrane–protein targets of small molecules, either in cell extracts or for in situ studies on intact cells [220]. Sacco et al. [221] systematically investigated signaling network perturbation after metformin treatment in the proteome and phosphoproteome of breast cancer cells via combining high-resolution MS and network analysis. They examined perturbations of 7875 proteins and 15,813 phosphosites after metformin treatment, and further found that metformin inhibited the p70S6K–rpS6 in a PP2A–phosphatase-dependent manner. This study demonstrated that deep proteomics and network analysis provided novel strategy for exploring both detailed and global mechanisms involving anticancer effects of metformin [221]. Compared with previous transcriptomics-based approaches, proteomics-based approaches provide more useful strategies for directly exploring both detailed and global molecular mechanisms of natural products at the human proteome level, not transcriptomics level.

Metabolomics

Metabolomics monitors the biochemical changes of many small molecules and their metabolites at various systems biology levels (e.g., cells, tissues, organs, etc.) via integrating high-throughput assays and bioinformatics analyses [9, 15, 222]. It provides a unique strategy to facilitate the identification of new drug targets of natural products or their metabolites [223–228]. Birkenstock et al. [223] used metabolic profiling to demonstrate that the antimicrobial agent triphenylbismuth dichloride (TPBC) blocked bacterial pyruvate catabolism in *Staphylococcus aureus* by quantifying the extracellular metabolites of bacterial cultures and helped to identify pyruvate dehydrogenase as a target for TPBC. Zhang et al. [224] used cellular metabolic profiling to quantitate six isoprenoid precursor metabolites. They demonstrated that fosmidomycin treatment blocked isoprenoid metabolism in two microbes (*Escherichia coli* and *Plasmodium falciparum*), suggesting that methylerythritol phosphate cytidylyltransferase (IspD) is a potential target for fosmidomycin. Bayet-Robert et al. [225] used metabolomics to elucidate the

MOA of three marine natural products (MNPs) against biochemical disorders induced in human MCF7 breast cancer cells. Metabolite profiling revealed that ascididemin involved enzyme blockade in citrate metabolism, and kahalalide F increased lipid membrane, suggesting new metabolic targets for these MNPs [225].

Owing to the dramatically increasing of multi-scale omics data, omics-based systems biology approaches offer powerful tools for drug–target identifications of natural products in drug discovery and development. In the future, new computational systems biology approaches to analyze and integrate new multi-scale omics data are urgently needed in this field.

15.4.4 Performance Evaluation of Computational Models

Several indicators are commonly used to evaluate the performance and robustness of computational models, including accuracy, precision, and recall. The details of these measurements are discussed in a previous study [9, 49]. Specifically, a receiver operating characteristic (ROC) curve is often drawn via measuring a series of true-positive rate and false-positive rate under different conditions. In addition, a precision–recall curve is often used via measuring an array of precision and recall values under different cutoff. Finally, the areas under these ROC curves or precision–recall curves will be yielded to evaluate the performance of various computational models.

The N-fold cross-validation is often used for evaluation of the performance of computational models, such as fivefold cross-validation and leave-one-out cross-validation. There are two differential cross-validation protocols for currently computational approaches or models: samples-based cross-validation versus links-based cross-validation. For example, during leave-one-out cross-validation in general ML-based models, each drug or target (not a drug–target link) in the entire drug–target network is extracted to be the test set in turn. The remnant dataset containing N-1 drugs or N-1 targets will be used as training set for building the models. For an N-fold cross-validation (e.g., fivefold cross-validation) in network-based link predictions [49], the entire links in drug–target network are randomly divided into five parts. One part is used as the test set in turn, which is not included in training set. The remnant network is used as the training set to build network-based model. However, N-fold cross-validations cannot measure the real generalization ability for most computational models [229]. It is crucial that all computational models are recommended to pass the external validation (model performance is evaluated by an external validation set that are not used to develop the model). Finally, a short list of predicted candidates with high-confidence score can be selected for further experimental validation (in vitro assays, in vivo animal models, or patient electronic medical records) to test the performance of computational models in practice.

References

1. Newman DJ, Cragg GM. Natural products as sources of new drugs over the 30 years from 1981 to 2010. J Nat Prod. 2012;75:311–35.
2. Ji HF, Li XJ, Zhang HY. Natural products and drug discovery: can thousands of years of ancient medical knowledge lead us to new and powerful drug combinations in the fight against cancer and dementia? EMBO Rep. 2009;10:194–200.
3. Harvey A. Natural products in drug discovery. Drug Discov Today. 2008;13:894–901.
4. Butler MS, Robertson AA, Cooper MA. Natural product and natural product derived drugs in clinical trials. Nat Prod Rep. 2014;31:1612–61.
5. Shen B. A new golden age of natural products drug discovery. Cell. 2015;163:1297–300.
6. Kingston DGI. Modern natural products drug discovery and its relevance to biodiversity conservation. J Nat Prod. 2011;74:496–511.
7. Chin YW, Balunas MJ, Chai HB, et al. Drug discovery from natural sources. AAPS J. 2006;8:E239–53.
8. Ho TT, Tran QT, Chai CL. The polypharmacology of natural products. Future Med Chem. 2018;10(11):1361–8.
9. Fang J, Liu C, Wang Q, et al. In silico polypharmacology of natural products. Brief Bioinform. 2018;19(6):1153–71.
10. Gu J, Gui Y, Chen L, et al. Use of natural products as chemical library for drug discovery and network pharmacology. PLoS One. 2013;8(4):e62839.
11. Kibble M, Saarinen N, Tang J, et al. Network pharmacology applications to map the unexplored target

space and therapeutic potential of natural products. Nat Prod Rep. 2015;32(8):1249–66.
12. Yildirim MA, Goh KI, Cusick ME, et al. Drug-target network. Nat. Biotech. 2007;25:1119–26.
13. Barneh F, Jafari M, Mirzaie M. Updates on drug-target network; facilitating polypharmacology and data integration by growth of DrugBank database. Brief Bioinform. 2016;17:1070–80.
14. DeCorte BL. Underexplored opportunities for natural products in drug discovery. J Med Chem. 2016;59:9295–304.
15. Harvey AL, Edrada-Ebel R, Quinn RJ. The re-emergence of natural products for drug discovery in the genomics era. Nat Rev Drug Discov. 2015;14:111–29.
16. Li JW, Vederas JC. Drug discovery and natural products: end of an era or an endless frontier? Science. 2009;325:161–5.
17. Fang J, Cai C, Wang Q, et al. Systems pharmacology-based discovery of natural products for precision oncology through targeting cancer mutated genes. CPT Pharmacometrics Syst Pharmacol. 2017;6:177–87.
18. Clardy J, Walsh C. Lessons from natural molecules. Nature. 2004;432:829–37.
19. Gu J, Zhang H, Chen L, et al. Drug-target network and polypharmacology studies of a Traditional Chinese Medicine for type II diabetes mellitus. Computational Biol Chem. 2011;35:293–7.
20. Vogt I, Mestres J. Drug-target networks Mol Informatics. 2010;29:10–4.
21. Mayr F, Möller G, Garscha U, et al. Finding new molecular targets of familiar natural products using in silico target prediction. Int J Mol Sci. 2020;21(19):7102.
22. Newman DJ, Cragg GM. Natural products as sources of new drugs from 1981 to 2014. J Nat Prod. 2016;79:629–61.
23. Koehn FE, Carter GT. The evolving role of natural products in drug discovery. Nat Rev Drug Discov. 2005;4:206–20.
24. Koeberle A, Werz O. Multi-target approach for natural products in inflammation. Drug Discov Today. 2014;19:1871–82.
25. Rodrigues T, Reker D, Schneider P, et al. Counting on natural products for drug design. Nat Chem. 2016;8:531.
26. Clemons PA, Bodycombe NE, Carrinski HA, et al. Small molecules of different origins have distinct distributions of structural complexity that correlate with protein-binding profiles. Proc Natl Acad Sci U S A. 2010;107:18787.
27. Wang XJ, Wei XM, Thijssen B, et al. Three-dimensional reconstruction of protein networks provides insight into human genetic disease. Nat Biotech. 2012;30:159–64.
28. Yang K, Ma WZ, Liang HH, et al. Dynamic simulations on the arachidonic acid metabolic network. PloS Computational Biol. 2007;3:523–30.
29. Zhang Y, Thiele I, Weekes D, et al. Three-dimensional structural view of the central metabolic network of thermotoga maritima. Science. 2009;325:1544–9.
30. Li QA, Li XD, Li CH, et al. A network-based multi-target computational estimation scheme for anticoagulant activities of compounds. PLoS One. 2011;6(3):e14774.
31. Hong JY. Role of natural product diversity in chemical biology. Curr Opinion Chem Biol. 2011;15:350–4.
32. Firn RD, Jones CG. Natural products – a simple model to explain chemical diversity. Nat Prod Reports. 2003;20:382–91.
33. Basso LA, da Silva LHP, Fett-Neto AG, et al. The use of biodiversity as source of new chemical entities against defined molecular targets for treatment of malaria, tuberculosis, and T-cell mediated diseases – a review. Memorias Do Instituto Oswaldo Cruz. 2005;100:575–606.
34. Quinn RJ, Carroll AR, Pham NB, et al. Developing a drug-like natural product library. J Nat Prod. 2008;71:464–8.
35. Feher M, Schmidt JM. Property distributions: differences between drugs, natural products, and molecules from combinatorial chemistry. J Chem Info Computer Sci. 2003;43:218–27.
36. Yongye AB, Waddell J, Medina-Franco JL. Molecular scaffold analysis of natural products databases in the public domain. Chem Biol & Drug Design. 2012;80:717–24.
37. Grabowski K, Baringhaus KH, Schneider G. Scaffold diversity of natural products: inspiration for combinatorial library design. Nat Prod Reports. 2008;25:892–904.
38. Lee ML, Schneider G. Scaffold architecture and pharmacophoric properties of natural products and trade drugs: application in the design of natural product-based combinatorial libraries. J Comb Chem. 2001;3:284–9.
39. Dobson CM. Chemical space and biology. Nature. 2004;432:824–8.
40. Rosen J, Gottfries J, Muresan S, et al. Novel chemical space exploration via natural products. J Medicinal Chem. 2009;52:1953–62.
41. Grabowski K, Schneider G. Properties and architecture of drugs and natural products revisited. Curr Chem Biol. 2007;1:115–27.
42. Henkel T, Brunne RM, Muller H, et al. Statistical investigation into the structural complementarity of natural products and synthetic compounds. Angewandte Chemie-International Edition. 1999;38:643–7.
43. Qiao XB, Hou TJ, Zhang W, et al. A 3D structure database of components from Chinese traditional medicinal herbs. J Chem Inf Comput Sci. 2002;42:481–9.
44. Doman TN, McGovern SL, Witherbee BJ, et al. Molecular docking and high-throughput screening

for novel inhibitors of protein tyrosine phosphatase-1B. J Medicinal Chem. 2002;45:2213–21.
45. Zhu F, Shi Z, Qin C, et al. Therapeutic target database update 2012: a resource for facilitating target-oriented drug discovery. Nucl Acids Res. 2012;40:D1128–36.
46. Fang J, Wu Z, Cai C, et al. Quantitative and systems pharmacology. 1. In silico prediction of drug-target interactions of natural products enables new targeted cancer therapy. J Chem Inf Model. 2017;57(11):2657–71.
47. Luo H, Mattes W, Mendrick DL, et al. Molecular docking for identification of potential targets for drug repurposing. Curr Top Med Chem. 2016;16:3636–45.
48. Ye H, Wei J, Tang K, et al. Drug repositioning through network pharmacology. Curr Top Med Chem. 2016;16:3646–56.
49. Cheng F, Liu C, Jiang J, et al. Prediction of drug-target interactions and drug repositioning via network-based inference. PLoS Comput Biol. 2012;8:e1002503.
50. Cheng F, Zhou Y, Li W, et al. Prediction of chemical-protein interactions network with weighted network-based inference method. PLoS One. 2012;7:e41064.
51. Li J, Lei K, Wu Z, et al. Network-based identification of microRNAs as potential pharmacogenomic biomarkers for anticancer drugs. Oncotarget. 2016;7:45584–96.
52. Li J, Wu Z, Cheng F, et al. Computational prediction of microRNA networks incorporating environmental toxicity and disease etiology. Sci Rep. 2014;4:5576.
53. Wu Z, Cheng F, Li J, et al. SDTNBI: an integrated network and chemoinformatics tool for systematic prediction of drug-target interactions and drug repositioning. Brief Bioinform. 2017;18:333–47.
54. Wu Z, Lu W, Wu D, et al. In silico prediction of chemical mechanism of action via an improved network-based inference method. Br J Pharmacol. 2016;173:3372–85.
55. Cheng F, Li W, Wang X, et al. Adverse drug events: database construction and in silico prediction. J Chem Inf Model. 2013;53:744–52.
56. Cheng F, Li W, Wu Z, et al. Prediction of polypharmacological profiles of drugs by the integration of chemical, side effect, and therapeutic space. J Chem Inf Model. 2013;53:753–62.
57. Lu W, Cheng F, Jiang J, et al. FXR Antagonism of NSAIDs contributes to drug-induced liver injury identified by systems pharmacology approach. Sci Rep. 2015;5:8114.
58. Campbell IB, Macdonald SJF, Procopiou PA. Medicinal chemistry in drug discovery in big pharma: past, present and future. Drug Discov Today. 2018;23:219–34.
59. Chen C, Huang H, Wu CH. Protein bioinformatics databases and resources. In: Chen C, Huang H, Wu CH, editors. Fundamentals of protein bioinformatics, vol. 1558. New York: Humana Press; 2017. p. 3–39.
60. Sliwoski G, Kothiwale S, Meiler J, et al. Computational methods in drug discovery. Pharmacol Rev. 2014;66:334–95.
61. Ashburn TT, Thor KB. Drug repositioning: identifying and developing new uses for existing drugs. Nat Rev Drug Discov. 2004;3:673–83.
62. Aronson JK. Old drugs—new uses. Br J Clin Pharmacol. 2007;64:563–5.
63. Hurle MR, Yang L, Xie Q, et al. Computational drug repositioning: from data to therapeutics. Clin Pharmacol Ther. 2013;93:335–41.
64. Cereto-Massagué A, Ojeda MJ, Valls C, et al. Tools for in silico target fishing. Methods. 2015;71:98–103.
65. Huang Y-W, Pineau I, Chang H-J, et al. Critical residues for the specificity of cofactors and substrates in human estrogenic 17β-hydroxysteroid dehydrogenase 1: variants designed from the three-dimensional structure of the enzyme. Mol Endocrinol. 2001;15:2010–20.
66. Sydow D, Burggraaff L, Szengel A, et al. Advances and challenges in computational target prediction. J Chem Inf Model. 2019;59:1728–42.
67. Kim S, Chen J, Cheng T, et al. PubChem 2019 update: improved access to chemical data. Nucleic Acids Res. 2018;47:D1102–9.
68. Casey SC, Amedei A, Aquilano K, et al. Cancer prevention and therapy through the modulation of the tumor microenvironment. Semin Cancer Biol. 2015;35:S199–223.
69. Liu Y, Cao X. Immunosuppressive cells in tumor immune escape and metastasis. J Mol Med (Berl). 2016;94:509–22.
70. Kotecha R, Takami A, Espinoza JL. Dietary phytochemicals and cancer chemoprevention: a review of the clinical evidence. Oncotarget. 2016;7:52517–29.
71. Jantan I, Ahmad W, Bukhari SN. Plant-derived immunomodulators: an insight on their preclinical evaluation and clinical trials. Front Plant Sci. 2015;6:655.
72. Calderon-Montano JM, Burgos-Moron E, Perez-Guerrero C, et al. A review on the dietary flavonoid kaempferol. Mini Rev Med Chem. 2011;11:298–344.
73. Kim SH, Hwang KA, Choi KC. Treatment with kaempferol suppresses breast cancer cell growth caused by estrogen and triclosan in cellular and xenograft breast cancer models. J Nutr Biochem. 2016;28:70–82.
74. Luo H, Rankin GO, Liu L, et al. Kaempferol inhibits angiogenesis and VEGF expression through both HIF dependent and independent pathways in human ovarian cancer cells. Nutr Cancer. 2009;61:554–63.
75. Goettert M, Schattel V, Koch P, et al. Biological evaluation and structural determinants of p38alpha mitogen-activated-protein kinase and c-Jun-N-terminal kinase 3 inhibition by flavonoids. Chembiochem. 2010;11:2579–88.
76. Arai Y, Endo S, Miyagi N, et al. Structure-activity relationship of flavonoids as potent inhibitors of carbonyl reductase 1 (CBR1). Fitoterapia. 2015;101:51–6.

77. Kasi PD, Tamilselvam R, Skalicka-Wozniak K, et al. Molecular targets of curcumin for cancer therapy: an updated review. Tumour Biol. 2016;37:13017–28.
78. Bar-Sela G, Epelbaum R, Schaffer M. Curcumin as an anticancer agent: review of the gap between basic and clinical applications. Curr Med Chem. 2010;17:190–7.
79. Chakraborty G, Jain S, Kale S, et al. Curcumin suppresses breast tumor angiogenesis by abrogating osteopontin-induced VEGF expression. Mol Med Rep. 2008;1:641–6.
80. Bhaumik S, Jyothi MD, Khar A. Differential modulation of nitric oxide production by curcumin in host macrophages and NK cells. FEBS Lett. 2000;483:78–82.
81. Surh YJ, Chun KS, Cha HH, et al. Molecular mechanisms underlying chemopreventive activities of anti-inflammatory phytochemicals: down-regulation of COX-2 and iNOS through suppression of NF-kappa B activation. Mutat Res. 2001;480–481:243–68.
82. Lu Y, Miao L, Wang Y, et al. Curcumin micelles remodel tumor microenvironment and enhance vaccine activity in an advanced melanoma model. Mol Ther. 2016;24:364–74.
83. Khan F, Niaz K, Maqbool F, et al. Molecular targets underlying the anticancer effects of quercetin: an update. Nutrients. 2016;8:529.
84. He D, Guo X, Zhang E, et al. Quercetin induces cell apoptosis of myeloma and displays a synergistic effect with dexamethasone in vitro and in vivo xenograft models. Oncotarget. 2016;7:45489–99.
85. Pratheeshkumar P, Budhraja A, Son YO, et al. Quercetin inhibits angiogenesis mediated human prostate tumor growth by targeting VEGFR- 2 regulated AKT/mTOR/P70S6K signaling pathways. PLoS One. 2012;7:e47516.
86. Kashyap D, Mittal S, Sak K, et al. Molecular mechanisms of action of quercetin in cancer: recent advances. Tumour Biol. 2016;37:12927–39.
87. Sternberg Z, Chadha K, Lieberman A, et al. Quercetin and interferon-beta modulate immune response(s) in peripheral blood mononuclear cells isolated from multiple sclerosis patients. J Neuroimmunol. 2008;205:142–7.
88. Hamalainen M, Nieminen R, Vuorela P, et al. Anti-inflammatory effects of flavonoids: genistein, kaempferol, quercetin, and daidzein inhibit STAT-1 and NF-kappaB activations, whereas flavone, isorhamnetin, naringenin, and pelargonidin inhibit only NF-kappaB activation along with their inhibitory effect on iNOS expression and NO production in activated macrophages. Mediat Inflamm. 2007;2007:45673.
89. Ruiz PA, Braune A, Holzlwimmer G, et al. Quercetin inhibits TNF-induced NF-kappaB transcription factor recruitment to proinflammatory gene promoters in murine intestinal epithelial cells. J Nutr. 2007;137:1208–15.
90. Lee J, Choi JW, Sohng JK, et al. The immunostimulating activity of quercetin 3-O-xyloside in murine macrophages via activation of the ASK1/MAPK/NF-kappaB signaling pathway. Int Immunopharmacol. 2016;31:88–97.
91. Spagnuolo C, Russo GL, Orhan IE, et al. Genistein and cancer: current status, challenges, and future directions. Adv Nutr. 2015;6:408–19.
92. Li F, Zhang J, Arfuso F, et al. NF-kappaB in cancer therapy. Arch Toxicol. 2015;89:711–31.
93. Xie J, Wang J, Zhu B. Genistein inhibits the proliferation of human multiple myeloma cells through suppression of nuclear factor-kappaB and upregulation of microRNA-29b. Mol Med Rep. 2016;13:1627–32.
94. Mir IA, Tiku AB. Chemopreventive and therapeutic potential of "naringenin," a flavanone present in citrus fruits. Nutr Cancer. 2015;67:27–42.
95. Qin L, Jin L, Lu L, et al. Naringenin reduces lung metastasis in a breast cancer resection model. Protein Cell. 2011;2:507–16.
96. Zhang F, Dong W, Zeng W, et al. Naringenin prevents TGF-beta1 secretion from breast cancer and suppresses pulmonary metastasis by inhibiting PKC activation. Breast Cancer Res. 2016;18:38.
97. Vanamala J, Leonardi T, Patil BS, Turner ND, et al. Suppression of colon carcinogenesis by bioactive compounds in grapefruit. Carcinogenesis. 2006;27:1257–65.
98. Lim W, Park S, Bazer FW, et al. Naringenin-induced apoptotic cell death in prostate cancer cells is mediated via the PI3K/AKT and MAPK signaling pathways. J Cell Biochem. 2017;118:1118–31.
99. Varoni EM, Lo Faro AF, Sharifi-Rad J, et al. Anticancer molecular mechanisms of resveratrol. Front Nutr. 2016;3:8.
100. Jang M, Cai L, Udeani GO, et al. Cancer chemopreventive activity of resveratrol, a natural product derived from grapes. Science. 1997;275:218–20.
101. MacCarrone M, Lorenzon T, Guerrieri P, et al. Resveratrol prevents apoptosis in K562 cells by inhibiting lipoxygenase and cyclooxygenase activity. Eur J Biochem. 1999;265:27–34.
102. Robb EL, Stuart JA. Resveratrol interacts with estrogen receptor-beta to inhibit cell replicative growth and enhance stress resistance by upregulating mitochondrial superoxide dismutase. Free Radic Biol Med. 2011;50:821–31.
103. Wang J, Guo Z, Fu Y, et al. Weak-binding molecules are not drugs? Toward a systematic strategy for finding effective weak-binding drugs. Briefing Bioinf. 2017;18:321–32.
104. Mestres J, Gregori-Puigjane E. Conciliating binding efficiency and polypharmacology. Trends Pharmacol Sci. 2009;30:470–4.
105. Jeong SK, Yang K, Park YS, et al. Interferon gamma induced by resveratrol analog, HS-1793, reverses the properties of tumor associated macrophages. Int Immunopharmacol. 2014;22:303–10.
106. Lai X, Pei Q, Song X, et al. The enhancement of immune function and activation of NF-kappaB by resveratrol-treatment in immunosuppressive mice. Int Immunopharmacol. 2016;33:42–7.

107. Jiao Y, Hannafon BN, Ding WQ. Disulfiram's anti-cancer activity: evidence and mechanisms. Anti Cancer Agents Med Chem. 2016;16:1378–84.
108. Chen D, Cui QC, Yang H, et al. Disulfiram, a clinically used anti-alcoholism drug and copper-binding agent, induces apoptotic cell death in breast cancer cultures and xenografts via inhibition of the proteasome activity. Cancer Res. 2006;66:10425–33.
109. Kim JY, Cho Y, Oh E, et al. Disulfiram targets cancer stem-like properties and the HER2/Akt signaling pathway in HER2-positive breast cancer. Cancer Lett. 2016;379:39–48.
110. Liu X, Wang L, Cui W, et al. Targeting ALDH1A1 by disulfiram/copper complex inhibits non-small cell lung cancer recurrence driven by ALDH-positive cancer stem cells. Oncotarget. 2016;7:58516–30.
111. Hatoum D, McGowan EM. Recent advances in the use of metformin: can treating diabetes prevent breast cancer? Biomed Res Int. 2015;2015:548436.
112. Gwak H, Kim Y, An H, et al. Metformin induces degradation of cyclin D1 via AMPK/GSK3beta axis in ovarian cancer. Mol Carcinog. 2017;56:349–58.
113. Gan RY, Li HB, Sui ZQ, et al. Absorption, metabolism, anticancer effect and molecular targets of epigallocatechin gallate (EGCG): an updated review. Crit Rev Food Sci Nutr. 2018;58(6):924–41.
114. Singh BN, Shankar S, Srivastava RK. Green tea catechin, epigallocatechin-3-gallate (EGCG): mechanisms, perspectives and clinical applications. BiochemPharmacol. 2011;82:1807–21.
115. Cheng CW, Shieh PC, Lin YC, et al. Indoleamine 2,3-dioxygenase, an immunomodulatory protein, is suppressed by (−)-epigallocatechin-3-gallate via blocking of gammainterferon-induced JAK-PKC-delta-STAT1 signaling in human oral cancer cells. J Agric Food Chem. 2010;58:887–94.
116. Shim JH, Choi HS, Pugliese A, et al. (-)-Epigallocatechin gallate regulates CD3-mediated T cell receptor signaling in leukemia through the inhibition of ZAP-70 kinase. J Biol Chem. 2008;283:28370–9.
117. Jin Y, Khadka DB, Cho WJ. Pharmacological effects of berberine and its derivatives: a patent update. Expert Opin Ther Pat. 2016;26:229–43.
118. Jabbarzadeh Kaboli P, Rahmat A, Ismail P, et al. Targets and mechanisms of berberine, a natural drug with potential to treat cancer with special focus on breast cancer. Eur J Pharmacol. 2014;740:584–95.
119. Wang N, Tan HY, Li L, et al. Berberine and Coptidis Rhizoma as potential anticancer agents: recent updates and future perspectives. J Ethnopharmacol. 2015;176:35–48.
120. Mantena SK, Sharma SD, Katiyar SK. Berberine, a natural product, induces G1-phase cell cycle arrest and caspase-3-dependent apoptosis in human prostate carcinoma cells. Mol Cancer Ther. 2006;5:296–308.
121. Ji C, Yang B, Yang YL, et al. Exogenous cell-permeable C6 ceramide sensitizes multiple cancer cell lines to Doxorubicin-induced apoptosis by promoting AMPK activation and mTORC1 inhibition. Oncogene. 2010;29:6557–68.
122. Lu JJ, Fu L, Tang Z, et al. Melatonin inhibits AP-2beta/hTERT, NF-kappaB/COX-2 and Akt/ERK and activates caspase/Cyto C signaling to enhance the antitumor activity of berberine in lung cancer cells. Oncotarget. 2016;7:2985–3001.
123. Ho YT, Yang JS, Li TC, et al. Berberine suppresses in vitro migration and invasion of human SCC-4 tongue squamous cancer cells through the inhibitions of FAK, IKK, NF-kappaB, u-PA and MMP-2 and -9. Cancer Lett. 2009;279:155–62.
124. Deng S, Hu B, Shen KP, et al. Inflammation, macrophage in cancer progression and Chinese herbal treatment. J Basic Clin Pharm. 2012;3:269–72.
125. Yildirim MA, Goh KI, Cusick ME, et al. Drug-target network. Nat Biotechnol. 2007;25:1119–26.
126. Bento AP, Gaulton A, Hersey A, et al. The ChEMBL bioactivity database: an update. Nucleic Acids Res. 2014;42:D1083–90.
127. Gilson MK, Liu T, Baitaluk M, et al. BindingDB in 2015: a public database for medicinal chemistry, computational chemistry and systems pharmacology. Nucleic Acids Res. 2016;44:D1045–53.
128. Wang Y, Bolton E, Dracheva S, et al. An overview of the PubChem BioAssay resource. Nucleic Acids Res. 2010;38:D255–66.
129. Law V, Knox C, Djoumbou Y, et al. DrugBank 4.0: shedding new light on drug metabolism. Nucleic Acids Res. 2014;42:D1091–7.
130. Yang H, Qin C, Li YH, et al. Therapeutic target database update 2016: enriched resource for bench to clinical drug target and targeted pathway information. Nucleic Acids Res. 2016;44:D1069–74.
131. Wagner AH, Coffman AC, Ainscough BJ, et al. DGIdb 2.0: mining clinically relevant drug-gene interactions. Nucleic Acids Res. 2016;44:D1036–44.
132. Kuhn M, Szklarczyk D, Pletscher-Frankild S, et al. STITCH 4: integration of protein-chemical interactions with user data. Nucleic Acids Res. 2014;42:D401–7.
133. Nickel J, Gohlke BO, Erehman J, et al. SuperPred: update on drug classification and target prediction. Nucleic Acids Res. 2014;42:W26–31.
134. Ye H, Ye L, Kang H, et al. HIT: linking herbal active ingredients to targets. Nucleic Acids Res. 2011;39:D1055–9.
135. Xue R, Fang Z, Zhang M, et al. TCMID: traditional Chinese Medicine integrative database for herb molecular mechanism analysis. Nucleic Acids Res. 2013;41:D1089–95.
136. Ru J, Li P, Wang J, et al. TCMSP: a database of systems pharmacology for drug discovery from herbal medicines. J Cheminform. 2014;6:13.
137. Mangal M, Sagar P, Singh H, et al. NPACT: naturally occurring plant-based anti-cancer compound-activity-target database. Nucleic Acids Res. 2013;41:D1124–9.

138. Tao W, Li B, Gao S, et al. CancerHSP: anticancer herbs database of systems pharmacology. Sci Rep. 2015;5:11481.
139. Bredel M, Jacoby E. Chemogenomics: an emerging strategy for rapid target and drug discovery. Nat Rev Genet. 2004;5:262–75.
140. Lamb J, Crawford ED, Peck D, et al. The Connectivity Map: using gene-expression signatures to connect small molecules, genes, and disease. Science. 2006;313:1929–35.
141. Duan Q, Flynn C, Niepel M, et al. LINCS Canvas Browser: interactive web app to query, browse and interrogate LINCS L1000 gene expression signatures. Nucleic Acids Res. 2014;42:W449–60.
142. Igarashi Y, Nakatsu N, Yamashita T, et al. Open TG-GATEs: a large-scale toxicogenomics database. Nucleic Acids Res. 2015;43:D921–7.
143. Ganter B, Snyder RD, Halbert DN, et al. Toxicogenomics in drug discovery and development: mechanistic analysis of compound/class-dependent effects using the DrugMatrix database. Pharmacogenomics. 2006;7:1025–44.
144. Lamb J. The connectivity map: a new tool for biomedical research. Nat Rev Cancer. 2007;7:54–60.
145. Adams JU. Genetics: big hopes for big data. Nature. 2015;527:S108–9.
146. Chatr-Aryamontri A, Breitkreutz BJ, Oughtred R, et al. The BioGRID interaction database: 2015 update. Nucleic Acids Res. 2015;43:D470–8.
147. Keshava Prasad TS, Goel R, Kandasamy K, et al. Human protein reference database–2009 update. Nucleic Acids Res. 2009;37:D767–72.
148. Mosca R, Ceol A, Aloy P. Interactome3D: adding structural details to protein networks. Nat Methods. 2013;10:47–53.
149. Szklarczyk D, Morris JH, Cook H, et al. The STRING database in 2017: quality-controlled protein-protein association networks, made broadly accessible. Nucleic Acids Res. 2017;45:D362–8.
150. Licata L, Briganti L, Peluso D, et al. MINT, the molecular interaction database: 2012 update. Nucleic Acids Res. 2012;40:D857–61.
151. Cheng F, Jia P, Wang Q, et al. Quantitative network mapping of the human kinome interactome reveals new clues for rational kinase inhibitor discovery and individualized cancer therapy. Oncotarget. 2014;5:3697–710.
152. Cheng F, Liu C, Lin CC, et al. A gene gravity model for the evolution of cancer genomes: a study of 3,000 cancer genomes across 9 cancer types. PLoS Comput Biol. 2015;11:e1004497.
153. Zhang C, Hong H, Mendrick DL, et al. Biomarker-based drug safety assessment in the age of systems pharmacology: from foundational to regulatory science. BiomarkMed. 2015;9:1241–52.
154. Cheng F, Murray JL, Zhao J, et al. Systems biology-based investigation of cellular antiviral drug targets identified by gene-trap insertional mutagenesis. PLoS Comput Biol. 2016;12:e1005074.
155. Cheng F, Liu C, Shen B, et al. Investigating cellular network heterogeneity and modularity in cancer: a network entropy and unbalanced motif approach. BMC Syst Biol. 2016;10(Suppl 3):65.
156. Cheng F, Zhao J, Hanker AB, et al. Transcriptome- and proteome-oriented identification of dysregulated eIF4G, STAT3, and Hippo pathways altered by PIK3CA H1047R in HER2/ER-positive breast cancer. Breast Cancer Res Treat. 2016;160:457–74.
157. Cheng F, Jia P, Wang Q, et al. Studying tumorigenesis through network evolution and somatic mutational perturbations in the cancer interactome. Mol Biol Evol. 2014;31:2156–69.
158. Cheng F, Zhao J, Zhao Z. Advances in computational approaches for prioritizing driver mutations and significantly mutated genes in cancer genomes. Brief Bioinform. 2016;17:642–56.
159. Vogelstein B, Papadopoulos N, Velculescu VE, et al. Cancer genome landscapes. Science. 2013;339:1546–58.
160. Hayes DN, Kim WY. The next steps in next-gen sequencing of cancer genomes. J Clin Invest. 2015;125:462–8.
161. Chin L, Andersen JN, Futreal PA. Cancer genomics: from discovery science to personalized medicine. Nat Med. 2011;17:297–303.
162. International Cancer Genome C, Hudson TJ, Anderson W, et al. International network of cancer genome projects. Nature. 2010;464:993–8.
163. Gao J, Aksoy BA, Dogrusoz U, et al. Integrative analysis of complex cancer genomics and clinical profiles using the cBioPortal. Sci Signal. 2013;6:pl1.
164. Forbes SA, Beare D, Gunasekaran P, et al. COSMIC: exploring the world's knowledge of somatic mutations in human cancer. Nucleic Acids Res. 2015;43:D805–11.
165. Nakagawa H, Wardell CP, Furuta M, et al. Cancer whole-genome sequencing: present and future. Oncogene. 2015;34:5943–50.
166. Chen X, Yan CC, Zhang X, et al. Drug-target interaction prediction: databases, web servers and computational models. Brief Bioinform. 2016;17:696–712.
167. Koutsoukas A, Simms B, Kirchmair J, et al. From in silico target prediction to multi-target drug design: current databases, methods and applications. J Proteome. 2011;74:2554–74.
168. Yue R, Shan L, Yang X, et al. Approaches to target profiling of natural products. Curr Med Chem. 2012;19:3841–55.
169. Jenkins JL, Bender A, Davies JW. In silico target fishing: predicting biological targets from chemical structure. Drug Discov Today Technol. 2007;3:413–21.
170. Harren J, Andrew RL. Structure-based drug discovery. Dordrecht: Springer; 2007. ISB ISBN: 978-1-4020-4406-9
171. Taboureau O, Baell JB, Fernandez-Recio J, et al. Established and emerging trends in computational drug discovery in the structural genomics era. Chem Biol. 2012;19:29–41.

172. Sakkiah S, Ng HW, Tong W, et al. Structures of androgen receptor bound with ligands: advancing understanding of biological functions and drug discovery. Expert Opin Ther Targets. 2016;20:1267–82.
173. Liu LJ, Leung KH, Chan DS, et al. Identification of a natural product-like STAT3 dimerization inhibitor by structure-based virtual screening. Cell Death Dis. 2014;5:e1293.
174. Zhong HJ, Lee BR, Boyle JW, et al. Structure-based screening and optimization of cytisine derivatives as inhibitors of the menin-MLL interaction. Chem Commun. 2016;52:5788–91.
175. Singh T, Gupta NA, Xu S, et al. Honokiol inhibits the growth of head and neck squamous cell carcinoma by targeting epidermal growth factor receptor. Oncotarget. 2015;6:21268–82.
176. Zhong HJ, Ma VP, Cheng Z, et al. Discovery of a natural product inhibitor targeting protein neddylation by structure-based virtual screening. Biochimie. 2012;94:2457–60.
177. Lee HM, Chan DS, Yang F, et al. Identification of natural product fonsecin B as a stabilizing ligand of c-myc G-quadruplex DNA by high-throughput virtual screening. Chem Commun. 2010;46:4680–2.
178. Ma D-L, Chan DS-H, Leung C-H. Molecular docking for virtual screening of natural product databases. ChemSci. 2011;2:1656–65.
179. Cierpicki T, Grembecka J. Challenges and opportunities in targeting the menin-MLL interaction. Future Med Chem. 2014;6:447–62.
180. Chen YZ, Zhi DG. Ligand-protein inverse docking and its potential use in the computer search of protein targets of a small molecule. Proteins. 2001;43:217–26.
181. Wang JC, Chu PY, Chen CM, et al. idTarget: a web server for identifying protein targets of small chemical molecules with robust scoring functions and a divide-and-conquer docking approach. Nucleic Acids Res. 2012;40:W393–9.
182. Lauro G, Romano A, Riccio R, et al. Inverse virtual screening of antitumor targets: pilot study on a small database of natural bioactive compounds. J Nat Prod. 2011;74:1401–7.
183. Lauro G, Masullo M, Piacente S, et al. Inverse virtual screening allows the discovery of the biological activity of natural compounds. Bioorg Med Chem. 2012;20:3596–602.
184. Vuong H, Cheng F, Lin CC, et al. Functional consequences of somatic mutations in cancer using protein pocket-based prioritization approach. Genome Med. 2014;6:81.
185. Zhao J, Cheng F, Wang Y, et al. Systematic prioritization of druggable mutations in approximately 5000 genomes across 16 cancer types using a structural genomics-based approach. Mol Cell Proteomics. 2016;15:642–56.
186. Lavecchia A. Machine-learning approaches in drug discovery: methods and applications. Drug Discov Today. 2015;20:318–31.
187. Yan X, Liao C, Liu Z, et al. Chemical structure similarity search for ligand-based virtual screening: methods and computational resources. Curr Drug Targets. 2016;17:1580–5.
188. Liu X, Ouyang S, Yu B, et al. PharmMapper server: a web server for potential drug target identification using pharmacophore mapping approach. Nucleic Acids Res. 2010;38:W609–14.
189. Yan X, Li J, Liu Z, et al. Enhancing molecular shape comparison by weighted Gaussian functions. J Chem Inf Model. 2013;53:1967–78.
190. Fang J, Yang R, Gao L, et al. Predictions of BuChE inhibitors using support vector machine and naive Bayesian classification techniques in drug discovery. J Chem Inf Model. 2013;53:3009–20.
191. Fang J, Yang R, Gao L, et al. Consensus models for CDK5 inhibitors in silico and their application to inhibitor discovery. Mol Divers. 2015;19:149–62.
192. Fang J, Li Y, Liu R, et al. Discovery of multitarget-directed ligands against Alzheimer's disease through systematic prediction of chemical-protein interactions. J Chem Inf Model. 2015;55:149–64.
193. Fang J, Pang X, Wu P, et al. Discovery of neuroprotective compounds by machine learning approaches. RSC Adv. 2016;6:9857.
194. Cheng F, Li W, Liu G, et al. In silico ADMET prediction: recent advances, current challenges and future trends. Curr Top Med Chem. 2013;13:1273–89.
195. Sprague B, Shi Q, Kim MT, et al. Design, synthesis and experimental validation of novel potential chemopreventive agents using random forest and support vector machine binary classifiers. J Comput Aided Mol Des. 2014;28:631–46.
196. Bilsland AE, Pugliese A, Liu Y, et al. Identification of a selective G1-phase benzimidazolone inhibitor by a senescence targeted virtual screen using artificial neural networks. Neoplasia. 2015;17:704–15.
197. Liu H, Sun J, Guan J, et al. Improving compound-protein interaction prediction by building up highly credible negative samples. Bioinformatics. 2015;31:i221–9.
198. Mousavian Z, Masoudi-Nejad A. Drug-target interaction prediction via chemogenomic space: learning-based methods. Expert Opin Drug Metab Toxicol. 2014;10:1273–87.
199. Yamanishi Y. Chemogenomic approaches to infer drug target interaction networks. Methods Mol Biol. 2013;939:97–113.
200. Zhao S, Li S. Network-based relating pharmacological and genomic spaces for drug target identification. PLoS One. 2010;5:e11764.
201. Yamanishi Y, Araki M, Gutteridge A, et al. Prediction of drug target interaction networks from the integration of chemical and genomic spaces. Bioinformatics. 2008;24:i232–40.
202. Bleakley K, Yamanishi Y. Supervised prediction of drug target interactions using bipartite local models. Bioinformatics. 2009;25:2397–403.
203. Keum J, Yoo S, Lee D, et al. Prediction of compound-target interactions of natural products

using large-scale drug and protein information. BMC Bioinformatics. 2016;17(Suppl 6):219.
204. Yu H, Chen J, Xu X, et al. A systematic prediction of multiple drug-target interactions from chemical, genomic, and pharmacological data. PLoS One. 2012;7:e37608.
205. Huang C, Zheng C, Li Y, et al. Systems pharmacology in drug discovery and therapeutic insight for herbal medicines. Brief Bioinform. 2014;15:710–33.
206. Sawada R, Kotera M, Yamanishi Y. Benchmarking a wide range of chemical descriptors for drug-target interaction prediction using a chemogenomic approach. Mol Inform. 2014;33:719–31.
207. Cheng F, Zhou Y, Li J, et al. Prediction of chemical-protein interactions: multitarget-QSAR versus computational chemogenomic methods. Mol BioSyst. 2012;8:2373–84.
208. Cheng F, Li W, Zhou Y, et al. Prediction of human genes and diseases targeted by xenobiotics using predictive toxicogenomic derived models (PTDMs). Mol BioSyst. 2013;9:1316–25.
209. Kitano H. Systems biology: a brief overview. Science. 2002;295:1662–4.
210. Kitano H. Computational systems biology. Nature. 2002;420:206–10.
211. Berg EL. Systems biology in drug discovery and development. Drug Discov Today. 2014;19:113–25.
212. Cheng F, Murray JL, Rubin DH. Drug repurposing: new treatments for Zika virus infection? Trends Mol Med. 2016;22:919–21.
213. Qu XA, Rajpal DK. Applications of connectivity map in drug discovery and development. Drug Discov Today. 2012;17:1289–98.
214. Hieronymus H, Lamb J, Ross KN, et al. Gene expression signature-based chemical genomic prediction identifies a novel class of HSP90 pathway modulators. Cancer Cell. 2006;10:321–30.
215. Wei G, Twomey D, Lamb J, et al. Gene expression-based chemical genomics identifies rapamycin as a modulator of MCL1 and glucocorticoid resistance. Cancer Cell. 2006;10:331–42.
216. Langley SR, Dwyer J, Drozdov I, et al. Proteomics: from single molecules to biological pathways. Cardiovasc Res. 2013;97:612–22.
217. Bensimon A, Heck AJ, Aebersold R. Mass spectrometry-based proteomics and network biology. Annu Rev Biochem. 2012;81:379–405.
218. Savitski MM, Reinhard FB, Franken H, et al. Tracking cancer drugs in living cells by thermal profiling of the proteome. Science. 2014;346:1255784.
219. Franken H, Mathieson T, Childs D, et al. Thermal proteome profiling for unbiased identification of direct and indirect drug targets using multiplexed quantitative mass spectrometry. Nat Protoc. 2015;10:1567–93.
220. Reinhard FB, Eberhard D. Thermal proteome profiling monitors ligand interactions with cellular membrane proteins. Nat Methods. 2015;12:1129–31.
221. Sacco F, Silvestri A, Posca D, et al. Deep proteomics of breast cancer cells reveals that metformin rewires signaling networks away from a pro-growth state. Cell Syst. 2016;2:159–71.
222. Kaddurah-Daouk R, Kristal BS, Weinshilboum RM. Metabolomics: a global biochemical approach to drug response and disease. Annu Rev Pharmacol Toxicol. 2008;48:653–83.
223. Birkenstock T, Liebeke M, Winstel V, et al. Exometabolome analysis identifies pyruvate dehydrogenase as a target for the antibiotic triphenylbismuthdichloride in multiresistant bacterial pathogens. J Biol Chem. 2012;287:2887–95.
224. Zhang B, Watts KM, Hodge D, et al. A second target of the antimalarial and antibacterial agent fosmidomycin revealed by cellular metabolic profiling. Biochemistry. 2011;50:3570–7.
225. Bayet-Robert M, Lim S, Barthomeuf C, et al. Biochemical disorders induced by cytotoxic marine natural products in breast cancer cells as revealed by proton NMR spectroscopy-based metabolomics. Biochem Pharmacol. 2010;80:1170–9.
226. Pulido MR, Garcia-Quintanilla M, Gil-Marques ML, et al. Identifying targets for antibiotic development using omics technologies. Drug Discov Today. 2016;21:465–72.
227. Zhao Y, Hu Q, Cheng F, et al. SoNar, a highly responsive NADþ/NADH sensor, allows high-throughput metabolic screening of anti-tumor agents. Cell Metab. 2015;21:777–89.
228. Kim HK, Wilson EG, Choi YH, et al. Metabolomics: a tool for anticancer lead-finding from natural products. Planta Med. 2010;76:1094–102.
229. Collins GS, de Groot JA, Dutton S, et al. External validation of multivariable prediction models: a systematic review of methodological conduct and reporting. BMC Med Res Methodol. 2014;14:40.

16 Polypharmacology and Polypharmacokinetics

Abstract

Polypharmacokinetics is a new concept in pharmacology. Polypharmacokinetics to polypharmacology is like pharmacokinetics to pharmacology. While pharmacokinetics normally deals with only a single compound with its liberation, absorption, distribution, metabolism, and excretion (LADME), polypharmacokinetics studies multiple compounds as a single therapeutic regimen, a single formulation, or a single pill/tablet. Clearly, polypharmacokinetics is enormously complex and intricate, yet being critical to ensure synergetic efficacy and improved safety with mitigated toxic effects of the clinical applications of polypharmacology in disease therapy. The present chapter summarizes the most updated studies on polypharmacokinetics with an emphasis on the metabolic approach for elucidating LADME of drug therapy and the influence of drug–drug interactions on polypharmacological therapy of human disease. Specifically, this chapter will start with an introduction to the general concept of pharmacokinetics and polypharmacokinetics and then turn to detailed descriptions of the metabolomics approach for studying polypharmacokinetics. Finally, the impact of drug–drug interaction on polypharmacokinetics will be discussed.

Keywords

Polypharmacokinetics (poly-PK) · Metabolomics · LADME · Combination drug therapy · Drug–drug interactions (DDIs)

16.1 General Concept of Pharmacokinetics and Polypharmacokinetics

16.1.1 Basics of Pharmacokinetics (PK)

Pharmacokinetics, often abbreviated as PK, is a branch of pharmacology dedicated to determining the fate of substances administered to a living organism [1, 2]. The substances of interest include any chemical xenobiotic such as pharmaceutical drugs, pesticides, food additives, cosmetics, etc. PK attempts to analyze chemical metabolism in human body and to discover the fate of a chemical from the moment that it is administered up to the point at which it is completely eliminated from the body. Whereas pharmacodynamics (PD) is the study of how the drug affects the organism, PK is the study of how an organism affects a drug. Both together influence dosing, benefit, and adverse effects of medicines. Hence, pharmacokinetics plays an important role in drug development [3–5].

Z. Wang, B. Yang, *Polypharmacology*, https://doi.org/10.1007/978-3-031-04998-9_16

PK describes how the body affects a specific xenobiotic/chemical after administration through the mechanisms of absorption, distribution, metabolism, and excretion [1, 2]. Clinical pharmacokinetics is the application of pharmacokinetic principles to the safe and effective therapeutic management of drugs in an individual patient, as well as the metabolic changes of the substance in the body (e.g., by metabolic enzymes such as cytochrome P450 or glucuronosyltransferase enzymes), and the effects and routes of excretion of the metabolites of the drug. Pharmacokinetic properties of chemicals are affected by the route of administration and the dose of administered drug. These may affect the absorption rate [6].

PK of a single compound is commonly referred to as the ADME scheme (absorption, distribution, metabolism, and excretion) or LADME if liberation is included as a separate step from absorption:

Liberation: the process of release of a drug from the pharmaceutical formulation [7, 8].

Absorption: the process of a substance entering the blood circulation.

Distribution: the dispersion or dissemination of substances throughout the fluids and tissues of the body.

Metabolism: also called biotransformation or inactivation. It is the recognition by the organism that a foreign substance is present and the irreversible transformation of parent compounds into daughter metabolites.

Excretion: the removal of the substances from the body. In rare cases, some drugs irreversibly accumulate in body tissue.

16.1.2 Differences Between Polypharmacokinetics and Regular Pharmacokinetics

For single-target drugs (STDs) of modern pharmacology (Single Agent on Single Target for Single Disease' Strategy; see Chaps. 1 and 2 for detail), PK depends on patient-related factors as well as on the chemical properties of a drug. Some patient-related factors (e.g., renal function, genetic makeup, sex, age, race, pregnancy, etc.) can be used to predict the pharmacokinetic parameters in populations [2, 9–16]. For example, the half-life of some drugs, especially those that require both metabolism and excretion, may be remarkably long in older people than in young men.

For combination drug therapy (CDT) (Multiple Agents on Multiple Targets for Single Disease or Combinational Drug Therapy) and fixed-dose combination (FDC) (Single Pill with Multiple Agents on Multiple Targets for Single Disease or Polypill; see Chaps. 1 and 2 for detail) of polypharmacology, however, PK or more precisely polypharmacokinetics (poly-PK) depends not only on patient-related factors and the chemical properties of drugs but also on interactions among multiple medications. In this perspective, traditional medicine (TM) that is considered natural multicomponent medicine resembles CDT and FDC in terms of its poly-PK.

Intriguingly, a multitarget drug (MTD) (Single Agent on Multiple Targets for Complex Disease) drug essentially follows the principles of regular pharmacokinetics as a STD despite that pharmacodynamically, it interacts with multiple therapeutic targets.

Poly-PK differs from regular (single-compound) pharmacokinetics in several aspects:

First, poly-PK involves multiple chemical components administered simultaneously or sequentially, and different components likely have distinct chemical properties, thereby distinct metabolism in the body [17–19].

Second, poly-PK involves complex metabolic and pharmacokinetic interactions among multiple co-applied drugs and between drugs and host cells/tissues, which are absent in regular pharmacokinetics.

Third, a multicomponent drug is often administered with different components at different concentrations/doses.

Finally, different components in a multitarget drug are likely metabolized or processed in different organs and excreted via different routes.

These differences can well make poly-PK more complicated and difficult to determine than regular pharmacokinetics.

Thus, the traditional approach to understand the pharmacology of a multicomponent agent is to study the effects of single components on single biological reactions, enzymes, genes, and so forth, and gradually assemble the findings into a whole picture [20]. However, assembling the results obtained from such a reductionist approach to achieve a system understanding of a concerted pharmacological intervention has proven impractical [20, 21]. The results of such attempts all failed to accurately capture the complex pharmacokinetic behavior of compound drugs and herbal medicines. Additionally, the PK of a given compound in a multicomponent assay may be significantly different from that in a single compound assay due to drug–drug interactions in human body.

At present, the most used method for poly-PK analysis is metabolomics approach. Automated blood sampling (ABS) in conjunction with liquid chromatography–mass spectrometry (LC–MS) analysis is another well-described technique. Additionally, in silico analysis of drug–drug interactions is also an important tool for poly-PK. This chapter will focus on these three methodologies.

16.2 Metabolomics Approach for Polypharmacokinetics

Lan and colleagues [20] developed a metabolomics approach coupled with multivariate statistical tools for comprehensive analysis of small molecules in biofluids to address poly-PK, the metabolism of multicomponent xenobiotic agents, such as compound drugs (CDT and FDC drugs), dietary supplements, and herbal medicines. They proposed a research framework that integrates the dynamic concentration profile of bioavailable xenobiotic molecules that result from in vivo absorption and hepatic and gut bacterial metabolism, as well as the human metabolic response profile. Notably, this framework also helps elucidate the pharmacological and molecular mechanisms of the polypharmacological compounds. As a matter of fact, the approaches and methodologies of poly-PK were primarily developed based on the analysis of herbal medicines that contain a vast number of components with dynamic pharmacodynamics and pharmacokinetics. Below is a brief introduction of their metabolomics approach and framework on herbal medicines [20].

16.2.1 Basic Concept of Metabolomics

Metabolomics is the study of the metabolome, the complete set of endogenous metabolites in a biological sample. In context of systems biology [22, 23], metabolomics is favored not only for decoding how a biological system responds to a stimulus by identifying the most significantly affected endogenous metabolites and their metabolic pathways but also resolving the relationships between endogenous and xenobiotic metabolic processes [24]. Metabolomics techniques have offered xenobiotic studies a novel way to differentiate the exogenous from the endogenous by untangling the interactive metabolic pathways for both. Capable of simultaneously analyzing hundreds and thousands of variables, metabolomics is uniquely suited to develop a new generation of PK platforms that can measure multiple nutraceutical components in vivo, as well as identify characteristic metabolic alterations in humans exposed to nutraceuticals [25, 26]. Along these lines, utilizing metabolomics to study herbal medicine efficacy and toxicity has been a key focus of recent herbal and pharmaceutical research [27–31]. Recently, we proposed an integrated metabolomic profiling strategy for PK and PD studies of multicomponent drugs using tandem mass spectrometry (MS/MS) [32] and conducted a proof-of-concept poly-PK study of Pu-erh tea intervention in humans based on metabolomics [33]. Metabolomics might assume the new mission to model the variations of both exogenous and endogenous metabolites through a given biological system, which has vital implications in both systems biology and multicomponent medications.

Distinct from the pharmacology of single compound drugs, the polypharmacology of multicomponent agents entails a "network" approach, in which multiple compounds interact in vivo with multiple targets to elicit interdependent activities and achieve an optimal effect [34, 35]. Due to the complexity of biological samples (e.g., blood, urine, tissues, and botanical samples), the analytical approaches to quantitatively measure the time-dependent concentration profiles of bioavailable molecules (i.e., poly-PK) are beyond the scope of traditional research. With such a complex network involving a large number of variables, it is technically challenging to identify each metabolite that changes significantly in the global metabolite pool, and to assess the human (endogenous) biochemical responses to exposure to these exogenous (xenobiotic) compounds.

The metabolic fate of chemicals in the human body is primarily determined by the chemical's structure and varies with dose, routes of exposure, inter- and intraindividual genetic differences, gut microbiota, diet, lifestyle, and environment, as well as other xenobiotics intentionally or unintentionally present. In order to simplify these intricate factors, a knowledge-based engine that sequentially manipulates each factor in silico for prediction, coupled with in vitro screening and identification, and in vivo validation is deployed for xenobiotic metabolism studies [36]. However, because cross-platform extrapolations are not always reliable or robust [37], the in vivo xenobiotic metabolite studies require special attention; these compounds are often intermixed against the background of endogenous metabolites in biological samples. Recent advances in liquid and gas chromatography (LC and GC) and capillary electrophoresis (CE) coupled with mass spectrometry (LC–MS, GC–MS, and CE–MS) and nuclear magnetic resonance (NMR) have made it possible to simultaneously detect various small-molecule metabolites [38].

16.2.2 Metabolomics in Xenobiotic Intervention Study

Metabolomics enables the profiling of a vast number of small-molecule metabolites in an organism or a biological system as a whole and helps understand how these metabolites respond to a stimulus, highlighting the most relevant potential biomarkers. When the stimulus is a xenobiotic intervention, metabolomics techniques can be employed to profile the metabolite pool with and without the xenobiotic and highlight the differential variables to identify the xenobiotic metabolites. The major difference between the two methods is the data processing procedure to differentiate exogenous metabolites against the endogenous metabolite background. The metabolomics strategy employs multivariate data pretreatment methods [39] (e.g., deconvolution, alignment, integration) coupled with multivariate analyses [26, 40] (e.g., principal component analysis (PCA), partial least squares discriminant analysis (PLS-DA)). In contrast, the classical strategy utilizes arbitrary endogenous background subtraction and/or knowledge-based mass defect filtering, which takes advantage of known mass changes between the parent ion and metabolite ions [41].

Such a metabolomics-based protocol was first proposed in 2003 by Plumb and colleagues in an attempt to find the in vivo metabolites of GSK-X, the structure of which was unknown to the investigators [42]. Shortly after, another metabolomics-based protocol identified several new metabolites of citalopram [43]. Consequently, Dr. Gonzalez from the National Cancer Institute at the National Institutes of Health expanded LC–MS-based metabolomics for xenobiotic metabolism studies by developing various strategies, including (a) identification of metabolites by metabolomic analysis with or without xenobiotic intervention [44], (b) validation of metabolites through metabolomic comparison between xenobiotic treatments with or without stable isotope label in order to highlight the xenobiotic metabolites in contrast to the endogenous ones altered by xenobiotic intervention, (c) identification of metabolic pathways through metabolomic comparison of wild-type and genetically modified mice, and (d) identification of polymorphisms of genes encoding human xenobiotic metabolism enzymes (XMEs) that may be responsible for adverse drug reactions [44]. Two recent reviews address the technical issues and progress in this field [45, 46].

Compared to the classical methods in xenobiotic intervention studies, metabolomics strategies have clear advantages [20]: first, in the capacity of handling a great number of variables—which allows a shift in purpose from targeting xenobiotic metabolites to profiling the complete set of metabolites in biological samples—and second, in the unbiased selection of variables that are significantly altered, which facilitates the discrimination of xenobiotic metabolites from endogenous ones. Due to these advantages of metabolomic strategies, a novel, evidence-based methodology that starts with mapping in vivo metabolites and leads to further in vitro characterization has been established, which is in stark contrast to the classic, knowledge-based methodology that starts with in vitro identification to in vivo characterization and is frequently muddled by inconsistent results [20]. Such a methodological shift obviates potential false-negative results that usually result from traditional empirical strategies, and significantly deepens our insight into the in vivo metabolic map of xenobiotics in at least two ways. First, metabolomics captures in vivo xenobiotic metabolites that are otherwise easily overwhelmed by the endogenous background. Examples of such metabolites are those with unexpected skeleton biotransformation or those at extremely trace levels with potential clinical significance and wide usage in the general population. Second, the consideration of genetic and environmental factors affecting the metabolism of xenobiotics in metabolomics studies facilitates the in vivo identification of metabolic pathways and factors affecting them.

16.2.3 Metabolomics-Based Strategy for Poly-PK Analysis

The first attempt of utilizing metabolomics strategy for poly-PK was the NMR-based metabolomics study on human nutritional intervention of grape/wine extract reported by van Velzen and colleagues in 2008 [47]. This study aimed to distinguish between-subject and within-subject variations in metabolomic data matrices using multivariate paired data analyses corresponding to a crossover design, which mainly concerned about the metabolites of polyphenols. Later, the same group developed the integration of PK modeling into metabolomics [48]. The proposed multivariate paired data analysis [49] and dynamic metabolomic data analysis [50], in combination with the study designs that tackle the challenges arising from dietary metabolite background and microbial–human co-metabolism in the gut [51], have collectively coined the term "nutrikinetics," a new discipline that combines cutting-edge technologies and new methodologies to get a complete picture of what happens to the consumed metabolites and supplements in the human body [52, 53].

The metabolomics-based strategies utilizing automated blood sampling in conjunction with high-performance liquid chromatography coupled to mass spectrometry (LC/MS) have clear advantages, as they can handle a great number of variables and large datasets, enable graphic representation of metabolism-related sample classification, and identify drug compounds and significant drug metabolites. In fact, several publications reported the use of metabolomics to capture both drug-derived and drug-induced metabolites [54] and the alterations of metabolites in urine after cocoa powder consumption [32, 55, 56].

Metabolomics is a "top-down" -omics technology that can be used for evaluating small-molecule metabolites as the ultimate downstream products of genomic, transcriptomic, and/or proteomic perturbations. Besides its phytochemical applications focusing on the quality assessments of botanic products [57–59], metabolomics in herbal medicine mostly addresses biomedical implications for the endogenous metabolism [60–64], while paying little attention to the metabolism of the xenobiotics, which is tightly linked to and has important implications for human metabolic networks.

Lan et al. [20] proposed a research framework to integrate the dynamic concentration profile of bioavailable xenobiotic molecules due to in vivo absorption and the hepatic and gut bacterial metabolism, as well as the human metabolic response profile. Their framework highlights the flux of small-molecule metabolites through the human body with gut as the major entrance and urinary lumen as the major exit points. Recent multicompartmental (plasma, urine, and cecal contents) metabolomic work depicted this framework in a smaller scale [65]. In our proposed model, the human body has evolved as a superorganism encompassing the genome, transcriptome, proteome, metabolome, and symbiotic gut microbiome, that latter coexisting with the host depending on genetic and environmental factors [66–70]. It is of special importance for the gut microbiome and the host to maintain ceaseless substance exchange, in which the small-molecule metabolites can permeate the biomembrane barrier. These metabolites can be divided into endogenous and exogenous metabolites according to their predicted sources; however, the distinction between endogenous and exogenous is not so clear due to the gut microbiome. As a result, metabolites with unclear origins can be classified into an intermediate group. For example, endogenous ethanol is not a metabolite of human metabolism, but a byproduct of gut microbes [71, 72], which is associated with nonalcoholic fatty liver disease [73].

The host–gut microbial metabolic networks for endogenous and exogenous metabolites are highly interactive but with subtle differences in metabolic fates. It is beyond doubt that the superorganism's metabolism has to strike a balance between the uptake of essential nutrients and the elimination of superfluous metabolites to maintain homeostasis, yielding a relatively stable internal environment of endogenous metabolites. The involvement of gut microbiota in endogenous metabolic pathways [74–78], as a matter of fact, has given rise to interactive host–microbiota metabolic, signaling, and immuneinflammatory axes that physiologically connect the gut, liver, muscle, and brain [79]. On the other hand, the superorganism cannot be too lenient with regard to exogenous metabolites, maintaining allostasis to achieve a stable internal environment by eliminating exogenous metabolites. Gut microbiota supplements the human metabolic system with xenobiotics [80] by directly taking part in their metabolism, especially in the metabolism of natural products such as flavonoids, saponins [81], polyphenols [51], and alkaloids [82, 83], as well as the metabolism of exogenous metabolites, such as the competitive sulfation of acetaminophen and p-cresol [84], and the regulation of XMEs [85, 86]. Based on the distinction between the metabolic fates of exogenous and endogenous metabolites, trend analysis was proposed as an alternative to metabolite discovery [87].

16.2.4 Other Metabolomics-Based Platforms for Poly-PK Analysis

To extend and expand the application of the metabolomics approach to poly-PK analysis, the research group led by Jia Wei [88–90] developed an R package polyPK, an automated data analysis platform for simultaneous monitoring of drug constituents and endogenous metabolites of multicomponent drugs and herbal medicines, which integrates information on both the diversity of the drug's chemical composition and its complex effects on the mammalian metabolic pathways [89]. polyPK provides ten functions for data preprocessing, compound identification and grouping, PK parameters calculation, multivariate statistical analysis, correlations, cluster analyses, and resulting visualization [89].

Ping et al. [91] validated a robust platform using Thermo Scientific™ Dionex™ UltiMate™ 3000 rapid separation LC, integrated with

Thermo Scientific™ Q Exactive™ MS (UPLC-HRMS) for acquiring, in a high-resolution full-scan mode, a global profile of all ionized components present in plasma after oral administration of a compound drugs. It offers a powerful tool for the pharmacokinetic screening and characterization of multiple components in CDT/FDC, medicinal herbs, and traditional Chinese medicine formulas, enabling the clarification of metabolic processes and fates of the compound drugs.

16.3 Drug–Drug Interactions (DDIs) and Polypharmacokinetics

For the treatment of complex disorders, patients often take multiple medications at the same time. CDT and FDC are two important and popular forms of multiple medications or polypharmacology. CDT refers to a combination of multiple theoretically single-targeted drugs and FDC is to encapsulate multiple active pharmaceutical ingredients into a single pharmaceutical formulation to produce a final medicinal product, usually in pill form (see Chaps. 1 and 2 for detail). The coexistence of multiple compounds can well lead to inextricable metabolic and pharmacokinetic interactions creating obstacles for pharmacological evaluation and nutraceutical/drug development [92].

Drug interactions can be categorized into three groups: interactions of drug with disease condition (drug–disease interaction), drugs with other drugs (drug–drug interaction, DDIs), and drugs with food (drug–food interaction). Most of the important drug interactions result from a change in the absorption, metabolism, or elimination of a drug. Among the three categories of drug interactions, DDIs have an important impact on pharmacokinetics. DDIs often occur when two or more drugs react with each other and may cause unexpected side effects, particularly when two drugs that have similar (additive) effects or opposite (canceling) effects on the body are administered together [93–95].

DDIs manifest in the mutual influence of drugs on their biotransformation, its slowdown, or acceleration, and leads to a change in the pharmacological action of drugs. Moreover, DDIs can be divided into three types: pharmaceutical, pharmacodynamic, and pharmacokinetic [92, 96]. Pharmaceutical DDIs may appear due to physical or chemical interactions, for example, when drugs are mixed in a syringe before infusion, and such DDIs are rare. Pharmacodynamic DDIs may occur when a pair or more co-administered drugs act on the same physiological system or target. Pharmacokinetic DDIs are very common and occur when one of the drugs ("violator" or "precipitant" drug) affects the absorption, distribution, metabolism, or excretion of another drug ("victim" or "object" drug). Such DDIs provoke an increase or a decrease in the exposure of an object drug and lead to a change in drug pharmacological action. In this study, we focused on the pharmacokinetic DDIs at the metabolism level (biotransformation), the so-called metabolic DDIs.

The most common drug-metabolizing enzymes (DMEs) in the first phase of xenobiotic metabolism in the human body are several isoforms of the cytochrome P450 superfamily. The US Department of Health and Human Services Food and Drug Administration Center for Drug Evaluation and Research (FDA CDER) requires determining which drug-metabolizing enzymes (CYP3A, CYP2D6, CYP2C19, CYP2C9, CYP2C8, CYP2B6, or CYP1A2) metabolize the investigational drug during in vitro studies of metabolic DDIs estimates [97].

In silico methods can help prioritize drug discovery efforts by guiding, but not replacing, in vitro and in vivo experiments [92, 98]. Most of such in silico methods predict DDIs indirectly. A recently presented machine learning (ML) method used different molecular fingerprints to classify compounds as inhibitors or noninhibitors of five major cytochrome P450 isoenzymes [99]. Ligand-based and structure-based methods deal with substrates, inhibitors [100, 101], and inducers [102] of particular DMEs. Results of prediction could help to determine possible DDIs.

However, such conclusions are not sufficiently reliable, as the pairs of substances that are substrates and inhibitors (or inducers) of DMEs may not exhibit DDIs. On the other hand, DDIs have often occurred between substances that could act as substrates, inducers, and inhibitors (that may act by various inhibition mechanisms); for example, this is a widespread case for cytochrome P450 CYP3A4 [103]. At best, a pair of potentially exhibiting DDI substances should be considered together during prediction as the whole entity. However, previously developed ligand-based and structure-based computational methods did not consider two substances in pairs simultaneously. Direct DDIs estimation methods for the pairs of substances include structure resemblance and functional similarities methods and literature-based DDIs prediction methods [92, 104–109]. These methods deal with the pairs of substances but require information about the pharmacokinetics and pharmacodynamics [0, 9, 14], interaction profile, target and side effects [92, 105, 108], and the phenotypic, therapeutic, chemical, and genomic properties [106] of substances or medical records [107]. It is clear that for new, not-yet-synthesized, and virtual substances, such information does not exist. The results of predictions of this group of methods [92, 104–109] have often been presented as datasets containing a bulk conglomerate of information about potential DDIs predicted between the existing drugs. Such examples include 430,128 [89, 105], 145,068 [92, 108], and over 250,000 [92, 109] records of unknown potential DDIs in the sets of predicted results. However, this bulk of information concerning drug pairs is provided without assessment of the possibility of DDIs manifestation.

Dmitriev et al. [92] created the computational structure–activity relationship (SAR) models to predict metabolic DDIs mediated by CYP1A2, CYP2B6, CYP2C19, CYP2C8, CYP2C9, CYP2D6, or CYP3A4. The same group has previously developed models for DDIs severity prediction [110, 111] that used the PASS (prediction of activity spectra for substances) program and PoSMNA (pairs of substances multilevel neighborhoods of atoms) substructural descriptors. These models are able to predict the classes of DDIs severity for pairs of molecules according to OpeRational ClassificAtion (ORCA). Using the same methods and descriptors, Dmitriev et al. [92] predicted whether two molecules would manifest metabolic DDIs mediated by the seven cytochromes mentioned above. Due to the limited possibilities of creating an appropriate training set, the stereochemical features of molecules were not taken into account in their descriptors. In addition, DDI predictions were obtained in qualitative mode ("YES" or "NO"). Unlike other ligand-based and structure-based methods [92, 99–103], their approach operated with two substances in pairs at once. This is reasonable for the DDI phenomenon, in which two substances interact simultaneously. It gives a direct indication of DDIs for the pairs of molecules without suggestions of the role of particular compounds, which is not always obvious (without consideration of inhibition or induction of a particular enzyme). In contrast to structure resemblance, functional similarities, and literature-based DDIs prediction methods [92, 104–109], our prediction method uses only structural formulas of compounds; it does not require any information about their biological activity. This implies that the method can be applied for not-yet-investigated, new, and virtual substances, in addition to providing a probabilistic assessment of possible DDIs and evaluating the possibility of DDIs manifestation for predicted pairs.

To avoid drug withdrawal from the market due to DDIs, pharmaceutical companies perform in vitro and in vivo studies. Physiologically based pharmacokinetic (PBPK) modeling is the in silico method of DDIs prediction that has already proved its applicability in the drug discovery and development process. It is clear that in silico methods will be used more intensively to reduce investigation costs [98].

The main procedures of in silico methods involve:

(a) *Information on DDIs and training set creation*: DDIs data can be collected from two sources of information. The first source was DrugBank Version 4.1 (University of Alberta and The Metabolomics Innovation Centre,

Edmonton, AB, Canada) [112] that contains information about interactions derived from public drug databases. The second source of DDIs data was the Fujitsu ADME Database (Chemistry & Life Science Group, Fujitsu, Tokyo, Japan) [113]. The final dataset includes information from both sources.

(b) *PASS*: The PASS software (Laboratory of Structure-Function Based Drug Design, Institute of Biomedical Chemistry, Moscow, Russia) [114] is based on the advanced naïve Bayes classifier and predicts the profiles of biological activity for drug-like compounds. The PASS algorithm creates a classification model of structure–activity relationships based on the training set with structures and known biological activities of known pharmaceutical agents. The PASS prediction results are presented as a ranked list of various biological activities with calculated probabilities P_a ("to be active") and P_i ("to be inactive"). The most probable activities are those predicted with the maximum value $\Delta P = P_a - P_i$. Currently, PASS predicts more than 8000 types of biological activities, including pharmacological effects, mechanisms of action, influences on gene expression, toxic and adverse effects, and interactions with metabolic enzymes and transporters. Biological activities for particular molecules in the PASS program are represented qualitatively as "active" or "inactive." The structural formulas of drug-like organic compounds are described by multilevel neighborhoods of atoms (MNA) descriptors.

(c) *Pairs of substances multilevel neighborhoods of atoms descriptors*: To describe the structures of drug pairs, we used PoSMNA descriptors instead of the MNA descriptors applied in the standard PASS software version [114]. PoSMNA descriptors can be used to predict various phenomena, e.g., synergistic effects of two drugs or the prediction of DDIs.

The main issue related to in silico methods is the study and use of the relationship of chemical compound structure and the phenomenon of metabolic DDIs mediated by the seven isoforms of cytochrome P450 most involved in drug metabolism [92]. The models created can be applied for virtual and not-yet-synthesized molecules using only their structural formulas. The implementation of PoSMNA descriptors and the PASS program algorithm for DDIs prediction at the level of cytochromes P450 makes it possible to consider a pair of molecules interacting as one entity without specifying the roles (substrate, inhibitor, or inducer, "object," or "precipitant" drug) of particular substances in the DDI process. Such an approach is unique and has already been used to create models for DDIs severity prediction [110, 111]. However, when predicting the DDIs severity without taking into account concrete pharmacokinetic or pharmacodynamic DDIs mechanisms, the accuracy of the prediction was not high enough, as compared to that obtained in the current study that considers only pharmacokinetic DDIs mediated by the seven cytochrome P450 isoforms (0.84 for three classes and 0.75 for five classes of severity vs 0.92 for DDIs prediction mediated by cytochrome P450 isoforms). Such a lower accuracy may be explained by the unclear separation of DDIs of these severity classes among themselves and the cases of DDIs in neighboring classes in the training set and by neglecting the DDIs mechanisms. According to Dmitriev et al. [92], the average accuracy of DDIs prediction at the level of cytochrome P450 isoforms is higher (0.92) due to the structural specificity of substances from the pairs that interact at a particular level of the cytochrome P450 isoform.

References

1. Benet LZ, Zia-Amirhosseini P. Basic principles of pharmacokinetics. Toxicol Pathol. 1995;23(2):115–23.
2. Johnson JA. Influence of race or ethnicity on pharmacokinetics of drugs. J Pharm Sci. 1997;86(12):1328–33.
3. Loisios-Konstantinidis I, Paraiso RLM, Fotaki N, et al. Application of the relationship between pharmacokinetics and pharmacodynamics in drug development and therapeutic equivalence: a PEARRL review. J Pharm Pharmacol. 2019;71(4):699–723.
4. Chen B, Dong JQ, Pan WJ, et al. Pharmacokinetics/pharmacodynamics model-supported early

drug development. Curr Pharm Biotechnol. 2012;13(7):1360–75.
5. Chaikin P, Rhodes GR, Bruno R, et al. Pharmacokinetics/pharmacodynamics in drug development: an industrial perspective. J Clin Pharmacol. 2000;40(12 Pt 2):1428–38.
6. Knights K, Bryant B. Pharmacology for health professionals. Amsterdam: Elsevier; 2002. ISBN 0-7295-3664-5
7. Koch HP, Ritschel WA. Liberation. Synopsis der Biopharmazie und Pharmakokinetik (in German). Landsberg, München: Ecomed. 1986:99–131. ISBN 3-609-64970-4
8. Ruiz-Garcia A, Bermejo M, Moss A, et al. Pharmacokinetics in drug discovery. J Pharm Sci. 2008;97(2):654–90.
9. Gleiter CH, Gundert-Remy U. Gender differences in pharmacokinetics. Eur J Drug Metab Pharmacokinet. 1996;21(2):123–8.
10. Schwartz JB. The influence of sex on pharmacokinetics. Clin Pharmacokinet. 2003;42(2):107–21.
11. Tasnif Y, Morado J, Hebert MF. Pregnancy-related pharmacokinetic changes. Clin Pharmacol Ther. 2016;100(1):53–62.
12. Swami R, Shahiwala A. Impact of physiochemical properties on pharmacokinetics of protein therapeutics. Eur J Drug Metab Pharmacokinet. 2013;38(4):231–9.
13. Avram MJ. Pharmacokinetic studies in pregnancy. Semin Perinatol. 2020;44(3):151227.
14. van Baak MA. Influence of exercise on the pharmacokinetics of drugs. Clin Pharmacokinet. 1990;19(1):32–43.
15. Alcorn J, McNamara PJ. Pharmacokinetics in the newborn. Adv Drug Deliv Rev. 2003;55(5):667–86.
16. Shi S, Klotz U. Age-related changes in pharmacokinetics. Curr Drug Metab. 2011;12(7):601–10.
17. Mather GG, Levy RH. Pharmacokinetics of polypharmacy: prediction of drug interactions. Epilepsy Res Suppl. 1996;11:113–21.
18. Honig PK, Cantilena LR. Polypharmacy. Pharmacokinetic perspectives Clin Pharmacokinet. 1994;26(2):85–90.
19. Löscher W, Wauquier A. Use of animal models in developing guiding principles for polypharmacy in epilepsy. Epilepsy Res Suppl. 1996;11:61–5.
20. Lan K, Xie G, Jia W. Towards polypharmacokinetics: pharmacokinetics of multicomponent drugs and herbal medicines using a metabolomics approach. Evid Based Complement Alternat Med. 2013;2013:819147.
21. Xue T, Roy R. Studying traditional Chinese medicine. Science. 2003;300(5620):740–1.
22. Lindon JC, Holmes E, Nicholson JK. Metabolomics: systems biology in pharmaceutical research and development. Curr Opinion Mol Ther. 2004;6(3):265–72.
23. Nicholson JK, Lindon JC. Systems biology: metabonomics. Nature. 2008;455(7216):1054–6.
24. Nicholson JK, Wilson ID. Understanding "global" systems biology: metabonomics and the continuum of metabolism. Nat Rev Drug Discov. 2003;2(8):668–76.
25. Wang Y, Tang H, Nicholson JK, et al. A metabonomic strategy for the detection of the metabolic effects of chamomile (Matricaria recutita L.) ingestion. J Agr Food Chem. 2005;53(2):191–6.
26. Sumner LW, Urbanczyk-Wochniak E, Broeckling CD. Metabolomics data analysis, visualization, and integration. Meth Mol Biol. 2007;406:409–36.
27. Xie G, Plumb R, Su M, et al. Ultra-performance LC/TOF MS analysis of medicinal Panax herbs for metabolomic research. J Separation Sci. 2008;31(6–7):1015–26.
28. Wu SC, Yen GC, Wang BS, et al. Antimutagenic and antimicrobial activities of pu-erh tea. LWT—Food Science and Technology. 2007;40(3):506–12.
29. Wang BS, Yu HM, Chang LW, et al. Protective effects of pu-erh tea on LDL oxidation and nitric oxide generation in macrophage cells. LWT—Food Science and Technology. 2008;41(6):1122–32.
30. Ni Y, Su M, Lin J, et al. Metabolic profiling reveals disorder of amino acid metabolism in four brain regions from a rat model of chronic unpredictable mild stress. FEBS Lett. 2008;582(17):2627–36.
31. Chen XQ, Ye Y, Cheng H, et al. Comparative analysis of theanine, caffein and polyphenolic constituents in green tea, black tea and puer tea. Food Res Dev. 2007;28(12):141–4.
32. Lan K, Jia W. An integrated metabolomics and pharmacokinetics strategy for multi-component drugs evaluation. Curr Drug Metab. 2010;11(1):105–14.
33. Xie G, Zhao A, Zhao L, et al. Metabolic fate of tea polyphenols in humans. J Proteome Res. 2012;11(6):3449–57.
34. Chan K. Progress in traditional Chinese medicine. Trends Pharmacol Sci. 1995;16(6):182–7.
35. Plumb GW, Garcia-Conesa MT, Kroon PA, et al. Metabolism of chlorogenic acid by human plasma, liver, intestine and gut microflora. J Sci Food Agr. 1999;79(3):390–2.
36. Nassar AEF, Talaat RE. Strategies for dealing with metabolite elucidation in drug discovery and development. Drug Discov Today. 2004;9(7):317–27.
37. Pelkonen O, Tolonen A, Korjamo T, et al. From known knowns to known unknowns: predicting in vivo drug metabolites. Bioanalysis. 2009;1(2):393–414.
38. Tolonen A, Turpeinen M, Pelkonen O. Liquid chromatography-mass spectrometry in in vitro drug metabolite screening. Drug Discov Today. 2009;14(3–4):120–33.
39. Wei X, Shi X, Kim S, et al. Data preprocessing method for liquid chromatography-mass spectrometry-based metabolomics. Anal Chem. 2012;84(18):7963–71.
40. Trygg J, Holmes E, Lundstedt T. Chemometrics in metabolomics. J Proteome Res. 2007;6(2):469–79.

41. Ma S, Chowdhury SK. Analytical strategies for assessment of human metabolites in preclinical safety testing. Analytical Chem. 2011;83(13):5028–36.
42. Plumb RS, Stumpf CL, Granger JH, et al. Use of liquid chromatography/time-of-flight mass spectrometry and multivariate statistical analysis shows promise for the detection of drug metabolites in biological fluids. Rapid Commun Mass Spectrometry. 2003;17(23):2632–8.
43. Idborg H, Edlund PO, Jacobsson SP. Multivariate approaches for efficient detection of potential metabolites from liquid chromatography/mass spectrometry data. Rapid Commun in Mass Spectrometry. 2004;18(9):944–54.
44. Chen C, Gonzalez FJ, Idle JR. LC-MS-based metabolomics in drug metabolism. Drug Metab Rev. 2007;39(2–3):581–97.
45. Patterson AD, Gonzalez FJ, Idle JR. Xenobiotic metabolism: a view through the metabolometer. Chem Res Toxicol. 2010;23(5):851–60.
46. Johnson CH, Patterson AD, Idle JR, et al. Xenobiotic metabolomics: major impact on the metabolome. Ann Rev Pharmacol Toxicol. 2012;52:37–56.
47. van Velzen EJJ, Westerhuis JA, van Duynhoven JPM, et al. Multilevel data analysis of a crossover designed human nutritional intervention study. J Proteome Res. 2008;7(10):4483–91.
48. van Velzen EJJ, Westerhuis JA, van Duynhoven JPM, et al. Phenotyping tea consumers by nutrikinetic analysis of polyphenolic end-metabolites. J Proteome Res. 2009;8(7):3317–30.
49. Smilde AK, Westerhuis JA, Hoefsloot HCJ, et al. Dynamic metabolomic data analysis: a tutorial review. Metabolomics. 2010;6(1):3–17.
50. Westerhuis JA, van Velzen EJJ, Hoefsloot HCJ, et al. Multivariate paired data analysis: multilevel PLSDA versus OPLSDA. Metabolomics. 2010;6(1):119–28.
51. van Duynhoven J, Vaughan EE, Jacob DM, et al. Metabolic fate of polyphenols in the human superorganism. Proc Nat Acad Sci U S A. 2011;108(suppl1):4531–8.
52. van Duynhoven JPM, van Velzena EJJ, Westerhuis JA. Nutrikinetics: concept, technologies, applications, perspectives. Trends Food Sci Technol. 2012;26(1):4–13.
53. Llorach R, Garcia-Aloy M, Tulipani S, et al. Nutrimetabolomic strategies to develop new biomarkers of intake and health effects. J Agr Food Chem. 2012;60(36):8797–808.
54. Croixmarie V, Umbdenstock T, Cloarec O, et al. Integrated comparison of drug-related and drug-induced ultra performance liquid chromatography/mass spectrometry metabonomic profiles using human hepatocyte cultures. Anal Chem. 2009;81(15):6061–9.
55. Llorach-Asunción R, Jauregui O, Urpi-Sarda M, et al. Methodological aspects for metabolome visualization and characterization. A metabolomic evaluation of the 24 h evolution of human urine after cocoa powder consumption. J Pharmaceutical Biomedical Analysis. 2010;51(2):373–81.
56. Lenz EM, Wilson ID. Analytical strategies in metabonomics. J Proteome Res. 2007;6(2):443–58.
57. Lan K, Zhang Y, Yang J, et al. Simple quality assessment approach for herbal extracts using high performance liquid chromatography-UV based metabolomics platform. J Chromatography A. 2010;1217(8):1414–8.
58. Wang Z, Hu H, Chen F, et al. Metabolic profiling assisted quality assessment of Rhodiola rosea extracts by high-performance liquid chromatography. Planta Med. 2012;78(7):740–6.
59. Duan L-X, Li M, Chen M, et al. Use of the metabolomics approach to characterize Chinese medicinal material Huangqi. Mol Plant. 2012;5(2):376–86.
60. Wang M, Lamers RJAN, Korthout HAAJ, et al. Metabolomics in the context of systems biology: bridging traditional Chinese medicine and molecular pharmacology. Phytother Res. 2005;19(3):173–82.
61. Zhang A, Sun H, Wang Z, et al. Metabolomics: towards understanding traditional Chinese medicine. Planta Med. 2010;76(17):2026–35.
62. Wang X, Sun H, Zhang A, et al. Potential role of metabolomics approaches in the area of traditional Chinese medicine: as pillars of the bridge between Chinese and Western medicine. J Pharmaceutical Biomedical Analysis. 2011;55(5):859–68.
63. Liu P, Liu S, Tian D, et al. The applications and obstacles of metabolomics in traditional Chinese medicine. Evidence-based Complementary Altern Med. 2012;2012:4.
64. Wu T, Yang M, Wei H-F, et al. Application of metabolomics in traditional Chinese medicine differentiation of deficiency and excess syndromes in patients with diabetes mellitus. Evidence-based Complementary Altern Med. 2012;2012:11.
65. Jové M, Serrano JCE, Ortega N, et al. Multicompartmental LC-Q-TOF-based metabonomics as an exploratory tool to identify novel pathways affected by polyphenol-rich diets in mice. J Proteome Res. 2011;10(8):3501–12.
66. Turnbaugh PJ, Ley RE, Hamady M, et al. The human microbiome project. Nature. 2007;449(7164):804–10.
67. Qin J, Li R, Raes J, et al. A human gut microbial gene catalogue established by metagenomic sequencing. Nature. 2010;464(7285):59–65.
68. Arumugam M, Raes J, Pelletier E, et al. Enterotypes of the human gut microbiome. Nature. 2011;473(7346):174–80.
69. Consortium THMP. A framework for human microbiome research. Nature. 2012;486(7402):215–21.
70. Consortium THMP. Structure, function and diversity of the healthy human microbiome. Nature. 2012;486(7402):207–14.
71. Rosabelle McManus I, Contag AO, Olson RE. Characterization of endogenous ethanol in the mammal. Science. 1960;131(3393):102–3.

72. Spinucci G, Guidetti M, Lanzoni E, et al. Endogenous ethanol production in a patient with chronic intestinal pseudo-obstruction and small intestinal bacterial overgrowth. Eur J Gastroenterol Hepatol. 2006;18(7):799–802.
73. Compare D, Coccoli P, Rocco A, et al. Gut-liver axis: the impact of gut microbiota on non alcoholic fatty liver disease. Nutrition, Metab Cardiovasc Dis. 2012;22(6):471–6.
74. Martin FPJ, Dumas ME, Wang Y, et al. A top-down systems biology view of microbiome-mammalian metabolic interactions in a mouse model. Mol Systems Biol. 2007;3:112.
75. Martin FPJ, Wang Y, Sprenger N, et al. Probiotic modulation of symbiotic gut microbial-host metabolic interactions in a humanized microbiome mouse model. Mol Systems Biol. 2008;4:157.
76. Martin FPJ, Sprenger N, Yap IKS, et al. Panorganismal gut microbiome-host metabolic crosstalk. J Proteome Res. 2009;8(4):2090–105.
77. Martin FPJ, Wang Y, Yap IKS, et al. Topographical variation in murine intestinal metabolic profiles in relation to microbiome speciation and functional ecological activity. J Proteome Res. 2009;8(7):3464–74.
78. Zheng X, Xie G, Zhao A, et al. The footprints of gut microbial-mammalian co-metabolism. J Proteome Res. 2011;10(12):5512–22.
79. Nicholson JK, Holmes E, Burcelin R, et al. Host-gut microbiota metabolic interactions. Science. 2012;336(6086):1262–7.
80. Wilson ID, Nicholson JK. The role of gut microbiota in drug response. Curr Pharmaceutical Design. 2009;15(13):1519–23.
81. Li H, Zhou M, Zhao A, et al. Traditional Chinese medicine: balancing the gut ecosystem. Phytother Res. 2009;23(9):1332–5.
82. Zhao YF, Song FR, Guo XH, et al. Studies on the biotransformation of aconitine in human intestinal bacteria using soft-ionization mass spectrometry. Chem J Chin Universities. 2008;29(1):55–9.
83. Wang X, Pi Z, Liu W, et al. Effect of pH on the metabolism of aconitine under rat intestinal bacteria and analysis of metabolites using HPLC/MS-MSn technique. Chin J Chem. 2010;28(12):2494–500.
84. Clayton TA, Baker D, Lindon JC, et al. Pharmacometabonomic identification of a significant host-microbiome metabolic interaction affecting human drug metabolism. Proc Nati Acad Sci U S A. 2009;106(34):14728–33.
85. Meinl W, Sczesny S, Brigelius-Flohé R, et al. Impact of gut microbiota on intestinal and hepatic levels of phase 2 xenobiotic-metabolizing enzymes in the rat. Drug Metab Dispos. 2009;37(6):1179–86.
86. Björkholm B, Bok CM, Lundin A, et al. Intestinal microbiota regulate xenobiotic metabolism in the liver. PLoS One. 2009;4(9):e6958.
87. Peters S, Janssen HG, Vivó-Truyols G. Trend analysis of time-series data: a novel method for untargeted metabolite discovery. Anal Chim Acta. 2010;663(1):98–104.
88. Jia W, Fan T-P, Wang NX, et al. The polypharmacokinetics of herbal medicines. Science. 2015;350:S76–9.
89. Li M, Wang S, Xie G, et al. polyPK: an R package for pharmacokinetic analysis of multi-component drugs using a metabolomics approach. Bioinformatics. 2018;34(10):1792–4.
90. Xie G, Wang S, Zhang H, et al. Poly-pharmacokinetic study of a multicomponent herbal medicine in healthy Chinese volunteers. Clin Pharmacol Ther. 2018;103(4):692–702.
91. Ping W, Tinglan Z, Guohua Y, et al. Polypharmacokinetic strategy-delineated metabolic fate of bioactive compounds in a traditional Chinese medicine formula, Yuanhu Zhitong tablets, using parallel reaction monitoring mode. Phytomedicine. 2019;53:53–61.
92. Dmitriev AV, Rudik AV, Karasev DA, et al. In silico prediction of drug-drug interactions mediated by cytochrome P450 isoforms. Pharmaceutics. 2021;13(4):538.
93. Niu J, Straubinger RM, Mager DE. Pharmacodynamic drug-drug interactions. Clin Pharmacol Ther. 2019;105(6):1395–406.
94. Carpenter M, Berry H, Pelletier AL. Clinically relevant drug-drug interactions in primary care. Am Fam Physician. 2019;99(9):558–64.
95. Roblek T, Vaupotic T, Mrhar A, et al. Drug-drug interaction software in clinical practice: a systematic review. Eur J Clin Pharmacol. 2015;71(2):131–42.
96. Kennedy C, Brewer L, Williams D. Drug Interactions Medicine. 2016;44:422–6.
97. In Vitro Drug Interaction Studies—Cytochrome P450 Enzyme- and Transporter-Mediated Drug Interactions Guidance for Industry. [(Accessed 27 Oct 2020)]; Available online: https://www.fda.gov/media/134582/download.
98. Dmitriev AV, Lagunin AA, Karasev DA, et al. Prediction of drug-drug interactions related to inhibition or induction of drug-metabolizing enzymes. Curr Top Med Chem. 2019;19:319–36.
99. Banerjee P, Dunkel M, Kemmler E, et al. SuperCYPsPred—A web server for the prediction of cytochrome activity. Nucleic Acids Res. 2020;48:W580–5.
100. Hochleitner J, Akram M, Ueberall M, et al. A combinatorial approach for the discovery of cytochrome P450 2D6 inhibitors from nature. Sci Rep. 2017;7:8071.
101. Kaserer T, Höferl M, Müller K, et al. In silico predictions of drug—drug interactions caused by CYP1A2, 2C9 and 3A4 inhibition—a comparative study of virtual screening performance. Mol Inform. 2015;34:431–57.
102. Torimoto-Katori N, Huang R, Kato H, et al. In silico prediction of hPXR activators using structure-

based pharmacophore modeling. J Pharm Sci. 2017;106:1752–9.
103. Fahmi OA, Maurer TS, Kish M, et al. A combined model for predicting CYP3A4 clinical net drug-drug interaction based on CYP3A4 inhibition, inactivation, and induction determined in vitro. Drug Metab Dispos. 2008;36:1698–708.
104. Takeda T, Hao M, Cheng T, et al. Predicting drug-drug interactions through drug structural similarities and interaction networks incorporating pharmacokinetics and pharmacodynamics knowledge. J Cheminform. 2017;9:16.
105. Vilar S, Uriarte E, Santana L, et al. Similarity-based modeling in large-scale prediction of drug-drug interactions. Nat Protoc. 2014;9:2147–63.
106. Cheng F, Zhao Z. Machine learning-based prediction of drug-drug interactions by integrating drug phenotypic, therapeutic, chemical, and genomic properties. J Am Med Inf Assoc. 2014;21:e278–86.
107. Duke JD, Han X, Wang Z, et al. Literature based drug interaction prediction with clinical assessment using electronic medical records: novel myopathy associated drug interactions. PLoS Comput Biol. 2012;8:e1002614.
108. Zhang P, Wang F, Hu J, et al. Label propagation prediction of drug-drug interactions based on clinical side effects. Sci Rep. 2015;5:12339.
109. Ferdousi R, Safdari R, Omidi Y. Computational prediction of drug-drug interactions based on drugs functional similarities. J Biomed Inf. 2017;70:54–64.
110. Dmitriev AV, Filimonov DA, Rudik AV, et al. Drug-drug interaction prediction using PASS. SAR QSAR Environ Res. 2019;30:655–64.
111. Dmitriev A, Filimonov D, Lagunin A, et al. Prediction of severity of drug-drug interactions caused by enzyme inhibition and activation. Molecules. 2019;24:3955.
112. DrugBank. [Accessed 27 Oct 2020]; Available online: https://go.drugbank.com/.
113. ADME Database. [Accessed 27 Oct 2020]; Available online: https://www.fujitsu.com/jp/group/kyushu/en/solutions/industry/lifescience/admedatabase/.
114. Poroikov VV, Filimonov DA, Borodina YV, et al. Robustness of biological activity spectra predicting by computer program PASS for noncongeneric sets of chemical compounds. J Chem Inf Comput Sci. 2000;40:1349–55.

17 General Process for Rational Design and Discovery of MTDs

Abstract

Drug discovery and development are a complex and costly process. Particularly, the single-target approach to drug development has encountered difficulties in predicting drugs that both are clinically efficacious and avoid toxicity. The recent ascent of polypharmacology as a new paradigm in drug development has many implications for disease therapy; however, it also adds more complexity to the process. From the present chapter on, this book will place an emphasis to the technical aspect of MTD development from several angles: strategies, technologies, methodologies, databases, and protocols. This chapter introduces the general steps in drug discovery and development and some specific points on MTD discovery: multitargets identification and validation as well as MTD hit/lead identification. The aim is to pave the way for better understanding of the contents in the subsequent chapters on the rational design of MTDs.

Keywords

Drug discovery · MTD discovery · Multitarget identification · Multitarget validation · MTD hit/lead identification

17.1 General Process for Drug Discovery and Development

New drug development is a highly regulated, complicated process that requires specialists and intense research and development (R&D) skill sets in the medical research community. All regulations and safety indications must be observed carefully, and human and animal clinical trials subjects treated professionally and with the utmost care. The goal of drug development is to prevent human and animal pain and suffering whenever possible and find and provide new drugs that we can depend on to improve our health and happiness.

Nowadays, it easily takes about 12–15 years and $1 billion for a new drug to get approved to market [1–4]. Such huge time and monetary investments are ascribed to the complexity in new drug development, which has increased manifolds over the past decades. Drug development process requires preclinical testing, investigational new drug (IND) applications, and completed clinical testing before marketing approval from the FDA [1]. Generally, new drug applications (NDAs) or biologics license applications (BLA) are reviewed comprehensively before approval, and then drug performance is resubmitted to regulatory agencies for post-

Z. Wang, B. Yang, *Polypharmacology*, https://doi.org/10.1007/978-3-031-04998-9_17

marketing studies. The overarching goal is to bring more efficient and safer treatments to the patients as quickly as possible after a thorough medical evaluation.

There are five critical steps in the US drug development process, including many phases and stages within each of them [1]. We will discuss these different phases and stages to develop an in-depth understanding of the entire process. The five steps are:

Step 1: Discovery and Development.
Step 2: Preclinical Research.
Step 3: Clinical Development.
Step 4: FDA Review.
Step 5: FDA Post-market Safety Monitoring.

17.1.1 Step 1: Discovery and Development

Drug discovery is how new medications are discovered. The drug discovery process underpins the entire pharmaceutical industry, encompassing the early stages of research from target discovery and validation right through to the identification of a drug candidate or lead compound. Initial identification of small therapeutic candidates comes about via a variety of streams. Research can lead to new insights into disease processes that highlight novel pathways for which drugs can be developed to intervene. Alternatively, companies conduct large-scale trial and error-based programs in order to identify molecular compounds that may be of interest. This is the process most often performed during initial lead discovery, with a view to take novel compounds right the way through to preclinical and clinical trials. Thoroughly calculated risk analysis at this point can increase the chances of success when investments into a lead are made.

Historically, drugs were mostly found by identifying active ingredients from traditional medicines or purely by chance. Afterward, classical pharmacology was used to investigate chemical libraries including small molecules, natural products, or plant extracts, and find those with therapeutic effects. Since human DNA was sequenced, reverse pharmacology has found remedies to existing diseases through testing.

Disease processes, molecular compound tests, existing treatments with unanticipated effects, and new technologies spur drug discovery through the cycle below.

Today, drug discovery involves screening hits, medicinal chemistry, and optimization of hits to reduce potential drug side effects (increasing affinity and selectivity). Efficacy or potency, metabolic stability (half-life), and oral bioavailability are also improved in this step of the drug development process.

Collectively, drug discovery process involves the following steps [1–3].

17.1.1.1 Target Identification and Validation

Target Identification

Target identification and validation kicks off the whole drug discovery process. During this first phase of drug development, in vitro research is performed to identify targets involved in specific diseases. Drugs fail in the clinic for two main reasons; the first is that they do not work and the second is that they are not safe. As such, one of the most important steps in developing a new drug is target identification and validation.

A target is a broad term which can be applied to a range of biological entities which may include proteins, genes, and RNA [3]. A good target needs to be efficacious and safe, meet clinical and commercial needs, and, above all, be "druggable." A "druggable" target is accessible to the putative drug molecule and its activity can be modulated by an exogenous compound, that is, a small molecule or larger biologicals and upon binding, can elicit a biological response which may be measured both in vitro and in vivo [3]. In addition, a good target is also required to meet clinical needs of the prospective patient. Naturally occurring cellular or modular structures that appear to play an important role in pathogenicity or disease progression are normally targets for therapeutics. It is now known that certain target classes are more amenable to small-molecule drug discovery, for example, G-protein-coupled receptors (GPCRs), whereas antibodies are good at blocking protein–protein interactions [3]. Good target identification and validation enables increased confidence in the relationship between target and disease and allows us to explore whether target modulation will lead to mechanism-based side effects.

Target Validation

After identifying and selecting a potential target, researchers must demonstrate that it is involved in the progression of a given disease and that its activity can be regulated. Conducting careful and precise target validation experiments is essential for the success of drug development in the following stages with a systematic validation approach. The approach itself depends on the therapeutic area but has a set of general principles that include disease association, bioactive molecules, protein interactions, signaling pathways analysis, preclinical evidence in cell-based models, preclinical evidence in intact systems (i.e., transgenic animals), and literature survey and competitor information.

17.1.1.2 Hit Identification and Validation

Hit Identification

Following target validation, compound screening assays are developed to identify the small-molecule leads that have the desired effect against the identified targets. There are a number of approaches by which hits can be identified, including high-throughput screening, knowledge-based approaches, and virtual screening.

Hit Validation

Validation of hits is required following initial screening, and again there are a few options to choose from.

Assay development and screening: Assays are test systems that evaluate the effects of the new drug candidate at the cellular, molecular, and biochemical levels.

High-throughput screening: High-throughput screening (HTS) uses robotics, data processing/control software, liquid handling devices, and sensitive detectors to rapidly conduct millions of pharmacological, chemical, and genetic tests, eliminating hours of painstaking testing by scientists. HTS identifies active compounds, genes, or antibodies that affect human molecules.

17.1.1.3 Lead Identification and Validation

After a number of hit series have been established, the aim at this point is the refinement of each hit series in order to produce more selective compounds—leads, the hit-to-lead (H2L) process. Multiple series should be worked on in tandem, as it is likely that some hit series will fail, often due to particular characteristics of the series. Focusing on multiple structurally different sets of hit series will help to offset this possibility. In the H2L process, small-molecule hits from an HTS are evaluated and optimized in a limited way into lead compounds.

Lead Identification

Lead compound identification is the process of identifying or creating a compound that can interact with the target previously selected. Researchers can conduct screening experiments to identify possible naturally occurring compounds that can be repurposed as drugs. Alternatively, synthetic compounds can be designed that will target the predicted target while not interfering with other cellular processes. In addition to testing the mechanism of action of the drug, initial safety tests are conducted in cell culture. Both the pharmacokinetics and pharmacodynamics of the drug are also tested—how it is metabolized and how it affects various bodily functions, respectively.

The lead compound then moves on to the lead optimization process.

Lead Optimization

Once a compound (or compounds) has been identified, they need to be optimized for efficacy and safety. The design of synthetic molecules can be altered to prevent off-target binding, making them less likely to interact with molecules other than the target. Additionally, the optimal dosage and introduction route (oral, injection) is tested on two- and three-dimensional cell culture platforms.

This stage also includes safety testing prior to introduction into multiple in vivo animal models in the following preclinical development stage. Animal models such as mice and rats can be used at this stage; however, some tests for safety are first conducted in vitro.

One specific example of a three-dimensional platform is a scaffold developed for the testing of new glaucoma drugs (Torrejon, KY, 2013). Glaucoma can result from increased intraocular pressure in the eye, which is directly related to the outflow of aqueous humor through a structure called the trabecular meshwork. The scaffold developed is a three-dimensional trabecular meshwork which effectively mimics the microenvironment around the eye. By screening drugs first in this 3D in vitro setting, animal models used in the next stage will have better chances of encountering a safe and effective drug.

17.1.1.4 Lead Optimization

In the lead optimization (LO) process, the lead compounds discovered in the H2L process are synthesized and modified to improve potency and reduce side effects. Lead optimization conducts experimental testing using animal efficacy models and ADMET tools, designing the drug candidate. At this stage, the aim is to maintain the desired properties of lead compounds while improving on possible deficiencies of their structures, with a view to produce a preclinical drug candidate. This stage can be used to find out whether the drug metabolizes in the right area of the body, or whether there are currently any side effects that are cause for concern. For this process, an integrated approach is recommended. The combination of specialists in computational chemistry, medical chemistry, drug metabolism, and other areas can provide unique insights into this late stage of the process.

17.1.1.5 Late-Stage Lead Optimization

Before progression to preclinical and clinical trials, late-stage optimization, in which further pharmacological safety of a lead compound is assessed, is a vital step. If this stage is overlooked, problems in efficacy, pharmacokinetics, and safety are more likely to occur later on in drug development. Safety optimization is a core stage; the aims are to identify and progress the leads with the best overall safety profile, remove the most toxic leads, and establish a well-characterized hazard and translational risk profile to enable further in vitro tests.

17.1.1.6 Active Pharmaceutical Ingredients

Active pharmaceutical ingredients (APIs) are biologically active ingredients in a drug candidate that produce effects. All drugs are made up of the API or APIs and excipients (inactive substances that deliver the drug into the human system). High potency active pharmaceutical ingredients (HP APIs) are molecules that are effective at much smaller dosage levels than standard APIs. They are classified based on toxicity, pharmacological potency, and occupational exposure limits, and used in complex drug development involving more than ten steps.

The drug discovery process ends when one lead compound is found for a drug candidate, and the process of drug development starts.

17.1.2 Step 2: Preclinical Research

Once a lead compound is found, drug development begins with preclinical research to determine the efficacy and safety of the drug. Researchers determine the following about the drug:

Absorption, distribution, metabolization, and excretion information.

Potential benefits and mechanisms of action.

Best dosage and administration route.

Side effects/adverse events.

Effects on gender, race, or ethnicity groups.

Interaction with other treatments.

Effectiveness compared to similar drugs.

Preclinical trials test the new drug on nonhuman subjects for efficacy, toxicity, and pharmacokinetic (PK) information. These trials are conducted by scientists in vitro and in vivo with unrestricted dosages.

17.1.2.1 Determine ADME

Absorption, distribution, disposition, metabolism, and excretion (ADME) is a PK process of measuring the ways the new drug affects the body. ADME involves mathematical descriptions of each effect.

17.1.2.2 Proof of Principle/Proof of Concept

Proof of principle (PoP) are studies that are successful in preclinical trials and early safety testing. Proof of concept (PoC) terminology is used almost interchangeably with PoP in drug discovery and development projects. Successful PoP/PoC studies lead to program advancement to the phase II studies of dosages.

17.1.2.3 In Vivo, In Vitro, and Ex Vivo Assays

These three types of studies are conducted on the whole, living organisms or cells, including animals and humans, or using nonliving organisms or tissue extract. In vivo, preclinical research examples are the development of new drugs using mice, rat, and dog models. In vitro is research conducted in a laboratory. Ex vivo uses animal cells or tissues from a nonliving animal. Examples of ex vivo research assays are finding effective cancer treatment agents, measurements of tissue properties (physical, thermal, electrical, and optical), and realistic modeling for new surgical procedures. In an ex vivo assay, a cell is always used as the basis for small explant cultures that provide a dynamic, controlled, and sterile environment.

17.1.2.4 In Silico Assays

In silico assays are test systems or biological experiments performed on a computer or via computer simulation. These are expected to become increasingly popular with the ongoing improvements in computational power, and behavioral understanding of molecular dynamics and cell biology.

17.1.2.5 Determine Drug Delivery Methods

New drug delivery methods include oral, topical, membrane, intravenous, and inhalation. Drug delivery systems are used for targeted delivery or controlled release of new drugs. Physiological barriers in animal or human bodies may prevent drugs from reaching the targeted area or releasing when they should. The goal is to prevent the drug from interacting with healthy tissues while still being effective.

Oral Application

Oral delivery of medications is reliable, cost-effective, and convenient for patients. Oral drug delivery may not monitor precise dosages to the desired area but is ideal for prophylactic vaccinations and nutritional regimens. Delayed action, stomach enzyme destruction, absorption inconsistencies, or patients with gastrointestinal issues or upset can occur, and patients must be conscious during administration.

Topical Application

Topical drug delivery involves ointments, creams, lotions, or transdermal patches that deliver a drug by absorption into the body. Topical delivery is more useful for patient skin or muscular conditions—it is preferred by patients due to noninvasive delivery and their ability to self-administer the medicine.

Parenteral (IM, SC, or LP Membrane) Administration

Parenteral drug delivery utilizes bodily membranes, including intramuscular (IM), intraperitoneal (IP), or subcutaneous or (SC). It is often used for unconscious patients and avoids epithelial barriers that are difficult for drugs to cross.

Parenteral (Intravenous): Intravenous injection is one of the fastest drug delivery absorption methods. IV injection ensures entire doses of drugs enter the bloodstream, and it is more effective than IM, SC, or LP membrane methods.

Parenteral (Inhalation): Inhalation drug delivery gets the drug rapidly absorbed into the mucosal lungs, nasal passages, throat, or mouth. Problems with inhalation delivery include difficulty delivering the optimum dosage due to small mucosal surface areas and patient discomfort. Pulmonary inhalation drug delivery uses fine drug powders or macromolecular drug solutions. Lung fluids resemble blood, so they can absorb small particles easily and deliver them into the bloodstream.

17.1.2.6 Optimizing Formulation and Improving Bioavailability

Formulation optimization is ongoing throughout preclinical and clinical stages. It ensures drugs are delivered to the proper place at the right time and in the right concentration. Optimization may include overcoming the solubility of compounds.

17.1.3 Step 3: Clinical Development

Once preclinical research is complete, researchers move on to clinical drug development, including clinical trials and volunteer studies to fine-tune the drug for human use.

17.1.3.1 Complexity of Study Design, Associated Cost, and Implementation Issues

The complexity of clinical trial design and its associated costs and implementation issues may affect trials carried out during this phase. Trials must be safe and efficacious and be completed under the drug development budget, using a methodology to ensure the drug works as well as possible for its intended purpose. This rigorous process must be set up correctly and enroll many volunteers to be effective.

17.1.3.2 Clinical Trials: Dose Escalation, Single Ascending, and Multiple Dose Studies

Proper dosing determines medication effectiveness, and clinical trial examine dose escalation, single ascending, and multiple dose studies to determine the best patient dosage.

Phase I: Study in Healthy Volunteers

This phase is the first time the drug is tested on humans; less than 100 volunteers will help researchers assess the safety and pharmacokinetics, absorption, metabolic, and elimination effects on the body, as well as any side effects for safe dosage ranges.

Phase II and Phase III: Studies in Patient Population

Phase II assesses drug safety and efficacy in an additional 100–500 patients, who may receive a placebo or standard drug previously used as treatment. Analysis of optimal dose strength helps create schedules while adverse events and risks are recorded. Phase III enrolls 1000–5000 patients, enabling medication labeling and instructions for proper drug use. Phase III trials require extensive collaboration, organization, and independent ethics committee (IEC) or institutional review board (IRB) coordination and regulation in anticipation of full-scale production following drug approval.

17.1.3.3 Biological Samples Collection, Storage, and Shipment

During clinical trials, biological samples are collected, stored, and shipped from testing sites according to global standards and regulations. Transport containers of biological samples may include dry ice packs or other temperature stabilizing methods. Different requirements apply to different types of biological samples.

17.1.3.4 Pharmacodynamic (PD) Biomarkers

PD biomarkers are molecular indicators of the drug's effects on the target human area, and link drug regimen and biological responses. This data can help select rational combinations of targeted agents and optimize drug regimens and schedules. Rationality and hypothesis-testing power are increased through the use of PD endpoints in human trials.

17.1.3.5 Pharmacokinetic Analysis

Pharmacokinetic analysis is an experimental trial that determines the theory of how a new drug behaves in the human body. The volume of distribution, clearance, and terminal half-life are defined through compartmental modeling.

17.1.3.6 Bioanalytical Method Development and Validation

Bioanalytical methods detect analytes and metabolites such as drug or biomarkers in biological or human samples to determine drug efficacy and safety. The complete bioanalytical assay consists of sample collection, clean-up, analysis, and detection.

17.1.3.7 Drug (Analyte) and Metabolite Stability in Biological Samples

Stability is important in determining human drug efficacy, and biological samples are required. Drug and drug metabolites are susceptible to degradation, which can lower drug concentration over the life of the drug.

17.1.3.8 Blood, Plasma, Urine, and Feces Sample Analysis for Drug and Metabolites

Biological samples used in clinical trials include blood, plasma, urine, and feces to determine and analyze various properties and effects of the drug and its metabolites on humans.

17.1.3.9 Patient Protection: GCP, HIPAA, and Adverse Event Reporting

Human patients must always be protected during clinical trials, and Good Clinical Practices (GCP), the Health Insurance Portability and Accountability Act (HIPAA), and adverse event reporting to IEC/IRB regulate and ensure their safety.

17.1.4 Step 4: FDA Review

Once the new drug has been formulated for its best efficacy and safety, and the results from clinical trials are available, it is advanced forward for wholistic FDA review. At this time, the FDA reviews and approves, or does not approve, the drug application submitted by the drug development company.

17.1.4.1 Regulatory Approval Timeline

The new drug regulatory approval timeline may be standard, fast track, breakthrough, accelerated approval, or priority review depending on its applications and necessity for patients. If standard or priority review is required, the approval timeline may be up to a year. Fast track, breakthrough, or accelerated approvals may occur sooner.

17.1.4.2 IND Application

IND applications are submitted to the FDA before starting clinical trials. If clinical trials are ready to be conducted, and the FDA has not responded negatively about the drug, developers may start the trials.

17.1.4.3 NDA/ANDA/BLA Applications

An NDA abbreviated new drug application (ANDA), or BLA, is submitted to the FDA after clinical trials demonstrate drug safety and efficacy. The FDA reviews study data and decides whether to grant approval or not. Additional research or an expert advisory panel may be required before a final decision is made.

17.1.4.4 Orphan Drug

An orphan drug is intended to treat disease so rare that financial sponsors are unwilling to develop it under standard marketing conditions. These drugs may not be approved quickly or at all.

17.1.4.5 Accelerated Approval

New drugs may be granted accelerated approval if there is strong evidence of positive impact on a surrogate endpoint instead of evidence of impact on actual clinical benefits the drug provides. Expedition of approval means the medication can help treat severe or life-threatening conditions.

17.1.4.6 Reasons for Drug Failure

New drug applications may fail for a variety of reasons, including toxicity, efficacy, PH properties, bioavailability, or inadequate drug performance.

Toxicity
If the toxicity of a new drug is too high in human or animal patients, the drug may be rejected due to safety concerns about its use following manufacture.

Efficacy
If a new drug's efficacy is not high enough or evidence is inconclusive, the FDA may reject it.

PK Properties or Bioavailability
PK properties or poor bioavailability due to low aqueous solubility, or high first-pass metabolism, may also cause a drug to fail FDA review. PK causes of drug failure include inadequate action duration and unanticipated human drug interactions.

Inadequate Drug Performance
If the new drug performs the desired function, but only at a shallow level, the FDA may reject the application in favor of a formulation that performs better.

17.1.5 Step 5: Post-market Monitoring

Following drug approval and manufacturing, the FDA requires drug companies to monitor the safety of its drug using the FDA Adverse Event Reporting System (FAERS) database. FAERS helps FDA implement its post-marketing safety surveillance program. Through this program, manufacturers, health professionals, and consumers report problems with approved drugs (Fig. 17.1).

17.2 Other Relevant Drug Development Concepts

17.2.1 Drug Master File

A Drug Master File (DMF) is a submission to the FDA used to provide confidential, detailed information about facilities, processes, or articles used in the manufacturing, processing, packaging, and storing of a human drug.

17.2.2 Drugs for Pediatric Use

Drugs for pediatric use are intended for use in children or youth, generally under the age of 21. In some cases, the American Academy of Pediatrics (AMA) may make exceptions if a pediatrician and family agree on an older age adult.

17.2.3 Drugs for Veterinary Use

Drugs for veterinary use are intended for use in animals, pets, and livestock. However, some veterinary drugs get their start in humans and then change to human and animal drugs.

17.2.4 Small Molecule vs Biologics

Small molecules have a variety of applications or biological functions. Large molecules (also called biologics) are proteins with a therapeutic effect. Small molecules work by cell signaling. Large molecule drugs are complex and may be composed of over 1300 amino acids. They are identical versions of human proteins.

17.2.5 Process Scale-Up Differences and Difficulties

Drug development involves generating progressively larger medicine batch sizes, and changes in processes for different-sized batches may cause unexpected difficulties. Use of the right pharmaceutical equipment can be helpful, as well as the discovery of parameters that affect critical process parameters (CPPs).

17.3 Multitargets Identification and Validation (MT-IV)

Identification of high-quality drug targets is at the heart of successful drug discovery and development but remains extremely challenging.

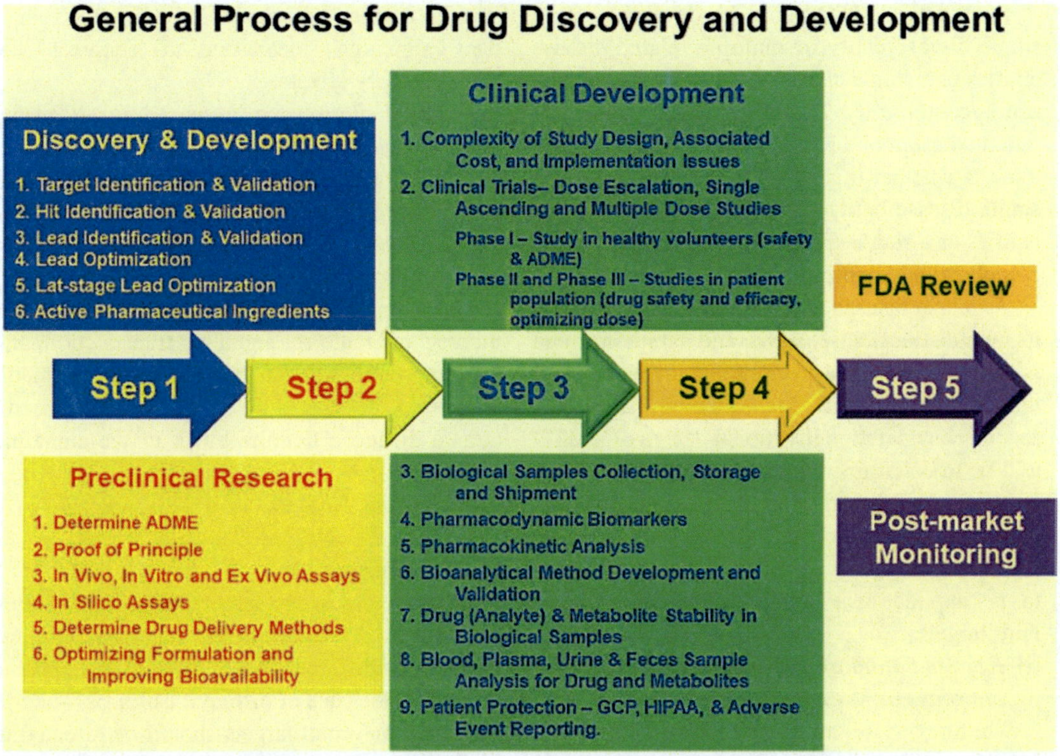

Fig. 17.1 Diagram elucidating the general procedures of drug discovery and development

Identification and validation of multiple targets for a MTD are even more difficult than for a STD. Multitargets in polypharmacology refers to multiple biomolecules that when deregulated and/or malfunctioning cause or critically modulate a single unique disease entity. The strategy and methodology for identification and validation of multitargets are essentially the same as for those of single-target or mono-target. However, it is natural that the identification and validation procedures are more complex for multitargets.

17.3.1 Identification of Multiple Targets

17.3.1.1 Principles for Multitargets Selection

In the drug discovery process, identification, deconvolution, and validation of therapeutic targets for a unique disease entity are of central importance as the process lays the groundwork for the subsequent development of novel drugs with potential interactions with these therapeutic targets. This is even more so for MTD discovery and development. As described in Chap. 1, multitargeticity, referring to the capability of a given drug to exhibit activity with respect to multiple molecular (binding site) targets, is the most fundamental element and core connotation of polypharmacology. The "multitargeticity" includes two different categories: designed multitargeticity and unintended multitargeticity, both of which are within the scope of polypharmacology for drug research and development. Designed multitargeticity is about the discovery and development of drugs as more effective and safer therapeutics, whereas unintended multitargeticity aims to identify and characterize drug promiscuity that essentially forms the source of off-target side effects or beneficial actions (such as the naturally occurring products and the synthetic compounds that happen to give rise to net beneficial effects).

In theory, a MTD could be developed for a single disease entity or multiple related disease entities or a single disease entity plus its complication/comorbidity. Therefore, in the process of rational design of a MTD, the first thing to determine is whether this MTD will be directed to a single disease entity or multiple related disease entities or a single specific disease entity plus its complication/comorbidity.

A unique disease entity is a real disorder with its own distinct cause, signs, and symptoms, and most human diseases are complex involving multiple etiologies and mechanisms. Thus, the second consideration in rational design of a MTD will be to determine which molecular targets are to be attacked, based upon the most updated scientific information on the disease and its major complication. In this step, several principles need to be applied as a guide to the selection of multitargets.

(i) *Key Mediators of Different Disease-Causing Pathways as Multitargets.*

Multitargets or multiple biomolecules for a MTD are generally identified based on multiple mediators of multiple different but related disease-causing signaling pathways relevant to a single disease entity. The valid targets should be the key molecules to their respective disease pathways. Key molecules are usually those that upon inhibition or activation could lead to definite and substantial blockade or activation of the disease-causing pathways and robust changes of phenotypes both in vitro and in vivo. Targeting multiple signaling mediators of different disease pathways could yield synergetic therapeutic effects on a disease. This explains why partial inhibition of a surprisingly small number of targets can often be more efficient than the complete inhibition of a single target.

(ii) *Signaling Mediators of Redundant Disease-Causing Pathways as Multitargets.*

One of the main causes for the low efficacy or even failure of many STDs in disease therapy can be ascribed to the redundancy of disease-causing signaling pathways: blockade of one disease-causing pathway could well be compensated by another disease-causing signaling pathway due to redundancy. In such a case, multitargets for a disease entity may be selected based on the redundant signaling mediators to suppress the compensatory effects.

(iii) *Key Mediators of a Unique Disease-Causing Pathway as Multitargets.*

Multitargets may also be selected from multiple mediators along the same disease-causing signaling pathway pertinent to a single disease entity in order to acquire additive actions. In this scenario, one of the multiple mediators should be the key molecule for signaling transduction, and the other target molecules may be those immediately upstream and/or downstream the key mediator to ensure a definite block of the signaling pathway.

(iv) *Multiple Isoforms of a Key Molecule in a Unique Disease-Causing Pathway as Multitargets.*

In many cases, the same protein has multiple isoforms, and these isoforms may have identical function or different functions or even opposing effects. In the case of different isoforms—identical function—multitargets should be directed to bind all or most isoforms to avoid redundancy. For the different isoforms—different functions or opposing effects scenario—multitargets should be designed to avoid binding to multiple isoforms.

(v) *Multiple Members of a Protein Family in a Unique Disease-Causing Pathway as Multitargets.*

Owing to high sequence similarity in the active sites within a protein family, small-molecule ligands often bind with high affinity to multiple members of that family [5]. This type of polypharmacology is particularly well established in kinase drug discovery, where privileged structures of certain hinge-binding motifs display low selectivity within the family, an effect that can lead to a poor therapeutic index (staurosporin) or be crucial for efficacy (aurora/FLT3) [5]. If the multiple members of a protein family are more or less involved in the regulation of a unique disease entity, then they can be considered multitargets for an MTD. If they have distinct or even countering pathophysiological roles, then it should be avoided having them incorporated into a multitargets chemical entity.

(vi) *High-Affinity-Specific Interactions with Different Protein Families.*

High-affinity-specific interactions can occur for ligands binding to different protein families despite no apparent binding site or sequence similarity [5]. This concept was recently highlighted by the discovery that the well-characterized potent kinase inhibitor, BI-2536 (PLK1, IC_{50} = 0.83 nM), binds with high affinity to the bromodomain family protein BRD4 (IC_{50} = 25–37 nM) [5]. Though it is presently unclear whether the interfamily activity represents an inherent conservation in protein structure for binding or is due to other attributes, one can take this advantage for identifying multitargets the interfamily activity is favorable for a unique disease entity or one should avoid this such interfamily activity when designing multitargets.

17.3.1.2 General Steps of Multitargets Identification (MT-I)

Multitargets identification (MT-I) involves in general the following basic steps:

(a) Formulating the strategy for MT-I.
(b) Choosing the technology/methodology for MT-I.
(c) Selecting the platform/database for MT-I.
(d) Conducting multitargets deconvolution.

These steps may take the following specific processes:

Data Mining

Data mining of available biomedical data has led to a significant increase in target identification. In this context, data mining refers to the use of a bioinformatics approach to help not only in identifying but also selecting and prioritizing potential disease targets [3, 5]. The data which are available come from a variety of sources but include publications and patent information, gene expression data, proteomics data, transgenic phenotyping, and compound profiling data. Identification approaches also include examining mRNA/protein levels to determine whether they are expressed in disease and if they are correlated with disease exacerbation or progression.

Genetic Associations

Another powerful approach is to look for genetic associations, for example, is there a link between a genetic polymorphism and the risk of disease or disease progression or is the polymorphism functional [3, 6]. For example, familial Alzheimer's disease (AD) patients commonly have mutations in the amyloid precursor protein or presenilin genes which lead to the production and deposition in the brain of increased amounts of the Abeta peptide, characteristic of AD [7]. There are also examples of phenotypes in humans where mutations can nullify or overactivate the receptor, for example, the voltage-gated sodium channel Nav1.7; both mutations incur a pain phenotype, insensitivity, or oversensitivity, respectively [8, 9].

Phenotypic Screening

An alternative approach is to use phenotypic screening to identify disease-relevant targets [3]. In an elegant experiment, Kurosawa et al. [10] used a phage display antibody library to isolate human monoclonal antibodies (mAbs) that bind to the surface of tumor cells. Clones were individually screened by immunostaining and those that preferentially and strongly stained the malignant cells were chosen. The antigens recognized by those clones were isolated by immunoprecipitation and identified by mass spectroscopy. Of 2114 mAbs with unique sequences, they identified 21 distinct antigens highly expressed on several carcinomas, some of which may be useful targets for the corresponding carcinoma therapy and several mAbs which may become therapeutic agents [3].

Biological Assays

The biological assay, such as RNAi, RNA sequencing, DNA microarray, and Gal4/UAS system, is also commonly used to identify the target in recent years (Fig. 17.2).

17.3.1.3 Multitargets Validation (MT-V)

Once identified, the target then needs to be fully prosecuted. Like MT-I, multitargets validation (MT-V) also involves four major steps:

(a) Formulating the strategy for MT-V.

(b) Choosing the technology/methodology for MT-V.
(c) Selecting the platform/database for MT-V.
(d) Performing MT-V.

According to Hughes et al. [3], target validation techniques range from in vitro tools through the use of whole animal models to modulation of a desired target in disease patients. While each approach is valid in its own right, confidence in the observed outcome is significantly increased by a multi-validation approach.

Antisense Technology
Antisense technology is a potentially powerful technique which utilizes RNA-like chemically modified oligonucleotides which are designed to be complimentary to a region of a target mRNA molecule [3, 11]. Binding of the antisense oligonucleotide to the target mRNA prevents binding of the translational machinery, thereby blocking synthesis of the encoded protein. A prime example of the power of antisense technology was demonstrated by researchers at Abbott Laboratories who developed antisense probes to the rat P2X3 receptor [12]. When given by intrathecal minipump, to avoid toxicities associated with bolus injection, the phosphorothioate antisense P2X3 oligonucleonucleotides had marked anti-hyperalgesic activity in the complete Freund's adjuvant model, demonstrating an unambiguous role for this receptor in chronic inflammatory states. Interestingly, after administration of the antisense oligonucleonucleotides was discontinued, receptor function and algesic responses returned. Therefore, in contrast to the gene knockout approach, antisense oligonucleotide effects are reversible, and a continued presence of the antisense is required for target protein inhibition [13]. However, the chemistry associated with creating oligonucleotides has resulted in molecules with limited bioavailability and pronounced toxicity, making their *in vivo* use problematic. This has been compounded by nonspecific actions, problems with controls for these tools, and a lack of diversity and variety in selecting appropriate nucleotide probes [11].

Transgenic Animal Models
Transgenic animals are an attractive validation tool as they involve whole animals and allow observation of phenotypic endpoints to elucidate the functional consequence of gene manipulation [3]. In the early days of gene targeting, animals were generated that lacked a given gene's function from inception and throughout their lives. This work yielded great insights into the *in vivo* functions of a wide range of genes [3]. One such example is through use of the P2X7 knockout mouse to confirm a role for this ion channel in the development and maintenance of neuropathic and inflammatory pain [14]. In mice lacking P2X7 receptors, inflammatory and neuropathic hypersensitivity is completely absent to both mechanical and thermal stimuli, while normal nociceptive processing is preserved. These transgenic animals were also used to confirm the mechanism of action for this ablation *in vivo* as the transgenic mice were unable to release the mature pro-inflammatory cytokine IL-1beta from cells although there was no deficit in IL-1beta mRNA expression. An alternative to gene knockouts is gene knock-ins, where a non-enzymatically functioning protein replaces the endogenous protein. These animals can have a different phenotype to a knockout, for example, when the protein has structural as well as enzymatic functions [15], and these mice should ostensibly mimic more closely what happens during treatment with drugs, that is, the protein is there but functionally inhibited.

Tissue-Restricted and/or Inducible Knockout Animal Models
More recently, the desire to be able to make tissue-restricted and/or inducible knockouts has grown. Although these approaches are technically challenging, the most obvious reason for this is the need to overcome embryonic lethality of the homozygous null animals. Other reasons include avoidance of compensatory mechanisms due to chronic absence of a gene-encoded function and avoidance of developmental phenotypes. However, the use of transgenic animals is expen-

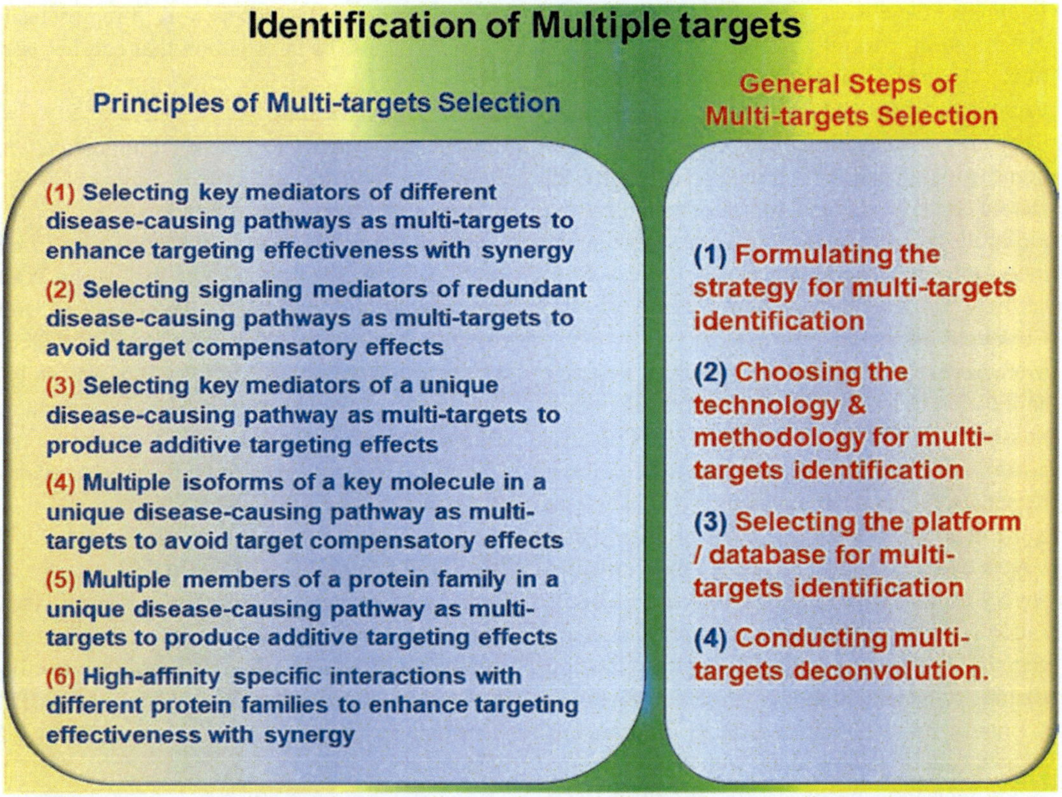

Fig. 17.2 Summary of the proposed principles and general steps of multiple targets (multitargets) selection

sive and time-consuming. So in order to circumvent some of these issues, the use of small interfering RNA (siRNA) has become increasingly popular for target validation. Double-stranded RNA (dsRNA) specific to the gene to be silenced is introduced into a cell or organism, where it is recognized as exogenous genetic material and activates the RNAi pathway. The ribonuclease protein Dicer is activated which binds and cleaves dsRNAs to produce double-stranded fragments of 21–25 base pairs with a few unpaired overhang bases on each end. These short double-stranded fragments are called siRNAs. These siRNAs are then separated into single strands and integrated into an active RNA-induced silencing complex (RISC). After integration into the RISC, siRNAs base-pair to their target mRNA and induce cleavage of the mRNA, thereby preventing it from being used as a translation template (reviewed by Castanotto and Rossi [16]). However, RNAi technology still has the major problem of delivery to the target cell, but many viral and nonviral delivery systems are currently under investigation (for review, see Whitehead et al. [17]).

Monoclonal Antibodies

Monoclonal antibodies are an excellent target validation tool as they interact with a larger region of the target molecule surface, allowing for better discrimination between even closely related targets and often providing higher affinity [3]. In contrast, small molecules are disadvantaged by the need to interact with the often more conserved active site of a target, while antibodies can be selected to bind to unique epitopes. This exquisite specificity is the basis for their lack of nonmechanistic (or "off-target") toxicity—a major advantage over small-molecule drugs [3].

However, antibodies cannot cross cell membranes restricting the target class mainly to cell surface and secreted proteins. One impressive example of the efficacy of a mAb *in vivo* is that of the function neutralizing anti-TrkA antibody MNAC13, which has been shown to reduce both neuropathic pain and inflammatory hypersensitivity [18], thereby implicating NGF in the initiation and maintenance of chronic pain. Finally, the classic target validation tool is the small bioactive molecule that interacts with and functionally modulates effector proteins.

Chemical Genomics

More recently, chemical genomics, a systemic application of tool molecules to target identification and validation, has emerged [3]. Chemical genomics can be defined as the study of genomic responses to chemical compounds [3]. The goal is the rapid identification of novel drugs and drug targets embracing multiple early phase drug discovery technologies ranging from target identification and validation, over compound design and chemical synthesis, to biological testing. Chemical genomics brings together diversity-oriented chemical libraries and high-information-content cellular assays, along with the informatics and mining tools necessary for storing and analyzing the data generated (reviewed in Zanders et al. [19]). The ultimate goal of this approach is to provide chemical tools against every protein encoded by the genome. The aim is to use these tools to evaluate cellular function prior to full investment in the target and commitment to a screening campaign.

17.3.1.4 An Example of MT-IV Procedures

Pinzi et al. [20] performed a large-scale analysis of the PDB by integrating different ligand-based and structure-based approaches, with the aim of identifying promising target associations for polypharmacology based on reported crystal structure information. First, the 2D and 3D similarity profiles of the crystallographic ligands were evaluated using different ligand-based methods.

Protein Data Bank (PDB)

The Protein Data Bank (PDB) provides experimentally validated structural information about targets and bound ligands as well. It represents a valuable source of information that can be used for:

Identifying active sites.
Understanding pharmacophore requirements.
Designing novel ligands.
Inferring structure–activity relationships.

The first step in the study reported by Pinzi et al. [20] was a large-scale analysis of the PDB by integrating different ligand-based and structure-based approaches, with the aim of identifying promising target associations for polypharmacology based on reported crystal structure information. First, the 2D and 3D similarity profiles of the crystallographic ligands were evaluated using different ligand-based methods.

Bioactivity Annotations Databases

Then, activity data of pairs of similar ligands binding to different targets were inspected by comparing structural information with bioactivity annotations reported in the ChEMBL, BindingDB, Binding MOAD, and PDBbind databases.

Molecular Docking

Afterward, extensive docking screenings of ligands in the identified cross-targets were made in order to validate and refine the ligand-based results [19].

Therapeutic Target Database (TTD)

Finally, the therapeutic relevance of the identified target combinations for polypharmacology was evaluated from comparison with information on therapeutic targets reported in the Therapeutic Target Database (TTD) [20]. The results led to the identification of several target associations with high therapeutic potential for polypharmacology.

17.3.1.5 An Example of MT-IV for Alzheimer's Disease (AD)

Alzheimer's disease (AD) is one of the most complicated progressive neurodegeneration diseases that involve many genes, proteins, and their complex interactions. No effective medicines or treatments are available yet to stop or reverse the progression of the disease due to its polygenic nature. To facilitate discovery of new AD drugs and better understand the AD neurosignaling pathways involved, Liu et al. [21] constructed an Alzheimer's disease domain-specific chemogenomics knowledgebase, AlzPlatform (www.cbligand.org/AD/) with cloud computing and sourcing functions.

AlzPlatform is implemented with powerful computational algorithms, including our established TargetHunter, HTDocking, and BBB Predictor for target identification and polypharmacology analysis for AD research. The platform has assembled various AD-related chemogenomics data records, including 928 genes and 320 proteins related to AD, 194 AD drugs approved or in clinical trials, and 405,188 chemicals associated with 1,023,137 records of reported bioactivities from 38,284 corresponding bioassays and 10,050 references [21].

Furthermore, the investigators described the application of the AlzPlatform in three case studies for identification of multitargets and polypharmacology analysis of FDA-approved drugs and also for screening and prediction of new AD active small chemical molecules and potential novel AD drug targets by our established TargetHunter and/or HTDocking programs [21]. The predictions were confirmed by reported bioactivity data and in vitro experimental validation. Overall, AlzPlatform can enrich our knowledge for AD target identification, drug discovery, and polypharmacology analyses and, also, facilitate the chemogenomics data sharing and information exchange/communications in aid of new anti-AD drug discovery and development.

17.3.2 Multitarget Lead Identification and Validation for MTDs

MTDs can in general be divided into three groups according to their sources or origin: (a) rationally designed MTDs, (b) naturally occurring MTDs, and (c) rationally designed single-target compounds or biologics that happen to possess unintended but desirable multitarget activities. Whereas the category (c) is an "opportunist" not truly subject to purposed design, MTDs belonging to the categories (a) and (b) can be developed by appropriate drug discovery strategies and technologies. The strategies for multitarget hit/lead identification and validation for rationally designed MTDs are distinct from those for naturally occurring MTDs. However, rational design of MTDs may well take the advantage of some naturally occurring pharmacophores or privileged scaffolds as core structures that possess multitarget activities. Moreover, in addition to privileged scaffold, molecular hybrid is another widely used strategy for generating multitarget compounds by taking advantage of structural commonalities to combine or overlap two distinct frameworks into a single chemical entity. Alternatively, MTDs could also be derived from "second-hand drugs" (currently used for a different disease entity or have-been-abandoned old drugs) by drug repositioning approach. The detailed information on MTD discovery has been presented in Chap. 12: "Polypharmacology in Drug Discovery and Development: Basis for Rational Design of Multitarget Drugs." In addition, the strategies, technologies, methodologies, and databases for MTDs discovery and development are also detailed in Chaps. 18, 19, 20, 21 and 22.

Polypharmacology studies are critical in MTD discovery and development. However, the exhaustive coverage of all targets by experimental methods is still an industry-wide challenge [22]. Apparently, the most competent approach would be to utilize the -omics (proteomics, cheminformatics, etc.) technologies [22–24]. The enormous molecular data generated in the post-genomic era has significantly accelerated the polypharmacology research [22]. There are a large number of public and private molecular databases available and are continuously growing in both size and number [22]. They integrate diverse information of molecular pathways, crystal structures, binding experiments, side effects, and drug targets. There are also a number of other small-molecule databases, including ZINC [25], PubChem [26], Ligand Expo [27], KEGG DRUG [28], etc., with enormous information about their disease relevance, chemical properties, and biological activities. These databases can be used not only to predict the protein targets of a small molecule but also to obtain insights of designing polypharmacological agents in a prospective manner [22]. Detailed introduction and discussion on these aspects will be given in the following chapters.

References

1. The Drug Development Process. US FDA. Retrieved on December 12, 2021. Available at: https://www.fda.gov/patients/learn-about-drug-and-device-approvals/drug-development-process.
2. Horizny C. The drug discovery process. Taconic Biosciences. Available at: https://www.taconic.com/taconic-insights/quality/drug-development-process.html
3. Hughes JP, Rees S, Kalindjian SB, et al. Principles of early drug discovery. Br J Pharmacol. 2011;162(6):1239–49.
4. Lansdowne LE. Exploring the drug development process. Technology networks: drug discovery. Published on March 13, 2020. Retrieved on December 10, 2021. Available at: https://www.technologynetworks.com/drug-discovery/articles/exploring-the-drug-development-process-331894.
5. Meyers J, Chessum NEA, Ali S, et al. Privileged structures and polypharmacology within and between protein families. ACS Med Chem Lett. 2018;9(12):1199–204.
6. Yang Y, Adelstein SJ, Kassis AI. Target discovery from data mining approaches. Drug Discov Today. 2009;14:147–54.
7. Bertram L, Tanzi RE. Thirty years of Alzheimer's disease genetics: the implications of systematic meta-analyses. Nat Rev Neurosci. 2008;9:768–78.
8. Yang Y, Wang Y, Li S. Mutations in SCN9A, encoding a sodium channel alpha subunit, in patients with primary erythermalgia. J Med Genet. 2004;41:171–4.
9. Cox JJ, Reimann F, Nicholas AK. An SCN9A channelopathy causes congenital inability to experience pain. Nature. 2006;444:894–8.
10. Kurosawa G, Akahori Y, Morita M, et al. Comprehensive screening for antigens overexpressed on carcinomas via isolation of human mAbs that may be therapeutic. Proc Natl Acad Sci U S A. 2008;105:7287–92.
11. Henning SW, Beste G. Loss-of-function strategies in drug target validation. Curr Drug Discov. 2002;May:17–21.
12. Honore P, Kage K, Mikusa J, et al. Analgesic profile of intrathecal P2X3 antisense oligonucleotide treatment in chronic inflammatory and neuropathic pain states. Pain. 2002;99:11–9.
13. Peet NP. What constitutes target validation? Targets. 2003;2:125–7.
14. Chessell IP, Hatcher JP, Bountra C, et al. Disruption of the P2X7 purinoceptor gene abolishes chronic inflammatory and neuropathic pain. Pain. 2005;114:386–96.
15. Abell AN, Rivera-Perez JA, Cuevas BD, et al. Ablation of MEKK4 kinase activity causes neurulation and skeletal patterning defects in the mouse embryo. Mol Cell Biol. 2005;25:8948–59.
16. Castanotto D, Rossi JJ. The promises and pitfalls of RNA-interference-based therapeutics. Nature. 2009;457:426–33.
17. Whitehead KA, Langer R, Anderson DG. Knocking down barriers: advances in siRNA delivery. Nat Rev Drug Discov. 2009;8:129–38.
18. Ugolini G, Marinelli S, Covaceuszach S, et al. The function neutralizing anti-TrkA antibody MNAC13 reduces inflammatory and neuropathic pain. Proc Natl Acad Sci U S A. 2007;104:2985–90.
19. Zanders ED, Bailey DS, Dean PM. Probes for chemical genomics by design. Drug Discov Today. 2002;7:711–8.
20. Pinzi L, Rastelli G. Identification of target associations for polypharmacology from analysis of crystallographic ligands of the protein data bank. J Chem Inf Model. 2020;60(1):372–90.
21. Liu H, Wang L, Lv M, et al. AlzPlatform: an Alzheimer's disease domain-specific chemogenomics knowledgebase for polypharmacology and target identification research. J Chem Inf Model. 2014;54(4):1050–60.
22. Reddy AS, Zhang S. Polypharmacology: drug discovery for the future. Expert Rev Clin Pharmacol. 2013;6(1):41–7.
23. Schmid A, Blank LM. Hypothesis-driven omics integration. Nat Chem Biol. 2010;6(7):485–7.
24. Joyce AR, Palsson BO. The model organism as a system: integrating 'omics' data sets. Nat Rev Mol Cell Biol. 2006;7(3):198–210.
25. Irwin JJ, Shoichet BK. ZINC – a free database of commercially available compounds for virtual screening. J Chem Inf Model. 2005;45(1):177–12.
26. Wheeler DI, Barrett T, Benson DA, et al. Database resources of the National Center for Biotechnology Information. Nucleic Acids Res. 2008;36(Database issue):D13–21.
27. Feng Z, Chen L, Maddula H, et al. Ligand depot: a data warehouse for ligands bound to macromolecules. Bioinformatics. 2004;20(13):2153–5.
28. Kanehisa M, Goto S, Sato Y, et al. KEGG for integration and interpretation of large-scale molecular data sets. Nucleic Acids Res. 2012;40(Database issue):D109–14.

18 General Strategies for Rational Design and Discovery of Multitarget Drugs

Abstract

The first step for drug discovery and development is to choose a right or optimal strategy to begin with. This is more so for rational design of multitarget drugs (MTDs). With the rapid advances in biomedical science and technology together with the emergence of interdisciplinary fields, numerous creative and innovative drug discovery strategies have been formulated. While different strategies may have different principles involving different theories and technologies, they may have some common and even overlapping applicability. Each of these strategies can be used as a stand-alone approach or in combination with other ones. This chapter provides a succinct overview on the frequently used strategies in drug discovery with a focus to their original concepts and possible applications to MTD discovery, along with some real examples. The contents aim to give a list of validated, well-established strategies that could be employed for the rational design of MTDs.

Keywords

Drug discovery strategies · Rational design of MTDs · Scaffold · Pharmacophores · Molecular hybrid

18.1 Target-Based Approach

18.1.1 Concept and Applicability of Target-Based Strategy in Drug Discovery

The target-based approach is a rational and conventional experimental approach for screening drug candidates targeting a biomolecule that causes diseases [1–3]. The starting point is a defined molecular target that is hypothesized to have an important role in disease. In recent years, with the advancement of high-throughput experimental technologies, the target-based approach has become used widely for drug discovery [1, 2, 4]. In target-based approach, the gene that codes for a protein target involved in the disease is first identified and compounds are screened to find for high-affinity binding. After this, the active compounds need to be checked in tissues and whole organisms. Target-based screening has been the method of choice in drug discovery for the past two decades.

By some measures, target-based drug discovery has been highly successful. Of 113 first-in-class drugs approved by the US FDA from 1999 to 2013, 70% were identified through target-based drug discovery [5, 6]. The approach is not limited to small-molecule drug discovery.

Z. Wang, B. Yang, *Polypharmacology*, https://doi.org/10.1007/978-3-031-04998-9_18

Antibody drugs and other protein biologics are generally discovered through target-based approaches, and the discovery of gene therapy and nucleic acid-based therapeutics is inherently target based as well [5].

One advantage of target-based approaches is that they are often simpler to execute than phenotypic approaches, operating with knowledge of a drug's molecular mechanisms from an earlier stage, and are generally faster, easier, and less costly to develop and run [5, 7]. With a molecular target in hand, drug discovery efforts can utilize crystallography, computational modeling, biochemistry, binding kinetics, molecular pharmacology, genomics, and mutational analysis to understand how a drug interacts with a target, enabling efficient structure–activity relationship (SAR) development, biomarker development, and the discovery of future generations of drugs acting at the target [5]. In contrast, optimizing compounds in the absence of a known molecular target, as may occur with a phenotypic strategy, can be challenging.

One of the most powerful tools in the identification of drug targets is human genetics [5]. When a human gene is closely linked to a disease, this provides strong evidence that a drug targeting that gene or gene product has a good chance to succeed. Some target classes have been particularly amenable to a target-based approach. Once thought to be extremely difficult to target, protein kinases are now targeted by increasing numbers of drugs, particularly in oncology [5, 8]. G-protein-coupled receptors (GPCRs) are the most common class of drug target, targeted by over 30% of approved drugs [5, 9]. Older drugs targeting GPCRs were often identified phenotypically based on pharmacological responses, while genomics and recombinant technology have enabled target-based GPCR drug discovery, leading to improved next-generation drugs [5]. Generally, as molecular understanding of a disease increases, so does the utility of a target-based approach [5].

However, in drug discovery, there are two serious problems, target deconvolution and polypharmacology, that cannot be solved by the target-based approach [1, 2, 5, 6]. Target deconvolution is used to identify a target biomolecule responsible for a phenotypic response. In addition, because the target-based approach is a method for searching for compounds that act directly on a known target biomolecule, it cannot be applied if the target biomolecule is unknown. Thus, target deconvolution is a bottleneck in the target-based approach to drug discovery. Several experimental methods have been developed in the chemical biology and chemical genetics research fields to address this problem [7]. In silico approaches have been also developed such as protein–ligand docking [8], virtual screening of small ligands [9], and inverted virtual screening methods [10]. These methods have a limitation in that they require available target protein structures.

The second critical issue in drug discovery with the target-based approach is multitarget polypharmacology [1, 6]. To discover multitargets and an MTD, it is necessary to achieve a target deconvolution for a plurality of target biomolecules that may affect MTD efficacy and toxicity. However, it is extremely difficult to perform target deconvolution experiments for multiple biomolecules from the viewpoint of labor and cost. Furthermore, even if target deconvolution for multitargets can be achieved, it is very hard to design and synthesize a chemical compound that interacts with multiple target molecules by experiment [1].

To overcome these two problems, i.e., target deconvolution and polypharmacology, Iwata et al. [1] developed a new in silico method using the probabilistic framework. This method is based on a machine learning technique that integrates data from compound–target protein interactions obtained from the target-based approach and data from compound–phenotype associations obtained from the phenotypic approach. It estimates a relevant network from compound to phenotype via target proteins. Specifically, the method consists of the following two steps. In the first step, a plurality of protein candidates is inferred a compound that can target by using a prediction model that is trained on the data of the compound–target protein interactions [1]. In the second step, the target proteins related to a phenotype are selected from protein candidates predicted in the first step model using a lasso model constructed by learning the data of compound–phenotype associations. Therefore, one can deduce a plurality of target proteins that can be interacted with by a specific compound and can be related to the phenotypic response [1]. In other words, this probabilistic framework integrates both data from the phenotypic and target-based approaches. The proposed method enables computational execution of target deconvolution for polypharmacology, which cannot be solved by conventional experimental approaches [1]. The method consists of two machine learning models. The first model predicts proteins targeted by a compound using the CGBVS model trained on the compound–target protein interaction data. The second model selects statistically significant proteins related to a phenotype from candidate proteins predicted by the CGBVS model. This is the first method for computationally executing target deconvolution considering polypharmacology, which has been unsolved by previous experimental approaches [1]. This method could provide keys for understanding the pathway and molecular mechanism from compound to phenotype, thus promoting drug discovery.

Another example of application of target-based approach to drug discovery was presented by the study reported by Swamidass et al. [11] who conducted a global analysis of public domain data, building a network of assays and connecting the assays if they shared non-promiscuous active molecules. Public domain data in PubChem alone represent more than 158 million measurements, 1.2 million molecules, and 4300 assays. This network spans both phenotypic and target-based screens, recapitulates known biology, and identifies new polypharmacology. Phenotypic screens are extremely important for drug discovery, contributing to the discovery of a large proportion of new drugs. Connections between phenotypic and biochemical, target-based screens can suggest strategies for repurposing both small-molecule and biologic drugs.

18.1.2 Assays for Validating Phenotype-Based Drug Discovery

18.1.2.1 Target-Based Screening

Target-based approach is a high-throughput method that uses defined/known molecular targets to find a drug that efficiently induces/inhibits it and is dependent on a specific molecular hypothesis formulated from prior knowledge, sometimes found from phenotypic screening [1]. It is also termed as reverse pharmacology in which scientists move in the opposite direction of the norm; instead of knowing the function and then finding the genomic component, the genomic study occurs first, and after several specific screenings, functional studies will be performed. Target-based screening is the use of recombinant technology and genomics for drug discovery [1].

18.1.2.2 Target-Based Assays

Target-based assays use isolated pharmacological targets to perform biochemical assays (e.g., G-coupled protein receptors, enzymes, interleukin/cytokine receptors, nuclear receptors, DNA, etc.). The assays are designed to provide insight into the interaction and mode of action of drugs designed to inhibit or activate pharmacological targets.

18.2 Phenotype-Based Approach

18.2.1 Concept and Applicability of Phenotype-Based Strategy in Drug Discovery

Phenotype-based drug discovery, in which drugs are tested for their efficacy on disease phenotypes rather than specific therapeutic targets, implies screening where the molecular mechanism of action is not assumed and does not require knowledge of the molecular target [1, 12, 13]. As such a phenotype-based approach is comparable to empirical screening, which was historically used in drug discovery before more target-based approaches became popular. Phenotypic screening is a type of screening used in biological research and drug discovery to identify substances such as small molecules, peptides, or RNAi that alter the phenotype of a cell or an organism in a desired manner. The phenotypic approach is an experimental approach involving techniques such as cell-based assay and in vivo assay for evaluating the phenotypic response of cells or tissues by chemical compounds [1, 12, 13]. These assays are used in drug discovery to search for active compounds that induce phenotypic responses that improve the disease state of cells or tissues.

Historically, phenotypic screening has a slight advantage when it comes to identifying first-in-class drugs, while target-based screening has yielded more best-in-class drugs [1, 12, 13]. In contrast to target-based drug discovery, phenotypic screening does not rely on knowing the identity of the specific drug target or its hypothetical role in the disease. A key benefit this approach has over target-based screening is its capacity to capture complex biological mechanisms that are not otherwise achievable. Increasing the understanding of the cell signaling networks involved in a disease process can decrease the amount of clinical failures by identifying factors contributing to unexpected activity, toxicities, and lack of efficacy [14].

When searching for biologically active molecules, phenotypic screening is often the most straightforward and intuitive way to discover relevant hits. The alternative is target-based screening in which a large number of compounds are screened against a single target protein, and subsequently, the active hits can be further optimized through medicinal chemistry efforts [15, 16]. However, according to a recent analysis of new molecular entities, target-based approaches are not as efficient as traditional phenotype-based methods in terms of generating first-in-class small-molecule drugs [15, 17]. One of the major limitations of target-based strategies is the fact that many compounds are found to interact with multiple targets, with most drug molecules interacting with six known molecular targets on average [15, 18]. Therefore, the one drug, "one target" paradigm, thought to be the cornerstone of target-based methods, often does not hold true for compounds identified using target-based methods. This deficiency has led to a paradigm shift that, when coupled with recent technological advances in proteomics and genomics methods, has resulted in a renaissance for phenotype-based screening methods.

One of the major advantages of phenotype-based approaches is that they provide an unbiased way to find active compounds in the context of complex biological systems [15]. Because phenotypic screening takes place in a physiologically relevant environment of cells or whole organism, the results from such screens provide a more direct view of the desired responses as well as highlight potential side effects. More importantly, phenotypic screens can lead to the identification of multiple proteins or pathways that may not have been previously linked to a given biological output [15]. Therefore, identifying the molecular targets of active hits from phenotypic screens is a crucial process that is required to understand underlying mechanisms and to further optimize active compounds [15]. Because target identification from phenotypic screens is expected to generate a spectrum of possible targets, the term "target deconvolution" was coined to more accurately define the process [15].

Over the last decade, a number of technologies from a wide range of fields have been explored to identify targets from phenotypic screens. In particular, proteomics- and genomics-based approaches have become more powerful when combined with whole-genome sequencing [15, 19]. High-throughput imaging platforms and computational analysis also have helped to find relevant pathways and proteins based on phenotype changes [15, 20]. Recent advances in quantitative mass spectrometry techniques have facilitated quantitative analysis of proteins and greatly enhanced the sensitivity of target detection [15, 21]. In this review, we will focus on the most recent examples of target deconvolution techniques in modern phenotypic profiling.

In the phenotypic approach, even if we successfully identify an active compound that affects a target phenotype, it is not possible to reveal the target biomolecule on which the hit compound directly acts and the relation between the target biomolecule and the phenotype [3, 17].

18.2.2 Assays for Validating Phenotype-Based Drug Discovery

In phenotype-based approach, animal models and cell-based assays are both strategies used to identify hits/leads.

18.2.2.1 In Vivo Assays

In vivo assays allow direct screening of compounds in preclinical disease models. A drawback is that these models are usually lower-throughput due to their intricacy. Higher-throughput large-scale in vivo phenotypic screens have been developed and could potentially be used for testing an array of diseases simultaneously [22], although in some cases these may not be commercially or ethically viable.

There is an alternative. Innovative automated phenotypic screening platforms have been developed to mirror the in vivo environment in an in vitro setting, ultimately capturing the insights typically gained from performing in vivo assays [23]. Having the ability to screen compounds across multiple human disease models in a high-throughput automated fashion is surely a "win-win." Safety, efficacy, potency, toxicity, and mechanism of action information can all be generated from the one platform, to ultimately determine the therapeutic potential in a system mirroring patient physiology [18].

18.2.2.2 Cell-Based (In Vitro) Assays

Cell-based (in vitro) assays present a significant advantage; namely, these can be easily adapted to a high-throughput format for automated phenotypic analysis [22]. Although there has been considerable advancement in phenotypic cell-based screening tools, there are still challenges associated with this approach. These include validation of "hits" and uncovering the molecular targets of these hits, also known as target deconvolution [10, 15].

18.3 Sequence-Based Approach

The sequence-based MTD approach has been discussed in detail in Chap. 12 (Polypharmacology in Drug Discovery and Development—Basis for Rational Design of Multitarget Drugs). The sequence-based strategy implies the discovery and application of therapeutic drugs that act by sequence-complementary mechanisms to bind the target DNAs (e.g., antisense oligonucleotides), mRNAs, ncRNAs (e.g., siRNA, miRNA inhibitor), decoy ODNs, or peptides/proteins (e.g., antibodies). It offers remarkable opportunities for the easy and precise design of MTDs able to act on multiple targets (multitargeticity) with preselected specificities (polyspecificity), representing the core and essence of polypharmacology [24–32]. This strategy can be best applied to gene therapy and epigenetic therapy and is virtually the only one that can afford exact polyspecificity or preselected specificities. (For detailed description on the sequence-based MTD approach, please see Chap. 12).

While the sequence-based approach takes DNAs and RNAs as the primary molecular targets, the structure-based approach to be introduced in the next subsection is mainly protein-directed therapeutic targeting strategy.

18.4 Structure-Based Drug Design in Drug Discovery

18.4.1 Concept and Applicability of Structure-Based Strategy in Drug Discovery

Drug design that relies on the knowledge of the three-dimensional structure of the biomolecular target is known as structure-based drug design (SBDD), which is powered by protein crystallography and computer graphics technologies [33–35]. The field of structure-based drug design is a rapidly growing area in which many successes have occurred in recent years, which is becoming an essential tool for faster and more cost-efficient lead discovery relative to the traditional method. Genomic, proteomic, and structural studies have provided hundreds of new targets and opportunities for future drug discovery. Perhaps, in the entire drug discovery paradigm, SBDD is the most powerful and efficient process.

Using the structure of the biological target, candidate drugs that are predicted to bind with high affinity and selectivity to the target may be designed using interactive graphics and the intuition of a medicinal chemist. Alternatively, various automated computational procedures may be used to suggest new drug candidates [33, 36]. If an experimental structure of a target is not available, it may be possible to create a homology model of the target based on the experimental structure of a related protein. Computational resources serve as an efficient technology for accelerating the drug discovery process, which includes various screening procedures, combinatorial chemistry, and calculations of such properties as absorption, distribution, metabolism, excretion, and toxicity (ADMET) [37].

As summarized by Batool and colleagues in their excellent review article [33], SBDD is an iterative process, and it proceeds through multiple cycles leading an optimized drug candidate to clinical trials [33]. Generally, a drug discovery process consists of four steps: the discovery phase, development phase, clinical trial phase, and registry phase [33]. In the first phase, a potential therapeutic target and active ligands are identified. The fundamental step involves cloning of the target gene followed by the extraction, purification, and 3D structure determination of the protein [33]. Many computer algorithms can be used to dock the huge databases of small molecules or fragments of compounds into the binding cavity of the target protein. These molecules are ranked according to a scoring system based on electrostatic and steric interactions with the binding site. Thorough investigation of electrostatic properties of the binding site, including the presence of cavities, clefts, and allosteric pockets, can be carried out using a 3D structure of the target molecule. Current SBDD methods consider the key features of the binding cavity of the therapeutic target to design efficient ligands [33, 38, 39]. In the second phase, the top hits are synthesized and optimized [40]. Furthermore, the top-ranked compounds with high affinity for selective modulation of the target protein are tested in vitro in biochemical assays. These ligands interfere with crucial cellular pathways, thereby leading to the development of drugs with a desired therapeutic and pharmacological effect [41]. Biological properties like efficacy, affinity, and potency of the selected compounds are evaluated by experimental methods [42]. The next step is to determine the 3D structure of the target protein in complex with the promising ligand obtained in the first phase [33]. The 3D structure provides detailed information about the intermolecular features that aid in the process of molecular recognition and binding of the ligand. Structural insights into the ligand–protein complex help with the analysis of various binding conformations; identification of unknown binding pockets and ligand–protein interactions; elucidation of conformational changes resulting from ligand binding; and detailed mechanistic studies [43]. Subsequently, multiple iterations increase the efficacy and specificity of the lead. The third phase includes clinical trials of the lead compounds. Those compounds that pass the clinical trials proceed to the fourth phase in which the drug is distributed in the market for clinical use.

SBDD is a computational technique widely used by pharmaceutical companies and scientists [33]. There are numerous drugs available on the market that have been identified by SBDD. Human immunodeficiency virus (HIV)-1-inhibiting FDA-approved drugs represent the foremost success story of SBDD [44]. Moreover, other drugs identified by the SBDD technique include a thymidylate synthase inhibitor, raltitrexed [34]; amprenavir, a potential inhibitor of HIV protease discovered by protein modeling and MD simulation [44, 45]; and the antibiotic norfloxacin [46]. Some of the failure cases have also been documented; for example, RPX00023 has been reported as an antidepressant that was claimed to have an agonistic activity toward receptor 5-HT1A, but it inhibited the receptor [47]. These failure cases are the reason for limitations in SBDD strategies.

18.4.2 Methods for Structure-Based Strategy in Drug Discovery

Current methods for structure-based drug design can be divided roughly into three main categories [48]. The first method is identification of new ligands for a given receptor by searching large databases of 3D structures of small molecules to find those fitting the binding pocket of the receptor using fast approximate docking programs. This method is known as virtual screening. A second category is de novo design of new ligands. In this method, ligand molecules are built up within the constraints of the binding pocket by assembling small pieces in a stepwise manner. These pieces can be either individual atoms or molecular fragments. The key advantage of such a method is that novel structures, not contained in any database, can be suggested [49–51]. A third method is the optimization of known ligands by evaluating proposed analogs within the binding cavity [48].

The completion of the human genome project has resulted in an increasing number of new therapeutic targets for drug discovery [52]. At the same time, high-throughput protein purification, crystallography, and nuclear magnetic resonance spectroscopy techniques have been developed and contributed to many structural details of proteins and protein–ligand complexes. These advances allow the computational strategies to permeate all aspects of drug discovery today [51–56], such as the virtual screening (VS) techniques [57] for hit identification and methods for lead optimization. Compared with traditional experimental high-throughput screening (HTS), VS is a more direct and rational drug discovery approach and has the advantage of low-cost and effective screening [58–60]. VS can be classified into ligand-based and structure-based methods. When a set of active ligand molecules is known and little or no structural information is available for targets, the ligand-based methods, such as pharmacophore modeling and quantitative structure–activity relationship (QSAR) methods, can be employed. As to structure-based drug design, molecular docking is the most common method which has been widely used ever since the early 1980s [61]. Programs based on different algorithms were developed to perform molecular docking studies, which have made docking an increasingly important tool in pharmaceutical research. Various excellent reviews on docking have been published in the past [56, 62–65], and many comparison studies were conducted to evaluate the relative performance of the programs [66–69].

18.4.2.1 Molecular Docking

Molecular docking is one of the most frequently used methods in structure-based drug design, due to its ability to predict the binding conformation of small-molecule ligands to the appropriate target binding site. Characterization of the binding behavior plays an important role in rational design of drugs as well as to elucidate fundamental biochemical processes [56, 70, 71]. Although it was originally developed to help in understanding the mechanisms of molecular recognition between small and large molecules, uses and applications of docking in drug discovery have heavily changed over the last years. In the field of molecular modeling, docking is a method which predicts the preferred orientation of one molecule to a second when bound to each other to form a stable complex [72]. Knowledge of the preferred orientation in turn may be used to predict the strength of association or binding affinity between two molecules using, for example, scoring functions. The associations between biologically relevant molecules such as proteins, peptides, nucleic acids, carbohydrates, and lipids play a central role in signal transduction. Furthermore, the relative orientation of the two interacting partners may affect the type of signal produced (e.g., agonism vs antagonism). Therefore, docking is useful for predicting both the strength and type of signal produced.

One can think of molecular docking as a problem of "lock-and-key," in which one wants to find the correct relative orientation of the "key" which will open up the "lock" (where on the surface of the lock is the keyhole, which direction to turn the key after it is inserted, etc.). Here, the protein can be thought of as the "lock," and the ligand can be thought of as a "key." Molecular docking may be defined as an optimization problem, which would describe the "best-fit" orientation of a ligand that binds to a particular protein of interest. However, since both the ligand and the protein are flexible, a "hand-in-glove" analogy is more appropriate than "lock-and-key" [73]. During the course of the docking process, the ligand and the protein adjust their conformation to achieve an overall "best-fit", and this kind of conformational adjustment resulting in the overall binding is referred to as "induced-fit" [74].

Molecular docking research focuses on computationally simulating the molecular recognition process. It aims to achieve an optimized conformation for both the protein and ligand and relative orientation between protein and ligand such that the free energy of the overall system is minimized. Two approaches are particularly popular within the molecular docking community:

One approach uses a matching technique that describes the protein and the ligand as complementary surfaces [75–77].

The second approach simulates the actual docking process in which the ligand–protein pairwise interaction energies are calculated [78].

The molecular docking method can be used to model the interaction between a small molecule and a protein at the atomic level, which allow us to characterize the behavior of small molecules in the binding site of target proteins as well as to elucidate fundamental biochemical processes [52, 79]. The docking process involves two basic steps: prediction of the ligand conformation as well as its position and orientation within these sites (usually referred to as *pose*) and assessment of the binding affinity. These two steps are related to sampling methods and scoring schemes, respectively, which will be discussed in the theory section [52].

Knowing the location of the binding site before docking processes significantly increases the docking efficiency. In many cases, the binding site is indeed known before docking ligands into it [52]. Also, one can obtain information about the sites by comparison of the target protein with a family of proteins sharing a similar function or with proteins co-crystallized with other ligands. In the absence of knowledge about the binding sites, cavity detection programs or online servers, e.g., GRID [80, 81], POCKET [82], SurfNet [83, 84], PASS [85], and MMC [86], can be utilized to identify putative active sites within proteins. Docking without any assumption about the binding site is called blind docking.

The early elucidation for the ligand–receptor binding mechanism is the lock-and-key theory proposed by Fischer [87], in which the ligand fits into the receptor like lock and key. The earliest reported docking methods [61] were based on this theory, and both the ligand and receptor were treated as rigid bodies accordingly. Then the "induced-fit" theory [88, 89] created by Koshland takes the lock-and-key theory a step further, stating that the active site of the protein is continually reshaped by interactions with the ligands as the ligands interact with the protein. This theory suggests that the ligand and receptor should be treated as flexible during docking. Consequently, it could describe the binding events more accurately than the rigid treatment [52].

Striking progresses have been made in structural and molecular biology along with advances in biomolecular spectroscopic structure determination methods. These methods have provided three-dimensional (3D) structures of more than 100,000 proteins [43]. In conjunction with the storage of (and organizing) such data, there has been much hype about the development of sophisticated and robust computational techniques. Completion of the Human Genome Project and advances in bioinformatics increased the pace of drug development because of the availability of a huge number of target proteins. The availability of 3D structures of therapeutically important proteins favors identification of binding cavities and has laid the foundation for structure-based drug design (SBDD). This is becoming a fundamental part of industrial drug discovery projects and of academic research [34]. SBDD is a more specific, efficient, and rapid process for lead discovery and optimization because it deals with the 3D structure of a target protein and knowledge about the disease at the molecular level [90]. Among the relevant computational techniques, structure-based virtual screening

(SBVS), molecular docking, and molecular dynamics (MD) simulations are the most common methods used in SBDD. These methods have numerous applications in the analysis of binding energetics and ligand–protein interactions and evaluation of the conformational changes occurring during the docking process [91]. In recent years, developments in the software industry have been driven by a massive surge in software packages for efficient drug discovery processes. Nonetheless, it is important to choose outstanding packages for an efficient SBDD process [92]. Briefly, automation of all the steps in an SBDD process has shortened the SBDD timeline [34]. Moreover, the availability of supercomputers, computer clusters, and cloud computing has sped up lead identification and evaluation. In this review, we offer an overview of the SBDD process and the methods being used in the present era. Moreover, we provide an in-depth discussion about the machine learning (ML) methods intended to speed up this process and big data handling.

18.4.2.2 Inverse Molecular Docking

Molecular docking of small molecules into the receptor active sites is one of the most widely used structure-based methods in the field of computer-aided drug discovery. It has been shown that current docking algorithms are able to predict docking poses of small molecules similar to experimental structures. Recent advances in high-throughput protein crystallography, nuclear magnetic resonance (NMR) spectroscopy, and electron microscopy have generated a large number of 3D protein structures in the Protein Data Bank (PDB). Availability of such structural data has allowed development of inverse docking methods where the primary aim is to dock a small molecule into binding sites of multiple targets for hit identification. Unlike traditional docking algorithms where small molecules are scored and ranked, in inverse docking target receptors are ranked according to their scores. In theory, all of the available docking algorithms should be able to carry out inverse docking. However, the major challenge is to handle a large amount of structural data of macromolecules and implement scoring functions that could accurately rank the targets instead of small molecules.

18.4.2.3 Binding-Site Structural Similarity (BsSS)

Application of structure-based approach is becoming more and more popular in drug discovery [93–95]. Nowadays, the majority of the methods are based on measuring the similarity of a query molecule against the hundreds of thousands of molecules for which pharmacological data on thousands of proteins are available in public sources. However, similarity-based methods are inherently biased by the chemical coverage offered by the active molecules present in those public repositories, which limits significantly their capacity to predict interactions with proteins structurally and functionally unrelated to any of the already known targets for drugs. It is in this respect that structure-based methods aiming at identifying similar binding sites may offer an alternative complementary means to ligand-based methods for detecting distant polypharmacology [95].

Detection of remote binding-site similarity in proteins is important for drug discovery and for drug repositioning and off-target effect prediction as well [96]. Various non-covalent interactions such as hydrogen bonds and van der Waals forces drive ligands' molecular recognition by binding sites in proteins. The increasing amount of available structures of protein–small-molecule complexes enabled the development of comparative approaches. Several methods have been developed to characterize and compare protein–ligand interaction patterns. Usually implemented as fingerprints, these are mainly used for post-processing docking scores and (off-)target prediction. In the latter application, interaction profiles detect similarities in the bound interactions of different ligands and thus identify essential interactions between a protein and its small-molecule ligands. Interaction pattern similarity correlates with binding-site similarity and is thus contributing to a higher precision in binding-site similarity assessment of proteins with distinct global structure [96].

The structure-based design of small-molecule inhibitors of protein–ligand and protein–protein interfaces is a key component of drug discovery. The underlying protein interactions can be regarded based on structural similarity of the secondary structure elements: similarities around the binding site ("ligand-sensing cores") or in the protein interface ("interface-sensing surfaces") in otherwise unrelated proteins can be useful in predicting polypharmacology and identifying new lead structures. Even small, conserved motifs can provide similar interaction patterns in proteins with a completely different fold and function [97]. The identification of these structural similarities can help in the design of new drugs by guiding further optimization.

18.4.2.4 Structure-Based Virtual Screening

Structure-based virtual screening is highly used in the early stages of drug discovery to identify new putative lead compounds for a given target [98]. However, when a small molecule elicits a biological effect, but its target is unknown, or the side effects it causes arise from its undesired interaction with unknown counterparts, the identification of its interacting targets represents an indispensable task. The computational procedure named inverse virtual screening, which relies on docking a molecule (or a small set of compounds) against panels of target proteins to select the most promising complexes, could be useful to overcome these issues. Panels can contain thousands of proteins, and they must be correctly prepared to assure the best docking performance. Therefore, the preparation of panels of proteins collected in the Protein Data Bank (www.rcsb.org), if manually performed, may be costly in terms of time and efforts, and this can limit the applicability of this approach in high-throughput virtual screening workflows. We here show an automated workflow to speed up panel preparation and development, and to test its performance, this protocol was initially applied to a panel of 628 viral proteins and, afterward, to a panel of transferase proteins (2789 entries) to perform a large inverse virtual screening study, testing a small set of compounds synthesized in our laboratory. Tankyrase 2 (PARP 5b) was selected as their preferred target of interaction, and the predicted binding was validated by means of surface plasmon resonance experiments. This protocol is useful for the rapid identification of the interacting target for a bioactive compound; accordingly, it facilitates the re-evaluation of the pharmacological activity of known active compounds, addressing the repurposing and the polypharmacology concepts [98].

18.4.3 An Overview of SBDD Process

Batool et al. [33] presented a comprehensive overview on a typical SBDD process, which greatly helps us better understand the SBDD approach in MTD discovery. Their procedures are composed of five steps.

18.4.3.1 Target Protein and Binding-Site Identification

The basic step in a typical SBDD process is target protein identification and validation [99]. The 3D structures of all therapeutically important proteins are determined experimentally by integrative structure biology techniques such as nuclear magnetic resonance (NMR), X-ray crystallography, or cryo-electron microscopy, but if a solution structure is not available, in silico methods are used to model the protein's 3D structure. There are three well-known structure prediction methods such as comparative modeling, threading, and ab initio modeling. Among them, homology modeling is one of the best and reliable approaches because it predicts the 3D structure of a target protein on the basis of the knowledge about the structure of homologous proteins with >40% similarity [100]. Once the 3D structure of the target is predicted, it is necessary to validate the model by checking the stereochemical properties in a Ramachandran plot. It shows the possible conformations of ψ and φ angles for all amino acid residues present in the protein structure [101]. There are many other methods for validation of the model [100, 102, 103].

After the structure of the target protein is resolved, the next step is to identify the binding pocket. This is a small cavity where ligands bind to the target to produce the desired effect.

Therefore, it is necessary to identify the appropriate site on the target protein. In spite of the protein's dynamic nature, there are a few methods capable of spotting the potential binding residues. These methods consider the knowledge about interaction energy and van der Waals (vdW) forces for binding-site mapping. Many methods have been developed for binding-site mapping by interaction energy calculations specifically for SBDD. This method identifies particular sites on the target protein which interact favorably with important functional groups on drug-like molecules [104]. These methods identify energetically favorable interactions of specific probes with the proteins. Q-SiteFinder [104] is an energy-based method commonly used for binding-site prediction. This method calculates vdW interaction energies of proteins with a methyl probe. Those with favorable energies are retained and clustered. These probe clusters are ranked based on their total interaction energies. In addition, interacting protein residues are functionally annotated to determine the binding site. The next step is hit discovery, which is done by docking of compound libraries into the binding cavity of the target protein. In the initial phases of lead discovery, it is important to choose a specific set of ligands that play a key part in the lead identification and optimization [105]. For hit hunting, SBDD integrates two divergent methods (i.e., virtual screening (VS) and de novo design).

18.4.3.2 Virtual Screening: A Lead Identification Approach

In medicinal chemistry, VS is a robust approach to lead identification [106]. In VS, databases of millions of drug-like or lead-like compounds are screened computationally against the target proteins with well-known 3D structures. The screening of compound libraries is accomplished by docking, where ligands are filtered based on their binding affinity [107, 108]. The top hits of the computational screening are then tested in vitro [106, 109]. VS is classified into two major types: ligand-based VS (LBVS) and SBVS. In LBVS, biological data are analyzed to separate inactive compounds from the active compounds. This information is then employed to identify highly active scaffolds on the basis of consensus pharmacophores [110], similarity, or various descriptors. In SBVS, the knowledge about the 3D structure of the target protein is necessary. The target protein is docked with the huge libraries of drug-like compounds, available commercially, via computer algorithms. A scoring function is executed to evaluate the binding force of the docked complex followed by experimental assays to validate the binding. The scoring of ligands is a critical step in SBVS. Unlike ligand-based methods, structure-based approaches do not rely on already available experimental data.

18.4.3.3 De Novo Drug Design

De novo drug design is a method of building novel chemical compounds starting from molecular units [33]. The gist of this approach is to develop chemical structures of the small molecules that bind to the target binding cavity with good affinity [33]. Generally, a stochastic approach is used for de novo design, and it is important to take the search space knowledge into consideration in the design algorithm. The two designs, positive and negative, are being used. In the former design, a search is restricted to the specific regions of chemical space with higher probability of finding hits having required features. In contrast, the search criteria are predefined in the negative mode, to prevent the selection of false positives. The chemical compound designing by computational techniques can be related to imitation of synthetic chemistry, while scoring functions perform binding assays. Critical assessment of candidates is crucial for the design process, and the scoring function is one of the assessment tools. Multiple scoring functions can be employed parallelly for multi-objective drug design [111], which considers multiple features at once.

Two methods—(i) ligand-based and (ii) receptor-based de novo drug design—can be used. The latter approach is more prevalent [33]. The quality of target protein structures and accurate knowledge about its binding site are important for receptor-based design because suitable small molecules are designed by fitting the fragments into the binding cavities of the receptors.

This could be either done by means of a computational program or by co-crystallization of the ligand with the receptor. There are two techniques for receptor-based design: by linking building blocks, either atoms or fragments such as single rings, amines, and hydrocarbons, together to form a complete chemical compound or simply by growing a ligand from a single unit. In the fragment-linking method, the binding site is identified to map the probable interacting points for different functional groups present in the fragments [33]. These functional groups are attached together to build an absolute compound. In the fragment-growing technique, the growth of fragments is accomplished within the binding site monitored by suitable search algorithms. These search algorithms involve scoring functions to assess the probability of growth. Fragment-based de novo design uses the whole chemical space to generate novel compounds. In case of the linking approach, the selection of linkers is critical. Fragment anchoring in the binding site can be performed by (i) the outside-in approach and (ii) the inside-out approach. In the former approach, the building blocks are primarily arranged at the periphery of the binding site, and it grows inward. In the course of the inside-out approach, building blocks are casually fitted into the binding site and built outward.

18.4.3.4 Molecular Docking

Docking is a technique of virtual simulation of molecular interactions [33]. Molecular docking predicts the conformation and binding of ligands within a target active site with high accuracy; therefore, it is the most popular technique in SBDD [52]. This method can be applied to study important molecular phenomena such as a ligand-binding pose and intermolecular interactions for stability of a complex. Moreover, docking algorithms predict binding energies and rank the ligands by means of various scoring functions. The appropriate ligand-binding conformation depends on two factors: (i) large conformational space defining possible binding poses and (ii) explicit prediction of binding energy correlating with each conformation [33, 52, 112]. Multiple iterations are performed, until the minimum energy state is attained, in which ligand binding is assessed by various scoring functions.

There are two types of molecular docking: flexible-ligand search docking and flexible-protein docking. In flexible-ligand search docking, three types of algorithms are designed to deal with the ligand flexibility. These algorithms are the stochastic method, systematic method, and simulation method [33]. The systematic algorithms are aimed at analyzing degrees of freedom. This task can be accomplished by the fragmentation method, one of the frequently used techniques. In this method, a ligand grows gradually inward in a binding cavity [33, 52]. In the conformational search technique, rotatable bonds of the molecule are rotated 360 ° systematically at a fixed-increment rate, or in the database approach, pregenerated libraries of conformational ensembles are utilized for ligand flexibility. In the stochastic algorithms, random modifications are applied to a single ligand or a group. These modifications are accepted or rejected depending upon probability functions such as genetic algorithm methods and the Monte Carlo (MC) method [33]. Lastly, MD simulation is a comprehensive technique for studying the dynamic behavior of macromolecules. Energy minimization is implemented as integration with simulations to achieve local minima. The algorithms available for energy minimization are the Newton–Raphson method, steepest descent, least squares methods, and a conjugate gradient [113]. Many biological systems show movements upon ligand binding; thus, in the flexible-protein docking method, the receptor remains flexible during the docking procedure to mimic the natural biological environment. In addition to the full protein movement, in a few cases, small motions are also noticed such as side chain rearrangement or movement of highly flexible loops. MD and MC methods are suitable for flexible-protein docking [114].

18.4.3.5 Scoring Functions

A scoring function helps a docking program to delve into the ligand-binding site. Once a significant binding conformation is identified, the scoring function calculates binding affinity. Accordingly, scoring functions are thought to have a substantial impact on docking. Scoring functions are trained on a training dataset of a similar class of compounds for which their experimental binding affinity is available. Scoring functions are divided into four general classes: force field, empirical, knowledge-based, and machine learning (ML) [115]. The force field is calculated by estimating the intermolecular interactions such as electrostatic and vdW forces between the binding partners. Empirical scoring functions are calculated based on the atom numbers in the ligand and target protein and are used for affinity and pose prediction [116]. The latter includes hydrophobic forces, hydrophilic forces, hydrogen bonding, and entropy. A statistical method called multiple linear regression is employed to fit scoring-function coefficients. A knowledge-based scoring function depends on statistical potentials of intermolecular interactions. This method is based purely on the assumption that frequently occurring functional groups or a certain type of atoms are energetically favorable and contribute to binding affinity [117]. In contrast to classical scoring functions, ML methods do not constrain analysis to a predefined functional form among structural features and binding affinity values. ML methods are dynamic techniques for construction and optimization of models to predict a binding pose and affinity. Lately, the development of novel scoring functions by ML is becoming popular [118]. These methods implicitly take into consideration the interactions between a ligand and target while ignoring error-prone interactions. Furthermore, different methods of the ML technique such as random forest (RF), support vector machine (SVM), and neural networks (NN) work with nonlinear dependence among binding interactions. Thus, ML-based scoring functions perform better than others do in case of binding energy calculations. Another scoring function known as consensus scoring employs collective scores to minimize the error rate in individual scores and to increase the possibility of true positive selection [113].

The efficiency of various scoring functions has been compared in many studies [33], regarding binding affinity prediction, reproducibility of a known binding conformation, and ranking of a library. All modern scoring functions have different accuracy rates under different conditions. Thus, none of the scoring functions can outperform the others. However, consensus scoring function can perform better than single-scoring approach and is widely used in various bioinformatics applications. Consensus scoring function compensates the limitations of single-scoring functions. It improves the hit rate by combining multiple scoring functions based on a simple cause: the true value tends to be closer to the mean value of replicated experiments. In case of single-scoring functions, a binding pose can be predicted accurately, but in terms of binding energy calculations, there is still a need to improve the performance of current scoring functions. Hence, a lot of efforts have been made to upgrade the abilities of the currently available scoring functions. Prevalent methods include the addition of certain features for calculation of entropic and solvation effects [119], development of a consensus scoring function to overcome the limitations of others, and calculations of quantum energy terms [120]. Targeted scoring functions are known to significantly enhance VS performance and might be a solution to the limitations of other scoring functions [121]. Such scoring functions generate output with higher probability of true hits and a decreased rate of false positives.

18.5 Scaffold-Based Approach

18.5.1 Concept and Applicability of Scaffold-Based Strategy in Drug Discovery

A molecular scaffold is a privileged structure or molecular framework with versatile binding properties, such that a single scaffold is able to provide potent and selective ligands for a range of different biological targets through modification of functional groups. In other words, privileged scaffolds refer to effective templates or core structures, as building blocks in the synthesis of chemically relevant organic compounds for drug discovery and design or construction of high-affinity ligands that can bind more than one receptor. Clearly, scaffold-based approach is one of the most relevant strategies for polypharmacology.

The term "privileged scaffold" was coined by Evans et al. [122, 123] as a core structure that can be adapted to different protein targets by functionalization, early examples of which were benzodiazepines, indoles, and 1,4-dihydropyridines [124]. This concept has been explored widely for chemically diverse scaffolds, especially flat heterocyclic systems. Benzodiazepines, otherwise known as allosteric enhancers at the γ-aminobutyric acid (GABA)A ionotropic receptor, can be functionalized to achieve high-affinity binding at various G-protein-coupled receptors (GPCRs) of interest such as cholecystokinin (CCK) receptors [122]. Notably, the term "privileged scaffold" has been used fairly liberally versus Evans' original conception of the term, in that the ability to bind multiple targets is less thoroughly employed as a strict criterion for membership versus the notion of multiple molecules of the same scaffold having bioactivity. [The concept of privileged scaffold has been introduced in Chap. 12, Sect. 12.5.

Over the past 15 years the privileged structure concept has emerged as a fruitful approach to the discovery of novel biologically active molecules. Privileged structures typically exhibit good drug-like properties, which in turn leads to more drug-like compound libraries and leads. The net result is the production of high-quality leads that provide a solid foundation for further development. Privileged structure-based libraries collect distinct target families (e.g., GPCRs, LGIC, enzymes/kinases) and have been successfully exploited across and within different target families and promises to be an effective approach to the discovery and optimization of novel bioactive molecules. Among the strategies that can lead to the discovery of new drugs, the identification and use of privileged structures that are able to interact with more than one target gained particular attention, in an attempt to find new drugs in a shorter time with respect to other strategies. These structures that have been identified mainly by empirical observations can target only a given protein family or can be able to interact with more, unrelated targets.

Medicinal chemistry has as its main mission the understanding of molecular reasons for the activity of a drug or drug candidate. In this understanding, a few structural subunits of a certain bioactive molecule may be more relevant to a specific pharmacologic activity, governing the main interactions with a receptor. Those are the pharmacophoric contributions or pharmacophoric molecular groupings. Not unusually, the structures of drugs or their precursors have several functional groups, as well as the pharmacophoric ones, and all of them are called auxophoric subunits. Evidently, they all contribute to the total free energy of the drug–receptor complex, distinctively influencing the activity. Therefore, some molecular scaffolds may have pharmacophoric characteristics for a certain type of receptor and not for others. Some scaffolds may have privileged characteristics, being recognized molecularly by distinctive receptors without being important pharmacophores. Privileged structures have been successfully exploited across and within different target families and promises to be an effective approach to the discovery and optimization of novel bioactive molecules.

Privileged chemical scaffolds can interact with a broad array of receptors, kinases, and other disease-causing factors as therapeutic targets. Understanding privileged chemical scaffolds can provide an understanding of the chemical vectors for rational polypharmacology. Below are brief introductions to some frequently used molecular scaffolds.

18.5.2 Examples of Privileged Scaffolds

18.5.2.1 Conformationally Locked Methanocarba Nucleosides

A single molecular scaffold can be adapted to interact with diverse targets, either separately or simultaneously [125]. Single molecular scaffold nucleosides and nucleotides in which ribose is substituted with bicyclo[3.1.0]hexane are an example of a versatile drug-like scaffold for increasing selectivity at their classical targets: kinases, polymerases, adenosine, and P2 receptors. Also, by applying structure-based functional group manipulations, rigidified adenosine derivatives can be repurposed to satisfy pharmacophoric requirements of various GPCRs, ion channels, enzymes, and transporters, initially detected as off-target activities. Recent examples include $5HT_{2B}$ serotonin receptor antagonists and novel dopamine transporter allosteric modulators. This directable target diversity establishes rigid nucleosides as privileged scaffolds.

Privileged structures, fragments, or scaffolds have been identified by diverse screening and by design, including combinatorial design of bicyclic structures [125–128]. Purine nucleobases were previously identified as privileged structures for medicinal chemistry [126, 127]. The studies from Jacobson et al. [125] and other research groups have modified nucleoside derivatives to expand the range of their target proteins and to shift their selectivity. It is now apparent that nucleosides can serve as privileged scaffolds to bind to diverse proteins, and the many nucleoside drugs approved for therapy testify to the pharmacological versatility of this scaffold [129].

18.5.2.2 N-Propargylamines

N-Propargylamines are one of the most useful and versatile precursors in the synthesis of various nitrogen heterocyclic compounds and complex natural products, such as pyrroles, pyridines, thiazoles, and oxazoles [130–132]. Thiazoles and their hydrogenated analogs are not only key structural units in a wide variety of natural products, but they also constitute important building blocks in medicinal chemistry. Therefore, the synthesis of these compounds using new protocols is always interesting. It is well-known that N-propargylamines can undergo a number of cyclization reactions to produce various nitrogen-containing heterocycles. The synthesis of thiazoles and their hydrogenated analogs from N-propargylamines offers several advantages, such as high functional group tolerance and high atom and step economy.

Thiazoles are an important class of azole compounds that have attracted considerable attention due to the fact that they exhibit a wide variety of pharmacological activities [130]. For example, abafungin is an antifungal drug marketed worldwide for the treatment of dermatomycoses. It works by inhibiting the enzyme sterol 24-C-methyltransferase. Febuxostat, also known by its brand name adenuric, is a xanthine oxidase inhibitor that helps to prevent gout flare-ups. Ritonavir (norvir) is an HIV protease inhibitor. It works by blocking the growth of HIV. Tiazofurin is a C-nucleoside analog with antineoplastic activity and acts by inhibition of the guanosine triphosphate (GTP) biosynthesis through a reduction of PI and PIP kinase activity. This compound class is also a crucial part of many natural products such as vitamin B1 (thiamine), epothilone, dolastatin, and many more. The hydrogenated thiazoles (thiazoline and thiazolidine derivatives) are also important structural motifs that are widely found in biologically active natural or synthetic products [130, 133]. Compounds containing these rings have widespread biological applications as anticancer [130, 134], anti-HIV, anti-inflammatory, antimicrobial, and specially antibiotic agents. Despite their great relevance in drug design, only very few synthetic methods toward these compounds have been reported to date. *N*-Propargylamines are one of the most specific class of alkynes having diverse reaction patterns. It is well-known that they can undergo a number of cyclization reactions to produce various *N*-heterocycles and complex natural products.

18.5.2.3 Quercetin

Quercetin, a naturally occurring flavonoid, inhibits multiple cancer cell lines and is used as an anticancer drug in phase I clinical trial [135]. Quercetin glycosides have also received some attention due to their high bioavailability and activity against various diseases including cancer. However, these have been studied to a lesser extent. In the study reported by Baby et al. [135], the structural basis of the multitarget inhibitory activity of quercetin and isoquercitrin, a glycoside derivative, on serine/threonine kinases using molecular modeling was explored. Structural analysis showed that both quercetin and isoquercitrin exhibited good binding energies and interacted with aspartate in the highly conserved Asp-Phe-Gly motif. The results indicate that isoquercitrin could be a more potent inhibitor of several members of the serine/threonine kinase family. In summary, their structural evaluation highlights the multitarget inhibitory property of quercetin and its potential to be a chemical platform for oncological polypharmacology.

18.5.2.4 1,2-Benzisoxazole

As stated by Rakesh et al. [136], heterocyclic compounds are an exceedingly important class of compounds, and they have attracted more attention for diverse biological studies. All available natural and synthetic heterocyclic compounds can participate in biochemical reactions in the human body [137]. The engineering and rationale behind drug design are closely related to the strategic incorporation of heterocyclic-like fragments with specific physicochemical properties. The potency and selectivity through bioisosteric replacements, lipophilicity, polarity, and aqueous solubility can ultimately be fine-tuned to the point of altering and conditioning the possible mechanisms of action of pharmaceutical drugs in an attempt to obtain molecularly targeted agents [138].

As heterocyclic compounds, the benzisoxazole analogs represent one of the privileged structures in medicinal chemistry, and there has been an increasing number of studies on benzisoxazole-containing compounds [136, 139]. The unique benzisoxazole scaffold also exhibits an impressive potential as antimicrobial, anticancer, anti-inflammatory, and anti-glycation agents and so on. Numerous benzisoxazole moieties have been found to possess a broad spectrum of pharmacological activities, which encouraged research activities in these areas. The presence of an amide bond, benzisoxazoles, chromans, and fluorine atom substituents can alter the chemical properties, disposition, and biological activities of drugs. Many fluorinated compounds, 1,2-benzisoxazole derivatives, and various amides are currently used in the treatment of diseases [140]. These include antidepressants, anti-inflammatory agents, antimalarial drugs, antipsychotics, antiviral agents, steroids, and anesthetics. The 1,2-benzisoxazole ring containing drugs zonisamide and risperidone has been used as anticonvulsants [141]. Fluorine substitution can also have a profound effect on drug disposition, in terms of distribution, drug clearance, route(s), and extent of drug metabolism [142].

Benzisoxazoles are currently the most important building blocks in drug discovery, with a high number of positive hits encountered in biological screens of this heterocycle and its derivatives [136]. The benzisoxazole template forms the molecular backbone, possesses versatile binding properties with a frequently occurring binding motif, and provides potent and selective ligands for a range of different biological targets in medicinal chemistry [136]. The benzisoxazole scaffold and its analogs are important pharmacophores that can be found in biologically active compounds across a number of different therapeutic areas as anti-HIV, antimicrobial, antipsychotic,11–14 anti-inflammatory, analgesic, dopamine and serotonin receptor, anticonvulsant, acetylcholine esterase, anticancer, antioxidation, antidiabetic agents, etc. [136].

1,2-Benxisoxazole derivatives are possessing a rich diversity of biological properties especially in the area of central nervous system (CNS) disorders and therefore considered privileged structures [143]. Uto presented a review article analyzing the most significant examples of 1,2-benzisoxazoles as privileged structures in terms of polypharmacology at the molecular level, specifically focusing on four 1,2-benzisoxazoles (zonisamide, risperidone, paliperidone, and iloperidone) which have been in clinical use with established therapeutic effectiveness. Furthermore, an updated and detailed account of the pharmacological properties of 1,2-benzisoxazole derivatives as therapeutics for CNS disorders is described. And finally, outlooks on current issues and future directions in this field are also provided. 1,2-Benzisoxazole was successfully employed in the discovery and development of zonisamide for the treatment of epilepsy and Parkinson's disease. 1,2-Benzisoxazole is also a significantly important structure for the development of atypical antipsychotics. 1,2-Benzisoxazole is therefore considered a good example of a privileged structure because it forms the centerpiece of small-molecule chemical entities with a wide range of pharmacological properties, especially in the area of CNS disorders [136].

18.5.2.5 Indazole

Heterocyclic compounds, also called heterocycles, are a major class of organic chemical compound that plays a vital role in the metabolism of all living cells [144]. The heterocyclic compound, indazole, belonging to a class of nitrogen containing chemical compounds composed of a pyrazole ring condensed with a benzene ring, has attracted more attention in recent years and is widely present in numerous commercially available drugs. Indazole-containing derivatives, representing one of the most important heterocycles in drug molecules, are endowed with a broad range of biological properties. A literature search was conducted by Qin et al. [144] in PubMed, Google Scholar, and Web of Science regarding articles related to indazole and its therapeutic application. Based on the versatile biological activities, the indazole-containing compounds can be divided into six groups: **(a)** anti-inflammatory, **(b)** antibacterial, **(c)** anti-HIV, **(d)** antiarrhythmic, **(e)** antifungal, and **(f)** antitumor. At least 43 indazole-based therapeutic agents were found to be used in clinical application or clinical trials.

The indazole scaffold is of great pharmacological importance as it forms the basic structure of a large number of compounds with potential therapeutic value [144–146]. Several review articles have summarized the recent advances in the development of indazole-based anticancer agents [146–148].

18.5.2.6 Imidazole

Imidazole is a heterocyclic moiety having immense biological importance. Since ancient times, the imidazole nucleus is considered to be a promising moiety in the field of chemistry [149–152]. Preliminary in vitro and in vivo studies have provided valuable scientific evidence for its use. Subsequently, imidazole constitutes a new class of compounds for new drug development as the presence of this nucleus in diverse therapeutic categories, viz., antimicrobial, anti-inflammatory, anticancer, immunomodulator, antiviral, etc., has made it an interesting moiety for the design and development of new pharmacological agents.

In the field of heterocyclic medicinal chemistry, especially five-membered ring structures containing a nitrogen atom, imidazole core is an imperative aromatic heterocycle which is usually present in naturally occurring products and synthetic bioactive molecules [151]. The occurrence of imidazole moiety in therapeutic compounds may be beneficial in terms of improving water-soluble properties due to its two nitrogen atoms which leads to the creation of hydrogen bonds. The imidazole nucleus has also been recognized as an important isostere of triazole, pyrazole, thiazole, tetrazole, oxazole, amide, etc. for the purpose of designing and development of various biologically active molecules. Moreover, imidazole core scaffold contains three carbon atoms and two nitrogen with electronic-rich characteristics that are responsible for readily binding with a variety of enzymes, proteins, and receptors compared to the other heterocyclic rings. Thus, imidazole core is an attractive binding site with diverse cations and anions as well as biomolecules through different reactions in the human biological system, thus displaying extensive biological activities [152]. This effort thoroughly provides a wide-ranging assessment in current drug discovery and developments of imidazole-based analogs in the entire series of synthetic medicinal chemistry as antibacterial and antifungal, anticancer, antitubercular, analgesic and anti-inflammatory, anti-neuropathic, antihypertensive, anti-allergic, anti-parasitic, antiviral, antidepressant, anti-obesity, and so on, altogether with their prospective approaches in diagnostic and pathological field [151].

Quinazoline Scaffold

Quinazoline derivatives are fused heterocyclic ring systems which have been explored for their inhibitory activity toward various protein kinase enzymes [153–155]. The chemistry of quinazoline derivatives has been a point of interest in the early 1950s with the elucidation of the structure of a natural quinazoline alkaloid, 3-[β-keto-gamma-(3-hydroxy-2-piperidyl)-propyl]-4-quinazolone, from an Asian plant known for its antimalarial properties. In a quest to find more antimalarial agents, various substituted quinazolines have been synthesized. Quinazoline derivatives occupy a crucial place in modern medicinal chemistry. Numerous studies have been performed on quinazoline and their derivatives for anticancer activity, and pharmacological importance of this scaffold has been well established. Several molecules containing quinazoline scaffold elicit anticancer activity through inhibition of epidermal growth factor receptor (EGFR) tyrosine kinases [153–155]. Alam et al. [153] developed alkylphosphocholines–gefitinib conjugates as multitarget anticancer agents. The antiproliferative activities of the newly synthesized compounds were evaluated against cell lines representing lung, breast, liver, and skin cancers. In addition, the capability of the newly synthesized compounds to inhibit Akt phosphorylation and EGFR tyrosine kinases was determined. The results emphasized the influence of the linkers' length on the elicited bioactivity. The long chain linkers possessing conjugates were more active regarding both of the elicited antiproliferative effect and inhibition of Akt phosphorylation while maintaining the ability to inhibit EGFR tyrosine kinases. Their cytotoxic activities were superior or comparable to erlotinib and miltefosine.

With special focus on quinazoline moiety, a review article [154] offers a detailed account of multiple mechanisms of action of various quinazoline derivatives: inhibition of the DNA repair enzyme system, inhibition of EGFR, thymidylate enzyme inhibition, and inhibitory effects for tubulin polymerization by which these derivatives have shown promising anticancer potential. Exhaustive literature survey indicated that quinazoline derivatives are associated with properties of inhibiting EGFR and thymidylate enzyme. It was also found to be involved in disturbing tubulin assembly. Furthermore, quinazoline derivatives have been found to inhibit critical targets such as DNA repair enzyme. These derivatives have shown significant activity against cancer. It appears that in cancer therapy, quinazoline derivatives seem to be quite promising and act through multiple mechanisms as their polypharmacological properties.

18.6 Pharmacophore-Based Approach

18.6.1 Concept and Applicability of Pharmacophore-Based Approach in Drug Discovery

The term "pharmacophores" is used by many to define functional or structural elements possessing biological activity. This does not conform to the official definition elaborated by an IUPAC working party and published in 1998 [156]: "A pharmacophore is the ensemble of steric and electronic features that is necessary to ensure the optimal supramolecular interactions with a specific biological target structure and to trigger (or to block) its biological response." Based on IUPAC definition, pharmacophores have the following properties:

(a) The pharmacophore describes the essential, steric, and electronic, function-determining points necessary for an optimal interaction with a relevant pharmacological target.
(b) The pharmacophore does not represent a real molecule or a real association of functional groups, but a purely abstract concept that accounts for the common molecular interaction capacities of a group of compounds toward their target structure.
(c) Pharmacophores are not specific functional groups (e.g., sulfonamides) or "pieces of molecules" (e.g., dihydropyridines, arylpiperazines).
(d) A pharmacophore is an abstract description of molecular features that are necessary for molecular recognition of a ligand by a biological macromolecule.

(e) A pharmacophore can be considered as the highest common denominator of a group of molecules exhibiting a similar pharmacological profile and which are recognized by the same site of the target protein.
(f) A pharmacophore model explains how structurally diverse ligands can bind to a common receptor site; in contrast, a molecular scaffold provides a core structure for the design or construction of a potent and selective ligand that can bind a range of different biological targets.
(g) Furthermore, pharmacophores are also used as the starting point for developing three-dimensional quantitative structure–activity relationship (3D-QSAR) models.

The retrospective analysis of the chemical structures of the various drugs used in medicine led medicinal chemists to identify some molecular motifs that are associated with high biological activity more frequently than other structures. Such molecular motifs were called privileged structures by Evans et al. [122], to represent substructures that confer activities to two or more different receptors. The implication was that the privileged structure provides the scaffold and that the substitutions on it provide the specificity for a particular receptor [157, 158]. Among the most popular privileged structures, historical representatives are arylethylamines (including indolylethylamines), diphenylmethane derivatives, tricyclic psychotropics, and sulfonamides. Dihydropyridines [159], benzodiazepines [122, 159], N-arylpiperazines, biphenyls, and pyridazines [160] are more recent contributions. A statistical analysis of NMR-derived binding data on 11 protein targets indicates that the biphenyl motif is a preferred substructure for protein binding [161]. Even before the advent of computer-aided drug design, simple pharmacophores were described in the literature and considered as tools for the design of new drug molecules.

Pharmacophores are often perceived to be the same as scaffolds. While it is somewhat difficult to distinguish the two in certain situations, they are different concepts. The major distinction between the two concepts is that a pharmacophore focuses on how structurally diverse ligands can bind to a common receptor site, whereas a scaffold centralizes to a core chemical structure that can be used as building block for the design or construction of a ligand through modification of functional groups to acquire improved potency and selectivity for binding multiple different molecular targets.

18.6.2 Criteria for a Satisfactory Pharmacophore Model in Drug Discovery

To be recognized as a useful tool, a pharmacophore model has to provide valid information for the medicinal chemist exploring SAR [162–165]. The criteria are comprehensively summarized by Wermuth [162].

18.6.2.1 Highlighting Functional Groups and Predictive Power

A pharmacophore model should highlight the functional groups involved in the interaction with the target, the nature of the non-covalent bonding, and the different intercharge distances. This means that worthless images of ribbon and spaghetti models [166], without indication of the molecular features of the interacting partners, have to be avoided. This is true also for many unnecessary and opaque theoretical digressions. The model also has to show some predictive power and lead to the design of new, more potent compounds or, even better, of totally novel chemical structures, not evidently deriving from the translation of structural elements from one active series into the other. An interesting aspect of pharmacophore-based analog design is referred to as scaffold hopping. It consists in the design of functional analogs by searching within large virtual compound libraries of isofunctional structures but based on a different scaffold. The objective is to escape from a patented chemical class in identifying molecules in which the central scaffold is changed, but the essential function-determining points are preserved and form the basis of a relevant pharmacophore [167].

18.6.2.2 Perfect Stereochemical Complementarity Between a Ligand and a Protein Target

The second criterion for a valid pharmacophore model is that it should discriminate stereoisomers. Stereospecificity is one of the principal attributes of pharmacological receptors, and a perfect stereochemical complementarity between the ligand and the binding-site protein is an essential criterion for high affinity and selectivity. A convincing example of enantiomeric discrimination was observed for GABA-A receptor antagonists [168].

18.6.2.3 Distinguishing Between Agonists and Antagonists

In a similar manner, the ideal model should distinguish between agonists and antagonists. This is relatively easy for the specific category of antagonists which, according to Ariëns et al. theory [169], derive from the agonists simply through the addition of some supplementary aromatic rings which play the role of additional binding sites (e.g., the passage from muscarinic agonists to muscarinic antagonists [170] or from GABA agonists to GABA antagonists [168]). The discrimination between the two categories becomes less evident when the passage from agonist to antagonist relies on relatively subtle changes such as one observes for glutamate, oxotremorine, and benzodiazepine antagonists.

18.6.2.4 Explaining Paradoxical Observations

Sometimes a good pharmacophore model can explain apparently paradoxical observations, e.g., the unexpected affinity reversal found in R- and S-enantiomers of the sulpiride series on changing N-ethyl to N-benzyl derivatives [36].

18.6.2.5 Generation of New Chemical Entities

A good pharmacophore model should be able to account for the lack of activity of certain analogs of the active structures. The knowledge of structural or electronic parameters leading to poorly active or inactive compounds is a cost-lowering factor that allows the number of compounds to be synthesized to be reduced.

It is highly probable that computer-driven combinations of two pharmacophores can lead to the design of new active entities combining in one molecule with the critical structural elements of two partners, MTDs.

18.6.3 The Process for Developing a Pharmacophore Model

The process for developing a pharmacophore model generally involves the following steps [162].

18.6.3.1 Select a Training Set of Ligands

A structurally diverse set of molecules will be used for developing the pharmacophore model. As a pharmacophore model should be able to discriminate between molecules with and without bioactivity, the set of molecules should include both active and inactive compounds.

18.6.3.2 Conformational Analysis

Generate a set of low-energy conformations that is likely to contain the bioactive conformation for each of the selected molecules.

18.6.3.3 Molecular Superimposition

Superimpose ("fit") all combinations of the low-energy conformations of the molecules. Similar (bioisosteric) functional groups common to all molecules in the set might be fitted (e.g., phenyl rings or carboxylic acid groups). The set of conformations (one conformation from each active molecule) that results in the best fit is presumed to be the active conformation.

18.6.3.4 Abstraction

Transform the superimposed molecules into an abstract representation. For example, superimposed phenyl rings might be referred to more conceptually as an "aromatic ring" pharmacophore element. Likewise, hydroxy groups could be designated as a "hydrogen-bond donor/acceptor" pharmacophore element.

18.6.3.5 Validation

A pharmacophore model is a hypothesis accounting for the observed biological activities of a set of molecules that bind to a common biological target. The model is only valid insofar as it is able to account for differences in biological activity of a range of molecules.

Typical pharmacophore features include hydrophobic centroids, aromatic rings, hydrogen bond acceptors or donors, cations, and anions. These pharmacophoric points may be located on the ligand itself or may be projected points presumed to be located in the receptor. The features need to match different chemical groups with similar properties, in order to identify novel ligands. Ligand–receptor interactions are typically "polar positive," "polar negative," or "hydrophobic." A well-defined pharmacophore model includes both hydrophobic volumes and hydrogen bond vectors.

18.7 Fragment-Based Approach

18.7.1 Concept and Applicability of Fragment-Based Approach in Drug Discovery

Fragments are small organic molecules which are small in size and low in molecular weight [172–175]. It is based on identifying small chemical fragments, which may bind only weakly to the biological target, and then growing them or combining them to produce a lead with a higher affinity. Fragment screening is a widely applied method for the discovery of lead molecules in the Fragment-Based Drug Design (FBDD) or Fragment-Based Lead Discovery (FBLD) [172–175]. FBDD/FBLD involves the selection of compounds with a very small molecular mass and weak binding affinities in the μM to mM range as a hallmark of fragments. The FBDD entails the screening of a small number of low-molecular-weight compounds in the hopes of finding even low affine but high ligand efficient fragments that have high probability to become drug candidates. Since 1996, when the first paper on FBDD was reported, the potentialities of this strategy became progressively more apparent as testified by the growing number of publications.

FBDD usually generates a compound starting from a chemical fragment with a low binding affinity to the target, low complexity in chemical structures, and low molecular weight (less than 300 Da) [172, 176, 177]. These starting hits are usually identified from a compound library using sensitive biophysical methods. The identified hit is then grown into drug-like molecules through different strategies. Although FBDD cannot replace high-throughput screening (HTS) campaigns in drug discovery, it has some attractive advantages such as saving experimental cost, offering diverse hits, and exhibiting multiple ways to develop novel compounds [172, 178]. These advantages have encouraged researchers to adopt this method to develop inhibitors of different types of targets. With the development of new approaches in screening and progress made in structural biology, FBDD has been readily carried on and playing important roles in target-based drug discovery [172, 179]. Several drugs such as vemurafenib—an inhibitor of oncogenic B-RAF kinase activity derived from fragment-based approach—have been approved by FDA [180].

18.7.2 Procedures of Fragment-Based Approach in Drug Discovery

Fragment-based drug discovery is mainly applicable to target-based drug discovery. Druggability of a target is always analyzed in target-based drug discovery projects and is utilized to predict possibility of developing drugs by identifying a pocket favoring binding to small-molecule compounds [172, 181, 182] (Aretz et al., 2016; Gee et al., 2016). In most cases, druggable targets are of great interests as the probability to develop small-molecule drugs is very high [183]. FBDD has been applied to quite a few druggable and undruggable targets. The procedures of FBDD approach in drug discovery generally involve the following steps.

18.7.2.1 Building Fragment Library

It is suggested that fragments should follow the "rules-of-three" in which compounds have (a) molecular weight < 300, (b) ClogP <3, and (c) the number of hydrogen bond donors and acceptors with each being <3 and the number of rotatable bonds should be <3 [172, 184, 185]. However, recent studies indicated that a fragment does not have to follow the rule-of-three as the fragment in screening utilizes simple organic compounds that can be modified efficiently [186]. Researchers usually build their own customized libraries in FBDD, and molecular weight of a fragment can be above 300 Da [172]. In a virtual screening, the fragment library can be expanded with increasing diversity as the screening can be accomplished in a short time [172]. As fragments can provide diverse compounds for optimization, the number of compounds in the fragment library is not a limitation factor. A potent compound was able to be developed through FBDD in which a library of about 800 fragments was utilized [187]. Only a few fragment libraries are commercially available [188]. Many fragment libraries were built based on empirical information and evidence.

18.7.2.2 Performing Fragment Screening

Binding affinities between fragments and their targets are normally very low in micromolar to millimolar range. Probing fragment and target interactions through biochemical methods which are based on spectrophotometric and fluorescence assays is very challenging. Therefore, other sensitive approaches that are able to determine low binding affinities are very useful in hit identification [172]. These approaches such as differential scanning fluorimetry (DSF), isothermal titration calorimetry (ITC), nuclear magnetic resonance (NMR), surface plasmon resonance (SPR), microscale thermophoresis (MST), and X-ray crystallography have been widely used in FBDD. (For detailed descriptions of the methods, please see the review article by Li Q [172].)

18.7.2.3 Conducting Compound Optimization

As fragments usually bind weakly to targets and exhibit no potent inhibitory effect on the activity of the targets, further chemical modification of the hits is required in hit-to-lead step. In this procedure, hits will be developed into leads which bind to the target with higher affinities and exhibit potent activities against the target [172]. Three strategies, namely, fragment growing, fragment hopping, and fragment linking are frequently utilized.

Fragment Growing

Fragment growing is the most commonly used strategy to grow fragments into compounds with higher molecular weights and higher potencies. Various chemical groups can be added to the building block (hit) to improve its potency. Co-crystal structures of hits with the target are very important for fragment growing. It has been noted that a fragment can be grown into a potent compound even without structural information.

Fragment Merging/Scaffold Hopping

Fragment merging or scaffold hopping is another strategy to grow fragments into potent compounds. This strategy is based on condition that the identified fragments have an overlapped binding site. Potent compounds can be developed by combining/merging chemical features of two or more fragments. To carry out fragment merging of the identified hits, binding modes of fragments should be determined through X-ray crystallography, NMR spectroscopy, or docking methods. This strategy is very useful for replacing non-drug-like core structures in the hits with suitable scaffolds to generate more drug-like molecules. In addition, it offers a chance to generate more patentable compounds.

Fragment Linking

Fragment linking is considered as the most powerful way to develop potent inhibitors from fragments. A lead compound can be developed by linking two or more fragments together. This is an attractive strategy as binding affinities can be improved dramatically. Based on the calculated binding free energy, linking two fragments with binding affinities in mM range will result in a compound with an affinity in nM range.

Fragment-based approaches are particularly suited for polypharmacology, owing to the inherent promiscuous nature of fragments [189]. Frett et al. [190] used fragment-based chemical screen and identified a novel dual kinase inhibitor, Pz-1, that demonstrated antitumor activity as an MTD. To systematically probe the potential of fragment-based multiple ligand discovery, Brunst et al. [174] have employed a large fragment library for comprehensive screening on five targets chosen from proteins for which multitarget ligands have been successfully developed previously (soluble epoxide hydrolase, leukotriene A4 hydrolase, 5-lipoxygenase, retinoid X receptor, farnesoid X receptor). Their results demonstrate the applicability of fragment-based approaches to identify starting points for polypharmacological compound development with certain limitations.

18.8 Molecular Hybridization Approach

18.8.1 Concept and Applicability of Molecular Hybridization Approach in Drug Discovery

"Molecular hybridization," also termed "molecular hybrids" or "hybrid compounds," is a new concept in drug design and development based on the combination of pharmacophoric moieties of different bioactive substances to produce a new hybrid compound with improved affinity and efficacy, when compared to the parent drugs [191–195]. This strategy can result in compounds presenting modified selectivity profile, different and/or dual modes of action, and reduced undesired side effects. Research on molecular hybridization is gaining momentum worldwide showing their promise in drug discovery. In many cases, molecular hybrids are also called as multitarget directed compounds, multifunctional compounds, dual-acting compounds, me-better drugs, multiple ligands combi-molecules, double drugs, combinational drugs, dual-drugs multiple-drugs, bifunctional drugs, bivalent ligands, drugs with two heads, etc. [196].

Molecular hybridization approach has the following strengths and weaknesses.

The advantages include:

- Designing and producing MTDs.
- Better affinity and efficacy when compared to the parent drugs.
- Enhanced specificity for a particular body target site.
- Reduction of undesirable side effects.
- Decreases in drug–drug interactions.
- Reduced emergence or spread of drug resistance in microorganisms and protozoans.
- Create new pharmacological properties with improved pharmacodynamics.

The disadvantages include:

- The chemical structures of the parent molecules have to be optimized before subjecting them for molecular hybridization.
- Altering the structure of parent molecules may affect their affinity, selectivity, and metabolism profiles.
- Molecular hybrids require activity in similar concentration range.
- Molecular hybrids with flexible structure may reduce the target binding efficacy.
- The high molecular weight of molecular hybrids may lower oral bioavailability.
- Molecular hybrids lose the ability to cross the blood–brain barrier.

18.8.2 Classification of Molecular Hybrids

18.8.2.1 Conjugates

Conjugates are the molecular hybrids that contain pharmacophores for each target and separated by a distinct linker group that is not found in either of the individual drugs. Most conjugates comprise of a metabolically stable linker.

18.8.2.2 Cleavable Conjugates

Cleavable conjugates are the molecular hybrids with a linker designed to be metabolized to release the two drugs that interact independently with each target.

18.8.2.3 Fused Hybrids

Fused hybrids are the non-super-posed molecular hybrids with the decreased size of the linker such that the scaffold of the pharmacophores just touches each other.

18.8.2.4 Merged Hybrids

Merged hybrids are the superposed molecular hybrids with merged frameworks by taking benefit of commonalities in the structures of the starting compounds, which give rise to smaller and simpler molecules.

Below are a few examples of molecular hybrids and their uses:

- Streptoniazid: streptomycin and isoniazid—antituberculár and antibiotics.
- Quinine acetyl salicylate: quinine and acetylsalicylic acid—antipyretic, analgetic, antimalarial.
- Acetamino salol: salicylic acid and acetamido phenol—analgetic.
- Guaicyl phenyl cinchoninate: guaiacol and cinchophen—expectorant.

18.8.3 Application of Molecular Hybrids

Recently, molecular hybridization approach has been recognized as an effective approach to design ligands able to modulate multiple targets of interest and has actually been successfully used by many research groups around the world and has had very promising results with diseases having multifactorial profiles, like Alzheimer's, Parkinson's disease, cancer, inflammation, and hypertension among others [194–199]. Hybrid compounds can be obtained by linking (presence of a linker) or framework integration (merging or fusing) strategies.

Although very promising to combat the multifactorial nature of complex diseases, the development of molecular hybrids faces the critical issues of selecting the right target combination and the achievement of a balanced activity toward them while maintaining drug-like properties.

In addition to molecular hybridization, there are at least other two molecular conjugation approaches available for drug discovery. Whereas molecular hybridization approach involves the association of two or more non-identical molecules through hybridization, "molecular addition approach" involves the association of two or more identical or non-identical molecules through weak forces of attraction, and "molecular replication approach" involves the association of two or more identical molecules through covalent bonding [196].

18.9 Activity-Based Proteomics Approach

18.9.1 Concept and Applicability of Activity-Based Proteomics Approach in Drug Discovery

Activity-based proteomics, or activity-based protein profiling (ABPP), is a functional proteomic technology that uses chemical probes that react with mechanistically related classes of enzymes critical in disease-causing pathways [20, 197]. A major advantage of ABPP is the ability to monitor the availability of the enzyme active site directly, rather than being limited to protein or mRNA abundance. With classes of enzymes such as the serine hydrolases [198] and metalloproteases [199] that often interact with endogenous inhibitors or that exist as inactive zymogens, this technique offers a valuable advantage over traditional techniques that rely on abundance rather than activity. In recent years ABPP has been combined with tandem mass spectrometry enabling the identification of hundreds of active enzymes from a single sample. This technique, known as ABPP-MudPIT (multidimensional protein identification technology), is especially useful for profiling inhibitor selectivity as the potency of an inhibitor can be tested against hundreds of targets simultaneously.

The basic unit of ABPP is the probe, activity-based probes (ABPs), which are small-molecule tools that can be used to monitor the activity of specific classes of enzymes. Over the past decade and a half, various ABPs have been designed to study proteases, hydrolases, phosphatases, histone deacetylases, and glycosidases [20, 200], and these probes have proven to be valuable in investigating enzyme-related disease mechanisms including cancer [201], microbial and parasitic pathogenesis [20], and metabolic disorders [20]. Typical activity-based protein profiling follows a similar overall workflow as affinity chromatography, including probe binding, protein separation, sequence analysis, and database searching. However, because most ABPs are designed to target a specific enzyme class, ABPs are particularly useful for phenotypic screening and lead optimization where a specific enzyme or enzyme family is suspected to be involved in a certain disease state or pathway [20].

ABPs have three components: a reactive electrophile for covalent modification of enzyme active site (RG, sometimes called a "warhead"), a linker or a specificity group for directing probes to specific enzymes, and a reporter or a tag for separating labeled enzymes [20]. Additionally, some probes may contain a binding group which enhances selectivity. The reactive group usually contains a specially designed electrophile that becomes covalently linked to a nucleophilic residue in the active site of an active enzyme. An enzyme that is inhibited or posttranslationally modified will not react with an activity-based probe. The tag may be either a reporter such as a fluorophore or an affinity label such as biotin or an alkyne or azide for use with the Huisgen 1,3-dipolar cycloaddition (also known as click chemistry) [202].

Therefore, a library of ABPs can be directly used for phenotypic screening and target identification simultaneously [203, 204]. Additionally, covalent enzyme inhibitors can be readily converted to ABPs by attaching a tagging group, and the combined use of covalent inhibitors and ABPs in phenotypic screening greatly facilitates target identification and also offers a powerful tool to study function and mechanism of identified proteins [205, 206]. For example, Hall et al. successfully identified a small molecule that blocks the process of host cell invasion by the *Toxoplasma gondii*. The selected inhibitor, WRR-086, was then converted to an ABP by attaching an alkyne group for click chemistry and used to identify TgDJ-1, a poorly characterized protein involved in oxidative stress response as a key player in host cell invasion [206].

18.9.2 Design of Activity-Based Probes for Drug Discovery

As mentioned above, a typical ABP contains three groups: a reactive group, a linking group or binding group, and a reporter tag. For probe design, the first factor to consider is the reactivity of compound. Most probes are based on bioactive small molecules. So far, many ABPs have utilized electrophilic reactive groups, including epoxides, Michael addition units, disulfides, lactones, β-lactams, and quinone compounds [207]. These groups can react with serine, tyrosine, or glutamine to modulate enzyme activity. However, there are many compounds that react with targets via non-covalent interactions. To overcome this problem, a more intuitive and unbiased strategy for identifying binding partners of unreactive NPs is to use photoaffinity labeling (PAL) [207]. PAL makes use of photoreactive moieties that are inert under standard synthetic-chemical and biological conditions but can be activated by UV light, generating highly reactive, transient species. Benzophenone and aliphatic and aromatic diazirines are the most commonly used PAL groups.

In the process of probe design, the choice of linking groups can also be critical. Linking groups can attach the reactive groups with the label groups together and reduce the impact of the label group on the reactive group [207]. The choice of linker group is also significant for reducing nonspecificity [207]. In this basic form, a linker can take the form of an extended alkyl or polyethylene glycol (PEG) spacer. Furthermore, of late, the design of cleavable linkers for protein enrichment has received much attention, especially for the isoTOP-ABPP strategy [207–209].

The other critical challenge in the process of probe design is the reporter group. The widely used reporters are the biotin–streptavidin system for pull-down assays and fluorescent reporters for imaging-based detection. Because of the existence of intrinsically biotinylated proteins, some nonspecific background can interfere with the identification of targets; however, fluorescent reporters can be used to avoid this problem. An increasing number of studies are combining these two reporters to identify targets [207, 210, 211].

In order to covalently attach ABPs to target proteins, an active site nucleophile such as cysteine or serine is required. However, not all enzymes have a nucleophile in their active site. One way to overcome this problem is to incorporate a photoreactive group, as has been demonstrated for probes of γ-secretase [212], metalloproteases [213], and the proteasome [214]. Alternatively, an electrophile can be introduced to react with any cysteine residue in close proximity to the site of probe binding [215, 216]. Incorporating a clickable group such as an alkyne or an azide on an ABP can also facilitate a direct link from phenotypic screening to target identification [206, 217, 218]. Additionally, to improve efficiency of two-step labeling, other tagging methods employing copper-free click chemistry [219], sulfo-click chemistry [220], Staudinger ligation [221], or Diels-Alder reaction [222] were developed as an alternative to conventional click chemistry.

ABPs are not only useful for target identification but also are powerful tools for the discovery of disease-related proteins. For example, a broad-spectrum cathepsin-C specific probe was used to show that dipeptidylpetidase1 (DPAP1) plays a significant role in malarial infection and is a potentially valuable drug target. In other examples, a broad-spectrum probe was used to link several serine hydrolases including retinoblastoma-binding protein 9 (RBBP9) [223], KIAA1363 [224], and monoacylglycerol lipase (MAGL) [225] to cancer progression. Furthermore, the broad-spectrum ABPs used for target validation are also useful to set up class-wide enzyme assays to identify new inhibitors or to test existing library of small molecules in phenotypic screening [226–228].

18.10 Omics-Based Approaches

Recent technical advances in sequencing, microarray, and mass spectometry (MS) technologies allow scientists to generate genomics, transcriptomics, proteomics, and other -omic data types at an unprecedented level of resolution [229]. Many studies have used these technologies to better understand the molecular mechanisms underlying complex diseases and to provide information on drug treatments. The resulting rich information data can be utilized to identify drug targets, to uncover the mechanism of action of drugs, and to assess (or infer) their side effects. Omics-based studies can also provide essential information to deliver personalized medicine. For example, it has been shown that genetic variations can help clinicians assess efficacy or toxicity of some targeted agents for specific subsets of molecularly profiled patients [229–232]. If systematically integrated, omics-driven molecular profiles of diseases and drug treatments/exposures could significantly accelerate drug discovery and development process [229].

Omics technologies have brought unprecedented abilities to screen biological samples at the gene, transcript, protein, metabolite, and their interaction network level in searching of novel targets [229–232]. In particular, genome-wide association studies (GWASs), whole-genome sequencing, and transcriptome analysis constitute essential tools to discover or validate new drug targets, since they can provide a systematic approach to evaluate their therapeutic efficacy and related side effects. Even though omics studies can provide information on the efficacy of drug targets, they can be exploited to better understand their mechanisms of action and, most importantly, to detect in advance drug-induced side effects. Omics technologies generate large amount of data from single experiments, and this information is often useless in their original formats.

18.10.1 Genomics in Drug Discovery

Traditionally, molecular targets for drug discovery were derived as a consequence of a lengthy "function to gene" process whereby a gene of interest derived from animal tissues was then cloned, expressed in a recombinant host, and functionally characterized. The tedious and time-consuming process, however, delivered specific targets whose functions were well characterized [231]. The successful completion of the human genome project in 2003 spurred on a revolution in biotechnological advances resulting in the diverse umbrella discipline of "Omics" as we know it today. Genomics have provided the earliest applications for drug target discovery. In particular, the application of DNA microarray and next-generation sequencing (NGS) technologies has enabled high-throughput analysis of genotype–phenotype relationships on human populations, opening a new era of genetics-informed drug discovery. DNA microarrays have been extensively used to conduct GWASs, which have helped scientists identify loci that harbor genetic variants (typically single nucleotide polymorphisms; SNPs) that are associated with risk for diseases and traits [233]. As stated by Paananen et al. [229], GWASs have helped to identify thousands of genes contributing to complex genetic traits. To date, almost 10,000 strong associations have been reported between genetic variants and one or more complex traits. Among these findings, there are several examples of disease-associated genes that have been identified as being effective drug targets. GWAS data have also been employed for the development of in silico methods for disease genes and drug target identification [234]. GWAS can also enable the discovery of biological pathways that confer susceptibility to diseases. However, these studies alone cannot elucidate how the variants affect downstream pathways and lead to a disease. Indeed, one of the most relevant post-GWAS challenges is to combine GWAS findings with additional omics data layers in order to shed the light on the biological systems underlying complex diseases and provide more effective drug targets. Indeed, GWAS-driven analyses have been often merged with transcriptomics data, which can help identify disease-related genes by comparing gene expression profiles between disease and control groups [229]. GWASs have also rapidly changed the landscape of pharmacogenomic research. Pharmacogenomics is the study of the impact of genetic variations of individuals on their drug response or drug metabolism. From a future perspective, it is envisaged that the most significant impact from genomics applications will be seen in the clinical phase of the drug development pipeline. Improving clinical trial design by using objective genetic inclusion and exclusion criteria will enable improved stratification of disease and thereby the definition of a more homogeneous cohort of patients. Similarly, genetic markers for efficacy and toxicity could be important predictive signatures capable of preselecting trial cohorts inherently susceptible to non-target effects, thereby improving the costly late-phase failures in the pipeline [231].

18.10.2 Transcriptomics in Drug Discovery

Genome-wide transcriptional profiling provides a global view of cellular state and how this state changes under different treatments (e.g., drugs) or conditions (e.g., healthy and diseased). In particular, drug-induced gene expression profiles in human cell lines and in vivo models can be used to elucidate the biological effects of putative targets and evaluate in advance therapeutic efficacy [229]. Transcriptomics signals are important for clinical candidate selection as they provide an evaluation of potential adverse effects of drug targets at an early stage in drug development. Indeed, in recent years, transcriptomics data have been intensively used in the toxicogenomics field. This has led to the development of large-scale public databases, such as DrugMatrix and Open TG-GATE, which collect compound-induced gene expression data with in vivo histopathological data, and Connectivity Map and the Library of Integrated Network-based Cellular Signatures L1000 dataset, which collect transcriptional drug perturbations for thousands of compounds tested on more than 70 cell lines. Other gene expression data following drug treatments can be retrieved from the Gene Expression Omnibus (GEO) repository [235] and the ArrayExpress Archive of Functional Genomics Data which are continuously updated. The data provided by these databases can be used in combination with genomic and network features to prioritize drugs with less likelihood of causing side effects. Similar computational strategies are currently proposed to support alternative methods for chemical risk assessment. Moreover, gene expression and transcriptome profiling can help researchers improve the design of clinical trials in phase I and phase II studies. Genome expression profiling can also be combined with the genotype of trait-associated variants using in vivo data, thus identifying target genes and the directionality of the effect of trait variants. Expression quantitative trait loci (eQTL) analyses are useful in this regard as they can provide genome-wide lists of genetic variants that associate with gene expression in a particular tissue. eQTL analysis can be used to identify causal and also to discover genetic networks that might play significant roles in drug resistance responses. Currently, many eQTL databases exist, and they can be queried to determine if a trait-associated variant (or variants in linkage disequilibrium) associates with the expression of a specific gene. One of the most used databases is the Genotype-Tissue Expression [236]. Over the past decade, the transcriptomics field has developed rapidly, thanks to the advent of NGS technologies. In particular, whole-transcriptome analysis with total RNA sequencing (RNA-seq) has become an indispensable tool for gene expression profiling. Total RNA-seq profiles provide an exceptional opportunity to study ncRNAs, including microRNAs and long non-coding RNAs. Similar to the protein-coding genes, ncRNAs can play critical roles in tumor progression and cancer therapy. Currently available large-scale cancer genome and pharmacogenomics projects, such as The Cancer Genome Atlas (TCGA), Cancer Cell Lines Encyclopedia (CCLE), and Genomics of Drug Sensitivity in Cancer (GDSC), can be used to systematically determine the regulatory roles of ncRNA in cancer drug response by combining RNA-seq data in conjunction with clinical and drug response data from thousands of tumor samples and cancer cell lines. These databases provide information on different omics data types, including mutation and copy number variation.

18.10.3 Proteomics in Drug Discovery

Proteomics refers to the analysis of the entire protein content of a cell, tissue, or organism under a specific condition. Several techniques have been developed to study the proteome of an organism, and among them MS has become the tool of choice. The three primary applications of MS to proteomics are cataloging protein expression, defining protein interactions, and identifying sites of protein modification. These sources of information have been extensively used as a DTD tool [229]. However, while transcriptome data cover the whole range of expressed genes, a typical untargeted mass spectrometry proteomics experiment can usually detect and quantify up to 5000 proteins, which is less than half of the expressed human proteome. Proteomics can also be used to address several steps of the drug development process, including identification and validation of drug targets, informing assay development for screening of leads, and generating in vitro and in vivo biomarkers as surrogate endpoints for efficacy, toxicology, and disease stratification. Most of the mass spectrometry-based proteomics research studies in drug discovery have been performed to characterize protein expression profiling, functional proteomics, and phosphoproteomics. These experiments aim to measure the protein expression levels, protein–protein complexes, and signal transduction relative to a control treatment.

However, in drug discovery it is very important to discover protein targets from phenotypic assays and to understand on- and off-target engagement of potential therapeutic compounds. This task can be addressed by using chemoproteomics [229, 237, 238]. Chemoproteomics refers to a new technology that facilitates large-scale study of proteins by combining chemical methods with mass spectrometry proteomics. In particular, it provides direct binding of small molecules with protein targets, helping one to quantify the amount of drug required to bind a target and subsequently produce a therapeutic effect and drug selectivity determination (through the assessment of off-target interactions) [237]. The term "chemical proteomics" is often used to define a specific focus area within the broader field of proteomics in which a small molecule is used to directly reduce the complexity of an entire proteome to focus only on proteins that interact with that target molecule. There are multiple approaches that can be employed in chemical proteomic workflows. These include small-molecule affinity- and activity-based probes that can be used to isolate targets and, more recently, label-free techniques to directly identify small-molecule binding proteins. Since, many reviews have covered the general principles of these approaches, we will focus only on the most recent examples of each technique.

Proteomics is also used in drug target identification by applying protein–protein interaction networks (PPINs). PPINs are typically modeled via graphs, whose nodes represent proteins and whose edges connect pairs of interacting proteins.

18.10.4 Metabolomics in Drug Discovery

Metabolomics is the study of the metabolome, i.e., all the metabolites present in a cell, tissue, or organism at a given time [229]. It provides an overview of the metabolic status and global biochemical events associated with a cellular or biological system. Unlike genes and proteins, whose functions are influenced by intriguing regulatory mechanisms such as epigenetic regulation and protein posttranslational modifications, metabolites provide direct signatures of biochemical activity and are therefore easier to correlate with phenotype. Metabolomic-based clinical applications and tests are now emerging [239] and can help to understand disease mechanisms from a new perspective. Metabolomics have contributed to identifying metabolic causes and biomarkers for chronic diseases such as diabetes, Alzheimer's disease, atherosclerosis, and cancer. Metabolites play an important role in tumor cell proliferation. Metabolomics can be used to implement effective precision medicine approaches such as personalized phenotyping and individualized drug response monitoring. Toxicity assessment of drug targets can be addressed by using metabolomics data. In particular, the information from metabolomic analysis can be used to determine the off-targets of a drug candidate and thus provide a mechanistic understanding of drug toxicity [240]. Comprehensive metabolomics databases include The Human Metabolome Database (HMDB) and MetaboLights [241].

18.10.5 Chemical Genomics or Chemogenomics

Chemical genomics or chemogenomics is a new term that describes the development of target-specific chemical ligands and the use of such chemical ligands to globally study gene and protein functions [242–244]. It begins to play a critical role in the genomic age of biological research and drug discovery. Chemogenomics aims toward the systematic identification of small molecules that interact with the products of the genome and modulate their biological function [242]. While historically the approach is based on efforts that systematically explore target gene families like kinases, today additional knowledge-based systematization principles are followed within early drug discovery projects which aim to biologically validate the targets and to identify starting points for chemical lead optimization. While the expectations of chemogenomics are very high, the reality of drug discovery is quite sobering with very high project attrition rates.

Chemogenomics involves the combination of a compound's effect on biological targets together with modern genomic technologies [244]. The merger of these two methodologies is creating a new way to screen for compound–target interactions, as well as map chemical and biological space in a parallel fashion. The challenge associated with mining complex databases has initiated the development of many novel in silico tools to profile and analyze data in a systematic way. The ability to analyze the combinatorial effects of chemical libraries on biological systems will aid the discovery of new therapeutic entities. Chemogenomics provides a tool for the rapid validation of novel targeted therapeutics, where a specific molecular target is modulated by a small molecule. Along with targeted therapies comes the ability to discovery pathway nodes where a single molecular target might be an essential component of more than one disease. Several disease areas will benefit directly from the chemogenomic approach, the most advanced being cancer. A genetic loss-of-function screen can be modulated in the presence of a compound to search for genes or pathways involved in the compound's activity. Several recent papers highlight how chemogenomics is changing with RNA interference-based screening and shaping the discovery of new targeted therapies [245, 246]. Together, chemical and RNA interference-based screens open the door for a new way to discover disease-associated genes and novel targeted therapies.

18.10.6 Microbiomics in Drug Discovery

Microbiomics is a fast-growing field in which all the microorganisms of a given community are investigated together. Human skin, mucosal surfaces, and the gut are colonized by microorganisms, including bacteria, viruses, and fungi, collectively known as the microbiota (and their genes constituting the microbiome) [247]. The human microbiome is enormously complex; for example, the gut contains roughly 100 trillion bacteria from 1000 different species [247]. There are substantial variations in microbiota composition between individuals resulting from seed during birth and development, diet and other environmental factors, drugs, and age [247, 248]. Many studies have implicated perturbations in gut bacteria in a variety of disorders, including diabetes, obesity, cancer, colitis, heart disease, and autism. The microbiome can be profiled by amplifying and then sequencing certain hypervariable regions of the bacterial 16S rRNA genes followed by clustering the sequences into operational taxonomic units. Shotgun metagenomics sequencing, in which total DNA is sequenced, can provide additional resolution for distinguishing genetically close microbial species. Several analytic tools have been developed for analyzing NGS data from targeted 16S or metagenomics analysis, such as QIIME (quantitative insights into microbial ecology) [247, 249]. These allow accurate quantitative determination of taxa that can be correlated with disease or other phenotypes of interest [247, 250]. Associated technologies include NGS application for 16S ribosomal abundance and metagenomics quantification [247, 251, 252].

18.10.7 Multi-omics in Drug Discovery

Each type of omics data provides important information highlighting differences between normal and disease conditions. These data can be utilized to discover diagnostic and prognostic markers and to give insight as to which biological processes are different between the disease and control samples. However, single-level omics data analysis is limited, reflecting reactive processes rather than causative ones [247]. Indeed, integrating multiple omics data types could reveal important molecular mechanisms that regulate for complex diseases and help scientists understand the dynamics that lead to disease manifestations. Moreover, a more detailed understanding of disease mechanisms would be beneficial when searching for novel, more effective biomarkers. Current DTD platforms do not provide tools for multi-omics data analysis mainly because of the absence of standardized analytical protocols. However, recent studies have proposed interesting approaches applied to the cancer drug discovery [253]. It is acknowledged that multi-omics cancer data have the potential in improvement of targeted therapy and the effectiveness of traditional therapies, in clarification of molecular mechanisms of cancer therapeutic resistance, and in the discovery of novel biomarkers and targeted drugs. For example, genomic and transcriptomic data and long-term clinical outcomes were recently analyzed to detect changes of gene expression based on somatic gene copy number aberrations. This analysis revealed important targeted therapeutic response-related events [254]. An integrative analysis of genomic and proteomic data demonstrated that aberrations of the PI3K pathway are particularly common in hormone receptor-positive breast cancer, which might be important in clinical selection of targeted therapies [255]. Another multi-omics data analysis, conducted in sorafenib-treated failure HCCs, combined quantitative proteomics and phosphoproteomics data to better understand the molecules targeted by this drug [256].

18.11 Network-Based Approach

18.11.1 Concept and Applicability of Network-Based Approach in Drug Discovery

Polypharmacology is network-based pharmacology. Systems pharmacology is an emerging area of pharmacology which utilizes network analysis of drug action as one of its approaches [257]. By considering drug actions and side effects in the context of the regulatory networks within which the drug targets and disease gene products function, network analysis promises to greatly increase our knowledge of the mechanisms underlying the multiple actions of drugs. Systems pharmacology can provide new approaches for drug discovery for complex diseases. The integrated approach used in systems pharmacology can allow for drug action to be considered in the context of the whole genome. Over the past decade, various computational methods were proposed to predict potential therapeutic targets and druggable compounds as well as drug–target interaction (DTI) with high efficiency and low costs. These methods can be roughly divided into several categories, such as structure-based molecular docking and pharmacophore-based, similarity-based, machine learning-based, and network-based methods [258]. Among them, network-based methods, which do not rely on three-dimensional structures of targets and negative samples, have shown great advantages over the others.

With the rapid development of high-throughput genomic technologies and the accumulation of genome-wide datasets for human disease, it has been shown that using only reductionistic principles has been difficult to capture the complex biological networks and design rational medication [259]. However, the emerging paradigm of "network-based approach" proposes to harness the power of networks to uncover relationships between various data types of interest for drug discovery. Recent advances include networks that encompass relationships between drugs, disease-related genes, therapeutic targets, and diseases. It is shown how network techniques can help in the investigation of the mechanism of action of existing drugs and new molecules or to identify novel disease genes and targets [258]. Network-based studies are becoming an increasingly important tool in understanding the relationships between drug action and disease susceptibility genes.

Network-based methods are one of the simplest methods, and it does not need any other additional information like structure or sequence for the prediction of drug–target interaction. Several other methods were also developed based on this method. Since these methods are not dependent on the three-dimensional structure of the protein, it can be useful for the targets whose 3D structures are unknown. The major advantage of these techniques includes simple, fast, and accurate. It is based on very simple physical processes such as resource diffusion, collaborative filtering, and random walk on networks [260, 261].

Sidders et al. [262] applied an in silico network pharmacology approach to identify small-molecule compounds with the potential to selectively disrupt the structure of a chronic pain-specific disease network, which we validate using a novel phenotypic screen that recapitulates key aspects of neuronal and pain biology by measuring changes in neuronal excitability in native sensory neurons. The combination of network pharmacology with a phenotypic screen is a powerful approach; we show that hit rates increase from 26% to 42%. This represents a rational approach to the discovery of compounds with a polypharmacology-based therapeutic value, which will be vital for the discovery of treatments for complex disease.

18.11.2 Drug–Target Network

In a review article, Janga and Tzakos [263] formalized the definition of a drug–target network by decomposing it into drug, target, and disease spaces and provide an overview of its structure and organizational principles. They suggest that drug–target networks can be decomposed to be studied at a variety of levels and argue that such network-based approaches have important implications in understanding disease phenotypes and in accelerating drug discovery.

Cellular proteins are the mediators of multiple organism functions being involved in physiological mechanisms and disease. By discovering lead compounds that affect the function of target proteins, the target diseases or physiological mechanisms can be modulated. Based on knowledge of the ligand–receptor interaction, the chemical structures of leads can be modified to improve efficacy and selectivity and reduce side effects. The global set of relationships between protein targets of all drugs and all disease gene products in the human protein–protein interaction or "interactome" network remains uncharacterized. Yildirim et al. [264] built a bipartite graph composed of US Food and Drug Administration-approved drugs and proteins linked by drug–target binary associations. The resulting network connects most drugs into a highly interlinked giant component, with strong local clustering of drugs of similar types according to Anatomical Therapeutic Chemical classification. Topological analyses of this network quantitatively showed an overabundance of "follow-on" drugs, that is, drugs that target already targeted proteins. By including drugs currently under investigation, they identified a trend toward more functionally diverse targets improving polypharmacology. To analyze the relationships between drug targets and disease gene products, they measured the shortest distance between both sets of proteins in current models of the human interactome network. Significant differences in distance were found between etiological and palliative drugs.

Cheng et al. [265] developed a computational approach, named bioactivity profile similarity search (BASS), for associating targets to small molecules by using the known target annotations of related compounds from public databases. To evaluate BASS, a bioactivity profile database was constructed using 4296 compounds that were commonly tested in the US National Cancer Institute 60 human tumor cell line anticancer drug screen (NCI-60). Each compound was used as a query to search against the entire bioactivity profile database, and reference compounds with similar bioactivity profiles above a threshold of 0.75 were considered as neighbor compounds of the query. Potential targets were subsequently linked to the identified neighbor compounds by using the known targets of the query compound. About 45% of the predicted compound–target associations were successfully verified retrospectively, suggesting the possible application of BASS in identifying the targets of uncharacterized compounds and thus providing insight into the study of promiscuity and polypharmacology. Furthermore, BASS identified a significant fraction of structurally diverse compounds with similar bioactivities, indicating its feasibility of "scaffold hopping" in searching novel molecules against the target of interest.

Tabei et al. [266] presented a novel method for systematic analyses of the underlying features characteristic of drug–protein interaction networks, which we call "drug-protein interaction signatures" from the integration of large-scale heterogeneous data of drugs and proteins. They developed a new efficient algorithm for extracting informative drug–protein interaction signatures from the integration of large-scale heterogeneous data of drugs and proteins, which is made possible by space-efficient representations for fingerprints of drug–protein pairs and sparsity-induced classifiers. The drug–protein interaction signatures consist of the associations between drug chemical substructures, adverse drug reactions, protein domains, biological pathways, and pathway modules.

18.11.3 Disease–Drug Networks

Hu and Agarwal [268] performed a systematic, large-scale analysis of genomic expression profiles of human diseases and drugs to create a disease–disease, drug–drug, and disease–drug network (to establish the links between different diseases and drugs), by directly matching their molecular profiles, in particular their transcriptomic profiles, thanks to the accumulation of whole-genome gene expression data available in the public domain. The main assumption of their approach is that gene expression profiles of many (but not all) diseases and drugs can characterize to some extent the effects of disease and drugs; therefore, these diseases and drugs can be related based on the similarity/dissimilarity of their induced expression profiles. In this study, a network of 170,027 significant interactions was extracted from the ~24.5 million comparisons between ~7000 publicly available transcriptomic profiles. The network includes 645 disease–disease, 5008 disease–drug, and 164,374 drug–drug relationships. At least 60% of the disease–disease pairs were in the same disease area as determined by the Medical Subject Headings (MeSH) disease classification tree. The remaining can drive a molecular level nosology by discovering relationships between seemingly unrelated diseases, such as a connection between bipolar disorder and hereditary spastic paraplegia and a connection between actinic keratosis and cancer. Among the 5008 disease–drug links, connections with negative scores suggest new indications for existing drugs, such as the use of some antimalaria drugs for Crohn's disease and a variety of existing drugs for Huntington's disease, while the positive scoring connections can aid in drug side effect identification, such as tamoxifen's undesired carcinogenic property. From the ~37 K drug–drug relationships, they discovered relationships that aid in target and pathway deconvolution, such as (1) KCNMA1 as a potential molecular target of lobeline and (2) both apoptotic DNA fragmentation and G2/M DNA damage checkpoint regulation as potential pathway targets of daunorubicin [268].

18.12 Computation-Based Approaches

Exhaustive polypharmacology studies in vitro or in vivo are not practical, computational methods of predicting unknown targets or side effects are being developed. Recent success in computational prediction of multitargeting drugs has established polypharmacology as a promising alternative approach to tackle some of the daunting complications in drug discovery. This will not only help discover more effective agents but also present tremendous opportunities to study novel target pharmacology and facilitate drug repositioning efforts in the pharmaceutical industry. Computer-aided drug design is a major workhorse of target-based drug discovery. Based on docking studies and virtual screening, drug candidates with optimal potency and selectivity can be predicted. With the recent renaissance of phenotype-based drug discovery, these in silico technologies have found an important new role in the process of target prediction. Over the last two decades, extensive information regarding activity, structures, and targets of small-molecule libraries has been deposited into public databases such as ChEMBL, DrugBank, and ChemBank [20]. There are also public web services such as TarFisDock and SEA (Similarity Ensemble Approach) for target prediction of small molecules. Using these tools, targets of active compounds can be predicted based on similarities in structure between an active hit and well-characterized drugs in these databases [20]. This computer-aided target prediction has been widely used to identify new targets of known drugs [20], to predict the targets of active hits from a library screening, and to investigate the mechanism of action of hits discovered from phenotypic screens. For example, Lounkine et al. used a public database ChEMBL and performed a large-scale "ligand-based similarity search" to predict target proteins, which lead to the identification of 73 unintended off-targets of 656 marketed drugs [269]. Additionally, recent advances in high-content screening platforms using automated imaging systems enable the establishment of phenotypic SAR, thus improving confidence levels in target prediction and providing important information regarding drug mode of action [20].

Computation-based approaches are not single method but a collection of many computer-aided strategies including structure-based, omics-based, network-based, sequence-based, and fragment-based approaches. Indeed, any computer-aided drug design (CADD) strategies belong to computation-based approaches [270–275]. The application of CADD can be cost-effective using experiments to compare predicted and actual drug activity, and the results from which can be used iteratively to improve compound properties [270]. Probably the most important implication of CADD is its applications to polypharmacology to develop MTDs. For polypharmacology, the two major areas are the discovery and prediction of multiple protein targets using the tools of computer-aided drug design and the use of these protein targets in predicting therapeutic potential in the context of biological networks [271–275].

18.13 Similarity Approaches

Molecular similarity is a key concept in drug discovery. It is based on the assumption that structurally similar molecules frequently have similar properties. Assessment of similarity between small molecules (ligand similarity) has been highly effective in the discovery and development of various drugs. Nonetheless, similarity approaches are not a single method but include diverse strategies covering several of the approaches described above such as binding-site structural similarity, expression profile/signature similarity, side effect similarity, and the like. Here only ligand-based similarity is discussed.

18.13.1 Concept and Applicability of Ligand-Based Similarity Approach in Drug Discovery

Ligand-based drug design (LBDD or indirect drug design) relies on knowledge of other molecules that bind to the biological target of interest. These other molecules may be used to derive a pharmacophore model that defines the minimum necessary structural characteristics a molecule must possess in order to bind to the target [276]. In other words, a model of the biological target may be built based on the knowledge of what binds to it, and this model in turn may be used to design new molecular entities that interact with the target. Alternatively, a quantitative structure–activity relationship (QSAR), with a correlation between calculated properties of molecules and their experimentally determined biological activity, may be derived. These QSAR relationships in turn may be used to predict the activity of new analogs [277].

LBDD is also a traditional method for the prediction of drug–target interaction [260, 278]. It purely depends on the assumption that similar drugs will have similar targets and vice versa. The similarity among two different drugs can be analyzed either using the structure of a drug or its profile, and the similarity between the targets can be identified using a sequence of the targets. The input data including the structure or profile of a drug or the target sequence should be given by the user. If the drug structure is given as an input data, the web server analyzes for the similarity between the given data and the data available in the databases which can be used to predict the DTI from known drug target to the unknown one [279]. Different types of similarity-based methods are available including 3D shape-based similarity [280], two-dimensional (2D) fingerprint-based similarity [281], and phenotypic-based similarity [282] methods. Many web applications including ChemMapper [283] and Similarity Ensemble Approach (SEA) [284] are available to which applies 2D and 3D similarity for the prediction of drug–target interaction.

18.13.2 Ligand-Based Interaction Fingerprint

Cao and Wang [285] described a ligand-based interaction fingerprint (LIFt) approach for target prediction. Together with physics-based docking and sampling methods, they assessed the performance systematically by modeling the polypharmacology of 12 kinase inhibitors in three stages. First, they examined the capacity of this approach to differentiate true targets from false targets with the promiscuous binder staurosporine, based on native complex structures. Second, they performed large-scale profiling of kinase selectivity on the clinical drug sunitinib by means of computational simulation. Third, they extended the study beyond kinases by modeling the cross-inhibition of bromodomain-containing protein 4 (BRD4) for ten well-established kinase inhibitors. On this basis, they made prospective predictions by exploring new kinase targets for the anticancer drug candidate TN-16, originally known as a colchicine site binder and microtubule disruptor. As a result, p38α was highlighted from a panel of 187 different kinases. The prediction was then validated by an in vitro kinase assay, which showed TN-16 as a low-micromolar p38α inhibitor.

18.13.3 Ligand-Based Similarity Ranking

Ligand-based in silico target fishing can be used to identify the potential interacting target of bioactive ligands, which is useful for understanding the polypharmacology and safety profile of existing drugs. The underlying principle of the approach is that known bioactive ligands can be used as reference to predict the targets for a new compound. Liu et al. [286] tested a pipeline enabling large-scale target fishing and drug repositioning, based on simple fingerprint similarity rankings with data fusion. A large library containing 533 drug relevant targets with 179,807 active ligands was compiled, where each target was defined by its ligand set. For a given query molecule, its target profile is generated by similarity searching against the ligand sets assigned to each target, for which individual searches utilizing multiple reference structures are then fused into a single ranking list representing the potential target interaction profile of the query compound. The proposed approach was validated by tenfold cross-validation and two external tests using data from DrugBank and Therapeutic Target Database (TTD). The use of the approach was further demonstrated with some examples concerning the drug repositioning and drug side effect prediction. The promising results suggest that the proposed method is useful for not only finding promiscuous drugs for their new usages but also predicting some important toxic liabilities. It appears that the simple ligand-based target fishing approach can assist future drug design and discovery from the rapidly increasing volume and diversity of data concerning drug-related targets and their ligands.

The elucidation of non-obvious binding-site similarities has provided useful indications for the establishment of polypharmacology, the identification of potential off-targets, or the repurposing of known drugs. The concept underlying all of these approaches is promiscuous binding or multitarget binding which can be analyzed from a ligand-based or a binding site-based perspective. Ehrt et al. [287] applied methods for the automated analysis and comparison of protein binding sites to study multitarget binding on a novel dataset of sites in complex with ligands sharing common shape and physicochemical properties. Their results showed the suitability of this dataset for the benchmarking of novel binding-site comparison methods by revealing promising directions for further in-depth analyses of promiscuity and druggability in a pocket-centered manner.

18.13.4 Ligand-Based Virtual Screening

Virtual screening methodologies have emerged as efficient alternatives for the discovery of new drug candidates. Meanwhile, ensemble methods are nowadays frequently used to overcome the limitations of employing a single model in ligand-based drug design. However, many applications of ensemble methods to this area do not consider important aspects related to both virtual screening and the modeling process. During the application of ensemble methods to virtual screening, the proper validation of the models in virtual screening conditions is often neglected. No analysis of the diversity of the ensemble members is performed frequently, or no considerations regarding the applicability domain of the base models are being made. In a review article [288], Helguera introduced basic concepts and definitions related to virtual screening and proposed a method based on genetic algorithms optimization for the generation of virtual screening tailored ensembles which address the previously identified problems in the current applications of ensemble methods to virtual screening. The proposed methodology was successfully applied to the generation of ensemble models for the ligand-based virtual screening of dual-target A2A adenosine receptor antagonists and MAO-B inhibitors as potential Parkinson's disease therapeutics.

Abdul Hameed et al. [289] studied polypharmacology using a ligand-based target fishing (LBTF) protocol. To implement the protocol, the investigators first generated a chemogenomic database that links individual protein targets with a specified set of drugs or target representatives. Target profiles were then generated for a given query molecule by computing maximal shape/chemistry overlap between the query molecule and the drug sets assigned to each protein target. The overlap was computed using the program ROCS (Rapid Overlay of Chemical Structures). They next validated this approach using the Directory of Useful Decoys (DUD). DUD contains 2950 active compounds, each with 36 property-matched decoys, against 40 protein targets. They chose a set of known drugs to represent each DUD target and carried out ligand-based virtual screens using datasets of DUD actives seeded into DUD decoys for each target. The researchers computed Receiver Operator Characteristic (ROC) curves and associated area under the curve (AUC) values. For the majority of targets studied, the AUC values were significantly better than for the case of a random selection of compounds. In a second test, the method successfully identified off-targets for drugs such as rimantadine, propranolol, and domperidone that were consistent with those identified by recent experiments. The results from their ROCS-based target fishing approach are promising and have potential application in drug repurposing for single and multiple targets, identifying targets for orphan compounds, and adverse effect prediction.

18.14 Label-Free Approach

18.14.1 Concept and Applicability of Label-Free Approach in Drug Discovery

Protein microarray detection techniques are often classified as being label-based and label-free. Most of the microarray applications have employed labeled detection such as fluorescent, chemiluminescent, and radioactive labeling. These labeling strategies have synthetic challenges and multiple label issues and may exhibit interference with the binding site. Before the target-directed approaches took the center stage three decades ago, drugs were mostly discovered through phenotypic assessment of isolated tissues and whole animals without labels [290]. With the rise of combinatorial chemistry, molecular biology, genomics, and pharmacology in the past decades, labeling techniques have become dominating in early drug discovery process [290]. Compared with traditional approaches, labeling techniques provide clear advantages in sensitivity, throughput, and molecular specificity. The ability to perform high-throughput screening (HTS) with high sensitivity empowers labeling techniques to rapidly identify maximally selective lead molecules for many targets of interest. The molecular specificity permits the application of molecular and chemical knowledge to investigate specific molecular hypotheses. However, labeling techniques have caveats associated with label interference, cell engineering, and target/mechanism of action constraint [290]. Clearly, early drug discovery is achieved mainly through two strategies, target-based and phenotypic approaches [291]. Target-based screens use high-throughput and label-dependent molecular assays to measure the effect of compounds on a specific target protein in vitro, while phenotypic screens use unbiased phenotypic assays to examine the effect of compounds on a specific phenotype of cells, tissues, or animals. Target-based approaches have been dominating early drug discovery in the past quarter of century, which is coincident with the continuous decline in productivity of pharmaceutical research and development [290–294].

According to Yang [290, 291], the problems of early strategies are associated with the following technical issues:

(a) The common use of engineered cells for screening precludes the investigation of drug action in native and physiologically relevant environments.
(b) The target/mechanism of action predetermined in a given screen may be not essential to disease pathogenesis but may induce undesired toxicity, and the molecular mechanism of action (MMOA) investigated may be irrelevant to the disease pathogenesis, non-druggable, or inefficacious. The MMOA describes the interaction between a drug and its target (or targets) that creates a specific response.
(c) Many drugs display clinically relevant polypharmacology—the specific binding of a drug to more than one target, suggesting that single target-based screen may de facto be ineffective.
(d) Molecular assays for target-based screens generally rely on the use of labels, which may cause artifacts in results. For example, fluorescent and luminescent labels may alter the functions of probe molecules, proteins, or cells, thus introducing bias or artifacts in results [295–297].
(e) Traditional phenotypic approaches suffer disadvantages associated with low-to-moderate throughput and difficulty in target deconvolution and in governing medicinal chemistry optimization.

Thus, there is high demand in implementing innovative technologies to discover drugs, in particular, for unmet therapeutic needs and unexploited biological mechanisms. The past decade has witnessed the great expansion of label-free platforms and methods with enhanced and/or totally new capabilities [290, 291].

Many label-free techniques such as SPR, carbon nanotubes (CNTs) and nanowires, nanohole arrays, interferometry, etc. have been successfully integrated with protein microarrays and are emerging rapidly as a potential complement to labeling methods. Label-free techniques have the following advantages:

(a) Label-free techniques do not require any chemical modification of an active compound, which can greatly facilitate the target identification process. This relatively new type of target identification strategy relies on changes in thermodynamic stability as the result of a protein–drug interaction. These methods are based on the concept that a protein has conformational flexibility in solution, making it more susceptible to proteolysis; however, once it binds to a small molecule, the overall complex will be more resistant to proteolysis [298]. One such label-free technique called DARTS (Drug Affinity Responsive Target Stability) has been used to successfully identify cellular targets of rapamycin, FK506, didemnin B, and resveratrol [299]. Similarly, a "pulse proteolysis" technique demonstrated that ligand-bound proteins are more stable upon protein denaturation and proteolysis compared to the samples without a ligand [300].
(b) Compared with labeling approaches, certain label-free techniques (e.g., resonant wave-

guide grating, RWG) have comparable or even better sensitivity and throughput for cell-based screening, while others (e.g., mass spectroscopy [MS]) may have better molecular specificity even at the proteome level. Nowadays, new-generation label-free techniques have been implemented in various stages of drug discovery process including hit identification, lead optimization, target engagement determination, drug safety assessment, and clinical diagnostics [290, 291].

(c) Owing to its noninvasive and holistic measure, label-free technique allows for multistep screen campaigns that can simultaneously identify different classes of ligands for the target of interest (e.g., agonists, biased agonists, antagonists, and allosteric modulators for a G-protein-coupled receptor) with relatively low-false positives and negatives [290, 291]. This is unlike conventional label-dependent screens which typically result in high false-negative rates (mostly due to the bias defined by the molecular hypothesis for screening) and high false-positive rates (mostly related to the chemo-physical properties [e.g., aggregation, colors, or fluorescence] of compounds and the presence of compensatory targets/pathways).

(d) Label-free techniques can also be used to assess the safety of lead-like molecules. First, almost all drugs display polypharmacology, the ability to bind to more than one target. Polypharmacology could be beneficial or harmful, depending on the targets bound by the drug. RWG has high sensitivity to detect functional receptome in living cells ranging from G-protein-coupled receptors to receptor tyrosine kinases, ion channels, transporter, immune receptors, cytoskeletal proteins, and kinases, enabling multitarget profiling [290, 291]. Label-free profiling with an enlarged panel of cell lines can dramatically increase the target coverage for polypharmacology assessment of drug molecules. MS can assess polypharmacology of drugs acting at the entire human proteome, thus facilitating the identification of primary target(s) and toxicity-inducing anti-target(s) for the drugs. Second, drug-induced toxicities in the liver, heart, kidney, and brain currently account for more than 70% of drug attrition and drug withdrawal. The toxicity potential of drug candidates prior to in vivo studies can be assessed in vitro using highly predictive cellular models.

(e) Label-free profiling with distinct probe- and gene-modulated cells provides a high-resolution self-referenced means to differentiate and classify lead-like molecules into different clusters, so leading candidates can be selected for in vivo testing.

(f) Combining label-free profiling in disease-relevant cell models with text mining, one can predict the biological functions of novel targets.

Taken together, label-free investigations are well positioned to facilitate early drug discovery. For follow-on drug discovery where the molecular mechanism of action is known, label-free can be implemented in a way similar to conventional label-dependent screens, and lead-like molecules can be optimized and advanced based on binding affinity, kinetics, and efficacy. For unproven targets where the desired molecular mechanism of action is unknown, multiple label-free and dependent techniques shall be combined to provide systematic comprehension of the druggability of the target and lead-like molecules.

18.14.2 Label-Free Techniques

Label-free technologies include SPR, NMR, high-throughput mass spectrometry, resonant waveguide plate-based screening, transmitted-light imaging, isothermal titration calorimetry, optical and impedance cell-based assays, and other biophysical methods. The technologies are used as screening technologies, high-content technologies, hit finding and hit validation strategies, mode of action and ADME/T, access to difficult target classes, cell-based receptor/ligand interactions particularly orphan receptors, and antibody and small-molecule affinity and kinetic analysis.

Central to drug discovery and development is to comprehend the target(s), potency, efficacy, and safety of drug molecules using pharmacological assays. Label-free techniques have been used to determine almost every important aspect of drug–target interactions starting from affinity to kinetics, structure, and thermodynamics, all of which are important to assess the druggability of compounds. All drugs need to bind to their target(s) to exert effects; the binding affinity of a drug to its primary target is a traditional surrogate for its in vivo efficacy prediction. Thus, measuring binding affinity is critical for understanding drug–target interactions and identifying efficacious leads. The following specific label-free techniques provide the opportunity.

18.14.2.1 Label-Free Biosensors

Label-free biosensors can measure the affinity of a drug binding to its target in different environments, which is essential to comprehend drug–target interactions in vivo where environments are diverse and complex. Surface plasmon resonance (SPR) and RWG can measure the affinity of drugs binding to purified receptors immobilized on the surface, while surface acoustic wave can measure the affinity of drugs binding to the receptor at the surface of living cells, and backscattering interferometry can determine drug–target binding affinity in homogenous solution or in vivo-like environment. Second, the kinetics, in particular, residence time, of a drug is believed to be more critical to its clinical efficacy and safety profiles than binding affinity. SPR is the standard technique to determine binding kinetics. RWG under microfluidics not only enables kinetic profiling with increased throughput but also manifests the effect of binding kinetics on receptor signaling in living cells. SPR imaging permits selectivity profiling of a specific drug against an array of targets, while biolayer interferometry is useful for kinetic profiling of biologic drugs with moderate throughput. Third, the structural features of drug binding are useful to determine druggability, that is, the likelihood of being able to modulate a target with a small-molecule drug. Structure-based techniques such as X-ray, NMR, and in silico docking can identify possible druggable sites of targets, enabling rational drug design. Fourth, the thermodynamic signatures including free energy, enthalpy, and entropy are essential to medicinal chemistry optimization and can be obtained using isothermal titration calorimetry.

18.14.2.2 Label-Free Binding Techniques

Label-free binding techniques, in particular microplate-based RWG, have been used in HTS for hit identification. NMR spectroscopy is particularly valuable for validating hits and ligand binding, as well as discovering active fragments including weak binders without preknowledge of binding interactions [12]. Unlike labeled techniques that are often limited to detect compounds competing with the same binding site of the labeled probe used, label-free can detect hits for all possible binding sites of the target protein. With the ever-reducing cost of producing high-quality and purified proteins, label-free binding based on screening may become affordable in the near future.

18.14.2.3 Label-Free Cell Phenotypic Profiling

Label-free cell phenotypic techniques (e.g., RWG and electric biosensor) have also been employed in HTS and even ultra-HTS using native cells for hit identification [301]. These techniques permit profiling and screening of drug molecules with a wide coverage in targets, pathways, cellular processes, and cell types. For label-free screens, hits can be selected from the library based on a desired size of response or based on similarity of the cell phenotypic profiles of compounds to reference probe molecules in a cell line or a panel of cell lines. Label-free techniques are powerful for lead and probe identification, in particular for poorly characterized targets. Using chemical similarity search based on hits, fragments, or substructures identified from the HTS campaign, one can quickly expand compound library and follow up with label-free profiling to discover potent probes, and even endogenous agonist(s), for the novel target.

Label-free cell phenotypic profiling techniques afforded by optical or electric biosensors offer clear advantages in rich information content, real-time kinetics, highly flexible assay formats, and high throughput, besides wide pathway coverage and ability in multitarget profiling and screening that are common to all phenotypic assays [301–303]. Optical biosensors such as resonant waveguide grating (RWG) measure drug-induced dynamic mass redistribution (DMR) signals, while electric biosensors measure drug-induced impedance signals [302].

18.14.2.4 Label-Free Imaging Techniques

Label-free imaging techniques are useful for chemical and phenotypic profiling of living cells [290, 291]. MS can be used to map out drug metabolites in living cells. In-cell NMR can provide novel insights into structural biology and drug–target interactions in the crowded native cellular environments. Nonlinear optical microscopies (e.g., two-photon fluorescence, second-harmonic generation, and digital holographic microscopy) enable chemical and morphological mapping of living cells and tissues. Coherent anti-Stokes Raman scattering and stimulated Raman scattering microscopies permit imaging specific molecules, macromolecular assembles, and organelles of living cells. Together, it is possible to probe many differ aspects of drug pharmacology in living cells and tissues that are otherwise difficult to be studied by conventional labeling techniques.

18.14.2.5 Label-Free Protein Quantification

Label-free protein quantification is a mass spectrometry-based method for identifying and quantifying relative changes in two or more biological samples instead of using a stable isotope-containing compound to label proteins. Unlike other methods for protein quantification, label-free quantification does not use a stable isotope-containing compound to chemically bind to and thus label the protein [304, 305]. Label-free quantification may be based on precursor signal intensity or on spectral counting. The first method is useful when applied to high precision mass spectra, such as those obtained using the new generation of time-of-flight (ToF), Fourier transform ion cyclotron resonance (FTICR), or Orbitrap mass analyzers. The high-resolution power facilitates the extraction of peptide signals on the MS1 level and thus uncouples the quantification from the identification process. In contrast, spectral counting simply counts the number of spectra identified for a given peptide in different biological samples and then integrates the results for all measured peptides of the protein(s) that are quantified.

18.15 Protein–Protein Interaction Approach

18.15.1 Concept and Applicability of Protein–Protein Interactions in Drug Discovery

Protein–protein interactions (PPIs) have a key role in regulating a wide range of cellular functions and pathophysiological processes and offer attractive opportunities for therapeutic intervention, as PPIs represent a highly populated class of targets for drug discovery [306–313]. Small-molecule inhibition of the direct PPIs that mediate many important biological processes is an emerging and challenging area in drug design [306–313]. Different PPI targets have been implicated in cancer, metabolic diseases, neurological disorders, and many other diseases [314, 315].

Conventional drug design has mainly focused on the inhibition of a single protein, usually an enzyme or receptor, since these proteins often contain a clearly defined ligand-binding site with which a small-molecule drug can be designed to interact. Designing a small molecule to bind to a PPI and subsequently inhibit the interaction poses several challenges, including the initial identification of suitable PPIs, the surface area of the interface (it is often large), and the location of "hot spots" (small regions suitable for drug binding) with the lack of well-defined binding pockets. Although efforts to modulate these interactions are more challenging and much less mature than work on conventional drug discovery pathways, significant progress has been made on several fronts [313]. There have recently been great advances in the development of peptide-based inhibitors, including side chain and backbone cross-linked agents, peptide scaffolds, and small-molecule inhibitors of protein–protein interactions, such as those containing terephthalate or bis-imidazole scaffolds [313].

Studies suggest that 15–40% of the human interactome is comprised of protein–peptide interactions [316, 317]. With an estimated 650,000 PPIs as part of the human interactome [316, 318], it is evident that these interactions play crucial roles in various cellular processes and pathways. Dysregulation in PPIs are often found to be the primary cause of several disease pathologies, making them attractive drug targets. However, the development of PPI inhibitors and stabilizers has been hindered because of the seemingly low druggability of PPI interfaces [316]. Extensive studies for PPI targets and modulators have been performed to understand these complex targets and identify distinct properties in its network, conformational structure, and ligand chemical space. PPI druggability continues to be poorly understood because these targets show substantial diversity as compared to protein families that have traditionally been targeted until now. Despite this, its relevance to multiple diseases and relative novelty in the drug discovery field encourage researchers to pursue these difficult targets. Moreover, the discovery of PPI inhibitors and therapeutics in the last few decades has ascertained that these targets are tractable and can be modulated by small-molecule compounds.

18.15.2 Development of PPI Inhibitors

Targeting the interfaces between proteins has huge therapeutic potential, but discovering small-molecule drugs that disrupt PPIs is an enormous challenge. Several recent success stories, however, indicate that protein–protein interfaces might be more tractable than has been thought. These studies discovered small molecules that bind with drug-like potencies to "hotspots" on the contact surfaces involved in protein–protein interactions. Remarkably, these small molecules bind deeper within the contact surface of the target protein, and bind with much higher efficiencies, than do the contact atoms of the natural protein partner. Some of these small molecules are now making their way through clinical trials, so this high-hanging fruit might not be far out of reach [311].

18.15.2.1 Small-Molecule Inhibitors of PPIs

Analysis of known PPI modulators showed that these compounds are larger and more hydrophobic and contain more multiple bonds and aromatic rings [316, 319, 320]. Interestingly, PPI ligands have a higher topological polar surface area (TPSA), even with its high hydrophobicity, as compared to traditional drugs, resulting in their tendency to form more hydrogen bonds in the PPI surface. This can be attributed to their larger sizes that allow for the attachment of more polar atoms [320]. More success is acquired for PPIs with grooves or small pockets, compared to globular interfaces. However, those with hot spot pockets that are spread out across a large interface are naturally more challenging. Because of this, PPI inhibitors also tend to have a more 3D conformation than the average inhibitor [321], especially for extended surfaces and disjointed pockets, due to its propensity to mimic the binding epitope of the partner protein and anchor into small pockets present in PPI interfaces [316, 322]. Based on these, it is evident that PPI inhibitors occupy a different chemical space as compared to typical small-molecule inhibitors and, hence, should be carefully evaluated in a different manner. Whereas the primarily hydrophobic interacting features of PPI inhibitors can be buried in hot spot regions, the rest of their structures are often solvent-exposed due to their binding location on the protein interface. These elements can be exploited to tailor pharmacokinetic properties without negatively influencing binding affinity [323].

18.15.2.2 Mimetics Targeting PPIs

The rational design of compounds that mimic key interactions at the protein–protein interface is another successful strategy for PPI inhibitor discovery [324, 325]. Compared with low-molecular-weight compounds, mimetics can be more selective and show lower toxicity. Modification and optimization of peptide mimetics can also improve structural stability, resulting in increased oral availability [326]. For example, Groß et al. designed and synthesized a soluble peptide mimicking CXCR4 to interrupt the gp-120 and CXCR4 interaction [327]. In this context, the combined utilization of mimicking strategies and virtual screening has broad application prospects for drug discovery. Based on this strategy, many mimetics targeting cancer-related PPIs were discovered using virtual screening, such as p53 mimetics for the MDM2–p53 interaction [328–330], BH3 mimetics for the Bcl-2–BH3 interaction [331], and SMAC mimetics for the IAP–SMAC interaction [332, 333]. Knowledge of the 3D structures of proteins allows the use of different approaches for mimetics design. For the interruption of PPIs, inhibitors may be larger in size than traditional drug compounds. Peptides and proteins mimetics are increasingly considered to be viable therapeutics for PPI inhibitor discovery [334].

18.15.2.3 Protein–Protein Docking

In order to apply structure-based design techniques to design PPI modulators, a three-dimensional structure of protein complex has to be available. In this context in silico approaches, in particular protein–protein docking, are a valuable complement to experimental methods for elucidating 3D structure of protein complexes. Protein–protein docking is easy to use and does not require significant computer resources and time (in contrast to molecular dynamics), and it results in 3D structure of a protein complex (in contrast to sequence-based methods of predicting binding interfaces) [335–338]. However, protein–protein docking cannot address all the aspects of protein dynamics, in particular the global conformational changes during protein complex formation [335–338]. In spite of this fact, protein–protein docking is widely used to model complexes of water-soluble proteins and less commonly to predict structures of transmembrane protein assemblies [335–338], including dimers and oligomers of G-protein-coupled receptors (GPCRs).

18.16 Big Data in Drug Discovery

The "big data" approach influences our daily life, and drug discovery is not an exception. By current computational techniques, molecular characteristics can be studied in a logical and systematic manner. The data collected from each compound can be subjected to analyses from different perspectives [33, 339]. In the modern era of technology, there has probably been an increase in the size of data generation. According to a recent estimate, the total size of stored data is approximately two zettabytes (10^{21}) with expected doubling every 2 years [340]. Hence, excavation of massively produced digital information offers a multitude of opportunities to increase productivity. Nevertheless, apart from the volume and production rate of big data, the variety and complexity of big data pose challenges for effective analysis [341]. Furthermore, sometimes generated data contain inconsistencies, such as missing or incomplete information, errors, and duplications, thereby affecting the outcomes of accurate simulation and analytical activities. Therefore, preliminary analysis and curation are required as advanced measures to ensure fairness, accuracy, and experimental efficacy [342]. On the other hand, precollection and curation measures vary among research communities, depending on preceding observations and experimental records. Yet, there is high demand for a simple, unified, and well-established curation protocol that ensures the quality of generated simulation and analytical datasets.

Several studies examined the impact of quality on research activities [343]. Several others recommend conducting a fair evaluation of the quality and impact of a particular work [344]. Hence, the existing standard of research continues to adhere to the "less-is-more" principle. Big data have played a vital role in medicinal and combinatorial chemistry, whereas HTS contributes to the generation of a huge amount of data over a short span of time. Big data dependency will likely increase as the perception of personalized medicine improves. Earlier, big data have been regarded as the beginning of computation-oriented medicinal chemistry (i.e., processing stacks of generated data, resulting in shortening of the time taken to complete a drug development cycle). For instance, a well-known global pandemic spanning more than 40 years, HIV, has infected more than 37 million people, where only 57% are being treated with antiviral agents (World Health Organization (WHO) 2018; https://www.who.int/news-room/spotlight/why-the-hiv-epidemic-is-not-over). In the past few years, many studies have addressed the inhibition of viral reverse transcriptase and/or integrase [345, 346]. Although this technique has proven effective enough, it comes with several shortcomings such as viral resistance and poor bioavailability.

In the early 1990s, the roles of chemokines and $CD4^+$ cells were described. Chemokine activity is associated with their G-protein-coupled receptors (GPCRs); in the CCR5 case, it is a "C-C" receptor with 75% homology to CCR2 [347]. With the emergence of CCR5 as an interesting and a druggable novel target to combat HIV, numerous pharmaceutical firms turned to their GPCR inhibitor libraries in search of a putative ligand for this protein. A strong lead, an imidazopyridine (UK107543), was identified by Pfizer, a well-known pharmaceutical company, using HTS [348]. Maraviroc (Selzentry), an antiretroviral drug, classified as an entry inhibitor was later declared as an approved drug for HIV-1 treatment by the FDA [349]. Such real-world use cases spotlight the significance of big data resources in medicinal chemistry. Therefore, among medicinal chemists, we are seeing a major demand for rational awareness of data-driven processes and for information-handling skills [350].

From this standpoint, the scientific communities started investing in the development of applications, tools, and software to handle massively generated and already stored data. Nevertheless, a major concern limiting the usability of these computational platforms includes security and privacy concerns for the users [351]. Aside from these factors, freely and publicly accessible resources provide a versatile collection, which can be manipulated beyond the pharmaceutical scope [352].

18.17 Artificial Intelligence and Machine Learning in Drug Discovery

Artificial intelligence (AI) mimics human behavior by simulating human intelligence by computer techniques [353]. ML, a subfield of AI, uses statistical methods for learning with or without being programmed [354]. In the drug development process, AI has shifted the mood from hype to hope [355]. Computational technologies and ML algorithms have revolutionized drug discovery in the pharmaceutical industry. Integration of ML algorithms in an automatic manner—to discover new compounds by analyzing, learning, and explaining pharmaceutical big data—is the application of AI to drug design [356]. Big Pharma is increasing investment in AI; this situation shows the truth behind the use of ML algorithms to identify and screen potential drug candidates. For instance, SYNSIGHT has introduced an AI-based integrated platform in combination with VS and molecular modeling to create huge biological models for drug development [357]. Many leading biopharmaceutical companies are collaborating to integrate AI and ML methods with their drug discovery pipelines. Pfizer has been collaborating with IBM since December 2016 to take advantage of their multicloud platform Watson [358] for immuno-oncology drug discovery [359]. Similarly, Exscientia Ltd., a UK-based world class AI-driven drug design company [360], is collaborating with Sanofi to find a cure for metabolic disorders [359], and Celgene, another leading pharmaceutical company, aims to accelerate drug discovery in the areas of autoimmunity and oncology [361]. Recently, Exscientia announced a success story in collaboration with GlaxoSmithKline (GSK), where they claimed the discovery of a highly potent lead molecule for the treatment of chronic obstructive pulmonary disease by means of AI-based drug discovery workflow [362].

ML success has been repeatedly demonstrated in classification, generative modeling, and reinforcement learning (RL). Different categories of ML are supervised learning, unsupervised learning, and RL. The subcategory of supervised learning, classification, and regression methods predicts the model on the basis of input and output data sources. Supervised ML is applicable to a disease in diagnostic methods, ADMET in a classification method's output, and to drug efficacy in regression methods [362]. SVMs with supervised ML algorithms use binary activity prediction to distinguish between a drug and nondrug [363, 364] or between specific and nonspecific compounds [365, 366]. SVM classification is performed in LBVS to rank the database compounds by decreasing activity probability. To minimize error in SVM ranking, optimized special ranking functions are used [367]. The clustering method for an unsupervised learning category can discover a disease subtype as outputs, while a feature-finding method can identify a target in a disease [368, 369]. Decision-making RL maximizes its performance in de novo drug design via modeling and quantum chemistry. RL is less dependent on dataset learning. With RL, the desired physical and biological properties of newly generated chemical structures can be biased [370]. ML exploits the relationship between a biological activity and chemical structure during drug design. Structure prediction of biological targets (protein structure, binding pocket, transmembrane regions, and phosphorylation and glycosylation sites) and quantitative structure–activity relationship (QSAR) models, pharmacophore models, molecular docking analysis, and ranking/scoring functions in similarity searches can be implemented and statistically validated by ML techniques [371]. Classifying a pharmacokinetic and toxicological (ADMET) profile, discovery or optimization of biologically active hit compounds, and the constructed model or biological activity of a new ligand can aid with a drug discovery process at several steps by ML techniques [372]. Multiple ML models can be used to drive multiparameter optimization. The output of ML methods depends on multiple parameters like diversity of the training dataset, an ability to handle imbalanced datasets of active and inactive compounds in the library, and defining precise parameters to cover full chemical space including active and inactive molecules [373]. Proficient ML models can be developed to screen huge libraries which generate few false positives and a good number of active compounds in the output. This goal can be attained using versatile training datasets comprising predicted inactive compounds [374, 375].

18.18 Deep Learning in Drug Design

Machine learning uses algorithms to parse data, learn from that data, and make informed decisions based on what it has learned. Deep learning structures algorithms in layers to create an "artificial neural network" that can learn and make intelligent decisions on its own. Deep learning (also known as deep structured learning) is part of a broader family of machine learning methods based on artificial neural networks with representation learning. Learning can be supervised, semi-supervised, or unsupervised. Deep learning is a branch of machine learning. It is an algorithm that abstracts data by using multiple processing layers composed of complex structures or multiple nonlinear transformations. Deep learning has revolutionized most areas of science and technology, including drug discovery.

Neural networks (NNs) represent a supervised neurology-inspired machine learning (ML) technique that is employed routinely and successfully to address such issues as speech and image recognition. Artificial neural networks (ANNs) are ML algorithms that operate as neurons in the brain: they receive numerous input signals and generate an activation response by calculating a weighted sum of the inputs through a nonlinear activation function and pass the output signal to subsequent connected neurons. The basic structure of an ANN consists of an input layer, hidden layer, and the output layer.

In the ANN, the processing nodes are either fully or partially connected. From input nodes, the input variables are taken and are transformed through hidden nodes into the output nodes where output values are calculated. By backpropagation methods, the ANN training is done in an iterative fashion to train the network. Due to over-fitting, a diminishing gradient, and other problems, the traditional ANN methods have not performed well and have been replaced by other ML algorithms like RF and SVM. The deep learning (DL) concept has originated from ANN's feedforward NNs with many hidden layers. DL's recent development has given the ANN a renaissance. With the rapid explosion of chemical "big data" from combinatorial synthesis and HTS, ML techniques have become an indispensable tool for drug designers to retrieve chemical information from large compound databases to design drugs rationally. Big data volume, velocity, variety, and veracity characterization are not possible via traditional QSAR approaches. ML techniques are more efficient than the physical model for scaling big datasets. DL, being the data-hungry ML algorithm for analyzing and exploring big data, is in high demand. As compared to other ML methods, the DL architecture is flexible. Atomwise, the first DL-based technology for structure-based small-molecule drug discovery has helped to design new potential drugs for 27 disease targets with accuracy and precision. A straightforward method with a fully connected deep neural network (DNN) is used for model building of compounds having the same number of molecular descriptors. To the Merck Kaggle challenge dataset, Ma et al. [376] applied a DNN and showed better performance as compared to RF on 13 of the total 15 targets. DNNs can handle thousands of descriptors without over-fitting and feature selection problems as in the traditional ANN, in an optimized manner, owing to the number of nodes and hidden layers. Mayr et al.'s multitasking DNN method won the Tox21 dataset challenge consisting of 12,000 compounds for 12 high-throughput toxicity assays. In this challenge the computational toxicity prediction of chemicals and drugs was given. The chemical structures and assay measurements from stress and nuclear receptor signaling pathway assays for 12 different toxic effects were available to the participants to check structure–activity relationships. Mayer et al. developed a DeepTox pipeline for toxicity prediction which uses deep learning algorithms. DeepTox normalizes the chemical structures followed by computation of the chemical descriptors. The computed descriptors are used in DL methods to predict the toxicity of chemicals. Later, these models are combined to ensembles. Statistically, a DNN outperforms other ML models such as SVM, RF, and others when applied to seven datasets selected from ChEMBL database. In variational autoencoder (VAE), an encoder NN generates a chemical structure via unsupervised learning to map chemical structures from a database onto a latent space. The trained VAE from the latent vector in the latent space transforms the molecular structure into a simplified molecular-input line-entry system (SMILES) string. Kadurin et al. [377] have generated new structures having specific anticancer properties by coupling the generative adversarial network (GAN) with VAE. In a GAN, two ANN models—the generator and discriminator—are trained simultaneously and generate a new molecule from scratch by optimizing a different and opposing objective function in a zero-sum game. A reinforced adversarial neural computer (RANC) with DL architecture, based on the GAN paradigm and RL, generates unique and adequate structures. The RANC uses the SMILES string dataset with key distribution of chemical features like molecular weight, log P, and topological polar surface area for de novo design of small molecules against different biological targets and pathways. Relevant to drug discovery, RANC trained on SMILES string representation outperforms other methods on several metrics. Segler et al. [378] and Yuan et al. [379] have used a recurrent neural network (RNN) for new structure generation acknowledging its success in natural language processing. RNN generates molecular structures by using the probability distribution learning on the SMILES string training set. Target-specific libraries were generated by Segler et al. [378] while exploring the RNNs. RNN together with deep Q-learning of the RL technology generates SMILES with desirable properties like quantitative estimate of drug-likeness (QED) and ClogP. Olivecrona et al. [380] overcame the incorporation of handwritten rules for undesirable structure penalties by tuning the pretrained RNN using the policy-based RL approach. Pereira et al. reported deep learning-based virtual screening method where they compared 95,316 decoys with 2950 ligands docked on 40 receptors and those ranked by the deep convolutional neural network showed better performance than other docking programs.

18.19 Expression-Based Approaches

Expression-based drug screening (e.g., transcriptomic-based drug screening) integrates an in silico computational drug-screening approach to select the drugs from transcriptomic-based drug screening. This approach looks at the drug-induced gene expression profiles to predict and subsequently characterize drug actions. Thus, in silico computational drug-screening approaches can be applied to selecting the drugs from transcriptomic-based drug screening. Expression cloning techniques utilize a library of cDNAs inserted into cloning vectors to express a library of proteins. A small-molecule–protein interaction can be detected by adding a tagged small molecule followed by affinity purification. In a sense, expression cloning techniques are similar to typical affinity purification because they also require chemical modification to attach a tag. However, when the target of interest is low abundance or is unstable, expression cloning can be an excellent alternative.

One method to express a large-scale library of proteins is phage display. Phage display is an affinity selection technique initially developed to identify antigen–antibody interactions and protein–protein interactions. A library of DNA sequences can be fused to a gene encoding a phage coat protein. Hence, the phage will display one unique protein on its surface per phage particle. Phage particles that bind to a small-molecule target with high affinity can be isolated. The isolated phage can be used for subsequent rounds of selection that can lead to further enrichment. Although traditional phage display techniques have identified molecular targets of many natural and synthetic ligands [381–383], more improved techniques have facilitated the process. In a recent example, Van Dorst et al. demonstrated that lytic (T7) cDNA phage display can be used as a fast, cost-effective alternative to conventional filamentous phage (M13) display and identified molecular targets of 17β estradiol [384].

With recent advances in robotics and high-content screening and analysis methods, transcriptional profiling can now be utilized as a comprehensive forward chemical genomics platform for drug discovery, lead selection, and lead optimization. It can be used to define the state of a cell on the basis of gene networks and to search for drugs that can shift cellular states in a manner predicted at the genome level to be therapeutically beneficial. The treatment of cells with compounds produces transcriptional "fingerprints" that reveal mechanism of action and enable discrimination between individual compounds based on drug behaviors important to all phases of drug discovery and development.

As an alternative to phage, mRNA display was introduced as a method in which proteins could be expressed as a fusion to their corresponding mRNAs [385]. This allows direct affinity screening and rapid identification of the target protein by sequencing of the corresponding cDNA tag. However, since the initial proof-of-concept experiment a decade ago [386], few actual examples of target identification have been reported. Most examples of mRNA display applications have been focused on protein–protein interaction and lead discovery.

Another alternative display method is the yeast three-hybrid screen. This is an adaptation of the commonly used two-hybrid system for identification of protein–protein interactions. The three-hybrid system makes use of the same elements but includes a "chemical dimerizer" that is used to link the small molecule of interest to the bait protein so that interaction with the prey domain can be measured. Although the idea was introduced nearly two decades ago [387], there have been only a few reports using three-hybrid system for target identification of small molecules [388]. Recently, Chidley et al. incorporated a SNAP-tag to covalently label drug molecules inside yeast and were able to identify previously unknown targets for clinically approved drugs [389].

Laenen et al. [390] proposed a network-based computational method for drug target prediction, applicable on a genome-wide scale. Their approach relies on the analysis of gene expression following drug treatment in the context of a functional protein association network. By diffusing differential expression signals to neighboring or correlated nodes in the network, genes are prioritized as potential targets based on the transcriptional response of functionally related genes. Different diffusion strategies were evaluated on 235 publicly available gene expression datasets for treatment with bioactive molecules having a known target. AUC values of up to more than 90% demonstrate the effectiveness of their approach and indicate the predictive power of integrating experimental gene expression data with prior knowledge from protein association networks.

The strategies described above are summarized in Table 18.1 for direct comparison.

Table 18.1 General strategies for rational design and discovery of multitarget drugs

Strategy	Basic characteristics
Target-based	Define and select a molecular that is known to have an important role in disease as the therapeutic target Screening drug candidates that have high affinity to the selected biomolecule target
Phenotype-based	Screening approach for discovery of drugs via testing their efficacy on disease phenotypes rather than specific therapeutic targets Knowledge of the molecular target and the molecular mechanism of action is not required
Sequence-based	The design of therapeutic drugs that act by sequence-complementary mechanisms to bind the target DNAs (e.g., antisense oligonucleotides), mRNAs, ncRNAs (e.g., siRNA, miRNA inhibitor), or peptides/proteins (e.g., antibodies) It is primarily applied to epigenetic MTDs, gene therapy MTDs, and macromolecule MTDs/biologics
Structure-based	Drug design that relies on the knowledge of the three-dimensional structure of the biomolecular target is known as structure-based drug design (SBDD), which is powered by protein crystallography and computer graphics technologies It is primarily applied to protein targets and maybe to RNA targets too Molecular docking and inverse molecular docking are the main techniques for structure-based strategy
Scaffold-based	This approach relies on employing privileged scaffolds as effective templates or core structures or building blocks in the synthesis of chemically relevant organic compounds for drug discovery and design or construct high-affinity ligands that can bind more than one receptor

(continued)

Table 18.1 (continued)

Strategy	Basic characteristics
Pharmacophore-based	It utilizes pharmacophore of the ensemble of steric and electronic features that is necessary to ensure the optimal supramolecular interactions with a specific biological target structure and to trigger (or to block) its biological response A pharmacophore can be considered as the highest common denominator of a group of molecules exhibiting a similar pharmacological profile and which are recognized by the same site of the target protein
Fragment-based	It is based on identifying small chemical fragments, which may bind only weakly to the biological target, and then growing them or combining them to produce a lead with a higher affinity Fragments have three "lows": Low binding affinity to the target, low complexity in chemical structures, and low molecular weight (less than 300 Da)
Molecular hybridization	Drug design and development based on the combination of pharmacophoric moieties of different bioactive substances to produce a new hybrid compound with improved affinity and efficacy, when compared to the parent drugs. This strategy can result in compounds presenting modified selectivity profile, different and/or dual modes of action, and reduced undesired side effects
Activity-based proteomics	A functional proteomic technology that uses chemical probes that react with mechanistically related classes of enzymes critical in disease-causing pathways
Omics-based	Omics technologies enable screening of biological samples at the gene, transcript, protein, metabolite, and their interaction network level in searching of novel targets Genome-wide association studies (GWASs) enable whole-genome sequencing and transcriptome analysis constituting essential tools to discover or validate new drug targets

(continued)

Table 18.1 (continued)

Strategy	Basic characteristics
Network-based	This approach encompasses relationships between drugs, disease-related genes, therapeutic targets, and diseases. It is one of the simplest methods and does not need any other additional information like structure or sequence for the prediction of drug–target interaction
Computation-based	It takes the advantages of public databases for drug design and applies computational models to predicting unknown targets, candidate drug actions, and side effects It is not a single method but a collection of many computer-aided strategies including structure-based, omics-based, network-based, sequence-based, fragment-based approaches, etc.
Similarity-based	It is based on the assumption that structurally similar molecules frequently have similar properties It is not a single method but includes diverse strategies such as binding-site structural similarity, expression profile/signature similarity, side effect similarity, and the like
Label-free	Label-free techniques do not require any chemical modification of an active compound, which can greatly facilitate the target identification process. It may offer better sensitivity and throughput for cell-based screening Owing to its noninvasive and holistic measure, label-free technique allows for multistep screen campaigns that can simultaneously identify different classes of ligands for the target of interest
Protein–protein interaction	Protein–protein interactions (PPIs) have a key role in regulating a wide range of cellular functions and pathophysiological processes, representing a highly populated class of targets for drug discovery. Small-molecule inhibition of PPIs is an emerging and challenging area in drug design

(continued)

Table 18.1 (continued)

Strategy	Basic characteristics
Big data for data mining	Big data are regarded as the beginning of computation-oriented medicinal chemistry (i.e., processing stacks of generated data, resulting in shortening of the time taken to complete a drug development cycle)
Artificial intelligence and machine learning	Integration of ML algorithms in an automatic manner—To discover new compounds by analyzing, learning, and explaining pharmaceutical big data—Is the application of AI to drug design
Deep learning	Machine learning uses algorithms to parse data, learn from that data, and make informed decisions based on what it has learned. deep learning structures algorithms in layers to create an "artificial neural network" that can learn and make intelligent decisions on its own. Deep learning has revolutionized most areas of science and technology, including drug discovery
Expression-based	High-content screening and analysis methods enable transcriptional profiling to provide a comprehensive forward chemical genomics platform for drug discovery, lead selection, and lead optimization. In silico computational drug-screening approaches can be applied to selecting the drugs from transcriptomic-based drug screening

References

1. Iwata H, Kojima R, Okuno Y. An in silico approach for integrating phenotypic and target-based approaches in drug discovery. Mol Inf. 2020;39(1–2):e1900096.
2. Croston GE. The utility of target-based discovery. Expert Opin Drug Discov. 2017;12(5):427–9.
3. Sams-Dodd F. Target-based drug discovery: is something wrong? Drug Discov Today. 2005;10(2):139–47.
4. Heilker R, Lessel U, Bischoff D. The power of combining phenotypic and target-focused drug discovery. Drug Discov Today. 2019;24(2):526–32.
5. Croston GE. The utility of target-based discovery. Expert Opin Drug Discovery. 2017;12(5):427–9.
6. Eder J, Sedrani R, Wiesmann C. The discovery of first-in-class drugs: origins and evolution. Nat Rev Drug Discov. 2014;13(8):577–87.
7. Zheng W, Thorne N, McKew JC. Phenotypic screens as a renewed approach for drug discovery. Drug Discov Today. 2013;18(21–22):1067–73.
8. Moffat JG, Rudolph J, Bailey D. Phenotypic screening in cancer drug discovery – past, present and future. Nat Rev Drug Discov. 2014;13(8):588–602.
9. Santos R, Urso O, Gaulton A, et al. A comprehensive map of molecular drug targets. Nat Rev Drug Discov. 2017;16(1):19–34.
10. Moffat JG, Vincent F, Lee JA, et al. Opportunities and challenges in phenotypic drug discovery: an industry perspective. Nat Rev Drug Discov. 2017;16(8):531–43.
11. Swamidass SJ, Schillebeeckx CN, Matlock M, et al. Combined analysis of phenotypic and target-based screening in assay networks. J Biomol Screen. 2014;19(5):782–90.
12. Peters J-U. Polypharmacology – foe or friend? J Med Chem. 2013;56(22):8955–71.
13. Terstappen GC, Schlüpen C, Raggiaschi R, et al. Target deconvolution strategies in drug discovery. Nat Rev Drug Discov. 2007;6(11):891–903.
14. Jones LH, Bunnage ME. Applications of chemogenomic library screening in drug discovery. Nat Rev Drug Discov. 2017;16(4):285–96.
15. Kumar A, Zhang KYJ. Hierarchical virtual screening approaches in small molecule drug discovery. Methods. 2015;71:26–37.
16. Chen YZ, Zhi DG. Ligand-protein inverse docking and its potential use in the computer search of protein targets of a small molecule. Proteins. 2001;43(2):217–26.
17. Swinney DC, Phenotypic vs. Target-based drug discovery for first-in-class medicines. Clin Pharmacol Ther. 2013;93(4):299–301.
18. Feng Y, Mitchison TJ, Bender A, et al. Multi-parameter phenotypic profiling: using cellular effects to characterize small-molecule compounds. Nat Rev Drug Discov. 2009;8(7):567–78.
19. Berg EL, Hytopoulos E, Plavec I, et al. Approaches to the analysis of cell signaling networks and their application in drug discovery. Curr Opin Drug Discov Dev. 2005;8(1):107–14.
20. Lee J, Bogyo M. Target deconvolution techniques in modern phenotypic profiling. Curr Opin Chem Biol. 2013;17(1):118–26.
21. Drews J. Drug discovery: a historical perspective. Science. 2000;287:1960–4.
22. Swinney DC, Anthony J. How were new medicines discovered? Nat Rev Drug Discov. 2011;10:507–19.
23. Mestres J, Gregori-Puigjane E, Valverde S, et al. The topology of drug-target interaction networks: implicit dependence on drug properties and target families. Mol BioSyst. 2009;5:1051–7.
24. Wang Z. MicroRNA-interference technologies. Heidelberg/New York: Springer-Verlag; 2009; ISBN-13: 978-3-642-00488-9

25. Lu Y, Xiao J, Lin H, et al. A single anti-microRNA antisense oligodeoxyribonucleotide (AMO) targeting multiple microRNAs offers an improved approach for microRNA interference. Nucleic Acids Res. 2009;37:e24–33.
26. Wang Z. New concepts of miRNAi technologies. In: MicroRNA interference technologies by Wang Z. Heidelberg: Springer-Verlag; 2009.
27. Wang Z. The concept of multiple-target anti-miRNA antisense oligonucleotides technology. Methods Mol Biol. 2011;676:51–7.
28. Wang Z. The concept of multiple-target anti-miRNA antisense oligonucleotides technology. In: Wu W, editor. MicroRNA and cancer. Totowa: Humana Press; 2011.
29. Allakhverdi Z, Allam M, Guimond A, et al. Multitargeted approach using antisense oligonucleotides for the treatment of asthma. Ann N Y Acad Sci. 2006;1082:62–73.
30. Zhang Y, Xie X, Ma W, et al. Multi-targeted antisense oligonucleotide delivery by a framework nucleic acid for inhibiting biofilm formation and virulence. Nano-Micro Lett. 2020;12(1):74.
31. Gao H, Xiao J, Yang B, et al. A single decoy oligodeoxynucleotides targeting multiple oncoproteins produces strong anti-cancer effects. Mol Pharmacol. 2006;70:1621–9.
32. Wang X, Liu Q, Hou B, et al. Concomitant targeting of multiple key transcription factors effectively disrupts cancer stem cells enriched in side population of human pancreatic cancer cells. PLoS One. 2013;8(9):e73942.
33. Batool M, Ahmad B, Choi S. A structure-based drug discovery paradigm. Int J Mol Sci. 2019;20(11):2783.
34. Anderson AC. The process of structure-based drug design. Chem Biol. 2003;10(9):787–97.
35. Reynolds CH, Merz KM, Ringe D, editors. Drug design: structure- and ligand-based approaches. 1st ed. Cambridge: Cambridge University Press; 2010.
36. Mauser H, Guba W. Recent developments in de novo design and scaffold hopping. Curr Opin Drug Discov Dev. 2008;11(3):365–74.
37. Batool M, Choi S. Proceedings of the 2017 IEEE Life Sciences Conference (LSC). In: Identification of druggable genome in staphylococcus aureus multidrug resistant strain. Sydney, NSW, Australia; 2017. p. 270–3.
38. Blaney J. A very short history of structure-based design: how did we get here and where do we need to go? J Comput Aided Mol Des. 2012;26:13–4.
39. Mandal S, Moudgil M, Mandal SK. Rational drug design. Eur J Pharm. 2009;625:90–100.
40. Wilson GL, Lill MA. Integrating structure-based and ligand-based approaches for computational drug design. Future Med Chem. 2011;3:735–50.
41. Urwyler S. Allosteric modulation of family c g-protein-coupled receptors: from molecular insights to therapeutic perspectives. Pharm Rev. 2011;63:59–126.
42. Fang Y. Ligand-receptor interaction platforms and their applications for drug discovery. Expert Opin Drug Discovery. 2012;7:969–88.
43. Ferreira LG, dos Santos RN, Oliva G, et al. Molecular docking and structure-based drug design strategies. Molecules. 2015;20:13384–421.
44. Wlodawer A, Vondrasek J. Inhibitors of HIV-1 protease: a major success of structure-assisted drug design. Annu Rev Biophys Biomol Struct. 1998;27:249–84.
45. Clark DE. What has computer-aided molecular design ever done for drug discovery? Expert Opin Drug Discovery. 2006;1:103–10.
46. Rutenber EE, Stroud RM. Binding of the anticancer drug zd1694 to *E. Coli* thymidylate synthase: assessing specificity and affinity. Structure. 1996;4:1317–24.
47. De Paulis T. Drug evaluation: Prx-00023, a selective 5-ht1a receptor agonist for depression. Curr Opin Investig Drugs. 2007;8:78–86.
48. Klebe G. Recent developments in structure-based drug design. J Mol Med. 2000;78(5):269–81.
49. Wang R, Gao Y, Lai L. LigBuilder: a multi-purpose program for structure-based drug design. J Mol Model. 2000;6(7–8):498–516.
50. Schneider G, Fechner U. Computer-based de novo design of drug-like molecules. Nat Rev Drug Discov. 2005;4(8):649–63.
51. Jorgensen WL. The many roles of computation in drug discovery. Science. 2004;303(5665):1813–8.
52. Meng X-Y, Zhang H-X, Mezei M, et al. Molecular docking: a powerful approach for structure-based drug discovery. Curr Comput Aided Drug Des. 2011;7(2):146–57.
53. Bajorath J. Integration of virtual and high-throughput screening. Nat Rev Drug Discov. 2002;1(11):882–94.
54. Walters WP, Stahl MT, Murcko MA. Virtual screening – an overview. Drug Discov Today. 1998;3:160–78.
55. Langer T, Hoffmann RD. Virtual screening: an effective tool for lead structure discovery? Curr Pharm Des. 2001;7(7):509–27.
56. Kitchen DB, Decornez H, Furr JR, et al. Docking and scoring in virtual screening for drug discovery: methods and applications. Nat Rev Drug Discov. 2004;3(11):935–49.
57. Gohlke H, Klebe G. Approaches to the description and prediction of the binding affinity of small-molecule ligands to macromolecular receptors. Angew Chem Int Ed Engl. 2002;41(15):2644–76.
58. Moitessier N, Englebienne P, Lee D, et al. Towards the development of universal, fast and highly accurate docking/scoring methods: a long way to go. Br J Pharmacol. 2008;153(Suppl 1):S7–26.
59. Shoichet BK, McGovern SL, Wei B, et al. Hits, leads and artifacts from virtual and high throughput screening. 2002. Mol Informatics: Confronting Complexity.

60. Bailey D, Brown D. High-throughput chemistry and structure-based design: survival of the smartest. Drug Discov Today. 2001;6(2):57–9.
61. Kuntz ID, Blaney JM, Oatley SJ, Langridge R, Ferrin TE. A geometric approach to macromolecule-ligand interactions. J Mol Biol. 1982;161(2):269–88.
62. Halperin I, Ma B, Wolfson H, et al. Principles of docking: an overview of search algorithms and a guide to scoring functions. Proteins. 2002;47(4):409–43.
63. Coupez B, Lewis RA. Docking and scoring--theoretically easy, practically impossible? Curr Med Chem. 2006;13(25):2995–3003.
64. Kontoyianni M, Madhav P, Suchanek E, et al. Theoretical and practical considerations in virtual screening: a beaten field? Curr Med Chem. 2008;15(2):107–16.
65. Brooijmans N, Kuntz ID. Molecular recognition and docking algorithms. Annu Rev Biophys Biomol Struct. 2003;32:335–73.
66. ten Brink T, Exner TE. Influence of protonation, tautomeric, and stereoisomeric states on protein-ligand docking results. J Chem Inf Model. 2009;49(6):1535–46.
67. Cross JB, Thompson DC, Rai BK, et al. Comparison of several molecular docking programs: pose prediction and virtual screening accuracy. J Chem Inf Model. 2009;49(6):1455–74.
68. Li X, Li Y, Cheng T, et al. Evaluation of the performance of four molecular docking programs on a diverse set of protein-ligand complexes. J Comput Chem. 2010;31(11):2109–25.
69. Plewczynski D, Lazniewski M, Augustyniak R, et al. Can we trust docking results? Evaluation of seven commonly used programs on PDBbind database. J Comput Chem. 2011;32(4):742–55.
70. Chen L, Morrow JK, Tran HT, et al. From laptop to benchtop to bedside: structure-based drug design on protein targets. Curr Pharm Des. 2012;18(9):1217–39.
71. Mostashari-Rad T, Arian R, Mehridehnavi A, et al. Study of CXCR4 chemokine receptor inhibitors using QSPR and molecular docking methodologies. J Theor Comput Chem. 2019;178(4)
72. Lengauer T, Rarey M. Computational methods for biomolecular docking. Curr Opin Struct Biol. 1996;6(3):402–6.
73. Jorgensen WL. Rusting of the lock and key model for protein-ligand binding. Science. 1991;254(5034):954–5.
74. Wei BQ, Weaver LH, Ferrari AM, et al. Testing a flexible-receptor docking algorithm in a model binding site. J Mol Biol. 2004;337(5):1161–82.
75. Goldman BB, Wipke WT. QSD quadratic shape descriptors. 2. Molecular docking using quadratic shape descriptors (QSDock). Proteins. 2000;38(1):79–94.
76. Meng EC, Shoichet BK, Kuntz ID. Automated docking with grid-based energy evaluation. J Comput Chem. 1992;13(4):505–24.
77. Morris GM, Goodsell DS, Halliday RS, et al. Automated docking using a Lamarckian genetic algorithm and an empirical binding free energy function. J Comput Chem. 1998;19(14):1639–62.
78. Feig M, Onufriev A, Lee MS, et al. Performance comparison of generalized born and Poisson methods in the calculation of electrostatic solvation energies for protein structures. J Comput Chem. 2004;25(2):265–84.
79. McConkey BJ, Sobolev V, Edelman M. The performance of current methods in ligand-protein docking. Curr Sci. 2002;83:845–55.
80. Goodford PJ. A computational procedure for determining energetically favorable binding sites on biologically important macromolecules. J Med Chem. 1985;28(7):849–57.
81. Kastenholz MA, Pastor M, Cruciani G, et al. GRID/CPCA: a new computational tool to design selective ligands. J Med Chem. 2000;43(16):3033–44.
82. Levitt DG, Banaszak LJ. POCKET: a computer graphics method for identifying and displaying protein cavities and their surrounding amino acids. J Mol Graph. 1992;10(4):229–34.
83. Laskowski RA. SURFNET: a program for visualizing molecular surfaces, cavities, and intermolecular interactions. J Mol Graph. 1995;13(5):323–330, 307–328.
84. Glaser F, Morris RJ, Najmanovich RJ, et al. A method for localizing ligand binding pockets in protein structures. Proteins. 2006;62(2):479–48.
85. Brady GP Jr, Stouten PF. Fast prediction and visualization of protein binding pockets with PASS. J Comput Aided Mol Des. 2000;14(4):383–401.
86. Mezei M. A new method for mapping macromolecular topography. J Mol Graph Model. 2003;21(5):463–72.
87. Fischer E. Einfluss der configuration auf die wirkung derenzyme. Ber Dt Chem Ges. 1894;27:2985–93.
88. Koshland DE Jr. Correlation of structure and function in enzyme action. Science. 1963;142:1533–41.
89. Hammes GG. Multiple conformational changes in enzyme catalysis. Biochemistry. 2002;41(26):8221–8.
90. Lionta E, Spyrou G, Vassilatis DK, et al. Structure-based virtual screening for drug discovery: principles, applications and recent advances. Curr Top Med Chem. 2014;14:1923–38.
91. Kalyaanamoorthy S, Chen YP. Structure-based drug design to augment hit discovery. Drug Discov Today. 2011;16:831–9.
92. Searls DB. Data integration: challenges for drug discovery. Nat Rev Drug Discov. 2005;4:45–58.
93. Scotti L, Mendonca Junior FJ, Ishiki HM, et al. Docking studies for multi-target drugs. Curr Drug Targets. 2017;18(5):592–604.
94. Dutta D, Das R, Mandal C, et al. Structure-based kinase profiling to understand the polypharmacological behavior of therapeutic molecules. J Chem Inf Model. 2018;58(1):68–89.

95. Jalencas X, Mestres J. Identification of similar binding sites to detect distant polypharmacology. Mol Inf. 2013;32(11–12):976–90.
96. Salentin S, Haupt VJ, Daminelli S, et al. Polypharmacology rescored: protein-ligand interaction profiles for remote binding site similarity assessment. Prog Biophys Mol Biol. 2014;116(2–3):174–86.
97. Koch O. Use of secondary structure element information in drug design: polypharmacology and conserved motifs in protein-ligand binding and protein-protein interfaces. Future Med Chem. 2011;3(6):699–708.
98. De Vita S, Lauro G, Ruggiero D, et al. Protein preparation automatic protocol for high-throughput inverse virtual screening: accelerating the target identification by computational methods. J Chem Inf Model. 2019;59(11):4678–90.
99. Grant MA. Protein structure prediction in structure-based ligand design and virtual screening. Comb Chem High Throughput Screen. 2009;12:940–60.
100. Song CM, Lim SJ, Tong JC. Recent advances in computer-aided drug design. Brief Bioinform. 2009;10:579–91.
101. Krieger E, Joo K, Lee J, et al. Improving physical realism, stereochemistry, and side-chain accuracy in homology modeling: four approaches that performed well in casp8. Proteins. 2009;77:114–22.
102. Bordoli L, Kiefer F, Arnold K, et al. Protein structure homology modeling using swiss-model workspace. Nat Protoc. 2009;4:1–13.
103. Potapov V, Cohen M, Inbar Y, et al. Protein structure modelling and evaluation based on a 4-distance description of side-chain interactions. BMC Bioinf. 2010;11:374.
104. Laurie AT, Jackson RM. Q-sitefinder: an energy-based method for the prediction of protein-ligand binding sites. Bioinformatics. 2005;21:1908–16.
105. Wunberg T, Hendrix M, Hillisch A, et al. Improving the hit-to-lead process: data-driven assessment of drug-like and lead-like screening hits. Drug Discov Today. 2006;11:175–80.
106. Lavecchia A, di Giovanni C. Virtual screening strategies in drug discovery: a critical review. Curr Med Chem. 2013;20:2839–60.
107. Shoichet BK. Virtual screening of chemical libraries. Nature. 2004;432:862–5.
108. Phatak SS, Stephan CC, Cavasotto CN. High-throughput and in silico screenings in drug discovery. Expert Opin Drug Discovery. 2009;4:947–59.
109. Reddy AS, Pati SP, Kumar PP, et al. Virtual screening in drug discovery—a computational perspective. Curr Protein Pept Sci. 2007;8:329–51.
110. Pedretti A, Mazzolari A, Gervasoni S, et al. Rescoring and linearly combining: a highly effective consensus strategy for virtual screening campaigns. Int J Mol Sci. 2019;20:2060.
111. Gillet VJ. New directions in library design and analysis. Curr Opin Chem Biol. 2008;12:372–8.
112. Lopez-Vallejo F, Caulfield T, Martinez-Mayorga K, et al. Integrating virtual screening and combinatorial chemistry for accelerated drug discovery. Comb Chem High Throughput Screen. 2011;14:475–87.
113. Sousa SF, Fernandes PA, Ramos MJ. Protein-ligand docking: current status and future challenges. Proteins. 2006;65:15–26.
114. Taylor RD, Jewsbury PJ, Essex JW. A review of protein-small molecule docking methods. J Comput Aided Mol Des. 2002;16:151–66.
115. Moitessier N, Englebienne P, Lee D, et al. Towards the development of universal, fast and highly accurate docking/scoring methods: a long way to go. Br J Pharm. 2008;153:7–26.
116. Guedes IA, Pereira FSS, Dardenne LE. Empirical scoring functions for structure-based virtual screening: applications, critical aspects, and challenges. Front Pharm. 2018;9:1089.
117. Li H, Peng J, Leung Y, et al. The impact of protein structure and sequence similarity on the accuracy of machine-learning scoring functions for binding affinity prediction. Biomol Ther. 2018;8:12.
118. David H, Gary BF. Computational intelligence methods for docking scores. Curr Comput Aided Drug Des. 2009;5:56–68.
119. Huang SY, Zou X. Inclusion of solvation and entropy in the knowledge-based scoring function for protein-ligand interactions. J Chem Inf Model. 2010;50:262–73.
120. Raub S, Steffen A, Kamper A, et al. Aiscore chemically diverse empirical scoring function employing quantum chemical binding energies of hydrogen-bonded complexes. J Chem Inf Model. 2008;48:1492–510.
121. Seifert MH. Targeted scoring functions for virtual screening. Drug Discov Today. 2009;14:562–9.
122. Evans BE, Rittle KE, Bock MG, et al. Methods for drug discovery: development of potent, selective, orally effective cholecystokinin antagonists. J Med Chem. 1988;31:2235.
123. Horton DA, Bourne GT, Smythe ML, et al. The combinatorial synthesis of bicyclic privileged structures or privileged substructures. Chem Rev. 2003;103:893–930.
124. Ioan P, Carosati E, Micucci M, et al. 1,4-Dihydropyridine scaffold in medicinal chemistry, the story so far and perspectives (part 1): action in ion channels and GPCRs. Curr Med Chem. 2011;18:4901–22.
125. Jacobson KA, Tosh DK, Toti KS, et al. Polypharmacology of conformationally locked methanocarba nucleosides. Drug Discov Today. 2017;22(12):1782–91.
126. Costantino L, Daniela B. Privileged structures as leads in medicinal chemistry. Front Med Chem. 2010;5:381–422.
127. Welsch ME, Snyder SA, Stockwell BR, et al. Privileged scaffolds for library design and drug discovery. Curr Opin Chem Biol. 2010;14:347–61.

128. Kombarov R, Altieri A, Genis D, et al. Bio cores: identification of a drug/natural product-based privileged structural motif for small-molecule lead discovery. Mol Divers. 2010;14:193–200.
129. Jordheim LP, Durantel D, Zoulim F, et al. Advances in the development of nucleoside and nucleotide analogues for cancer and viral diseases. Nat Rev Drug Discov. 2013;12:447–64.
130. Arshadi S, Vessally E, Edjlali L, et al. N-Propargylamines: versatile building blocks in the construction of thiazole cores. Beilstein J Org Chem. 2017;13:625–38.
131. Zindo FT, Joubert J, Malan SF. Propargylamine as functional moiety in the design of multifunctional drugs for neurodegenerative disorders: MAO inhibition and beyond. Future Med Chem. 2015;7(5):609–29.
132. Lauder K, Toscani A, Scalacci N, et al. Synthesis and reactivity of propargylamines in organic chemistry. Chem Rev. 2017;117(24):14091–200.
133. Gaumont A-C, Gulea M, Levillain J. Overview of the chemistry of 2-thiazolines. Chem Rev. 2009;109:1371–401.
134. Havrylyuk D, Kovach N, Zimenkovsky B, et al. Synthesis and anticancer activity of isatin-based pyrazolines and thiazolidines conjugates. Arch Pharm. 2011;344:514–22.
135. Baby B, Antony P, Al Halabi W, et al. Structural insights into the polypharmacological activity of quercetin on serine/threonine kinases. Drug Des Devel Ther. 2016;10:3109–23.
136. Rakesh KP, Shantharam CS, Sridhara MB, et al. Benzisoxazole: a privileged scaffold for medicinal chemistry. MedChemComm. 2017;8(11):2023–39.
137. Eguch S. Bioactive Heterocycles II. In: Topics in Heterocyclic Chemistry. 2007; Vol 8(XII): p 249.
138. Gomtsyan A. Heterocycles in drugs and drug discovery. Chem Heterocycl Compd. 2012;48:7–10.
139. Kirk KL, Filler R. Biomedical frontiers of fluorine chemistry, Symposium series, vol. 639. Washington, DC: American Chemical Society; 1996. p. 1–24.
140. Gelders YG, Heylen SLE, Vander BG, et al. Pilot clinical investigation of risperidone in the treatment of psychotic patients. J Pharmacopsychiatry. 1990;23:206–11.
141. Dollery C. Therapeutic drugs. Edinburgh: Churchill Livingstone; 1999.
142. Park BK, Kitteringham NR. Effects of fluorine substitution on drug metabolism: pharmacological and toxicological implications. Drug Metab Rev. 1994;26:605–43.
143. Uto Y. 1,2-Benzisoxazole: a privileged structure with a potential for polypharmacology. Curr Pharm Des. 2016;22(21):3201–11.
144. Qin J, Cheng W, Duan YT, et al. Indazole as a privileged scaffold: the derivatives and their therapeutic applications. Anti Cancer Agents Med Chem. 2021;21(7):839–60.
145. Denya I, Malan SF, Joubert J. Indazole derivatives and their therapeutic applications: A patent review (2013-2017). Expert Opin Ther Pat. 2018;28(6):441–53.
146. Wan Y, He S, Li W, et al. Indazole derivatives: promising anti-tumor agents. Anti Cancer Agents Med Chem. 2018;18(9):1228–34.
147. Dong J, Zhang Q, Wang Z, et al. Recent advances in the development of indazole-based anticancer agents. ChemMedChem. 2018;13(15):1490–507.
148. Zhang SG, Liang CG, Zhang WH. Recent advances in indazole-containing derivatives: synthesis and biological perspectives. Molecules. 2018;23(11):2783.
149. Sharma A, Kumar V, Kharb R, et al. Imidazole derivatives as potential therapeutic agents. Curr Pharm Des. 2016;22(21):3265–301.
150. Chopra B, Dhingra AK, Prasad DN. Imidazole: an emerging scaffold showing its therapeutic voyage to develop valuable molecular entities. Curr Drug Res Rev. 2020;12(2):103–17.
151. Chopra PN, Sahu JK. Biological significance of imidazole-based analogues in new drug development. Curr Drug Discov Technol. 2020;17(5):574–84.
152. Alghamdi SS, Suliman RS, Almutairi K, et al. Imidazole as a promising medicinal scaffold: current status and future direction. Drug Des Dev Ther. 2021;15:3289–312.
153. Alam MM, Hassan AHE, Kwon YH, et al. Design, synthesis and evaluation of alkylphosphocholine-gefitinib conjugates as multitarget anticancer agents. Arch Pharm Res. 2018;41(1):35–45.
154. Das R, Mehta DK, Dhanawat M. Bestowal of quinazoline scaffold in anticancer drug discovery. Anti Cancer Agents Med Chem. 2021;21(11):1350–68.
155. Ismail RSM, Ismail NSM, Abuserii S. Recent advances in 4-aminoquinazoline based scaffold derivatives targeting EGFR kinases as anticancer agents. Future J Pharmaceut Sci. 2016;2(1):9–19.
156. Wermuth CG, Ganellin CR, Lindberg P, et al. Glossary of terms used in medicinal chemistry (IUPAC recommendations 1997). Annu Rep Med Chem. 1998;33:385–95.
157. Trainor G. Privileged structures – an update. Annu Rep Med Chem. 2000;35:289–98.
158. Sheridan RP, Miller MD. A method for visualizing recurrent topological substructures in sets of active molecules. J Chem Inf Comput Sci. 1998;38:915–24.
159. Thompson LA, Ellman JA. Synthesis and applications of small molecule libraries. Chem Rev. 1966;96:555–600.
160. Wermuth CG. Search for new lead compounds: the example of the chemical and pharmacological dissection of aminopyridazines. J Heterocyclic Chem. 1998;35:1091–100.
161. Hajduk PJ, Bures M, Praestgaard J, et al. Privileged molecules for protein binding identified from NMR-based screening. J Med Chem. 2000;43:3443–7.
162. Wermuth CG. Pharmacophores: historical perspective and viewpoint from a medicinal chemist. In: Langer T, Hoffmann RD, editors. Pharmacophores and pharmacophore searches. Weinheim:

WILEY-VCH Verlag GmbH & Co. KGaA; 2006. ISBN3-527-31250-1.
163. Wermuth CG, Langer T. Pharmacophore identification. In: Kubinyi H, editor. 3D QSAR in drug design. Theory methods and applications. Leiden: ESCOM; 1993. p. 117–36.
164. Duarte CD, Barreiro EJ, Fraga CAM. Privileged structures: a useful concept for the rational design of new lead drug candidates. Mini Rev Med Chem. 2007;7(11):1108–19.
165. Pradeepkiran JA. Pharmacophore-based models for therapeutic drugs against phosphorylated tau in Alzheimer's disease. Drug Discov Today. 2019;24(2):616–23.
166. Wermuth CG. The impact of QSAR and CADD methods in drug design. In: Hoeltje HD, Sippl W, editors. Rational approaches to drug design. Barcelona: Prous Science; 2001. p. 3–20.
167. Schneider G, Giller T, Neidhart W, et al. "Scaffold-hopping" by topological pharmacophore search: a contribution to virtual screening. Angew Chem Int Ed. 1999;38:2894–6.
168. Rognan D, Boulanger T, Hoffmann R, et al. Structure and molecular modeling of GABAA antagonists. J Med Chem. 1992;35:1969–77.
169. Ariëns EJ, Rodrigues de Miranda JF, Simonis AM. The pharmacon-receptor–effector concept: a basis for understanding the transmission of information in biological systems. In: O'Brien RD, editor. The receptors. New York: Plenum Press; 1979. p. 33–91.
170. Wermuth CG. Aminopyridazines–an alternative route to potent muscarinic agonists with no cholinergic syndrome. Farmaco. 1993;48:253–74.
171. Rognan D, Sokoloff P, Mann A, et al. Optically active benzamides as predictive tools for mapping the dopamine D2 receptor. Eur J Pharmacol Mol Pharmacol Sect. 1990;3:59–70.
172. Li Q. Application of fragment-based drug discovery to versatile targets. Front Mol Biosci. 2020;7:180.
173. Price AJ, Howard S, Cons BD. Fragment-based drug discovery and its application to challenging drug targets. Essays Biochem. 2017;61(5):475–84.
174. Brunst S, Kramer JS, Kilu W, et al. Systematic assessment of fragment identification for multitarget drug design. ChemMedChem. 2021;16(7):1088–92.
175. Sancineto L, Massari S, Iraci N, et al. From small to powerful: the fragments universe and its "chem-appeal". Curr Med Chem. 2013;20(11):1355–81.
176. Murray CW, Rees DC. The rise of fragment-based drug discovery. Nat Chem. 2009;1:187–92.
177. Doak BC, Norton RS, Scanlon MJ. The ways and means of fragment-based drug design. Pharmacol Ther. 2016;167:28–37.
178. Erlanson DA, Fesik SW, Hubbard RE, et al. Twenty years on: the impact of fragments on drug discovery. Nat Rev Drug Discov. 2016;15:605–19.
179. Harner MJ, Frank AO, Fesik SW. Fragment-based drug discovery using NMR spectroscopy. J Biomol NMR. 2013;56:65–75.
180. Erlanson DA. Introduction to fragment-based drug discovery. In: Davies TG, Hyvönen M, editors. Fragment-based drug discovery and X-ray crystallography. Berlin: Springer; 2012. p. 1–32.
181. Gee CT, Arntson KE, Urick AK, et al. Protein-observed 19F-NMR for fragment screening, affinity quantification and druggability assessment. Nat Protoc. 2016;11:1414.
182. Aretz J, Kondoh Y, Honda K, et al. Chemical fragment arrays for rapid druggability assessment. Chem Commun. 2016;52:9067–70.
183. Dang CV, Reddy EP, Shokat KM, et al. Drugging the 'undruggable' cancer targets. Nat Rev Cancer. 2017;17:502–8.
184. Tounge BA, Parker MH. Designing a diverse high-quality library for crystallography-based FBDD screening. Fragment-based drug design – tools, practical approaches, and examples. Methods Enzymol. 2011;493:3–20.
185. Congreve M, Carr R, Murray C, et al. A 'rule of three' for fragment-based lead discovery? Drug Discov Today. 2003;8:876–7.
186. Jhoti H, Williams G, Rees DC, et al. The 'rule of three' for fragment-based drug discovery: where are we now? Nat Rev Drug Discov. 2013;12:644.
187. Sabbah M, Mendes V, Vistal RG, et al. Fragment-based design of mycobacterium tuberculosis InhA inhibitors. J Med Chem. 2020;63:4749–61.
188. Singh M, Tam B, Akabayov B. NMR-fragment based virtual screening: a brief overview. Molecules. 2018;23:233.
189. Bottegoni G, Favia AD, Recanatini M, et al. The role of fragment-based and computational methods in polypharmacology. Drug Discov Today. 2012;17(1–2):23–34.
190. Frett B, Carlomagno F, Moccia ML, et al. Fragment-based discovery of a dual pan-RET/VEGFR2 kinase inhibitor optimized for single-agent polypharmacology. Angew Chem Int Ed Engl. 2015;54(30):8717–21.
191. Viegas-Junior C, Danuello A, da Silva BV, et al. Molecular hybridization: a useful tool in the design of new drug prototypes. Curr Med Chem. 2007;14(17):1829–52.
192. de Oliveira PM, Duarte da Cruz RM, de Oliveira VJ, et al. Hybrid compounds as direct multi-target ligands: A review. Curr Top Med Chem. 2017;17(9):1044–79.
193. Ivasiv V, Albertini C, Gonçalves AE, et al. Molecular hybridization as a tool for designing multitarget drug candidates for complex diseases. Curr Top Med Chem. 2019;19(19):1694–711.
194. Bérubé G. An overview of molecular hybrids in drug discovery. Expert Opin Drug Discovery. 2016;11(3):281–305.
195. Decker M, editor. Design of hybrid molecules for drug development. Elsevier; 2017. ISBN:978-0-08-101011-2

196. Sunil RJ, Sarbani P, Jayashree A. Molecular hybridization – an emanating tool in drug design. Med Chem (Los Angeles). 2019;9:6.
197. Berger AB, Vitorino PM, Bogyo M. Activity-based protein profiling: applications to biomarker discovery, in vivo imaging and drug discovery. Am J Pharmacogenomics. 2004;4(6):371–81.
198. Liu Y, Patricelli MP, Cravatt BF Activity-based protein profiling: the serine hydrolases. Proc Natl Acad Sci U S A 1999; 96(26): 14694–9.
199. Saghatelian A, Jessani N, Joseph A, et al. Activity-based probes for the proteomic profiling of metalloproteases. Proc Natl Acad Sci U S A. 2004;101(27):10000–5.
200. Evans MJ, Cravatt BF. Mechanism-based profiling of enzyme families. Chem Rev. 2006;106:3279–301.
201. Nomura DK, Dix MM, Cravatt BF. Activity-based protein profiling for biochemical pathway discovery in cancer. Nat Rev Cancer. 2010;10:630–8.
202. Speers AE, Adam GC, Cravatt BF. Activity-based protein profiling in vivo using a copper(i)-catalyzed azide-alkyne [3 + 2] cycloaddition. J Am Chem Soc. 2003;125(16):4686–7.
203. Puri AW, Lupardus PJ, Deu E, et al. Rational design of inhibitors and activity-based probes targeting Clostridium difficile virulence factor TcdB. Chem Biol. 2010;17:1201–11.
204. Deu E, Leyva MJ, Albrow VE, et al. Functional studies of Plasmodium falciparum dipeptidyl aminopeptidase I using small molecule inhibitors and active site probes. Chem Biol. 2010;17:808–19.
205. Arastu-Kapur S, Ponder EL, Fonovic UP, et al. Identification of proteases that regulate erythrocyte rupture by the malaria parasite Plasmodium falciparum. Nat Chem Biol. 2008;4:203–13.
206. Hall CI, Reese ML, Weerapana E, et al. Chemical genetic screen identifies Toxoplasma DJ-1 as a regulator of parasite secretion, attachment, and invasion. Proc Natl Acad Sci U S A. 2011;108:10568–73.
207. Wang S, Tian Y, Wang M. Advanced activity-based protein profiling application strategies for drug development. Front Pharmacol. 2018;9:353.
208. Leriche G, Chisholm L, Wagner A. Cleavable linkers in chemical biology. Bioorg Med Chem. 2012;20(2):571–82.
209. Rudolf GC, Heydenreuter W, Sieber SA. Chemical proteomics: ligation and cleavage of protein modifications. Curr Opin Chem Biol. 2013;17(1):110–7.
210. Liao LX, Song XM, Wang LC, et al. Highly selective inhibition of IMPDH2 provides the basis of antineuroinflammation therapy. Proc Natl Acad Sci U S A. 2017;114(29):E5986–94.
211. Nasheri N, Joyce M, Rouleau Y, et al. Modulation of fatty acid synthase enzyme activity and expression during hepatitis C virus replication. Chem Biol. 2013;20(4):570–82.
212. Crump CJ, am Ende CW, Ballard TE, et al. Development of clickable active site-directed photoaffinity probes for gamma-secretase. Bioorg Med Chem Lett. 2012;22:2997–3000.
213. Sieber SA, Niessen S, Hoover HS, et al. Proteomic profiling of metalloprotease activities with cocktails of active-site probes. Nat Chem Biol. 2006;2:274–81.
214. Geurink PP, Florea BI, Van der Marel GA, et al. Probing the proteasome cavity in three steps: bio-orthogonal photo-reactive suicide substrates. Chem Commun (Camb). 2010;46:9052–4.
215. Barglow KT, Saikatendu KS, Bracey MH, et al. Functional proteomic and structural insights into molecular recognition in the nitrilase family enzymes. Biochemistry. 2008;47:13514–23.
216. Pace NJ, Pimental DR, Weerapana E. An inhibitor of glutathione S-transferase omega 1 that selectively targets apoptotic cells. Angew Chem Int Ed Engl. 2012;51:8365–8.
217. Wright AT, Song JD, Cravatt BF. A suite of activity-based probes for human cytochrome P450 enzymes. J Am Chem Soc. 2009;131:10692–700.
218. Saario SM, McKinney MK, Speers AE, et al. Clickable, photoreactive inhibitors to probe the active site microenvironment of fatty acid amide hydrolase. Chem Sci. 2012;3:77–83.
219. van der Linden WA, Li N, Hoogendoorn S, et al. Two-step bioorthogonal activity-based proteasome profiling using copper-free click reagents: a comparative study. Bioorg Med Chem. 2012;20:662–6.
220. Kulkarni SS, Hu X, Doi K, et al. Screening of protein-protein interaction modulators via sulfo-click kinetic target-guided synthesis. ACS Chem Biol. 2011;6:724–32.
221. Berry AF, Heal WP, Tarafder AK, et al. Rapid multilabel detection of geranylgeranylated proteins by using bioorthogonal ligation chemistry. Chembiochem. 2010;11:771–3.
222. Willems LI, Verdoes M, Florea BI, et al. Two-step labeling of endogenous enzymatic activities by Diels-Alder ligation. Chembiochem. 2010;11:1769–81.
223. Shields DJ, Niessen S, Murphy EA, et al. RBBP9: a tumor-associated serine hydrolase activity required for pancreatic neoplasia. Proc Natl Acad Sci U S A. 2010;107:2189–94.
224. Chiang KP, Niessen S, Saghatelian A, et al. An enzyme that regulates ether lipid signaling pathways in cancer annotated by multidimensional profiling. Chem Biol. 2006;13:1041–50.
225. Nomura DK, Long JZ, Niessen S, et al. Monoacylglycerol lipase regulates a fatty acid network that promotes cancer pathogenesis. Cell. 2010;140:49–61.
226. Bachovchin DA, Brown SJ, Rosen H, et al. Identification of selective inhibitors of uncharacterized enzymes by high-throughput screening with fluorescent activity-based probes. Nat Biotechnol. 2009;27:387–94.
227. Deu E, Yang Z, Wang F, et al. Use of activity-based probes to develop high throughput screening assays that can be performed in complex cell extracts. PLoS One. 2010;5:e11985.
228. Bachovchin DA, Mohr JT, Speers AE, et al. Academic cross-fertilization by public screening

yields a remarkable class of protein phosphatase methylesterase-1 inhibitors. Proc Natl Acad Sci U S A. 2011;108:6811–6.
229. Paananen J, Fortino V. An omics perspective on drug target discovery platforms. Brief Bioinform. 2020;21(6):1937–53.
230. Mun J, Choi G, Lim B. A guide for bioinformaticians: 'omics-based drug discovery for precision oncology. Drug Discov Today. 2020;2020:S1359.
231. Matthews H, Hanison J, Nirmalan N. "Omics"-informed drug and biomarker discovery: opportunities, challenges and future perspectives. Proteomes. 2016;4(3):28.
232. Simon R, Roychowdhury S. Implementing personalized cancer genomics in clinical trials. Nat Rev Drug Discov. 2013;12(5):358–69.
233. Cannon ME, Mohlke KL. Deciphering the emerging complexities of molecular mechanisms at GWAS Loci. Am J Hum Genet. 2018;103(5):637–53.
234. Uenaka T, Satake W, Cha P-C, et al. *In silico* drug screening by using genome-wide association study data repurposed dabrafenib, an anti-melanoma drug, for Parkinson's disease. Hum Mol Genet. 2018;27:3974–85.
235. Clough E, Barrett T. The gene expression omnibus database. Methods Mol Biol. 2016;1418:93–110.
236. GTEx Consortium. Human genomics. The genotype-tissue expression (GTEx) pilot analysis: multitissue gene regulation in humans. Science. 2015;348(6235):648–60.
237. Jones LH, Neubert H. Clinical chemoproteomics-opportunities and obstacles. Sci Transl Med. 2017;9(386):eaaf7951.
238. Nguyen C, West GM, Geoghegan KF. Emerging methods in chemoproteomics with relevance to drug discovery. Methods Mol Biol. 2017;1513:11–22.
239. Wishart DS. Emerging applications of metabolomics in drug discovery and precision medicine. Nat Rev Drug Discov. 2016;15(7):473–84.
240. Wang P, Shehu AI, Ma X. The opportunities of metabolomics in drug safety evaluation. Curr Pharmacol Rep. 2017;3(1):10–5.
241. Kale NS, Haug K, Conesa P, et al. MetaboLights: an open-access database repository for metabolomics data. Curr Protoc Bioinformatics. 2016;53:14.13.1–14.13.18.
242. Jacoby E. Chemogenomics: drug discovery's panacea? Mol BioSyst. 2006;2(5):218–20.
243. Zheng XF, Chan TF. Chemical genomics: a systematic approach in biological research and drug discovery. Curr Issues Mol Biol. 2002;4(2):33–43.
244. Gaither LA. Chemogenomics approaches to novel target discovery. Expert Rev Proteomics. 2007;4(3):411–9.
245. Hsieh YY, Huang TC, Lo HL, et al. Systematic discovery of drug action mechanisms by an integrated chemical genomics approach: identification of functional disparities between azacytidine and decitabine. Oncotarget. 2016;7(19):27363–78.
246. Bajorath J. Computational approaches in chemogenomics and chemical biology: current and future impact on drug discovery. Expert Opin Drug Discovery. 2008;3(12):1371–6.
247. Hasin Y, Seldin M, Lusis A. Multi-omics approaches to disease. Genome Biol. 2017;18(1):83.
248. Org E, Parks BW, Joo JW, et al. Genetic and environmental control of host-gut microbiota interactions. Genome Res. 2015;25:1558–69.
249. Caporaso JG, Kuczynski J, Stombaugh J, et al. QIIME allows analysis of high-throughput community sequencing data. Nat Methods. 2010;7:335–6.
250. Org E, Mehrabian M, Lusis AJ. Unraveling the environmental and genetic interactions in atherosclerosis: central role of the gut microbiota. Atherosclerosis. 2015;241:387–99.
251. Klindworth A, Pruesse E, Schweer T, et al. Evaluation of general 16S ribosomal RNA gene PCR primers for classical and next-generation sequencing-based diversity studies. Nucleic Acids Res. 2013;41:e1.
252. Scholz MB, Lo CC, Chain PS. Next generation sequencing and bioinformatic bottlenecks: the current state of metagenomic data analysis. Curr Opin Biotechnol. 2012;23:9–15.
253. Lu M, Zhan X. The crucial role of multiomic approach in cancer research and clinically relevant outcomes. EPMA J. 2018;9(1):77–102.
254. Curtis C, Shah SP, Chin SF, et al. The genomic and transcriptomic architecture of 2,000 breast tumours reveals novel subgroups. Nature. 2012;486(7403):346–52.
255. Stemke-Hale K, Gonzalez-Angulo AM, Lluch A, et al. An integrative genomic and proteomic analysis of PIK3CA, PTEN, and AKT mutations in breast cancer. Cancer Res. 2008;68(15):6084–91.
256. Dazert E, Colombi M, Boldanova T, et al. Quantitative proteomics and phosphoproteomics on serial tumor biopsies from a sorafenib-treated HCC patient. Proc Natl Acad Sci U S A. 2016;113(5):1381–6.
257. Berger SI, Iyengar R. Network analyses in systems pharmacology. Bioinformatics. 2009;25(19):2466–72.
258. Wu Z, Li W, Liu G. Network-based methods for prediction of drug-target interactions. Front Pharmacol. 2018;9:1134.
259. Li P, Fu Y, Yonghua WY. Network based approach to drug discovery: a mini review. Mini Rev Med Chem. 2015;15(8):687–95.
260. Jayachandran P, Ilango S, Paital B, et al. Approaches for network based drug discovery. Front Biosci (Schol Ed). 2021;13(1):30–43.
261. Engin HB, Gursoy A, Nussinov R, et al. Network-based strategies can help mono- and polypharmacology drug discovery: a systems biology view. Curr Pharm Des. 2014;20(8):1201–7.
262. Sidders B, Karlsson A, Kitching L, et al. Network-based drug discovery: coupling network pharmacology with phenotypic screening for neuronal excitability. J Mol Biol. 2018;430(18 Pt A):3005–15.

263. Janga SC, Tzakos A. Structure and organization of drug-target networks: insights from genomic approaches for drug discovery. Mol BioSyst. 2009;5(12):1536–48.
264. Yildirim MA, Goh KI, Cusick ME, et al. Drug-target network. Nat Biotechnol. 2007;25(10):1119–26.
265. Cheng T, Li Q, Wang Y, et al. Identifying compound-target associations by combining bioactivity profile similarity search and public databases mining. J Chem Inf Model. 2011;51(9):2440–8.
266. Tabei Y, Kotera M, Sawada R, et al. Network-based characterization of drug-protein interaction signatures with a space-efficient approach. BMC Syst Biol. 2019;13(Suppl 2):39.
267. Tang J, Aittokallio T. Network pharmacology strategies toward multi-target anticancer therapies: from computational models to experimental design principles. Curr Pharm Des. 2014;20(1):23–36.
268. Hu G, Agarwal P. Human disease-drug network based on genomic expression profiles. PLoS One. 2009;4(8):e6536.
269. Lounkine E, Keiser MJ, Whitebread S, et al. Large-scale prediction and testing of drug activity on side-effect targets. Nature. 2012;486:361–7.
270. Hung CL, Chen CC. Computational approaches for drug discovery. Drug Dev Res. 2014;75(6):412–8.
271. Chaudhari R, Fong LW, Tan Z, et al. An up-to-date overview of computational polypharmacology in modern drug discovery. Expert Opin Drug Discovery. 2020;15(9):1025–44.
272. Chaudhari R, Tan Z, Huang B, et al. Computational polypharmacology: a new paradigm for drug discovery. Expert Opin Drug Discovery. 2017;12(3):279–91.
273. Plake C, Schroeder M. Computational polypharmacology with text mining and ontologies. Curr Pharm Biotechnol. 2011;12(3):449–57.
274. Wathieu H, Issa NT, Byers SW, et al. Harnessing polypharmacology with computer-aided drug design and systems biology. Curr Pharm Des. 2016;22(21):3097–108.
275. Lauria A, Bonsignore R, Bortolotta R, Perricone U, Martorana A, Gentile C. Drugs polypharmacology by in silico methods: new opportunities in drug discovery. Curr Pharm Des. 2016;22(21):3073–81.
276. Guner OF. Pharmacophore perception, development, and use in drug design. La Jolla: International University Line; 2000. ISBN 978-0-9636817-6-8
277. Tropsha A. QSAR in Drug Discovery. In: Reynolds CH, Merz KM, Ringe D, editors. Drug design: structure- and ligand-based approaches. 1st ed. Cambridge: Cambridge University Press; 2010. p. 151–64.
278. Willett P, Barnard JM, Downs GM. Chemical similarity searching. J Chem Inform Comput Sci. 1998;38:983–96.
279. Wang C, Kurgan L. Survey of similarity-based prediction of drug-protein interactions. Curr Med Chem. 2019;27:5856–86.
280. Hu G, Kuang G, Xiao W, et al. Performance evaluation of 2D fingerprint and 3D shape similarity methods in virtual screening. J Chem Inform Model. 2012;52:1103–13.
281. Willett P. Similarity-based virtual screening using 2D fingerprints. Drug Discov Today. 2006;11:1046–53.
282. Campillos M, Kuhn M, Gavin AC, et al. Drug target identification using side-effect similarity. Science. 2008;321:263–6.
283. Gong J, Cai C, Liu X, et al. ChemMapper: a versatile web server for exploring pharmacology and chemical structure association based on molecular 3D similarity method. Bioinformatics. 2013;29:1827–9.
284. Keiser MJ, Roth BL, Armbruster BN, et al. Relating protein pharmacology by ligand chemistry. Nat Biotechnol. 2007;25:197–206.
285. Cao R, Wang Y. Predicting molecular targets for small-molecule drugs with a ligand-based interaction fingerprint approach. ChemMedChem. 2016;11(12):1352–61.
286. Liu X, Xu Y, Li S, et al. In Silico target fishing: addressing a "Big Data" problem by ligand-based similarity rankings with data fusion. J Cheminform. 2014;6:33.
287. Ehrt C, Brinkjost T, Koch O. Binding site characterization – similarity, promiscuity, and druggability. MedChemCommun. 2019;10(7):1145–59.
288. Helguera AM, Perez-Castillo Y, Cordeiro MNDS, et al. Ligand-based virtual screening using tailored ensembles: a prioritization tool for dual A2A adenosine receptor antagonists/monoamine oxidase B inhibitors. Curr Pharm Des. 2016;22(21):3082–96.
289. Abdul Hameed MD, Chaudhury S, Singh N, et al. Exploring polypharmacology using a ROCS-based target fishing approach. J Chem Inf Model 2012; 52(2): 492–505.
290. Fang Y. Are label-free investigations the best approach to drug discovery? Future Med Chem. 2015;7(12):1561–4.
291. Fang Y. Label-free drug discovery. Front Pharmacol. 2014;5:52.
292. Scannell JW. Diagnosing the decline in pharmaceutical R&D efficiency. Nat Rev Drug Discov. 2012;11:191–200.
293. Pammolli F, Magazzini L, Riccaboni M. The productivity crisis in pharmaceutical R&D. Nat Rev Drug Discov. 2011;10:428–38.
294. Paul SM, Mytelka DS, Dunwiddie CT, et al. How to improve R&D productivity: the pharmaceutical industry's grand challenge. Nat Rev Drug Discov. 2010;9:203–14.
295. Pacholec M, Bleasdale JE, Chrunyk B, et al. SRT1720, SRT2183, SRT1460, and resveratrol are not direct activators of SIRT1. J Biol Chem. 2010;285:8340–51.
296. Hu H, Deng H, Fang Y. Label-free phenotypic profiling identified D-luciferin as a GPR35 agonist. PLoS One. 2012;7:e34934.
297. Beher D, Wu J, Cumine S, et al. Resveratrol is not a direct activator of SIRT1 enzyme activity. Chem Biol Drug Des. 2009;74:619–24.

298. Fontana A, de Laureto PP, Spolaore B, et al. Probing protein structure by limited proteolysis. Acta Biochim Pol. 2004;51:299–321.
299. Lomenick B, Hao R, Jonai N, et al. Target identification using drug affinity responsive target stability (DARTS). Proc Natl Acad Sci U S A. 2009;106:21984–9.
300. Chang Y, Schlebach JP, Verheul RA, et al. Simplified proteomics approach to discover protein-ligand interactions. Protein Sci. 2012;21:1280–7.
301. Fang Y. Troubleshooting and deconvoluting label-free cell phenotypic assays in drug discovery. J Pharmacol Toxicol Methods. 2013;67(2):69–81.
302. Fang Y. Live cell optical sensing for high throughput applications. Adv Biochem Eng Biotechnol. 2010;118:153–63.
303. Wang J, Zhang X, Fang Y, et al. Label-free cell phenotypic assays for assessing drug polypharmacology. Curr Pharm Des. 2016;22(21):3190–200.
304. Bantscheff M, Schirle M, Sweetman G, et al. Quantitative mass spectrometry in proteomics: a critical review. Anal Bioanal Chem. 2007;389(4):1017–31.
305. Asara JM, Christofk HR, Freimark LM, et al. A label-free quantification method by MS/MS TIC compared to SILAC and spectral counting in a proteomics screen. Proteomics. 2008;8(5):994–9.
306. White AW, Westwell AD, Brahemi G. Protein-protein interactions as targets for small-molecule therapeutics in cancer. Expert Rev Mol Med. 2008;10:e8.
307. Arkin M, Wells JA. Small-molecule inhibitors of protein-protein interactions: progressing towards the dream. Nat Rev Drug Discov. 2004;3:301–17.
308. Fry DC, Vassilev LT. Targeting protein-protein interactions for cancer therapy. J Mol Med. 2005;83:955–63.
309. Yin H, Hamilton AD. Strategies for targeting protein-protein interactions with synthetic agents. Angew Chem Int Ed Engl. 2005;44:4130–63.
310. Lindsley CW. Small molecule inhibition of protein-protein interaction: an emerging paradigm in drug design. Curr Top Med Chem. 2007;7:921.
311. Wells JA, McClendon CL. Reaching for high-hanging fruit in drug discovery at protein-protein interfaces. Nature. 2007;450:1001–9.
312. Fry DC. Protein-protein interactions as targets for small molecule drug discovery. Biopolymers. 2006;84(6):535–52.
313. Zhao L, Chmielewski J. Inhibiting protein-protein interactions using designed molecules. Curr Opin Struct Biol. 2005;15(1):31–4.
314. Sun JC, Zhao ZM. A comparative study of cancer proteins in the human protein-protein interaction network. BMC Genomics. 2010;11:S5.
315. Zhao YJ, Aguilar A, Bernard D, et al. Small-molecule inhibitors of the mdm2-p53 protein-protein interaction (mdm2 inhibitors) in clinical trials for cancer treatment. J Med Chem. 2015;8:1038–52.
316. Macalino SJY, Basith S, Clavio NAB, et al. Evolution of in silico strategies for protein-protein interaction drug discovery. Molecules. 2018;23(8):1963.
317. Petsalaki E, Russell RB. Peptide-mediated interactions in biological systems: new discoveries and applications. Curr Opin Biotechnol. 2008;19:344–50.
318. Stumpf MP, Thorne T, de Silva E, et al. Estimating the size of the human interactome. Proc Natl Acad Sci U S A. 2008;105:6959–64.
319. Bourgeas R, Basse MJ, Morelli X, et al. Atomic analysis of protein-protein interfaces with known inhibitors: the 2p2i database. PLoS One. 2010;5:e9598.
320. Sperandio O, Reynes CH, Camproux AC, et al. Rationalizing the chemical space of protein-protein interaction inhibitors. Drug Discov Today. 2010;15:220–9.
321. Basse MJ, Betzi S, Bourgeas R, et al. 2p2idb: a structural database dedicated to orthosteric modulation of protein-protein interactions. Nucleic Acids Res. 2013;41:D824–7.
322. Labbe CM, Laconde G, Kuenemann MA, et al. Ippi-db: a manually curated and interactive database of small non-peptide inhibitors of protein-protein interactions. Drug Discov Today. 2013;18:958–68.
323. Laraia L, McKenzie G, Spring DR, et al. Overcoming chemical, biological, and computational challenges in the development of inhibitors targeting protein-protein interactions. Chem Biol. 2015;22:689–703.
324. Wu KJ, Lei PM, Liu H, et al. Mimicking strategy for protein-protein interaction inhibitor discovery by virtual screening. Molecules. 2019;24(24):4428.
325. Mullard A. Protein–protein interaction inhibitors get into the groove. Nat Rev Drug Discov. 2012;11:173–5.
326. Mason JM. Design and development of peptides and peptide mimetics as antagonists for therapeutic intervention. Future Med Chem. 2010;2:1813–22.
327. Groß A, Möbius K, Haußner C, et al. Mimicking protein–protein interactions through peptide–peptide interactions: HIV-1 gp120 and CXCR4. Front Immunol. 2013;4:257.
328. Carry J-C, Garcia-Echeverria C. Inhibitors of the p53/hdm2 protein–protein interaction—path to the clinic. Bioorg Med Chem Lett. 2013;23:2480–5.
329. Rew Y, Sun D, Gonzalez-Lopez De Turiso F, et al. Structure-based design of novel inhibitors of the MDM2–p53 interaction. J Med Chem. 2012;55:4936–54.
330. Vassilev LT, Vu BT, Graves B, et al. In vivo activation of the p53 pathway by small-molecule antagonists of MDM2. Science. 2004;303:844–8.
331. Opydo-Chanek M, Gonzalo O, Marzo I. Multifaceted anticancer activity of BH3 mimetics: current evidence and future prospects. Biochem Pharmacol. 2017;136:12–23.
332. Cekay MJ, Roesler S, Frank T, et al. Smac mimetics and type II interferon synergistically induce necroptosis in various cancer cell lines. Cancer Lett. 2017;410:228–37.

333. Bai L, Smith DC, Wang S. Small-molecule SMAC mimetics as new cancer therapeutics. Pharmacol Ther. 2014;144:82–95.
334. Modell AE, Blosser SL, Arora PS. Systematic targeting of protein–protein interactions. Trends Pharmacol Sci. 2016;37:702–13.
335. Kaczor AA, Bartuzi D, Stępniewski TM, et al. Protein-protein docking in drug design and discovery. Methods Mol Biol. 2018;1762:285–305.
336. Grosdidier S, Fernández-Recio J. Protein-protein docking and hot-spot prediction for drug discovery. Curr Pharm Des. 2012;18(30):4607–18.
337. Grosdidier S, Fernández-Recio J. Identification of hot-spot residues in protein-protein interactions by computational docking. BMC Bioinf. 2008;9:447.
338. Barradas-Bautista D, Rosell M, Pallara C, et al. Structural prediction of protein-protein interactions by docking: application to biomedical problems. Adv Protein Chem Struct Biol. 2018;110:203–49.
339. Prieto-Martínez FD, López-López E, Eurídice Juárez-Mercado K, et al. Chapter 2: Computational drug design methods—current and future perspectives. In: Roy K, editor. In silico drug design. Cambridge, MA: Academic; 2019. p. 19–44.
340. Akoka J, Comyn-Wattiau I, Laoufi N. Research on big data—a systematic mapping study. Comput Stand Interfaces. 2017;54:105–15.
341. Secchi P. On the role of statistics in the era of big data: a call for a debate. Stat Probab Lett. 2018;136:10–4.
342. Cox DR, Kartsonaki C, Keogh RH. Big data: some statistical issues. Stat Probab Lett. 2018;136:111–5.
343. Bornmann L. Measuring the societal impact of research. EMBO Rep. 2012;13:673.
344. Mårtensson P, Fors U, Wallin S-B, et al. Evaluating research: a multidisciplinary approach to assessing research practice and quality. Res Policy. 2016;45:593–603.
345. Singh AK, Das K. Insights into HIV-1 Reverse Transcriptase (RT) Inhibition and Drug Resistance from Thirty Years of Structural Studies. Viruses. 2022;14(5):1027.
346. Ghosh AK, Osswald HL, Prato G. Recent progress in the development of HIV-1 protease inhibitors for the treatment of hiv/aids. J Med Chem. 2016;59:5172–208.
347. Barmania F, Pepper MS. C-c chemokine receptor type five (ccr5): an emerging target for the control of HIV infection. Appl Transl Genom. 2013;2:3–16.
348. MacArthur RD, Novak RM. Reviews of anti-infective agents: Maraviroc: the first of a new class of antiretroviral agents. Clin Infect Dis. 2008;47:236–41.
349. Kuritzkes D, Kar S, Kirkpatrick P. Maraviroc. Nat Rev Drug Discov. 2008;7:15.
350. Lusher SJ, McGuire R, van Schaik RC, et al. Data-driven medicinal chemistry in the era of big data. Drug Discov Today. 2014;19:859–68.
351. Ebejer JP, Fulle S, Morris GM, et al. The emerging role of cloud computing in molecular modelling. J Mol Graph Model. 2013;44:177–87.
352. Kissin I. What can big data on academic interest reveal about a drug? Reflections in three major us databases. Trends Pharm Sci. 2018;39:248–57.
353. Mak KK, Pichika MR. Artificial intelligence in drug development: present status and future prospects. Drug Discov Today. 2019;24:773–80.
354. Bishop CM. Model-based machine learning. Philos Transact A Math Phys Eng Sci. 2013;371:20120222.
355. Duch W, Swaminathan K, Meller J. Artificial intelligence approaches for rational drug design and discovery. Curr Pharm Des. 2007;13:1497–508.
356. Probst C, Schneider S, Loskill P. High-throughput organ-on-a-chip systems: current status and remaining challenges. Curr Opin Biomed Eng. 2018;6:33–41.
357. IBM Ibm Watson. Accessed 1 May 2019; Available online: https://www.ibm.com/watson.
358. Smalley E. Ai-powered drug discovery captures pharma interest. Nat Biotechnol. 2017;35:604–5.
359. Exscientia At the forefront of small molecule drug discovery. Accessed 1 May 2019; Available online: https://www.exscientia.co.uk/.
360. Fleming N. How artificial intelligence is changing drug discovery. Nature. 2018;557:55–7.
361. Exscientia Celgene and exscientia enter 3-year AI drug discovery collaboration focused on accelerating drug discovery in oncology and autoimmunity. Accessed 1 May 2019; Available online: https://www.exscientia.co.uk/news.
362. Exscientia Exscientia achieves molecule discovery milestone as part of gsk collaboration. Accessed 1 May 2019; Available online: https://www.exscientia.co.uk/news.
363. Guncar G, Kukar M, Notar M, et al. An application of machine learning to haematological diagnosis. Sci Rep. 2018;8:411.
364. Byvatov E, Fechner U, Sadowski J, et al. Comparison of support vector machine and artificial neural network systems for drug/nondrug classification. J Chem Inf Comput Sci. 2003;43:1882–9.
365. Zernov VV, Balakin KV, Ivaschenko AA, et al. Drug discovery using support vector machines. The case studies of drug-likeness, agrochemical-likeness, and enzyme inhibition predictions. J Chem Inf Comput Sci. 2003;43:2048–56.
366. Warmuth MK, Liao J, Ratsch G, et al. Active learning with support vector machines in the drug discovery process. J Chem Inf Comput Sci. 2003;43:667–73.
367. Jorissen RN, Gilson MK. Virtual screening of molecular databases using a support vector machine. J Chem Inf Model. 2005;45:549–61.
368. Koohy H. The rise and fall of machine learning methods in biomedical research. F1000Res. 2012;2017:6.
369. Young JD, Cai C, Lu X. Unsupervised deep learning reveals prognostically relevant subtypes of glioblastoma. BMC Bioinf. 2017;18:381.
370. Popova M, Isayev O, Tropsha A. Deep reinforcement learning for de novo drug design. Sci Adv. 2018;4:eaap7885.

371. Lo YC, Rensi SE, Torng W, et al. Machine learning in chemoinformatics and drug discovery. Drug Discov Today. 2018;23:1538–46.
372. Lima AN, Philot EA, Trossini GH, et al. Use of machine learning approaches for novel drug discovery. Expert Opin Drug Discovery. 2016;11:225–39.
373. Ma XH, Jia J, Zhu F, et al. Comparative analysis of machine learning methods in ligand-based virtual screening of large compound libraries. Comb Chem High Throughput Screen. 2009;12:344–57.
374. Han LY, Ma XH, Lin HH, et al. A support vector machines approach for virtual screening of active compounds of single and multiple mechanisms from large libraries at an improved hit-rate and enrichment factor. J Mol Graph Model. 2008;26:1276–86.
375. Liu XH, Ma XH, Tan CY, et al. Virtual screening of abl inhibitors from large compound libraries by support vector machines. J Chem Inf Model. 2009;49:2101–10.
376. Ma J, Sheridan RP, Liaw A, et al. Deep neural nets as a method for quantitative structure-activity relationships. J Chem Inf Model. 2015;55:263–74.
377. Kadurin A, Nikolenko S, Khrabrov K, et al. Drugan: an advanced generative adversarial autoencoder model for de novo generation of new molecules with desired molecular properties in silico. Mol Pharm. 2017;14:3098–104.
378. Segler MHS, Kogej T, Tyrchan C, et al. Generating focused molecule libraries for drug discovery with recurrent neural networks. ACS Cent Sci. 2018;4:120–31.
379. Yuan W, Jiang D, Nambiar DK, et al. Chemical space mimicry for drug discovery. J Chem Inf Model. 2017;57:875–82.
380. Olivecrona M, Blaschke T, Engkvist O, et al. Molecular de-novo design through deep reinforcement learning. J Cheminform. 2017;9:48.
381. Van Dorst B, Mehta J, Rouah-Martin E, et al. cDNA phage display as a novel tool to screen for cellular targets of chemical compounds. Toxicol In Vitro. 2010;24:1435–40.
382. Jung HJ, Shim JS, Lee J, et al. Terpestacin inhibits tumor angiogenesis by targeting UQCRB of mitochondrial complex III and suppressing hypoxia-induced reactive oxygen species production and cellular oxygen sensing. J Biol Chem. 2010;285:11584–95.
383. Takakusagi Y, Takakusagi K, Sugawara F, et al. Use of phage display technology for the determination of the targets for small-molecule therapeutics. Expert Opin Drug Discovery. 2010;5:361–89.
384. Van Dorst B, Mehta J, Rouah-Martin E, et al. The identification of cellular targets of 17beta estradiol using a lytic (T7) cDNA phage display approach. Toxicol In Vitro. 2011;25:388–93.
385. Roberts RW, Szostak JW. RNA-peptide fusions for the in vitro selection of peptides and proteins. Proc Natl Acad Sci U S A. 1997;94:12297–302.
386. McPherson M, Yang Y, Hammond PW, et al. Drug receptor identification from multiple tissues using cellular-derived mRNA display libraries. Chem Biol. 2002;9:691–8.
387. Licitra EJ, Liu JO. A three-hybrid system for detecting small ligand-protein receptor interactions. Proc Natl Acad Sci U S A. 1996;93:12817–21.
388. Becker F, Murthi K, Smith C, et al. A three-hybrid approach to scanning the proteome for targets of small molecule kinase inhibitors. Chem Biol. 2004;11:211–23.
389. Chidley C, Haruki H, Pedersen MG, et al. A yeast-based screen reveals that sulfasalazine inhibits tetrahydrobiopterin biosynthesis. Nat Chem Biol. 2011;7:375–83.
390. Laenen G, Thorrez L, Börnigen D, et al. Finding the targets of a drug by integration of gene expression data with a protein interaction network. Mol BioSyst. 2013;9(7):1676–85.

19 Databases for Rational Design and Discovery of Multitarget Drugs

Abstract

Rational drug design and discovery, particularly rational multitarget drug (MTD) design and discovery, heavily relies on computational approaches exploring the Internet data resources. Internet resources can be classified into two classes (Potemkin V, Potemkin A, Grishina M, Curr Top Med Chem 18:1955–1975, 2018). The first class of resources accumulates information about drugs, drug candidates, compounds, and bioassays, which is a starting point in drug discovery and design. The second class of Internet resources includes web portals performing online computations for drug discovery and design. Here in this chapter, we intend to classify drug discovery Internet resources into three categories: (**a**) spaces that contain theoretically infinite number of data or information in particular fields; (**b**) databases that collect curated sets of disease-, target-, or drug-related data and information and mostly have the power to predict drug candidates or druggability; and (**c**) online algorithms/web servers or programs/software that can be used for target and/or drug candidate predictions. This chapter will give brief introductions to these Internet resources in terms of their applicability to and strengths/weaknesses in drug discovery. It should be noted that many excellent databases that could be useful to polypharmacology-based drug design and discovery are unfortunately not included in this chapter due to the limited page space.

Keywords

Polypharmacological resources · Spaces · Databases · Online algorithms/web servers · Programs/software

19.1 Information Spaces

Rational drug design and discovery, particularly rational multitarget drug (MTD) design and discovery, heavily rely on both tangible and intangible resources of various sorts, including knowledge, data, and tools of multiple disciplines (chemical, biological, pharmacological, pathological, etc.) and multiple levels (molecular, cellular, tissue, organism). Probably, the most invaluable resource is the Internet resources. Nowadays Internet resources have become an indispensable requirement for drug discovery. Internet resources can be classified into two classes [1]. The first class of resources accumulates information about drugs, drug candidates, compounds, and bioassays [1]. This information is a starting point in drug discovery and design. It is necessary for a training dataset composition. The data found at this step are needed in the search for the rules predicting a biological activity or recognizing active compounds among other molecules. The second

Z. Wang, B. Yang, *Polypharmacology*, https://doi.org/10.1007/978-3-031-04998-9_19

class of Internet resources includes web portals performing online computations for drug discovery and design [1]. The web portals perform (**a**) modeling of molecular structure such as geometry optimization and molecular docking; (**b**) online computations of various descriptors and physical-chemical and ADMET properties influencing the bioprocesses occurring in a living organism along the road of the drug therapeutic action; (**c**) quantitative structure–activity relationship (QSAR) and quantitative structure–property relationship (QSPR) studies; (**d**) prognosis of bioactivities of compounds; and (**e**) design of new drug candidates [1]. Here in this chapter, we intend to classify drug discovery Internet resources into three categories: (**a**) spaces that contain theoretically infinite number of data or information in particular fields; (**b**) databases that collect curated sets of disease-, target-, or drug-related data and information and mostly have the power to predict drug candidates or druggability; and (**c**) online algorithms/web servers or programs/software that can be used for target and/or drug candidate predictions. This chapter will give brief introductions to these Internet resources in terms of their applicability to and strengths/weaknesses in drug discovery.

19.1.1 Chemical Space

Chemical space is a concept in cheminformatics referring to the property space spanned by all possible molecules and chemical compounds adhering to a given set of construction principles and boundary conditions [2, 3]. Chemical space contains every possible chemical compound. It includes every drug and material we know and every one we can find in the future. More specifically, chemical space is the ensemble of all possible molecules, which is believed to contain at least 10^{60} organic molecules below 500 Da of possible interest for drug discovery. In layman's terms, chemical space is the conceptual territory inhabited by all possible compounds, and it is where scientists have found every known medicine and material.

19.1.1.1 Theoretical Spaces

An upper estimate says it contains 10^{180} compounds, more than twice the magnitude of the number of atoms in the universe. In cheminformatics, chemical space is restricted to only potential pharmacologically active molecules, and in this sense, its size is estimated to be in the order of 10^{60} molecules. The assumptions used for estimating the number of potential pharmacologically active molecules, however, use Lipinski's rules, in particular the molecular weight limit of 500. The estimate also restricts the chemical elements used to be "CHONS"—carbon, hydrogen, oxygen, nitrogen, and sulfur. It further makes the assumption of a maximum of 30 atoms to stay below 500 Daltons, allows for branching and a maximum of 4 rings, and arrives at an estimate of 10^{63}. In addition to the drug-like space and lead-like space that are, in part, defined by Lipinski's rule of five, the concept of known drug space (KDS), which is defined by the molecular descriptors of marketed drugs, has also been introduced [4–6]. KDS can be used to help predict the boundaries of chemical spaces for drug development by comparing the structure of the molecules that are undergoing design and synthesis to the molecular descriptor parameters that are defined by the KDS.

19.1.1.2 Empirical Spaces

As of July 2009, there were 49,037,297 organic and inorganic substances registered with the Chemical Abstracts Service, indicating that they have been reported in the scientific literature [7].Chemical libraries used for laboratory-based screening for compounds with desired properties are examples for real-world chemical libraries of small size (a few hundred to hundreds of thousands of molecules).

Éliás et al. [8] reported the systematic elaboration and thorough evaluation of a data matrix generated by sampling the chemical space around a common 5,6-fused bicyclic heteroaromatic template applying characteristic pharmacophore elements of central nervous system relevant aminergic G-protein-coupled receptor ligands, aiming to address the application of chemical space to polypharmacology.

The Known Chemical Space[a]

database	description	size[a]	Web address	ref
PubChem	known molecules from various public sources	32.5 M	http://pubchem.ncbi.nlm.nih.gov	(11)
Chemspider	online resource from the Royal Society of Chemistry	26.0 M	http://www.chemspider.com/	(12)
ZINC	commercially available small molecules	21.0 M	http://zinc.docking.org	(13)
NCI Open	anticancer and AIDS compounds with screening data	0.25 M	http://cactus.nci.nih.gov/ncidb2.1	(14)
ChemDB	commercially available small molecules	4.1 M	http://cdb.ics.uci.edu	(15)
BindingDB	bioactive molecules with binding affinity data	0.36 M	http://www.bindingdb.org	(16)
ChemBank	small molecules annotated with screening data	1.2 M	http://chembank.broadinstitute.org/	(17)
ChEMBL	small molecules annotated with experimental data	1.1 M	https://www.ebi.ac.uk/chembldb	(18)
CTD	comparative toxicogenomics database	0.17 M	http://ctdbase.org	(19)
HMDB	human metabolome database	0.0085 M	http://www.hmdb.ca	(20)
SMPDB	small molecule pathway database	0.001 M	http://www.smpdb.ca	(21)
DrugBank	experimental and approved small molecule drugs	0.0065 M	http://www.drugbank.ca	(22)

19.1.2 Pharmacologic Space

Pharmacologic space, by the same logic of chemical space, represents the collection of all pharmacologic compounds and molecules, which is constructed based on the global relationships between chemical structure and biological targets [9–12]. The global mapping of pharmacological space by the integration of several vast sources of medicinal chemistry structure–activity relationships (SARs) data enables confident identification of the human targets for which chemical tools and drugs have been discovered to date. The integration of SAR data from diverse sources by unique canonical chemical structure, protein sequence, and disease indication enables the construction of a ligand–target matrix to explore the global relationships between chemical structure and biological targets. Using the data matrix, we are able to catalog the links between proteins in chemical space as a polypharmacology interaction network. Paolini et al. [12] demonstrated that probabilistic models can be used to predict pharmacology from a large knowledge base. The relationships between proteins, chemical structures, and drug-like properties provide a framework for developing a probabilistic approach to drug discovery that can be exploited to increase research productivity.

Indeed, a precise mapping of the biological target space is essential in drug discovery for many reasons [11]. First, it enables to properly describe targets of existing drugs and clinical candidates and therefore estimate the druggability of major target families. Second, it permits to orient early drug discovery toward previously uncharted target space (e.g., protein–protein interactions, epigenetic targets) with the hope of discovering selective and safe drug candidates. Last, it provides an ideal framework to better understand molecular reasons for promiscuity and polypharmacology [11].

The analysis of pharmacological space is becoming highly relevant in light of the emerging polypharmacology paradigm, that is, the increasing evidence that many drugs elicit therapeutic effects and adverse drug reactions through interactions with multiple targets. Nisius and Bajorath [13] provided an overview of computational approaches for analyzing pharmacological space and put their opportunities and limitations in perspective. Insights into computational approaches for the study of target–ligand interactions and polypharmacology are provided and put into scientific context.

19.1.3 Biological Space

Petit-Zeman [14] defines biological space as the area(s) of chemical space that possess biologically active ("bioactive") compounds for a specific target or target class. As such, they can modulate a given biological system and subsequently influence disease development and progression. As such, they can modulate a given biological system and subsequently influence disease development and progression. In literature, this space has also been called "biologically relevant chemical space" [15].

Only a small percentage of the vast chemical space has been estimated to be biologically active and is thus relevant for drug development, as randomly searching bioactive compounds in chemical space with no prior information resembles the search for "the needle in a haystack." Hence, bioactive molecules are often used as a starting point in in silico explorations of biological space. The plethora of in silico methods for this task includes similarity and pharmacophore searching methods for novel compounds, scaffold-hopping approaches to derive novel chemotypes, or the development of quantitative structure–activity relationships (QSARs) to explore the interplay between the 3D chemical structure and its biological activity toward a specific target. The biological space is comprised of small molecules which are active on specific targets [14]. If researchers want to explore the role of targets in a given biological system experimentally, they can use small molecules which are potent and selective toward a specific target (thus confided to a particular area in chemical space).

The structural comparison of protein binding sites is increasingly important in drug design; identifying structurally similar sites can be useful for techniques such as drug repurposing, and also in a polypharmacological approach to deliberately affect multiple targets in a disease pathway, or to explain unwanted off-target effects. Once similar sites are identified, identifying local differences can aid in the design of selectivity. Siragusa et al. [16] developed a semiautomated approach, called BioGPS, that is based on mapping of biological space incorporating the software FLAP which combines GRID Molecular Interactions Fields (MIFs) and pharmacophoric fingerprints. BioGPS comprises the automatic preparation of protein structure data, identification of binding sites, and subsequent comparison by aligning the sites and directly comparing the MIFs. Chemometric approaches are included to reduce the complexity of the resulting data on large datasets, enabling focus on the most relevant information. Individual site similarities can be analyzed in terms of their Pharmacophoric Interaction Field (PIF) similarity, and importantly the differences in their PIFs can be extracted. Here we describe the BioGPS approach and demonstrate its applicability to rationalize off-target effects (ERα and SERCA), to classify protein families and explain polypharmacology (ABL1 kinase and NQO2), and to rationalize selectivity between subfamilies (MAP kinases p38α/ERK2 and PPARδ/PPARγ) [16].

19.1.4 Genomic Space

The genomic space is an area of chemical space that contains all genomic data available, which are coming from the fortuitous confluence of technological advances in protein and DNA analysis as well as imaging advances in cell biology [17, 18]. The genomic space comprises four main layers [18].

19.1.4.1 Genomic Data Production

In the Genomic Data Production layer, physical DNA is extracted from its biological environs and abstracted into a computational equivalent that can be stored on machines. Recall that the entirety of the DNA found in an organism constitutes its whole genome. This post details the types of genetic data that can be produced from a biological sample since the whole genome is not always sequenced [18].

19.1.4.2 Genomic Data Storage and Management

This layer comprises data companies that help others store and process great quantities of genetic data. Several general cloud storage and analytics service companies including Amazon Web Services, Google Cloud, and Databricks are found here, together with more omics-focused cloud companies (e.g., DNA Nexus, Seven Bridges) that strive to facilitate and optimize established genomic pipelines while adhering to best practices. Furthermore, some of the companies here also control repositories of genomic data to further empower their platform [18].

19.1.4.3 Genomic Data Analysis

This layer is where the most progress remains to be achieved, and the greatest value will be unlocked. It has traditionally been the domain of academic research groups and in-house analytics teams at pharmaceutical, biotechnology, and agricultural firms. Increasingly, as the sheer importance of the problem and size of the opportunity here begin to be appreciated by more people together with the increased availability of data and computational access, many new ventures are joining this layer. Some can be extensions of existing, general data analytics companies [18].

19.1.4.4 Implementation of Genomic Data Insights to Build Products and Services

This layer is incomparably diverse and comprises three main segments: biotechnology/pharmaceutical companies, diagnostics companies, and provider groups (e.g., hospitals and health systems). The burgeoning predictive health segment also belongs here. Together, they constitute groups that utilize genetic insights produced in layer III to make medical products or render health services. Often, this process involves engaging in drug and diagnostics development, as well as patient care and health management. Health insurance companies would fit into layer IV as well, as they have demonstrated increased interest in genomic data.

19.1.5 Chemogenomic Space

19.1.5.1 Basics of Chemogenomic Space

Chemogenomics, or chemical genomics, is the systematic screening of targeted chemical libraries of small molecules against individual drug target families (e.g., GPCRs, nuclear receptors, kinases, proteases, etc.) with the ultimate goal of identification of novel drugs and drug targets [19]. Chemogenomics attempts to combine genomic data, structural biological data, classical dendrograms, and selectivity data to explore, define, and classify the medicinally relevant kinase space. By identifying screening hits that modulate the activity of the less well-characterized members of the target family, the function of these novel targets can be elucidated. Furthermore, the hits for these targets can be used as a starting point for drug discovery. The completion of the human genome project has provided an abundance of potential targets for therapeutic intervention. Chemogenomics strives to study the intersection of all possible drugs on all potential targets [20].

A common method to construct a targeted chemical library is to include known ligands of at least one and preferably several members of the target family. Since a portion of ligands that were designed and synthesized to bind to one family member will also bind to additional family members, the compounds contained in a targeted chemical library should collectively bind to a high percentage of the target family [21].

Chemogenomics integrates target and drug discovery by using active compounds, which function as ligands, as probes to characterize proteome functions. The interaction between a small compound and a protein induces a phenotype. Once the phenotype is characterized, we could associate a protein to a molecular event. Compared with genetics, chemogenomics techniques are able to modify the function of a protein rather than the gene. Also, chemogenomics is able to observe the interaction as well as reversibility in real time. For example, the modification of a phenotype can be observed only after addition of a specific compound and can be interrupted after its withdrawal from the medium.

19.1.5.2 Approaches of Chemogenomic Space

Currently, there are two experimental chemogenomic approaches: forward (classical) chemogenomics and reverse chemogenomics. Forward chemogenomics attempts to identify drug targets by searching for molecules which give a certain phenotype on cells or animals, while reverse chemogenomics aims to validate phenotypes by searching for molecules that interact specifically with a given protein [22]. Both of these approaches require a suitable collection of compounds and an appropriate model system for screening the compounds and looking for the parallel identification of biological targets and biologically active compounds. The biologically active compounds that are discovered through forward or reverse chemogenomic approaches are known as modulators because they bind to and modulate specific molecular targets, thus they could be used as "targeted therapeutics" [19].

Forward Chemogenomics

In forward chemogenomics, which is also known as classical chemogenomics, a particular phenotype is studied and small compound interacting with this function is identified. The molecular basis of this desired phenotype is unknown. Once the modulators have been identified, they will be used as tools to look for the protein responsible for the phenotype. For example, a loss-of-function phenotype could be an arrest of tumor growth. Once compounds that lead to a target phenotype have been identified, identifying the gene and protein targets should be the next step [23]. The main challenge of forward chemogenomics strategy lies in designing phenotypic assays that lead immediately from screening to target identification.

Reverse Chemogenomics

In reverse chemogenomics, small compounds that perturb the function of an enzyme in the context of an in vitro enzymatic test will be identified. Once the modulators have been identified, the phenotype induced by the molecule is analyzed in a test on cells or on whole organisms. This method will identify or confirm the role of the enzyme in the biological response [23]. Reverse chemogenomics used to be virtually identical to the target-based approaches that have been applied in drug discovery and molecular pharmacology over the past decade. This strategy is now enhanced by parallel screening and by the ability to perform lead optimization on many targets that belong to one target family.

19.1.5.3 Applications of Chemogenomic Space

Chemogenomics expedites the discovery of therapeutically relevant targets from phenotypic screens. However, the vast majority of proteins in the proteome lack selective pharmacological modulators, necessitating the development of new technologies that significantly expand chemogenomic space. Chemoproteomics has emerged as a robust platform to map small molecule–protein interactions in cells using functionalized chemical probes in conjunction with mass spectrometry analysis [24]. Exploration of the ligandable proteome in this manner has led to the development of new pharmacological modulators of diverse proteins.

Determining Mode of Action

Chemogenomics has been used to identify mode of action (MOA) for traditional Chinese medicine (TCM) and Ayurveda. Compounds contained in traditional medicines are usually more soluble than synthetic compounds, have "privileged structures" (chemical structures that are more frequently found to bind in different living organisms), and have more comprehensively known safety and tolerance factors. Therefore, this makes them especially attractive as a resource for lead structures when developing new molecular entities. With databases containing chemical structures of compounds used in alternative medicine along with their phenotypic effects, in silico analysis may be of use to assist in determining MOA, for example, by predicting ligand targets that were relevant to known phenotypes for traditional medicines [25]. In a case study for TCM, the therapeutic class of "toning and replenishing medicine" was evaluated. Therapeutic actions (or phenotypes) for that class include anti-inflammatory, antioxidant, neuroprotective, hypoglycemic activity, immunomodulatory, antimetastatic, and hypotensive. Sodium–glucose transport proteins and PTP1B (an insulin signaling regulator) were identified as targets which link to the hypoglycemic phenotype suggested. The case study for Ayurveda involved anticancer formulations. In this case, the target prediction program enriched for targets directly connected to cancer progression such as steroid-5-alpha-reductase and synergistic targets like the efflux pump P-gp. These target–phenotype links can help identify novel MOAs.

Beyond TCM and Ayurveda, chemogenomics can be applied early in drug discovery to determine a compound's mechanism of action and take advantage of genomic biomarkers of toxicity and efficacy for application to phase I and II clinical trials [26].

Identifying New Drug Targets

Chemogenomics profiling can be used to identify totally new therapeutic targets, for example, new antibacterial agents [27]. The study capitalized on the availability of an existing ligand library for an enzyme called murD that is used in the peptidoglycan synthesis pathway. Relying on the chemogenomics similarity principle, the researchers mapped the murD ligand library to other members of the mur ligase family (murC, murE, murF, murA, and murG) to identify new targets for the known ligands. Ligands identified would be expected to be broad-spectrum Gram-negative inhibitors in experimental assays since peptidoglycan synthesis is exclusive to bacteria. Structural and molecular docking studies revealed candidate ligands for murC and murE ligases.

Identifying Genes in Biological Pathway

Thirty years after the posttranslationally modified histidine derivative diphthamide was determined, chemogenomics was used to discover the enzyme responsible for the final step in its synthesis [28]. Diphthamide is a posttranslationally modified histidine residue found on the translation elongation factor 2 (eEF-2). The first two steps of the biosynthesis pathway leading to dipthine have been known, but the enzyme responsible for the amidation of dipthine to diphthamide remained a mystery. The researchers capitalized on *Saccharomyces cerevisiae* co-fitness data. Co-fitness data is data representing the similarity of growth fitness under various conditions between any two different deletion strains. Under the assumption that strains lacking the diphthamide synthetase gene should have high co-fitness with strain lacking other diphthamide biosynthesis genes, they identified ylr143w as the strain with the highest co-fitness to all other strains lacking known diphthamide biosynthesis genes. Subsequent experimental assays confirmed that YLR143W was required for diphthamide synthesis and was the missing diphthamide synthetase.

The ChEMBLSpace graphical explorer enables the identification of compounds from the ChEMBL database, which exhibit a desirable polypharmacology profile. This profile can be predefined or created iteratively, and the tool can be extended to other data sources [29].

Chemogenomics comprises a systematic relationship between targets and ligands that are used as target modulators in living systems such as cells or organisms. In recent years, data on small molecule–bioactivity relationships have become increasingly available and consequently so have the number of approaches used to translate bioactivity data into knowledge. Bende et al. [30] conducted chemogenomic data analysis to predict small-molecule targets and generate biological fingerprints that reflect a new awareness of biological networks and polypharmacology.

Chemogenomics data generally refers to the activity data of chemical compounds on an array of protein targets and represents an important source of information for building in silico target prediction models. The increasing volume of chemogenomics data offers exciting opportunities to build models based on big data. Preparing a high-quality dataset is a vital step in realizing this goal, and this work aims to compile such a comprehensive chemogenomics dataset. ExCAPE-DB comprises over 70 million SAR data points from publicly available databases (PubChem and ChEMBL) including structure, target information, and activity annotations as a chemogenomics resource reflecting industry-scale data not only for building predictive models of in silico polypharmacology and off-target effects but also for the validation of cheminformatics approaches in general [31].

19.2 Polypharmacology Databases

19.2.1 Polypharmacology Browser (PPB)

The polypharmacology browser (PPB) is a web-based platform developed by Reymond's research group [22–25], which predicts possible targets for small molecules by searching for nearest neighbors using ten different fps describing composition, substructures, molecular shape, and pharmacophores. PPB searches through 4613 groups of at least 10 same target annotated bioactive molecules from ChEMBL and returns a list of predicted targets ranked by consensus voting scheme and p value. A validation study across 670 drugs with up to 20 targets showed that combining the predictions from all 10 fps gives the best results, with on average 50% of the known targets of a drug being correctly predicted with a hit rate of 25%. Furthermore, when profiling a new inhibitor of the calcium channel TRPV6 against 24 targets taken from a safety screen panel, we observed inhibition in 5 out of 5 targets predicted by PPB and in 7 out of 18 targets not predicted by PPB. The rate of correct (5/12) and incorrect (0/12) predictions for this compound by PPB was comparable to that of other web-based prediction tools.

Several web-based tools have been reported recently which predict the possible targets of a small molecule by similarity to compounds of known bioactivity using molecular fingerprints; however, predictions in each case rely on similarities computed from only one or two fps. Considering that structural similarity and therefore the predicted targets strongly depend on the method used for comparison, it would be highly desirable to predict targets using a broader set of molecular fingerprints simultaneously. PPB offers a versatile platform for target prediction based on multi-fingerprint comparisons.

By screening a focused library of kinase inhibitor analogs in a phenotypic co-culture assay for angiogenesis inhibition, Reymond and colleagues identified an aminotriazine that acts as a cytostatic nanomolar inhibitor. However, this aminotriazine was found to be completely inactive in a whole-kinome profiling assay. To decipher its mechanism of action, a study used the online target prediction tool PPB2 [32–35], which suggested lysophosphatidic acid acyltransferase β (LPAAT-β) as a possible target for this aminotriazine as well as several analogs identified by structure–activity relationship profiling. LPAAT-β inhibition ($IC_{50} \approx 15$ nm) was confirmed in a biochemical assay and by its effects on cell proliferation in comparison with a known LPAAT-β inhibitor. These experiments illustrate the value of target prediction tools to guide target identification for phenotypic screening hits and significantly expand the rather limited pharmacology of LPAAT-β inhibitors.

PPB website: http://gdbtools.unibe.ch:8080/PPB/

PPB2 website: https://ppb2.gdb.tools/

19.2.2 Polypharma

Polypharma database is a unique and specific database for the multitargeted ligands and targets [36]. The highly accurate structural and activity (both binary and continuous) could be used to develop various ligand-based (e.g., QSAR) to predict off-targets or design polypharmacological agents. It consists of FDA-approved drugs and ligands from Ligand Expo complexed with multiple target structures in PDB that belong to distinct target families along with their biological activities. It incorporates both ligand-based prediction of targets, which correlates the ligand properties to multitarget activity data, and structure-based prediction of targets using docking/inverse docking approaches. All compounds curated have been co-crystallized with more than one unique protein with intensive reports of their multitargeting activities. Polypharma was built as a dedicated database specifically designed for polypharmacology studies by providing accurate, experimentally validated structural and activity data of multitargeting ligands.

Polypharma integrates a variety of online resources to build a novel database specifically for multitargeting ligands, along with their modulated targets and quantitative bioactivities (e.g., binding affinity), in particular for polypharmacology modeling. To date, Polypharma includes 953 ligands that are complexed with two or more protein structures belonging to distinct target families. It also provides other information such as molecular properties of ligands and their targets. A set of query functions has been implemented to search the database, and molecular networks can be constructed to depict ligand–target interactions. The query results can also be visualized with integrated molecular visualization tools.

Polypharma website: http:/imdlab.org/polypharma/

19.2.3 Computational Analysis of Novel Drug Repurposing Opportunities (CANDO)

The Computational Analysis of Novel Drug Repurposing Opportunities (CANDO) platform for multi-scale therapeutic discovery, initially funded by a NIH Director's Pioneer Award, was created by Samudrala's team upon basic protein and proteome structure, function, interaction, evolution, and design research to overcome the limitations of traditional single target/disease approaches [37–41]. The platform screens and ranks drugs/compounds for every disease/indication by considering their interactions to every biological protein target, as well as integration of extensive, multi-scale data, from comprehensive well-curated libraries. The interactions are determined using multiple fast bioanalytic docking and hierarchical fragment-based docking with dynamics protocols, among others. The platform continually uses machine learning protocols to iterate on new relevant data from preclinical and clinical studies and improve performance. CANDO not only enables discovery of drugs tailored holistically to their relevant environments but also provides insight into disease system etiologies and mechanisms via analytics of drug–protein interactions at multiple scales.

CANDO was built based on a hypothesis that drugs function by interacting with multiple protein targets to create a molecular interaction signature that can be exploited for therapeutic repurposing and discovery. The researchers compiled a library of compounds that are human ingestible with minimal side effects, followed by an "all-compounds" vs "all-proteins" fragment-based multitarget docking with dynamics screen to construct compound–proteome interaction matrices that were then analyzed to determine similarity of drug behavior. The proteomic signature similarity of drugs is then ranked to make putative drug predictions for all indications in a shotgun manner. They have previously applied this platform with success in both retrospective benchmarking and prospective validation, and to understand the effect of druggable protein classes on repurposing accuracy. They have also used the CANDO platform to analyze and determine the contribution of multitargeting (polypharmacology) to drug repurposing benchmarking accuracy. Their results indicate that a large number of protein structures with diverse fold space and a specific polypharmacological interactome are necessary for accurate drug predictions using the CANDO database [38].

CANDO website: http://protinfo.compbio.buffalo.edu/cando/

19.2.4 QuartataWeb

QuartataWeb is a user-friendly server developed for polypharmacological and chemogenomics analyses [42]. It uses known (experimentally verified) chemical–target interactions (CTIs) from DrugBank and STITCH in a probabilistic matrix factorization algorithm to predict new CTIs. The engine parameters have been optimized to ensure high CTI prediction accuracy. Users can easily obtain information on experimentally verified (known) and computationally predicted (new) interactions between 5494 drugs and 2807 human proteins in DrugBank and between 315,514 chemicals and 9457 human proteins in the STITCH database. In addition, QuartataWeb links targets to KEGG pathways and GO annotations, completing the bridge from drugs/chemicals to function via protein targets and cellular pathways. It allows users to query a series of chemicals, drug combinations, or multiple targets, to enable multidrug, multitarget, multipathway analyses, toward facilitating the design of polypharmacological treatments for complex diseases.

QuartataWeb database website: http://quartata.csb.pitt.edu

19.2.5 SAMNet

The rapid development of high-throughput biotechnologies has led to an onslaught of data describing genetic perturbations and changes in mRNA and protein levels in the cell. Because each assay provides a one-dimensional snapshot of active signaling pathways, it has become desirable to perform multiple assays (e.g., mRNA expression and phosphoproteomics) to measure a single condition. However, as experiments expand to accommodate various cellular conditions, proper analysis and interpretation of these data have become more challenging. Fraenkel's research team developed a novel approach called SAMNet, for Simultaneous Analysis of Multiple Networks, that is able to interpret diverse assays over multiple perturbations [43, 44]. The algorithm uses a constrained optimization approach to integrate mRNA expression data with upstream genes, selecting edges in the protein–protein interaction network that best explain the changes across all perturbations. The result is a putative set of protein interactions that succinctly summarizes the results from all experiments, highlighting the network elements unique to each perturbation. The group evaluated SAMNet in both yeast and human datasets [43]. The yeast dataset measured the cellular response to seven different transition metals, and the human dataset measured cellular changes in four different lung cancer models of epithelial–mesenchymal transition (EMT), a crucial process in tumor metastasis. SAMNet was able to identify canonical yeast metal-processing genes unique to each commodity in the yeast dataset, as well as human genes such as β-catenin and TCF7L2/TCF4 that are required for EMT signaling but escaped detection in the mRNA and phosphoproteomics data. Moreover, SAMNet also highlighted drugs likely to modulate EMT, identifying a series of less canonical genes known to be affected by the BCR-ABL inhibitor imatinib (Gleevec), suggesting a possible influence of this drug on EMT.

SAMNet was built upon a network strategy that takes into account the polypharmacology of small molecules in order to generate hypotheses for their broader mode of action. It has also been tested via a screen for kinase inhibitors that increase the efficacy of gemcitabine, the first-line chemotherapy for pancreatic cancer [44]. Eight kinase inhibitors emerge that are known to affect 201 kinases, of which only 3 kinases have been previously identified as modifiers of gemcitabine toxicity. The authors used the SAMNet algorithm to identify pathways linking these kinases and genetic modifiers of gemcitabine toxicity with transcriptional and epigenetic changes induced by gemcitabine using DNaseI-seq and RNA-seq. SAMNet uses a constrained optimization algorithm to connect genes from these complementary datasets through a small set of protein–protein and protein–DNA interactions. The resulting network recapitulates known pathways including DNA repair, cell proliferation, and the epithelial-to-mesenchymal transition.

SAMNet website: http://www.github.com/sgosline/SAMNet

19.2.6 Multiple Target Ligand Database (MTLD)

Chen et al. [45] constructed a free, Internet-accessible database called the Multiple Target Ligand Database (MTLD) through data mining of the Protein Data Bank (PDB; to be introduced in a subsequent section). The MTLD initially contained 1732 multitarget ligands (MTLs) which bind to 14,996 binding sites extracted from 12,759 PDB structures. Among the MTLs, 222 entries are approved drugs and 1334 entries are drug-like compounds. The MTLD can also provide information on the side effects of drugs through predicting drug promiscuity. The MTLD database was later updated [46] to contain 2444 MTLs that bind with 21,424 binding sites from 18,231 crystal structures. Of the MTLs, 304 entries are approved drugs, and 1911 entries are drug-like compounds. Additionally, new functions such as multiple conditional search and linkage visualization were added.

MTLD website: www.mtdcadd.com

19.2.7 Multitarget Ligand Finder ePlatton

The ePlatton web-based platform was constructed by Du and Shi [47] by integrating disease, disease pathways, drug adverse reaction, and the relationship of targets and ligand clusters in the ligand cluster-based protein network (LCBN) and sequence-based protein database (SBN). The database aims to assist compound design, polypharmacology exploration, and drug combination formulation. It contained 234,591 protein–ligand interactions extracted from ChEMBL. By the 2D structure similarity, 13,769 ligands emerged from 156,151 distinct ligands which were recognized by 1477 proteins. ePlatton considerably reduced the redundancy of the ligand set of targets and made it easy to deduce the possible relationship between compounds and targets, pathways, and side effects. ePlatton behaved reliably in validation experiments and also fast in virtual screening and information retrieval.

ePlatton website: http://www.megabionet.org/eplatton

19.2.8 Promiscuous 2.0

Promiscuous 2.0 provides a powerful yet easy to use resource, which enables experts and non-experts alike to create complex interaction networks in an intuitive and interactive way and potentially infer new indications for existing compounds [48]. Moreover, it is possible to search for existing drugs or even submit new molecular structures and be presented with suggested application areas or, vice versa, get potential drug candidates for disease indications of interest, a completely unique feature across all published databases.

Promiscuous 2.0 website: http://bioinformatics.charite.de/promiscuous2

19.2.9 The Drug Gene Pathway (DRUGPATH) for Mapping Polypharmacology

The Drug Gene Pathway (DRUGPATH) metadatabase was created as a response to the complex treatment required for various diseases including Gulf War illness (GWI) and post-traumatic stress disorder (PTSD), where therapy involves using multiple drugs in combination [49]. The objective was to develop a database that maps the interactions between drugs, genes, pathways, and targets for use in the treatment of complex diseases, including the prediction of off-target interactions, otherwise known as side effects. Using MATLAB and Python scripts, interactions between known drugs, genes, targets, and pathways amalgamated from numerous expert-curated sources such as PharmGKB, DrugBank, DGIdb, ConsensusPathDB, Guide to PHARMACOLOGY, HUGO Gene Nomenclature Committee, Toxin and Toxin-Target Database, repoDB, the FDA's National Drug Code database, etc. were mapped together. The raw data was first downloaded from its source and subsequently cleaned, where extraneous information such as data from non-humans, internal identifiers, timestamps, etc. were removed. The remaining information was then integrated into an SQLite database. DRUGPATH currently contains a total of 2,632,516 unique entries, and of these, there are 54,757 unique genes, 2,632,242 unique pathways, and 31,042 unique drugs. DRUGPATH allows researchers and clinicians to discern which pathways are affected by each drug, reducing the likelihood of an adverse drug reaction occurring. The incorporation of drug, gene, target, and pathway information makes DRUGPATH a powerful resource for predicting potential side effects when designing or refining a given drug combination therapy. Not only that, it also contains the FDA status, half-life, and indication for each drug whenever possible for clinical applications of this database.

Table 19.1 provides an incomplete list of polypharmacology databases and related resources.

19.3 Pharmacology Databases

19.3.1 DrugBank

The DrugBank database is a comprehensive, freely accessible, online database containing information on drugs and drug targets created and maintained by the University of Alberta and The Metabolomics Innovation Centre located in Alberta, Canada [50]. As both a bioinformatics and a cheminformatics resource, DrugBank combines detailed drug (i.e., chemical, pharmacological, and pharmaceutical) data with comprehensive drug target (i.e., sequence, structure, and pathway) information [50]. The DrugBank Online website is available to the public as a free-to-access resource. However, use and redistribution of content from DrugBank Online or the underlying DrugBank Data, in whole or part, and for any purpose require a license. Academic users can apply for a free license for certain use cases, while all other users require a paid license.

The latest release of the database (version 5.0) [51, 52] contains 9591 drug entries including 2037 FDA-approved small molecule drugs, 241 FDA-approved biotech (protein/peptide) drugs, 96 nutraceuticals, and over 6000 experimental drugs. Additionally, 4270 non-redundant protein (i.e., drug target/enzyme/transporter/carrier) sequences are linked to these drug entries. Each DrugCard entry contains more than 200 data fields with half of the information being devoted to drug/chemical data and the other half devoted to drug target or protein data [52].

Four additional databases, HMDB [53], T3DB [54], SMPDB [55], and FooDB, are also part of a general suite of metabolomic/cheminformatic databases. HMDB contains equivalent information on more than 40,000 human metabolites, T3DB contains information on 3100 common toxins and environmental pollutants, SMPDB contains pathway diagrams for nearly 700 human metabolic pathways and disease pathways, while FooDB contains equivalent information on ~28,000 food components and food additives.

DrugBank [51] collects comprehensive molecular information about drugs, their mechanisms, interactions, and targets. It is primarily focused on providing data mining tools needed to facilitate target discovery and drug development. Latest versions of the DrugBank also provide information on the effect of hundreds of drugs on metabolite levels (pharmacometabolomics data), gene expression (pharmacotranscriptomics data), and protein expression (pharmacoprotoemics data). New data have also been added on the status of hundreds of new drug clinical trials and existing drug repurposing trials. Data in DrugBank are provided in an XML format. This makes data downloads and development of data extraction routines simpler and faster for programmers and database developers.

DrugBank combines detailed drug (i.e., chemical, pharmacological, and pharmaceutical) data with comprehensive drug target (i.e., sequence, structure, and pathway) information. The database contains 6711 drug entries including 1447 FDA-approved small molecule drugs, 131 FDA-approved biotech (protein/peptide) drugs, 85 nutraceuticals, and 5080 experimental drugs [52].

The knowledge base consists of proprietary authored content describing clinical level information about drugs such as side effects and drug interactions, as well as molecular level data such as chemical structures and what proteins a drug interacts with. DrugBank offers a suite of products powered by the DrugBank Platform and has customers located around the world crossing multiple industries including precision medicine, electronic health records, drug development, and regulatory agencies. DrugBank also provides DrugBank Online as a free-to-access resource for academic research and is used by millions of pharmacists, pharmacologists, health professionals, and pharmaceutical researchers every year.

Table 19.1 List of polypharmacology databases (in blue) and related resources (in black)

Database	Description	Website
Polypharmacology Browser (PPB)	PPB predicts possible targets for small molecules by searching for nearest neighbors using ten different fps describing composition, substructures, molecular shape, and pharmacophores	http://gdbtools.unibe.ch:8080/PPB/ https://ppb2.gdb.tools/
Polypharma	Polypharma integrates a variety of online resources to build a novel database specifically for multitargeting ligands, along with their modulated targets and quantitative bioactivities (e.g., binding affinity), in particular for polypharmacology modeling It incorporates both ligand-based prediction of targets, which correlates the ligand properties to multitarget activity data, and structure-based prediction of targets using docking/inverse docking approaches	http:/imdlab.org/polypharma/
CANDO	The platform screens and ranks drugs/compounds for every disease/indication by considering their interactions to every biological protein target, as well as integration of extensive, multi-scale data, from comprehensive well-curated libraries It was built based on a hypothesis that drugs function by interacting with multiple protein targets to create a molecular interaction signature that can be exploited for therapeutic repurposing and discovery	http://protinfo.compbio.buffalo.edu/cando/
QuartataWeb	It allows users to query a series of chemicals, drug combinations, or multiple targets, to enable multidrug, multitarget, multi-pathway analyses, toward facilitating the design of polypharmacological treatments for complex diseases	http://quartata.csb.pitt.edu
SAMNet	It was built upon a network strategy that takes into account the polypharmacology of small molecules in order to generate hypotheses for their broader mode of action	http://www.github.com/sgosline/SAMNet
Multiple Target Ligands Database (MTLD)	MTLD database contains 2444 multitarget ligands that bind with 21,424 binding sites from 18,231 crystal structures	www.mtdcadd.com
Multi-Target Ligand Finder ePlatton	The database aims to assist compound design, polypharmacology exploration, and drug combination formulation	http://www.megabionet.org/eplatton
Promiscuous 2.0	It enables the creation of complex interaction networks in an intuitive and interactive way and infer new indications for existing compounds	http://bioinformatics.charite.de/promiscuous2
The Drug Gene Pathway (DRUGPATH	It was created for various diseases where therapy involves using multiple drugs in combination It can be used to map the interactions between drugs, genes, pathways, and targets for use in the treatment of complex diseases, including the prediction of off-target interactions	https://drugpath.app/
DrugBank	Provides chemical, pharmacological, and pharmaceutical data with comprehensive drug target (i.e., sequence, structure, and pathway) information. Database contains 6711 drug entries including 1447 FDA-approved small molecule drugs, 131 FDA-approved biotech (protein/peptide) drugs, 85 nutraceuticals, and 5080 experimental drugs	http://www.drugbank.ca/
STITCH	Contains interactions for between 300,000 small molecules and 2.6 million proteins from 1133 organisms. Chemicals are linked to other chemicals and proteins by evidence derived from experiments, databases, and the literature	http://stitch.embl.de/
BindingDB	Database of measured binding affinities, focusing chiefly on the interactions of protein considered to be drug targets with small, drug-like molecules. It contains 832,773 binding data for 5765 protein targets and 362,123 small molecules	http://www.bindingdb.org/

(continued)

Table 19.1 (continued)

Database	Description	Website
SuperTarget	An extensive web resource for analyzing 332,828 drug–target interactions	http://insilico.charite.de/supertarget/
IUPHAR-DB	Detailed pharmacological, functional, and pathophysiological information on G-protein-coupled receptors, voltage-gated ion channels, ligand-gated ion channels, and nuclear hormone receptors	http://www.iuphar-db.org/
PubChem's BioAssay Database	The public repository of bioassay database contains 500,000 descriptions of assay protocols, covering 5000 protein targets and 30,000 gene targets and providing over 130 million bioactivity outcomes	http://pubchem.ncbi.nlm.nih.gov/
WOMBAT	Contains 331,872 entries (268,246 unique SMILES), representing 1966 unique targets captured from 15,320 papers published in medicinal chemistry journals between 1975 and 2009	http://www.sunsetmolecular.com/
ChEMBL	Contains 2D structures, calculated properties (e.g., logP, molecular weight, Lipinski parameters, etc.) and abstracted bioactivities (e.g. binding constants, pharmacology, and ADMET data)	https://www.ebi.ac.uk/chembl/

All data in DrugBank is derived from public non-proprietary sources. Nearly every data item is fully traceable and explicitly referenced to the original source. DrugBank data is available through a public web interface.

Users may query DrugBank in a number of ways:

(a) Simple text queries of the entire textual component of the database are supported. Clicking on the Browse button generates a tabular synopsis of DrugBank's content. This view allows users to casually scroll through the database or re-sort its contents.
(b) Clicking on a given DrugCard button brings up the full data content for the corresponding drug. A complete explanation of all the DrugCard fields and sources is given there.
(c) The PharmaBrowse button allows users to browse through drugs as grouped by their indication. This is particularly useful for pharmacists and physicians but also for pharmaceutical researchers looking for potential drug leads.
(d) The ChemQuery button allows users to draw (using ChemAxon applets) or write (as a SMILES string) a chemical compound and to search DrugBank for chemicals similar or identical to the query compound.
(e) The TextQuery button supports a more sophisticated text search (partial word matches, case sensitive, misspellings, etc.) of the text portion of DrugBank.
(f) The SeqSearch button allows users to conduct BLAST (protein) sequence searches of the 18,000 sequences contained in DrugBank. Both single and multiple sequence (i.e., whole proteome) BLAST queries are supported.
(g) The Data Extractor button opens an easy-to-use relational query search tool that allows users to select or search over various combinations of subfields. The Data Extractor is the most sophisticated search tool for DrugBank.

Network pharmacology elucidates the relationship between drugs and targets. As the identified targets for each drug increase, the corresponding drug–target network (DTN) evolves from solely reflection of the pharmaceutical industry trend to a portrait of polypharmacology. A study evaluated the potentials of DrugBank database in advancing systems pharmacology [56]. The investigators constructed and analyzed DTN from drug and target associations in the DrugBank 4.0 database. Our results showed that in bipartite DTN, increased ratio of identified targets for drugs augmented density and connectivity of drugs and targets and decreased modular structure. To clear up the details in the network structure, the DTNs were projected into two networks, namely, drug similarity network (DSN) and target similarity network (TSN). In DSN, various classes of Food and Drug Administration-approved drugs with distinct therapeutic categories were linked together based on shared targets. Projected TSN also showed complexity because of promiscuity of the drugs. By including investigational drugs that are currently being tested in clinical trials, the networks manifested more connectivity and pictured the upcoming pharmacological space in the future years. Diverse biological processes and protein–protein interactions were manipulated by new drugs, which can extend possible target combinations. The authors concluded that network-based organization of DrugBank 4.0 [56] data not only reveals the potential for repurposing of existing drugs but also allows generating novel predictions about drugs off-targets, drug–drug interactions, and their side effects. Similar application of DrugBank as an in silico model for interpreting polypharmacology in drug–target networks has also been documented [57].

DrugBank websites: http://www.drugbank.ca/

19.3.2 DGidb

The DGidb database is an open-access database and web interface with collection of drug–gene relationships and gene druggability information from 30 distinct repositories, including papers, databases, and web resources [58]. It has been the first drug target discovery platform providing tools to capture and prioritize genes that are known to be targeted by existing drugs, especially targeted drugs rather than broad chemotherapeutics. Drug–gene interactions have been mined from existing databases and literature to populate DGIdb. Similarly, genes have been categorized as potentially druggable according to membership in selected pathways, molecular functions, and gene families. In the latest version, druggable genes from GWASs are also included. All data from DGIdb are available as tab-delimited data downloads and also through a web services API.

For the version 3.0, there has been a substantial expansion of the content of the database through the addition of new sources and the updating of existing sources. Six new sources have been added, which brings the total number of sources represented by the DGIdb to 30. Three of these new sources provide drug–gene interactions from prominent expert-curated databases of clinically actionable variants similar to the CIViC and Clearity Foundation sources, already included in the DGIdb [59]. These include the Precision Oncology Knowledge Base (OncoKB), cancer genome interpreter (CGI), and The Jackson Laboratory Clinical Knowledgebase (CKB) [60, 61]. From these three new sources, 2822 new interaction claims were added. These sources were added as antineoplastic interaction sources, resulting in an 36.7% increase in antineoplastic drugs. Drug–gene interactions were also added from the United States Food and Drug Administration (FDA) Pharmacogenomics website and the National Cancer Institute (NCI) Cancer Gene Index, resulting in an additional 276 and 6231 interaction claims, respectively. In addition to these sources, the existing druggable gene category sources were expanded with data extracted from a new druggable genome paper that used computational approaches to identify druggable genes from genome-wide association studies [62]. The inclusion of this source into the DGIdb results in an additional 2300 potentially druggable genes, a 58% increase from 2.0. Notably, this independent definition of the druggable genome almost completely encapsulates many other gene categories (e.g., kinases, G-protein-coupled receptors) that are expected to be good drug targets.

DGIdb website: www.dgidb.org

19.3.3 The IUPHAR/BPS Guide to PHARMACOLOGY or IUPHAR Database (IUPHAR-DB)

The IUPHAR/BPS Guide to PHARMACOLOGY is an open-access website, acting as a portal to information on the biological targets of licensed drugs and other small molecules [63, 64]. The Guide to PHARMACOLOGY (with GtoPdb being the standard abbreviation) is developed as a joint venture between the International Union of Basic and Clinical Pharmacology (IUPHAR) and the British Pharmacological Society (BPS). This replaces and expands upon the original 2009 IUPHAR Database (standard abbreviation IUPHAR-DB). The Guide to Pharmacology aims to provide a concise overview of all pharmacological targets, accessible to all members of the scientific and clinical communities and the interested public, with links to details on a selected set of targets. The information featured includes pharmacological data, target, and gene nomenclature, as well as curated chemical information for ligands. Overviews and commentaries on each target family are included, with links to key references.

The target groups currently included on the Guide to PHARMACOLOGY are:

Catalytic receptors
Enzymes
G-protein-coupled receptors
Ion channels
Kinases
Nuclear receptors
Transporters
Other protein targets including fatty acid-binding proteins, sigma receptors, and adiponectin receptors

Information for each target group is subdivided into families based on classification, with a separate data page for each family. Within each page, targets are arranged into lists of tables, with each table including the protein and gene nomenclature for the target with links to gene nomenclature databases and listing selected ligands with activity at the target, including agonists, antagonists, inhibitors, and radioligands. Pharmacological data and references are given, and each ligand is hyperlinked to a ligand page displaying nomenclature and a chemical structure or peptide sequence, along with synonyms and relevant database links. The Guide to PHARMACOLOGY also includes a list of all ligand molecules included on the site, subdivided into categories including small organic molecules (including mammalian metabolites, hormones, and neurotransmitters), synthetic organic molecules, natural products, peptides, inorganic molecules, and antibodies. A complete list of all the approved drugs included on the website is also available via the ligand list. The Guide to PHARMACOLOGY is being expanded to include clinical information on targets and ligands, in addition to educational resources. Search features on the website include quick and advanced search options and receptor and ligand searches, including support for ligand structures using chemical structures. Other features include "Hot topic" news items and a recent receptor–ligand pairing list.

The Guide to PHARMACOLOGY portal offers a remit to:

(a) Provide access to data on all known biological targets.
(b) Enable students and scientists in academia and industry, working in areas related to pharmacology and drug/target research, to exploit the full potential of the considerable amount of information on drug action available in the published literature.
(c) Provide an entry point into the pharmacological literature for basic and clinical scientists from other disciplines.
(d) Provide an integrated educational resource with access to high-quality training in the principles of basic and clinical pharmacology and techniques.
(e) Foster innovative drug discovery.

IUPHAR-DB websites: http://www.iuphar-db.org/; https://www.guidetopharmacology.org/

19.4 Therapeutic Target Database

19.4.1 Therapeutic Target Database (TTD)

Therapeutic Target Database (TTD) is a pharmaceutical and medical repository [65, 66] constructed by the Innovative Drug Research and Bioinformatics Group (IDRB) at Zhejiang University, China, and the Bioinformatics and Drug Design Group at the National University of

Singapore. It provides information about known and explored therapeutic protein and nucleic acid targets, the targeted disease, pathway information, and the corresponding drugs directed at each of these targets [67]. Detailed knowledge about target function, sequence, 3D structure, ligand-binding properties, enzyme nomenclature and drug structure, therapeutic class, and clinical development status is crucial [67].

This database currently contains 3419 therapeutic targets (461 successful, 1191 clinical trial, 207 patented, and 1560 research targets) and 37,316 drugs (2649 approved, 9465 clinical trial, 5059 patented, and 20,143 experimental drugs). The targets and drugs in TTD cover 583 protein biochemical classes and 958 drug therapeutic classes, respectively [68]. The latest version of the International Classification of Diseases (ICD-11) codes released by WHO are incorporated in TTD to facilitate the clear definition of disease/disease class.

TTD also provides information about drug resistance mutations, gene expression profiles in the disease-relevant drug-targeted tissue of the patients and healthy individuals, and target combinations of multitarget drugs and drug combinations. The database is organized through five main panels which authorize to browse it by advanced search, patient data, target or drug groups, or model data. It also permits to download various datasets.

The TTD can be used for the following purposes.

19.4.1.1 Validation of Primary Therapeutic Target

Target validation normally requires the determination that the target is expressed in the disease-relevant cells/tissues, that it can be directly modulated by a drug or drug-like molecule with adequate potency in biochemical assay, and that target modulation in cell and/or animal models ameliorates the relevant disease phenotype. Therefore, TTD collects three types of target validation data [69]:

(a) Experimentally determined potency of drugs against their primary target or targets.
(b) Evident potency or effects of drugs against disease models (cell lines, ex vivo, in vivo models) linked to their primary target or targets.
(c) Observed effects of target knockout, knockdown, RNA interference, transgenetic, antibody, or antisense treated in vivo models.

19.4.1.2 Mutation and Expression Profile of Target

Extensive efforts have been directed at the discovery, investigation, and clinical monitoring of targeted therapeutics. These efforts may be facilitated by the convenient access of the genetic, proteomic, interactive, and other aspects of the therapeutic targets. The related data are provided in TTD [67]:

(a) 2000 drug resistance mutations in 83 targets and 104 target/drug regulatory genes
(b) Differential expression profiles of 758 targets in the disease-relevant drug-targeted tissue and the non-targeted tissues.
(c) 1008 target combinations of 1764 drugs and the 1604 target combinations of 664 multi-target drugs.

19.4.1.3 Clinical Trial and Patent Protected Target

Increasing numbers of proteins, nucleic acids, and other molecular entities have been explored as therapeutic targets, hundreds of which are targets of approved, clinical trial, and patent-protected drugs. Knowledge about these targets and corresponding drugs, particularly those in clinical uses and trials and patented, is highly useful for facilitating drug discovery. The latest version of TTD provided [68]:

(a) Clinical trial agents and their targets together with their structures and experimental activity values.
(b) Patented agents and their targets together with their structures and experimental activity values.

19.4.1.4 Regulator and Signaling Pathway of Target

Knowledge of therapeutic targets and early drug candidates is useful for improved drug discovery.

Particularly, the data of target regulators and affiliated signaling pathways can facilitate the research regarding druggability, systems pharmacology, new trends, molecular landscapes, and the development of drug discovery tools. Some transporter databases are provided. TTD is thus developed to provide such information about [68]:

(a) Target-regulating microRNAs and transcription factors and target-interacting proteins.
(b) Comprehensive target–pathway pairs fully referenced by multiple pathway databases.
(c) The convenient access of the multiple targets and drugs cross-linked to each of these pathway entries.

TTD website: http://db.idrblab.net/ttd/

19.4.2 Drug Target Commons (DTC)

Drug Target Commons (DTC) [70] is a crowdsourcing web platform that aims to standardize the collection, management, curation, and annotation of the notoriously heterogeneous compound–target bioactivity data to facilitate drug discovery, target identification, and drug repurposing. It integrates different, publicly available bioactivity data, which are generated using various assays, on compound–target interactions. Multiple data sources are systematically used for discovery of new indications for drugs (i.e., the selection of compound affecting specific proteins or biological pathways). DTC also provides access to QSAR models that can be used to extend target spaces for drugs. Besides, the drug-related bioactivity data are combined with chemical proteomics data in order to characterize biological pathways that are affected by certain drugs, to identify markers for drug monitoring, and to determine drug cocktails. These data can be used to characterize potential safety issues associated with certain drugs (e.g., in vivo absorption, distribution, metabolism, excretion, and toxicity properties) [71].

DTC website: http://drugtargetcommons.fimm.fi

19.4.3 Adverse Drug Reaction Classification System–Target Profile (ADReCS-Target)

ADReCS-Target [72] provides target profiles for aiding drug safety research and application by collecting data about adverse drug reactions (ADRs) caused by drug interactions with protein, gene, and genetic variation. ADReCS-Target contains more than 66,000 association pairs with over 2200 standard ADR terms manually curated from text mining of the public scientific literatures. All the terms are standardized by using ADReCS ontology and represented as a connected network or systematic fashion. The user can download selected records via the embedded download function in six formats such as JSON, XML, CSV, TXT, SQL, and MS-Excel. ADReCS-Target also allows batch data retrieval.

ADReCS-Target website: http://bioinf.xmu.edu.cn/ADReCS-Target

19.5 Drug–Target Interaction Databases

19.5.1 BindingDB

BindingDB is a public, web-accessible database of measured binding affinities, focusing chiefly on the interactions of protein considered to be drug-targets with small, drug-like molecules [73]. As of February 7, 2021, BindingDB contains 41,328 entries, each with a DOI, containing 2,114,159 binding data for 8202 protein targets and 928,022 small molecules. There are 2823 protein–ligand crystal structures with BindingDB affinity measurements for proteins with 100% sequence identity and 8263 crystal structures allowing proteins to 85% sequence identity.

The purpose of BindingDB is to support medicinal chemistry and drug discovery via literature awareness and development of structure–activity relations (SAR and QSAR); validation of computational chemistry and molecular modeling approaches such as docking, scoring, and free energy methods; chemical biology and chemical genomics; and basic studies of the physical chemistry of molecular recognition.

The data collection derives from a variety of measurement techniques, including enzyme inhibition and kinetics, isothermal titration calorimetry, NMR, and radioligand and competition assays. BindingDB includes data extracted from the scientific literature by the BindingDB project, selected PubChem confirmatory BioAssays, and ChEMBL entries for which a well-defined protein target is provided. BindingDB's web interface provides a range of browsing, query, and data download tools. These include browsing by the name of a protein target or by journal citation, query by chemical similarity and substructure, and downloads by target or query result.

BindingDB website: http://www.bindingdb.org/

19.5.2 SuperTarget

In 2008, the first version of SuperTarget was developed with the intention to accentuate drug–target interactions themselves and to provide references to other resources for more elaborate analysis [74]. SuperTarget is a database developed in the first place to collect information about drug–target relations and an extensive web resource for analyzing 332,828 drug–target interactions. The aim of SuperTarget is to offer an at most comprehensive datasets. It integrated drug–target relations from different well-known databases as DrugBank, BindingDB, and SuperCyp.

SuperTarget consists mainly of three different types of entities: drugs, proteins, and side effects. It also contains information about pathways and ontologies. The analysis of drug–target interactions can play a crucial role to improve the process of drug design and admission. SuperTarget provided a variety of drug–target interactions and affected biological pathways in a user-friendly manner. This second release of SuperTarget contains a core dataset of ~330,000 drug–target interactions, of which about 310,000 interactions have binding affinity data. We consider a drug–target relation as a specific interaction of a small chemical compound, which could be used to treat or diagnose a disease. Thus, SuperTarget now enables scientists to carry out not only qualitative but also quantitative analysis of drug–target interactions. To provide users with further information on drug–target interactions, SuperTarget provides links to physicochemical properties and further structural information of drugs. Proven or potential target proteins are represented as stored in UniProtKB, by functional annotations extracted from GO, and by related pathway information provided by KEGG.

SuperTarget website: http://insilico.charite.de/supertarget/

19.6 Genomic Databases

19.6.1 DisGeNET

The DisGeNET platform [75] aims to overcome the fragmentation and heterogeneity of available genomic data for mining gene–disease associations. As one of the largest available collections of genes and variants involved in human diseases, DisGeNET integrates data from expert curated repositories, GWAS catalogues, animal models, and the scientific literature. DisGeNET data are homogeneously annotated with controlled vocabularies and community-driven ontologies. Additionally, several original metrics are provided to assist the prioritization of genotype–phenotype relationships. The information is accessible through a web interface, a Cytoscape App, an RDF SPARQL endpoint, scripts in several programming languages, and an R package. DisGeNET is a versatile platform that can be used for different research purposes including the investigation of the molecular underpinnings of specific human diseases and their comorbidities, the analysis of the properties of disease genes, the generation of hypothesis on drug therapeutic action and drug adverse effects, the validation of computationally predicted disease genes, and the evaluation of text-mining methods performance.

DisGeNET features a score based on the supporting evidence to prioritize gene–disease associations. These scores rely on data retrieved from databases of curating genetic association studies

(the GWAS Catalog and the Genetic Association Database) and genomic information extracted from animal models. DisGeNET data are available for downloading in several formats: as SQLite database, as tab-separated files, and as dump files, serialized in RDF/Turtle.
DisGeNET website: http://www.disgenet.org

19.6.2 Open Targets Platform (OTP)

The Open Targets Platform (OTP) [76] is a platform for therapeutic target identification and validation, providing either a target-centric workflow to identify diseases that may be associated with a specific target or a disease-centric workflow to identify targets that may be associated with a specific disease. Users can easily transition between these target- and disease-centric workflows. It is also designed to support identification and prioritization of biological targets for follow-up. Coverage includes genetic associations, somatic mutations, known drugs, gene expression, affected pathways, literature mining, and animal models. Each drug target is linked to a disease using integrated genome-wide data from a broad range of data sources. The platform provides either a target-centric workflow to identify diseases that may be associated with a specific target or a disease-centric workflow to identify targets that may be associated with a specific disease. The latest version of Open Target provides information regarding the tractability of a target, which measures the "ligandability" of putative drug targets [77], and safety risk information associated with selected targets. OTP allows programmatic retrieval of data via a set of REST services or, alternatively, the access to dump files.
OPT website: https://www.targetvalidation.org

19.6.3 Pharos

The "druggable genome" encompasses several protein families, but only a subset of targets within them have attracted significant research attention and thus have information about them publicly available. The Illuminating the Druggable Genome (IDG) program was initiated in 2014 and has the goal of developing experimental techniques and a Knowledge Management Center (KMC) that would collect and organize information about protein targets from four families, representing the most common druggable targets with an emphasis on understudied proteins [78]. Two resources were developed by the KMC [78]: the Target Central Resource Database (TCRD) which collates many heterogeneous gene/protein datasets and Pharos, a multimodal web interface that presents the data from TCRD. The aim of Pharos is to encourage "serendipitous browsing," whereby related, relevant information is made easily discoverable.

Pharos [78] provides a web interface for data collected by the Illuminating the Druggable Genome initiative. It incorporates text-mined bibliometric associations and statistics from the biomedical and patent literature, mRNA and protein expression data, disease and phenotype associations, bioactivity data, drug–target interactions, and omics-driven data imported from the Harmonizome. It also integrates with the functionality of the Drug Central and DTO resources. The Drug Target Ontology is a database providing tools to classify and integrate drug discovery data based on formalized and standardized classifications and annotations of druggable protein targets. DTO integrates phylogenecity, function, target development level, disease association, tissue expression, chemical ligand and substrate characteristics, and target family specific characteristics. Protein classes are linked to tissue and disease via different levels of confidence. DTO also contains drug target development classifications, a large collection of cell lines from the LINCS project and relevant cell–disease and cell–tissue relations. DTO is modeled in OWL2-DL to enable further classification by inference reasoning and SPARQL queries. DTO is implemented following a modularization approach. DTO will serve as the organizational framework for drug targets in the IDG PHAROS User Interface Portal.
Pharos website: https://pharos.nih.gov

19.6.4 Comparative Toxicogenomics Database (CTD)

The Comparative Toxicogenomics Database (CTD) [79] is a premier public resource that provides information for literature-based, manually curated associations and interactions between chemicals, gene products, phenotypes, diseases, environmental exposures, and pharmaceuticals and their relationships to diseases. Professional biocurators manually curate the scientific literature, transforming text, tables, figures, and supplemental files into annotated data that are seamlessly integrated and available through CTD's public web application (PWA). Different data sources for toxicogenomics, phenotypes, diseases, environmental exposures, and pharmaceuticals are considered in order to build drug–gene and drug–disease interactions. Overall, CTD includes over 38 million toxicogenomic relationships for analysis and hypothesis development. This information is organized through community-accepted controlled vocabularies and ontologies with accession identifiers, in order to ensure that CTD's content is cohesive, manageable, and computable, as well as adhering to the FAIR principle. CTD's vocabularies and content are described and made freely available for users to download in a variety of formats. Additionally, all CTD interactions are embellished with taxa identification (enabling data to be compared across species, from model laboratory organisms to humans) and are directly linked to the original source article (providing full transparency and traceability). Furthermore, CTD maintains good data stewardship with the digital informatics community by facilitating semantic standards for the environmental health science community, complying with reporting standards set by the BioSharing Information Resources, and registering with BioDBcore (https://biosharing.org/biodbcore-000173) and the NAR Molecular Biology Database Collection (http://www.oxfordjournals.org/our_journals/nar/database/summary/1188).
CTD website: http://ctdbase.org/

19.7 Protein Databases

19.7.1 Protein Data Bank (PDB)

The Protein Data Bank (PDB) [80] is a database for the three-dimensional structural data of large biological molecules, such as proteins and nucleic acids. The data, typically obtained by X-ray crystallography, NMR spectroscopy, or, increasingly, cryo-electron microscopy and submitted by biologists and biochemists from around the world, are freely accessible on the Internet via the websites of its member organizations (PDBe, PDBj, RCSB, and BMRB). The PDB is overseen by an organization called the Worldwide Protein Data Bank, wwPDB.

The PDB is a key in areas of structural biology, such as structural genomics. Most major scientific journals and some funding agencies now require scientists to submit their structure data to the PDB. Many other databases use protein structures deposited in the PDB. For example, SCOP and CATH classify protein structures, while PDBsum provides a graphic overview of PDB entries using information from other sources, such as Gene Ontology.

The Protein Data Bank (PDB) provides experimentally validated structural information about targets and bound ligands, representing a valuable source of information to help in (**a**) identifying active sites, (**b**) understanding pharmacophore requirements, (**c**) designing novel ligands, and (**d**) inferring structure–activity relationships.

The structure files may be viewed using one of several free and open source computer programs, including Jmol, Pymol, VMD, and Rasmol. Other non-free, shareware programs include ICM-Browser, MDL Chime, UCSF Chimera, Swiss-PDB Viewer, StarBiochem (a Java-based interactive molecular viewer with integrated search of protein databank), Sirius, and VisProt3DS (a tool for Protein Visualization in 3D stereoscopic view in anaglyph and other modes) and Discovery Studio. The RCSB PDB website contains an extensive list of both free and commercial molecule visualization programs and web browser plugins.

In a study, Pinzi et al. [81] performed a large-scale analysis of the PDB by integrating different ligand-based and structure-based approaches, with the aim of identifying promising target associations for polypharmacology based on reported crystal structure information. First, the 2D and 3D similarity profiles of the crystallographic ligands were evaluated using different ligand-based methods. Then, activity data of pairs of similar ligands binding to different targets were inspected by comparing structural information with bioactivity annotations reported in the ChEMBL, BindingDB, Binding MOAD, and PDBbind databases. Afterward, extensive docking screenings of ligands in the identified cross-targets were made in order to validate and refine the ligand-based results. Finally, the therapeutic relevance of the identified target combinations for polypharmacology was evaluated from comparison with information on therapeutic targets reported in the Therapeutic Target Database (TTD). The results led to the identification of several target associations with high therapeutic potential for polypharmacology.
PDB website: http://www.wwpdb.org/

19.7.2 Protein–Protein Interaction (PPI) Databases

Protein–protein interactions (PPIs) [82] are physical contacts of high specificity established between two or more protein molecules as a result of biochemical events steered by interactions that include electrostatic forces, hydrogen bonding, and the hydrophobic effect. Many are physical contacts with molecular associations between chains that occur in a cell or in a living organism in a specific biomolecular context.

Proteins rarely act alone as their functions tend to be regulated. Many molecular processes within a cell are carried out by molecular machines that are built from numerous protein components organized by their PPIs. These physiological interactions make up the so-called interactomics of the organism, while aberrant PPIs are the basis of multiple aggregation-related diseases, such as Creutzfeldt–Jakob and Alzheimer's diseases.

PPIs have been studied with many methods and from different perspectives: biochemistry, quantum chemistry, molecular dynamics, and signal transduction, among others. All this information enables the creation of large protein interaction networks—similar to metabolic or genetic/epigenetic networks—that empower the current knowledge on biochemical cascades and molecular etiology of disease, as well as the discovery of putative protein targets of therapeutic interest.

Over the past few years, the number of known protein–protein interactions has increased substantially. To make this information more readily available, a number of publicly available databases have set out to collect and store protein–protein interaction data [83]. Protein–protein interactions have been retrieved from six major databases and integrated and the results compared. The six databases (the Biological General Repository for Interaction Datasets [BioGRID], the Molecular INTeraction database [MINT], the Biomolecular Interaction Network Database [BIND], the Database of Interacting Proteins [DIP], the IntAct molecular interaction database [IntAct], and the Human Protein Reference Database [HPRD]) differ in scope and content; integration of all datasets is nontrivial owing to differences in data annotation. With respect to human protein–protein interaction data, HPRD seems to be the most comprehensive. To obtain a complete dataset, however, interactions from all six databases have to be combined. To overcome this limitation, meta-databases such as the Agile Protein Interaction Database (APID) offer access to integrated protein–protein interaction datasets, although these also currently have certain restrictions [83].

Large-scale identification of PPIs generated hundreds of thousands of interactions, which were collected together in specialized biological databases that are continuously updated in order to provide complete interactomes. The first of these databases was the Database of Interacting Proteins (DIP) [84].

Primary databases collect information about published PPIs via small-scale or large-scale experimental methods, for example, DIP,

Biomolecular Interaction Network Database (BIND), Biological General Repository for Interaction Datasets (BioGRID), Human Protein Reference Database (HPRD), IntAct Molecular Interaction Database, Molecular Interactions Database (MINT), MIPS Protein Interaction Resource on Yeast (MIPS-MPact), and MIPS Mammalian Protein–Protein Interaction Database (MIPS-MPPI).

Meta-databases normally result from the integration of primary databases information but can also collect some original data.

Prediction databases include many PPIs that are predicted using several techniques (main article), for example, Human Protein–Protein Interaction Prediction Database (PIPs) [85], Interlogous Interaction Database (I2D), Known and Predicted Protein–Protein Interactions (STRING-db), and Unified Human Interactive (UniHI).

DIP websaite: http://dip.doe-mbi.ucla.edu
BioGRID website: http://www.thebiogrid.org
MINT website: http://mint.bio.uniroma2.it/mint
BIND website: http://bond.unleashedinformatics.com
IntAct website: http://www.ebi.ac.uk/intact
HPRD website: http://www.hprd.org

19.7.3 Binding MOAD

Binding MOAD (Mother of All Databases) [86] is a database of 9836 protein–ligand crystal structures. All biologically relevant ligands are annotated, and experimental binding affinity data is reported when available. Binding MOAD has almost doubled in size since it was originally introduced in 2004, demonstrating steady growth with each annual update. Several technologies, such as natural language processing, help drive this constant expansion. Along with increasing data, Binding MOAD has improved usability. The website now showcases a faster, more featured viewer to examine the protein–ligand structures. Ligands have additional chemical data, allowing for cheminformatics mining. Lastly, logins are no longer necessary, and Binding MOAD is freely available to all at http://www.BindingMOAD.org.

The goal of Binding MOAD is to provide users with a dataset focused on high-quality X-ray crystal structures that have been solved with biologically relevant ligands bound. Where available, experimental binding affinities (K_a, K_d, K_i, IC_{50}) are provided from the primary literature of the crystal structure. The database has been updated regularly since 2005, and this most recent update has added nearly 7000 new structures (growth of 21%). MOAD currently contains 32,747 structures, composed of 9117 protein families and 16,044 unique ligands.

Binding MOAD website: http://www.BindingMOAD.org

19.7.4 Protein-Drug Interaction Database (PDID)

Protein-Drug Interaction Database (PDID) [87] addresses incompleteness of these data by providing access to putative protein–drug interactions that cover the entire structural human proteome. It covers 9652 structures from 3746 proteins and houses 16,800 putative interactions generated from close to 1.1 million accurate, all-atom structure-based predictions for several dozens of popular drugs. The predictions were generated with three modern methods: ILbind, SMAP, and eFindSite. They are accompanied by propensity scores that quantify likelihood of interactions and coordinates of the putative location of the binding drugs in the corresponding protein structures. PDID complements the current databases that focus on the curated interactions and the BioDrugScreen database that relies on docking to find putative interactions. Moreover, we also include experimentally curated interactions which are linked to their sources: DrugBank, BindingDB, and Protein Data Bank. Our database can be used to facilitate studies related to polypharmacology of drugs including repurposing and explaining side effects of drugs.

PDID website: http://biomine.ece.ualberta.ca/PDID/

19.7.5 Protein Pharmacology Interaction Network (PhiN)

The protein pharmacology interaction network database (PhIN) [88] aims to assist multitarget drug discovery by providing comprehensive and flexible network pharmacology analysis. Overall, PhIN contains 1,126,060 target–target interaction pairs in terms of shared compounds and 3,428,020 pairs in terms of shared scaffolds, which involve 12,419,700 activity data, 9414 targets, 314 viral targets, 652 pathways, 1,359,400 compounds, and 309,556 scaffolds. Using PhIN, users can obtain interacting target networks within or across human pathways, between human and virus, by defining the number of shared compounds or scaffolds under an activity cutoff. We expect PhIN to be a useful tool for multitarget drug development.

PhIN website: http://cadd.pharmacy.nankai.edu.cn/phin/

19.7.6 eMatchSite

Detecting similarities between ligand-binding sites in the absence of global homology between target proteins has been recognized as one of the critical components of modern drug discovery. Local binding site alignments can be constructed using sequence order-independent techniques; however, to achieve a high accuracy, many current algorithms for binding site comparison require high-quality experimental protein structures, preferably in the bound conformational state. This, in turn, complicates proteome scale applications, where only various quality structure models are available for the majority of gene products. To improve the state of the art, Brylinski [89] developed eMatchSite, a new method for constructing sequence order-independent alignments of ligand-binding sites in protein models. Large-scale benchmarking calculations using adenine-binding pockets in crystal structures demonstrate that eMatchSite generates accurate alignments for almost three times more protein pairs than SOIPPA. More importantly, eMatchSite offers a high tolerance to structural distortions in ligand-binding regions in protein models. For example, the percentage of correctly aligned pairs of adenine-binding sites in weakly homologous protein models is only 4–9% lower than those aligned using crystal structures. This represents a significant improvement over other algorithms, e.g., the performance of eMatchSite in recognizing similar binding sites is 6% and 13% higher than that of SiteEngine using high- and moderate-quality protein models, respectively. Constructing biologically correct alignments using predicted ligand-binding sites in protein models opens up the possibility to investigate drug–protein interaction networks for complete proteomes with prospective systems-level applications in polypharmacology and rational drug repositioning.

eMatchSite website: http://www.brylinski.org/ematchsite

Table 19.2 provides an incomplete list of protein databases.

19.8 Non-coding RNA Databases

19.8.1 MicroRNA Databases

As part of epigenetic mechanism, non-coding RNAs (ncRNAs) play critical roles in regulating nearly every biological function and pathological processes in humans. More importantly, these ncRNAs themselves may be therapeutic targets and/or therapeutic agents for many human diseases. Databases of ncRNAs therefore are important resources for multitarget drug (MTD) design and discovery. The categories of ncRNAs that have been extensively and intensively studied mainly include microRNAs (miRNAs), long non-coding RNAs (lncRNAs), and circular RNAs (circRNAs). The microRNA database and microRNA target databases provided below are a compilation of databases and web portals and servers used for microRNAs and their targets.

19.8.1.1 MicroRNA Databases (Table 19.3a)

19.8.1.2 MicroRNA Target Gene Databases (Table 19.3b)

19.8.2 Long Non-coding RNA Databases (Table 19.4)

19.8.3 Circular RNA Databases (Table 19.5)

19.9 Human Disease Databases

19.9.1 MalaCards: The Human Disease Database

MalaCards: The Human Disease Database is a robust database designed to enable genomic and genetic researchers, investigators, and scholars to efficiently navigate the universe of human genes, genetic variants, proteins, cells, and biological pathways related to various human diseases. Available since 2013, MalaCards provides detailed data on thousands of human diseases and their relationships with respective genes. Rappaport et al. describe the MalaCards project as an "attempt to generate a complete lexicon of all human diseases" [130, 131]. MalaCards is part of the LifeMap Sciences suite of products that includes the GeneCards Suite product line and contains relevant, integrated content from its genetics product line, such as GeneCards and GeneAnalytics.

Each disease is arranged in MalaCards under two distinct categories: anatomical and global. The anatomical disease section includes 18 major categories with more than 26,000 diseases representing all areas of the body, including blood, bone, immune, muscle, and reproductive diseases. The global disease section contains 6 categories, cancer, fetal, genetic, infectious, metabolic, and rare diseases, and almost 18,000 specific diseases.

The disease names have been compiled from 10 primary disease databases with another 11 databases as secondary sources. Overall, the site includes over 19,000 disease entries, which correspond with more than 13,000 genes from approximately 70 prominent databases and websites. Unique entries number close to 13,000 [131]. MalaCards uses several medical classification schemes, language systems, and related ontologies. Genetic information, variations, and associated information related to the disease are included along with annotations, matrices, and mapped tables.

MalaCards website: https://www.malacards.org/

19.9.2 DisGeNET: A Database of Gene–Disease Associations

DisGeNET is a discovery platform containing one of the largest publicly available collections of genes and variants associated with human diseases [75, 132, 133]. DisGeNET integrates data from expert curated repositories, GWAS catalogues, animal models, and the scientific litera-

Table 19.2 List of protein databases

Database	Description	Website
Repository for 3D biological macromolecule structures		
Protein Data Bank	Macromolecule structure coordinates archive	http://www.wwpdb.org
Structure resources		
RCSB Protein Data Bank	Macromolecular structure deposition, search, and analysis tools (USA)	http://www.rcsb.org
PDB in Europe (PDBe)	Macromolecular structure deposition, search, and analysis tools (Europe)	http://www.pdbe.org
PDB in Japan (PDBj)	Macromolecular structure deposition, search, and analysis tools (Japan)	http://www.pdbj.org
BioMagResBank (BMRB)	NMR dataset deposition, search, and validation tools	http://www.bmrb.wisc.edu
EMDataBank	Cryo-EM data deposition, search, and analysis tools	http://www.emdatabank.org
Nucleic Acid Database (NDB)	Nucleic acid structure search and analysis tools	http://ndbserver.rutgers.edu
Biosis	SAS data deposition	http://www.bioisis.net/
Protein Ensemble Database (pe-DB)	NMR and SAXS data deposition for intrinsically disordered proteins	http://pedb.vib.be
Structural annotation resources		
SCOP suite (SCOPe & SCOP2)	Structural classification	http://scop.mrc-lmb.cam.ac.uk/scop/
CATH	Structural classification	http://www.cathdb.info
SUPERFAMILY	Structural alignment and annotation	http://supfam.cs.bris.ac.uk/SUPERFAMILY/
Gene3D	Structural alignment and extended annotation	http://gene3d.biochem.ucl.ac.uk/Gene3D/
Genome3D	Structural alignment and extended annotation	http://genome3d.eu/
ArchDB	Structural classification of loops	http://sbi.imim.es/archdb/
MobiDB	Intrinsically disordered protein database	http://mobidb.bio.unipd.it
DisProt	Intrinsically disordered protein database	http://www.disprot.org
Other structural annotation databases		
PDBePISA	Intermolecular interactions and quaternary structure	http://pdbe.org/pisa
Dictionary of Interfaces in Proteins (DIP)	Protein–protein interactions	http://dip.doe-mbi.ucla.edu/dip/Main.cgi
ProtBuD	Known and predicting protein interactions	http://dunbrack.fccc.edu/ProtBuD/
3did	Domain–domain and domain–peptide interactions	http://3did.irbbarcelona.org/
Membrane Proteins of Known Structure (mpstruc)	Membrane protein structure search and annotation	http://blanco.biomol.uci.edu/mpstruc/listAll/list
Protein Data Bank of Transmembrane Proteins (PDBTM)	Membrane protein structure search and annotation	http://pdbtm.enzim.hu/
GPCRDB	G-protein-coupled receptor protein structure search and annotation	http://www.gpcr.org/7tm/
Transporter Classification Database	Classification of membrane transport proteins	http://www.tcdb.org
Data aggregators		
Structural Biology Knowledgebase (SBKB)	Structure-centric data aggregator and search tool for sites below	http://sbkb.org
PSI Protein Model Portal	Theoretical model data aggregator	http://www.proteinmodelportal.org

(continued)

Table 19.2 (continued)

Database	Description	Website
TargetTrack	Protein Structure Initiative structural target registration and experimental history database	http://sbkb.org/tt/
KB-Rank Search tool	Biomedical annotation structure search	http://protein.tcmedc.org
PSI Technology Portal	Catalog of PSI laboratory technologies and methods	http://technology.sbkb.org/portal/
PSI Publications Portal	Catalog of articles published by PSI	http://olenka.med.virginia.edu/psi/
Community-driven annotation		
Proteopedia	Community annotation of structures	http://proteopedia.org
TOPSAN	Community annotation of structural genomics structures	http://www.topsan.org

Table 19.3a MicroRNA databases

Name	Description	Website	References
deepBase	deepBase is a database for annotating and discovering small and long ncRNAs (microRNAs, siRNAs, piRNAs, etc.) from high-throughput deep sequencing data	https://rna.sysu.edu.cn/deepBase/	[90]
miRBase	miRBase database is a searchable database of published miRNA sequences and annotation	https://www.mirbase.org/	[91]
microRNA.org	microRNA.org is a database for experimentally observed microRNA expression patterns and predicted microRNA targets and target downregulation scores	http://www.microrna.org/microrna/getExprForm.do	[92]
miRGen 4.0	DIANA-miRGen v4: indexing promoters and regulators for more than 1500 microRNAs	https://diana.e-ce.uth.gr/mirgenv4	[93]
miRNAMap	miRNAMap: genomic maps of microRNA genes and their target genes in mammalian genomes	http://mirnamap.mbc.nctu.edu.tw/	[94]
TargetScan	TargetScan7.0 classifies microRNAs according to their level of conservation (i.e., species-specific, conserved among mammals, or broadly conserved among vertebrates) and aggregates them into families based upon their seed sequence. It also annotates conserved isomiRs using small RNA sequencing datasets	http://www.targetscan.org/vert_80/	[95]
VIRmiRNA	*VIRmiRNA* is the first dedicated resource on experimental *viral miRNA* and their *targets*. This resource also provides inclusive knowledge about anti-viral miRNAs known to play role in antiviral immunity of host	http://crdd.osdd.net/servers/virmirna/	[96]

ture. DisGeNET data are homogeneously annotated with controlled vocabularies and community-driven ontologies. Additionally, several original metrics are provided to assist the prioritization of genotype–phenotype relationships.

The current version of DisGeNET (v7.0) contains 1,134,942 gene–disease associations (GDAs), between 21,671 genes and 30,170 diseases, disorders, traits, and clinical or abnormal human phenotypes, and 369,554 variant–disease associations (VDAs), between 194,515 variants and 14,155 diseases, traits, and phenotypes.

The information in DisGeNET can be accessed in several ways:

Table 19.3b MicroRNA target gene database

Name	Description	Website	References
StarBase	starBase is designed for decoding miRNA–lncRNA, miRNA–mRNA, miRNA–circRNA, miRNA–pseudogene, miRNA–sncRNA, protein–lncRNA, protein–sncRNA, protein–mRNA, and protein–pseudogene interactions and ceRNA networks from 108 CLIP-Seq (HITS-CLIP, PAR-CLIP, iCLIP, CLASH) datasets. It also provides Pan-Cancer Analysis for microRNAs, lncRNAs, circRNAs, and protein-coding genes from 6000 tumor samples	https://web.archive.org/web/20110222111721/http://starbase.sysu.edu.cn/	[97, 98]
StarScan	StarScan is developed for scanning small RNA (miRNA, piRNA, siRNA)-mediated RNA cleavage events in lncRNA, circRNA, mRNA, and pseudogenes from degradome sequencing data	http://mirlab.sysu.edu.cn/starscan/	[99]
Cupid	Cupid is a method for simultaneous prediction of miRNA–target interactions and their mediated competing endogenous RNA (ceRNA) interactions. It is an integrative approach which significantly improves on miRNA–target prediction accuracy as assessed by both mRNA and protein level measurements in breast cancer cell lines. Cupid is implemented in three steps: Step 1: re-evaluate candidate miRNA binding sites in 3' UTRs. Step2: interactions are predicted by integrating information about selected sites and the statistical dependency between the expression profiles of miRNA and putative targets. Step 3: Cupid assesses whether inferred targets compete for predicted miRNA regulators. *Only the source code for step 3 is provided	http://cupidtool.sourceforge.net/	[100]
TargetScan	Predicts biological targets of miRNAs by searching for the presence of sites that match the seed region of each miRNA. In flies and nematodes, predictions are ranked based on the probability of their evolutionary conservation. In zebrafish, predictions are ranked based on site number, site type, and site context, which includes factors that influence target site accessibility. In mammals, the user can choose whether the predictions should be ranked based on the probability of their conservation or on site number, type, and context. In mammals and nematodes, the user can choose to extend the predictions beyond conserved sites and consider all sites	http://www.targetscan.org/vert_80/	[95]
TarBase	A comprehensive database of experimentally supported animal microRNA targets	http://diana.imis.athena-innovation.gr/DianaTools/index.php?r=tarbase/index	[101]
Diana--microT	DIANA-microT 3.0 is an algorithm based on several parameters calculated individually for each microRNA, and it combines conserved and non-conserved microRNA recognition elements into a final prediction score	https://web.archive.org/web/20101208180159/http://diana.cslab.ece.ntua.gr/microT/	[102]
miRecords	An integrated resource for microRNA–target interactions	http://c1.accurascience.com/miRecords/	[103]
PicTar	PicTar is combinatorial microRNA target predictions	http://pictar.bio.nyu.edu/	[104]
PITA	PITA incorporates the role of target site accessibility, as determined by base-pairing interactions within the mRNA, in microRNA target recognition	https://genie.weizmann.ac.il/pubs/mir07/mir07_data.html	[105]
RepTar	A database of inverse miRNA target predictions, based on the RepTar algorithm that is independent of evolutionary conservation considerations and is not limited to seed pairing sites	https://web.archive.org/web/20131214044623/http://reptar.ekmd.huji.ac.il/	[106]

Name	Description	Website	References
RNA22	The first link (predictions) provides RNA22 predictions for all protein-coding transcripts in human, mouse, roundworm, and fruit fly. It allows you to visualize the predictions within a cDNA map and also find transcripts where multiple miRs of interest target. The second website link (custom) first finds putative microRNA binding sites in the sequence of interest and then identifies the targeted microRNA	https://cm.jefferson.edu/rna22/ https://cm.jefferson.edu/rna22/	[107]
miRTarBase	The experimentally validated microRNA–target interaction database. As a database, miRTarBase has accumulated more than three hundred and sixty thousand miRNA–target interactions (MTIs), which are collected by manually surveying pertinent literature after NLP of the text systematically to filter research articles related to functional studies of miRNAs. Generally, the collected MTIs are validated experimentally by reporter assay, Western blot, microarray, and next-generation sequencing experiments. While containing the largest amount of validated MTIs, the miRTarBase provides the most updated collection by comparing with other similar, previously developed databases	http://mirtarbase.mbc.nctu.edu.tw/	[108]
miRwalk	Aggregates and compares results from other miRNA-to-mRNA databases	http://zmf.umm.uni-heidelberg.de/apps/zmf/mirwalk2/	[109]
MBSTAR	Multiple-instance approach for finding out true or functional microRNA binding sites	https://www.isical.ac.in/~bioinfo_miu/MBStar30.htm	[110]

Table 19.4 Long non-coding RNA (lncRNA) database

Name	Description		References
deepBase	Identification, expression, evolution, and function of long non-coding RNAs (LncRNAs), small RNAs, and circular RNAs from deep-sequencing data	https://rna.sysu.edu.cn/deepBase/	[111]
LNCipedia	A comprehensive compendium of human long non-coding RNAs	https://lncipedia.org/	[112, 113]
lncRNAdb	The reference database for functional long non-coding RNAs	http://lncrnadb.com/	[114]
LncRNAWiki	A wiki-based, publicly editable, and open-content platform for community curation of human long non-coding RNAs (lncRNAs)	https://ngdc.cncb.ac.cn/lncrnawiki1/index.php/Main_Page	[115]
LncBook	A comprehensive collection of 270,044 human lncRNAs and systematic curation of lncRNAs' annotation by multi-omics data integration, function annotation, and disease association	https://ngdc.cncb.ac.cn/lncbook/index	[116]
MONOCLdb	The MOuse NOnCode lung database provides the annotations and expression profiles of mouse long non-coding RNAs (lncRNAs) involved in influenza and SARS-CoV infections	https://www.monocldb.org/	[117]
NONCODE	An integrated knowledge database dedicated to ncRNAs, especially lncRNAs	http://www.noncode.org/	[118]
lncRNome	A comprehensive searchable biologically oriented knowledge base for long non-coding RNAs in humans	http://genome.igib.res.in/lncRNome/	
NRED	A database of long non-coding RNA expression	http://nred.matticklab.com/cgi-bin/ncrnadb.pl	[119]
C-It-Loci	A tool to explore and to compare the expression profiles of conserved loci among various tissues in three organisms	http://c-it-loci.uni-frankfurt.de/	[120]
MiTranscriptome	A catalog of human long poly-adenylated RNA transcripts derived from computational analysis of high-throughput RNA-seq data from over 6500 samples, spanning diverse cancer and tissue types	https://mitranscriptome.org/	[121]
slncky Evolution Browser	This site contains alignments and evolutionary metrics of conserved lncRNAs		[122]
Cancer LncRNA Census (CLC)	Database of long non-coding RNAs causally implicated in cancer through in vivo, in vitro, and other evidence	https://www.gold-lab.org/clc	[123]
BIG Transcriptome	High confidence of coding and non-coding transcriptomes assembled with hundreds of pseudo-stranded and stranded RNA-seq datasets		[124]

Table 19.5 Circular RNA (circRNA) databases

Name	Description	Website	References
circRNABase	circRNABase is designed for decoding miRNA–circRNA interaction networks from thousands of circRNAs and 108 CLIP-Seq (HITS-CLIP, PAR-CLIP, iCLIP, CLASH) datasets	http://web.archive.org/20130922084530/starbase.sysu.edu.cn/mirCircRNA.php	[98]
circBase	circBase explores public circRNA datasets and downloads the custom python scripts needed to discover circRNAs in your own (ribominus) RNA-seq data	http://www.circbase.org	[125]
CircNet	CircNet records public and novel circRNAs and putative circRNA–miRNA interactions. All the information is presented in an interactive user interface and available for download	https://awi.cuhk.edu.cn/~CircNet	[126]
Circ2Traits	Circ2Traits is a comprehensive database for circular RNA potentially associated with disease and traits	https://github.com/shaoli86/circ2Traits	[127]
circRNADb	circRNADb 1.0, circular RNA (or circRNA) database version 1.0, is a comprehensive database of circular RNA molecules in humans. It is freely available for non-commercial use. The latest version of this circRNA database contains 32,914 annotated exonic circRNAs and can be a valuable resource for large-scale studies of circRNA, especially in humans	http://reprod.njmu.edu.cn/cgi-bin/circrnadb/circRNADb.php	[128]
CIRCpedia v2	CIRCpedia v2 is a comprehensive database containing circRNA annotations from over 180 RNA-seq datasets across 6 different species. This atlas allows users to search, browse, and download circRNAs with expression characteristics/features in various cell types/tissues, including disease samples. It incorporates conservation analysis of circRNAs between humans and mice. Finally, the web interface also contains computational tools for the comparison of circRNA expression between samples	http://www.picb.ac.cn/rnomics/circpedia/	[129]

- The web interface, through the Search and Browse functionalities
- The DisGeNET REST API
- The Resource Description Framework (DisGeNET-RDF) representation via the SPARQL endpoint and the Faceted Browser
- The DisGeNET Cytoscape App
- Scripts in the most commonly used programming languages
- The disgenet2r package
- The SQLite database.
- Tab separated files. See downloads section

DisGeNET is a versatile platform that can be used for different research purposes including the investigation of the molecular underpinnings of human diseases and their comorbidities, the analysis of the properties of disease genes, the generation of hypothesis on drug therapeutic action and drug adverse effects, the validation of computationally predicted disease genes, and the evaluation of text-mining method performance.
DisGeNET website: https://www.disgenet.org/

19.9.3 KEGG DISEASE Database

KEGG DISEASE is a collection of disease entries focusing only on the perturbants, for the details of molecular networks are unknown for most diseases. Each entry is identified by the H number and contains a list of known genetic factors (disease genes), environmental factors, pathogens, and therapeutic drugs (see, e.g., the disease entry of chronic myeloid leukemia H00004). Disease entries are classified in the following BRITE hierarchy files.

Human diseases [+ gene]
Infectious diseases [+ genome]
Human diseases in ICD-11 classification

The KEGG DISEASE entry only indicates membership information, i.e., association of genes and molecules to diseases, but may reflect the underlying molecular network. For single-gene diseases, perturbed pathway maps are not drawn, but causative genes are mapped to normal pathway maps through disease entries (see, e.g., the disease entry of congenital disorders of glycosylation type I H00118 and the normal pathway map of N-glycan biosynthesis hsa00510).

19.9.3.1 Disease Pathway Maps

The Human Diseases category of the KEGG PATHWAY database is a collection of disease pathway maps.
KEGG Pathway Maps: Human Diseases

It contains multifactorial diseases such as cancers, immune system diseases, neurodegenerative diseases, cardiovascular diseases, and metabolic diseases where known disease genes are marked in red. It also contains infectious diseases where interacting molecular networks of both pathogens and humans are depicted.

19.9.3.2 Network Variation Maps

Network variation maps in the KEGG NETWORK database indicate how perturbed molecular networks caused by gene variants, viruses, and other factors are associated with selected classes of diseases.
KEGG DISEASE Database website: https://www.genome.jp/kegg/disease/

19.9.4 DNetDB: The Human Disease Network Database

Yang et al. [134] assessed disease similarities based on gene expression data by using differential coexpression analysis, a recently emerging method, which has been proved to be more potential to capture dysfunctional regulation mechanisms than differential expression analysis. They constructed the human disease network database (DNetDB) based on the disease relationships they identified together with the relevant information. Benefiting from the use of differential coexpression analysis, the potential common dysfunctional regulation mechanisms shared by disease pairs (i.e., disease relationships) were extracted and presented. Statistical indicators, common disease-related genes, and drugs shared by disease pairs were also included in DNetDB. In total, 1326 disease relationships among 108 diseases, 5598 pathways, 7357 disease-related genes, and 342 disease drugs are recorded in

DNetDB, among which 3762 genes and 148 drugs are shared by at least two diseases. DNetDB is the first database focusing on disease similarity from the viewpoint of gene regulation mechanism. It provides an easy-to-use web interface to search and browse the disease relationships and thus helps to systematically investigate etiology and pathogenesis, perform drug repositioning, and design novel therapeutic interventions.
DNetDB website: http://app.scbit.org/DNetDB/#.

19.9.5 HuVarBase: The Human Variant Database

The HuVarBase (HUmanVARiantdataBASE) [135] assimilates publicly available human variant data at protein level and gene level into a comprehensive resource. Protein level data such as amino acid sequence, secondary structure of the mutant residue, domain, function, subcellular location, and posttranslational modification are integrated with gene level data such as gene name, chromosome number & genome position, DNA mutation, mutation type origin, and rs ID number. Disease class has been added for the disease-causing variants. HuVarBase database includes 702,048 disease-causing variants, of which 652,399 are "missense," 10,174 are "nonsense," 8850 are "insertion," and 30,625 are "deletion" variants. There are 72,815 neutral variants (i.e., limited or no evidence on the pathogenic role of the mutation) of which 66,191 are "missense," 2885 are "nonsense," 259 are "insertion," and 3480 are "deletion" variants. In total, there are 774,863 variants reported from 18,318 proteins.

The comprehensive HuVarBase facilitates searching for a variant and obtaining variant-related sequence and structural information for viewing or downloading. The HuVarBase includes small-scale variant types such as "missense," "insertion," and "deletion" along with the type "non-sense" in the human genome. The applications of HuVarBase range from analysis, disease prediction to personalized medicine as exemplified by earlier efforts [136, 137]. Analysis of the variants can be performed at sequence and structural level, in order to understand the effects of mutations leading to disease. Prediction algorithms requiring a comprehensive variant dataset can make use of the vast dataset available with the HuVarBase. The database will be updated periodically. The updates will be on a quarterly basis, and the update information will be reflected in the "What's New" section of the web server.
HuVarBase website: https://www.iitm.ac.in/bioinfo/huvarbase

19.9.6 ChEMBL-Neglected Tropical Disease (ChEMBL-NTD)

ChEMBL-NTD is a repository for Open Access primary screening and medicinal chemistry data directed at neglected diseases—endemic tropical diseases of the developing regions of Africa, Asia, and the Americas. The primary purpose of ChEMBL-NTD is to provide a freely accessible and permanent archive and distribution center for deposited data. ChEMBL-NTD is a subset of the data in the free medicinal chemistry and drug discovery database ChEMBL. We actively encourage download and use of the data.
ChEMBL-NTD website: https://chembl.gitbook.io/chembl-ntd/

19.10 Biological Databases

19.10.1 Gene Ontology

The Gene Ontology (GO) is a major bioinformatics initiative to unify the representation of gene and gene product attributes across all species [138]. More specifically, the project aims to (**a**) maintain and develop its controlled vocabulary of gene and gene product attributes; (**b**) annotate genes and gene products and assimilate and disseminate annotation data; and (**c**) provide tools for easy access to all aspects of the data provided by the project and to enable functional interpretation of experimental data using the GO, for example, via enrichment analysis. GO is part of a larger classification effort, the Open Biomedical

Ontologies, being one of the Initial Candidate Members of the OBO Foundry.

Whereas gene nomenclature focuses on gene and gene products, the Gene Ontology focuses on the function of the genes and gene products. The GO also extends the effort by using markup language to make the data (not only of the genes and their products but also of curated attributes) machine readable and to do so in a way that is unified across all species (whereas gene nomenclature conventions vary by biological taxon).
GO website: http://geneontology.org/

19.10.2 KEGG: Kyoto Encyclopedia of Genes and Genomes

KEGG (Kyoto Encyclopedia of Genes and Genomes) is a suite of databases and associated software for understanding and simulating higher-order functional behaviors of the cell or the organism from its genome information [139]. First, KEGG computerizes data and knowledge on protein interaction networks (PATHWAY database) and chemical reactions (LIGAND database) that are responsible for various cellular processes. Second, KEGG attempts to reconstruct protein interaction networks for all organisms whose genomes are completely sequenced (GENES and SSDB databases). Third, KEGG can be utilized as reference knowledge for functional genomics (EXPRESSION database) and proteomics (BRITE database) experiments.

KEGG is a collection of databases dealing with genomes, biological pathways, diseases, drugs, and chemical substances. KEGG is utilized for bioinformatics research and education, including data analysis in genomics, metagenomics, metabolomics and other omics studies, modeling and simulation in systems biology, and translational research in drug development.

The KEGG database project was initiated in 1995 by Minoru Kanehisa, Professor at the Institute for Chemical Research, Kyoto University, under the then ongoing Japanese Human Genome Program [140]. Foreseeing the need for a computerized resource that can be used for biological interpretation of genome sequence data, he started developing the KEGG PATHWAY database. It is a collection of manually drawn KEGG pathway maps representing experimental knowledge on metabolism and various other functions of the cell and the organism. Each pathway map contains a network of molecular interactions and reactions and is designed to link genes in the genome to gene products (mostly proteins) in the pathway. This has enabled the analysis called KEGG pathway mapping, whereby the gene content in the genome is compared with the KEGG PATHWAY database to examine which pathways and associated functions are likely to be encoded in the genome.

19.10.2.1 Systems Information

PATHWAY—pathway maps for cellular and organismal functions
MODULE—modules or functional units of genes
BRITE—hierarchical classifications of biological entities

19.10.2.2 Genomic Information

GENOME—complete genomes
GENES—genes and proteins in the complete genomes
ORTHOLOGY—ortholog groups of genes in the complete genomes
Chemical information
COMPOUND, GLYCAN—chemical compounds and glycans
REACTION, RPAIR, RCLASS—chemical reactions
ENZYME—enzyme nomenclature

19.10.2.3 Chemical Information

KEGG COMPOUND—small molecules
KEGG GLYCAN—glycans
KEGG REACTION/RCLASS—reactions and reaction class
KEGG ENZYM—enzyme nomenclature

19.10.2.4 Health Information

DISEASE—human diseases
DRUG—approved drugs
ENVIRON—crude drugs and health-related substances

The most unique data object in KEGG is the molecular networks—molecular interaction, reaction, and relation networks representing systemic functions of the cell and the organism. KEGG and REACTOME pathway data describe known molecular interaction and reaction networks. These data are integrated with chemicals, genes, and diseases in CTD to provide insights into molecular networks that may be affected by chemicals and possible mechanisms underlying environmental diseases. You can browse pathways, or use them to formulate gene or chemical–gene interaction queries. Pathway information is provided for chemical, gene, and disease detail pages. Pathways that are statistically enriched among genes/proteins that interact with a chemical are displayed for each chemical.

Experimental knowledge on such systemic functions is captured from literature and organized in the following forms:

Pathway map—in KEGG PATHWAY (see Pathway maps)

Brite hierarchy and table—in KEGG BRITE (see Brite hierarchies)

Membership (logical expression)—in KEGG MODULE

Membership (simple list)—in KEGG DISEASE

These databases constitute the reference knowledge base for biological interpretation of genomes and high-throughput molecular datasets through the process of KEGG mapping (see KEGG mapping).

KEEG website: http://www.genome.jp/kegg/; http://www.genome.ad.jp/kegg/

19.10.3 PubChem's BioAssay Database

PubChem's BioAssay database has served as a public repository for small-molecule and RNAi screening data since 2004 providing open access of its data content to the community [141, 142]. PubChem accepts data submission from worldwide researchers at academia, industry, and government agencies. PubChem also collaborates with other chemical biology database stakeholders with data exchange. With over a decade's development effort, it becomes an important information resource supporting drug discovery and chemical biology research. To facilitate data discovery, PubChem is integrated with all other databases at NCBI. The PubChem BioAssay database currently contains 500,000 descriptions of assay protocols, covering 5000 protein targets and 30,000 gene targets and providing over 130 million bioactivity outcomes [141, 142]. PubChem's bioassay data are integrated into the NCBI Entrez information retrieval system, thus making PubChem data searchable and accessible by Entrez queries.

PubChem mostly contains small molecules but also larger molecules such as nucleotides, carbohydrates, lipids, peptides, and chemically modified macromolecules. We collect information on chemical structures, identifiers, chemical and physical properties, biological activities, patents, health, safety, toxicity data, and many others.

PubChem's BioAssay database website: https://pubchem.ncbi.nlm.nih.gov

19.11 Chemical Databases

19.11.1 PubChem

PubChem is a database of chemical molecules and their activities against biological assays. The system is maintained by the National Center for Biotechnology Information (NCBI), a component of the National Library of Medicine, which is part of the US National Institutes of Health (NIH). PubChem can be accessed for free through a web user interface. Millions of compound structures and descriptive datasets can be freely downloaded via FTP. PubChem contains multiple substance descriptions and small molecules with fewer than 100 atoms and 1000 bonds. More than 80 database vendors contribute to the growing PubChem database [143].

PubChem is taken as a source of polypharmacology [144]. In PubChem, 71,303 molecules have been identified as ligands that have two or more biological targets, and more than 30,000 ligands were found to be active against more than 400 targets [145]. A bibliometric analysis of

PubChem applications is presented by reviewing 1132 research articles [146]. The massive volume of chemical structure and bioactivity data in PubChem and its online services have been used globally in various fields including chemical biology, medicinal chemistry, and informatics research. PubChem supports drug discovery in many aspects such as lead identification and optimization, compound–target profiling, polypharmacology studies, and unknown chemical identity elucidation. PubChem has also become a valuable resource for developing secondary databases, informatics tools, and web services. The growing PubChem resource with its public availability offers support and great opportunities for the interrogation of pharmacological mechanisms and the genetic basis of diseases, which are vital for drug innovation and repurposing.

PubChem website: https://pubchem.ncbi.nlm.nih.gov/

19.11.2 ChEMBL

ChEMBL [147] is a well-established resource in the fields of drug discovery and medicinal chemistry research. It curates and stores standardized bioactivity, molecule, target, and drug data extracted from multiple sources, including the primary medicinal chemistry literature. Moreover, the ChEMBL database includes data typically generated in the preclinical and clinical phases of drug discovery, specifically drug metabolism and disposition data. These data help researchers better understand the key aspects of successful drug discoveries. ChEMBL data can be downloaded in a number of standard formats that can be automatically imported to external applications for further analyses. Alternatively, ChEMBL provides a REST API-based service, allowing the remote retrieval of ChEMBL data and its integration into other applications.

ChEMBL is an open large-scale bioactivity database, initially described in the 2012 Nucleic Acids Research Database Issue. Since then, a variety of new data sources and improvements in functionality have contributed to the growth and utility of the resource. In particular, more comprehensive tracking of compounds from research stages through clinical development to market is provided through the inclusion of data from United States Adopted Name applications; a new richer data model for representing drug targets has been developed; and a number of methods have been put in place to allow users to more easily identify reliable data. Finally, access to ChEMBL is now available via a new Resource Description Framework format, in addition to the web-based interface, data downloads, and web services.

ChEMBL website: https://www.ebi.ac.uk/chembl/

19.11.3 STITCH

STITCH is a database of known and predicted interactions between chemicals and proteins. The interactions include direct (physical) and indirect (functional) associations; they stem from computational prediction, from knowledge transfer between organisms, and from interactions aggregated from other (primary) databases. STITCH contains interactions for between 300,000 small molecules and 2.6 million proteins from 1133 organisms. Chemicals are linked to other chemicals and proteins by evidence derived from experiments, databases, and the literature.

STITCH website: http://stitch.embl.de/

19.11.4 UniChem

UniChem is a very simple, large-scale nonredundant database of pointers between chemical structures and EMBL-EBI chemistry resources. Its purpose is to optimize the efficiency with which structure-based hyperlinks may be built and maintained between chemistry-based resources and is particularly suitable for creating such links “on the fly” (by use of REST web services). Primarily, this service has been designed to maintain cross references between EBI chemistry resources. These include primary chemistry resources (ChEMBL and ChEBI), and other resources where the focus is not small molecules, but which may nevertheless contain some small molecule information (e.g., Gene Expression Atlas, PDBe).

The advantage of UniChem is that it implements an easily scalable and maintainable system of chemical structure integration within the EBI-Search infrastructure, allowing future new chemistry resources within the EBI to be included with minimal effort. Interestingly, centralized mapping for external resources, such as PubChem, ZINC, etc., is also straightforward and economical and provides a wider cross-referencing service. The recent exposure of UniChem externally to EMBL-EBI provides these same mapping benefits to resources outside of EMBL. Recent open-access papers listed on the citation page describe further background.

Downloading of data from source databases and the uploading and registration into UniChem is highly automated. Because the configuration and resources of different source databases vary widely, it is necessary to employ a variety of data downloading procedures (such as ftp, web services, oracle calls, etc.). The downloading procedures adopted for each source are summarized on the summary page for each source, accessed via the sources page. Since some data is unsuitable for use in UniChem, or may damage the data integrity within the database, a series of rules are automatically applied when loading data. These rules are summarized on the Rules for Loading page.

UniChem website: https://www.ebi.ac.uk/unichem/

19.11.5 ChemProt

ChemProt is a disease chemical biology database, which is based on a compilation of multiple chemical–protein annotation resources, as well as disease-associated protein–protein interactions (PPIs) [148, 149]. It is a collection of more than 700,000 unique chemicals with biological annotation for 30,578 proteins, as well as over 2-million chemical–protein interactions, which were integrated in a quality scored human PPI network of 428,429 interactions. The PPI network layer allows for studying disease and tissue specificity through each protein complex. ChemProt can assist in the in silico evaluation of environmental chemicals, natural products, and approved drugs, as well as the selection of new compounds based on their activity profile against most known biological targets, including those related to adverse drug events. Results from the disease chemical biology database associate citalopram, an antidepressant, with osteogenesis imperfect and leukemia and bisphenol A, an endocrine disruptor, with certain types of cancer, respectively.

ChemProt website: http://www.cbs.dtu.dk/services/ChemProt/

19.11.6 ChemMapper

ChemMapper is an online platform to predict polypharmacology effect and mode of action for small molecules based on 3D similarity computation [149]. ChemMapper collects >350,000 chemical structures with bioactivities and associated target annotations (as well as >3,000,000 non-annotated compounds for virtual screening). Taking the user-provided chemical structure as the query, the topmost similar compounds in terms of 3D similarity are returned with associated pharmacology annotations. ChemMapper is designed to provide versatile services in a variety of chemogenomics, drug repurposing, polypharmacology, novel bioactive compounds identification, and scaffold-hopping studies.

ChemMapper is a versatile web-based tool for exploring target pharmacology and chemical relationships against any given small molecules via SHAFTS, a fast 3D similarity method in which the 3D similarity calculation is driven by the hybrid information of molecular shape and chemotype features [150]. ChemMapper assembles a large repertoire of bioactive chemical database annotated with target information and multiple screening databases from different catalogs of the chemical vendors. It facilitates identification of potential targets that may play a pivotal role in biological response to the drugs and other active chemicals, as well as the discovery of new hit compounds with similar pharmacology profiles but novel scaffolds. Moreover, ChemMapper can also be useful in the application of drug repurposing, as well as investigation of potential side effects related to the drugs and other active chemicals.

ChemMapper website: http://lilab.ecust.edu.cn/chemmapper/

19.12 Prediction Programs

19.12.1 FastTargetPred Program

Several web-based tools predict the putative targets of a small molecule query compound by similarity to molecules with known bioactivity data using molecular fingerprints. In numerous situations, it would however be valuable to be able to run such computations on a local computer. Chaput et al. [151] built a new program for the prediction of protein targets for small molecule queries, which they named FastTargetPred. Structural similarity computations rely on a large collection of confirmed protein–ligand activities extracted from the curated ChEMBL 25 database. The program allows to annotate an input chemical library of ~100 k compounds within a few hours on a simple personal computer. FastTargetPred is written in Python 3 (≥3.7) and C languages. The program can be run on Linux, MacOS, and Windows operating systems. Precompiled versions are available at https://github.com/ludovicchaput/FastTargetPred. FastTargetPred is licensed under the GNU GPLv3. The program calls some scripts from the free chemistry toolkit MayaChemTools.

Contact for more info: bruno.villoutreix@inserm.fr

19.12.2 SEABED

The SEABED web server integrates a variety of docking and QSAR techniques in a user-friendly environment [152]. SEABED goes beyond the basic docking and QSAR web tools and implements extended functionalities like receptor preparation, library editing, flexible ensemble docking, hybrid docking/QSAR experiments, or virtual screening on protein mutants. SEABED is not a monolithic workflow tool but Software as a Service platform.

SEABED website: http://www.bsc.es/SEABED

19.12.3 LigBuilder V3 Drug Design Program

LigBuilder V3 [153] represents the first de novo MTD design program, which can be used to design ligands to target multiple receptors, multiple binding sites of one receptor, or various conformations of one receptor. LigBuilder V3 is generally applicable in de novo MTD design and optimization, especially for the design of concise ligands for protein targets with large difference in binding sites. To demonstrate the utility of LigBuilder V3, the inventors of the program have used it to design dual-functional inhibitors targeting HIV protease and HIV reverse transcriptase with three different strategies, including multitarget de novo design, multitarget growing, and multitarget linking. The designed compounds were computational validated by MM/GBSA binding free energy estimation as highly potential multitarget inhibitors for both HIV protease and HIV reverse transcriptase.

LigBuilder V3 website: http://www.pkumdl.cn/ligbuilder3/

19.12.4 C-SPADE

The advent of polypharmacology paradigm in drug discovery calls for novel chemoinformatic tools for analyzing compounds' multitargeting activities. Such tools should provide an intuitive representation of the chemical space through capturing and visualizing underlying patterns of compound similarities linked to their polypharmacological effects. Most of the existing compound-centric chemoinformatics tools lack interactive options and user interfaces that are critical for the real-time needs of chemical biologists carrying out compound screening experiments. Toward that end, Ravikumaret al [154]. introduced C-SPADE, an open-source exploratory web tool for interactive analysis and visualization of drug profiling assays (biochemical, cell-based, or cell-free) using compound-centric similarity clustering. C-SPADE allows the users to visually map the chemical diversity of a screening panel, explore investigational compounds in terms of their similarity to the screening panel, perform polypharmacological analyses, and guide drug–target interaction predictions. C-SPADE requires only the raw drug profiling data as input, and it automatically retrieves the structural information and constructs the compound clusters in real-time, thereby reducing the time required for manual analysis in drug development or repurposing applications. The web tool provides a customizable visual workspace that can either be downloaded as figure or Newick tree file or shared as a hyperlink with other users.

C-SPADE website: http://cspade.fimm.fi/

19.12.5 DrugGenEx-Net (DGE-NET)

The DrugGenEx-Net (DGE-NET) [155] is a system polypharmacology platform which can be used to predict empirical drug–target (DT) interactions, integrates interaction pairs into a multi-tiered network analysis, and ultimately predicts disease-specific drug polypharmacology through systems-based gene expression analysis. Incorporation of established biological network annotations for protein target–disease, signaling pathway, molecular function, and protein–protein interactions enhances predicted DT effects on disease pathophysiology. Over 50 drug–disease and 100 drug–pathway predictions are validated. For example, the predicted systems pharmacology of the cholesterol-lowering agent ezetimibe corroborates its potential carcinogenicity. When disease-specific gene expression analysis is integrated, DGE-NET prioritizes known therapeutics/experimental drugs as well as their contraindications. Proof of concept is established for immune-related rheumatoid arthritis and inflammatory bowel disease, as well as neurodegenerative Alzheimer's and Parkinson's diseases. Thus, DGE-NET is a novel computational method in predicting drug therapeutic and counter-therapeutic indications by uniquely integrating systems pharmacology with gene expression analysis. DGE-NET correctly predicts various drug–disease indications by linking the biological activity of drugs and diseases at multiple tiers of biological action and is therefore a useful approach to identifying drug candidates for re-purposing.

References

1. Potemkin V, Potemkin A, Grishina M. Internet resources for drug discovery and design. Curr Top Med Chem. 2018;18(22):1955–75.
2. Reymond J-L, Awale M. Exploring chemical space for drug discovery using the chemical universe database. ACS Chem Neurosci. 2012;3(9):649–57.
3. Kirkpatrick P, Ellis C. Chemical space. Nature. 2004;432(7019):823–65.
4. Mirza A, Desai R, Reynisson J. Known drug space as a metric in exploring the boundaries of drug-like chemical space. Eur J Med Chem. 2009;44(12):5006–11.
5. Bade R, Chan HF, Reynisson J. Characteristics of known drug space. Natural products, their derivatives and synthetic drugs. Eur J Med Chem. 2010;45(12):5646–52.
6. Matuszek AM, Reynisson J. Defining known drug space using DFT. Mol Inform. 2016;35(2):46–53.
7. Ruddigkeit L, van Deursen R, Blum LC, et al. Enumeration of 166 billion organic small molecules in the chemical universe database GDB-17. J Chem Inf Model. 2012;52(11):2864–75.
8. Éliás O, Kovács Z, Wágner G, et al. Charting the chemical space around the (iso)indoline scaffold, a comprehensive approach towards multi-target directed ligands. Bioorg Med Chem Lett. 2016;26(17):4211–115.
9. Sidorov P, Gaspar H, Marcou G, et al. Mappability of drug-like space: towards a polypharmacologically competent map of drug-relevant compounds. J Comput Aided Mol Des. 2015;29(12):1087–108.
10. Hopkins AL. Pharmacological space. In: Wermuth C, Aldous D, Raboisson P, Rognan D, editors. The practice of medicinal chemistry. Academic Press; 2015. p. 395–408.
11. Peter KP. Exploring pharmacological space. Nat Rev Drug Discov. 2006;5:719.
12. Paolini GV, Shapland RH, van Hoorn WP, et al. Global mapping of pharmacological space. Nat Biotechnol. 2006;24(7):805–15.

13. Nisius B, Bajorath J. Mapping of pharmacological space. Expert Opin Drug Discovery. 2011;6(1):1–7.
14. Petit-Zeman S. Biological Space—a starting point in in-silico drug design and in experimentally exploring biological systems. Oxford Protein Informatics Group. https://www.blopig.com/blog/2017/07/biological-space-a-starting-point-in-silico-drug-design-and-in-experimentally-exploring-biological-systems/
15. Koch MA, Schuffenhauer A, Scheck M, et al. Charting biologically relevant chemical space: a structural classification of natural products (SCONP). Proc Natl Acad Sci U S A. 2005;102:17272–7.
16. Siragusa L, Cross S, Baroni M, et al. BioGPS: navigating biological space to predict polypharmacology, off-targeting, and selectivity. Proteins. 2015;83(3):517–32.
17. Vukmirovic OG, Tilghman SM. Exploring genome space. Nature. 2000;405:820–2.
18. Adanve B. A brief overview of the genomic space. https://blog.geneticintelligence.com/overview-of-the-genomic-space/#:~:text=The%20genomic%20space%20comprises%20four,to%20build%20products%20and%20services
19. Bredel M, Jacoby E. Chemogenomics: an emerging strategy for rapid target and drug discovery. Nat Rev Genet. 2004;5(4):262–75.
20. Namchuk M. Finding the molecules to fuel chemogenomics. Targets. 2002;1(4):125–9.
21. Caron PR, Mullican MD, Mashal RD, et al. Chemogenomic approaches to drug discovery. Curr Opin Chem Biol. 2001;5(4):464–70.
22. Ambroise Y. Chemogenomic techniques. Archive. Today. Archived from the original on 23 August 2013. Retrieved 28 Nov 2021.
23. Wuster A, Madan BM. Chemogenomics and biotechnology. Trends Biotechnol. 2008;26(5):252–8.
24. Jones LH. Expanding chemogenomic space using chemoproteomics. Bioorg Med Chem. 2019;27(15):3451–3.
25. Mohd Fauzi F, Koutsoukas A, Lowe R, et al. Chemogenomics approaches to rationalizing the mode-of-action of traditional Chinese and Ayurvedic medicines. J Chem Inf Model. 2013;53(3):661–73.
26. Engelberg A. Iconix Pharmaceuticals, Inc.—removing barriers to efficient drug discovery through chemogenomics. Pharmacogenomics. 2004;5(6):741–4.
27. Bhattacharjee B, Simon RM, Gangadharaiah C, et al. Chemogenomics profiling of drug targets of peptidoglycan biosynthesis pathway in Leptospira interrogans by virtual screening approaches. J Microbiol Biotechnol. 2013;23(6):779–84.
28. Cheung-Ong K, Song KT, Ma Z, et al. Comparative chemogenomics to examine the mechanism of action of DNA-targeted platinum-acridine anticancer agents. ACS Chem Biol. 2012;7(11):1892–901.
29. Fechner N, Papadatos G, Evans D, et al. ChEMBLSpace—a graphical explorer of the chemogenomic space covered by the ChEMBL database. Bioinformatics. 2013;29(4):523–4.
30. Bender A, Young DW, Jenkins JL, et al. Chemogenomic data analysis: prediction of small-molecule targets and the advent of biological fingerprint. Comb Chem High Throughput Screen. 2007;10(8):719–31.
31. Sun J, Jeliazkova N, Chupakin V, et al. ExCAPE-DB: an integrated large scale dataset facilitating Big Data analysis in chemogenomics. J Cheminform. 2017;9:17.
32. Awale M, Reymond JL. The polypharmacology browser: a web-based multi-fingerprint target prediction tool using ChEMBL bioactivity data. J Cheminform. 2017;9:11.
33. Awale M, Reymond JL. Polypharmacology Browser PPB2: target prediction combining nearest neighbors with machine learning. J Chem Inf Model. 2019;59(1):10–7.
34. Awale M, Reymond JL. Web-based tools for polypharmacology prediction. Methods Mol Biol. 2019;1888:255–72.
35. Poirier M, Awale M, Roelli MA, et al. Identifying lysophosphatidic acid acyltransferase beta (LPAAT-beta) as the target of a nanomolar angiogenesis inhibitor from a phenotypic screen using the polypharmacology browser PPB2. ChemMedChem. 2019;14(2):224–36.
36. Reddy AS, Tan Z, Zhang S. Curation and analysis of multitargeting agents for polypharmacological modeling. J Chem Inf Model. 2014;54(9):2536–43.
37. Minie M, Chopra G, Sethi G, et al. CANDO and the infinite drug discovery frontier. Drug Discov Today. 2014;19:1353–63.
38. Chopra G, Samudrala R. Exploring polypharmacology in drug discovery and repurposing using the CANDO platform. Curr Pharm Des. 2016;22(21):3109–23.
39. Mangione W, Falls Z, Chopra G, et al. cando.py: open source software for predictive bioanalytics of large scale drug-protein-disease data. J Chem Inf Model. 2020;60:4131–6.
40. Falls Z, Mangione W, Schuler J, et al. Exploration of interaction scoring criteria in the CANDO platform. BMC Res Notes. 2019;12:318.
41. Mangione W, Samudrala R. Identifying protein features responsible for improved drug repurposing accuracies using the CANDO platform: implications for drug design. Molecules. 2019;24:167.
42. Li H, Pei F, Taylor DL, et al. QuartataWeb: integrated chemical-protein-pathway mapping for polypharmacology and chemogenomics. Bioinformatics. 2020;36(12):3935–7.
43. Sara JC, Gosline S, Spencer J, et al. SAMNet: a network-based approach to integrate multi-dimensional high throughput datasets. Integr Biol. 2012;11:1415–27.
44. Ursu O, Gosline SJC, Beeharry N, et al. Network modeling of kinase inhibitor polypharmacology reveals pathways targeted in chemical screens. PLoS One. 2017;12(10):e0185650.

45. Chen C, He Y, Wu J, Zhou J. Creation of a free, internet-accessible database: the multiple target ligand database. J Cheminform. 2015;7:14.
46. Chen C, Wu M, Cen S, et al. MTLD, a database of multiple target ligands, the updated version. Molecules. 2017;22(9):1375.
47. Du Y, Shi T. Ligand cluster-based protein network and ePlatton, a multi-target ligand finder. J Cheminform. 2016;8:23.
48. Gallo K, Goede A, Eckert A, et al. PROMISCUOUS 2.0: a resource for drug-repositioning. Nucleic Acids Res. 2021;49(D1):D1373–80.
49. Jaundoo R, Craddock TJA. DRUGPATH: the drug gene pathway meta-database. Int J Mol Sci. 2020;21(9):3171.
50. Wishart DS, Knox C, Guo AC, et al. DrugBank: a comprehensive resource for in silico drug discovery and exploration. Nucleic Acids Res. 2006;34:D668–72.
51. Wishart DS, Feunang YD, Guo AC, et al. DrugBank 5.0: a major update to the DrugBank database for 2018. Nucleic Acids Res. 2018;46:D1074–82.
52. Law V, Knox C, Djoumbou Y, et al. DrugBank 5.0: shedding new light on drug metabolism. Nucleic Acids Res. 2014;42:D1091–7.
53. Wishart DS, Guo AC, Eisner R, et al. HMDB: a knowledgebase for the human metabolome. Nucleic Acids Res. 2009;37:D603–10.
54. Lim E, Pon A, Djoumbou Y, et al. T3DB: a comprehensively annotated database of common toxins and their targets. Nucleic Acids Res. 2010;38:D781–6.
55. Jewison T, Su Y, Disfany FM, et al. Small molecule pathway database. Nucleic Acids Res. 2014;42:D478–84.
56. Barneh F, Jafari M, Mirzaie M. Updates on drug-target network; facilitating polypharmacology and data integration by growth of DrugBank database. Brief Bioinform. 2016;17(6):1070–80.
57. Takigawa I, Tsuda K, Mamitsuka H. An in-silico model for interpreting polypharmacology in drug-target networks. Methods Mol Biol. 2013;993:67–80.
58. Cotto KC, Wagner AH, Feng Y-Y, et al. DGIdb 3.0: a redesign and expansion of the drug–gene interaction database. Nucleic Acids Res. 2018;46:D1068–73.
59. Griffith M, Spies NC, Krysiak K. CIViC is a community knowledgebase for expert crowdsourcing the clinical interpretation of variants in cancer. Nat Genet. 2017;49(2):170–4.
60. Chakravarty D, Gao J, Phillips SM, et al. OncoKB: a precision oncology knowledge base. JCO Precis Oncol. 2017;2017:PO.17.00011.
61. Patterson SE, Liu R, Statz CM. The clinical trial landscape in oncology and connectivity of somatic mutational profiles to targeted therapies. Hum Genomics. 2016;10:4.
62. Finan C, Gaulton A, Kruger FA. The druggable genome and support for target identification and validation in drug development. Sci Transl Med. 2017;9(383):eaag1166.
63. Armstrong JF, Faccenda E, Harding SD, et al. The IUPHAR/BPS Guide to PHARMACOLOGY in 2020: extending immunopharmacology content and introducing the IUPHAR/MMV Guide to MALARIA PHARMACOLOGY. Nucleic Acids Res. 2020;48(D1):D1006–21.
64. Alexander SP, Benson HE, Davenport A, et al. GuideToPharmacology.org—an update. Brit J Pharmacol. 2012;167(4):697–8.
65. Chen X, Ji ZL, Chen YZ. TTD: therapeutic target database. Nucleic Acids Res. 2002;30:412–5.
66. Zhou Y, Zhang YT, Lian XC, et al. Therapeutic target database update 2022: facilitating drug discovery with enriched comparative data of targeted agents. Nucleic Acids Res. 2022;50(D1):1398–407.
67. Li YH, Yu CY, Li XX, et al. Therapeutic target database update 2018: enriched resource for facilitating bench-to-clinic research of targeted therapeutics. Nucleic Acids Res. 2018;46(D1):D1121–7.
68. Wang Y, Zhang S, Li F, et al. Therapeutic target database 2020: enriched resource for facilitating research and early development of targeted therapeutics. Nucleic Acids Res. 2019;48(D1):D1031–41.
69. Zhu F, Shi Z, Qin C, et al. Therapeutic target database update 2012: a resource for facilitating target-oriented drug discovery. Nucleic Acids Res. 2012;40:D1128–36.
70. Tanoli Z, Alam Z, Vähä-Koskela M, et al. Drug Target Commons 2.0: a community platform for systematic analysis of drug-target interaction profiles. Database (Oxford). 2018;2018:1–13.
71. Ravikumar B, Aittokallio T. Improving the efficacy-safety balance of polypharmacology in multi-target drug discovery. Expert Opin Drug Discovery. 2018;13:179–92.
72. Huang L-H, He Q-S, Liu K, et al. ADReCS-Target: target profiles for aiding drug safety research and application. Nucleic Acids Res. 2018;46:D911–7.
73. Liu T, Lin Y, Wen X, et al. BindingDB: a web-accessible database of experimentally determined protein-ligand binding affinities. Nucleic Acids Res. 2007;35:D198–201.
74. Günther S, Kuhn M, Dunkel M, et al. SuperTarget and Matador: resources for exploring drug-target relationships. Nucleic Acids Res. 2008;36:D919–22.
75. Piñero J, Bravo À, Queralt-Rosinach N, et al. DisGeNET: a comprehensive platform integrating information on human disease-associated genes and variants. Nucleic Acids Res. 2017;45(D1):D833–9.
76. Koscielny G, An P, Carvalho-Silva D, et al. Open targets: a platform for therapeutic target identification and validation. Nucleic Acids Res. 2017;45:D985–94.
77. Bauer U, Breeze AL. "Ligandability" of drug targets: assessment of chemical tractability via experimental and *in silico* approaches. Lead Generation. 2016;2016:35–62.
78. Nguyen D-T, Mathias S, Bologa C, et al. Pharos: collating protein information to shed light on

the druggable genome. Nucleic Acids Res. 2017;45:D995–1002.
79. Davis AP, Grondin CJ, Johnson RJ, et al. The Comparative Toxicogenomics Database: update 2019. Nucleic Acids Res. 2019;47:D948–54.
80. wwPDB, Consortium. Protein Data Bank: the single global archive for 3D macromolecular structure data. Nucleic Acids Res. 2019;47(D1):520–8.
81. Pinzi L, Rastelli G. Identification of target associations for polypharmacology from analysis of crystallographic ligands of the protein data bank. J Chem Inf Model. 2020;60(1):372–90.
82. Herce HD, Deng W, Helma J, et al. Visualization and targeted disruption of protein interactions in living cells. Nat Commun. 2013;4:2660.
83. Lehne B, Schlitt T. Protein-protein interaction databases: keeping up with growing interactomes. Hum Genomics. 2009;3(3):291–7.
84. Xenarios I, Rice DW, Salwinski L, et al. DIP: the database of interacting proteins. Nucleic Acids Res. 2000;28(1):289–91.
85. McDowall MD, Scott MS, Barton GJ. PIPs: human protein-protein interaction prediction database. Nucleic Acids Res. 2009;37:D651–6.
86. Smith RD, Clark JJ, Ahmed A, et al. Updates to binding MOAD (mother of all databases): polypharmacology tools and their utility in drug repurposing. J Mol Biol. 2019;431(13):2423–33.
87. Wang C, Hu G, Wang K, et al. PDID: database of molecular-level putative protein-drug interactions in the structural human proteome. Bioinformatics. 2016;32(4):579–86.
88. Wang Z, Li J, Dang R, et al. PhIN: a protein pharmacology interaction network database. CPT Pharmacometrics Syst Pharmacol. 2015;4(3):e00025.
89. Brylinski M. eMatchSite: sequence order-independent structure alignments of ligand binding pockets in protein models. PLoS Comput Biol. 2014;10(9):e1003829.
90. Yang JH, Shao P, Zhou H, et al. deepBase: a database for deeply annotating and mining deep sequencing data. Nucleic Acids Res. 2010;38:D123–30.
91. Griffiths-Jones S, Saini HK, van Dongen S, et al. miRBase: tools for microRNA genomics. Nucleic Acids Res. 2008;36:D154–8.
92. Betel D, Wilson M, Gabow A, et al. The microRNA.org resource: targets and expression. Nucleic Acids Res. 2007;36:D149–53.
93. Perdikopanis N, Georgakilas GK, Grigoriadis D, et al. DIANA-miRGen v4: indexing promoters and regulators for more than 1500 microRNAs. Nucleic Acids Res. 2021;49:D151–9.
94. Hsu PW, Huang HD, Hsu SD, et al. miRNAMap: genomic maps of microRNA genes and their target genes in mammalian genomes. Nucleic Acids Res. 2006;34:D135–9.
95. Agarwal V, Bell GW, Nam JW, et al. Predicting effective microRNA target sites in mammalian mRNAs. elife. 2015;4:e05005.
96. Abid Q, Nishant T, Isha M, et al. VIRmiRNA: a comprehensive resource for experimentally validated viral miRNAs and their targets. Database: J Biol Database Curation. 2014;2014:bau103.
97. Yang J-H, Li J-H, Shao P, et al. StarBase: a database for exploring microRNA-mRNA interaction maps from Argonaute CLIP-Seq and Degradome-Seq data. Nucleic Acids Res. 2010;39:D202–9.
98. Li JH, Liu S, Zhou H, et al. starBase v2.0: decoding miRNA-ceRNA, miRNA-ncRNA and protein-RNA interaction networks from large-scale CLIP-Seq data. Nucleic Acids Res. 2014;42(1):D92–7.
99. Liu S, Li JH, Wu J, et al. StarScan: a web server for scanning small RNA targets from degradome sequencing data. Nucleic Acids Res. 2015;43(W1):W480–6.
100. Chiu HS, Llobet-Navas D, Yang X, et al. Cupid: simultaneous reconstruction of microRNA-target and ceRNA networks. Genome Res. 2015;25(2):257–67.
101. Sethupathy P, Corda B, Hatzigeorgiou AG. TarBase: a comprehensive database of experimentally supported animal microRNA targets. RNA. 2006;12(2):192–7.
102. Maragkakis M, Alexiou P, Papadopoulos GL, et al. Accurate microRNA target prediction correlates with protein repression levels. BMC Bioinform. 2009;10:295.
103. Xiao F, Zuo Z, Cai G, et al. miRecords: an integrated resource for microRNA-target interactions. Nucleic Acids Res. 2009;37:D105–10.
104. Krek A, Grün D, Poy MN, et al. Combinatorial microRNA target predictions. Nat Genet. 2005;37(5):495–500.
105. Kertesz M, Iovino N, Unnerstall U, et al. The role of site accessibility in microRNA target recognition. Nat Genet. 2007;39(10):1278–84.
106. Elefant N, Berger A, Shein H, et al. RepTar: a database of predicted cellular targets of host and viral miRNAs. Nucleic Acids Res. 2011;39:D188–94.
107. Miranda KC, Huynh T, Tay Y, et al. A pattern-based method for the identification of MicroRNA binding sites and their corresponding heteroduplexes. Cell. 2006;126(6):1203–17.
108. Hsu SD, Lin FM, Wu WY, et al. miRTarBase: a database curates experimentally validated microRNA-target interactions. Nucleic Acids Res. 2011;39:D163–9.
109. Dweep H, Sticht C, Pandey P, et al. miRWalk-database: prediction of possible miRNA binding sites by "walking" the genes of three genomes. JBI. 2011;44(5):839–47.
110. Bandyopadhyay S, Ghosh D, Mitra R, et al. MBSTAR: multiple instance learning for predicting specific functional binding sites in microRNA targets. Sci Rep. 2015;5:8004.
111. Zheng LL, Li JH, Wu J, et al. deepBase v2.0: identification, expression, evolution and function of small RNAs, LncRNAs and circular RNAs from deep-sequencing data. Nucleic Acids Res. 2016;44:D196–202.

112. Volders PJ, Helsens K, Wang X, et al. LNCipedia: a database for annotated human lncRNA transcript sequences and structures. Nucleic Acids Res. 2012;41:D246–51.
113. Volders PJ, Verheggen K, Menschaert G, et al. An update on LNCipedia: a database for annotated human lncRNA sequences. Nucleic Acids Res. 2015;43:D174–80.
114. Amaral PP, Clark MB, Gascoigne DK, et al. LncRNAdb: a reference database for long noncoding RNAs. Nucleic Acids Res. 2010;39:D146–51.
115. Ma L, Li A, Zou D, et al. LncRNAWiki: harnessing community knowledge in collaborative curation of human long non-coding RNAs. Nucleic Acids Res. 2015;43:D187–92.
116. Ma L, Cao J, Liu L, et al. LncBook: a curated knowledgebase of human long non-coding RNAs. Nucleic Acids Res. 2019;47:D128–34.
117. Josset L, Tchitchek N, Gralinski LE, et al. Annotation of long non-coding RNAs expressed in collaborative cross founder mice in response to respiratory virus infection reveals a new class of interferon-stimulated transcripts. RNA Biol. 2014;11(7):875–90.
118. Bu D, Yu K, Sun S, et al. NONCODE v3.0: integrative annotation of long noncoding RNAs. Nucleic Acids Res. 2011;40:D210–5.
119. Dinger ME, Pang KC, Mercer TR, et al. NRED: a database of long noncoding RNA expression. Nucleic Acids Res. 2009;37:D122–6.
120. Weirick T, David J, Stefanie D, et al. C-It-Loci: a knowledge database for tissue-enriched loci. Bioinformatics. 2015;31(21):3537–43.
121. Iyer MK, Niknafs YS, Malik R, et al. The landscape of long noncoding RNAs in the human transcriptome. Nat Genet. 2015;47(3):199–208.
122. Chen J, Shishkin AA, Zhu X, et al. Evolutionary analysis across mammals reveals distinct classes of long non-coding RNAs. Genome Biol. 2016;17:19.
123. Carlevaro-Fita J, Lanzós A, Feuerbach L, et al. Cancer LncRNA census reveals evidence for deep functional conservation of long noncoding RNAs in tumorigenesis. Commun Biol. 2020;3:56.
124. You B-H, Yoon S-H, Nam JW. High-confidence coding and noncoding transcriptome maps. Genome Res. 2017;27(6):1050–62.
125. Memczak S, Jens M, Elefsinioti A, et al. Circular RNAs are a large class of animal RNAs with regulatory potency. Nature. 2013;495(7441):333–8.
126. Liu Y-C, Li J-R, Sun C-H, et al. CircNet: a database of circular RNAs derived from transcriptome sequencing data. Nucleic Acids Res. 2016;44(D1):D209–15.
127. Ghosal S, Das S, Sen R, et al. Circ2Traits: a comprehensive database for circular RNA potentially associated with disease and traits. Front Genet. 2013;4:283.
128. Chen X, Han P, Zhou T, et al. circRNADb: a comprehensive database for human circular RNAs with protein-coding annotations. Sci Rep. 2016;6:34985.
129. Dong R, Ma X-K, Li G-W, et al. CIRCpedia v2: an updated database for comprehensive circular RNA annotation and expression comparison. Genomics Proteomics Bioinformatics. 2018;16(4):226–33.
130. Espe S. Malacards: the human disease database. J Med Libr Assoc. 2018;106(1):140–1.
131. Rappaport N, Twik M, Plaschkes I, et al. MalaCards: an amalgamated human disease compendium with diverse clinical and genetic annotation and structured search. Nucleic Acids Res. 2017;45(D1):D877–87.
132. Piñero J, Queralt-Rosinach N, Bravo À, et al. DisGeNET: a discovery platform for the dynamical exploration of human diseases and their genes. Database (Oxford). 2015;2015:bav028.
133. Piñero J, Ramírez-Anguita JM, Saüch-Pitarch J. The DisGeNET knowledge platform for disease genomics: 2019 update. Nucleic Acids Res. 2020;48(D1):D845–55.
134. Yang J, Wu S-J, Yang S-Y. DNetDB: the human disease network database based on dysfunctional regulation mechanism. BMC Syst Biol. 2016;10(1):36.
135. Ganesan K, Kulandaisamy A, Binny Priya S, et al. HuVarBase: a human variant database with comprehensive information at gene and protein levels. PLoS One. 2019;14(1):e0210475.
136. Shen J, Song K, Slater AJ, et al. STOPGAP: a database for systematic target opportunity assessment by genetic association predictions. Bioinformatics. 2017;33(17):2784–6.
137. Gosalia N, Economides AN, Dewey FE, et al. MAPPIN: a method for annotating, predicting pathogenicity and mode of inheritance for nonsynonymous variants. Nucleic Acids Res. 2017;45(18):10393–402.
138. The Gene Ontology Consortium. The Gene Ontology project in 2008. Nucleic Acids Res. 2008;36:D440–4.
139. Kanehisa M. The KEGG database. Novartis Found Symp. 2002;247:91–101.
140. Kanehisa M, Goto S. KEGG: Kyoto Encyclopedia of Genes and Genomes. Nucleic Acids Res. 2000;28(1):27–30.
141. Wang Y, Xiao J, Suzek TO, et al. PubChem's BioAssay database. Nucleic Acids Res. 2012;40:D400–12.
142. Wang Y, Bryant SH, Cheng T, et al. PubChem BioAssay: 2017 update. Nucleic Acids Res. 2017;45(D1):D955–63.
143. PubChem Source Information. The PubChem Project. USA: National Center for Biotechnology Information.
144. Chen B, Wild D, Guha R. PubChem as a source of polypharmacology. J Chem Inf Model. 2009;49(9):2044–55.
145. Hu Y, Gupta-Ostermann D, Bajorath J. Exploring compound promiscuity patterns and multi-target activity spaces. Comput Struct Biotechnol J. 2014;9:e201401003.

146. Cheng T, Pan Y, Hao M, et al. PubChem applications in drug discovery: a bibliometric analysis. Drug Discov Today. 2014;19(11):1751–6.
147. Mendez D, Gaulton A, Bento AP, et al. ChEMBL: towards direct deposition of bioassay data. Nucleic Acids Res. 2019;47:D930–40.
148. Taboureau O, Nielsen SK, Audouze K, et al. ChemProt: a disease chemical biology database. Nucleic Acids Res. 2011;39:D367–72.
149. Kringelum J, Kjaerulff SK, Brunak S, et al. ChemProt-3.0: a global chemical biology diseases mapping. Database (Oxford). 2016;2016:bav123.
150. Gong J, Cai C, Liu X, et al. ChemMapper: a versatile web server for exploring pharmacology and chemical structure association based on molecular 3D similarity method. Bioinformatics. 2013;29(14):1827–9.
151. Chaput L, Guillaume V, Singh N, et al. FastTargetPred: a program enabling the fast prediction of putative protein targets for input chemical databases. Bioinformatics. 2020;36(14):4225–6.
152. Fenollosa C, Otón M, Andrio P, et al. SEABED: small molEcule activity scanner weB servicE baseD. Bioinformatics. 2015;31(5):773–5.
153. Yuan Y, Pei J, Lai L. LigBuilder V3: a multi-target de novo drug design approach. Front Chem. 2020;8:142.
154. Ravikumar B, Alam Z, Peddinti G, et al. C-SPADE: a web-tool for interactive analysis and visualization of drug screening experiments through compound-specific bioactivity dendrograms. Nucleic Acids Res. 2017;45(W1):W495–500.
155. Issa NT, Kruger J, Wathieu H, et al. DrugGenEx-Net: a novel computational platform for systems pharmacology and gene expression-based drug repurposing. BMC Bioinform. 2016;17(1):202.

Methods for Rational Design and Discovery of Multitarget Drugs

20

Abstract

Choosing a proper strategy is the first step or prerequisite for drug design/discovery, particularly rational design of multitarget drugs (MTDs), which has been discussed in Chap. 18. And using appropriate web-based data resources for initial analyses and predictions is the next step to success in drug design/discovery, and this information has been conveyed in Chap. 19. Yet to be able to put the strategy and data resource into a real-world process for hit identification and validation as well as hit-to-lead development, application of proper and efficient methods is indispensable. The present chapter provides detailed descriptions on step-by-step procedures of a few most commonly and frequently used methods in drug design and discovery, including high-throughput screening assays, molecular docking methods, target and compound similarity-based methods, X-ray crystallography, machine learning, and information mining (data mining and text maiming).

Keywords

Polypharmacological methodologies · High-throughput screening · Molecular docking · Similarity analysis · Information mining

20.1 High-Throughput Screening Assays

High-throughput screening (HTS) is up to now the most frequently applied method for scientific experimentation in drug discovery and other relevant fields of biology and chemistry [1]. Using robotics, data processing/control software, liquid handling devices, and sensitive detectors, high-throughput screening allows a researcher to quickly conduct millions of chemical, genetic, or pharmacological tests. Through this process, active compounds, antibodies, or genes that modulate a particular biomolecular pathway can be rapidly identified. The results of these experiments provide starting points for drug design and discovery. HTS is a relatively recent innovation, made feasible largely through modern advances in robotics and high-speed computer technology. It still takes a highly specialized and expensive screening laboratory to run an HTS operation, so in many cases, a small- to moderate-size research institution will use the services of an existing HTS facility rather than set up one for itself.

Z. Wang, B. Yang, *Polypharmacology*, https://doi.org/10.1007/978-3-031-04998-9_20

HTS in general involves the following steps:

20.1.1 Step 1: Assay Plate Preparation

The key labware or testing vessel of HTS is the microtiter plate: a small container, usually disposable and made of plastic, that features a grid of small, open divots called wells. In general, microplates for HTS have either 96, 192, 384, 1536, 3456, or 6144 wells. These are all multiples of 96, reflecting the original 96-well microplate with spaced wells of 8 × 12 with 9 mm spacing. Most of the wells contain test items, depending on the nature of the experiment. These could be different chemical compounds dissolved, e.g., in an aqueous solution of dimethyl sulfoxide (DMSO). The wells could also contain cells or enzymes of some type. (The other wells may be empty or contain pure solvent or untreated samples, intended for use as experimental controls.)

A screening facility typically holds a library of stock plates, whose contents are carefully catalogued, and each of which may have been created by the lab or obtained from a commercial source. These stock plates themselves are not directly used in experiments; instead, separate assay plates are created as needed. An assay plate is simply a copy of a stock plate, created by pipetting a small amount of liquid (often measured in nanoliters) from the wells of a stock plate to the corresponding wells of a completely empty plate.

Assay plate is often handled by a robot arm.

20.1.2 Step 2: Reaction Observation

To prepare for an assay, the researcher fills each well of the plate with some biological entity that they wish to conduct the experiment upon, such as a protein, cells, or an animal embryo. After some incubation time has passed to allow the biological matter to absorb, bind to, or otherwise react (or fail to react) with the compounds in the wells, measurements are taken across all the plate's wells, either manually or by a machine. Manual measurements are often necessary when the researcher is using microscopy to (for example) seek changes or defects in embryonic development caused by the wells' compounds, looking for effects that a computer could not easily determine by itself. Otherwise, a specialized automated analysis machine can run a number of experiments on the wells (such as shining polarized light on them and measuring reflectivity, which can be an indication of protein binding). In this case, the machine outputs the result of each experiment as a grid of numeric values, with each number mapping to the value obtained from a single well. A high-capacity analysis machine can measure dozens of plates in the space of a few minutes like this, generating thousands of experimental data points very quickly.

Depending on the results of this first assay, the researcher can perform follow-up assays within the same screen by "cherrypicking" liquid from the source wells that gave interesting results (known as "hits") into new assay plates, and then rerunning the experiment to collect further data on this narrowed set, confirming, and refining observations.

20.1.3 Step 3: Automation Systems

Automation is an important element in HTS's usefulness. Typically, an integrated robot system consisting of one or more robots transports assay–microplates from station to station for sample and reagent addition, mixing, incubation, and finally readout or detection. An HTS system can usually prepare, incubate, and analyze many plates simultaneously, further speeding the data collection process. HTS robots that can test up to 100,000 compounds per day currently exist [2]. Automatic colony pickers pick thousands of microbial colonies for high-throughput genetic screening [3]. The term uHTS or ultrahigh-throughput screening refers (circa 2008) to screening in excess of 100,000 compounds per day [3].

20.1.4 Step 4: Experimental Design and Data Analysis

With the ability of rapid screening of diverse compounds (such as small molecules or siRNAs) to identify active compounds, HTS has led to an explosion in the rate of data generated in recent years [4]. Consequently, one of the most fundamental challenges in HTS experiments is to glean biochemical significance from mounds of data, which relies on the development and adoption of appropriate experimental designs and analytic methods for both quality control and hit selection [5]. HTS research is one of the fields that have a feature described by John Blume, Chief Science Officer for Applied Proteomics, Inc., as follows: soon, if a scientist does not understand some statistics or rudimentary data-handling technologies, he or she may not be considered to be a true molecular biologist and, thus, will simply become "a dinosaur."

20.1.4.1 Quality Control

High-quality HTS assays are critical in HTS experiments. The development of high-quality HTS assays requires the integration of both experimental and computational approaches for quality control (QC). Three important means of QC are (a) good plate design, (b) the selection of effective positive and negative chemical/biological controls, and (c) the development of effective QC metrics to measure the degree of differentiation so that assays with inferior data quality can be identified [6]. A good plate design helps to identify systematic errors (especially those linked with well position) and determine what normalization should be used to remove/reduce the impact of systematic errors on both QC and hit selection [5].

Effective analytic QC methods serve as a gatekeeper for excellent quality assays. In a typical HTS experiment, a clear distinction between a positive control and a negative reference such as a negative control is an index for good quality. Many quality-assessment measures have been proposed to measure the degree of differentiation between a positive control and a negative reference. Signal-to-background ratio, signal-to-noise ratio, signal window, assay variability ratio, and Z-factor have been adopted to evaluate data quality [5, 7]. Strictly standardized mean difference (SSMD) has recently been proposed for assessing data quality in HTS assays [8].

20.1.4.2 Hit Selection

A compound with a desired size of effects in an HTS is called a hit. The process of selecting hits is called hit selection. The analytic methods for hit selection in screens without replicates (usually in primary screens) differ from those with replicates (usually in confirmatory screens). For example, the z-score method is suitable for screens without replicates, whereas the t-statistic is suitable for screens with replicates. The calculation of SSMD for screens without replicates also differs from that for screens with replicates [5].

For hit selection in primary screens without replicates, the easily interpretable ones are average fold change, mean difference, percent inhibition, and percent activity. However, they do not capture data variability effectively. The z-score method or SSMD, can capture data variability based on an assumption that every compound has the same variability as a negative reference in the screens [9]. However, outliers are common in HTS experiments, and methods such as z-score are sensitive to outliers and can be problematic. As a consequence, robust methods such as the z*-score method, SSMD*, B-score method, and quantile-based method have been proposed and adopted for hit selection [5, 10].

In a screen with replicates, we can directly estimate variability for each compound; as a consequence, we should use SSMD or t-statistic that does not rely on the strong assumption that the z-score and z*-score rely on. One issue with the use of t-statistic and associated p-values is that they are affected by both sample size and effect size. They come from testing for no mean difference, and thus are not designed to measure the size of compound effects. For hit selection, the major interest is the size of effect in a tested compound. SSMD directly assesses the size of effects [11]. SSMD has also been shown to be better than other commonly used effect sizes. The population value of SSMD is comparable across experiments, and, thus, we can use the same cutoff for the population value of SSMD to measure the size of compound effects.

20.1.5 Techniques for Increased Throughput and Efficiency

Unique distributions of compounds across one or many plates can be employed either to increase the number of assays per plate or to reduce the variance of assay results, or both. The simplifying assumption made in this approach is that any N compounds in the same well will not typically interact with each other, or the assay target, in a manner that fundamentally changes the ability of the assay to detect true hits.

For example, imagine a plate wherein compound A is in wells 1–2-3, compound B is in wells 2–3-4, and compound C is in wells 3–4-5. In an assay of this plate against a given target, a hit in wells 2, 3, and 4 would indicate that compound B is the most likely agent while also providing three measurements of compound B's efficacy against the specified target. Commercial applications of this approach involve combinations in which no two compounds ever share more than one well, to reduce the (second-order) possibility of interference between pairs of compounds being screened.

Automation and low volume assay formats were leveraged by scientists at the NIH Chemical Genomics Center (NCGC) to develop quantitative HTS (qHTS), a paradigm to pharmacologically profile large chemical libraries through the generation of full concentration–response relationships for each compound. With accompanying curve fitting and cheminformatics software, qHTS data yields half maximal effective concentration (EC50), maximal response, and Hill coefficient (nH) for the entire library enabling the assessment of nascent structure–activity relationships (SAR) [12].

In March 2010, research was published demonstrating an HTS process allowing 1000 times faster screening (100 million reactions in 10 hours) at one-millionth the cost (using 7–10 times the reagent volume) than conventional techniques using drop-based microfluidics [12]. Drops of fluid separated by oil replace microplate wells and allow analysis and hit sorting while reagents are flowing through channels.

In 2010, researchers developed a silicon sheet of lenses that can be placed over microfluidic arrays to allow the fluorescence measurement of 64 different output channels simultaneously with a single camera [13]. This process can analyze 200,000 drops per second.

In 2013, researchers have disclosed an approach with small molecules from plants. In general, it is essential to provide high-quality proof-of-concept validations early in the drug discovery process. Here technologies that enable the identification of potent, selective, and bioavailable chemical probes are of crucial interest, even if the resulting compounds require further optimization for development into a pharmaceutical product. Nuclear receptor RORα, a protein that has been targeted for more than a decade to identify potent and bioavailable agonists, was used as an example of a very challenging drug target. Hits are confirmed at the screening step due to the bell-shaped curve. This method is very similar to the quantitative HTS method (screening and hit confirmation at the same time), except that using this approach greatly decreases the data point number and can screen easily more than 100,000 biological relevant compounds [14].

Whereby traditional HTS drug discovery uses purified proteins or intact cells, very interesting recent development of the technology is associated with the use of intact living organisms, like the nematode *Caenorhabditis elegans* and zebrafish (*Danio rerio*) [15].

In 2016–2018, plate manufacturers began producing specialized chemistry to allow for mass production of ultralow adherent cell repellent surfaces which facilitated the rapid development of HTS amenable assays to address cancer drug discovery in 3D tissues such as organoids and spheroids, a more physiologically relevant format.

20.1.6 PubChem BioAssay on HTS

PubChem BioAssay [16] repository stores the results of biochemical HTS target oriented assays. Based on the PubChem BioAssay data model, results of HTS target oriented assays, in simple terms, could be interpreted as a subset of small molecules reported to inhibit (in rare cases activate) the targeted protein. The abundance of this information (> 1000 independent assays) from the PubChem BioAssay repository could be integrated on a large scale to derive binding spectra for approximately half of the million molecules across several hundred proteins [17, 18]. Let us denote KA to be the number of active molecules (molecules exhibiting activity in the assay at relevant concentrations) in the HTS UPS (UBL) assay and as KB the number of inactive molecules. For each off-target protein "Z," we count the number (*kA*) of the active molecules (from the UPS assay) which target "Z" (have been reported in the other HTS assay as active versus "Z") and *kB* is the number of inactive molecules which target Z. Odds for the molecules (to be active) targeting "Z" is (*kA/kB*), while the same odds for the molecules which do not target "Z" is ((*KA-kA*)/(KB-kB)). The odds ratio is (*kA/kB*)/((*KA-kA*)/(*KB-kB*)) and indicates the increase/decrease of odds for a molecule to target "Z" while targeting the UPS–UBL protein. Significance of the odds ratio is computed using X^2-distribution, and the adjustment of p-values for multiple testing (each target one hypothesis) is done using the FDR procedure [19–21].

20.2 Molecular Docking Methods

Molecular docking of small molecules into the receptor active sites is one of the most widely used structure-based methods in the field of computer-aided drug discovery. It has been shown that current docking algorithms are able to predict docking poses of small molecules similar to experimental structures. In the field of molecular modeling, docking is a method which predicts the preferred orientation of one molecule to a second when a ligand and a target are bound to each other to form a stable complex [22]. Knowledge of the preferred orientation in turn may be used to predict the strength of association or binding affinity between two molecules using, for example, scoring functions. The associations between biologically relevant molecules such as proteins, peptides, nucleic acids, carbohydrates, and lipids play a central role in signal transduction. Furthermore, the relative orientation of the two interacting partners may affect the type of signal produced (e.g., agonism vs antagonism). Therefore, docking is useful for predicting both the strength and type of signal produced.

Molecular docking is one of the most frequently used methods in structure-based drug design, due to its ability to predict the binding conformation of small-molecule ligands to the appropriate target binding site. Characterization of the binding behavior plays an important role in rational design of drugs as well as to elucidate fundamental biochemical processes [23].

20.2.1 Molecular Docking

20.2.1.1 Docking Approaches

Two approaches are particularly popular within the molecular docking community. One approach uses a matching technique that describes the protein and the ligand as complementary surfaces [24, 25]. The second approach simulates the actual docking process in which the ligand–protein pairwise interaction energies are calculated. Both approaches have significant advantages as well as some limitations. These are outlined below.

Shape Complementarity

Geometric matching/shape complementarity methods describe the protein and ligand as a set of features that make them dockable [26]. These features may include molecular surface/complementary surface descriptors. In this case, the receptor's molecular surface is described in terms of its solvent-accessible surface area, and the ligand's molecular surface is described in terms of its matching surface description. The complementarity between the two surfaces amounts to the shape matching description that may help finding the complementary pose of docking the target and the ligand molecules. Another approach is to describe the hydrophobic features of the protein using turns in the main-chain atoms. Yet another approach is to use a Fourier shape descriptor technique. Whereas the shape complementarity-based approaches are typically fast and robust, they cannot usually model the movements or dynamic changes in the ligand/protein conformations accurately, although recent developments allow these methods to investigate ligand flexibility. Shape complementarity methods can quickly scan through several thousand ligands in a matter of seconds and actually figure out whether they can bind at the protein's active site and are usually scalable to even protein–protein interactions. They are also much more amenable to pharmacophore-based approaches, since they use geometric descriptions of the ligands to find optimal binding.

Simulation

Simulating the docking process is much more complicated. In this approach, the protein and the ligand are separated by some physical distance, and the ligand finds its position into the protein's active site after a certain number of "moves" in its conformational space. The moves incorporate rigid body transformations such as translations and rotations, as well as internal changes to the ligand's structure including torsion angle rotations. Each of these moves in the conformation space of the ligand induces a total energetic cost of the system. Hence, the system's total energy is calculated after every move. The obvious advantage of docking simulation is that ligand flexibility is easily incorporated, whereas shape complementarity techniques must use ingenious methods to incorporate flexibility in ligands. Also, it more accurately models reality, whereas shape complementary techniques are more of an abstraction.

Clearly, simulation is computationally expensive, having to explore a large energy landscape. Grid-based techniques, optimization methods, and increased computer speed have made docking simulation more realistic.

20.2.1.2 Mechanics of Docking

To perform a docking screen, the first requirement is a structure of the protein of interest. Usually, the structure has been determined using a biophysical technique such as X-ray crystallography, NMR spectroscopy, or cryo-electron microscopy (cryo-EM), but can also derive from homology modeling construction. This protein structure and a database of potential ligands serve as inputs to a docking program. The success of a docking program depends on two components: the search algorithm and the scoring function.

Search Algorithm

The search space in theory consists of all possible orientations and conformations of the protein paired with the ligand. However, in practice with current computational resources, it is impossible to exhaustively explore the search space—this would involve enumerating all possible distortions of each molecule (molecules are dynamic and exist in an ensemble of conformational states) and all possible rotational and translational orientations of the ligand relative to the protein at a given level of granularity. Most docking programs in use account for the whole conformational space of the ligand (flexible ligand), and several attempt to model a flexible protein receptor. Each "snapshot" of the pair is referred to as a pose.

A variety of conformational search strategies have been applied to the ligand and to the receptor. These include:

(a) Systematic or stochastic torsional searches about rotatable bonds.
(b) Molecular dynamics simulations.
(c) Genetic algorithms to "evolve" new low-energy conformations and where the score of each pose acts as the fitness function used to select individuals for the next iteration.

Ligand Flexibility

Conformations of the ligand may be generated in the absence of the receptor and subsequently docked or conformations may be generated on-the-fly in the presence of the receptor-binding cavity [27], or with full rotational flexibility of every dihedral angle using fragment-based docking. Force field energy evaluations are most often used to select energetically reasonable conformations, but knowledge-based methods have also been used [28].

Peptides are both highly flexible and relatively large-sized molecules, which makes modeling their flexibility a challenging task. A number of methods were developed to allow for efficient modeling of flexibility of peptides during protein–peptide docking [29].

Receptor Flexibility

Computational capacity has increased dramatically over the last decade making possible the use of more sophisticated and computationally intensive methods in computer-assisted drug design. However, dealing with receptor flexibility in docking methodologies is still a thorny issue [30]. The main reason behind this difficulty is the large number of degrees of freedom that have to be considered in this kind of calculations. Neglecting it, however, in some of the cases may lead to poor docking results in terms of binding pose prediction [31].

Multiple static structures experimentally determined for the same protein in different conformations are often used to emulate receptor flexibility [23]. Alternatively rotamer libraries of amino acid side chains that surround the binding cavity may be searched to generate alternate but energetically reasonable protein conformations [32, 33].

Scoring Function

Docking programs generate a large number of potential ligand poses, of which some can be immediately rejected due to clashes with the protein. The remainder are evaluated using some scoring function, which takes a pose as input and returns a number indicating the likelihood that the pose represents a favorable binding interaction and ranks one ligand relative to another.

Most scoring functions are physics-based molecular mechanics force fields that estimate the energy of the pose within the binding site.

Alternative approaches use modified scoring functions to include constraints based on known key protein–ligand interactions [34], or knowledge-based potentials derived from interactions observed in large databases of protein–ligand structures (e.g., the Protein Data Bank) [35].

There are a large number of structures from X-ray crystallography for complexes between proteins and high-affinity ligands, but comparatively fewer for low affinity ligands as the later complexes tend to be less stable and therefore more difficult to crystallize. Scoring functions trained with this data can dock high-affinity ligands correctly, but they will also give plausible docked conformations for ligands that do not bind. This gives a large number of false-positive hits, i.e., ligands predicted to bind to the protein that actually do not when placed together in a test tube.

One way to reduce the number of false positives is to recalculate the energy of the top scoring poses using (potentially) more accurate but computationally more intensive techniques such as generalized Born or Poisson-Boltzmann methods.

20.2.1.3 Docking Assessment

The interdependence between sampling and scoring function affects the docking capability in predicting plausible poses or binding affinities for novel compounds. Thus, an assessment of a docking protocol is generally required (when experimental data is available) to determine its predictive capability. Docking assessment can be performed using different strategies, such as:

(a) Docking accuracy (DA) calculation.
(b) The correlation between a docking score and the experimental response or determination of the enrichment factor (EF)
(c) The distance between an ion-binding moiety and the ion in the active site.
(d) The presence of induce-fit models.

Docking Accuracy

Docking accuracy represents one measure to quantify the fitness of a docking program by rationalizing the ability to predict the right pose of a ligand with respect to that experimentally observed [36].

Enrichment Factor

Docking screens can also be evaluated by the enrichment of annotated ligands of known binders from among a large database of presumed nonbinding, "decoy" molecules. In this way, the success of a docking screen is evaluated by its capacity to enrich the small number of known active compounds in the top ranks of a screen from among a much greater number of decoy molecules in the database. The area under the receiver operating characteristic (ROC) curve is widely used to evaluate its performance.

Prospective
Resulting hits from docking screens are subjected to pharmacological validation (e.g., IC_{50}, affinity, or potency measurements). Only prospective studies constitute conclusive proof of the suitability of a technique for a particular target. In the case of G-protein-coupled receptors (GPCRs), which are targets of more than 30% of marketed drugs, molecular docking led to the discovery of more than 500 GPCR ligands [37].

Benchmarking
The potential of docking programs to reproduce binding modes as determined by X-ray crystallography can be assessed by a range of docking benchmark sets.

For small molecules, several benchmark datasets for docking and virtual screening exist, e.g., Astex Diverse Set consisting of high-quality protein–ligand X-ray crystal structures [38] or the Directory of Useful Decoys (DUD) for evaluation of virtual screening performance [39].

An evaluation of docking programs for their potential to reproduce peptide binding modes can be assessed by Lessons for Efficiency Assessment of Docking and Scoring (LEADS-PEP) [40].

20.2.1.4 Applications

A binding interaction between a small-molecule ligand and an enzyme protein may result in activation or inhibition of the enzyme. If the protein is a receptor, ligand binding may result in agonism or antagonism. Docking is most commonly used in the field of drug design—most drugs are small organic molecules, and docking may be applied to:

(a) *Hit identification*: Docking combined with a scoring function can be used to quickly screen large databases of potential drugs in silico to identify molecules that are likely to bind to protein target of interest (see virtual screening). Reverse pharmacology routinely uses docking for target identification.
(b) *Lead optimization*: Docking can be used to predict in where and in which relative orientation a ligand binds to a protein (also referred to as the binding mode or pose). This information may in turn be used to design more potent and selective analogs.
(c) *Bioremediation*: Protein–ligand docking can also be used to predict pollutants that can be degraded by enzymes.

20.2.2 Inverse Docking

Recent advances in high-throughput protein crystallography, nuclear magnetic resonance (NMR) spectroscopy, and electron microscopy have generated a large number of 3D protein structures in the Protein Data Bank (PDB). Availability of such structural data has allowed development of inverse docking methods where the primary aim is to dock a small molecule into binding sites of multiple targets for hit identification. Unlike traditional docking algorithms where small molecules are scored and ranked, in inverse docking, target receptors are ranked according to their scores [41].

Reverse or inverse docking is proving to be a powerful tool for drug repositioning and drug rescue. It involves docking a small-molecule drug/ligand in the potential binding cavities of a set of clinically relevant macromolecular targets [42]. Inverse docking, first proposed in 2001 by Chen et al. [43], refers to computationally docking a specific small molecule of interest to a library of receptor structures. The technique may be used to identify new potential biological targets of known compounds [44–46], or to identify targets for compounds among a family of related receptors [47]. The technique has shown success in distinguishing between homology models of receptors [47]. The technique may also be used to generate a compound's predicted pharmacological profile [48], or to generate a virtual selectivity profile that characterizes the promiscuity of the inhibitors [49]. Given the multifaceted nature of a pharmacologically active compound's biological effects, inverse docking is especially helpful, because it may generate new hypotheses for the action mechanism.

In theory, all of the available docking algorithms should be able to carry out inverse docking. However, the major challenge is to handle a large amount of structural data of macromolecules and implement scoring functions that could accurately rank the targets instead of small molecules [41]. It has been found that traditional scoring functions used in current docking algorithms are not sufficient to rank targets accurately and robustly [41, 50, 51]. Therefore, in recent years, significant efforts have been put to develop modified scoring functions specific for target ranking [41], such as INVDOCK [52], TarFisDock [53], and idTarget [54]. Chen et al. developed the first automated inverse docking method INVDOCK [52].

In the work conducted by Grinter et al. [42], the investigators used the inverse docking approach to screen for potential molecular targets of PRIMA-1. The objective is to guide future assays of the inhibitors of these predicted targets for their efficacy in inhibiting tumor cell proliferation, as such results may lead to potential cancer treatments, as well as provide clues regarding PRIMA-1's action mechanism. Below are brief descriptions of the steps taken in their study.

20.2.2.1 In Silico Screening

The authors [42] used our protein–ligand docking software package MDock [55, 56] (zoulab.dalton.missouri.edu/software.htm) to dock PRIMA-1 into many potential drug targets. Although MDock is sufficiently computationally efficient for PDB-wide database screening, we chose to start with the well-characterized Potential Drug Target Database (PDTD), which at the time of use contained about 1100 experimentally determined structures of 830 actual or suspected drug targets (http://www.dddc.ac.cn/pdtd) [57]. They also used the PDTD's binding-site definitions, which in most cases are based on the set of amino acid residues that are within 6.5 Å of the bound ligand. OMEGA Version 2.2.1 was used to generate conformations of PRIMA-1 for flexible-ligand docking (OpenEye Scientific Software Inc., Santa Fe, NM) with the *rms* parameter set to 0.1 Å, *maxconfs* to 1,000,000, *maxconfgen* to10000000, and *ewindow* to 10. As PRIMA-1 has few rotatable bonds, this only resulted in 42 generated ligand conformations. Each of these conformations was docked to each protein as a rigid body.

For each protein in the database, a molecular surface of the binding site was generated, along with the associated sphere points representing potential initial positions for ligand atom centers. Ligand atoms were matched to these sphere points, and orientations were sampled and ranked by our knowledge-based scoring function, ITScore [55, 56]. All of MDock's default parameters were used in this work, with the exception of write_score_total, which was set to 1 so that only the highest-scoring orientation is recorded when each protein/PRIMA-1 conformation pair is docked as a rigid body. We then ranked each protein according to the lowest ITScore (corresponding to the highest predicted affinity) recorded for any of the 42 PRIMA-1 conformations that were docked to it. Because PDTD contains redundant experimental structures of the same protein [57], we clustered the resulting docked structures into groups sharing 90% sequence identity. They then ran a BLAST search in order to map the PDTD proteins, which come from various species to human gene sequences. Inhibitors of the top human or human analog proteins were considered candidate anticancer agents for assay.

20.2.2.2 Cell Viability Assay

The investigators used the sulforhodamine B (SRB) assay [58] to evaluate the effect of the OSC-inhibitor Ro 48–8071 on the viability of breast cancer cells. This cell protein dye-binding assay determines the protein content in surviving cells as an index to determine cell growth, viability, and survival. The effects of PRIMA-1 on cell growth and viability were assessed.

20.3 Target and Compound Similarity-Based Methods

Molecular similarity is a key concept in drug discovery. It is based on the assumption that structurally similar molecules frequently have similar properties. Assessment of similarity between small molecules has been highly effective in the discovery and development of various drugs [59]. Especially, two-dimensional (2D) similarity approaches have been quite popular due to their simplicity, accuracy, and efficiency. Recently, the focus has been shifted toward the development of methods involving the representation and comparison of three-dimensional (3D) conformation of small molecules [59]. Among the 3D similarity methods, evaluation of shape similarity is now gaining attention for its application not only in virtual screening but also in molecular target prediction, drug repurposing, and scaffold hopping. A wide range of methods have been developed to describe molecular shape and to determine the shape similarity between small molecules. The most widely used methods include atom distance-based methods, surface-based approaches such as spherical harmonics and 3D Zernike descriptors, and atom-centered Gaussian overlay-based representations. Several of these methods demonstrated excellent virtual screening performance not only retrospectively but also prospectively. In addition to methods assessing the similarity between small molecules, shape similarity approaches have been developed to compare shapes of protein structures and binding pockets. Additionally, shape comparisons between atomic models and 3D density maps allowed the fitting of atomic models into cryo-electron microscopy maps. This review aims to summarize the methodological advances in shape similarity assessment highlighting advantages, disadvantages, and their application in drug discovery.

According to Kumar and Zhang [59], there are two components of molecular similarity analysis: (a) structural representations and (b) quantitative measurements of similarity between two structural representations. Many types of structural representations have been suggested to measure the similarity between two molecules. These include physiochemical properties, topological indices, molecular graphs, pharmacophore features, molecular shapes, molecular fields, etc. Further, there are various methods to quantify the similarity between two structural representations, e.g., Tanimoto coefficient, Dice index, cosine coefficient, Euclidean distance, Tversky index, etc. Among these, Tanimoto coefficient [60] is the most popular and widely used similarity measure. Based on the structural representation, molecular similarity approaches can be broadly classified into 2D or 3D similarity methods. The 2D similarity methods rely only on the 2D structural information and are among the fastest, efficient, and most popular similarity search methods. Moreover, they do not rely on structural alignments for estimating the similarity between two molecules. These methods include substructure search, fingerprint similarity search, and 2D descriptor-based methods. However, most of these methods are limited in their ability to enable scaffold hopping and provide no structural and mechanistic insights. To deal with the limitations associated with 2D similarity methods, several approaches were developed that account for 3D conformations of a molecule while performing similarity search. These methods include pharmacophore modeling, shape similarity, molecular field-based methods, and 3D fingerprints, among others. In recent years, ligand 3D shape-based similarity analysis has become a method of choice in increasing number of virtual screening campaigns. Several successful applications of shape similarity to discover new molecules have been published in the literature. The major advantage with shape-based virtual screening methods is that scaffold hopping can be conveniently accomplished and scaffolds other than the query can be identified. Below is a brief description on 3D shape similarity methods and their applications to drug discovery, essentially based on the review article by Kumar and Zhang [59] (for detailed information, please refer to the article by Kumar and Zhang [59]: Kumar A, Zhang KYJ. Advances in the development of shape similarity methods and their application in drug discovery. Front Chem. 2018; 6: 315).

20.3.1 3D Shape Similarity Methods

The 3D shape has been widely recognized as a key determinant for the activity of small molecules and other biomolecules [61, 62]. The shape complementarity between ligand and receptor is necessary for bringing the receptor and ligand sufficiently close to each other so they can form critical interactions necessary for binding [59]. Two molecules with similar shape are likely to fit the same binding pocket and thereby exhibiting similar biological activity. Shape comparison methods could be broadly classified as (a) alignment-free or non-superposition methods and (b) alignment- or superposition-based methods. Both of these methods have their own advantages and disadvantages [59]. Alignment-free methods are independent of the position and orientations of molecules. As such, they are much faster and could be used to screen large compound databases. Alignment-based methods rely on finding the optimal superposition between the compounds. Alignment-based methods are highly effective in identifying shape similarities among the molecular structures, but they are computationally expensive. These methods enable comparison of the surface properties such as hydrophobicity and polarity. Visualization is one of the advantages with the alignment-based methods and the similarity between two molecules can be displayed. This information is useful in the design of new molecules and to guide further optimization. However, a subpar molecular alignment may lead to errors in comparing two molecules. Apart from this broad classification, shape similarity methods could be classified based on the underlying representation of molecular shape. The similarity between these shape representations is evaluated by employing various similarity metrics [59].

20.3.1.1 Atomic Distance-Based Descriptors

These methods are based on the assumption that the shape of a molecule can be described by the relative positions of its atoms [59]. The similarity between molecules can be then calculated by comparing the corresponding distributions of atomic distances. As these descriptors only require the computation of interatomic distances in compounds, these methods are faster compared to other shape comparison methodologies. Additionally, these methods do not require the alignment between two molecules for shape comparison. One of the earlier atomic distance-based shape comparison methods was based on atom triplet distances [63]. This method considered each molecule as a collection of three atom submolecules. The atom triplet triangle perimeters were used to generate shape histograms which were then utilized to compare the shape of molecules. This method however has a few limitations. It is difficult to select bin size suitable for all molecules. Each molecule typically generates 300–500 atom triplets and storing them requires large space especially when comparing a large database of molecules. To deal with this limitation, another atom triplet-based molecular shape comparison method was developed where a 2048 bits long single condensed triplet shape signature was employed to represent the entire set of triplets in each molecule [64]. A signature of the query molecule is first compared with the already stored signatures of database molecules. Then only the compounds with adequately similar signatures are compared in detail by generating all triplets. Although this method was efficient, there was a risk of missing similar compounds due to the use of highly reduced signature representation [59]. Another group developed molecular descriptors based on atom triplet triangles, angular information from surface point normal and local curvature to facilitate shape comparisons [65]. However, these descriptors have limited discriminating power and require large disk space for storage [59].

20.3.1.2 Atom-Centered Gaussian-Based Shape Similarity Methods

Among many methods of describing the molecular shape of a molecule, hard sphere [66] and Gaussian sphere [67] are two most widely adopted models [59]. Both of these models describe the shape in terms of the volume of a molecule. Two molecules will possess similar shape if they have similar volume. Hard sphere model represents a molecule by a set of merged spheres where each sphere serves as an atom with its van der Waals radius. The volume of a molecule can be calculated by a formula that describes the union of a number of sets and their intersection. Although the analytical expression of the volume and its derivatives is reported in the original publication [68], it is not easy to implement as the formulas become very complicated with increasing number of intersections [59]. Gaussian sphere model [67] represents a molecule using a set of overlapping Gaussian spheres and measures the integral volume over all overlapping Gaussians. In this model, each intersection is expressed as the integral of a set of overlapping atom-centered Gaussian spheres, and the volume of a molecule is described based on the inclusion–exclusion principle. Analytical expression for the volume calculation is given in the original publication which describes highly accurate volume calculation up to sixth-order intersections [67]. Several methods based on Gaussian overlays were developed to measure the shape similarity between two molecules. Among these, Rapid Overlay of Chemical Structures (ROCS) is undoubtedly the most widely used method that utilizes Gaussian functions to measure the shape similarity between two molecules [69, 70].

20.3.1.3 Surface-Based 3D Shape Similarity Comparison Methods

Molecular surface is another way of depicting the shape of a molecule. Comparison of molecular surfaces based on their shapes can reveal similarity in their physical and biological properties [59]. There are many ways to describe the surface of a molecule. Precise definitions such as surface based on quantum mechanical wave functions are not practical especially for large molecules [71]. Surface definitions such as solvent-accessible surface and van der Waals surface are more practical and much easier to calculate. Some studies employed alpha shapes which is a coarse representation of Connolly surface [71] to describe the shape of a molecule [72]. Alpha shapes of a set of points “S” are generalization of convex hull and utilize a parameter, α to describe the shape with varying levels of details. For large α values, the alpha shape is equivalent to convex hull and shape feature details such as concavities and voids started to appear with decrease in α value. The alpha shape method has been applied to represent and compare shapes of 3D molecules [72].

Shape signatures or shape histograms offer another representation of molecular shape that can be used to explore 3D volume of a molecule confined by the solvent-accessible surface. Shape signatures are probability distribution histograms borrowed from a computer graphics technique, ray tracing. In this method, a ray is initiated within a molecule bound by its solvent-accessible surface. Propagation of a ray trace inside of the triangulated solvent-accessible surface is recorded as probability distribution histograms.

Several methods employed local surface shape similarity to align and estimate the similarity between molecules. One such method applied subgraph isomorphism to molecular surface comparison [73]. In this method, molecular surface was represented by patches of the same shape. Alignment between two molecules was obtained by using a clique-detection algorithm to obtain overlapping patches. Quadratic shape descriptors [74] exploited a similar concept where molecular surface was divided into a series of patches. Each patch was represented by geometrically invariant descriptors such as the normal, the shape index, and the principle curvatures which were then used to identify similar patches. SURFCOMP [75] further applied several filters such as surrounding shape and physicochemical properties to identify corresponding patches on surfaces of two molecules.

20.3.2 Application of Shape Similarity Methods in Drug Discovery

20.3.2.1 Application in Virtual Screening

Shape similarity attempts to quantify the resemblance between two molecules utilizing several descriptions of molecular shape as described previously. This approach has been successfully utilized as a virtual screening tool to identify molecules similar to a given query from the library of chemicals [59]. Several retrospective studies have been published demonstrating the utility of shape-based similarity methods over 2D and other 3D similarity methods [59]. Some other studies also presented computational approaches to improve the performance and efficiency of shape comparison methods. One study recommended the selection of a suitable query and incorporation of chemical information such as pharmacophoric features of the query molecule to improve the performance of shape-based virtual screening. Another study demonstrated that the application of a machine learning method, support vector machine (SVM), to shape comparisons can significantly improve virtual screening efficiency [76]. The need of automation was further suggested specially to carry out multiple query searches which ensure a diverse hit list [59].

20.3.2.2 Applications in Protein Structure Comparison

Evaluation of structural similarity between protein structures has many applications including but not limited to classification of protein structures, evolutionary relationship between protein structures, identification of templates for homology modeling, functional annotation, protein–protein interactions, etc. Conventional methods for protein structure comparison are based on the alignment of protein atoms or residues. These methods require extensive rotational and translational sampling, thereby limiting their utility for large-scale protein structure comparisons. Several methods have been developed that utilize shape similarity to detect global or local similarity between protein structures. Classification of these methods also follows the previously described classification including Gaussian overlay-based methods, surface-based methods using spherical harmonic descriptors, 3D Zernike descriptors, etc. Among these, surface-based methods were developed previously to measure similarity between protein structures. Only later they were applied to the small-molecule area. Several methods of protein structure comparison employed SH to represent shapes of protein structures [59]. Like SH, 3D Zernike-based moments are also suitable to compare shapes of protein structures. Not only they were suitable to estimate the similarity between two proteins but also their rotation–translation invariant nature allows fast real-time search of similar proteins in structural databases such as PDB [77]. A Gaussian mixture model-based protein shape similarity method [78] also allows large-scale comparisons of proteins with data from PDB and EMDB. This method has been implemented as Omokage search in PDB Japan. The server compares global shapes of proteins and results are obtained reasonably fast within 1 min after submission of a query. Large-scale comparison of protein structures based on shape is useful in functional annotation, selection of templates for comparative modeling, etc.

20.3.3 Similarity Ensemble Approach (SEA)

The similarity ensemble approach relates proteins based on the set-wise chemical similarity among their ligands [79]. It can be used to rapidly search large compound databases and to build cross-target similarity maps. Keiser and colleagues used the similarity ensemble approach (SEA) [80, 81] in a large scale to predict the activity of 656 marketed drugs on 73 unintended "side effect" targets and confirmed half of the predictions with the IC_{50} activity values ranging from 1 nM to 30 μM [81]. Among these new associations was the finding that the abdominal pain side effect of chlorotrianisene was due to its newly discovered inhibition of cyclooxygenase-1 (COX-1). Oprea et al. [82, 83] also reported results from their text mining of 7684 approved drugs and mapped the "adverse reactions" of 988 unique drugs onto 174 side effects. They were then clustered with principal component analysis into a self-organizing map and integrated it into a Cytoscape network. They expected that this type of data could streamline drug repurposing [83]. An exemplified large-scale study where the similarity ensemble approach was used to predict the activity of 656 marketed drugs on 73 unintended "side effect" targets [81]. Experimental evaluations confirmed that half of the predictions were correct with the activity values ranging from 1 nM to 30 μM [81].

20.3.4 Bioactivity Profile Similarity Search

Cheng et al. [84] developed a computational approach, named bioactivity profile similarity search (BASS), for associating targets to small molecules by using the known target annotations of related compounds from public databases. To evaluate BASS, a bioactivity profile database was constructed using 4296 compounds that were commonly tested in the US National Cancer Institute 60 human tumor cell line anticancer drug screen (NCI-60). Each compound was used as a query to search against the entire bioactivity profile database, and reference compounds with similar bioactivity profiles above a threshold of 0.75 were considered as neighbor compounds of the query. Potential targets were subsequently linked to the identified neighbor compounds by using the known targets of the query compound. About 45% of the predicted compound–target associations were successfully verified retrospectively, suggesting the possible application of BASS in identifying the targets of uncharacterized compounds and thus providing insight into the study of promiscuity and polypharmacology. Furthermore, BASS identified a significant fraction of structurally diverse compounds with similar bioactivities, indicating its feasibility of "scaffold hopping" in searching novel molecules against the target of interest.

20.3.5 Binding-Site Similarity

Binding-site similarity-based search is frequently used for target prediction. It is based on the assumption that structurally similar proteins have similar molecular function, and thus it is likely that they bind to structurally similar compounds [41]. This phenomenon can be explained by the "lock and key" model that was postulated by Emil Fischer in 1894. Even differing in their overall sequences or structures, proteins may share local similarities in the form of binding sites, enabling them to bind similar ligands. A majority of these tools utilize some variations of geometric patterns where the 3D coordinates of proteins are converted into simplified patterns that can be compared [41]. The geometric matching of triplets is the most commonly used strategy, and the triplets are aligned based on the measurement of the distances between each triplet.

The Fingerprints for Ligands and Proteins (FLAP) algorithm was developed recently by combining the GRID Molecular Interaction Fields (GRID-MIFs) and pharmacophoric features [85]. In the initial steps, various chemical probes that cover all types of chemical interactions are used to generate MIFs. These MIFs are then converted into pharmacophoric features using the weighted energy-based and space-coverage functions. All possible combinations of pharmacophoric features are used to generate four-point pharmacophore fingerprints, which are used to find ligand–ligand, ligand–receptor, and receptor–receptor similarities. The FLAP algorithm has been implemented in BioGPS, an automated method that calculates binding-site similarities [86]. This method works in two steps. First, it uses the Fixpdb algorithm for automatic protein structure preparation. In the second step, it employs the FLAPsite algorithm for binding-site detection using 16 different GRID probes. As described earlier, the FLAP protocol then calculates MIFs for the binding sites. These MIFs can then be superposed against all binding-site MIFs in the database. Such binding-site similarity tools can be used for drug repurposing and hit identifications. By comparing binding sites, it can be used in lead optimization process as well.

20.4 X-Ray Crystallography

20.4.1 Basics of X-Ray Crystallography

Structural information has paved the way for structure-guided drug discovery and laid the foundations for structural bioinformatics. Knowledge of three-dimensional structures of biological macromolecules is essential for a complete understanding of many biological processes. Macromolecular X-ray crystallography has been the primary methodology for determining the three-dimensional structures of proteins, nucleic acids, and viruses. X-ray crystallography is the most widely used technique in structural biology and can provide highly detailed structures of proteins, nucleic acids, or macromolecular complexes without any size limit. With the advent of structural biology in the drug discovery process, medicinal chemists gained the opportunity to use detailed structural information in order to progress screening hits into leads or drug candidates [87–91].

X-ray crystallography has proven to be an invaluable tool in this respect, as it is able to provide exquisitely comprehensive structural information about the interaction of a ligand with a pharmacological target. As fragment-based drug discovery emerged in the recent years, X-ray crystallography has also become a powerful screening technology, able to provide structural information on complexes involving low-molecular-weight compounds, despite weak binding affinities. Given the low numbers of compounds needed in a fragment library, compared to the hundreds of thousand usually present in drug-like compound libraries, it now becomes feasible to screen a whole fragment library using X-ray crystallography, providing a wealth of structural details that will fuel the fragment to drug process [88]. In the past decade, several recent advances in biological crystallography and automation of data collection and structure solution allowed extraordinary progresses, and now more than 93,000 crystal structures have been deposited into the Protein Data Bank [90]. This wealth of structural data significantly helped the elucidation of many biological processes and led to the development of new drugs.

However, X-ray crystallography still has a few fundamental limitations, some of which may be overcome and complemented using emerging methods and technologies in other areas of structural biology. The technical challenges and time constraints have traditionally limited its use to lead optimization. The application of X-ray crystallography is now being extended beyond structure determination into new approaches for lead discovery. Structure–activity relationships by nuclear magnetic resonance have been widely used to detect ligand binding and to give some indication of the location of the binding site. X-ray crystallography has the advantage of defining ligand-binding sites with greater certainty. High-throughput approaches make this method applicable to screening to identify molecular fragments that bind protein targets, and to defining precisely their binding sites. X-ray crystallography can then be used as a rapid technique to guide the elaboration of the fragments into larger molecular weight compounds that might be useful leads for drug discovery [92].

To allow readers of this book to get some impression of the real steps in the application of X-ray crystallography, a typical workflow of X-ray crystallography is described below primarily based on the excellent review article presented by Maveyraud and Mourey [88] (for detailed information, please refer to the article by Maveyraud and Mourey [88]: Maveyraud L, Mourey L. Protein X-ray crystallography and drug discovery. Molecules. 2020; 25(5): 1030).

20.4.2 Workflow of X-Ray Crystallography

The classical workflow of a rational drug design project is usually initiated with the structure determination of the—supposedly validated—pharmacological target. X-ray crystallography is the more common tool for this endeavor, but nuclear magnetic resonance (NMR) and, more recently, cryo-electron microscopy (cryo-EM) are also important players in the field. In the absence of experimental structural information, modeling of the target's structure can also provide useful insights, at the condition that at least one structure of a homologous protein exists at the Protein Data Bank (PDB) [93]. Besides the structure of the pharmacological target, at least one ligand has to be identified in order to initiate the rational drug design process.

20.4.2.1 Ligand Identification

Ligand identification can be addressed using several technologies [88]:

High-Throughput Screening (HTS)

High-throughput screening (HTS) allows one to experimentally screen large chemical libraries, if an appropriate assay is available, resulting in a collection of potential ligands. Preliminary confirmation of direct binding to the protein target using biophysical methods will help to select the most appropriate compounds for structural studies.

Knowledge-Based Drug Discovery

It is quite common that in the course of the characterization of a protein target, some knowledge is gained on potential ligands, such as substrate or cofactor analogs, which are often potent inhibitors.

Virtual Screening Methods

Success in target structure determination opens access to virtual screening methods, which aim at identifying potential ligands of a target using computational methods. This step serves often as a preliminary filter to identify a few hundred potential binders among huge virtual libraries, in order to reduce the experimental work needed to confirm the interaction. Experimentally confirmed ligands are then used in structural studies.

Fragment-Based Drug Discovery (FBDD)

Recently, fragment-based drug discovery (FBDD), also known as fragment-based drug design, or fragment-based ligand/lead discovery (FBLD) has emerged as a powerful tool to identify ligands, albeit with poor affinity given the reduced molecular weight of fragments (about 150–250 Da, containing 10 to 20 non-hydrogen atoms) [88]. With such compounds, detection of binding requires sensitive techniques, and X-ray crystallography has proved to be very effective at identifying weak binders, with affinity as low as a few mM. In most favorable cases, fragment libraries of a few hundreds of compounds can be screened using X-ray crystallography, delivering dozens of structures of complexes [94, 95]. Although FBDD was initially used as an alternative approach to HTS, i.e., for targets without hits, it appears that these two methodologies are more and more often used in parallel. Despite numerous fragment-based screening technologies existing, X-ray crystallography and protein-based NMR are getting more and more popular since they provide direct and experimental structural information on fragment mode of binding. More recently, the concept of fragment-based screening has been extended to crystallographic screening of ultralow-molecular-weight compounds, called MiniFrags and typically containing five to seven non-hydrogen atoms. Given the low affinity of these compounds, this new crystal-soaking methodology requires working at very high concentrations, typically 1 M. As of October 2018, more than 40 FBLD-derived drugs have been in clinical development, three of them having been approved for clinical use [88].

Knowledge of the structure of a target with a first bound compound, the hit, provides structural information allowing the medicinal chemist to design chemical modifications susceptible to improve the interaction with the target, and therefore the potency [88]. However, improving the interaction is not the sole consideration that should guide the so-called "hit2lead" process, as ADME–Tox properties should also be improved in order for the final drug to be used in humans. Moreover, as efficient binding to the target in vitro does not necessarily translate into an efficient drug in vivo, biological evaluation of drug potency has to be introduced in the process as early as possible. Multiple chemical evolutions of the hit(s) are usually undertaken, and structure–activity relationships (SARs) have to be deciphered. This hit2lead process can therefore be seen as an iterative process where each cycle delivers improved compounds. Structure determination at each cycle of the process is of crucial importance for the chemist in order to validate the expected binding mode of the modified compound and helps to rationalize SAR [88].

Structure determination of a complex formed of a protein and a small-molecular-weight compound does not present any specific difficulties from the crystallographic point of view, especially if the structure of the protein is already known [88]. However, in the case of a crystallographic fragment screening campaign, the process is far more complicated because of the amount of data that is generated. Indeed, hundreds of crystals are to be prepared hopefully resulting in as many datasets and structures to be solved, refined, and analyzed. With such an amount of data to be processed and analyzed within a short time, automation is of paramount importance, both for crystal preparation and handling, and for computing tasks. Besides, a properly dimensioned computing environment should be available, as terabytes of data will have to be processed.

20.4.2.2 Crystallizing the Target

Whether engaged in a crystallographic screening or in a hit2lead process, many crystallographic structures of complexes will have to be obtained. It is therefore useful to carefully optimize the crystallization procedure of the target to ensure that it reproducibly delivers a suitable crystal form diffracting to the highest possible resolution, preferentially better than 2.5 Å. In some extreme cases, this step will require to optimize the protein construct and/or the expression and purification steps [88]. Seeding, that is the introduction of microcrystals or crystal fragments into an equilibrated crystallization drop, might be considered in order to attain suitable reproducibility. Crystallization conditions allowing for cooling of the crystal to 100 K without the need of prior cryoprotection will definitely facilitate the crystal harvesting steps. The targeted binding site of the protein should not be blocked by crystal contacts with symmetry mates, nor should it be occupied by a molecule coming from the crystallization solution since such compound could compete with ligand binding. Tolerance to the presence of small amounts of the organic solvent used to solubilize putative ligands to be studied should be ensured. From the crystallographic point of view, in order to facilitate data collection, a crystal form with a high-symmetry space group should be preferred, i.e., triclinic and monoclinic space groups would necessitate more data to be collected.

20.4.2.3 Preparation of Crystals of Complexes

Soaking

Soaking experiments are easy to set up and require only small amounts of ligand. A single crystal is immersed in a 1 μL droplet of a 10–100 mM solution of ligand, for a period of time ranging from a few seconds to several days. Soaking is well suited when multiple ligands are planned to be investigated, since the production of protein crystals is well controlled and does not have to be optimized for each assay. The solubility of the ligand is often a limiting factor. Despite the fact that it is difficult to determine the ligand concentration, adding the ligand in the solid form or from saturated solution directly to the crystal may be successful. For soluble ligands, their concentration should be such that 90% of the protein is bound to the ligand, meaning that a ligand concentration at least ten times the K_d value is required. Because the binding affinity is often not known, ligands are used at concentrations ranging from 0.1 to 1 mM and from 20 to 50 mM for high-molecular-weight ligands and fragments, respectively. Soaking may allow/accommodate small conformation changes, but co-crystallization should be preferred in case large changes are expected upon ligand binding. It is not uncommon that soaking experiments render the crystal fragile. In that case, cross-linking of the crystal can be attempted prior to soaking. Although this method often results in the apparition of cracks on the crystals, detrimental for diffraction, it has proved to be instrumental in several cases [96]. Soaking crystals when many compounds are to be evaluated might be a tedious process, as the addition of the concentrated ligand solution on the crystal or the transfer of each crystal in a concentrated droplet of ligand is manually performed. However, automation of the soaking procedure has recently been developed using, for instance, acoustic dispensing of nanodrops of concentrated ligands into crystal containing droplets, where several hundreds of soaking experiments can be performed within minutes [97].

Co-crystallization

For co-crystallization experiments, the protein has to be conditioned in the presence of the ligand prior to crystallization. As the presence of the ligand itself might alter the crystallization conditions, screening of new crystallization conditions might be necessary, in the worst case for each ligand. Precoating of crystallization plates with ligands prior to crystallization, followed or not by in situ X-ray diffraction, provides a very practical approach elegantly adapted for fragment screening in 96-well format [98, 99]. As for soaking experiments, the ligand concentration that should be used depends on the binding affinity and should be several times greater than the K_d. In the case of ligands with poor solubility, incubation of diluted protein and ligand followed by concentration may be necessary. Temperature and time of incubation prior to launching crystallization assays may vary from 4 °C to room temperature and from 1 h to overnight, respectively.

Crystal Mounting and Data Collection

X-ray diffraction data collection, whether at a synchrotron or on a home source, usually requires to harvest crystals from the crystallization drop, using a loop or mesh, and to cool them at 100 K in order to limit radiation damages. If performed manually, these steps are quite time-consuming, especially if an additional cryoprotection step is necessary [88]. Various automation procedures have been developed [100, 101]. Some synchrotron beamlines now offer automated sample loading and centering coupled to unattended data collection and processing. An average throughput of 120 crystals/day has been reported at the MASSIF-1 beamline at the European Synchrotron Radiation Facility (ESRF), which includes crystal mounting, centering, data collection, and data processing with multiple pipelines. An elegant alternative to this data collection strategy is the in situ data collection, where diffraction data are collected at room temperature directly from crystallization drops in 96-well plates [98]. When coupled to the use of precoated plates with dried ligands, this method eliminates the need of crystal harvesting and cooling. However, high-quality crystals are needed as X-ray scattering from the crystal is contaminated with the scattering by the plate and the crystallization medium. Additionally, a complete rotation of the crystal in the X-ray beam is not possible. The available rotation range is restricted to about 90°, as the 96-well plate has to remain more or less perpendicular to the beam. Combination of multiple datasets is therefore commonly required, and high-symmetry space groups are privileged. The use of Mylar-based film sandwich plates has recently been described that allows room temperature and cryogenic high-throughput in situ data collection [102].

Data Processing, Structure Determination, and Refinement

In the hit–lead process, structures of complexes allow one to guide the next chemical optimization cycle. Hence, at each cycle, a few structures will be solved and low to medium throughput is expected. The use of automated data processing and refinement pipelines will be helpful, but standard, manual crystallographic procedures are still applicable. In contrast, in a crystallographic fragment screening project, where several hundreds of structures need to be determined and analyzed within a few hours or days, automation is mandatory [88].

The data processing step corresponds to the analysis of X-ray diffraction images for the extraction of diffraction spots intensities. The derived observed structure factors are the actual experimental data that will allow structure determination and refinement [88]. Most synchrotron beamlines offer automatic data processing using various widely spread pipelines, such as AutoProc [103], XDSAPP [104], xia2 [105], or locally developed ones, such as GRENADES at ESRF

[106] or XDSme at SOLEIL (Legrand P. (2017), GitHub repository https://github.com/legrandp/xdsme "XDSME: XDS Made Easier," Synchrotron SOLEIL, Gif-sur-Yvette, France). These pipelines rely on the use of one of the standard data processing software: XDS [107], iMosflm [108], or Dials [109]. Each software uses a different assumption and algorithm, and the best processed dataset has to be selected for structure determination and refinement. Structure determination is usually straightforward as the structure of the target is available, and the crystallization procedure has been optimized beforehand to reproducibly deliver a selected crystal form. In some cases, when the presence of the ligand induces large cell parameter modifications, or even space group changes, molecular replacement might be needed, but usually a few cycles of rigid body refinement suffice.

In a screening project, full refinement of all generated structures is not desirable since only a few structures will actually display a ligand. The objective is to refine sufficiently the structure to the point that the absence of the ligand is reasonably certain, so that the refinement can be abandoned [88]. The efforts can therefore be concentrated on the refinement of the few actual complexes. In a fragment screening project, a hit rate of 5–10% is expected. This corresponds to 50 to 100 structures that should be fully refined for a 1000-fragment library. Full refinement of non-bound structures is time-consuming and will produce no usable information, and should be avoided as much as possible. Automatic refinement pipelines include the phenix.ligand_pipeline [110], dimple from the CCP4 package [111], and Buster/TNT [112]. It has been suggested that up to a quarter of the ligands could remain undetected if refinement is not pushed to full completion. Indeed, presence of the ligand is easily confirmed for nearly fully occupied tightly bound compounds, as clear positive peaks in the Fourier difference maps will pop up even at an early stage of refinement. On the contrary, low occupancy fragments remain easily overlooked unless specific detection procedures are used. It should be stressed that the interpretation of the electron density maps is often performed by a human. In an extreme case, careful examination of the electron density maps resulted in the retraction of the original structure of the complex. Less dramatic errors can result in bad positioning of part of the ligand, especially as X-ray crystallography does not allow for the distinguishing of carbon, nitrogen, and oxygen atoms in the electron density maps. However, such distinction may confidently rely on the interpretation of the possible hydrogen bond network with surrounding protein atoms or solvent molecules. Different procedures have been described in order to detect those weakly bound ligands. The Pipedream pipeline (Sharff A, Keller P, Vonrhein C, Smart O, Womack T, Flensburg C, Paciorek C and Bricogne G (2011). Pipedream, version 1.2.3, Global Phasing Ltd., Cambridge, United Kingdom) analyzes individual structures independently and treats the expected ligand-binding site specifically, for instance, with respect to bulk solvent correction, in order not to obscure the corresponding electron density. A more global approach has been elaborated in the PanDDA procedure [113]: an average electron density map is computed from multiple isomorphous apo-structures, after proper superimposition, and significant deviations from the mean in individual datasets are flagged as events. Each structure of complexes is then refined as an ensemble of bound and unbound states. This procedure is able to identify weakly bound compounds, when the classical Fourier difference analysis fails to detect any.

Refining structures of complexes that include low-molecular-weight compounds is not straightforward as these compounds are usually not recognized by refinement programs. Thus, a proper dictionary has to be elaborated in order to describe the ideal stereochemical parameters of any "new" molecule. Most refinement programs include dedicated tools that are usually suited for case-by-case generation of suitable restraints, at least in their basic academic version (Grade for BUSTER/TNT, libcheck for CCP4, REEL for PHENIX). Dedicated servers are also available, such as PRODRG2, which are able to generate proper restraints for various refinement programs. These tools may present some limitations in the type of atoms they can properly handle, e.g., metals, or in the way they can automatically process specific chemical groups, e.g., ionizable moieties. Most of these tasks are easily handled when only a few structures have to be determined. However, when many structures have to be handled, like in a crystallographic fragment screening project, dedicated data management and/or workflow tools are needed. XChemExplorer [114] was specifically designed to handle crystallographic screening projects from the diffraction data processing step up to the identification of complexes with PanDDA and deposition of refined structures within the PDB. Although XChemExplorer has been tailored to manage data collected at the diamond light source, it is possible to adapt it to data collected at any source. This is not the first platform designed to handle such cases: this task may also be performed using the AutoSolve platform from the proprietary program Astex [115].

X-ray crystallography has been used for discovery of dual inhibitors of tyrosine and phosphoinositide kinases as targeted polypharmacology [116]. Apsel et al. [116] reported the systematic discovery of molecules that potently inhibit both tyrosine kinases and phosphatidylinositol-3-OH kinases, two protein families that are among the most intensely pursued cancer drug targets. Through iterative chemical synthesis, X-ray crystallography, and kinome-level biochemical profiling, the authors identified compounds that inhibit a spectrum of new target combinations in these two families. Crystal structures revealed that the dual selectivity of these molecules is controlled by a hydrophobic pocket conserved in both enzyme classes and accessible through a rotatable bond in the drug skeleton. They showed that one compound, PP121, blocks the proliferation of tumor cells by direct inhibition of oncogenic tyrosine kinases and phosphatidylinositol-3-OH kinases. These molecules demonstrate the feasibility of accessing a chemical space that intersects two families of oncogenes.

Gall et al. [117] disclosed the unexpected discovery of a peptidic ligand 1 by X-ray crystallography, which was auto-tailored by the therapeutic target MMP-13 through partial self-degradation and subsequent structure-based optimization to a highly potent and selective β-sheet peptidomimetic inhibitor derived from the endogenous tissue inhibitors of metalloproteinases (TIMPs). The incorporation of non-proteinogenic amino acids in combination with a cyclization strategy proved to be key for the de novo design of TIMP peptidomimetics. The optimized cyclic peptide 4 (ZHAWOC7726) is membrane permeable with an IC_{50} of 21 nm for MMP-13 and an attractive selectivity profile with respect to a polypharmacology approach including the anticancer targets MMP-2 (IC_{50}: 170 nm) and MMP-9 (IC_{50}: 140 nm).

20.5 Machine Learning

20.5.1 Basics of Machine Learning Technologies

The advancements of information technology and related processing techniques have created a fertile base for progress in many scientific fields and industries. Machine learning (ML) is the study of computer algorithms that can improve automatically through experience and by the use of data [118]. It is seen as a part of artificial intelligence. Machine learning algorithms build a model based on sample data, known as training data, in order to make predictions or decisions without being explicitly programmed to do so. Machine learning algorithms are used in a wide variety of applications, such as in medicine, email filtering, speech recognition, and computer vision, where it is difficult or unfeasible to develop conventional algorithms to perform the needed tasks.

The methods for designing drug targets and novel drug discovery now routinely combine ML and deep learning algorithms to enhance the efficiency, efficacy, and quality of developed outputs [119–121]. The generation and incorporation of big data, through technologies such as high-throughput screening and high-throughput computational analysis of databases used for both lead and target discovery, has increased the reliability of the ML and deep learning incorporated techniques. The use of these virtual screening and encompassing online information has also been highlighted in developing lead synthesis pathways. ML approaches provide a set of tools that can improve discovery and decision-making for well-specified questions with abundant, high-quality data. Opportunities to apply ML occur in all stages of drug discovery [119–121]. Examples include target validation, identification of prognostic biomarkers, and analysis of digital pathology data in clinical trials. Applications have ranged in context and methodology, with some approaches yielding accurate predictions and insight.

The popularity of machine learning (ML) across drug discovery continues to grow, yielding impressive results. It should be noted that while as their use increases, so do their limitations become apparent. Such limitations include their need for big data, their sparsity in data, and their lack of interpretability. It has also become apparent that the techniques are not truly autonomous, requiring retraining even post deployment.

In this subsection, application of artificial intelligence (AI) in drug design, ML methods to drug discovery, and ML in drug design applications are introduced. The contents are essentially adopted from the excellent comprehensive review article by Dara et al. [119] (for detailed information, please refer to the article by Dara et al. [119]: Dara S, Dhamercherla S, Jadav SS, et al. Machine learning in drug discovery: A Review. Artif Intell Rev. 2021: 1–53).

20.5.2 Machine Learning Methods to Drug Discovery

Artificial intelligence (AI) innovation has a high priority in drug design through the enhancement of ML approaches and the collection of pharmacological data. AI does not rely upon any hypothetical improvements, but it has more essence in transforming medical information into studies like reusable methods. In general, there are different approaches such as random forest, naïve Bayesian classification (NBC), multiple linear regression (MLR), logistic regression (LR), linear discriminant analysis (LDA), probabilistic neural networks (PNN), multilayer perceptron (MLP), support vector machine (SVM), etc. are considered in the context of ML [119]. In order to gain capability in feature extraction and feature generalization, AI advancements are specifically used as a deep learning technique toward drug design.

Many designed features in traditional ML models are performed manually, but deep learning approaches will accelerate various features through available initialized data automatically because multilayer feature extraction techniques are used to convert straightforward features into complex features. One advantage of using deep learning approaches was presence of low quantity generalization blunders, so it recovers more exact results. CNN, RNN, Auto Encoder, DNN, and RBN are considered as different deep learning techniques [119].

20.5.2.1 Support Vector Machines

SVM model is a supervised learning algorithm basically utilized in predicting the class labeled data, i.e., binary data. The SVM is widely used in drug discovery using its various kernels [122].

20.5.2.2 Random Forest Algorithm

The random forest algorithm was a supervised algorithm. It is a way of creating a forest from various perspectives to make it random. The significant advantage of the random forest algorithm is its relevance to both regression and classification issues. Random forests can apply trained algorithmic techniques, i.e., bagging. The selection of molecular descriptors is seen as an important step in virtual screening to identify bioactive molecules during the drug development process, because this choice of descriptors shows predictions with lower accuracy. Hence, the random forest technique was used to improve prediction and then select naturally trained molecular descriptors for kinase ligands, hormone receptors, enzymes, etc. [119].

20.5.2.3 Multilayer Perception (MLP)

The multilayer perception model is also known as a feedforward neural network. MLP provides an outcome based on a set of input sources. For training any sort of information, the backpropagation approach is utilized. This model is similar to a directed graph because of the essence of multiple layers as input nodes and output nodes are associated with some weights. After processing the data, the perceptron can fluctuate each connected weight in the network. In this way, the presence of error in actual output can be compared with the expected output. In general, MLP can be used very easily and very quickly, but fulfilling its duties in training is very difficult, and MLP also does not offer any guarantee of global minimum performance.

20.5.2.4 Deep Learning

Deep learning is a part of machine learning, having the capability to extract a greater level of features through utilization of multiple layers from input data [123]. Deep learning is an immense field that is creating massive premiums nowadays. In general, deep learning is the same neural network architecture that consists of several layers, and data can be transformed between these layers.

20.5.3 ML in Drug Design Applications

The drug design techniques rely on the databases which are in turn developed based on the different ML algorithms. The precise training, validation, and application of ML algorithms in the drug discovery era provide an enthusiastic outcome by easing the complicated error-prone protocols. The ML techniques are introduced in most of the drug design processes to reduce the time as well as manual interference. Specifically, ML has the following application related to drug design.

20.5.3.1 Homology Modeling/Prediction of Protein Folding

The folding of secondary structure like β-sheets and α-helices, which is formed by the interaction of side-chain amino acid residues, is very critical to regulating the smooth functioning of three-dimensional proteins. An accurate protein folding along with its prerogative active ligand site can be experimentally obtained by X-ray crystallography, NMR spectroscopy, and cryogenic electron microscopic techniques (cryo-EM).

20.5.3.2 Target Identification

The target identification for new chemical entity (NCE; a drug that contains an active moiety that has never been approved by the FDA or marketed in the United States) is an extreme task due to lack of knowledge on their off-targets such as enzymes, ion channels, proteins, or receptors. The binding-site recognition for the NCEs is another key task for computational/bioinformatics experiments where more than one active site exist in the protein. In the above cases, the predefined most popular web servers (FTMap), as well as specific modules such as "Sitemap" developed with the help of algorithms, can define the preferential binding site to speed up the drug discovery process [119]. A few other online programs like GHECOM, POCASA, Pocketome, SURFNET, ConCavity, LIGSITE, Q-SiteFinder, Fpocket, and PASS predict the feasible binding sites located within the provided protein templates. The metaPocket 2.0 program utilizes the above platforms to afford the most reliable ligand-binding sites present on templates. Further, AI models like FD/DCA can also predict the druggable sites in the provided biological macromolecules. Recently, the DeepDTnet as a new target identifier in drug repurposing has been tested. The DeepDTnet strategy is developed by amalgamating the multi-disease cellular targets, pathogenic genes (genomics), and drugs (chemical spaces) being utilized for their treatment [119].

20.6 Prediction of Protein Folding

20.6.1 Prediction of Protein–Protein Structure

20.6.1.1 Prophecy of Protein–Protein Interactions

In recent years, the PPI-based drug discovery programs are experimentally produced a hopeful pharmacological substance. Yet, the time and cost for experimental prediction of PPI are considered as rate-limiting barriers. In this regard, the different databases hosted the web servers (few are publicly available) framed by targeting PPI which are prevailing as preliminary PPI identification tools to accelerate the medicinal chemistry research.

20.6.1.2 Hit Discovery

In drug discovery, small molecules are considered as hits for target binding to identify the best-altered functions. The detection of hit by diverse algorithms is currently prevailing as a robust technique in the current drug discovery paradigm. An application of multivariate parameters (K-nearest neighbors (K-NN) and support vector machine (SVM)) on high-content screening (HCS) analysis can produce a variety of hits against neurological complications.

20.7 Drug Repurposing

20.7.1 Virtual Screening

20.7.1.1 High-Throughput Virtual Screening and Scoring in Molecular Docking Techniques

Activity Scoring

In virtual scoring, the scoring function is a fundamental component in molecular docking for assessing binding affinities toward target. In machine learning, mapping ability features can yield great accomplishment to extract physical, geometric, and chemical features to retrieve scores. Based on scores, data-driven black box models which are considered to predict interactions in binding affinities and furthermore avoiding few concepts in docking like physical function are very hard to study. Random forest and SVM concepts identified with AI utilization for better performance in the scoring function. For instance, an SVM model can be utilized instead of a linear additive method related to the energy terms concept, since an SVM can characterize the relationship between experimental binding affinities and own energy terms, i.e., can be extracted from docking program eHiTS. Thus, data gives better execution in scoring power and screening power [119].

20.7.1.2 Hit to Lead

It is also referred to as lead generation in the beginning phases of drug discovery. It locates small molecules referred to as hits from the high-throughput screening (HTS) through deficient optimization to locate promising lead compounds.

20.8 QSAR Analysis

20.8.1 De Novo Drug Architecture

20.8.1.1 Lead Optimization

The lead optimization is an essential step of the drug discovery process in which the best medicinally active fragment hits are considered leads to extend the medicinal chemistry projects. The main aim of the lead optimization is to eliminate the side effects/notorious effects of the existing active analogs by a minimal structural modification to yield a better and safer scaffold.

Chemical and Physical Properties

In the drug discovery pipeline, physical and chemical properties have been utilized to reduce significant failures. At that point, deep learning models are used as lead optimization techniques to improve unique methodologies.

Pharmacokinetics: Absorption, Distribution, Metabolism, and Excretion

Many AI strategies adopted a huge amount of drug metabolism information to predict unique metabolic enzymes like UDP-glucuronosyltransferases (UGTs), cytochrome P450s, etc.

20.8.2 Toxicity

20.8.2.1 ML in e-Resources for Drug Discovery

The AI and ML algorithms prevailed as the main computational scoring functions for evaluation when a predicted value was added as a parameter, which is involved in the basic drug discovery paradigm. The detailed applications of the ML algorithms specified in the e-resource are described below.

ML in Pan-Assay Interference Screening (PAINS)
The precise information about hits can be obtained from primary or secondary biological screening assays of purchasable/commercially available databases which are the most important parameters before starting medicinal chemistry projects. Thus, elimination of the compounds present in different cellular biological assays considered as pan-assay derived hits could reduce the cost and time of the medicinal chemists. The pan-assay information can be accessed from the PAINS database on request. Therefore, the Hit Dexter 2.0 web server has been launched compiled from PubChem library and screening assays.

ML in Drug Metabolite and Metabolic Site Prediction
The identification of metabolic site for any kind of drug or new chemical entity is very essential before its administration into the human body. The prediction of drug metabolism can be done by animal models (preclinical studies) which was a rate-limiting step as well as costly, and it is mandatory to retrieve therapeutic approval of new chemical entities. The site of metabolism can be predicted by several modules among which "ADMET Predictor" of SimulationsPlus tools has gained attention and is pure works on the models compiled by the artificial intelligence algorithms. The FAME3 is one of the online servers which predicts the region for the given drug/compound which undergoes metabolism validated databases gathering phase I/phase II metabolic parameters associated with several databases validated. The online services such as GLORY/GLORYx provides the precise information about the possibilities of new metabolite and their relevant formation data with respect to mitochondrial cytochrome P450 enzyme and conjugations [124].

ML in Skin Sensitivity Parameter Prediction
The prediction of skin sensitivity is one of the essential criteria for assessing safety parameters of the new drugs/compounds and it is patient-to-patient specifications. In this regard, the AI models such as random forest-based MACCS (RF_MACCS) and support vector machine (SVM)-based PaDEL (SVM_PaDEL) algorithms trained with approximately 1400 ligands linked with local lymph node assay (LLNA) information can be utilized.

ML in Natural Product Identification
The ML trained with 265,000 natural product (NP) isolates and synthetic libraries validated by MCC is being used as a basic predictive model NP Scout online server will reveal the probable identity of the newly discovered analogs. The application of NP Scout in the prediction of sources for the query molecule might provide information about their natural product sources and could become a part of natural product-based drug discovery.

20.9 Information Mining

In the "mega data/big data" era as we are in now, information is the most precious and largest gold mine. Information can take various forms, and data and text are the two most common forms. For science in general and pharmaceutical science in particular, datasets are probably the most valuable information [125–127]. Big data is a field that treats ways to analyze, systematically extract information from, or otherwise deal with datasets that are too large or complex to be dealt with by traditional data processing application software. Data with many fields (columns) offer greater statistical power, while data with higher complexity (more attributes or columns) may lead to a higher false discovery rate. Big data analysis challenges include capturing data, data storage, data analysis, search, sharing, transfer, visualization, querying, updating, information privacy, and data source. The analysis of big data presents challenges in sampling, and thus previously allowing for only observations and sampling. Therefore, big data often includes data with sizes that exceed the capacity of traditional software to process within an acceptable time and value.

The advances of genomics, sequencing, and high-throughput technologies have led to the creation of large volumes of diverse datasets for drug discovery. Analyzing these datasets to better understand disease and discover new drugs is becoming more common [125–127]. Recent open data initiatives in basic and clinical research have dramatically increased the types of data available to the public. The past few years have witnessed successful use of big data in many sectors across the whole drug discovery pipeline. Proper and efficient interpretation of big data in the drug discovery community should enhance project timelines and reduce clinical attrition through improved early decision-making. Yet, the issues we encounter start with the sheer volume of data and how we first ingest it before building an infrastructure to house it to make use of the data in an efficient and productive way. There are many problems associated with the data itself including general reproducibility, but often, it is the context surrounding an experiment that is critical to success. Assistance in the forms of artificial intelligence (AI) and machine learning is required to understand and translate the context.

Therefore, information mining becomes the first step and most critical act for using the information. The two most important perspectives of information mining are data mining and text mining [127] (For detailed information, please refer to the article by Kumar and Zhang [59]: Kumar A, Zhang KYJ. Advances in the development of shape similarity methods and their application in drug discovery. Front Chem. 2018; 6: 315).

20.9.1 Data Mining

Data mining is the act of computationally extracting new information from large amounts of data, and the biological sciences are generating enormous quantities of data, ushering in the era of "big data." Stephens et al. [128] state that sequencing data alone constitutes ~35 petabases/year and will grow to ~1 zettabase/year by 2025. This creates a large opportunity for the development and deployment of novel mining algorithms, and two recent reviews on data and text mining in the era of big data are found in Che et al. [129] and Herland et al. [130]. A wide variety of methods for extracting value from different types and models of data fall under the umbrella of "data mining." Classification algorithms (decision trees, naïve Bayesian classification, and other classifiers), frequent pattern algorithms (association rule mining, sequential pattern mining, and others), clustering algorithms (including methods to cluster continuous and categorical data), and graph and network algorithms have all evolved to present a diverse landscape for research and an arsenal to deploy against the toughest data challenges. Most researchers consider some other areas, including text mining, as being under the data mining umbrella. For example, Piatetsky-Shapiro [131] state: "Data Mining in my opinion includes: text mining, image mining, web mining, predictive analytics, and much of the techniques we use for dealing with massive datasets, now known as Big Data." The methods applied to text mining, however, are specialized to such a degree that it is common to view it as a separate area of specialty. Data mining courses do not usually include any text-mining material, but rather there are separate courses dedicated to it, and the same applies to textbooks.

Kernel Methods in Computational Biology by Schölkopf, Tsuda, and Vert [132] covers methods specific to computational biology. *Introduction to Data Mining* [133] *and Data Mining: Concepts and Techniques, third edn* [134] are two popular textbooks in data mining and give an excellent overview of the field. A more concise presentation can be found in the paper by Xindong Wu et al., Top 10 algorithms in data mining [135], which were identified in December 2006 as C4.5, k-means, SVM, Apriori, EM, PageRank, AdaBoost, kNN, naïve Bayes and CART, covering clustering, classification and association analysis, which are among the most important topics in data mining research:

According to Jain et al. [136] in "Data clustering: a review," "Clustering is the unsupervised classification of patterns (observations, data items, or feature vectors) into groups (clusters)."

Classification is akin to clustering because it segments data into groups called classes, but unlike clustering, classification analyses require knowledge and specification of how classes are defined.

Statistical learning theory seeks "to provide a framework for studying the problem of inference that is, of gaining knowledge, making predictions, making decisions or constructing models from a set of data" states Bousquet et al. [137]. A textbook on statistical learning expands on these notions [138].

Association analysis facilitates the unmasking of hidden relationships in large datasets. The discovered associations are then expressed as rules or sets of items that frequently occur together. Challenges to association analysis methods include that discovering such patterns can be computationally expensive given a large input dataset and that there could potentially be many spurious associations "discovered" that simply occur by chance. A well-known introduction to the topic is found in [139], and in particular, a seminal paper on mining association rules from clinical databases is found in Stilou et al. [140].

Link analysis analyzes hyperlinks and the graph structure of the web for the ranking of web search results. PageRank is perhaps the best-known algorithm for link analysis [141].

In a notable transition showing the power of new algorithms and data, data mining approaches are now being used to learn, not just the primary features but also context-specific features. For example, initial data mining approaches that constructed gene–gene networks built a single network [142]. In contrast, recent approaches learn multiple context-specific networks, allowing the construction of process-specific [143] and tissue-specific networks [144, 145]. An individual is made up of a personalized combination of such context-specific networks, so we anticipate that continued advances in the context specificity of data mining approach will play an important role in the broad implementation of precision medicine.

20.9.2 Text Mining

Text mining is a subfield of data mining that seeks to extract valuable new information from unstructured (or semi-structured) sources [146]. Text mining extracts information from within those documents and aggregates the extracted pieces over the entire collection of source documents to uncover or derive new information. This is the preferred view of the field that allows one to distinguish text mining from natural language processing (NLP). Thus, given as input a set of documents, text-mining methods seek to discover novel patterns, relationships, and trends contained within the documents. Aiding the overall goal of discovering new information are NLP programs that go from the relatively simple text processing tasks at the lexical or grammatical levels (such as a tokenizing or a part-of-speech tagger), to relatively complex information extraction algorithms [like named entity recognition (NER) to find concepts such as genes or diseases, normalization to map them to their unique identifiers, or relationship extraction and sentiment analysis systems, among others]. The greater the complexity of the task, the more likely it is to integrate methods from data mining (such as classification or statistical learning).

Text-mining subareas include the following:

(a) Information retrieval deals with the problem of finding relevant documents in response to a specific information need (query).
(b) NER is at the core of the automatic extraction of information from text and deals with the problem of finding references to entities (mentions) such as genes, drugs, and diseases present in natural language text and tagging them with their location and type. NER is also referred to as "entity tagging" or "concept extraction." This is a basic building block for almost all other extraction tasks. NER in the biomedical domain is generally considered to be more difficult than other domains, such as geography or news reports. This is owing to inconsistency in how known entities, such as symptoms or drugs, are named (e.g., nonstandard abbreviations and new ways of referring to them). An open-source NER engine, BANNER [147], with

models to recognize genes and diseases mentioned in biomedical text, is currently available for gene and disease NER, and LINNAEUS is available for species [148]. Rebholz-Schuhmann et al. [149] present an overview of the NER solutions for the second CALBC task, including protein, disease, chemical (drug), and species entities. Campos et al. [150] discuss a recent survey of tools for biomedical NER. A system assigning text to a wide range of semantic classes using linguistic rules is presented in [151], illustrating a slightly different than standard NER because classes potentially overlap. Verspoor et al. [152] use the CRAFT corpus to improve the evaluation of gene NER (and some lower-level tasks like part-of-speech and sentence segmentation).

(c) Named Entity Identification allows the linkage of objects of interest, such as genes, to information that is not detailed in a publication (such as their Entrez Gene identifier) [153]. Two open-source systems using largely dictionary-based approaches to normalize gene names appear in [153, 154].

(d) Association extraction is one of the higher-level tasks still considered purely an information extraction application. It uses the output from the prior subtasks to produce a list of (binary or higher) associations among the different entities of interest. Catalysts for advances in this area have been the Biocreative and BioNLP shared tasks, with excellent teams from around the world putting their systems to the test against carefully annotated datasets. A survey of submissions to Biocreative III [155] and BioNLP [156, 157] shows a good overview of approaches responsive to the respective shared tasks. Putting together associations into networks of molecular interactions that can explain complex biological processes is the next logical step, and one that still is considered the "holy grail" of automatic biomolecular extraction. Ananiadou et al. [158] and Li et al. [159] discuss comprehensive surveys of methods for the extraction of network information from the scientific literature and the evaluation of extraction methods against reference corpora. Semantic-based approaches such as [160] will make their mark in the coming years.

(e) Event extraction is similar to association extraction but instead of separately extracting various relations between different entities in text, this task focuses on identifying specific events and the various players involved in it (arguments). For instance, the arguments of a transport event will include the molecule being transported, the cell to which it is being transported, and the cell from which it is being transported. Event extraction was a key component of the BioNLP Shared Tasks both in 2011 [157] and 2013 [161], challenging the biomedical community to expand and cultivate their approaches in this area and leading to steadily improving results.

(f) Pathway extraction is a budding branch of biomedical text mining closely following the footsteps of event extraction. It involves the automated construction of biological pathways through the extraction and ordering of pathway-related events from text. However, the majority of researchers in this domain have been focusing their efforts on supporting pathway curation through event extraction, rather than entirely automating the process. Tari et al. [162] were able to achieve promising results for the automated synthesis of pharmacokinetic pathways by applying an automated reasoning-based approach for event ordering. The first shared task on pathway curation was organized by BioNLP in 2013 [161] to establish the current state-of-the-art performance level for extracting pathway-relevant events such as phosphorylation and transport.

In the end, a set of the different subtask solutions are used in a pipeline that allows information to be integrated and analyzed toward knowledge discovery. However, this multiplies the effects of errors down the pipeline, leaving systems highly vulnerable.

An overarching challenge for biomedical text mining is to incorporate the many knowledge resources that are available to us into the NLP pipeline. In the biomedical domain, unlike the general text-mining domain, we have access to large numbers of extensive, well-curated ontologies and knowledge bases. Biomedical ontologies provide an explicit characterization of a given domain of interest. The quality of data mining efforts would likely increase if existing ontologies (e.g., UMLS [163] and BioPortal [16]) were used as sources of terms in building lexicons, for figuring out what concept subsumes another, and as a way of normalizing alternative names to one identifier. For example, using ontologies as described enabled the use of unstructured clinical notes for generating practice-based evidence on the safety of a highly effective, generic drug for peripheral vascular disease [164].

20.9.3 Information-Mining Methods

The dataset for information-mining comprises PubMed articles [165] and can be summarized as follows: Step 1: Identify all diseases, genes, and mutations in PubMed abstracts; Step 2: Associate mutations with a given disease from all articles in PubMed about that disease; and Step 3: Link the proteins or genes with each mutation using an aggregation of information extracted from PubMed, the Web (using Bing search engine features), and sequence analysis.

Step 1: Identifying Disease, Gene, and Mutation Mentions.

To identify disease, gene, and mutation mentions, Singhal et al used the following entity recognition tools for these tasks:

GNormPlus GNormPlus extracts gene/proteins from a given text in two steps. In the first step, it recognizes the gene/protein entity mention using a novel CRF++ based module in combination with the species recognition module SR4GN. In the second step, the gene/protein mention is normalized using GenNorm in combination with the composite mention simplification tool SimConcept and an abbreviation resolution tool Ab3P.

tmVar tmVar uses a conditional random field (CRF) model to detect mutation mentions in text. The CRF model identifies the mutation type, reference amino acid or nucleotide, mutant nucleotide or amino acid, and the mutation position as the states for the CRF, i.e., each component of a mutation is treated as an individual state in the CRF and thus sequentially retrieved. The information about these components is learned from a curated tmVar corpus. Some additional rare mutation mentions are handled in the post-processing module.

DNorm DNorm segments the given text into sentences and identifies disease mentions using BANNER. The mentions are then normalized to disease concepts in MeSH by finding the best match using a learning-to-rank technique. Finally, the disambiguation related to primary disease or a disease synonym is resolved using a disease concept hierarchy. The output is a tuple of disease mention text span and corresponding concept (MeSH ID) [165].

Step 2: Disease–Mutation Relationship Extraction

Mutations associated with the queried disease are identified using an ML-based classification algorithm trained to detect disease-related mutations using a novel feature set derived by mining the text in PubMed abstracts [165]. The feature set captures information such as mutation and disease mention proximity, disease mention frequency, same-sentence mention of mutation, and disease and text sentiment (polarity) to distinguish between disease-related and non-related mutation mentions in text.

In the second step of the disease-mutation identification process, the PMID-mutation pairs for the disease are converted into an ordered list. Unique mutations are ranked in descending order by the number of PMIDs in which each mutation appears, and a corresponding list of referencing PMIDs is provided for each mutation [165].

Step 3: Gene–Variant Relationship Extraction

Once the disease-mutations are identified (Step 1), the next step is to extract the gene associated with each mutation [165]. Performing this step within the local textual and structural information of individual abstracts is challenging and error-prone because (a) automatic gene recognition remains quite difficult, and (b) in many cases, abstracts mention multiple genes and mutations. Singhal et al [165] designed a three-step approach that takes advantage of global information beyond a single abstract.

(a) *Gene extraction from a PubMed corpus (PubMed Rank)*: Given a mutation and a referencing PMID list, all genes (normalized to Entrez Gene ID) mentioned in the PubMed abstracts of corresponding PMIDs are collected and then ranked based on the aggregate score of their mentions in the local abstract as well as in all other abstracts [165].
(b) *Gene extraction from a Web corpus (Bing Rank)*: Parallel to the process described above, a *Bing Rank list* of candidate genes is also generated from the Web using the Bing search engine's free API service. The most frequent mention for each entry from the Mutation list (obtained from the referencing abstracts) is given as the input query to the Bing search engine. Gene mentions for each query are mined from the title and text snippets of the top-20 Bing search results [165].
(c) Rank *Rank aggregation* [165].
(d) *Sequence validation* [165].

Machine learning and data mining often employ the same methods and overlap significantly, but while machine learning focuses on prediction, based on known properties learned from the training data, data mining focuses on the discovery of (previously) unknown properties in the data (this is the analysis step of knowledge discovery in databases). Data mining uses many machine learning methods, but with different goals; on the other hand, machine learning also employs data mining methods as "unsupervised learning" or as a preprocessing step to improve learner accuracy. Much of the confusion between these two research communities (which do often have separate conferences and separate journals, ECML PKDD being a major exception) comes from the basic assumptions they work with: in machine learning, performance is usually evaluated with respect to the ability to reproduce known knowledge, while in knowledge discovery and data mining (KDD), the key task is the discovery of previously unknown knowledge. Evaluated with respect to known knowledge, an uninformed (unsupervised) method will easily be outperformed by other supervised methods, while in a typical KDD task, supervised methods cannot be used due to the unavailability of training data.

References

1. Macarron R, Banks MN, Bojanic D, et al. Impact of high-throughput screening in biomedical research. Nat Rev Drug Discov. 2011;10(3):188–95.
2. Hann MM, Oprea TI. Pursuing the leadlikeness concept in pharmaceutical research. Curr Opin Chem Biol. 2004;8(3):255–63.
3. Michael S, Auld D, Klumpp C, et al. A robotic platform for quantitative high-throughput screening. ASSAY Drug Develop Technol. 2008;6(5):637–57.
4. Howe D, Costanzo M, Fey P, et al. Big data: the future of biocuration. Nature. 2008;455(7209):47–50.
5. Zhang XHD. Optimal high-throughput screening: practical experimental design and data analysis for genome-scale RNAi research. Cambridge University Press; 2011. ISBN 978-0-521-73444-8
6. Zhang XH, Espeseth AS, Johnson EN, et al. Integrating experimental and analytic approaches to improve data quality in genome-scale RNAi screens. J Biomol Screen. 2008;13(5):378–89.
7. Zhang JH, Chung TD, Oldenburg KR. A simple statistical parameter for use in evaluation and validation of high throughput screening assays. J Biomol Screen. 1999;4(2):67–73.
8. Zhang XHD. A pair of new statistical parameters for quality control in RNA interference high-throughput screening assays. Genomics. 2007;89(4):552–61.
9. Zhang XHD. A new method with flexible and balanced control of false negatives and false positives for hit selection in RNA interference high-throughput screening assays. J Biomol Screen. 2007;12(5):645–55.
10. Brideau C, Gunter G, Pikounis B, et al. Improved statistical methods for hit selection in high-throughput screening. J Biomol Screen. 2003;8(6):634–47.
11. Zhang XHD. A method for effectively comparing gene effects in multiple conditions in RNAi and expression-profiling research. Pharmacogenomics. 2009;10(3):345–58.
12. Inglese J, Auld DS, Jadhav A, et al. Quantitative high-throughput screening (qHTS): A titration-based approach that efficiently identifies biological activities in large chemical libraries. Proc Natl Acad Sci U S A. 2006;103(31):11473–8.

13. Schonbrun E, Abate AR, Steinvurzel PE, et al. High-throughput fluorescence detection using an integrated zone-plate array. Lab on a Chip Royal Society of Chemistry. 2010;10(7):852–6.
14. Helleboid S, Haug C, Lamottke K, et al. The identification of naturally occurring neoruscogenin as a bioavailable, potent, and high-affinity agonist of the nuclear receptor RORα (NR1F1). J Biomol Screen. 2014;19(3):399–6406.
15. Atanasov AG, Waltenberger B, Pferschy-Wenzig EM, et al. Discovery and resupply of pharmacologically active plant-derived natural products: a review. Biotechnol Adv. 2015;33(8):1582–614.
16. Whetzel PL, Noy NF, Shah NH, et al. BioPortal: enhanced functionality via new web services from the National Center for biomedical ontology to access and use ontologies in software applications. Nucleic Acids Res. 2011;39:W541–5.
17. Amelio I, Landré V, Knight RA, et al. Polypharmacology of small molecules targeting the ubiquitin-proteasome and ubiquitin-like systems. Oncotarget. 2015;6(12):9646–56.
18. Wang Y, Bolton E, Dracheva S, et al. An overview of the PubChem BioAssay resource. Nucleic Acids Res. 2010;38:D255–66.
19. Antonov AV. BioProfiling. de: analytical web portal for high-throughput cell biology. Nucleic Acids Res. 2011;39:W323–7.
20. Dietmann S, Lee W, Wong P, et al. CCancer: a bird's eye view on gene lists reported in cancer- related studies. Nucleic Acids Res. 2010;38:W118–23.
21. Antonov AV, Mewes HW. Complex functionality of gene groups identified from high-throughput data. J Mol Biol. 2006;363:289–96.
22. Lengauer T, Rarey M. Computational methods for biomolecular docking. Curr Opin Struct Biol. 1996;6(3):402–6.
23. Kitchen DB, Decornez H, Furr JR, et al. Docking and scoring in virtual screening for drug discovery: methods and applications. Nat Rev Drug Discov. 2004;3(11):935–49.
24. Goldman BB, Wipke WT. QSD quadratic shape descriptors. 2. Molecular docking using quadratic shape descriptors (QSDock). Proteins. 2000;38(1):79–94.
25. Morris GM, Goodsell DS, Halliday RS, et al. Automated docking using a Lamarckian genetic algorithm and an empirical binding free energy function. J Comput Chem. 1998;19(14):1639–62.
26. Shoichet BK, Kuntz ID, Bodian DL. Molecular docking using shape descriptors. J Comput Chem. 2004;13(3):380–97.
27. Friesner RA, Banks JL, Murphy RB, et al. Glide: a new approach for rapid, accurate docking and scoring. 1. Method and assessment of docking accuracy. J Med Chem. 2004;47(7):1739–49.
28. Klebe G, Mietzner T. A fast and efficient method to generate biologically relevant conformations. J Computer-Aided Mol Des. 1994;8(5):583–606.
29. Ciemny M, Kurcinski M, Kamel K, et al. Protein-peptide docking: opportunities and challenges. Drug Discov Today. 2018;23(8):1530–7.
30. Antunes DA, Devaurs D, Kavraki LE. Understanding the challenges of protein flexibility in drug design. Expert Opin Drug Discov. 2015;10(12):1301–13.
31. Cerqueira NM, Bras NF, Fernandes PA, et al. MADAMM: a multistaged docking with an automated molecular modeling protocol. Proteins. 2009;74(1):192–206.
32. Totrov M, Abagyan R. Flexible ligand docking to multiple receptor conformations: a practical alternative. Curr Opin Struct Biol. 2008;18(2):178–84.
33. Hartmann C, Antes I, Lengauer T. Docking and scoring with alternative side-chain conformations. Proteins. 2009;74(3):712–26.
34. Arcon JP, Turjanski AG, Martí MA, et al. Biased docking for protein–ligand pose prediction. In: Ballante F, editor. Protein-ligand interactions and drug design, Methods in molecular biology, vol. 2266. New York: Springer; 2021. p. 39–72. ISBN 978-1-0716-1209-5.
35. Gohlke H, Hendlich M, Klebe G. Knowledge-based scoring function to predict protein-ligand interactions. J Mol Biol. 2000;295(2):337–56.
36. Ballante F. Protein-ligand docking in drug design: performance assessment and binding-pose selection. In: Rational drug design: methods in molecular biology, vol. 2018; 1824. p. 67–88.
37. Ballante F, Kooistra AJ, Kampen S, et al. Structure-based virtual screening for ligands of g protein–coupled receptors: what can molecular docking do for you? Pharmacol Rev. 2021;73(4):527–65.
38. Hartshorn MJ, Verdonk ML, Chessari G, et al. Diverse, high-quality test set for the validation of protein-ligand docking performance. J Med Chem. 2007;50(4):726–41.
39. Huang N, Shoichet BK, Irwin JJ. Benchmarking sets for molecular docking. J Med Chem. 2006;49(23):6789–801.
40. Hauser AS, Windshügel B. A benchmark data set for assessment of peptide docking performance. J Chem Info Model. 2015;56(1):188–200.
41. Chaudhari R, Tan Z, Huang B, et al. Computational polypharmacology: a new paradigm for drug discovery. Expert Opin Drug Discov. 2017;12(3):279–91.
42. Grinter SZ, Liang Y, Huang Y-U, et al. An inverse docking approach for identifying new potential anti-cancer targets. J Mol Graph Model. 2011;29(6):795–9.
43. Chen YZ, Zhi DG. Ligand-protein inverse docking and its potential use in the computer search of protein targets of a small molecule. Proteins. 2001;43:217–26.
44. Do QT, Renimel I, Andre P, et al. Reverse pharmacognosy: application of selnergy, a new tool for lead discovery. The example of epsilon-viniferin. Curr Drug Discov Technol. 2005;2:161–7.
45. Muller P, Lena G, Boilard E, et al. In silico-guided target identification of a scaffold-focused library:

1,3,5-triazepan-2,6-diones as novel phospholipase A2 inhibitors. J Med Chem. 2006;49:6768–78.
46. Zahler S, Tietze S, Totzke F, et al. Inverse in silico screening for identification of kinase inhibitor targets. Chem Biol. 2007;14:1207–14.
47. Schapira M, Abagyan R, Totrov M. Nuclear hormone receptor targeted virtual screening. J Med Chem. 2003;46:3045–59.
48. Rollinger JM. Accessing target information by virtual parallel screening—the impact on natural product research. Phytochem Lett. 2009;2:53–8.
49. Bissantz C, Logean A, Rognan D. High-throughput modeling of human g-protein coupled receptors: amino acid sequence alignment, three-dimensional model building, and receptor library screening. J Chem Inf Comput Sci. 2004;44:1162–76.
50. Kellenberger E, Rodrigo J, Muller P, et al. Comparative evaluation of eight docking tools for docking and virtual screening accuracy. Proteins Struct Funct Genet. 2004;57:225–42.
51. Schomburg KT, Bietz S, Briem H, et al. Facing the challenges of structure-based target prediction by inverse virtual screening. J Chem Inf Model. 2014;54(6):1676–86.
52. Chen YZ, Ung CY. Computer automated prediction of potential therapeutic and toxicity protein targets of bioactive compounds from Chinese medicinal plants. Am J Chin Med. 2002;30(1):139–54.
53. Li H, Gao Z, Kang L, et al. TarFisDock: a web server for identifying drug targets with docking approach. Nucleic Acids Res. 2006;34:W219–24.
54. Wang JC, Chu PY, Chen CM, et al. idTarget: a web server for identifying protein targets of small chemical molecules with robust scoring functions and a divide-and-conquer docking approach. Nucleic Acids Res. 2012;40:W393–9.
55. Huang S-Y, Zou X. An iterative knowledge-based scoring function to predict protein–ligand interactions: I. Derivation of interaction potentials. J Comput Chem. 2006;27:1865–75.
56. Huang S-Y, Zou X. An iterative knowledge-based scoring function to predict protein–ligand interactions: II. Validation of the scoring function. J Comput Chem. 2006;27:1876–82.
57. Gao Z, Li H, Zhang H, et al. PDTD: a web-accessible protein database for drug target identification. BMC Bioinform. 2008;9(104):1–7.
58. Rubinstein LV, Shoemaker RH, Paull KD. Comparison of in vitro anticancer-drug-screening data generated with a tetrazolium assay versus a protein assay against a diverse panel of human tumor cell lines. J Natl Cancer Inst. 1990;82:1113–8.
59. Kumar A, Zhang KYJ. Advances in the development of shape similarity methods and their application in drug discovery. Front Chem. 2018;6:315.
60. Rogers DJ, Tanimoto TT. A computer program for classifying plants. Science. 1960;132:1115–8.
61. Zauhar RJ, Moyna G, Tian L, et al. Shape signatures: a new approach to computer-aided ligand- and receptor-based drug design. J Med Chem. 2003;46:5674–90.
62. Kortagere S, Krasowski MD, Ekins S. The importance of discerning shape in molecular pharmacology. Trends Pharmacol Sci. 2009;30:138–47.
63. Bemis GW, Kuntz ID. A fast and efficient method for 2D and 3D molecular shape description. J Comput Aided Mol Des. 1992;6:607–68.
64. Nilakantan R, Bauman N, Venkataraghavan R. New method for rapid characterization of molecular shapes: applications in drug design. J Chem Inf Comput Sci. 1993;33:79–85.
65. Good AC, Ewing TJ, Gschwend DA, et al. New molecular shape descriptors: application in database screening. J Comput Aided Mol Des. 1995;9:1–12.
66. Connolly ML. Computation of molecular volume. J Am Chem Soc. 1985;107:1118–24.
67. Grant JA, Pickup BT. A gaussian description of molecular shape. J Phys Chem. 1995;99:3503–10.
68. Masek BB, Merchant A, Matthew JB. Molecular shape comparison of angiotensin II receptor antagonists. J Med Chem. 1993;36:1230–8.
69. Rush TS III, Grant JA, Mosyak L, et al. A shape-based 3-D scaffold hopping method and its application to a bacterial protein-protein interaction. J Med Chem. 2005;48:1489–95.
70. Hawkins PCD, Skillman AG, Nicholls A. Comparison of shape-matching and docking as virtual screening tools. J Med Chem. 2007;50:74–82.
71. Mezey PG. Molecular surfaces. In: Lipkowitz KB, Boyd DB, editors. Reviews in computational chemistry. New York: VCH Publishers; 2007. p. 265–94.
72. Wilson JA, Bender A, Kaya T, et al. Alpha shapes applied to molecular shape characterization exhibit novel properties compared to established shape descriptors. J Chem Inf Model. 2009;49:2231–24.
73. Cosgrove DA, Bayada DM, Johnson AP. A novel method of aligning molecules by local surface shape similarity. J Comput Aided Mol Des. 2000;14:573–91.
74. Goldman BB, Wipke WT. Qadratic shape descriptors. 1. Rapid superposition of dissimilar molecules using geometrically invariant surface descriptors. J Chem Inf Comput Sci. 2000;40:644–58.
75. Hofbauer C, Lohninger H, Aszódi A. SURFCOMP: a novel graph-based approach to molecular surface comparison. J Chem Inf Comput Sci. 2004;44:837–47.
76. Sato T, Yuki H, Takaya D, et al. Application of support vector machine to three-dimensional shape-based virtual screening using comprehensive three-dimensional molecular shape overlay with known inhibitors. J Chem Inf Model. 2012;52:1015–26.
77. Kihara D, Sael L, Chikhi R, et al. Molecular surface representation using 3D zernike descriptors for protein shape comparison and docking. Curr Protein Pept Sci. 2011;12:520–30.
78. Kawabata T. Multiple subunit fitting into a low-resolution density map of a macromolecular complex using a gaussian mixture model. Biophys J. 2008;95:4643–58.

79. Reddy AS, Zhang S. Polypharmacology: drug discovery for the future. Expert Rev Clin Pharmacol. 2013;6(1) https://doi.org/10.1586/ecp.12.74.
80. Keiser MJ, Roth BL, Armbruster BN, et al. Relating protein pharmacology by ligand chemistry. Nat Biotechnol. 2007;25(2):197–206.
81. Lounkine E, Keiser MJ, Whitebread S, et al. Large-scale prediction and testing of drug activity on side-effect targets. Nature. 2012;486(7403):361–7.
82. Oprea TI, Mestres J. Drug repurposing: far beyond new targets for old drugs. AAPS J. 2012;14(4):759–63.
83. Hopkins AL. Network pharmacology. Nat Biotechnol. 2007;25(10):1110–1.
84. Cheng T, Li Q, Wang Y, et al. Identifying compound-target associations by combining bioactivity profile similarity search and public databases mining. J Chem Inf Model. 2011;51(9):2440–8.
85. Cross S, Baroni M, Carosati E, et al. FLAP: GRID molecular interaction fields in virtual screening. Validation using the DUD data set. J Chem Inf Model. 2010;50(8):1442–50.
86. Siragusa L, Cross S, Baroni M, et al. BioGPS: navigating biological space to predict polypharmacology, off-targeting, and selectivity. Proteins. 2015;83(3):517–32.
87. Shi Y. A glimpse of structural biology through X-ray crystallography. Cell. 2014;159(5):995–1014.
88. Maveyraud L, Mourey L. Protein X-ray crystallography and drug discovery. Molecules. 2020;25(5):1030.
89. Zheng H, Handing KB, Zimmerman MD, et al. X-ray crystallography over the past decade for novel drug discovery – where are we heading next? Expert Opin Drug Discov. 2015;10(9):975–89.
90. Ennifar E. X-ray crystallography as a tool for mechanism-of-action studies and drug discovery. Curr Pharm Biotechnol. 2013;14(5):537–50.
91. Carvalho AL, Trincão J, Romão MJ. X-ray crystallography in drug discovery. Methods Mol Biol. 2009;572:31–56.
92. Blundell TL, Patel S. High-throughput X-ray crystallography for drug discovery. Curr Opin Pharmacol. 2004;4(5):490–6.
93. Du H, Brender JR, Zhang J, et al. Protein structure prediction provides comparable performance to crystallographic structures in docking-based virtual screening. Methods. 2015;71:77–84.
94. McIntyre PJ, Collins PM, Vrzal L, et al. Characterization of three druggable hot-spots in the Aurora-A/TPX2 interaction using biochemical, biophysical, and fragment-based approaches. ACS Chem Biol. 2017;12:2906–14.
95. Kirsch P, Hartman AM, Hirsch AKH, et al. Concepts and core principles of fragment-based drug design. Molecules. 2019;24:4309.
96. Andersen OA, Schonfeld DL, Toogood-Johnson I, et al. Cross-linking of protein crystals as an aid in the generation of binary protein-ligand crystal complexes, exemplified by the human PDE10a-papaverine structure. Acta Crystallogr D Biol Crystallogr. 2009;65:872–4.
97. Collins PM, Ng JT, Talon R, et al. Gentle, fast and effective crystal soaking by acoustic dispensing. Acta Crystallogr D Struct Biol. 2017;73:246–55.
98. le Maire A, Gelin M, Pochet S, et al. In-plate protein crystallization, in situ ligand soaking and X-ray diffraction. Acta Crystallogr D Biol Crystallogr. 2011;67:747–55.
99. Gelin M, Delfosse V, Allemand F, et al. Combining 'dry' co-crystallization and in situ diffraction to facilitate ligand screening by X-ray crystallography. Acta Crystallogr D Biol Crystallogr. 2015;71:1777–87.
100. Deller MC, Rupp B. Approaches to automated protein crystal harvesting. Acta Crystallogr F Struct Biol Commun. 2014;70:133–55.
101. Zander U, Hoffmann G, Cornaciu I, et al. Automated harvesting and processing of protein crystals through laser photoablation. Acta Crystallogr D Struct Biol. 2016;72:454–66.
102. Broecker J, Morizumi T, Ou WL, et al. High-throughput in situ X-ray screening of and data collection from protein crystals at room temperature and under cryogenic conditions. Nat Protoc. 2018;13:260–92.
103. Vonrhein C, Flensburg C, Keller P, et al. Data processing and analysis with the autoPROC toolbox. Acta Crystallogr D Biol Crystallogr. 2011;67:293–302.
104. Sparta KM, Krug M, Heinemann U, et al. XDSAPP2.0. J Appl Crystallogr. 2016;49:1085–92.
105. Winter G. Xia2: an expert system for macromolecular crystallography data reduction. J Appl Crystallogr. 2010;43:186–90.
106. Monaco S, Gordon E, Bowler MW, et al. Automatic processing of macromolecular crystallography X-ray diffraction data at the ESRF. J Appl Crystallogr. 2013;46:804–10.
107. Kabsch W. Xds. Acta Crystallogr D Biol Crystallogr. 2010;66:125–32.
108. Battye TG, Kontogiannis L, Johnson O, et al. iMOSFLM: a new graphical interface for diffraction-image processing with MOSFLM. Acta Crystallogr D Biol Crystallogr. 2011;67:271–81.
109. Winter G, Waterman DG, Parkhurst JM, et al. DIALS: implementation and evaluation of a new integration package. Acta Crystallogr D Struct Biol. 2018;74:85–97.
110. Echols N, Moriarty NW, Klei HE, et al. Automating crystallographic structure solution and refinement of protein-ligand complexes. Acta Crystallogr D Biol Crystallogr. 2014;70:144–54.
111. Winn MD, Ballard CC, Cowtan KD, et al. Overview of the CCP4 suite and current developments. Acta Crystallogr D Biol Crystallogr. 2011;67:235–42.
112. Blanc E, Roversi P, Vonrhein C, et al. Refinement of severely incomplete structures with maximum likelihood in BUSTER-TNT. Acta Crystallogr D Biol Crystallogr. 2004;60:2210–21.
113. Pearce NM, Krojer T, Bradley AR, et al. A multi-crystal method for extracting obscured crystallo-

graphic states from conventionally uninterpretable electron density. Nat Commun. 2017;8:15123.
114. Krojer T, Talon R, Pearce N, et al. The XChemExplorer graphical workflow tool for routine or large-scale protein-ligand structure determination. Acta Crystallogr D Struct Biol. 2017;73:267–78.
115. Mooij WT, Hartshorn MJ, Tickle IJ, et al. Automated protein-ligand crystallography for structure-based drug design. ChemMedChem. 2006;1:827–38.
116. Apsel B, Blair JA, Gonzalez B, et al. Targeted polypharmacology: discovery of dual inhibitors of tyrosine and phosphoinositide kinases. Nat Chem Biol. 2008;4(11):691–9.
117. Gall FM, Hohl D, Frasson D, et al. Drug design inspired by nature: crystallographic detection of an auto-tailored protease inhibitor template. Angew Chem Int Ed Engl. 2019;58(12):4051–5.
118. Mitchell T. Machine learning. New York: McGraw Hill; 1997. isbn:0-07-042807-7.
119. Dara S, Dhamercherla S, Jadav SS, et al. Machine learning in drug discovery: a review. Artif Intell Rev. 2021:1–53.
120. Vamathevan J, Clark D, Czodrowski P, et al. Applications of machine learning in drug discovery and development. Nat Rev Drug Discov. 2019;18(6):463–77.
121. Patel L, Shukla T, Huang X, et al. Machine learning methods in drug discovery. Molecules. 2020;25(22):5277.
122. Smola AJ, Schölkopf B. A tutorial on support vector regression. Stat Comput. 2004;14(3):199–222.
123. Deng L, Dong Y. Deep learning: methods and applications. Found Trends Sign Process. 2014;7(3–4):197–387.
124. de Bruyn KC, Stork C, Šícho M, et al. Glory: generator of the structures of likely cytochrome p450 metabolites based on predicted sites of metabolism. Front Chem. 2019;7:402.
125. Brown N, Cambruzzi J, Cox PJ, et al. Big data in drug discovery. Prog Med Chem. 2018;57(1):277–356.
126. Chen B, Butte AJ. Leveraging big data to transform target selection and drug discovery. Clin Pharmacol Ther. 2016;99(3):285–97.
127. Gonzalez GH, Tahsin T, Goodale BC, et al. Recent advances and emerging applications in text and data mining for biomedical discovery. Brief Bioinform. 2016;17(1):33–42.
128. Stephens ZD, Lee SY, Faghri F, et al. Big data: astronomical or genomical? PLoS Biol. 2015;13:e1002195.
129. Che D, Safran M, Peng Z. From big data to big data mining: challenges, issues, and opportunities. Database Syst Adv Appl Lect Notes Comput Sci. 2013;7827:1–15.
130. Herland M, Khoshgoftaar TM, Wald R. A review of data mining using big data in health informatics. J Big Data. 2014;1:2.
131. Piatetsky-Shapiro G. The journey of knowledge discovery. In: Gaber M, editor. Journeys to data mining. Berlin/Heidelberg: Springer; 2012. p. 173–96.
132. Schölkopf B, Tsuda K, Vert J-P. Kernel methods in computational biology. Cambridge: The MIT Press; 2004.
133. Tan PN, Steinbach M, Kumar V. Introduction to data mining. Pearson Education; 2007.
134. Han J, Kamber M, Pei J. Data mining: concepts and techniques. Waltham: Morgan Kaufmann; 2012.
135. Wu X, Kumar V, Ross Quinlan J, et al. Top 10 algorithms in data mining. Knowl Inf Syst. 2007;14:1–37.
136. Jain AK, Murty MN, Flynn PJ. Data clustering: a review. ACM Comput Surv. 1999;31:264–323.
137. Bousquet O, Boucheron S, Lugosi G. Introduction to statistical learning. In: Advanced lectures on machine learning. Berlin/Heidelberg: Springer; 2004. p. 169–207.
138. Friedman J, Hastie T, Tibshirani R. The elements of statistical learning. New York: Springer; 2001.
139. Tan PN, Steinbach M, Kumar V. Association analysis: basic concepts and algorithmstitle. In: Introduction to data mining. Pearson Education; 2007. p. 327–414.
140. Stilou S, Bamidis PD, Maglaveras N, et al. Mining association rules from clinical databases: an intelligent diagnostic process in healthcare. Stud Health Technol Inform. 2001;84:1399–403.
141. Manning CD, Raghavan P, Schütze H. Introduction to information retrieval. New York: Cambridge University Press; 2008.
142. Lee I, Date SV, Adai AT, et al. A probabilistic functional network of yeast genes. Science. 2004;306:1555–8.
143. Myers CL, Troyanskaya OG. Context-sensitive data integration and prediction of biological networks. Bioinformatics. 2007;23:2322–30.
144. Goya J, Wong AK, Yao V, et al. FNTM: a server for predicting functional networks of tissues in mouse. Nucleic Acids Res. 2015;43:W182–7.
145. Greene CS, Krishnan A, Wong AK, et al. Understanding multicellular function and disease with human tissue-specific networks. Nat Genet. 2015;47:569–76.
146. Feldman R, Sanger J. The text mining handbook: advanced approaches in analyzing unstructured data. Cambridge University Press; 2007.
147. Leaman R, Gonzalez G. BANNER: an executable survey of advances in biomedical named entity recognition. Pac Symp Biocomput. 2008:652–63.
148. Gerner M, Nenadic G, Bergman CM. LINNAEUS: a species name identification system for biomedical literature. BMC Bioinform. 2010;11:85.
149. Rebholz-Schuhmann D, Jimeno Yepes A, Li C, et al. Assessment of NER solutions against the first and second CALBC silver standard corpus. J Biomed Semantics. 2011;2(Suppl 5):S11.
150. Campos D, Matos S, Oliveira JL. Biomedical named entity recognition: a survey of machine-learning tools. In: Sakurai S, editor. Theory and applications for advanced text mining. InTech Open Access; 2012.

151. Cohen KB, Christiansen T, Baumgartner WA Jr, et al. Fast and simple semantic class assignment for biomedical text. In: Proceedings of the 2011 workshop on biomedical natural language processing. Stroudsburg, PA, USA: Association for Computational Linguistics (ACL)-HLT 2011; 2011. p. 38–45.
152. Verspoor K, Cohen KB, Lanfranchi A, et al. A corpus of full-text journal articles is a robust evaluation tool for revealing differences in performance of biomedical natural language processing tools. BMC Bioinform. 2012;13:207.
153. Hakenberg J, Plake C, Leaman R, et al. Interspecies normalization of gene mentions with GNAT. Bioinformatics. 2008;24:i126–32.
154. Wermter J, Tomanek K, Hahn U. High-performance gene name normalization with GENO. Bioinformatics. 25:815–21.
155. Krallinger M, Vazquez M, Leitner F, et al. The protein-protein interaction tasks of BioCreative III: classification/ranking of articles and linking bio-ontology concepts to full text. BMC Bioinform. 2011;12(Suppl 8):S3.
156. Kim J-D, Ohta T, Pyysalo S, et al. Overview of BioNLP'09 shared task on event extraction. In: Proceedings of the BioNLP shared task 2009 workshop. Madison: Omnipress, Inc; 2009. p. 1–9.
157. Kim J-D, Pyysalo S, Ohta T, et al. Overview of BioNLP shared task 2011. In: Proceedings of the BioNLP shared task 2011 workshop. Madison: Omnipress, Inc; 2011. p. 1–6.
158. Ananiadou S, Pyysalo S, Tsujii J, et al. Event extraction for systems biology by text mining the literature. Trends Biotechnol. 2010;28:381–90.
159. Li C, Liakata M, Rebholz-Schuhmann D. Biological network extraction from scientific literature: state of the art and challenges. Brief Bioinform. 2014;15:856–77.
160. Faiz R, Amami M, Elkhlifi A. Semantic event extraction from biological texts using a Kernel-based method. Adv Knowl Discov Manag. 2014;527:77–94.
161. Nédellec C, Bossy R, Kim J-D, et al. Overview of BioNLP shared task 2013. In: Proceedings of the BioNLP shared task 2013 workshop. Madison: Omnipress, Inc; 2013. p. 1–7.
162. Tari L, Anwar S, Liang S, et al. Synthesis of pharmacokinetic pathways through knowledge acquisition and automated reasoning. Pac Symp Biocomput. 2010:465–76.
163. Bodenreider O. The unified medical language system (UMLS): integrating biomedical terminology. Nucleic Acids Res. 2004;32:267D–270.
164. Leeper NJ, Bauer-Mehren A, Iyer SV, et al. Practice-based evidence: profiling the safety of cilostazol by text-mining of clinical notes. PLoS One. 2013;8:e63499.
165. Singhal A, Simmons M, Lu Z. Text mining genotype-phenotype relationships from biomedical literature for database curation and precision medicine. PLoS Comput Biol. 2016;12(11):e1005017.

21 Exemplary Protocols of Rational Design of Multitarget Drugs

Abstract

The overall goal of this monograph is to promote the development of polypharmacology as a new sub-discipline of pharmacology and polypharmacology-based multitarget drug (MTD) discovery for human disease therapy. This chapter is the final one of the monograph. Previous chapters have given the fundamental knowledge and clinical applications of polypharmacology, as well as the strategies, resources, and methodologies for MTD discovery. While these contents weight on the theoretical accounts of the polypharmacology-guided MTD discovery, here this final chapter will provide some "real-life" examples in this regard. It is not to give a panoramic view of the new field of rational design of MTDs with only a few examples, but to shed some light on the topic to create certain impression. These examples present several representative protocols based on contemporary state-of the-art strategies and technologies, which would be of practical value for understanding the procedures of MTD design and discovery. The protocols selected for discussion include vector analysis for identification of MTDs, fragment-based identification of MTDs, and scaffold-based molecular hybrid approaches for rational design of MTDs for Alzheimer's disease.

Keywords

Polypharmacological protocols · Rational design of MTDs · Vector analysis · Fragment-based MTD design · Scaffold-based molecular hybridization

21.1 Creative Vector Analysis for Identifying MTDs

The design of MTDs is an essential area of research in medicinal chemistry and the first step to MTD discovery and development for the management of complex human diseases. For the design of MTDs, the first issue to resolve is to measure and define "multitargeticity" in a quantitative way, i.e., how multitargeted a ligand is. There is no doubt that multitarget metrics could help multitarget drug design by providing comparable and workable parameters for drugs and ligands. Yet there has not been any uniform criterion in this regard. Sánchez-Tejeda et al. [1] proposed a vector analysis for quantitatively measuring and defining "multitargeticity." The authors developed terms, such as order and force of a ligand, to generate two critical parameters for the discrimination of MTDs: multitarget indexes 1 and 2 (MTI-1 and MTI-2). They constructed several training sets to test the usefulness of the indexes: an experimental training set with

Z. Wang, B. Yang, *Polypharmacology*, https://doi.org/10.1007/978-3-031-04998-9_21

real affinities, a docking training set within theoretical values, and an extensive database training set. The indexes MTI-1 and MTI-2 proved to be useful, as they were successfully used independently in silico and experimental data, identifying actual multitarget compounds and even selective ligands in most of the training sets. The investigators then applied these indexes to evaluate a virtual library of potential ligands for targets related to multiple sclerosis (MS), identifying ten compounds that are likely leads for the development of MTDs based on their in silico behavior.

Originally, this project by Sánchez-Tejeda et al. [1] started with the purpose of designing dual-target ligands for neurodegenerative diseases (e.g., a dual cholinesterase–monoamine oxidase-B (MAO-B) inhibitor). They designed 211 ligands and decided to figure out a parameter that could summarize or identify the ligands that had the most potency toward the two targets (the nature of the ligands and the targets will be explained later). To analyze the data, they plotted the docking score of the ligands of one target against the docking score of the second one. In the plot, a ligand can be described by the coordinates or docking scores of both targets. Since they are coordinates, the ligand describes an *arrow* or vector, starting from the origin. The angle described by the vector is the selectivity; in fact, the formula of the tangent is the formula for selectivity.

The following contents of this subsection of Chapter 21 are to re-present the steps Sánchez--Tejeda et al. [1] were taking for the rational design of MTDs in a quantitative manner. This protocol involves multiple strategies for rational MTD design, which are outlined in Chap. 18, including structure-based approach/molecular docking, scaffold-based approach, virtual screening, cheminformatics, etc. For the sake of simplicity, the equations and calculations formulated by the authors are omitted, and for detailed information and better understanding of this protocol, please read the original article by these authors, available online at https://pubmed.ncbi.nlm.nih.gov/32232029/.

21.1.1 Step 1: Construction of an Experimental Training Set: Multi-kinase Ligands

Sánchez-Tejeda et al. [1] selected ten known FDA-approved drugs labeled multi-kinase-directed drugs as models of MTDs [2]. Additionally, two non-multi-kinase drugs were included as negative controls. The binding affinity (defined in terms of K_i) of each drug to its target was searched in the BindingDB. The targets of each ligand were selected according to the FDA-approved information. For the negative controls, tyrosine-protein kinase ABL1 was included in the analysis. The K_i was transformed into pM units and linearized. The objective of this analysis was to test if the multitarget index could correctly classify drugs using experimental values. The analysis was done, as stated in section Construction of the Multitarget Index.

21.1.2 Step 2: Construction of a Docking Training Set: Multi-kinase Ligands

The authors then selected ten known FDA-approved drugs labeled multi-kinase-directed drugs as models of multitarget drugs [2], along with three non-multi-kinase drugs as negative controls. The 13 ligands were docked on the tyrosine-protein kinase KIT (PDB id: 4HVS), vascular endothelial growth factor receptor 2 (PDB id: 3VO3), and platelet-derived growth factor receptor beta (PDB id: 1SHA). The docking studies were carried out in Molegro Virtual Docker version 6.0 using the standard protocol suggested by the manufacturer. All waters, cofactors, and non-active ligands were removed from the workspace. MolDock optimizer was used as the running algorithm with 25 runs per ligand. A sphere with a radius of 15 Å was constructed around the active sites of the three proteins and selected as the search site. The poses with the lowest MolDock score were used for further analysis. The objective of this analysis was to determine if the multitarget index could correctly classify drugs using theoretical values. The data were processed, as stated in section Construction of the Multitarget Index.

21.1.3 Step 3: Construction of an Experimental Evaluation Set: DUD

Next, they downloaded the database of DUD (Directory of Useful Decoys), containing the energy scores of nearly 98,258 molecules against 40 targets; some targets had a smaller number of calculated energies but were included in the analysis, as this would challenge the indexes. The package was cleaned so that only the negative energies of each ligand were analyzed. The objective of this analysis was to determine if the multitarget index could filter an extensive database. The data were processed, as stated in section Construction of the Multitarget Index.

21.1.4 Step 4: Virtual Library of Ligands for MS

21.1.4.1 Selection and Construction of the Virtual Library

For ligand construction, PAR-1 antagonists with demonstrated activity were searched. Vorapaxar is a commercially available platelet antiaggregant whose mechanism of action is PAR-1 antagonism; therefore, it was used as a reference ligand. F16357 and SCH79797 are molecules whose antagonism has been previously studied, and thus, they were used as starting points for the design of multitarget molecules. Four possible scaffolds were selected for PAR-1, of which scaffolds W and X were obtained by scaffold hopping from F16357 and vorapaxar, respectively. F16357 was used as scaffold Y, and scaffold Z was an annular modification of SCH79797.

For the selection of KLK-6 ligands, benzamidine isosteres were designed, because this compound is known to be a serine protease inhibitor. The selected benzamidine isosteres were aminopyridine (A), aminopyridine with carboxylic acid (B), aminopyridine with alcohol (C), 2-aminopirimidine aminoquinoline (E), aminoisoquinoline (F), aminoquinazoline (G), and benzylamine (H). Currently, there are no commercially available drugs whose mechanism of action is selective inhibition of KLK-6. However, it was found that 0HM, a benzylamine derivative, was previously determined as a compound with high binding energy to KLK-6 and thus was used as a reference ligand for this enzyme [3].

Finally, 211 compounds were constructed from a combination of both types of ligands with the help of Marvin Sketch 16.2.22.0 and saved in *.smiles format. The three-dimensional geometry was optimized with Spartan '14 (1.1.4) using MMFF and HF 6–31 G*.

21.1.4.2 Molecular Docking

The crystalline structures of PAR-1 complexed with vorapaxar (PDB id: 3VW7), human S1PR1(PDB id: 3V2Y), and human KLK-6 with 0HM (PDB id: 4D8N) were downloaded. The water molecules and co-crystallized ligands were removed from the work area. The docking procedure was carried out in the same manner used for multi-kinase ligands.

21.1.4.3 Cheminformatics Analysis

The smiles codes of the 211 ligands were placed in admetSAR in order to predict some of their pharmacokinetic properties [4]. The following probability scores were obtained: permeability of the blood–brain barrier (BBB), human intestinal absorption (HIA), glycoprotein P substrate (PGP-substrate), carcinogen, acute oral toxicity (AOT), and inhibition of hERG (human Ether-à-go-go-Related Gene). A coefficient of +1 was assigned to all values that fulfilled the following conditions: BBB +, HAI +, PGP-non-substrate, non-carcinogen, AOT III or IV, and weak hERG inhibitor. Otherwise, a negative coefficient was assigned in such a way that the desirable properties were considered positive. With these coefficients, an average of the chemoinformatic properties was calculated, which was called Chemoinformatic Score (CIS).

21.1.5 Step 5: Construction of the Multitarget Index

Vector analysis, mentioned in the introduction, was used as the mathematical basis for the index construction. Besides, vector analysis allowed new interpretations of ligands, concepts, and parameters that may have a significant impact on multitarget drug design.

21.1.5.1 Order of a Ligand

The core idea of the index is to interpret ligands as vectors. The theoretical affinity or score for a target may be interpreted as a coordinate within this vector. This interpretation treats targets as independent variables that are orthogonal to each other.

21.1.5.2 Force of a Ligand

This parameter corresponds to the norm of the vector, which is a metric that combines all the affinities of the ligands into a single value. It is generally understood as magnitude, meaning that greater values correspond to ligands whose particular affinities are large. It is a useful parameter when trying to compare combined affinities. However, this metric is also sensible to extreme values.

21.1.5.3 Binding Capacity and Total Multitarget Capacity

Because each coordinate is a vector, the cross product of all the targets will give a new vector. This new vector is an indirect measure of the binding capacity of a ligand. This metric is more sensible than the force because it is a multiplication of affinities, and considerable differences between affinities have greater repercussions. The authors interpreted the metric as the binding capacity, a measure of a ligand's tendency to bind to more targets.

The average is interpreted as a ligand's affinity to all the targets. The average simulates a drug that has equal affinities for all the proteins. Therefore, the average is defined as the total multitarget capacity of the ligand (MTc).

Finally, the quotient of the binding capacity and the total multitarget capacity gives a proportion of how much of that multitarget capacity is being used. If the binding capacity equals the total multitarget capacity, then the ligand is a true multitarget ligand. By itself, this quotient is an index of "multitargeticity" that can be expressed in percentage for easier reading and read as "used multitarget capacity."

21.1.5.4 Index Standardization, Definition, and Interpretation

Since the idea behind the index is to compare different ligands, it is necessary to standardize the index. The following scheme was proposed: the ligand should be its own reference for standardization.

This simplifies the standardized, used multitarget capacity to a simple geometric mean ranging from 0 to 1, effectively making it an index or measurement of "multitargeticity."

The interpretation is the one originally described in the introduction: how similar the hyperrectangle described by the ligand is similar to a hypercube. Alternatively, in a less abstract way, it is an efficiency measurement: how much "multitarget capacity" is being used.

21.1.5.5 A Second Multitarget Index

A second parameter, the standard deviation, was calculated. With this, another index was constructed that could measure the dispersion of the affinities: the bigger the value, the less variation among the targets. This is a more sensitive parameter that ranges from 1 to negative values. This index also encodes selectivity: smaller values, even negative, indicate more selectivity.

21.1.5.6 Defining of a Multitarget Ligand

With the two indexes, the authors proposed the following values for classifying a ligand as a multitarget:

The ligand is multitarget if MTI-1 $\geq$ 0.84 and MTI-2 $\geq$ 0.60.

These values correspond to an ~20% deviation from the mean affinity. It is worth mentioning that, although this gives a quantitative definition of multitargeticity, ligands that do not fulfill the criteria should not be discarded. These indexes quantify the dispersion and variation of the affinities and do not indicate in any way the potency.

21.1.5.7 Multitarget Potency and Selectivity

Equally low affinities will give high MT index values. For this reason, a final critical parameter was introduced: the multitarget potency. This value is the product of the force times, both multitarget indexes, which is the equivalent of calculating how much of that force is due to the multitargeticity of the ligand.

The multitarget potency is a metric for drug design since the highest values of potency represent a possible multitarget hit or even multitarget lead. In the same line of thought, the next parameter that we propose is an attempt to identify selective ligands.

Both parameters maintain the desired properties (higher values indicate higher potency and selectivity) and are useful for identifying possible multitarget and selective ligands. These metrics, as with any other, have their benefits and drawbacks and will be discussed further on.

21.2 Systematic Assessment of Fragment Identification for Multitarget Drug Design

Fragment-based strategies have been postulated as a versatile avenue to the discovery of MTD leads. Brunst et al. [5] employed a large fragment library for comprehensive screening on five targets chosen from proteins for which multitarget ligands have been successfully developed previously (soluble epoxide hydrolase, leukotriene A4 hydrolase, 5-lipoxygenase, retinoid X receptor, farnesoid X receptor). Differential scanning fluorimetry served as primary screening method before fragments hitting at least two targets were validated in orthogonal assays. In this way, they obtained valuable fragment leads with dual-target engagement for six out of ten target combinations. Their results demonstrate the applicability of fragment-based approaches to identify starting points for polypharmacological compound development with certain limitations. This subsection gives a detailed account of the protocols presented in their study [5] with some minor modifications. For detailed information and better understanding of their protocols, readers are referred to the original article which is available at: https://www.ncbi.nlm.nih.gov/pmc/articles/PMC8049054/.

There are a couple of ways to create MTDs, molecular hybrid, pharmacophore fusion, etc. Molecular hybridization could render MTDs with poor ADME properties resulting from high molecular weights as a consequence of pharmacophore linkage as the simplest strategy to obtain MTDs by joining of pharmacophores via a molecular linker. Thus, pharmacophore fusion for two (or more) targets in a common molecular framework or identification of a merged pharmacophore is a more attractive approach to design MTDs with favorable profile. Morphy and Rankovic proposed the concept of fragment-based design of DMLs with merged pharmacophore [6]. This strategy implies that a low-molecular-weight fragment binding to the desired multiple targets is optimized for potency simultaneously on the target proteins. Several studies have demonstrated the feasibility of this approach [7–9]. The key step of this concept is the identification of a suitable molecular fragment to serve as a starting point. Studies by Hann et al. [10] and Hopkins et al. [11] imply that there is a high probability to identify such fragments from screening due to the fact that binding promiscuity increases with lower molecular weight.

Based on these reasonings, Brunst et al. [5] aimed to systematically probe the feasibility of identifying fragment hits for MTD discovery. For this, they chose five proteins that have been successfully targeted by MTDs previously to ensure that multitarget ligands for these proteins are possible. Their target choice covered the enzymes 5-lipoxygenase (5-LOX), soluble epoxide hydrolase (sEH), and leukotriene A4 hydrolase (LTA4H), as well as the nuclear receptors farnesoid X receptor (FXR) and retinoid X receptor (RXR). For several combinations of these proteins, sEH/5-LOX, sEH/LTA4H, and sEH/FXR, feasibility of a DML has been demonstrated.

Despite their common feature of binding lipids and fatty acid mimetics, the protein fold, catalyzed reaction, as well as binding site shape and residues strongly differ in all five targets. 5-LOX is an iron-dependent enzyme catalyzing the epoxidation of arachidonic acid to leukotriene A4 (LTA4), which is subsequently converted to LTB4 by zinc-dependent LTA4H. sEH, by converting fatty acid epoxides to their corresponding diols, performs a different hydrolysis reaction of polyunsaturated fatty acid epoxides. FXR and RXR belong to the family of nuclear receptors and are activated by bile acids and fatty acids, respectively.

21.2.1 Step 1: Pre-evaluation of the Chemical Ligand Space of the Protein Targets

In order to computationally pre-evaluate the chemical ligand space of the protein targets and select a suitable fragment collection, the investigators trained a self-organizing map (SOM) on the known modulators of all five targets with an IC_{50} or Ki <10 μM retrieved from the ChemblDB v.24 as well as on the Prestwick Chemical library (Prestwick Chemical, Illkirch, France) containing off-patent approved drugs. The FragFP descriptor, a substructure-based fingerprint, was calculated for all compounds using OSIRIS DataWarrior v.5.0.0 (www.openmolecules.org) and employed for training of a SOM with 50 × 50 neurons. Analysis of the SOM revealed that the active compound for the individual target proteins occupies distinct clusters, whereas the approved drugs were widely distributed between the clusters, suggesting that a fragment library derived from these compounds is suitable to discover actives on all five targets.

21.2.2 Step 2: Primary Screening of the Fragment Library

Based on this observation, the authors selected the core set of the Prestwick Drug-Fragment Library (PDFL) comprising 480 compounds for the fragment screening approach [5]. This structurally diverse fragment library was generated by virtual fragmentation of approved drugs and provides a broad distribution of chemical motifs and functional groups present in bioactive compounds.

For the primary screen of this fragment library, they employed differential scanning fluorimetry (DSF, also known as thermal shift). DSF is generally applicable to a large panel of protein targets and provides a robust, low-cost screening technology for fragment-based approaches. All five proteins of interest are soluble and were recombinantly expressed in *Escherichia coli*. They performed an initial fragment screen on all five targets by DSF. The use of a uniform screening method for all targets ensured consistent data and provided the opportunity to identify false-positive hits which interfered with the screening technology. The conditions of the DSF assay were optimized for the desired screening and validated with reference compounds. Compounds causing a positive thermal shift $\Delta T_m \geq 1.0\,°C$ were considered as active. Of note, the reference 5-LOX inhibitor did not exhibit protein stabilization observed by a thermal shift, but the shape of the resulting melting curve was considered to be sufficient for screening, and a lower cutoff ($\Delta T_m \geq 0.9\,°C$) was applied for 5-LOX.

DSF measurement of the PDFL (single concentration of 500 μM, duplicates, 96-well format) on the 5 proteins retrieved 19 fragment hits on sEH, 28 on LTA4H, 12 on 5-LOX, 16 on RXRα, and 14 on FXR, respectively. Among the hits, three compounds gave a positive signal on all five proteins likely qualifying them as pan assay interference compounds (PAINS). Fragments binding to multiple targets were identified for seven out of ten target combinations. Among them, fragments **4**, **6**, **7**, and **9** match previously identified privileged scaffolds for fatty acid mimetics. Substructure search among the active compounds revealed that 4-benzylphenol **4** as well as 4-hydroxybiphenyl **9** have been successfully incorporated in ligands of all five targets. Active ligands containing *N*-phenylbenzamide **6** and the *N*-phenylbenzylamine **7** have been described for four of the protein targets.

Furthermore, indole, which has been characterized as a privileged heterocycle for fatty acid mimetics, is present in fragments **5** and **10**. Dual 5-LOX/RXR ligand **1** was present in 5-LOX inhibitors, while dual FXR/LTA4H modulator **3** is found among known FXR ligands. Fragments **2** (sEH/5-LOX) and **8** (LTA4H/RXR) are novel scaffolds not appearing in known ligands of any of the five targets.

21.2.3 Step 3: Hit Validation

For hit validation, Brunst et al. [5] employed secondary assays with orthogonal readouts to confirm fragment activities. Biochemical activity assays were chosen for the enzymes sEH, LTA4H, and 5-LOX. Compound activity on sEH and LTA4H was observed using recombinant protein and fluorogenic substrates, while activity on recombinant 5-LOX was determined by HPLC-based detection of product formation (5-HETE). RXRα and FXR modulation was assessed in cell-based hybrid reporter gene assays. All ten compounds were tested at a single concentration of 500 μM in the sEH and LTA4H activity assays and with a concentration of 100 μM on 5-LOX, RXRα, and FXR. Fragments showing an inhibition greater than 50% (sEH-H and LTA4H) or greater than 70% (5-LOX) as well as the hits determined in the DSF experiments were further characterized in full concentration–response curves between 0.1 and 1000 μM. Toxicity at concentrations of 100 μM and above in the cell-based reporter gene assays prevented reliable characterization on FXR and RXRα.

As alternative cell-free assay, they also employed isothermal titration calorimetry (ITC) to determine binding affinity of the fragments toward the LBDs of both nuclear receptors. ITC was performed with a protein concentration of 50 μM and compound concentrations of 250 μM. DSF hits **2** (sEH/5-LOX), **4** (LTA4H/FXR), **9** (sEH/FXR), and **10** (5-LOX/LTA4H) were confirmed active in the orthogonal experiments. Ligand efficiency (LE) calculated from the respective IC_{50} or K_d values was found favorable for the confirmed hits (LE ≥ 0.3 kcal/mol) rendering them as valuable starting points for multitarget design.

21.2.4 Step 4: Substructure Search to Validate Multitargeting Property

Next, the investigators performed a substructure search with each hit fragment in the Prestwick Chemical library and retrieved eight approved drugs (**11–18**) that contained a motif of a multitarget active fragment from the screening (Scheme 2). Drugs **11–18** were evaluated in the enzyme activity assays for sEH, LTA4H, and 5-LOX as well as in the cellular reporter gene assays for FXR and RXR. None of the approved drugs 11–20 exhibited dual target activity, and their potencies were less favorable considering their molecular weights being higher than those of the validated fragment hits. Most interestingly, calcium channel blocker **15** (bepridil), a derivative of fragment **7**, inhibited 5-LOX with an IC_{50} of 3.2 μM while completely losing activity toward sEH. Thus, simple expansion of drug-derived fragment hits to the respective drugs is not sufficient to discover leads for multitarget design.

The systematic fashion of our approach also allows for comparison to previous attempts to identify dual ligands by means of rational design of a starting fragment for subsequent optimization. The biphenyl fragment **9** was identified as a dual sEH/FXR ligand. Substructure search of ChemblDB revealed that the biphenyl fragment was previously incorporated in potent FXR partial agonist **19** and sEH inhibitor **20**. Rationally designed dual fragment **21**, which enabled development of potent dual sEH/FXR modulators, was not present in the screening library. The most similar (Tanimoto coefficient on FragFP) fragment **22** displayed no activity in the DSF screen. The same holds true for the dual 5-LOX/sEH fragment inhibitor **23**, which was identified previously by virtual screening, while its closest neighbor **24** displayed no activity. This observation leads to the assumption that a fragment-based screening approach is complementary to rational design of multitarget ligands and vice versa.

21.2.5 Important Notes from This Systematic Model Study

According to Brunst et al. [5], several interesting points and lessons for MTD design can be deduced from their systematic model study.

21.2.5.1 Fragment-Based Approaches Can Be Used for MTD Design

For six out of ten target combinations, fragment hits were successfully obtained from library screening, and for four out of ten, the hits could be validated in an orthogonal assay system. Most previous multitarget-optimization studies have relied on a rational combination of the pharmacophores. This approach is in many cases highly efficient, yet it requires deep understanding of the pharmacophores and the underlying structure–activity relationship for the targets of interest, which is not always available. Furthermore, linking, merging, or fusing pharmacophores mostly yields previously described chemical scaffolds or combinations thereof, whereas a systematic fragment-based approach is unbiased by previously described ligands.

21.2.5.2 The Size of the Compounds Matters

This systematic experimental fragment screening fully supports the theoretical assumptions by Hann et al. [10] and Hopkins et al. [11]—the target promiscuity of chemical compounds decreases with increasing molecular weight. While the fragments **1–10** were able to hit multiple targets, their larger approved drug counterparts **11–18** exhibited only weak selective activity with respect to their molecular weight. Therefore, a promising starting point for a multitarget ligand discovery campaign is a low-molecular-weight hit, exactly as for the discovery of single-target drugs.

21.2.5.3 The Size and the Composition of the Library Matters

The hit rates for fragment-based screening are generally described in the range of 2–8% [12, 13]. In this study, hit rates of 2.5–5.8% were reached on the individual target. Furthermore, six out of ten dual target hits represent privileged scaffolds for fatty acid mimetics. The concept of using focused fragment libraries was successfully applied for kinases or metal-binding proteins and could be potentially adopted for multitarget fragment screening.

21.2.5.4 The Screening Technology Matters

Several studies have evaluated the applicability of different screening technologies for fragment identification. In this study, DSF screening was employed as a universal low-cost method for primary screening. However, follow-up characterization of the fragments using fluorescence-based activity assays for sEH and LTA4H revealed that even compounds without a positive shift in DSF can exhibit inhibitory activity, which should certainly be handled with care. Furthermore, given the low probability to identify a multitarget fragment hit, a more permissive primary screening technique would offer advantages. Screening by X-ray crystallography or NMR spectroscopy would additionally provide valuable structural information which is indispensable for fragment-based approaches, especially in the design of multitarget compounds.

21.3 Scaffold-Based Molecular Hybrid Approaches for Rational Design of MTDs for Alzheimer's Disease

Alzheimer's disease (AD) is a complex neurodegenerative disorder with a multifaceted pathogenesis. The multiple factors involved in AD include amyloid aggregation to form insoluble neurotoxic plaques of Aβ, hyperphosphorylation of tau protein, oxidative stress, calcium imbalance, mitochondrial dysfunction, and deterioration of synaptic transmission. A recent approach for the rational design of new drug candidates, also called multitarget-directed ligands (MTDLs) or simply multitarget drugs (MTDs) approach, has gained increasing attention by many research groups and led to development of a variety of hybrid compounds acting simultaneously on diverse biological targets. Dias et al. [14] presented an excellent review on some recent advances and examples of the exploitation of MTD approach in the rational design of novel drug candidate prototypes for the treatment of AD. The examples described in this review article are some typical representations of molecular hybrids derived from privileged scaffolds. This subsection provides a brief description to each of the examples essentially by following the discussions in the paper by Dias et al. [14] with some rearrangement and modifications.

21.3.1 Hydroquin: Benzenediol–Berberine Hybrids

Since 2005, the literature has shown several results from applying this innovative approach of drug design [14]. Drugs such as donepezil, tacrine, and rivastigmine [15, 16] have been used as structural models for molecular hybridization with bioactive synthetic and natural products such as curcumin [17, 18], berberine [19, 20], and 8-hydroxyquinolines [21], among others in the search for new chemical entities with multiple bioactive properties useful for the treatment of AD.

As one example of this strategy, berberine was used as a basic skeleton for the construction of a series of hybrid compounds with molecular subunits that include melatonin and ferulic acid. The goal was to obtain new derivatives with antioxidant properties and anti-Aβ aggregation, but with reduced inhibitory potency of acetylcholinesterase (AChE). The hybrid compound hydroquin demonstrated the best ability for suppression of Aβ aggregation, plus excellent antioxidant effect and reasonable ability to inhibit acetylcholinesterase (AChE) and butyrylcholinesterase (BuChE). In addition, another derivative was identified as a potent inhibitor of AChE, with strong antioxidant properties, confirming the multitargeting potential of these substances in comparison to berberine, the original prototype [20].

21.3.2 Memoquin: Benzoquinone–Polyamine Hybrid

Memoquin, a 1,4-benzoquinone-polyamine hybrid compound, was reported by Cavalli and co-workers [18] as a promising drug prototype for AD. The structural design of compound was based on a polyamine core, derived from coproctamine, a dimeric agent with anticholinesterasic and antimuscarinic. Aiming to add to these properties, the ability to neutralize ROS (reactive oxygen species) and neuroprotection, authors proposed the insertion of a 1,4-benzoquinone, from coenzyme Q10 (CoQ10), a fragment with a recognized potent mitochondrial antioxidant property and a protector of the hippocampus neurons against $A\beta_{1-40}$-induced neurotoxicity. The inner polymethylene chain was replaced by the benzoquinone nucleus. The insertion of the benzoquinone nucleus contributes to increase the neuroprotective activity, because a hydrophobic and planar π system is generated, which is able in principle to bind Aβ [18].

Memoquin exhibited the ability to inhibit AChE (IC_{50} = 1.55 nM), almost 15-fold more potent than the reference compound (coproctamine). Memoquin also inhibited AChE-induced $A\beta_{1-40}$ aggregation with an IC_{50} of 28.3 μM and the self-induced $A\beta_{1-42}$ aggregation with an IC_{50} of 5.93 μM. In addition, it is also capable to inhibit BACE-1 (β secretase) activity in range nanomolar concentrations (IC_{50} = 108 nM) and showed antioxidant properties, reducing the formation of free radicals by 44.1%. It was demonstrated that the 1,4-benzoquinone moiety can be reduced in vivo to the 1,4-dihydroquinone form, increasing its antioxidant potential and scavenging properties. Finally, in vivo assays showed the potential of memoquin to restore the cholinergic deficit and successfully reduce Aβ expression and accumulation, also reducing hyperphosphorylation of tau protein.

Considering all these multitarget profile of memoquin (**8**), Bolognesi and co-workers [22] reported in 2009 the design of new hybrid lipoic acid–memoquin (**11**) derivatives inspired on the memoquin structure and lipoic acid structure which is a potent antioxidant. In addition to the observed biological properties of memoquin, compound **12** also exhibited a good neuroprotective effect against oxidative stress, greater than the reference control lipoic acid (**10**).

21.3.3 Curcumin

Curcumin is the major constituent of *Curcuma longa*, a plant species used in Indian Traditional Medicine, with Aβ aggregation inhibitory properties have been used as a structural model for molecular to generate new analogs of memoquin with enhanced neuroprotective AChE and BuChE inhibitory properties. Among the novel memoquin hybrid derivatives, two compounds showed the best multiple-target action profile with selective inhibitory activities of AChE and BuChE, with modulatory effects on the amyloid aggregation and inhibition of neurofibril formation [23].

21.3.4 Tacrine

Tacrine is the first approved ChE inhibitor by the FDA for the treatment of AD. However, due to some undesirable side effects, this drug has its therapeutic use restricted and, therefore, the search for more effective and secure tacrine derivatives is still of interest. In recent years, many studies have focused on combining the potent ChE inhibitory properties of tacrine with additional biological properties to search for new MTDLs.

21.3.4.1 Tacrine–Benzylamino-3-Pyridyl Hybrid

Contelles and co-workers [24] synthesized a new series of 4H-pyrano [2,3-b] quinolone derivatives, planned by molecular hybridization of tacrine fused to a benzylamino-3-pyridyl ring moiety, aiming to obtain new AChEIs with an innovative molecular scaffold and to explore its pharmacophoric contributions to additional neuroprotective properties and modulatory effects on voltage-dependent Ca^{2+} channels. In general, the new tacrine hybrid compounds showed to be better selective for AChE inhibition than tacrine. It seems that the substitution at C-20 position of the 3-pyridyl ring system was deleterious for AChE inhibition.

21.3.4.2 Tacrine Dimer: Tacrine–Tacrine Hybrid

The dimer bis-(7)-tacrine was one of the first homodimers reported in the literature with increased AChE affinity, exhibiting a 1000-fold higher inhibitory potency than tacrine. This best inhibitory profile is due to a dual simultaneous interaction with active and peripheral sites of AChE. Further studies demonstrated that this compound was also capable to inhibit the AChE-induced Aβ aggregation with an IC_{50} of 41.7 μM. Because of this symbiotic profile, dimer bis-(7)-tacrine has been described as an important structural model for rational design of novel MTDLs [25].

21.3.4.3 BW284c51 and Ambenonium: Tacrine Dimer–Chelating Group Hybrid

Bolognesi and co-workers [16] proposed modifications on the structure of bis-(7)-tacrine with an insertion of a spacer subunit with functional groups capable to chelate metals, which are involved in the degenerative process. For this purpose, compounds BW284c51 and ambenonium were selected as model prototypes due to their cholinesterase inhibitory activity and their singular structural feature, carrying carbonyl and oxalamide functionalities which could lead to the desired metal chelation property. The multifunctional compounds maintained a potent AChEI in the nanomolar range and showed additional inhibitory effect of AChE-induced amyloid-β aggregation. The results showed the ability for chelation of Cu^{2+} and Fe^{3+} ions and suggest that these compounds might act against AD by a chelation mechanism [16].

21.3.4.4 Tacrine Dimer–Cystamine Hybrid

Minarini and co-workers [25] also explored the structure of bis-(7)-tacrine, in an attempt to obtain the hybrid derivative cystamine bis-(7)-tacrine, with an insertion of the cystamine moiety as a linker between two acrydine subunits of the bis-tacrine framework. Cystamine has shown important biological properties such as antioxidant and neuroprotective. This study revealed the hybrid dimer cystamine-tacrine was able to inhibit AChE (IC_{50} = 5.04 nM), BuChE (IC_{50} = 4.23 nM), self-Aβ aggregation (IC_{50} = 24.2 μM), and AChE-induced Aβ aggregation (52,6%) in the same range of the reference compound, with additional neuroprotective effect on SH-SY5Y cell line against H_2O_2-induced oxidative injury, with low toxicity [25].

21.3.4.5 Tacrine–Ferulic Acid Hybrid

Fang and co-workers [26] reported in 2008 the synthesis and pharmacologic evaluation of a new series of tacrine–ferulic acid hybrid derivatives. They planned a new structural pattern by the connection of ferulic acid to the structure of tacrine via an alkylenediamine as a linker side chain. This approach was used aiming to obtain new multitarget hybrid compounds which could conjugate the AChE inhibitory activity originating from the tacrine template and the antioxidant activity from the ferulic acid moiety. The compounds showed lower antioxidant activity than ferulic acid, but all hybrids tested showed moderate to good antioxidant activity. All compounds effectively inhibited AChE in in vitro assays. Particularly, a compound showed a tenfold higher AChE inhibitory activity (IC_{50} = 4.4 nM) than tacrine. Enzymatic kinetics studies suggest that this compound possess high-affinity binding to the peripheral anionic site (PAS) of AChE and, therefore, could inhibit Aβ-peptide fibril formation [26].

21.3.4.6 Tacripyrines: Tacrine–Dihydropyridine Hybrid

In another approach, tacripyrines were designed by combining tacrine with a calcium antagonist such as nimodipine [27]. In this study, the AChE inhibitory activity of the nimodipine–tacrine hybrids is potentiated, with IC_{50} in the nanomolar range for AChE. Besides acting as an inhibitor of AChE, these compounds also showed significant neuroprotective effects. In particular, one compound (IC_{50} = 105 ± 15 nM) proved to be moderately active in models AChE-induced and Aβ-self aggregation models (30 and 35% inhibition, respectively). Further kinetic studies and molecular modeling data suggest that this compound could interact with the PAS of the AChE, exerting its inhibitory properties by noncompetitive mode. Additionally, most of the compounds showed moderate Ca^{2+} channel blocking effect, and therefore the compounds are neuroprotective agents [27].

21.3.4.7 Tacrine–Caffeic Acid Hybrids

A novel series of tacrine–caffeic acid hybrids were designed by combining the structure of caffeic acid and tacrine as multitarget-directed ligands against Alzheimer's disease. In vitro studies showed that most of the target molecules exhibited significant antioxidant activities on oxygen radical absorbance capacity method (ORAC). The hybrid molecules showed to be good metal chelators and also exhibited higher antioxidant capacity than the caffeic acid with values of 2.75–9.37 μM. In particular, a compound exhibited the greatest ability in inhibition of self- or AChE-induced β-amyloid$_{1-40}$ aggregation, as well as showed potent neuroprotective effects against H_2O_2- and glutamate-induced cell death, with low toxicity in HT22 cells. Further molecular modeling and enzymatic kinetics studies clarified that these inhibitors may act in two binding sites of AChE, which could explain the inhibitory effect exerted over Aβ aggregation [28].

21.3.4.8 Tacrine–Phenylthiazole Hybrids

All ten compounds derived from tacrine–phenylthiazole hybrids were potent inhibitors of cholinesterases with IC_{50} values ranging from 5.78 ± 0.05 to 7.14 ± 0.01 nM for AChE and from 5.75 ± 0.03 to 10.35 ± 0.15 nM for BuChE. Moreover, a structure–activity relationship study was conducted to study the influence on the type of middle linker and substitutions at

40-position of 4-phenyl-2-aminothiazole. The results were indicative that the length of middle linker affected the AChE inhibitory potency. The derivative, with the largest linker subunit, exhibited the greater potency in AChE inhibition (IC_{50} = 7.14 ± 0.01 nM). Data from kinetic studies have shown that this compound is a mixed-type inhibitor and could bind simultaneously at the catalytic and the peripheral sites of AChE. Most of phenylthiazole–tacrine hybrids showed a good inhibitory potency on $A\beta_{1-42}$ self-aggregation, although the activity was lower when compared with subunit 4-phenyl-2-aminothiazole. Additionally, the compound displayed a blockade effect on Ca^{2+} overload in the primary cultured cortical neurons [29].

21.3.5 Galanthamine–Memantine Hybrid

According to the studies of Simoni and co-workers [30], a series of hybrid derivatives of galanthamine, an AChE inhibitor, and memantine, a noncompetitive NMDA receptor antagonist, showed potent AChE inhibitory activities with multiple neuroprotective effects in vitro. Some of these hybrid compounds were also capable to prevent glutamate-induced neurotoxicity by moderately blocking glutamate receptor NMDA subtype and neuronal cell death via NMDA receptor-dependent effect at nanomolar concentrations (e.g., memagal, IC_{50} = 0.28 nM). These molecules were the first examples reported of dual AChE/NR2B drug prototype candidates [30].

21.3.6 Bivalent Β-Carboline Derivatives

A new series of 30 bivalent β-carboline derivatives was synthesized by Rook and co-workers [31] based on monovalent β-carbolines (as potent dual AChE/BuChE inhibitors). The researchers also found several of these compounds being also potent NMDA receptor blockers, in addition to the potent inhibitory properties of both AChE/BuChE enzymes. A structure–activity study revealed that the inhibitory activity of quaternary N9-bivalent β-carbolines was increased compared to their monovalent analogues. One of the compounds showed the higher AChE inhibitory activity (IC_{50} = 0.5 nM) and proved to be a potent inhibitor of the transient induced-glutamate Ca^{2+} with an IC_{50} = 1.4 μM, which is fourfold more potent than memantine (IC_{50} = 5 5.6 μM) in the same assay conditions [31].

21.3.7 Benzofuran-Based Hybrid

In another approach for drug design of MTDLs, Rizzo and co-workers [32, 33] reported a new series of hybrid molecules based on the structure of the AChE inhibitor *N*-methyl-*N*-benzylamine and compound SKF-64346, a benzofuran derivative with good inhibitory properties of Aβ-fibril formation, using a heptyloxy moiety as a linker subunit. All compounds evaluated have shown to be good AChE and Aβ-fibril formation inhibitors, reducing the Aβ neurotoxicity. The compound showed an IC_{50} of 1.82 μM in BuChE inhibition, being 5.7-fold more selective toward BuChE than AChE (IC_{50} = 10.5 μM). Furthermore, this compound was also able to inhibit Aβ self-aggregation with an IC_{50} of 13 μM and showed a neuroprotective effect of 63% at 30 mM [32, 33]. Some years later, aiming to improve its pharmacological profile, the researchers modified the original structure of the compound by variation of the spacer subunit between the 2-arylbenzofuran and the *N*-methyl-*N*-benzylamine moieties. They proposed a variation in the length of the spacer subunit, modification in the substituents in 3 position of the benzofuran scaffold, and change of the *N*-methyl-*N*-benzylamine heptyloxy side chain from the *para* to *meta* position [33]. These structural modifications led to a new series that exhibited an increase in activity for most of the modified compounds.

21.3.8 AChE Inhibitor–MAO Inhibitor Hybrids

21.3.8.1 Donepezil–PF9601N Hybrid

A new family of multitarget molecules able to interact with both AChE and BuChE, as well as with monoamine oxidase (MAO) A and B, has been synthesized by Samadi and collaborators [34]. Rational structural pattern of these compounds considered conjunctive approach by combination of benzylpiperidine and *N*-propargylamine moieties present in the AChE inhibitor donepezil and the MAO inhibitor PF9601N, respectively, connected by a central pyridine or naphthyridine ring system. The most promising derivative showed a potent inhibitory activity of AChE (IC_{50} = 37 nM), but less potent than donepezil. This compound also showed a moderate inhibitory potency of MAO-A (IC_{50} = 41 μM). Moreover, molecular modeling showed that these inhibitors probably act in two binding sites of AChE and that the length of the spacer subunit is of particular relevance in the modulation of AChE inhibition, being crucial for the dual interaction of these molecules with both CAS and PAS sites of the enzyme [34].

21.3.8.2 Donepezil–N-(5benzyloxy-1-Methyl-1H-Indol-2-Yl) Methyl-N-Methylprop-2-Yn-1-Amine

A recent publication by Bolea and collaborators [35] reported the synthesis and pharmacological evaluation of a new family of multifunctional molecules capable to interact simultaneously with the AD-related enzymes AChE, BuChE, MAO-A, and MAO-B. The structural architecture of the series was based on the molecular hybridization of the benzylpiperidine moiety of donepezil and the indolyl-propapylamino subunit present in the structure of the MAO inhibitor N-(5benzyloxy-1-methyl-1H-indol-2-yl)methyl-N-methylprop-2-yn-1-amine, connected by a central oxy-methylene chain [35]. The compound was identified as the most promising ligand, showing a good selectivity and high potency in the inhibition of MAO-A with an IC_{50} = 5.2 nM ($IC_{50\ MAO\text{-}B}$ = 43.0 nM), but less potent and with poor selectivity for the AChE (IC_{50} = 0.35 μM) and BuChE (IC_{50} = 0.46 μM). Further molecular modeling and enzymatic kinetics studies clarified that these inhibitors act in two binding sites of AChE, which could explain the inhibitory effect over Aβ-self-induced (47.8 ± 2.1%) and AChE-dependent aggregation (32.4 ± 7%) [35].

21.3.8.3 Donepezil–Benzylidene-Indanone Pharmacophore Hybrid

Taking the structure of donepezil as a model, a group [36] reported a new family of derivatives based on the structure of benzylidene–indanone pharmacophoric subunit with the insertion of set of substituted benzene ring systems. In vitro evaluation showed that most of the compounds were potent inhibitors of MAO-B (IC_{50} of 7.5–40.5 μM), besides antioxidant (ORAC-FL value of 2.75–9.37 μM) and metal chelator properties. In addition, they showed a great ability for inhibition of self-induced β-amyloid aggregation (10.5–80.1%, 20 μM). The compound was identified as the most potent inhibitor agent of $A\beta_{1-42}$ aggregation (80.1%) and MAO-B (IC_{50} = 7.5 μM). Moreover, it proved to be an excellent antioxidant and metal chelator agent, being able to inhibit Cu^{2+}-induced $A\beta_{1-42}$ aggregation and disassembling the Aβ fibrils.

21.3.9 Resveratrol–Catechol Pharmacophore

Resveratrol, a potent phenolic natural plant metabolite, with remarkable antioxidant and anti-inflammatory properties, has been recently described as an inhibitor of Aβ aggregation. Thus, its molecular scaffold has been also explored for assembly of innovative MTDLs with antioxidant, anti-inflammatory, and inhibitory of Aβ aggregation properties. Two compounds were the lead compounds identified among a series of resveratrol-derived molecular hybrids, planned by the combination of the catechol pharmacophore with the well-known metal chelator clioquinol [37]. These two hybrid compounds showed high potency in the inhibition of self-induced Aβ aggregation (IC_{50} = 7.56 μM and IC_{50} = 6.51 μM, respectively) in comparison to resveratrol (IC_{50} = 15.11 μM). Furthermore, antioxidant assays performed by the ORAC method showed the ability of the compounds in decreasing the generation of ROS in 4.72 and 4.70 trolox equivalent, respectively. These compounds were also capable to disassemble the highly structured Aβ fibrils generated by self- and Cu (II)-induced Aβ aggregation and exhibited significant inhibitory effect on MAO-A and MAO-B with a moderate AChE inhibition and low neurotoxicity [37].

References

1. Sánchez-Tejeda JF, Juan F, Sánchez-Ruiz JF, Salazar JR, et al. A definition of "multitargeticity": identifying potential multitarget and selective ligands through a vector analysis. Front Chem. 2020;8:176. Available online at: https://pubmed.ncbi.nlm.nih.gov/32232029/
2. Li YH, Wang PP, Li XX, et al. The human kinome targeted by fda approved multi-target drugs and combination products: a comparative study from the drug-target interaction network perspective. PLoS One. 2016;11:e0165737.
3. Liang G, Chen X, Aldous S, et al. Human kallikrein 6 inhibitors with a para-amidobenzylanmine P1 group identified through virtual screening. Bioorg Med Chem Lett. 2012;22:2450–5.
4. Cheng F, Li W, Zhou Y, et al. admetSAR: a comprehensive source and free tool for assessment of chemical ADMET properties. J Chem Inf Model. 2012;52:3099–105.
5. Brunst S, Kramer JS, Kilu W, et al. Systematic assessment of fragment identification for multitarget drug design. ChemMedChem. 2021;16(7):1088–92. Available online at: https://www.ncbi.nlm.nih.gov/pmc/articles/PMC8049054/
6. Morphy R, Rankovic Z. Drug Discov Today. 2007;12:156–60.
7. Artis DR, Lin JJ, Zhang C, et al. Scaffold-based discovery of indeglitazar, a PPAR pan-active anti-diabetic agent. Proc Natl Acad Sci U S A. 2009;106:262–7.
8. Zhang C, Ibrahim PN, Zhang J, et al. Design and pharmacology of a highly specific dual FMS and KIT kinase inhibitor. Proc Natl Acad Sci U S A. 2013;110:5689–94.
9. Achenbach J, Klingler F-M, Blöcher R, et al. Exploring the chemical space of multitarget ligands using aligned self-organizing maps. ACS Med Chem Lett. 2013;4:1169–72.
10. Hann MM, Leach AR, Harper G. Molecular complexity and its impact on the probability of finding leads for drug discovery. J Chem Inf Comput Sci. 2001;41:856–64.
11. Hopkins AL, Mason JS, Overington JP. Can we rationally design promiscuous drugs? Curr Opin Struct Biol. 2006;16:127–36.
12. Keserű GM, Erlanson DA, Ferenczy GG, et al. Design principles for fragment libraries: maximizing the value of learnings from pharma fragment-based drug discovery (FBDD) programs for use in academia. J Med Chem. 2016;59:8189–206.
13. Schuffenhauer A, Ruedisser S, Marzinzik A, et al. Library design for fragment-based screening. Curr Top Med Chem. 2005;5:751–62.
14. Dias KS, Viegas C Jr. Multi-target directed drugs: a modern approach for design of new drugs for the treatment of Alzheimer's disease. Curr Neuropharmacol. 2014;12(3):239–55.
15. Samadi A, Valderas C, de los Ríos C, et al. Cholinergic and neuroprotective drugs for the treatment of Alzheimer and neuronal vascular diseases II. Synthesis biological assessment and molecular modelling of new tacrine analogues from highly substituted 2-aminopyridine-3-carbonitriles. Bioorg Med Chem. 2011;19:122–33.
16. Bolognesi ML, Cavalli A, Luca Valgimigli L, et al. Multi-target-directed drug design strategy: from a dual binding site acetylcholinesterase inhibitor to a trifunctional compound against Alzheimer's disease. J Med Chem. 2007;50:6446–9.
17. Chen S-Y, Chen Y, Li Y-P, et al. Design, synthesis, and biological evaluation of curcumin analogues as multifunctional agents for the treatment of Alzheimer's disease. Bioorg Med Chem. 2011;19:5596–604.
18. Cavalli A, Bolognesi ML, Capsoni S, et al. A small molecule targeting the multifactorial nature of Alzheimer's disease. Angew Chem Int. 2007;46:3689–92.
19. Shan W-J, Huang L, Zhou Q, et al. Synthesis, biological evaluation of 9-N-substituted berberine derivatives as multi-functional agents of antioxidant, inhibitors of acetylcholinesterase, butyrylcholinesterase and amyloid-ß aggregation. Eur J Med Chem. 2011;46:5885–93.
20. Jiang H, Wang X, Huang L, et al. Benzenediol-berberine hybrids: multifunctional agents for Alzheimer's disease. Bioorg Med Chem. 2011;19:7228–35.
21. Fernandez-Bachiller MI, Pérez C, González-Muñoz GC, et al. Novel Tacrine-8-hydroxyquinoline hybrids as multifunctional agents for the treatment of Alzheimer's disease, with neuroprotective, cholinergic, antioxidant, and copper-complexing properties. J Med Chem. 2010;53:4927–37.
22. Bolognesi ML, Cavalli A, Bergamini C, et al. Toward a rational design of multitarget-directed antioxidants: merging memoquin and lipoic acid molecular frameworks. J Med Chem. 2009;52(23):7883–6.
23. Bolognesi ML, Bartolini M, Tarozzi A, et al. Multitargeted drugs discovery: balancing anti-amyloid and anticholinesterase capacity in a single chemical entity. Bioorg Med Chem Lett. 2011;21(9):2655–8.
24. Contelles JM, Leo R, Lopez MG, et al. Synthesis and biological evaluation of new 4H-pyrano[2,3-b] quinolone derivatives that block acetylcholinesterase and cell calcium signals, and cause neuroprotection against calcium overload and free radicals. Eur J Med Chem. 2006;41:1464–9.
25. Minarini A, Milelli A, Tumiatti V, et al. Cystamine-tacrine dimer: a new multi-target-directed ligand as

potential therapeutic agent for Alzheimer's disease treatment. Neuropharmacology. 2012;62:997–100.
26. Fang L, Kraus B, Lehmann J, et al. Design and synthesis of tacrine-ferulic acid hybrids as multi-potent anti-Alzheimer drug candidates. Bioorg Med Chem Lett. 2008;18:2905–9.
27. Contelles JM, Leon L, de los Rios C, et al. Tacripyrines the first tacrine-dihydropyridine hybrids as multitarget-directed ligands for the treatment of Alzheimer's Disease. J Med Chem. 2009;52:2724–32.
28. Chao X, He X, Yang Y, et al. Design synthesis and pharmacological evaluation of novel tacrine-caffeic acid hybrids as multi-targeted compounds against Alzheimer's disease. Bioog Med Chem Lett. 2012;22:6498–502.
29. Wang Y, Wang F, Yu JP, et al. Novel multipotent phenylthiazole-tacrine hybrids for the inhibition of cholinesterase activity, b-amyloid aggregation and Ca^{2+} overload. Bioorg Med Chem. 2012;20:6513–22.
30. Simoni E, Daniele S, Bottegoni G, et al. Combining Galantamine and Memantine in multitargeted new chemical entities potentially useful in Alzheimer's disease. J Med Chem. 2012;55:9708–21.
31. Rook Y, Schmidtke KU, Gaube F, et al. Bivalent β-carbolines as potential multitarget anti-Alzheimer agents. J Med Chem. 2010;53:3611–7.
32. Rizzo S, Riviere C, Piazzi L, et al. Benzofuran-based hybrid compounds for the inhibition of cholinesterase activity β amyloid aggregation and aβ neurotoxicity. J Med Chem. 2008;51:2883–6.
33. Rizzo S, Tarozzi A, Bartolini M, et al. A. 2-Arylbenzofuran-based molecules as multipotent Alzheimer's disease modifying agentes. Eur J Med Chem. 2012;58:519–32.
34. Samadi A, Chioua M, Bolea I, et al. Synthesis biological assessment and molecular modeling of new multipotente MAO and cholinesterase inhibitors as potential drugs for the treatment of Alzheimer's disease. Eur J Med Chem. 2011;46:4665–8.
35. Bolea I, Jimenez JJ, de los Rios C, et al. Synthesis biological evaluation, and molecular modeling of donepezil and N-[(5-(Benzyloxy)-1-methyl-1H-indol-2-yl) methyl]-N-methylprop-2-yn-1-amine hybrids as new multipotent cholinesterase/monoamine oxidase inhibitors for the treatment of Alzheimer's Disease. J Med Chem. 2011;54:8251–70.
36. Minarini A, Milelli A, Tumiatti V, et al. Multitarget-directed benzylideneindanone derivatives anti-ß-amyloid (aβ) aggregation antioxidant metal chelation and monoamine oxidase b (mao-b) inhibition properties against Alzheimer's disease. Neuropharmacology. 2012;62:997–1003.
37. Lu C, Guo Y, Yan J, et al. Design synthesis and evaluation of multitarget-directed resveratrol derivatives for the treatment of Alzheimer's disease. J Med Chem. 2013;56:5843–59.

Index

Z. Wang, B. Yang, *Polypharmacology*, https://doi.org/10.1007/978-3-031-04998-9

O

T

U